CEREBRAL CORTEX

Volume 12
Extrastriate Cortex in Primates

CEREBRAL CORTEX

Series Editors: Edward G. Jones and Alan Peters

Volume 1: Cellular Components of the Cerebral Cortex

Volume 2: Functional Properties of Cortical Cells

Volume 3: Visual Cortex

Volume 4: Association and Auditory Cortices

Volume 5: Sensory-Motor Areas and Aspects of Cortical Connectivity

Volume 6: Further Aspects of Cortical Function, Including Hippocampus

Volume 7: Development and Maturation of Cerebral Cortex

Volume 8A: Comparative Structure and Evolution of Cerebral Cortex, Part I

Volume 8B: Comparative Structure and Evolution of Cerebral Cortex, Part II

Volume 9: Normal and Altered States of Function

Volume 10: Primary Visual Cortex in Primates

Volume 11: The Barrel Cortex of Rodents

Volume 12: Extrastriate Cortex in Primates

Series Editors: Edward G. Jones and Alan Peters

CEREBRAL CORTEX

Volume 12
Extrastriate Cortex in Primates

Edited by

KATHLEEN S. ROCKLAND
The University of Iowa
Iowa City, Iowa

JON H. KAAS
Vanderbilt University
Nashville, Tennessee

and

ALAN PETERS
Boston University School of Medicine
Boston, Massachusetts

Springer Science+Business Media, LLC

The Library of Congress cataloged the first volume of this title as follows:

Main entry under title:

Cerebral cortex.

Includes bibliographical references and index.
Contents: v. 1. Cellular components of the cerebral cortex.
1. Cerebral cortex—Collected works. I. Peters, Alan, 1929– II. Jones, Edward G., 1939– . [DNLM: 1. Cerebral cortex—Anatomy and histology. 2. Cerebral cortex—Physiology. WL 307 C4136]

QP383.C45 1984 612'.825 84-1982

ISBN 978-1-4757-9627-8 ISBN 978-1-4757-9625-4 (eBook)
DOI 10.1007/978-1-4757-9625-4

Originally published by Plenum Press, New York in 1988
Softcover reprint of the hardcover 1st edition 1988

http://www.plenum.com

10 9 8 7 6 5 4 3 2 1

Contributors

Jean Bullier	Cerveau et Vision INSERM 371, 69675 Bron Cedex, France
Stephanie Clarke	Institut de Physiologie, Université de Lausanne, CH-1005 Lausanne, Switzerland
Catherine G. Cusick	Department of Anatomy SL49, Tulane School of Medicine, New Orleans, Louisiana 70112
Charles G. Gross	Department of Psychology, Princeton University, Princeton, New Jersey 08544
Balázs Gulyás	Division of Human Brain Research, Department of Neuroscience, Karolinska Institute, S-17177 Stockholm, Sweden
Jon H. Kaas	Department of Psychology, Vanderbilt University, Nashville, Tennessee 37240
Tirin Moore	Department of Brain and Cognitive Sciences, Massachusetts Institute of Technology, Cambridge, Massachusetts 02139
Lionel G. Nowak	Section of Neurobiology, Yale University School of Medicine, New Haven, Connecticut 06510
Guy A. Orban	Laboratorium voor Neuro- en Psychofysiologie, KU Leuven, Medical School, Campus Gasthuisberg, B-3000 Leuven, Belgium
Esther Peterhans	Department of Neurology, University Hospital Zurich, CH-8091 Zurich, Switzerland

Heather L. Read Center for Molecular and Behavioral Neuroscience, Rutgers University, Newark, New Jersey 07102

Kathleen S. Rockland Department of Neurology, 2007 RCP, College of Medicine, University of Iowa, Iowa City, Iowa 52242-1053

Hillary R. Rodman Department of Psychology and Yerkes Regional Primate Research Center, Emory University, Atlanta, Georgia 30322

Anna Wang Roe Section in Neurobiology, Yale University School of Medicine, New Haven, Connecticut 06510

Marcello G. P. Rosa Vision, Touch and Hearing Research Centre, Department of Physiology and Pharmacology, The University of Queensland, Queensland 4072, Australia

Jeffrey D. Schall Vanderbilt Vision Research Center, Department of Psychology, Vanderbilt University, Nashville, Tennessee 37240

Peter H. Schiller Department of Brain and Cognitive Sciences, Massachusetts Institute of Technology, E25-634 MIT, Cambridge, Massachusetts 02139

Ralph M. Siegel Center for Molecular and Behavioral Neuroscience, Rutgers University, Newark, New Jersey 07102

Keiji Tanaka Brain Science Institute, RIKEN, 2-1 Hirosawa, Wako-shi, Saitama, 351-01, Japan

Daniel Y. Ts'o Department of Neurobiology, Rockefeller University, New York, New York 10021

S. Zeki Wellcome Department of Cognitive Neurology, University College London, London WC1E 6BT, England

Karl Zilles C. & O. Vogt-Institute of Brain Research, University of Düsseldorf, D-40225 Düsseldorf, Germany

Preface

Over the last twenty-five years, there has been an extensive effort, still growing for that matter, to explore and understand the organization of extrastriate cortex in primates. We now recognize that most of caudal neocortex is visual in some sense and that this large visual region includes many distinct areas. Some of these areas have been well defined, and connections, neural properties, and the functional consequences of deactivations have been studied. More recently, non-invasive imaging of cortical activity patterns during visual tasks has led to an expanding stream of papers on extrastriate visual cortex of humans, and results have been related to theories of visual cortex organization that have emerged from research on monkeys. Against this backdrop, the time seems ripe for a review of progress and a glance at the future.

One caveat important to emphasize at the very onset is that the reader may be puzzled or confused by the use of different terminologies. Individual investigators commonly tend to favor different terminologies, but in general some prove more advantageous than others. As discussed by Rowe and Stone (1977) as well as by others, there is an unfortunate tendency for role-indicating names to lead to fixed ideas about function, in contrast to those that are more neutral and adaptable to new findings. Terminology differences may often seem trivial in that they reflect no basic disagreements (the names "MT," "V5," and "motion area" designate the same area; "blobs," "patches," and "puffs" designate the same subdivisions in area V1). Other differences reflect unresolved issues and disparate viewpoints. For example, is the region called "ventral V3" part of a larger visual area, V3, or is it a separate visual area, VP? Likewise, "dorsal V3" has been conceptualized as part of another area, DM, rather than as a component of V3. As another example, it makes a great deal of difference for comparisons across cortical areas and across species if the sublayer in V1 that projects to "MT" or "V5" is considered as layer IIIC or IVB.

Terminology also reflects and, even more, guides ways of thinking about the visual system as a whole. Does the "extrastriate" rubric include visually related cortex in the parietal lobe? In the frontal lobe? Probably the most popular conceptualization of visual cortex has been a fixed hierarchy of processing areas. However useful this approach may be, ambiguities exist in the assignment of individual areas to levels, as graphically portrayed in any number of M. C. Escher prints (for example, the 1960 lithograph *Ascending and Descending*, which is discussed by D. Hofstadter [1979, p. 13] in the context of strange loops and tangled hierarchies). Ultimately what is "ascending" or "descending" may be less than certain. Moreover, the terminology of hierarchy does not easily incorporate other aspects of visual processing, such as temporal factors or the possibility—which the field is just beginning to consider—of dynamic interactions and groupings.

The present volume covers a range of relevant topics. The first chapter, by Charles Gross, reviews the historical development of ideas and concepts about extrastriate visual cortex. As we approach the symbolic milestone of the next millennium, this chapter provides a gracious tribute to the antecedents of current views and a reminder of the future progress we can expect. The following five chapters review and examine general features of cortical organization. Peter Schiller discusses the implications of conceptualizing processing at the level of single neurons and further considers the question of how functional specificity might emerge from neurons that are multifunctional within dynamically adjustable rather than dedicated systems. Chapters by Jon Kaas and Marcello Rosa consider theories of how cortex is divided into areas, criteria for identifying fields, and the nature of cortical representations. Lionel Nowak and Jean Bullier cover the important and somewhat neglected topic of the temporal characteristics of cortical processing (see also Guy Orban, Chapter 9). These authors summarize evidence for fast and slow pathways as a compilation of earlier distinctions between serial and parallel processing. Kathleen Rockland reviews recent advances in understanding of the microarchitecture of cortical connections, including features of spatial convergence and divergence, as well as laminar patterns of terminations, and discusses how these data might relate to theories of hierarchical organization.

Chapters 7–13 focus on features of specific visual areas or groups of related visual areas. The second visual area, V2, is reviewed in separate chapters by Anna Roe and Dan Ts'o and by Esther Peterhans. Roe and Ts'o discuss modular specialization with V2, the remapping of visual space within the different classes of modules, and interaction between modules. Peterhans stresses the complex response properties of neurons in V2 and relates response features to aspects of perception. The structural and functional characteristics of area MT/V5 and associated areas are thoughtfully considered by Guy Orban, who also develops the issue of homologous areas in monkeys and humans. "Higher order" visual areas in the temporal lobe, including the region of the "superior temporal polysensory area," and subdivisions of classical inferotemporal cortex are covered in separate chapters by Catherine Cusick and Keiji Tanaka. Ralph Siegel and Heather Read consider the functions of visual and visuomotor areas of posterior parietal cortex, while Jeffrey Schall extensively reviews visuomotor areas of especially the frontal lobe.

A single chapter, by Hillary Rodman and Tirin Moore, addresses the normal development of connections and function in extrastriate cortex, as well as the reorganization that occurs in extrastriate cortex as the result of injury or experience.

The last group of chapters, 15–17, concentrates on the organization of extrastriate cortex in humans. The scope and detail contained in these chapters, largely based on research over the last few years, testify to the gathering momentum of research on visual cortex of humans, both from functional mapping and from anatomical studies. From this perspective, the authors again treat the difficult issues of how to identify cortical areas, functional specializations, and network interactions. Results in human experiments benefit from and even require the conceptual context that has emerged from earlier studies in monkeys and other mammals, and the importance of continuing investigations in nonhuman mammals is clear. The closing chapter by Semir Zeki questions whether there is "one overall consciousness or many" while considering the possible contributions of several visual areas or regions to human experience, such as those concerned with color or motion perception. (An additional invited chapter, "Topography and Function of Human Visual Cortex," had been prepared by Dr. Edgar A. DeYoe, but owing to unforeseeable circumstances could not be included in this volume.)

Volume 12 of *Cerebral Cortex* on extrastriate cortex of primates complements Volume 10, *Primary Visual Cortex in Primates.* Volume 3 in this series, entitled *Visual Cortex,* covered aspects of organization and function in several different species. Important features of visual cortex development were discussed in Volume 7, entitled *Development and Maturation of Cerebral Cortex.* Overall, the authors of the present volume have provided an extensive and detailed review of extrastriate cortex of primates. Some emphasis has been placed on features that are likely to exist in both monkey and human brains. Several relevant topics could have been considered more fully in this already large volume, and more points of view could have been expressed. However, some comparative and evolutionary aspects of visual cortex organization have been addressed in Volume 8 of this series, and future chapters and volumes on visual cortex can be anticipated.

Kathleen S. Rockland
Jon H. Kaas
Alan Peters

Iowa City, Nashville, and Boston

References

Hofstadter, D. R., 1979, *Gödel, Escher, Bach: An Eternal Golden Braid,* Basic Books, New York.
Rowe, M. H., and Stone, J., 1977, Naming of neurons, *Brain Behav. Evol.* **14:**185–216.

Contents

Color Plates ... xxiii

Chapter 1

From Imhotep to Hubel and Wiesel: The Story of Visual Cortex

Charles G. Gross

1. Introduction ... 1
2. Ancient Egyptian Surgery and Medicine ... 2
   2.1. The First Written Mention of the Brain ... 2
   2.2. The Legacy of Egyptian Medicine ... 3
3. Greek Philosopher-Scientists and the Beginning of Brain Science ... 4
   3.1. Miletus, Cradle of Science ... 4
   3.2. Alcmaeon of Croton, the First Neuroscientist ... 6
   3.3. The Hippocratic Doctors ... 7
   3.4. The Alexandrians and the Beginning of Neuroanatomy ... 8
   3.5. Galen of Pergamon ... 10
4. The Medieval "Cell" Doctrine of Brain Function ... 11
5. The Rebirth of Brain Science ... 14
   5.1. Vesalius Restarts Neuroanatomy ... 14
   5.2. Cerebral Cortex: Gland or Rind? ... 15
   5.3. Thomas Willis Turns toward Cortex ... 16
   5.4. Emanuel Swedenborg: Unknown Neurogenius ... 20
   5.5. Von Haller and the Insensitivity of Cortex ... 21
   5.6. Gennari and His Stripe ... 22
6. The Beginning of the Modern Era of Cortical Localization ... 23
   6.1. Gall and Phrenology ... 23
   6.2. Flourens Attacks Gall but the Cortex (Re)emerges as a "Higher" Structure ... 24
   6.3. Broca "Confirms" Gall ... 25
   6.4. Evolution and Brain Function ... 26
   6.5. The Search for Sensory Areas in the Cerebral Cortex ... 29

7. The Discovery of a Visual Center in the Cerebral Cortex 31
7.1. Bartolomeo Panizza: The First Claim 31
7.2. The Battle for Visual Cortex: Ferrier versus Munk and Schäfer 31
7.3. Striate Cortex Is Visual Cortex 36
7.4. Neurophysiology of Striate Cortex Begins 38
7.5. The Microelectrode Comes to the Visual System 40
8. Beyond Striate Cortex 42
8.1. The Surrounding Zones 42
8.2. Temporal Association Cortex as a Visual Area 44
8.3. Parietal Association Cortex as a Visual Area 48
9. Coda: Two or More Cortical Visual Systems 51
10. References 52

Chapter 2

Past and Present Ideas About How the Visual Scene Is Analyzed by the Brain

Peter H. Schiller

1. Introduction 59
2. The Specificity of Neurons 60
3. The Specificity of Brain Areas 63
4. Countercurrents to Neuronal Specificity 67
5. Countercurrents to Areal Specificity 75
6. Brain Topography and Visual Representation 80
7. Summary 85
8. References 87

Chapter 3

Theories of Visual Cortex Organization in Primates

Jon H. Kaas

1. Introduction 91
2. What Are We Mapping? 92
3. Early Maps 94
4. Early Modern Theories 98
5. Recent Maps of Visual Cortex 100
6. V2 and MT, Areas Common to All Primates 104
7. Regions of Contention: Visual Areas of the Third Tier versus V3 and V4 108
8. Area V3 and Dorsal V3 108
9. Area DM 110
10. Area M 111
11. Area DI 113

12. Area V4 113
13. DLr and DLc 114
14. Upper Temporal Lobe Areas: MTc, FSTd, FSTv, and MST 115
15. VP and VA 116
16. Parietal Visual Areas: LIP, VIP, PIP, and VPP 117
17. Summary and Conclusions 118
18. References 119

Chapter 4

Visuotopic Organization of Primate Extrastriate Cortex

Marcello G. P. Rosa

1. Introduction 127
2. Study of Cortical Visuotopy 130
   2.1. Methods Used in the Definition of Visuotopic Maps 131
   2.2. Visuotopically Organized versus Nonvisuotopically Organized Areas 135
   2.3. Receptive Field Size, Scatter, and Point-Image Size 136
   2.4. Types of Representation and Topographic Transitions between Areas 140
3. A Survey of Visuotopic Maps and Cortical Areas in Different Primate Species 144
   3.1. The Second Visual Area 144
   3.2. The Middle Temporal Area 150
   3.3. Visual Areas Surrounding MT 154
   3.4. The Dorsomedial Complex 158
   3.5. The Lateral and Ventral Areas 165
   3.6. The Occipitotemporal Transition (TEO) 172
   3.7. Summary 173
4. Why Have Different Visuotopic Maps? 176
5. Large Receptive Fields, Scatter, and Spatial Resolution 181
6. Dynamics of Receptive Field Position 186
7. Conclusions 189
8. References 191

Chapter 5

The Timing of Information Transfer in the Visual System

Lionel G. Nowak and Jean Bullier

1. Introduction 205
2. Structural Constraints: Axonal Conduction and Integration Times 206
   2.1. Axonal Conduction Time 206
   2.2. Neuronal Integration Time 215

3. Timing of Information Transfer to and between Visual Cortical Areas ... 220
3.1. Latency in the Retina and Lateral Geniculate Nucleus ... 221
3.2. Visual Latencies in Cat Cortex ... 224
3.3. Visual Latencies in Primate Cortex ... 225
4. Conclusion: How Does Cortex Process Information So Rapidly? ... 231
4.1. What Connects Together Fires Together ... 231
4.2. How Rapidly Is Response Selectivity Established? ... 232
4.3. A Functional Model of the Slow Brain ... 233
5. References ... 233

Chapter 6

Elements of Cortical Architecture: Hierarchy Revisited

Kathleen S. Rockland

1. Introduction ... 243
2. Elements of a "Computational" Architecture ... 245
2.1. Laminar Terminations of Single Axons ... 245
2.2. Spatial Configuration of Single Axons ... 250
2.3. Postsynaptic Population ... 254
2.4. Connectional Weight ... 258
2.5. Some Quantitative Parameters ... 264
2.6. Subtypes of Connections? ... 275
2.7. Intrinsic Connections ... 277
3. Hierarchy Revisited ... 279
3.1. Laminar Patterns ... 280
3.2. Reciprocity ... 281
3.3. Collateralized Connections ... 281
3.4. Other Connections ... 283
4. Summary: Cortical Architectures ... 284
5. Concluding Remarks ... 286
6. References ... 287

Chapter 7

The Functional Architecture of Area V2 in the Macaque Monkey: Physiology, Topography, and Connectivity

Anna Wang Roe and Daniel Y. Ts'o

1. Introduction ... 295
2. Receptive Field Properties with Respect to V2 Stripes ... 297
3. Modular Organization within Stripes ... 303
4. Topography ... 305
4.1. Multiple Maps ... 307

4.2. Cortical Magnification, Anisotropy, Modular Maps, and Receptive Field Size 311
4.3. Point Image versus Point Set 312
4.4. What Makes V2 a Single Cortical Area? 313
5. Connectivity 314
5.1. Color, Form, and Motion: How Segregated Are They? 314
5.2. Serial or Parallel? 320
6. Relationship of Structure and Circuitry to Perception 322
6.1. The Role of V2 in Perceptual Binding across Modalities 322
6.2. Role of V2 in Contour Perception: An Implementation 323
7. Why Another Stage of Cortical Processing Apart from V1? 326
8. References 328

Chapter 8

Functional Organization of Area V2 in the Awake Monkey

Esther Peterhans

1. Introduction 335
2. Methodological Considerations 336
2.1. Animal Training 337
2.2. Visual Stimulation 338
2.3. Relating Anatomy and Physiology 338
3. Visual Contours 338
3.1. Contrast Borders 340
3.2. Color Borders 342
4. Mechanisms of Figure–Ground Segregation 343
4.1. Occlusion Cues 343
4.2. Motion Cues 348
4.3. Stereoscopic Cues 352
5. Conclusions 353
6. References 354

Chapter 9

Visual Processing in Macaque Area MT/V5 and Its Satellites (MSTd and MSTv)

Guy A. Orban

1. Introduction 359
1.1. Definition of Areas 360
1.2. Connections 363
1.3. Nature of the Input to Area MT/V5 365
2. Physiological Exploration of Macaque Area MT/V5 and Its Satellites 367
2.1. General Response Properties 367

2.2. Exploration of the Classical RF: Simple Parameter Selectivities 376
2.3. Exploration of the Classical RF: Responses to Complex Stimuli 388
2.4. The Antagonistic Surround........ 398
3. Behavioral Exploration of Macaque Area MT/V5 and Its Satellites.. 404
3.1. Motion Perception........ 405
3.2. Guidance of Eye Movements........ 412
3.3. Guidance of Body Movements 418
3.4. Extraction of 2D and 3D Shape from Motion 418
4. Functional Imaging of Area MT/V5 and Satellites in Humans 419
5. Conclusions 425
6. References 425

Chapter 10

The Superior Temporal Polysensory Region in Monkeys

Catherine G. Cusick

1. Definition, Location, and Extent of Superior Temporal Polysensory Cortex in Monkeys........ 435
1.1. Early Views of the Organization of Cerebral Cortex 435
1.2. Early Studies of the Location and Extent of the Superior Temporal Polysensory Area of Monkeys 437
1.3. Initial Revision of the Extent of the STP Region: Visual Areas of the Upper Bank of the STS........ 438
2. Alliances of Superior Temporal "Polysensory" Cortex with Dorsal and Ventral Visual Processing Streams 440
2.1. Dorsal Stream Cortical Connections of STP Cortex........ 440
2.2. Pulvinar Connections 441
2.3. Cortical versus Subcortical Influences on STP Cortex........ 441
2.4. Behavioral Studies of Visuospatial and Object Recognition Functions after Ablation of STP Cortex 442
2.5. Neurophysiological Evidence for Functional Heterogeneity within Polysensory Cortex of the Upper Bank of the STS 443
3. Subparcellation of the STP Region: Alternative Views........ 445
3.1. Caudal and Rostral STP........ 445
3.2. PP and STP Proper........ 445
3.3. Connectional/Architectonic Subdivisions of "Area TPO" (TPO-1–TPO-4)........ 446
3.4. Other Proposals: STPp, STPa........ 447
4. A Revision of the STP Region Based on Chemoarchitecture and Connections with Other High-Order Cortical Areas........ 448
4.1. Revision of Architectonic Subdivisions of the Upper Bank and Fundus of the STS 448
4.2. Area TPO Can Be Identified by Distinctive Connections with "High-Order" Association Cortex 452

4.3. Proposed Physiological Correspondences of Subdivisions of Area TPO 460
4.4. Are Subdivisions within the STP Region (Area TPO) Separate Cortical Areas? 461
5. Concluding Remarks 462
6. References 463

Chapter 11

Columnar Organization in the Inferotemporal Cortex

Keiji Tanaka

1. Introduction 469
2. Stimulus Selectivity of Cells in TE 470
3. Position, Orientation, Size, and Attributes Defining the Contours ... 473
4. Columnar Organization in TE 477
5. Organization of Afferents to TE 477
6. Optical Imaging of the Columnar Organization 486
7. Changeability of the Selectivity in the Adult 489
8. Functions of the TE Columns 490
9. Binding Activities in Distant Columns 492
10. "Face Neurons" in the Anterior Part of the Superior Temporal Sulcus 493
11. Object Recognition by Activities Distributed over the Brain 494
12. References 495

Chapter 12

Construction and Representation of Visual Space in the Inferior Parietal Lobule

Ralph M. Siegel and Heather L. Read

1. Introduction 499
2. Neurological Findings 500
3. The Macaque Model 501
4. The Connections of the Macaque Inferior Parietal Lobule 502
5. Behavioral Methods 504
6. Electrophysiological Studies 505
6.1. Visual Properties 505
6.2. Extraretinal Signals 514
7. A Mathematical Model for Spatial Representation 516
8. Conclusion 519
9. References 520

Chapter 13

Visuomotor Areas of the Frontal Lobe

Jeffrey D. Schall

1. Introduction 527
2. Postarcuate Premotor Cortex 530
 2.1. Ventral Premotor Cortex 531
 2.2. Dorsal Premotor Cortex 535
3. Frontal Eye Field 540
 3.1. Overview of the Oculomotor System 540
 3.2. Location of FEF in Nonhuman Primates 542
 3.3. Location of FEF in Humans 545
 3.4. FEF Connectivity 548
 3.5. FEF Role in Saccade Production 552
 3.6. FEF Sensory Function 569
 3.7. FEF Role in Slow Pursuit Eye Movements 580
 3.8. FEF Function in Human 582
4. Supplementary Eye Field 584
 4.1. Supplementary Motor Area 584
 4.2. Connectivity of SEF 585
 4.3. Neuronal Properties 594
 4.4. Role of SEF in Saccade Production 601
5. Prefrontal Cortex 604
 5.1. Dorsolateral Prefrontal Cortex—Areas 46 and 9 605
 5.2. Ventrolateral Prefrontal Cortex—Area 12 610
 5.3. Inferior Orbital Prefrontal Cortex—Area 13 612
6. Cingulate Cortex 614
7. Conclusions 616
8. References 616

Chapter 14

Development and Plasticity of Extrastriate Visual Cortex in Monkeys

Hillary R. Rodman and Tirin Moore

1. Introduction and Overview 639
2. Normal Development of Extrastriate Visual Cortex 640
 2.1. Prenatal Specification of Visual Cortical Areas 640
 2.2. Development of Connectivity of Extrastriate Areas 641
 2.3. Development of Neuronal Activity in Extrastriate Areas 646
 2.4. Development of Patterns of Neurochemical Markers 652
3. Plasticity of Extrastriate Visual Cortex in Adulthood 653
 3.1. Damage-Induced Plasticity 654
 3.2. Experience or Learning-Induced Plasticity 658
4. Plasticity of Extrastriate Visual Cortex in Development 660
 4.1. Effects of Early Lesions of Extrastriate Cortex 660
 4.2. Effects of Early Lesions of Primary Visual Cortex 660

5. Comparison with Development and Plasticity of Extrastriate Cortex in Other Mammals 662
5.1. Specification and Development 662
5.2. Parallel Activation and Plasticity of Extrastriate Areas........ 663
6. Concluding Discussion........ 663
6.1. General Principles 663
6.2. Sequence of Development and Functional Streams........ 664
6.3. Plasticity and Recovery of Function after Early or Late Brain Damage........ 665
7. References 666

Chapter 15

Architecture, Connectivity, and Transmitter Receptors of Human Extrastriate Visual Cortex: Comparison with Nonhuman Primates

Karl Zilles and Stephanie Clarke

1. Historical Concepts and Early Maps of the Human Visual Cortex... 673
1.1. Bipartition of the Visual Cortex or "Visuosensory" versus "Visuopsychic" Regions........ 674
1.2. Tripartition of the Visual Cortex and the First Hierarchical Concepts........ 675
1.3. Further Cyto-, Myelo-, and Pigmentoarchitectonic Parcellations........ 682
1.4. Subdivisions of the Visual Cortex Derived from Myelogenetic Studies 685
2. Extrastriate Visual Cortex in Prosimians and New and Old World Monkeys 685
2.1. Maps of Visual Cortical Areas 686
2.2. Architecture of Extrastriate Areas........ 688
3. Connectivity of Extrastriate Areas in Nonhuman Primates 692
3.1. Thalamic Connections 692
3.2. Intrahemispheric Connections 693
3.3. Interhemispheric Connections 693
4. Transmitter Receptors in Extrastriate Areas of Nonhuman Primates 695
4.1. Transmitter Receptors in Area V2........ 695
4.2. Transmitter Receptors in Other Extrastriate Areas........ 699
5. Human Extrastriate Visual Cortex 699
5.1. Functional Specialization within Human Extrastriate Cortex .. 699
5.2. Cyto- and Myeloarchitecture of Human Extrastriate Areas.... 704
5.3. Cytochrome Oxidase Staining as a Marker for Visual Areas and Intraareal Compartments........ 706
6. Connectivity of Human Extrastriate Cortex........ 710
6.1. Thalamic Connections 710
6.2. Interhemispheric Connections 713
6.3. Visual Areas as Derived from the Pattern of Callosal Connections........ 717

6.4. Intrahemispheric Connections 718
6.5. Intrinsic Connections 721
7. Transmitter Receptors in Extrastriate Areas of the Human Brain ... 723
7.1. Receptors in V2 724
7.2. Receptors in Other Extrastriate Areas 726
8. Anatomical Studies as a Key to the Functional Organization of the Human Extrastriate Cortex 727
8.1. Concept of Visual Areas 727
8.2. Modular Organization of Visual Areas 728
8.3. Extrastriate Neuronal Network 728
9. References 729

Chapter 16

Functional Organization of Human Visual Cortical Areas

Balázs Gulyás

1. Introduction 743
2. Techniques Applicable to the Functional Exploration of Human Visual Cortex 744
3. What the Monkey Brain Tells Us about the Human Visual Cortex .. 747
4. Human Visual Cortical Areas and Human Visual Cortex 749
5. Topographic Organization of Early Visual Cortical Areas 752
6. Cortical Magnification Factors 758
7. Functional Properties of Early Visual Cortical Areas 761
8. Information Processing in Human Visual Cortex: Basic Visual Cues 762
8.1. Color-Sensitive Regions 762
8.2. Visual Motion-Sensitive Regions 764
8.3. Disparity 765
8.4. Visual Form 765
8.5. Form Perception versus Visuospatial Operations in the Visual Cortex? 767
9. Convergence and Divergence in the Human Visual System 769
10. References 770

Chapter 17

The Color and Motion Systems as Guides to Conscious Visual Perception

S. Zeki

1. Introduction 777
2. Residual Vision 778
2.1. Characteristics of Residual Vision 779
2.2. The Neural Basis of Residual Motion Vision 781
3. The Reciprocal Connections between V1 and V5 787

4. Dynamic Parallelism 791
4.1. The Parallel Pathways to V1 and V5 791
4.2. The Direct Measurement of Latencies in Visual Cortex 793
5. The Color System 794
5.1. The Color Pathways in the Brain 795
5.2. Carbon Monoxide Poisoning 797
5.3. V4, Orientation Selectivity, and Form Analysis 799
5.4. The Human Color Center 801
5.5. The Effects of Lesions in the Human Color Center 802
5.6. The Implications of the Loss of Color Constancy Mechanisms 802
6. Visual Consciousness as a Distributed Process: The Direct Contribution of Individual Areas to Conscious Vision 803
7. References 804

Index 811

Color Plates

Chapter 13, Figure 8. Location of frontal and supplementary eye movement fields in human frontal cortex. The images are statistical maps of the difference between the blood flow measured when subjects made self-paced large-amplitude (40°) saccade in complete darkness and the blood flow measured when subjects were in a resting condition making no eye movements. The top panel shows a coronal section and two associated horizontal sections. Color intensity reflects statistical reliability of the difference in blood between the saccade and fixation conditions. The locations of FEF and SEF are indicated. The bottom panel shows a horizontal and two coronal sections illustrating the location of the central focus of FEF identified in eight PET studies (reviewed in Paus, 1996) of oculomotor (red spots) and manual (yellow spots) control. CS, Central Sulcus; PreCs, precentral sulcus; SFS, superior frontal sulcus.

Chapter 15, Figure 8. (A, B) Regional distribution of AMPA and kainate receptors respectively (demonstrated with [^{3}H]AMPA and [^{3}H]kainate binding; K. Zilles, unpublished observations), in adjacent parasagittal sections through the macaque cortex. The color scales show the receptor densities in fmol/mg protein. A considerable regional heterogeneity of the receptor distribution is visible, which allows the delineation of numerous extrastriate areas, including V1, V2v, V2d, VP, V3, V3A, V4, V4d/V4t, V5/MT, MST, and the dorsomedial zone, DMZ. Further delineations are possible in frontal, prefrontal, and temporal regions. Abbreviations: ces, central sulcus; lus, lunate sulcus; M1, primary motor cortex; SS1, primary somatosensory cortex; sts, superior temporal sulcus. (C–F) The regional distribution of muscarinic (mainly) M1, α_1, kainate, and $GABA_A$ receptors, respectively (labeled with [^{3}H]pirenzepine, [^{3}H]prazosin, [^{3}H]kainate, and [^{3}H]muscimol; K. Zilles, unpublished observations) in adjacent coronal sections (C: M1; 8D: α_1) and an oblique (between coronal and horizontal cutting planes) section (E: kainate receptor) through the human occipital and parietal cortex. (D) Distribution of $GABA_A$ receptors at higher resolution. The color scales show the respective receptor densities in fmol/mg protein. Crossed arrows indicate the different orientations of the sections (d, dorsal; dr, rostrodorsal; l, lateral; m, medial; v, ventral; vc, ventrocaudal). A considerable regional heterogeneity of the receptor distribution is visible in all cases, which allows the delineation of various extrastriate areas. Abbreviations: cas, calcarine sulcus; cols, collateral sulcus; fus, fusiform gyrus; ils, intraligual sulcus; ips, intraparietal sulcus; los, lateral occipital sulcus; ls, lingual sulcus; otls, occipitotemporal sulcus; p-os, parietooccipital sulcus. Thin arrows in panel E point to layer IVB. For further details see text.

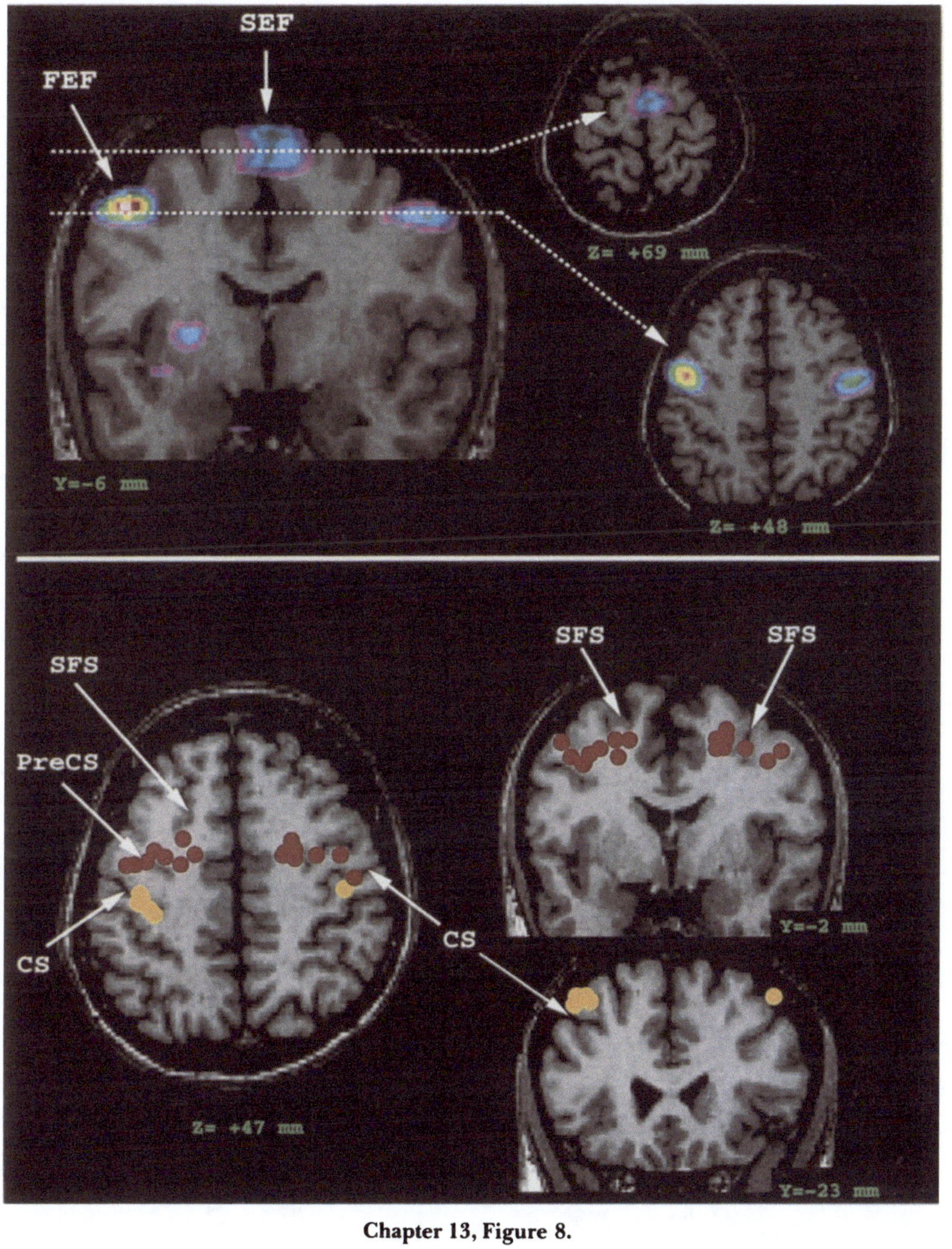

Chapter 13, Figure 8.

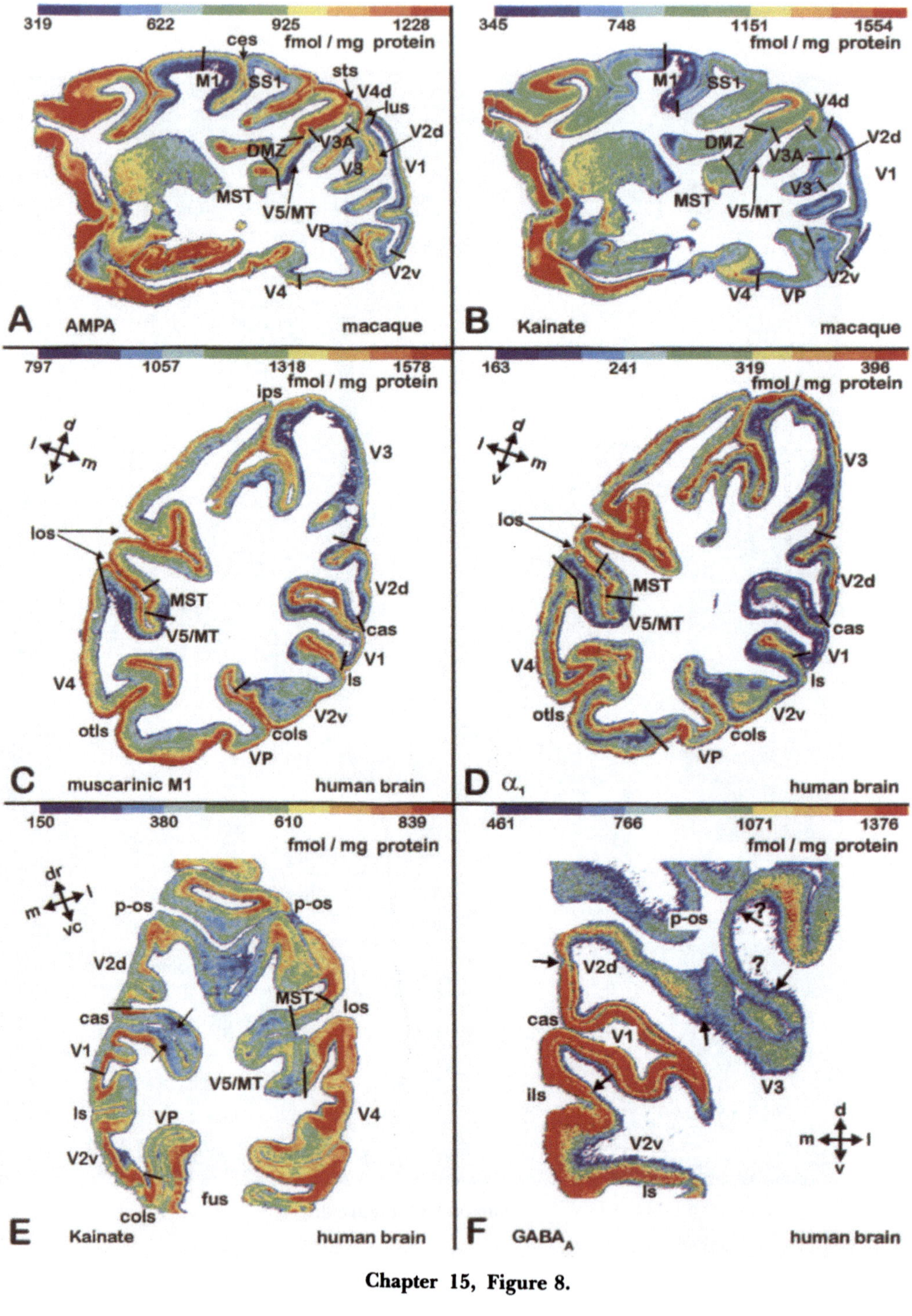

Chapter 15, Figure 8.

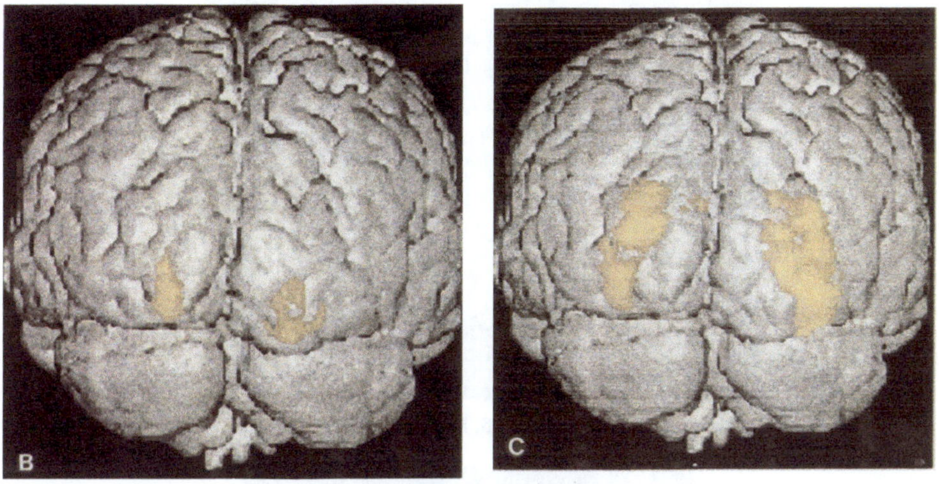

Chapter 16, Figure 2.

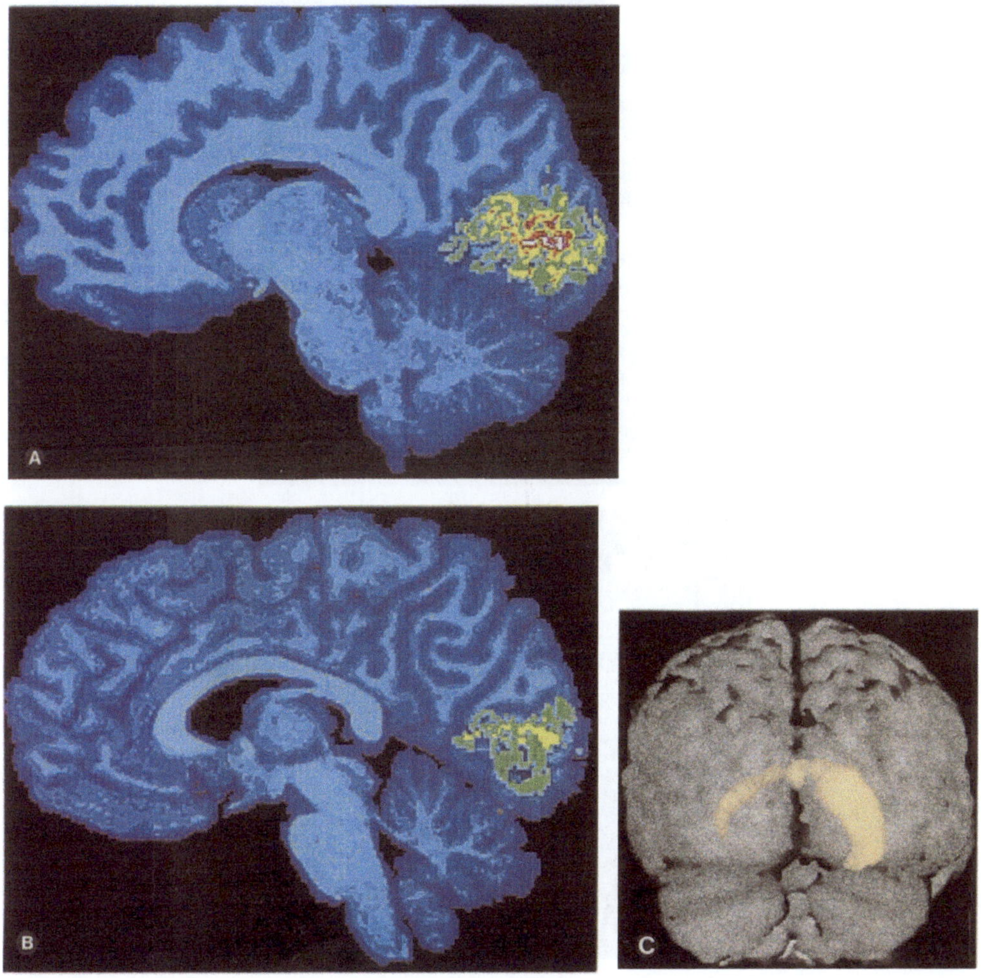

Chapter 16, Figure 4.

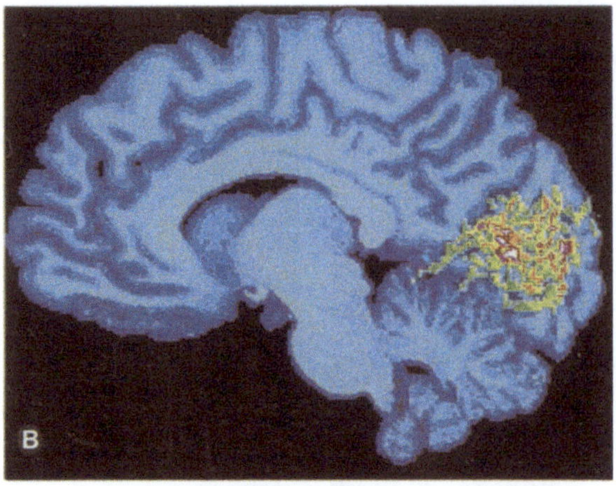

Chapter 16, Figure 5.

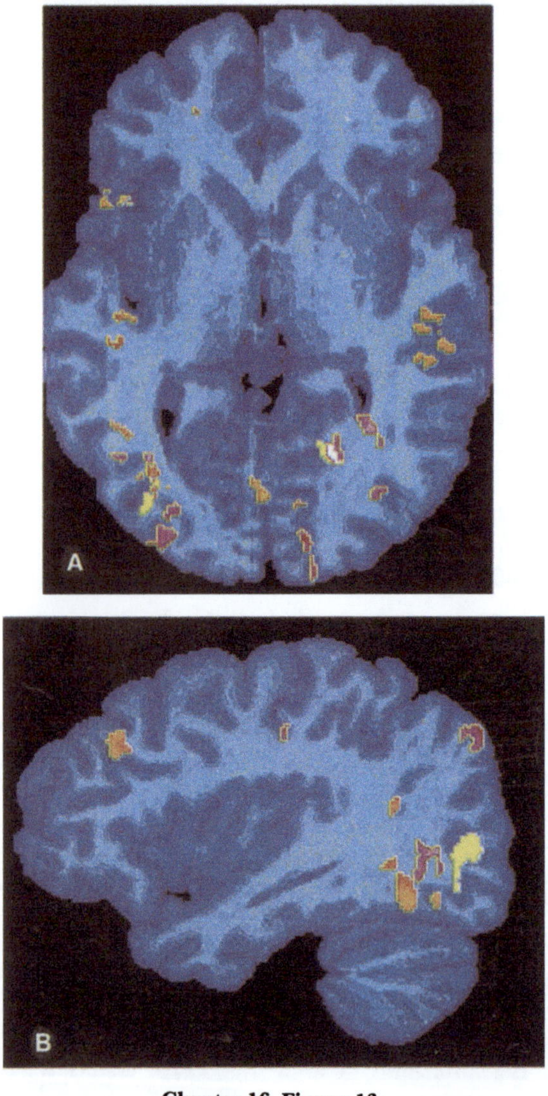

Chapter 16, Figure 13.

Chapter 16, Figure 2. The location of the primary visual cortex (area V1) (B) and area V2 (C) on the posterior surface of the brain. The reconstruction is based on quantitative cytoarchitectonic data of five normal postmortem brains, the colored regions represent 50% or more overlap levels and are projected onto the normal MRI brain of the Human Brain Atlas (Roland *et al.*, 1994). Original data by Karl Zilles and Katrin Amunts; reconstruction made by Jonas Larsson.

Chapter 16, Figure 4. The anatomical variability of a visual cortical area: striate cortex. Five normal postmortem brains were anatomically standardized with the help of the Human Brain Atlas (Roland *et al.*, 1994). The color-coded images show overlaps between areas V1 of the five anatomically standardized brains. (A) Parasagittal slice made 10 mm from the midline in the left hemisphere (i.e., $x = -0$ mm), showing the largest overlaps. (B) Parasagittal slice made 5 mm from the midline in the left hemisphere ($x = -5$ mm), i.e., showing the overlaps approximately on the medial surface of the hemisphere. Color code indicating number of overlaps out of five subjects' area V1: white: 5, red: 4, yellow: 3, green: 2, blue: 1. (Original cytoarchitectonic data were kindly provided by Karl Zilles and Torsten Schorman.) (C) The anatomical variability of the area V1/V2 border. The colored region projected onto the posterior surface of the brain represents the overlap between those parts of areas V1 and V2 that are not shown in Fig. 2C, i.e., that represent at the 50% overlap level the border between areas V1 and V2. Reconstruction made by Jonas Larsson.

Chapter 16, Figure 5. The anterior sector of the putative foveal representation in area V1 (population average). Note that this region is anterior to the occipital pole wherein the foveal representation in the striate cortex is suggested by numerous studies. (B) This discrepancy can be explained by the fact that in an overlap image of the anatomically standardized striate cortex the maximum overlap region is also anterior to the occipital pole (for explanation, see Fig. 6; parasagittal slice, $x = -7$ mm; white region: five overlaps). This is mainly due to the fact that sulcal variability increases away from the center of the brain, i.e., more occipital sectors of the foveal representations in the striate cortex will result in fewer overlaps.

Chapter 16, Figure 13. Parallel processing of visual form information generated by various visual cues. (A) Horizontal slice, $z = 2$ mm. HBA convention, as in Fig. 9 (Roland *et al.*, 1994). (B) Parasagittal slice, $x = -35$ mm. Yellow: regions processing visual form generated by motion cues; orange: regions processing visual form generated by texture cues; red: regions processing visual form generated by luminance cues. Note that there is only one overlap region in the left occipito-parietal sulcus; for the rest the regions are nonoverlapping. From the experimental series described in Gulyás *et al.* (1994a, b).

1

From Imhotep to Hubel and Wiesel

The Story of Visual Cortex

CHARLES G. GROSS

1. Introduction

This chapter traces the origins of our current ideas about visual cortex. We begin, in Section 2, long before the beginning of science, in the 30th century BCE, with the earliest description of the cerebral cortex. In Section 3 we consider the views of Greek philosopher-scientists on the functions of the brain. Section 4 concerns the long period in which there were virtually no advances in Europe in understanding the brain. In Section 5 we describe how even after Western brain research was well underway again, the cerebral cortex tended to be ignored. Section 6 considers the beginning of the modern study of the cerebral cortex and the localization therein of psychological functions. Our focus narrows in Section 7 and we consider how a specifically visual area of the cortex was delineated. The final section brings us to the theme of the entire volume, the extrastriate visual cortices. This review ends in the early 1980s about the time of the award of the Nobel Prize to Hubel and Wiesel for their discoveries on the visual cortex.

CHARLES G. GROSS • Department of Psychology, Princeton University, Princeton, New Jersey 08544.

Cerebral Cortex, Volume 12, edited by Rockland *et al.* Plenum Press, New York, 1997.

2. Ancient Egyptian Surgery and Medicine

2.1. The First Written Mention of the Brain

The first written reference to the cortex, indeed to any part of the brain, occurs in the *Edwin Smith Surgical Papyrus*. Although written about 1700 BCE, this papyrus is a copy of a much older surgical treatise dating back to the Pyramid age of the Old Kingdom (about 30th century BCE). The papyrus was bought in 1862 by an early American Egyptologist, Edwin Smith, from a local in Luxor, probably one of those hereditary tomb robbers who inhabited a nearby village. It eventually found its way to the great American Egyptologist James H. Breasted (Breasted, 1930).

The publication of Breasted's translation in 1930 made an enormous impact on medical historians and Egyptologists (Sarton, 1959; Sigerist, 1951). Previously, Egyptian medicine had been thought to be a jumble of incantations, amulets, and superstition. Rational medicine was supposed to have begun only with the Greeks. Yet the Edwin Smith papyrus is clear evidence of a scientific observer attempting to understand the human body and to treat its injury rationally.

The papyrus consists of a coolly empirical description of 48 cases starting from the head and working down to the shoulders, where the copyist stops in the middle of a sentence. For each case, the author systematically describes the examination, the diagnosis, and whether treatment was feasible. Each diagnosis comes to one of three conclusions: that the patient should be told that this is "an ailment which I will treat," "an ailment that I will try to treat," or "an ailment that I will not treat."

The word for brain first comes up in Case 6, a case of a skull fracture:

> (*Title*) Instructions concerning a gaping wound in his head, penetrating to the bone, smashing his skull, (and) rending open the brain of his skull.
>
> (*Examination*) If thou examinest a man having a gaping wound in his head, penetrating to the bone, smashing his skull, and rending open the brain of his skull, thou shouldst palpate his wound. Shouldst thou find that smash which is in his skull [like] those corrugations which form in molten copper, (and) something therein throbbing (and) fluttering under thy fingers, like the weak place of an infant's crown before it becomes whole . . . (and) he [the patient] discharges blood from both his nostrils, (and) he suffers with stiffness in his neck.
>
> (*Diagnosis*) [you say] "an ailment not to be treated."

Indeed, the "corrugations" which form in molten copper during the smelting process such as that used in early Egypt really do look like cerebral cortex.

In several cases, the author notes the relation of the laterality of the injury to the laterality of the symptom. For example, in Case 5, the patient "walks shuffling with his sole on the side of him having that injury which is in his skull" (presumably, a contra-coup injury).

The author was clearly aware that the site of injury determines the locus and nature of the symptoms. Thus, in Case 31, "It is a dislocation of a vertebra of the neck extending to this backbone which causes him to be unconscious of his two arms and legs." Elsewhere, the membranes covering the brain (meninges) and the fluid bathing it (the cerebrospinal fluid) are mentioned and aphasia ("he speaks not to thee") and seizures ("he shudders exceedingly") are described.

Although the document is startling in its rationality and empiricism and in the virtual absence of superstition and magic, Breasted (1930) did tend to overinterpret the papyrus, for example, writing, "this recognition of the localization of function in the brain . . . shows an astonishing early discernment which has been more fully developed by modern surgeons only within the present generations." Perhaps Breasted's greatest flight of fancy was the suggestion that the papyrus was written by Imhotep, a famous physician who flourished about the time that the original of the papyrus was written. Although Imhotep does date from about the time of the composition of the Edwin Smith papyrus, there is absolutely no evidence that he wrote it. It is in fact very unlikely that he did so, since the papyrus deals largely with battle wounds, and in the very hierarchically ordered world of Egyptian medicine, Imhotep was certainly not a battlefield surgeon.*

2.2. The Legacy of Egyptian Medicine

By about the Middle Kingdom (2000 BCE) there was a gradual decline in the artistic, architectural, and intellectual creativity and vibrancy that characterized the earlier dynasties. The society became more rigid and hierarchical, intellectual life more dominated by priests, the sculpture largely copies of earlier works, and the building more gigantic and grandiose. The rational and empirical spirit of medical practice that suffuses the Edwin Smith papyrus largely gives way to mysticism, religion, and elaborate speculations on the next world (Sarton, 1959; Sigerist, 1951). Yet the fame of ancient Egyptian medicine lived on, in the *Odyssey,* in the *Old Testament,* among the Presocratic physicians, in Galen, and in the Cabala, and continues today, in any New Age boutique or "health food" store.

It is important to view the correlations between brain injury and paralysis in the Edwin Smith papyrus in the context of ancient Egyptian theory and practice. We know that the Egyptians thought that the heart was the most important organ in the body, the seat of the mind and the center of intellectual activities. This is clear from their philosophical and religious writings and is emphasized by their practice of mummification. Both Herodotus' descriptions of the process of embalming and the examination of mummies show the contrast between the degrees of importance they attributed to the heart and brain. The first step in mummification was to scoop the brain out through the nostrils with an iron bar. By contrast, the heart (and most other internal organs) was either elaborately wrapped and replaced in the body or carefully stored in canopic jars near the body. As indicated in the *Book of the Dead,* the ancient Egyptians considered it essential that the dead body be preserved and all the important organs be re-

*Imhotep is fascinating in his own right. He was the grand vizier of the 3rd dynasty Pharaoh Zoser (2700–2650 BCE). He is described in a contemporary inscription as "chancellor of the king of Lower Egypt, the first after the King of Upper Egypt, administrator of the great palace, hereditary noble, high priest of Heliopolis, the builder, the sculptor." He is credited with designing the Step pyramid of Sakkara: the Tomb of Zoser, the first pyramid and the first example of large-scale dressed stone architecture. He was also a priest, astrologer, and magician. Yet his fame as a physician seems to have impressed his contemporaries and later generations most of all. Miniature statues of him were used as amulets to ward off disease and eventually he was deified as the Egyptian God of Medicine, an unusual honor even for a successful physician (Sarton, 1959; Sigerist, 1951; Guthrie, 1945).

tained so that in the afterlife the body would be in a suitable condition for resurrection when the soul returned to it. The dead Pharaohs were prepared for their next life with everything but a brain (Sarton, 1959; Sigerist, 1951).

The idea of the heart as the sensory and intellectual center of the body seems to have been universal not only in Egyptian writings, but also in other Near Eastern as well as Indian civilizations. It was even held much later, by the greatest of all Greek savants, Aristotle (Gross, 1995).

3. Greek Philosopher-Scientists and the Beginning of Brain Science

The approach to head injury of the Edwin Smith surgical papyrus stands out as a rock of empiricism in the sea of mysticism and superstition that characterized biological and medical writings in the Near East for about the next 24 centuries. But one could hardly call the papyrus scientific. Science is not just craft or technique or knowledge. Medical science is not just description of symptoms or treatment and it is not just the absence of superstition or magic. Rather, science, or perhaps we should say formal, self-conscious science, is the assumption that the world is something that can be understood by man's reason—a mechanism that works in some consistent way with a regularity governed by a limited set of rules. In this world view that defines science, the universe is not the playground of gods and ghosts acting in a capricious fashion moved by passion and whim. Science is public: it demands rational, critical debate; it involves observation, description, and measurement; it carries the assumption that there are underlying principles or "laws" potentially accessible by these methods.

This idea of formal science begins, at least in the West, with a group of Greek thinkers known as the Presocratic philosophers (Sarton, 1959; Schrödinger, 1954; Freeman, 1954; Longrigg, 1993; Farrington, 1944). They used the term "physiologia" to describe their enterprise, which is perhaps best translated as natural philosophy rather than physiology, physics, or just philosophy.

3.1. Miletus, Cradle of Science

It all started in the 6th century BCE in Ionia, a set of Greek city-states on the west shore of Turkey, or Anatolia (Fig. 1). The leading city-state, from which the earliest Presocratics came, was Miletus. What was special about this time and place that made it the cradle of science? The Ionians were a Greek people deriving from Crete. They were pioneers living in a new land and creating a new set of political institutions. The Bronze Age was giving way to the Iron Age, enabling the cheap production of tools and weapons, and thus these city-states could maintain themselves, at least for a while, in the face of the empires to their east. By the 6th century BCE, Miletus was a great port city that had established trading colonies throughout the Mediterranean and Black seas. It was a place where the sea lanes used by Greek, Phoenician, and Egyptian traders met the overland caravans from the east, from as far away as India and China. Its wealth derived both from its merchant ships and from local industries such as textiles and pottery. With its rich ferment of races, cultures, and ideas, Miletus was an interface between East and West (Sarton, 1959; Farrington, 1944).

At about this time, the rule of the landed aristocracy was breaking up and power was going to the merchant classes. They had the wealth to support speculation on the nature of the universe and they had the desire for new techniques, particularly in mathematics and astronomy. The development of alphabetic writing had taken away the monopoly on written knowledge from the class of scribe-priests that had characterized the cuneiform and hieroglyphic civilizations. The Ionian philosophers were neither prophets nor priests, but usually inventors, engineers, traders, or politicians and often several of these. Slavery had not yet become so pervasive that the ruling classes regarded manual labor with contempt. Finally, in these new city-states there was debate about the nature of society and about the best form of government. These freedoms to question the nature of social institutions seem to have been part of a spirit of inquiry that extended to the physical and biological world. All this ferment bubbled up into the beginning of the systematic inquiry into the universe that we call science.

Thales (ca. 583 BCE) was the first of the Presocratic philosophers, and thus is traditionally named the first (Western) scientist. He visited Egypt, came back with a number of geometric facts, and then applied them to new practical problems such as measuring the height of a building and the distance of a ship at sea. He seems to have been the first to conceive of the value of a general proposition or theorem in geometry. He is credited with such proofs as that the base angles

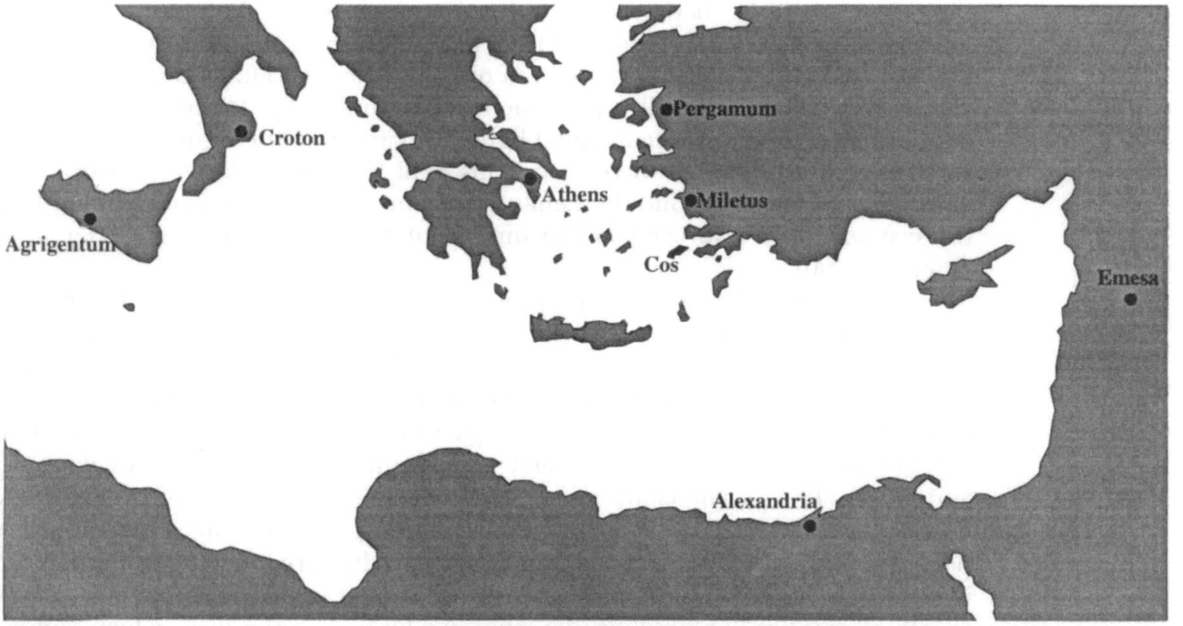

Figure 1. Some of the important centers of classical medicine and biology.

of an isosceles triangle are equal and a circle is bisected by its diameter (Freeman, 1954; Sarton, 1959; Lloyd, 1970).

Thales is most famous for his idea that *water* was the basic and original substance. He thought that the earth is a flat disc floating on water, that there is water all around us, and that the heavenly bodies are water vapor. What is new or scientific about this? After all, the Egyptians, the Babylonians, and all peoples have cosmologies about how things began, and water cosmologies are particularly common. For example, in one Babylonian legend, the creator is Marduk—and "All the lands were sea . . . Marduk bound a rush mat upon the fact of the waters, he made dirt and piled it on the rush mat" (Farrington, 1944). Thales' cosmology was fundamentally different from the Babylonian and other prescientific ones for two reasons. First, he left gods such as Marduk out of his scheme. Second, he sought a common element underlying all phenomena.

3.2. Alcmaeon of Croton, the First Neuroscientist

By the middle of the fifth century BCE there were three major centers of Greek medical science: Croton, in what is now southern Italy, Agrigentum on the south coast of modern Sicily, and Cos, an island off modern Turkey (Fig. 1) (Sigerist, 1961; Sarton, 1959; Longrigg, 1993). The oldest of these medical centers was in Croton, and its most famous member was Alcmaeon. Alcmaeon (ca. 450 BCE) was the first writer to champion the brain as the site of sensation and cognition (Lloyd, 1975; Edelstein, 1943; Longrigg, 1993; Theophrastus, 1917). He also seems to have been the first practitioner of anatomic dissection as a tool of intellectual inquiry. His most detailed dissections and theories were on the senses, particularly vision. Alcmaeon described the optic nerves and noted that they came together "behind the forehead" (which is why, he opined, the eyes move together) and suggested that they were "light-bearing paths" to the brain. He removed and dissected the eye and observed that it contained water. Observations of what are now called phosphenes following a blow to the eye led him to conclude that the eye also contained light (fire) and that this light was necessary for vision. This idea that the eye contains light became the basis of theories of vision that persisted beyond the Renaissance. Indeed, Alcmaeon's idea of light in the eye was only disproved in the middle of the 18th century (Grusser and Hagner, 1990).

Among the other Presocratic philosopher-scientists who adopted and expanded on Alcmaeon's view of the functions of the brain were Democritus, Anaxagoras, and Diogenes (all ca. 425 BCE) (Longrigg, 1993; Theophrastus, 1917; Freeman, 1954; Beare, 1906). Democritus developed a version that became very influential because of its impact on Plato. Specifically, Democritus taught that everything in the universe is made up of atoms of a particular size and shape. The psyche (soul, mind, vital principle) is made up of the lightest, most spherical, and fastest moving atoms. Although the psychic atoms are dispersed among other atoms throughout the body, they are much more numerous in the brain. Slightly cruder atoms are concentrated in the heart, making it the center of emotion, and still cruder ones are located in the liver, which consequently is the seat of lust and appetite. This trichotomy developed into Plato's

(428–347 BCE) hierarchy of the parts of the soul in which there is no question about the supremacy of the brain. As he put in the *Timaeus* (Plato, 1959), "It is the divinest part of us and lords over all the rest." Then, much later, in Galen's medical theorizing, the three soul-parts became the three pneumas of humoral physiology which dominated medical thought for so many centuries (Singer, 1957; Galen, 1968).

Alcmaeon's view of the hegemony of the brain was not universal among the Presocratic philosopher scientists. For example, Empedocles (ca. 445 BCE), the leading member of the medical center at Agrigentum, the second great center of Greek medicine, taught that the blood was the medium of thought, and that the degree of intelligence depended on the composition of the blood (Freeman, 1954; Beare, 1906). Thus, for him, the heart was the central organ of intellect and the seat of mental disorder.

3.3. The Hippocratic Doctors

The third great center for the teaching and practice of medicine in the 5th century BCE was the island of Cos, and its most famous member was Hippocrates. The first large body of Western scientific writings that have survived is the Hippocratic corpus. Although there is no question that Hippocrates (ca. 425 BCE) was a real historical figure, it is not clear which of the writings called Hippocratic were actually written by him (Lloyd, 1978). The Hippocratic corpus consists of over 60 treatises which vary enormously in style and technical level and which were not written by one author, or even in one period. Perhaps this set of medical writings is just the remaining part of the medical library at Cos.

Unlike Alcmaeon and the Croton school, the Hippocratic doctors did not practice dissection, and their knowledge of anatomy was slight. However, like the Presocratic thinkers in general, they rejected supernatural causes of disease and sought natural explanations through observation and extended case studies (Sigerist, 1961; Sarton, 1959; Longrigg, 1993). Similarly detailed accounts of disease processes were rare until after the Renaissance and even then tended to be advertisements for the skill of the physician rather than empirical studies.

The Hippocratic work of greatest relevance to brain function is the famed essay "*On the sacred disease*" (Hippocrates, 1950), epilepsy. The work, probably designed as a lecture for laymen, opens with an homage to reason and the rejection of superstition:

> I do not believe that the "Sacred Disease" is any more divine or sacred than any other disease, but, on the contrary, has specific characteristics and a definite cause . . .
>
> It is my opinion that those who first called this disease "sacred" were the sort of people we now call witch-doctors, faith-healers, quacks, and charlatans. These are exactly the people who pretend to be very pious and to be particularly wise. By invoking a divine element they were able to screen their own failure to give suitable treatment and so called this a 'sacred' malady to conceal their ignorance of its nature.

The author has no doubt that "the brain is the seat of this disease." As to the general functions of the brain, he is equally clear:

> It ought to be generally known that the source of our pleasure, merriment, laughter, and amusement, as of our grief, pain, anxiety, and tears, is none other than the

> brain. It is specially the organ which enables us to think, see, and hear, and to distinguish the ugly and the beautiful, the bad and the good, pleasant and unpleasant . . . It is the brain too which is the seat of madness and delirium, of the fears and frights which assail us, often by night, but sometimes even by day; it is there where lies the cause of insomnia and sleep-walking, of thoughts that will not come, forgotten duties, and eccentricities.

Furthermore, he states that neither the diaphragm nor the heart has any mental functions as some have claimed: "Neither of these organs takes any part in mental operations, which are completely undertaken by the brain."

What then is the cause of epilepsy, the so-called sacred disease? It attacks only the phlegmatic, those with an excess of phlegm or mucus.

> Should . . . [the] . . . routes for the passage of phlegm from the brain be blocked, the discharge enters the blood-vessels . . . this causes aphonia, choking, foaming at the mouth, clenching of the teeth and convulsive movements of the hands; the eyes are fixed, the patient becomes unconscious and, in some cases, passes a stool . . . All these symptoms are produced when cold phlegm is discharged into the blood which is warm, so chilling the blood and obstructing its flow.

These extracts from "On the Sacred Disease" typify Hippocratic medicine: absence of superstition, accurate clinical description, ignorance of anatomy, and a physiology that is largely an absurd mixture of false analogy, speculation, and humoral theory. Perhaps the entire history of medicine can be viewed as the narrowing of the gap between the medical empiricism characteristic of the School of Cos and the knowledge of structure and mechanism sought by the School of Croton.

3.4. The Alexandrians and the Beginning of Neuroanatomy

Neither the Presocratics nor the Hippocratic doctors referred specifically to the convolutions of the cerebral cortex. The first to do so were Praxagoras of Cos (ca. 300 BCE) and his student Philotimos (Steckerl, 1958). Praxagoras' primary claim to fame was that he was among the first to clearly distinguish arteries and veins and to use the pulse as a major diagnostic technique. He referred to the "long flexuosities and winding and folding of the convolutions" of the brain. However, consideration of the functions of these convolutions did not occur until the beginning of the systematic study of human anatomy in the museum at Alexandria.

The museum was founded at the end of the fourth century BCE by Ptolemy I, the first Greek ruler of Egypt. It was a vast state-supported institute for research, perhaps like some combination of the National Institutes of Health and the Institute for Advanced Study or All Souls College. Over 100 professors lived communally and had their salaries and expenses paid. The museum included lecture and study rooms, an astronomical observatory, a zoo, a botanical garden, and dissecting and operating rooms (Farrington, 1949; Fraser, 1972). Its huge library was considered one of the wonders of the ancient world.

It was here that Herophilus (ca. 270 BCE) and then Erasistratus (ca. 260 BCE) began the systematic study of the structure of the human body (Von Staden, 1989; Dobson 1926–27; Longrigg, 1988). The immediate cause of this extraordinary surge in anatomical interest in second century Alexandria was that it was the first time and place where open dissection of the human body could be carried out. Previously, anatomical dissections had been done only on animals. The Greek reverence (and dread) of the dead human body had made its dissection quite impossible. What made Alexandria different? A number of factors seem to have come together (Fraser, 1972; Von Staden, 1989; Longrigg, 1988; Edelstein, 1943). One was that Herophilus and Erasistratus had the full support of a totalitarian regime determined to glorify itself through the achievement of its scientists. As absolute rulers in a foreign land, the Ptolemys brought few inhibitions with them. A second factor must have been that dissection of the human body for the purposes of mummification had been practiced in Egypt for centuries, and thus the general cultural background of Egypt undoubtedly helped make human dissection possible. However, it is very unlikely that the Greek anatomists had any contact with the Egyptian embalmers, as the social gap between the Greeks in Alexandria and the "natives" surrounding them seems to have been enormous (Fraser, 1972). Another factor may have been the changes in philosophical attitudes toward dying and the human corpse that were becoming common by this time: Aristotle had taught that after death the body was no more than a physical frame without feeling or rights. The uniqueness of anatomy at Alexandria is revealed by the fact that not only was human dissection practiced first in Alexandria, but Alexandria was the first and virtually the only place where human vivisection was carried out for scientific or medical purposes (until the Germans and Japanese in World War II).

Both Herophilus and Erasistratus were particularly interested in the brain. They provided the first detailed, accurate descriptions of the human brain including the ventricles (Von Staden, 1989; Dobson, 1926; Longrigg, 1988). They, like Alcmaeon and the Hippocratic doctors before them, had no question about the dominant role of the brain in sensation, thought, and movement.*

Herophilus claimed that the fourth ventricle was the "command center," a view later rejected by Galen who instead stressed the importance of brain tissue itself (Von Staden, 1989). Herophilus compared the cavity in the posterior floor of the fourth ventricle to the cavity in the pens in use in Alexandria at that time, and it is still called the *calamus scriptorius,* for "reed pen," or, sometimes, *calamus Herophili.*

Erasistratus likened the convolutions of the brain to the coils of the small intestine, a comparison that continued into the 19th century, often with Erasistratus's name still attached (Dobson, 1926–27). Indeed in the 19th century the cerebral convolutions were often called the "enteroid processes" and many drawings of the cortex looked more like the small intestine than like the brain (Schiller, 1965; Clarke and Dewhurst, 1966) (see below, Figs. 6 and 8). Erasistratus

*There is the egregious and curious case of one Greek philosopher-scientist, Aristotle (384–322 BCE), indeed the greatest of them all, who denied any hegemonic role for the brain and did so in a detailed set of arguments. As discussed in detail elsewhere (Gross, 1995), Aristotle adduced anatomical, embryological, comparative, experimental, and introspective arguments for why the heart, not the brain, was the center of sensation, mentation, and movement. But at this time the crucial line of evidence for brain function was the clinical, the observation of brain-injured humans. However, clinical medicine was virtually the only field of knowledge in which Aristotle took no interest. By contrast, Alcmaeon and the other Greek natural philosophers who did realize the functions of the brain were practicing physicians.

compared the brain convolutions of a number of animals including hares and stags with those of humans. From these comparisons, he attributed the greater intelligence of humans to their more numerous convolutions. Galen (1968) later ridiculed Erasistratus's correlation between intelligence and the number of convolutions by noting that a donkey had more brain convolutions than humans. However, Galen and possibly Erasistratus were referring to the convolutions of the cerebellum rather than those of the cerebrum. In any case, Galen's sarcasm about any relationship between the convolutions of the brain and intelligence had an extraordinarily pervasive influence and was repeatedly quoted over the next 1500 years, by Vesalius, writing in 1543, among others. It seemed to have inhibited any serious interest in the convolutions until Thomas Willis in the 17th century.

Erasistratus traced both sensory and motor nerves into the brain and was reported to have made experiments on the living brain to determine its functions, but no accounts of these survive. He certainly had a good feeling for the nature of research, as reflected in this quote from his work *On Paralysis*:

> Those who are completely unused to inquiry are in the first exercise of their mind, blinded and dazed and straightway leave off the inquiry from mental fatigue and an incapacity that is no less than that of those who enter races without being used to them. But the man who is used to inquiry tries every possible loophole as he conducts his research and turns in every direction and so far from giving up the inquiry in the space of a day, does not cease his search throughout his life. (Lloyd, 1973)

After Herophilos and Erasistratus there was a rapid decline of Alexandrian medicine into various schools sterilely warring over medical theory (Von Staden, 1989; Longrigg, 1988, 1993). Dissection virtually disappears in the West until the rise of the medieval universities in the 12th century, when it begins again, initially for forensic purposes (Singer, 1957).

3.5. Galen of Pergamon

Galen (A.D. 129–199) was the most important figure in ancient medical science and our best source of information about it. He represents its peak and it is through his eyes that the medieval world saw the human body and that we today see the panorama of classical anatomy, physiology, and medicine (Sarton, 1954; Siegel, 1968, 1970; Lloyd, 1973). His main extant works that deal extensively with the brain are *On the usefulness of the parts of the body* (Galen, 1968), *On antomical procedures* (Galen, 1956, 1962), and *On the doctrines of Plato and Hippocrates* (Galen, 1978–84).

Galen provided an accurate and detailed account of the anatomy of the brain. Indeed, quite how accurate was not appreciated until recently when it was realized that his descriptions fit the brain of the then much more available ox than they do the human brain (Woolam, 1958; Spillane, 1981). He described the ventricles in considerable detail because they were crucial in his physiological system, being responsible for the transformation of vital pneuma into psychic pneuma. The latter was the active principle of both sensory and motor nerves and the central nervous system.

On the basis of his extensive clinical experience at the gladiatorial school in Pergamon, Galen distinguished sensory and motor nerves. He believed the former traveled to the anterior part of the brain and the latter came from the posterior part, a clear anticipation of Muller's Doctrine of Specific Nerve Energy (the idea that the functions of a nerve are determined by its central connections). He viewed sensation as a central process, since from his clinical data as well from his animal experiments he knew that sensation could be impaired by brain injury even when the sense organs were intact.

Although ventricles, particularly the anterior ventricle, were important as a source of psychic pneuma, Galen located the soul not in the ventricles, but in the solid portions of the brain. Among the arguments for this was his demonstration that when brain lesions penetrated to the ventricles, death did not invariably result even if both sensation and movement were lost. Thus, he placed both the soul and higher cognitive function in the solid portions of the brain. In spite of his localization of cognitive function in the substance rather than the spaces of the brain, Galen ridiculed Erasistratus's correlation between intelligence and the number of brain convolutions, with amazingly long lasting effects, as mentioned above.

Galen described the optic chiasm and tract and noted that the tract was "intimately and firmly connected . . . with a part of the brain of a peculiar kind, different in boundaries and circumference from the other parts," presumably the lateral geniculate body. He thought that the optic nerves originated in "the anterior part of the lateral ventricles" and noted that experimental pressure on this region of the anterior ventricle resulted in blindness.

At about the time of Galen's death in 199, Greek science and Greek medicine died. People preferred to believe rather than to discuss, critical faculty gave way to dogma, interest in this world declined in favor of the world to come, and worldly remedies were replaced by prayer and exorcism.

4. The Medieval "Cell" Doctrine of Brain Function

The central feature of the medieval view of the brain was the localization of the mental faculties in the ventricles of the brain. In its basic form, the faculties of the mind (derived from Aristotle) were distributed among the spaces within the brain, (the ventricles described by Galen). The lateral ventricles were collapsed into one space, the first "cell" or small room. It received input from all the sense organs and was the site of the "sensus communis" or common sense which integrated across the modalities. The sensations yielded images, and thus fantasy and imagination were often located here, too. The second or middle cell was the site of cognitive processes: reasoning, judgement, and thought. The third cell or ventricle was the site of memory. See Fig. 2.

Although the basic doctrine remained intact for about 1200 years, there were some minor developments. By the 10th century there was a shift from the original static localization to a more dynamic process analogous to digestion. The sensory inputs were made into images in the first cell and then they were transferred to the second cell whose central location made it warmer, appropri-

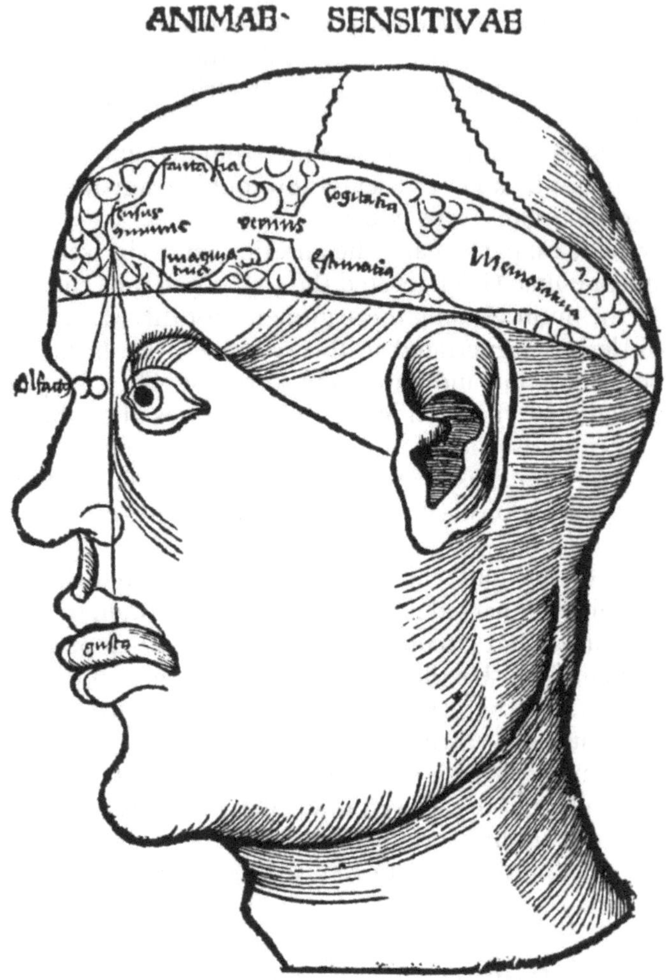

Figure 2. The organs of the "sensitive soul" (*animae sensitivae*) from G. Reisch, 1503, *Margarita Philosophica* (*Pearls of Philosophy*), one of the first modern encyclopedias. This illustration of ventricular doctrine was literally copied by many subsequent authors. Messages from the organs of smell, taste, seeing, and hearing are united in the "common sense" (*sensus communis*) in the first ventricle in which fantasy (*fantasia*) and imagination (*imaginativa*) also reside. The first ventricle communicates with the second by the vermis; thought (*cogitativa*) and judgement (*estimativa*) are located there. Memory (*memoria*) is in the third ventricle. The curlicues around the ventricles may represent cerebral convolutions. As described in the text, Vesalius ridiculed this particular figure.

ate for the further processing (i.e., digestion) into cognition. Leftover thoughts were then transferred to the third cell for storage. These "transfers of information" occurred through the passages between the ventricles that had been described by Galen. There was also a change in the quality of the drawings of the heads in which the ventricles lay: from the crude medieval conceptual representations to the sophisticated pictorial representations of such Renaissance masters as Durer and Leonardo (Clarke and Dewhurst, 1996; Pagel, 1958; Clarke, 1962; Woolam, 1957).

How did the cell doctrine arise and why was it so attractive to the medieval and early renaissance mind? The cell doctrine developed out of a curious amalgam of Greek medical theory and observation and the ideological concerns of the early church fathers. Although Galen had described the ventricles in great detail, he localized the mental faculties in the solid portions of the cerebrum. The 4th century Byzantine Poseidonus further developed this idea (Lewy, 1847). He seems to have been the first to report in detail on the effects of localized brain damage in humans. He said that lesions of the anterior brain substance impaired imagination and lesions of the posterior brain impaired memory, but damage to the middle ventricle produced deficits in reasoning.

The early church fathers were very much concerned with the nonmaterial nature of the soul. Therefore, rather than localize the soul, they localized Aristotle's classification of its functions, namely the functions of the mind such as sensation and memory. Furthermore, they felt that brain tissue was too earthy, too dirty to act as an intermediary between the body and the soul, so they put the mental faculties in the empty space in the brain—the ventricles. Besides the desire for a suitable intermediary between the body and noncorporeal soul, another contribution to the doctrine of three brain cells may have been the parallel with the Trinity. Thus Nemesius, Bishop of Emesia (ca. 390), put all the faculties of the soul into the ventricles following the same anterior–posterior localization as his contemporary Poseidonus, but making the localization of the mental faculties entirely ventricular (Nemesius, 1955).

The three stages of processing postulated for the three cells was influenced or at least rationalized by a comparison with the spatial division of function in classical law courts, as in the following quote from the *Anatomia Nicolai Physici,* a 12th century text derived from an Islamic synthesis of Nemesius and Poseidonus with Greek humoral and pneumatic physiology:

> On the account of the three divisions of the brain the ancient philosophers called it the temple of the spirit, for the ancients had three chambers in their temples: first the vestibulum, then the consistorium, finally the apotheca. In the first the declarations were made in law-cases; in the second the statements were sifted; in the third final sentence was laid down. The ancients said that the same processes occur in the temple of the spirit, that is, the brain. First we gather ideas into the cellular phantisca, in the second cell we think them over, in the third we lay down our thought, that is we commit to memory. (Corner, 1927)

The specific localization in the anterior and posterior cells clearly derives from Galen. As noted above, Galen had put sensory processing in the soft and impressionable anterior regions. Galen had thought the posterior portions were motor in function and therefore hard in order to be able to move muscles. The early church fathers chose this hard region as a good one for the safe storage of valuable brain goods, that is, memories.

"Empirical support" for the cell doctrine was not lacking, as in this quote from Andre du Laurens (ca. 1597), Professor of Medicine and Chancellor, Montpellier University, and physician to Henry IV:

> If we will (saith Aristotle in his Problemes) enter into any serious and deepe conceit we knit the browes and draw them up: if we will call to mind and remember anything, wee hang downe the head, and rub the hinder part, which sheweth very well that the imagination lieth before and the memorie behinde . . . in the diverse pettie chambers in the braine, which the Anatomists call ventricles . . ." (Hunter and Macalpine, 1963)

5. The Rebirth of Brain Science

5.1. Vesalius Restarts Neuroanatomy

Andreas Vesalius of Padua (1514–1564) was the greatest of the Renaissance anatomists, the first to rekindle anatomical science and break the stranglehold of Galen. He is often paired with Copernicus as an initiator of the scientific revolution (Hall, 1954). In his *De Humani Corporis Farica (On the Structure of the Human Body)* published in 1543 the study of nature, particularly human nature, begins in the West again, and, by implication, the dependence on Church-sanctified authority for knowledge is rejected.

Figure 3 shows one of Vesalius's famous and beautiful drawings of a horizontal dissection of the human brain. Vesalius ridiculed the ventricular doctrine of brain function, writing in regard to Reisch's representation of ventricular localization (Fig. 2), "Such are the inventions of those who never look into our

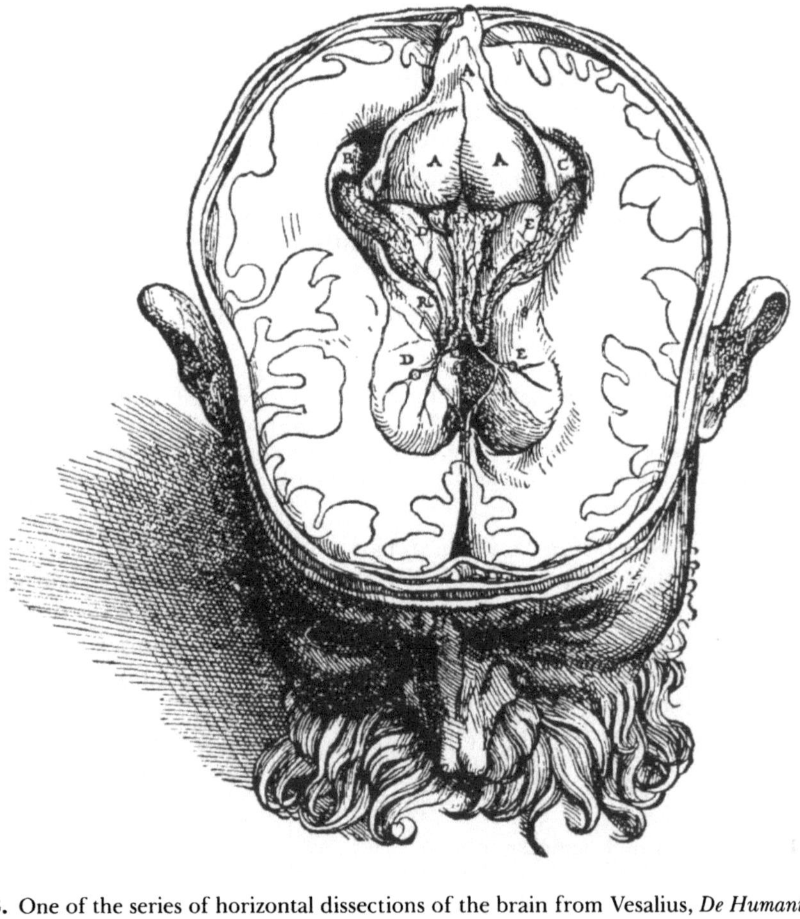

Figure 3. One of the series of horizontal dissections of the brain from Vesalius, *De Humani Corporis Fabrica* (1543). The fornix (A) has been retracted. Note how the various ventricular structures have been drawn and labeled in detail, but the cortex is drawn in the most rudimentary fashion.

maker's ingenuity in the building of the human body" (Singer, 1952). His principal argument against the localization of the functions of the soul in the ventricles was that many animals have ventricles similar to those in humans and yet they are denied a reigning soul. As to the true functions of the ventricles, he commented, "I believe nothing ought to be said of the locations of the faculties of the principal soul in the brain—even though they are so assigned by those who today rejoice in the name of theologians" (Singer, 1952). In spite of this skepticism, note that Vesalius drew and labeled the ventricular structures in much more detail and with much more care than the cortex.

Ventricular localizations continued among both scientific and lay writers (see below, Fig. 7). Perhaps the most recent attribution of important cognitive function to the ventricles by a major scientist was Sir Richard Owen's attempt, in the middle of the 19th century, to find the uniqueness of humans in their supposed unique ventricular structures, particularly the hippocampus minor (Gross, 1993a, b). The most famous lay mention of the ventricles was certainly Shakespeare's in *Love's Labour's Lost*: "A foolish extravagant spirit, full of forms, figures, shapes, objects, ideas, apprehensions, motions, revolutions. They are begat in the ventricle of memory, nourished in the womb of pia mater."

Turning to the convolutions, Vesalius pointed out, you "may learn [about the convolutions of the brain] . . . by observing the brain of some animals at breakfast or at dinner." He agreed that Galen was correct in rejecting Erasistratus's correlating their number with intelligence, and believed their true function was to allow the blood vessels to bring nutriment to the deeper parts of the brain: "nature impressed those sinuous foldings throughout the substance of the brain, so that the thin membrane, folded with numerous vessels, could insert itself into the substance of the brain and so very dexterously administer nourishment."

5.2. Cerebral Cortex: Gland or Rind?

The first clear distinction between the cerebral cortex and white matter was by Archiangelo Piccolomini (1526–1586), Professor of Anatomy in Rome, who succeeded in separating the two in gross dissection. He called the former "cerebrum" and the latter "medulla" and noticed "certain lines" in the cerebrum (Meyer, 1971; Clarke and O'Malley, 1996). The terms "cortex" (or rind), "substantia cineretia" (or brown substance), convolutions, and cerebrum continued to be used interchangeably into the 19th century. Medullary substance also continued into the 19th century as a synonym for white matter. As reflected in the term "rind," most workers attributed little importance to the cortex.

Marcello Malpighi (1628–1694), Professor in Bologna, the founder of microscopic anatomy and discoverer of capillaries, was the first to examine microscopically the cortex. He saw it as made up of little glands with attached ducts (Fig. 4). Similar "globules" were reported by Leeuwenhoek and many subsequent microscopists (Meyer, 1971; Clarke and O'Malley, 1996). Perhaps they were observing pyramidal cells (Nordenskiold, 1928). At least in Malpighi's case artifact is a more likely possibility, since his globules were more prominent in boiled than fresh tissues (Clarke and Bearn, 1968). Malpighi's view of the brain

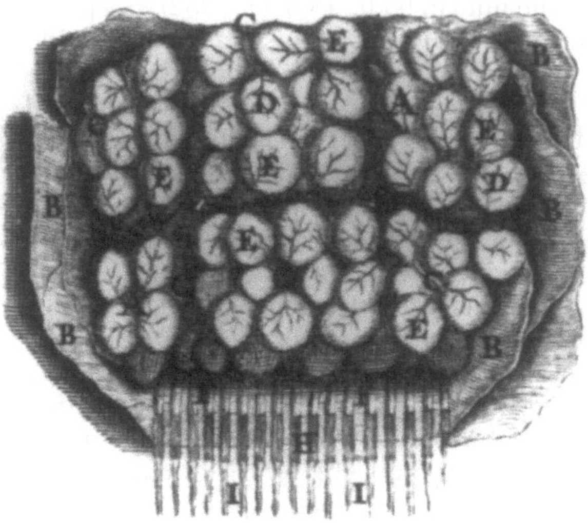

Figure 4. Malpighi's cortical glands from his *De Cerebri Cortice* (1666) with their attached fibers. Although he may have seen brain cells, these drawings are likely to be of artifacts (Clarke and Bearn, 1968). Swedenborg used these supposed cortical elements to build an elaborate theory of brain function that has similarities with the neuron doctrine (Gross, 1996).

as a glandular organ was commonly held in the 17th and 18th centuries, perhaps because it fit with the much earlier, but still persisting views of Aristotle that the brain was a cooling organ and of the Hippocratic doctors that it was the source of phlegm (Meyer, 1971). The other common view was that the cortex is largely made up of blood vessels. One of the earliest advocates of this was Frederik Ruysch (1628–1731), Professor of Anatomy in Amsterdam, who noted: "the cortical substance of the cerebrum is not glandular, as many anatomists have described it, nay have positively asserted, but wholly vascular" (Clarke and O'Malley, 1996). On this view the convolutions were viewed as mechanisms for protecting the delicate blood vessels of the cortex. Representative of this view was Thomas Bartholin, (1616–1680) Professor of Anatomy in Copenhagen and discoverer of the lymphatic system. After yet again rejecting Erasistratus's association of the convolutions with intelligence, Bartholin (Schiller, 1965; Meyer, 1971) indicated that their true purpose was "to make the cerebral vessels safe by guiding them through these tortuosities and so protect them against danger of rupture from violent movements, especially during full moon when the brain swells in the skull."

5.3. Thomas Willis Turns toward Cortex

Prior to Gall and the development of his phrenological system at the beginning of the 19th century there were only a very few isolated figures who advocated significant functions for the cerebral cortex. The first and most important of these was Thomas Willis (1621–1675), who has been called the founder of British neurology and was one of the most important figures in brain science since Galen (Dow, 1940; Dewhurst, 1982; Meyer and Hierons, 1965). Willis was educated at Oxford, quite early gained the Chair of Natural Philosophy there as well as a very lucrative private practice, and was one of the founders of the Royal Society. His *Cerebral Anatomy* (1664) was the first monograph on the brain and dealt with brain physiology, brain chemistry, and clinical neurology as well as brain anatomy. Many of its illustrations were by the great architect Sir Christopher Wren, then Professor of Astronomy at Oxford (Fig. 5).

Willis implicated the "cortical and grey part of the cerebrum" in the functions of memory and will. In his scheme, sensory signals came along the sensory pathways into the corpus striatum where the common sense was located. They were then elaborated into perceptions and imagination in the overlying white matter (then called the corpus callosum or hard body since it was harder than the cortex) and then passed to the cerebral cortex where they were stored as memories. In his words:

> As often as a sensible impression, such as a visual stimulus, arrives from the periphery it turns inwards like an undulation of water and is transferred to the corpora striata where the sensation received from outside becomes a perception of internal sense. If,

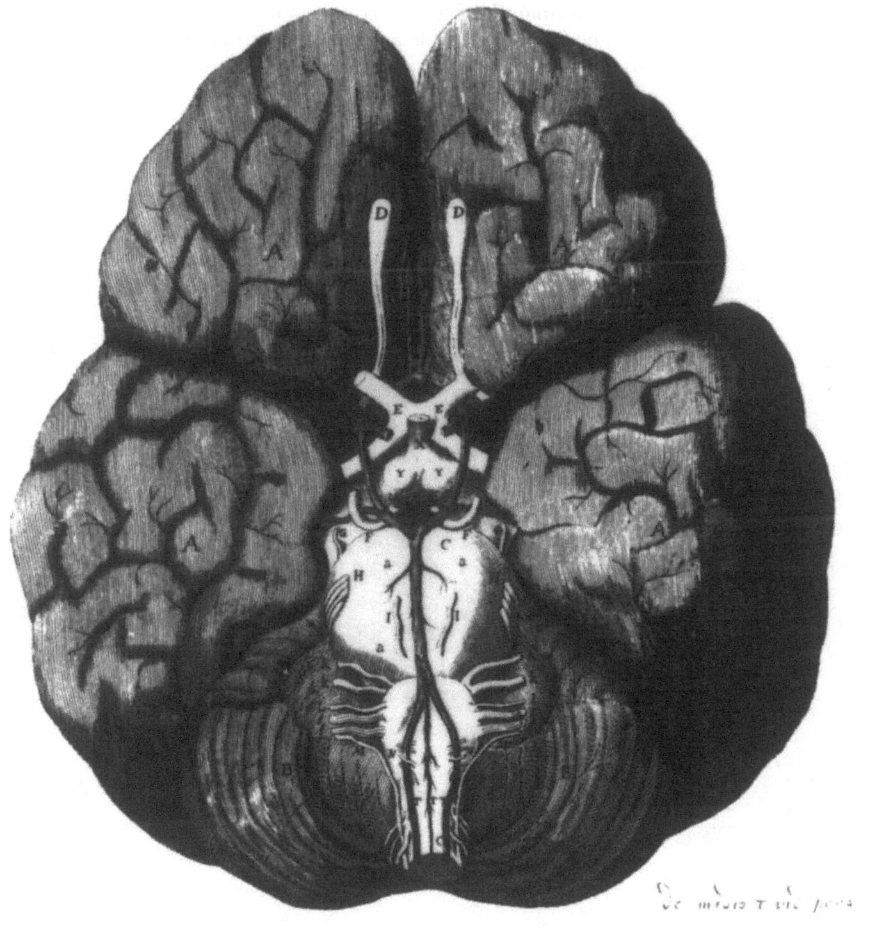

Figure 5. Ventral view of the brain from Willis, *Cerebral Anatomy* (1664), drawn by the great architect Christopher Wren. Note the detailed drawing and labeling of the cranial nerves and basal brain structures (including the circle of Willis) in contrast to the vague and partially obscured representation of the cerebral cortex, all of which has the single designation "A." This schematic and stylized treatment of the cortex was characteristic of all of Willis's illustrations, although he took relatively more interest in the cortex than did most others in the surrounding centuries.

however, this impression is carried further and penetrates the corpus callosum, imagination takes the place of sensation. If after this the same undulation of the spirits strikes against the cortex, as it were the outermost banks, it imprints there a picture or character of the object which, when it is later reflected from there revives the memory of the same thing. (Willis, 1664)

According to Willis, the cortex initiates voluntary movement, whereas the cerebellum is involved only in involuntary movement.

Willis's ideas on brain function came not only from his dissections, but also from his experiments on animals and his correlation of symptoms and pathology in humans. Willis noticed that whereas the cerebellum was similar in a variety of different mammals, the complexity of the cerebral convolutions varied greatly among different animals and this variation was correlated with intellectual capacity:

Hence these folds or convolutions are far more numerous and rarer in man than in any other animal because of the variety and number of acts of the higher Faculties, but they are varied by a disordered and almost haphazard arrangement so that the operations of the animal function might be free, changeable and not limited to one. Those gyri are fewer in quadrupeds, and in such as the cat, they are found to have a particular shape

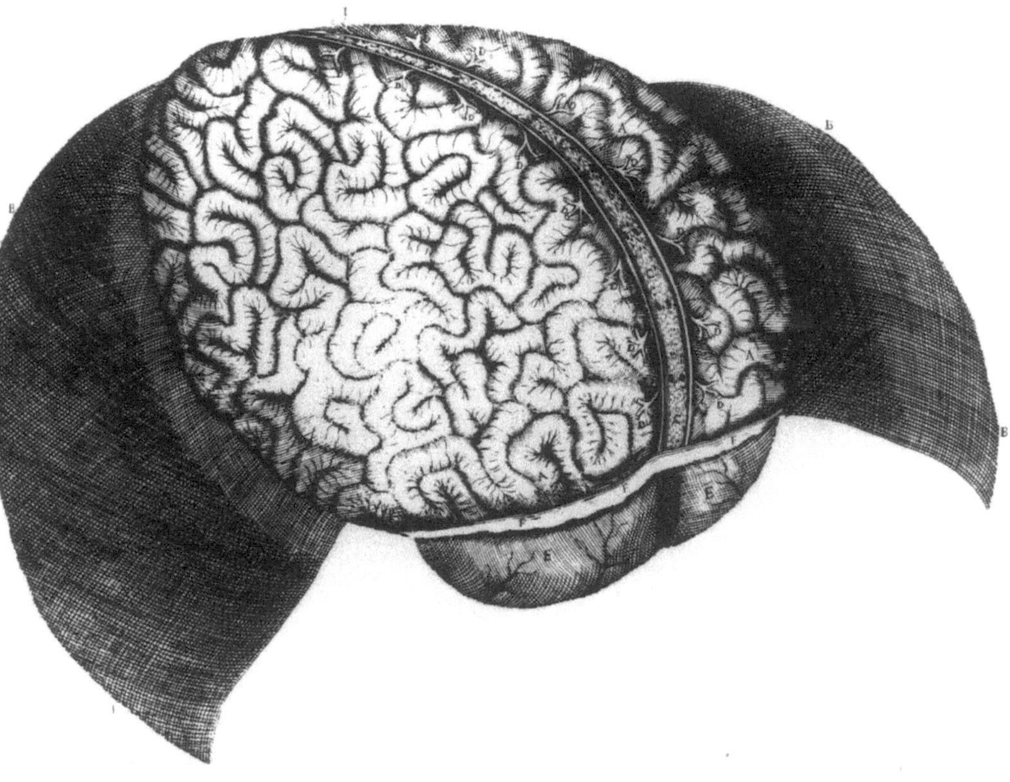

Figure 6. The depiction of the cerebral convolutions by Raymond de Viessens of Montpellier, a leading neuroanatomist of the late 17th century, in his *Neurographia Universalis* (1685). The convolutions are not differentiated in any way and, still following Erasistratus, look like intestines.

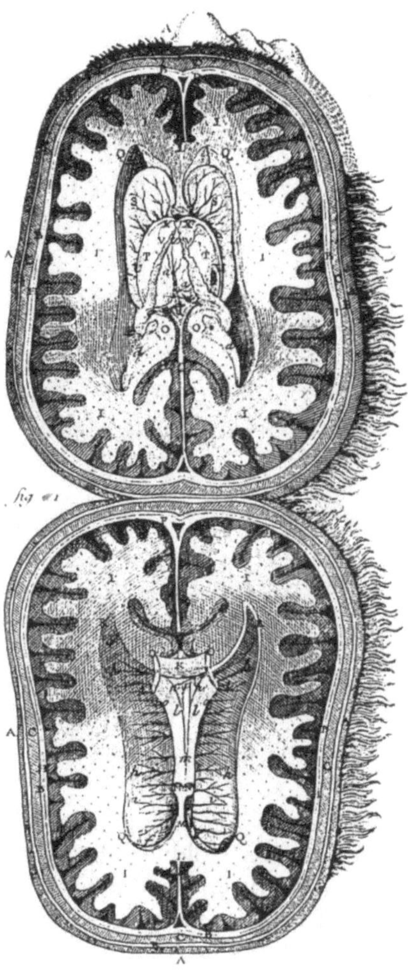

Figure 7. In this drawing from the article on anatomy in Diderot's *Encyclopedia,* note how the ventricular structures are drawn and labeled in detail, whereas the cortical convolutions are represented very schematically and hardly labeled, reminiscent of Vesalius's drawing (Fig. 3) 200 years earlier. Reproduced from M. Diderot and M. D'Alembert, *Encyclopedie ou Dictionairre Raisonné des Sciences, des Arts et des Metiers* (1751), courtesy Wellcome Institute Library, London.

> and arrangement so that this beast considers or recalls scarcely anything except what the instincts and demands of nature suggest. In the smaller quadrupeds, and also in birds and fish, the surface of the brain is flat . . . Hence it is that animals of this sort understand or learn few things. (Willis, 1664)

In spite of the importance of the cerebral cortex in Willis's schema, there is no adequate drawing of the cortex in his work: he apparently never asked Wren or anybody else to produce one. In fact, for another 150 years the cortex continued to be drawn as Erasistratus had first suggested: as coils of the small intestine (Figs. 6 and 8; see also Fig. 7).

Although Willis was a major figure in his time and beyond, his ideas on the importance of the cerebral cortex soon disappeared and the view of the cortex as a glandular, vascular, or protective rind returned to its original dominance. There were two figures who did challenge this view. The first was Francois Pourfour du Petit (1644–1741), a French army surgeon (Neuburger, 1981; Kruger, 1963). He carried out a series of systematic experiments on the effects of

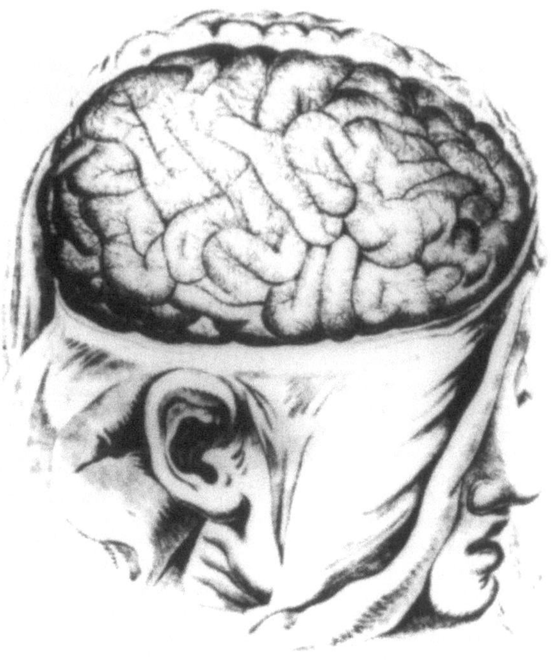

Figure 8. In Vicq d'Azyr's *Traité d'Anatomie* (1786) the convolutions still are drawn looking like intestines, but now some relation to the actual morphology of the brain is discernible.

cortical lesions in dogs and related them to his clinicopathological observations on wounded soldiers. From these studies he realized that the cerebral cortex plays a critical role in normal movement and that this influence is a contralateral one. However, his observations were totally ignored until they were rediscovered much later. One reason was that he did not hold an academic post and published his account in a very limited edition. Yet his observation that the cortex was insensitive to touch was repeatedly cited in support of the views of von Haller, who, as discussed below, was the dominant physiologist of the day. Thus, du Petit's work demonstrating motor functions of cortex was probably ignored largely because of the anti-cortex ideology of the time, not because it was published in a minor journal.

5.4. Emanuel Swedenborg: Unknown Neurogenius

The second major figure advocating the importance of the cortex between Willis and Gall was Emanuel Swedenborg (1688–1772), the founder and mystical prophet of the "New Jerusalem" or Swedenborgian Church (which is still active in the United States and Great Britain). Before having the spiritual epiphany that led to the founding of the Church, Swedenborg made significant contributions to astronomy, geology, metallurgy, and physics. His interest in the nature of the soul led him to the anatomy and physiology of the brain. Although he did some empirical work on the brain, most of his ideas came from reviewing the literature of the day (Gross, 1996).

Malpighi, the first microscopist, saw when he examined the cerebral cortex (Fig. 4). In an extraordinary anticipation of neuron theory, Swedenborg argued that these globules or cerebellula ("little brains") were functionally independent units which were connected to each other and to the periphery by nerve fibers. Their operations, he argued, were the basis of sensation, mentation, and movement. He localized all higher function in these units of the cerebral cortex:

> . . . the brain is the sensorium commune, with respect to its cortical substance . . . since to it are referred the impressions of the external sense organs as if to their one and only internal centre. The cortical substance is also the motorium commune volutarium, for whatever actions are mediated by the nerves and muscles are determined beforehand by the will, i.e., initiated by the cortex. (Swedenborg, 1740–41)

Furthermore, he localized the cerebella with different functions in different cortical regions: "The cerebella are . . . wisely arranged in order that they might correspond exactly to the various external sensations. For these cerebellula are compacted into units . . . In this way, every type of sensation—vision, hearing, taste, olfaction—can be received distinctly . . ." (Swedenborg, 1740–41).

Remarkably, Swedenborg had the idea of the somatotopic organization of motor function in the cerebral cortex. He correctly localized control of the foot in the dorsal cortex (he called it the "highest lobe"), the trunk in an intermediate site, and the face and head in the ventral cortex (his third lobe): "the muscle and actions which are in the ultimates of the body or in the soles of the feet depend more immediately upon the highest parts; upon the middle lobe the muscles which belong to the abdomen and thorax, and upon the third lobe those which belong to the face and head; for they seem to correspond to one another in an inverse ratio (Swedenborg, 1882). There is no other suggestion of the somatotopic organization of motor cortex until the experiments of Fritsch and Hitzig in 1870.

Overall, Swedenborg was not a particularly obscure figure. His writings on religion and spiritualism had an enormous impact on European and American writers and artists. Blake, Yeats, Balzac, the Brownings, Baudelaire, and Strindberg, for example, all claimed to be particularly influenced by him (Gross, 1996). In 19th century America his influence was strong among those interested in spiritualism and in transcendentalism. Although substantial portions of his writings on the brain were published in Latin in the middle of the 18th century and in English at the end of the 19th century and again in the early 20th century, they went unnoticed by mainstream neuroscience (Swedenborg, 1740–41, 1882–87, 1938; Gross, 1996). Thus, his extraordinary anticipations of modern neuroscience had no discernible impact. Swedenborg never taught, and as a neuroscientist, he never had students or followers. His prescient publications on brain function were finally brought to the attention of the scientific world at the beginning of this century by the historian of medicine Max Neuberger (1901). By that time they were only of "historical" interest.

5.5. Von Haller and the Insensitivity of Cortex

The space we have just given to Willis, du Petit, and Swedenborg, figures who thought the cortex was a crucial brain structure, is somewhat anachronistic since the opposite view was very much the dominant one throughout the 2000 years between Erasistratus and Gall. Much more representative and influential was Albrecht von Haller (1708–1777), Professor at Tubingen and later Bern, who dominated physiology in the middle of the 18th century and was also famous as a botanist, poet, novelist, and politician (Neuberger, 1981). In his monumental *Elementa Physiologiae Corporis Humani* (1757–1765, in eight volumes) and his *Icones Anatomicae* (1743–1756) he divided the organs of the body as well as parts of the nervous system into those "irritable" (such as muscle) and those "sensible" (such as the sense organs and nerves). He tested sensibility with mechanical stimuli, e.g., picking with a scalpel, puncturing with a needle, and pinching with forceps, and with chemical stimuli, e.g., silver nitrate, sulfuric acid, and alcohol. With these methods he found the cortex completely insensitive. By contrast, he reported the white matter and subcortical structures such as the thalamus and medulla to be highly sensitive; their stimulation, he reported, produced expressions of pain, attempts to escape, or convulsions. The following is a typical confirmation of his views by one of his students, J. G. Zinn (1728–1795), of "Zonule of Zinn" fame, Professor of Medicine at Gottingen:

> Having cut out a small circular piece of the cranium of a dog with a trephine . . . I pierced the exposed dura mater, touched it with a blade of a scalpel, and poured a solution of mercury sublimate on it; the animal, however, gave no signs of pain and suffered no convulsions. Since I thought the dog ought to have become apoplectic, I irritated the reflected skin and he showed that he felt pain by giving out a loud cry . . . Having incised the dura mater, I cut the cortex into pieces, pierced it, irritated it, but the animal showed no sign of pain. [By contrast, when he thrust a trocar through the skull, corpus callosum and corpus striatum to the base of the brain in another dog (as confirmed at autopsy) the dog howled pitifully . . . vomited repeatedly . . . and died the next day]. (Neuberger, 1981)

From observations such as these Haller concluded that all parts of the cortex were equivalent (because stimulation had the same negative effect) and that all subcortical regions were also equivalent because their stimulation had similar positive effects. Because of his prestige and many students and followers, Haller's views of the insensitivity and equipotentiality of cortex suppressed the observations and ideas on the importance of cortex of Willis, Swedenborg, and du Petit and persisted well into the next century (Clarke and Jacyna, 1987; Neuberger, 1981). As to the cortex itself, von Haller was of the cortex-as-blood-vessels school: "the greater part of it consists of mere vessels . . . as to glandules making the fabric . . . that notion has been discarded; nor has there been any opinion received with less probability than this" (Neuberger, 1981).

5.6. Gennari and His Stripe

A few years after Swedenborg died, there was an event particularly central to this volume, namely the discovery of the stripe of Gennari, which we now know marks in primates the location of striate cortex, the primary visual cortex. Francesco Gennari (1752–1797), then a medical student, in the course of examining frozen sections of an unstained human brain, observed and reported on a white line in the cortex that was especially prominent and sometimes double in the posterior part of the brain (See Fig. 9). This was the first evidence that the

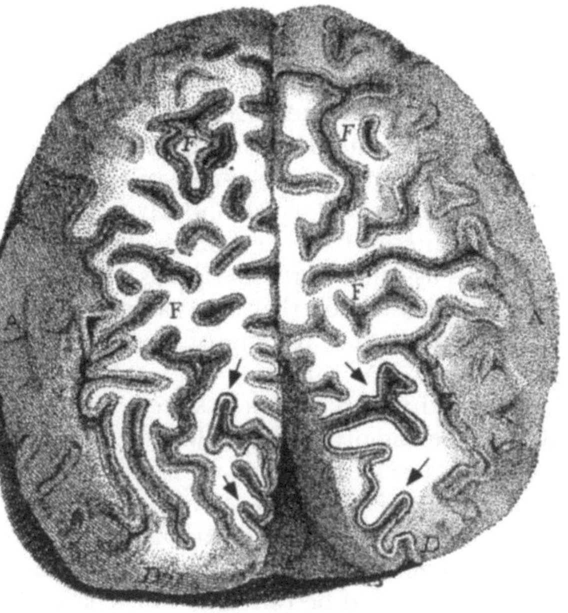

Figure 9. In his *De peculiari structura cerebri nonnullisque eius morbis* (1782), Gennari was the first to describe regional variations in the structure of the cerebral cortex. Specifically, he noticed a white line in the cortex which is more prominent in the medial and posterior portions of a frozen human brain (arrows) and which is now known as the stripe of Gennari.

cerebral cortex was not uniform in structure. The more famous Vicq d'Azyr* (1748–1794) rediscovered the stripe and for a while it was known as "the stripe of Vicq d'Azyr" until priority was sorted out and the name reverted to Gennari (Clarke and O'Malley, 1996; Glickstein and Rizzolatti, 1984). As to its function, Gennari commented, "Just as the use of so many other things is as yet concealed from us, so I do not know the purpose for which this substance was created" (Clarke and O'Malley, 1996).

Gennari never published again, and died young, penniless, and a compulsive gambler (Glickstein and Rizzolatti, 1984).

6. The Beginning of the Modern Era of Cortical Localization

6.1. Gall and Phrenology

The localization of different psychological functions in different regions of the cerebral cortex begins with Franz Joseph Gall (1758–1828) and his collabora-

*Although he never held an academic post, Vicq d'Azyr was renowned as an anatomist (the mammillothalmic tract bears his name) and as a physician (Marie Antoinette being one of his more famous patients).

tor J. C. Spurzheim (1776–1832), the founders of phrenology.* Prior to the development of his phrenological system, Gall (with Spurzheim) had made a number of major neuroanatomical discoveries that would have fixed him in the history of neuroscience, even if he had never begun his phrenological project of correlating the morphology of the cranium with psychological faculties. Among his significant anatomical findings were the recognition that the grey matter was functioning neural tissue connected to the underlying white matter (to which he attributed a conductive function), the first description of postembryonic myelinization, proof of the decussation of the pyramids, the first clear description of the commissures, the demonstration that the cranial nerves originate below the cerebrum, and the realization that folding of the brain was to conserve space (Temkin, 1953).

The central idea of the phrenological system was that the brain was an elaborately wired machine for producing behavior, thought, and emotion and the cerebral cortex was a set of organs with different functions (Gall and Spurzheim, 1835). They postulated about "35 affective and intellectual faculties" and assumed (a) these were localized in specific organs of the cerebral cortex, (b) the development or prominence of these faculties was a function of their activity and the amount of activity would be reflected in the size of the cortical organ, and (c) the size of each cortical organ was indicated by the prominence of the overlying skull, i.e., in cranial bumps. The primary method of data collection by Gall and Spurzheim was examining the skulls of a great variety of people, from lunatics and criminals to the eminent and accomplished. Neuropsychological and animal experimental data, even gathered by themselves, were viewed as only minor and ancillary evidence (Gross, 1987; Young, 1970; Temkin, 1953).

Phrenology had wide popular appeal particularly in England and the United States and among many leading intellectuals, such as Honoré de Balzac, A. R. Wallace, Horace Mann, and George Eliot (Young, 1970; Cooter, 1985; Clarke and Jacyna, 1987). However, it met considerable opposition from the religious, political, and scientific establishments of the day. For example, Gall's public lectures were banned in Austria because they led to materialism and opposed religion and morality. His works were placed on the Index of the Catholic Church for similar reasons. In 1908 the French Institute, later the Academy of Science, under the leadership of the great Cuvier, totally rejected even the anatomical parts of a paper he submitted for admission.

6.2. Flourens Attacks Gall but the Cortex (Re)emerges as a "Higher" Structure

In the scientific world the most important and influential critique of Gall came form Pierre Flourens (1794–1867), later Professor of Natural History at the Sorbonne (Young, 1970; Clarke and Jacyna, 1987; Clarke and O'Malley, 1996). A technically brilliant experimenter, Flourens quickly rose in the French

*Gall preferred the term "organology" over Spurzheim's "phrenology." For differences between the more cautious Gall and the more popularizing Spurzheim see Clarke and Jacyna (1987) and Zola-Morgan (1995).

scientific establishment and at the age of 35 was elected to the Academy that had rejected Gall. Starting in the 1820s and continuing for over 20 years Flourens carried out a series of experiments on the behavioral effects of brain lesions, particularly with pigeons. Flourens reported that lesions of the cerebrum had devastating effects on willing, judging, remembering, and perceiving. However, he found that the site of the lesion was irrelevant: all regions of the cerebrum contributed to these functions. The only exception was vision, in that a unilateral lesion produced only contralateral blindness, but again there was no localization within the hemisphere. These holistic results tended to eclipse Gall's ideas of punctate localization, but only in scientific circles and only temporarily.

Flourens's finding of cognitive losses after cerebral lesions was actually a confirmation of Gall's emphasis on the cognitive role of the cortex, a viewpoint that had been virtually absent before Gall. This change in the attitude toward the cortex was reflected in the midcentury textbooks that attributed intellectual function to the cortex. William Carpenter, in his authoritative *Principles of Human Physiology,* noted that the convolutions of the cerebrum were: "the centre of intellectual action . . . the site of ideas . . . restricted to intellectual operations . . . the sole instrument of intelligence . . . it is probably by them alone that ideas . . . of surrounding objects are acquired . . . and that these ideas are made the groundwork of mental operations . . . that would also seem to be the exclusive seat of Memory . . . and Will" (Carpenter, 1845).

The cortex was termed a "superadded" structure lying hierarchically and physically above the highest sensory structure, the thalamus and the highest motor structure, the corpus striatum. The general idea that the thalamus had major sensory functions and the corpus striatum major motor functions was generally accepted by the middle of the 19th century on the basis of a number of studies involving tracing sensory and motor tracts from the periphery and making experimental lesions in animals (Meyer, 1971).

This view of the "higher" functions of the cortex, common for the period, combines Haller's view of the insensitivity of cortex and both Gall's and Flourens's attribution of "higher faculties," but neither sensory nor motor functions, to the cortex.

6.3. Broca "Confirms" Gall

In spite of the bitter attacks of Flourens, Gall's idea of punctate localization and even many of his specific localizations such as language in the frontal lobes and sexuality in the cerebellum continued to be actively debated in the middle of the 19th century (Young, 1970; Clarke and Jacyna, 1987). At least in the scientific community, the supposed correlations between skull and brain morphology were quickly recognized as erroneous. Yet Gall's ideas stimulated the search for correlations between the site of brain injury and specific psychological deficits in patients as well as in experimental animals. Reports of such correlations were published in both the phrenological and mainstream neurological literature and the question of the localization of psychological function in the brain was hotly debated at scientific meetings.

Thus, in 1848, J. B. Bouillard (1796–1881), Professor at La Charite, Paris, and a powerful figure in the medical establishment,* offered a cash prize for a patient with major frontal lobe damage who did *not* have a language deficit. The debate about localization reached a climax at a series of meetings of the Paris Société d'Anthropologie in 1861. At the April meeting, Paul Broca (1824–1880), Professor of Pathology at the Sorbonne and founder of the society, announced that he had a critical case on this issue. A patient with longstanding language difficulties, nicknamed "Tan" because that was all that he could say, had just died. The next day Broca displayed his brain at the meeting and indeed it had widespread damage in the left frontal lobe. He presented several similar cases in the next few months. Not only did Broca's cases finally establish the principle of discrete localization of psychological function in the brain, but the discovery itself was hailed as a vindication of Gall. Broca himself had regarded Gall's work as "the starting point for every discovery in cerebral physiology in our century" (Young, 1970; Clarke and Jacyna, 1987).

6.4. Evolution and Brain Function

Another factor in the growing interest in the cerebral cortex was the ideas about organic evolution that were in the air in the decades before the publication of Darwin's *Origin of Species* (1861). In J.-B. Lamarck's (1809) theory of evolution, the first coherent one, evolution involved continuous upward progress, an inevitable transformation of lower into upper forms. The anonymous and widely influential best seller *Vestiges of the Natural History of Creation* (1844) took a similar "progressive" view of evolution (Gross, 1993a).

6.4.1. Herbert Spencer

Spencer (1820–1903) was the first and most important figure to apply evolutionary ideas to the nervous system and psychology (Young, 1970; Boakes, 1984). Although precocious, Spencer had virtually no formal education, but read widely in the sciences as a boy. A seminal experience was hearing a lecture on phrenology by Spurzheim at the age of 11 and it was decades before he decisively parted from phrenology. Before he did so he published in phrenological journals and invented a new, more accurate device for measuring skull bumps. After a few years as a railway engineer he drifted into political journalism, where he came into contact with T. H. Huxley, George Henry Lewis (and, to use a modern but particularly apt expression, his partner, George Eliot), Thomas Carlyle, and the scientific and political issues of the day.

In his first book, *Social Statics* (1851), Spencer set out a quasi-Lamarckian "progressive" theory of evolution. He argued that it justified "survival of the fittest" in human society (a phrase which Darwin later adopted). This led him to oppose such things as government help for the poor, public health, and public education. These views were the theoretical bases of the ultraindividualist and conservative ideology that later became known as "Social Darwinism," although

*Bouillard was known as the "red dean" for his participation in the Revolution of 1848. He reportedly prided himself on being the model for Balzac's Dr. Horace Bianchon (Schiller, 1979).

"Spencerism" would have been a more appropriate designation (Hofstadter, 1955). Spencer's social views were particularly welcome among the elites in post-bellum America. As John D. Rockefeller put it in a Sunday School address: "The growth of a large business is merely survival of the fittest . . . The American beauty rose can be produced in the splendor and fragrance which bring cheer to its beholder only by sacrificing the early buds which grow up around it. This is not an evil tendency in business. It is merely the working-out of a law of nature and a law of God" (Hofstadter, 1955).

In his next work, *Principles of Psychology* (1855), Spencer combined association psychology with evolutionary theory to produce a new "evolutionary associationism." From evolution he took the idea of a progressive increase in the complexity of the nervous system both phylogenetically and ontogenetically. This led to the conception of the cortex as the newest, highest, and most important level of the nervous system. Furthermore, Spencer argued that there must be localization of function in the cortex just as there clearly is in lower nervous structures:

> But no physiologist who calmly considers the question . . . can long resist the conviction that different parts of the cerebrum subserve different kinds of mental action. Localization of function is the law of all organization whatever: separateness of duty is universally accompanied with separateness of structure: and it would be marvellous were an exception to exist in the cerebral hemispheres. Let it be granted that the cerebral hemispheres are the seat of higher psychical activities; let it be granted that among these higher psychical activities there are distinctions of kind . . . more or less distinct kinds of psychical activity must be carried out in more or less distinct parts of the cerebral hemispheres . . . It is proved experimentally, that every bundle of nerve fibers and every ganglion, has a special duty; and that each part of every bundle of nerve fibers and every such ganglion, has a duty still more special. Can it be, then, that in the great hemispherical ganglia alone, this specialization of duty does not hold? (Spencer, 1855)

When the *Origin of Species* was published in 1859, Spencer became a enthusiastic follower of Darwin. He set out to unify all knowledge along the principles of Darwinian evolution and produced his massive, multivolumed *Principles of Synthetic Philosophy* which attempted to do so. Today, his synthetic philosophy is all but forgotten, whereas the disastrous consequences of his social views are still reverberating. However, Spencer did make one permanent and major contribution to the development of modern neuroscience. That was the profound influence of his views of the evolution of the nervous system on John Hughlings Jackson.

6.4.2. John Hughlings Jackson

John Hughlings Jackson (1835–1911) is the perennial holder of the title, "Father of English Neurology." As a medical student in Yorkshire, Jackson was so enthralled with Spencer's writings that he almost abandoned medicine to pursue their study full time. Instead he spent 45 years as a clinical neurologist at the National Hospital, Queen Square, London, applying Spencer's ideas on the "evolution and dissolution" of the nervous system. Indeed many of his over 300 papers began with sentiments like, "I should say that a very great part of this

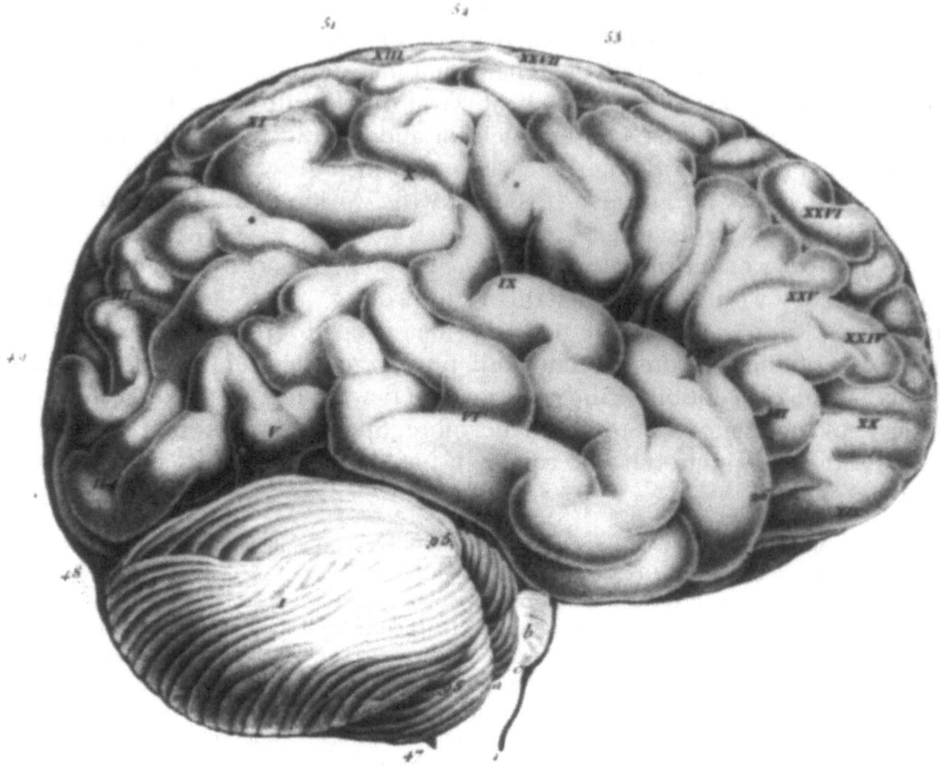

Figure 10. This drawing of the cortical convolutions in Gall and Spurzheim's *Anatomie et Physiologie du Systeme Nerveux* (1810) is quite accurate, one of the first to be so. Perhaps it was necessary to consider the cortex a very important structure in order to portray it accurately. The numbers refer to different phrenological organs.

paper is nothing more than the application of certain of Herbert Spencer's principles."

Spencer had taught that evolution implied a continuity of nervous organization from spinal cord to cerebral cortex and therefore, as Jackson put it, "If the doctrine of evolution be true, all nervous centers [including the cortex] must be of sensori-motor constitution," that is, they must have both sensory and motor functions (Jackson, 1958; Young, 1970). It was the combination of Spencer's notion of cortex as a sensorimotor structure, Spencer's insistence on cerebral localization of function, and Jackson's many observations of seizures (including his wife's) that led him to the brilliant clinical inference that the seizures we now call "Jacksonian" reflect a somototopically organized cortical motor mechanism.

Jackson's ideas on the motor mechanisms of the cerebral cortex were dramatically confirmed in 1870 by Fritsch and Hitzig's demonstration of specific movements from electrical stimulation of the cortex of the dog. These authors were not reticent about the more general implications of their results, as shown by the final lines of their paper: "It further appears, from the sum of all our experiments . . . certainly some psychological functions and perhaps all of them . . . need certain circumscribed centers of the cortex" (Fritsch and Hitzig, 1870).

In summary, in spite of their temporary eclipse under the shadow of Flourens's experiments, Gall's general ideas of punctate localization in the cortex were essentially vindicated by the third quarter of the 19th century. By that time, they were viewed as confirmed by Broca's demonstration in 1861 (see Broca, 1960) of an association between damage to the frontal lobe and aphasia, were spread in a more palatable fashion by Spencer and Jackson, and were seen as confirmed again in 1870 by Fritsch and Hitzig's experiments on stimulation of motor cortex. Gall's ideas on the localization of mental function had a deep and lasting influence through stressing (a) that the human mind could be subdivided into specific functions, (b) that these specific functions were mediated by discrete brain structures, and (c) that the cerebral cortex was crucially important in mental activity. It is interesting to note that one of the first accurate drawings of the cerebral cortex was by Gall and Spurzheim (Fig. 10). Before them the cortex was often portrayed as a pile of intestines (Figs. 6 and 8) or in a crude, schematic way with no attention to detail (Fig. 7). It was as if it was necessary to believe a structure had important functions before going to the trouble to portray it accurately.

6.5. The Search for Sensory Areas in the Cerebral Cortex

The last quarter of the 19th century saw an intense search for the localization of "sensory centers" in the cortex. In addition to the increasing interest in the cortex from the work of Gall, Flourens, Spencer, Jackson, and Fritsch and Hitzig, a major spur to the search for sensory centers was Johannes Müller's (1838) Doctrine of Specific Nerve Energies. Müller (1801–1858), Professor of Anatomy and Physiology at Berlin, dominated mid-19th century physiology through his personality, his many influential students, and his massive *Handbuch der Physiologie* (1833–40). Müller's doctrine had three essential elements. The first and most fundamental asserted that sensation was the awareness of the states of sensory nerves, not of the outer world itself. The second element was that when a given nerve type or nerve "energy" was excited, the same type of experience is produced no matter what the stimulus. Thus photic, mechanical, and electrical stimulation of the eye all produce visual sensations. Müller, following Aristotle, assumed that there were five nerve types or "nerve energies." We would call them qualities or modalities. The third element of the doctrine was that the same physical stimulus applied to different sense organs gives rise to different sensations. Thus a blow to the eye and one to the ear produce visual and auditory sensations, respectively. Müller was unsure of the locus of the nerve "specificity." As he put it, "It is not known whether the essential cause of the peculiar 'energy' of each nerve of sense is seated in the nerve itself or in the parts of the brain and spinal cord with which it is connected."

Müller's student, however, the great Hermann von Helmholtz (1863), philosopher, physicist, and psychologist, located the specificity squarely in the nerve terminations.

The idea that the specificity of nerves derived from their central connections was not new. On the basis of his clinical practice among gladiators in Pergamon, Galen had distinguished between sensory and motor nerves and argued that sensory nerves were connected to the anterior part of the brain and motor nerves to the posterior. Charles Bell (1774–1842), the codiscoverer of the Law of Spinal Roots (or, rather, the sensory half of it*), extended the idea of specificity inherent in that law to the five senses to yield in 1811 an account of nerve specificity essentially identical to Müller's later published one. As Bell put it, "the nerves of sense depend for their attributes on the organs of the brain to which they are severally attached . . . the properties of the nerves are derived from their connections with the parts of the brain."

Bell's (1811) account was in a "tiny little" pamphlet of 36 pages, each with 4.5 × 2.5 inches of text per page, entitled *Idea of a New Anatomy of the Brain,* 100 copies of which were privately printed for his friends. It did not seem to be noticed in Europe although it was published *in toto* in an American medical journal at the time (Bell, 1812).

It is important to note that for both Bell and Müller it was not the terminations in the cerebral cortex that conveyed specificity on the sensory nerves. Rather, for both of them, and, as noted above, more generally for almost all the physiologists and anatomists of the first half of the 19th century, the cortex still had no sensory (or motor) functions. The main argument for this view was still von Haller's that since the cortex was "insensitive" to touch, it could hardly be sensory. Rather, the cortex was believed to be the site of the highest intellectual functions. This idea was often supported both by the phylogenetic correlation of cortical complexity with intelligence and reports of intellectual deficits after cortical lesions. It was clearly also heavily influenced by Gall's ideas. Note that of all the 35 faculties that Gall had put into the cerebral cortex, none were sensory or motor. Some of Gall's faculties do have "sensory sounding" names, but on examination are actually cognitive. For example, as to the faculty of Color, Gall notes "I do not mean the simple faculty of seeing or perceiving colors . . .[but rather] distinguishing the relations of colors: the talent for painting" (Gall and Spurzheim, 1835).

What turned Müller's doctrine and everybody else's attention toward the possible sensory functions of the cerebral cortex was Fritsch and Hitzig's (1870) discovery of motor cortex by electrical stimulation. This unambiguously demonstrated that the cerebral cortex had more than just "higher functions."

Müller's Doctrine of Specific Nerve Energies now became directed toward cortex as the locus of "specific energies." Thus, under its influence, in the later part of the 19th century, (a) neural pathways were traced from the sense organs into the brain to find the specific regions in which they ended, (b) the cortex was divided up into separate centers or organs on the basis of the pattern of its structure, thereby yielding the techniques of cytoarchitectonics and myeloarchitectonics, (c) cortical lesions were made in animals to find the sensory

*There was a long and bitter priority controversy between Sir Charles Bell and the great French physiologist Francois Magendie (1783–1855) over the discovery of the Law of Spinal Roots, i.e., that the ventral spinal roots are motor and the dorsal ones sensory. In fact, (a) Bell proposed only the sensory functions of the dorsal roots, (b) there is no reason to believe that Magendie knew of Bell's claims before he carried out and published his own experiments, and (c) both halves of the law were experimentally demonstrated by Magendie, whereas Bell's consideration of the functions of the dorsal roots were largely anatomically based; see Cranefield (1974).

"centers," and (d) in close parallel, attempts were made to correlate sensory losses in humans with the site of cortical damage.

7. The Discovery of a Visual Center in the Cerebral Cortex

7.1. Bartolomeo Panizza: The First Claim

The first discrete localization of visual function in the cortex on the basis of systematic investigations seems to have been that of Bartolomeo Panizza (1785–1867) Professor of Anatomy at Pavia, and a follower of Gall (Panizza, 1855; Mazzarello and Della Sala, 1993). After examining the brains of several patients who were blind after strokes, he attributed vision to the posterior cortex. He then tested this idea by lesions and enucleations in a number of species and concluded that the occipital region was the crucial one for vision. He also studied the anatomical and behavioral effects of monocular enucleation as a function of age and concluded that the effects on the brain were more profound in adults than infants. Panizza's work seems to have been totally ignored at the time. One reason for this may have been because he only published in local journals, those of the Royal Institute of Lombardy of Science, Arts and Letters; however, these journals were exchanged with those of the Royal Society and presumably other scientific societies.

A more likely reason for the lack of impact of Panizza's work was the prevailing theoretical view of the relative role of cortex and subcortex. As we have indicated, at that time it was thought that the thalamus was the highest sensory center and the basal ganglia the highest motor center. By contrast, the cortex was believed to be concerned not with sensation or movement, but with intellectual operations (Carpenter, 1845; Walshe, 1958). The importance of Panizza's work was only realized after there was good evidence for a cortical visual area from the work of Ferrier, Munk, and Schäfer as described in the next section (Tamburini, 1880; Manni and Petrosini, 1994).

7.2. The Battle for Visual Cortex: Ferrier versus Munk and Schäfer

Immediately after Fritsch and Hitzig's publication, the English physiologist David Ferrier (1843–1928), working at the West Riding Lunatic Asylum and at Kings College, London, confirmed their work, first in dogs and then monkeys (Ferrier, 1873, 1875a). He then applied their electrical stimulation methods to search for the sensory cortices. He found that stimulation of the "angular gyrus" (area 7 in posterior parietal cortex) in the monkey produced conjugate eye movements and interpreted this as indicating that this area was the seat of the perception of visual impressions; by contrast, he found that stimulation of the occipital lobe or other regions did not have these effects (Ferrier, 1873, 1875a). He further tested this idea by making angular gyrus lesions (Fig. 11) and reported that unilateral lesions produced temporary blindness in the contralateral eye

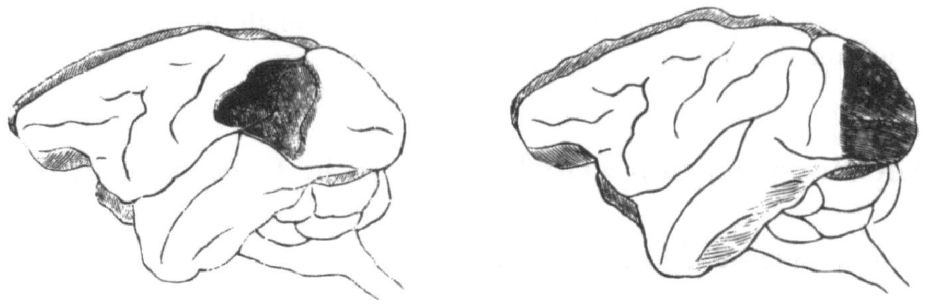

Figure 11. Right: Ferrier (1886) found these occipital lesions not to cause any visual deficits. Note that the lesion spares what we now know to be the representation of the fovea in striate cortex. Left: The lesion of the "angular gyrus" that Ferrier first claimed produced blindness (Ferrier, 1876), and later, only temporary blindness (Ferrier, 1886). Both were bilateral lesions made in macaques.

and bilateral lesions produced permanent blindness in both eyes (Ferrier, 1875b, 1876). However, the animals were only observed for a few days before he sacrificed them, the operations having been done without antiseptic techniques. Summarizing the results on four animals with angular gyrus lesions, he wrote:

> The loss of visual perception is the only result of this lesion, the other senses and the powers of voluntary movement being retained so long as the lesion remains confined to the angular gyrus itself. By the term visual perception I wish to indicate the consciousness of visual impressions, and to distinguish this from mere impressions on the optical apparatus and reactions which are only of a reflex nature . . . (Ferrier, 1875b)

By contrast, monkeys with large occipital lesions (Fig. 11) showed no visual disturbances at all unless their lesions encroached upon the angular gyrus (Ferrier, 1875b). Indeed the only effects of occipital lesions were a temporary loss of appetite. From this he speculated that the occipital lobes were related to the "organic sensibilities and are the anatomical substrata of the correlated feelings which form a large part of our personality and subjectivity" (Ferrier, 1878).

How did Ferrier account for the finding that the visual disturbances were evident only through the contralateral eye? Isaac Newton in 1704 (see Newton, 1952) had described the partial decussation of the optic pathways and its significance for binocular vision clearly in his *Opticks* (Fig. 12), and several other 18th century figures held similar views. Indeed, homonymous hemianopia after unilateral brain damage was explained in terms of partial decussation as early as 1723 (Polyak, 1957). Ferrier was aware of both the partial decussation and its possible relation to "hemiopia" as he called it. However, he thought that the uncrossed fibers crossed to the opposite hemisphere at some level beyond the chiasm so that the cortex of each hemisphere received from the entire contralateral eye. Thus, he thought that only subcortical lesions produced homonymous hemianopia (Ferrier, 1878).

Soon after these first studies by Ferrier, Hermann Munk (1839–1912), Professor of Physiology at the Veterinary Hochschule in Berlin, reported very different results on the effects of occipital lesions in dogs and monkeys (1881), and the battle began, a battle that was not to be resolved for more than a decade. Munk's surgical and aseptic techniques were much better than those of Ferrier's and he was able to study his animals for many months. He described two types of

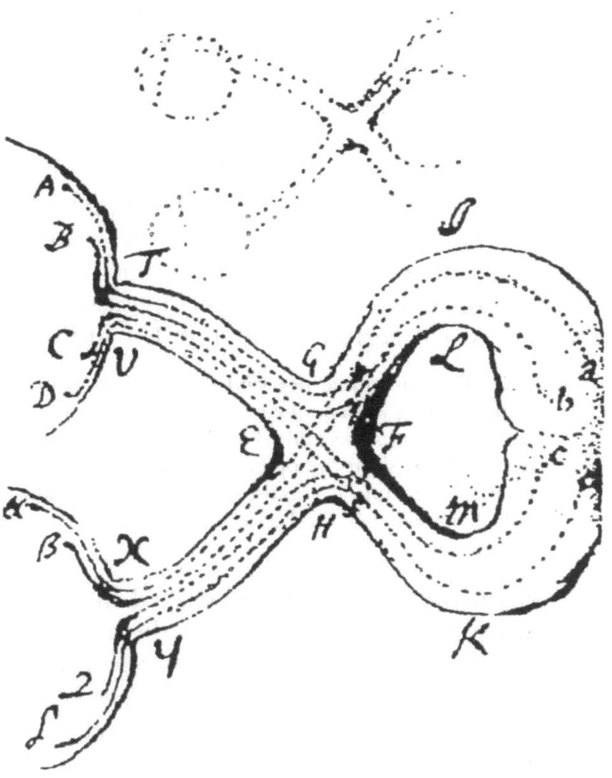

Figure 12. Newton (1704) was the first to suggest, in his *Opticks,* partial decussation at the optic chiasm which resulted in binocular convergence. This is clearly and elegantly illustrated in this sketch that Grusser found in Newton's manuscript pages of the *Opticks* (Grusser and Landis, 1991). Note that Newton thought that binocular fusion occurred in the chiasm itself. There is no reason to believe that Newton had any direct anatomical evidence for his model.

blindness after occipital lesions. The first type he called *Seelenblindheit* or "psychic blindness," and he reported it to follow limited occipital lesions in dogs. The dogs saw objects and avoided bumping into them but did not recognize their meaning:

> No abnormalities of hearing, taste, smell, motricity or sensation. The dog walks freely about the room without bumping into objects. If one blocks his path, he avoids or adroitly jumps over obstacles. But within they psychic domain of vision a distinctive defect exists: he pays no attention to water or food, even if he is hungry and thirsty. He seems indifferent to everything he sees; threats do not frighten him. One can bring a match up to his eyes without him backing away. Seeing his master or seeing other dogs leaves him impassive . . . he no longer knows or recognizes what he sees.

Although these results were never replicated by others and Munk's interpretation was disputed, the term "psychic blindness" caught on, in part, because the concept fit the associationist theories of the period (Hecaen and Albert, 1978; James, 1890). Munk's observations on psychic blindness were brought to a wide audience by William James (1890), who discussed them in detail in his *Principles of Psychology.* Thus when Lissauer (1890) published the first detailed anatomic-clinical report of a human visual recognition deficit in the absence of sensory losses, he adopted Munk's term and went on to distinguish two types of psychic blindness, namely, "apperceptive" and "associative." Later, Sigmund Freud (1891) coined the term "visual agnosia" to replace "psychic blindness." However, the term psychic blindness continued in use and, as we shall see, was immediately applied by Heinrich Klüver and Paul Bucy to describe the behavior of their temporal-lobectomized monkeys.

The second type of blindness which Munk distinguished he called "*Rindenblindheit*" or "cortical blindness." It was a total absence of any vision and he found it to follow complete removal of the occipital cortex in both dogs and monkeys. With his monkeys, Munk realized that complete unilateral occipital lesions produce blindness not in the opposite eye, but in half of each retina. Presumably the fact that only half the retinal fibers cross in the monkey but about 80% of them cross in the dog made this phenomenon of homonymous hemianopsia much easier to detect by casual observation in the monkey than in the dog. As for Ferrier, Munk (1881) had this to say:

> In my first communication on the physiology of the cortex . . . I did not say anything about Ferrier's work on the monkey because there was nothing good to say about it . . . [Ferrier's] statements and what followed from them . . . are worthless and gratuitous constructions since the operated animals were examined by Mr. Ferrier in quite an insufficient manner . . . as the experiments show now I have said at that time rather too little than too much, Mr. Ferrier had not made one correct guess, all his statements have turned out to be wrong.

About this time, Lister described his techniques for aseptic surgery, and soon after, Ferrier and Yeo used them in a new series of cortical lesions in monkeys (Ferrier and Yeo, 1884; Ferrier, 1886). Now the animals could be studied for several months after operation and Ferrier modified his previous views as to the permanence of the blindness after angular gyrus lesions: "Formerly, I localized the visual centres in the angular gyrus, to the exclusion of the occipital lobes. This being a partial truth is an error . . ." (Ferrier, 1886); and "Complete destruction of the angular gyri on both sides causes for a time total blindness, succeeded by a lasting visual impairment in both eyes. The only lesion which causes complete and permanent blindness is total destruction of the occipital lobes and angular gyri on both sides" (Ferrier and Yeo, 1884).

In spite of this retreat, Ferrier (1886) still insisted that Munk's conclusions on the location of the visual area were "entirely erroneous" and "vitiated by the occurrence of secondary encephalitis." We will return later to Ferrier's observations on angular gyrus lesions in our discussion of the discovery of visual areas in the parietal lobe.

Now Edward Albert Schäfer (1850–1935), Professor of Physiology at University College, London, and later at Edinburgh, entered the fray. In his first experiments, carried out with his student Victor Horsley (the coinventor of the stereotaxic instrument), they obtained results opposite from those of Ferrier, namely more eye movements from stimulation of the occipital lobe than of the angular gyrus and much greater visual deficits from occipital lesions than from angular gyrus ones (Horsley and Schäfer, 1888). Then Schäfer carried out a series of further experiments with an American neurologist, Sanger Brown, in which the occipital lesions were more complete than anybody had made previously (Fig. 13), and several of the animals were studied in detail for several months (Brown and Schäfer, 1888; Schäfer, 1888a). They convincingly showed that total removal of the occipital lobe produced permanently blind animals, but only if the lesion extended on the ventral surface into the temporal lobe (Schäf-

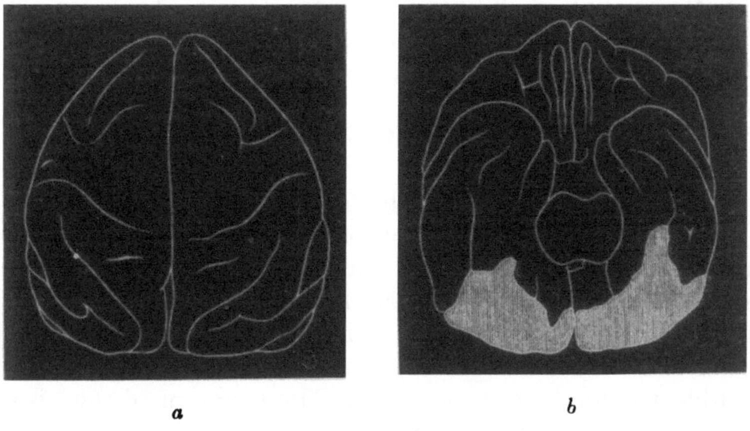

Figure 13. The occipital lesions that Brown and Schäfer (1888) reported to cause blindness in macaques. Note that the lesions include what we now know to be the entire representation of the visual field in striate cortex, including both the representation of the fovea on the lateral surface (left, dorsal view) and in the far anterior of the calcarine sulcus (right, ventral view).

er, 1888a, b). Angular gyrus involvement, however, was neither necessary nor sufficient to produce such blindness.

Ferrier and Schäfer continued to quarrel over whether the occipital lobe or the angular gyrus was the visual area, both in the journals and at various national and international meetings to which they brought their critical monkeys as demonstrations and to be examined by special committees (Ferrier, 1888; Schäfer, 1888a, b, 1900). In 1890, William James in his influential *The Principles of Psychology*, after complaining of all this internecine warfare, came down unambiguously for a visual area in the occipital lobes. The battle was virtually over by then.

Today, the bases for the apparent contradictions between Ferrier and Munk and Schäfer in the location of the visual area are understandable. From his descriptions and drawings (Fig. 11), it is clear that Ferrier removed the occipital lobes by an incision parallel to and about a half an inch or more posterior to the lunate sulcus. This site was chosen to make sure that the entire angular gyrus, his supposed visual center, was entirely spared. By his estimates this would remove "at least two thirds of the occipital lobes." Today we know that such a lesion would leave the representation of about the peripheral 30° of the visual field in striate cortex intact and, more importantly, about a few degrees of the entire representation of the lower half of the vertical meridian as well (Gattass *et al.*, 1981). This is enough residual striate cortex to account for the visually guided behavior described by Ferrier after his occipital lesions.

By contrast, Schäfer's occipital lesions not only included all the striate cortex on the lateral surface (by making his lobectomy through the floor of the lunate sulcus), but in the only animal totally and permanently blind, the bilateral lesion extended on the ventral surface far enough forward to have included all the buried striate cortex in the anterior calcarine fissure (Fig. 13). Munk provided less detailed information on the site of his lesions, but they certainly included more of striate cortex than did Ferrier's and included at least some of the striate cortex in the calcarine sulcus on the medial surface (Munk, 1890).

7.3. Striate Cortex Is Visual Cortex

By the turn of the century, with the resolution of the Ferrier–Schäfer–Munk debate, anatomical, clinicopathological, and experimental data were converging as to the identity of the visual area in the cortex of humans and monkeys. The French anatomist Gratiolet's (1854) identification of the optic radiation (initially called "Gratiolet's radiation") extending from the geniculate to the posterior cortex had been very important as the first demonstration of a sensory pathway extending to the cortex. The terminus of this visual pathway was more accurately delimited in the developmental myeloarchitectonic studies of Paul Flechsig (1847–1929), Professor at Leipzig, beginning in the 1870s. On the basis of the time of myelination of human cortex, he divided cortex into three zones, "projection," myelinating at birth, "intermediate," myelinating at 1 month, and "terminal," myelinating later. The intermediate and terminal areas taken together he termed "association cortex" (Fig. 14). By 1896, Flechsig could identify the target of the visual radiations with the most posterior projection zone and he realized it was the region of the stripe of Gennari. This region was soon named by Elliot Smith (1907) as "area striata."

During the 1880s, studies of human brain damage by Hermann Wilbrand in Hamburg, M. Allan Starr at Columbia, Henry Hun in Albany, and others (see references in Polyak, 1957) were identifying blindness with damage to the occipital cortex. Finally, the Swedish neuropathologist Salomon Henschen (1893) of the Karolinska Institute in Stockholm collected over 160 cases of blindness and hemianopia after cortical lesions which led him to identify "the centre of vision" or "cortical retina" with the calcarine cortex and later with all of striate cortex. Final experimental proof of the identification of striate cortex with vision came with Minkowski's (1911) behavioral and anatomical studies on animals.

The term "calcarine sulcus" had been coined by T. H. Huxley (1825–1895) in the course of his bitter dispute with Richard Owen (1804–1892) over the hippocampus minor and man's place in nature (Gross, 1993a, b). Owen had claimed that a unique feature of the human brain was that only humans had a "hippocampus minor," also known as the calcar avis. This structure is a ridge in the floor of the posterior horn of the lateral ventricle. To prove Owen wrong, Huxley and his allies set out to demonstrate the existence of the hippocampus minor in a variety of primates. In the course of his study of the brain of the spider monkey for this purpose, Huxley (1861) provided the first accurate description of the calcarine sulcus. He coined the name "calcarine" for it because its indentation into the lateral ventricle is what forms the calcar avis.

As the localization problem was being solved, the next issue was how striate cortex was organized. The great Arab visual scientist ibn al-Haytham (965–1039), known in Europe as Alhazen, had proposed a point-to-point projection of the retinal image onto the brain (Gross, 1981; Lindberg, 1976). This idea was well known in Europe through the 1572 translation of his work *De Aspectibus,* the standard textbook on physiological optics until Kepler and beyond. Depictions

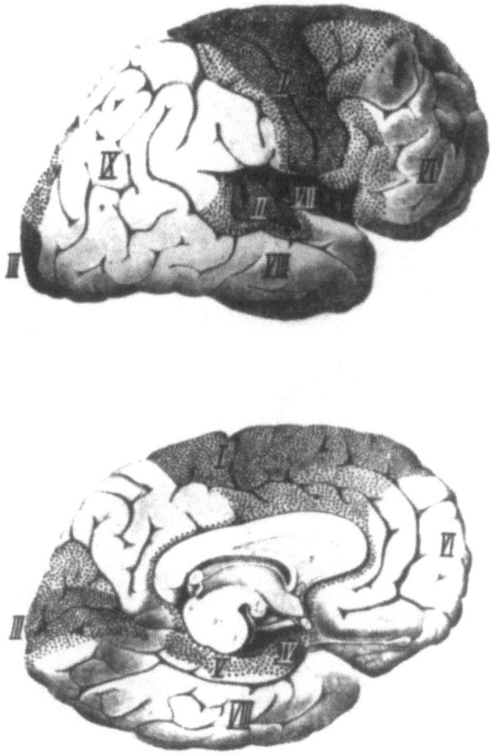

Figure 14. Flechsig's (1896) parcellation of the brain based on time of myelinization. The densely stippled areas are the projection zones surrounded by the marginal or intermediate zones. The terminal areas are unstippled. Association cortex is made up of the intermediate and terminal zones.

of the visual pathways from the Renaissance onward typically show a point-to-point projection from eye to brain whether fanciful as in Descartes (Fig. 15) or remarkably prescient as in Newton (Fig. 12). This idea of a topographic projection seemed to have derived from the theoretical considerations of Alhazen, rather than from any empirical evidence.

Salomon Henschen (1893), with his large number of cases, made a good start at empirically decoding the topography of striate cortex. He correctly placed the representation of the upper visual field in the lower bank of the calcarine sulcus and that of the lower one in the upper bank, but he reversed the center–periphery organization. This error was hardly surprising, given how large and diffuse many of his lesions were. As Glickstein and Whitteridge (1987) have pointed out, it was the subsequent introduction of high-velocity bullets in the Russo-Japanese War that produced discrete lesions and often small entry and exit wounds and thus made it possible to plot the locus of destroyed brain and correlate it with visual field defects. In that war the Japanese ophthalmologist Tatsuji Inouye (1909; Glickstein and Whitteridge, 1987) produced the first reasonably accurate scheme of how the retina is mapped on striate cortex, including the magnification of the representation of the fovea, which had not been observed before. In World War I there were a large number of studies (Polyak, 1957) with similar results, but the most widely known is that of the British neurologist Sir Gordon Holmes (1918a), perhaps because his easy-to-understand schematic diagram was reproduced in so many textbooks (Fig. 16).

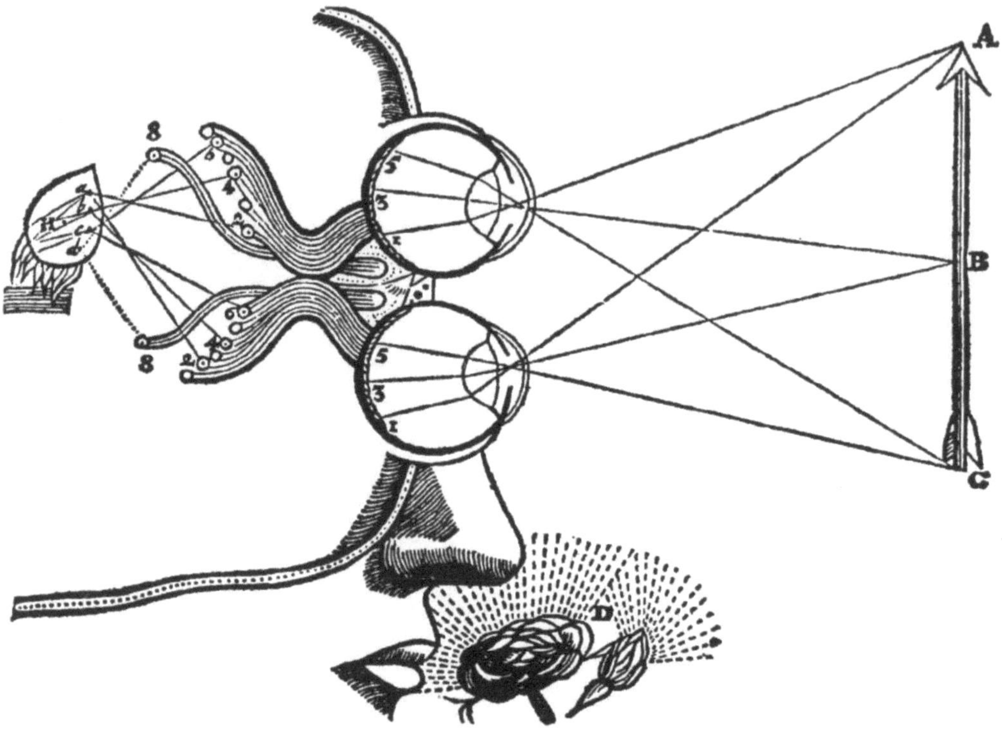

Figure 15. Several aspects of Descartes theory of sensory processing are illustrated in this figure from his physiology textbook, *De Homine* (1662). Light from the arrow enters the eye; the lens throws an inverted but topographically ordered image onto the retina. The message then travels in the hollow optic nerves from each eye by way of the animal spirits to the central pineal gland, where the information from the two eyes is united in a corresponding fashion to yield a single upright image. Olfactory messages from the flower also travel to the pineal body, but the strength of the visual signal (due to attention) suppresses this olfactory input.

7.4. Neurophysiology of Striate Cortex Begins

In 1886, Adolf Beck began to work for his doctorate at the University of Kracow (Brazier, 1988). This was not only the period of intense searching for sensory centers in the cortex, but also the beginning of electrophysiology. I. M. Sechenov and his students had recorded electrical changes in the spinal cord and brain of a frog after stimulation of its leg. Beck then set out in his doctoral thesis research to use this method to try to localize the different sensory systems. He wrote, "The question arises, are there any currents in the nervous centers of the brain and spinal cord? If so, are there changes in these currents during activity? And would the localizing of such changes be of any help in demonstrating a state of activity of a focal nature in the central nervous system?"

This was a period before amplifiers and before any knowledge of electrical activity in the intact organism, except for the "injury potential" (Brazier, 1988). After a series of experiments on frogs in which he thought he found spontaneous electrical activity, Beck turned to the cortex of rabbits and dogs. He placed

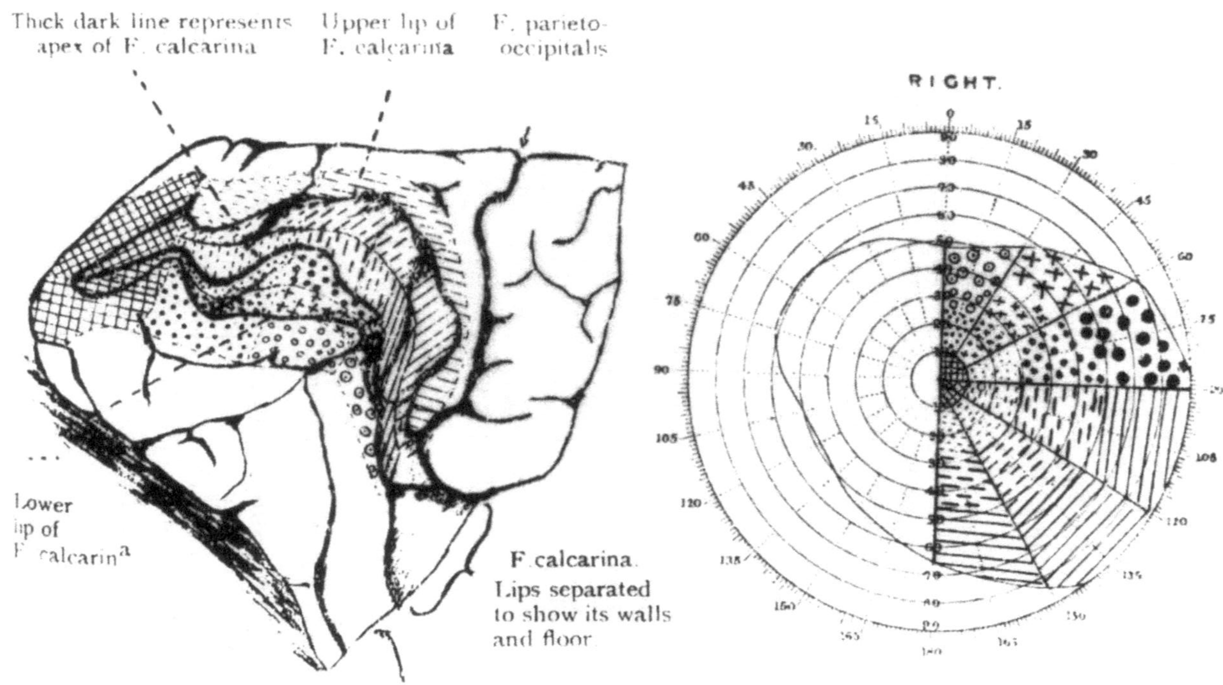

Figure 16. The representation of the retina in striate cortex according to Gordon Holmes: "A diagram of the probable representation of the different portions of the visual fields in the calcarine cortex. On the left is a drawing of the mesial surface of the left occipital lobe with the lips of the calcarine separated so that its wall and floor are visible. The markings on the various portions of the visual cortex which is thus exposed correspond with those shown in the chart of the right half of the field of vision. This diagram does not claim to be in any respect accurate; it is merely a scheme" (Holmes, 1918).

pairs of electrodes in various cortical regions and presented visual, auditory, and tactile stimuli. He found an oscillating potential difference in the occipital region in the case of visual stimuli and used it to plot the extent of the visual cortex. As his thesis was in Polish, he published a three-page summary in the leading physiology journal of the day, *Centralblatt für Physiologie* (Beck, 1890). The importance of his demonstration of sensory evoked responses was immediately recognized; indeed it stimulated a flood of letters claiming priority. One of these letters was from Richard Caton of Liverpool, who had indeed published similar, if less extensive experiments earlier (Caton, 1875). However, not only had they not been noticed in Poland, but they were totally ignored in England. The Establishment there had thought Caton's "weak electric currents" quite irrelevant.

Beck went on to a distinguished academic career in Poland, including the rectorship of the University of Lwow. When he was 80, the Germans came to take him because he was a Jew, and he swallowed the cyanide capsule supplied by his son, a doctor, and escaped the gas chamber (Brazier, 1988).

The American psychologist S. Howard Bartley (1934) was the first to carry out a systematic study of the visual evoked response of cerebral cortex and did so in rabbits. In the early 1940s, at Johns Hopkins, S. A. Talbot and Wade Marshall (e.g., 1941) used visual evoked responses to carry out their pioneering studies of the visual topography of striate cortex first in cats, then macaques, and then, with Clinton Woolsey and others, in a variety of other mammals (reviewed in Woolsey, 1971). Particularly in gyrocephalic animals these maps tended to be incomplete since the macroelectrodes used confined the recording to the surface of the cortex. Subsequently, as described in the next section, using single-neuron recording, Daniel and Whitteridge (1961) in the monkey and Hubel and Wiesel (1962) in the cat, followed by many other studies, confirmed and extended these electrophysiological maps of the visuotopic organization of striate cortex.

7.5. The Microelectrode Comes to the Visual System

7.5.1. From Lord Adrian to Stephen Kuffler

The analysis of visual processing by single neurons begins with the work of E. A. Adrian (1889–1977). Indeed, virtually all of modern neurophysiology begins with Adrian. Among his other achievements were the establishment of the all-or-none law, the first recording from single neurons, the concepts of labeled line and rate coding, the first recording of spontaneous activity from cerebral and cerebellar cortex neurons, and the confirmation of the existence of brain waves, the EEG (Adrian, 1928, 1947). Titles and awards accrued: he was made a Baron, awarded the Order of Merit and the Nobel Prize (1932), and elected Master of Trinity College and Professor of Physiology in the University of Cambridge and President of the Royal Society and of the British Association for the Advancement of Science.

In 1927, Adrian and Bryan Matthews recorded spike trains from the optic nerve of the conger eel and noted that the rate of firing increased and the latency decreased as the intensity of the light increased. Following this up, H.

Keffer Hartline (1903–1983) dissected out single optic fibers first in *Limulus* (Hartline and Graham, 1932) and then in the frog, where he distinguished on, off, and on/off responses for the first time and introduced the concept of a receptive field (Hartline, 1938). Hartline spent most of his career at the University of Pennsylvania and Rockefeller University and shared the Nobel Prize with George Wald and Ragnar Granit in 1967.

The next major development was that of Stephen Kuffler (1913–1980), then at The Johns Hopkins University. Working with cats, he developed for the first time a technique for recording from the retina without having to remove the cornea and lens as had been done previously. This maintained the normal optics of the eye and enabled him to focus light on the portion of the retina from which he was recording. With these techniques he discovered the center–surround on–off antagonistic organization of the receptive fields of retinal ganglion cells (Kuffler, 1953). Horace Barlow (1953), then working in Kuffler's laboratory, made similar observations in the frog. They noted that this receptive field organization made the cells much more sensitive to edges and contours than to diffuse light. (Barlow even called one of the classes of cells he described a "bug detector.") We now know that this receptive field structure is fundamental to the organization of the entire visual system. It was the extension of Kuffler's work from retina to cortex by Hubel and Wiesel that formed the basis of the current study of visual cortex.

7.5.2. Hubel and Wiesel

In 1959 David Hubel, a Canadian, and Torsten Wiesel, a Swede, came to Kuffler's laboratory at Johns Hopkins as post doctoral fellows. Visual physiology, and indeed all of sensory physiology and psychology, were never the same again. Through the brilliant use of single-neuron physiology they revealed the functional architecture of striate cortex and showed how it begins the binocular analysis of form and color (Hubel and Wiesel, 1959, 1962). This research promised the possibility of understanding perception in terms of neurons and became the model for subsequent explorations of visual neurons inside and outside of striate cortex and indeed for all of contemporary neurophysiology. Subsequently,they moved to Harvard with Kuffler and in 1981 they shared the Nobel Prize with Roger Sperry. Their remarkable achievements (Hubel, 1988), which extended into visual neuroanatomy and neural development, have been extensively reviewed in this and other volumes of *Cerebral Cortex* and will not concern us in detail here except for two historical notes.

The first is their description of their first observation of an orientation-selective neuron in a cat, perhaps the opening wedge in revealing the secrets of striate cortex:

> We had been doing experiments for about a month . . . and were not getting very far; the cells simply would not respond to our spots and annuli. [The stimuli that had been used by Kuffler to reveal the properties of retinal ganglion cells.] One day we made an especially stable recording . . . For 3 or 4 hours we got absolutely nowhere. Then gradually we began to elicit some vague and inconsistent responses by stimulating somewhere in the midperiphery of the retina. We were inserting [a] glass slide with a black spot into a [projecting ophthalmoscope] when suddenly over the audiomonitor the cell went off like a machine gun. After some fussing and fiddling we found out what

> was happening. The response had nothing to do with the black dot. As the glass slide was inserted its edge was casting onto the retina a faint but sharp shadow, a straight dark line on a light background. That was what the cell wanted, and it wanted it, moreover, in just one narrow range of orientations. (Hubel, 1982)

A few years later they realized that cells with similar orientation selectivity and cells with similar ocular dominance were arranged in orientation and ocular dominance columns, respectively (Hubel and Wiesel, 1962). This discovery, presumably, must have been facilitated by the proximity at Hopkins of Vernon Mountcastle, who had discovered columnar organization in somatosensory cortex (Mountcastle, 1957).

The second historical point is that Hubel and Wiesel were by no means the first to record from single neurons in striate cortex. In 1952, the Freiburg group starring R. Jung, G. Baumgartener, O. Creutzfeldt, and O. J. Grusser had begun a systematic program of research on the visual activity of single neurons in striate cortex of the cat (June, 1992). Although their techniques were technically sophisticated, their central finding for about the first 10 years was actually that striate neurons showed little visual responsiveness: 50% of the many cells sampled showed no responses and the responses of many of the others, by subsequent standards, were rather feeble. As Jung (1992) later candidly admitted, a primary reason for their failure to activate striate cells was that their elaborate apparatus (which took 2 years to build) was too inflexible to vary the orientation of the visual stimulus. As he put it, "We missed the orientation specificity . . . [because of] . . . premature quantification and a too rigid methodological restriction."

This completes our story of research on striate cortex. We now turn to visual cortex outside of striate cortex.

8. Beyond Striate Cortex

Whereas striate cortex has remained within the borders delimited for it at the start of this century, visual cortex has flowed forward out of the occipital lobe, deep into the temporal, parietal, and even frontal lobes, and now consists of more than two dozen discrete areas making up more than half the primate cerebral cortex. We first consider the discovery of the second and third visual areas adjacent to striate cortex in the occipital lobe and then the discovery of visual cortex in the temporal and parietal lobes.

8.1. The Surrounding Zones

As mentioned previously, Flechsig had divided the cortex on the basis of onset of myelination into three zones: projection, intermediate, and terminal, the latter two together being termed "association" cortex (Fig. 14). He chose the term "association" because he believed that these regions became myelinated in humans at about the age he thought that children began to associate the different senses with each other and with movement. In his scheme, only the primary areas received sensory input from below the cortex and this input was confined to one modality. The output of the modality-specific primary areas was to the intermediate zones surrounding each of the primary areas; they remained unimodal and further elaborated the sensations into perceptions which might then be "associated" with the other sensory modalities by way of long association fibers in the terminal zone. At this time psychology was dominated by British associationism (typified by John Stuart Mill and Alexander Bain), and association cortex was quickly assigned the tasks of the "association of sensations into perceptions, images, ideas and memories" (Boring, 1950). The idea that sensations were fundamentally different from perceptions was much older, deriving from Aristotle ("It is not the bell that enters the ear, but the energy of the sound that leaves the bell") and emphasized by David Hume (1739) and, in our century, by E. B. Titchener (1910).

Inspired by these theoretical conceptions, the intermediate visual zone was soon termed "visuopsychic" to distinguish it from the primary visual or "visuosensory" area now identified with the stripe of Gennari. On "histological" grounds (both cytoarchitectonic and myeloarchitectonic, in later terminology), the visuopsychic area was defined as a strip of 1.3–2 cm surrounding the visuosensory area, by J. S. Bolton (1900) and Alfred Campbell (1905), both working in the Rainhill Asylum, Lancashire. As the latter put it, the visuosensory center is "a primary station . . . for the receipt of impressions derived from . . . both retinal fields [and the visuopsychic center] . . . is an area wherein visual impressions are further dealt with in the process of elaboration and intellectual interpretations."

The bases for the functions attributed to the visuopsychic area as opposed to the visuosensory one were strictly anatomical, embryological, and theoretical: Campbell was pessimistic of finding patients with sufficiently discrete lesions to distinguish the two types of areas.

G. Elliot Smith (1907), Professor of Anatomy at the Egyptian School of Medicine in Cairo, then divided the Bolton–Campbell visuopsychic area into a zone adjacent to striate cortex, which he named parastriate cortex, and an outer zone, which he named peristriate cortex. These areas corresponded to Brodmann's (1909) areas 18 and 19 designated at about the same time, and to the ones later named by von Economo (1929) as areas OB and OA. The outer zone, peristriate or 19, was assumed to have more complex psychic functions than the inner zone, parastriate or 18. Collectively these two "visual association areas" were later called prestriate cortex.

Originally, on the basis of anatomical studies, it had been assumed that only striate cortex was a "cortical retina," i.e. was topographically organized, and that parastriate and peristriate cortex (areas 18 and 19) were not topographically organized, being concerned with "perceptual and association functions" and not "sensory" function. However, with the rise of electrical mapping of sensory cortex, quite the contrary became clear.

In the 1940s, using evoked responses, Adrian reported a second somatosensory area and then Talbot and Woolsey found second auditory and visual areas (references in Woolsey, 1971). Soon after, Woolsey and his many students at the University of Wisconsin begun their decades-long program to use evoked responses to map sensory and motor cortex in a variety of mammals (Woolsey, 1971). In 1965 Hubel and Wiesel, using microelectrode recording of single-neuron activity in the cat, confirmed the second visual area of Talbot and Marshall (now called V2) and found an adjacent third visual area (now called V3). In current terminology, in the primate, area 18 (parastriate cortex, OB) corresponds roughly to areas V2 and V3, and area 19 (peristriate cortex, OA) corresponds roughly to much of visual area V4.

A new phase of visual physiology began in the 1970s when Jon Kaas and John Allman, initially in Woolsey's lab, began their now classic series of studies of extrastriate visual cortex in the owl monkey (Allman and Kaas, 1971; Kaas, this volume). They used single-neuron recording rather than evoked potentials and soon freed themselves of the assumption of only two or three visual areas, rapidly finding an amazing multiplicity of visuotopically organized visual areas extending well into the temporal and parietal lobes. Similar extrastriate visual areas were then mapped in the macaque by a number of laboratories, including those of S. Zeki, D. Van Essen, L. Ungerleider, and R. Desimone and myself, eventually filling much of "posterior association cortex" with over two dozen visual areas. These developments are considered in subsequent chapters of this volume.

8.2. Temporal Association Cortex as a Visual Area

8.2.1. Brown and Schäfer's Lost Observations

In the confusion and controversy over the location of the visual and auditory centers at the end of the 19th century, two important observations were lost (or ignored) and had to be rediscovered decades later. When these observations were eventually repeated and extended, one led to the discovery of the critical role of inferior temporal cortex in object recognition, the other to the discovery of the critical role of posterior parietal cortex in spatial vision. We consider the first development in this section and the second one later.

The first lost observation was made in the article by Brown and Schäfer (1888) which, as described above, played such a definitive role in establishing the visual area in the occipital lobes. Brown and Schäfer opposed Ferrier's views not only on the location of the visual area, but also on his localization of the auditory area in the superior temporal gyrus. One of their strongest pieces of evidence against an auditory center in the temporal lobe was an animal who had received large bilateral temporal lobe lesions. This animal was demonstrated at several meetings and its lack of deafness attested to by an examining committee at the 1887 meeting of the Neurological Society (Schäfer, 1888b). More interesting (to this author, at least) than its lack of deafness were the changes in postoperative behavior that Brown and Schäfer described thus:

> A remarkable change is manifested in the disposition of the monkey. Prior to the operations he was very wild and even fierce, assaulting any person who teased or tried to handle him. Now he voluntarily approaches all persons indifferently, allows himself to be handled, or even to be teased or slapped without making any attempt at retaliation or endeavoring to escape. His memory and intelligence seem deficient. He gives evidence of hearing, seeing and of his senses generally, but it is clear that he no longer clearly understands the meaning of the sights, sounds and other impressions that reach him . . . He appears no longer to discriminate between the different kinds of food . . .

He still however possesses the sense of taste, for when given a raisin which has been partially filled with quinine he shows evident signs of distaste, and refuses to eat the fruit.

The authors concluded that, "On localization of functions the experiment throws no direct light . . . a general depression of the intellectual faculties has resulted, and a condition resembling idiocy produced."

Schäfer repeatedly used this animal's normal hearing as an argument against the existence of a temporal lobe auditory center. However, he never again even mentioned the strange effects of removal of the temporal lobes except to say that it produced "idiocy" (Schäfer, 1888a, b, 1900). Given the disputes over the location of the basic cortical sensory areas and the fact that even Schäfer himself never mentioned again the effects of temporal lobe removal, it is not surprising that these observations were lost until well after Klüver and Bucy repeated them in 1938. Why Klüver and Bucy did so is described in the next section of this paper.

8.2.2. Discovery of the Klüver–Bucy Syndrome

Heinrich Klüver was a Professor at the University of Chicago for over 50 years and responsible for the development of many of the modern techniques for the study of visual cognition in monkeys (Klüver, 1933). He was also quite interested in the effects of mescaline on perception (particularly his own) and wrote a little book called *Mescal, The Divine Plant* (Klüver, 1926, 1928). Klüver thought that the hallucinations (aura) that sometimes precede a temporal lobe seizure in patients resembled a mescaline-induced hallucination, and on that basis he speculated that the temporal lobes might be the site of action of mescaline. To test this possibility he arranged for the neurosurgeon Paul Bucy to remove the temporal lobes of several monkeys in order to see whether, postoperatively, mescaline still induced behaviors suggesting mescaline hallucination (Klüver and Bucy, 1938, 1939; Klüver, 1948).

The effects of mescaline seem to have been the same after the temporal lobe removals as before, but the undrugged temporal lobectomized monkeys showed a strange and intriguing set of behavioral changes. This set of symptoms became known as the "Klüver–Bucy" syndrome. Klüver (1948) summarized them as follows:

1. . . . psychic blindness or visual agnosia . . . the ability to recognize and detect the meaning of objects on visual criteria alone seems to be lost although the animal exhibits no or at least no gross defects in the ability to discriminate visually.

2. . . . strong oral tendencies in the sense that the monkey insists on examining all objects by mouth.

3. . . . hypermetamorphosis [after Wernicke]. There is an excessive tendency to attend and react to every visual stimulus.

4. . . . profound changes in emotional behavior, and there may even be a complete loss of emotional responses in the sense that . . . anger and fear are not exhibited. All expressions of emotions . . . may be completely lost.

5. . . . striking increase in the amount and diversity of sexual behavior. It exhibits forms of autosexual, heterosexual and homosexual behavior rarely or never seen in normal monkeys.

6. . . . a remarkable change in dietary habits . . . [particularly] . . . a tendency to consume large quantities of . . . [meat].

Ten years after his original study, Klüver (1948) came upon the 1881 Brown and Schäfer paper with the description of the monkey with the bilateral temporal lobectomy, and wrote:

> It is evident that the observations of Sanger Brown and Schäfer agree fully with ours, particularly with reference to the picture of psychic blindness, the oral tendencies, the 'hypermetamorphosis' and the striking changes in emotional behavior. It is evident that the bilateral lesions, for example in monkey 6, were comparable in extent to the lesions produced in our experimental animals.

8.2.3. Fractionation of the Klüver–Bucy Syndrome

Klüver and Bucy (1939) had reported that they could not produce their syndrome or any of its components by smaller lesions and suggested that the syndrome was a unitary one which could not be fractionated. However, about 10 years later, fractionation is exactly what began to happen at the Yerkes Laboratory of Primate Biology in Orange Park, Florida. By 1942 this first U.S. primate center was a joint Harvard–Yale enterprise, Harvard providing the director, Karl Lashley, and Yale providing the laboratory, originally founded by Robert M. Yerkes.

Until moving to Orange Park, Lashley had spent much of his experimental career failing to find evidence for cortical localization of complex behavior in rats (Lashley, 1933, 1950). He now set out to study the effects of lesions of "visual association" cortex in monkeys. He tested the monkeys on a extensive battery of visual tests which included measures of sensory status, learning, and memory (Lashley, 1948). He had planned to destroy Brodmann's area 18 and 19, but as detailed in his classic attack on cytoarchitectonics (Lashley and Clark, 1946), he could not identify them reliably. Therefore, he made "no pretense of dealing with a distinct anatomic or functional area" and instead removed "a band of cortex encircling the striate area which he called "the prestriate region."

His prestriate lesions included "not less than 60% of Brodmann's areas 18 and 19" (his estimate) and not much more (my estimate). They included some limited encroachment into posterior parietal and posterior temporal cortex. He found that these lesions "did not produce any trace of object agnosia . . . [or] . . . any significant amnesia for . . . color or form . . ." Lashley concluded that "interruption of transcortical associative connections is therefore not the cause of visual agnosia [and] comparison of the experimental and clinical evidence indicates that visual agnosia cannot be ascribed to uncomplicated loss of prestriate tissue."

The apparently conclusive antilocalization implications of Lashley's findings were reversed soon after the arrival in Orange Park of two graduate students, Josephine Semmes (Blum) from Yale and Kao Liang Chow from Harvard, joined by Karl Pribram, who had trained in neurosurgery with Bucy at Chicago. In a joint study they made lesions much larger than Lashley's (1948) and encroached further into the parietal lobe and much more anteriorly and ventrally into the temporal lobe. These lesions succeeded in producing "psychic blindness," i.e., severe deficits in visual discrimination learning (as well as some tactile learning deficits which will be discussed later). Since the locus common to their lesions and Klüver and Bucy's was temporal cortex, they surmised that temporal cortex must be crucial for visual learning (Blum *et al.*, 1950). Chow (1951) then went on to show that, indeed, "lesions of the lateral surface of the temporal lobe and especially the middle temporal gyrus, produced visual deficits" whereas lesions of adjacent areas did not.

In these studies the cortical lesions that produced the visual deficits did not produce any of the other components of the Klüver–Bucy syndrome such as tameness and changes in sexual and eating behavior. Rather, as Pribram and Bagshaw (1953) showed, these components, without the visual difficulties, followed lesions of the amygdala and adjacent cortex. Thus the original Brown–Schäfer/Klüver–Bucy syndrome was fractionated into visual discrimination difficulties following temporal cortical damage and other, "motivational" symptoms following amygdala damage.

Pribram continued his studies of cortical function in monkeys first in John Fulton's laboratory at Yale and then in a new laboratory established for him at the Institute of Living, a private mental hospital in Connecticut. He was joined by Mortimer Mishkin, a graduate student from McGill; they further localized the visual components of the Klüver–Bucy syndrome to the middle and inferior temporal gyri (Mishkin and Pribram, 1954; Mishkin, 1954). This region became known as inferotemporal cortex, or, more recently, inferior temporal (IT) cortex. In the mid 1950s, research on the behavioral effects of IT lesions was also continuing at Orange Park and in Harry Harlow's laboratory at the University of Wisconsin. By the end of the decade the behavioral effects of IT cortex lesions were being studied in laboratories all across the country (e.g., those of Chow at the University of Chicago, William and Martha Wilson at Bryn Mawr, Mishkin at NIMH, and Donald Meyer at Ohio State University) as well as in England (Larry Weiskrantz at Cambridge University, and George Ettlinger at Queen's Square, London). Much of this effort was directed toward characterizing the visual learning deficit that followed IT lesions. This research is reviewed in Gross (1973, 1994) and Dean (1976).

8.2.4. Inferior Temporal Cortex Joins the Visual System

There were three major advances which led to the current period of research on IT cortex. The first was a set of developments which brought IT cortex into anatomical relation with the rest of the visual system. It had been supposed from degeneration and strychnine neuronography studies that IT cortex received visual information from the pulvinar (to which both the superior colliculus and striate cortex project) and from striate cortex over a corticocortical multisynaptic pathway (Chow and Hutt, 1953). The puzzle was that both large lesions of the pulvinar and attempted interruption of the putative corticocortical pathway repeatedly failed to produce the IT deficit on visual learning (e.g., Chow, 1961).

In 1966, Mishkin solved this conundrum by disconnecting IT and striate cortex by a crossed lesion paradigm: he combined a striate lesion in one hemisphere, an IT lesion in the other hemisphere, and section of the forebrain commissures. The combination of all three lesions, but not any two, cut the pathway between striate and IT cortex and by deafferenting IT cortex produced a severe impairment on visual learning.

At this time little was known about the occipitotemporal region lying between (and connecting) striate and IT cortex. In the 1970s, as the result of the pioneering studies initiated by Allman and Kaas, it began to be filled with a set of visuotopically organized areas. These areas were to be grouped in the 1980s as the occipitotemporal or "ventral" cortical visual stream specialized for the analysis of form and color.

The second major advance was the functional subdivision of IT cortex into a more anterior area, corresponding with cytoarchitectonic area TE and a more posterior area corresponding to area TEO. On the basis of the differential effects of lesions of the two areas Iwai and Mishkin (1968) and Cowey and Gross (1970) suggested that area TE was more involved in mnemonic processes and area TEO more involved in perceptual ones. Subsequent work revealed differences in the neuronal properties and connections of these two areas. These developments are discussed in detail elsewhere (Gross, 1973, 1994; Gross *et al.*, 1993; Mishkin, 1972).

The third major advance was the beginning of the study of the visual properties of single neurons in IT cortex. In 1967, Peter Schiller, Carolyn Wells, George Gerstein, and the author reported that IT neurons responded to visual but not auditory stimuli (Gross *et al.*, 1967). The author, Carlos Eduardo Rocha-Miranda, and David Bender (Gross *et al.*, 1969) then went on to show that these neurons had visual receptive fields that were relatively large, always included the fovea, often extended across the midline into the ipsilateral visual field, and were not visuotopically organized. Furthermore, many of these cells responded only or best to highly complex stimuli. The most striking of these cells were selective for faces. Finally, the responses of these cells could be modified by attentional and mnemonic parameters (Gross *et al.*, 1979). The intellectual milieu that led to these discoveries is described in Gross (1994).

By the early 1980s, the end of the time window of this chapter, the existence of face-selective neurons in IT cortex as well as their other properties had been replicated in other laboratories (Perrett *et al.*, 1982; Richmond and Wurtz, 1982). The next decade saw the beginning of systematic studies of the role of experience, attention, and development in the properties of IT neurons. These developments are discussed in this volume (see chapter by Rodman and Moore) and elsewhere (Tanaka, 1996; Miyashita, 1993; 1996; Desimone *et al.*, 1995).

8.3. Parietal Association Cortex as a Visual Area

8.3.1. Ferrier's Lost Observations

We now turn to the second set of findings that were "lost" after the battle over the location of the "cortical visual center" was over. These were observations on the effect of lesions of the angular gyrus (area 7, posterior parietal cortex). As described above, Ferrier had argued that the visual center was located in the angular gyrus. Here is what Ferrier and Yeo (1884) had to say about the monkey's recovery from this lesion:

> On the fourth day there were some indications of returning vision. A piece of orange was held before it, whereupon it came forward in a groping manner and tried

> to lay hold . . . but missed repeatedly . . . On the fifth day . . . it was evidently able to see its food, but constantly missed laying hold of it, putting its hand beyond it or short of it . . . On the sixth day . . . it was able to pick up grains of rice scattered on the floor, but always with uncertainty as to their exact position . . . Four weeks after the operation . . . the same want of precision was still seen as regards its power of putting its hands on objects it wished to pick up . . . At this date when it was walking on a table it tumbled off, having come too near the edge without seeming to be aware of the fact.

From this description it seems clear that, rather than being blind, this monkey was having problems with localizing stimuli and reaching for them. However, since Ferrier had clearly been wrong in localizing the "visual center" in the parietal lobe, these observations of what he called "blindness" were usually attributed to some artifact such as damage to the optic radiations or transient interference with occipital function and then were almost forgotten. In fact Ferrier's opponents Brown and Schafer had made an essentially similar, and even more forgotten, observation, noting that after bilateral posterior parietal lesions, their monkey "would evidently see and run up to [a raisin], but then often fail to find it . . ."

8.3.2. Parietal Lesions in Humans

Thirty years later, at the end of World War I, Sir Gordon Holmes (1918b) described a set of veterans with disturbances of visual and spatial orientation after bilateral posterior parietal lesions. Their symptoms included deficits in reaching and pointing to visual targets, avoiding obstacles, learning and remembering routes, judging distance and size, recognizing spatial relations, fixating a target, and following a moving stimulus. By contrast, their object recognition ability and other cognitive functions were essentially normal. They had all suffered bilateral penetrating missile wounds of the posterior parietal cortex. Holmes' careful observations established the critical role of posterior parietal cortex in visuospatial functions. His paper was also one of the very few on parietal damage in humans or monkeys that cited Ferrier's early observations on visuospatial deficits after parietal lesions in monkeys. The next major development was the realization by W. R. Brain (1941), Lord Brain, the long-time editor of the journal *Brain,* that unilateral posterior parietal lesions, particularly of the right hemisphere, can produce dramatic neglect of the contralateral side of space. By the time of Malcolm Critchley's (1953) classic monograph, *The Parietal Lobes,* it was clear that a variety of complex visual, visuospatial, and visuomotor symptoms followed damage to parietal cortex in human.

8.3.3. Parietal Cortex Lesions in the Monkey: 20th Century

The first 20th century investigation of posterior parietal cortex function in monkeys was by T. C. Ruch and J. F. Fulton (e.g., Ruch *et al.,* 1938). They found lesions of this area to produce impairments on roughness and weight discrimination learning. They interpreted these results as indicating that posterior parietal cortex was a somatosensory association area following the idea that the regions adjacent to each primary sensory area (somatosensory cortex in this case) had the function of developing sensory information into perceptions. Further support for this idea came from studies in Lashley's laboratory at the Yerkes Primate

Center and then in Pribram's laboratory at the Institute of Living (e.g., Blum *et al.*, 1950; Blum, 1951; Wilson, 1957; Wilson *et al.*, 1960). These studies continued to emphasize the role of posterior parietal cortex in tactile learning, and indeed, viewed this cortex as playing an analogous role for tactile learning as inferior temporal cortex did for visual learning.

The attitude of dismissing the visuospatial effects of posterior parietal lesions as "side effects" altered with the arrival at Pribram's laboratory of George Ettlinger from the Institute of Neurology, Queen's Square, London. He had studied human parietal patients, and was aware of Ferrier's earlier observations on the effects of parietal lesions in monkeys. Ettlinger stressed that posterior parietal lesions in monkeys produced visual orientation and reaching deficits and increased sensitivity to stimulus–response separation. He argued that these deficits were the bases of the difficulties in tactile discrimination learning (e.g., Bates and Ettlinger, 1960; Ettlinger and Kahlsbeck, 1962; Ettlinger *et al.*, 1966). By the late 1970s several other workers had moved away from the stress on a tactile learning deficit as the main effect of posterior parietal lesions, to view the deficit as a supramodal spatial one (Semmes, 1965; Milner *et al.*, 1977). With the beginning of microelectrode studies of parietal cortex this controversy was replaced by a new one about the functions of posterior parietal cortex.

8.3.4. The Microelectrode Arrives at Posterior Parietal Cortex

Single-neuron analysis of posterior parietal cortex began with the work of Juhani Hyvarinen and his colleagues in Helsinki starting in 1970 (Hyvarinen, 1973; reviewed in Hyvarinen, 1982). They found cells that fired "when a sensory stimulus which interested the animal was placed in a specific location in space where it became the target of the monkey's gaze or manual reaching, tracking or manipulation . . . Some cells were clearly related to eye movements whereas others appeared to discharge in response to visual sensory stimuli."

They interpreted their results as indicating that posterior parietal cortex was involved in the visuospatial guidance of movement, thereby providing an explanation for the visuospatial deficit after parietal lesions. However, their procedures were somewhat naturalistic: eye and arm movement and position were not measured or controlled and stimulus presentation was often informal.

Soon after, Vernon Mountcastle and his colleagues at Johns Hopkins explored similar phenomena and did so under greater experimental control. They found cells specifically related to projection of the arm, to hand manipulation, to visual tracking, and to visual fixation as well as cells with more complex properties. (Mountcastle, 1976; Mountcastle *et al.*, 1975; Lynch *et al.*, 1977). At least in their early papers, they claimed that very few, if any, of their cells were "directly activated" by visual stimuli. They described their parietal cells as "function[ing] in a command fashion, directing visual attention to and exploration of the immediately surrounding extrapersonal space (Mountcastle *et al.*, 1975), and as "execut[ing] a matching function between the neural signals of the nature of objects and the internal drive states of the organism, and contain[ing] a neural apparatus for the direction of visual attention to objects of interest and for shifting the focus of attention from one to another" (Lynch *et al.*, 1977).

Michael Goldberg and David Robinson (1977, 1980; Robinson *et al.*, 1978) at the National Institutes of Health then demonstrated that Mountcastle *et al.* were quite wrong: many posterior parietal neurons did indeed respond to visual stimuli and did so in the absence of movement, thereby unambiguously falsifying Mountcastles's "motor command" hypothesis. They did find that the visual responses could often be enhanced by the animal's attention, but this did not necessarily involve any eye movement at all (Bushnell *et al.*, 1981). Goldberg and Robinson had returned posterior parietal cortex to the visual system where Ferrier had first put it.

In the next decade there were several further significant developments on these issues. As they are outside of our time frame, we can merely list two of them. First, Andersen *et al.* (1985) showed that the visual responses of posterior parietal neurons can be modulated by eye position and therefore an ensemble of such cells carry information about visual targets in spatial coordinates. Second, it became clear that there are a number of different visual areas in the posterior parietal cortex which appear to play different roles in a variety of visual and visuospatial functions such as the analysis of stimulus movement, the coding of space, and the visual control and monitoring of eye and hand movement (e.g., Colby and Duhamel, 1991; Gross and Graziano, 1995; Duhamel *et al.*, 1992).

9. Coda: Two or More Cortical Visual Systems

By the early 1980s and the end of the period covered by this review over a dozen cortical visual areas had been described, extending visual cortex from the occipital lobe well into the temporal and parietal lobes. In 1982, Leslie Ungerleider and Mortimer Mishkin proposed a powerful and highly influential theory that brought considerable order into this plethora of visual areas. They proposed that the various extrastriate visual areas could be grouped into two processing streams. The first was a "dorsal" pathway from striate cortex, through a set of dorsal prestriate areas, to posterior parietal cortex, forming a system for spatial vision. The second was a "ventral" pathway from striate cortex through a set of ventral prestriate areas to inferior temporal cortex, forming a system for object vision. Their theory had two major roots. The first was the suggestion by Schneider (1967) and Trevarthen (1968) that visual functions could be divided, spatially and functionally, into *what?* or identity functions and *where?* or localization functions. The second basis of Ungerleider and Mishkin's (1982) proposal was their own confirmation and extension of the two "lost" 19th century observations detailed above, that temporal lesions produce a psychic blindness for recognizing objects and that posterior lesions produce a deficit in spatial vision.

Today, the number of cortical visual areas in monkeys exceeds two dozen and they now comprise more than half of the entire cortical mantle. Ungerleider and Mishkin's scheme continues to be a powerful and effective way of organizing these areas. Not surprisingly, it has been modified and developed in a variety of ways. One development has been the extension of both the dorsal and ventral systems into the frontal lobes. Another development has been the recognition that each system may be divided into subsystems or streams. For example, within the dorsal system, a subsystem specialized for motion may be distinguished consisting of areas MT, MST, FST, and STP. A third development has been the much greater stress on the visuomotor functions of the dorsal system. The remainder of this volume is concerned with these and other advances in our understanding of extrastriate visual cortex.

ACKNOWLEDGMENTS. I thank the following for their help in the preparation of this paper: Mairi Benson (Sherrington Collection in the History of Neuroscience, Oxford University), Greta Berman, Linda Chamberlin (Princeton University Library), Helen Chung, Mazyar Fallah, Michael Graziano, George Krauthamer, Hilary Rodman, Maida Rosengarten, Wellcome Institute Library; I also acknowledge support from NIH Grant EY 11347–25.

10. References

Adrian, E. D., 1928, *The Basis of Sensation, The Action of the Sense Organs,* Christophers, London.
Adrian, E. D., 1947, *The Physical Background of Perception,* Clarendon Press, Oxford.
Adrian, E. D., and Matthews, B., 1927, The action of light on the eye, I: The discharge of impulses in the optic nerve and its relation to the electric changes in the retina, *J. Physiol.* **63:**378–414.
Allman, J. M., and Kaas, J. H., 1971, A representation of the visual field in the caudal third of the middle temporal gyrus of the owl monkey (*Aotus trivirgatus*), *Brain Res.* **31:**85–101.
Anderson, R., Essick, G., and Siegel, R., 1985, Encoding of spatial location by posterior parietal neurons, *Science* **230:**456–458.
Barlow, H. B., 1953, Summation and inhibition in the frog's retina, *J. Physiol.* **119:**69–88.
Bartley, S. H., 1934, Relation of intensity and duration of brief retinal stimulation by light to the electrical response of the optic cortex of the rabbit, *Am. J. Physiol.* **108:**397–408.
Bates, J., and Ettlinger, G., 1960, Posterior biparietal ablations in the monkey, *Arch. Neurol.* **3:**177–192.
Beare, J. I., 1906, *Greek Theories of Elementary Cognition from Alcmaeon to Aristotle,* Clarendon Press, Oxford.
Beck, A., 1890, Die Bestimmung der Localisation der Gehirn- und Ruckenmarkfunctionen vermittelst der elektrischen Erscheinungen, *Centralbl. Physiol.* **4:**473–476.
Bell, C., 1811, *Idea of a New Anatomy of the Brain,* Warren, Winchester.
Bell, C., 1812, Idea of a new anatomy of the brain, *Balt. Med. Phil. Lycaeum* **1:**303–318.
Blum, J., 1951, Cortical organization in somesthesis: Effects of lesions in posterior associative cortex on somatosensory function in *Macaca mulatta, Comp. Psychol. Monogr.* **20:**219–249.
Blum, J., Chow, K., and Pribram, K., 1950, A behavioral analysis of the organization of the parieto-temporor-preoccipital cortex, *J. Comp. Neurol.* **93:**53–100.
Boakes, R., 1984, *From Darwin to Behaviorism: Psychology and the Minds of Animals,* Cambridge University Press, Cambridge.
Bolton, J., 1900, The exact histological localization of the visual area of the human cerebral cortex, *Phil. Trans. R. Soc. Lond.* **193:**165–222.
Boring, E. G., 1950, *A History of Experimental Psychology,* Appleton-Century-Crofts, New York.
Brain, W., 1941, Visual disorientation with special reference to lesions of the right cerebral hemisphere, *Brain* **64:**244–272.
Brazier, M. A. B., 1988, *A History of Neurophysiology in the Nineteenth Century,* Raven Press, New York.
Breasted, J. H., 1930, *The Edwin Smith Surgical Papyrus,* University of Chicago Press, Chicago.
Broca, P., 1960, Remarks on the seat of the faculty of articulate language, followed by an observation of aphemia [1861], in: *Some Papers on the Cerebral Cortex* (G. Von Bonin, trans.), Thomas, Springfield, IL, pp. 49–72.
Brodmann, K., 1909, *Vergleichende Lokalisationslehre der Grosshirnrinde in ihren Prinzipien dargestellt auf Grund des Zellenbaues,* Barth, Leipzig.
Brown, S., and Schäfer, E., 1888, An investigation into the functions of the occipital and temporal lobes of the monkey's brain, *Phil. Trans. R. Soc. Lond.* **179:**303–327.

Bushnell, M., Goldberg, M., and Robinson, D., 1981, Behavioral enhancement of visual responses in monkey cerebral cortex, I. Modulation in posterior parietal cortex related to selective visual attention, *J. Neurophysiol.* **46:**755–772.

Campbell, A. W., 1905, *Histological Studies on the Localization of Cerebral Function,* Cambridge University Press, Cambridge.

Carpenter, W., 1845, *Principles of Human Physiology,* Lea and Blanchard, Philadelphia.

Caton, R., 1875, The electric currents of the brain, *Br. Med. J.* **2:**278.

Chow, K.-L., 1951, Effects of partial extirpations of the posterior association cortex on visually mediated behavior, *Comp. Psychol. Monogr.* **20:**187–217.

Chow, K.-L., 1961, Anatomical and electrographical analysis of temporal neocortex in relation to visual discrimination learning in monkeys, in: *Brain Mechanisms and Learning* (J. Delafresnay, ed.), Blackwell, Oxford, pp. 507–526.

Chow, K.-L., and Hutt, P., 1953, The "association cortex" of *Macaca mulatta*: A review of recent contributions to its anatomy and functions, *Brain* **76:**625–677.

Clarke, E., 1962, The early history of cerebral ventricles, *Trans. Stud. Coll. Physicians Philadelphia* **30:**85–89.

Clarke, E., and Bearn, J., 1968, The brain 'glands' of Malpighi elucidated by practical history, *J. Hist. Med. Allied Sci.* **23:**309.

Clarke, E., and Dewhurst, K., 1996, *An Illustrated History of Brain Function,* Norman, San Francisco.

Clarke, E., and Jacyna, L. S., 1987, *Nineteenth-Century Origins of Neuroscientific Concepts,* University of California Press, Berkeley.

Clarke, E., and O'Malley, C. D., 1996, *The Human Brain and Spinal Cord. A Histological Study Illustrated by Writings from Antiquity to the Twentieth Century,* Norman, San Francisco.

Colby, C., and Duhamel, J., 1991, Heterogeneity of extrastriate visual areas and multiple parietal areas in the macaque monkey, *Neuropsychologia* **29:**517–537.

Cooter, R. S., 1985, *The Cultural Meaning of Popular Science: Phrenology and the Organization of Consent in Nineteenth-Century Britain,* Cambridge University Press, Cambridge.

Corner, G., 1927, *Anatomical Texts of the Earlier Middle Ages: A Study in the Transmission of Culture,* Carnegie Institution of Washington, Washington, DC.

Cowey, A., and Gross, C., 1970, Effects of foveal prestriate and inferotemporal lesions on visual discrimination by rhesus monkeys, *Exp. Brain Res.* **11:**128–144.

Cranefield, P. F., 1974, *The Way In and the Way Out: Francois Magendie, Charles Bell and the Roots of the Spinal Nerves,* Futura, Mount Kisco, NY.

Critchley, M., 1953, *The Parietal Lobes,* Arnold, London.

Daniel, P., and Whitteridge, D., 1961, The representation of the visual field on the cerebral cortex in monkeys, *J. Physiol.* **159:**203–221.

Dean, P., 1976, Effects of inferotemporal lesions on the behavior of monkeys, *Psychiatr. Bull.* **83:**41–71.

Descartes, R., 1972, *Treatise of Man* [1662] (T. S. Hall, trans.), Harvard University Press, Cambridge, MA.

Desimone, R., Miller, E., Chelazzi, L. and Lueschow, A., 1995, Multiple memory systems in the visual cortex, in: *The Cognitive Neurosciences* (M. Gazzaniga, ed.), MIT Press, Cambridge, MA, pp. 475–486.

Dewhurst, K. E., 1982, Thomas Willis and the foundations of British neurology, in: *Historical Aspects of the Neurosciences* (F. Rose and W. Bynum, ed.), Raven Press, New York, pp. 327–346.

Dobson, J. F., 1926–27, Erasistratus, *Proc. R. Soc. Med.* **20:**825–832.

Dow, R. S., 1940, Thomas Willis (1621–1675) as a comparative neurologist, *Ann. Med. Hist.,* 2nd ser., **2:**181–194.

Duhamel, J., Colby, C., and Goldberg, M., 1992, The updating of the representation of visual space in parietal cortex by intended eye movements, *Science* **255:**90–92.

Edelstein, L., 1943, The history of anatomy in antiquity, in: *Ancient Medicine, Selected Papers of Ludwig Edelstein* (O. Temkin and C. Temkin, eds.), Johns Hopkins University Press, Baltimore, MD, pp. 247–301.

Elliot Smith, G., 1907, New studies on the folding of the visual cortex and the significance of the occipital sulci in the human brain, *J. Anat.* **41:**198–207.

Ettlinger, G., and Kahlsbeck, J. E., 1962, Changes in tactual discrimination and in visual reaching after successive and simultaneous bilateral posterior parietal ablations in the monkey, *J. Neurol. Neurosurg. Psychiatr.* **25:**256–268.

Ettlinger, G., Morton, H., and Moffett, E., 1966, Tactile discrimination performance in the monkey. The effect of bilateral posterior parietal and lateral frontal ablations, and of callosal section, *Cortex* **2**:5–29.

Farrington, B., 1944, *Thales to Aristotle,* Penguin, Harmondsworth, UK.

Farrington, B., 1949, *Theopharastus to Galen,* Penguin, Harmondsworth, UK.

Ferrier, D., 1873, Experimental researches in cerebral physiology and pathology, *West Riding Lunatic Asylum Med. Rep.* **3**:30–96.

Ferrier, D., 1875a, On the brain of monkeys, *Proc. R. Soc. Lond.* **23**:409–432.

Ferrier, D., 1875b, Experiments on the brain of monkeys, *Phil. Trans. R. Soc. Lond.* **165**:433–488.

Ferrier, D., 1876, *Functions of the Brain,* Smith, Elder, London.

Ferrier, D., 1878, *The Localisation of Cerebral Disease,* Smith, Elder, London.

Ferrier, D., 1886, *The Functions of the Brain,* Smith, Elder, London.

Ferrier, D., 1888, Shafer on the temporal and occipital lobes, *Brain* **11**:7–30.

Ferrier, D., and Yeo, G., 1884, A record of the experiments on the effects of lesions of different regions of the cerebral hemispheres, *Phil. Trans. R. Soc. Lond.* **175**:479–564.

Flechsig, P., 1886, *Gehirn und Steele,* Veit, Leipzig.

Fraser, P., 1972, *Ptolemaic Egypt,* Oxford University Press, Oxford.

Freeman, K., 1954, *The Pre-Socratic Philosophers,* Blackwell, Oxford.

Freud, S., 1953, *On Aphasia: A Critical Study* [1891], International University Press, New York.

Fritsch, G. J., and Hitzig, E., 1870, Uber die elektrische Erregbarkeit des Grosshirrns, *Arch. Anat. Physiol.* **1870**:300–332.

Galen, 1956, *On anatomical procedures, the surviving books* (2nd c.), Oxford University Press, Oxford.

Galen, 1962, *On anatomical procedures, the later books* (2nd c.), Cambridge University Press, Cambridge.

Galen, 1968, *On the usefulness of the parts of the body* (2nd c.), Cornell University Press, Ithaca, NY.

Galen, 1978–84, *On the Doctrines of Hippocrates and Plato* (2nd c.), Akademie-Verkag, Berlin.

Gall, F., and Spurzheim, J., 1835, *On the function of the brain and each of its parts: with observations on the possibility of determining the instincts, propensities and talents, or the moral and intellectual dispositions of men and animals, by the configuration of the brain and head* [1810] (W. Lewis, Jr., trans.), Marsh, Capen and Lyon, Boston.

Gattass, R., Gross, C. G., and Sandell, J. 1981, Visual topography of V2 in the macaque, *J. Comp. Neurol.* **201**:519–539.

Glickstein, M., and Rizzolatti, G., 1984, Francesco Gennari and the structure of the cerebral cortex, *Trends Neurosci.* **7**:464–467.

Glickstein, M., and Whitteridge, D., 1987, Tatsuji Inouye and the mapping of the visual fields on the human cerebral cortex, *Trends Neurosci.* **10**:350–353.

Goldberg, M., and Robinson, D., 1977, Visual responses of neurons in monkey inferior parietal lobule: The physiologic substrate of attention and neglect, *Neurology* **27**:350.

Goldberg, M., and Robinson, D., 1980, The significance of enhanced visual responses in posterior parietal cortex, *Behav. Brain Sci.* **3**:503–505.

Gratiolet, P., 1854, Note sur les expansions des racines cérébrales du nerf optique et sur leur terminaison dans une region determinée de l'écorde des hémisphères, *C. R. Acad. Sci.* **39**:274–278.

Gross, C. G., 1973, Visual functions of inferotemporal cortex, in *Handbook of Sensory Physiology, Vol. VII/3B* (H. Autrum, R. Jung, W. Lowenstein, D. McKay, and H. Teuber, eds.), Springer, Berlin, pp. 451–482.

Gross, C. G., 1981, Ibn-al-Haytham on eye and brain, vision and perception, *Bull. Islamic Med.* **1**:309–312.

Gross, C. G., 1987, Phrenology, in: *Encyclopedia of Neuroscience* (G. Adelman, ed.), Birkhauser, Boston, pp. 948–950.

Gross, C. G., 1993a, Hippocampus minor and man's place in nature: A case study in the social construction of neuroanatomy, *Hippocampus* **3**:403–415.

Gross, C. G., 1993b, Huxley vs. Owen: The hippocampus minor and evolution, *Trends Neurosci.* **16**:493–497.

Gross, C. G., 1994, How inferior temporal cortex became a visual area, *Cerebral Cortex* **4**:455–469.

Gross, C. G., 1995, Aristotle on the brain, *Neuroscientist* **1**:245–250.

Gross, C. G., 1996, Emanuel Swedenborg: A neuroscientist before his time, *Neuroscientist,* in press.

Gross, C. G., and Graziano, M. S. A., 1995, Multiple representations of space in the brain, *Neuroscientist* **1**:43–49.
Gross, C. G., Schiller, P. H., Wells, C., and Gerstein, G., 1967, Single-unit activity in temporal association cortex of the monkey, *J. Neurophysiol.* **30**:833–843.
Gross, C. G. Bender, D. B., and Rocha-Miranda, C., 1969, Visual receptive fields of neurons in inferotemporal cortex of the monkey, *Science* **166**:1303–1306.
Gross, C. G., Bender, D., and Gerstein, G. L., 1979, Activity of inferior temporal neurons in behaving monkeys, *Neuropsychology* **17**:215–229.
Gross, C. G., Rodman, H. R., Gochin, P., and Colombo, M., 1993, Inferior temporal cortex as a pattern recognition device, in: *Computational Learning and Cognition: Proceedings of the Third NEC Research Symposium* (E. B. Baum, ed.), SIAM, Philadelphia, PA, pp. 44–72.
Grusser, O., and Hagner, M., 1990, On the history of deformation phosphenes and the idea of internal light generated in the eye for the purpose of vision, *Ophthalmology* **74**:57–85.
Grusser, O., and Landis, T., 1991, Visual agnosias and other disturbances of visual perception and cognition, in: *Vision and Visual Dysfunction* (J. Cronly-Dillon, ed.), MacMillan, New York.
Guthrie, D., 1945, *A History of Medicine,* T. Nelson, London.
Hall, A. R., 1954, *The Scientific Revolution 1500–1800. The Formation of the Modern Scientific Attitude,* Beacon, Boston.
Hartline, H., 1938, The response of single optic nerve fibers of the vertebrate retina, *Am. J. Physiol.* **113**:59–60.
Hartline, H., and Graham, C., 1932, Nerve impulses from single receptors in the eye, *J. Cell. Comp. Physiol.* **1**:277–295.
Hecaen, H., and Albert, M., 1978, *Human Neuropsychology,* Wiley, New York.
Henschen, S., 1893, On the visual path and centre, *Brain* **16**:170–180.
Hippocrates, 1950, On the sacred disease, in: *The Medical Works of Hippocrates* (5th c. BCE) (J. Chadwick, ed.), Blackwell, Oxford, pp. 179–189.
Hofstadter, R., 1955, *Social Darwinism in American Thought,* Beacon, Boston.
Holmes, G. M., 1918a, Disturbance of vision by cerebral lesions, *Br. J. Ophthalmol.* **2**:353–384.
Holmes, G. M., 1918b, Disturbances of visual orientation, *Br. J. Ophthalmol.* **2**:449–468, 506–516.
Horsley, V. A. H., and Schäfer, E., 1888, A record of experiments upon the functions of the cerebral cortex, *Phil. Trans. R. Soc. Lond.* **179**:1–45.
Hubel, D., 1982, Evolution of ideas on the primary visual cortex, 1955–1978: A biased historical account, *Biosci. Rep.* **2**:435–469.
Hubel, D., 1988, *Eye, Brain and Vision,* Freeman, New York.
Hubel, D., and Wiesel, T., 1959, Receptive fields of single neurons in the cat's striate cortex, *J. Physiol.* **148**:574–591.
Hubel, D., and Wiesel, T., 1962, Receptive fields, binocular interaction and functional architecture in the cat's visual cortex, *J. Physiol.* **160**:106–154.
Hubel, D., and Wiesel, T., 1965, Receptive fields and functional architecture in two non-striate area visual areas (18 and 19) of the cat, *J. Physiol.* **195**:215–243.
Hume, D., 1948, An enquiry concerning human understanding, in: *Philosophical Essays Concerning Human Understanding* [1739] (L. Selby-Bigge and P. Nidditch, eds.), Clarendon, Oxford.
Hunter, R., and Macalpine, I., 1963, *Three Hundred Years of Psychiatry, 1535–1860,* Oxford University Press, Oxford.
Huxley, T. H., 1861, On the brain of *Ateles paniscus, Proc. Zool. Soc.* **1861**:247–260.
Hyvarinen, J., 1973, Functional properties of cells in the parietal, somatosensory and association cortices of the monkey, *Acta Physiol. Scand. Suppl.* **396**:36.
Hyvarinen, J., 1982, *The Parietal Cortex of Monkey and Man,* Springer-Verlag, Berlin.
Inouye, T., 1909, *Die Sehstorungen bei Schussverletzungen der kortikalen Sehsphare, nach Beobachtungen an Verwundeten der letzten japanischen Kriege,* W. Englemann, Leipzig.
Iwai, E., and Mishkin, M., 1968, Two visual foci in the temporal lobe of monkey, in: *Neurophysiological Basis of Learning and Memory* (N. Yoshii and N. Buchwald, eds.), Osaka University Press, Osaka, pp. 23–33.
Jackson, J. H., 1958, *Selected Writings of John Hughlings Jackson,* Basic Books, New York.
James, W., 1890, *Principles of Psychology,* Dover, New York.
Jung, R., 1992, Sensory research in historical perspective: Some philosophical foundations of per-

ception, in: *Handbook of Physiology: The Nervous System, Vol. III: Sensory Processes, Part 1* (J. Brookhart and V. Mountcastle, eds.), American Physiological Society, Bethesda, MD, pp. 1–73.
Kluver, H., 1926, Mescal visions and the eidetic types, *Am. J. Psychol.* **37:**502–515.
Kluver, H., 1928, *Mescal, The Divine Plant,* Paul, Trent and Teubner, London.
Kluver, H., 1933, *Behavior Mechanisms in Monkeys,* Chicago University Press, Chicago.
Kluver, H., 1948, Functional differences between the occipital and temporal lobes with special reference to the interrelations of behavior and extracerebral mechanisms, in: *Cerebral Mechanisms in Behavior* (L. Jeffress, ed.), Wiley, New York.
Kluver, H., and Bucy, L., 1938, An analysis of certain effects of bilateral temporal lobectomy in the rhesus monkey, with special reference to "psychic blindness," *J. Psychol.* **5:**33–54.
Kluver, H., and Bucy, L., 1939, Preliminary analysis of functions of the temporal lobes in monkeys, *Arch. Neurol. Psychiatr.* **42:**979–1000.
Kruger, L., 1963, Francois Pourfour du Petit 1664–1741, *Exp. Neurol.* **7:**2–5.
Kuffler, S., 1953, Discharge patterns and functional organization of mammalian retina, *J. Neurophysiol.* **16:**37–68.
Lashley, K. S., 1933, Integrative functions of the cerebral cortex, *Physiol. Rev.* **13:**1–42.
Lashley, K. S., 1948, The mechanism of vision. XVIII. Effects of destroying the visual "associative areas" of the monkey, *Genet. Psychol. Monogr.* **37:**107–166.
Lashley, K. S., 1950, In search of the engram, *Symp. Exp. Biol.* **4:**45–48.
Lashley, K. S., and Clark, G., 1946, The cytoarchitecture of the cerebral cortex of *Ateles:* A critical examination of the architectonic studies, *J. Comp. Neurol.* **85:**223–306.
Lewy, A., 1847, Ueber die Bedeutung des Antyllus, Philagrius and Posidonius in der Geschichte der Heilkunde, *Janus Leipzig* **3:**166–184.
Lindberg, D., 1976, *Theories of Vision from Al-Kindi to Kepler,* University of Chicago Press, Chicago.
Lissauer, H., 1890, Ein Fall von Seelenblindheit nebst einem Bertrage zur Theorie derselben, *Arch. Psychiatr. Nervenkr.* **21:**22–270.
Lloyd, G., 1970, *Early Greek Science: Thales to Aristotle,* Norton, New York.
Lloyd, G., 1973, *Greek Science After Aristotle,* Norton, New York.
Lloyd, G., 1975, Alcmaeon and the early history of dissection, *Sudhoff's Arch. Geschichte Med.* **59:**113-147.
Lloyd, G., 1978, The Hippocratic question, *Class Qt.* **28:**202–222.
Longrigg, J., 1988, Anatomy in Alexandria in the third century BC, *Br. J. Hist. Sci.* **21:**455–488.
Longrigg, J., 1993, *Greek Rational Medicine: Philosophy and Medicine from Alcmaeon to the Alexandrians,* Routledge, London.
Lynch, J., Mountcastle, V., Talbot, W., and Yin, T., 1977, Parietal lobe mechanisms for directed visual attention, *J. Neurophysiol.* **40:**362–389.
Manni, E., and Petrosini, L., 1994, Contributions by Bartolomeo Panizza to the anatomy and physiology of some cranial nerves, *J. Hist. Neurosci.* **3:**187–197.
Mazzarello, P., and Della Sala, S., 1993, The demonstration of the visual area by means of the atrophic degeneration method in the work of Bartolomeo Panizza (1955), *J. Hist. Neurosci.* **2:**315–322.
Meyer, A., 1971, *Historical Aspects of Cerebral Anatomy,* Oxford University Press, Oxford.
Meyer, A., and Hierons, R., 1965, On Thomas Willis's concepts of neurophysiology: Part II, *Med. Hist.* **9:**142–155.
Milner, A., Ockleford, E. M., and Dewar, W., 1977, Visuo-spatial performance following posterior parietal and lateral frontal lesions in stumptail macaques, *Cortex* **13:**350–460.
Minkowski, M., 1911, Zur Physiologie Der Sehsphare, *Pflugers Arch.* **141:**171–327.
Mishkin, M., 1954, Visual discrimination performance following partial ablations of the temporal lobe: II. Ventral surface vs. hippocampus, *J. Comp. Physiol. Psychol.* **47:**187–193.
Mishkin, M., 1966, Visual mechanisms beyond the striate cortex, in: *Frontiers in Physiological Psychology* (R. W. Russell, ed.), Academic Press, New York, pp. 93–118.
Mishkin, M., 1972, Cortical visual areas and their interactions, in: *The Brain and Human Behavior* (A. Karczmar and J. Eccles, eds.), Springer-Verlag, Berlin, pp. 187–202.
Mishkin, M., and Pribram, K., 1954, Visual discrimination performance following partial ablations of the temporaral lobe. I. Ventral vs. lateral, *J. Comp. Physiol. Psychol.* **47:**14–20.
Miyashita, Y., 1993, Inferior temporal cortex: Where visual perception meets memory, *Annu. Rev. Neurosci.* **16:**245–263.

Mountcastle, V., 1957, Modality and topographic properties of single neurons of cat's somatic sensory cortex, *J. Neurophysiol.* **20:**408–434.

Mountcastle, V., 1976, The world around us: Neural command functions of selective attention, *Neurosci. Res. Program Bull. Suppl.* **14:**1–47.

Mountcastle, V., Lynch, J., Georgopoulos, A., Sakata, H., and Acuna, C., 1975, Posterior parietal association cortex of the monkey: Command functions for operations within extrapersonal space, *J. Neurophysiol.* **38:**871–908.

Muller, J., 1838, On the specific energies of nerves, in: *A Sourcebook of the History of Psychology* (R. Herrnstein and E. Boring, ed.), Harvard University Press, 1965, Cambridge, MA, Vol. 2, pp. 26–33.

Munk, H., 1960, On the functions of the cortex [1881] in: *Some Papers on the Cerebral Cortex* (G. Von Bonin, trans.), Thomas, Springfield, IL, pp. 97–117.

Munk, H., 1890, Of the visual area of the cerebral cortex and its relation to eye movements, *Brain* **13:**450–469.

Nemesius, 1955, On the nature of man, in: *Cyril of Jerusalem and Nemesius of Emesa* (W. Telfer, ed.), Westminster Press, Philadelphia, pp. 224–453.

Neuburger, M., 1901, Swedenborg's Beziehungen zur Gehirnphysiologie, *Wien. Med. Wochenschr.* **44:**165.

Neuburger, M., 1981, *The Historical Development of Experimental Brain and Spinal Cord Physiology Before Flourens,* Johns Hopkins University Press, Baltimore, MD.

Newton, I., 1952, *Opticks: Or a treatise of the reflections, refractions, inflexions and colours of light* [1704], Dover, New York.

Nordenskiold, E., 1928, *The History of Biology,* Tudor, New York.

Pagel, W., 1958, Medieval and Renaissance contributions to the knowledge of the brain and its functions, in: *The History and Philosophy of Knowledge of the Brain and Its Functions* (F. Poytner, ed.), Thomas, Springfield, IL, pp. 95–114.

Panizza, B., 1855, Osservazioni sul nervo ottico, *Gior. Roy. Ist Lombardo Lett. Art Bibl. Ital.* **7:**237–252.

Perrett, D., Rolls, E., and Caan, W., 1982, Visual neurons responsive to faces in the monkey temporal cortex, *Exp. Brain Res.* **47:**329–342.

Plato, 1959, *Timaeus* (3rd c. BCE), Bobbs-Merrill, New York.

Polyak, H., 1957, *The Vertebrate Visual System,* University of Chicago Press, Chicago.

Pribram, K., and Bagshaw, M., 1953, Further analysis of the temporal lobe syndrome utilizing fronto-temporal ablations, *J. Comp. Neurol.* **99:**347–375.

Richmond, B., and Wurtz, R., 1982, Inferotemporal cortex in awake monkeys, in: *Changing Concepts of the Nervous System* (A. R. Morrison and P. Strick, eds.), Academic Press, New York, pp. 411–422.

Robinson, D., Goldberg, M., and Stanton, G., 1978, Parietal association cortex in the primate: Sensory mechanisms and behavioral modulations, *J. Neurophysiol.* **41:**910–936.

Ruch, T., Fulton, J., and German, W., 1938, Sensory discrimination in monkey, chimpanzee and man after lesions of the parietal lobe, *Arch. Neurol. Psychiatr.* **39:**919–938.

Sarton, G., 1954, *Galen of Pergamon,* University of Kansas Press, Lawrence, KA.

Sarton, G., 1959, *A History of Science, Vol. 1, Ancient Science Through the Golden Age of Greece,* Harvard University Press, Cambridge, MA.

Schäfer, E., 1888a, Experiments on special sense localisations in the cortex cerebri of the monkey, *Brain* **10:**362–380.

Schäfer, E., 1888b, On the functions of the temporal and occipital lobes: A reply to Dr. Ferrier, *Brain* **11:**145–161.

Schäfer, E., 1900, The cerebral cortex, in: *Textbook of Physiology* (E. Schäfer, ed.), Pentland, Edinburgh.

Schiller, F., 1965, The rise of the 'enteroid processes' in the 19th century: Some landmarks in cerebral nomenclature, *Bull. Hist. Med.* **41:**515–538.

Schiller, F., 1979, *Paul Broca: Founder of French Anthropology, Explorer of the Brain,* University of California Press, Berkeley.

Schneider, G. E., 1967, Contrasting visuomotor functions of tectum and cortex in the golden hamster, *Psychol. Forschung.* **31:**52–62.

Schrödinger, E., 1954, *Nature and the Greeks,* Cambridge University Press, Cambridge.

Semmes, J., 1965, A non-tactual factor in astereognosis, *Neuropsychologia* **3:**295–315.

Siegel, R., 1968, *Galen's System of Physiology and Medicine,* Karger, New York.

Siegel, R., 1970, *Galen on Sense Perception,* Karger, New York.

Sigerist, H., 1951, *A History of Medicine, Vol. I, Primitive and Archaic Medicine,* Oxford University Press, Oxford.

Sigerist, H., 1961, *A History of Medicine, Vol. II, Early Greek, Hindu and Persian Medicine,* Oxford University Press, Oxford.

Singer, C., 1952, *Vesalius on the Human Brain,* Oxford University Press, Oxford, pp. 18–20.

Singer, C., 1957, *A Short History of Anatomy and Physiology from the Greeks to Harvey,* Dover, New York.

Spencer, H., 1885, *Principles of Psychology,* Williams and Norgate, London.

Spillane, J. D., 1981, *The Doctrine of the Nerves, Chapters in the History of Neurology,* Oxford University Press, Oxford.

Steckerl, F., 1958, *The Fragments of Praxagoras of Cos and His School,* Brill, Leiden.

Swedenborg, E., 1740–41, *Oeconomia regni animalis* (A. Clissold, trans.), Franc, London.

Swedenborg, E., 1882–87, *The Brain Considered Anatomically, Physiologically and Philosophically* (R. L. Tafal, trans.), Speirs, London.

Swedenborg, E., 1938, *Three Transactions on the Cerebrum* (A. Acton, trans.), Swedenborg Scientific Association, Philadelphia.

Talbot, S., and Marshall, W., 1941, Physiological studies on neural mechanisms of visual localization and discrimination, *Am. J. Ophthalmol.* **24:**1255–1263.

Tamburini, A., 1880, Rivendicazione al panizza della scoperta delcentro visivo cordicale, *Rev. Sper. Frent. Med. Leg.* **6:**152–154.

Tanaka, K., 1996, Inferotemporal cortex and object vision, *Annu. Rev. Neurosci.* **19:**109–140.

Temkin, O., 1953, Remarks on the neurology of Gall and Spurzheim, in: *Science, Medicine and History* (E. Underwood, ed.), Oxford University Press, Oxford, Vol. II, pp. 282–290.

Theophrastus, 1917, On the senses, in: *Theophrastus and the Greek Physiological Psychology Before Aristotle* (2nd c. BCE) (G. M. Stratton, trans.), Allen and Unwin, London, pp. 67–151.

Titchener, E., 1910, *A Textbook of Psychology,* MacMillan, New York.

Trevarthen, C. B., 1968, Two mechanisms of vision in primates, *Psychol. Forschung.* **31:**299–337.

Ungerleider, L., and Mishkin, M., 1982, Two cortical visual systems, in: *Analysis of Visual Behavior* (D. Ingle, M. Goodale, and R. Mansfield, eds.), MIT Press, Cambridge, MA, pp. 549–586.

Von Economo, C., 1929, *The Cytoarchitectonics of the Human Cerebral Cortex,* Oxford University Press, Oxford.

Von Helmholtz, H., 1863, *On Sensation of Tone as Physiological Basis for the Theory of Music,* Dover, New York.

Von Staden, H., 1989, *Herophilus: The Art of Medicine in Early Alexandria,* Cambridge, University Press, Cambridge.

Walshe, F., 1958, Some reflections upon the opening phase of the physiology of the cerebral cortex 1850–1900, in: *The History and Philosophy of Knowledge of the Brain and Its Function* (F. Poynter, ed.), Thomas, Springfield, IL, pp. 223–234.

Willis, T., 1664, *Cerebral Anatomy,* Martyn and Allestry, London.

Wilson, M., 1957, Effects of circumscribed cortical lesions upon somesthetic and visual discrimination in the monkey, *J. Comp. Physiol. Psychol.* **50:**630–635.

Wilson, M., Stamm, J., and Pribram, K., 1960, Deficits in roughness discrimination after posterior parietal lesions in monkeys, *J. Comp. Physiol. Psychol.* **53:**535–539.

Woolam, D., 1957, The historical significance of the cerebrospinal fluid, *Med. Hist.* **1:**91–114.

Woolam, D., 1958, Concepts of the brain and its functions in classical antiquity, in: *The History and Philosophy of Knowledge of the Brain and Its Function* (F. Poynter, ed.), Thomas, Springfield, IL, pp. 5–18.

Woolsey, C., 1971, Comparative studies on cortical representation of vision, *Vis. Res. Suppl.* **3:**365–382.

Young, R. M., 1970, *Mind, Brain and Adaptation in the Nineteenth Century,* Oxford University Press, Oxford.

Zola-Morgan, S., 1995, Localization of brain function: The legacy of Franz Joseph Gall (1758–1828), *Annu. Rev. Neurosci.* **18:**359–383.

2

Past and Present Ideas About How the Visual Scene Is Analyzed by the Brain

PETER H. SCHILLER

1. Introduction

The dominant view during the past 40 years has been that the visual system analyzes the visual scene by breaking it down into basic attributes such as color, form, motion, depth, and texture. It was proposed that individual, dedicated neurons and specific visual areas are devoted to the analysis of each of these attributes. Current research has challenged these views by emphasizing that neurons, especially in the cortex, have multifunctional properties and therefore serve as general-purpose analyzers rather than feature detectors. Consequently it appears that most extrastriate visual areas, rather than each being devoted to the analysis of a specific basic visual attribute, perform several different tasks and thereby engage in more advanced and complex analyses than had been realized.

PETER H. SCHILLER • Department of Brain and Cognitive Sciences, Massachusetts Institute of Technology, E25-634 MIT, Cambridge, Massachusetts 02139.

Cerebral Cortex, Volume 12, edited by Rockland *et al.* Plenum Press, New York, 1997.

The aim of this chapter is to examine the basic assumptions and ideas about the manner in which the visual scene is analyzed by the visual system. The topics central to the chapter are the specificity of neuronal analysis, the specificity of visual areas, and the representation of the visual scene in the primary visual cortex.

2. The Specificity of Neurons

In 1906 Santiago Ramón y Cajal traveled to Stockholm to receive the Nobel Prize in Medicine and Physiology he was to share with Camillo Golgi. At the ceremonies Cajal gave a gracious address, the central focus of which was his enunciation of the neuron doctrine. He was followed on the podium by Camillo Golgi, who proceeded to attack Cajal viciously while promulgating his own reticular theory. Ironically, these two radically different views of nervous tissue organization, which formed one of the hottest debates of the time, were derived from the same reduced silver-stain method fortuitously discovered by Golgi; looking at such tissue through the microscope, Golgi saw a continuous network, whereas Cajal perceived individual neurons whose processes were in close apposition to, but not continuous with, other neuronal processes (Ramón y Cajal, 1989). After a protracted battle the neuron doctrine won out and went on to have a major influence on our views of nervous tissue function.

One of the central ideas that has emerged from this effort is the specificity of individual neuronal analysis. This idea had become deeply ingrained in Cajal's thinking. A specific example comes from his work on the retina and pertains to the inferences he made about the rod and cone pathways. That there are two different kinds of photoreceptors—the rods and the cones—was discovered by Schultze in the 1860s; he conjectured that the rods are for crepuscular vision, having observed that they do not exist in the fovea, where night vision is poor. Cajal, with excellent anatomical methods at his disposal, wanted to understand the manner in which the rods and cones made their central connections. He reasoned as follows:

> Since the impression received by the rod is different from that taken up by the cone, it is necessary from every point of view that each of these specific impressions should be conveyed through the retina by a separate channel.
>
> When we reason with common sense and lift the war club determined upon vigorous action, nature ultimately hears us. Knowing what I was looking for, I began to explore eagerly . . . ; and finally, as the reward of my faith, there deigned to appear most clearly and brilliantly those two types of bipolar cells demanded by theory and guessed by reason . . . Observe how one variety of bipolar makes a contact . . . with the body of a certain giant ganglionic neuron . . . The bipolar cell for the cone . . . comes into relation with the . . . medium and small ganglionic cells. (Ramón y Cajal, 1989)

Here we have what appears to be the first compelling evidence in the visual system for the creation of two parallel pathways, one for the cones and the other for the rods, which Cajal believed to course independently to the central nervous system via separate sets of ganglion cells and perform different analyses. This arrangement is depicted in Fig. 1A. Cajal's view, as we shall see, has recently been effectively challenged.

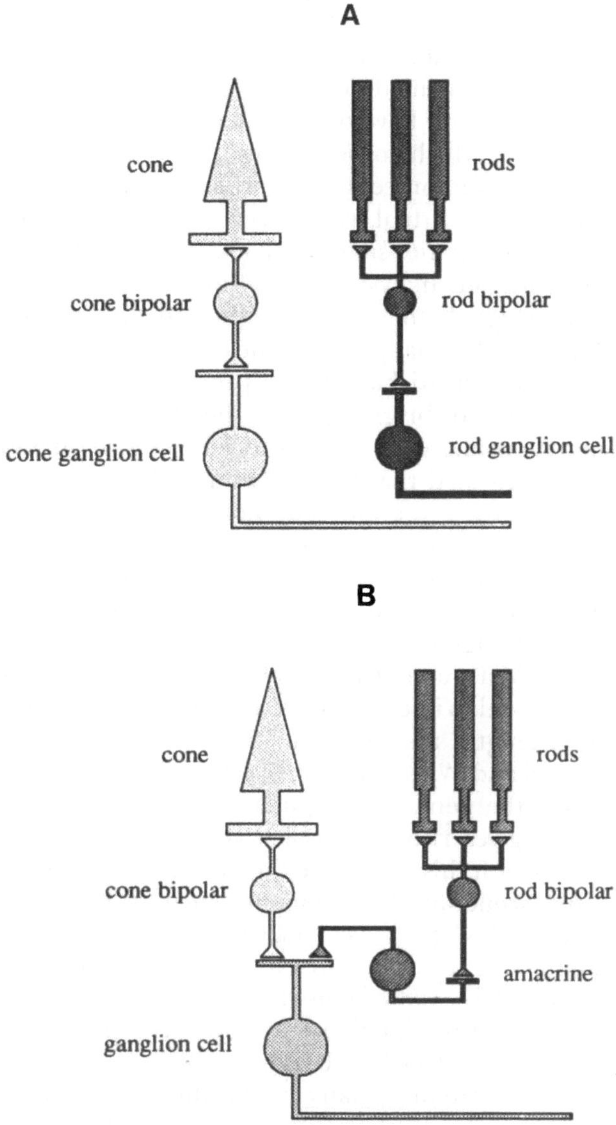

Figure 1. The basic inferred connections of the rods and cones in the primate retina. (A) Cajal's model according to which the rods and cones connect, via the rod and cone bipolars, to rod and cone ganglion cells, yielding thereby separate pathways to the central nervous system. (B) Connections of the rods and cones as determined in recent studies showing that ganglion cells receive convergent input from the rods and the cones; the cone bipolars make direct connections with the ganglion cells, while the rod bipolars make their connections via amacrine cells. For a more detailed account of these connections for the ON and OFF channels and for various other classes of retinal ganglion cells see Waessle and Boycott (1991) and Schiller (1995).

The neurocentric thought that ascribes highly selective functions to single neurons reached its zenith with the advent of single-cell recordings. Much as the atom had at one time been thought to be the basic unit of matter, the neuron had become the basic functional unit of the brain. Neurons were thought of as features detectors that extracted specific attributes from the environment (Barlow, 1972). Driven by convictions much like Cajal's when he conceived of the rod and cone pathways, neuroscientists soon discovered high levels of specificity in many parts of the visual system.

The first recordings from single cells in the visual system were made by the Nobel laureate Halden Keffer Hartline, who arrived at this great achievement by painstakingly dissecting the fibers of the frog optic nerve (Hartline, 1938). His effort yielded immediate payoff by revealing that retinal ganglion cells are of three different types: ON, OFF, and ON/OFF, a discovery that became all the more remarkable when it was determined subsequently that the ON and OFF systems are formed not at the level of the photoreceptors, which all hyperpolarize to light and have only graded potentials, but at the level of the bipolar cells, one set of which conserves and the other of which inverts the signals it receives from the photoreceptors (Werblin and Dowling, 1969). Thus, at the level of the bipolar cells, a dual system is created from a single one. That nature has gone to such complexities suggests that great benefits must be reaped from this arrangement for the analysis of visual information. The initial ideas were that the ON system signals the onset of light and the OFF system the termination of light. Another idea that gained some degree of acceptance, suggested initially by Hartline as well as Richard Jung (Jung, 1961), is that the ON system gives rise to the perception of brightness and the OFF system to the perception of darkness. A third consideration was advanced by more physiologically minded investigators, who suggested that the ON and OFF systems have evolved to provide for the antagonistic center/surround organization of receptive fields in retinal ganglion cells and in the cortex (Schiller, 1995).

Further advances and inferences about neuronal function soon followed: the studies of Barlow and of Lettvin and his collaborators identified bug-detector and edge-detector neurons in the frog retina (Barlow, 1953; Lettvin *et al.*, 1959; Maturana *et al.*, 1960). In the mammalian cortex the Nobel laureates Hubel and Wiesel discovered neurons that responded selectively to the orientation, length, and width of line segments or to specific directions of motion (Hubel and Wiesel, 1962a, b; 1968). A hierarchic principle emerged according to which the geniculate input to cortex first forms simple cells, an aggregate of which gives rise to complex cells that in turn create hypercomplex neurons. Along with this idea came the notion that the visual system first breaks down the visual scene into fine elements, such as the orientation of line segments, spatial frequencies, specific colors, directions of motion, and so on, and then somehow combines these attributes to yield a perception of the world (Livingstone and Hubel, 1987, 1988). This view has led to the idea of the pontifical neuron, which, when it fires, signals such highly complex attributes as the faces of specific personnages (Barlow, 1972).

Subsequently, many of the discoveries of visual system organization were interpreted with these ideas of neuronal specificity in mind. One of the major findings following on the heels of striate cortex organization was that several parallel pathways, in addition to the rod and cone systems and the ON and OFF channels, originate in the retina (Enroth-Cugell and Robson, 1966; Schiller, 1986; Stone, 1983). This was interpreted as further proof that the visual scene is indeed broken down into specific subelements in the visual system. Among the parallel pathways identified were the cat X, Y, and W channels, which subsequently acquired several different names, as they were found in just about every species that had been subjected to study. In the primate, the putative analogue to the X and Y systems was identified as the midget and the parasol retinal ganglion cells—midget because of the size of the cells, parasol because of the umbrella-shaped arbors (Waessle and Boycott, 1991). Driven by the prevalent view of the times, these two systems were believed to extract different attributes from the visual scene. What exactly these attributes were was not fully agreed upon;

according to one view, the midget system, coursing through the parvocellular portions of the lateral geniculate nucleus toward visual cortex, extracts color, and the parasol system, coursing through the magnocellular layers of the lateral geniculate nucleus, extracts luminance information. Subsequently new complexities were realized; the midget system was thought to extract both color and form, and the parasol system both motion and stereopsis (Livingstone and Hubel, 1987; Schiller, 1986). Thus the idea that the visual scene is broken down into specific basic subunits persisted; what varied were the specific subunits to which the various systems responded.

Just what the basic elements are of visual perception and how many elements there might be remain unsettled issues. Driven by philosophical presuppositions and psychological insight, however, it was proclaimed that visual perception is indeed broken down into logical subunits; implicit in such thought was the assumption that such subunits should have their counterparts in the brain. The basic elements of visual perception were conceived to be color, form, depth, motion, and texture. The assumed neural isomorphism demanded that there be neurons and brain areas that perform these analyses. Even more problematic, if often ignored, was the problem these assumptions created: once broken down into various categories, how does the brain put the image together again? One of my students, Karl Zipser, aptly refers to this as the Humpty Dumpty problem:

> Humpty Dumpty sat on a wall,
> Humpty Dumpty had a great fall;
> All the King's horses, and all the King's men
> Cannot put Humpty Dumpty together again.

Until recently, then, the prevalent view that has occupied center stage in systems neuroscience is that individual neurons perform dedicated, unique, and highly specific analyses of a basic nature.

3. The Specificity of Brain Areas

The specificity of neural tissue had, of course, been extensively debated well before Cajal carved his indelible impressions onto the history pages of biology. Francis Joseph Gall, the father of phrenology, postulated not only that different brain areas perform different jobs, but that the qualities in which an individual excels are the product of a greater than average amount of brain tissue devoted to that particular quality, which, in turn, is reflected in the protrusions of the skull. He and John Gaspar Spurzheim systematized these qualities and their putative locations in 1809 (Gall and Spurzheim, 1989). Many textbooks to this day are graced by a fanciful picture of the human head with inscriptions designating the "mental faculties" that were believed to reside in the various brain regions. The fact that scientific thought is strongly driven by the outlook, the values, and the politics of the times was clearly reflected in the selection of these qualities: of Spurzheim's 35 areas, 21 were devoted to such propensities and sentiments as amativeness, cautiousness, benevolence, veneration, wonder, and ideality, qualities that in today's world, brain scientists—or anyone else for that matter—would allocate only limited space in the brain at best.

scorned, the idea of areal specificity persisted, soon to be supported by much more respectable efforts. Studying patients with brain injuries, Broca and Wernicke identified centers for the processing of language, and, after a few false starts, Hermann Munk and others established that the primary visual cortex is located in the occipital pole (Glickstein, 1985). These findings led, on the one hand, to the elusive task of specifying all brain areas and, on the other, to the denial of specificity for many functions. In the occipital lobe, Brodmann, in the early part of the century, histologically differentiated areas 17, 18, and 19, and proposed that each has a distinct visual function (Brodmann, 1909). Documented first in the visual cortex by Talbot and Marshall (1941) using evoked potentials, it was established that in most visual areas the retinal surface—and hence the visual field—is laid out in an orderly fashion. Physiological mapping studies took this discovery several steps further by uncovering many more visual cortical areas, in most of which an orderly re-representation of the visual field was demonstrated (Felleman and Van Essen, 1991; Kaas, 1978, 1982).

Fueled in part by the pervasive belief in neuronal specificity, research in many brain areas using single-cell recordings ushered in a new age of phrenology. As more and more brain regions were identified and investigated, in most stunning specificity was documented. That each of these areas has evolved for the purpose of analyzing a specific visual attribute became the most popular and most widely embraced view. In 1970 Hubel and Wiesel published a paper in *Nature* declaring area 18 (V2) to be the region that is uniquely involved in the analysis of stereoscopic depth perception (Hubel and Wiesel, 1970). Shortly thereafter Zeki claimed that area V4 is the region devoted to color analysis and area V5 (MT) to the analysis of motion, giving rise thereby to the one-area, one-function hypothesis (Zeki, 1973, 1974). Among the most celebrated were the neurons recorded in the inferotemporal cortex of the monkey that responded selectively to faces and toilet brushes (Gross, 1973). Here, at last, seemed to be a region where pontifical cells resided for the perception of complex visual attributes.

The topographic order seen in the visual cortex, coupled with the orderly arrangement of the left and right eye inputs and the orientation selectivity of V1 neurons, led to specific hypotheses about the functional organization, or cytoarchitecture, of this area and subsequently of extrastriate regions. Hubel and Wiesel proposed an elegant model, referred to by some as the "ice cube" model, as shown at the top of Fig. 2, according to which the striate cortex is composed of hypercolumns, each of which represents a small region of the visual field within a full complement of orientations and the left and right eye inputs reside, thus comprising the basic module necessary to perform the local calculations—whatever they might be—for perceptual analysis (Hubel and Wiesel, 1977).

A new level of complexity in the organization of extrastriate cortex was realized when Margaret Wong-Riley discovered the cytochrome oxidase puffs in area V1, an orderly distribution of small, metabolically active ovoid regions of about 250 by 150 μ in diameter, which were shown to form neat rows centered within each of the ocular dominance columns (Horton and Hubel, 1980; Hubel and Wiesel, 1969; Wong-Riley *et al.*, 1978). This new finding suggested that another, yet-unrealized specialization exists in the striate cortex. In hot pursuit of this curious new discovery, it was soon claimed that the puffs contained cells that were unoriented and color-selective (Livingstone and Hubel, 1984). These that were unoriented and color-selective (Livingstone and Hubel, 1984). These claims were taken as further proof that the visual scene is indeed broken down into the basic attributes of color, form, motion, and depth. Specific projections were identified from V1 to other extrastriate areas in support of the specificity of areas for the analysis of color, motion, and stereopsis. Revised models followed, one of which is shown in the middle panel of Fig. 2, to which one might refer as the "radial model."

Further intricacies in the organization of the striate cortex were uncovered when, after arduous and clever efforts, optical recording methods succeeded. Reexamination of striate cortex organization now revealed curious discon-

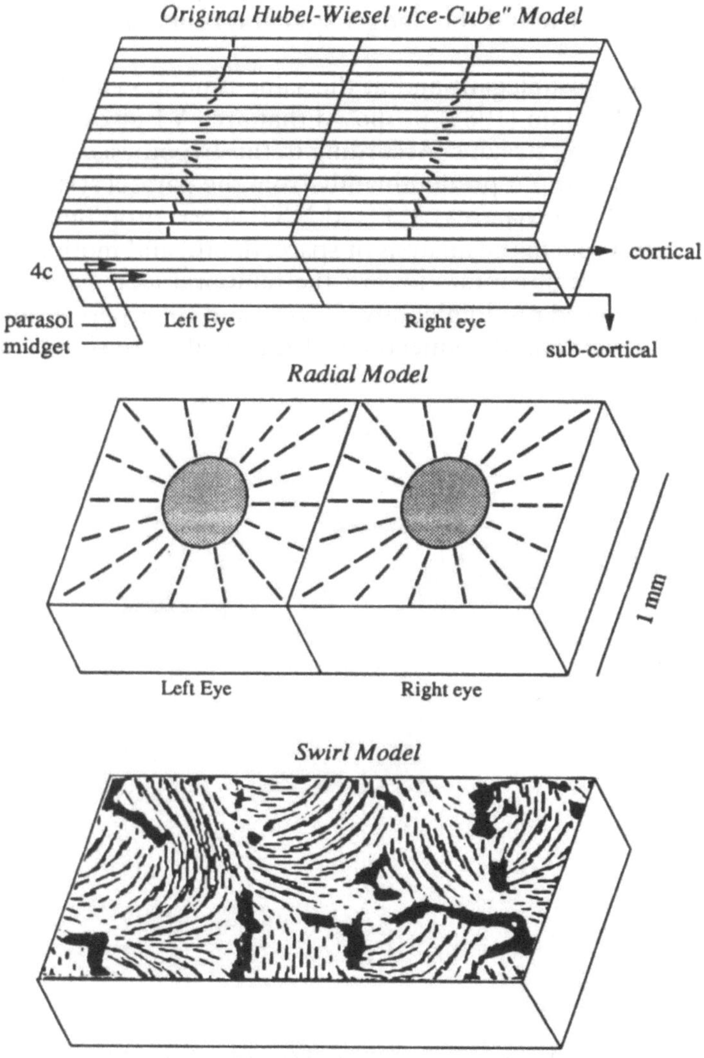

Figure 2. Three models of visual cortex organization—the original "ice cube" model, the radial model, and the swirl model, which is based on optical recording results. Also shown are (1) the inputs to layers 4c alpha and beta from the parvocellular and magnocellular portions of the lateral geniculate nucleus that receive their inputs from the midget and parasol cells of the retina and (2) the inference that the upper layers project to cortical and the deeper layers to subcortical regions of the brain.

tinuities or fractures. The stunning pictures generated from this work yielded additional impressions of the spatial layout of striate cortex (Blasdel, 1992a, b). An example of this is shown in the bottom panel of Fig. 2, entitled here the "swirl model," which shows the spatial arrangement of orientation selectivities in a patch of cortex; the dark areas designate regions of discontinuity in the layout of orientations.

These relatively modest modifications in the perceived layout of the striate cortex did not question the central principles of columnar and laminar organization. Anatomical studies disclosed well-defined laminations and established that the inputs from the midget and parasol systems coursing through the lateral geniculate nucleus terminate in separate cortical layers, with the parasol system terminating in layer 4cα and the midget system terminating in 4cβ (Hubel and Wiesel, 1972).

In one of the most elegant and at the same time most ambitious schemes, Ungerleider and Mishkin (1982) proposed that area V1 gives rise to two major information-processing streams. According to this hypothesis, the so-called dorsal stream receives its input predominantly from the parasol system and courses toward the parietal lobe via areas MT and MST; it was proposed that this pathway plays a central role in the analysis of space, depth, and motion. The second, so-called ventral stream, courses toward the temporal lobe via area V4; it was proposed to be involved predominantly in shape and color perception. The basic structures and hypothesized connections of this model are shown in Fig. 3.

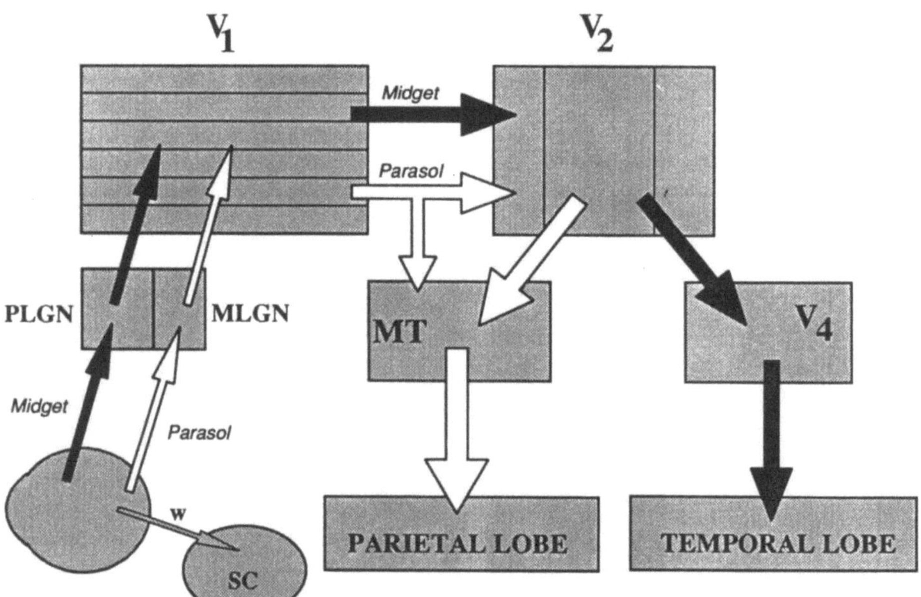

Figure 3. Basic hypothesized connections of the midget, parasol, and W-cells of the retina. The midget cells project to the parietal lobe via the parvocellular layers of the lateral geniculate nucleus (PLGN), areas V1 (striate cortex), V2, and MT. The parasol cells project to the temporal lobe via the magnocellular layers of the lateral geniculate nucleus (MLGN), areas V1, V2, and V4. The W-cells project to the superior colliculus. For sake of clarity, interconnections among these structures and the projections of the parasol and W-cells to other subcortical regions are omitted.

Following on the heels of the discovery and the specification of the visual areas in mammals were lesion studies that lent additional support to the one-area, one-function hypothesis. Lesions of area V4 led to deficits in color vision, especially color constancy, and lesions of area MT led to deficits in motion perception (Newsome and Wurtz, 1988; Wild *et al.*, 1985). Perhaps most dramatic were studies that showed highly selective deficits in humans such as achromatopsia following lesions of regions that putatively included area V4 (Heywood and Cowey, 1987; Heywood *et al.*, 1992; Zeki, 1990).

Coupled with the idea of dedicated, specific neuronal analysis, the dominant thought until recently has been that visual areas are highly specific and perform unique visual analyses.

4. Countercurrents to Neuronal Specificity

Emblematic of events to come that are reshaping our ideas about neuronal specificity today was the discovery that Cajal's impressions about the connections of the rods and cones were not quite correct. While the rods and cones do indeed have their own bipolars, it has now been established that they do not feed into separate pathways coursing toward the central nervous system (Schiller, 1995; Waessle and Boycott, 1991). Instead, in those portions of the retina where rods and cones coexist, ganglion cells receive a convergent input from them and do so via different intraretinal pathways as depicted in Fig. 1b. Consequently, when ganglion cells are driven by the cones, they can contribute to color vision, whereas at night, when driven by the rods, they subserve achromatic, crepuscular vision. Thus the same ganglion cell can process two rather different kinds of signals. Also discovered was the fact that both the midget and the parasol retinal ganglion cells receive convergent input from the rods and cones (Schiller *et al.*, 1990; Wiesel and Hubel, 1966). These findings seriously question the idea of labeled lines and highlight the multifunctional properties of neurons even at early stages of visual processing.

Further studies provided additional evidence regarding the multiplicity of messages carried by the ganglion cells. With the discovery of the midget and parasol systems, which were seen as pathways that broke down the visual scene into different components, it has been suggested that the midget system carries wavelength information and the parasol system luminance information. This attractive idea, however, was short-lived. Even the earliest single-cell recording studies—the very ones that had shown that the rod and cone input converges on ganglion cells—established that both the midget and the parasol systems are capable of carrying luminance information (Wiesel and Hubel, 1966). Studies in which the visual capacities of animals were examined following selective disruption of either the midget or the parasol system showed that most basic visual functions, such as brightness, texture, pattern, shape, motion, flicker, and depth perception, as well as scotopic and photopic vision, are carried out by both systems (Schiller and Logothetis, 1990; Schiller *et al.*, 1990). The major difference is that the midget system can process information up to higher spatial frequencies, whereas the parasol system can do so up to higher temporal frequencies (see Fig. 7). The exception is differential wavelength analysis, which appears to be done exclusively by the midget system. These findings led to the idea that the midget and parasol systems have evolved not for the purpose of analyzing different basic visual attributes, but to extend the range of vision, with the midget system extending it in the spatial and wavelength domains and the parasol system extending it in the temporal domain (Schiller and Logothetis, 1990). This inference applied also to the rods and cones of the retina, whose emergence has resulted in extending the range of vision in the light-intensity domain.

Early ideas about the functions of the ON and OFF systems that originate in the retina have also given way to different views largely brought about, as is so often the case, by technical advances. One of these was the use of a newly created artificial neurotransmitter analogue, 2-amino-4-phosphonobutyrate (APB), which was shown to render inoperative the ON bipolar cells without significantly affecting the OFF bipolars (Schiller, 1995; Slaughter and Miller, 1981). This made it possible to examine the consequences of reversibly blocking the ON system on the activity of single cells in various parts of the visual system and also to determine how this procedure affects visual perception (Schiller, 1982, 1995). These studies established that after ON channel block with APB, OFF cells for the most part retain their center/surround organization and cortical cells retain their direction and orientation specificities. It appears, therefore, that the ON and OFF channels did not evolve to create center/surround organization or other receptive field attributes such as orientation and direction specificity in the cortex. Studies in which APB was used to block the ON channel in animals trained to perform on a variety of visual tasks showed that the prime deficit lay in the ability to detect visual stimuli made visible by virtue of light increment, but not by light decrement. These and related findings suggest, therefore, that the reason the ON and OFF channels have evolved might have been to optimize information transfer by having excitatory processes for both increments and decrements in illumination. The rapidity with which events in the visual world need to be processed in the animal kingdom is at a premium. A fish in the ocean, in order to survive, must be able to respond with minimum delay both to a bird looming dark against the sky as it swoops down and to a predatory fish that rises from the dark depths illuminated from above; the OFF system provides the best signal for the approaching bird and the ON system for the fish.

Surprisingly, visual acuity using high-contrast stimuli is little affected by blocking the ON channel with APB as a result of which half of the bipolar cells in the retina are rendered inoperative. This finding is less baffling now because refined anatomical studies have established that in the central retina of the primate, each cone gives rise to at least two midget bipolar cells and two sets of retinal ganglion cells whose dendritic processes arborize in different sublamina in the inner plexiform layer. In the retina and also in the lateral geniculate nucleus the ON and OFF channels in the midget and parasol systems remain separate (Nelson *et al.*, 1978; Schiller, 1995; Waessle and Boycott, 1991). However, in the striate cortex many cells receive convergent input from all of these systems, indicating once again the consequent multifunctional properties of neurons.

Attesting to the complexity of the visual system is the fact that in the retina there is a class of ganglion cells that falls into the grab-bag category of W-cells which receive a convergent input from ON and OFF bipolar cells, thereby forming cells of the sort already noted by Hartline (1938): the ON/OFF ganglion cells. The W-cells as well as the parasol cells project to several visual nuclei in the primate that include the thalamus, the pretectum, and the superior colliculus (Hendry and Yoshioka, 1994; Malpeli *et al.*, 1981; Schiller, 1986; Schiller and Logothetis, 1990; Stone, 1983). The midget ganglion cells, however, send their axons only to the lateral geniculate nucleus (LGN).

It is noteworthy that the ON/OFF organization seen in the retina has its counterpart in the temperature receptors of the somatosensory system, one set of which is excited by increases and the other by decreases in temperature; both are thereby signaled to the central nervous system by an excitatory process.

Since retinal ganglion cells and LGN cells are capable of carrying numerous messages, it should come as no surprise that cells in the striate cortex, rather than extracting a single feature from the visual scene, have greatly expanded multifunctional properties. Striate cortex cells, by virtue of extensive interconnections, transform the thalamic input in a variety of ways. While it is well known that cells in the striate cortex show selectivity for orientation, direction, spatial frequency, wavelength, and depth, often overlooked is the fact that many or all of these properties are commonly found in a single cell. Thus many single cells, rather than being selective for a single specific feature of the visual scene, such as orientation of line segments or direction of motion or a specific color, respond to many different features. That cortical cells have such multipurpose properties was already shown in early papers by Hubel and Wiesel (1959, 1962, 1968). It is evident, then, that a given cell will respond to motion when motion is present in the visual scene and will, in addition, respond to line segments of certain orientation in the absence of motion, to small stimuli that contain little information about orientation, to a range of colors when the cones are active, and to achromatic stimuli under dark-adapted conditions. In addition, the cell's response may be modulated by stereoscopic depth and spatial frequency. These cells, therefore, can convey information about many different aspects of the visual scene; they are multifunctional.

While it was recognized early on that in the striate cortex there is convergent input to single cells from the ON and OFF channels and from the two eyes, it was assumed that such convergence does not take place for the midget and parasol systems. Instead, these systems were believed to remain separate in the striate cortex and beyond, as depicted in Fig. 3. This issue was examined in a series of studies in which either the parasol or the midget system was selectively blocked at the level of the lateral geniculate nucleus while recording from single neurons in the striate cortex (Malpeli *et al.*, 1981). The results established that while some cells are indeed driven selectively by the midget or the parasol system, there are numerous cells which receive a convergent input from these two systems. Figure 4 depicts this state of affairs to highlight the multiprocessing properties of individual striate cortex cells.

Early alternatives to the cortical feature detector idea, which resulted from the discovery of spatial frequency selectivity in the striate cortex, were models in which cells acted as spatial and temporal filters. Rather than extracting specific features, neurons were thought to form classes within each of which cells re-

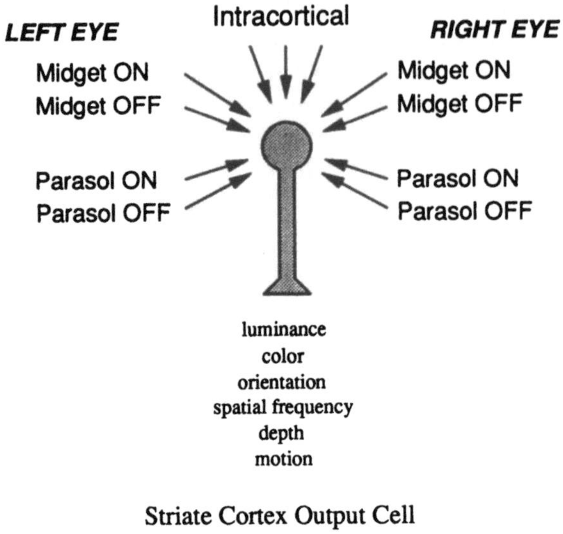

Figure 4. Schematic representation of the inputs and the analyses performed by a single multipurpose neuron in the striate cortex.

sponded optimally to different portions of a continuum such as spatial frequency (De Valois and De Valois, 1980). This was an idea akin to the way Young and Helmholtz (1924) thought about the processing of wavelength information.

The neuronal specificity implied by the feature detector model gives rise to another major problem: it would require an inordinately large number of neurons in the brain to cope with the incredible richness and variety present in the visual scene, numbers which it fortunately does not possess. Such high specificity would also mean that most neurons would be silent most of the time, a rather uneconomical state of affairs. Yet economy in living organisms is at a premium. The more you can do with less, the better. Size, weight, and metabolic requirements must be kept to a minimum. To accomplish this, sensory analysis has to be optimized using every possible trick. An example of how the nervous system goes about economizing can be seen in the processing of wavelength information. While we can perceive hundreds of colors, there are only three types of cone photoreceptors. Contrary to "common sense," for these three photoreceptors to do a good job in recording the color image they need to be broadly rather than narrowly tuned; as a result, typically all three photoreceptors are activated by most stimuli, but the ratio of their activity differs even for small differences in the wavelength composition of objects. To facilitate this process further and to increase sensitivity, the output of the photoreceptors is combined in an "opponent fashion" forming just two major systems in the midget retinal ganglion cells, red/green and blue/yellow (Lennie, 1984). How these signals are combined in the visual cortex to give rise to the hundreds of colors we are capable of discerning is only marginally known. It would seem unlikely, however, that each color experience would be a product of singular neuronal activity.

Similar considerations apply to other sensory analyses, such as stereoscopic depth perception and motion. Physiological studies of stereoscopic depth perception initially led to the idea that there is a separate complement of neurons for each disparity in each hypercolumn that represents a tiny segment of the visual field. Once again, such an arrangement would require an inordinate number of neurons. Instead, it appears now that stereopsis follows principles similar to those of color; only a few distinct disparities are coded (Poggio, 1972; Poggio and Poggio, 1984). The gamut of stereoscopic depths the organism is capable of perceiving is the product of computational processes that are derived from the relative activity of the different neuronal groups. Thus is appears that there is no specific neuron for each depth, just as there is no specific neuron for each color.

Similarly, the neural analysis of motion not long ago was thought to involve highly specific neurons. Logically investigators have broken down the way we analyze motion into several subcategories. These include planar, radial, and circular motion, object motion, and self-motion which gives rise to optic flow (Duffy and Wurtz, 1991; Roy and Wurtz, 1990). Having created such categories, researchers surmised that each of them is represented in the brain by a different set of neurons or by different subregions of areas in the parietal lobe. Recent studies have shown, however, that the distinctions made by the mind do not necessarily have their counterparts in the brain. Wurtz and his collaborators have shown that most neurons in area MST, for example, coprocess several of these attributes. For example, of the planar, radial, and circular types of motion tested in one set of studies, they showed that 40% of the neurons carry information about all three components, 30% carry information about two, and only 20% carry information about just one component, indicative of a continuum of specificity.

Another prevalent view that has recently come under scrutiny is the idea that striate cortical neurons have very small receptive fields whose analysis is consequently confined to very small regions of the visual field. Presaged by work on the retina showing surprisingly extensive lateral interactions (McIlwain, 1966), current work has shown that in the striate cortex the responses of neurons can be dramatically influenced by events relatively distant from their receptive fields. Furthermore, such influence can be quite general in nature. Thus it has been shown that the responses of single cells can be modulated by a variety of stimulus conditions: borders that are made visible by texture differences, by luminance, depth, motion, and color, can all influence a single cell. An example of this from the work of Zipser *et al.* (1994) appears in Fig. 5. The data shown in this figure establish, first of all, that the presence of a border quite distant from the classical receptive field of the cell notably affects its response. Second, it shows that this effect occurs in the cell irrespective of the means by which the border is made visible; borders created by differences in texture composition, depth, motion, and luminance all can modify the response of a given neuron. The response modification is most evident during the latter portions of the discharge, implicating possible feedback systems in giving rise to the effect. Thus it appears that many cells in the striate cortex, instead of being dedicated, have rather general properties that allow them to process several kinds of visual information.

Given these considerations, one might ask why earlier work placed so much emphasis on the dedicated, feature detection properties of single cells in the visual system. The answer lies, in part, in the limitation of single-cell recordings from which the majority of the inferences were derived. Most cells can be studied for only brief periods of time before they disappear from the tip of the microelectrode; during this brief period, their responses can be examined for only a rather limited set of stimulus conditions and the stability of the response over time cannot be assessed at all. These techniques are akin to showing a Martian a photograph of a deciduous tree taken in the winter. Based on such a picture the Martian would be most unlikely to gain the idea of leaves, of growth, of damage to the branches due to storms, and of the wide variety of species. Thus when it comes to single-cell physiology, we are much like Martians; we are restricted to brief glimpses of a small sample of neurons. And the sample is limited indeed. During a typical vertical electrode penetration through the 2 mm of cortical gray matter, investigators are lucky to isolate 4 or 5 cells out of the more than 100 that the electrode encounters in its path. Since much of the work has been carried out on paralyzed animals, it is difficult to imagine that the crucial role of the multifarious inputs that include feedback from higher cortical

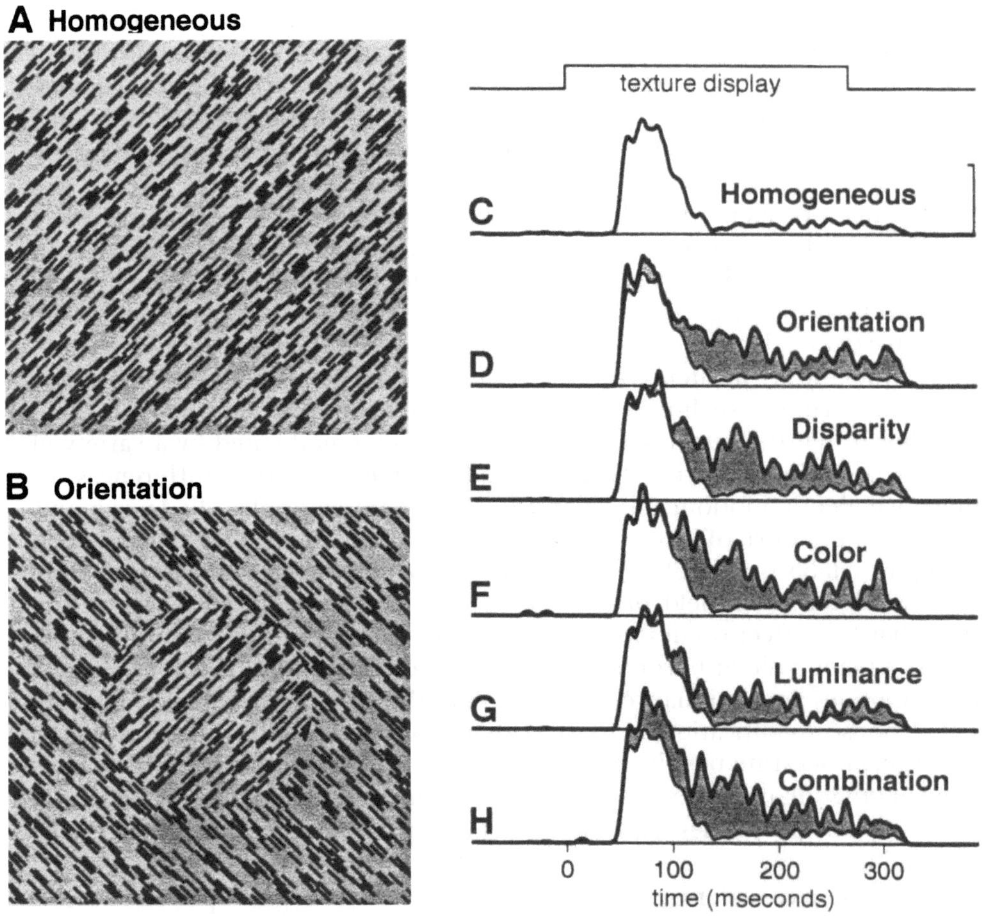

Figure 5. The responses of a striate cortex neuron to texture surfaces in which a shape is made visible by a variety of cues. (A, B) Shape segmentation using differential texture cues. The response of a cell is shown for conditions in which a shape is made visible by a variety of means as indicated on the right side of each histogram.

centers would operate normally. In this case of the analysis of visual stimuli, it must also be recognized that the retinal image of even small objects will drive thousands of cells, the response of each of which may be only subtly modified and would be difficult to detect when studied singly.

When Alan Turning conceived of the computer, his ideas were modeled on how he thought the brain worked; the brain, he believed, was a general-purpose instrument (Hodges, 1983). Now that computers have reached high levels of sophistication and have become part of everyday life, it is not uncommon to reverse this thought process and use computers as metaphors for the brain. The major input device to the early general-purpose computers, however, was a dedicated keyboard not unlike the classic Underwood typewriter on which any key, when punched, produced a specific, invariant letter on the paper. By contrast, today's computers have been given keyboards with conditional properties; what the outcome is of pressing any given key depends on the program that is running and on what other keys have concurrently been activated. The situation is similar, but of course far more complicated and elaborate, even at early stages, in the visual system. Already the responses of retinal ganglion cells and certainly of striate cortex cells is determined not only by the specific stimulus that appears in the center of the cell's receptive field, but, as just shown, on what other stimuli appear in the surrounding parts of the visual field and on the state of the system.

According to these new views, then, most neurons in the visual system are not dedicated to the analysis of a singular, highly specific basic feature of the visual scene. Instead, they have multipurpose properties and can integrate information over a considerable area; each neuron has the capacity to perform quite a number of different, if specific tasks. Thus, rather than constituting dedicated systems, neurons, especially in the cortex, have multipurpose properties and engage in parallel, integrated processing. It appears, therefore, that much as the atom is no longer regarded as the basic unit of matter, the neuron is no longer to be regarded as the basic information processing element of nervous tissue.

If specific sensations and perceptions are not signaled by the activity of dedicated sets of neurons, how else could this be accomplished? How can multipurose neurons yield the necessary specificity? Several hypotheses have been advanced, according to most of which coding involves populations of neurons that produce different spatiotemporal signals for different percepts (McClurkin *et al.,* 1994). A straightforward, if simple-minded scheme exemplifying such ideas appears in Fig. 6, in which all connections are excitatory. Each horizontal trace depicts the activity of a different neuron, with the vertical lines representing action potentials generated during a short time period, say 500 msec. In the first area, area V, circumscribed by the dotted rectangle, say the striate cortex, the activity of nine hypothetical neurons is shown whose receptive fields cover three adjacent locations, A, B, and C. At each location the three neurons shown have different direction specificities, 1, 2, and 3. The cells respond to both stationary, flickering stimuli and to motion. In the upper display in Fig. 6 (FLICKER) responses elicited by a stationary flickering stimulus are depicted: all the cells whose receptive fields are stimulated by this stimulus discharge to each flash (cells B1, B2, and B3); the other cells remain silent, since the stimulus is not within their receptive fields. In the lower display (MOTION) the responses elicited in these same cells by a moving stimulus are shown: only those cells

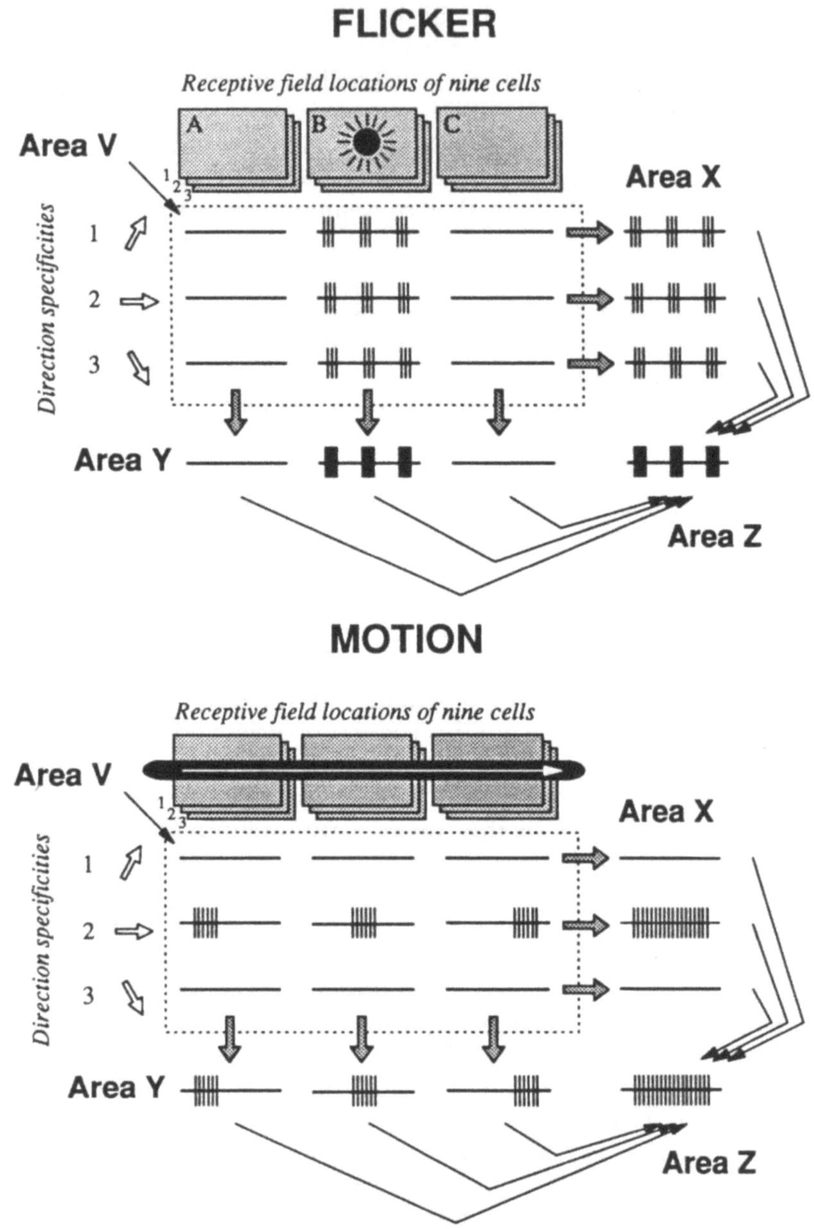

Figure 6. Hypothetical scheme of the responses and connections of a set of neurons in which action potentials of different spatiotemporal patterns are elicited by flickering and by moving stimuli.

selective to the particular direction of motion extant are activated (cells A2, B2, and C2).

Next it may be assumed that these cells make a variety of connections with other areas in the visual system, say areas MT and V4. In one case the information in this hypothetical scheme is pooled for space, but not for direction (area X); in the other case information is pooled for direction, but not for space (area

Y). As can be seen, cells in both of these areas continue to respond to the flashing and the moving stimuli, but do so very differently. Even if the outputs of stages X and Y are combined (at area Z), the temporal patterning of activity is distinct, thereby providing the opportunity to yield different inscriptions for flicker and for motion.

The scheme shown here is exceedingly simple. The majority of neurons, already in the striate cortex, have more than two specificities; furthermore, the interactions among neurons involve both excitatory and inhibitory circuitry. Increasing complexity by adding just one additional factor, such as color, for example, would necessitate a three-dimensional representation in the scheme shown in Fig. 6. Additional capacities in multifunctional cortical cells, such as selectivity for spatial frequency or for stereoscopic depth, would add further dimensions to the scheme, and would yield so many possible interconnections and spatiotemporal response profiles that it would boggle the mind. Yet examination of the connectivity of cells within and between the various visual areas reveals a complexity that far exceeds what is seen in most models that have been proposed (Felleman and Van Essen, 1991; Lund, 1988; McClurkin *et al.*, 1994).

5. Countercurrents to Areal Specificity

Adherents to the specialization of cortical function, spearheaded by Gall and Spurzheim in the beginning of the 19th century, encountered powerful opposition 100 years later. In 1927, Heinrich Kluver, in summarizing the visual functions of occipital cortex, concluded that it is impossible to reduce visual perception to simple capacities that are separately localized (Kluver, 1927). A more general argument was made by Henry Head (1926), who said, "No function is 'localized' strictly in any part of the cortex and no form of activity, somatic or physical, is built up into a mosaic of elementary processes which become evident when it is disturbed by a lesion of the brain."

These views subsequently received experimental support from Lashley, whose influential lesion studies in rats showed that deficits in learning were a product not of what region of cortex, but of how much of it had been removed (Lashley, 1963).

The ensuing anatomical and physiological discoveries during the past 40 years swung the pendulum back, reasserting the specificity of cortical areas. Now, as a result of many recent discoveries, the pendulum is once again on the reverse swing. A series of careful studies revealed that there are more than 30 visual areas in the primate cortex that make more than 300 interconnections (DeYoe and Van Essen, 1988; Felleman and Van Essen, 1991). Most of the extrastriate areas are not uniformly organized as one might expect were each devoted to the execution of a single, unique function; instead, each area contains modules within which neurons perform a great variety of analyses. Even the claimed specificity within identified modules has come under fire recently. In the striate cortex, where modules defined by the cytochrome oxidase label had been thought to contain unoriented, color-selective cells and regions outside these areas orientation-selective neurons, recent studies have failed to find such clear-cut specialization (Leventhal *et al.*, 1995; Purves *et al.*, 1992; Swindale, 1990). In V2, the characteristics of neurons in the various anatomically identified modules have also been shown to exhibit less specialization than past studies had postulated (Levitt *et al.*, 1994).

These and related studies have raised questions about the specificity of laminar and columnar organization in the visual cortex, which were central to the idea of neuronal and areal specificity in the visual system. In addition to the studies questioning the selective nature of neuronal analysis in the various cortical modules, it was shown that the inputs and outputs of the striate cortex are not quite as specific as had been believed. The inputs from the lateral geniculate nucleus were shown to terminate not only in layers 4cα and 4cβ (see Fig. 2), but in many other layers as well. The outputs from the deeper layers of the cortex were shown to project not only to subcortical areas, but also to other cortical areas (Lund, 1988). With the discovery of several other specific analyses by neurons, each assumed to be done within a specific little module, such as the analysis of spatial frequency, depth, direction, and color, just how all of these might be laid out in the cortex became quite a conceptual problem. It was established, furthermore, that simple cells, rather than being confined to layers that lie in close apposition to 4c, reside in most cortical layers. Studies failed to find compelling evidence for columnar or laminar organization for direction selectivity; in fact, many simple cells were shown to have subfields with opposing selectivity for light and dark edges (Schiller *et al.*, 1976). No obvious spatial organization has been demonstrated for spatial frequency and stereopsis in the primate. The segregation of the left and right eye inputs, while exhibiting clear-cut columnar organization in layer 4c, turned out to be less obvious in other layers, where most cells were shown to be binocularly activated. These new considerations highlight the multifunctional properties of neurons and regions in the striate cortex and suggest that a new, more dynamic model superseding those shown in Fig. 2 may be just around the corner.

It was also discovered that the inputs to some of the extrastriate areas are not as specific as the one-area, one-function hypothesis would demand. Neurophysiological studies in which the midget or the parasol inputs to cortical cells were reversibly blocked at the level of the lateral geniculate nucleus revealed that many neurons in area V4 receive convergent input from both the midget and parasol systems (Ferrera *et al.*, 1992). Area MT, on the other hand, appears more specialized; it receives a predominant, although not unique, input from cells in V1 and V2 that are driven by parasol cells (Maunsell *et al.*, 1990). In fact, as noted in the previous section, many neurons already in V1 receive convergent input from the midget and parasol systems much like the retinal ganglion cells that receive convergent input from the rods and cones (Malpeli *et al.*, 1981). It has been estimated, in fact, that a single cell in area V4 receives input from several thousand V1 cells; each of those V1 cells is in turn influenced by input not only from many LGN cells, but also from lateral and feedback processes from thousands of other V1 and extrastriate cells (Desimone and Schein, 1987).

Recent studies that have examined the characteristics of single cells in area V4 have shown that this area is not uniquely devoted to the analysis of color (Schiller, 1994; Desimone and Schein, 1984; Schein and Desimone, 1990). Many cells process information about shape, depth, and even motion. More strikingly, studies carried out in trained, behaving monkeys show that the responses of many of these cells are modulated by such factors as attention and the state of the animal; inputs through modalities other than visual can also have an influence (Haenny and Schiller, 1988; Haenny *et al.*, 1988; Moran and Desimone, 1985; Schiller, 1994). Thus area V4 appears to have properties that in the past have been attributed to regions called "association cortex." Even in area MT, which for the most part continues to be perceived as a region devoted to the analysis of motion (Albright, 1984; Albright *et al.*, 1984), it is clear that most single cells, in addition to exhibiting directional preference, are also orientation-specific; the cells respond vigorously to stationary stimuli that are flashed on and off or that impinge on the retina in the course of saccadic eye movements, suggesting that they participate in the analysis of the visual scene in the absence of motion information.

Current experiments that have examined the consequences of brain lesions for the most part lend strong support to a more generalist view of information processing in the visual cortex. It can be safely asserted that selective lesions of areas V4 and MT in monkeys fail to "knock out" highly specific basic visual functions (Schiller, 1993; Schiller and Lee, 1991). After MT lesions monkeys show deficits in motion and flicker perception, but do so only when the tasks introduced by the experimenter are made quite difficult. A surprising degree of motion processing is retained; furthermore, after MT lesions there is notable recovery with time. V4 lesions result in only mild to moderate deficits on color perception tasks and on pattern perception tasks. More pronounced deficits are evident, however, when tasks that are used necessitate higher level analyses, such as the recognition of objects that have undergone various transformations (size and perspective changes, partial occlusions, etc.) visual learning, and the selection of objects in the visual scene that are less salient (smaller or lesser contrast) than other objects with which they appear.

It has also been established that paired lesions of areas V4 and MT do not produce deficits that are more than the sum of the deficits produced by the lesions separately. Such paired lesions fail to manifest deficits in brightness perception, stereopsis, and coarse shape perception. These findings are not in consonance with the two-stream hypothesis, according to which one stream passes through V4 and the other through MT, and indicate that additional pathways must be involved in visual information processing that bypass these two areas.

Figure 7 provides a summary of deficit magnitudes on a broad range of visual capacities following lesions of the parvocellular and magnocellular portions of the lateral geniculate nucleus that selectively block the midget and parasol systems and of areas V4 and MT (Schiller, 1993; Schiller and Logothetis, 1990).

Clinical studies of patients have been another significant source of support of areal specificity (Heywood *et al.*, 1992; Meadows, 1974; Schiller, 1994; Zeki, 1990). It has been demonstrated, for example, that some patients with damage to the lingual and fusiform gyri of the occipital lobe have major deficits in color vision (achromatopsia). Although damage to visual cortex in most afflicted patients is not sharply confined, and typically involves both white and gray matter, these reports have been hailed as major support for areal specificity. However, it must be noted that clinical studies suffer from two serious fallacies. The first is a

Deficit magnitude following PLGN, MLGN, V4 and MT lesions

VISUAL CAPACITY		*PLGN*	*MLGN*	*V_4*	*MT*	*V_4+ MT*
color vision		severe	none	mild	none	mild
texture perception		severe	none	mild	none	mild
pattern perception		severe	none	mild	none	mild
shape perception	fine	severe	none	mild	none	mild
	coarse	mild	none	none	none	none
brightness perception		none	none	none	none	none
coarse scotopic vision		none	none	none	none	none
contrast sensitivity	fine	severe	none	mild	mild	mild
	coarse	mild	none	none	mild	mild
stereopsis	fine	severe	none	none	none	none
	coarse	pronounced	none	none	none	none
motion perception		none	moderate	none	moderate	moderate
flicker perception		none	severe	none	pronounced	pronounced
choice of "lesser" stimuli		severe	none	severe	none	severe
visual learning		*not tested*	*not tested*	severe	none	severe
perception of transformed objects		*not tested*	*not tested*	pronounced	*not tested*	*not tested*

Figure 7. Summary results of deficit magnitude on visual tests following lesions of the parvocellular (PLGN) and magnocellular (MLGN) portions of the lateral geniculate nucleus, of area V4, and area MT. The choice of "lesser" stimuli refers to conditions in which monkeys perform on an oddity task in which one stimulus, the target, is different from the other stimuli, which are all identical; this target stimulus is smaller, has less contrast, or contains dots that move at a slower rate than do the other stimuli in the array. When the situation is reversed so that the target is the "greater" stimulus, the deficit seen is minimal in all cases. The perception of transformed objects refers to a condition in which the monkey performs on a match-to-sample paradigm and has to make matches to samples that have been transformed in a variety of ways. The transforms include changes in the size of the match stimuli, in occluding them to various degrees, and in varying the amount of contour information provided. The other visual tasks listed should be self-explanatory. For details see Schiller *et al.* (1990).

sampling fallacy. Published clinical studies for the most part report on only interesting cases. Of the thousands of humans that suffer brain damage, only a handful make it into published reports. The plethora of unreported cases are likely to have deficits that are complex, diffuse, and uninterpretable, hardly the stuff of a good report. What we have, therefore, is a heavily biased sample in favor of specificity that provides a simple and probably rather incorrect view of brain specialization. The second fallacy lies in the limited number of tests administered to the patient and the choice of these tests, typically based on preconceived notions. As an extreme example, if one hypothesizes that area V4 processes color and then proceeds to administer only tests for color vision, a deficit seen would, by exclusion, seem specific for this capacity. To assert specificity convincingly, a detailed battery of tests assessing many visual capacities is essential. These considerations suggest, therefore, that caution must be exercised in inferring areal specificity from clinical reports.

During the past few years some of the most dramatic results raising questions about neuronal and areal specificity have come from studies examining the plasticity of the nervous system. What these studies reveal is that in many, but not all parts of the brain dramatic reorganization can take place. Several mechanisms have been identified. The rapidity with which some of these changes can occur suggests that dynamic, in-place networks are involved. Longer term changes involving structural changes have also been documented. Some regions of course are more subject to such reorganization than are others. It is surprising that such changes have been reported to also occur near both the sensory and motor ends (Merzenich and Sameshima, 1993).

Figure 8 shows some of the modifications that need to be made in the initial scheme that was presented in Fig. 3. Since the interconnections and areas involved are far too numerous to be readily assimilated, only those pathways and structures that are most salient to the arguments I have made are depicted. Three major pathways are shown to originate in the retina, although several more have been identified. The midget system projects only to the lateral geniculate nucleus, whereas the parasol and W-systems project to several other regions besides the lateral geniculate nucleus, of which only the superior colliculus is shown (Hendry and Yoshioka, 1994; Schiller and Malpeli, 1977; Schiller and Logothetis, 1990). Cells projecting from the striate cortex to other regions may be driven by the midget, the parasol, or both systems. Area MT and the superior colliculus receive inputs predominantly from the parasol system, whereas area V4 receives a mixed input from these two pathways. Little is known about the role of the W pathway in the geniculostriate system. In addition to

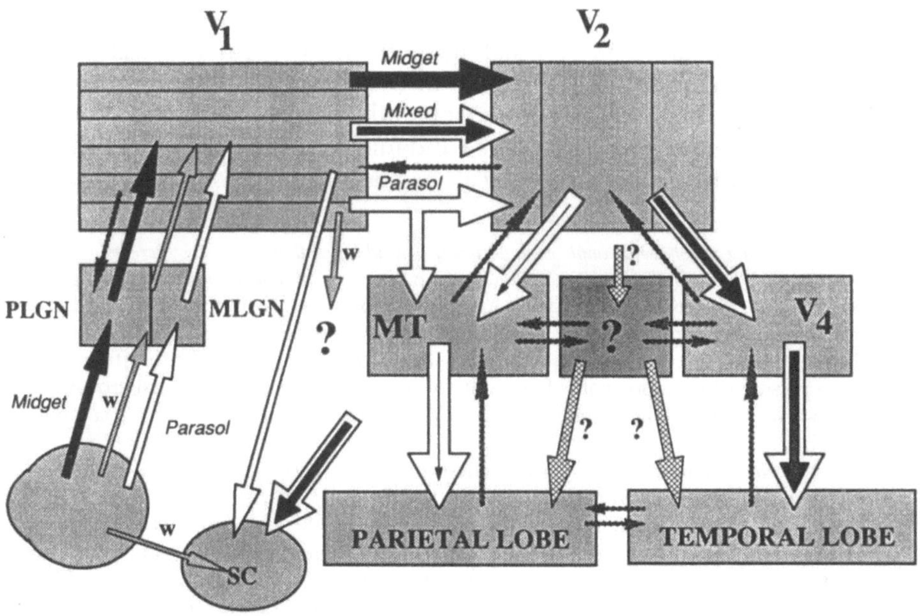

Figure 8. Modified and expanded schematic of major structures and connections in the visual system as considered in the text.

inputs from area, MT and V4, the parietal and temporal lobes receive inputs from other, not yet clearly identified extrastriate regions. The superior colliculus receives projections from most extrastriate regions as well as from the temporal, parietal, and frontal (not shown) lobes, which in turn most likely receive inputs from all three systems that originate in the retina. Most of the areas shown make extensive interconnections.

According to current views, then, it appears that each extrastriate visual area, rather than performing a unique, one-function analysis, is engaged, as are most neurons in the visual system, in many different tasks. Visual areas in posterior cortex make profuse interconnections and receive input from many different systems that originate early in the visual pathways. Consequently, they carry out interactive analyses of many different aspects of the visual scene. It is therefore entirely possible that the process of visual analysis does not truly involve the breaking down of the visual scene into basic elements. It appears, therefore, that Humpty (Vision) Dumpty may not have fallen and broken up after all. He is still sitting up there, whole and healthy, wondering just what the process really is that makes him visible.

6. Brain Topography and Visual Representation

I now turn to the thorniest problem in vision, about which so little is known even at this advanced stage of our study of the visual system: the representation and recognition of visual objects.

One of the early ideas of object perception uses the metaphor of photography; the impression of an object imaged on the retinal surface is sent to the brain, where it is represented relatively unchanged on the topographically ordered visual cortical surface. Cajal was one of the early investigators who thought about this problem. He did so in relation to the manner in which the optic fibers make their central connections, which at the time was under considerable debate. Cajal thought that "the fundamental crossing of the optic tracts is necessarily bound up with the physical mechanism of vision." Cajal wrote optimistically:

> *Everything will have a simple explanation if it is admitted that the correct perception of an object implies the congruence of the cerebral surfaces of projection, that is those representing each point in space.* Hence, in order that the mental perception may be unified and may agree exactly with the external reality, or, in other words, in order that the image conveyed through the right eye may be continuous with that conveyed through the left eye, the intercrossing of the optic paths from side to side is quite necessary; a total crossing in animals with panoramic vision, a partial crossing in animals endowed with a common visual field.
>
> The first diagram [Fig. 9A] shows the form and direction which the mental visual image would have had on the supposition that the optic nerves had not crossed. The incongruity of the two images is evident: that projected through the right eye does not fit with that from the left eye, and it would be impossible for the animal to synthesize the two images into a continuous representation. The horizon would be presented as a panormaic view formed from two photographs, the right and left ones being laterally inverted.
>
> Let us now examine the mental image resulting from the intercrossing of the optic nerves, an intercrossing adopted by nature in case of lenticular eyes. [Figure 9B] shows

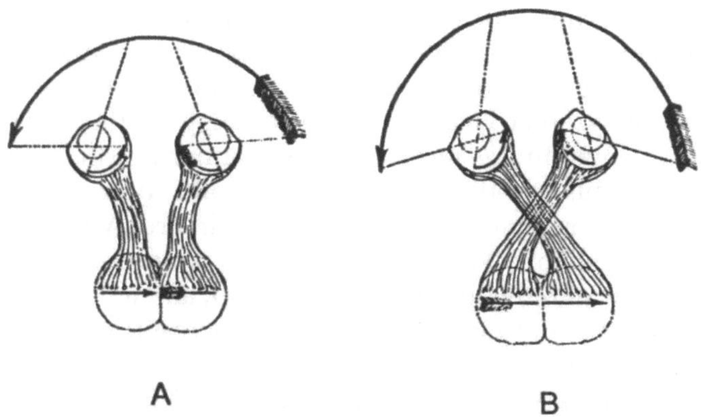

Figure 9. Drawings by Cajal showing the consequences of uncrossed (A) and completely crossed (B) fibers at the optic chiasma on the hypothesized representation of the visual field in the brain. The "central image" of the arrow in the visual field is depicted in each of the drawings. The representation of the arrow is continuous only with the crossed fibers. Irrelevant parts of the original figures have been deleted.

> with the greatest clearness, that, thanks to the crossing, the two images, right and left, correspond and make a continuous panorama, the later inversion disappearing. (Ramón y Cajal, 1989).

Cajal went on to acknowledge the fact that in mammals the fibers from the temporal hemiretinae remain uncrossed, while those from the nasal hemiretinae cross: "Things take place somewhat differently in the mammals, in which the double visual projection reproduces the same region of space. In these animals there exists the uncrossed tract, through which the duplication of visual images is ingeniously avoided, while the advantages of the crossing are retained."

Clearly, the major underlying idea here is that the shape of objects is represented in the brain as the result of a reproduction of the original, much like a photograph, which makes an imprint on the cortical surface and does so in a continuous fashion across the two hemsipheres. In order for the object to be perceived, one implicitly assumes that this display is viewed by the mind. Skeptics refer to this as the homunculus theory of object perception.

Modern-day reincarnation and elaboration of this idea has resulted in elegant mathematical formulations that accurately specify the rules of transformation whereby retinal images are translated to form cortical imprints in area V1 (Schwartz, 1994). These models take into account what is known about the organization of the striate cortex, such as the differential spatial representation of the central and peripheral portions of the visual field (generally discussed in terms of the magnification factor), ocular dominance columns, and cytochrome oxidase patches. On the basis of this it has been established that the mapping process is conformal. This means that although the cortical impression of objects is transformed because much more cortical space is allocated for foveal than for peripheral vision, the relationship of nearby points is maintained by preserving angular relationships among them, thereby providing locally equal magnification in all directions. This is true, however, only within each hemifield, as will be considered further below.

While these formulations are most satisfying and yield spectacular computer-generated displays, they do not really provide major new insights about how the brain goes about seeing and recognizing objects. What we now know about the topographic layout in the cortex suggests that Cajal's idea about continuous representation was incorrect. The manner in which the optic fibers project to the thalamus and then to the cortex results in having corresponding points in the two eyes reach corresponding points in area V1, where the inputs from the two eyes combine. While this projection pattern is probably essential for the realization of stereoscopic depth perception, among other things, it does not actually make for a continuous representation across the hemispheres as had been proposed by Cajal.

To make these points evident, the manner in which simple objects are laid out on the surface of the lissencephalic portion of the rhesus monkey striate cortex (central 7–8 deg of the visual field) is shown in Fig. 10 and 11. In Fig. 10 the cortical representation is shown for seven identical arrows under two conditions. In the first the arrows are confined to the right visual hemifield, which projects to the occipital cortex of the left hemisphere. Because of the magnification factor, the images of the arrows on the cortical surface are different in size, have various curvatures, and are laid out quite differently from the way they appear in the visual field. The difference in representation becomes especially notable when the arrows are centered on the vertical meridian, as depicted on the right side of Fig. 10. In this case the images are split between the hemispheres and, notably, do not form any sort of continuum across the left and right occipital cortices. This point is made more tellingly in Fig. 11, where the cortical representation of three identical disks is depicted under similar conditions when the disks are confined to the right hemifield and when they are presented along the midline. While in the former cases the shape of the two disks appearing more peripherally is reasonably well preserved, the central circle forms an oval and its cortical representation is much larger than that of the peripheral inscription. When the disks are centered along the vertical meridian, their representation in each hemisphere forms half-circles without any evidence for continuous representation. One may wonder, indeed, how such imprints lead to the perception of a solid disk.

Add to this curious state of affairs the fact that when we peruse the visual field we typically make two to four saccadic eye movements per second. With each glance even a small object activates thousands of cells in the cortex. The result is a series of impressions in the striate cortex that would border on the incomprehensible were there persistence to them. Of course one of the major features of image processing is that the neural activity produced by each image does not persist within the retina, the lateral geniculate nucleus, and the striate cortex. The images are transformed further in higher cortical areas where one presumes the different retinal impressions that arise during eye and body movements are discounted to allow for a stable representation of the world. Part of this process, one might assume, is reflected in the progressive enlargement of receptive fields in higher visual areas and in the loss of topographic representation. As a result, it has been conjectured that the system eventually ends up with cell groups that produce the same pattern of discharge to specific shapes irrespective of where they appear in the visual field. This is evident already, on a

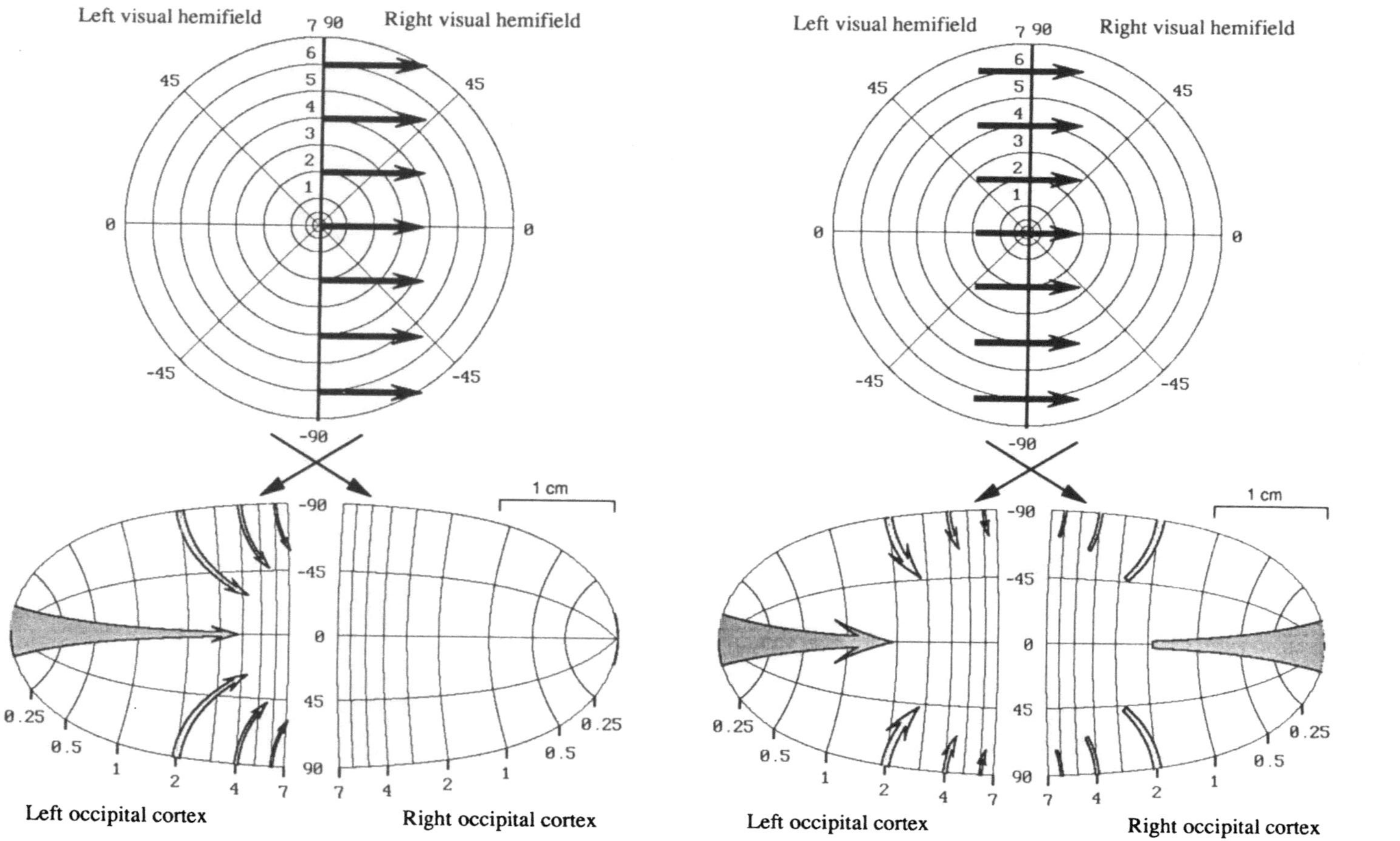

Figure 10. The images formed in the striate cortex of the rhesus monkey by seven identical arrows in the visual field. In the left portion of the figure the arrows are confined to the right visual hemifield. On the right the arrows are centered on the vertical meridian. Due to the magnification factor and the layout of the cortex, the images formed are quite different from what appears in the visual field, and do not form a continuous representation when the figures fall along the vertical meridian.

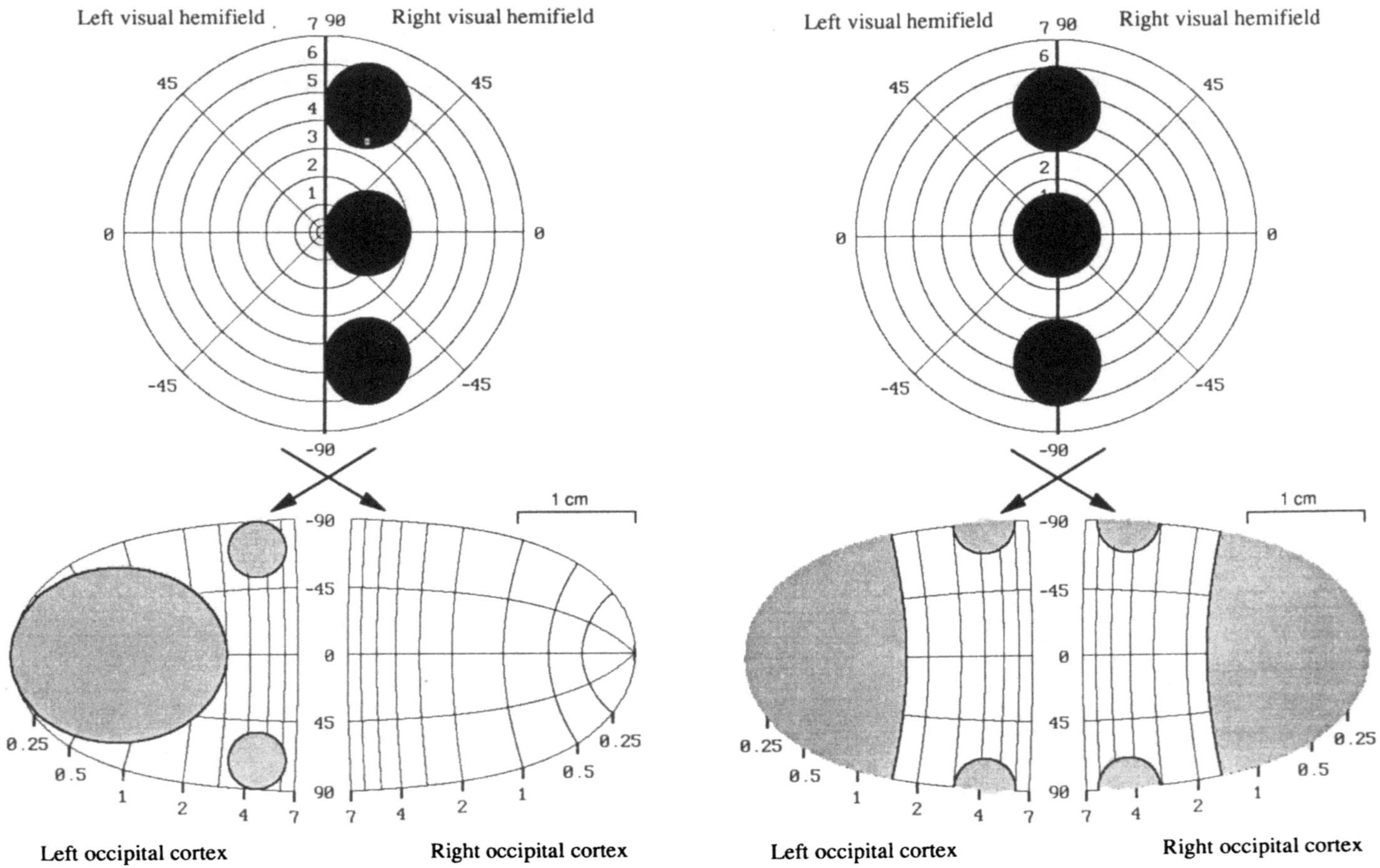

Figure 11. Images formed by three identical disks on the cortex when they appear in the right visual field and when centered along the vertical meridian. In both cases the shading shown for the cortical image depicts the fact that cortical neurons respond much more vigorously at the borders of figures.

small scale, in the complex cells of the striate cortex that respond to a range of stimuli, such as small line segments of certain orientations, irrespective of where they appear within their receptive fields.

This line of inference leads to the assumption that somewhere in higher cortical areas the left and right visual hemifields are combined to form a unified representation. Whether that is a necessary condition for image processing remains to be determined. If this does happen, say in inferotemporal cortex, where the receptive fields of most cells cross the vertical meridian, one is confronted with yet another problem: Since there are two inferotemporal cortices, one on each side of the brain, is the representation duplicated or are different things represented within each?

One of the essential requirements of object perception is the storage of images that can be retrieved and compared with new incoming images to yield recognition. Remarkably, we are capable of storing millions of sensory images for long time periods and recalling them in detail. We have an almost infinite capacity for this. We can recognize faces, pictures, and just about every visual scene, not to mention other modalities, even when we have not encountered them for years. Similar capacities exist in most species, including fish that return to their spawning grounds, birds that migrate, and sandwasps that find their digs through repeated circling above the immediate area before they fly off. That such storage is not a product of reverberating circuits (Hebb, 1949) has been made amply clear by clinical data. When under certain surgical conditions the heart is bypassed, the blood is replaced with serum, and body temperature is lowered to a level where the EEG goes flat, indicating no neural activity at all for several hours, following recovery patients still have all their memories. Thus the storage involves the laying down of a more or less permanent trace even for visual images seen just once or perhaps a few times. Because of the inordinate memory capacity reflected in this process, it is likely to involve processes that can best be understood at the molecular rather than the cellular level.

7. Summary

Neuroscientists have made remarkable progress during this century. It is a field studded with discoveries and Nobel prizes; we have learned more about the organization and functions of the nervous system during this period than in all of the time preceding it. It is a vital, exciting field. Consequently it is not surprising that it also abounds with controversy. As depicted in this chapter it may appear that those investigators in the field of neuroscience who study the visual system form two irreconcilable groups, one of which is committed to neuronal and areal specificity and the other to general, multipurpose systems. In reality, most investigators nowadays realize that in the brain both dedicated and general-purpose systems exist and that there is a complex interplay between such systems.

One may conjecture that the multifunctionality seen in the nervous system arose out of necessity rather than pure choice for creating the best analyzers. Dedicated systems, after all, tend to perform a better job. Dedicated tools, such as pliers, hammers, and screwdrivers, do better than multipurpose tools, which we buy mostly out of necessity, the necessity having often to do with spatial restrictions; one cannot readily carry around a well-equipped tool chest in the glove compartment. A multipurpose tool will fit and will do a passable job. Similar considerations may have arisen for the evolution of the nervous system. Due to space limitations, the number of neurons, which nature had difficulty miniaturizing, needed to be kept to a minimum. Thus, to increase computing power, they were turned into more complex, multipurpose devices where possible. That this is not an ideal solution can be seen from the fact that in most species dedicated systems remain. In the visual system numerous examples can be seen. One of these is the accessory optic system, which begins with a specialized set of ganglion cells in the retina, the displaced cells of Dogiel, that project to the terminal nuclei and from there through the dorsal cap of Kooy of the inferior olive to the cerebellum (Simpson, 1984). The circuit is completed by projections to the vestibular nuclei and from there to the oculomotor complex. This system is dedicated to the stabilization of the eye to the visual scene. Any retinal slip is recognized by the specialized, directionally selective ganglion cells, whose signals are converted in the circuit to activate the eye muscles to cancel the retinal slip. This is an effective dedicated system which in the rabbit utilizes only 6000–7000 ganglion cells out of a population of more than 350,000.

The advantages of maintaining some degree of specialization can also be seen in the geniculostriate system of primates. Whereas in most regions of the retina, presumably for the sake of economy, rods and cones converge on the ganglion cells, thereby giving rise to pathways that can carry different messages under different illumination conditions, this is not the case in the fovea. The fovea is rod-free. The main reason for this appears to be the necessity to optimize acuity. By eliminating the rods and reducing the diameter of the cones, high packing density can be realized, which thereby provides for the analysis of fine detail. The cortical cells to which the fovea projects, however, comprise a multipurpose system, many cells of which are capable of analyzing several different attributes of the visual scene.

The considerations presented in this chapter bring us to the following conclusions:

1. In the nervous system there are both dedicated and multipurpose systems; multipurpose systems are evident already at early stages of processing.

2. Neurons become increasingly multifunctional as one ascends from peripheral to central structures in the nervous system; this is an especially notable property of cortical neurons.

3. Neuronal multifunctionality represents a major evolutionary trend. Thus, rather than creating more and more dedicated systems to meet the ever-rising demands organisms had to face over the eons, general-purpose systems have evolved that not only are able to perform a number of analyses concurrently, but, instead of being "hard wired," are also modifiable by experience.

4. The process of *perceptual analysis* in the visual system is not to "break down the visual scene" into basic components. Such analysis is performed interactively by areas and neurons with multipurpose properties.

5. In the course of evolution, the numerous extrastriate visual areas did not arise for the purpose of separately analyzing basic visual attributes such as color,

motion, pattern, and depth. Even relatively primitive systems lacking multiple visual cortical areas, or lacking cortex entirely for that matter, had to perform such analyses. Nature did not have the freedom to rework everything completely and to thereby relegate basic analyses to newly created structures. Instead, it would seem reasonable to assume that these areas have evolved for entirely different reasons; among these was the need to improve on existing analyses, to meet new contingencies, and to allow for selective modification as the result of experience.

6. With each glance of the eye any one object in the visual scene activates many thousands of cells in the brain. The momentary neural signature of an object is inscripted at numerous brain sites, which when combined with images held in store, leads to object recognition. Because of the continual shifts in gaze, many different locations are activated in rapid succession in the striate cortex. One might assume that as the result of further processing such shifts in gaze are discounted, leading to a stable representation of the visual scene in higher cortical areas. Although the mapping of visual objects in the striate cortex is conformal, there is no continuous representation for objects whose images fall across the vertical meridian. Under these latter conditions object representation is split. Whether in higher areas this problem is rectified by creating a unified representation of objects still needs to be determined.

ACKNOWLEDGMENTS. The research by the author and his collaborators as described in this chapter was supported by NIH Grant EY00676. Figures 10 and 11 were constructed by Warren M. Slocum.

8. References

Albright, T. D., 1984, Direction and orientation selectivity of neurons in visual area MT of the macaque, *J. Neurophysiol.* **48:**338–351.

Albright, T. D., Desimone, R., and Gross, C. G., 1984, Columnar organization of directionally selective cells in visual area MT of the macaque, *J. Neurophysiol.* **51:**16–31.

Barlow, H. B., 1953, Summation and inhibition in the frog's retina, *J. Physiol.* (Lond.) **119:**69–88.

Barlow, H. B., 1972, Single units and sensation: A neuron doctrine for perceptual psychology? *Perception* **1:**371–395.

Blasdel, G. G., 1992a, Orientation selectivity, preference, and continuity in monkey striate cortex, *J. Neurosci.* **12:**3139–3161.

Blasdel, G. G., 1992b, Differential imaging of ocular dominance and orientation selectivity in monkey striate cortex, *J. Neurosci.* **12:**3115–3138.

Brodmann, K. 1909, *Vergleichende Lokalisationslehre der Grosshirnrinde,* Barth, Leipzig.

Desimone, R., and Schein, J., 1987, Visual properties of neurons in area V4 of the macaque: Sensitivity to stimulus form, *J. Neurophysiol.* **57:**835–868.

De Valois, R. L., and De Valois, K. K., 1980, Spatial vision, *Annu. Rev. Psychol.* **31:**309–341.

De Yoe, E. A., and Van Essen, D. C., 1988, Concurrent processing streams in monkey visual cortex, *Trends Neurosci.* **11:**219–226.

Duffy, C. J., and Wurtz, R. H., 1991, Sensitivity of MST neurons to optic flow stimuli. I. A continuum of response selectivity to large-field stimuli, *J. Neurophysiol.* **65:**1329–1345.

Enroth-Cugell, C., and Robson, J. G., 1966, The contrast sensitivity of retinal ganglion cells of the cat, *J. Physiol.* (Lond.) **18:**517–552.

Felleman, D. J., and Van Essen, D. C., 1991, Distributed hierarchical processing in the primate cerebral cortex, *Cerebral Cortex* **1:**1–47 (1991).

Ferrera, V. P., Nealy, T. A., and Maunsell, J. H. R., 1992, Mixed parvocellular and magnocellular geniculate signals in visual area V4, *Nature* **358:**756–758.

Gall, F. J., and Spurzheim, G., 1989, Research on the nervous system in general and on that of the brain in particular, in: *Brain and Behavior* (K. H. Pribram, ed.), Penguin, Middlesex, U.K., pp. 20–26.

Glickstein, M. E., 1985, Ferrier's mistake, *Trends Neurosci.* **8:**341–344.

Gross, C. G., 1973, Inferotemporal cortex and vision, in: *Progress in Physiological Psychology,* Vol. 5 (E. Stellar and J. M. Sprague, eds.), Academic Press, New York, pp. 77–123.

Haenny, P. E., and Schiller, P. H., 1988, State dependent activity in monkey visual cortex. I. Single cell activity in V1 and V4 on visual tasks, *Exp. Brain Res.* **69:**225–244.

Haenny, P. E., Maunsell, J. H. R., and Schiller, P. H., 1988, State dependent activity in monkey visual cortex. II. Retinal and extraretinal factors in V4, *Exp. Brain Res.* **69:**245–259.

Hartline, H. K., 1938, The responses of single optic nerve fibers of the vertebrate eye to illumination of the retina, *Am. J. Physiol.* **121:**400–415.

Head, H., 1926, *Aphasia and Kindred Disorders of Speech,* Macmillan, New York.

Hebb, D. O., 1949, *The Organization of Behavior,* Wiley, New York.

Helmholtz, H., 1924, *Physiological Optics* (J. P. C. Southall, ed.), Optical Society of America, Rochester, New York.

Hendry, S. H. C., and Yoshioka, T., 1994, A neurochemically distinct third channel in the macaque dorsal lateral geniculate nucleus, *Science* **264:**575–577.

Heywood, C. A., and Cowey, A., 1987, On the role of cortical area V4 in the discrimination of hue and pattern in macaque monkeys, *J. Neurosci.* **7:**2601–2617.

Heywood, C. A., Gadotti, A., and Cowey, A., 1992, Cortical area V4 and its role in the perception of color, *J. Neurosci.* **12:**4056–4065.

Hodges, A., 1983, *Alan Turing: The Enigma,* Simon & Schuster, New York.

Horton, J. C., and Hubel, D. H., 1980, Cytochrome oxidase stain preferentially labels intersection of ocular dominance and vertical orientation columns in macaque striate cortex, *Soc. Neurosci. Abstr.* **6:**315.

Hubel, D. H., and Wiesel, T. N., 1959, Receptive fields of single neurones in the cat's striate cortex, *J. Physiol.* (Lond.) **148:**574–591.

Hubel, D. H., and Wiesel, T. N., 1962, Receptive fields, binocular interaction and functional architecture in the cat's visual cortex, *J. Physiol.* (Lond.) **160:**106–154.

Hubel, D. H., and Wiesel, T. N., 1968, Receptive fields and functional architecture of monkey striate cortex, *J. Physiol.* (Lond.) **195:**215–243.

Hubel, D. H., and Wiesel, T. N., 1969, Anatomical demonstration of columns in the monkey striate cortex, *Nature* **225:**41–42.

Hubel, D. H., and Wiesel, T. N., 1970, Cells sensitive to binocular depth in area 18 of the macaque cortex, *Nature* **225:**41–42.

Hubel, D. H., and Wiesel, T. N., 1972, Laminar and columnar distribution of geniculo-cortical fibers in the macaque monkey, *J. Comp. Neurol.* **158:**267–294.

Hubel, D. H., and Wiesel, T. N., 1977, Functional architecture of macaque visual cortex, *Proc. R. Soc. B* **198:**1–59.

Jung, R., 1961, Korrelationen von Neuronentatigkeit und Sehen, in: *Neurophysiologie und Psychophysik des visuellen Systems* (R. Jung and H. H. Kornhuber, eds.), Springer-Verlag, New York, pp. 410–435.

Kaas, J. H., 1978, The organization of visual cortex in primates, in: *Sensory Systems of Primates* (C. Noback, ed.), Plenum Press, New York, pp. 151–179.

Kaas, J. H., 1982, The segregation of function in the nervous system: Why do sensory systems have so many subdivisions? in: *Contributions to Sensory Physiology,* Vol. 7 (W. D. Neef, ed.), Academic Press, New York, 1982, pp. 201–240.

Kluver, H., 1927, Visual disturbances after cerebral lesions, *Psychol. Bull.* **24:**316–358.

Lashley, K. S., 1963, *Brain Mechanisms and Intelligence,* Dover, New York.

Lennie, P., 1984, Recent developments in the physiology of color vision, *Trends Neurosci.* **7:**243–248.

Lettvin, J. Y., Maturana, W. S., McCullogh, W. S., and Pitts, W. H., 1959, What the frog's eye tells the frog's brain, *Proc. Inst. Radio. Eng.* **47:**1940–1951.

Leventhal, A. G., Thompson, K. G., Liu, D., Zhou, Y., and Ault, S. J., 1995, Concomitant sensitivity to orientation, direction and color of cells in layers 2, 3, and 4 of monkey striate cortex, *J. Neurosci.* **15:**1808–1818.

Levitt, J. B., Kiper, D. C., and Movshon, J. A., 1994, Receptive fields and functional architecture of macaque V2, *J. Neurophysiol.* **71:**2517–2542.

Livingstone, M. S., and Hubel, D. H., 1984, Anatomy and physiology of a color system in the primate visual cortex, *J. Neurosci.* **4:**309–356.

Livingstone, M. S., and Hubel, D. H., 1987, Psychophysical evidence for separate channels for the perception of form, color, movement, and stereopsis, *J. Neurosci.* **7:**3416–3468.

Livingstone, M. S., and Hubel, D. H., 1988, Segregation of form, color, movement, and depth: Anatomy, physiology, and perception, *Science* **240:**740–749.

Lund, J. S., 1988, Anatomical organization of macaque monkey striate visual cortex, *Annu. Rev. Neurosci.* **11:**253–288.

Malpeli, J. G., Schiller, P. H., and Colby, C. L., 1981, Response properties of single cells in monkey striate cortex during reversible inactivation of individual geniculate laminae, *J. Neurophysiol.* **49:**1102–1119.

Maturana, J. R., Lettvin, J. Y., McCullogh, W. S., and Pitts, W. H., 1960, Anatomy and physiology of vision in the frog (*Rana pipiens*), *J. Gen. Physiol.* **43:**129–175.

Maunsell, J. H. R., Nealey, T. A., and De Priest, D. D., 1990, Magnocellular and parvocellular contributions to responses in the middle temporal visual area (MT) of the macaque monkey, *J. Neurosci.* **10:**3323–3334.

McClurkin, J. W., Zarbock, J. A., and Optican, L. M., 1994, Temporal codes for colors, patterns and memories, in: *Cerebral Cortex,* Vol. 10, *Primary Visual Cortex in Primates* (A. Peters and K. S. Rockland, eds.), Plenum Press, New York.

McIlwain, J. T., 1966, Some evidence concerning the physiological bases of the periphery effect in the cat's retina, *Exp. Brain Res.* **1:**265–271.

Meadows, J. C., 1974, Distributed perception of colours associated with localized cerebral lesions, *Brain* **97:**615–632.

Merzenich, M. M., and Sameshima, K., 1993, Cortical plasticity and memory, *Curr. Opin. Neurobiol.* **3:**187–196.

Moran, J., and Desimone, R., 1985, Selective attention gates visual processing in the extrastriate cortex, *Science* **229:**782–784.

Nelson, R., Famiglietti, E., and Kolb, H., 1978, Intracellular staining reveals different levels of stratification for On- and Off-center ganglion cells in the cat retina, *J. Neurophysiol.* **41:**472–483.

Newsome, W. T., and Wurtz, R. H., 1988, Probing visual cortical function with discrete chemical lesions, *Trends Neurosci.* **11:**394–400.

Poggio, G., 1972, Spatial properties of neurons in striate cortex of unanesthetized macaque monkey, *Invest. Ophthalmol.* **11:**368–377.

Poggio, G. F., and Poggio, T., 1984, The analysis of stereopsis, *Annu. Rev. Neurosci.* **7:**379–412.

Purves, D., Riddle, R. R., and La Mantia, A. S., 1992, Iterated patterns of brain circuitry (or how the cortex gets its spots), *Trends Neurosci.* **15:**362–368.

Ramón y Cajal, S., 1989, *Recollections of My Life* (E. Horne Craigie, trans.), MIT Press, Cambridge, MA.

Roy, J.-P. and Wurtz, R. H., 1990, The role of disparity-sensitive cortical neurons in signalling the direction of self-motion, *Nature* **348:**160–162.

Schein, S. J., and Desimone, R., 1990, Spectral properties of V4 neurons on the macaque, *J. Neurosci.* **10:**3369–3389.

Schiller, P. H., 1982, Central connections of the retinal On and Off pathways, *Nature* **297:**580–583.

Schiller, P. H., 1986, The central visual system, *Vision Res.* **26:**1351–1386.

Schiller, P. H., 1993, The effects of V4 and middle temporal (MT) area lesions on visual performance in the rhesus monkey, *Visual Neurosci.* **10:**717–746.

Schiller, P. H., 1994, Area V4 of the primate visual cortex, *Curr. Dir.* **3:**89–92.

Schiller, P. H., 1995, The ON and OFF channels of the mammalian visual system, in: *Retinal and Eye Research,* Vol. 15, No. 1 (N. N. Osborne and G. J. Chader, eds.), Pergamon Press, Elsevier Science Inc., New York.

Schiller, P. H., and Lee, K., 1991, The role of the primate extrastriate area V4 in vision, *Science* **251:**1251–1253.

Schiller, P. H., and Logothetis, N. K., 1990, The color-opponent and broad-band channels of the primate visual system, *Trends Neurosci.* **13:**392–398.

Schiller, P. H., and Malpeli, J., 1977, The properties and tectal projections of monkey retinal ganglion cells, *J. Neurophysiol.* **40:**428–445.

Schiller, P. H., Finlay, B., and Volman, S., 1976, Quantitative studies of single-cell properties in monkey striate cortex, *J. Neurophysiol.* **39:**1288–1374.

Schiller, P. H., Logothetis, N. K., and Charles, E. R., 1990, Role of color-opponent and broad-band channels in vision, *Visual Neurosci.* **5:**321–346.

Schwartz, E. L., 1994, Computational studies of the spatial architecture of primate visual cortex: Columns, maps, and protomaps, in: *Cerebral Cortex,* Vol. 10, *Primary Visual Cortex in Primates* (A. Peters and K. S. Rockland, eds.), Plenum Press, New York.

Simpson, J. I., 1984, The accessory optic system, *Annu. Rev. Neurosci.* **7:**13–41.

Slaughter, M. M., and Miller, R. F., 1981, 2-Amino-4-phosphonobutyric acid: A new pharmacological tool for retina research, *Science* **211:**182–184.

Stone, J., 1983, *Parallel Processing in the Visual System,* Plenum Press, New York.

Swindale, N. V., 1990, Is the cerebral cortex modular? *Trends Neurosci.* **13:**487–492.

Talbot, S. A., and Marshall, W. H., 1941, Physiological studies on neural mechanisms of visual localization and discrimination, *Am J. Ophthalmol.* **24:**1255–1264.

Ungerleider, L. G., and Mishkin, M., 1982, Two cortical visual systems, in: *Analysis of Visual Behavior* (D. J. Ingle, M. A. Goodale, and R. J. W. Mansfield, eds.), MIT Press, Cambridge, MA, pp. 549–586.

Waessle, H., and Boycott, B. B., 1991, Functional architecture of the mammalian retina, *Physiol. Rev.* **71:**447–479.

Werblin, F. S., and Dowling, J. E., 1969, Organization of the retina of the mudpuppy, *Nectorus maculosus.* II. Intracellular recording, *J. Neurophysiol.* **32:**339–355.

Wiesel, T. N., and Hubel, D. H., 1966, Spatial and chromatic interactions in the lateral geniculate body of the rhesus monkey, *J. Neurophysiol.* **29:**1115–1156.

Wild, H. M., Butler, S. R., Carden, D. J., and Kulikowski, J. J., 1985, Primate cortical area V4 important for color constancy but not wavelength discrimination, *Nature* **313:**133–135.

Wong-Riley, M. T. T., Merzenich, M. M., and Leake, P. A., 1978, Changes in endogenous enzymatic activity to DAB induced by neuronal inactivity, *Brain Res.* **141:**185–192.

Zeki, S. M., 1973, Colour coding in rhesus monkey prestriate cortex, *Brain Res.* **53:**422–427.

Zeki, S. M., 1974, Functional organization of a visual area in the posterior bank of the superior temporal sulcus of the rhesus monkey, *J. Physiol.* (Lond.) **236:**549–573.

Zeki, S. M., 1990, A century of cerebral achromatopsia, *Brain* **113:**1721–1777.

Zipser, K., Lee, T.-S., Lamme, V. A. F., and Schiller, P. H., 1994, Spatial extent of V1 extra-RF mechanisms in macaque, *Soc. Neurosci. Abstr.* **20:**1477.

3

Theories of Visual Cortex Organization in Primates

JON H. KAAS

1. Introduction

The present review outlines and evaluates theories of how visual cortex is divided into areas in primates. Maps of cortical areas have long been used as guides for further research and they clearly have implications for how information is processed in the visual system. Early maps such as those of Brodmann (1909) and Von Economo (1929) have had great impact on current theories of visual cortex organization, and parts of these early theories remain in use. Yet early investigators disagreed on how extrastriate cortex is subdivided, and the usefulness of the architectonic methods used to formulate early proposals has been repeatedly questioned (e.g., Lashely and Clark, 1946). Current proposals are more complex and include many visual areas. In principle, current proposals should be more accurate because they are based on additional sorts of information, especially patterns of cortical connections and retinotopic organization. Indeed there is widespread agreement on the locations and extent of some proposed fields such as V2 and MT (V5). However, our maps of cortex also differ in many ways, suggesting that the supporting evidence is ambiguous and limited enough to allow different interpretations. As a reflection of this uncertainty, Felleman and Van Essen (1991), after an extensive review and synthesis, conclude that of 32 proposed visual areas, only five rate a high confidence level of 1 on a scale of 1–3. Possibly one might take an even more conservative view, since only three areas (V1, V2, and MT) are components of most proposals. In any case, it seems useful to review the progression from early to recent theories of cortical organization in an effort to see how they evolved and influenced each

JON H. KAAS • Department of Psychology, Vanderbilt University, Nashville, Tennessee 37240.
Cerebral Cortex, Volume 12, edited by Rockland *et al.* Plenum Press, New York, 1997.

other, as well as determine both reliable features and those that require further study and evaluation.

A remaining and important goal will be to determine what visual areas are common to all primates, and how primate taxa differ. Most likely, a number of visual areas, perhaps 10 or more, have been retained from the first primates (Kaas and Preuss, 1993; Northcutt and Kaas, 1995). The premise that visual areas have been retained from the common ancestor allows us to relate concepts of visual cortex organization based on experimental results in monkeys to humans. Research on monkeys more likely applies to humans if the research involves subdivisions of the brain that are widely shared among primates, especially if they seem to have varied little across taxa. Thus, comparative studies have the value of providing information about consistencies in brain organization across primates that likely apply to humans (Kaas, 1993). In addition, comparative studies provide another level of validation of theories of brain organization for given taxa, since the differences in the theories should be explainable in terms of the evolution of variations from a common plan.

In considering proposals and evidence, it is useful to remember that there are over 200 species of primates. Their brains undoubtedly differ in many ways, some of which are quite obvious. Yet primates just as clearly share a number of visual areas. Since we often make assumptions about the visual cortex of humans from studies on monkeys, it is the shared features of visual cortex organization that are of special interest. Difficulties arise, however, since comparative information is extremely limited, and proposals are largely based on data collected from a few species of either New or Old World monkeys. While it is tempting to conclude that differences in maps of cortex based on data from either New or Old World monkeys reflect real phyletic specializations, there is little reason to do so. Instead, interpretations of data seem most influenced by the theoretical framework of investigators. This review concentrates on theories of visual cortex organization that have emerged from experimental studies on monkeys, regardless of species, with an effort to determine common features.

2. What Are We Mapping?

As maps of the earth can reflect different physical and politically determined variables, maps of the brain can be based on various characteristics (Dykes and Ruest, 1986). The most important characteristic for the present review is that the neocortex is subdivided into cortical areas or fields. The concept of the cortical area is old, stemming back to early arguments about localization of function in the brain (Gross, this volume; Finger, 1994, for review). Localizationists have clearly prevailed and we now conceive of the cortex much as Brodmann (1909) did: that is, as a patchwork of areas, each participating in a distinct set of functions as a result of a unique pattern of inputs, outputs, and intrinsic circuitry. Cortical areas are generally assumed to have sharp boundaries, much as primary visual cortex clearly has, although historically many investigators held the view that some functional areas could be highly overlapping (e.g., Herrick,

1891). However, there is no clear evidence for gradually merging fields, and the assumption of sharp boundaries largely prevails.

Investigators have repeatedly addressed the issue of how to distinguish cortical areas. Early researchers largely had to depend on histological differences to distinguish fields. The basic assumption was that cortical areas, or the functionally distinct "organs of the brain" (Brodmann, 1909), would be structurally distinct. As Herrick (1891) noted, "function and structure are intimately connected, and difference in function implies difference in structure." In a similar manner, Ramón y Cajal (1899) referred to the "structural plurality" of the cortex in support of the doctrine that "diversities of function result from the particular structure of each cortical area." The difficulty with the traditional architectonic approach, unfortunately, has been that structural differences between areas are often subtle. Thus, few areas could be identified with certainty, and uncertainties led investigators to generate quite different proposals (see below). Fortunately, other methods have become increasingly used to supplement architectonic procedures, and architectonic approaches became more varied.

Patterns of connections have been used extensively to define cortical areas. At first, patterns of thalamocortical connections were used to subdivide cortex because they could be revealed by retrograde degeneration after cortical lesions (e.g., Walker, 1938). Presently, corticocortical connection patterns provide the bulk of the evidence for the existance of many visual areas. Electrophysiological mapping methods, especially with microelectrodes, also proved to be extremely useful (e.g., Adrian, 1941). Regional differences in the receptive field properties of neurons provide further information (e.g., Van Essen and Zeki, 1978; Baker *et al.*, 1981; Peterson *et al.*, 1988), but largely after a visual area has been defined otherwise. Likewise, lesions can be used to demonstrate a functional role for a postulated field in support of its validity. However, because all methods have strengths and weaknesses, most investigators agree that cortical areas are most reliably defined by the congruence of evidence based on different methods, that is, areas are best defined by multiple criteria (Kaas, 1982, 1989b; Van Essen, 1985).

Despite widespread agreement on how to define visual areas, difficulties in making valid distinctions remain. Cortical areas are often quite similar to each other histologically. Areas can also be internally heterogeneous in structure. The portions of primary visual cortex devoted to binocular and monocular vision, for example, look different. This led Brodmann (1909) to misidentify the monocular portion of area 17 of squirrels as area 18. Adjoining areas typically have quite similar connections (Young *et al.*, 1995), and thus they can be confounded. Furthermore, different portions of an area, especially regions devoted to central or peripheral vision (Krubitzer and Kaas, 1990), can have different magnitudes or targets of connections, and of course modules within areas can have quite different connections (see below). In addition, most cortical areas representing a sensory surface have congruent borders (Allman and Kaas, 1975). For visual areas, this means that very similar parts of the visual hemifield are represented in adjoining locations across the borders of fields, making borders difficult to distinguish. Finally, the difficulties in defining visual areas increase for "higher order" areas where visual field maps may not be simple in organization, receptive fields may be larger, and neuronal responsiveness is more depressed by anesthesia (Colby and Duhamel, 1991).

Another and particularly vexing problem is distinguishing between cortical areas and the modules that subdivide them. Research over the last 20 years or so, especially more recently, has indicated that some areas are heterogeneous in organization so that two or more maps interdigitate within the field. Primary somatosensory cortex of monkeys, for example, has separate territories for thalamic inputs related to the slowly and rapidly adapting classes of cutaneous receptors (Sur *et al.*, 1981). In visual cortex of primates, primary visual cortex, V1, is divided into "blob" and "interblob" zones, each with different chemoarchitectonics, connections, and functional characteristics (Casagrande and Kaas, 1994). The second visual area, V2, has three sets of band-shaped modules that alternate as they cross the narrow width of the field, and each set of bands can be considered a separate map of the visual field (Rosa *et al.*, 1988; Roe and Ts'o, 1995). There is evidence for modular organization within other visual fields as well. Many and perhaps most cortical areas contain a mosaic of small regions of different circuitry. Such modules repeat, often with considerable regularity (Kaas, 1990; Krubitzer, 1995; Purves *et al.*, 1992). The patchy nature of connections throughout cortex (e.g., Goldman and Nauta, 1977) is evidence that modularity is a fundamental feature of cortical areas (Kaas, 1990). Thus, in producing maps of the areas of visual cortex, we have the practical problems of both distinguishing areas from each other and discriminating between areas and modules.

In visual cortex, the only reliable distinction between areas and modules is that we presume areas represent the complete (or nearly complete) visual hemifield, while each module represents only part of the hemifield. Thus, evidence on the completeness of the map should help distinguish modules from small visual areas. However, regions of cortex have been proposed as areas even when it is clear that they represent only part of the visual hemifield (see below). In addition, separate modules within a field may overlap in representation and include large portions of the hemifield. Elsewhere, we have argued that a series of such proposed visual areas along the lateral border of V1 of rats and other rodents are actually modules of V2 (Kaas *et al.*, 1989a).

3. Early Maps

Early views on the extent and subdivisions of visual cortex considered only the occipital region and divided visual cortex into only two or three areas (Fig. 1). Striate or primary visual cortex was easily identified, and its significance as a visual structure readily appreciated, since lesions produced dramatic impairments which Munk (1890) referred to as "psychical blindness." Lesions of adjoining extrastriate cortex also changed visual performance, but in less dramatic ways. Thus, cortical processing was thought to start with striate cortex as the site of sensation, relay through one or two higher levels of visual processing concerned with perception, and then relay to regions of association cortex for higher level abilities (Kaas, 1988). Campbell (1905) identified striate cortex as the visuosensory area, while including much of the remaining occipital lobe in a visuopsychic area. G. E. Smith (1906) developed a similar but more elaborate scheme, with two extrastriate processing fields, peristriate and parastriate areas, as well as bimodal visuosomatosensory and visuoauditory bands. Brodmann (1909) expanded the visual sphere with a large occipital region, area 18, and a

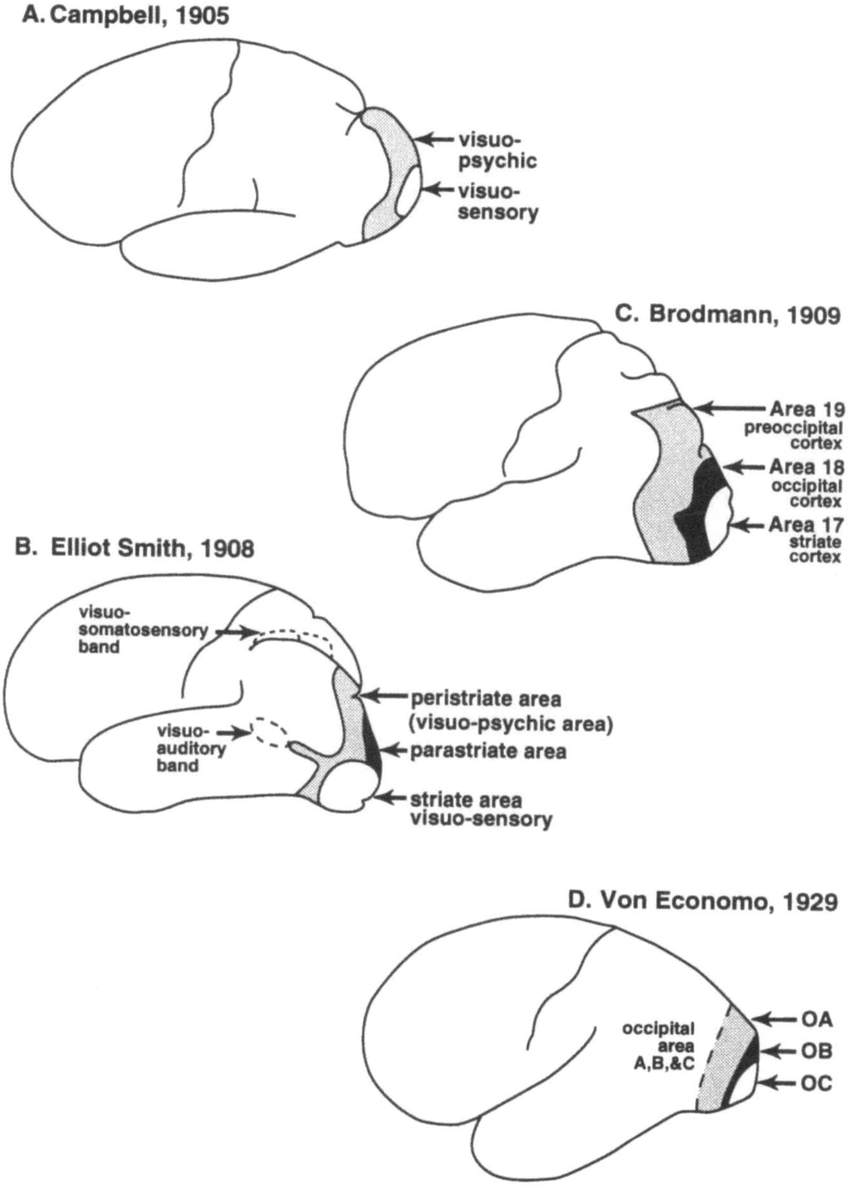

Figure 1. Early views on how visual cortex in human brains is subdivided into functional areas. Investigators postulated only two or three extrastriate visual areas. Elliot Smith included bimodal association fields. See text for further explanation.

larger preoccipital region, area 19. Von Economo (1929) divided the visual territory into extrastriate occipital areas A and B and striate cortex or occipital area C. All of these early maps were based on interpretations of cortical architecture, influenced by clinical data, and were similar in having few subdivisions. Nevertheless, there was no agreement on the extents and locations of extrastriate areas.

The parcellations of Campbell and Elliot Smith have been abandoned, while those of Von Economo and especially Brodmann continue to have great influence. The impact of Brodmann's subdivisions can be appreciated by the widespread use of his terminology. Two notable examples of Brodmann's impact on subsequent interpretations of early experimental data are provided here. McCulloch (1944), for example, attempted to relate new experimental results on cortical connections (based on recordings after local strychninization) to Brodmann's subdivisions and produced a flattened, surface-view scheme of the major connections of all neocortex and the basal ganglia (Fig. 2). McCulloch portrayed

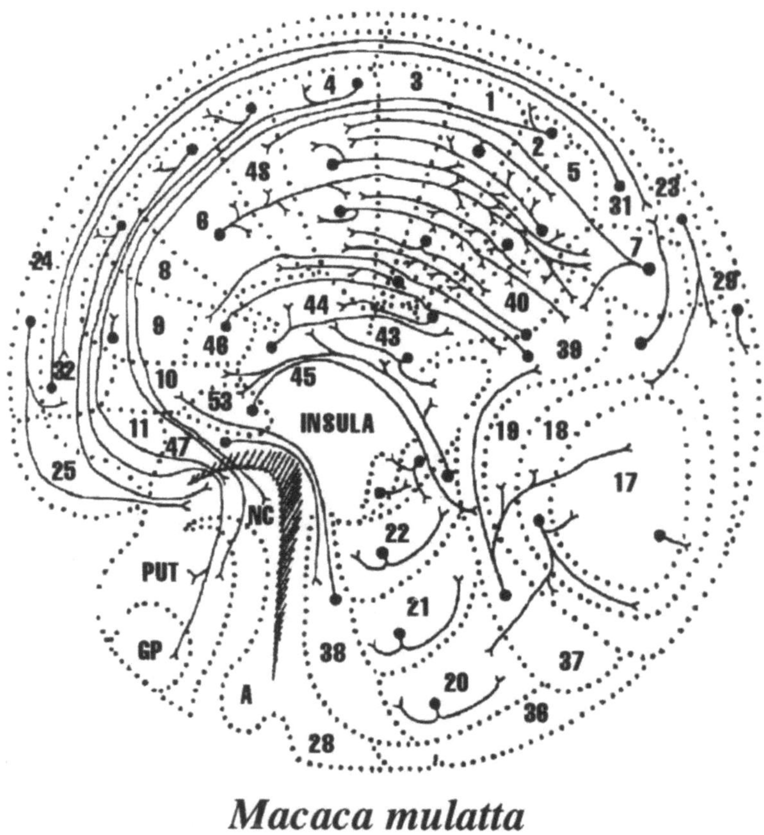

Figure 2. A summary diagram from McCulloch (1944) showing his view of how Brodmann's areas are arranged in the flattened cortex of macaque monkeys. The basal ganglia are also included in this diagram. The illustrated connections were based on experimental results of recordings after local strychninization ("physiological neuronography"). Area 17 is surrounded by ringlike areas 18 and 19, and these three fields form three successive stages of processing.

areas 18 and 19 as successive rings completely surrounding area 17 (this was somewhat uncertain from Brodmann's maps; see Jouandet *et al.*, 1989), and areas 17, 18, and 19 were seen as serially connected processing fields. In addition, parts of temporal, parietal, and limbic cortex were thought to be involved in visual processing. Although McCulloch (1944) and co-workers used the terms 18 and 19, they depended on physiological evidence of connections to define the fields. They noted that it was nearly impossible to identify architectonic boundaries, and that "area 18 has one ill-defined and one well-defined boundary," while area 19 "has two ill-defined boundaries" (Bonin *et al.*, 1942). Area 18, as defined by McCulloch, was larger than area 18 of Brodmann and included all cortex activated by the strychinization of area 17.

In another influential summary, Konorski (1967) provided a "conceptual" map of flattened human neocortex with functional names replacing Brodmann's numbers (Fig. 3). Yet, the influence of Brodmann's scheme remained clear. Following the reports of Hubel and Wiesel (1965) in which areas 17, 18, and 19 in cats were identified as three visual representations VI, VII, and VIII, Konorski proposed that areas 17, 18, and 19 of Brodmann correspond to VI, VII, and VIII in humans. Higher order visual fields were related to other Brodmann areas. Thus, V-Sn (visual sign field) corresponded to area 7b; V-MO (visual for manipulable objects) was 7b; V-VO (visual for visual objects) was area 39; and V-F (for visual faces) and V-AO (visual animated objects) together comprised area 37. The serial flow of processing from VI to higher areas was part of the proposal. The important point here is that both these and other influential schemes of brain organization and function retained the basic framework of Brodmann. Yet, contemporaries of Brodmann, and other investigators using the architectonic method disagreed on how cortex is subdivided, and such differ-

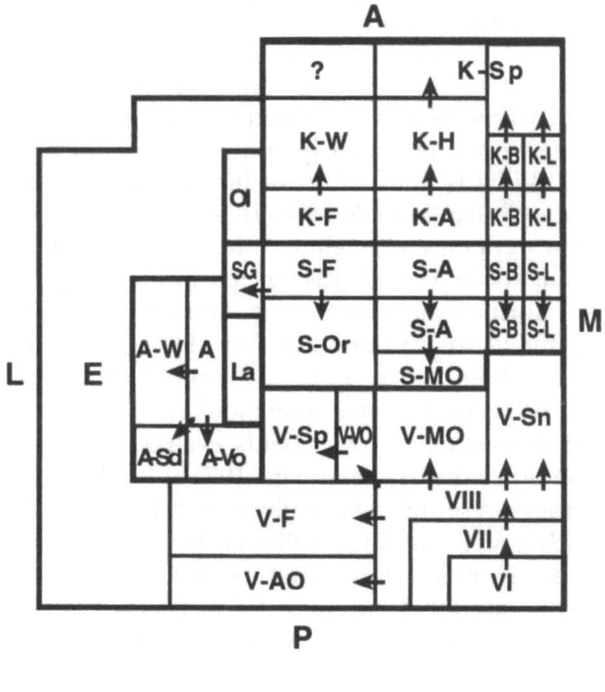

Figure 3. A conceptual map of the subdivisions of the human cerebral cortex from Konorski (1967). Areas VI, VII, and VIII correspond to Brodmann's fields 17, 18, and 19. Other specialized visual fields (V-F, etc.) were related to other Brodmann areas. Auditory (A), somatosensory (S), and kinesthetic (K) fields are also included. See Konorski (1967) for further details.

ences in opinion led many early workers (e.g., Lashley and Clark, 1946; LeGros Clark, 1952; Sholl, 1956) to question the validity of proposals based solely on architecture.

4. Early Modern Theories

The 1960s through the early 1970s was a period when new degeneration methods for more accurately revealing the details of cortical connection patterns came into widespread use (referred to as "the Nauta revolution" by Ebbesson, 1984). In addition, for the first time, microelectrode and surface recordings were used to reveal the retinotopic organizations of regions of cortex. The studies of Hubel and Wiesel (1965) in cats were extremely important and influential in that they showed (1) that each of the three architectonic areas 17, 18, and 19 of "Brodmann" (although Brodmann did not delineate these fields in cats; see Otsuka and Hassler, 1962, and Sanides and Hoffmann, 1969, for these fields) contains a systematic representation of the contralateral visual hemifield; (2) area 17 projects to area 18, and area 18 to area 19; and (3) receptive fields become larger across this sequence with receptive field properties becoming more complex. It was natural that this elegant elaboration of Brodmann's basic scheme was readily applied to primates.

While general connection patterns were revealed by degeneration after large cortical lesions in monkeys (e.g., Kuypers *et al.,* 1965; Jones and Powell, 1970), Myers (1965) seems to have been the first to use the details of connection patterns to define extrastriate visual areas and determine their retinotopy. Myers based his conclusions (Fig. 4A) on his studies of corpus callosum connections (Myers, 1962) and projections of striate cortex (Myers, 1965). Callosal connections ringed striate cortex, demonstrating a band of tissue devoted to the zero vertical meridian. Myers also noted that projections from the middle of V1, corresponding to the zero horizontal meridian, project to the outer margin of V2. Myers thereby described projection patterns, both callosal and from V1, that have been confirmed in subsequent studies (see below) and accurately outline V2. What is unusual about his proposal was the retention, but redefinition, of each of Brodmann's areas 18 and 19 into two separate fields. Cortex we would now define as V2 included the callosal connection zone as juxtastriate area 18, while the rest of V2 was termed striate-receptive area 19. Area 18 and 19 "proper" formed additional rings outside of V2 in this proposal. This scheme also included areas 21 and 20 of Brodmann as larger halves of another ringlike field.

Subsequent proposals of Zeki (1969) and Cragg (1969) also were based on V1 projections in macaque monkeys. Zeki (Fig. 4B) redefined areas 18 and 19 of Brodmann as narrower fields, each containing a mirror-image representation of the other. Later, Zeki (e.g., 1971b) used the terms V2 and V3 for these fields. Cragg presented a basically similar, but somewhat more elaborate view (Fig. 4C), with a VII and VIII, followed by a large, third ringlike field with convergent input from V1 (horizontal meridian plus vertical meridian). According to both proposals, any location in V1 projects to separate locations in VII or "area 18" and VIII or "area 19." As discussed in following pages, projections from V1 to

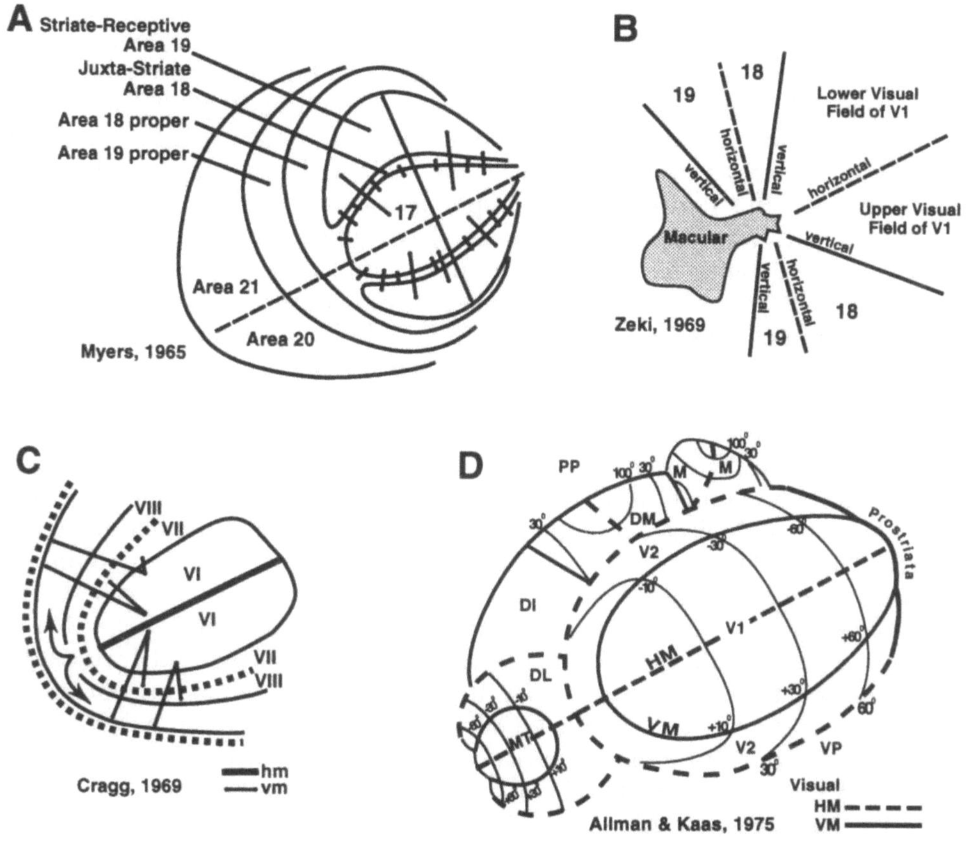

Figure 4. Early modern portrayals of the postulated organization of extrastriate cortex in monkeys. Each scheme shows cortical fields from a surface-view of flattened cortex, with caudal to the right and medial above. Schemes A–C were based on connection patterns in macaque monkeys, whereas scheme D was based on microelectrode recodings in owl monkeys. (A) In the proposal of Myers (1965), lines mark the projections of area 17 or V1 to what we now term V2. The juxtastriate area 18 not only received inputs from the border region of area 17, thus providing evidence for the representation of the zero vertical meridian, but also it received callosal connections related to the vertical meridian (Myers, 1962). The dashed line marks the zero horizontal meridian. Since the middle of area 17, corresponding to the horizontal meridian, projects to the outer border of striate-receptive area 19, this cortex was also seen as representing the horizontal meridian. (B) The scheme of Zeki (1969) was based on the projection pattern of V1 (area 17). V1 was found to project to V2 (area 18), as found by Myers (1965), and to more rostral cortex termed area 19 or V3. The proposed projections to ventral "area 19" have not been demonstrated, and the projections to dorsal area 19 have been reinterpreted as to DM (Krubitzer and Kaas, 1993). (C) The scheme of Cragg (1969) was also based on V1 projections (lines). Again, projections are summarized as forming two ringlike patterns to V2 (VII) and V3 (VIII). However, as noted above, the pattern of projections from V1 to a ventral V3 has not been found. Cragg also illustrated a third ringlike projection zone of converging inputs from parts of V1 representing the horizontal and vertical meridians. Subsequent investigations indicate that these projections must have been to the middle temporal visual area, MT, since V1 projections have not been found to other parts of this proposed ring. In addition, the projection from different parts of V1 to MT (e.g., Weller and Kaas, 1983) do not converge as proposed. (D) A scheme for visual cortex organization in owl monkeys based on Allman and Kaas (1975). Besides V1 and V2, proposed areas include the dorsolateral (DL), the dorsointermediate (DI), and the dorsomedial (DM) visual areas as well as the medial (M), middle temporal (MT), ventroposterior (VP), and posterior parietal (PP) areas.

the region of a ventral V3 have not been discovered (Weller and Kaas, 1983; Van Essen *et al.*, 1986; Krubitzer and Kaas, 1993). In addition, the outer ring of convergent input of Cragg (Fig. 4C) does not exist, but is was probably postulated after observing projections to a small part of the ring. Projections from V1 to the region now known as the middle temporal visual area, MT or V5, were initially described as convergent (see Zeki, 1971a). These maps of the 1960s contained elements that have been supported by subsequent investigation (the projection pattern to V2) and features that have not (the projections to a ventral VIII and convergently to all parts of a more distant ring). In all three schemes, the influence of Brodmann is clear. Other investigators at about that time also described connections between "areas 18 and 19" as two successive stages of processing (Rockland and Pandya, 1979; Tigges *et al.*, 1981).

A quite different plan of extrastriate cortex organization was generated from microelectrode mapping studies in New World owl monkeys (Fig. 4D). Instead of V3 on the outer border of V2, Allman and Kaas (1975) divided this cortex into five distinct visual areas, and allowed for the possibility of others. This plan followed initial experiments on the organization of the middle temporal visual area, MT, where cortex between MT and V1 was first divided into the traditional "areas 18 and 19," since limited recordings in cortex between MT and V2 suggested two mirror-image representations (Allman and Kaas, 1971a), much like V2 and V3 of cats. However, subsequent analysis and further data collection led to the conclusion that at least five representations border V2, including the dorsolateral area, DL (Allman and Kaas, 1974b), the dorsomedial area, DM, the dorsointermediate area, DI (Allman and Kaas, 1975), the medial area, M (Allman and Kaas, 1976) and the incompletely explored ventroposterior area, VP (Newsome and Allman, 1980). Thus, the concept of a single visual area, V3 or area 19, along the outer border of V2 was replaced with the idea of a third tier of several visual areas.

5. Recent Maps of Visual Cortex

With the advent of more experimental data, recent maps of visual cortex in primates have become more elaborate. For the most part, proposals have expanded by modifying one or the other of two basic frameworks (Fig. 4A–C), one with a V3 and one without (Fig. 4D). Since the two types of maps emerged from studies of different primate taxa, one possible interpretation is that visual cortices of New and Old World monkeys are subdivided quite differently. However, both basic frameworks have now been applied to New and Old World monkeys (Figs. 5 and 6), suggesting instead that differences in cortical organization between these two lines of primate evolution may not be the explanation for the different schemes.

Recent versions of how visual cortex is subdivided in macaque monkeys continue to differ. The summary (Fig. 5A) of Van Essen and Maunsell (1983) was based on findings from several laboratories, including early collaborations of Van Essen and Zeki (1978), and later experimental work of Van Essen and coworkers (e.g., Van Essen *et al.*, 1981, 1986). Interesting new features of this

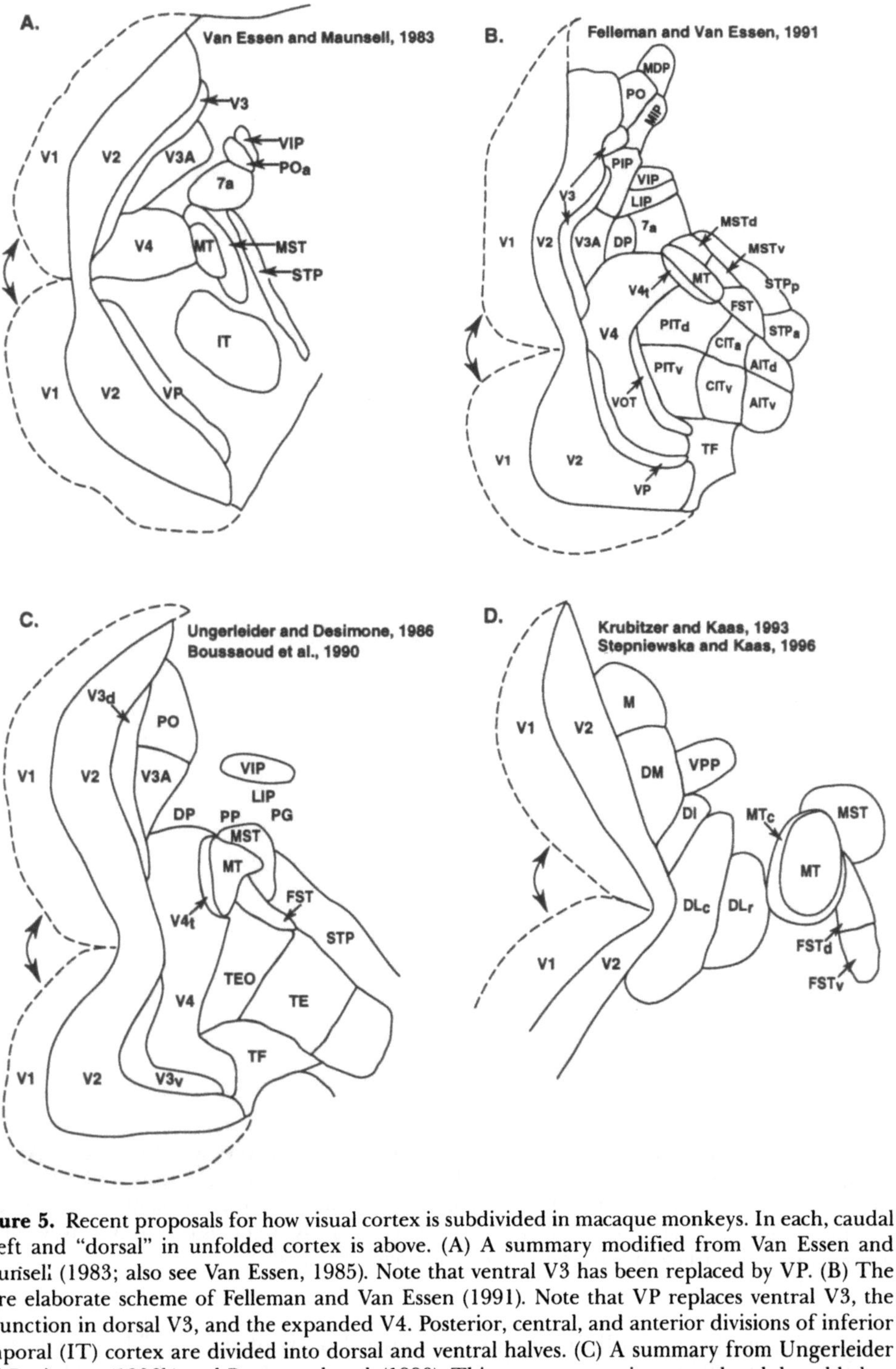

Figure 5. Recent proposals for how visual cortex is subdivided in macaque monkeys. In each, caudal is left and "dorsal" in unfolded cortex is above. (A) A summary modified from Van Essen and Maunsell (1983; also see Van Essen, 1985). Note that ventral V3 has been replaced by VP. (B) The more elaborate scheme of Felleman and Van Essen (1991). Note that VP replaces ventral V3, the disjunction in dorsal V3, and the expanded V4. Posterior, central, and anterior divisions of inferior temporal (IT) cortex are divided into dorsal and ventral halves. (C) A summary from Ungerleider and Desimone (1986b) and Boussaoud *et al.* (1990). This summary retains ventral and dorsal halves of V3 (Fig. 4B, C), which is split centrally by a broad expanse of V4. In addition, V2 is bordered by PO and TF [temporal field "F" after the terminology of Von Economo (1929)]. (D) A proposal for how some of extrastriate cortex is organized in macaque monkeys that postulates similarities between New and Old World monkeys. Compare with Fig. 6A.

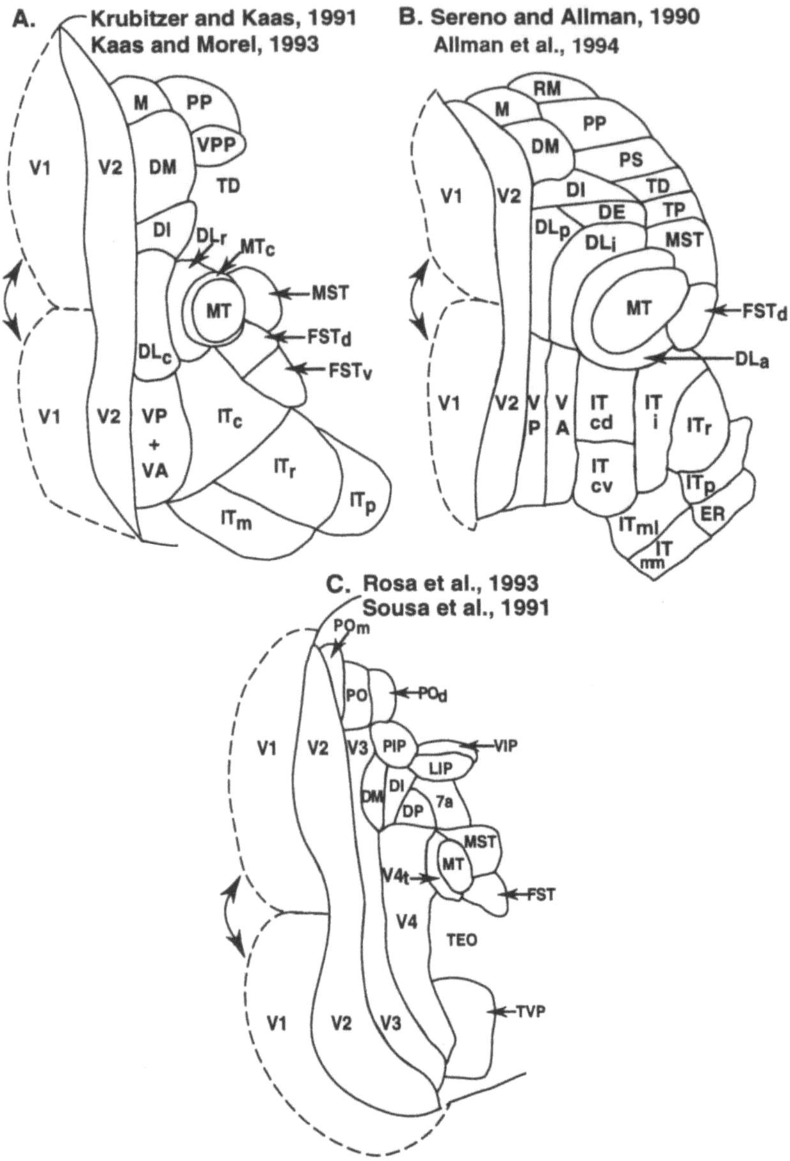

Figure 6. Recent proposals for how visual cortex is subdivided in New World monkeys. (A) The organization of extrastriate cortex in owl monkeys (based on Krubitzer and Kaas, 1991; Kaas and Morel, 1993; also see Kaas, 1994). (B) The organization of visual cortex in owl monkeys according to Sereno and Allman (1990) and Allman *et al.* (1994). (C) Subdivisions of visual cortex in New World cebus monkeys according to Rosa *et al.* (1993) and Sousa *et al.* (1991).

proposal include (1) a split in dorsal and ventral parts of the "V3" belt to accommodate a V4, (2) the conversion of ventral V3 to another visual area, VP, while dorsal V3 remains as an incomplete visual area representing only the lower visual quadrant, and (3) an estimate of the complete outer boundary of V4 of Zeki (1971b). Other visual areas have been added, including MT (Allman and

Kaas, 1971a), MST (Maunsell and Van Essen, 1983), VP (see Fig. 4D; Newsome and Allman, 1980), VIP (Maunsell and Van Essen, 1983), POa (Pandya and Seltzer, 1982), and V3a (Zeki, 1978). A later summary (Fig. 5B) by Felleman and Van Essen (1991) added quite a few visual areas, but also portrayed a greatly expanded V4 and a V3 broken into two parts. In contrast to earlier views on the extent of V3 (Fig. 4), five or more visual areas border the rostral margin of V2 in this interpretation, and V3 is reduced to a narrow strip along only part of the dorsal half of V2. The enlargement of V4, also part of the plan of Ungerleider and Desimone (1986b) (Fig. 5C), reflects the conclusion of Gattass *et al.* (1988), based on microelectrode mapping studies, that V4 extends far into the temporal lobe and disproportionately represents the upper visual quadrant.

In contrast to Van Essen and collaborators, the interpretation of Ungerleider and Desimone (Fig. 5C) retains a ventral V3. They added a transitional zone, V4t, between V4 and MT, area PO of Colby *et al.* (1988), a fundal area of the superior temporal sulcus, FST, and a ventral intraparietal area (VIP), subsequently described more fully by Colby *et al.* (1993). Area V3A has been retained from Zeki (1978). V4t was thought to be an incomplete area representing only the lower visual quadrant.

Our interpretation of the organization of visual cortex in macaque monkeys (Fig. 5D) was guided by earlier conclusions based on the microelectrode maps in owl monkeys (Fig. 4D) and later anatomical results in macaques that seem consistent with observations on New World monkeys. In brief, we propose that cortex in the region of dorsal V3 and V3A is DM, cortex medial to DM is a homologue of M (Colby *et al.*, 1988, speculated that PO may be M, and we agree and use the older term), and the V4 region includes two dorsolateral fields, DLc and DLr, with DLr about half the size of DLc. MTc is comparable to part of V4t, but we postulate a larger field representing the upper as well as the lower visual quadrant, as in owl monkeys (Kaas and Morel, 1993). The term MTc is based on the earlier architectonic description of a ring of dense patches of cytochrome oxidase around MT in owl monkeys (Tootell *et al.*, 1985). It also seems likely that FST is divided into dorsal (FSTd) and ventral (FSTv) fields, as in owl monkeys (Kaas and Morel, 1993), since the region has two patterns of connections with V2 in macaques (Stepniewska and Kaas, 1996). Our proposed ventral posterior parietal field, VPP, is the major projection target of DM, and it may correspond to VIP or LIP (see below).

Our summary of the organization of visual cortex in New World monkeys (Fig. 6A) reflects both our extensive research efforts in New World monkeys and the advantages provided by monkeys with more cortical areas accessible on the exposed brain surface. Supporting evidence for dividing DL (Fig. 4D) into DLc, DLr, and MTc comes from the anatomical studies of Cusick and Kaas (1988), Morel and Kaas (1993) and Steele *et al.* (1991). The caudal, rostral, medial, and polar subdivisions of inferior temporal cortex (ITc, ITr, ITm, ITp) are from Weller and Kaas (1985, 1987).

The proposal of Sereno and Allman (1990) is basically similar, although more elaborate (Fig. 6B). Some areas (M, DM, DI, VP, MT) are from earlier mapping studies (Fig. 4D), and others are based on interpretations of anatomical studies (e.g., Weller and Kaas, 1987) and subsequent mapping studies that are largely unpublished (Allman *et al.*, 1994). An important point of agreement is that the former territory of DL is divided into three areas (DLp, DLi, and DLa), much like DLc, DLr, and MTc (Fig. 6A). Thus, both connection patterns and the results of microelectrode mapping studies provide evidence for dividing the cortex between V2 and MT into three areas.

A somewhat different concept of visual cortex organization resulted from studies of connections of V1 (Sousa *et al.*, 1991) and microelectrode mapping (Rosa *et al.*, 1993) in cebus monkeys. These New World monkeys have a large brain, and the disadvantage in mapping studies of having as many fissures in visual cortex as Old World monkeys. Interestingly, the scheme for cebus monkeys (Fig. 6C) resembles some of the proposals for macaque monkeys in that there is a long V3 forming most of the outer border of V2, and a large V4 extending far into the temporal lobe. However, DM is recognized as a separate area, but this DM does not border V2 as in other proposals (Fig. 5D and 6A, B). Instead, DM occupies the position of V3A (Fig. 5A–C). The investigators conclude that V3A is DM. More recent interpretations of visual cortex organization in New World marmosets by Rosa and Schmid (1995) do not include a V3.

In summary, current theories of extrastriate visual cortex organization in primates have a few fields in common, V2 and MT in particular, but obvious differences. When early proposals were based on research restricted to either New or Old World monkeys, it seemed possible that many of the differences simply reflected real variations in cortical organization across primate taxa (Van Essen and Zeki, 1978). Now it seems that a more likely explanation for the major differences, especially for early stages of extrastriate cortex, is that ambiguous data are interpreted within the conceptual framework of the investigators. Our early interpretations of research on New World monkeys influenced our later interpretations of cortical organization in macaque monkeys, whereas some of the researchers working on New World cebus monkeys had previously studied macaque monkeys (e.g., Gattass *et al.*, 1988) and they adopted concepts of cortical organization from proposals based on macaques. Superficially, the present situation resembles the time of the early architectonic studies, where proposals for cortical organization simply differed, and no resolution was forthcoming. However, the present situation is more promising in that data have been and can be generated from an array of approaches. Thus, we can expect a resolution of most of the points of disagreement, especially as critical questions are brought into focus.

The next portions of this review briefly consider V2 and MT, visual areas that are now well established, to see what features allow visual areas to achieve widespread credibility. We then consider the evidence for the organization of the disputed region of V3–V4 or DM–DL.

6. V2 and MT, Areas Common to All Primates

The second visual area, V2, and the middle temporal visual area, MT, have been identified in all primates that have been appropriately studied (Fig. 7). The second visual area may be an area that emerged with the primary area in reptiles (Reiner, 1993) or early mammals, with both areas being retained in all or nearly

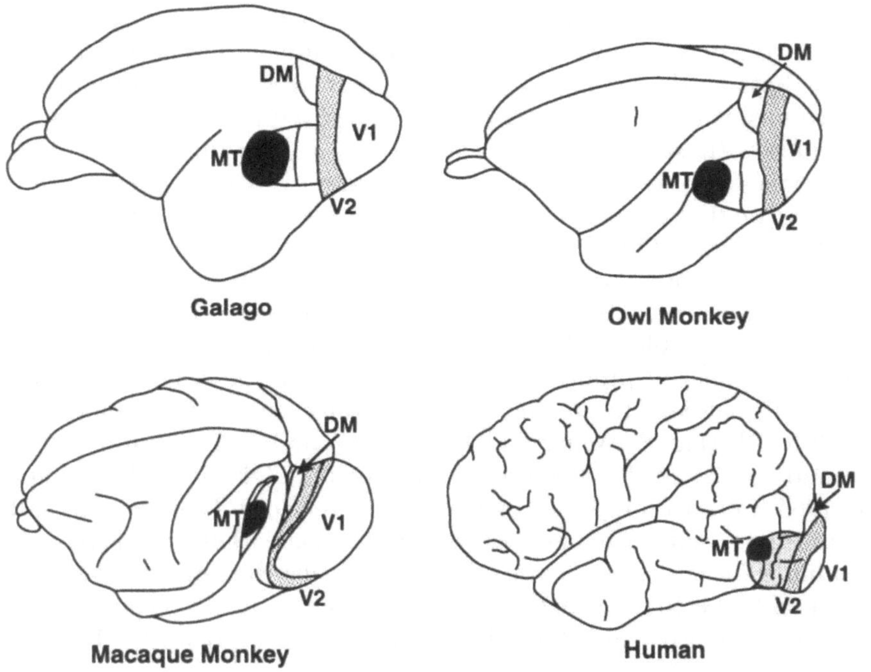

Figure 7. The proposed locations of visual areas V2, MT, and DM in prosimian galagos, New World owl monkeys, Old World macaque monkeys, and humans.

all extant mammals (Kaas, 1995a; Northcutt and Kaas, 1995). MT, in contrast, seems to be a more recently evolved visual area, since MT is found in all primates, but it has not been identified with certainty in any nonprimate (Kaas and Preuss, 1993; Northcutt and Kaas, 1995). MT may have evolved from a visual area bordering V2 that was subsequently displaced into temporal cortex.

Both V2 and MT can be identified by a number of characteristics, including distinctive architecture, retinotopic maps, and direct inputs from V1. V2 resembles the first band of extrastriate cortex of early reports (Fig. 1), but V2 does not precisely correspond to any of these fields. However, Brodmann (1909) defined area 18 differently in different species, so that V2 closely corresponds to the area 18 Brodmann (1909) described for marmosets, but V2 is much narrower than the area 18 described for Old World monkeys and humans. The concept of V2 or VII stems from the early discovery of a second representation of the visual hemifield in cortex lateral to striate cortex in cats (Talbot and Marshall, 1941), and the term VII was applied to indicate that the area was second in order of discovery, rather than to imply that it constitutes a second level of processing (Woolsey and Fairman, 1946).

The exposed dorsolateral portion of V2 was first mapped in a primate by Cowey (1964), who showed that the anterior border of dorsolateral V2 in squirrel monkeys corresponds to the zero horizontal meridian, a conclusion soon supported by anatomical evidence that central parts of V1 project to the outer margins of V2 in macaque monkeys (Fig. 4). Allman and Kaas (1974a) provided the first complete electrophysiological map of V2 in primates, and they described V2 as a "split" or second-order transformation of the visual hemifield bordering 90% of V1 in owl monkeys. Subsequently, V2 has been fully mapped in cebus (Rosa *et al.*, 1988) and macaque (Gattass *et al.*, 1981) monkeys with similar results. Partial maps of V2 have been derived in other primates, including marmosets (Rosa and Schmid, 1995) and galagos (Allman *et al.*, 1973). In addition, imaging of visually induced activity patterns has provided considerable information on the retinotopy of V2 in humans (DeYoe *et al.*, 1995; Sereno *et al.*, 1995; Shipp *et al.*, 1995).

The modular organization of V2 was first revealed by uneven patterns of connections. Most notably, patchy terminations of V1 projections in V2 were described by Wong-Riley (1979; also see Weller and Kaas, 1983; Lin *et al.*, 1982) and bandlike zones of input from the pulvinar were described by Curcio and Harting (1978; also see Levitt *et al.*, 1995). Bandlike arrays of neurons projecting from V2 to DL (Fig. 8; Weller and Kaas, 1985) or V4 (Shipp and Zeki, 1985; DeYoe and Van Essen, 1985) and to MT (Shipp and Zeki, 1985; DeYoe and Van Essen, 1985) were soon reported. A banding pattern of V2 connections with MT also exists in prosimian primates (Krubitzer and Kaas, 1990). Band-shaped modules within V2 seem to be a basic characteristic of primates, one that probably emerged with the first primates (Kaas and Preuss, 1993). Types of modular organization also occur in V2 of other mammals (Cusick *et al.*, 1984; Kaas *et al.*, 1989).

An important advance in understanding extrastriate cortex organization in primates occurred when it became apparent that V2 can be delineated by a banding pattern in sections processed for cytochrome oxidase (CO). In CO preparations, V2 is divided into a series of CO dark bands separated by paler interbands (Livingstone and Hubel, 1982; Tootell *et al.*, 1983). The interbands stain more densely for myelin (Krubitzer and Kaas, 1989). One set of CO dark bands projects to MT, the other to DL–V4, while the interbands also project to DL–V4. The CO dark bands have been distinguished as belonging to two classes, thick or thin, with the thicker bands in macaques projecting to MT. However, thickness or thinness does not seem to be a reliable way to distinguishing the bands across primates (Fig. 9, see Krubitzer and Kaas, 1990). Fortunately, the antibody Cat-301 selectively marks the so-called thick bands in V2 of macaque monkeys (DeYoe *et al.*, 1990). The banding pattern of V2 has been identified in a

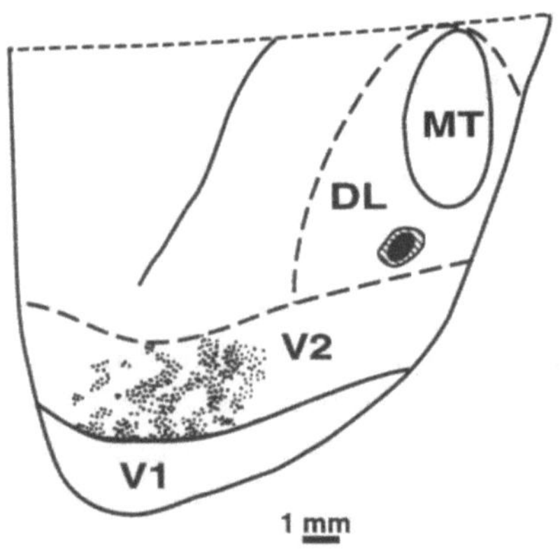

Figure 8. Bands of neurons in V2 projecting to the dorsolateral visual area in an owl monkey. The dorsal view of the caudal third of the right hemisphere shows where neurons were labeled in V2 (dots) by an injection (dark oval) in DL. Visual areas MT, DL, VII (V2), and V1 are indicated. From Weller and Kaas (1985).

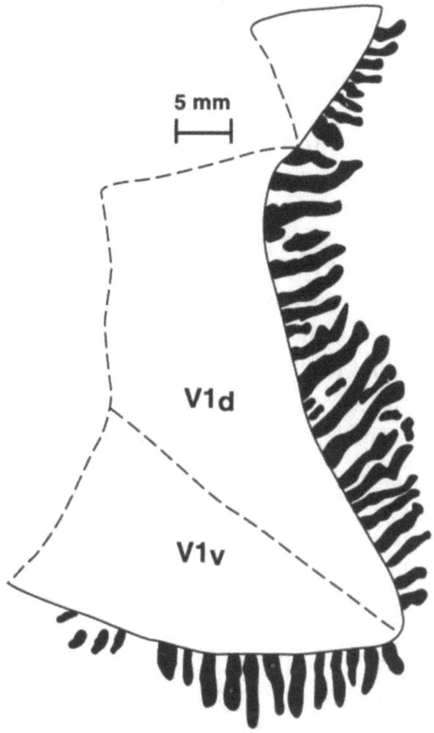

Figure 9. A reconstruction of the pattern of dark cytochrome oxidase (CO) bands in V2 of a macaque monkey based on sections cut parallel to the surface of manually flattened cortex. The reconstruction of the pattern in the dorsal half of V2 (above) appears to be nearly complete, resulting in approximately 30 CO bands or 60 for V2 if both halves are symmetrical. Brain sections were prepared by L. A. Krubitzer.

variety of anthropoid primates (*Macacca, Homo, Saimiri, Callithrix, Cebus, Aotus*), but the bands are faint in prosimian primates (Condo and Casagrande, 1990; Krubitzer and Kaas, 1990; Preuss *et al.*, 1993; Preuss and Kaas, 1996). The relevance of the banding pattern for theories of cortical organization is that the bands can be used to identify the boundaries of V2 precisely, especially in sections from flattened cortex. Before such preparations, it was possible to identify V2 architectonically (Allman and Kaas, 1974a), but the distinctions between V2 and rostral areas were slight (Allman and Kaas, 1974a). The presence of the CO-banding pattern as a marker of V2 would allow a precise delineation of all parts of the field in humans (Wong-Riley *et al.*, 1993), although this has not been done yet. Another advantage provided by the banding pattern in V2 is that it can be used as a scale by counting bands from the center of V2 to compare portions of V2 or locate areas along the V2 border across individuals. The dorsomedial half of V2 in macaques appears to have about 30 CO-dense bands (Fig. 9), suggesting a total of around 60 bands.

The other extrastriate visual area that is included in all current theories of visual cortex organization is area MT or V5. Area MT contains a first-order (undivided) representation of the visual hemifield. Such a representation has now been mapped in detail in owl monkeys (Allman and Kaas, 1971a), galagos (Allman *et al.*, 1973), macaques (Gattass and Gross, 1981), and marmosets (Rosa and Schmid, 1995). Most importantly, MT can be identified by its dense myelination (e.g., Allman and Kaas, 1971a) and dark appearance in CO preparations (e.g., Tootell *et al.*, 1985). Thus, MT has been readily identified by architectonic features in many primates, including humans (Tootell and Taylor, 1995; also see Tootell *et al.*, 1995) and endangered prosimian species (Preuss and Kaas, 1996) after natural death. MT can also be revealed by a topographic pattern of input from V1 (Spatz, 1977; Symonds and Kaas, 1978; for review see Weller and Kaas, 1983; Preuss *et al.*, 1993), although the existence of topographic connections was once in question (Cragg, 1969; Zeki, 1971a). MT is characterized by inputs dominated by the magnocellular geniculate relay to V1 (Maunsell *et al.*, 1990), and a high percentage of neurons that are selective for the direction of moving stimuli (Zeki, 1971c, 1980; Baker *et al.*, 1981; Felleman and Kaas, 1984).

As for V2, there is considerable evidence that MT has a modular organization, although the nature of the modules is not well understood. The connections of MT with MT of the other hemisphere and with other fields are quite patchy (e.g., Krubitzer and Kaas, 1990, 1993; Kaas and Morel, 1993), as is the staining for myelin (Krubitzer and Kaas, 1990) and cytochrome oxidase (Tootell *et al.*, 1985). Albright *et al.* (1984) suggested that MT contains alterations of axis-of-motion columns of about 0.5 mm in width, but these columns would be larger than the patches of connections and myelin. Another possibility is that the patches relate to the segregation of global and local processing modules (Born and Tootell, 1992) or shape and motion modules (Malonek *et al.*, 1994).

While some differences in the response properties of MT neurons in macaques and owl monkeys once led to questions about the homology of the area across species (Zeki, 1980), there now seems to be little doubt that MT is an area common to all primates (e.g., Rosa *et al.*, 1993). Thus, V2 and MT are landmarks that can serve as guides for the relative locations of other visual areas across primates. Because V2 and MT can be accurately located by architectonic criteria, it is possible to measure and compare the sizes of these fields across primates in studies of adaptation and brain evolution (e.g., Pessoa *et al.*, 1992).

7. Regions of Contention: Visual Areas of the Third Tier versus V3 and V4

Presently, there are no extrastriate visual areas besides V2 and MT that one can clearly identify as having emerged early in primate evolution and having been widely retained. This is partly the consequence of the rather ambiguous evidence for most visual areas. The ambiguousness of the data is recognized by most investigators of extrastriate visual cortex, but uncertainties and alternative interpretations are not always stressed in experimental reports. Here we review evidence for different interpretations of the organization of cortex just rostral to V2. This cortex has been described as containing either a third tier of visual areas or areas V3 and V4.

8. Area V3 and Dorsal V3

V3 in macaque monkeys was originally conceived as a mirror-image representation of V2 (Fig. 4B, C), much as is V3 of cats (Hubel and Wiesel, 1965). The representation of the zero horizontal meridian was thought to separate V2 and V3, with the zero vertical meridian forming the outer border of V3. The original evidence was the finding that dorsal V1 projects to both V2 and cortex rostral to dorsal V2 in a manner suggesting two parallel, mirror-image patterns (Fig. 4). Furthermore, dorsal V2 was found to relay to adjacent locations along its outer border (Zeki, 1971b). Given these patterns, a reasonable assumption was that the projections of ventral V1 and V2 are symmetrical to those of dorsal V1 and V2, and that a complete third representation borders V2. Additional evidence came from studies of corpus callosum connections that were already known to ring V1 along the representation of the vertical meridian (Fig. 4A). Callosal terminations in cortex rostral to V2 were also thought to mark the representation of the vertical meridian and the rostral border of V3 (Zeki, 1977; Van Essen and Zeki, 1978). Later, microelectrode mapping results were interpreted as consistent with the concept of V3 in that regions of the lower visual quadrant were found to be represented along the dorsal border of V2, while regions of the upper visual quadrant were found along the ventral border of V2 (Gattass *et al.,* 1988).

While many findings were consistent with the traditional concept of V3 in macaques, it soon became apparent that some observations were not. Most notably, projections were found from dorsal V1 to cortex rostral to dorsal V2, but not from ventral V1 to cortex rostral to ventral V2 (Weller and Kaas, 1983). Van Essen *et al.* (1986) stressed this difference in the projection pattern, and also noted a dorsoventral asymmetry in the myeloarchitecture. Cortex along part of the dorsal border of V2 was densely myelinated, while cortex along the ventral border of V2 was not. Thus, Van Essen *et al.* (1986) proposed that only dorsal cortex was V3, and the ventral cortex was not (Fig. 5A). Instead, part of the ventral cortex was considered to be the ventral posterior area (Newsome *et al.,* 1986), one of the third-tier areas of owl monkeys (Fig. 4D). Neurons in the dorsal and ventral zones of cortex were also found to have quite different response properties (Burkhalter and Van Essen, 1986; Burkhalter *et al.,* 1986), providing further evidence that dorsal and ventral "V3" regions are not parts of the same field.

Another component of the evidence for V3 is also questionable. Terminations of callosally projecting neurons have been used repeatedly to define the rostral border of "V3" (Fig. 4; also see Clarke and Miklossy, 1990). Corpus callosum terminations do not form a narrow band along the postulated outer border of V3 as they do for V1. Surface views of the callosal pattern revealed broad rather than narrow zones of terminations rostral to V2 (Van Essen and Zeki, 1978; Newsome and Allman, 1980; Cusick *et al.,* 1984). While these callosal terminations may in part coincide with the representation of the zero vertical meridian, they are too broad and scattered to be useful indicators of the boundaries of extrastriate visual areas (Cusick and Kaas, 1986; Kaas, 1995c). For example, 90% of the outer border of MT, which is a well-defined visual area, corresponds to the vertical meridian, and yet this border is not outlined by callosal connections. Instead, callosal connections include both border and nonborder regions of MT (Cusick *et al.,* 1984; Maunsell and Van Essen, 1987; Krubitzer and Kaas, 1990). Even for locations where callosal connections are concentrated along the representation of the zero vertical meridian, they do not necessarily distinguish between different theories of extrastriate visual cortex organization.

One would expect any visual area to have some representation of the vertical meridian.

Overall, the original concept of V3 seems hard to justify. We also question the usefulness and validity of proposing a greatly modified V3, one that is just one of four or more areas bordering V2, and an area that represents only the lower visual field, and even then, possibly in a discontinuous manner (Fig. 5B). We have referred to such incomplete representations as "improbable areas" (Kaas, 1993), since it seems improbable that any visual area would represent only the lower or upper visual field. While the paracentral lower field has an expanded representation in some areas such as MT of macaques (Maunsell and Van Essen, 1987), we know of no visual abilities that exist for only lower field vision, although disproportionate representation and connections may lead to enhanced performance for tasks within the lower field (Previc, 1980; Rubin *et al.*, 1996).

9. Area DM

The dorsomedial area, DM, was one of the first extrastriate visual areas mapped with microelectrodes in owl monkeys (Allman and Kaas, 1975). DM represents both the upper and lower visual quadrants in a densely myelinated region of cortex along the rostral border of V2 at the medial margin of the hemisphere. Early studies also revealed major connections with more rostral posterior parietal cortex and MT (Wagor *et al.*, 1975) as well as inputs from different parts of V2 (Kaas and Lin, 1977) and V1 (Lin *et al.*, 1982). Input patterns from both parts of V1 and V2 devoted to upper or lower vision were consistent with the concept of a complete hemifield representation in DM, rather than only a portion of lower quadrant representation as in V3. A recent reinvestigation of the area in owl monkeys (Krubitzer and Kaas, 1993) provided additional evidence for DM as a representation that includes both upper and lower visual quadrants in a densely myelinated region of cortex. Surface view patterns of connections were more extensively studied, and inputs to DM from portions of V1 and V2 that represent the upper as well as lower visual quadrant were revealed. Other connections of DM were with a ventral portion of the posterior parietal cortex (termed VPP; see Fig. 6A), M, MT, MST, dorsal FST, DLr, the region of VP plus VA, and frontal visuomotor cortex. Most of the connections of DM place the field within the dorsal visual stream of processing of Ungerleider and Mishkin (1982) that converges in posterior parietal cortex. Yet, surprisingly, neurons in V1 that project to DM were largely within the cytochrome oxidase blobs, processing modules considered to be key elements of the ventral stream of processing (Casagrande and Kaas, 1994). Overall, DM in owl monkeys seems to be one of the better identified visual areas, having a retinotopic map, distinctive myeloarchitecture, and a unique pattern of connections.

Krubitzer and Kaas (1993) also provided evidence that DM is an area that is more generally found in primates. Injections in V1 revealed interconnections with densely myelinated cortex on the border of V2 that was judged to be DM in two additional species of New World monkeys, three species of Old World monkeys, and prosimian galagos. Other evidence for the widespread existence of DM in primates has accumulated. As already noted, a DM, as well as a narrow dorsal V3, has been described in cebus monkeys (Fig. 6B). More recently, Rosa and Schmid (1995) described a DM without a dorsal V3 in marmosets. Rosa and Schmid (1995) also provided details on the representation of the visual field in DM that seem to apply more broadly to other diurnal primates. Their results indicate that the representation of central vision is lateral in DM, peripheral vision is medial, and the upper quadrant is split into rostrolateral and medial portions. A similar map of the representation of central and paracentral vision in presumptive DM of macaque monkeys has been determined from connection patterns after the injection of up to six distinguishable tracers in dorsolateral V2 (Stepniewska and Kaas, 1996). Recently, microelectrode mapping in prosimian galagos has revealed a representation with the features of DM (Rosa *et al.*, 1997). Finally, retinotopic activation patterns in the human brain (Sereno *et al.*, 1995) reveal activity related to the upper visual quadrant, as well as the lower visual quadrant, in the expected location of DM (Kaas, 1995b), suggesting that DM exists in humans as well. Perhaps more importantly, injections in putative DM of macaque monkeys label the same targets as DM injections in owl monkeys, including preferentially the CO blobs of V1 (Beck and Kaas, 1995). We conclude from this evidence that DM is common to all New and Old World monkeys, and probably all primates. The location or expected location of DM for prosimian, simian, and human primates is shown in Fig. 7.

10. Area M

The medial area, M, was first described in owl monkeys as a small representation of the visual hemifield along the outer border of V2 just "medial" to DM (Fig. 4D; Allman and Kaas, 1976). M was found to have proportionately little representation of central compared to peripheral vision. Connections of area M in owl monkeys have only partially been described, but they include projections to parts of V2 representing both upper and lower quadrants (Graham *et al.*, 1978), a pattern expected for a small, complete visual area, but not for a restricted part of a dorsal V3. Recently, parts of M have been remapped in owl monkeys, and these results provide further evidence for the validity of the area (Krubitzer and Kaas, 1993). A similar map of the hemifield in the same relative region has been termed PO in New World cebus monkeys (Neuenschwander *et al.*, 1994; also see Fig. 6C) after the term used in macaque monkeys (Covey *et al.*, 1982). Somewhat surprisingly, the authors argued that "area PO" in cebus monkeys is homologous with PO of macaques but not M of owl monkeys. This conclusion raises a phylogenetic issue. Why would a New World monkey and an Old World monkey have the same area, while two New World monkeys do not? A more plausible alternative is that PO of macaques is a homologue of M in owl monkeys, as suggested by Colby *et al.* (1988), and it is because of the strength of this argument that we use the term M in both New and Old World monkeys (Fig. 10B). Area M or PO appears to have similar connections, including those with paracental parts of V1, in both New World (Martinez-Millán and Höllander, 1975; Sousa *et al.*, 1991) and Old World monkeys (Colby *et al.*, 1988). However, differences in architecture have been proposed, and other visual areas in the region may exist to complicate identification (Fig. 6C) (Neuenschwander *et al.*, 1994). Interestingly, Rosa and Schmid (1995) propose that both M and a narrow medial PO (POm) exist in marmosets. By its position, their area M appears to comprise much of area M of owl monkeys, whereas POm is shown as a sliver of cortex wedged between V2 and M (and adjoining DM) that represents only a part of the upper visual quadrant. Thus, a number of interpretations of how the region is subdivided seem plausible. Nevertheless, all of these current proposals indicate that a small visual area, rather than part of V3, is located along a medial portion of the border of V2. Furthermore, there is no compelling reason to postulate a major difference in the organization of this region of cortex between New and Old World monkeys.

Finally, given the adjacency of M and DM, or PO and V3d, and the possibility of several other small visual areas in the region, observations on the connections of any specific field may have confounded results across areas. As noted

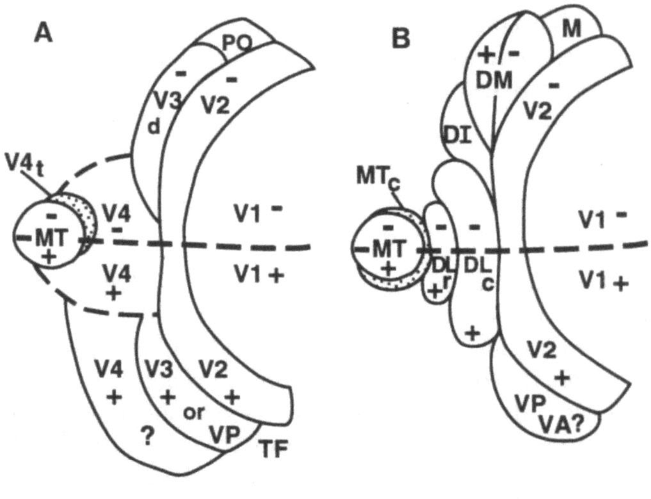

Figure 10. A schematic of major differences in theories of the organization of caudal extrastriate cortex in monkeys. For some fields, representations of the upper (+) and lower (−) visual quadrants are indicated. A horizontal dashed line divides some fields into parts representing upper or lower quadrants. (A) This scheme reflects the major features of most proposals based on Old World macaque monkeys. Differences in views are also expressed in the figure. Most investigators now include a V4t and an expanded V4 with an enlarged representation of the upper visual quadrant (see Fig. 5B, C). Dashed lines indicate dorsal and ventral boundaries of V4 from early proposals (Fig. 5A). V4 is also seen as extended ventrally. The scheme also allows for V3 as having dorsal and ventral halves (Fig. 5C), or being only dorsal to a ventral region, VP (Fig. 5B). PO is from Fig. 5B, C, and is different from POa (Fig. 5A), but possibly homologous to M (see below and text). (B) A scheme that has emerged from our studies on New and Old World monkeys. Dorsal V3 is replaced by DM and DI, M replaces PO because M was described earlier, V4t is enlarged into an MTc that represents both upper and lower quadrants, and the remaining V4 region is divided into two dorsolateral fields, DLr and DLc. VP and VA are presumptive "areas" that largely or completely relate to the upper visual quadrant (see text).

above, injections in DM of macaque as well as other monkeys selectively labeled output neurons that are coextensive with the CO blobs of V1 (Beck and Kaas, 1995; Krubitzer and Kaas, 1993). More medial injections in macaques, possibly in M, labeled V1 neurons in layer 3C [4B in some schemes (Casagrande and Kaas, 1994)] outside the blobs (P. D. Beck and J. H. Kaas, unpublished observations). Similar projections of V1 have been described as terminating in dorsal V3 (Felleman *et al.*, 1996; Burkhalter *et al.*, 1986). Thus, the described laminar and modular differences in the origins of inputs to the DM–V3 region may correspond to projections to different visual areas.

11. Area DI

The dorsointermediate area, DI, occupies a region of cortex between DL and DM in owl monkeys that responds to visual stimuli and has retinotopic organization (Allman and Kaas, 1975, 1976). The retinotopy of DI in owl monkeys has not been described in detail. Microelectrode mapping data have been reported for marmosets by Rosa and Schmid (1995), and they describe an area with central vision represented caudally along the V2 border, more peripheral parts of the visual field in progressively more rostral parts of DI, and the upper quadrant in cortex medial to the lower quadrant. The connection pattern with V2 suggests a similar organization for DI in macaque monkeys, at least for the representation of central and paracentral vision of the lower visual quadrant (Stepniewska and Kaas, 1996). Although the evidence for DI remains sparse, the DI region is not easily incorporated into the retinotopic maps of either DM medially or DL laterally, so that it seems reasonable to continue to propose the existence of this small area. Again, there is evidence for DI in both New and Old World monkeys, and little reason to propose different organizations in the region of cortex for these taxa.

12. Area V4

The V4 region of cortex on the anterior bank of the lunate sulcus of macaque monkeys was defined (Zeki, 1971b) as a target field of projections from areas 18 (V2) and 19 (V3). The original proposal was that the region contains two visual areas or two target regions, V4 and V4a, but this suggestion has been largely ignored or dismissed. Originally, the V4 region was thought to have little retinotopic organization, but microelectrode mapping evidence (Gattass *et al.*, 1988) has been used to demonstrate retinotopy in the region with the lower quadrant represented dorsal to the upper quadrant, which was shown as greatly expanded into the temporal lobe. The expansion from the original territory can be appreciated by comparing Fig. 5A and 5B. Earlier claims for two separate representations of at least the lower visual quadrant in the region of dorsal V4 (Baizer and Maquire, 1983; Maguire and Baizer, 1984) did not seem to be supported by the study of Gattass *et al.*, (1988), and this was probably an important factor in subsequent portrayals of the V4 region as a single field (Fig. 5B, C). As a further development, V4 was strongly associated with the processing of color information (Zeki, 1973; see Cowey and Heywood, 1995, for a critical review). Largely because patients with achromatopsia have ventromedial occipital brain damage and attention to color activates the same region, some proposals place V4 of humans in the region of the lingual gyrus (e.g., Zeki *et al.*, 1991) in a position relative to MT and central V2 that is quite different from the location of the original V4 in macaques. Thus, it seems quite unlikely that this cortex in humans is homologous with any of the original V4 region in macaques (Merigan, 1993; Ungerlieder and Haxby, 1994; Cowey and Heywood, 1995, Kaas, 1995b). Finally, there is no evidence that any of the proposed ventral extension of V4 is a target of the topographic pattern of V2 projections that occur in the original V4 region (Stepniewska and Kaas, 1996).

13. DLr and DLc

The dorsolateral cortex of owl monkeys was once thought to contain a single visual area, DL, between V2 and MT (Fig. 4D; Allman and Kaas, 1974b). Subsequent studies on patterns of connections led us to divide the area into two parallel representations, the caudal and rostral dorsolateral areas (DLc and DLr, Fig. 6A), and a narrow MT crescent wrapping around MT (see below). The results of microelectrode remapping of the region provided further evidence for three such fields (Sereno and Allman, 1990), although somewhat different terms were used (Fig. 6B). These areas represent the lower quadrant medially and the upper quadrant laterally and the representation of the zero horizontal meridian bisects both areas as it crosses from V1 to MT. In New World owl and squirrel monkeys, the region now identified as DLc receives the major feedforward output of V2 (Tigges *et al.*, 1974; Weller and Kaas, 1985; Cusick and Kaas, 1988; Steele *et al.*, 1991) and provides the major source of activation of the caudal subdivisions of inferior temporal cortex. Thus, DLc is a critically important relay in the ventral stream of visual processing (Ungerleider and Miskin, 1982; Weller, 1988). The smaller DLr has much less input from V2, and a more compressed representation of the visual hemifield. DLr has connections with MT and DM (Fig. 11), as well as caudal inferior temporal cortex (Steele *et al.*, 1991; Weller *et al.*, 1991). Thus, DLr contributes to both the dorsal and ventral streams, but it is more involved with the dorsal stream than DLc. DLc is further distinguished from DLr by being slightly more myelinated (Steele *et al.*, 1991) and darker in brain sections processed for cytochrome oxidase (Cusick and Kaas, 1988).

The organization of dorsolateral visual cortex in macaque monkeys appears to be quite similar to that in New World monkeys. The connection pattern revealed by injecting up to six different tracers in V2 of macaques identified DLc as an area in the caudal half of the cortex between MT and V2 with major V2 connections (Fig. 10B; Stepniewska and Kaas, 1996). DLc occupies about 30 mm of cortex along the border of central V2. Thus, DLc is much smaller than the currently postulated size of V4 (Fig. 5B, C). The region of DLc in macaques provides the major output to the adjoining caudal portion of inferior temporal

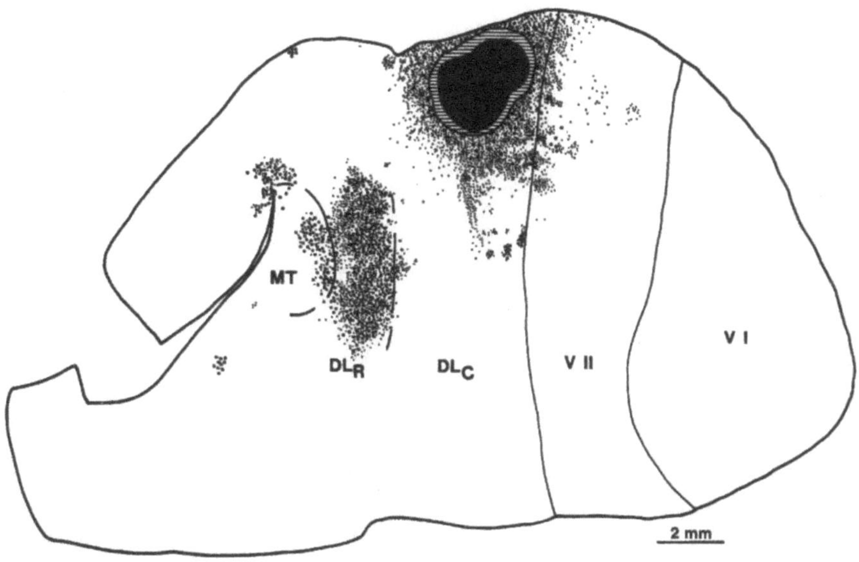

Figure 11. Labeled neurons (large dots) and terminations (small dots) after an injection of a tracer into presumptive DM of a squirrel monkey. The figure is of part of caudal cortex that was flattened and cut parallel to the surface. Areas V1, V2, DLc, DLr, and MT are identified. Note the dense label in DLr but not DLc. Modified from Fig. 8 of Weller *et al.* (1991).

cortex (Distler *et al.,* 1993; Morel and Bullier, 1990). The rostral division, DLr, is less densely interconnected with V2, and is much smaller, extending possibly 15 mm dorsoventrally (Stepniewska and Kaas, 1996). DLr may have modular subdivisions, since single injections in V2 typically reveal separate rostral and caudal patches of labeled connections. The DLr region of macaques also appears to have connections with both ventral stream areas in inferior temporal cortex and dorsal stream areas in posterior parietal cortex (Morel and Bullier, 1990; Baizer *et al.,* 1991; Tanaka *et al.,* 1990).

14. Upper Temporal Lobe Areas: MTc, FSTd, FSTv, and MST

The fundal area of the superior temporal sulcus, FST, was first distinguished by Ungerleider and Desimone (1986b) as the ventral part of a broader projection pattern of MT to the medial superior temporal region, MST (Maunsell and Van Essen, 1983). MST remains in most current maps (Fig. 5 and 6) as a smaller field that may be subdivided into dorsal and ventral fields, MSTd and MSTv (Fig. 5B). Our investigations of this cortex in owl monkeys led us to divide this region into four areas, dorsal (FSTd) and ventral (FSTv) fundal areas, an MT crescent, MTc (Fig. 12; Kaas and Morel, 1993), and an MST that is smaller than our original "ST" projection zone of MT in owl monkeys, which contained dorsal (MST) and ventral (FST) parts (Weller *et al.,* 1984). In owl monkeys and other primates, FST extends ventrally from MT as a moderately dark region of cortex in brain sections stained for myelin (Ungerleider and Desimone, 1986a; Krubitzer and Kaas, 1990). Somewhat surprisingly, we found that dorsal and

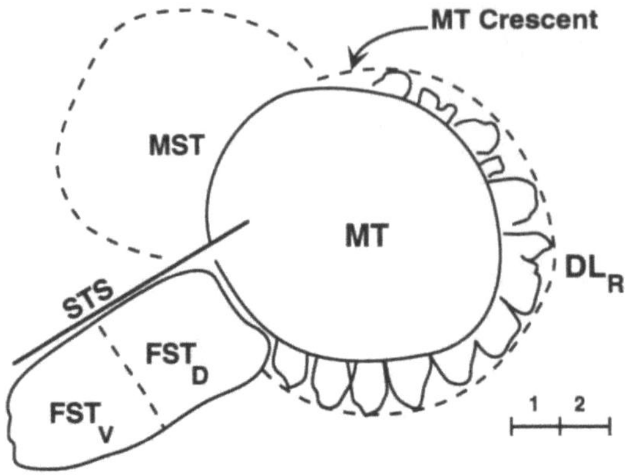

Figure 12. A surface view of FSTv, FSTd, and MTc in owl monkeys. The patchy appearance of MTc reflects the cytochrome oxidase- and myelin-dense modules with FSTv connections (see text). Modified from Fig. 1 of Kaas and Morel (1993). Scale bar = 2 mm.

ventral halves of FST had quite different connection patterns with other regions of cortex. Injections of tracers into the dorsal half of FST of owl monkeys revealed, as expected (e.g., Ungerleider and Desimone, 1986b), dense interconnections with MT. FSTd connections included both dorsal and ventral halves of V2, demonstrating that the small region contains representations of both the upper and lower quadrants. The majority of FSTd connections were with areas of the dorsal stream (MT, MST, DM, VPP), but they also included DI and VP. The connections of the ventral half of FST differed in several ways. First, FSTv injections labeled a series of patches of tissue that formed a ring or crescent around most of MT (Fig. 12). These patches are dark in myelin and cytochrome oxidase (CO) preparations, and they are separated by myelin and CO light septa. Because a crescent of CO-dark patches was first described in relation to MT by Tootell *et al.* (1985), we termed the area the MT crescent. MTc includes the V4 transitional (V4t) zone of Desimone and Ungerleider (1986), but MTc extends around more of MT. Other major cortical connections of FSTv included subdivisions of inferior temporal cortex, identifying the field as part of the ventral stream of visual processing. FSTv was sparsely connected with V2, and MT and MST were basically free of FSTv connections.

Clear evidence for FSTd and FSTv as separate fields is currently lacking in macaque monkeys. However, V2 injections label separate dorsal and ventral foci in FST (Stepniewska and Kaas, 1996), as one would expect for two fields. In squirrel monkeys, Weller and Steele (1992) designated an area as the dorsal division of caudal inferior temporal cortex (ITcd). This field appears to correspond, at least in part, to FST.

15. VP and VA

Cortex along the ventral border of V2 remains a region that has been incompletely studied, and the organization of this region is still poorly understood. In both New World owl monkeys (Newsome and Allman, 1980) and macaque monkeys (Newsome *et al.*, 1986), recordings just rostral to ventral V2 reveal a rerepresentation of part of the upper visual quadrant in a pattern that mirrors ventral V2. This region has been referred to as the ventroposterior "area," although as such it is an incomplete and "improbable" area (Kaas, 1993) representing only the upper visual quadrant. Many investigators still consider the region to be part of V3 (Figs. 5C and 10A), but differences in connections, neuronal response properties, and architecture have been cited in arguments against this interpretation (e.g., Burkhalter *et al.*, 1986; Burkhalter and Van Essen, 1986; Newsome *et al.*, 1986; Felleman *et al.*, 1997). Cortex just anterior to VP also responds to visual stimulation and it may form another representation of the upper visual quadrant (Allman *et al.*, 1994). Thus, this "ventral anterior area," VA, apparently does not complete a hemifield representation by joining a lower field representation to the upper field representation in VP. Yet neurons responsive to stimuli in the lower visual quadrant exist in the region (Gattass *et al.*, 1988; Boussaoud *et al.*, 1991). Other interpretations may be possible. Some investigators, for example, include the VA region in V4 (Fig. 5B, C), and Rosa (this volume) suggests that DLc and DLr continue ventrally as VP and VA. But this interpretation results in visual areas that greatly overrepresent the upper visual quadrant, and does not account for the presence of neurons responsive to stimuli in the lower visual quadrant.

16. Parietal Visual Areas: LIP, VIP, PIP, and VPP

A number of other visual areas have been proposed. Most notably there is clear evidence for several distinct visual areas in posterior parietal cortex (see Colby and Duhamel, 1991, for review). However, these complex higher order fields are more difficult to define precisely. Early designations may have been based on a single characteristic, and different terms have been used to refer to the same area. For example, Maunsell and Van Essen (1983) introduced the term ventral intraparietal area, or VIP, for the projection zone of MT in posterior parietal cortex of macaque monkeys (also see Colby *et al.*, 1993). A comparable region was previously called focus 5 of MT projections in marmosets by Spatz and Tigges (1972), and was identified as a major target of DM (Wagor *et al.*, 1975; Kaas *et al.*, 1977; Weller *et al.*, 1991; Krubitzer and Kaas, 1993) as well as MT (Weller *et al.*, 1984; Krubitzer and Kaas, 1990) in owl and squirrel monkeys. We currently refer to the region in New World monkeys as the ventral posterior parietal area, VPP (Fig. 6A; Kaas and Morel, 1993), but it seems likely that VIP and VPP are the same area. Other proposed areas in posterior parietal cortex include a lateral intraparietal area, LIP, originally defined as a region of posterior parietal cortex projecting to frontal cortex (Andersen *et al.*, 1985). Possibly VIP and LIP are parts of a larger area (however, see Colby *et al.*, 1993). Another possibility is that LIP contains separate ventral and dorsal areas, LIPv and LIPd (Blatt *et al.*, 1990). More medially in posterior parietal cortex, connections were used to define a medial intraparietal area, MIP, as well as a posterior intraparietal area, PIP, between V3 and PO (Colby *et al.*, 1988). Felleman and Van Essen

(1991) place PIP somewhat differently (see Fig. 5B). Thus, uncertainties remain, and these and other proposed fields require further study.

17. Summary and Conclusions

For nearly the last 100 years, many neuroscientists have attempted to determine how neocortex is subdivided into areas of functional significance. Accurate concepts of how cortex is subdivided into areas are obviously important as guides for further study and for the interpretation of previous results. More relevant experimental data have been gathered for visual cortex of primates than possibly any other cortical system and current proposals are much more valid than earlier proposals. However, areas can be difficult to define, especially by a single feature. Thus, modern maps of visual cortex still differ in many ways, and we do not have a comprehensive outline that includes different primate taxa. Nevertheless, some conclusions are well supported, and further progress can be expected as issues come into focus.

The present review has a few main points. (1) While early concepts of how visual cortex is subdivided were inconsistent with each other, and have been shown to be incompatible with recently collected experimental data, these concepts continue to have impact and influence. In particular, the concept of V3 has persisted as an extension of Brodmann's portrayal of area 19. It now seems clear that the region of cortex along the outer border of V2 contains a number of visual areas in primates, none of which are like area 19 of Brodmann (1909). Proposals that retain a V3 generally do so in a greatly modified form. Rather than submit V3 to such modifications, the concept of V3 as an equivalent of area 19 should be abandoned. (2) Of the proposed extrastriate visual areas, only two, V2 and MT, are recognized by all investigators regardless of primate taxa. Nevertheless, other areas are likely to be basic to all primates, and DM is one of these fields. A number of other visual areas have been reasonably well defined in some monkeys, but the designations have not been fully accepted. Further study is important and can be expected to lead rapidly to widespread agreement for some of these fields. (3) The importance of identifying areas and modules held in common across taxa has not been fully appreciated. The significance of having evidence for areas in several primate taxa is that these fields are likely to be retained in branches of evolution, such as humans, where experimental approaches are limited. In addition, theories of cortical organization for different species should be compatible with each other within the framework of evolutionary change. Thus, comparisons across species serve as a further means of evaluating theories for particular species. (4) While it once seemed that New World and Old World monkeys differed greatly in how extrastriate visual cortex is organized, this conclusion has not been supported by more recent comparative studies. Primate taxa undoubtedly differ in many ways, but presently we have a better understanding of how primates are similar rather than different in extrastriate cortex organization.

ACKNOWLEDGMENTS. I thank Pam Beck, Cassie Cusick, Sherre Florence, Leah Krubitzer, Marcie Pospical, Kathy Rockland, Jeff Schall, Iwona Stepniewska, and

Roz Weller for helpful comments on the manuscript, and Barbara Martin for illustrations. Research from the author's laboratory has been supported by Grant EY-02686 from the National Eye Institute.

18. References

Adrian, E. D., 1941, Afferent discharges to the cerebral cortex from peripheral sense organs, *J. Physiol.* (Lond) **100:**159–191.

Albright T. D., Desimone, R., and Gross, C. G., 1984, Columnar organization of directionally selective cells in visual area MT of the macaque, *J. Neurophysiol.* **51:**16–31.

Allman, J. M., and Kaas, J. H., 1971a, A representation of the visual field in the caudal third of the middle temporal gyrus of the owl monkey (*Aotus trivirgatus*), *Brain Res.* **31:**85–105.

Allman, J. M., and Kaas, J. H., 1971b, Representation of the visual field in striate and adjoining cortex of the owl monkey (*Aotus trivirgatus*), *Brain Res.* **35:**89–106.

Allman, J. M., and Kaas, J. H., 1974a, The organization of the second visual area (V-II) in the owl monkey: A second-order transformation of the visual hemifield, *Brain Res.* **76:**247–265.

Allman, J. M., and Kaas, J. H., 1974b, A crescent-shaped cortical visual area surrounding the middle temporal area (MT) in the owl monkey (*Aotus trivirgatus*), *Brain Res.* **81:**199–213.

Allman, J. M., and Kaas, J. H., 1975, The dorsomedial cortical visual area: A third tier area in the occipital lobe of the owl monkeys (*Aotus trivirgatus*), *Brain Res.* **100:**473–487.

Allman, J. M., and Kaas, J. H., 1976, Representation of the visual field in the medial wall of the occipital-parietal cortex in the owl monkey, *Science* **191:**572–576.

Allman, J., Jeo, R., and Sereno, M., 1994, The functional organization of visual cortex in owl monkeys, in: *Aotus: The Owl Monkey* (J. F. Baer, R. E. Weller, and I. Kakoma, eds.), Academic Press, Orlando, FL. pp. 287–320.

Allman, J. M., Kaas, J. H., and Lane, R. H., 1973, The middle temporal visual area (MT) in the bush baby (*Galago senegalensis*), *Brain Res.* **57:**197–202.

Andersen, R. A., Asanuma, C., and Cowan, W. M., 1985, Callosal and prefrontal associational projecting cell populations in area 7a of the macaque monkey: A study using retrogradely transported fluorescent dyes, *J. Comp. Neurol.* **232:**493–455.

Baizer, J. S., and Maguire, W. M., 1983, Double representation of lower visual quadrant in prelune-ate gyrus of rhesus monkey, *Invest. Opthalmol. Vis. Sci.* **24:**1436–1439.

Baizer, J. S., Ungerleider, L. G., and Desimone, R., 1991, Organization of visual inputs to the inferior temporal and posterior parietal cortex in macaques, *J. Neurosci.* **11:**168–190.

Baker, J. F., Petersen, S. E., Newsome, W. T., and Allman, J. M., 1981, Visual response properties of neurons in four extrastriate visual areas of the owl monkey (*Aotus trivirgatus*): A quantitative comparison of medial, dorsomedial, dorsolateral, and middle temporal areas, *J. Neurophysiol.* **45:**397–416.

Beck, P. D., and Kaas, J. H., 1995, Evidence for the presence of the dorsomedial visual area (DM) in five primate species, *Soc. Neurosci. Abstr.* **21:**1275.

Blatt, G. J., Andersen, R. A., and Stoner, G. R., 1990, Visual receptive field organization and corticocortical connections of the lateral intraparietal area (area LIP) in the macaque, *J. Comp. Neurol.* **299:**421–445.

Bonin, G. V., Garol, H. W., and McCulloch, W. S., 1942, The functional organization of the occipital lobe, *Biol. Symp.* **1:**165–192.

Born, R. T., and Tootell, R. B. H., 1992, Segregation of global and local motion processing in primate middle temporal area, *Nature* **357:**497–499.

Boussaoud, D., Ungerleider, L. G., and Desimone, R., 1990, Pathways for motion analysis: Cortical connections of the medial superior temporal and fundus of the superior temporal visual areas in the macaque, *J. Comp. Neurol.* **296:**462–495.

Boussaoud, D., Desimone, R., and Ungerleider, L. G., 1991, Visual topography of area TEO in the macaque, *J. Comp. Neurol.* **306:**554–575.

Brodmann, K., 1909, *Vergleichende Lokalisationslehre der Grosshirnrinde,* Barth, Leipzig.

Burkhalter, A., and Van Essen, D. C., 1986, Processing of color, form and disparity information in visual areas VP and V2 of ventral extrastriate cortex in the macaque monkey, *J. Neurosci.* **6:**2327–2351.

Burkhalter, A., Felleman, D. J., Newsome, W. T., and Van Essen, D. C., 1986, Anatomical and physiological asymmetries related to visual areas V3 and VP in macaque extrastriate cortex, *Vision Res.* **26:**63–80.

Campbell, A. W., 1905, *Histological Studies on the Localization of Cerebral Function,* Cambridge University Press, Cambridge.

Casagrande, V. A., and Kaas, J. H., 1994, The afferent, intrinsic, and efferent connections of primary visual cortex in primates, in: *Cerebral Cortex,* Volume 10, *Primary Visual Cortex in Primates* (A. Peters and K. Rockland, eds.), Plenum Press, New York, pp. 201–259.

Clarke, S., and Miklossy, J., 1990, Occipital cortex in man: Organization of callosal connections, related myelo- and cytoarchitecture, and putative boundaries of functional visual areas, *J. Comp. Neurol.* **208:**188–214.

Colby, C. L., and Duhamel, J.-R., 1991, Heterogeneity of extrastriate visual areas and multiple parietal areas in macaque monkey, *Neuropsychologia* **29:**517–537.

Colby, C. L., Gattass, R., Olson, C. R., and Gross, C. G., 1988, Topographical organization of cortical afferents to extrastriate area PO in the macaque: A dual tracer study, *J. Comp. Neurol.* **269:**392–413.

Colby, C. L., Duhamel, J.-R., and Goldberg, M. E., 1993, The ventral intraparietal area (VIP) of the macaque: Anatomical location and visual response properties, *J. Neurophysiol.* **69:**902–914.

Condo, G. J., and Casagrande, V. A., 1990, Organization of cytochrome oxidase staining in the visual cortex of nocturnal primates (*Galago crassicaudatus* and *Galago senegalensis*), *J. Comp. Neurol.* **293:**632–645.

Covey, E., Gattass, R., and Gross, C. G., 1982, A new visual area in the parietooccipital sulcus of the macaque, *Soc. Neurosci. Abstr.* **8:**861.

Cowey, A., 1964, Projection of the retina on to striate and prestriate cortex in the squirrel monkey, *Saimiri sciureus, J. Neurophysiol.* **27:**366–393.

Cowey, A., and Heywood, C. A., 1995, There's more to colour than meets the eye, *Behav. Brain Res.* **71:**89–100.

Cragg, B. G., 1969, The topography of the afferent projections in the circumstriate visual cortex of the monkey studied by the Nauta method, *Vision Res.* **9:**733–747.

Curcio, C. A., and Harting, J. K., 1978, Organization of pulvinar afferents to area 18 in the squirrel monkey: Evidence for stripes, *Brain Res.* **143:**155–161.

Cusick, C. G., and Kaas, J. H., 1986, Interhemispheric connections of cortical, sensory and motor maps in primates, in: *Two Hemispheres: One Brain* (F. Lepore, M. Pitto, and H. H. Jasper, eds.), Liss, New York, pp. 83–102.

Cusick, C. G., and Kaas, J. H., 1988, Cortical connection of area 18 and dorsolateral visual cortex in squirrel monkeys, *Visual Neurosci.* **1:**211–237.

Cusick, C. G., Gould III, H. J., and Kaas, J. H., 1984, Interhemispheric connections of visual cortex of owl monkeys (*Aotus trivirgatus*), marmosets (*Callithrix jacchus*) and galagos (*Galago crassicaudatus*), *J. Comp. Neurol.* **230:**311–336.

Desimone, R., and Ungerleider, L. G., 1986, Multiple visual areas in the caudal superior temporal sulcus of the macaque, *J. Comp. Neurol.* **248:**164–189.

DeYoe, E. A., and Van Essen, D. C., 1985, Segregation of efferent connections and receptive field properties in visual area V2 of the macaque, *Nature* **317:**58–61.

DeYoe, E. A., Hockfield, S., Garren, H., and Van Essen, D. C., 1990, Antibody labeling of functional subdivisions in visual cortex. Cat-301 immunoreactivity in striate and extrastriate cortex of the macaque monkey, *Visual Neurosci.* **5:**67–81.

DeYoe, E. A., Carman, G., Bandettini, P., Glickman, S. J., Wieser, J., Cox, R., Miller, D., and Neitz, J., 1995, Mapping striate and extrastriate visual areas in human cerebral cortex, *Proc. Natl. Acad. Sci. USA* **93:**2382–2386.

Distler, C., Boussaoud, D., Desimone, R., and Ungerleider, L. G., 1993, Cortical connections of inferior temporal area TEO in macaque monkeys, *J. Comp. Neurol.* **334:**125–150.

Dykes, R. W., and Ruest, A., 1986, What makes a map in somatosensory cortex? in: *Cerebral Cortex,* Vol. 5, *Sensory-Motor Areas and Aspects of Cortical Connectivity* (E. G. Jones and A. Peters, eds.), Plenum Press, New York,pp. 1–29.

Ebbesson, S. O. E., 1984, Evolution and ontogeny of neural circuits, *Behav. Brain Sci.* **7**:321–366.

Felleman, D. J., and Kaas, J. H., 1984, Receptive field properties of neurons in the middle temporal visual area (MT) of owl monkeys, *J. Neurophysiol.* **52**:488–513.

Felleman, D. J., and Van Essen, D. C., 1991, Distributed hierarchical processing in the primate cerebral cortex, *Cerebral Cortex* **1**:1–47.

Felleman, D. J., Burkhalter, A., and Van Essen, D. C., 1997, Cortical connections of areas V3 and VP of macaque monkey extrastriate visual cortex, *J. Comp. Neurol.* **379**:21–47.

Finger, S., 1994, *Origins of Neuroscience,* Oxford University Press, Oxford.

Gattass, R., and Gross, C. G., 1981, Visual topography of the striate projection zone in the posterior superior temporal sulcus (MT) of the macaque, *J. Neurophysiol.* **46**:521–538.

Gattass, R., Gross, C. G., and Sandell, J. H., 1981, Visual topography of V2 in the macaque, *J. Comp. Neurol.* **201**:519–530.

Gattass, R., Sousa A. P. B., and Gross, C. G., 1988, Visuotopic organization and extent of V3 and V4 of the macaque, *J. Neurosci.* **8**:1831–1845.

Goldman, P. S., and Nauta, W. J. H., 1977, Columnar distribution of cortico-cortical fibers in the frontal association, limbic, and motor cortex of the developing rhesus monkey, *Brain Res.* **122**:393–414.

Graham, J., Wall, J. T., and Kaas, J. H., 1978, Cortical projections of the medial visual area in the owl monkey (*Aotus trivirgatus*), *Neurosci. Lett.* **15**:109–114.

Herrick, C. J., 1891, The problems of comparative neurology, *J. Comp. Neurol.* **1**:93–105.

Hubel, D. H., and Wiesel, T. N., 1965, Receptive fields and functional architecture in two nonstriate visual areas (18 and 19) of the cat, *J. Neurophysiol.* **30**:1561–1573.

Jones, E. G., and Powell, T. P. S., 1970, An anatomical study of converging sensory pathways within the cerebral cortex of the monkey, *Brain* **93**:793–820.

Jouandet, M. I., Tramo, M. J., Herron, D. M., Hermann, A., Loftus, W. C., Barzell, J., and Gazzaniga, M. S., 1989, Brain prints: Computer-generated two-dimensional maps of the human cerebral cortex *in vivo, J. Cognitive Neurosci.* **1**:88–117.

Kaas, J. H., 1982, The segregation of function in the nervous system: Why do the sensory systems have so many subdivisions? *Contrib. Sens. Physiol.* **7**:201–240.

Kaas, J. H., 1988, Changing concepts of visual cortex organization in primates, in: *Neuropsychology of Visual Perception* (J. W. Brown, ed.), Erlbaum, Hillsdale, NJ, pp. 1–32.

Kaas, J. H., 1989a, The evolution of complex sensory systems in mammals, *J. Exp. Biol.* **146**:165–176.

Kaas, J. H., 1989b, Why does the brain have so many visual areas? *J. Cognitive Neurosci.* **1**:121–135.

Kaas, J. H., 1990, Processing areas and modules in sensory-perceptual cortex, in: *Signal and Sense: Local and Global Order in Perceptual Maps* (G. M., Edelman, W. E. Gall, and W. M. Cowan, eds.), Wiley, New York, pp. 67–82.

Kaas, J. H., 1993, The organization of visual cortex in primates: Problems, conclusions, and the use of comparative studies in understanding the human brain, in: *The Functional Organization of the Human Visual Cortex* (B. Gulyás, D. Ottoson, and P. E. Roland, eds.), Pergamon Press, Oxford, pp. 1–11.

Kaas, J. H., 1994, The organization of sensory and motor cortex in owl monkeys, in: *Aotus: The Owl Monkey,* Academic Press, Orlando, FL, pp. 331–351.

Kaas, J. H., 1995a, The evolution of isocortex, *Brain Behav. Evol.* **46**:187–196.

Kaas, J. H., 1995b, Human visual cortex; Progress and puzzles, *Curr. Biol.* **5**:1126–1128.

Kaas, J. H., 1995c, The organization of callosal connections in primates, in: *Epilepsy and the Corpus Callosum II* (A. G. Reeves and D. W. Roberts, eds.), Plenum Press, New York, pp. 15–27.

Kaas, J. H., and Lin, C. S., 1977, Cortical projections of area 18 in owl monkeys, *Vision Res.* **17**:739–741.

Kaas, J. H., and Morel, A., 1993, Connections of visual areas of the upper temporal lobe of owl monkeys: The MT crescent and dorsal and ventral subdivisions of FST, *J. Neurosci.* **13**:534–546.

Kaas, J. H., and Preuss, T. M., 1993, Archontan affinities as reflected in the visual system, in: *Mammal Phylogeny* (F. Szalay, M. Novacek, and M. McKenna, eds.), Springer-Verlag, New York, pp. 115–128.

Kaas, J. H., Lin, C.-S., and Wagor, E., 1977, Cortical projections of posterior parietal cortex in owl monkeys, *J. Comp. Neurol.* **177**:387–408.

Kaas, J. H., Krubitzer, L. A., and Johanson, K. L., 1989, Cortical connections of areas 17 (V-I) and 18 (V-II) of squirrels, *J. Comp. Neurol.* **281**:426–446.

Konorski, M. D., 1967, *Integrative Activity of the Brain,* University of Chicago Press, Chicago.
Krubitzer, L. A., 1995, The organization of neocortex in mammals: Are species differences really so different? *TINS* **18:**408–417.
Krubitzer, L. A., and Kaas, J. H., 1989, Cortical integration of parallel pathways in the visual system of primates, *Brain Res.* **478:**161–165.
Krubitzer, L. A., and Kaas, J. H., 1990, Cortical connections of MT in four species of primates: Areal, modular, and retinotopic patterns, *Visual Neurosci.* **5:**165–204.
Krubitzer, L. A., and Kaas, J. H., 1993, The dorsomdeial visual area of owl monkeys: Connections, myeloarchitecture, and homologies in other primates, *J. Comp. Neurol.* **334:**497–528.
Kuypers, H. G. J. M., Szwarcbart, M. K., Mishkin, M., and Rosvold, H. E., 1965, Occipitotemporal corticocortical connections in the rhesus monkey, *Exp. Neurol.* **11:**245–262.
Lashley, K. S., and Clark, G., 1946, The cytoarchitecture of the cerebral cortex of *Ateles:* A critical examination of architectonic studies, *J. Comp. Neurol.* **85:**223–305.
LeGros Clark, W. E., 1952, A note on cortical cyto-architectonics, *Brain* **75:**96–104.
Levitt, J. B., Yoshioka, T., and Lund, J. S., 1995, Connections between the pulvinar complex and cytochrome oxidase-defined compartments in visual area V2 of macaque monkey, *Exp. Brain Res.* **104:**419–430.
Lin, C.-S., Weller, R. E., and Kaas, J. H., 1982, Cortical connections of striate cortex in the owl monkey, *J. Comp. Neurol.* **211:**165–176.
Livingstone, M. S., and Hubel, D. H., 1982, Thalamic inputs to cytochrome oxidase-rich regions in monkey visual cortex, *Proc. Natl. Acad. Sci. USA* **79:**6098–6101.
Malonek, D., Tootell, R. B. H., and Grinvald, A., 1994, Optical imaging reveals the functional architecture of neurons processing shape and motion in owl monkey area MT, *Proc. R. Soc. Lond. B* **258:**109–119.
Maquire, W. M., and Baizer, J. S., 1984, Visuotopic organization of the prelunate gyrus in rhesus monkey, *J. Neurosci.* **4:**1690–1704.
Martinez-Millán, L., and Holländer, H., 1975, Cortico-cortical projections from striate cortex of the squirrel monkey (*Saimiri sciureus*): A radioautographic study, *Brain Res.* **83:**405–417.
Maunsell, J. H. R., and Van Essen, D. C., 1983, The connections of the middle temporal visual area (MT) and their relationship to a cortical hierarchy in the macaque monkey, *J. Neurosci.* **3:**2563–2586.
Maunsell, J. H. R., and Van Essen, D. C., 1987, Topographic organization of the middle temporal visual area in the macaque monkey: Representational biases and the relationship to callosal connections and myeloarchitectonic boundaries, *J. Comp. Neurol.* **266:**535–555.
Maunsell, J. H. R., Nealey, T. A., and DePriest, D. D., 1990, Magnocellular and parvocellular contributions to responses in the middle temporal visual area (MT) of the macaque monkey, *J. Neurosci.* **10:**3323–3334.
McCulloch, W. S., 1944, Functional organization of cerebral cortex, *Physiol. Rev.* **24:**390–407.
Merigan, W. H., 1993, Human V4? *Curr. Biol.* **3:**226–229.
Morel, A., and Bullier, J., 1990, Anatomical segregation of two cortical visual pathways in the macaque monkey, *Visual Neurosci.* **4:**555–578.
Munk, H., 1890, Of the visual area of the cerebral cortex, and its relation to eye movements, *Brain* **13:**45–70.
Myers, R. E., 1962, Commissural connections between occipital lobes of the monkey, *J. Comp. Neurol.* **118:**1–16.
Myers, R. E., 1965, Organization of visual pathways, in: *Functions of the Corpus Callosum* (E. G. Ettlinger, ed.), Churchill, London, p. 133.
Neuenschwander, S., Gattas, S. R., Sousa, A. P. B., and Pinon, M. C. G. P., 1994, Identification and visuotopic organization of area PO and Pod in *Cebus* monkey, *J. Comp. Neurol.* **340:**65–86.
Newsome, W. T., and Allman, J. M., 1980, Interhemispheric connections of visual cortex in the owl monkey, *Aotus trivirgatus,* and the bushbaby, *Galago senegalensis, J. Comp. Neurol.* **194:**209–233.
Newsome, W. T., Maunsell, J. H. R., and Van Essen, D. C., 1986, Ventral posterior visual area of the macaque: Visual topography and areal boundaries, *J. Comp. Neurol.* **252:**139–153.
Northcutt, R. G., and Kaas, J. H., 1995, The emergence and evolution of mammalian neocortex. *TINS* **18:**373–379.
Otsuka, R., and Hassler, R., 1962, On the construction and organization of the cortical visual region in the cat, *Arch Psychiatr. Neurol.* **203:**213–234.

Pandya, D. N., and Seltzer, B., 1982, Intrinsic connections and architectonics of posterior parietal cortex in the rhesus monkey, *J. Comp. Neurol.* **204:**196–210.

Pessoa, V. F., Abrahao, J. C. H., Pacheco, R. A., Pereira, L. C. M., Magalhaes-Castro, B., and Saraiva, P. E. S., 1992, Relative sizes of cortical visual areas in marmosets: Functional and phylogenetic implications, *Exp. Brain Res.* **88:**459–462.

Petersen, S. E., Miezen, F. M., and Allman, J. M., 1988, Transient and sustained responses in four extrastriate visual areas of the owl monkey, *Exp. Brain Res.* **70:**55–60.

Preuss, T. M., Kaas, J. H., 1996, Cytochrome oxidase "blobs" and other characteristics of primary visual cortex in a lemuriform primate, *Cheriogaleus medius, Brain Behav. Evol.* **47:**103–112.

Preuss, T. M., Beck, P. D., and Kaas, J. H., 1993, Areal, modular, and connectional organization of visual cortex in a prosimian primate, the slow loris (*Nycticebus coucong*), *Brain Behav. Evol.* **2:**237–251.

Previc, F. H., 1980, Functional specialization in the upper and lower visual fields in man: Origins and implications, *Behav. Brain Sci.* **13:**519–575.

Purves, D., Riddle, D. R., and LaMantia, A. S., 1992, Iterated patterns of brain circuitry (or how the cortex gets its spots), *TINS* **15:**362–368.

Ramón y Cajal, S., 1899, *Comparative Study of the Sensory Areas of the Human Cortex,* Clark Union Press, Worcester, MA, pp. 311–359.

Reiner, A., 1993, Neurotransmitter organization and connections of turtle cortex: Implications for the evolution of mammalian isocortex, *Comp. Biochem. Physiol.* **104A:**735–748.

Rockland, K. S., and Pandya, D. N., 1979, Laminar origins and terminations of cortical connections of the occipital lobe in the rhesus monkey, *Brain Res.* **179:**3–20.

Roe, A. W., and Ts'o, D. Y., 1995, Visual topography in primate V2: Multiple representations across functional stripes, *J. Neurosci.* **15:**3684–3715.

Rosa, M. G. P., and Schmid, L. M., 1995, Visual areas in the dorsal and medial extrastriate cortices of the marmoset, *J. Comp. Neurol.* **359:**272–299.

Rosa, M. G. P., Sousa, A. P. B., and Gattass, R., 1988, Representation of the visual field in the second visual area in the *Cebus* monkey, *J. Comp. Neurol.* **275:**326–345.

Rosa, M. G. P., Soares, J. G. M., Florani, Jr., M., and Gattass, R., 1993, Cortical afferents of visual area MT in the *Cebus* monkey: Possible homologies between New and Old World monkeys, *Visual Neurosci.* **10:**827–855.

Rosa, M. G. P., Casagrande, V. A., Preuss, T., and Kaas, J. H., 1997, Visual field representation in striate and prestriate cortices of a prosimian primate (Galago garnetti), *J. Neurophysiol.* **77:**3193–3217.

Rubin, N., Nakayama, K., and Shapley, R., 1996, Enhanced perception of illusory contours in the lower versus upper visual hemifields, *Science* **271:**651–653.

Sanides, F., and Hoffman, J., 1969, Cyto- and myloarchitecture of the visual cortex of the cat and of the surrounding integration cortices, *J. Hirnforsch.* **11:**79–104.

Sereno, M. I., and Allman, J. M., 1990, Cortical visual areas in mammals, in: *The Neural Basis of Visual Function,* Vol. 4 (A. G. Leventhal, ed.), Macmillan, London, pp. 160–172.

Sereno, M. I., Pale, A. M., Reppas, J. B., Kwong, K. K., Belliveau, J. W., Brady, T. J., Rosen, B. R., and Totell, R. B. H., 1995, Borders of multiple visual areas in humans revealed by functional magnetic resonance imaging, *Science* **268:**889–893.

Shipp, S., and Zeki, S., 1985, Segregation of pathways leading from area V2 to areas V4 and V5 of macaque monkey visual cortex, *Nature* **315:**322–325.

Shipp, S., Watson, J. D. G., Frackowiak, R. S. J., and Zeki, S., 1995, Retinotopic maps in human prestriate visual cortex: The demarcation of areas V2 and V3, *Neuroimage* **2:**125–132.

Sholl, P. A., 1956, *The Organization of the Cerebral Cortex,* Methuen, London.

Smith, G. E., 1906, A new topographic survey of human cerebral cortex, being an account of the distribution of the anatomically distinct cortical areas and their relationship to the cerebral sulci, *J. Anat. Physiol.* **42:**237–254.

Sousa, A. P. B., Pinon, M. C. G. P., Gattass, R., and Rosa, M. G. P., 1991, Topographic organization of cortical input to striate cortex in the cebus monkey: A fluorescent tracer study, *J. Comp. Neurol.* **308:**665–682.

Spatz, W. B., 1977, Topographically organized reciprocal connections between areas 17 and MT (visual area of superior temporal sulcus) in the marmoset *Callithrix jacchus, Exp. Brain Res.* **27:**559–572.

Spatz, W. B., and Tigges, J., 1972, Experimental-anatomical studies on the "middle temporal visual area (MT)" in primates. I. Efferent cortico-cortical connections in the marmoset, *Callithrix jacchus, J. Comp. Neurol.* **146:**451–464.

Steele, G. E., Weller, R. E., and Cusick, C. G., 1991, Cortical connections of the caudal subdivision of the dorsolateral area (V4) in monkeys, *J. Comp. Neurol.* **306:**495–520.

Stepniewska, I., and Kaas, J. H., 1996, Topographic patterns of V2 cortical connections in macaque monkeys, *J. Comp. Neurol.* **371:**129–152.

Sur, M., Wall, J. T., and Kaas, J. H., 1981, Modular segregation of functional cell classes within the postcentral somatosensory cortex of monkeys, *Science* **212:**1059–1061.

Symonds, L. L., and Kaas, J. H., 1978, Connections of striate cortex in the prosimian, *Galago senegalensis, J. Comp. Neurol.* **181:**477–512.

Talbot, S. A., and Marshall, W. H., 1941, Physiological studies on neural mechanisms of visual localization and discrimination, *Am. J. Ophthalmol.* **24:**1255–1264.

Tanaka, J., Lindsey, E., Lausmann, S., and Creutzfeldt, O. D., 1990, Afferent connections of the prelunate visual association cortex (areas V4 and DP), *Anat. Embryol.* **181:**19–30.

Tigges, J., Spatz, W. B., and Tigges, M., 1974, Efferent cortico-cortical fiber connections of area 18 in the squirrel monkey (*Saimiri*), *J. Comp. Neurol.* **158:**219–236.

Tigges, J., Tigges, M., Anschel, S., Cross, N. A., Letbetter, W. D., and McBride, R. L., 1981, Areal and laminar distribution of neurons interconnecting the central visual cortical areas 17, 18, 19 and MT in squirrel monkey (*Saimiri*), *J. Comp. Neurol.* **202:**539–560.

Tootell, R. B. H., and Taylor, J. B., 1995, Anatomical evidence for MT and additional cortical visual areas in humans, *Cerebral Cortex* **1:**39–55.

Tootell, R. B. H., Silverman, M. S., DeValois, R. L., and Jacobs, G. H., 1983, Functional organization of the second cortical area of primates, *Science* **220:**737–739.

Tootell, R. B. H., Hamilton, S. L., and Silverman, M. S., 1985, Topography of cytochrome oxidase activity in owl monkey cortex, *J. Neurosci.* **5:**2786–2800.

Tootell, R. B. H., Reppas, J. B., Kwong, K. K., Malach, R., Born, R. T., Brady, T. J., Rosen, B. R., and Belliveau, J. W., 1995, Functional analysis of human MT and related visual cortical areas using magnetic resonance imaging, *J. Neurosci.* **15:**3215–3230.

Ungerleider, L. G., and Desimone, R., 1986a, Projections to the superior temporal sulcus from the central and peripheral field representations of V1 and V2, *J. Comp. Neurol.* **248:**147–163.

Ungerleider, L. G., and Desimone, R., 1986b, Cortical projections of visual area MT in the macaque, *J. Comp. Neurol.* **248:**190–222.

Ungerleider, L. G., and Haxby, J. V., 1994, "What" and "where" in the human brain, *Curr. Opin. Neurobiol.* **4:**157–165.

Ungerleider, L. G., and Mishkin, M., 1979, The striate projection zone in the superior temporal of *Macaca mulatta:* Location and topographic organization, *J. Comp. Neurol.* **188:**347–366.

Underleider, L. G., and Mishkin, M., 1982, Two cortical visual systems, in: *Analysis of Visual Behavior* (D. J. Ingle, M. A. Goodale, and R. L. W. Mansfield, eds.), MIT Press, Cambridge, MA, pp. 544–586.

Van Essen, D. C., 1985, Functional organization of primate visual cortex, in: *Cerebral Cortex,* Vol. 3, *Visual Cortex* (A. Peters and E. G. Jones, eds.) Plenum Press, New York, pp. 259–329.

Van Essen, D. C., and Maunsell, J. H. R., 1983, Hierarchical organization and the functional streams in the visual cortex, *TINS* **6:**370–375.

Van Essen, D. C., and Zeki, S. M., 1978, The topographical organization of rhesus monkey prestriate cortex, *J. Physiol.* **277:**193–226.

Van Essen, D. C., Maunsell, J. H. R., and Bixby, J. L., 1981, The middle temporal visual area in the macaque: Myeloarchitecture, connections, functional properties and topographic organization, *J. Comp. Neurol.* **199:**293–326.

Van Essen, D. C., Newsome, W. T., Maunsell, J. H. R., and Bixby, J. L., 1986, The projections from striate cortex (V1) to areas V2 and V3 in the macaque monkey: Asymmetries, areal boundaries and patchy connections, *J. Comp. Neurol.* **244:**451–480.

Von Economo, C., 1929, *The Cytoarchitectonics of the Human Cortex,* Oxford University Press, Oxford.

Wagor, E., Lin, C.-S., and Kaas, J. H., 1975, Some cortical projections of the dorsomedial visual area (DM) of association cortex in the owl monkey (*Aotus trivirgatus*), *J. Comp. Neurol.* **163:**227–250.

Walker, A. E., 1938, *The Primate Thalamus,* University of Chicago Press, Chicago.

Weller, R. E., 1988, Two cortical visual systems in Old and New World primates, *Prog. Brain Res.* **75**:293–306.

Weller, R. E., and Kaas, J. H., 1983, Retinotopic patterns of connections of area 17 with visual areas V-II and MT in macaque monkeys, *J. Comp. Neurol.* **220**:253–279.

Weller, R. E., and Kaas, J. H., 1985, Cortical projections of the dorsolateral visual area in owl monkeys: The prestriate relay to inferior temporal cortex, *J. Comp. Neurol.* **234**:35–59.

Weller, R. E., and Kaas, J. H., 1987, Subdivisions and connections of inferior temporal cortex in owl monkeys, *J. Comp. Neurol.* **256**:137–172.

Weller, R. E., and Steele, G. E., 1992, Cortical connections of subdivisions of inferior temporal cortex in squirrel monkeys, *J. Comp. Neurol.* **324**:37–66.

Weller, R. E., Wall, J. T., and Kaas, J. H., 1984, Cortical connections of the middle temporal visual area (MT) and the superior temporal cortex in owl monkeys, *J. Comp. Neurol.* **228**:81–104.

Weller, R. E., Steele, G. E., and Cusick, C. G., 1991, Cortical connections of dorsal cortex rostral to VII in squirrel monkeys, *J. Comp. Neurol.* **306**:521–537.

Wong-Riley, M. I. T., 1979, Columnar cortico-cortical interconnections within the visual system of the squirrel and macaque monkeys, *Brain Res.* **162**:201–207.

Wong-Riley, M. I. T., Hevner, R. F., Cuttan, R., Earnest, M., Egan, R., Frost, J., and Ngugen, T., 1993, Cytochrome oxidase in the human visual cortex: Distribution in the developing and adult brain, *Visual Neurosci.* **10**:41–58.

Woolsey, C. N., and Fairman, D., 1946, Contralateral, ipsilateral, and bilateral representations of cutaneous receptors in somatic areas I and II of the cerebral cortex of pig, sheep, and other mammals, *Surgery* **19**:684–702.

Young, M. P., Scannell, J. W., and Burns, G., 1995, *The Analysis of Cortical Connectivity,* Landes, Austin, TX.

Zeki, S. M., 1969, Representation of central visual fields in prestriate cortex of monkey, *Brian Res.* **14**:271–291.

Zeki, S. M., 1971a, Convergent input from the striate cortex (area 17) to the cortex of the superior temporal sulcus in the rhesus monkey, *Brain Res.* **29**:338–340.

Zeki, S. M., 1971b, Cortical projections from two prestriate areas in the monkey, *Brain Res.* **34**:19–35.

Zeki, S. M., 1971c, Response properties and receptive fields of cells in an anatomically defined region of the superior temporal sulcus of the monkey, *Brain Res.* **35**:528–532.

Zeki, S. M., 1973, Colour coding in rhesus monkey prestriate cortex, *Brain Res.* **53**:422–427.

Zeki, S. M., 1977, Simultaneous anatomical demonstration of the representations of the vertical and horizontal meridians in areas V2 and V3 of rhesus monkey visual cortex, *Proc. R. Soc. Lond. B* **195**:517–523.

Zeki, S. M., 1978, The third visual complex of rhesus monkey prestriate cortex, *J. Physiol.* **277**:245–272.

Zeki, S., 1980, The response properties of cells in the middle temporal area (area MT) of owl monkey visual cortex, *Proc. R. Soc. Lond. B* **207**:239–248.

Zeki, S., Watson, J. P. G., Lueck, C. D., Friston, K., Kennard, C., and Frackowjak, R. S. J., 1991, A direct demonstration of functional specialization in human visual cortex, *J. Neurosci.* **11**:641–649.

4

Visuotopic Organization of Primate Extrastriate Cortex

MARCELLO G. P. ROSA

1. Introduction

One of the fundamental aspects of the organization of extrastriate visual cortex in mammals is the presence of multiple representations of the visual field. Although the existence of these visuotopic maps has been known for more than 50 years (Talbot, 1942), there are still many open issues regarding their organization. Descriptions of the number, boundaries, and visuotopic organization of cortical visual areas vary not only among primates, but also between studies of single species by different groups (e.g., Fig. 1; see Table I for abbreviations). It is not clear whether these differences reflect real individual or interspecies variability, or whether they merely reflect the need for more study or better criteria for the definition of visual areas. The aim of this chapter is to review the current evidence related to the precision, extent, and topological characteristics of visuotopic maps in extrastriate areas of primates. A critical evaluation of the published evidence on these subjects reveals that some of the present points of contention are the result not only of the complexity of the problem, but also of the scarcity of the data available for interpretation. Response properties, architecture, connections, and modular patterns can also be used to lend or deny support to specific hypotheses generated on the basis of visuotopy, but these criteria will not be reviewed in detail here. In addition, studies of nonprimate

MARCELLO G. P. ROSA • Vision, Touch and Hearing Research Centre, Department of Physiology and Pharmacology, The University of Queensland, Queensland 4072, Australia.

Cerebral Cortex, Volume 12, edited by Rockland *et al.* Plenum Press, New York, 1997.

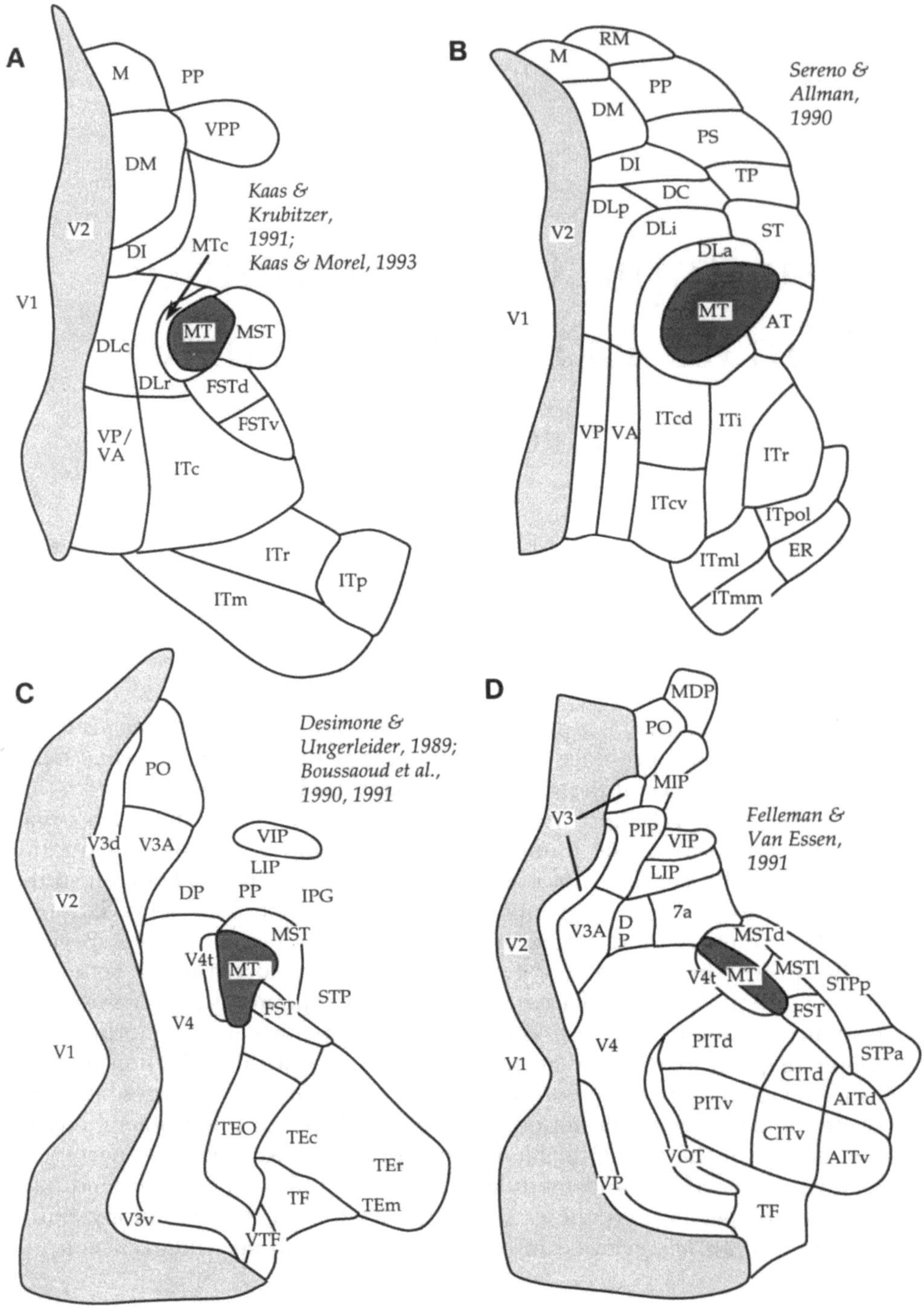

Figure 1. Different models of the organization of extrastriate cortex in the owl monkey (A, B) and macaque monkey (C, D), according to recent studies. Each diagram was drawn as a flat reconstruction of a right hemisphere (dorsal upward, caudal to the left). The location of area V2 is shown in light gray, and that of MT in dark gray. For abbreviations, see Table I.

Table I. Abbreviations for Cortical Areas

AITd	Anterior inferior temporal, dorsal subdivision
AITv	Anterior inferior temporal, ventral subdivision
AL	Anterolateral
AT	Anterior temporal
CITd	Central inferior temporal, dorsal subdivision
CITv	Central inferior temporal, ventral subdivision
D	Dorsal
DA	Dorsoanterior
DC	Dorsocaudal
DI	Dorsointermediate
DL	Dorsolateral
DLa	Dorsolateral anterior
DLc	Dorsolateral, caudal subdivsion
DLi	Dorsolateral intermediate
DLp	Dorsolateral posterior
DLr	Dorsolateral, rostral subdivision
DM	Dorsomedial
DP	Dorsal prelunate
DSR	Direction/size/rotation
DZ	Dorsal zone
ER	Entorhinal
FST	Fundus of superior temporal
FSTd	Fundus of superior temporal, dorsal subdivision
FSTv	Fundus of superior temporal, ventral subdivision
IPG	Inferior parietal, gyral subdivision
IT	Inferior temporal
ITc	Inferior temporal, caudal subdivision
ITcd	Caudal inferior temporal, dorsal division
ITcv	Caudal inferior temporal, ventral division
ITi	Inferior temporal, intermediate subdivision
ITm	Inferior temporal, medial subdivision
ITml	Medial inferior temporal, lateral division
ITmm	Medial inferior temporal, medial division
ITp/ITpol	Inferior temporal polar division
ITr	Inferior temporal, rostral subdivision
LIP	Lateral intraparietal
M	Medial
MDP	Medial dorsal parietal
MIP	Medial intraparietal
MST	Medial superior temporal
MSTd	Medial superior temporal, dorsal subdivision
MSTl	Medial superior temporal, lateral subdivision
MT	Middle temporal
MTc	Middle temporal crescent
MTp	Middle temporal peripheral
PG	Cytoarchitectonic area PG[a]
PIP	Posterior intraparietal
PITd	Posterior inferotemporal, dorsal subdivision
PITv	Posteror inferotemporal, ventral subdivision
PM	Posteromedial
PO	Parietooccipital
POd	Parietooccipital dorsal
POm	Parietooccipital medial
PP	Posterior parietal
PS	Posterior sylvian
RM	Rostromedial
ST	Superior temporal
STP	Superior temporal polysensory
STPa	Superior temporal polysensory, anterior subdivision
STPp	Superior temporal polysensory, posterior subdivision
TE	Cyotarchitectonic area TE[a]
TEc	Caudal subdivision of cytoarchitectonic area TE[a]
TEm	Medial subdivision of cytoarchitectonic area TE[a]
TEr	Rostral subdivision of cytoarchitectonic area TE[a]
TEO	Cytoarchitectonic area TEO[a]
TP	Temporoparietal
V1	Primary visual area (visual area 1)
V2	Second visual area (visual area 2)
V3	Visual area 3
V3A	Visual area 3A
V3d	Visual area 3, dorsal subdivision
V3v	Visual area 3, ventral subdivision
V4	Visual area 4
V4t	Visual area 4, transitional
V4v	Visual area 4, ventral subdivision
V5	Visual area 5
V6	Visual area 6
VA	Ventral anterior
VIP	Ventral intraparietal
VOT	Ventral occipitotemporal
VP	Ventral posterior
VPP	Ventral posterior parietal
VTF	Visual subdivision of cytoarchitectonic area TF[a]

[a] Von Bonin and Bailey (1947).

species will be considered only insofar as they clarify the organization in primates.

The first part of this chapter will define some key concepts that are important for the understanding of visuotopic representation in primate cortex. The current evidence regarding the visuotopic maps in specific regions of cortex will then be reviewed, and points of controversy will be identified. Finally, some functional aspects of the visuotopic maps, including the issues of spatial resolution, neuronal diversity, and plasticity of visuotopic representations, will be discussed.

2. Study of Cortical Visuotopy

The study of visuotopic organization is concerned with issues such as the location of the neurons which respond to the stimulation of a given point of the visual field, how neurons that represent different points of the visual field in one functional area are located relative to each other, how representations in different areas are interrelated, and the precision of the representations in different areas. Thus, visuotopy is basically a matter of spatial transformations: the way in which neural pathways distort, filter, cut, transpose, or multiply the retinal maps in order to form different cortical representations.

If one considers the hypothetical functions of extrastriate areas from the purely computational point of view, there are alternative ways to organize neurons in cortex which may not result in visuotopic maps (Sutherland, 1973; Barlow, 1986). Yet, empirical observations in diverse mammals have demonstrated that neurons in caudal extrastriate cortex respond to visual stimulation within defined receptive fields and that, in many areas, the position of neuronal receptive fields in visual space varies gradually as a function of neuronal position in the cortex (e.g., Cowey, 1964; Hubel and Wiesel, 1965, 1969; Dubner and Brown, 1968; Wright, 1969; Kaas *et al.*, 1970, 1972; Allman and Kaas, 1971, 1974a; Dubner and Zeki, 1971; Hall *et al.*, 1971; Montero *et al.*, 1973; Zeki, 1974). In these areas, by sampling a number of points in the cortex, it is possible to interpolate a "map" of visual space that fits the individual observations, and that predicts the position of receptive fields of other, nonstudied neurons.

The study of visuotopic maps may give clues to the function of a given area, to the computational strategies employed by the brain, or to the "instructions" that constrain brain development. One of the main reasons for studying the visuotopy of extrastriate cortex is the possibility of identifying visual areas and locating their boundaries. In the situation where each area contains a well-defined topographic map of the visual field, the delimitation of such a map allows one to determine the boundaries of a visual area even if other connectional, architectonical, and functional data are not available. For example, if there is evidence that portions of the visual field are rerepresented in a topographically systematic fashion within an undifferentiated architectonic region, it can be argued that more than one functional area is present within this region. Classical applications of this criterion include the demonstrations of multiple visual areas in primate cytoarchitectural areas 18 and 19 by Allman and Kaas (1971, 1974a,b, 1975, 1976) and by Zeki (1969, 1978a; Van Essen and Zeki, 1978). However, in practice, the delimitation of visual areas on a purely topographic criterion is not straightforward. As we will see below (Sections 2.2 and 2.3), among other possible complexities, local rerepresentations of portions of the visual field can occur within single areas as a consequence of segregation of function between columns (Hubel and Wiesel, 1977; Van Essen and Zeki, 1978; Blasdel and Fitzpatrick, 1984; Rosa *et al.,* 1988b; Tootell *et al.,* 1988; Roe and Ts'o, 1995, and this volume). Therefore, it is important to consider the evidence for rerepresentation in the light of the local precision of the map. In addition, there can be many ways to combine partial representations to form meaningful "maps" of the visual field, and the uncertainty of the proposed boundaries increases with the sparseness of the sampling. An unambiguous solution is more likely to be reached if the sample is extensive enough to cover the entire area in question, as well as adjacent areas. Finally, the number of meaningful interpretations of a given set of data can be reduced by the parallel analysis of mapping parameters other than receptive field position, such as receptive field size and response properties.

In many ways, these problems parallel those that were faced by somatosensory physiologists in defining body maps that were meaningful in a functional context (Dykes and Ruest, 1986; Krubitzer *et al.,* 1995). As pointed out by these authors, two different maps can both be "right" in the sense that each represents the data accurately, and the differences may simply reflect the fact that each of them was drawn to represent a different set of mapping variables (see also Phillips *et al.,* 1994, and Schreiner, 1995, on the primary auditory cortex). Discrepancies require more studies, and only when the topographic data are correlated with the cellular anatomy and connections is it likely that one of the several "solutions" will prove to be the most parsimonious in a functional context (e.g., Burton *et al.,* 1995, and Krubitzer *et al.,* 1995, in somatosensory cortex).

2.1. Methods Used in the Definition of Visuotopic Maps

Although ideally the definition of an area should take into account the combined evidence obtained through different approaches, the method that offers the most complete and precise idea of visuotopic organization is that of electrophysiological mapping of neural responses to visual stimulation. In a typical experiment (e.g., Fig. 2), an electrode is moved to different points of the brain, and the receptive fields corresponding to each point are determined by moving simple visual patterns such as rectangles, spots, or gratings in front of the animal while monitoring the activity of single units, unit clusters, or field potentials.* This method saw its first application to primate extrastriate areas

*In electrophysiological mapping studies, in order to describe receptive field position, the visual field is imagined as represented on a "panoramic" screen, and spherical coordinates are assigned relative to axes of reference, which are usually the vertical meridian (a vertical line through the center of the fovea, dividing the visual field into right and left hemifields, which roughly corresponds to the center of the line of decussation in the retina) and the horizontal meridian (a perpendicular to the ventrical meridian through the center of the fovea, which further subdivides the hemifields into upper and lower quadrants). Although different systems of coordinates can be used to define points on the surface of a sphere, polar coordinates centered on the fovea are the most commonly used in primate research.

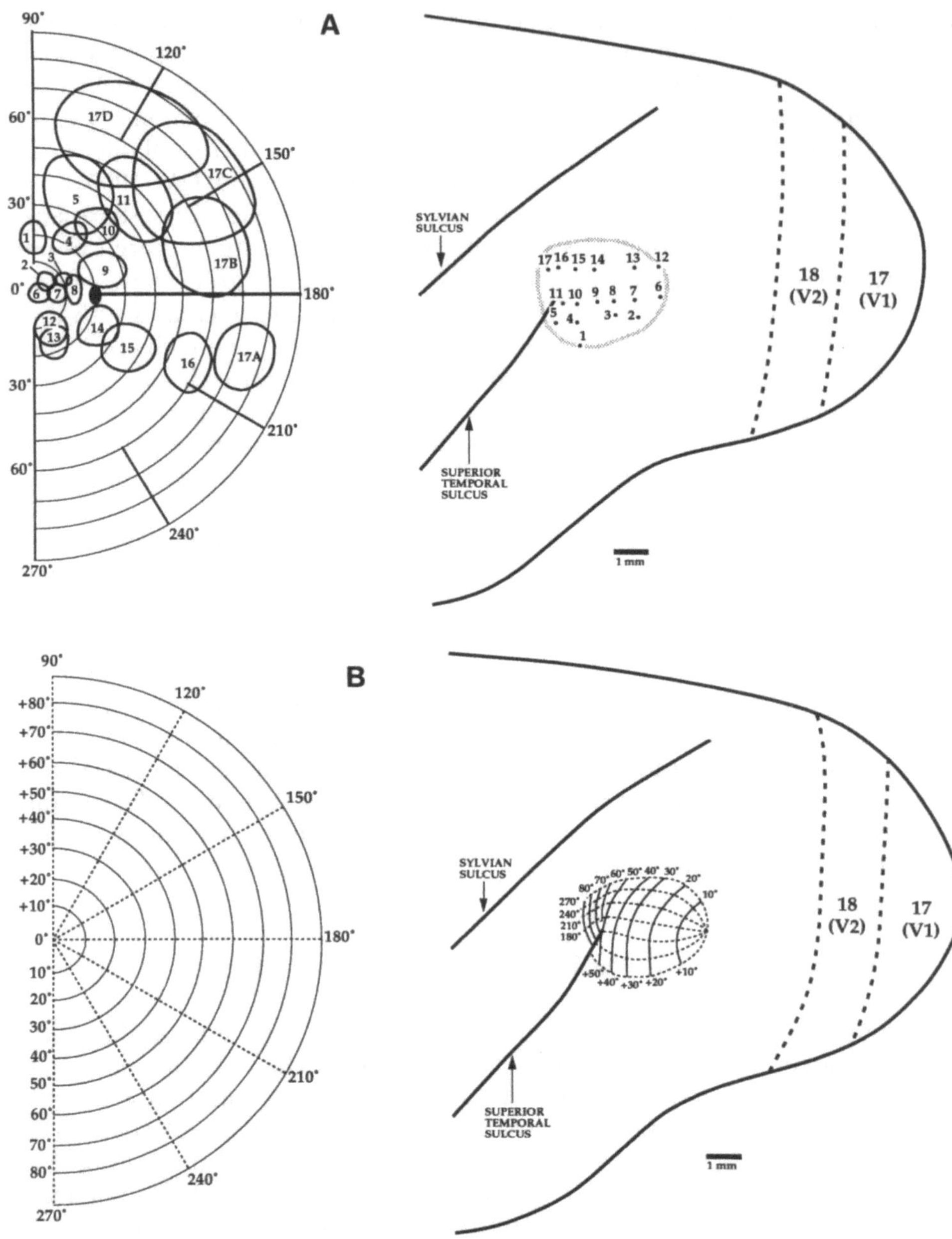

Figure 2. First demonstration of visuotopic organization in area MT of the owl monkey, redrawn from Figs. 9 and 11 of Allman and Kaas (1971). (A) Data from one experiment, in which neurons at different recording sites in area MT yielded receptive fields restricted to different portions of the visual field. The perimeter chart shown at the upper left represents the contralateral hemifield of the animal, and shows locations of receptive fields (ovals) relative to the horizontal meridian (180° polar angle line) and the vertical meridian (90°–270° line). Receptive field eccentricity (distance in degrees from the fovea) is also indicated by semicircles (e.g., 0°, 30°, 60°). The diagram at the upper right is a dorsolateral view of the caudal half of the left hemisphere of the brain of an owl monkey (rostral to the left, dorsal up), showing the location of recording sites in MT (dots), and the architectonic boundaries of MT (gray line), V1, and V2 (dashed lines). In penetration 17, four receptive fields were recorded down the medial bank of the superior temporal sulcus, at successively deeper sites (17A, B, C, D). (B) Summary diagram of the representation of the visual field in area MT of the owl monkey, based on data from several experiments. The coordinates of the contralateral hemifield were superimposed onto the cortex, to generate an idealized view of the visuotopic organization in MT.

more than 30 years ago (Cowey, 1964; Hubel and Wiesel, 1970; Allman and Kaas, 1971; Dubner and Zeki, 1971), but, with modern improvements, it remains very useful. Notably, it is now possible to sample much larger numbers of neurons in a single animal, therefore reducing the error involved in pooling topographical data obtained in different cases into a single "representative" map. This is due both to advances in surgical technique (which allow the maintenance of primates in optimal condition for extended periods of anesthesia and neuromuscular block) and to the introduction of awake, behaving preparations (Dow *et al.*, 1981; Maguire and Baizer, 1984; Komatsu and Wurtz, 1988; Erickson *et al.*, 1989).

In principle, it is possible to combine high-resolution topographic mapping of visual responses with single-unit analysis of response properties, thereby yielding a more reliable definition of boundaries between areas. In practice, however, one is still confronted with the opposing alternatives of mapping a few simple parameters (such as receptive field size and position) across a large extent of cortex, or conducting quantitative analyses of response properties of cells at fewer sites. Long-term chronic preparations, while allowing the study of large populations of single units in detail, suffer from limited precision regarding the location of the recording sites, especially within sulci. As high-resolution *in vivo* neuroimaging techniques become more accessible to research laboratories, the precise three-dimensional reconstruction of recording sites during experiments may become possible. This development will allow the combination of topographic and functional analyses of large populations of neurons in single animals.

In the last decade, techniques have also been developed for automatic interpretation of electrophysiological data, providing "bias-free" maps of cortical visuotopy. One such technique involves the use of computers to interpolate visuotopic coordinates between points sampled by electrode penetrations within a single area, based on digitized grids of recording sites and receptive field coordinates (Maunsell and Van Essen, 1987; Fiorani *et al.*, 1989). This technique allows a quantification of certain visuotopic parameters (such as biases toward the representation of specific sectors of the visual field) that do not become evident through the analysis of individual receptive fields. However, it does not address the possibility of ambiguous boundaries between areas, and its optimal use requires that the boundaries of the area under study are determined prior to the application of the interpolation algorithm (Neuenschwander *et al.*, 1994). Otherwise, contamination of the data matrix with samples taken from adjacent areas may result in an artifactual impression of complex topography, involving duplications or discontinuities. In contrast, the recently described method of Sereno *et al.* (1994) does not require predetermination of areal boundaries. This technique, which is based on the interpolation of coordinates across the entire data set, uses local visual field sign as a criterion for automatically parsing the cortex into areas. Visual field sign is a local measure of the chirality: whether the visuotopic map, as seen from the cortical surface, forms a mirror or non-mirror image of the animal's visual field. The assumption behind this technique is that all of the cortex belonging to a single area represents the visual field according to the same field sign; therefore, transitions in field sign should correspond to borders between areas. The main merit of this approach is that it generates an impartial hypothesis regarding both the areal boundaries and the visuotopy, based on the same data set. However, the "human factor" has not been totally removed, as the exact final configuration of the maps is dependent on parameters (used by the interpolation algorithm) that are subjectively chosen by the experimenter. Moreover, fine-scale transitions of visual field sign have been reported within a single area, V2 (Roe and Ts'o, 1995), as a consequence of segregation of function by stripelike compartments. Another potential problem, which is common to all of the current "bias-free" methods, is that the algorithms assume local continuity of the maps. As explained below (Section 2.4), recent studies suggest that map discontinuities may be a real feature of the visuotopic map of some areas. Thus, by interpolating visual field coordinates across a map discontinuity, there is the risk of generating artifactual "representations" of portions of the visual field. Because the "bias-free" methods are recent, it is still unclear to what extent the caveats mentioned above will prove to be practical problems. Even though the interpolation techniques are useful aids for data analysis, the hypotheses generated on the basis of these methods should currently be interpreted with the same caution as any electrophysiological mapping data in isolation.

The visuotopic organization within one area can also be deduced by tracing the connections from (or to) another area for which the visuotopy has already been determined (e.g., Cragg and Ainsworth, 1969; Zeki, 1969; Spatz *et al.*, 1970; Ungerleider and Mishkin, 1979; Maunsell and Van Essen, 1983; Weller and Kaas, 1983; Weller *et al.*, 1984; Newsome *et al.*, 1986; Perkel *et al.*, 1986; Van Essen *et al.*, 1986; Colby *et al.*, 1988; Cusick and Kaas, 1988; Rosa *et al.*, 1988b, 1993b; Krubitzer and Kaas, 1990; Sousa *et al.*, 1991; Steele *et al.*, 1991). This method relies on the observation that corticocortical connections *preferentially* link cells that represent the same region of the visual field (Cowey, 1964). One of the most important tests of a hypothesis regarding the borders of a cortical area is the demonstration of a coherent pattern of connections with other parts of the brain: different portions of the same area should receive similar inputs and outputs. However, for the purposes of a first definition of visuotopic organization, the picture provided by studying connections alone is at best coarse. This imprecision stems in part from the fact that corticocortical connections may also integrate representations of portions of the visual field that are not exactly matching (for a review, see Salin and Bullier, 1995). In addition, because adjacent areas often share inputs (or outputs), the possibilities of erroneous interpretation are many. Although the amount of information obtained in each animal can be increased by the use of multiple tracers, topographic patterns of connections are still best judged in the context of electrophysiologically defined areas.

The study of interhemispheric connections has also contributed to the determination of the visuotopy of primate cortex. As we will see below (Section 3.5), the pattern of interhemispheric connections of visuotopically organized areas is highly reproducible between animals (Myers, 1962; Zeki, 1970, 1977; Newsome and Allman, 1980), and the patches of origin or termination of callosal axons tend to concentrate around the representations of the vertical meridian of the visual field and the fovea (Van Essen and Zeki, 1978; Van Essen *et al.*, 1982), at least when they are assessed with anterograde degeneration methods. Even though the introduction of more sensitive tracers has shown that representations of portions of the visual field located away from the midline may also connect with the contralateral hemisphere (Maunsell and Van Essen, 1983; Cusick *et al.*, 1984), mapping of interhemispheric connections still provides a coarse global view of some aspects of extrastriate visuotopy.

Finally, metabolic mapping methods such as 2-deoxyglucose, *in vivo* imaging of intrinsic signals, and functional nuclear magnetic resonance imaging can be used to study visuotopic representations. These methods have seen little application as a tool for analysis of visuotopy beyond the primary visual area, V1 (Tootell *et al.*, 1982, 1988; DeYoe *et al.*, 1994; Frostig, 1994; Grinvald *et al.*, 1994), but they hold the promise of allowing the simultaneous determination of visuotopy, functional response properties, and connections in extrastriate cortex without the need of prohibitively long experiments (for recent applications, see Tootell and Hamilton, 1989; Malach *et al.*, 1994; Malonek *et al.*, 1994; Sereno *et al.*, 1995).

2.2. Visuotopically Organized versus Nonvisuotopically Organized Areas

As we will see in Section 3, visuotopic representations form a large portion of the visual cortex, and, indeed, of the primate brain. According to one of the most widely used models of organization of visual cortex, visual areas are thought to form two main parallel pathways. These pathways, which flow from the primary visual area (V1) into the posterior parietal and temporal cortices, have been referred to as the "dorsal" and "ventral" streams, respectively (Ungerleider and Mishkin, 1982). Within each pathway, the precision of visuotopic maps is progressively eroded, without a sharp limit between visuotopically organized and nonorganized cortex. In general, in all primates the more rostral an area is in the cortex, the less precise is its visual topography. It is unclear to what extent the processing of visual information along the dorsal and ventral streams is hierarchical (in the sense that "higher" areas in the hierarchy depend on inputs from the "lower" areas); indeed, recent studies strongly argue against such a framework (for a review, see Bullier and Nowak, 1995). However, the hierarchical levels of processing defined on the basis of laminar patterns of origin and termination of connections (Rockland and Pandya, 1979; Tigges *et al.*, 1981; Maunsell and Van Essen, 1983; Felleman and Van Essen, 1991) are useful for the present analysis, since they correlate reasonably well with some of the properties of visuotopic maps, such as their precision and the receptive field sizes.

Although it is hard to draw a precise line between topographically organized and nontopographically organized visual cortices, the visuotopy can be described as crude at best in the caudal subdivision of the inferior temporal cortex (Fig. 1; area ITc in New World monkeys and prosimian primates, also called area TEO in Old World monkeys based on the cytoarchitectonic map of von Bonin and Bailey, 1947) and in some of the visual areas surrounding the middle temporal area (MT). Visuotopic analyses of these ares in simian primates show some segregation between the representations of the upper and lower visual quadrants, or between central and peripheral space, but little further evidence of topographic order (Maunsell and Van Essen, 1983; Weller *et al.*, 1984; Desimone and Ungerleider, 1986; Komatsu and Wurtz, 1988; Fiorani *et al.*, 1989; Boussaoud *et al.*, 1990, 1991; Krubitzer and Kaas, 1990; Steele *et al.*, 1991; Rosa *et al.*, 1993b). In prosimian primates, a degree of topography also exists in area ITc and in the cortex surrounding MT rostrally and ventrally (*Cheirogaleus*, Paolini *et al.*, 1994; *Otolemur*, M. Rosa, V. Casagrande, T. Preuss, and J. Kaas, unpublished observations). As far as the current data indicate, these areas may be seen as the rostral limits of the topographically organized visual cortex in the ventral and dorsal streams, respectively. So far there is so no clear evidence of visuotopy in the intermediate and rostral portions of the inferior temporal cortex, in the superior temporal polysensory area (STP), or in the posterior parietal cytoarchitectonic area PG, areas which lie at the next levels of processing defined by connectional patterns (Gross *et al.*, 1969; Robinson *et al.*, 1978; Desimone and Gross, 1979; Motter and Mountcastle, 1981; Bruce *et al.*, 1982, 1986; Motter *et al.*, 1987; Boussaoud *et al.*, 1991). Interestingly, as the visuotopic order vanishes, the excitatory receptive fields in these areas become essentially bilateral: many fields are centered on the vertical meridian, and include a representation of the ipsilateral hemifield which may be as extensive as the contralateral representation. These properties contrast sharply with those observed at earlier stages of processing, where the excitatory receptive fields are largely restricted to the contralateral hemifield, and this may provide a criterion for delimiting topographically organized visual cortex. However, a note of caution is necessary: it should be remembered that some of the early reports on the organization of visual areas such as MT also failed to observe visuotopy, due to the samples being restricted to small portions of the area under study. Thus, future investigations of high-order visual areas in the parietal and temporal lobes may still observe remnants of a visuotopic order, even if it is "rather crude and essentially quadrantic" (e.g., Dubner and Zeki, 1971).

2.3. Receptive Field Size, Scatter, and Point-Image Size

Receptive fields of neurons in extrastriate areas are often quite extensive, if one considers the "silent" peripheries that can modulate the responses to stimuli presented at the center of the receptive field (Zeki, 1983a,b; Allman *et al.*, 1985; Tanaka *et al.*, 1986; Desimone and Schein, 1987; Gaska *et al.*, 1987; Schein and Desimone, 1990; Born and Tootell, 1992; Olavarria *et al.*, 1992; Desimone *et al.*, 1993). However, there is usually a restricted central excitatory region ("classical" receptive field; Allman *et al.*, 1985), within which stimuli reliably produce excitatory responses. For the sake of simplicity, it is common to represent visuotopic maps as point-to-point representations of the retina in which imaginary projection lines crossing the field of vision map into lines crossing the surface of a cortical area. In reality, factors such as the size of the excitatory neuronal receptive fields and a degree of local variation of receptive field position and extent (receptive field scatter) define the grain of a visuotopic map.

Because receptive fields of extrastriate neurons have finite sizes, even in the hypothetical situation in which the visuotopic representation was perfectly linear (with each small displacement of electrode position in the cortex yielding a smooth, predictable displacement of neuronal receptive field centers in the visual field), "points"* in the visual field would project to areas in the visuotopic map. The cortical *point-image size* corresponds to the diameter of the cortical region activated by a point in the visual field, measured parallel to the cortical layers. This linear measurement also corresponds to the distance one needs to travel parallel to the cortical layers in order to obtain completely nonoverlapping receptive fields (e.g., Hubel and Wiesel, 1974; McIlwain, 1975). In contrast with the idealized situation, the existence of receptive field scatter means that the cortical point-image is unlikely to have precisely defined boundaries, and is better described in probabilistic terms (Dow *et al.*, 1981; Albright and Desimone, 1987). Determining the receptive field size and representational scatter are of obvious importance for the analysis of extrastriate visuotopy. For example, there is little sense in analyzing the position of receptive field centers in an area in which each cell responds to stimulation through most of the visual field. In this case, the small variations of receptive field center position observed from one neuron to the next may simply reflect the greater imprecision in defining the receptive field boundaries, the receptive field scatter, or the difference between the geometrical center and the functional center (point generating maximal response) of the receptive field (Maguire and Baizer, 1984). The last factor, in particular, becomes more important as the receptive field size increases, as the error involved in taking the geometrical center of the receptive field as an estimate of the point of maximal response is likely to increase (McIlwain, 1983).

The diameter of classical receptive fields in extrastriate cortex can vary from a fraction of a degree (in the foveal representation of the second visual area, V2; Van Essen and Zeki, 1978; Gattass *et al.*, 1981; Rosa *et al.*, 1988b) to close to the entire visual field (in some of the superior temporal and caudal parietal areas; Gross *et al.*, 1979; Motter and Mountcastle, 1981; Bruce *et al.*, 1982; Saito *et al.*, 1986). However, this variation in receptive field size seems to conform to some well-defined rules. Within a given area, receptive field size depends on which portion of the visual field is represented, increasing markedly with increasing eccentricity following linear (Gattass *et al.*, 1981, 1985; Desimone and Ungerleider, 1986; Newsome *et al.*, 1986; Albright and Desimone, 1987; Komatsu and Wurtz, 1988; Rosa *et al.*, 1988b) or logarithmic (Maunsell and Van Essen, 1987; Fiorani *et al.*, 1989; Rosa and Schmid, 1995) relationships (Fig. 3). In addition, even within a given cell column (in which all neurons share the representation of a region of visual field), receptive field size will vary according to laminar position, with neurons in layer 4 usually having the smallest receptive fields. Finally, different areas show different relationships between receptive field size and eccentricity (Gattass *et al.*, 1981, 1985, 1988; Burkhalter and Van Essen, 1986; Desimone and Ungerleider, 1986; Felleman and Van Essen, 1987; Rosa and Schmid, 1995). In the context of anatomically defined hierarchical

*Geometrically, a "point" has no dimension, and cannot be seen. Thus, references to "points in the visual field" should be taken as equivalent to "very small portions of the visual field." A plausible working definition is equating the dimension of a "point" to the minimum angle of resolution at optimal contrast at each eccentricity.

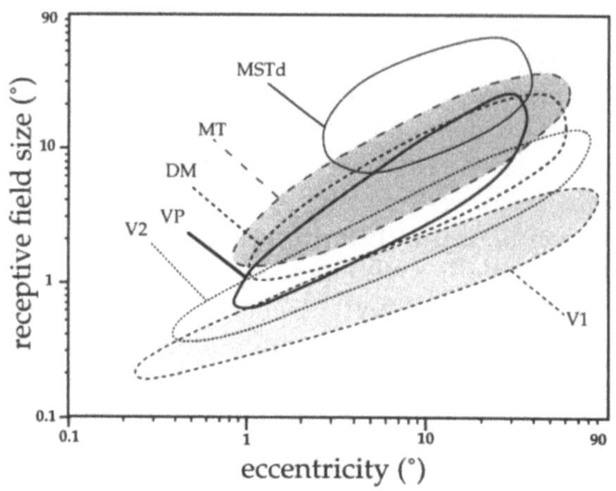

Figure 3. Distributions of multiunit receptive field size (estimated as square root of receptive field area) as a function of eccentricity in different visual areas. Each ellipse was drawn so as to encompass all measurements of receptive field size observed in a given cortical area of a capuchin monkey (*Cebus apella*). In all areas, the average receptive field size increases with increasing eccentricity, following a near-logarithmic relationship. In addition, for a given eccentricity, the average receptive field size increases at successively higher hierarchical levels of visual cortex (for example, from V1, to V2, to MT, to MSTd). Measurement of V1, V2, DM, and VP receptive fields are all from a single animal, and those of MT and MSTd receptive fields from another animal (data from Fiorani *et al.*, 1989, and unpublished experiment by M. G. P. Rosa, G. Gattass, and A. P. B. Sousa).

levels, a general principle is that the "higher" an area is, the larger will be the receptive fields for a given eccentricity. For example, neurons in V2 of both macaque and capuchin monkeys have the smallest receptive fields in extrastriate cortex (Van Essen and Zeki, 1978; Gattass *et al.*, 1985; Rosa *et al.*, 1988b), and receptive fields of neurons in MT are on average smaller than those of neurons at the next levels, the medial superior temporal and fundus of superior temporal areas (MST and FST; Desimone and Ungerleider, 1986; Komatsu and Wurtz, 1988; Erickson *et al.*, 1989; Fiorani *et al.*, 1989; see Fig. 3). It is interesting to observe that the gradual increase in receptive field size at successively "higher" stages of visual processing is not a universal characteristic of sensory systems. For example, the auditory pathway seems to use different strategies, as frequency-tuning curves of single neurons show a much smaller degree of variation at different levels of the pathway (e.g., Calford *et al.*, 1983).* Moreover, even within a single auditory area, neurons along the same isofrequency band may show a systematic variation of frequency bandwidth according to position in the cortex (Schreiner, 1991, 1995; Schreiner and Sutter, 1992), whereas in visual areas most of the variation of receptive field size for the representation of a given eccentricity is laminar.

As receptive field size increases from one area to the next, it has been observed that the receptive field scatter also increases (Van Essen and Zeki, 1978; Maguire and Baizer, 1984; Albright and Desimone, 1987; Fiorani *et al.*, 1989). In particular, a comparison between V1 and MT suggests that receptive field scatter varies in a manner proportional to receptive field size (Albright and Desimone, 1987). The possibility of a proportional relationship between receptive

*In fact, considering the relative frequency of some categories of cells (such as those which show a nonmonotonic or multipeaked relationship between response and stimulus intensity) the population tuning curves may actually become sharper at some of the "higher" stages (Sutter and Schreiner, 1991; Clarey *et al.*, 1992; Irvine, 1992; Calford and Semple, 1995), especially due to lateral inhibition at high sound pressure levels.

field size and scatter has not been investigated in other primate visual areas. However, evidence from studies of cat visual cortex suggests that the amount of local scatter (within single cellular columns) may vary less steeply than the receptive field size as one moves from V1 into extrastriate cortex (Sherk and Mulligan, 1992). It is important to note that, in addition to the local scatter, a degree of apparent visuotopic disorder may be introduced in the global map of a visual area if the representation includes topographic fractures, duplications and biases (Van Essen *et al.,* 1981; Sherk and Mulligan, 1992, 1993; see Section 2.4). Ideally, local and global "scatter" in representations should be considered separately. Local scatter may reflect processing strategies of the columnar processing module, such as the parallel encoding of multiple properties (Ballard, 1987) or population coding of spatial location by interpolation (Section 5). In fact, local scatter may simply be a reflection of biological constraints on the development of a visuotopic map. The exact quantification of local scatter requires electrode penetrations that are parallel to the cortical columns. The global scatter in the visual field representation of an area may reflect, in addition, the interdigitation of columns which encode different groups of properties within a given point set* (e.g., Newsome *et al.,* 1986; Rosa *et al.,* 1988b; Gattass *et al.,* 1990). Multiple interdigitating representations may result in larger point-image size without degrading the precision of representation for a given class of neurons or for a given set of image parameters (Swindale, 1991). In practice, however, isolating the local scatter from the other sources of topographic imprecision is technically difficult unless transitions of response properties or cortical architecture accompany the transitions between functional columns within a given point set (e.g., Blasdel and Fitzpatrick, 1984; Roe and Ts'o, 1995). It should also be noted that the view of strict segregation of function in parallel compartments, which has emerged as a result of studies in the last decade or so (e.g., Livingstone and Hubel, 1984; Hubel and Livingstone, 1987), has been increasingly challenged by more recent work which emphasizes gradual variability of neuronal response properties in both V1 and V2 (e.g., Girard and Bullier, 1989; DeBruyn *et al.,* 1993; Peterhans and von der Heydt, 1993; Levitt *et al.,* 1994a; Leventhal *et al.,* 1995). Perhaps a model in which cells responding to different combinations of visual parameters are generated as a result of the overlap of smooth gradients of distribution of afferents (Ballard, 1987; Malach, 1994) will prove to be more realistic than the view of multiple interdigitating compartments. In this case, the precision of representation and point-image size will vary in parallel as a function of the number of significant combinations of parameters.

In conclusion, the relationship between receptive field size and scatter in different areas is an interesting issue that demands more study. A constant relationship between these variables would suggest that a similar "canonical" circuit defines the visuotopic precision in all areas. Moreover, the establishment of such a relationship could allow an objective criterion for the determination of

*In order to avoid confusion with anatomically defined "modules" such as cytochrome oxidase blobs and pyramidal cell modules (Peters and Sethares, 1991), I will use the term point set (introduced by Roe and Ts'o, 1995) to designate a region of cortex large enough to encompass columns selective for a full cycle of the stimulus variables, representing a point in the visual field. For example, in V1 a point set would encompass orientation, ocular dominance, spatial frequency, and color hypercolumns.

nonvisuotopically organized areas, i.e., those in which the aggregate receptive field† of each point set includes the entire hemifield.

2.4. Types of Representation and Topographic Transitions between Areas

The existence of finite receptive fields and of receptive field scatter demonstrates the basic fact that, throughout visual cortex, point-to-point transformations do not exist. In addition, the "zone-to-zone" representations of the visual field in extrastriate areas may present further complexities in the form of discontinuities, incongruencies, and multiplication of representations of specific sectors. As a guide for discussion, some of the theoretical possibilities involved in converting a map of the contralateral visual hemifield onto a cortical representation are illustrated in Fig. 4, reproduced from Van Essen *et al.* (1981). When this figure was first published, the authors had to resort to cortical areas belonging to other sensory systems and subcortical nuclei to exemplify the occurrence of some of these transformations. Fifteen years later, we can limit our examples to the visual cortex.

Early in the story of cortical mapping, it was realized that not all visual areas represented the hemifield according to the same rules. A formal distinction between first-order transformations and second-order transformations of the visual field was proposed by Allman and Kaas (1974a), based on the concept of topologically equivalent and nonequivalent structures. Imagining the cortex within a visual area as a two-dimensional* representation of the contralateral hemifield, *first-order transformations* (Fig. 4B) are the most straightforward: adjacent points of the hemifield always project to single adjacent points in the cortex, and adjacent points in the cortex always represent adjacent points in the visual field. More realistically, the principle behind first-order transformations can be restated as "finite small sectors of the visual field project to single, partially overlapping regions of cortex." Note that these definitions imply nothing about the shape or local distortion of the map. If the map is imagined as drawn on an elastic surface, it is possible to stretch or bend the surface at will and still have a first-order transformation provided that no tears are introduced and that nonadjacent points along the border are not made adjacent. First-order representations are rare. The quintessential example, in every mammal studied so far, is still V1; in spite of different sizes and shapes observed in different species, this area always preserves the image continuity if the aggregate fields of adjacent columns are used as the mapping parameter (see Rosa *et al.*, 1993a, for a review). In extrastriate cortex, the only established first-order transformation is the middle temporal area (MT) in prosimians and New World monkeys (Allman and Kaas, 1971; Allman *et al.*, 1973; Spatz, 1977; Fiorani *et al.*, 1989). In Old World

†The aggregate receptive field of a cortical column, hypercolumn, or point set is the region of the visual field which includes all receptive fields of single units sharing that cortical locus; this definition takes into account both the unitary field size and the field scatter (e.g., Dow *et al.*, 1981).

*For the sake of simplicity, it can be assumed that all receptive fields located in a cellular column share the representation of a region of the visual field, thereby converting the three-dimensional cortex into a two-dimensional surface.

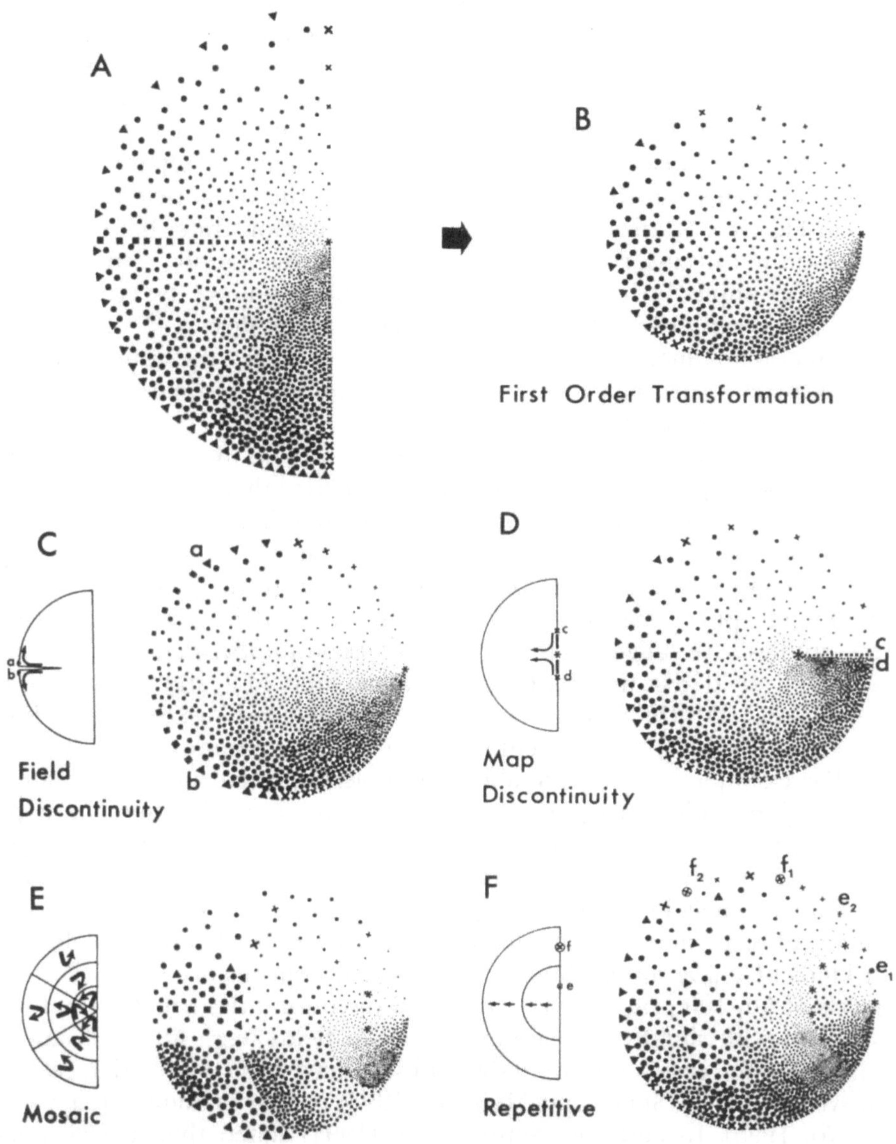

Figure 4. Different transformations of the visual hemifield (A) into hypothetical circular cortical areas (B–F). In A, the size of the dots reflects eccentricity in the visual field, and the dot density varies with polar angle: the highest density indicates the portions of the visual field near the vertical meridian in the lower quadrant, and the lowest density those near the vertical meridian in the upper quadrant. The center of gaze is represented by an asterisk, the vertical meridian by Xs, the horizontal meridian by squares, and the visual field periphery by triangles. The different classes of transformation (B–F) are explained in the main text. Reproduced with permission from Van Essen *et al.* (1981).

monkeys, the visuotopy of MT is reported to be more complex (Van Essen *et al.*, 1981; Maunsell and Van Essen, 1987) and, as discussed in detail below (Section 3.2), it is unclear if it fits the definition.

Second-order transformations are of several types, all of which result in topologically nonequivalent representations. The first category, *field discontinuities* (Fig. 4C), result in adjacent points of the visual field not always being represented at adjacent points in cortex. The first example of a "field discontinuity" second-order transformation to be described in primates was V2, where the map of the visual field is split along a line that corresponds approximately to the horizontal meridian (Allman and Kaas, 1974a; Gattass *et al.,* 1981). As a result of this particular transformation, regions of the visual field located immediately above and below the horizontal meridian may be separated by large expanses of cortex, which represent other parts of the visual field (e.g., points a and b in Fig. 4C). Studies in nonprimate species demonstrate that field discontinuities do not necessarily correspond to the representation of the horizontal meridian; examples include areas with discontinuities along the visual field periphery (Tusa *et al.,* 1979; Albus and Beckmann, 1980; Rosa *et al.,* 1994) and the vertical meridian (Sherk and Mulligan, 1993). In primates, the visuotopic map in most extrastriate areas may include field discontinuities, in some cases combined with other sources of topographic complexity.

Map discontinuities occur whenever adjacent sites on the cortical map represent nonadjacent points in the visual field. In the hypothetical case shown in Fig. 4D, the region of foveal representation is internalized in the map; as a result, two points in the visual field located along the vertical meridian (c and d) are represented at adjacent sites on the cortical map. Although there have been no descriptions of "pure" map discontinuities of this kind in primate extrastriate cortex, studies have suggested that such map discontinuities may occur in combination with field discontinuities in macaque area MT and marmoset dorsomedial area (DM) (Van Essen *et al.,* 1981; Maunsell and Van Essen, 1987; Rosa and Schmid, 1995). To what extent real map discontinuities exist is still an open issue. For example, if one were able to sample *every* neuron located along an electrode penetration crossing a map discontinuity, would there be a point where completely nonoverlapping receptive fields are observed for adjacent neurons? Or would the line of "map discontinuity" simply correspond to a region of extremely low cortical magnification factor,* but one where receptive fields of adjacent neurons nonetheless always overlap? There is evidence for both kinds of topographic abnormalities. When neurons are sampled at short intervals of 50–100 μm, most putative map discontinuities prove to be formed by neurons with slightly overlapping receptive fields (e.g., Maunsell and Van Essen, 1987). Yet, cases of outright discontinuity have been observed in cat lateral suprasylvian cortex (Sherk and Mulligan, 1993) and in macaque area V2 (Roe and Ts'o, 1995). In addition, Sereno *et al.* (1994) stated that in owl monkey extrastriate cortex they "very rarely observed large jumps in receptive field centers (a large jump being one that results in a completely nonoverlapping receptive field)." For most practical purposes (e.g., when one is trying to assess the orderliness of a map or to define the boundary between two areas) both types of transition may be treated as map singularities. Further study of this topic may be interesting for both theoretical and pragmatic reasons. For example, some theories of formation of visuotopic maps during development emphasize the role of gradients of chemical markers in the target area as determinants of the

*The cortical magnification factor (Daniel and Whitteridge, 1961) corresponds to the ratio between the distance separating two points in the cortex (measured parallel to the cortical layers) and the angular separation between the corresponding receptive field centers.

map (e.g., Sperry, 1963; Stuermer, 1991), whereas others suggest that epigenetic mechanisms, such as activity-driven plasticity, may impose substantial changes on the chemically specified map (Edelman, 1981; Schmidt, 1985). The existence of true map discontinuities may require sharp transitions in the distribution of chemoaffinity markers rather than smooth gradients, or suggest that the chemically specified map is modifiable to a large extent. On the practical side, the existence of map discontinuities may introduce spurious results when current algorithms for interpolation of visuotopic maps are applied, since they all assume map continuity (Maunsell and Van Essen, 1987; Fiorani *et al.*, 1989; Sereno *et al.*, 1994). If further study confirms the existence of map discontinuities, the precision of these algorithms may be improved by routines that avoid interpolation between pairs of recording sites yielding distant receptive fields.

Two other categories of transformation are the *mosaic* (Fig. 4E) and the *repetitive* (Fig. 4F) maps. In both cases, the visual field is subdivided into small sectors, which are mapped more or less independently. In a mosaic the representation of each sector, considered separately, is ordered, but the borders between pieces of the mosaic are not congruent. Thus, local field discontinuities and map discontinuities are scattered throughout the map. In a repetitive representation, each piece of the mosaic is represented more than once. Both of these transformations are still compatible with the existence of a global visuotopy in the area. For example, in parts E and F of Fig. 4, the centroperipheral axis and the upper/lower quadrant segregation are preserved. Thus, sparse grids of recording sites (e.g., one sample taken from each piece of the mosaic) may not be able to detect the presence of mosaics or repetitive maps. An example can be seen from a fine-grained analysis of primate V1. In this area, within layer 4C, the same portion of the visual field is rerepresented, once for each eye, in adjacent ocular dominance stripes (Hubel and Wiesel, 1977; Tootell *et al.*, 1982; Blasdel and Fitzpatrick, 1984). Nonetheless, this repetition is not apparent in the global visuotopy of V1 if the aggregate receptive fields of columns are taken as a mapping parameter. In V2 of the capuchin monkey a repetitive local representation has been correlated with the presence of cytochrome oxidase stripes (Rosa *et al.*, 1988b), and in the parietooccipital area (PO) a mosaic organization has been proposed by Neuenschwander *et al.* (1994). Therefore, both mosaic and repetitive transformations reflect local properties of the map, and can be observed in areas that are either first- or second-order transformations globally.

The above types of transformation all refer to the topographic organization within a single visual area. What happens at borders between areas? It is often observed that at the transition between two visuotopically organized areas, groups of neurons on either side of the border represent partially overlapping portions of the visual field. This situation defines a *congruent* border. This arrangement results in a supra-areal visual topography, which may have practical consequences: for example, a brain lesion involving parts of adjacent visual areas is likely to affect the representations of overlapping sectors of the visual field in these areas. Congruent borders are not universal, however. A few examples of noncongruent borders (in which the position of receptive fields changes abruptly as the boundary between areas is crossed) have been reported in primates. Studying the dorsal extrastriate cortex of the prosimian *Galago,* Allman *et al.* (1979) found abrupt topographic transitions at the border between V2 and the "dorsal visual area." For example, in Fig. 14, a peripheral receptive field (21) was recorded from a site in the dorsal area adjacent to the central lower quadrant representation in V2. In addition, Gattass *et al.* (1987) observed noncongruent borders between V1 and V2 corresponding to the representation of the visual field periphery. Again, finer analysis may prove that these borders are formed by regions of rapid change of receptive field position rather than topographic fractures. An interesting observation by Payne and Siwek (1990) suggests another type of topographic perturbation which may contribute to bridging the representational gap across noncongruent borders (and map discontinuities). In the caudal extrastriate cortex of cats, they found unusually large receptive fields at the zone of confluence of areas with noncongruent representations.

3. A Survey of Visuotopic Maps and Cortical Areas in Different Primate Species

This section includes a comparative description of the visuotopic organization of extrastriate areas in different species of primates. The main objective here is to identify what is known, and what is controversial about the visuotopy of primate extrastriate cortex. Even a brief look at Table I of abbreviations is sufficient to demonstrate the point that there are many more names than there are possibly visual areas in the brain. Comparisons between species are made cumbersome by the variety of names that have been used to designate the same areas, or parts of the same areas, by different investigators. Proposing a solution for the issue of nomenclature is beyond the scope of this chapter, but whenever the homologies are clear, I will try to list all the different designations.

3.1. The Second Visual Area

Even though a second visually responsive and topographically organized area was demonstrated in the cat in 1942 (Talbot, 1942), 20 years passed before visual responses in a corresponding area were recorded in a primate. In 1964, Cowey first reported on a topographically organized belt ("prestriate area") in front of the squirrel monkey's V1. This belt was defined by a dual approach: cortical responses were evoked in the prestriate area both by stimulating the visual field with spots of light and by electrically stimulating different points of V1 (Fig. 5A).* In this way, many important aspects of the organization of the "prestriate area" adjacent to V1, which is now known as the second visual area (V-II or V2), were established. For example, it was demonstrated that (a) the representations of the upper and lower quadrants in V2 are located in ventral and dorsal extrastriate cortex, respectively, (b) the receptive fields of V2 neurons

*Curiously, both Daniel and Whitteridge (1961) and Cowey (1964) observed no visual responses in the "prestriate" area of macaque monkeys. However, the existence of a similarly organized area in the macaque was suggested by recording responses in the lunate sulcus while electrically stimulating V1 (Cowey, 1964).

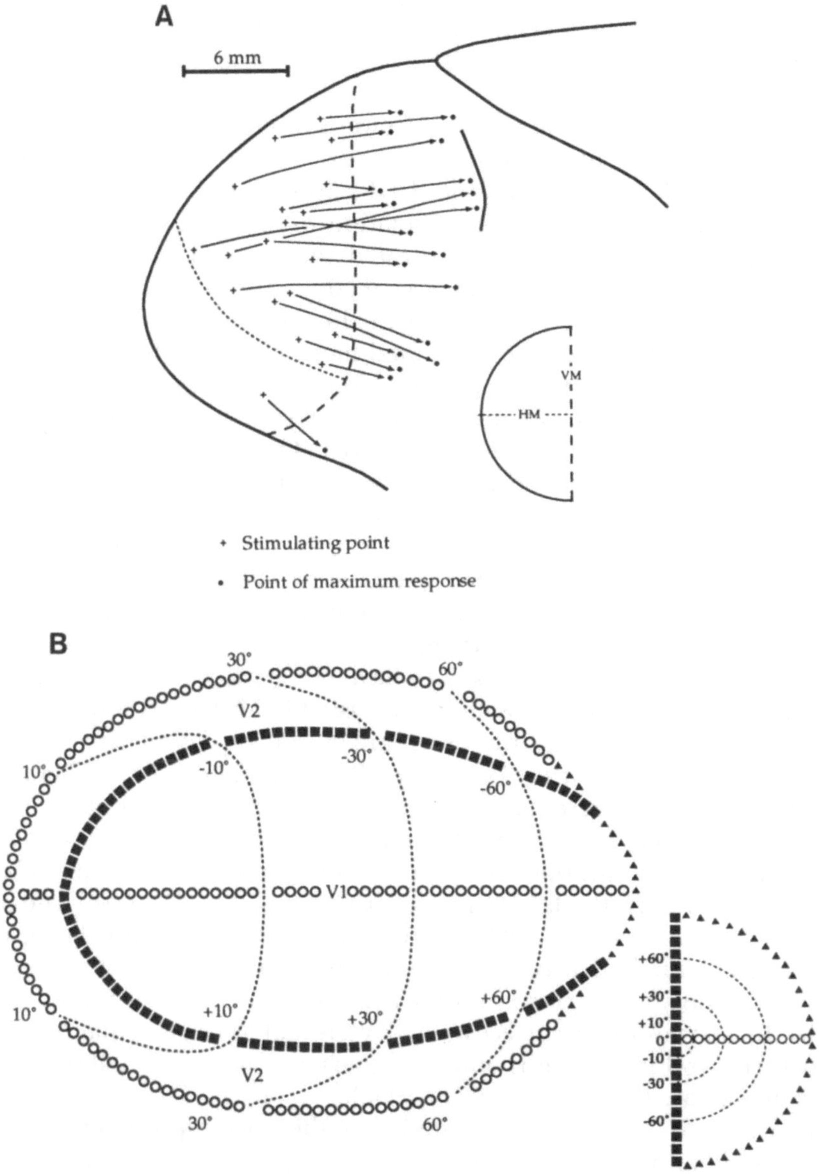

Figure 5. Early studies of visuotopic organization of V2 in simian primates. (A) Lateral view of the right occipital lobe of the squirrel monkey, showing the relationship between points of electrical stimulation in V1 (+) and points of maximal evoked response in the prestriate cortex (V2, dots). Stimulation of points in V1 near the representation of the horizontal meridian (dotted line) yielded maximal responses along the rostral border of V2, while stimulation of points near the representation of the vertical meridian (dashed line) yielded maximal responses in immediately adjacent parts of V2. A mediolateral gradient of visuotopy was also demonstrated, with the most lateral part of V1 (representing the center of the fovea) projecting to lateral V2, and medial parts of V1 (representing paracentral vision) projecting to V2 near the midline. Redrawn from Fig. 8 of Cowey (1964). (B) The Allman–Kaas model of visuotopic organization of V2 in the owl monkey (redrawn from Fig. 6 of Allman and Kaas, 1974a). This diagram represents an "unfolded" view of V1 and V2 in the left cerebral hemisphere, and illustrates the second-order transformation of the contralateral hemifield in this area. The insert (lower right) summarizes the symbols used for the representation of the visual field coordinates (vertical meridian: squares, horizontal meridian: circles; periphery: triangles; isoeccentricity lines: dashed lines). In V2, the horizontal meridian of the visual field is represented both dorsally (lower quadrant representation) and ventrally (upper quadrant representation).

move from the fovea to the periphery as the electrode is moved from lateral to medial in the cortex, (c) the receptive fields of V2 neurons are large in comparison with those of V1 neurons, and (d) the visual responses in V2 depend on the integrity of V1.

For the next 10 years, refinement of Cowey's model for the organization of V2 relied entirely on anatomical tracing of V1 projections (Cragg and Ainsworth, 1969; Zeki, 1969; Spatz *et al.*, 1970). Studies employing degeneration techniques provided independent confirmation of several aspects of the proposed organization of V2, and demonstrated that a similarly organized area exists in the macaque. However, they also made clear that, unlike in V1, the representation of the central visual field in this area includes a field discontinuity. The representation of the horizontal meridian in V2 forms its anterior boundary with other extrastriate areas; as a consequence, the upper and lower quadrant representations in V2 are separated except for a small connecting "bridge" which represents the center of the fovea. This model was formalized by Allman and Kaas (1974a), who collated the results of many of their own electrophysiological recording experiments in the owl monkey with other lines of evidence into a global model of the visuotopy of primate V2 (Fig. 5B).* A similar organization has since been confirmed in macaques (Van Essen and Zeki, 1978; Gattass *et al.*, 1981), capuchins (Rosa *et al.*, 1988b), marmosets (Rosa and Schmid, 1995; Fritsches, 1995), and humans (Sereno *et al.*, 1995). These diurnal primates differ from the nocturnal owl monkey in having a greater emphasis on central representation and in the exact point in the visual field that corresponds to the field discontinuity. Whereas in the owl monkey the representation of central vision may be continuous for the central 5–7° (Allman and Kaas, 1974a), in the diurnal primates the representation of the horizontal meridian at the rostral border of V2 is already split within the central 0.5–1°. In addition, it has been observed that the magnification factor in V2 does not hold a constant relationship with that in V1. In the large diurnal primates, V2 is about as large as V1 (Gattass *et al.*, 1981; Rosa *et al.*, 1988b; Sereno *et al.*, 1995). The foveal magnification factor is similar in both areas, but in some species there is a relative compression of the periphery that accounts for the slightly smaller area of V2 (Rosa *et al.*, 1988b). In the marmoset the surface area of V2 is half as large as that of V1, and in the nocturnal owl monkey it is only one-quarter as large (Tootell *et al.*, 1985; Krubitzer and Kaas, 1990; Pessoa *et al.*, 1992).

Allman and Kaas (1974a) suggested that the split representation of the horizontal meridian in V2 may have functional significance, by optimizing the configuration of axonal connections between this area and V1. On teleological grounds, it was suggested that the map of the visual field in V2 sacrifices internal continuity in order to allow connections of uniform length between V1 and V2. A corollary of this hypothesis is that, for V2 function, interaction with V1 is more "important" than interactions by intrinsic connections across the horizon-

*Although the results of Cowey (1964) suggested this organization (e.g., Fig. 5A), it was never made explicit by this author, who summarized his observations on the organization of V2 as indicating that "the first few degrees of the central retina are represented in prestriate cortex in a manner which resembles a *mirror image* of the striate map." In addition, responses in V2 were observed only on the dorsal and lateral surfaces of the occipital lobe, and attempts to record from a putative peripheral representation on midline cortex were not successful (Cowey, 1964).

tal meridian (Allman and Kaas, 1974a). While visual processing in primate V2 is highly dependent on inputs from V1 (Schiller and Malpeli, 1977; Bullier *et al.*, 1994), this is unlikely to be the only important factor. In some nonprimate mammals the visual reposes of neurons in the presumed homologue of V2 are much less dependent on V1 (e.g., cat: Dreher and Cottee, 1975; Sherk, 1978; Dreher, 1986; Dreher *et al.*, 1992; see also Olavarria and Torrealba, 1978, for evidence in rats). Areas V1 and V2 seem to work largely in parallel in cats, where multiple field discontinuities are nonetheless present in V2 (Tusa *et al.*, 1979; Albus and Beckmann, 1980). It is likely that the timing of neuronal firing in V1 and V2 is crucial, and that the map of V2 has a configuration that increases the chance of this firing being synchronous (Nelson *et al.*, 1992; Frien *et al.*, 1994; Singer, 1994). In this case, minimizing the variance of the distribution of axonal length of cells interconnecting V1 and V2 may be important. Whatever the function of the field discontinuity in primate V2, it is striking that intrinsic pathways in this area seem to be largely unable to integrate information across the horizontal meridian.* A possible strategy used by the visual system to circumvent this obstacle is the duplication of the representation of the portion of the visual field around the horizontal meridian (Van Essen and Zeki, 1978; Gattass *et al.*, 1981; Rosa *et al.*, 1988b). It is possible that a neural representation of a sector of the visual field surrounding the horizontal meridian is fed simultaneously to neurons in both dorsal and ventral V2, perhaps by means of a V1 projection involving neurons with bifurcating axons.

While there is a reasonable consensus abut the major aspects of the topographic organization of primate V2, there are still points which deserve further study, including the precision and continuity of the topographic map at a local level. A characteristic of V2 is the presence of a modular structure consisting of stripes, which can be revealed by cytochrome oxidase histochemistry. Cytochrome oxidase stripes have been observed in all simians studied (e.g., Livingstone and Hubel, 1982, 1984; Tootell *et al.*, 1983; Rosa *et al.*, 1988a, 1991; Krubitzer and Kaas, 1989, 1990; Wong-Riley *et al.*, 1993; Tootell and Taylor, 1995), as well as some (Preuss *et al.*, 1993; Paolini *et al.*, 1994), but not all prosimians (Condo and Casagrande, 1990). In most species these stripes run orthogonal to the border between V1 and V2, although in some macaques this relationship is not perfect (Van Essen *et al.*, 1990). At least in New World and Old World monkeys, three types of stripes have been reported, each containing neurons that have different response properties and connections (Livingstone and Hubel, 1982; Tootell *et al.*, 1983; DeYoe and Van Essen, 1985; Hubel and Livingstone, 1987; Krubitzer and Kaas, 1989; Tootell and Hamilton, 1989; see Peterhans and von der Heydt, 1993, and Levitt *et al.*, 1994a, for a somewhat different view).

The organization of V2 in functional stripes has direct implications for the

*An early study suggested the existence of connections between dorsal and ventral portions of V2 in squirrel monkey (Tigges *et al.*, 1974), but this has not been confirmed by other studies. More recent work reveals connections between the upper and lower quadrant representations in V2 only within the foveal representation (e.g., Cusick and Kaas, 1988; Stepniewska and Kaas, 1996). Because there was no histological control of the lesion extent relative to the rostral border of V2, the ventral focus of degeneration observed by Tigges *et al.* (1974) may represent projections from the upper field representation in area DM to V2.

visuotopy of this area. For example, analysis of the magnification factor has demonstrated that the visuotopic map in V2 is anisotropic, with the magnification factor measured parallel to the border of V2 (i.e., across the cytochrome oxidase stripes) being much larger than that perpendicular to this border (Rosa *et al.*, 1988b; Roe and Ts'o, 1995). It has been proposed (Rosa *et al.*, 1988b) that each point of the visual field projects to a compartment in V2 (equivalent to the point set) that is elongated, bridging adjacent portions of different stripes (Fig. 6E). This arrangement allows each point of the visual field to be represented multiple times (once for each type of visual parameter), and the apposition of such elongated "units" of representation results in an overall anisotropy of the map. Connectional studies have also demonstrated that circular lesions in V1 lead to elongated foci of degeneration in V2 (Spatz and Tigges, 1972), and that tracer injections within V2 label intrinsic connections that extend further parallel to the V1/V2 border than perpendicular to it (Rockland, 1985; Cusick and Kaas, 1988; Lund *et al.*, 1993; Levitt *et al.*, 1994b). These observations can be explained in the light of the marked anisotropy of cortical magnification factor within V2.

More direct evidence of a relationship between topographic irregularities in V2 and its modular organization emerged when it was observed in the capuchin monkey that receptive field eccentricity changes in a discontinuous way as V2 is crossed from lateral to medial (Fig. 6). Instead of the smooth centroperipheral displacement expected if V2 contained a continuous map, electrophysiological recordings suggested that the receptive field centers in this area are mapped in a discontinuous fashion, according to "isoeccentricity bands" 3–4 mm wide which cross V2 perpendicular to the V1/V2 border (Rosa *et al.*, 1988b; see also Fig. 7 of Van Essen and Zeki, 1978, for similar observations in the macaque). Within a band, receptive field eccentricity changes slowly and unpredictably, but at the border between bands, it changes abruptly. The mediolateral extent of the isoeccentricity bands matched closely the dimension of a full cycle of cytochrome oxidase stripes in capuchins (Rosa *et al.*, 1988a,b). These observations were based on a relatively coarse sampling, and do not allow one to distinguish between a

Figure 6. Representation bands in V2. (A–C) Recordings from V2 in the lateral bank of the calcarine sulcus of a capuchin monkey (based on Fig. 17 of Rosa *et al.*, 1988b). (A) Parasagittal section (caudal to the right, dorsal up) through the caudal half of the brain, showing the location (box) of the recording sites illustrated in B. (B) Five electrode penetrations, near tangential to the peripheral representation of the upper quadrant in V2, with recording sites indicated by circles. The corresponding receptive fields are illustrated in C. As expected from the topographic organization of V2, receptive fields of neurons change from the vertical meridian (fields 1, 13) to the horizontal meridian (fields 12, 20) as the lateral bank of the calcarine sulcus is crossed from dorsal to caudal. However, a smooth centroperipheral displacement of receptive fields is not observed as V2 is crossed from caudal to rostral. Instead, the receptive fields of the caudal 3 penetrations (light gray) rerepresent the same range of eccentricities. Moving rostrally, a new sector (dark gray, rostral penetrations) is reached. (D, E) Model of visuotopic organization of central representation in V2, based on Rosa *et al.* (1988b; Figs. 2A and 16). (D) View of the central 10° of the lower visual quadrant, with sectors of the visual field indicated in different shades of gray. The center of gaze is indicated by a star, the horizontal meridian by circles, and the vertical meridian by squares. (E) View of the corresponding sector of V2 (caudal to the right, medial to the top), with representation bands indicated. The white sectors separating the bands may be either regions of low magnification factor or topographic discontinuities. In the central representation, a point in the visual field projects to a sector of V2 that is 4 mm long parallel to the V1/V2 border and only 2–2.5 mm long perpendicular to this border.

model in which V2 is formed by rigidly segregated representational modules and another model in which the visual topography is continuous, but contains alternating regions of high and low magnification factor. Roe and Ts'o (1995) suggested that fine-scale maps also exist across cytochrome oxidase stripes, and that receptive fields may "jump back" in space at the borders between stripes. Because this result implies a rigid segregation between visual field representation within adjacent cytochrome oxidase stripes, it may be an important topic for further study, especially in view of the accumulating evidence that the anatomical and physiological segregation between stripes is not clear-cut (Peterhans and

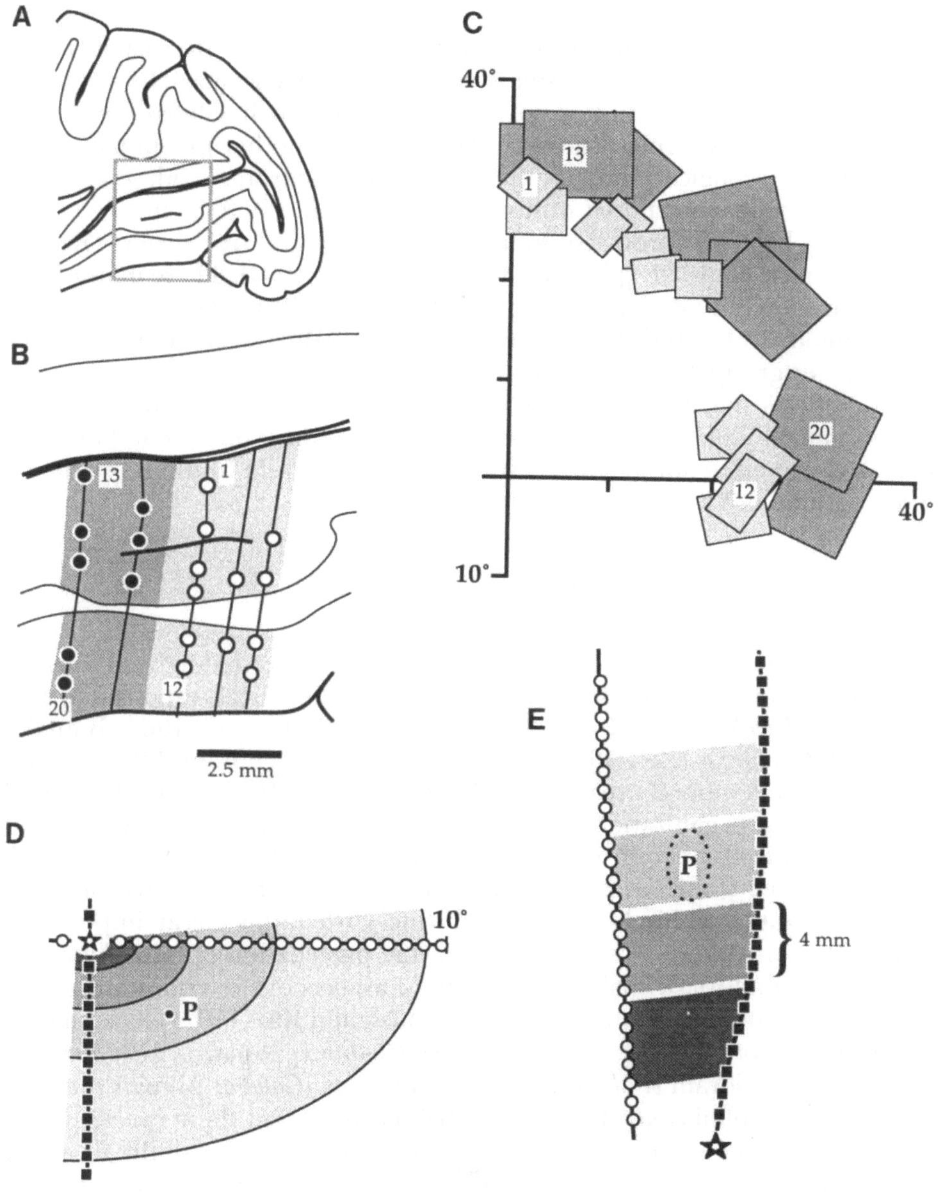

Figure 6.

von der Heydt, 1993; Levitt *et al.*, 1994a,b; Malach *et al.*, 1994; Peterhans, this volume). Differences between species also need to be taken into consideration: the anatomical distinctness of cytochrome oxidase stripes varies between species of monkey (e.g., Tootell *et al.*, 1985; Shipp and Zeki, 1989; Krubitzer and Kaas, 1990), and recordings in the marmoset monkey (Rosa *et al.*, 1997b) suggest that the degree of segregation of receptive fields by representation bands in this species is not as marked as that in capuchins.

Another topic that needs further study is the organization of V2 in prosimian primates. Isolated rows of penetrations crossing this area in the lorisiform *Galago* (e.g., Allman *et al.*, 1973, 1979) suggest an organization not unlike that observed in simian primates. However, there has been a brief report suggesting that in this species V2 is formed by two physically separate subareas, segregated by an interposed area which represents both quadrants of the visual field (Sereno and Allman, 1991). This result is striking, since V2 seems to be a continuous area in every monkey studied so far, a result which has been confirmed by cytochrome oxidase architecture. In the slow loris, another prosimian, there are no electrophysiological data, but the pattern of cytochrome oxidase stripes gives no hint of an area interposed within V2 (Preuss *et al.*, 1993). Similarly, a continuous V2 has been demonstrated in both flying foxes and tree shrews (Kaas *et al.*, 1972; Sesma *et al.*, 1984; Jain *et al.*, 1994; Rosa *et al.*, 1994). These mammals have been considered by different scientists (flying fox: Pettigrew *et al.*, 1989; Rosa and Schmid, 1994; tree shrew: Kaas and Preuss, 1993) as members of the primates' sister group. Supposing that the organization of V2 in flying fox and tree shrew reflects that present in early primates, the eventual confirmation of a discontinuous V2 in *Galago* would suggest an autapomorphous condition. Moreover, a recent electrophysiological study in *Galago garnetti* found no evidence for a discontinuous V2 (Rosa *et al.*, 1997a).

3.2. The Middle Temporal Area

By the late 1960s, several studies had demonstrated that in both New World and Old World monkeys visual stimulation can elicit neural responses in cortical areas rostral to V1 and V2 (Woolsey *et al.*, 1955; Doty *et al.*, 1964; Bignall and Singer, 1967; Gross *et al.*, 1967, 1969). However, it was not until the description of the middle temporal area (MT; Allman and Kaas, 1971) that it became clear that visuotopic organization is present through much of the caudal neocortex. This study (Fig. 2) not only described the existence of a map of the visual field in an area remote from V1, but also demonstrated that the neurons forming this map were located within histologically distinct cortex; this was an important step toward demonstrating a rational basis for the parcellation of "visual association cortex" into multiple visual areas. The basic aspects of the visuotopic organization of MT in owl monkeys reported by Allman and Kaas (1971) have since been confirmed in other New World monkeys (*Callithrix:* Spatz, 1977; Rosa *et al.*, 1995b; *Cebus:* Fiorani *et al.*, 1989) and prosimians (*Galago:* Allman *et al.*, 1973; *Cheirogaleus:* Paolini *et al.*, 1994). In global terms, MT in these species is a first-order transformation of the entire contralateral hemifield, with the fovea represented caudally, the periphery rostrally, the lower field dorsally, and the upper field ventrally. As in V1, the vertical meridian representation forms most of the outer border of MT. Interspecies variation of the visuotopy of MT is observed mainly in terms of the relative emphasis on central representation, with nocturnal primates having proportionally less of the surface area of MT dedicated to foveal and parafoveal vision. In addition, variability occurs between individuals of the same species with regard to details such as the relative extent of upper versus lower quadrant representation and the anatomical orientation of the representation of the horizontal meridian (Fiorani *et al.*, 1989). In spite of this variability and the large receptive fields, the visuotopic organization of MT in both New World monkeys and prosimians is very clear. This is reflected, for example, in the visuotopic patterns of connections between MT, V1, and V2 (Spatz, 1977; Lin *et al.*, 1982; Wall *et al.*, 1982; Weller *et al.*, 1984; Cusick and Kaas, 1988; Krubitzer and Kaas, 1990; Sousa *et al.*, 1991) and in the prominent "ring" of callosal connections that can be used to locate this area (Newsome and Allman, 1980; Cusick *et al.*, 1984).

In contrast, in spite of many studies, the details of the visuotopic organization of area MT in Old World monkeys* are still somewhat controversial. Early investigations of MT in the macaque (Dubner and Zeki, 1971; Zeki, 1974) reported that the receptive fields of nearby neurons overlapped, and that neurons recorded along penetrations in different portions of this area represented distant parts of the visual field. Although recognizing that these results suggested a visuotopic organization, Zeki (1974) also emphasized the relative disorderliness introduced by the large receptive fields. Yet, because the borders of the macaque "superior temporal motion area" were not delimited, the issues of whether there was a global visuotopic map in this area and of the homology with New World monkey area MT were only clarified by later studies (Ungerleider and Mishkin, 1979; Gattass and Gross, 1981; Van Essen *et al.*, 1981; Weller and Kaas, 1983). The main controversy regarding the visuotopy of MT in Old World monkeys refers to whether the visuotopic map departs significantly from a "first-order transformation." Analyses of receptive fields recorded along single-electrode penetrations always demonstrate a considerable degree of visuotopic order (e.g., Fig. 3 of Gattass and Gross, 1981; Van Essen *et al.*, 1981; Albright and Desimone, 1987; Maunsell and Van Essen, 1987). Yet, two-dimensional analyses of the global visuotopy in MT (by analyzing the data obtained in multiple-electrode penetrations) may suggest either a continuous first-order transformation (similar to, though more distorted than that observed in New World monkeys and prosimians; Gattass *et al.*, 1981; Desimone and Ungerleider, 1986), or the presence of additional irregularities, such as field and map discontinuities. For example, Van Essen and collaborators have found that in many cases the representation of the vertical meridian in MT is not located around the perimeter of this area. They also present evidence for double representation of parts of the visual field in distant portions of MT (Van Essen *et al.*, 1981; Maunsell and Van Essen, 1987). Maunsell and Van Essen (1987) argued that the differences between results obtained by different groups are more apparent than real, and that the visuotopic map in MT may present local order, but global disorder. Thus, evidence for disorderliness may not emerge when adjacent sites along single-electrode

*Area MT in Old World monkeys is sometimes referred to as V5 (visual area 5).

penetrations are compared, only becoming apparent when the entire pattern of the visuotopic map is considered.

Given the evidence of complex patterns of representation elsewhere in the sensory cortices, there is no *a priori* reason why field or map discontinuities should not occur in MT. However, it is unlikely that this is the entire explanation for the different results. Among the hundreds of electrode penetrations that have been made in MT by different investigators, at least some clear examples of map discontinuities, field discontinuities, and duplications would be likely to appear along single tracks. Yet, there are no published examples of (single penetration) field sequences that double onto themselves to rerepresent the same region of the visual field at points separated by a distance greater than the point-image size of MT, or that yield nonoverlapping receptive fields at sites less than 500 μm apart. Another possibility is that the representation within macaque MT is a first-order transformation similar to that observed in other primates, and that some of the evidence for disorderliness may be a consequence of the inclusion of data from adjacent areas (for a detailed discussion, see Fiorani *et al.*, 1989). Because the myeloarchitectonical definition of MT in macaques involves errors of up to 2.5 mm (Desimone and Ungerleider, 1986; Maunsell and Van Essen, 1987), and because the correspondence between MT and the zone containing direction-selective cells is not perfect (Komatsu and Wurtz, 1988; Erickson *et al.*, 1989; Fiorani *et al.*, 1989), this possibility cannot be discounted. In particular, the inclusion of parts of the area located dorsolateral to MT (see Section 3.3) could explain why the representation of the lower-field vertical meridian in macaque MT was sometimes observed well within the presumed borders of this area (as shown in Fig. 4 of Gattass and Gross, 1981, and Fig. 13 of

Figure 7. Topographic organization of visual area MT in the macaque. Recording sites (A) and receptive fields (B) in macaque area MT, redrawn from Fig. 3 of Maunsell and Van Essen (1987). (A) An unfolded reconstruction of the caudal bank of the superior temporal sulcus, with the myeloarchitectonic boundaries of area MT indicated. Seven sequences of recording sites, labeled A–G from ventral to dorsal, are indicated (ticks indicate the location of individual recording sites). (B) The location of corresponding receptive field centers and selected receptive field borders in the contralateral hemifield. The visuotopic organization of MT in the macaque appears to be much more complex than in New World monkeys. For example, the representation of the lower quadrant is many times larger than that of the upper quadrant, and the neurons with receptive fields nearest to the vertical meridian in some sequences (e.g., D, E) are found far from the border of MT. (C) Reanalysis of the same data with the visual field sign criterion, which distinguishes MT from surrounding areas in New World monkeys (Sereno *et al.*, 1994). The sector of mirror-image representation is shaded light gray, and the sector of non-mirror-image representation dark gray. If MT is defined as the sector of non-mirror-image representation, a simple and relatively more ordered first-order representation is observed in this area. The dotted line to the right of MT indicates an estimate of MTp, or the sector of peripheral representation in MT (Ungerleider and Desimone, 1986). Similar to MT, MTp receives direct projections from V1, but it is less densely myelinated. (D) Pattern of callosal connections in the same animal, redrawn from Fig. 11 of Maunsell and Van Essen (1987). In this bidimensional reconstruction, the dashed line indicates the fundus of the superior temporal sulcus, the continuous line the myeloarchitectural borders of MT, and the gray fill the pattern of interhemispheric connections of MT and surrounding areas, as evaluated with the Wiitanen (1969) method. Dark gray indicates dense degeneration, intermediate gray moderate degeneration, and light gray sparse degeneration. A callosal-free ring is observed within MT, similar to New World monkeys (Newsome and Allman, 1980).

Maunsell and Van Essen, 1987), resulting in a virtual double representation of the lower quadrant with mixed visual field sign (Fig. 7). Thus, although more work is needed to settle this question, it is possible that some of the observed complexity in macaque MT comes from including cells recorded in the adjacent areas into the visuotopic map of this area, and from excluding parts of MT that are not as densely meylinated (such as MTp, or "peripheral MT," of Ungerleider and Desimone, 1986). If one analyzes the Old World monkey area MT using a set of topographic and anatomical criteria similar to those used to define MT in New World monkeys (Fig. 7), a similar organization emerges for both groups of simians.

In macaques, the magnification factor of MT seems to vary in parallel with that of V1. In both areas, approximately half of the surface is dedicated to the representation of the central 10° of the visual field (Van Essen *et al.*, 1981; Desimone and Ungerleider, 1986; Komatsu and Wurtz, 1988; Erickson *et al.*,

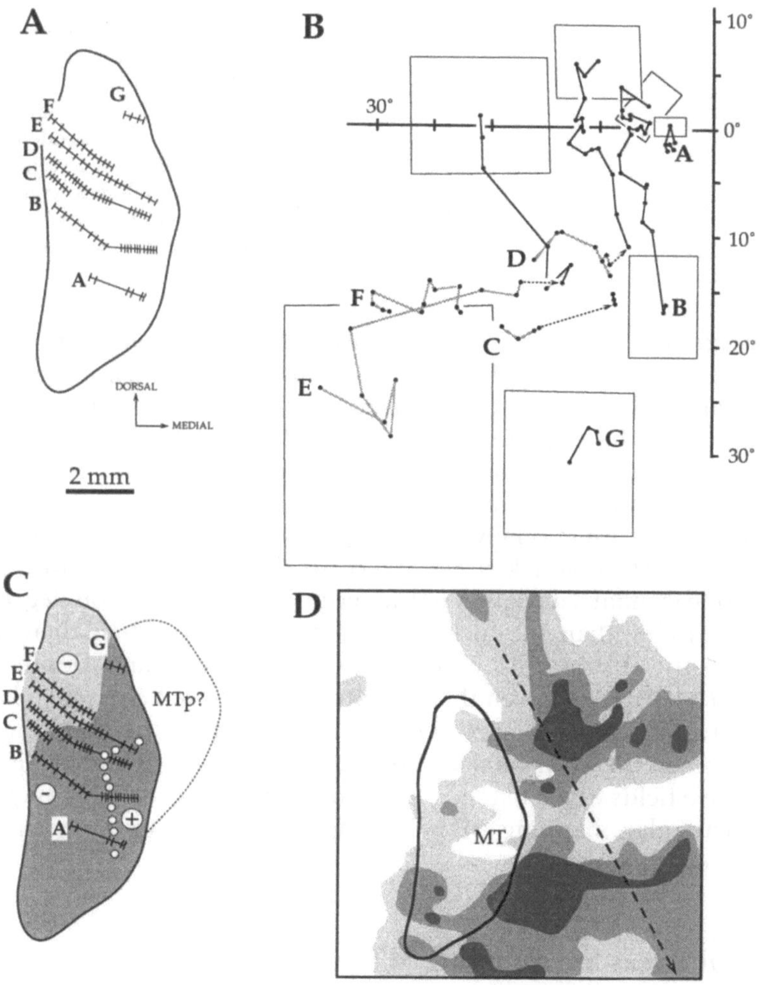

Figure 7.

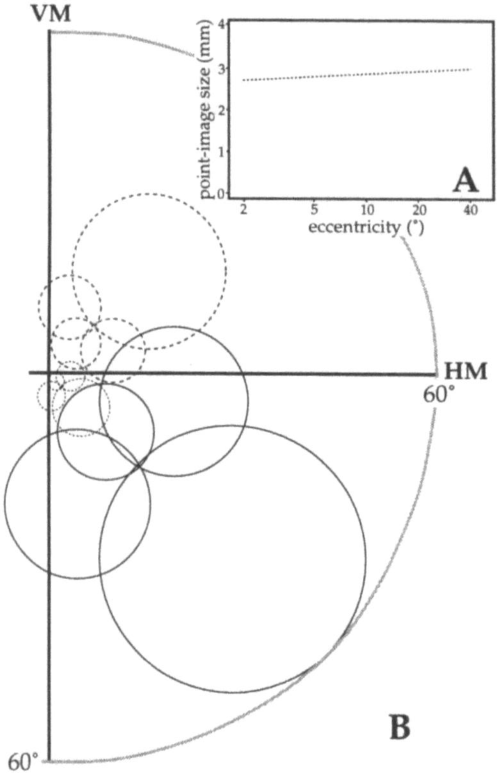

Figure 8. Coarse grain of the visuotopic map of area MT (redrawn from Fig. 13 of Fiorani *et al.*, 1989). (A) Point-image size as a function of eccentricity in area MT of *Cebus apella,* based on multiunit recordings. Any point of the visual field projects to a compartment of MT that is 2.5–3 mm in diameter, independent of eccentricity. (B) Schematic representations of aggregate receptive fields in MT (circles) plotted in a representation of the monkey's contralateral hemifield. The thick black lines indicate the horizontal (HM) and vertical (VM) meridians, and the thick gray line the border of the binocular field of vision. Dotted aggregate fields all include a point at 3° eccentricity, dashed fields include a point at 10° eccentricity, and continuous fields include a point at 20° eccentricity. Independent of eccentricity, points in the visual field located at a line equidistant from the HM and VM are included by aggregate fields which extend to both the HM and VM.

1989). By contrast, in New World monkeys, there seems to exist relatively less foveal representation in MT than in V1 (Fiorani *et al.,* 1989), but again this needs more careful study, ideally in conscious animals trained to foveate moving targets. In addition, the magnification factor in MT has been shown to vary locally, with single penetrations crossing patches of alternately rapid and slow receptive field displacement (Van Essen *et al.,* 1981). This variability may be imposed by the existence of a functionally heterogeneous local architecture in MT (Born and Tootell, 1992; Malonek *et al.,* 1994), In both New World and Old World monkeys, one standard deviation of the receptive field scatter (in degrees) in MT corresponds to less than one-fifth of the average receptive field diameter, which is similar to the relationship observed in V1 (Albright and Desimone, 1987; Fiorani *et al.,* 1989). Yet, given that receptive fields in MT are much larger than those in V1, this scatter is large enough that a point of the visual field located halfway between the horizontal and vertical meridians is included within aggregate receptive fields which extend to both meridians (Fig. 8). This may explain in part the irregular mapping of polar angle in MT (Van Essen *et al.,* 1981; Erickson *et al.,* 1989).

3.3. Visual Areas Surrounding MT

Visual responses in the cortex surrounding MT have been observed for the past quarter of a century (e.g., Allman and Kaas, 1971, 1974b; Gattass and

Gross, 1981; Van Essen *et al.*, 1981; Desimone and Ungerleider, 1986), and models of visual areas in this region have been the subject of a slow process of refinement as more data have been obtained (see Komatsu and Wurtz, 1988, and Kaas and Morel, 1993, for historical reviews). The data on the visuotopy of this region show several points of convergence between studies of different species (Komatsu and Wurtz, 1988; Fiorani *et al.*, 1989; Kaas and Morel, 1993; Sereno *et al.*, 1994). These points are as follows, and are summarized in Fig. 9:

(a) Most or all of the lower quadrant representation in MT is bordered by a topographically organized area in which neurons have receptive fields about as large as those of neurons in MT (Allman and Kaas, 1974b; Gattass and Gross, 1981; Schein *et al.*, 1982; Maguire and Baizer, 1984; Gattass *et al.*, 1985, 1988; Desimone and Ungerleider, 1986; Fiorani *et al.*, 1989; Rosa *et al.*, 1993b; Sereno *et al.*, 1994). The representation of the lower visual field in this area (Fig. 10B, sites 1–6) forms a thin strip approximately mirror-symmetrical to that in MT, with the vertical meridian represented near the border (see, however, the above discussion on visuotopy of MT in macaques). A representation of the upper quadrant may exist in this area, ventral to the foveal representation (Boussaoud *et al.*, 1991; Kaas and Morel, 1993; Rosa *et al.*, 1993b; M. Sereno, personal communication). Proposed names include transitional V4 (V4t, Desimone and Ungerleider, 1986; Gattass *et al.*, 1988), dorsal zone (DZ, Fiorani *et al.*, 1989), middle temporal crescent (MTc, Kaas and Morel, 1993; Tootell and Taylor,

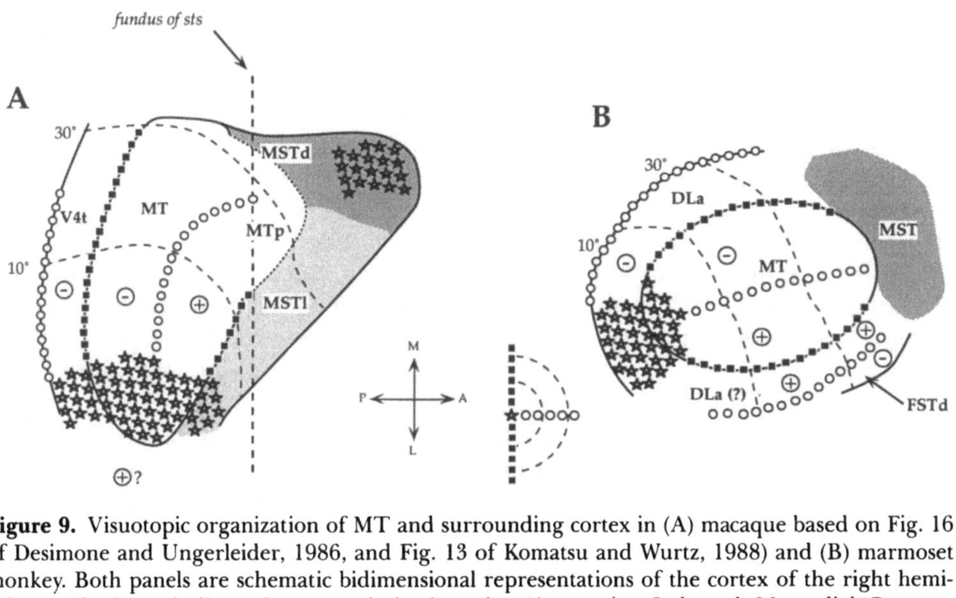

Figure 9. Visuotopic organization of MT and surrounding cortex in (A) macaque based on Fig. 16 of Desimone and Ungerleider, 1986, and Fig. 13 of Komatsu and Wurtz, 1988) and (B) marmoset monkey. Both panels are schematic bidimensional representations of the cortex of the right hemisphere. The insets indicate the anatomical orientation (A, anterior; L, lateral; M, medial; P, posterior) and the symbols used to represent visual field coordinates (squares: vertical meridian; circles: horizontal meridian; dashed lines: isoeccentricity lines; stars: fovea, corresponding to the central 2° of visual field). A non-topographically organized area with very large receptive fields (MST or MSTd; shown in dark gray) is located anterior to MT. The "+" and "−" signs indicate representations of the upper and lower quadrants, respectively, and "sts" is an abbreviation for superior temporal sulcus. No segregation between upper and lower quadrants has been reported for the macaque's lateral subdivision of MST (MSTl).

1995), and dorsolateral anterior area (DLa, Sereno and Allman, 1991; Sereno *et al.*, 1994).

(b) The peripheral representation of MT is bordered by an area in which topographic organization is not readily apparent. Some of the receptive fields of neurons in this region are large enough to include an entire quadrant of the visual field, even though they usually include the fovea. Proposed names include DSR (direction/size/rotation area of Saito *et al.*, 1986, a name coined to reflect the response properties of neurons recorded in this region), medial superior temporal area (MST; Desimone and Ungerleider, 1986; Fiorani *et al.*, 1989; Krubitzer and Kaas, 1990), and dorsal subdivision of MST (MSTd; Komatsu and Wurtz, 1988).

(c) The upper quadrant representation in MT is bordered by visuotopically organized cortex in which the centroperipheral axis is mapped in parallel with MT. In Old World monkeys (Fig. 9A), a single area has been proposed in this region, named either FST (fundus of superior temporal area; Desimone and Ungerleider, 1986) or MST1 (lateral subdivision of MST; Komatsu and Wurtz, 1988). In contrast, in New World monkeys (Fig. 9B), two areas have been recognized in this region, which are different in terms of cytochrome oxidase architecture and connections (Kaas and Morel, 1993) as well as visuotopy (Sereno and Allman, 1991). The cortex ventral to the caudal half of MT contains a representation of the upper quadrant, likely to belong to DLa (or MTc, or V4t). More rostrally, a different area, named AT (anterior temporal area; Sereno and Allman, 1991) or FSTd (dorsal subdivision of the fundus of superior temporal area; Kaas and Morel, 1993) seems to exist (Fig. 9B).

A point that deserves more study is the nature of the border between FSTd and DLa; this will hinge on the question of whether nor not DLa has a complete upper quadrant representation. The difficulty involved in the determination of this boundary is illustrated in Fig. 10. In the marmoset, the caudal two-thirds of MT is bordered ventrally by a strip representing the central 30–40° of the upper quadrant (Fig. 10B, sites 19–26), which seems to complement the lower quadrant representation observed dorsal and caudal to MT, in DLa. Exploring cortex lateral to this strip reveals large and diffuse receptive fields (Fig. 10B, sites 27–29), which are unlikely to belong to DLa. Moving rostrally in the cortex (Fig. 10A), the receptive field sequences corresponding to sites ventral to MT are displaced to represent more peripheral portions of the visual field up to 80–90°, as would be expected if DLa surrounded MT in this region (Fig. 10A, sites 5–7). Yet receptive fields in this region often cross the horizontal meridian, to represent the lower visual field periphery (Fig. 10A, sites 8–9). Thus, taken alone, the representation of the upper quadrant ventral to MT agrees with the proposal that DLa surrounds most of MT (Allman and Kaas, 1974b). However, the existence of neurons with relatively small receptive fields representing parts of the lower quadrant in this region argues against a simple model in which the lower quadrant representation of DLa borders the lower quadrant representation of MT, and the upper quadrant representation of DLa borders the upper quadrant representation of MT. For these reasons, the border between DLa and FSTd is still unclear. In fact, it is possible that DLa and FSTd are both portions of the same area, and that the line of field discontinuity in this area does not coincide exactly with the horizontal meridian, such as in area V2 of flying foxes (Rosa *et*

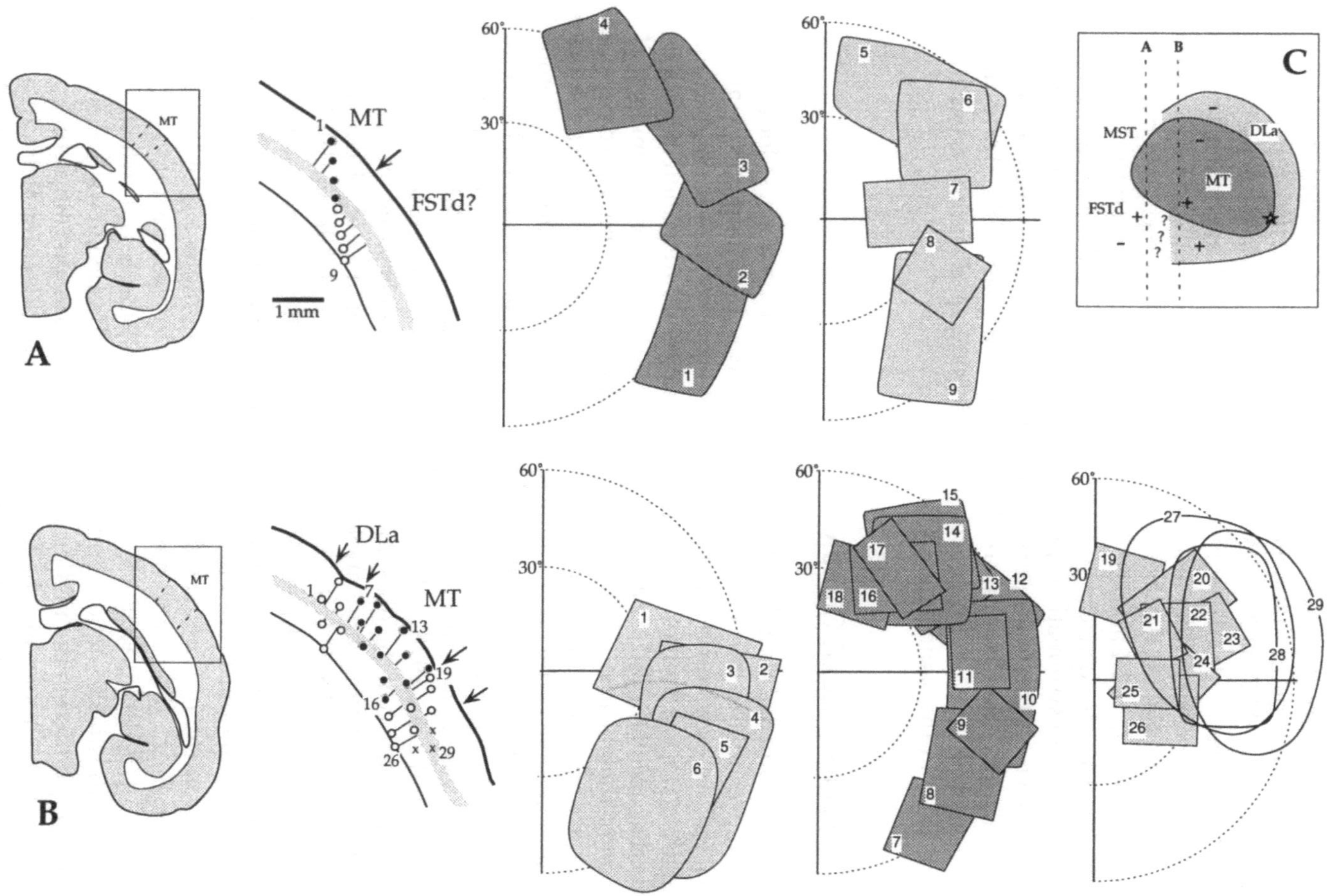

Figure 10. Visuotopic organization of cortex surrounding the peripheral representation of MT in the marmoset. (A, B) Sequences of recording sites in the cortex and receptive fields of neurons in MT and surrounding areas. The location of recording sites in the brain is indicated in coronal sections (left) in which the boundaries of MT are shown by dashed lines. The regions of cortex within the boxes are magnified in the middle panels, which indicate the location of individual recording sites in MT (black dots), areas surrounding MT (white circles), and an area not adjacent to MT (crosses) relative to layer 4 (gray). Recording sites have been sequentially numbered according to their projection to layer 4. The right panels illustrate the receptive fields recorded in MT (dark gray) and surrounding areas (light gray and outline) in schematic representations of the contralateral hemifield (continuous lines indicate the vertical and horizontal meridians, dashed lines indicate selected isoeccentricity lines). At an anterioposterior levei close to the anterior end of MT (A), receptive fields in both quadrants are observed ventral to MT. At this level, cortex dorsal to MT shows large and diffuse receptive fields. More caudally (B), a pattern that resembles DLa is seen both dorsal and ventral to MT. Receptive fields 27–29, recorded lateral to the putative upper quadrant representation in DLa, are larger than those in either MT or DLa, and are based on less brisk cellular responses. Insert C (upper right) indicates the level of sections A and B relative to the borders of MT, in a surface view of the left hemisphere (rostral is to the left). The uncertainty regarding the border between DLa and FSTd is indicated by question marks.

al., 1994). In the marmoset, both presumed FSTd and DLa are small areas, in which the visual field is represented with the same field sign (mirror-image in the convention of Sereno *et al.*, 1994), and which have neurons with receptive fields of similar size. A correlation of topographic mapping with functional response properties or anatomical connections may be needed to answer this question. Another point that needs to be addressed is whether any of the areas found ventral to MT in New World monkeys corresponds to the macaque's MST1. Receptive fields of neurons in macaque MST1 are much larger than those of MT neurons (Komatsu and Wurtz, 1988), whereas in the marmoset there is a substantial overlap between the distribution of receptive field sizes of neurons in MT and DLa/FSTd. Even though the representation of the centroperipheral axis in cortex ventral to MT is similar in both New World monkeys and macaques, no systematic representation of upper and lower quadrants has been described in the macaque's MST1. Thus, more study will be needed to establish if the cortex ventral to MT in New World and Old World monkeys contains homologous areas, functional analogues, or different areas that just happen to be in the same general location.

3.4. The Dorsomedial Complex

By "dorsomedial complex" I refer to a group of visual areas located rostral to V2 and caudal to the posterior parietal areas, including cortex along the medial wall. In New World monkeys, this complex encompasses the dorsomedial visual area (DM) and surrounding areas, including DI (dorsointermediate) and M (medial; Allman and Kaas, 1975, 1976; see Fig. 1 for a summary of the location of these areas). In Old World monkeys, the dorsomedial complex comprises the dorsal part of visual area 3 (V3d) and other areas in the annectent gyrus and midline cortex, including visual area 3A (V3A) and the parietooccipital area (PO). The boundaries and visuotopic organizations of visual areas in this region have been particularly controversial, partly because these areas have been regarded as essentially different in New World and Old World monkeys (e.g., Van Essen and Zeki, 1978). Although there have been suggestions of homology between some areas in the dorsomedial complex, mainly based on the pattern of anatomical connections and myeloarchitecture (Wagor *et al.*, 1975; Lin *et al.*, 1982; Gattass *et al.*, 1985; Colby *et al.*, 1988; Sereno and Allman, 1991; Weller *et al.*, 1991; Krubitzer and Kaas, 1993; Rosa *et al.*, 1993b), specific hypotheses to explain the different descriptions of topographic organization among species have been lacking until recently.

Studies in the owl monkey described two topographically organized areas, DM and M, in the dorsal and medial extrastriate cortices rostral to V2 (Allman and Kaas, 1975, 1976; Krubitzer and Kaas, 1993). Both DM (which is located partially on the dorsal surface, and partially along the medial wall) and M (which is completely buried in the midline cortex) seem to contain complete representations of both the upper and the lower quadrants of the visual field. In addition, the visuotopic maps in both DM and M have been described as relatively simple second-order transformations, each including a field discontinuity corresponding to the representation of the horizontal meridian (Fig. 11). However, they

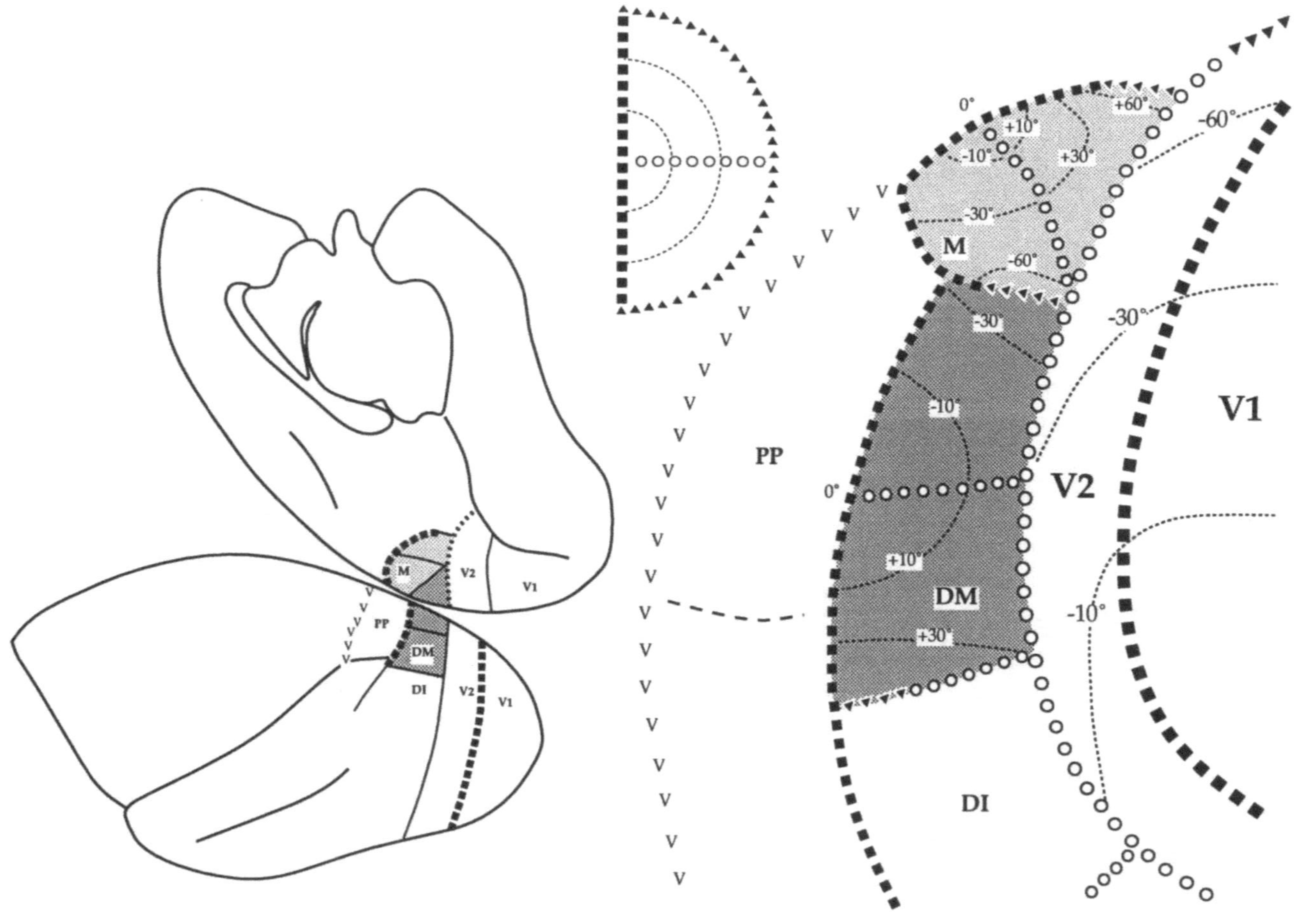

Figure 11. Visuotopy of dorsal and medial extrastriate cortices in the owl monkey. The diagram on the right illustrates a two-dimensional view of dorsal cortex in the owl monkey, based on Fig. 2 of Allman and Kaas (1975). The topographic organization of DM (dark gray) and M (light gray) is indicated according to the conventions shown in the inset. Numbers (0°, +10°, +30°, etc.) refer to isoeccentricity lines, with positive values indicating the upper quadrant representations and negative values the lower quadrant representations. The Vs indicate the rostral border of visually responsive cortex (PP, posterior parietal area). The diagram on the left shows the locations of DM and M in lateral and medial views of the owl monkey brain (based on Fig. 1 of Baker *et al.*, 1981; rostral to the left). The boundaries of V2 are also indicated.

differ in the relative emphasis on representation of central vision: whereas DM, like most visual areas, has an expanded central representation, M is unique in that no preferential magnification of sectors of the visual field seems to exist (Allman and Kaas, 1976). A third area, DI, located lateral to DM, was proposed for the owl monkey (Allman and Kaas, 1975; Krubitzer and Kaas, 1990), but its visual topography has never been fully described.

Results of work in the marmoset have revealed areas that correspond to those described in owl monkeys (Rosa and Schmid, 1995). For example, areas DM and M were identified, each of which represents the entire visual field and occupies approximately the same position relative to V2 as in the owl monkey. However, the visuotopic organization of the dorsomedial cortex in the marmoset also differs in several aspects from that observed in the owl monkey (Fig. 12). In the marmoset, DM is an elongated area, which forms a complex second-order representation of the visual field. As in the owl monkey, the lower quadrant representation is continuous, with central vision represented laterally, peripheral vision medially, the horizontal meridian caudally, and the vertical meridian rostrally. In contrast,

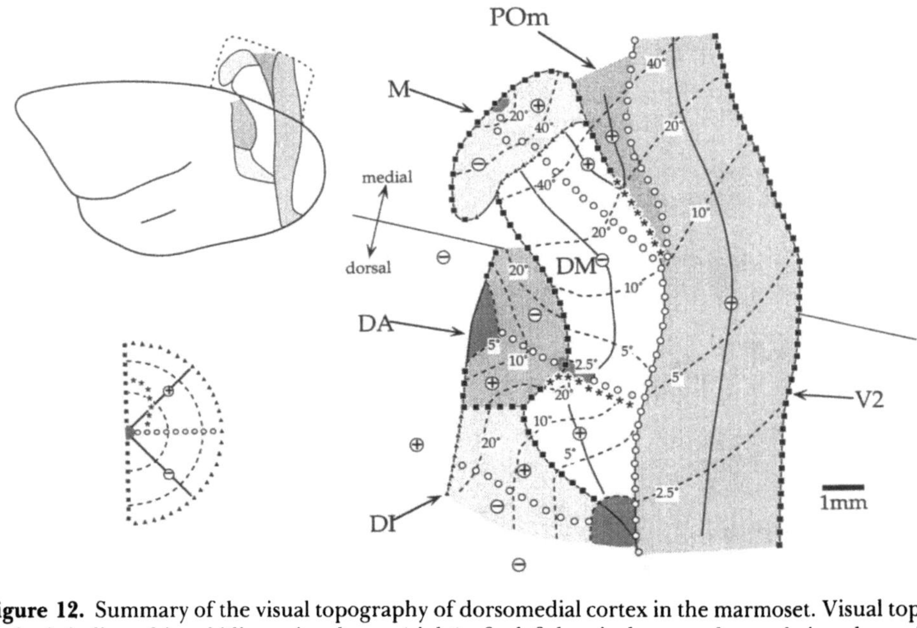

Figure 12. Summary of the visual topography of dorsomedial cortex in the marmoset. Visual topography is indicated in a bidimensional map (right) of a left hemisphere, each area being shown in a different tone of gray. The position of these areas on the brain is indicated in the top left inset, in which cortex located along the midline has been artistically "unfolded" and projected dorsally. The visuotopy is indicated according to the symbols illustrated in the bottom left inset. In the flat reconstruction, the transition between cortex located on the dorsal and medial surfaces is indicated. Foveal representations (<2° eccentricity) are indicated in dark gray, horizontal meridian representations are indicated by circles, vertical meridian by squares, visual periphery by triangles, isoeccentricity lines by dashed lines, and the 45° polar angle lines by continuous lines. A line across the upper quadrant, which defines a field and map discontinuity, is indicated by the asterisks. Modified from Fig. 17 of Rosa and Schmid (1995).

the upper quadrant representation is split, with the central portion represented at the lateral edge of DM on the dorsal surface, and the periphery along the midline. Thus, a field discontinuity runs obliquely through the upper quadrant representation, rather than along the representation of the horizontal meridian. More striking is the fact that, in the marmoset, DM also includes a map discontinuity. At some points of DM, as one crosses from the representation of the central upper quadrant into that of the central lower quadrant, there is a sudden shift of receptive field eccentricity (Fig. 13). This arrangement allows a congruent border between DM and V2 throughout, something that would not be possible if DM in the marmoset were organized like that in the owl monkey (Rosa and Schmid, 1995). The study of marmosets also revealed a relatively simple, but highly compressed representation of both quadrants of the visual field lateral to DM, which presumably corresponds to DI (Fig. 12).

In the macaque, the cortex rostral to V2 in the dorsal and medial extrastriate cortices was initially assigned to the dorsal half of the "third visual area" (V3; Zeki and Sandeman, 1976; Zeki, 1977, 1978a; Van Essen and Zeki, 1978; see Fig. 1C, D), which would also encompass an upper field representation, rostral to V2 in the ventral extrastriate cortex. Physiological and anatomical asymmetries between dorsal and ventral cortex forced a revision of this view, and "V3" (or "dorsal V3," V3d) was later redefined as an area representing only the lower visual quadrant (Van Essen *et al.*, 1982, 1986; Burkhalter *et al.*, 1986). In addition, electrophysiological mapping studies of cortex rostral to dorsal V2 in the macaque demonstrated that the visual topography of this region was incompatible with a simple, continuous representation of the lower quadrant as proposed by Zeki and collaborators (e.g., Zeki and Sandeman, 1976; Van Essen and Zeki, 1978). Instead, assuming that V3d formed a single representation at the lower quadrant, Gattass *et al.* (1988) proposed that the representation of the vertical meridian in this area was formed by multiple, spatially segregated islands (e.g., Fig. 17B). Physiological recording studies also revealed the existence of other areas, which represented both quadrants of the visual field, rostral and medial to V3. These were named V3A and PO, respectively (Van Essen and Zeki, 1978; Gattass *et al.*, 1985, 1988; Colby *et al.*, 1988; see also Fig. 17B).

More recently, it has been proposed that V3 is not a valid visual area. Instead, it has been suggested that an area homologous to DM in the macaque encompasses both a lower field representation originally assigned to "V3" and an upper quadrant representation originally assigned to "V3A" (Krubitzer and Kaas, 1993). In fact, in both the marmoset and the owl monkey, the lower field representation in DM is topographically similar to "V3d" of the macaque (Rosa and Schmid, 1995). In addition, it is probable that a split upper quadrant representation also exists in the macaque's DM, similar to that in the marmoset. This hypothesis is based on the observation that, in both the macaque and the capuchin (a New World monkey), "V3" has been portrayed as being bordered medially by area "PO" (Colby *et al.*, 1988; Rosa *et al.*, 1993b; Neuenschwander *et al.*, 1994). Recent studies of PO in these species (Neuenschwander *et al.*, 1994) concluded that in this area the lower quadrant is represented dorsally and the upper quadrant ventrally along the midline, and that increasing eccentricities are represented from caudal to rostral. This is exactly the pattern observed in the

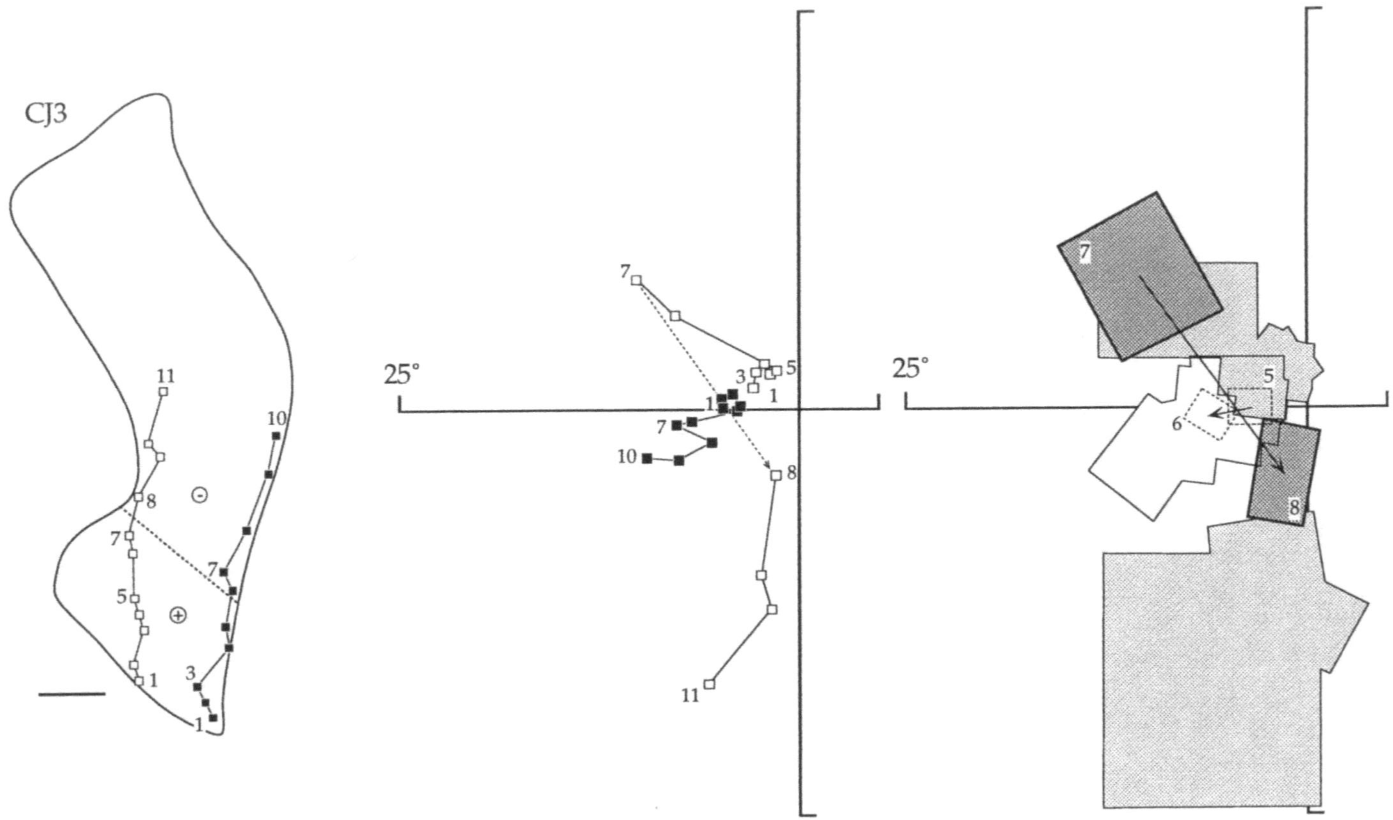

Figure 13. Topographic evidence for a map discontinuity in the representation of the visual field within the marmoset's area DM (modified from Fig. 6 of Rosa and Schmid, 1995). Left: Bidimensional reconstruction of DM in one animal, with selected sequences of recording sites crossing DM from lateral to medial indicated by filled squares (caudal sequence, parallel to the V2/DM border) and open squares (rostral sequence, across the topographic discontinuity within DM). A dotted line separates the central upper (+) and lower (−) field representations. Arrows indicate the anatomical orientation (R, rostral; M, medial), and the scale bar indicates 1 mm. Middle: Sequences of centers of receptive fields of neurons at the sites indicated in the left column. The dotted line with arrow indicates the transition between the central upper and lower field representations at the rostral border of DM. Right: Borders of the corresponding receptive fields. The region of the visual field encompassed by the receptive fields of neurons recorded in the caudal row of recording sites is indicated in white, and that corresponding to the rostral row in gray. In the rostral row, receptive fields recorded on either side of the discontinuity do not overlap (dark gray and arrow). Elsewhere in DM, receptive fields of neurons recorded at sites less than 1 mm apart always overlap. One example of a pair of receptive fields (5 and 6) recorded from the caudal part of DM, outside the region of map discontinuity, is indicated by the dotted outlines and arrows.

portion of the marmoset's DM which is buried along the midline cortex (Fig. 12). Thus, both "dorsal V3" and "PO" (together, perhaps, with parts of "V3A") are likely to form a single area, which is the homologue of DM. This hypothesis is further supported by the connections and myeloarchitecture of the dorsomedial areas (Krubitzer and Kaas, 1993; Rosa and Schmid, 1995). It may also explain why, in the original descriptions, the macaque's "dorsal V3" represented only the central 30° of the lower visual field, and "PO" did not contain the representation of the central part of the visual field (Gattass *et al.*, 1985, 1988; Colby *et al.*, 1988; Neuenschwander *et al.*, 1994). Finally, the visuotopy of a visual area located dorsal and rostral to PO, named visual area 6 (V6; Galletti *et al.*, 1990) or parietooccipital dorsal area (POd; Rosa *et al.*, 1993b; Neuenschwander *et al.*, 1994), has been shown to be similar in all respects to that of the owl monkey's area M, including a deemphasis on foveal representation (Rosa and Schmid, 1995).

In *Galago,* a topographically organized dorsal visual area representing both the upper and the lower visual quadrants was found between V2 and the posterior parietal cortex (Fig. 14). Although its anatomical location is equivalent to that of DM, in visuotopic terms it resembles the owl monkey's area M, because the upper quadrant representation is more medial than the lower quadrant representation (Allman *et al.*, 1979). However, hodological studies suggest that the dorsal area is homologous to DM (Krubitzer and Kaas, 1990, 1993). Given the differences in the visuotopic organization of DM between the owl monkey and marmoset, it is not inconceivable that the dorsal area is, indeed, homologous to DM, but that the relative position of the representation of the two quadrants in this area follows a third distinct pattern in *Galago* (Rosa *et al.*, 1997a).

The cortex located rostral to DM and M in owl monkeys has been described as non-topographically organized. It has been reported that neurons in this region have very large receptive fields and are poorly responsive in anesthetized preparations (ventral posterior parietal area, VPP, of Krubitzer and Kaas, 1993). However, Sereno and Allman (1991) have suggested that visuotopically organized cortex in the owl monkey may extend further rostrally. This has been supported by work in the marmoset (Rosa and Schmid, 1995), where at least one area rostral to DM (DA, or dorsoanterior area) has relatively small receptive fields and exhibits a topographic organization (Fig. 12). Likewise, in the macaque, one area located in the caudal parietal cortex shows evidence of visual topography (lateral intraparietal area, LIP; Blatt *et al.*, 1990); this may correspond to DA. In both New World and Old World monkeys, the representations of the upper and lower quadrants in this area are segregated medially and laterally, respectively, and the representation of central vision is rostral to that of peripheral vision. In contrast, electrophysiological recordings in the prosimian *Otolemur* indicate that the putative homologue of DM directly abuts a parietal visual area containing neurons with large receptive fields, encompassing one-quarter of the visual field or more, where no visuotopy is apparent (Rosa *et al.*, 1997a). This organization is also apparent in flying foxes; together, these observations suggest that, in early primates, DM was the only topographically organized area between V2 and the posterior parietal cortex.

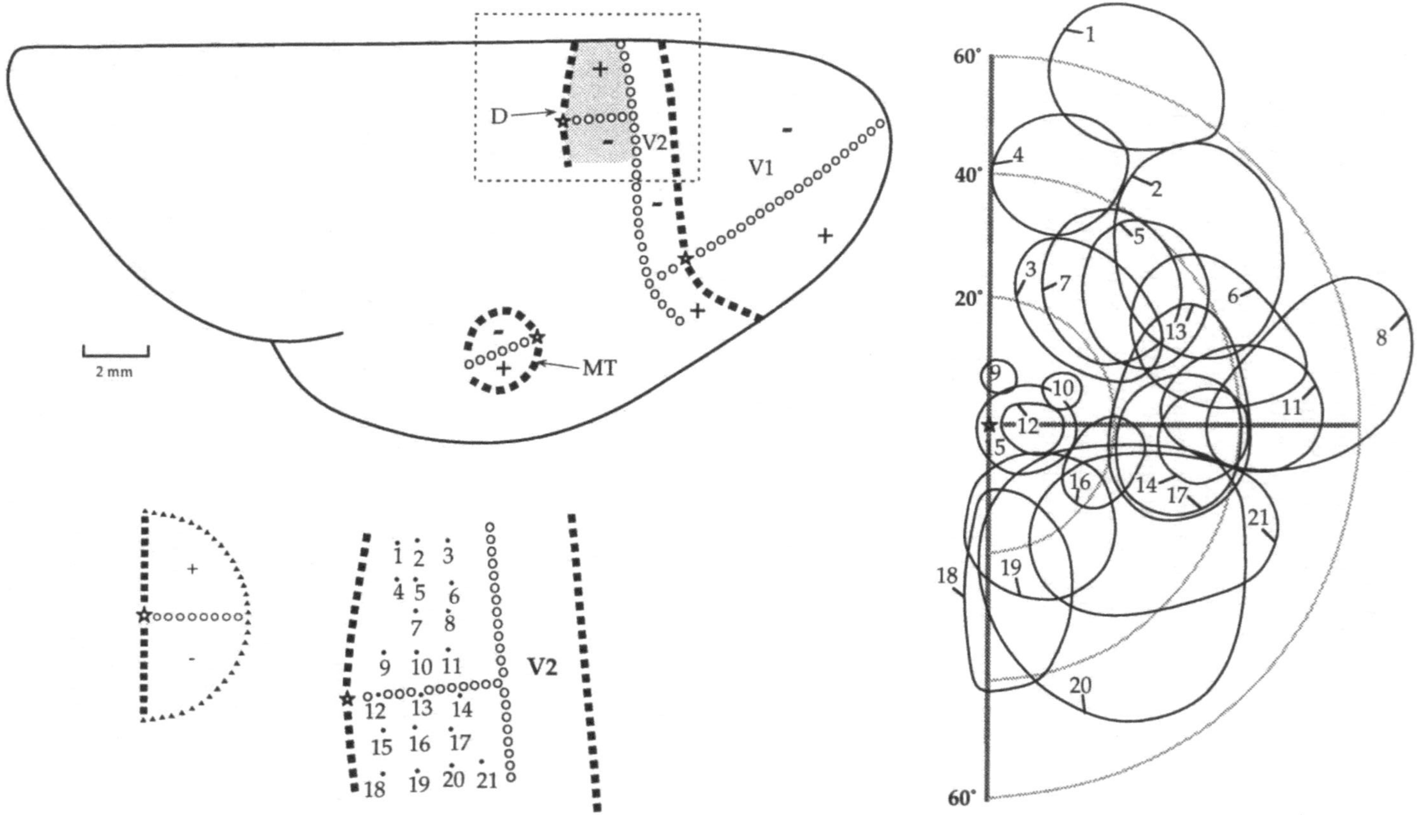

Figure 14. Visuotopic organization of the dorsal visual area (D) in the prosimian *Galago*. Upper left: Dorsal view of the left hemisphere of the brain (rostral to the left, medial upward) showing the location of area D (gray) relative to V1, V2, and MT. The representation of the visual field in these areas is indicated according to the inset (bottom left): stars indicate representations of the *area centralis*, circles representations of the horizontal meridian, squares representations of the vertical meridian, "+" signs representations of the upper quadrant, and "−" signs representations of the lower quadrant. The dorsal cortex within the dotted box is shown magnified at the bottom, where recording sites within area D are indicated (dots). The corresponding receptive fields (1–21) are illustrated in the right diagram. Modified from Fig. 1 of Allman *et al.* (1979).

For convenience of presentation, I will review in this section the evidence regarding the topographic organization of visual areas located in the lateral cortex, between the representation of central vision in V2 and area MT, and areas rostral to V2 in ventral extrastriate cortex (Fig. 1). According to previous investigations, these include the ventral anterior and ventral posterior areas (VA and VP), subdivisions of the owl monkey's dorsolateral area (DL), and subdivisions of the macaque's fourth visual area (V4). As initially defined, some of these areas formed incomplete representations of the contralateral visual field. In fact, as I will argue below, the representations of the upper quadrant in the ventral extrastriate cortex can be reasonably seen as continuations of the representations of the lower quadrant in lateral extrastriate cortex, and it is likely that complete representations also exist here.

The initial work on the visuotopy of lateral and ventral cortex in New World monkeys suggested the existence of a single area, named DL, occupying the cortex between V2 and MT (Allman and Kaas, 1974b), and two areas in the ventral cortex, named VP and VA (ventral posterior and ventral anterior areas; Newsome and Allman, 1980; Fig. 15A). In this initial model, DL was described as a "crescent-shaped" area containing a second-order map of the entire visual field with strong emphasis on foveal representation, whereas VP and VA were described as elongated parallel strips representing predominantly, if not exclusively, the upper quadrant. As discussed above (Section 3.3), later work indicated that different areas bordered MT dorsally and ventrally, therefore refuting the notion of DL as an area which completely surrounded MT. In addition, further studies of visuotopic organization of the lateral cortex (Sereno and Allman, 1991; Sereno *et al.*, 1987, 1994), based on finer sampling, demonstrated the existence of three parallel representations of the lower quadrant, where the single lower quadrant representation of DL was expected to exist. These have been named the dorsolateral posterior (DLp), dorsolateral intermediate (DLi), and dorsolateral anterior (DLa) areas,* which form a series of approximately mirror-symmetrical maps of the lower visual field (Fig. 15B; Sereno *et al.*, 1987). The demonstration of multiple representations in DL, and specifically of a representation of the vertical meridian which forms the border between DLp and DLi, solved a long-standing puzzle regarding the interhemispheric connections of extrastriate cortex (Fig. 16, upper right): while the interhemispheric connections of the owl monkey's areas V1, V2, DM, MT, and VP were, as expected, concentrated along the representation of the vertical meridian (Newsome and Allman, 1980; Cusick *et al.*, 1984), those of DL, as originally described by Allman and Kaas (1974b), did not seem to bear any special relationship to the representation of this meridian.

The question that remained, however, was whether the subdivisions of DL were restricted to the representation of the lower visual field. The study of

*Other nomenclature, which emerged from studies of anatomical connections, labels the areas which are probable homologues of DLp and DLi the caudal and rostral subdivisions of the dorsolateral area (DLc and DLr), and the probable homologue of DLa the middle temporal crescent (MTc) (Cusick and Kaas, 1988; Steele *et al.*, 1991; Kaas and Morel, 1993).

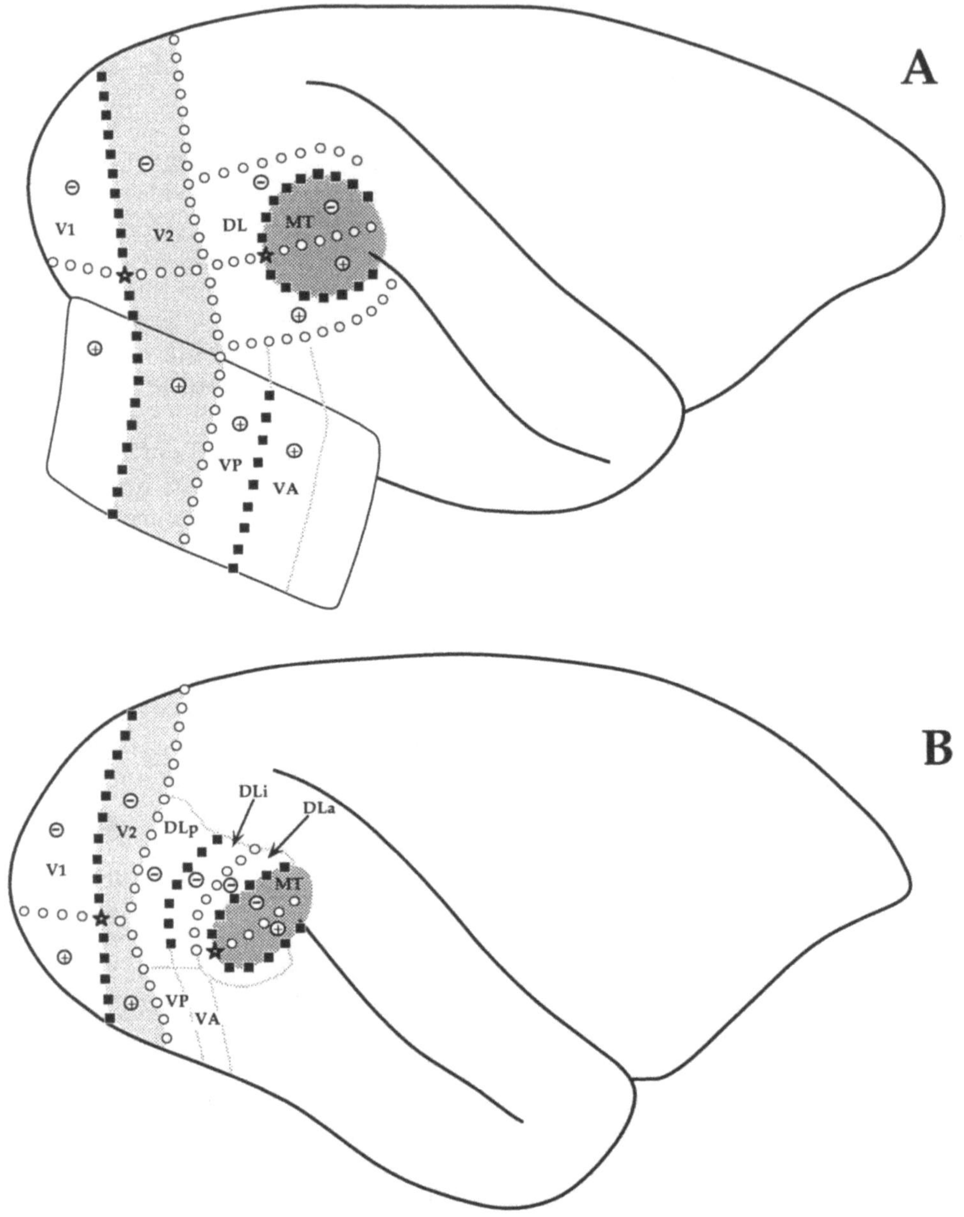

Figure 15. Changing concepts of organization of lateral and ventral extrastriate cortex in New World monkeys. (A–C) Three different views of the visuotopy of extrastriate cortex of the right hemisphere (A, B, owl monkey; C, marmoset monkey; rostral to the right). In all diagrams, V2 is shown in light gray, and MT in dark gray. In both A and C, ventral extrastriate cortex has been "unfolded" to allow the illustration of its visuotopic organization. The visuotopy is indicated according to the inset (bottom right): representations of the center of the fovea are indicated by a star, vertical meridian by squares, horizontal meridian by circles, isoeccentricity lines by dashed lines, and upper and lower quadrants by "+" and "−", respectively. (A) The Allman–Kaas model of organization of owl monkey cortex (based on Figs. 8 and 9 of Newsome and Allman, 1980). DL was defined as a "crescent-shaped" visual area interposed between V2 and MT (Allman and Kaas, 1974b). Boundaries and visual topography of ventral areas (VA and VP) are later modifications proposed by

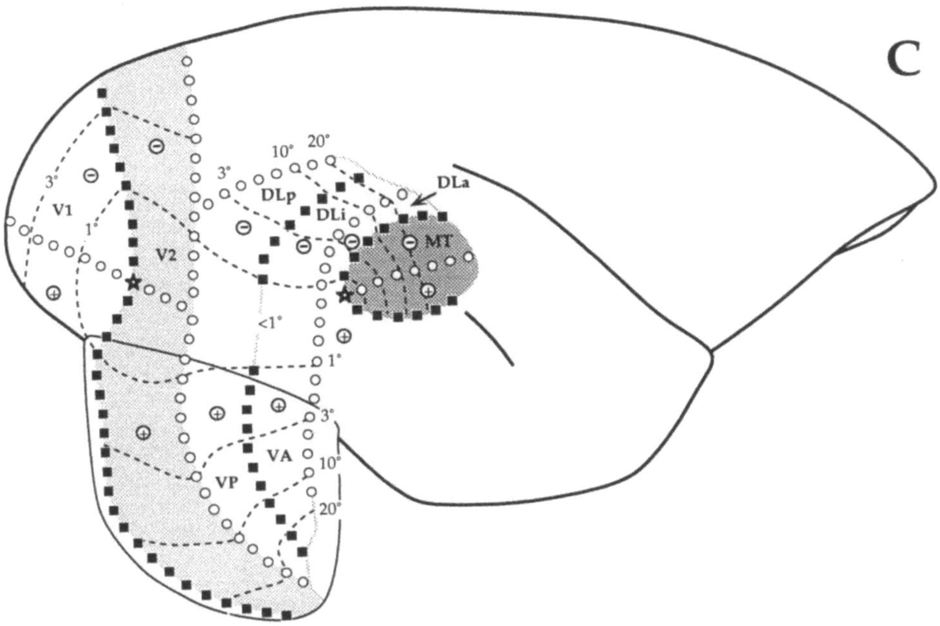

Figure 15. (*Continued*) Newsome and Allman (1980). (B) The demonstration, by Sereno and collaborators, of three subdivisions within DL in the owl monkey. The boundaries of DLp, DLi, DLa, MT, V2, VA, and VP are based on Fig. 7.3 of Sereno and Allman (1990); the visuotopic organization of DLp, DLi, and DLa is based on Sereno *et al.* (1987, 1994). The representation of the vertical meridian that separates DLp from DLi corresponds precisely to a band of callosal connections (region 5 in Fig. 16, upper right). (C) Model of the organization of lateral and ventral cortex in the marmoset, based on experiments by M. G. P. Rosa, G. N. Elston, and K. Fritsches (in preparation). The existence of three areas separating V2 from MT was confirmed in this species. DLp and VP together form one complete representation of the visual field (caudally), and DLi and VA together form another complete representation (rostrally).

Allman and Kaas (1974b) suggested that at least some portions of the lateral cortex immediately ventral to DLp and DLi represented the central portion of the upper visual field. Furthermore, anatomical studies support the idea that the lower field representations in DLp and DLi continue ventrally into upper quadrant representations (e.g., Tigges *et al.*, 1974; Cusick and Kaas, 1988; Sousa *et al.*, 1991; Steele *et al.*, 1991; Weller and Steele, 1992; Rosa *et al.*, 1993b). However, a preliminary report by Sereno *et al.* (1987) rejected this hypothesis, reporting instead the existence of "multiple, small representations of parts of both upper and lower fields" ventral to the lower field representations in DLp and DLi. More recently, we have been recording in the lateral and ventral portions of marmoset extrastriate cortex (Rosa *et al.*, 1997b). Our data are consistent with the notion of three parallel representations of the lower quadrant, located between the lower quadrant representations in V2 and MT. As in the studies of DL in the owl monkey, we also found that the representations in DLp and DLi are relatively irregular (Sereno *et al.*, 1987). However, unlike the results in the owl monkey, we found no evidence of multiplication of representations of the two quadrants as the recording sites moved into the ventrolateral and ventral extra-

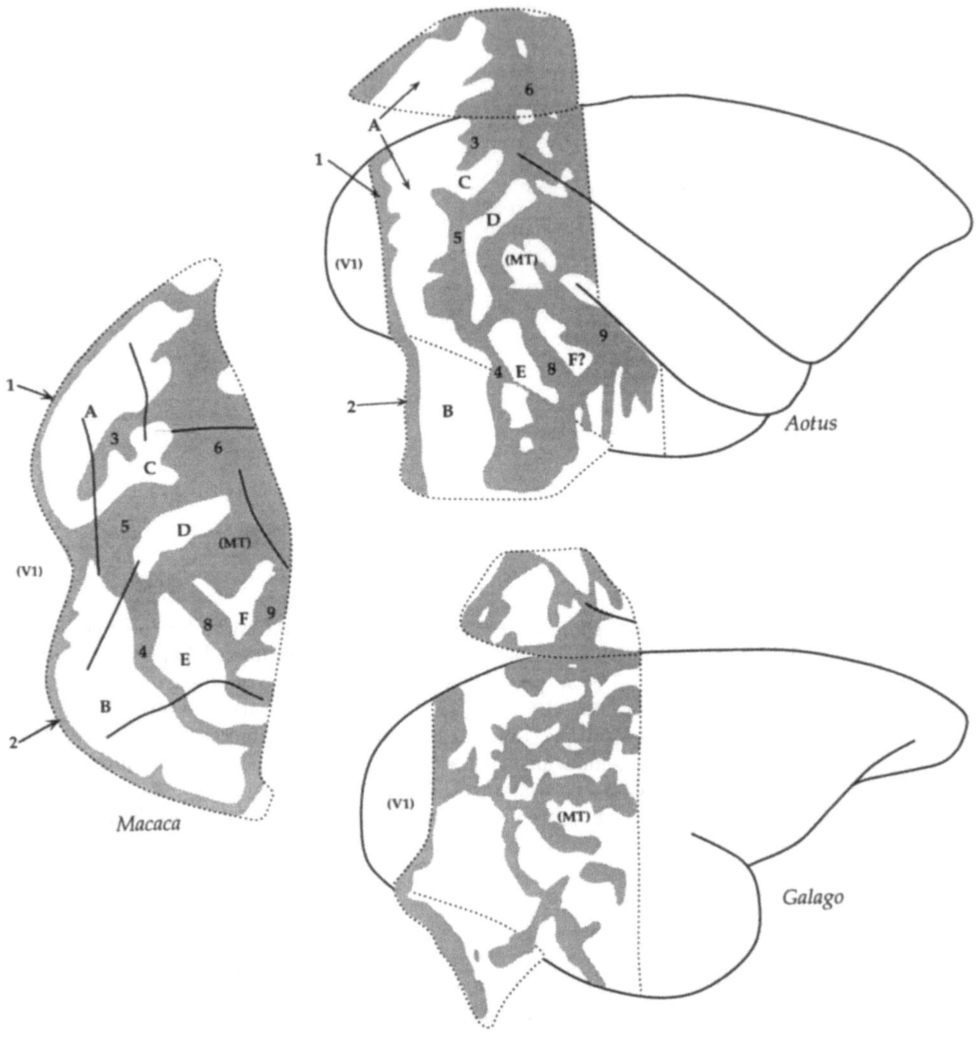

Figure 16. Patterns of interhemispheric connections of caudal neocortex in three primates, as revealed by anterograde degeneration techniques. Regions showing degenerating terminals after sectioning of the corpus callosum are indicated in gray, regions free of degenerating terminals in white. Left: "Unfolded" view of the pattern of interhemispheric connections of caudal neocortex in the macaque (redrawn from Fig. 3 of Van Essen *et al.*, 1982). The fundi of sulci are shown as continuous lines. The callosal-dense regions were designated by numbers (1–9) and the callosal-free regions by letters (A–F), according to Van Essen *et al.* (1982). Upper right: A comparable view of the owl monkey cortex, based on Figs. 8 and 9 of Newsome and Allman (1980). Dorsal and ventral "flaps" of cortex, corresponding to the cortex buried along the midline and on the ventral surface, respectively, were graphically added to a lateral view of the brain to produce an overall representation of the callosal connectivity of visual areas in this species. Note that, due to the effect of perspective, this technique of representation is subject to more distortion than that produced by the full bidimensional reconstruction. Letters and numbers similar to those used to designate the regions of callosal connection in the macaque (left) were placed to guide the comparison between the patterns in macaque and owl monkey. Lower right: Pattern of callosal connections in the prosimian *Galago,* based on Figs. 12 and 13 of Newsome and Allman (1980). The position of area MT, which is architectionally distinct in all three species, is also indicated.

striate cortices. Instead, as shown in Fig. 15C, our observations are compatible with a simpler model in which the lower quadrant representation in DLp continues ventrally, without rerepresentation, into the upper quadrant representation of VP, and DLi continues into VA. Moreover, receptive fields recorded at the representations of corresponding eccentricities in the lower (DLp, DLi) and upper (VP, VA) quadrants were similar in size. Purely from the point of view of visual topography, these results imply that VP and DLp, on one hand, and VA and DLi, on the other hand, may be parts of elongated areas that represent both quadrants of the visual field (see also Rosa *et al.*, 1993b, for anatomical data that support this view). Thus, the current evidence suggests that all three subdivisions of the ventrolateral cortex form second-order transformations, with field discontinuities. The enormous expansion of the representation of the fovea in these areas in the marmoset increases the difficulty of determining the borders at the lateral edge of the occipital cortex, which corresponds to the transition between upper and lower quadrants; here small variations of receptive field polar angle may be easier to demonstrate in the nocturnal owl monkey. Still, we have not yet observed any clear duplications of sectors of the visual field in this region in the marmoset, other than those expected from the normal receptive field scatter and point-image size. In support of a model whereby the dorsolateral and ventral cortices include parts of the same areas, the modular organization of cortex revealed by cytochrome oxidase histochemistry in New World monkeys also shows similarities between these regions. Although the patterns in these regions vary more between individuals than those in areas V1 and V2, in favorable preparations both the caudal parts of DL (Cusick and Kaas, 1988; Gattass *et al.*, 1990) and VA/VP (Tootell *et al.*, 1985) reveal thick "slabs" of high cytochrome oxidase reactivity separated by thin bridges of less reactive cortex.

In the macaque (Fig. 17A) and other Old World simians, there is also evidence for multiple areas between V2 and MT (e.g., Zeki, 1971; Gattass *et al.*, 1985; Sereno *et al.*, 1995). One of these (V4t, described in Section 3.3) is likely to correspond to New World monkeys' DLa. In addition, the organization of ventral cortex in the macaque is very similar to that in New World monkeys. A representation of the upper quadrant anterior to V2 has been identified as the homologue of VP (Burkhalter *et al.*, 1986; Newsome *et al.*, 1986; Gattass *et al.*, 1988),* and a visuotopic organization similar to that observed in New World monkeys' VA has been described in the ventral part of the V4 complex (V4v; Gattass *et al.*, 1985, 1988; see Sereno *et al.*, 1995, for evidence of VA and VP in humans). The main difference seems to be that much of the macaque dorsolateral cortex, between V2 and V4t, has been regarded as part of a single lower quadrant representation, corresponding to the dorsal half of area V4 (Gattass *et al.*, 1985; Van Essen, 1985). Yet a more complex pattern was observed by Van Essen and Zeki (1978), and Maguirre and Baizer (1984) described two parallel representations of the lower quadrant across the anterior bank of the lunate sulcus, prelunate gyrus, and posterior bank of the superior temporal sulcus, which they named the posteromedial (PM) and anterolateral (AL) areas.† These

*In macaques, VP is also known as ventral V3, or V3v.
†There were earlier reports of anatomical (Zeki, 1971) and physiological (Zeki, 1983c) differences between the rostral and caudal parts of the V4 complex.

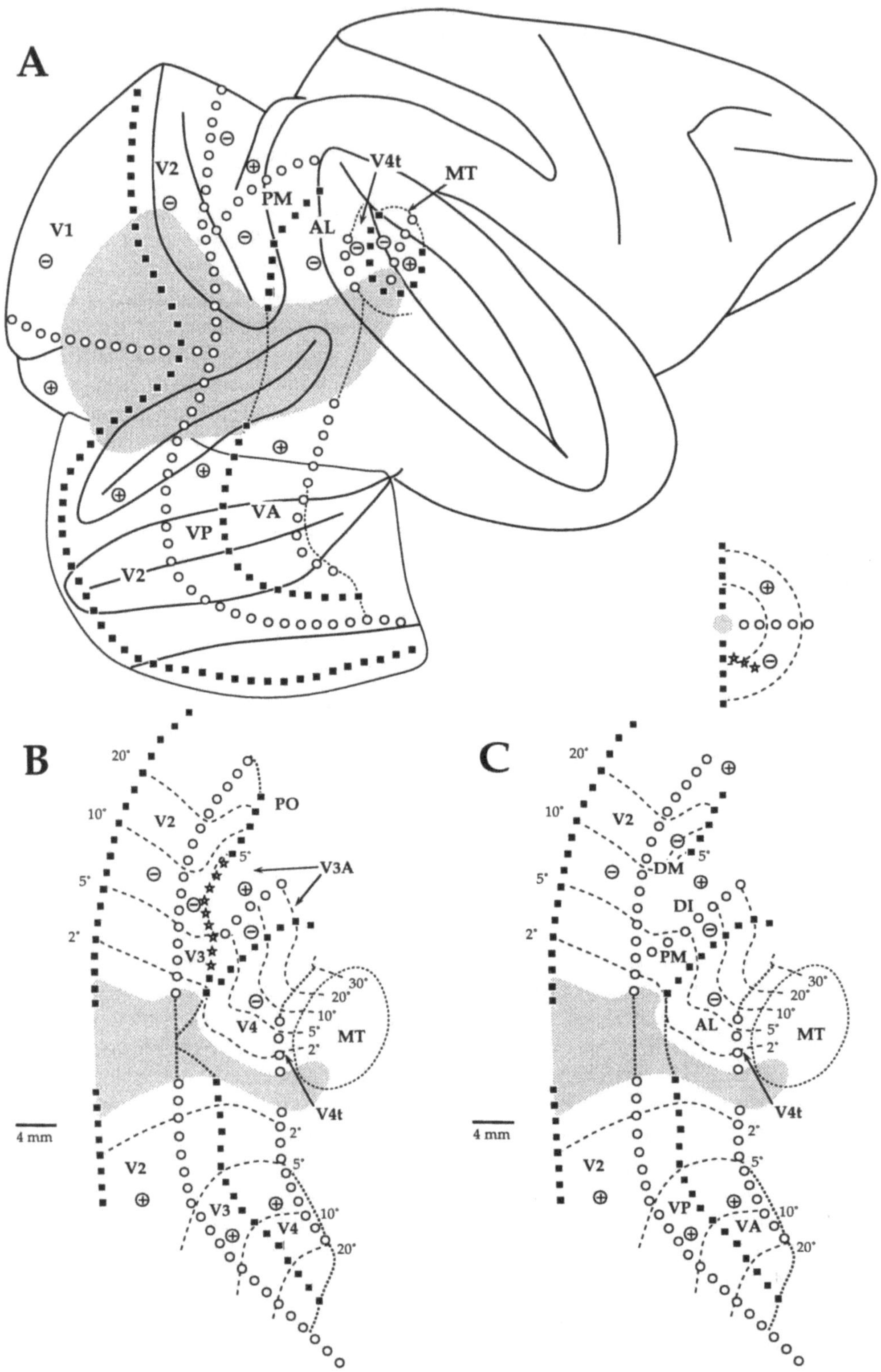
A
V1
V2
PM
AL
V4t
MT
VA
VP
V2
B
20°
10°
5°
2°
V2
PO
V3A
5°
V3
V4
MT
30°
20°
10°
5°
2°
V4t
4 mm
V2
2°
5°
V3
V4
10°
20°
C
20°
10°
5°
2°
V2
DM
5°
DI
PM
AL
MT
30°
20°
10°
5°
2°
V4t
4 mm
V2
2°
5°
VP
VA
10°
20°

two representations are separated by a representation of the vertical meridian which runs obliquely from the lunate to the superior temporal sulcus, a pattern that is strongly reminiscent of New World monkeys' DLp and DLi. The data of Gattass *et al.* (1988; see Fig. 17B) can be also interpreted in this way (Fig. 17C). Thus, although more study is clearly needed, it is possible that, in the macaque, area PM of Maguirre and Baizer (1984) forms the lower quadrant representation of VP, and that AL forms the lower quadrant representation of VA (Fig. 17A). This hypothesis is supported by the fact that "V4 and VP are very similar in their physiological response properties for all parameters that have been extensively tested in both areas" (Burkhalter and Van Essen, 1986). Moreover, the pattern of callosal connections of the dorsolateral and ventral cortices of the macaque is strikingly similar to that of the owl monkey (Fig. 16, left), suggesting that the representations of the vertical meridian are similarly located in New World and Old World monkeys.

The available information on the visuotopy of lateral and ventral visual cortices of prosimians is much more limited. In *Cheirogaleus,* a small lemur, Paolini *et al.* (1994) reported that a single visual area exists in the cortex between V2 and MT. In contrast, in *Galago* the pattern of interhemispheric connections suggests a much more complex topography of lateral cortex, with at least one additional representation of the vertical meridian interposed between the rostral border of V2 and the caudal border of MT (Fig. 16, bottom right; Newsome and Allman, 1980). There has been a brief report indicating that galagos may have three visuotopically organized areas between V2 and MT (Sereno and Allman, 1991). The ventral areas of prosimians, including VP, seem to be similar to those

Figure 17. Model of visuotopic organization of lateral and ventral cortices in the macaque monkey. (A) Lateral view of the right hemisphere of a macaque brain in which the sulci and the ventral surface were partially "unfolded" to allow visualization of the cortical ares. The visuotopic organization is indicated according to the insert (right): the foveal representation (<2°) is shaded gray, vertical meridian representations are indicated by squares, horizontal meridian representations by circles, upper quadrant representations by "+," and lower quadrant representations by "−." The visuotopic organization of V1 and V2 is based on Gattass *et al.* (1981), and the organization of MT and V4t is based on Gattass *et al.* (1988). The visuotopies of the ventral areas (VP and VA) are also based on Gattass *et al.* (1988); note, however, that these authors designated these areas as V3v and V4v in their paper. The areas of the anterior bank of the lunate sulcus and prelunate gyrus (AL and PM) are based on Maguirre and Baizer (1984). The key points of this illustration are that PM and VP are likely to form together a complete representation of the visual field, and that AL and VA together form another representation. This being the case, the organization of lateral and ventral areas in the macaque strongly resembles that observed in New World monkeys (Fig. 15C). (B) Original interpretation of boundaries of visual areas in the macaque dorsal cortex, by Gattass *et al.* (1988). This diagram represents a bidimensional reconstruction of the macaque caudal extrastriate cortex, with the visual topography of V2, V3, V3A, V4, and V4t indicated (gray area represents the central 1° of the visual field, dashed lines are isoeccentricity lines, and stars represent an oblique line crossing the lower quadrant; see insert for other symbols). The fine dotted lines indicate boundaries of areas which were interpolated based on myeloarchitectural evidence. Redrawn from Fig. 5 of Gattass *et al.* (1988), with the exception of the organization of V3A, which was based on Figs. 3, 8, 11, and 13 of the same reference. (C) A reinterpretation of the same data, based on the studies of New World monkeys and on the studies of Maguirre and Baizer (1984) in the macaque. Area V4 corresponds well to AL and VA together; however, the current evidence is that a striplike "V3" bordering V2 anteriorly is not a valid visual area. It is likely that a lower quadrant representation previously assigned to "V3A" corresponds to Maguirre and Baizer's area PM, forming the continuation of VP into the anterior bank of the lunate sulcus and prelunate gyrus.

of monkeys (Allman and McGuiness, 1983), a conclusion that is also supported by the pattern of callosal connections (Fig. 16).

3.6. The Occipitotemporal Transition (TEO)

In the macaque, there have been several reports of visuotopy in cortex located between the ventral anterior upper-quadrant representation (VA) and the non-topographically organized IT cortex. The current evidence points toward the existence of two areas in this region, with only the most lateral of these being visuotopically organized.

Representations of both quadrants were found in the area that occupies the lateral and ventrolateral aspects of the occipitotemporal transition. This area was named TEO (Gattass *et al.*, 1988; Boussaoud *et al.*, 1991) because it roughly corresponded to the cytoarchitectonic area TEO of Von Bonin and Bailey (1947). The most complete data on this region come from the study of Boussaoud *et al.* (1991), who correlated the topographic organization of TEO with its myeloarchitecture and pattern of connections (Fig. 18). Complete representations of both visual quadrants were observed, although the magnification factor of the upper quadrant representation was found to be greater than that of the lower quadrant representation. In order to explain this latter observation, the authors suggested that part of the upper quadrant representation they observed may in fact belong to V4t (see Section 3.3), but no clear topographic border was observed. The visuotopic map of TEO is hard to classify in terms of the types of representation shown in Fig. 4. The receptive fields in this area are often large enough to encompass portions of both the vertical and the horizontal meridians. In addition, although some large shifts of receptive field center position are observed at the transition between the representations of the upper and lower quadrant (Fig. 18), these are masked by the large receptive fields. Thus, even though the evidence is suggestive, it is still impossible to ascertain whether there are any real field or map discontinuities in TEO. At least provisionally, the visual map in this area may be regarded as a gradient: eccentricities grow from lateral to medial on the ventrolateral surface, with the upper quadrants represented caudally and the lower quadrants rostrally. Other investigators have suggested that the caudal part of TEO should be considered an autonomous visual area, VOT (ventral occipitotemporal area; Van Essen *et al.*, 1990; Felleman and Van Essen, 1991), which represents only the upper quadrant. However, there have been very few electrophysiological data published to back up this hypothesis.

The area located medial to TEO, adjacent to the rostral tip of the calcarine fissure, is less understood. In a brief report, Gattass *et al.* (1985) found receptive fields representing the lower quadrant in the ventromedial aspect of the caudal temporal lobe, in a region which they named VF (by analogy with cytoarchitectonic area TF, which encompassed the recording sites). Later studies, by Boussaoud *et al.* (1991), subdivided TF into visually responsive (VTF) and non-responsive sectors. Although some of the receptive fields recorded in this region by both studies were relatively small, no evidence of visual topography was found. This may reflect in part the difficulties in sampling associated with the location and the small size of this area.

Evidence of visuotopy in the occipitotemporal transition of New World monkeys comes mainly from anatomical tracer studies (Sousa *et al.*, 1991; Weller and Steele, 1992; Rosa *et al.*, 1993b). In *Cebus* monkeys, the occipitotemporal transition has been subdivided into two areas on the basis of anatomical connections: a medial area (temporal ventral posterior area, TVP) that projects to both V1 and MT, and a lateral area (ITc or TEO) that does not (Sousa *et al.*, 1991; Rosa *et al.*, 1993b). The connections between TVP (which may correspond to VTF in the macaque), V1, and MT follow a centroperipheral gradient of eccentricities, from lateral to medial. In addition, a segregation between upper and lower quadrant representations has been observed in ITc of squirrel monkeys (Weller and Steele, 1992).

Finally, there has been a brief report suggesting that, in prosimians, the caudal portions of IT cortex (which may correspond to TEO) present a clearer visuotopic organization than that observed in monkeys (Paolini *et al.*, 1994).

3.7. Summary

The above account of visuotopy of extrastriate areas highlights several important points that deserve more intensive study. Nonetheless, even at this stage some conclusions can be drawn. For example:

(a) The evidence, although fragmentary in several aspects, suggests a highly homologous organization of extrastriate areas in New World and Old World monkeys. In spite of the different names that have been given, and some variation in the exact placement of boundaries in different species, the published evidence demonstrates that the visuotopic organization of New World and Old World monkey cortices is more similar than previously thought. This is reflected, for example, in the similar pattern of callosal connections (Fig. 16).

(b) The case for areas with incomplete representations of the visual field is not strong. In each example where an area was originally defined as representing only part of the visual field, there is evidence (either from later work in the same species or from comparative studies) that the same data can be interpreted in alternative ways which preserve the completeness of the visuotopic map. One consequence of this revision is that the actual number of visual areas in primate cortex is probably significantly smaller than the 32 presently recognized in some models (e.g., Felleman and Van Essen, 1991). On the other hand, more nonvisuotopic areas may be discovered as a result of further study and subdivision of the inferior temporal and parietal cortices.

(c) There is a large variety of transformations among the visuotopic maps of extrastriate cortex. Simple maps of the visual field, either in the form of first-order or second-order transformations with field discontinuities corresponding to the horizontal meridian, as described by early studies (Allman and Kaas, 1974a,b, 1975, 1976; Van Essen and Zeki, 1978; Gattass *et al.*, 1981), may be the exception rather than the rule. For example, there may be map discontinuities in DM, and additional field discontinuities that do not correspond to the horizontal

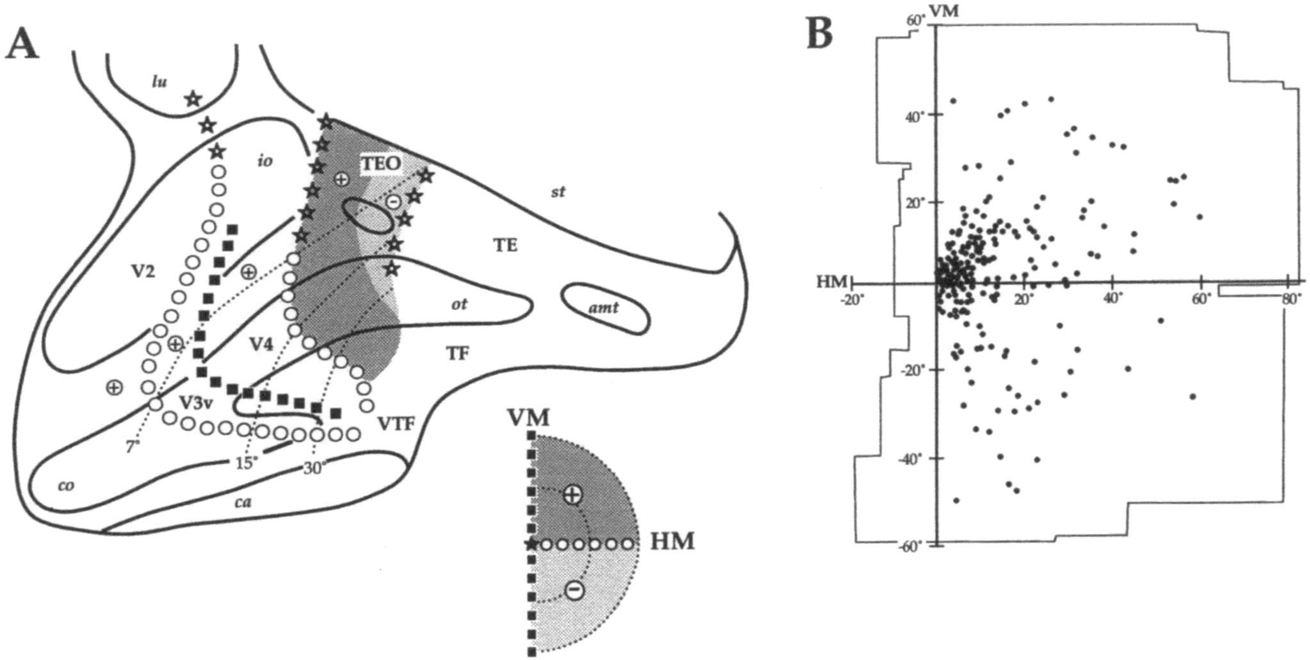

Figure 18. Organization of caudal inferotemporal cortex, area TEO, in the macaque monkey (redrawn from Figs. 7, 8, 12, and 15 of Boussaoud *et al.* (1991). (A) Summary map of the visuotopic organization of TEO, shown in a bidimensional reconstruction of ventral extrastriate cortex of a right hemisphere (rostral to the right, lateral to the top). The symbols used to represent visual field coordinates are according to the inset (stars indicate the foveal representations, squares indicate vertical meridian representations, circles indicate horizontal meridian representations, dotted lines indicate isoeccentricity lines). The lips of sulci are indicated in continuous lines (ca, calcarine; co, collateral; io, inferior occipital; lu, lunate; ot, occipitotemporal; st, superior temporal). The positions of cytoarchitectonic areas TE and TF are also indicated. (B) Evidence that TEO represents the entire contralateral hemifield. Each dot represents the center of a receptive field, and the thin line indicates the outer boundaries of all receptive fields of a neuron recorded from TEO cortex. (C, D) Sequences of recording sites (shown in parasagittal sections, top) and corresponding receptive fields of neurons (bottom) crossing TEO from caudal to rostral. The representation within the upper quadrant (white circles) and lower quadrant (black circles) sectors of TEO each seem to be coherent, in the sense that adjacent neurons tend to represent adjacent portions of the visual field. However, sudden topographic transitions are observed as the horizontal meridian is crossed from the upper to the lower quadrant.

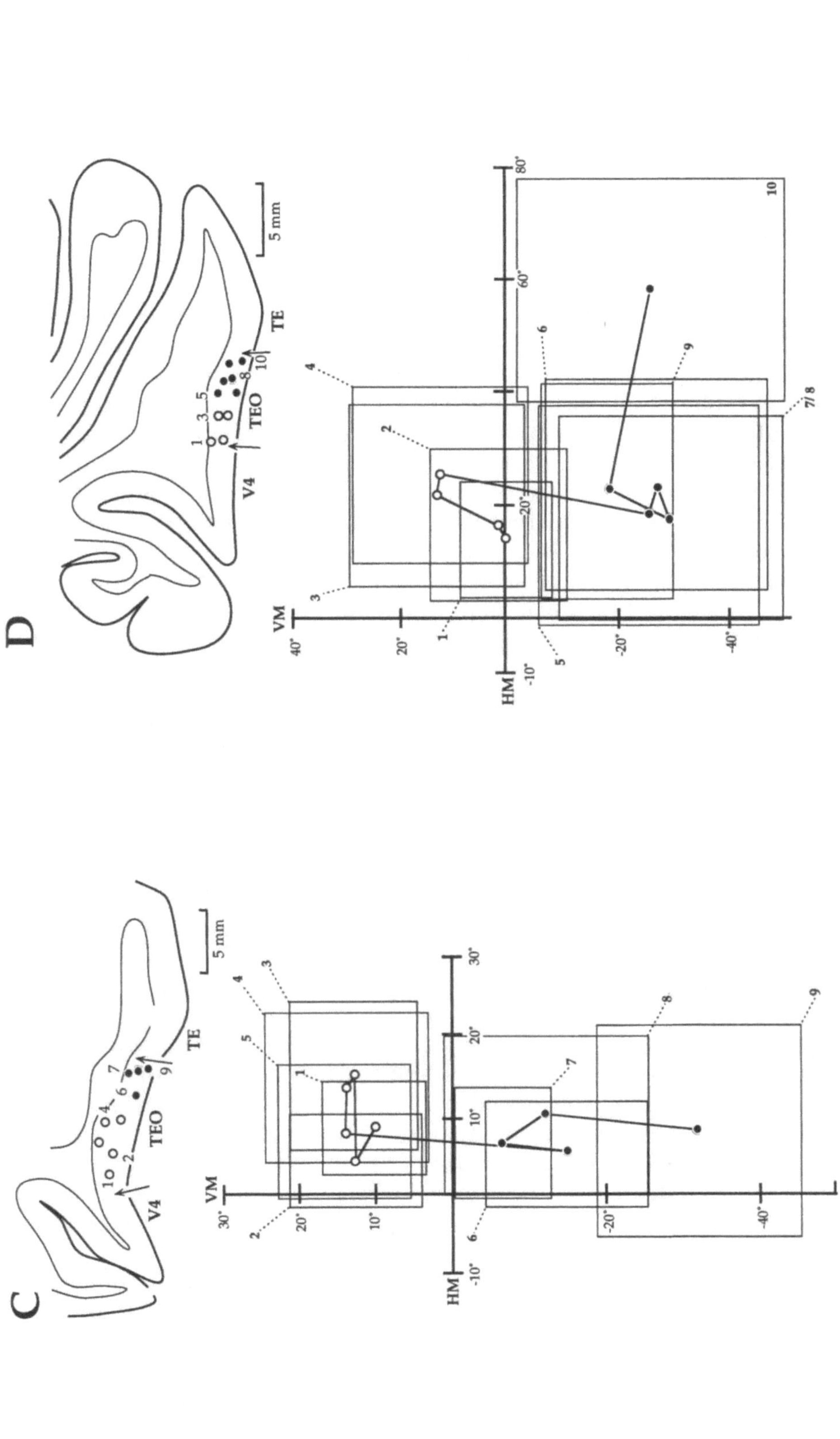

Figure 18. *(Continued).*

meridian in the dorsolateral areas. In some species, even in V2 the fine organization may be more complex than initially implied, by virtue of a repetitive or mosaic organization. In caudal parietal and temporal areas, the large receptive fields result in maps that may be better described as visuotopic gradients.

4. Why Have Different Visuotopic Maps?

This is a basic question that comes to mind when one is confronted with the complexity of the topographic organization of extrastriate cortex. Nonetheless, I find that studying the characteristics of the visuotopic maps is much easier than finding good solutions to such a fundamental problem. There have been many papers addressing the question of why multiple visual areas exist at all (e.g., Cowey, 1981; Edelman, 1981; Barlow, 1986; Kaas, 1989), and the intention here is not to review the different theories in detail. However, it is interesting to observe the extent to which current data on visuotopy of extrastriate areas support the various hypotheses. In general, explanations for the existence of visuotopic maps in mature animals can be grouped in two general categories, which may be referred to as "functional" and "developmental." The former emphasize the aspect of functional significance, that is, that a particular arrangement of the neurons representing different portions of the visual field in an area confers a specific advantage to the processing of visual information. Hypotheses grouped in the latter category see the existence and variety of visuotopic maps through much of visual cortex as epiphenomena, or natural consequences of the rules for establishment of connections during development. The dichotomy is not absolute, and "functional" and "developmental" hypotheses intersect at many points. For example, alternative genomes coding hypothetical developmental rules which result in a less functional visual cortex would be disfavored by natural selection, and therefore there is ultimately a "functional" aspect to any visuotopy. Yet, the hypotheses grouped under the "functional" and "developmental" categories differ in their predictions of the extent to which visuotopic maps can be modified during development, and in their implications for understanding visual function. A purely "developmental" approach implies that the visual topography of a given area is susceptible to a much larger degree of variation than does a purely "functional" context; for example, as a consequence of experimental manipulations during the period of formation of connections. Moreover, "developmental" hypotheses suggest that the function of a given area could still be performed in spite of a large amount of variability (natural or otherwise) of visuotopic maps, whereas in "functional" terms one would expect that phenotypes which are distant from the ideal would result in less-than-optimal visual performance.

"Developmental" hypotheses based on modulation of synaptic connections by correlated activity were presented, among others, by Kaas (1989), who argued that cortical visuotopic maps develop as a consequence of the action of Hebbian mechanisms during development. It was argued that selection for correlated activity would tend to create and preserve visuotopic maps, because both during development and later, in adult life, adjacency in retinal (and visual) space is a powerful determinant of correlated activity. Thus, a reinforcement of synaptic interactions is likely to occur between neurons which map nearby points of visual space. Since many synaptic interactions are local, short-range projections, a visuotopic map could gradually emerge even if only random connectivity or a coarse map existed initially (Schmidt, 1985). The principle of experience-dependent synaptic modification also lies behind explanations for map formation (Edelman, 1981) and plasticity (Pearson *et al.*, 1987) based on the neuronal group selection theory. Along these lines, Maunsell and Van Essen (1987) have suggested that visually guided movements of the hand, which occur mainly in the lower visual field, are correlated with a specific bias toward a larger representation of the lower quadrant in comparison with the upper quadrant in macaque area MT. They suggested that regions of the visual field that form the animal's praxic space (Hughes, 1977) may be relatively emphasized in visual areas as a consequence of preferential stimulation both during development and adult life (see, however, Fiorani *et al.*, 1989).

"Functional" hypotheses imply that, during the evolutionary process, a certain type of map has been selected as the one that best fits the function of an area in visual processing. Thus, the diversity of visuotopic organization that is observed in present-day primates may reflect the end result of a process of optimization by natural selection, and each type of map should bear a relationship with function. One such example, described above (Section 3.1), is the idea that the visual map of V2 reflects the importance of interactions with V1 (Allman and Kaas, 1974a). Schwartz (1977) argued that the visuotopic organizations of V1, V2, and some of the extrastriate areas are functionally significant in that they may provide a degree of size invariance, thereby simplifying processing of shape at further stations of the visual pathway. In addition, Gattass *et al.* (1990) and Neuenschwander *et al.* (1994) have suggested that the apparent local disorder of visuotopic organization in area PO may result from a mapping strategy that facilitates the processing of optical flow, by allowing neurons which represent different eccentricities along a given meridian to be located at nearby points in cortex. Finally, some interpretations of the variability of topographic precision and receptive field size suggest that the particular parameters observed in different areas optimize correlations across visual field compartments of different sizes (e.g., Gross and Mishkin, 1977; Zeki, 1983a,b; Barlow, 1986).

A more general question is, "why do visuotopic maps exist in the cortex at all?" Cowey (1981) suggested that the organization of cortex in visuotopic maps may optimize the process of feature extraction by allowing efficient lateral inhibition. In this model, the error involved in processing image features such as object borders, color, orientation, disparity, and spatial frequency is reduced through inhibition of cells that signal other, less-optimal descriptions of the stimulus. A visuotopic map is the best solution in a model that is heavily dependent on lateral inhibition, because "if there were no map and the cells concerned with a particular part of space were scattered throughout a visual area, the average length of their interconnecting processes would have to be much longer, and the problem of specifying them during development would be enormous" (Cowey, 1981). Thus, this theory also implies that the visuotopic order may have the functional role of facilitating the formation of appropriate connections (as opposed to being a consequence of the interactions which occur during development). In contrast, Gattass *et al.* (1990) proposed that visuotopic organization is preserved through much of the visual cortical pathways in order to facilitate interactions between cells that process *different* stimulus features. They argued that, if all cells in areas located along a hierarchical pathway processed the same stimulus parameter (at successively more global levels), only one visuotopic stage would be required, and that further processing could be better achieved by nonvisuotopic mapping strategies such as those suggested by Barlow (1986). According to this theory, the fact that some global visuotopy is observed for many stages of processing reflects a strategy to minimize connections between modules which process different stimulus parameters in visual areas; these are often organized in interdigitating mosaics such as the different types of cytochrome oxidase stripes in V2. This may, in turn, facilitate cooperative processing, in which one type of visual "cue" influences processing of other parameters (e.g., defining shape from the relative motion of components at different points of the scene). In addition, Swindale (1991) suggested that the deviation of visuotopic order from a strictly topographic representation, as observed in cortical visual areas, reflects the need to provide coverage at the same locus of visual space for neurons processing different types of information in parallel. All of these models share the implication that visuotopy has the functional value of simplifying or enhancing computation by local, intrinsic connections. Finally, it should be pointed out that there have been suggestions that the assumption of continuity, or "smoothness" of image processing gradients, may significantly facilitate the unambiguous resolution of computational problems such as determination of direction of motion of edges, optical flow, and color by neural networks (Poggio *et al.,* 1985). The assumption of continuity may also simplify the completion of contours that fall onto scotomas, such as those corresponding to the optic nerve head and large retinal vessels (Fiorani *et al.,* 1992; Tripathy *et al.,* 1995). A visuotopic representation may be the best way to implement "smoothness" in a neural system.

If optimization of intra-areal lateral interactions were the sole determining factor of visuotopic organization, either as an interactive process during development or as a prespecified "target," one would expect first-order transformations to predominate in the cortex; however, this is not observed experimentally (Section 3). As reviewed above, Allman and Kaas (1974a,b) also proposed that visuotopic maps (and particularly second-order representations) may be an efficient way to optimize the connections *between* areas. Fine-grained mapping of cortical visuotopy in different individuals of the same species reveals that in some areas, such as primate V2 and DM, the position of discontinuities in the representation is remarkably constant, both in terms of visual field coordinates and location in the cortex (Rosa *et al.,* 1988b, 1994; Rosa and Schmid, 1995). This may be interpreted as indirect evidence that the location of field and map discontinuities has functional significance. Even though there may not be a specific functional gain related to a discontinuity at a given location, it is possible that such interruptions will tend to occur in regions where they result in minimal functional loss. In other areas, the locations of field discontinuities may be more variable, suggesting that they result from stochastic, rather than deterministic processes during development. One classic example is the configuration of the border between areas V2 and V3 in the cat (Tusa *et al.,* 1979; Albus and Beckmann, 1980), which was found to vary considerably between individuals by virtue of the existence of additional field discontinuities. The border between subdivisions of the dorsolateral complex in primates may be similarly variable (Sereno *et al.*, 1987). One model that tries to account for these observations relates the existence of different degrees of field discontinuity to the degree of elongation of cortical areas (Wolf *et al.*, 1994). Given a predetermined shape and a few simple assumptions related to the degrees of emphasis on central vision and radial symmetry, the simulation of an activity-driven (Hebbian) developmental process results in maps that replicate in many ways the classes of field discontinuity observed in mammalian visual areas. In addition, different replications of the activity-driven process may result in different final configurations of the map (Fig. 19), especially for areas with extreme degrees of elongation. The latter observation fits nicely with the evidence for individual variability in areas such as cat V2/V3, and those in the primate dorsolateral/ventral complex. It is interesting to note that, in addition to visual topography, self-organizing feature map algorithms can also model the formation of representational mosaics (see Schreiner, 1995, for a recent review with emphasis on auditory cortex). According to these simulations, the emergence of "modules" defined by units sharing similar response properties can be seen as a result of mapping several independently varying parameters onto a single surface in such a way as to minimize connectional length for each unit class (Mitchison, 1995).

As pointed out by Wolf *et al.* (1994), development of visuotopic maps may be further constrained if there is a necessity for maximally congruent borders between areas. It has been suggested, along a "developmental" line, that the discontinuities in visuotopic maps observed in adults may represent the result of a process that drives the cortex toward a configuration of minimal disruption of the supra-areal topographic organization (Rosa and Schmid, 1995). According to this hypothesis, the formation of visuotopic maps depends on lateral interactions, both with other portions of the same area and with portions of adjacent areas. If there is a "protomap" that specifies to some extent the boundaries of each visual area in relation to the others (Rakic, 1988), and if visuotopic maps are formed in a way that is dependent on the representation in adjacent cortex (both within and outside an area), one would expect that the maps would be more regular in visual areas that mature earlier in development. In fact, the maps in V1 and MT, which develop early in relation to other extrastriate areas (Watson *et al.*, 1993), are perhaps the only first-order transformations in visual cortex.* The other areas, which become myelinated later, all form second-order transformations, perhaps because their locations are predetermined by the protomap, and they have to "fit" to a supra-areal topography already constrained by V1 and

*Primary (or primary-like) sensory areas tend to myelinate early during development, and it is intriguing that MT belongs, with V1, to the group that is already myelinated at birth (Flechsig, 1920). Perhaps MT and V1 should both be seen as "primary" cortical visual areas for the retinotectal and retinogeniculate pathways, respectively. This interpretation has found some support from neurophysiological and clinical studies (Raiguel *et al.*, 1989; Rodman *et al.*, 1989, 1990; Girard *et al.*, 1992; Beckers and Zeki, 1995). The importance of a specific telencephalic destination for the tectofugal pathway was first pointed out by Karten (e.g., Karten, 1979; Shimizu *et al.*, 1995) from studies in avian species, and it now appears that this is a general feature of the visual system of amniote vertebrates.

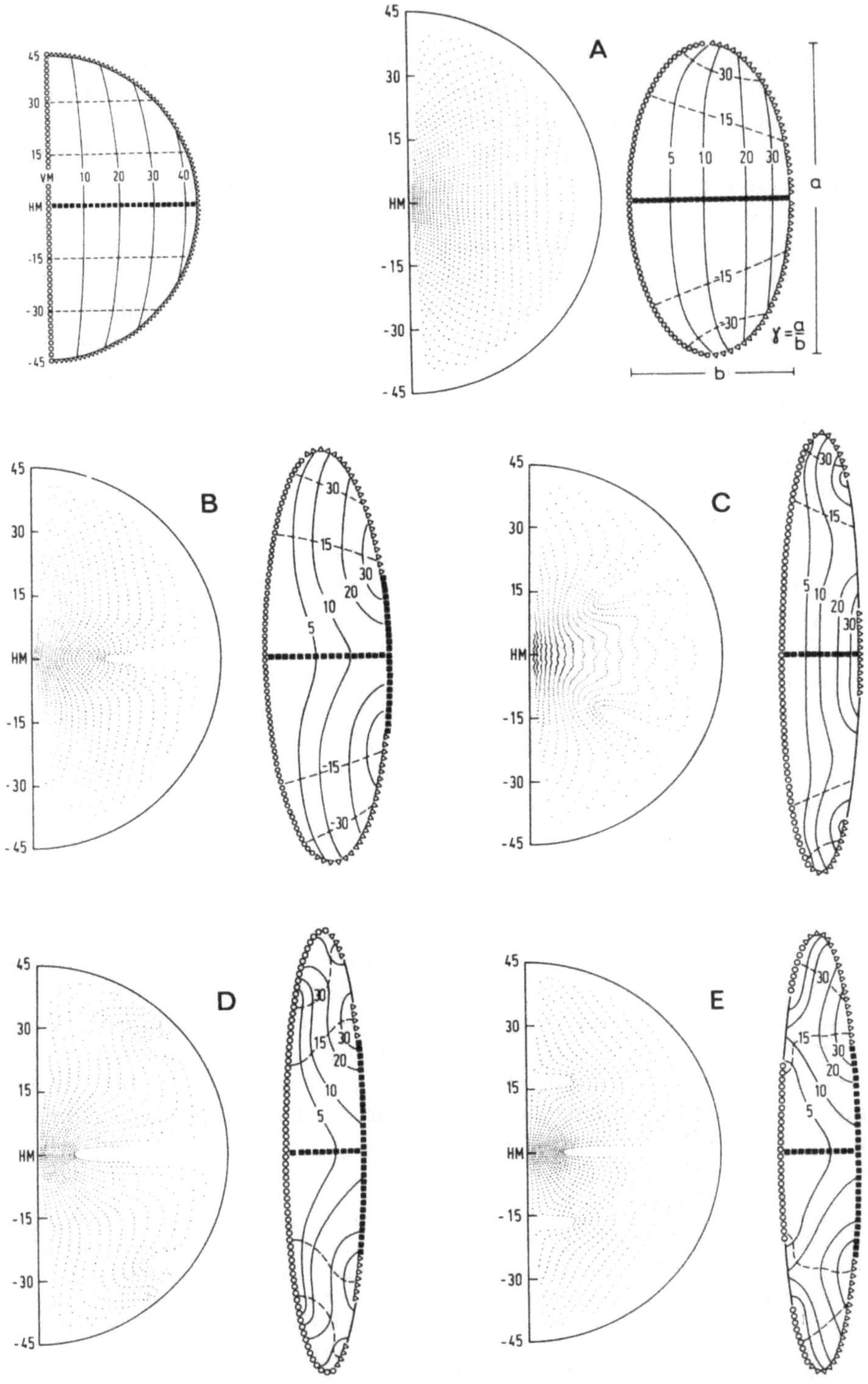
45
30
15
VM
10
20
30
40
HM
-15
-30
-45
A
B
C
D
E
5
$\gamma = \frac{a}{b}$
a
b

MT. This model suggests that the different visuotopic organizations are not determined by natural selection in order to confer specific advantages in processing; instead, they arise in development as a consequence of the cortex attaining a configuration of maximal stability (Lee and Malpeli, 1994) which results in the least disruption of visuotopic continuity both *between* and *within* areas.

5. Large Receptive Fields, Scatter, and Spatial Resolution

Large receptive field size and scatter are normal properties of the visuotopic representation in extrastriate cortex, which certainly do not facilitate the task of the experimenter who tries to make sense of the topographic relationships between receptive fields. From the functional point of view, however, their significance is obscure. One possible interpretation is that the increase in receptive field size and scatter at different anatomical-hierarchical levels of visual cortex is a functional correlate of a low-spatial-resolution system. For example, it has been suggested that the main functions of extrastriate neurons include tasks that demand comparison of stimulus features across large expanses of the retina (such as the processing of motion and color information). As a tradeoff for this capacity, extrastriate neurons might lack fine spatial resolution (Zeki, 1978b; Barlow, 1981, 1986; Creutzfeldt, 1988). In one model, the extrastriate areas were seen as specialized "feature extractors," with neurons in different areas processing different stimulus parameters (Zeki, 1978b). The image features would, at some later stage, be combined in "synthetic" areas such as IT cortex, posterior parietal area 7a, or STP (Desimone and Gross, 1979; Gross *et al.*, 1981). However, in the realm of such a distributed-hierarchical model, the hypothesis that links large receptive fields with lack of spatial resolution carries within itself a contradiction. Since the main projections to the "synthetic" areas arise from areas that have relatively large neuronal receptive fields themselves (Desimone *et al.*, 1980; Andersen *et al.*, 1990), by the time the image features become available for "synthesis," fine detail has been lost. One postulated solution to this problem

Figure 19. Visuotopic maps in hypothetical visual areas of elliptic shape, generated by Wolf *et al.* (1994) using the self-organizing feature map algorithm (Kohonen, 1989). The maps A–E all have the same stimulus nonuniformity (ratio of central density to peripheral density = 100), but the hemifields are mapped onto ellipses of different degrees of elongation (elongation is defined by the parameter γ, or ratio of the two axes of the ellipse). (A) First-order transformation, generated by mapping of the hemifield onto a cortical area of $\gamma = 2$. On the left, the central 45° of a visual hemifield is represented, with receptive field center positions corresponding to a uniform sampling grid in the cortex indicated by dots. On the right, the resulting visuotopic map is shown, using the symbols defined in the inset (upper left; VM, vertical meridian; HM, horizontal meridian). (B) Same as in A, but with the stimuli being mapped onto an ellipse of $\gamma = 3.5$. A field discontinuity corresponding to the representation of the horizontal meridian is generated. (C–E) As above, but for an ellipse of $\gamma = 6$; results of three separate replicates, each "trained" with different sequence of stimuli. In C and D, multiple field discontinuities are formed along the peripheral representation; in E, discontinuities are formed along both vertical and horizontal meridian representations. Redrawn from Wolf *et al.* (1994).

is to abandon the hierarchical aspect, such that processing occurs simultaneously both "upstream" and "downstream" from extrastriate cortex. Resolution of fine detail would be dependent on area V1 or some other hypothetical region containing neurons with small receptive fields, and interconnections between different brain areas would allow information about diverse stimulus features to be combined with spatial detail (e.g., Creutzfeldt, 1988). A similar concept, outside the realm of information-processing models, is that of an early-stage "spatial map" to which "global maps" can refer by reentrant mechanisms. This has been proposed as part of the neuronal group selection theory (Edelman, 1981).

Although more recent research suggests that the view of extrastriate areas as parallel feature extractors may be an oversimplification (e.g., Desimone *et al.*, 1985), the extent to which parallel or serial processes contribute to the analysis of the visual image and the degree of subdivision of function in extrastriate cortex is not clear. An entirely different issue, however, is the extent to which large excitatory receptive fields necessarily imply that neurons in extrastriate cortex form low-spatial-resolution systems. There is direct evidence that the spatial resolution of neurons in different visual areas is not inversely proportional to receptive field size; while receptive fields of neurons in V2 are 2–4 times larger (linearly) than those of neurons in V1, the sharpness of the focus of activity elicited in both areas by the same stimulus is similar, as evaluated by 2-deoxyglucose mapping (Tootell and Hamilton, 1989). In contrast, there is no evidence for a linear relationship between neuronal receptive field size in visual cortex and spatial resolution. The image of an average receptive field of V1 neurons for a given eccentricity (i.e., the projection of the contour of the receptive field onto an idealized visuotopic map; McIlwain, 1975) is at least 10 times larger than the minimal angle of resolution and 50 times larger than the hyperacuity threshold (Barlow, 1981). By considering both receptive field size and scatter, Dow *et al.* (1981) calculated that the point-image size in the macaque V1 is 30 times larger than the cortical projection of the minimal angle of resolution of humans. The cortical projection of the minimal angle of resolution in the central representation (50–100 μm) roughly corresponds to twice the separation between pyramidal cell somata (or pyramidal cell modules; Peters and Sethares, 1991) in V1. Hyperacuity thresholds, which are at least 5 times smaller and change more steeply with eccentricity than classical acuity thresholds (Westheimer, 1982),* may depend on the separation between granule cells (Barlow, 1981) or interaction between arrays of pyramidal cells.

McIlwain (1976) has also argued that loss of spatial detail is not an inevitable consequence of each cell having a large receptive field. He argued that high

*Because there has been some controversy, it is still unclear whether there is some direct relationship between classical acuity and magnification factor. Contradicting the results of earlier work (Rolls and Cowey, 1970; Cowey and Rolls, 1974), Levi *et al.* (1985) have suggested that only the variation of hyperacuity thresholds is proportional to the magnification factor in V1, and that classical acuity is limited by the retinal sampling array. However, this hypothesis involves comparison between psychophysical data in humans and physiological data in macaques, and the assumption that the cortical magnification factor varies similarly with eccentricity in both species. Because this assumption has been challenged (Sereno *et al.*, 1995), and because more recent studies have suggested that the variations of cortical magnification factor and ganglion cell density with eccentricity are nearly parallel (Wässle *et al.*, 1990; Wilder *et al.*, 1996), it is still too early to make make any definitive statements.

resolution can be achieved by population coding, i.e., through the consideration of responses of a large neural population, rather than single cells. Larger receptive field sizes allow more neurons to participate in feature extraction or localization processes at any visual field location in comparison with the hypothetical situation in which areas of different sizes all have small receptive fields and low scatter. If the latter were true, the number of processing units for each point of the visual field would necessarily become smaller for smaller visual areas. Processes such as spatial interpolation (Barlow, 1979, 1981; Crick *et al.*, 1981) may be optimized by the use of large numbers of neurons monitoring each point of the visual field. Thus, number of neurons, rather than receptive field size, is likely to be the limiting factor for resolution.

In theory, because the maximal resolvable frequency is a function of the number of sampling channels, it is still true that V1, which has more neurons than any other area, has the potential to resolve much finer detail than extrastriate areas. The point that should be made is that, based on the currently available data, the neural machinery in extrastriate areas may still allow a degree of spatial selectivity that can hardly be described as "coarse." As an illustration, the magnification factor in the central representation in macaque area MT is at least 2 mm/° (Albright and Desimone, 1987; Komatsu and Wurtz, 1988) and perhaps twice that (Erickson *et al.*, 1989). This is not an atypically high value; other areas such as V2 and those in the dorsolateral complex may have even higher foveal magnification factors (Gattass *et al.*, 1981, 1988; Maguirre and Baizer, 1984; Rosa *et al.*, 1988b). Considering the value of 2 mm/° given by Albright and Desimone (1987) for the peak magnification in MT, and assuming that, as in V1, the theoretical "classical" resolution limit roughly corresponds to the separation of pyramidal cell modules (Peters and Sethares, 1991; Peters *et al.*, 1994), there may be sufficient "room" for 60 functional units within the representation of the central 1°. Even this simplified calculation, which employs a linear array of modules, allows a theoretical resolution as small as 2′ of arc if the resolution threshold requires one "silent" pyramidal cell module between two excited ones; an even higher resolution could be achieved by a hexagonal array, which may be biologically more realistic (e.g., Hughes, 1977). Thus, in theory, the neuronal population in macaque MT is able to provide sufficient acuity to specify the location of a stimulus in the central visual field at a level higher than the peak resolution of the visual system of cats (Bisti and Maffei, 1974; Hughes, 1977), animals that are often considered to have developed vision. Such acuity may be perfectly adequate for the specification of the location of moving stimuli, without reference to another map.

A further point to be considered is that, in the behaving primate, unlike in controlled laboratory conditions, the size of receptive fields of extrastriate neurons may be effectively smaller. Even when simple visual stimuli such as a flashing spot or bar are employed, receptive fields of extrastriate units demonstrate a spatial heterogeneity. For a flashed stimulus of fixed size, responses are maximal within relatively restricted functional centers (Maguire and Baizer, 1984; Hubel and Livingstone, 1985). In addition, single neurons often yield a maximal response for stimuli which are many times smaller than the total extent of the receptive field (Baizer *et al.*, 1977; Hubel and Livingstone, 1985; Desimone and Schein, 1987). Because multiple changing visual stimuli are present throughout the visual field, mechanisms such as lateral inhibition between neurons representing adjacent retinal loci may modulate neuronal responsiveness, especially to stimulation of the "fringes" of receptive fields, effectively resulting in smaller excitatory areas. There is a rich network of GABAergic neurons throughout visual cortex (Jones, 1993), and the response magnitude of V1 neurons is modulated by pharmacological manipulations of GABAergic transmission (see Sillito, 1984, for review). Some of the intrinsic inhibitory axons may extend for as far as 2 mm parallel to the cortical surface (Albus and Wahle, 1994). In the somatosensory cortex, there is ample evidence for receptive field shaping by both within-field and surround inhibition, and GABA has been implicated in this function (Mountcastle and Powell, 1959; Laskin and Spencer, 1979; Hicks and Dykes, 1983; Dykes *et al.,* 1984; Alloway *et al.,* 1989). Unlike in visual mapping experiments, in which typically only a single stimulus is presented at a time, in the "real world" the multiplicity of competing patterns activating different points of the retina may cause strong inhibitory interactions between neurons located at different points of the visuotopic map of extrastriate areas, and perhaps a substantial shrinkage of excitatory receptive fields toward their functional center. Thus, lateral inhibition may contribute to the production of sharp peaks of neural activity as a function of location in the visuotopic map, even if each neuron has the potential to respond to stimulation across a wide sector of the visual field (see Kitano *et al.,* 1994, for evidence in support of this model). Although specific tests regarding the possibility of modulation of excitatory receptive field size by intra-areal inhibitory interactions are still lacking, especially in primate extrastriate cortex, there have been demonstrations of other phenomena that indicate the

Figure 20. Effect of receptive field scatter, receptive field spatial heterogeneity, and lateral inhibition on the distribution of neural activity across the cortex. Each panel (A–D) includes a highly schematic representation of seven cortical columns (with three neurons each; bottom), of the corresponding receptive fields and field centers (middle), and of the cellular activity evoked in the brain (top) by a stimulus (large black circle overlying the receptive fields). The dotted horizontal line ("sp." in panel A) represents the spontaneous activity. (A) Hypothetical model of an area in which the visuotopy is geometrically precise. All receptive fields of neurons located in a given column share exactly the same center, and their sizes are similar. In addition, receptive fields are spatially homogeneous (i.e., stimulation of any point of the receptive field yields a response of similar strength). In this case, any stimulus produces a focus of activity that has the same linear dimension as the point-image size, and neural activity varies as a function of position in the cortex according to a step function. (B) Model incorporating receptive field scatter. The receptive fields within a column are now randomly distributed around geometrical centers, which correspond to the same points represented in panel A. A maximal receptive field scatter (distance between the individual receptive field center and average center of all fields of the column) equivalent to one-third of the average receptive field size was allowed. Receptive field shape was also varied, but the average size of receptive fields was kept the same as in A. By virtue of the statistical distribution of receptive field centers in a column, the step of activity as a function of position in cortex (A) is converted into a near normal distribution with a defined peak. (C) As in B, but with heterogeneous receptive fields. In this model, the strength of neural response of a given neuron decays linearly as a function of distance of the stimulus from the receptive field center. (D) Model incorporating features of B and C, plus lateral inhibition between adjacent columns. In this case, a peak of activity many times narrower than the point-image size is produced by the stimulus. With larger numbers of columns and neurons, the precision of determination of the peak of activity increases. It is possible that in the real situation two just separable stimuli are coded by two "active" pyramidal cell modules separated by a single "inhibited" module.

malleability of the visual receptive fields of extrastriate neurons (see Section 6). Thus, the receptive fields of extrastriate visual neurons mapped in anesthetized animals, or in behaving animals performing tasks in which the attention is not directed to the stimulus, need to be seen as "probability" fields: they may include the many possible instantaneous spatial profiles of the receptive field that can

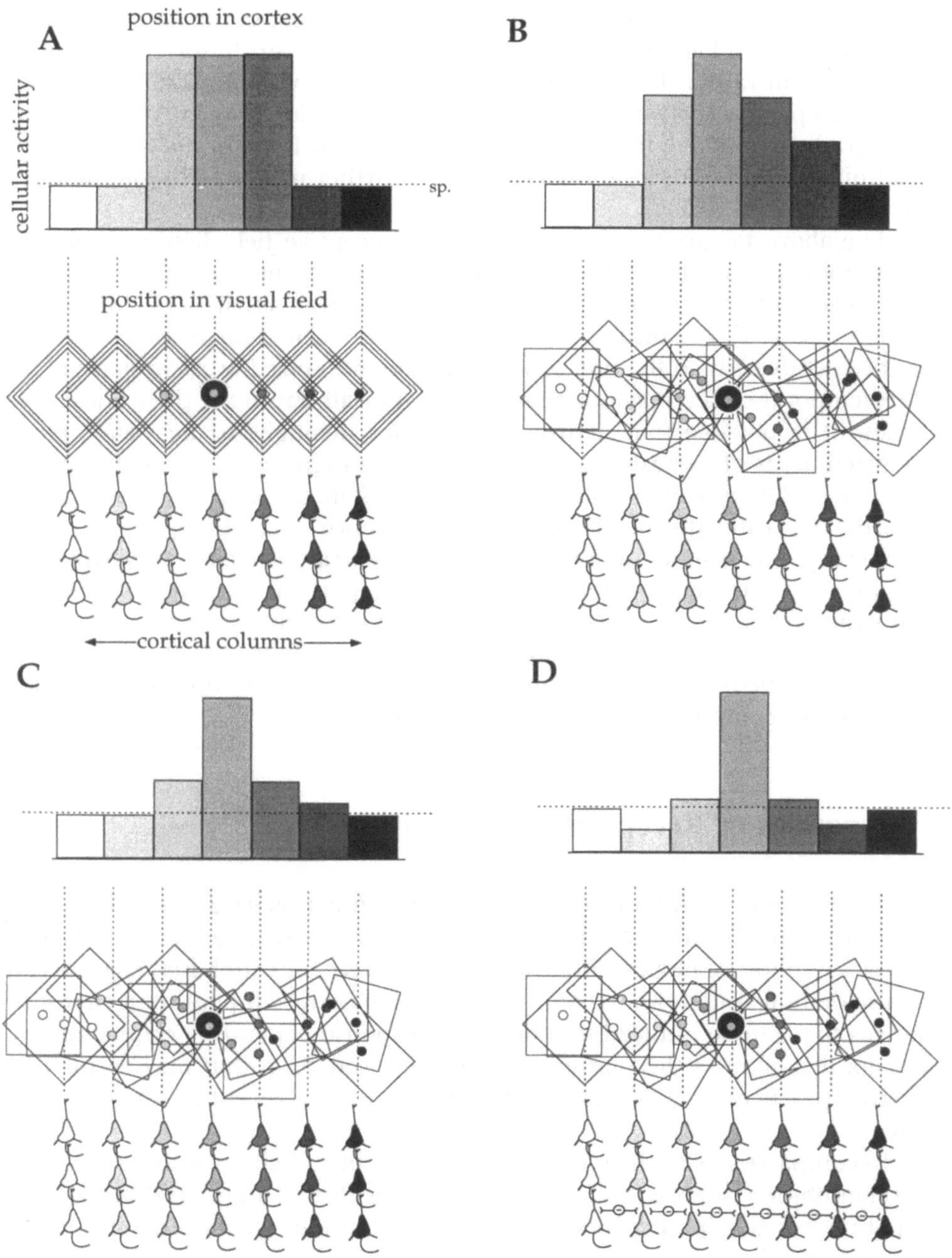

Figure 20.

result from presentation of different stimulus configurations. Another possibility is that the receptive field mapped under anesthesia reflects the optimal or "least-resistance" pathway from retina to cortex.

From a purely mechanistic point of view, it can also be argued that the heterogeneity of receptive fields observed at any given point of extrastriate cortex in terms of scatter and size may actually result in a sharper peak of neuronal activity across the surface of a cortical area, rather than degrading the spatial resolution of the system. Consider, for example, the hypothetical situations illustrated in Fig. 20. Because receptive fields are stochastically distributed around a point rather than following a point-to-point visuotopic map, the hypothetical "step" of activity as a function of cortical location (Fig. 20A) is converted into a gradient with a defined peak (Fig. 20B, C), which might reduce ambiguity in defining the stimulus location, especially if further narrowed by lateral inhibition (Fig. 20D).

The above factors (spatial interpolation, receptive field heterogeneity, lateral inhibition, and probabilistic sampling) are not mutually exclusive, and may add to each other to produce sharp peaks of neural activity as a function of location in cortex. In conclusion, the neural representations of the visual field that project directly to the non-topographically organized areas in parietal and inferotemporal cortex may not be so coarse after all. From the theoretical point of view, it is possible that neuronal ensembles throughout the visuotopically organized areas of extrastriate cortex preserve a high degree of spatial resolution in parallel with their role in the first steps in the extraction of stimulus features, obviating the need for a fine "reference" map except perhaps for hyperacuity. The preservation of high acuity in extrastriate cortex might help to explain, for example, how cells in the anterior part of the superior temporal cortex can respond in dramatically different ways to such slight changes of visual image as a change in the direction of gaze of a face (Perrett *et al.*, 1985, 1990). Even though the representations in this region are nonvisuotopic, it seems likely that some of its inputs convey very precise local signals.

6. Dynamics of Receptive Field Position

To this point, I have considered the possible functional significance of extrastriate visuotopy from a mechanistic point of view, based on geometry and probabilistic sampling strategies within single visual areas. Research in the last decade has also demonstrated two dynamic properties of cortical topographic maps: the changes that can be induced in a neuron's response depending on the animal's behavioral state, and the plasticity of the representation that may result from selective practice or deactivation of portions of the receptor surface or the cortex.

In comparison with the modulation of a receptive field's size and position by lateral inhibition or cross-columnar facilitation within a single area, the modulation of neuronal responses by selective covert attention, eye movements, or tasks involving comparison between sensory modalities is likely to depend on "extrinsic" influences arising from other cortical and subcortical structures (Desimone *et al.*, 1990). Change of attentional focus in behaving animals produces an effect in the macaque's area V4 that may be interpreted "as if the receptive field had contracted around the attended stimulus" (Moran and Desimone, 1985; Desimone *et al.*, 1990). Visuotopic changes on an even larger scale were demonstrated in neurons in the macaque's area LIP. When mapped in anesthetized monkeys, the responses of neurons in LIP show evidence of visuotopy (Blatt *et al.*, 1990), and relatively restricted receptive fields can be mapped in the behaving animal with a simple fixation task (Colby *et al.*, 1993a,b). However, in a paradigm in which the monkey had to execute a saccade bringing a visual stimulus to the center of the neuron's receptive field, it was observed that the response would start before the actual eye movement (Duhamel *et al.*, 1992). Neurons in LIP therefore demonstrate that instantaneous change of visuotopy occurs in some portions of the visual system in anticipation of eye movements, by responding to stimuli both in a classical receptive field and in the (previously silent) anticipated retinal location.

A surprising result, suggesting a similar phenomenon on a smaller scale, was that of Motter and Poggio (1990), who demonstrated that V1 receptive field positions are not strictly fixed to retinal coordinates. The receptive fields of V1 cells may remain stable in space while the monkey fixates a target, even though microsaccadic eye movements occur (Fig. 21). This compensation of receptive field position for eye movements occurring at such an early stage of the cortical visual pathway has the functional implication that the neural representation that is relayed to extrastriate cortex is already a stabilized version of that captured by the retinae, with the image captured by both eyes brought into register independent of the error introduced by microsaccades. It has been proposed that this property depends on shifter circuits involving several layers of relay neurons and inhibitory switches that can "slide" the entire monocular image parallel to the cortical surface, redirecting the monocular visuotopic maps (Anderson and Van Essen, 1987). The calculated maximal "shift" necessary for binocular registration (2–3 mm) is equivalent to the point-image size in foveal V1 of anesthetized monkeys (Hubel and Wiesel, 1974; Dow *et al.*, 1981; Gattass *et al.*, 1987). Perhaps the aggregate receptive fields of cell columns (or multiunit receptive fields), which are used for the calculation of point-image size, reflect not only the "instantaneous" size of unitary receptive fields, but also the probabilistic nature of center location (in retinal coordinates) as a function of time (Rosa *et al.*, 1991). Because of the logarithmic decay of magnification factor in most visual areas, the error introduced by microsaccades decays markedly relative to receptive field size with increasing eccentricity; the point-image size calculated on the basis of multiunit recordings of aggregate single-unit fields is also markedly reduced, although not in direct proportion (Maguirre and Baizer, 1984; Gattass *et al.*, 1987; Rosa *et al.*, 1988b).

The different scales of the immediate shifts of receptive field position that are possible in V1 and LIP are an expected consequence of the compression of the visuotopic map in the latter area relative to the former, if one argues that a similar neural circuit regulates the effective extent of the excitatory receptive field position in both areas. By the same token, similar processes of receptive field "shifting" may occur throughout extrastriate cortex, and this would be a plausible immediate substrate for shrinkage, expansion, or change of functional

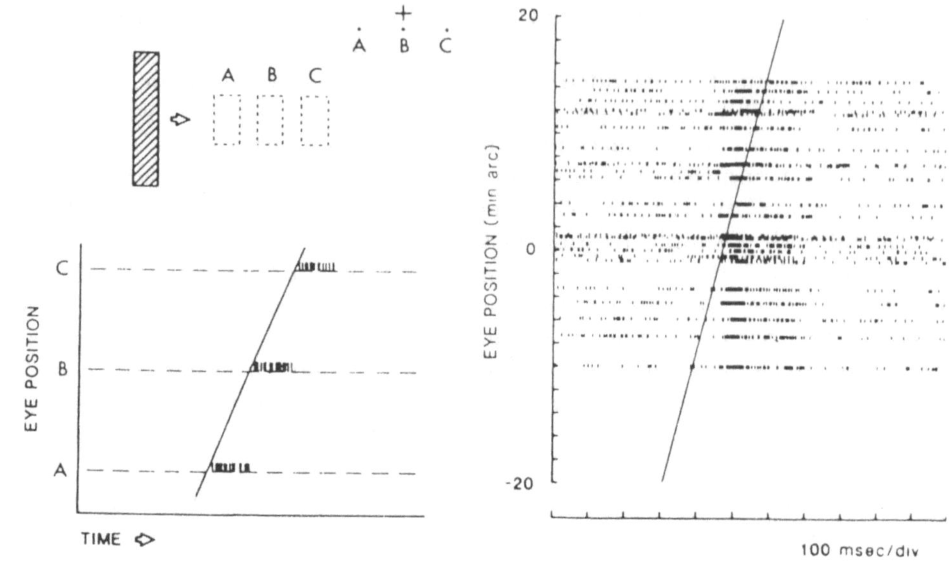

Figure 21. Immediate retinotopic plasticity of receptive field position in monkey visual cortex. This analysis is based on a trial-by-trial analysis of response onset to a moving stimulus as a function of position of the eye in the orbit. The left panel illustrates the hypothetical situation in which the receptive field of a neuron is "locked" to retinotopic coordinates. Upper left: While the monkey fixates the point defined by the cross-hairs, the fovea drifts around this point by virtue of microsaccadic eye movements. In different trials, with the fovea pointing to three slightly different points in space (A, B, C), the moving target will intersect the receptive field of the neuron (dotted rectangles) at different times. The expected relationship between eye position and response onset is illustrated at the bottom left: in some trials (e.g., A) the response will be advanced, and in others delayed (e.g., C). The right panel illustrates the data obtained from a neuron in monkey V1. The oblique line indicates the expected relationship between eye position and response onset for this neuron. In contrast to the prediction, the raster diagrams obtained in different trials indicate response onsets at similar times. This experiment reveals that the representation of the visual field in primate cortex is not rigidly locked to retinal coordinates. Instead, the visual system can compensate for small eye movements, therefore stabilizing the representation relative to the intended point of fixation in space. Reproduced with permission from Fig. 3 of Motter and Poggio (1990).

center of receptive fields in these areas relative to retinal coordinates. The question that remains, however, is what mechanisms control the shift. Alternatives include subcortical structures (Crick, 1984; Koch and Ullman, 1985; Desimone *et al.,* 1990; McIlwain, 1995), interlaminar projections, feedback from areas in the later stages of the visual pathway (Anderson and Van Essen, 1987; Van Essen and Anderson, 1990; Vidyasagar, 1996), or combinations of these.

In view of the above evidence that receptive field plasticity may be embedded in the normal mode of operation of visual cortex, it is not surprising that manipulation of the inputs to a cortical area may cause rearrangement of its visuotopic map. This has been reported by studies in V1 of both monkey and cat (Kaas *et al.,* 1990; Heinen and Skavenski, 1991; Gilbert and Wiesel, 1992; Darian-Smith and Gilbert, 1995; Rosa *et al.,* 1995a; Schmid *et al.,* 1995, 1996), but monkey extrastriate cortex has not yet received similar attention. In monkeys with retinal lesions extending from the center of gaze to 1.5° eccentricity, Heinen and Skavenski (1991) found an unresponsive sector in V1 extending from the lateral "apex" (where the center of the fovea is represented) to the point where the 0.5° isoeccentricity line was mapped prior to the lesion. Because the silent sector did not extend to the original 1.5° isoeccentricity line,* this result effectively suggests that visuotopic shifts across several millimeters of cortex are possible shortly after a retinal lesion (see also Fiorani *et al.,* 1992; Schmid *et al.,* 1995). This may be another consequence of rerouting of the visual pathway after some of the possible competing inputs are silenced. However, a more important question is whether the receptive field shifts can extend beyond the normal scatter of receptive field position that is observed in control animals (which, as speculated above, may represent the range of the shifter circuit). In the cat, if the normal point-image size is taken into account (Schmid *et al.,* 1995, 1996), few neurons with reorganized receptive fields are observed in "reorganized" cortex, and their responses are notably weaker than normal. Thus, the current evidence suggests that, within the first 3 months after a retinal lesion, the visuotopic reorganization beyond the normal point-image size is not extensive or complete. Long-term recoveries, however, may allow a functionally more significant reestablishment of responses in denervated sectors (Darian-Smith and Gilbert, 1995). In the context of possible functions of visuotopic maps, it is interesting to note that visuotopic reorganization may help to avoid the existence of sharp transitions of neural activity across cortical areas. Such discontinuities could disturb the computation of features that depend on the assumption of continuous gradients (or "smoothness") by analogue neural networks (Poggio *et al.,* 1985; see Section 4).

Because of the large receptive fields in primate extrastriate cortex, proof of visuotopic reorganization in these areas is likely to be difficult to obtain using the retinal lesion model. Yet, the neural elements necessary for producing functional changes of the visuotopic map are clearly present in these areas, as demonstrated by other paradigms. One example was the demonstration by Wurtz *et al.* (1990) that, after local lesions of area MT, the receptive fields of many neurons located around the lesioned sector expand to 2–3 times their original size (linearly) and then contract back to normal size (Fig. 22A, B). This process is reminiscent of the changes that occur in somatosensory cortex after digit amputation, denervation, or local anesthesia (Calford and Tweedale, 1988; Kolarik *et al.,* 1994). It remains to be seen whether, upon contraction back to normal size, the receptive fields of some of these neurons have their centers displaced toward the region of the visual field originally represented at the lesion site. This process could contribute toward the recovery of function within the motion scotoma that is observed in the weeks following restricted lesions of MT (Newsome *et al.,* 1985; Wurtz *et al.,* 1990).

7. Conclusions

1. One of the significant conclusions that should be drawn from this review is that there is still much to be learned about the limits, borders, and topographic

*That the lesion was effective up to 1.5° in the retina was confirmed by Heinen and Skavenski by measuring the monkey's shift in fixation relative to the prelesion condition.

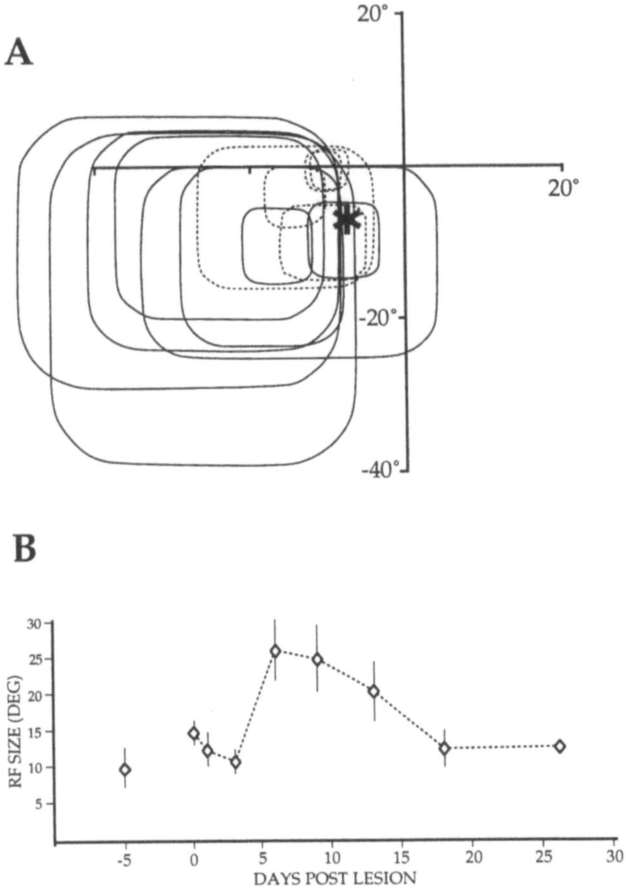

Figure 22. Effects of a small cortical lesion on the representation of the visual field in macaque area MT (redrawn from Fig. 4 of Wurtz *et al.* 1990). (A) Location of receptive fields of single neurons of area MT, recorded through the same guide tube on different days. The dotted fields are prelesion controls, showing the normal variability of receptive field size of MT neurons at a given eccentricity. The asterisk indicates the visuotopic locus (average of receptive field centers) of a point in MT where an electrolytic lesion was placed. A few days after the lesion, receptive fields of many neurons bordering the lesion (continuous outline), recorded through the same guide tube as the prelesion recordings, are enlarged relative to the control situation. (B) Time course of the enlargement of receptive fields. The receptive field sizes (diamonds indicate the mean size, bars one standard deviation) are significantly increased for most of the second week postlesion, but return to normal levels after 15 days. The expansion of receptive fields of neurons adjacent to lesioned cortex may help avoid the formation of a permanent motion scotoma after MT lesions.

organization of primate extrastriate visual areas. One of the main obstacles for a better comprehension of the structure and function of these areas has been the fact that studies of different species have yielded very different models of their organization. Still, the data are sufficient to allow the formulation of a number of specific hypotheses regarding the homologies of areas in different primates, which emphasize the overall similarity of visuotopic organization in spite of differences in brain size, morphology, and habits. This is reassuring, for if there were major differences between species of monkeys, the task of understanding the human brain on the basis of animal studies (e.g., Sereno *et al.*, 1995) would be much more difficult. Thus, through a comparative approach, we may be approaching a unified model of primate visual areas. While more work will be needed to clarify whether all simians share the same basic layout of areas in the topographically organized portion of visual cortex, as I propose, the data are certainly not sufficient to reject this hypothesis. Likewise, the data suggest that, similar to the somatosensory and auditory areas (Imig *et al.*, 1982; Dykes and

Ruest, 1986; Morel *et al.*, 1993; Krubitzer *et al.*, 1995), primate visual areas are organized in a series of reproducible maps which include the entire receptor surface.

2. This review also highlights the desirability of moving toward a unified nomenclature of primate visual areas. In many cases, areas that are probably homologues have been given three or four distinct names by different researchers, and it is also likely that different portions of single areas have been named differently. Because of the many points of uncertainty that need to be clarified by further research, the adoption of such a nomenclature may be premature, but will nonetheless be of great advantage in the future.

3. Finally, although the data certainly vindicate the use of visuotopy as a powerful criterion for identification and delimitation of cortical areas, it is still unclear to what extent the different characteristics of maps can be linked to function. Studies which modeled the development of visuotopy have demonstrated that maps remarkably similar to those in extrastriate areas both in terms of their constancy and variability can be generated based on a set of simple rules and multiple iterations with incoming stimuli. Yet, given the current status of research in this area, where quantitative determinations of the entire visuotopic organization of areas are still the exception, expecting to understand the reason for the diversity of visuotopic organizations would be perhaps expecting too much. Clearly, further work is necessary toward establishing such a quantitative database, as well as toward understanding the developmental mechanisms of map formation in extrastriate cortex and modeling inter-areal and intra-areal interactions between neurons.

ACKNOWLEDGMENTS. The preparation of this paper would not have been possible without the patient work of Rowan Tweedale, who reviewed successive drafts for style, grammar, spelling, and scientific consistency. I am also indebted to Mike Calford, Janine Clarey, Bogdan Dreher, Leah Krubitzer, Jack Pettigrew, and David Vaney for numerous suggestions which improved the manuscript. This work was supported by a project grant and an R. Douglas Wright fellowship from the Australian National Health and Medical Research Council (NH&MRC). Additional support was received from the Australian Research Council, through a Special Research Centre Grant to the Vision, Touch & Hearing Research Centre, The University of Queensland.

8. References

Albright, T. D., and Desimone, R., 1987, Local precision of visuotopic organization in the middle temporal area (MT) of the macaque, *Exp. Brain Res.* **65:**582–592.

Albus, K., and Beckmann, R., 1980, Second and third visual areas of the cat: Interindividual variability in retinotopic arrangement and cortical location, *J. Physiol. (Lond.)* **299:**247–276.

Albus, K., and Wahle, P., 1994, The topography of tangenital inhibitory connections in the postnatally developing and mature striate cortex of the cat, *Eur. J. Neurosci.* **6:**779–792.

Allman, J. M., and Kaas, J. H., 1971, A representation of the visual field in the caudal third of the middle temporal gyrus of the owl monkey (*Aotus trivirgatus*), *Brain Res.* **31:**85–105.

Allman, J. M., and Kaas, J. H., 1974a, The organization of the second visual area (V II) in the owl monkey: A second order transformation of the visual hemifield, *Brain Res.* **76:**247–265.

Allman, J. M., and Kaas, J. H., 1974b, A crescent-shaped cortical visual area surrounding the middle temporal area (MT) in the owl monkey (*Aotus trivirgatus*), *Brain Res.* **81:**199–213.

Allman, J. M., and Kaas, J. H., 1975, The dorsomedial cortical visual area: A third tier area in the occipital lobe of the owl monkey (*Aotus trivirgatus*), *Brain Res.* **100:**473–487.

Allman, J. M., and Kaas, J. H., 1976, Representation of the visual field on the medial wall of occipital-parietal cortex in the owl monkey, *Science* **191:**572–575.

Allman, J. M., and McGuinness, E., 1983, The organization of cortical visual areas in a strepsirhine primate, *Galago senegalensis, Soc. Neurosci. Abstr.* **9:**957.

Allman, J. M., Kaas, J. H., and Lane, R. H., 1973, The middle temporal area (MT) in the bushbaby, *Galago senegalensis, Brain Res.* **57:**197–202.

Allman, J., Campbell, C. B. G., and McGuinness, E., 1979, The dorsal third tier area in *Galago senegalensis, Brain Res.* **179:**355-361.

Allman, J., Miezin, F., and McGuinness, E., 1985, Stimulus specific responses from beyond the classical receptive field: Neurophysiological mechanisms for local–global comparisons in visual neurons, *Annu. Rev. Neurosci.* **8:**407–430.

Alloway, K. D., Rosenthal, P., and Burton, H., 1989, Quantitative measurements of receptive field changes during antagonism of GABAergic transmission in primary somatosensory cortex of cats, *Exp. Brain Res.* **78:**514–532.

Andersen, R. A., Asanuma, C., Essick, G., and Siegel, R. M., 1990, Corticocortical connections of anatomically and physiologically defined subdivisions within the inferior parietal lobule, *J. Comp. Neurol.* **296:**65–113.

Anderson, C. H., and Van Essen, D. C., 1987, Shifter circuits: A computational strategy for dynamic aspects of visual processing, *Proc. Natl. Acad. Sci. USA* **84:**6297–6301.

Baizer, J. S., Robinson, D. L., and Dow, B. M., 1977, Visual responses of area 18 neurons in awake, behaving monkey, *J. Neurophysiol.* **40:**1024–1037.

Baker, J. F., Petersen, S. E., Newsome, W. T., and Allman, J. M., 1981, Visual response properties of neurons in four extrastriate visual areas of the owl monkey (*Aotus trivirgatus*): A quantitative comparison of medial, dorsomedial, dorsolateral and middle temporal areas, *J. Neurophysiol.* **45:**397–416.

Ballard, D. H., 1987, Cortical connections and parallel processing: Structure and function, in: *Vision, Brain and Cooperative Computation* (M. A. Arbib and A. R. Hanson, eds.), MIT Press, Cambridge, MA, pp. 563–621.

Barlow, H. B., 1979, Three theories of cortical function, in: *Neurobiology of Vision* (R. D. Freeman, ed.), Academic Press, New York, pp. 1–16.

Barlow, H. B., 1981, Critical limiting factors in the design of the eye and visual cortex, *Proc. R. Soc. Lond. B* **212:**1–34.

Barlow, H. B., 1986, Why have multiple cortical areas? *Vision Res.* **26:**81–90.

Beckers, G., and Zeki, S., 1995, The consequences of inactivating areas V1 and V5 on visual motion perception, *Brain* **118:**49–60.

Bignall, K. E., and Singer, P., 1967, Auditory, somatic and visual input to association and motor cortex of the squirrel monkey, *Exp. Neurol.* **18:**300–312.

Bisti, S., and Maffei, L., 1974, Behavioural contrast sensitivity of the cat in various visual meridians, *J. Physiol.* (Lond.) **241:**201–210.

Blasdel, G. G., and Fitzpatrick, D., 1984, Physiological organization of layer 4 in macaque striate cortex, *J. Neurosci.* **4:**880–895.

Blatt, G. J., Andersen, R. A., and Stoner, G. R., 1990, Visual receptive field organization and cortico-cortical connections of the lateral intraparietal area (area LIP) in the macaque, *J. Comp. Neurol.* **299:**421–445.

Born, R. T., and Tootell, R. B. H., 1992, Segregation of global and local motion processing in primate middle temporal visual area, *Nature* **357:**497–499.

Boussaoud, D., Ungerleider, L. G., and Desimone, R., 1990, Pathways for motion analysis: Cortical connections of the medial superior temporal and fundus of the superior temporal visual areas in the macaque, *J. Comp. Neurol.* **296:**462–495.

Boussaoud, D., Desimone, R., and Ungerleider, L. G., 1991, Visual topography of area TEO in the macaque, *J. Comp. Neurol.* **306:**554–575.

Bruce, C. J., Desimone, R., and Gross, C. G., 1982, Visual properties of neurons in a polysensory area in superior temporal sulcus of the macaque, *J. Neurophysiol.* **46:**369–384.

Bruce, C. J., Desimone, R., and Gross, C. G., 1986, Both striate cortex and superior colliculus contribute to visual properties of neurons in superior temporal polysensory area of macaque monkey, *J. Neurophysiol.* **55:**1057–1075.

Bullier, J., and Nowak, L. G., 1995, Parallel versus serial processing: New vistas on the distributed organization of the visual system, *Curr. Opin. Neurobiol.* **5:**497–503.

Bullier, J., Girard, P., and Salin, P.-A., 1994, The role of area 17 in the transfer of information to extrastriate visual cortex, in: *Cerebral Cortex,* Volume 10, *Primary Visual Cortex in Primates* (A. Peters and K. S. Rockland, eds.), Plenum Press, New York, pp. 301–331.

Burkhalter, A., and Van Essen, D. C., 1986, Processing of color, form and disparity information in visual areas VP and V2 of ventral extrastriate cortex in the macaque monkey, *J. Neurosci.* **6:**2327–2351.

Burkhalter, A., Felleman, D. J., Newsome, W. T., and Van Essen, D. C., 1986, Anatomical and physiological asymmetries related to visual areas V3 and VP in macaque extrastriate cortex, *Vision Res.* **26:**63–80.

Burton, H., Fabri, M., and Alloway, K., 1995, Cortical areas within the lateral sulcus connected to cutaneous representations in area 3b and area 1: A revised interpretation of the 2nd somatosensory area in macaque monkeys, *J. Comp. Neurol.* **355:**539–562.

Calford, M. B., and Semple, M. N., 1995, Monaural inhibition in cat auditory cortex, *J. Neurophysiol.* **73:**1876–1891.

Calford, M. B., and Tweedale, R., 1988, Immediate and chronic changes in responses of somatosensory cortex in adult flying-fox after digit amputation, *Nature* **332:**446–448.

Calford, M. B., Webster, W. R., and Semple, M. S., 1983, Measurement of frequency selectivity of single neurons in the central auditory pathway, *Hearing Res.* **11:**395–401.

Clarey, J. C., Barone, P., and Imig, T. J., 1992, Physiology of thalamus and cortex, in: *The Mammalian Auditory Pathway: Neurophysiology* (A. N. Popper and R. R. Fay, eds.), Springer-Verlag, Berlin, pp. 232–334.

Colby, C. L., Gattass, R., Olson, C. R., and Gross, C. G., 1988, Topographic organization of cortical afferents to extrastriate area PO in the macaque: A dual tracer study, *J. Comp. Neurol.* **238:**1257–1299.

Colby, C. L., Duhamel, J.-R., and Goldberg, M. E., 1993a, Ventral intraparietal area in the macaque: Anatomic location and visual response properties, *J. Neurophysiol.* **69:**902–914.

Colby, C. L., Duhamel, J.-R., and Goldberg, M. E., 1993b, The analysis of visual space by the lateral intraparietal area of the monkey: The role of extraretinal signals, *Prog. Brain Res.* **95:**307–316.

Condo, G. J., and Casagrande, V. A., 1990, Organization of cytochrome oxidase staining in the visual cortex of nocturnal primates (*Galago crassicaudatus* and *Galago senegalensis*), *J. Comp. Neurol.* **293:**632–645.

Cowey, A., 1964, Projection of the retina on to striate and prestriate cortex in the squirrel monkey, *Saimiri sciureus, J. Neurophysiol.* **27:**366–393.

Cowey, A., 1981, Why are there so many visual areas? in: *The Organization of the Cerebral Cortex* (F. O. Schmitt, F. G. Worden, G. Adelman, and S. G. Dennis, eds.), MIT Press, Cambridge, MA, pp. 395–413.

Cowey, A., and Rolls, E. T., 1974, Human cortical magnification factor and its relation to visual acuity, *Exp. Brain Res.* **21:**447–454.

Cragg, B. G., and Ainsworth, A., 1969, The topography of the afferent projections in the circumstriate visual cortex of the monkey studied by the Nauta method, *Vision Res.* **9:**733–747.

Creutzfeldt, O. D., 1988, Extrageniculo-striate visual mechanisms: Compartmentalization of visual functions, *Prog. Brain Res.* **75:**307–320.

Crick, F., 1984, The function of the thalamic reticular complex: The searchlight hypothesis, *Proc. Natl. Acad. Sci. USA* **81:**4586–4590.

Crick, F. H. C., Marr, D. C., and Poggio, T., 1981, An information-processing approach to understanding the visual cortex, in: *The Organization of the Cerebral Cortex* (F. O. Schmitt, F. G. Worden, G. Adelman, and S. G. Dennis, eds.), MIT Press, Cambridge, MA, pp. 505–533.

Cusick, C. G., and Kaas, J. H., 1988, Cortical connections of area 18 and dorsolateral visual cortex in squirrel monkeys, *Visual Neurosci.* **1:**211–237.

Cusick, C. G., Gould, H. J., and Kaas, J. H., 1984, Interhemispheric connections of visual cortex of owl monkeys (*Aotus trivirgatus*), marmosets (*Callithrix jacchus*), and galagos (*Galago crassicaudatus*), *J. Comp. Neurol.* **230:**311–336.

Daniel, P. M., and Whitteridge, D., 1961, The representation of the visual field on the cerebral cortex in monkeys, *J. Physiol.* (Lond.) **159:**203–221.

Darian-Smith, C., and Gilbert, C. D., 1995, Topographic reorganization in the striate cortex of the adult cat and monkey is cortically mediated, *J. Neurosci.* **15:**1631–1647.

DeBruyn, E. J., Casagrande, A., Beck, P. D., and Bonds, A. B., 1993, Visual resolution and sensitivity of single cells in the primary visual cortex (V1) of a nocturnal primate (bush baby): Correlations with cortical layers and cytochrome oxidase patterns, *J. Neurophysiol.* **69:**3–18.

Desimone, R., and Gross, C. G., 1979, Visual areas in the temporal cortex of the macaque, *Brain Res.* **178:**363–380.

Desimone, R., and Schein, S. J., 1987, Visual properties of neurons in area V4 of the macaque: Sensitivity to stimulus form, *J. Neurophysiol.* **57:**835–868.

Desimone, R., and Ungerleider, L. G., 1986, Multiple visual areas in the caudal superior temporal sulcus of the macaque, *J. Comp. Neurol.* **248:**164–189.

Desimone, R., Fleming, J., and Gross, C. G., 1980, Prestriate afferents to inferior temporal cortex: An HRP study, *Brain Res.* **184:**41–55.

Desimone, R., Schein, S. J., Moran, J., and Ungerleider, L. G., 1985, Contour, color and shape analysis beyond the striate cortex, *Vision Res.* **25:**441–452.

Desimone, R., Wessinger, M., Thomas, L., and Schneider, W., 1990, Attentional control of visual perception: Cortical and subcortical mechanisms, *Cold Spring Harbor Symp. Quant. Biol.* **55:**963–971.

Desimone, R., Moran, J., Schein, S. J., and Mishkin, M., 1993, A role for the corpus callosum in visual area V4 of the macaque, *Visual Neurosci.* **10:**159–171.

DeYoe, E. A., and Van Essen, D. C., 1985, Segregation of efferent connections and receptive field properties in visual area V2 of the macaque, *Nature* **317:**58–61.

DeYoe, E. A., Bandettini, P., Neitz, J., Miller, D., and Winans, P., 1994, Functional magnetic resonance imaging (FMRI) of the human brain, *J. Neurosci. Meth.* **54:**171–187.

Doty, R. W., Kimura, D. S., and Mogenson, G. J., 1964, Photically and electrically elicited responses in the central visual system of the squirrel monkey, *Exp. Neurol.* **10:**19–51.

Dow, B. M., Snyder, A. Z., Vautin, R. G., and Bauer, R., 1981, Magnification factor and receptive field size in foveal striate cortex of the monkey, *Exp. Brain Res.* **44:**213–228.

Dreher, B., 1986, Thalamocortical and corticocortical interconnections in the cat visual system: Relation to the mechanisms of information processing, in: *Visual Neuroscience* (J. D. Pettigrew, K. J. Sanderson, and W. R. Levick, eds.), Cambridge University Press, Cambridge, pp. 290–314.

Dreher, B., and Cottee, L. J., 1975, Visual receptive field properties of cells in area 18 of the cat's cerebral cortex before and after acute lesions in area 17, *J. Neurophysiol.* **38:**735–750.

Dreher, B., Michalski, A., Cleland, B. G., and Burke, W., 1992, Effects of selective pressure block of Y-type optic nerve fibers on the receptive field properties of neurons in area 18 of the visual cortex of the cat, *Visual Neurosci.* **9:**65–78.

Dubner, R., and Brown, F. J., 1968, Response of cells to restricted visual stimuli in an association area of cat cerebral cortex, *Exp. Neurol.* **20:**70–86.

Dubner, R., and Zeki, S. M., 1971, Response properties and receptive fields of cells in an anatomically defined region of the superior temporal sulcus in the monkey, *Brain Res.* **35:**528–532.

Duhamel, J.-R., Colby, C. L., and Goldberg, M. E., 1992, The updating of the representation of visual space in parietal cortex by intended eye movements, *Science* **255:**90–92.

Dykes, R. W., and Ruest, A., 1986, What makes a map in somatosensory cortex, in: *Cerebral Cortex,* Volume 5, *Sensory-Motor Areas and Aspects of Cortical Connectivity* (E. G. Jones, and A. Peters, eds.), Plenum Press, New York, pp. 1–29.

Dykes, R. W., Landry, P., Metherate, R., and Hicks, T. P., 1984, Functional role of GABA in cat primary somatosensory cortex: Shaping receptive fields of cortical neurons, *J. Neurophysiol.* **52:**1066–1093.

Edelman, G. M., 1981, Group selection as the basis for higher brain function, in: *The Organization of the Cerebral Cortex* (F. O. Schmitt, F. G. Worden, G. Adelman, and S. G. Dennis, eds.), MIT Press, Cambridge, MA, pp. 535–563.

Erickson, R. G., Dow, B. M., and Snyder, A. Z., 1989, Representation of the fovea in the superior temporal sulcus of the macaque monkey, *Exp. Brain Res.* **78:**90–112.

Felleman, D. J., and Van Essen, D. C., 1987, Receptive field properties of neurons in area V3 of macaque monkey extrastriate cortex, *J. Neurophysiol.* **57:**889–920.

Felleman, D. J., and Van Essen, D. C., 1991, Distributed hierarchical processing in primate cerebral cortex, *Cerebral Cortex* **1:**1–47.
Fiorani, M., Gattass, R., Rosa, M. G. P., and Sousa, A. P. B., 1989, Visual area MT in the *Cebus* monkey: Location, visuotopic organization, and variability, *J. Comp. Neurol.* **287:**98–118.
Fiorani, M., Rosa, M. G. P., Gattass, R., and Rocha-Miranda, C. E., 1992, Dynamic surrounds of receptive fields in primate striate cortex: A physiological basis for perceptual completion? *Proc. Natl. Acad. Sci. USA* **89:**8547–8551.
Flechsig, P., 1920, *Anatomie des menschlichen Gehirns und Ruckenmarks,* Thieme Press, Leipzig.
Frien, A., Eckorn, R., Bauer, R., Woelbern, T., and Kehr, H., 1994, Stimulus-specific fast oscillations at zero phase between visual areas V1 and V2 of the awake monkey, *NeuroReport* **5:**2273–2277.
Fritsches, K., 1995, Visuotopic organization in the primary and second visual areas of the marmoset, Diplomarbeit Thesis, Technischen Hochschule Darmstadt.
Frostig, R. D., 1994, What does *in vivo* optical imaging tell us about the primary visual cortex in primates? in: *Cerebral Cortex,* Volume 10, *Primary Visual Cortex in Primates* (A. Peters and K. S. Rockland, eds.), Plenum Press, New York, pp. 331–358.
Galletti, C., Battaglini, P. P., and Fattori, P., 1990, Functional properties of neurons in the anterior bank of the parieto-occipital sulcus of the macaque monkey, *Eur. J. Neurosci.* **3:**452–461.
Gaska, J. P., Jacobson, L. D., and Pollen, D. A., 1987, Response suppression by extending sine-wave gratings within the receptive fields of neurons in visual cortical area V3a of the macaque monkey, *Vision Res.* **27:**1687–1692.
Gattass, R., and Gross, C. G., 1981, Visual topography of striate projection zone (MT) in posterior superior temporal sulcus of the macaque, *J. Neurophysiol.* **46:**621–638.
Gattass, R., Gross, C. G., and Sandell, J. H., 1981, Visual topography of V2 in the macaque, *J. Comp. Neurol.* **201:**519–539.
Gattass, R., Sousa, A. P. B., and Covey, E., 1985, Cortical visual areas of the macaque: Possible substrates for pattern recognition mechanisms, in: *Pattern Recognition Mechanisms* (C. Chagas, R. Gattass, and C. G. Gross, eds.), Pontificial Academy of Sciences Press, Vatican City, pp. 1–20.
Gattass, R., Sousa, A. P. B., and Rosa, M. G. P., 1987, Visual topography of V1 in the *Cebus* monkey, *J. Comp. Neurol.* **259:**529–548.
Gattass, R., Sousa, A. P. B., and Gross, C. G., 1988, Visuotopic organization and extent of V3 and V4 of the macaque, *J. Neurosci.* **8:**1831–1845.
Gattass, R., Rosa, M. G. P., Sousa, A. P. B., Piñon, M. C. G. P., Fiorani, M., and Neuenschwander, S., 1990, Cortical streams of visual information processing in primates, *Brazil. J. Med. Biol. Res.* **23:**375–393.
Gilbert, C. D., and Wiesel, T. N., 1992, Receptive field dynamics in adult primary visual cortex, *Nature* **356:**150–152.
Girard, P., and Bullier, J., 1989, Visual activity in area V2 during reversible inactivation of area 17 in the macaque monkey, *J. Neurophysiol.* **62:**1287–1302.
Girard, P., Salin, P. A., and Bullier, J., 1992, Response selectivity of neurons in area MT of the macaque monkey during reversible inactivation of area V1, *J. Neurophysiol.* **67:**1437–1446.
Grinvald, A., Lieke, E. E., Frostig, R. D., and Hildesheim, 1994, Cortical point-spread function and long-range lateral interactions revealed by real-time optical imaging of macaque monkey primary visual cortex, *J. Neurosci.* **14:**2545–2568.
Gross, C. G., and Mishkin, M., 1977, The neural basis of stimulus equivalence across retinal translation, in: *Lateralization in the Nervous System* (S. Harnad, R. Doty, J. Jaynes, L. Goldstein, and G. Krauthamer, eds.), Academic Press, New York, pp. 109–122.
Gross, C. G., Schiller, P. H., Wells, C., and Gerstein, G. L., 1967, Single-unit activity in temporal association cortex of the monkey, *J. Neurophysiol.* **30:**833–843.
Gross, C. G., Bender, D. B., and Rocha-Miranda, C. E., 1969, Visual receptive fields of neurons in inferotemporal cortex of the monkey, *Science* **166:**1303–1306.
Gross, C. G., Bruce, C. J., Desimone, R., Fleming, J., and Gattass, R., 1981, Cortical visual areas of the temporal lobe: Three areas in the macaque, in: *Cortical Sensory Organization,* Volume 2, *Multiple Visual Areas* (C. N. Woolsey, ed.), Humana Press, Clifton, NJ, pp. 187–216.
Hall, W. C., Kaas, J. H., Killackey, H., and Diamond, I. T., 1971, Cortical visual areas in the grey squirrel (*Sciurus carolinensis*): A correlation between cortical evoked potential maps and architectonic subdivisions, *J. Neurophysiol.* **34:**437–452.

Heinen, S. J., and Skavenski, A. A., 1991, Recovery of visual responses in foveal V1 neurons following bilateral foveal lesions in adult monkey, *Exp. Brain Res.* **83:**670–674.

Hicks, T. P., and Dykes, R. W., 1983, Receptive field size for certain neurons in primary somatosensory cortex is determined by GABA-mediated intracortical inhibition, *Brain Res.* **274:**160–164.

Hubel, D. H., and Livingstone, M. S., 1985, Complex-unoriented cells in a subregion of primate area 18, *Nature* **315:**325–327.

Hubel, D. H., and Livingstone, M. S., 1987, Segregation of form, color, and stereopsis in primate area 18, *J. Neurosci.* **7:**3378–3415.

Hubel, D. H., and Wiesel, T. N., 1965, Receptive fields and functional architecture in two non-striate visual areas (18 and 19) of the cat, *J. Neurophysiol.* **28:**229–289.

Hubel, D. H., and Wiesel, T. N., 1969, Visual area of the lateral suprasylvian gyrus (Clare-Bishop area) of the cat, *J. Physiol.* (Lond.) **202:**251–260.

Hubel, D. H., and Wiesel, T. N., 1970, Cells sensitive to binocular depth in area 18 of the macaque monkey cortex, *Nature* **225:**41–42.

Hubel, D. H., and Wiesel, T. N., 1974, Uniformity of monkey striate cortex: A parallel relationship between field size, scatter, and magnification factor, *J. Comp. Neurol.* **158:**295–306.

Hubel, D. H., and Wiesel, T. N., 1977, Functional architecture of macaque visual cortex, *Proc. R. Soc. London. B* **198:**1–59.

Hughes, A., 1977, The topography of vision in mammals of contrasting life style: Comparative optics and retinal organization, in: *Handbook of Sensory Physiology,* Volume VII/5, *The Visual System in Vertebrates* (F. Crescitelli, ed.), Springer-Verlag, Berlin, pp. 613–756.

Imig, T. J., Reale, R. A., and Brugge, J. F., 1982, The auditory cortex: Patterns of corticocortical projections related to physiological maps in the cat, in: *Cortical Sensory Organization,* Volume 3, *Multiple Auditory Areas* (C. N. Woolsey, ed.), Humana Press, Clifton, NJ, pp. 1–41.

Irvine, D. R. F., 1992, Physiology of the auditory brainstem, in: *The Mammalian Auditory Pathway: Neurophysiology* (A. N. Popper, and R. R. Fay, eds.), Springer-Verlag, Berlin, pp. 153–231.

Jain, N., Preuss, T. M., and Kaas, J. H., 1994, Subdivisions of the visual system labeled with the Cat-301 antibody in tree shrews, *Visual Neurosci.* **11:**731–741.

Jones, E. G., 1993, GABAergic neurons and their role in cortical plasticity in primates, *Cerebral Cortex* **3:**361–372.

Kaas, J. H., 1989, Why does the brain have so many visual areas?, *J. Cognitive Neurosci.* **1:**121–135.

Kaas, J. H., and Morel, A., 1993, Connections of visual areas of the upper temporal lobe of owl monkeys: The MT crescent and dorsal and ventral subdivisions of FST, *J. Neurosci.* **13:**534–546.

Kaas, J. H., and Preuss, T. M., 1993, Archontan affinities as reflected in the visual system, in: *Mammalian Phylogeny* (F. Szalay, M. Novacek, and M. McKenna, eds.), Springer-Verlag, New York, pp. 115–128.

Kaas, J., Hall, W. C., and Diamond, I. T., 1970, Cortical visual areas I and II in the hedgehog: The relation between evoked potential maps and architectonic subdivisions, *J. Neurophysiol.* **33:**595–615.

Kaas, J. H., Hall, W. C., Killackey, H., and Diamond, I. T., 1972, Visual cortex of the tree shrew (*Tupaia glis*): Architectonic subdivisions and representation of the visual field, *Brain Res.* **42:**491–496.

Kaas, J. H., Krubitzer, L. A., Chino, Y. M., Langston, A. L., Polley, E. H., and Blair, N., 1990, Reorganization of retinotopic cortical maps in adult mammals after lesions of the retina, *Science* **248:**229–231.

Karten, H. J., 1979, Visual lemniscal pathways in birds, in: *Neural Mechanisms of Behavior in the Pigeon* (A. M. Granda and J. H. Maxwell, eds.), Plenum Press, New York, pp. 409–430.

Kitano, M., Niiyama, K., Kasamatsu, T., Sutter, E. E., and Norcia, A. M., 1994, Retinotopic and nonretinotopic field potentials in cat visual cortex, *Visual Neurosci.* **11:**953–977.

Koch, C., and Ullman, S., 1985, Shifts in selective visual attention: Towards the underlying neural circuitry, *Hum. Neurobiol.* **4:**219–227.

Kohonen, T., 1989, *Self-Organization and Associative Memory,* 3rd ed., Springer-Verlag, Berlin.

Kolarik, R. C., Rasey, S. K., and Wall, J. T., 1994, The consistency, extent, and locations of early-onset changes in cortical nerve dominance aggregates following injury of nerves to primate hands, *J. Neurosci.* **14:**4269–4288.

Komatsu, H., and Wurtz, R. H., 1988, Relation of cortical areas MT and MST to pursuit eye movements. I. Localization and visual properties of neurons, *J. Neurophysiol.* **60:**580–603.

Krubitzer, L. A., and Kaas, J. H., 1989, Cortical integration of parallel pathways in the visual system of primates, *Brain Res.* **478:**161–165.

Krubitzer, L. A., and Kaas, J. H., 1990, Cortical connections of MT in four species of primates: Areal, modular, and retinotopic patterns, *Visual Neurosci.* **5:**165–204.

Krubitzer, L. A., and Kaas, J. H., 1993, The dorsomedial visual area of owl monkeys: Connections, myeloarchitecture, and homologies in other primates, *J. Comp. Neurol.* **334:**497–528.

Krubitzer, L. A., Clarey, J., Tweedale, R., Elston, G., and Calford, M., 1995, A redefinition of somatosensory areas in the lateral sulcus of macaque monkeys, *J. Neurosci.* **15:**3821–3839.

Laskin, S. E., and Spencer, W. A., 1979, Cutaneous masking. II. Geometry of excitatory and inhibitory receptive fields of single units in somatosensory cortex of the cat, *J. Neurophysiol.* **42:**1061–1082.

Lee, D., and Malpeli, J. G., 1994, Global form and singularity: Modeling the blind spot's role in lateral geniculate morphogenesis, *Science* **263:**1292–1294.

Leventhal, A. G., Thompson, K. G., Liu, D., Zhou, Y., and Ault, S. J., 1995, Concomitant sensitivity to orientation, direction, and color of cells in layers 2, 3, and 4 of monkey striate cortex, *J. Neurosci.* **15:**1808–1818.

Levi, D. M., Klein, S. A., and Aitsebaomo, A. P., 1985, Vernier acuity, crowding and cortical magnification, *Vision Res.* **25:**963–977.

Levitt, J. B., Kiper, D. C., and Movshon, J. A., 1994a, Receptive fields and functional architecture of macaque V2, *J. Neurophysiol.* **71:**2517–2542.

Levitt, J. B., Yoshioka, T., and Lund, J. S., 1994b, Intrinsic cortical connections in macaque visual area V2: Evidence for interactions between different functional streams, *J. Comp. Neurol.* **342:**551–570.

Lin, C. S., Weller, R. E., and Kaas, J. H., 1982, Cortical connections of striate cortex in owl monkeys, *J. Comp. Neurol.* **211:**165–176.

Lvingstone, M. S., and Hubel, D. H., 1982, Thalamic input to cytochrome oxidase-rich regions in monkey visual cortex, *Proc. Natl. Acad. Sci. USA* **79:**6098–6101.

Livingstone, M. S., and Hubel, D. H., 1984, Anatomy and physiology of a color system in the primate visual cortex, *J. Neurosci.* **4:**309–356.

Lund, J. S., Yoshioka, T., and Levitt, J. B., 1993, Comparison of intrinsic connectivity in different areas of macaque monkey cerebral cortex, *Cerebral Cortex* **3:**148–162.

Maguire, W. M., and Baizer, J. S., 1984, Visuotopic organization of the prelunate gyrus in rhesus monkey, *J. Neurosci.* **4:**1690–1704.

Malach, R., 1994, Cortical columns as devices for maximizing neuronal diversity, *Trends Neurosci.* **17:**101–104.

Malach, R., Tootell, R. B. H., and Malonek, D., 1994, Relationship between orientation domains, cytochrome oxidase stripes, and intrinsic horizontal connections in squirrel monkey area V2, *Cerebral Cortex* **4:**151–165.

Malonek, D., Tootell, R. B. H., and Grinvald, A., 1994, Optical imaging reveals the functional architecture of neurons processing shape and motion in owl monkey area MT, *Proc. R. Soc. Lond. B* **258:**109–119.

Maunsell, J. H. R., and Van Essen, D. C., 1983, The connections of the middle temporal visual area (MT) and their relationship to a cortical hierarchy in the macaque monkey, *J. Neurosci.* **3:**2563–2586.

Maunsell, J. H. R., and Van Essen, D. C., 1987, Topographic organization of the middle temporal visual area in the macaque monkey: Representational biases and the relationship to callosal connections and myeloarchitectonic boundaries, *J. Comp. Neurol.* **266:**535–555.

McIlwain, J. T., 1975, Visual receptive fields and their images in superior colliculus of the cat, *J. Neurophysiol.* **38:**219–230.

McIlwain, J. T., 1976, Large receptive fields and spatial transformations in the visual system, *Int. Rev. Physiol.* **10:**223–248.

McIlwain, J. T., 1983, Representation of the visual streak in visuotopic maps of the cat's superior colliculus: Influence of the mapping variable, *Vision Res.* **23:**507–516.

McIlwain, J. T., 1995, Lateral geniculate lamination and the corticogeniculate projection: A potential role in binocular vision in the quadrants, *J. Theor. Biol.* **172:**329–333.

Mitchison, G., 1995, A type of duality between self-organizing maps and minimal wiring, *Neural Computation* **7:**25–35.

Montero, V. M., Rojas, A., and Torrealba, F., 1973, Retinotopic organization of striate and prestriate visual cortex in the albino rat, *Brain Res.* **53:**202–207.

Moran, J., and Desimone, R., 1985, Selective attention gates visual processing in the extrastriate cortex, *Science* **229:**782–784.

Morel, A., Garraghty, P. E., and Kaas, J. H., 1993, Tonotopic organization, architecture fields, and connections of auditory cortex in macaque monkeys, *J. Comp. Neurol.* **335:**437–459.

Motter, B. C., and Mountcastle, V. B., 1981, The functional properties of the light-sensitive neurons of the posterior parietal cortex studied in waking monkeys: Foveal sparing and opponent vector organization, *J. Neurosci.* **1:**3–26.

Motter, B. C., and Poggio, G. F., 1990, Dynamic stabilization of receptive fields of cortical neurons (VI) during fixation of gaze in the macaque, *Exp. Brain Res.* **83:**37–43.

Motter, B. C., Steinmetz, M. A., Duffy, C. J., and Mountcastle, V. B., 1987, Functional properties of parietal visual neurons: Mechanisms of directionality along a single axis, *J. Neurosci.* **7:**154–176.

Mountcastle, V. B., and Powell, T. P. S., 1959, Neural mechanisms subserving cutaneous sensibility, with special reference to the role of afferent inhibition in sensory perception and discrimination, *Bull. Johns Hopkins Hosp.* **105:**201–232.

Myers, R. E., 1962, Commissural connections between occipital lobes of the monkey, *J. Comp. Neurol.* **118:**1–16.

Nelson, J. I., Salin, P. A., Munk, M. H. J., Arzi, M., and Bullier, J., 1992, Spatial and temporal coherence in corticocortical connections: A cross-correlation study in area 17 and area 18 in the cat, *Visual Neurosci.* **9:**21–37.

Neuenschwander, S., Gattass, R., Sousa, A. P. B., and Piñon, M. C. G. P., 1994, Identification and visuotopic organization of areas PO and POd in *Cebus* monkey, *J. Comp. Neurol.* **340:**65–86.

Newsome, W. T., and Allman, J. M., 1980, Interhemispheric connections of visual cortex in the owl monkey, *Aotus trivirgatus,* and the bushbaby, *Galago senegalensis., J. Comp. Neurol.* **194:**209–233.

Newsome, W. T., Wurtz, R. H., Dursteler, M. R., and Mikami, A., 1985, Deficits in visual motion processing following ibotenic acid lesions of the middle temporal visual area of the macaque monkey, *J. Neurosci.* **5:**825–840.

Newsome, W. T., Maunsell, J. H. R., and Van Essen, D. C., 1986, Ventral posterior visual area of the macaque: Visual topography and areal boundaries, *J. Comp. Neurol.* **252:**139–153.

Olavarria, J., and Torrealba, F., 1978, The effect of acute lesions of the striate cortex on the retinotopic organization of the lateral peristriate cortex in the rat, *Brain Res.* **151:**386–391.

Olavarria, J. F., DeYoe, E. A., Knierim, J. J., Fox, J. M., and Van Essen, D. C., 1992, Neural responses to visual texture patterns in middle temporal area of the macaque monkey, *J. Neurophysiol.* **68:**164–181.

Paolini, M., Sereno, M. I., Jeo, R., Dobbins, A., and Allman, J. A., 1994, Organization of extrastriate cortex in the primitive primates, *Cheirogaleus* and *Lemur, Soc. Neurosci. Abstr.* **20:**427.

Payne, B. R., and Siwek, D. F., 1990, Receptive fields of neurons at the confluence of cerebral cortical areas 17, 18, 20a, and 20b in the cat, *Visual Neurosci.* **4:**475–479.

Pearson, J. C., Finkel, L. H., and Edelman, G. M., 1987, Plasticity in the organization of adult cerebral cortical maps: A computer simulation based on neuronal group selection, *J. Neurosci.* **7:**4209–4223.

Perkel, D. J., Bullier, J., and Kennedy, H., 1986, Topography of the afferent connectivity of area 17 in the macaque monkey: A double-labelling study, *J. Comp. Neurol.* **253:**374–402.

Perrett, D. I., Smith, P. A. J., Potter, D. D., Mistlin, A. J., Head, A. S., Milner, A. D., and Jeeves, M. A., 1985, Visual cells in the temporal cortex sensitive to face view and gaze direction, *Proc. R. Soc. Lond. B* **223:**293–317.

Perrett, D. I., Harries, M. H., Mistlin, A. J., Hietanen, J. K., Benson, P. J., Bevan, R., Thomas, S., Oram, M. W., Ortega, J., and Brierley, K., 1990, Social signals analyzed at the single cell level: Someone is looking at me, something touched me, something moved! *Int. J. Comp. Psychol.* **4:**25–55.

Pessoa, V. F., Abrahão, J. C. H., Pacheco, R. A., Pereira, L. C. M., Magalhães-Castro, B., and Saraiva, P. E. S., 1992, Relative size of cortical visual areas in marmosets: Functional and phylogenetic implications, *Exp. Brain Res.* **88:**459–462.

Peterhans, E., and von der Heydt, R., 1993, Functional organization of area V2 in the alert macaque, *Eur. J. Neurosci.* **5:**509–524.

Peters, A., and Sethares, C., 1991, Organization of pyramidal neurons in area 17 of monkey visual cortex, *J. Comp. Neurol.* **306:**1–23.
Peters, A., Payne, B. R., and Budd, J., 1994, A numerical analysis of the geniculocortical input to striate cortex in the monkey, *Cerebral Cortex* **4:**215–229.
Pettigrew, J. D., Jamieson, B. G. M., Robson, S. K., Hall, L. S., McAnally, K. I., and Cooper, H. M., 1989, Phylogenetic relations between microbats, megabats and primates (Mammalia: Chiroptera and Primates), *Phil. Trans. R. Soc. Lond. B* **325:**489–559.
Phillips, D. P., Semple, M. N., Calford, M. B., and Kitzes, L. M., 1994, Level-dependent representation of stimulus frequency in cat primary auditory cortex, *Exp. Brain Res.* **102:**210–226.
Poggio, T., Torre, V., and Koch, C., 1985, Computational vision and regularization theory, *Nature* **317:**314–319.
Preuss, T. M., Beck, P. D., and Kaas, J. H., 1993, Areal, modular, and connectional organization of visual cortex in a prosimian primate, the slow loris (*Nycticebus coucang*), *Brain Behav. Evol.* **42:**321–335.
Raiguel, S. E., Lagae, L., Gulyas, B., and Orban, G. A., 1989, Response latencies of visual cells in macaque areas V1, V2 and V5, *Brain Res.* **493:**155–159.
Rakic, P., 1988, Specification of cerebral cortical areas, *Science* **241:**170–176.
Robinson, D. L., Goldberg, M. E., and Stanton, G. B., 1978, Parietal association cortex in the primate: Sensory mechanisms and behavioral modulations. *J. Neurophysiol.* **41:**910–932.
Rockland, K. S., 1985, A reticular pattern of intrinsic connections in primate area V2 (area 18), *J. Comp. Neurol.* **235:**467–478.
Rockland, K. S., and Pandya, D. N., 1979, Laminar origin and terminations of cortical connections of the occipital lobe in the rhesus monkey, *Brain Res.* **179:**3–20.
Rodman, H. R., Gross, C. G., and Albright, T. D., 1989, Afferent basis of visual response properties in area MT of the macaque. I. Effects of striate cortex removal, *J. Neurosci.* **9:**2033–2050.
Rodman, H. R., Gross, C. G., and Albright, T. D., 1990, Afferent basis of visual response properties in area MT of the macaque. II. Effects of superior colliculus removal, *J. Neurosci.* **10:**1154–1164.
Roe, A. W., and Ts'o, D. Y., 1995, Visual topography in primate V2: Multiple representations across functional stripes, *J. Neurosci.* **15:**3689–3715.
Rolls, E. T., and Cowey, A., 1970, Topography of the retina and striate cortex and its relationship to visual acuity in rhesus monkeys and squirrel monkeys, *Exp. Brain Res.* **10:**298–310.
Rosa, M. G. P., and Schmid, L. M., 1994, Topography and extent of visual-field representation in the superior colliculus of the megachiropteran *Pteropus*, *Visual Neurosci.* **11:**1037–1057.
Rosa, M. G. P., and Schmid, L. M., 1995, Visual areas in the dorsal and medial extrastriate cortices of the marmoset, *J. Comp. Neurol.* **359:**272–299.
Rosa, M. G. P., Gattass, R., and Fiorani, M., 1988a, Complete pattern of ocular dominance stripes in V1 of a New World monkey, *Cebus apella*, *Exp. Brain Res.* **72:**645–648.
Rosa, M. G. P., Sousa, A. P. B., and Gattass, R., 1988b, Representation of the visual field in the second visual area in the *Cebus* monkey, *J. Comp. Neurol.* **275:**326–345.
Rosa, M. G. P., Gattass, R., and Soares, J. G. M., 1991, A quantitative analysis of cytochrome oxidase-rich patches in the primary visual cortex of *Cebus* monkeys: Topographic distribution and effects of late monocular enucleation, *Exp. Brain Res.* **84:**195–209.
Rosa, M. G. P., Schmid, L. M., Krubitzer, L. A., and Pettigrew, J. D., 1993a, Retinotopic organization of the primary visual cortex of flying foxes (*Pteropus poliocephalus* and *Pteropus scapulatus*). *J. Comp. Neurol.* **335:**55–72.
Rosa, M. G. P., Soares, J. G. M., Fiorani, M., and Gattass, R., 1993b, Cortical afferents of visual area MT in the *Cebus* monkey: Possible homologies between New and Old World monkeys, *Visual Neurosci.* **10:**827–855.
Rosa, M. G. P., Schmid, L. M., and Pettigrew, J. D., 1994, Organization of the second visual area in the megachiropteran bat *Pteropus*, *Cerebral Cortex* **4:**52–68.
Rosa, M. G. P., Schmid, L. M., and Calford, M. B., 1995a, Responsiveness of cat area 17 after monocular inactivation: Limitation of topographic plasticity in adult cortex, *J. Physiol.* (Lond.) **482:**589–608.
Rosa, M. G. P., Schmid, L. M., and Clarey, J. C., 1995b, Visual areas in the extrastriate cortex of the marmoset, *Soc. Neurosci. Abstr.* **21:**903.
Rosa, M. G. P., Casagrande, V. A., Preuss, T., and Kaas, J. H., 1997a, Visual field representation in striate and prestriate cortices of a prosimian primate (*Galago garnetti*), *J. Neurophysiol.* **77:**3193–3217.

Rosa, M. G. P., Fritsches, K. A., and Elston, G. N., 1997b, The second visual area in the marmoset monkey: Magnification factors, architectural boundaries, and modularity, *J. Comp. Neurol.*, in press.

Saito, H., Yukie, M., Tanaka, K., Hikosaka, K., Fukada, Y., and Iwai, E., 1986, Integration of direction signals of image motion in the superior temporal sulcus of the macaque monkey, *J. Neurosci.* **6:**145–157.

Salin, P.-A., and Bullier, J., 1995, Corticocortical connections in the visual system: Structure and function, *Physiol. Rev.* **75:**107–154.

Schein, S. J., and Desimone, R., 1990, Spectral properties of V4 neurons in the macaque, *J. Neurosci.* **10:**3369–3389.

Schein, S. J., Marrocco, R. T., and DeMonasterio, F. M., 1982, Is there a high concentration of color-selective cells in area V4 of monkey visual cortex? *J. Neurophysiol.* **47:**193–213.

Schiller, P. H., and Malpeli, J. G., 1977, The effect of cooling striate cortex on area 18 cells in the monkey, *Brain Res.* **126:**366–369.

Schmid, L. M., Rosa, M. G. P., and Calford, M. B., 1995, Retinal detachment induces massive immediate reorganization in visual cortex, *NeuroReport* **6:**1349–1353.

Schmid, L. M., Rosa, M. G. P., Calford, M. B., and Ambler, J. S., 1996, Visuotopic reorganisation in the primary visual cortex of adult cats following monocular and binocular retinal lesions, *Cerebral Cortex* **6:**388–405.

Schmidt, J. T., 1985, Formation of retinotopic connections: Selective stabilization by an activity-dependent mechanism, *Cell. Mol. Neurobiol.* **5:**65–84.

Schreiner, C. E., 1991, Functional topographies in the primary auditory cortex of the cat, *Acta Otolaryngol. Suppl.* **491:**7–16.

Schreiner, C. E., 1995, Order and disorder in auditory cortical maps, *Curr. Opin. Neurobiol.* **5:**489–496.

Schreiner, C. E., and Sutter, M. L., 1992, Topography of excitatory bandwidth in cat primary auditory cortex: Single-neuron versus multiple-neuron recordings, *J. Neurophysiol.* **68:**1487–1502.

Schwartz, E. L., 1977, Spatial mapping in the primate sensory projection: Analytic structure and relevance to perception, *Biol. Cybernet.* **25:**181–194.

Sereno, M. I., and Allman, J. M., 1991, Cortical visual areas in mammals, in: *Vision and Visual Dysfunction,* Volume 4, *The Neural Basis of Visual Function* (A. G. Leventhal, ed.), Macmillan, London, pp. 160–172.

Sereno, M. I., McDonald, C. T., and Allman, J. M., 1987, Multiple visual areas between V2 and MT in the owl monkey, *Soc. Neurosci. Abstr.* **13:**625.

Sereno, M. I., McDonald, C. T., and Allman, J. M., 1994, Analysis of retinotopic maps in extrastriate cortex, *Cerebral Cortex* **4:**601–620.

Sereno, M. I., Dale, A. M., Reppas, J. B., Kwong, K. K., Belliveau, J. W., Brady, T. J., Rosen, B. R., and Tootell, R. B. H., 1995, Borders of multiple visual areas in humans revealed by functional magnetic resonance imaging, *Science* **268:**889–893.

Sesma, M. A., Casagrande, V. A., and Kaas, J. H., 1984, Cortical connections of area 17 in tree shrews, *J. Comp. Neurol.* **230:**337–351.

Sherk, H., 1978, Area 18 cell responses in the cat during reversible inactivation of area 17, *J. Neurophysiol.* **41:**204–215.

Sherk, H., and Mulligan, K., 1992, Retinotopic order is surprisingly good within cell columns in the cat's lateral suprasylvian cortex, *Exp. Brain Res.* **91:**46–60.

Sherk, H., and Mulligan, K. A., 1993, A reassessment of the lower visual field map in striate-recipient lateral suprasylvian cortex, *Visual Neurosci.* **10:**131–158.

Shimizu, T., Cox, K., and Karten, H. J., 1995, Intratelencephalic projections of the visual wulst in pigeons (*Columba livia*), *J. Comp. Neurol.* **359:**551–572.

Shipp, S., and Zeki, S., 1989, The organization of connections between area V5 and V2 in macaque monkey visual cortex, *Eur. J. Neurosci.* **1:**333–354.

Sillito, A. M., 1984, Functional considerations of the operation of GABAergic inhibitory processes in the visual cortex, in: *Cerebral Cortex,* Volume 2, *Functional Properties of Cortical Cells* (E. G. Jones, and A. Peters, eds.), Plenum Press, New York, pp. 91–117.

Singer, W., 1994, Coherence as an organizing principle of cortical functions, *Int. Rev. Neurobiol.* **37:**153–183.

Sousa, A. P. B., Piñon, M. C. G. P., Gattass, R., and Rosa, M. G. P., 1991, Topographic organization of cortical input to striate cortex in the *Cebus* monkey: A fluorescent tracer study, *J. Comp. Neurol.* **308:**665–682.

Spatz, W. B., 1977, Topographically organized reciprocal connections between areas 17 and MT (visual area of the superior temporal sulcus) in the marmoset *Callithrix jacchus, Exp. Brain Res.* **27:**559–572.

Spatz, W. B., and Tigges, J., 1972, Species difference between Old World and New World monkeys in the organization of the striate–prestriate association, *Brain Res.* **43:**591–594.

Spatz, W. B., Tigges, J., and Tigges, M., 1970, Subcortical projections, cortical associations, and some intrinsic interlaminar connections of the striate cortex in the squirrel monkey (*Saimiri*), *J. Comp. Neurol.* **140:**155–174.

Sperry, R. W., 1963, Chemoaffinity in the orderly growth of nerve fiber patterns and connections, *Proc. Natl. Acad. Sci. USA* **50:**703–709.

Steele, G. E., Weller, R. E., and Cusick, C. G., 1991, Cortical connections of the caudal subdivision of the dorsolateral area (V4) in monkeys, *J. Comp. Neurol.* **306:**495–520.

Stepniewska, I., and Kaas, J. H., 1996, Topographic patterns of V2 cortical connections in macaque monkeys, *J. Comp. Neurol.* **371:**129–152.

Stuermer, C. A. O., 1991, The formation of topographically ordered connections during development and regeneration of the vertebrate visual system, in: *Vision and Visual Dysfunction,* Volume 11, *Development and Plasticity of the Visual System* (J. Cronly-Dillon, ed.), Macmillan, London, pp. 88–111.

Sutherland, N. S., 1973, Object recognition, in: *Handbook of Perception,* Volume III, *Biology of Perceptual Systems* (E. C. Carterette, and M. P. Friedman, eds.), Academic Press, New York, pp. 157–185.

Sutter, M. L., and Schreiner, C. E., 1991, Physiology and topography of neurons with multipeaked tuning curves in cat primary auditory cortex, *J. Neurophysiol.* **65:**1207–1226.

Swindale, N. V., 1991, Coverage and the design of striate cortex, *Biol. Cybernet.* **65:**415–426.

Talbot, S. A., 1942, A lateral localization in the cat's visual cortex, *Fed. Proc.* **1:**84.

Tanaka, K., Hikosaka, K., Saito, H., Yukie, M., Fukada, Y., and Iwai, E., 1986, Analysis of local and wide-field movements in the superior temporal visual areas of the macaque monkey, *J. Neurosci.* **6:**134–144.

Tigges, J., Spatz, W. B., and Tigges, M., 1974, Efferent cortico-cortical fibre connections of area 18 in the squirrel monkey (*Saimiri*), *J. Comp. Neurol.* **158:**219–236.

Tigges, J., Tigges, M., Anschel, S., Cross, N. A., Letbetter, W. D., and McBride, R. L., 1981, Areal and laminar distribution of neurons interconnecting the central visual cortical areas 17, 18, 19 and MT in squirrel monkey (*Saimiri*), *J. Comp. Neurol.* **202:**539–560.

Tootell, R. B. H., and Hamilton, S. L., 1989, Functional anatomy of the second visual area (V2) in the macaque, *J. Neurosci.* **9:**2620–2644.

Tootell, R. B. H., and Taylor, J. B., 1995, Anatomical evidence for MT and additional cortical visual areas in humans, *Cerebral Cortex* **5:**39–55.

Tootell, R. B. H., Silverman, M. S., Switkes, E., and DeValois, R. L., 1982, Deoxyglucose analysis of retinotopic organization in primate striate cortex, *Science* **218:**902–904.

Tootell, R. B. H., Silverman, M. S., De Valois, R. L., and Jacobs, G. H., 1983, Functional organization of the second cortical visual area of primates, *Science* **220:**737–739.

Tootell, R. B. H., Hamilton, S. L., and Silverman, M. S., 1985, Topography of cytochrome oxidase activity in the owl monkey cortex, *J. Neurosci.* **5:**2786–2800.

Tootell, R. B. H., Switkes, E., Silverman, M. S., and Hamilton, S. L., 1988, Functional anatomy of macaque striate cortex. II. Retinotopic organization, *J. Neurosci.* **8:**1531–1568.

Tripathy, S., Levi, D. M., Ogmen, H., and Harden, C., 1995, Perceived length across the physiological blind spot, *Visual Neurosci.* **12:**385–402.

Tusa, R. J., Rosenquist, A. C., and Palmer, L. A., 1979, Retinotopic organization of areas 18 and 19 in the cat, *J. Comp. Neurol.* **185:**657–678.

Ungerleider, L. G., and Desimone, R., 1986, Projections to the superior temporal sulcus from the central and peripheral field representations of V1 and V2, *J. Comp. Neurol.* **248:**147–163.

Ungerleider, L. G., and Mishkin, M., 1979, The striate projection zone in the superior temporal sulcus of *Macaca mulatta:* Location and topographic organization, *J. Comp. Neurol.* **188:**347–366.

Ungerleider, L. G., and Mishkin, M., 1982, Two cortical systems, in: *Analysis of Visual Behavior* (D. J. Ingle, M. A. Goodale, and R. J. W. Mansfield, eds.), MIT Press, Cambridge, MA, pp. 549–586.

Van Essen, D. C., 1985, Functional organization of primate visual cortex, in: *Cerebral Cortex,* Volume 3, *Visual Cortex* (A. Peters, and E. G. Jones, eds.), Plenum Press, New York, pp. 259–329.

Van Essen, D. C., and Anderson, C. H., 1990, Information processing strategies and pathways in the primate retina and visual cortex, in: *An Introduction to Neural and Electronic Networks* (S. F. Zornetzer, J. L. Davis, and C. Lau, eds.), Academic Press, New York, pp. 43–72.

Van Essen, D. C., and Zeki, S. M., 1978, The topographic organization of rhesus monkey prestriate cortex, *J. Physiol.* (Lond.) **277:**193–226.

Van Essen, D. C., Maunsell, J. H. R., and Bixby, J. L., 1981, The middle temporal visual area in the macaque: Myeloarchitecture, connections, functional properties and topographic organization, *J. Comp. Neurol.* **199:**293–326.

Van Essen, D. C., Newsome, W. T., and Bixby, J. L., 1982, The pattern of interhemispheric connections and its relationship to extrastriate visual areas in the macaque monkey, *J. Neurosci.* **2:**265–283.

Van Essen, D. C., Newsome, W. T., Maunsell, J. H. R., and Bixby, J. L., 1986, The projections from striate cortex (V1) to areas V2 and V3 in the macaque monkey: Asymmetries, areal bounderies, and patchy connections, *J. Comp. Neurol.* **244:**451–480.

Van Essen, D. C., Felleman, D. J., DeYoe, E. A., Olavarria, J., and Knierim, J., 1990, Modular and hierarchical organization of extrastriate visual cortex in the macaque monkey, *Cold Spring Harbor Symp. Quant. Biol.* **55:**679–696.

Vidyasagar, T. R., 1996, Attentional searchlight gates responses of neurones in macaque primary visual cortex, *Proc. Aust. Neurosci. Soc.* **7:**36.

Von Bonin, G., and Bailey, P., 1947, *The Neocortex of Macaca mulatta,* University of Illinois Press, Urbana, IL.

Wagor, E., Lin, C. S., and Kaas, J. H., 1975, Some cortical projections of the dorsomedial visual area (DM) of association cortex in owl monkey, *Aotus trivirgatus, J. Comp. Neurol.* **163:**227–250.

Wall, J. T., Symonds, L. L., and Kaas, J. H., 1982, Cortical and subcortical projections of the middle temporal area (MT) and adjacent cortex in Galagos, *J. Comp. Neurol.* **211:**193–214.

Wässle, H., Grünet, U., Röhrenbeck, J., and Boycott, B., 1990, Retinal ganglion cell density and cortical magnification factor in the primate, *Vision Res.* **30:**1897–1911.

Watson, J. D. G., Myers, R., Frackowiak, R. S. J., Hajnal, J. V., Woods, R. P., Mazziotta, J. C., Shipp, S., and Zeki, S. M., 1993, Area V5 of the human brain: Evidence from a combined study using positron emission tomography and magnetic resonance imaging, *Cerebral Cortex* **3:**79–94.

Weller, R. E., and Kaas, J. H., 1983, Retinotopic patterns of connections of area 17 with areas V-II and MT in macaque monkeys, *J. Comp. Neurol.* **220:**253–279.

Weller, R. E., and Steele, G. E., 1992, Cortical connections of subdivisions of inferior temporal cortex in squirrel monkeys, *J. Comp. Neurol.* **324:**37–66.

Weller, R. E., Wall, J. T., and Kaas, J. H., 1984, Cortical connections of the middle temporal visual area (MT) and the superior temporal cortex in owl monkeys, *J. Comp. Neurol.* **228:**81–104.

Weller, R. E., Steele, G. E., and Cusick, C. G., 1991, Cortical connections of dorsal cortex rostral to V-II in squirrel monkeys, *J. Comp. Neurol.* **306:**521–537.

Westheimer, G., 1982, The spatial grain of the perifoveal visual field, *Vision Res.* **22:**157–162.

Wiitanen, J. T., 1969, Selective silver impregnation of degenerating axons and axon terminals in the central nervous system of the monkey (*Macaca mulatta*), *Brain Res.* **14:**546–548.

Wilder, H. D., Grünert, U., Lee, B. B., and Martin, P. R., 1996, Topography of ganglion cells and photoreceptors in the retina of the New World marmoset monkey *Callithrix jacchus, Visual Neurosci.* **13:**335–352.

Wolf, F., Bauer, H.-U., and Geisel, T., 1994, Formation of field discontinuities and islands in visual cortical maps, *Biol. Cybernet.* **70:**525–531.

Wong-Riley, M., Hevner, R. F., Cutlan, R., Earnest, M., Egan, R., Frost, J., and Nguyen, T., 1993, Cytochrome oxidase in the human visual cortex: Distribution in the developing and adult brain, *Visual Neurosci.* **10:**41–58.

Woolsey, C. N., Akert, K., Benjamin, R. M., Leibowitz, H., and Welker, W. I., 1955, Visual cortex of the marmoset, *Fed. Proc.* **14:**166.

Wright, M. J., 1969, Visual receptive fields of cells in a cortical area remote from striate cortex in the cat, *Nature* **223:**973–975.

Wurtz, R. H., Yamasaki, D. S., Duffy, C. J., and Roy, J. P., 1990, Functional specialization for visual motion processing in primate cerebral cortex, *Cold Spring Harbor Symp. Quant. Biol.* **55:**717–727.

Zeki, S. M., 1969, Representation of central visual fields in prestriate cortex of monkey, *Brain Res.* **14:**271–291.

Zeki, S. M., 1970, Interhemispheric connections of prestriate cortex in monkey, *Brain Res.* **19:**63–75.

Zeki, S. M., 1971, Cortical projections from two prestriate areas in the monkey, *Brain Res.* **34:**19–35.

Zeki, S. M., 1974, Functional organization of a visual area in the posterior bank of the superior temporal sulcus of the rhesus monkey, *J. Physiol.* (Lond.) **236:**549–573.

Zeki, S. M., 1977, Simultaneous anatomical demonstration of the representation of the vertical and horizontal meridians in areas V2 and V3 of rhesus monkey visual cortex, *Proc. R. Soc. Lond. B* **195:**517–523.

Zeki, S. M., 1978a, The third visual complex of rhesus monkey prestriate cortex, *J. Physiol.* (Lond.) **277:**245–272.

Zeki, S. M., 1978b, Uniformity and diversity of structure and function in rhesus monkey prestriate cortex, *J. Physiol.* (Lond.) **277:**273–290.

Zeki, S. M., 1983a, Colour coding in the cerebral cortex: The reaction of cells in monkey visual cortex to wavelengths and colours, *Neuroscience* **9:**741–765.

Zeki, S. M., 1983b, Colour-coding in the cerebral cortex: The responses of wavelength-selective and colour-coded cells in monkey visual cortex to changes in wavelength composition, *Neuroscience* **9:**767–781.

Zeki, S. M., 1983c, The distribution of wavelength and orientation selective cells in different areas of monkey visual cortex, *Proc. R. Soc. Lond.* **217:**449–470.

Zeki, S. M., and Sandeman, D. R., 1976, Combined anatomical and electrophysiological studies on the boundary between the second and third visual areas of rhesus monkey cortex, *Proc. R. Soc. Lond. B* **194:**555–562.

5

The Timing of Information Transfer in the Visual System

LIONEL G. NOWAK and JEAN BULLIER

1. Introduction

The mammalian visual cortex is composed of a constellation of cortical areas that are interconnected by a dense network of corticocortical connections. Among those connections, it is usual to distinguish between feedforward and feedback connections. Feedforward connections carry information away from area V1 toward the parietal and the temporal lobes, whereas feedback connections carry impulses in the reverse direction (Salin and Bullier, 1995). It is becoming increasingly apparent that, despite its complexity, the visual system processes information very rapidly. The delay imposed by neuronal processing in the correction of visually guided movements is of the order of 100 msec (Rossetti, 1997). Recent results also suggest that visual recognition of complex scenes is possible within 100–200 msec (Thorpe *et al.*, 1996).

The question we address is how such a complicated network of interconnected cortical areas can process information so rapidly. We first review the evidence concerning two major constraints for processing speed by neural net-

LIONEL G. NOWAK • Section of Neurobiology, Yale University School of Medicine, New Haven, Connecticut 06510. JEAN BULLIER • Cerveau et Vision INSERM 371, 69675 Bron Cedex, France

Cerebral Cortex, Volume 12, edited by Rockland *et al.* Plenum Press, New York, 1997.

works: the neuronal integration time and the axonal delays. We then present evidence for the rapid responses of neurons to visual stimulation at different stages of the visual system. We conclude by discussing how different models of the visual system are compatible with such a rapid processing time.

2. Structural Constraints: Axonal Conduction and Integration Times

2.1. Axonal Conduction Time

2.1.1. Minimizing the Sizes of Cortical Axons

In large-brained animals, white matter takes up a large proportion of the brain volume. For example, in humans this proportion is estimated at 40–60% (Hofman, 1985; Murre and Sturdy, 1995). In order to maintain brain volume within reasonable bounds, it is therefore important to minimize the space taken by axons. This is particularly relevant for corticocortical axons since they make up 99% of the white matter underlying the cortex (Abeles, 1991).

Minimizing axonal volume is probably the major constraint that pushes highly interconnected neurons to group together in the same cortical areas and modules, since these are highly efficient strategies to conserve axonal volume (M. E. Nelson and Bower, 1990; Mitchison, 1992; Murre and Sturdy, 1995). Another way to reduce the volume of white matter is to reduce axon diameter. This probably explains why, with a few exceptions noted below, corticocortical axons are generally small compared to axons of the peripheral nervous system. Within the gray matter of the visual cortex, most axons are smaller than 1 μm in diameter; for example, in cat and monkey area 17, the mode of the distribution of axon diameter is 0.6–0.7 μm (Haug, 1968; Peters and Sethares, 1996). The small size of cortical axons is also maintained in the white matter. This can be observed in the corpus callosum, which provides a good sample of corticocortical axons. In the macaque monkey, the mean diameter of callosal axons is less than 1 μm, and in the rostral part of the corpus callosum, all axons have diameters less than 1 μm (LaMantia and Rakic, 1990). This contrasts with the monkey optic nerve, in which most axons are larger than 1 μm (Reese and Ho, 1988).

A small number of giant axons (larger than 2.5 μm) is observed in the part of the corpus callosum containing axons between the extrastriate areas and parietal cortex. In contrast, the region of the corpus callosum containing axons linking together both temporal cortices does not appear to contain giant axons (LaMantia and Rakic, 1990). This suggests that giant axons are limited to specific networks for which high conduction velocity is necessary. We will see below that indeed processing speed is higher in dorsal extrastriate areas and parietal cortex than in temporal cortex.

Since there is a linear relationship between conduction velocity and fiber diameter (see (Section 2.1.2), minimizing axon size for conserving space has important consequences on delays due to conduction along axons.

2.1.2. Methods for Estimating Conduction Delays

There are several ways of estimating delays due to conduction of the action potential along the axon: calculating it from the distribution of axon sizes, measuring the position of the peak with respect to the origin of the time difference axis in cross-correlation histograms (CCH), measuring the latencies of excitatory postsynaptic potentials (EPSPs) recorded with spike-triggered averaging, or measuring latencies of spikes evoked by antidromic activation using electrical stimulation. In addition, in some cases, it is possible to use latencies of spikes or EPSPs evoked by orthodromic electrical stimulation.

The morphological method consists in estimating the distribution of fiber diameters and deducing that of axonal conduction velocities. The diameter of the axon with its myelin sheath is called the fiber diameter. Axon and fiber diameters are linearly related with a coefficient of 0.77 (Waxman and Bennett, 1972) (0.64–0.87 for rabbit callosal axons; Waxman and Swadlow, 1976). Using a coefficient of 5.5 between conduction velocity and fiber diameter (Hursh, 1939; Waxman and Bennett, 1972), one gets 7.14 as the coefficient between conduction velocity and axon diameter. Note, however, that this linear relationship applies only for myelinated fibers. For nonmyelinated axons, conduction velocity varies as the square root of the axon diameter (Rushton, 1951).

Deducing conduction velocity from axon diameter has the advantage of giving an estimate of the conduction delays of all the axons of a given pathway. However, to get a proper estimate of axon diameters, it is necessary to measure them on electron micrograph (EM) images. The EM technique is best used on well-identified tracts such as the interhemispheric commissures, which can be cut in a plane orthogonal to the direction of the axons. In the case of axons between two cortical areas, for example, it is necessary to stain fibers with axonal tracers such as biocytin or *Phaseolus vulgaris* leucoagglutin. These techniques usually produce a small number of well-stained axons and it is difficult to know to what extent they do not favor the largest axons. Also, estimation for axon diameter from EM images is more difficult to use in that case, because of the uncertainty concerning the angle between the section plane and the axis of the axon. Finally, it is known that large differences in diameter can exist between the main axon trunk and its cortical ramifications (Houzel *et al.*, 1994), which are often nonmyelinated. Measuring only the axon trunk diameter may therefore lead to a serious underestimate of the conduction time unless the entire axon is reconstructed.

Cross-correlation is a technique that gives the temporal relationship between the firing times of two neurons (A and B) recorded extracellularly. The action potentials of neuron A are used as time markers to trigger the averaging of the activity of the other neuron (B). The cross-correlation histogram is a measure of the probability of observing a spike in neuron B as a function of the time before or after the occurrence of an action potential in neuron A (Perkel *et al.*, 1967). It can also be seen as an estimate of the firing rate of neuron B as a function of the time before or after a spike in neuron A (Abeles, 1982). When the firing times of the two neurons are independent, the cross-correlation histogram is flat. If the neuron A activates neuron B with a delay t, the cross-correlation histogram shows a peak with a delay t, indicating that the firing probability of the second neuron increased after neuron A fired an action potential. When the two neurons are simultaneously activated by a third group of neurons, the cross-correlation histogram presents a peak centered at the origin of time (due to the common input, the discharge probability of neuron B increases at the same time as that of neuron A).

The cross-correlation method has the major advantage that it gives the actual average delay between the production of one spike in a neuron and a resultant spike in the target neuron, including axon conduction time and synaptic delay. Furthermore, measures derived from cross-correlation histograms are made from the activity of neurons activated by relatively "natural" stimuli, in contrast to electrical stimulation that artificially synchronizes the spikes of a large number of neurons. Calculating cross-correlation histograms is therefore the method of choice for measuring delays between two neurons. Its use is limited, however, by the difficulty of isolating more than a small number of pairs of interconnected neurons in two different structures.

Spike-triggered averaging is related to the calculation of cross-correlation histograms. Two neurons are recorded, one using intracellular and the other extracellular recording. The membrane potential of the intracellular neuron is averaged using the spikes of the other neuron as triggering events. In the averaged intracellular records, one can sometimes identify small synaptic potentials that are interpreted as resulting from the monosynaptic connection between the two neurons. Measuring the EPSP latency provides a good estimate of conduction time in the axon and synaptic delay between the two neurons. Another method with an even lower yield consists in recording intracellularly two neurons at the same time and recording the latency of the EPSP evoked in one neuron by stimulation of the other neuron. The advantages (reliability of measures and lack of artificial synchronization) and the disadvantage (low encounter rate) of these techniques are similar to those of cross-correlation.

Electrical stimulation can be used to stimulate axons in the orthodromic or antidromic direction. When orthodromic activation is used, latencies can be measured between the stimulus and the resulting spike in the postsynaptic neuron in case of extracellular recording, or the EPSP if the recording is intracellular. This technique should be used with caution with neural populations which are interconnected by reciprocal connections. This is due to the fact that electrical stimulation activates axons and not cell bodies (Nowak and Bullier, 1997a,b) and that, because of the presence of recurrent collaterals, it is difficult to know whether one neuron is activated by true orthodromic activation or whether electrical stimulation elicited a spike which traveled in the antidromic direction and provided a synaptic input through recurrent collaterals of neighboring neurons.

These difficulties disappear when antidromic activation is used. Measuring antidromic latencies with electrical stimulation is relatively easy and gives conduction times between structures that can be identified with electrophysiological recordings. Besides the possibility that electrical stimulation may be biased toward large neurons, which are easier to record from, and toward large axons, which are more easily activated by electrical stimuli (Ranck, 1975), one of the difficulties related to this method is that of knowing the actual site of stimulation. Unless small stimulating currents are used, it is difficult to avoid the possibility of activating axons in the white matter below the stimulated area. One way to reduce this risk is by verifying that the laminar distribution of activated neurons conforms to that predicted by the anatomy (Swadlow and Weyand, 1981).

Assessing the validity of the electrical stimulation method can be done by comparing the distribution of conduction velocities calculated from antidromic latencies and fiber length and that calculated from the axon sizes. When this is done on callosal connections of the macaque monkey using a coefficient of 7.14 between conduction velocity and axon size (Swadlow *et al.*, 1978; LaMantia and Rakic, 1990), the match is actually quite good, suggesting that antidromic latencies to electrical stimulation give a proper estimate of the conduction times of all the axons between two structures.

2.1.3. Specificity of Conduction Times in Different Connections and Species

Each set of connections possesses a characteristic pattern of conduction delays. This is well illustrated in Fig. 1, which presents the distributions of antidromic latencies of neurons in cortical area V1 of the awake rabbit following electrical stimulation in several afferent structures (Swadlow and Weyand, 1981). The figure shows that there are important differences between the conduction times in corticogeniculate, corticotectal, callosal, and association corticocortical axons.

It is remarkable that the ordering of conduction times between different targets of cortical neurons (corticotectal short, association intermediate, and corticothalamic long) is conserved in different species such as rabbit, monkey, and cat (see below). It is also observed for different sensory modalities: one can find a similar ordering of latencies for the efferent connections of area V1, area S1 and the motor cortex of the rabbit (Swadlow, 1992, 1994). This suggests that speed of information transfer in different corticofugal pathways is subjected to the same constraints across different sensory modalities and different species.

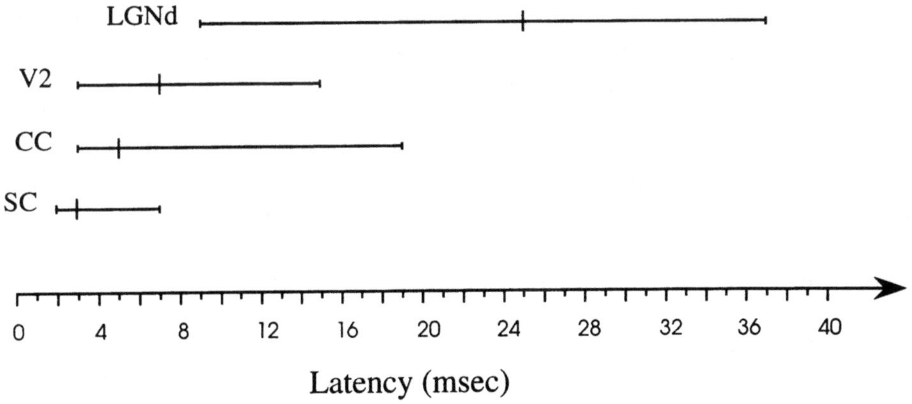

Figure 1. Distributions of antidromic latencies in rabbit area V1 after electical stimulation in the LGNd, area V2, visual cortex of the other hemisphere (CC), and superior colliculus (SC). The small vertical lines represent the 10%, 50% (median), and 90% centiles. Redrawn from Swadlow and Weyand (1981).

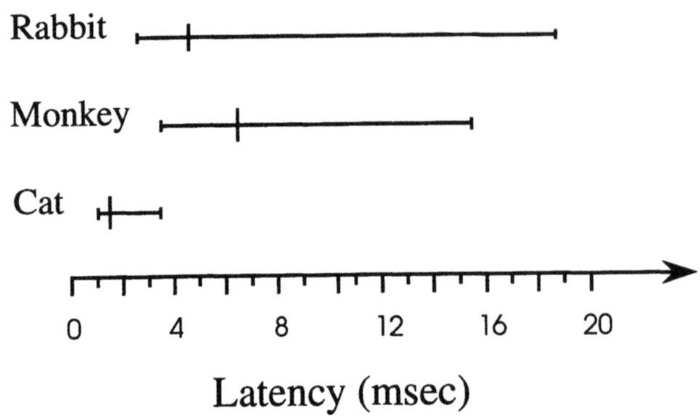

Figure 2. Distributions of antidromic latencies in area V1 after electrical stimulation in the opposite cerebral hemisphere in the rabbit (Swadlow and Weyand, 1981), monkey (Swadlow *et al.*, 1978), and cat (Innocenti, 1980). The small vertical lines represent the 10%, 50% (median), and 90% centiles.

Contrary to what one would expect, conduction time along a given fiber tract is not governed mainly by fiber length. Comparison of antidromic latencies of callosal axons in different species (Fig. 2) shows that conduction delays are much shorter in the cat than in rabbit, despite the fact that they cover a longer distance. Conduction times are similar in rabbit and monkey despite the much longer paths of callosal axons in monkey. In the corticotectal pathway, a similar interspecies order can be observed, with axons in the cat displaying shorter conduction times than in the rabbit or the monkey (which do not differ significantly; see references in Section 2.1.4f). Higher conduction speed is also found in the retinogeniculostriate pathway of the cat compared to that of the monkey (Stone, 1983). This suggests that brains of different species function at different speeds: cat brains are faster than monkey and rabbit brains, at least for the visual system.

2.1.4. Axon Conduction Delays in Different Sets of Connections

2.1.4a. Three Classes of Axons in the Retinogeniculate Pathways. The relatively large sizes of axons in the optic nerve and the optic tract in cats and monkeys (Hughes and Wässle, 1976; Reese and Guillery, 1987; Reese and Ho, 1988) suggest that transfer time between the retina and the cortex must be rapid. Approximately half of the conduction delay is due to the slow unmyelinated intraretinal part of the axon (Stanford, 1987). From the optic disk to the dorsal lateral geniculate nucleus (LGNd), axons are myelinated and conduct much faster. There appears to be a compensation of the intraretinal delay by the conduction velocity of the myelinated part of the axon so that the conduction delay between the retinal ganglion cell and the LGN does not depend on the position of the ganglion cell in the retina (Stanford, 1987).

Conduction times between the retina and the LGN in the cat have been measured by cross-correlation studies and shown to be around 4 msec on average, with a range of 1.8–7 msec for X and Y cells (Cleland *et al.*, 1971; Mastronarde, 1987). Delays between the LGN and the cortex appear shorter, on the order of 1–6 msec as found in cross-correlation studies (Tsumoto *et al.*, 1978; Tanaka, 1983). A heroic cross-correlation experiment between retina and cortex performed by Lee *et al.* (1977) showed that the delay between a spike in a retinal ganglion cell and its cortical target is of the order of 4.5–7.5 msec for X or Y cells. Such a short conduction time is remarkable given the long distance between the eye and the cortex of area 17. It is interesting to note that it takes a similar amount of time to get from the eye to area 17 than between two adjacent cortical areas (see Section 2.1.4c).

Since the early recordings of George Bishop in the 1930s, it has been known that the transfer of information in optic nerves of mammalian species is made along three groups of axons with different conduction velocities (Stone, 1983). These have been called X, Y, and W in the cat (Stone, 1983) and M, P, and K in the monkey (Casagrande, 1994; Hendry and Yoshioka, 1994). Differences between these groups in axon conduction times from retina to cortex can be sizable because there is a positive correlation between the conduction speeds of axons of geniculate cells and those of their retinal afferents (Cleland *et al.*, 1976; Marrocco, 1976; Wilson *et al.*, 1976; Schiller and Malpeli, 1978). In the cat, for example, if we add the retinogeniculate conduction times (Kirk *et al.*, 1975) to the geniculocortical delays (Wilson *et al.*, 1976), we find that, on the average, Y axons transfer information to cortex 3.5 msec faster than X axons, and X axons are faster by 6 msec than W axons. In the monkey, a similar calculation can be made for M and P pathways by adding conduction times from the retina to the optic chiasm (Gouras, 1969; Schiller and Malpeli, 1977b), from the chiasm to the LGN (Dreher *et al.*, 1976; Schiller and Malpeli, 1978), and from the LGN to cortex (Schiller and Malpeli, 1978; Bullier and Henry, 1980). This shows that, on average, M axons take 5.5 msec less time to transfer impulses to cortex than P axons. In keeping with that conclusion, the study of Mitzdorf and Singer (1979) shows that, after electrical stimulation of the optic nerve or optic chiasm, the onset latency of synaptic potentials in area V1 is shorter by 6 msec in layer 4Cα, which receive M inputs, than in layer 4Cβ, which is activated by P afferents.

2.1.4b. Intra-Area Axonal Conduction Times. Once afferent messages have reached the cortex of area 17, they are processed by interneuronal communication within the area and between different cortical areas. Little is known concerning the axon conduction delays in these two groups of pathways. For intra-area conduction, most data come from intracellular recordings in cortical slices of rat visual or somatosensory cortex, using spike-triggered averaging of membrane potential or double intracellular recordings. The values are remarkably long (several milliseconds), given that the recorded neurons are distant by only a few hundred micrometers (Thomson *et al.*, 1988; Mason *et al.*, 1991; Nicoll and Blakemore, 1993). These correspond to very slow conduction velocities (of the order of a few tenths of m/sec). With antidromic activation techniques, the axon conduction velocity for axons of intrinsic connections ranged between 0.26 and 0.47 m/sec within rat area 17 (Nowak *et al.*, 1997) and averaged 0.28 m/sec in rat area 18a (Lohmann and Rörig, 1994).

Conduction delays were also measured in cat visual cortex with spike-triggered averaging (Komatsu *et al.*, 1988). These are also long (e.g., 2.8 msec for 720 μm distance), corresponding to conduction velocities of the order of 0.3 m/sec, after subtraction of the synaptic delay. Interestingly, the conduction velocity for the intrinsic collaterals of an axon may be very different from that of the main trunk projecting elsewhere. Kang *et al.* (1988) used spike-triggered averaging to measure the latencies of intracortical EPSPs evoked by pyramidal tract cells in cat motor cortex. When these latencies are compared with the antidromic latencies evoked by stimulation of the pyramidal tract, it appears that conduction delays within cortex are of the same order of magnitude (a few milliseconds) as transfer time in the pyramidal tract axon, despite the enormous difference in conduction distances (a few hundred micrometers in intracortical connections versus several centimeters for pyramidal tract axon). This confirms the slow nature of conduction in intracortical collaterals of cortical axons.

These very slow conduction velocities for intra-area communication in rat and cat cortex fall below the range usually associated with myelinated axons (down to 1 or 2 m/sec; Rushton, 1951; Waxman and Bennett, 1972). This suggests that intrinsic collaterals of pyramidal cell axons that make up most local connections are usually not myelinated along their whole length. It is also in keeping with the observation that many cortical axons are not myelinated in the cortical gray matter (Peters and Sethares, 1996).

The low speed of information transfer within a given cortical area raises interesting questions concerning the temperal aspects of intra-area information processing. Local ramifications of cortical collaterals can reach several millimeters (Gilbert and Wiesel, 1981; Martin and Whitteridge, 1984). With conduction velocities of the order of 0.3 m/sec, this means that spikes will reach the target neurons located 3 mm away 10 msec later. This may explain the observation in *in vivo* intracellular recordings of cat cortical neurons of long delays (10–30 msec) of synaptic activity triggered by visual stimulation in peripheral parts of the receptive field (Fregnac and Bringuier, 1996). Slow conduction velocities of intrinsic connections may also be the reason for the slow (0.1–0.2 m/sec) spread of synaptic activity across monkey visual cortex reveled by optical imaging (Grinvald *et al.*, 1994).

2.1.4c. Association Corticocortical Axons. Corticocortical connections between areas of the same hemisphere are usually reciprocal and it is possible to identify feedforward and feedback connections (Felleman and Van Essen, 1991; Salin and Bullier, 1995). Besides anatomical and functional differences (Salin and Bullier, 1995), it is possible that these two types of connections show differences in conduction delays. Such differences have been found in the thalamocortical connections between LGN and cortex. For example, in the cat, most LGN axons conduct impulses in less than 4 msec in the feedforward direction (Wilson *et al.*, 1976), whereas numerous corticogeniculate axons take more than 4 msec to transfer information in the feedback direction (Harvey, 1980; Grieve and Sillito, 1995).

For corticocortical connections, this has been directly addressed in the rat between areas 17 and 18a using antidromic activation of single neurons. The results led to the conclusion that conduction times are similar in feedforward and feedback connections (Nowak *et al.*, 1997). Antidromic latencies of feedforward axons ranged between 3.7 and 7.9 msec and those of feedback between 4.1 and 8.9 msec. Differences in mean values of latencies (5.7 msec for feedforward and 6 msec for feedback) and conduction velocities (0.41 and 0.38 m/sec) were not statistically significant. In the monkey visual cortex, comparison of the diameter of axons in feedforward and feedback connections between monkey areas

V1 and V2 (Rockland and Virga, 1989, 1990) also leads to the conclusion that, in contrast to thalamocortical connections, feedforward and feedback corticocortical connections involve similar conduction times.

Whether measured with antidromic activation or by the onset latency of evoked potentials (Nowak *et al.*, 1997), it appears that conduction times in axons between areas 17 and 18a in the rat are long (several msec), given the short distance between these areas. The onset latency of synaptic responses obtained in rat extrastriate cortex after electrical stimulation in area 17 ranged between 3 and 10 msec, as measured by voltage-sensitive dyes *in vivo* (Orbach and Van Essen, 1993).

Long conduction times were also observed in the rabbit visual cortex. According to Swadlow and Weyand (1981), the majority of antidromic latencies of area 17 neurons are between 2 and 12 msec (median at 6–7 msec) after electrical stimulation in area 18. The corresponding conduction velocities range between 0.23 and 5.74 m/sec (median at 0.64 m/sec), values that are only marginally larger than in the rat.

Very little is known concerning the conduction times in connections between cortical areas in the cat and monkey. Recording of antidromically activated neurons in cat area 17 after stimulation in areas 18 and 19 shows that some feedforward corticocortical axons conduct rapidly (less than 2 msec delay; Toyama *et al.*, 1974; Bullier *et al.*, 1988), while others are slow (2–12 msec delays; Bullier *et al.*, 1988). Feedback connections from these areas onto V1 tend to involve small axons (below 1 μm; Henry *et al.*, 1991) and are therefore likely to conduct impulses slowly. Orthodromic and antidromic latencies of area 17 neurons activated by electrical stimulation in PMLS tend to be shorter (mostly below 3 msec; G. H. Henry, personal communication) than those from areas 18 or 19. This is an interesting observation given that PMLS is often considered to be the homologue of area MT in the macaque (Shipp and Grant, 1991; Payne 1993) and that connections between areas 17 and MT in the macaque involve very rapidly conducting axons (see below).

In the monkey, by subtracting the latencies of the current sinks evoked in ares V1 and V2 by electrical stimulation of the optic radiation (Mitzdorf and Singer, 1979), it is possible to deduce that it takes several milliseconds to transfer information between these two areas. The short distance between areas V1 and V2 (a few millimeters) and the fact that axon diameters are of the order of 1 μm (Rockland and Virga 1990) suggest that axon conduction times are of the order of 1 msec. Therefore, most of the delay between areas V1 and V2 probably corresponds to synaptic delays and integration times.

In contrast to the connections between areas V1 and V2, at least some of the neurons projecting from V1 to MT are characterized by the presence of giant axons, 3 μm or larger (Rockland, 1995). This predicts that very short latencies should be measured between areas V1 and MT. Indeed, despite the large separation (greater than 1 cm) between corresponding regions in areas V1 and MT, measurements of antidromic latencies of area V1 neurons after electrical stimulation in area MT gave very low values, between 1.1 and 1.7 msec (Movshon and Newsome, 1996). Beyond MT, there are several areas of the parietal cortex, such as MST and VIP, that show dense myelination in the gray matter. If we assume that this is a characteristic feature of terminals of rapidly conducting axons by analogy with area MT, we can conclude that the pathway to parietal cortex through areas MST and VIP must involve very short conduction delays. This is in keeping with the very short latencies to visual stimulation that have been reported in this region (see Section 3).

There are no direct measurements of conduction delays in other corticocortical connections in the monkey. Measurement of axon size in the connection between V2 and V4 (Rockland, 1992) and between TEO or V4 to V2 and V1 (Rockland *et al.*, 1994) gave values between 0.5 and 1.5 μm. Because of the deep sulci, the distance between interconnected regions in V2 and V4 must be of the order of 2 cm. This suggests that it takes several milliseconds to conduct impulses from V2 to V4.

More information is needed concerning the conduction delays between cortical areas in the monkey in order to understand the large range of latencies to visual stimulation (see Section 3). From what we know, it appears that most interarea axonal transfers take several milliseconds, with the exception of the connection between V1 and MT, which appears remarkably fast.

2.1.4d. Callosal Axons. Conduction times in the callosal pathway of rabbits has been extensively studied by Swadlow and his collaborators using electrical stimulation. The shortest antidromic latency was 2.4 msec, but most latencies were long, with a maximum value of 39.8 msec (Swadlow, 1974). Swadlow and Weyand (1981) obtained similar results, with conduction velocities ranging between 0.5 and 8.25 m/sec (median 2.8 m/sec). The large range of conduction velocities and the very low values for some axons correspond to the morphological characteristics of the callosal axons in the rabbit. In the posterior part of the splenium, 45% of the fibers are not myelinated, with diameters between 0.08 and 0.6 μm; the myelinated axons range between 0.3 and 1.85 μm in diameter (Waxman and Swadlow, 1976). A few very large fibers (2–3 μm diameter) were also observed.

Conduction times in cat callosal axons have been measured by several groups (Toyama *et al.*, 1974; Harvey, 1980; Innocenti, 1980; McCourt *et al.*, 1990). All these studies returned short conduction times, most of them being between 1 and 5 msec, corresponding to relatively rapid conduction velocities (9–17 m/sec; mean 13 m/sec). This is clearly much faster than in the rabbit (Fig. 2). The difference can be explained by the high proportion of large-diameter myelinated fibers in the cat corpus callosum (Naito *et al.*, 1971).

Fewer studies have been made of interhemispheric conduction delays in the macaque monkey. As mentioned above, Swadlow and his colleagues (Swadlow *et al.*, 1978) concluded that callosal axons of neurons in the prelunate gyrus of the monkey are relatively fast conducting (median 7.4 m/sec), leading to conduction delays between areas V4 of both hemispheres of 2.8–22.5 msec (median 7 msec). To estimate delays in other regions of the corpus callosum, one can use the comprehensive morphological study of LaMantia and Rakic (1990). Most axons must conduct impulses more slowly than 7 m/sec and very few giant axons conduct faster than 20 m/sec. Assuming an average axon length of 50 mm, this correspond to interhemispheric delays longer than 7 msec for most axons and below 2.5 msec for the fastest axons. In the rostral part of the corpus callosum, it is likely that the numerous nonmyelinated axons (LaMantia and Rakic, 1990) conduct more slowly than 1 m/sec, which would give interhemispheric delays larger than 50 msec.

2.1.4e. Slow Corticogeniculate Axons. As illustrated in Fig. 1, corticothalamic axons are characterized by long conduction delays due to very low conduction velocities (mean 0.64 m/sec in the rabbit). Studies in the cat have also identified a population of slow corticogeniculate axons (Harvey, 1980; Grieve and Sillito, 1995). On average, it takes several milliseconds to transfer impulses from LGN to cortex and 10 msec to return to the LGN. It is therefore likely that the initial 20 msec of responses of rapidly activated neurons in the LGN (Y cells) is not affected by cortical feedback. Such slow conduction of corticothalamic axons suggests that either this pathway is used to modify the long-term membrane properties of relay neurons (McCormick and Van Krosigk, 1992) or that it acts preferentially on neurons that respond late to visual stimulation, such as lagged cells (Mastronarde, 1987; Saul and Humphrey, 1990).

2.1.4f. A Fast Cortical Pathway to the Superior Colliculus. Latencies measured in the cortex following electrical stimulation of the tectum are short (Fig. 1), due to the high conduction velocities of corticotectal axons (average 8.4 m/sec in the rabbit). Similar rapid transfer times in the corticotectal pathway have been demonstrated in the cat (Palmer and Rosenquist, 1974; Toyama *et al.*, 1974; Harvey, 1980; Thalluri and Henry, 1989; Casanova, 1993) and in the monkey (Finlay *et al.*, 1976).

The superior colliculus is innervated by W and Y types of retinal axons in the cat (Hoffmann, 1973) and the monkey (Schiller and Malpeli, 1977b). It also receives in addition a cortical input from neurons located in layer 5 of various extrastriate cortical areas (Gilbert and Kelly, 1976; Fries, 1984). This indirect input appears to be under the control of the M system in the monkey (Schiller and Malpeli, 1978) and the Y system in the cat (Hoffmann, 1973). It is interesting to note that the fastest set of output axons from visual cortex is the corticotectal pathway (Fig. 1). There seems therefore to be a fast corticotectal circuit that is driven by fast retinal axons and provides rapid activation to the superior colliculus. Such an organization gives support to the idea that fast conducting axons are associated with specific roles requiring rapid reactions, such as activating eye movements.

2.2. Neuronal Integration Time

The axonal conduction delay is not the sole parameter that determines the time it takes to transfer information from one structure to another. Once the spikes have reached the axon terminals, one must consider the length of time it takes for a depolarizing event evoked by the incoming spike to bring the membrane potential of the target cell to spike threshold. This amount of time, which we will name the "integration time," can be relatively long and can vary depending on the state (e.g., the membrane potential) of the neuron.

2.2.1. Chronaxie Measurements

In order to characterize quantitatively the integration time in neurons, the simplest experiment situation consists in injecting depolarizing current pulses into intracellularly recorded neurons. It is then possible to determine when the threshold for action potential is reached as a function of the current intensity. Such studies have been performed in cat motor cortex neurons *in vivo* and rat visual cortex *in vitro*. The results showed what can be expected, namely that the threshold is reached faster for higher current intensities. A good linear fit of the relationship can be obtained by plotting the inverse of the firing time as a function of current intensity, such that $I = a/t + b$, with I the current intensity and t the time of spike occurrence. This relationship was established by Weiss (1901) for the extracellular stimulation of frog and toad motor nerves.

The relationship $I = a/t + b$ can be written as $I = I_0C/t + I_0$ (e.g., Ranck, 1975). The constant terms of the equation can be determined from the linear regression between $1/t$ and I. The parameter I_0 correspond to the rheobase, which is the current for which a spike would be obtained at a time equal to infinity (when $1/t \rightarrow 0$). The constant C is the well-known chronaxie. It corresponds to the latency of the spike that is obtained with a current intensity set to twice the rheobase value.

The average values of chronaxies obtained in cat motor cortex neurons are 9.8 msec for fast pyramidal tract neurons and 27.1 msec for slow pyramidal tract neurons (Koike *et al.*, 1970). In rat visual cortex, the average value obtained for neurons of the supragranular layers is 15 msec (Nowak and Bullier, 1997a). In other words, the delay between the onset of a depolarization and the spike that results from this depolarization varies from 1 to several tens of msec when the current intensity is twice the threshold value. This is much larger than the chronaxie values of axons, which are usually measured in tenths of a millisecond (Ranck, 1975; Nowak and Bullier, 1997a).

The chronaxies of neurons are likely to be influenced by their membrane properties. This includes the resting membrane potential, the passive time constant, and different conductances that are activated in the subthreshold range of membrane potential. Such mechanisms include the potassium current I_A, which acts as a brake and delays the time at which the threshold for the action potential is reached (Segal and Barker, 1984; Eder *et al.*, 1991), or the noninactivating sodium current $I_{Na(p)}$, which induces an anomalous rectification in cortical neurons which increases the apparent time constant (Stafstrom *et al.*, 1985; Sutor and Hablitz, 1989). Thus, the chronaxie, and therefore the integration time, is likely to be influenced by passive and active membrane properties, to an extent that remains to be established. In addition, it is possible that both network activity and neuromodulatory substances acting on these properties affect the integration time of cortical neurons. Integration time is therefore a parameter that is determined by multiple factors.

2.2.2. Integration Time at Resting Membrane Potential

It is difficult to extrapolate the results of chronaxie measurements to the integration time of cortical neurons when they are activated by natural stimuli. In these conditions, single EPSPs are too small to depolarize neurons from rest to spike threshold. Neuronal depolarization requires the temporal summation of synaptic potentials. The number and the degree of synchronization of these synaptic potentials determine the speed of the temporal summation. Altogether, the shape of the underlying synaptic current differs from the square pulse used for chronaxie measurements. Nonetheless, chronaxie measurements tell us that integration times in neurons must be of the order of several to several tens of milliseconds.

The integration time that can be measured for natural stimulations corresponds to the difference between the latency of a spike and the onset latency of the triggering compound synaptic potential. Despite its fundamental importance in understanding the temporal aspect of cortical processing, integration time has been measured only in a small number of studies, because such measurements require the use of the difficult technique of intracellular recordings *in vivo*.

Creutzfeldt and Ito (1968) used small visual stimuli flashed inside the receptive field of neurons intracellularly recorded in cat area 17. They found that the onset latency of the EPSPs was between 30 and 40 msec after stimulus onset. The spike latency was between 35 and 100 msec. In other words, the delay between the onset of the synaptic potentials and the action potential it elicited (the integration time) was between 5 and 60 msec. Integration time was also reported recently by Volgushev *et al.* (1995) in cat area 17 using whole-cell patch-clamp recording. The stimuli were bars flashed in the receptive fields of the neurons. The integration time was 5–12 msec for optimally oriented stimuli (mean 7.6 msec) and 6–15 msec (mean 109.6 msec) for orientation close to the optimal (nonoptimal orientation, but sufficiently close to it to elicit action potentials).

These two studies show that the integration time is much longer than what would be predicted from the results of orthodromic electrical activation of the retinothalamic pathway. Several studies showed that there is only a fraction of a millisecond between the arrival of afferent spikes in geniculate axons and the triggering of a spike in first-order neurons (Singer *et al.,* 1975; Bullier and Henry, 1979). These results indicates that cortical neurons *can* react with extremely fast integration times in the artificial situation of electrical stimulation of their afferents, when resulting EPSPs are presumably very numerous and synchronized by the electrical pulse. The relationship between current intensity and chronaxie (see above) suggests that such fast integration times require very high synaptic currents that are probably outside the physiological range.

However, as mentioned above, in response to visual stimuli, neurons reach their firing threshold quite slowly. This slowness suggests that either a small number of functional inputs contribute to the beginning of the response, and/or there is a rather coarse synchronization of the afferent activity. This would be the consequence of the small number of individual synapses established between a thalamocortical axon and a given cortical neuron (Freund *et al.,* 1985, 1989;

Peters *et al.*, 1994), combined with the latency scatter reported for thalamic relay cells in response to visual stimulation (see below).

Matsumura (1979) recorded intracellularly from the motor cortex of awake behaving monkeys. He noticed that the visual cue used to trigger the behavioural response of the monkey elicited an EPSP with a latency between 80 and 180 msec. This EPSP preceded the movement by 70 to 180 msec. More interestingly, the delay between the onset of this EPSP and the first spike was long: 39 ± 29 msec (mean ± S.D.) in pyramidal tract neurons, 77 ± 63 in nonpyramidal tract neurons. This again points to the fact that, in normal conditions, action potentials of cortical neurons may appear only very late after the first EPSPs. In that case, however, it is possible that the visually elicited synaptic response required the addition of synaptic potentials provided by other sources of afferents to lead to spike threshold.

2.2.3. How EPSPs Are Transformed in Spikes when Neurons Are Close to the Firing Threshold

The preceding section concerned the time it takes for the membrane potential of a neuron to increase from rest to firing threshold. This part of the discussion deals with the way and the time it takes for an EPSP to be transformed into a spike when the membrane potential is already close to firing threshold. For this, we need a tool that gives the delay between the onset of an EPSP and the onset of the resultant spike. One way is to relate the shape of the cross-correlation histogram (CCH) peak to that of the underlying EPSP.

In the case of a monosynaptic connection, the shape of a cross-correlation histogram peak indicates at what time and with what dispersion the spike was triggered from the underlying EPSP. In a modelization study, Knox (1974) established that, theoretically, the shape of a CCH peak must reflect that of the rising phase of the EPSP. This is easy to understand: the rising part of the EPSP is that part of the EPSP that will cross the firing threshold. Several studies have tested the prediction of Knox's modelization in the spinal cord (Kirkwood and Sears, 1978; Fetz and Gustafsson, 1983; Gustafsson and McCrea, 1983; Cope *et al.*, 1987). One general outcome from these studies was that the shape of a CCH peak corresponds to the temporal derivative of the underlying EPSP when the EPSP is of large amplitude. In other words, the time window during which an EPSP is effective to trigger an action potential (which corresponds to the integration time when the neuron is sufficiently close to firing threshold) corresponds to the duration of the rising part of the EPSP. However, for EPSPs of small amplitude, and for EPSP in a noisy neuron, the relationship between CCH shape and EPSP derivative no longer holds. In these conditions, the time window during which the EPSP would be effective in triggering an action potential can be as long as the whole EPSP duration.

In neocortex, the relationship between the shape of an EPSP and the timing of the action potential it triggers may be more complex: due to the presence of an active regenerative conductance ($I_{\mathrm{Na(p)}}$) that is activated below spike threshold (Stafstrom *et al.*, 1985), the time lag between the onset of an EPSP and the resulting spike can be long, up to several tens of milliseconds (Reyes and Fetz, 1993a,b). This "delayed crossing" effect is less prominent for larger EPSPs.

Among *in vivo* studies of neocortical neurons, none has directly established the relationship between EPSP shape and spike output timing. However, comparing the rise time of EPSPs reported from *in vitro* studies and the width of CCH peaks obtained for monosynaptic connections reported from *in vivo* cross-correlation study would allow one to determine if, indeed, the integration time is bounded by the duration of the EPSP rise time. The rise time of monosynaptic EPSPs has been determined in several studies relying on spike-triggered averaging or on dual intracellular recordings in brain slices. The reported rise times range from less than 1 msec to a few, at most 5 msec (Komatsu *et al.*, 1988; Thomson *et al.*, 1988, 1993; Mason *et al.*, 1991; Nicoll and Blakemore, 1993; Thomson and West, 1993). The width of cross-correlation histogram peaks corresponding to monosynaptic connections has been reported in some studies for neurons recorded within the same cortical area. Eggermont (1992) reported peak widths less than 5 msec. Toyama *et al.* (1981) found peak durations of 5–7 msec, with some departure in a small number of cases (up to 15 msec). Michalski *et al.* (1983) found peak widths between 0.5 and 6 msec (2.22 msec on average). In the study of Gochin *et al.* (1991) the majority of monosynaptic peaks have a width close to 1 msec.

Therefore, there seems to be a good match between CCH peak width and EPSP rise time, although some CCH peaks are wider than expected. In any case, the CCH peak width is less than that of the EPSP width measured at half-height (an underestimate of the real EPSP duration), which is between 4 and 20 msec (Komatsu *et al.*, 1988; Thomson *et al.*, 1988; Mason *et al.*, 1991; Nicoll and Blakemore, 1993).

2.2.4. Conclusion: Coincidence Detection and Temporal Summation

The amount of time required to trigger an action potential in a cortical neuron corresponds to the sum of the axonal delay and the integration time. The integration time itself depends on the state of the neuron. When the membrane potential of a cell is close to firing threshold, the firing of an action potential can occur as soon as the EPSP has appeared and can still be observed a few milliseconds after EPSP onset. In that case, the integration time appears to be very short and the time it takes for information to be transferred from one part of the brain to another would be slightly longer than the axon conduction delay.

If the cell is *not* already depolarized (i.e., if its membrane potential can be considered at rest or if it has been hyperpolarized by inhibitory inputs), a supplementary delay of *at least* 5 msec must be added. This situation applies to the onset latency of responses to visual stimulation. In that case, it would be a mistake to equate the physiological transfer time to the axonal delay that can be determined either by antidromic activation or by cross-correlation. Hence, knowing axonal conduction delay and integration time, it is possible to predict the difference in onset latency between two different structures in the brain. In the case of areas V1 and V2 in the monkey, it can be expected that most of the axonal conduction delays would be a few milliseconds. Adding a minimum of 5 msec to this conduction delay, it is possible to predict a latency difference between area V1 and V2 close to 10 msec. A value of 10 msec was indeed reported in two studies comparing visual latencies in areas V1 and V2 under the same conditions (Raiguel *et al.*, 1989; Nowak *et al.*, 1995). This 10-msec latency difference is much larger than expected on the basis of axonal conduction delays only.

Experimental data show that the integration time when a neuron has to be depolarized from rest to firing threshold is long (5 to several tens of msec). This sluggishness can result from several factors: the small size of the incoming EPSPs, and the possibility that only a fraction of them arrive synchronously at the target cell, because of the dispersion of axonal velocities and visual response latencies in the afferents. The duration of the integration time is close to the membrane time constant. It is also close to the mean interspike interval observed in cortical neurons in response to visual stimulation and, therefore, according to the definition of Konig *et al.* (1996), it can be considered that neurons in that regime behave as integrators.

However, when neurons are close to their firing threshold, their integration time appears to be very short, since it can be reduced to a fraction of the rise time (1–5 msec) of an EPSP. Under these conditions, the larger and faster rising EPSPs are more likely to trigger action potentials than small EPSPs with a slow rising phase. Therefore, the neurons would be sensitive to the temporal coincidence of EPSPs within a narrow time window (a few milliseconds) and behave as coincidence detectors under these conditions (following the definition of König *et al.*, 1996).

Whether a neuron functions as integrator or as coincidence detector would be determined by the amount of time it can be considered as close to firing threshold during a physiological stimulation. This has not been quantitatively established. However, the published intracellular recordings show that, in response to visual stimulation, neurons are not simply depolarized in a sustained fashion during the passage of the bar stimulus. Instead, the membrane potential undergoes marked fluctuations due to the mixture of incoming EPSPs and inhibitory postsynaptic potentials (IPSPs) (Douglas and Martin, 1991; Ferster and Jagadeesh, 1992; Pei *et al.*, 1994). Whether these large fluctuations of membrane potential are compatible with a coincidence detector model remains to be established (for recent reviews see Shadlen and Newsome, 1994; Softky, 1995; Konig *et al.*, 1996).

In addition, it remains to be established to what extent neurons that show synchronized firing also activate their target at the same time: this might be the case when the distance between source and target cells is short, but a certain degree of asynchrony might be introduced in the case of long-range connections like the callosal pathway, given the variability in axonal size and conduction velocity that has been reported in visual cortex (see Section 2.1.4d).

3. Timing of Information Transfer to and between Visual Cortical Areas

With the knowledge of the neuronal integration time and axonal delay, it is possible to interpret the latencies of neurons at different stages of the visual system to the flashed presentation of a visual stimulus. We will examine successively the latencies of retinal, thalamic, and cortical neurons.

3.1. Latency in the Retina and Lateral Geniculate Nucleus

3.1.1. Latencies in Cat Retina and LGN

Latencies of responses of retinal ganglion cells to visual stimulation provide an estimate of the processing time of the retinal network. In the report by Cleland and Enroth-Cugell (1970), the shortest response latency reported was less than 30 msec. In that of Levick (1973), the visual latency determined for 10 ganglion cells at the highest luminance tested (32 times above threshold) was between 20 and 35 msec. These results were obtained from unidentified cell types.

Bolz *et al.* (1982) measured the latencies of cat retinal ganglion cells with different types of visual stimulation. They found that the shortest latencies were obtained when the stimulus had the same size as the center region of the receptive field. Such stimuli were also those that yielded the maximal discharge rate. With this type of stimulation, Bolz *et al.* observed that the latency of X cells was longer than that of Y cells by 10–15 msec. The absolute value of latency depended on the contrast between the stimulus and the background, but the latency difference between X and Y cells remained whatever the contrast. For high-contrast stimulation, the absolute values of latencies were between 31 and 44 msec for brisk sustained (X) cells and between 25 and 38 msec for brisk transient (Y) cells.

Three important conclusions can be drawn from these studies (1) The visual latency of retinal ganglion cells is not less than 20 msec with the optimal stimulation at the highest contrast. This shows that the intraretinal processing is slow. For comparison, it is roughly twice the time needed for somatosensory information to reach the cortex (Welker *et al.,* 1993)! (2) The *average* latency of X cells is longer than that of Y cells when activated by their respective optimal stimuli. (3) Despite the difference of average latency, there is a scatter of response latency, such that there is an *overlap* of the latency distribution of X and Y cells.

Given these results, it is not unexpected that many of the studies in which visual response latencies have been measured in the cat LGN showed shorter latencies for Y cells compared to X cells. The latency difference that has been reported in different studies is between 6 and 20 msec (Orban *et al.,* 1985; Sestokas and Lehmkuhle, 1986; Saul and Humphrey, 1990; Hartveit and Heggelund, 1992; Humphrey and Saul, 1992; Mastronarde, 1992). However, although average response latencies differ for X and Y cells, the distributions of latencies show a large overlap for the two cell types, similar to what has been found for the retina. In other word, the likelihood that a Y input reaches the cortex after an X input is not negligible.

The visual latencies in cat LGN are even more widespread when one considers the categories of lagged cells that have been reported in several studies. The range of latencies for such cells is between 100 and 1000 msec (Saul and Humphrey, 1990). In the report of Humphrey and Saul (1992), the X lagged cell latency was on average 171 msec. Hartveit and Heggelund (1992) found a mean latency of 101 msec for Y lagged cells and 109 msec for X lagged cells. Such values are extremely long compared to those of cortical neurons (see Section 3.2).

Sestokas and Lehmkuhle (1986) established how visual latencies in the dorsal lateral geniculate nucleus of the cat are affected by the spatial frequency and contrast of the grating used for stimulation. The onset and peak latencies were found to vary as a function of spatial frequency for both X and Y cells. The shortest onset and peak latencies for X cells were obtained with a spatial frequency of 0.5–0.75 c/deg, and they increased at higher or lower spatial frequencies. For Y cells the shortest onset and peak latencies were obtained at around 0.17 c/deg (the lowest spatial frequency tested) and increased for higher spatial frequencies. The shortest onset latency obtained for X cells with the optimal spatial frequency was 60 msec on average, and about 50 msec for Y cells. Again, with their respective optimal stimuli, the onset latency of X cells was longer than that of Y cells by 10 msec. At low spatial frequency, the onset latency of Y cells was less than that of X cells. At a spatial frequency of about 0.75–1 c/deg, the latencies were similar. At high spatial frequency the *onset* latency of X cells was shorter than that of Y cells, but the *peak* latency of Y cells was never longer than that of X cells.

The latency also depends on the contrast of the stimulus. When tested with their optimal spatial frequency, both X and Y cells display their shortest latency at the highest contrast. The relationship between contrast and latency was monotonic. There was no range of contrast for which X-cell latency was shorter than Y latency.

In conclusion, there appears to be both a difference in the average value and an overlap of the latencies of X and Y cells in the cat retina and LGN.

3.1.2. Latencies in Primate Retina and LGN

The visual latencies have not been extensively investigated in the primate retina and LGN. One study performed in the retina (Gouras and Link, 1966) showed that, similar to what has been reported for the cat retina, the intraretinal processing is slow, visual latencies of ganglion cells being rarely less than 20 msec even at the highest contrast. Unfortunately, this study was conducted on a small sample of unidentified cell types, making it impossible to determine differences between different subtypes of primate ganglion cells.

Visual response latencies have been determined in the macaque lateral geniculate nucleus by Marrocco (1976), who found that the median latency of color-opponent LGN cells (presumably parvocellular neurons of type 1 and 2 according to the classification of Wiesel and Hube, 1966; see also Dreher *et al.*, 1976) is longer than that of non-color-opponent sustained cells (probably type 3 parvocellular cells) by 10 msec and longer by 18 msec than that of transient nonopponent cells (likely to correspond to magnocellular type 3 cells). Notice that this latency difference between visual latencies of magno- and parvocellular neurons is much longer than would be expected on the basis of axon conduction times only (see Section 2.1.4a). This suggests that, similar to what has been reported for the cat, most of the difference in latency originates from different

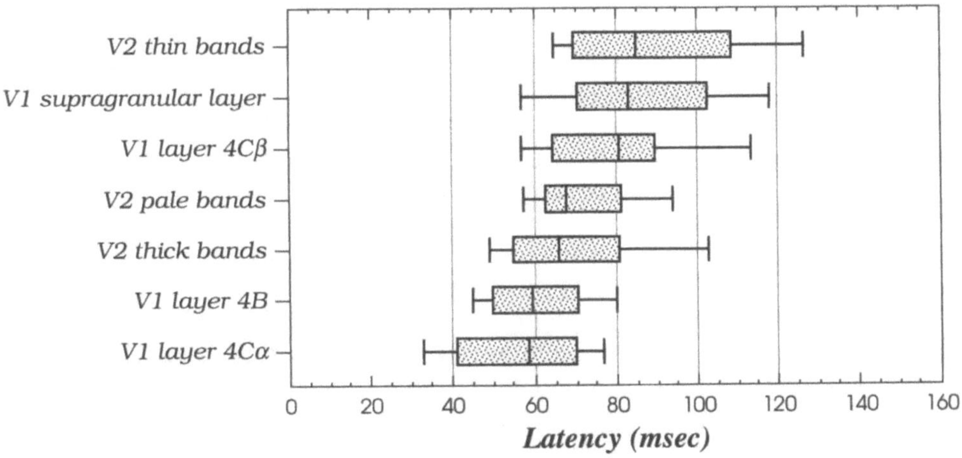

Figure 3. Distributions of latencies to visual stimulation in different modules of areas V1 and V2 of the macaque monkey. The central box corresponds to the 25–75% centile (vertical line is the median). The small vertical bars on either side of the box correspond to the 10% and 90% centiles. Redrawn from Bullier and Nowak (1995). Thin, thick, and pale bands refer to the cytochrome oxidase bands in area V2.

processing times within the retina. On the other hand, there appears to be an overlap in the distribution of latencies for the different cell types. A similar tendency, concerning both shorter latencies for magnocellular neurons and overlap of latencies for magnocellular and parvocellular neurons, has been reported by Blakemore and Vital-Durand (1986).

Interestingly, Marrocco (1976) observed a good correlation between the latency to visual stimulation and the latency of orthodromic responses after electrical stimulation of the optic chiasm. Neurons with the longest orthodromic latencies are probably those that are innervated by the smallest axons. Since there is a relationship between axon size and soma size (Wässle *et al.*, 1975), retinal ganglion cells that drive these neurons are likely to be small and to have smaller dendritic arbors than other neurons of the same class. Therefore, they may not sample as many inputs as larger neurons, which would lead to slower temporal summation, leading to a longer latency.

The visual latency of neurons in the different subdivisions of the LGN has been established by Irvin *et al.* (1986) for the galago. The mean latencies they found were 54 msec for neurons of the magnocellular layers, 68 msec for those of the parvocellular layers, 79 msec for those of the koniocellular layers, and 82 msec for the neurons of the interlaminar zone. These last two groups of cells are likely to be the equivalent of the cat W cells.

These studies, therefore, show that a latency difference of 10–20 msec exists between the parvocellular and magnocellular layer neurons. This latency difference is similar to that obtained between X and Y cells in the cat lateral geniculate nucleus and retina. It remains to be established how the latency of primate parvo- and magnocellular neurons is affected by different visual stimuli.

3.1.3. Conclusion

In cat and probably in monkey, there is a difference in the intraretinal processing time, such that the ganglion cells that display the fastest conduction velocity (Y cells or magnocellular neurons) are also those that display the shortest visual latencies for a broad range of visual stimulus parameters.

This latency difference must also appear in the cortex. Indeed, with square stimuli of 1–2 deg width, a latency difference of 20 msec has been observed between layer 4Cα (recipient of magnocellular inputs) and layer 4Cβ neurons (recipient of parvocellular inputs) of macaque monkey area V1 (Nowak *et al.*, 1995). This latency difference is larger than the latency difference that would be expected on the basis of difference in axonal conduction delay only (5–6 msec; see above). To establish the difference of *functional* delays that can be expected when the different inputs reach the cortex, the intraretinal processing time must be taken into account.

3.2. Visual Latencies in Cat Cortex

Visual response latencies in cat area 17 have been measured in a number of studies. Among the first studies to mention response latencies is that of Creutzfeldt and Ito (1968), who reported visual response latencies for spikes between 35 and 100 msec. These relatively long latencies to visual stimulation can be contrasted with the very short latencies obtained for EPSPs or spikes elicited by electrical stimulation of the subcortical afferent or spike-triggered averaging (see Section 2.1.4a). These long visual latencies have to be related to the long intraretinal processing time, combined with the integration time and the polysynaptic nature of some of the cortical responses.

The observation of long visual latencies in area 17 (compared with other sensory modalities) has been confirmed in a number of studies using extracellular recordings of action potentials. The peak response latencies reported by Ikeda and Wright (1975) averaged 47 msec (range 22–67 msec) for their transient cells and 57 msec (range 36–83 msec) for their sustained cells. Interestingly, Ikeda and Wright reported that simple and complex cells could not be distinguished on the basis of their visual latency (mean latencies 50.4 and 51.3 msec, respectively), another argument to be added to the numerous other arguments showing that, in cat primary visual cortex, simple and complex cells do not represent successive steps in a hierarchy of information processing, but most likely function as two parallel systems. Eschwiler and Rauscheker (1993) found that the onset latency (contrast 82%) ranges between 30 and 170 msec (median 65 msec). The peak latency is between 43 and 245 msec (median 85 msec). This indicates a large scatter of response latencies.

Best *et al.* (1986) measured response latencies in the different layers of cat area 17. For the whole of area 17, the latency distribution also showed a large scatter. Part of this scatter reflects different latencies in different cortical layers. Indeed, the average latency for the supragranular layers was longer by 20 msec than the average latency in layer 4. This large difference could reflect different pathways for the transmission of X and Y inputs, or a long integration time for polysynaptic responses, or both.

The visual latencies in the LGN and different cortical areas (17, 18, 19, PMLS, and area 7) of cat cortex have been determined by Dinse and Krüger (1994). The interesting aspect of that study is that the latencies in different areas were determined under similar experimental conditions. This gives the possibility to determine the interareal transmission delays with visual stimulation. As one progresses from the LGN to area 7, the mean latency is found to increase. In parallel to an increase in mean latency, there is also an increase in latency scatter. However, it is in area 18 that the shortest response latencies were obtained (40.4 msec on average). Those in area 17 were longer by 12 msec (52.7 msec). This latency difference is quite close to the latency difference reported between X (the main input to area 17) and Y (the main input to area 18) cells in cat retina and LGN (see Section 3.1.1).

It has been proposed that cat cortical areas can be organized in a hierarchy according to the pattern of reciprocal connections between them (Felleman and Van Essen, 1991). The hierarchy proposed by Felleman and Van Essen suggests that the order of activation should be 17-18-19-PMLS. The order revealed by the mean latencies obtained by Dinse and Krüger (1994) is clearly different: as already mentioned, the earliest latencies are observed in area 18, followed by area 17, then by area PMLS and area 19. Therefore, if visual cortical areas of the cat processed information in a serial manner, as suggested by the hierarchical arrangement, the functional order would be 18-17-PMLS-19. This suggests that the rank of a given area in the anatomical hierarchy does not necessarily correspond to that of the order of activation. Other examples will be found among the cortical areas of the primate visual system in Section 3.3.

3.3. Visual Latencies in Primate Cortex

3.3.1. Latencies in Primate Areas V1 and V2

Visual response latencies have been reported for area V1 in a number of studies in anesthetized and in awake, behaving monkeys. These studies also differed with respect to the visual stimuli activating the neurons and to the methods used to determine latency of the response. For these reasons, the mean or median response latencies can differ by as much as 40 msec from one study to another. At one of the two extreme ends, one can find the latencies of Bartlett and Doty (1974) and those of Maunsell and Gibson (1992), both using flashed visual stimuli (median for single units, 45 msec). At the other end, Raiguel *et al.* (1989) obtained a median latency of 85 msec with moving bar stimuli.

The latency distribution for some of the studies performed in area V1 is presented in Fig. 4 below, which summarizes the results of a number of latency studies in monkey visual cortex using flashing stimuli (with the exception of the work of Kawano *et al.*, who used sudden onset of fast-moving stimuli). Even among studies using flashing stimuli in area V1 in awake animals (all the reports except that of Nowak *et al.*, 1995), there is a large variability in absolute latency. The median latency varies between 45 and 60 msec. Even larger interstudy variability is seen in the earliest and latest latencies: the 10% centile of the distribution varies between 27 and 48 msec and the 90% centile varies between 65 and 120 msec.

The latency scatter (difference between the 90% and the 10% centile) in area V1 varies between 30 and 80 msec in different reports. It is larger in anesthetized animals than in awake, behaving animals. This might be related to the influence of anesthesia, which can modify the timing of retinal to geniculate information transfer for some cells (Mukerjee and Kaplan, 1995).

The latency scatter in V1 can be caused by several factors: the first possibility is that the scatter observed in the LGN is reproduced in the cortex. This can be determined by comparing the latency scatters in layers 4Cα and 4Cβ (Nowak *et al.*, 1995), the main recipients of the M and P channels, to those observed for presumed magnocellular and parvocellular neurons of the LGN (Marrocco, 1976); both studies were performed on anesthetized monkeys. It clearly appears, then, that the latency scatter observed for the presumed magnocellular neurons (37.7 msec) is similar to the one obtained in layer 4Cα (45 msec). The latency scatter observed in layer 4Cβ (60 msec) appears slightly larger than that obtained for color opponent cells (43 msec) and for nonopponent, sustained cells (47.5 msec). Notice also that part of the increase in latency scatter in layer 4 could be due to some polysynaptic transmission as well: neurons within these layers establish contacts with each other and receive in addition a dense innervation from layer 6 neurons (Levitt *et al.*, 1996).

Another factor that can account for the latency scatter is the difference in response onset for neurons driven by the magnocellular pathway compared to those driven by the parvocellular pathway. Knowing the difference in latency between M and P neurons in the LGN, it is of interest to investigate whether such a difference persists at the cortical level. This has been done by comparing the latencies of neurons in layers 4Cα and 4Cβ (Nowak *et al.*, 1995). A value of 20 msec was found, in keeping with the latency difference between M and P neurons in the LGN (see Section 3.1.2). Combining this 20-msec latency difference to the latency scatter in layer 4Cβ gives already a total latency scatter of 80 msec for the whole of layer 4. Another possibility is that some neurons are driven by koniocellular cells, and, if they exist in the monkey, by lagged cells of the LGN (as mentioned above, these neurons have the longest latencies in the lateral geniculate nucleus).

The third factor that may account for the latency scatter observed in area V1 is the polysynaptic nature of some of the responses. Neurons at the top of the supragranular layers, for example, receive their inputs not directly from layer 4, but after an intervening relay in the lowest part of the supragranular layers (Levitt *et al.*, 1996). Hence, these neurons are distant by two synapses from layer 4. Assuming an integration time between 5 and 10 msec and a negligible axonal delay, we have that these neurons would be activated only 10–20 msec after neurons in layer 4. The latency scatter now would be close to 100 msec. This is in fact the latency range obtained by Nowak *et al.* (1995) for ON responses (for OFF responses, a few neurons with exceptionally long latencies were observed). Finally, another path at the origin of polysynaptic responses in area V1, leading to long visual latency, could be one that involves feedback connections from extrastriate cortex.

Beyond the thalamorecipient layers in V1, the M and P channels appear to converge in many neurons (Malpeli *et al.,* 1981; Nealey and Maunsell, 1994; Levitt *et al.,* 1996) and it is therefore difficult to interpret differences in latencies of V1 neurons as reflecting different types of thalamic inputs. In area V2, it is known that the thick cytochrome oxidase bands are mainly innervated by layer 4B (the M pathway; Livingstone and Hubel, 1987). This may explain why latencies of neurons in that subdivision are among the shortest recorded in V2 (Fig. 3; Munk *et al.,* 1995). The pale cytochrome oxidase bands receive inputs from the interblobs in area V1 (Livingstone and Hubel, 1984). Short latencies were also observed in the pale bands, in keeping with the convergence of M and P inputs in the interblobs (Nealey and Maunsell, 1994; Yoshioka *et al.,* 1994).

Although the thin cytochrome oxidase bands in V2 are driven by the cytochrome oxidase blobs in V1, which themselves receive converging inputs from all channels (Casagrande, 1994; Hendry and Yoshioka, 1994; Nealey and Maunsell, 1994; Yoshioka *et al.,* 1994), their latencies are substantially longer than those recorded in pale and thick bands (Fig. 3; Munk *et al.,* 1995). One possible reason for these latencies is that the blobs, in contrast to layer 4B and to the interblobs, do not receive *direct* inputs from layer 4Cα. Instead, they receive their magnocellular inputs indirectly, after a relay in layer 4B (Lachica *et al.,* 1992; Yoshioka *et al.,* 1994). This additional relay may introduce a temporal delay of several milliseconds due to the integration time.

The long latencies of neurons in thin cytochrome oxidase bands of V2 may also be due to the predominance of nonoriented color-selective cells in that subdivision (although this is a matter of debate; see Salin and Bullier, 1995, for review). Since cytochrome oxidase blobs in V1 also contain a majority of nonoriented color-coded cells (Livingstone and Hubel, 1984; Yoshioka and Dow, 1996), it is possible that neurons in that subdivision are mainly under the influence of the parvocellular system, hence responding with longer visual latencies. Such long latencies would also be consistent with the projection of the late-activated K channel in the cytochrome oxidase blobs (Casagrande, 1994, Hendry and Yoshioka, 1994). A third explanation is that the blobs–thin bands channel may have response latencies delayed by strong inhibitory blockade from the M channel (L. G. Nowak, M. H. J. Munk, P. Girard, and J. Bullier, unpublished; see also Sato *et al.,* 1994).

The thick cytochrome oxidase bands in V2 project to area MT (Shipp and Zeki, 1985). The short latencies in this subdivision suggest that MT neurons are activated rapidly, a conclusion in keeping with published work (see Section 3.3.3) and with the very fast conducting axons between V1 and MT (see Section 2.1.4c). Area V4, the other major output of area V2, receives some short latency activity through the pale bands and long latency activity through the thin bands. Thin and pale bands might process different aspects (color in thin bands and orientation of borders in pale bands) of the visual scene. There is evidence that a segregation between color and orientation modules also exists in V4 (Tanaka *et al.,* 1986; Yoshioka and Dow, 1996). This suggests that in V4, some modules are activated earlier than others. A possible interpretation of such a segregation will be given in Section 4.3.

3.3.2. Transmission of Visual Signals May Rely on One Neuronal Layer per Area

In several studies the visual latencies have been determined with the same method in two different areas. Under these conditions it is possible to determine the interarea information transfer time. The latency difference that has been reported between areas V1 and V2 is 11 msec in Raiguel *et al.* (1989) and 10 msec for single-unit recordings in Nowak *et al.* (1995). The latency difference that has been reported between areas V1 and MT, which are reciprocally and monosynaptically connected, is 9 msec in the study by Raiguel *et al.* (1989) and 11 msec in the results of Maunsell (1987). Hence, for cortical areas that are directly connected, the latency difference appears to be close to 10 msec (see also Thorpe and Imbert, 1989).

Maunsell (1987) also reported a 22-msec latency difference between the transient cells in area V1 and neurons in V4. Areas V1 and V4 are connected through a relay in V2. The 22-msec difference between the latencies in V1 and V4 corresponds approximately to twice the interarea transfer time of 10 msec. Similar reasoning can be applied to the results obtained by Vogels and Orban (1990, 1994), who reported visual latencies in areas V1 and TE of macaque monkeys with the same experimental paradigm (in one monkey, recording was done in both areas). The median latency in TE was found to be larger than the median latency in V1 by 40 msec. This 40 msec corresponds exactly to what can be expected if the latency difference between two directly connected areas is 10 msec, because there are four relays between V1 and TE (V2, V4, TEO, TE).

As already mentioned, an interareal delay of 10 msec corresponds roughly to the sum of the axonal conduction delay between two cortical areas and the integration time. To explain a 22-msec latency difference between areas V1 and V4, or a 40-msec latency difference between areas V1 and TE, it must be assumed that there is *only one synaptic layer per cortical area and only one action potential per neuron* (Thorpe and Imbert, 1989). This is unexpected since neurons in layer 4 that receive information from a lower stage usually do not possess axons projecting in other cortical areas. However, the basal dendrites of neurons in lower layer 3 can enter in layer 4. In addition, except in monkey area V1, neurons of the infragranular layer could be contacted on their apical dendrites as they pass through layer 4. Furthermore, it is known that feedforward axons often terminate not only in layer 4, but also in the lower part of layer 3 and in the infragranular layers. In other words, the feedforward inputs can directly contact output neurons. For example, terminals of feedforward inputs from area V1 onto area MT are found in layer 4 and the lower portion of layer 3 (Rockland, 1989). It is also in the lower portion of layer 3 that neurons projecting from area MT to the frontal eye field are concentrated (Schall *et al.*, 1995).

This 10 msec per area rule is of course valid only on average for the short-latency neurons. For the activation of long-latency neurons, there may be some paths which involve several synaptic stages in a given area, combined with information transferred through feedback connections.

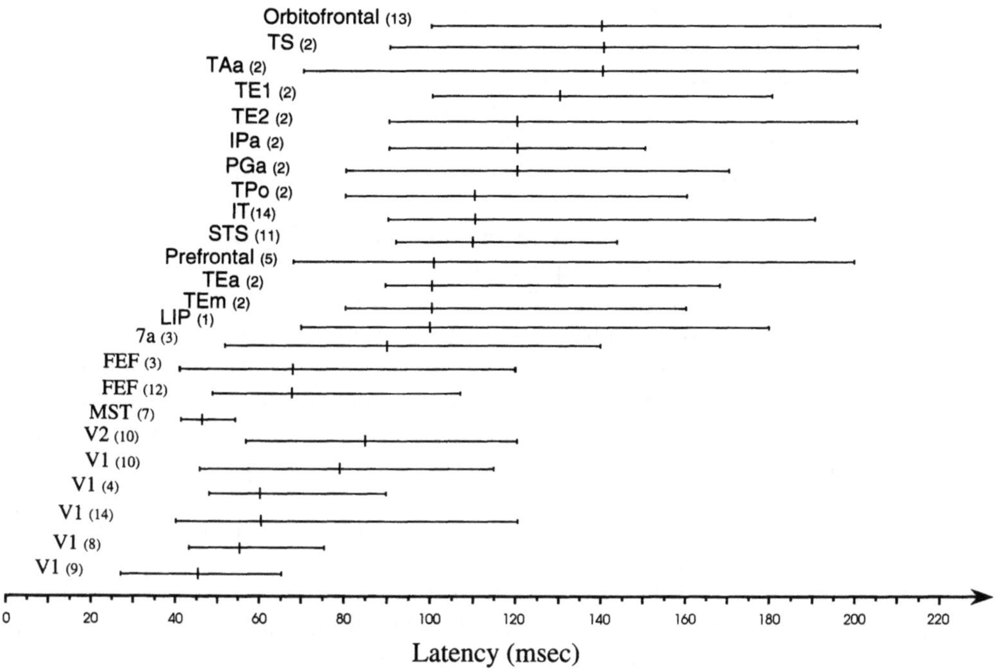

Figure 4. Distributions of latencies to visual stimulation in different areas of the visual system of the macaque monkey. Behaving monkey in all cases except case 10, Nowak *et al.* (1995). Small flashed stimulus in all cases except case 7, Kawano *et al.* (1994), in which sudden onset of a fast-moving visual pattern was used as the stimulus. Same presentation as Figs. 1 and 2 (bars represent 10%, 50%, and 90% centiles). (1) Barash *et al.* (1991); (2) Baylis *et al.* (1987); (3) Bushnell *et al.* (1991); (4) Celebrini *et al.* (1993); (5) Funahashi *et al.* (1990); (6) Goldberg and Bushnell (1991); (7) Kawano *et al.* (1994); (8) Knierim and Van Essen (1992); (9) Maunsell and Gibson (1992); (10) Nowak *et al.* (1995); (11) Perrett *et al.* (1982); (12) Thompson *et al.* (1996); (13) Thorpe *et al.* (1983); (14) Vogels and Orban (1994).

3.3.3. The Fast and the Slow Brain

Figure 4 presents a summary of a number of reports of latencies to flashed visual stimuli recorded in different cortical areas of awake monkeys. Note that there are three exceptions: the studies of Nowak *et al.* and Munk *et al.* were done in anesthetized animals and the study by Kawano *et al.* (1994) was done with stimuli consisting of large array of dots suddenly set in motion at high speed.

What is immediately evident in Fig. 4 is the large scatter of latencies. This scatter is such that latencies in adjacent cortical areas overlap substantially. There are two major groups of areas in Fig. 4: areas with median latencies between 40 and 80 msec, which constitute what we call the fast brain, and areas with median latencies between 100 and 150 msec, which make up what we call the slow brain.

The fast brain corresponds in Fig. 4 to areas V1, V2, MST, and the frontal eye field (FEF). In addition, we saw earlier that areas MT and V4 can also be included in this group because their latencies are displaced with respect to V1 by only 10 or 20 msec. Short latencies of MT and MST neurons to rapidly moving stimuli have also been reported by Orban's group (Lagae *et al.*, 1994). Given the short latencies in areas MT and MST, it is somewhat surprising that latencies in areas LIP and 7a are not shorter. In the case of LIP (Barash *et al.*, 1991), the methods used to measure latencies tend to overestimate them by 20 msec.

The fast brain is dedicated to an early processing of visual information in the case of V1 and occipital prestriate cortex and to a processing of information related to visuomotor integration in the case of the parietal cortex and frontal eye field. An early activation of parietal cortex is in keeping with its major input from the M channel in the parietal cortex via area MT. Most cortical areas involved in this pathway are heavily myelinated (areas MT, MST, VIP), suggesting that speed of transfer is an important parameter in this cortical region. We also know that, at least in the case of the V1 to MT pathway, axons are rapidly conducting. It is likely that rapidly conducting axons also link areas MT, MST, VIP, and FEF. The presence of rapidly conducting axons between MT and FEF would explain the early activation of FEF neurons by visual stimuli (Bushnell *et al.*, 1991; Bichot *et al.*, 1996; Thompson *et al.*, 1996), despite its long distance from area MT, which probably constitutes the major relay of visual information toward frontal cortex.

Another potential explanation for the early latencies in areas of the fast brain is the large number of bypass connections in the dorsal occipitoparietal stream. Examples of such bypasses are found in the direct projections from V1 to V3 and MT, which bypass a relay through V2 (Felleman and Van Essen, 1991). Another example is given by the direct projection from area V1 to area PO, which also receives indirect inputs through area MT (Colby *et al.*, 1988).

The slow brain groups most areas of the temporal lobe and some areas of the frontal cortex located rostral to the frontal eye field. The median latencies are in the 100–150-msec range and many neurons respond as late as 200 msec after the presentation of the stimulus. It is unclear at the moment which factors are most important in determining the long latencies of neurons of the slow brain. One could be that axons in the temporal cortex are slowly conducting. In addition, the presence of bypass connections appears to be less frequent in the ventral occipitotemporal stream, which seems to be organized in a more serial fashion. For example, to reach TE from V1, information has to relay in areas V2, V4, and TEO (although a few bypass connections also exist; Nakamura *et al.*, 1993).

3.3.4. Latencies and the Hierarchical Model of Cortical Organization

As was the case for the cat visual cortex, latencies to visual stimulation in monkey are not ordered as expected from the hierarchical organization based on anatomical grounds (Felleman and Van Essen, 1991). The FEF region, which is at a position close to the top of the hierarchy, is activated simultaneously with neurons of areas V1, which belong to the first level (Bichot *et al.*, 1996; Thompson *et al.*, 1996). The latency difference between neurons in these two areas does not leave much time for whatever processing is supposed to occur at each stage of the hierarchy. Another example that was cited earlier (Bullier and Nowak, 1995) is that of areas V2, MT, and V4. The reasoning goes like this: The median latency is longer by 10 msec in area V2 than in V1 when the latencies are determined under similar experimental conditions. A 10-msec latency difference has also been reported between area V1 and area MT (Maunsell, 1987; Raiguel *et al.*, 1989). In other words, if one uses latency to order cortical areas, then area MT should leave the fifth level it occupies in the anatomical hierarchy, to join area V2 at the second level (Raiguel *et al.*, 1989). In the anatomical hierarchy, areas MT and V4 occupy the same level. However, the latency difference between areas V1 and V4 is 12 msec longer than that between area V1 and area MT (Maunsell, 1987; Raiguel *et al.*, 1989). In terms of latency, therefore, it looks as if areas MT and V4 occupy two different hierarchical levels.

Thus, if one assumes that latencies to visual stimuli provide a reasonable estimate of the order of activation of cortical areas, it appears that this order does not follow the one suggested by the anatomical hierarchy of cortical areas. The major factor that influences the speed of activation of cortical areas appears to be whether or not a cortical area belongs to the dorsal stream that is almost exclusively activated by the M channel.

4. Conclusion: How Does Cortex Process Information So Rapidly?

4.1. What Connects Together Fires Together

Latencies in adjacent areas differ by 10 msec on the average. The latency distributions in different areas show dispersions much larger than 10 msec (Fig. 4). In addition, neurons of the visual cortex give responses lasting usually more than 100 msec even to the shortest stimuli (Rolls and Tovee, 1994; Kovacs *et al.*, 1995). Because of the large latency scatter and the sustained responses of visual neurons, neurons in adjacent cortical areas emit spikes during the same post-stimulus period. In the case of areas V1 and V2, this overlap of latency leads one to conclude that much of the processing in these two areas is done in parallel and not in a sequential fashion, as suggested by the results of reversible inactivation of V1 (Schiller and Malpeli, 1977a; Girard and Bullier, 1989). Given the simultaneity of firing of neurons, it is no surprise that many pairs of neurons in adjacent areas give synchronized responses (J. I. Nelson *et al.*, 1992; Nowak *et al.*, 1994). Neighboring areas are also often strongly linked with reciprocal connections (Felleman and Van Essen, 1991; Young, 1992). It is tempting to suggest that the synchronization of firing of neurons in adjacent areas is due to the dense network of reciprocal connections between them.

Beyond a certain distance on the cortical surface, connections become rare or absent. This is particularly the case in the temporal lobe, in which connections appear to link a small set of neighboring areas (Desimone *et al.*, 1980). It is interesting to remark that neurons in V1 and V2 which are not directly connected to areas of the temporal cortex also display much shorter latencies. On the contrary, despite their large separation on the cortical surface, neurons of areas MT/MST and FEF are interconnected and display similar latencies to visual stimulation. In other words, paraphrasing an earlier statement (Löwel and Singer, 1992), "what connects together fires together."

4.2. How Rapidly Is Response Selectivity Established?

Given the rapid activation of neurons in most cortical areas, an interesting question, discussed earlier by Thorpe and Imbert (1989), is to determine whether the early part of the response is already selective to different stimuli or whether the response is first broadly tuned and selectivity is acquired progressively as the response develops.

Face-selective cells in the STS have visual latencies as short as 80 msec (Perrett *et al.*, 1982). Response selectivity for different views of the face and head is already present only 5 msec after the onset of the response, even in the case of rapidly responding cells (Oram and Perrett, 1992). A fast acquisition of response selectivity has also been observed by Celebrini *et al.* (1993), who found that orientation selectivity in area V1 is already present at the onset of the response of even the most rapidly responding neurons. In their intracellular study of V1 neurons, Volgushev *et al.* (1995) showed that the subthreshold synaptic response is initially broadly tuned to orientation, but acquire its definitive tuning *before* action potentials are fired.

The very fast acquisition of response selectivity has an important consequence for the neurobiological basis of receptive field construction: at least for the rapidly responding neurons, the receptive field properties must be determined by the pattern of feedforward inputs (Thorpe and Imbert, 1989). If, in addition, inhibition is required in the sculpturing of response selectivity (Sillito *et al.*, 1980; Douglas and Martin, 1991), this implies that inhibitory neurons must have very short visual latencies and a very rapid integration time. Whether this is truly the case remains to be established for the visual cortex, but results consistent with this view have been obtained in the barrel field of rat somatosensory cortex (Welker *et al.*, 1993).

What happens to selectivity beyond the early response of the first activated neurons? Using an analysis based on information theory, Tovee *et al.* (1993) observed that the quantity of information reaches a high value very early after the onset of the response. A fast rise in information content of neuronal response has also been reported for areas V1, V2, and V4 by McClurkin and Optican (1996). However, the information content has also been found to increase during later parts of the response (Richmond and Optican, 1990; Heller *et al.*, 1995; McClurkin and Optican, 1996). Similarly, if the results obtained by Oram and Perrett (1992) showed a very fast acquisition of response selectivity, it is only 80 msec after response onset that the strongest difference between responses is reached. Therefore, it looks as if there is a very rapid acquisition of receptive field properties, followed by a further refinement of the selectivity. Other evidence that there are properties that develop only after the response onset can be found in the time course of responses of V1 neurons to pop-out stimuli (Knierim and Van Essen, 1992; Lamme, 1995).

The arguments presented above indicate that the rapid acquisition of response selectivity relies only on the feedforward transmission of information across cortical areas. However, the further refinement that follows that initial response could be related to the overlap of visual response latencies between different cortical areas that could allow bidirectional interactions involving the feedback connections.

Seeing is not just detecting the orientation of a bar stimulus. There are many situations in which ambiguities must be resolved rapidly in order to identify the different elements of the visual scene. We present here a model of the part of the visual system involved in pattern recognition, the slow brain. The model contains several of the features that were stressed above: (1) The wave of incoming information reaching the cortex is not synchronous; it is composed of at least three and possibly four (M, P, K, and lagged inputs) waves that are activated up to 200 msec after the stimulus onset. (2) Neurons function mainly as coincidence detectors and signal the coincidence of properly timed waves of activity. (3) Feedback connections are used to carry different hypotheses concerning the elements of the visual scene to be tested against the incoming waves of afferent activity (Ullman, 1995).

The model is an elaboration of that presented by Ullman (1995) to explain pattern recognition with a bidirectional flow of information. The idea is that the M channel acts as an ignition device to activate representations stored in the temporal cortex. The poor temporal resolution and the lack of color selectivity of the M channel leads to a sketchy view of the scene. Although limited in precision, this is the first information that reaches the temporal cortex. Different potential interpretations of an object lead to activation of different neurons that send activity down through feedback connections. As proposed in the model of Ullman (1995), these waves of descending activity meet the subsequent waves of feedforward activity (P, K, lagged), giving rise either to a reinforcement or to an extinction by a mechanism of coincidence detection between properly timed feedforward and feedback activity waves. The proper interpretation of the image is the one that leads to reinforced activity at all levels of the visual cortex.

If we integrate such a model of the slow brain with the existence of the fast brain, we arrive at the conclusion that analysis of a visual scene proceeds at different speeds depending on which aspects are privileged. A very fast analysis system appears to be devoted to the processing of localization and visuomotor interactions. It corresponds to the fast brain, is under the influence of the M system, and may function mostly in the feedforward mode. A relatively fast processing system, mostly under the influence of the M system in the slow brain, achieves a coarse analysis of broad features in the feedforward mode. Finally, a slow processing system deals with the fine detail analysis and the precise recognition using P and K systems and feedback connections.

5. References

Abeles, M., 1982, *Local Cortical Circuits*, Springer-Verlag, Berlin.

Abeles, M., 1991, Corticonics: Neural Circuits of the Cerebral Cortex, Cambridge University Press, Cambridge.

Barash, S., Bracewell, R. M., Fogassi, L., Gnadt, J. W., and Andersen, R. A., 1991, Saccade-related activity in the lateral intraparietal area. I. Temporal properties; comparison with area 7a, *J. Neurophysiol.* **66:**1095–1108.

Bartlett, J. R., and Doty, R. W., 1974, Response of units in striate cortex of squirrel monkeys to visual and electrical stimuli, *J. Neurophysiol.* **37:**621–641.

Baylis, G. C., Rolls, E. T., and Leonard, C. M., 1987, Functional subdivisions of the temporal lobe neocortex, *J. Neurosci.* **7:**330–342.

Best, J., Reuss, S., and Dinse, H. R. O., 1986, Lamina-specific differences of visual latencies following photic stimulation in the cat striate cortex, *Brain Res.* **385:**356–360.

Bichot, N. P., Schall, J. D., and Thompson, K. G., 1996, Visual feature selectivity in frontal eye fields induced by experience in mature macaques, *Nature* **381:**697–699.

Blakemore, C., and Vital-Durand, F., 1986, Organization and postnatal development of the monkey's lateral geniculate nucleus, *J. Physiol.* **380:**453–491.

Bolz, J., Rosner, G., and Wässle, H., 1982, Response latency of brisk-sustained (X) and brisk-transient (Y) cells in the cat retina, *J. Physiol.* **328:**171–190.

Bullier, J., and Henry, G. H., 1979, Ordinal position of neurons in cat striate cortex, *J. Neurophysiol.* **42:**1251–1263.

Bullier, J., and Henry G. H., 1980, Ordinal position and afferent input of neurons in monkey striate cortex, *J. Comp. Neurol.* **193:**913–935.

Bullier, J., and Nowak, L. G., 1995, Parallel versus serial processing: New vistas on the distributed organization of the visual system, *Curr. Opin. Neurobiol.* **5:**497–503.

Bullier, J., McCourt, M. E., and Henry, G. H., 1988, Physiological studies on the feedback connection to the striate cortex from cortical areas 18 and 19 of the cat, *Exp. Brain Res.* **70:**90–98.

Bushnell, M. C., Goldberg, M. E., and Robinson, D. L., 1991, Behavioral enhancement of visual responses in monkey cerebral cortex: I. Modulation in posterior parietal cortex related to selective visual attention, *J. Neurophysiol.* **46:**755–772.

Casagrande, V. A., 1994, A third visual pathway to primate area V1, *Trends Neurosci.* **17:**305–310.

Casanova, C., 1993, Response properties of neurons in area 17 projecting to the striate-recipient zone of the cat's lateralis posterior-pulvinar complex: Comparison with cortico-tectal cells, *Exp. Brain Res.* **96:**247–259.

Celebrini, S., Thorpe, S., Trotter, Y., and Imbert, M., 1993, Dynamics of orientation coding in area V1 of the awake primate, *Visual Neurosci.* **10:**811–825.

Cleland, B. G., and Enroth-Cugell, C., 1970, Quantitative aspects of gain and latency in the cat retina, *J. Physiol.* **206:**73–91.

Cleland, B. G., Dubin, M. W., and Levick, W. R., 1971, Simultaneous recording of input and output of lateral geniculate neurons, *Nature (New Biol.)* **231:**191–192.

Cleland, B. G., Levick, W. R., Morstyn, H., and Wagner, H. G., 1976, Lateral geniculate relay of slowly conducting retinal afferents to cat visual cortex, *J. Physiol.* **255:**299–320.

Colby, C. L., Gattass, R., Olson, C. R., and Gross, C. G., 1988, Topographical organization of cortical afferents to extrastriate visual area PO in the macaque: A dual tracer study, *J. Comp. Neurol.* **269:**392–413.

Cope, T. C., Fetz, E. E., and Matsumara, M., 1987, Cross-correlation assessment of synaptic strength of single Ia fibre connections with triceps surae motoneurones in cats, *J. Physiol.* **390:**161–188.

Creutzfeldt, O., and Ito, M., 1968, Functional synaptic organization of primary visual cortex neurons in the cat, *Exp. Brain Res.* **6:**324–352.

Desimone, R., Fleming, J., and Gross, C. G., 1980, Prestriate afferents to inferior temporal cortex: An HRP study, *Brain Res.* **184:**41–55.

Dinse, H. R., and Krüger, K., 1994, The timing of processing along the visual pathway in the cat, *NeuroReport* **5:**893–897.

Douglas, R. J., and Martin, K. A., 1991, A functional microcircuit for cat visual cortex, *J. Physiol.* (Lond.) **440:**735–769.

Dreher, B., Fukada, Y., and Rodieck, R. W., 1976, Identification, classification and anatomical segregation of cells with X-like and Y-like properties in the lateral geniculate nucleus of old-world primates, *J. Physiol.* **258:**433–452.

Eder, C., Ficker, E., Günde, L. J., and Heinemann, U., 1991, Outward currents in rat entorhinal cortex stellate cells studied with conventional and perforated patch recordings, *Eur. J. Neurosci.* **3:**1271–1280.

Eggermont, J. J., 1992, Neural interaction in cat primary auditory cortex. Dependence on recording depth, electrode separation, and age, *J. Neurophysiol.* **68:**1216–1228.

Eschwiller, G. W., and Rauschecker, J. P., 1993, Temporal integration in visual cortex of cats with surgically induced strabismus, *Eur. J. Neurosci.* **5:**1501–1509.

Felleman, D. J., and Van Essen, D. C., 1991, Distributed hierarchical processing in the primate cerebral cortex, *Cerebral Cortex* **1:**1–47.

Ferster, D., and Jagadeesh, B., 1992, EPSP–IPSP interactions in cat visual cortex studied with *in vivo* whole-cell patch recording, *J. Neurosci.* **12:**1262–1274.

Fetz, E. E., and Gustafsson, B., 1983, Relation between shapes of post-synaptic potentials and correlated changes in firing probability of cat motoneurones, *J. Physiol.* **341:**387–410.

Finlay, B. L., Schiller, P. H., and Voman, S. F., 1976, Quantitative studies of single-cell properties in monkey striate cortex. IV. Corticotectal cells, *J. Neurophysiol.* **39:**1352–1361.

Fregnac, Y., and Bringuier, V., 1996, Spatio-temporal dynamics of synaptic integration in cat visual cortical receptive fields, in: *Biological Basis and Computational Theory of Vision* (A. Aertsen and V. Braitenberg, eds.), Elsevier, Amsterdam, pp. 1–57.

Freund, T. F., Martin, K. A., Somogyi, P., and Whitteridge, D., 1985, Innervation of cat visual areas 17 and 18 by physiologically identified X- and Y-type thalamic afferents. II. Identification of postsynaptic targets by GABA immunocytochemistry and Golgi impregnation, *J. Comp. Neurol.* **242:**275–291.

Freund, T. F., Martin, K. A. C., Soltesz, I., Somogyi, P., and Whitteridge, D., 1989, Arborisation pattern and postsynaptic targets of physiologically identified thalamocortical afferents in striate cortex of the macaque monkey, *J. Comp. Neurol.* **289:**315–336.

Fries, W., 1984, Cortical projections to the superior colliculus in the macaque monkey: A retrograde study using horseradish peroxidase, *J. Comp. Neurol.* **230:**55–76.

Funahashi, S., Bruce, C. J., and Godman-Rakic, P. S., 1990, Visuospatial coding in primate prefrontal neurons revealed by oculomotor paradigms, *J. Neurophysiol.* **63:**814–831.

Gilbert, C. D., and Kelly, J. P., 1976, The projections of cells in different layers of the cat's visual cortex, *J. Comp. Neurol.* **163:**81–106.

Gilbert, C. D., and Wiesel, T. N., 1981, Morphology and intracortical projections of functionally characterised neurons in the cat visual cortex, *Nature* **280:**120–125.

Girard, P., and Bullier, J., 1989, Visual activity in area V2 during reversible inactivation of area 17 in the macaque monkey, *J. Neurophysiol.* **62:**1287–1302.

Gochin, P. M., Miller, E. M., Gross, C. G., and Gerstein, G. L., 1991, Functional interactions among neurons inferior temporal cortex of the awake macaque, *Exp. Brain Res.* **84:**505–516.

Goldberg, M. E., and Bushnell, M. C., 1991, Behavioral enhancement of visual responses in monkey cerebral cortex. II. Modulation in frontal eye fields related to saccades, *J. Neurophysiol.* **46:**773–787.

Gouras, P., 1969, Antidromic responses of orthodromically identified ganglion cells in monkey retina, *J. Physiol.* **204:**407–419.

Gouras, P., and Link, K., 1966, Rod and cone interaction in dark-adapted monkey ganglion cells, *J. Physiol.* **184:**499–510.

Grieve, K. L., and Sillito, A. M., 1995, Differential properties of cells in the feline primary visual cortex providing corticofugal feedback to the lateral geniculate nucleus and visual claustrum, *J. Neurosci.* **15:**4868–4874.

Grinvald, A., Lieke, E. E., Frostig, R. D., and Hidelsheim, R., 1994, Cortical point-spread function and long-range interactions revealed by real-time optical imagining of macaque primary visual cortex, *J. Neurosci.* **14:**2545–2568.

Gustafsson, B., and McCrea, D., 1983, Influence of stretch-evoked synaptic potentials on firing probability of cat spinal motoneurones, *J. Physiol.* **347:**431–451.

Hartveit, E., and Heggelund, P., 1992, The effect of contrast on the visual response of lagged and nonlagged cells in the cat lateral geniculate nucleus, *Visual Neurosci.* **9:**515–525.

Harvey, A. R., 1980, A physiological analysis of subcortical and commissural projections of areas 17 and 18 of the cat, *J. Physiol.* **302:**507–524.

Haug, H., 1968, Quantitative elektronenmikroskopische Untersuchungen über den markfaseraufbau in der sehrinde der katze, *Brain Res.* **11:**65–84.

Heller, J., Hertz, J. A., Kjaer, T. W., and Richmond, B. J., 1995, Information flow and temporal coding in primate pattern vision. *J. Comput. Neurosci.* **2:**175–193.

Hendry, S. H. C., and Yoshioka, T. Y., 1994, A neurochemically distinct third channel in the macaque dorsal lateal geniculate nucleus, *Science* **264:**575–577.

Henry, G. H., Salin, P. A., and Bullier, J., 1991, Projections from area 18 and 19 to cat striate cortex: Divergence and laminar specificity, *Eur. J. Neurosci.* **3:**186–200.

Hoffmann, K. P., 1973, Conduction velocity pathways from retina to superior colliculus in the cat: A correlation with receptive field properties, *J. Neurophysiol.* **36:**409–424.

Hofman, M. A., 1985, Size and shape of the cerebral cortex in mammals. I. The cortical surface, *Brain Behav. Evol.* **27:**28–40.

Houzel, J.-C., Milleret, C., and Innocenti, G., 1994, Morphology of callosal axons interconnecting areas 17 and 18 of the cat, *Eur. J. Neurosci.* **6:**898–917.

Hughes, A., and Wässle, H., 1976, The cat optic nerve: Fibre total count and diameter spectrum, *J. Comp. Neurol.* **169:**171–184.

Humphrey, A. L., and Saul, A. B., 1992, Action of brain stem reticular afferents on lagged and nonlagged cells in the cat lateral geniculate nucleus, *J. Neurophysiol.* **68:**673–691.

Hursh, J. B., 1939, Conduction velocity and diameter of nerve fibers, *Am. J. Physiol.* **127:**131–139.

Ikeda, H., and Wright, M. J., 1975, Retinotopic distribution, visual latency and orientation tuning of sustained and transient cortical neurons in area 17 of the cat's visual cortex, *Exp. Brain Res.* **22:**385–398.

Innocenti, G. M., 1980, The primary visual pathway through the corpus callosum: Morphological and functional aspects in the cat, *Arch. Ital. Biol.* **118:**124–188.

Irvin, G. E., Norton, T., Sesma, M. A., and Casagrande, V. A., 1986, W-like response properties of interlaminar zone cells in the lateral geniculate nucleus of a primate (*Galago crassicaudatus*), *Brain Res.* **362:**254–274.

Kang, Y., Endo, K., and Araki, T., 1988, Excitatory synaptic actions between pairs of neighboring pyramidal tract cells in the motor cortex, *J. Neurophysiol.* **59:**636–647.

Kawano, K., Shidara, M., Watanabe, Y., and Yamane, S., 1994, Neural activity in cortical area MST of alert monkey during occular following responses, *J. Neurophysiol.* **71:**2305–2324.

Kirk, D. L., Cleland, B. G., and Levick, W. R., 1975, Axonal conduction latencies of cat retinal ganglion cells, *J. Neurophysiol.* **38:**1395–1402.

Kirkwood, P. A., and Sears, T. A., 1978, The synaptic connections to intercostal motoneurones as revealed by the average common excitation potential, *J. Physiol.* **275:**287–314.

Knierim, J. J., and Van Essen, D. C. V., 1992, Neuronal responses to static texture patterns in area V1 of the alert macaque monkey, *J. Neurophysiol.* **67:**961–980.

Knox, C. K., 1974, Cross-correlation function for a neuronal model, *Biophys. J.* **14:**567–582.

Koike, H., Mano, N., Okada, Y., and Oshima, T., 1970, Repetitive impulses generated in fast and slow pyramidal tract cells by intracellularly applied current steps, *Exp. Brain Res.* **11:**263–281.

Komatsu, Y., Nakajima, S., Toyama, K., and Fetz, E. E., 1988, Intracortical connectivity revealed by spike-triggered averaging in slice preparations of cat visual cortex, *Brain Res.* **442:**359–362.

Konig, P., Engel, A. K., and Singer, W., 1996, Integrator or coincidence detector—The role of the cortical neuron revisited, *Trends Neurosci.* **19:**130–137.

Kovacs, G., Vogels, R., and Orban, G. A., 1995, Cortical correlate of backward masking, *Proc. Natl. Acad. Sci. USA* **92:**5587–5591.

Lachica, E. A., Beck, P. D., and Casagrande, V. A., 1992, Parallel pathways in macaque monkey striate cortex: Anatomically defined columns in layer III, *Proc. Natl. Acad. Sci. USA* **89:**3566–3570.

Lagae, L., Maes, H., Raiguel, S., Xiao, D.-K., and Orban, G. A., 1994, Responses of macaque STS neurons to optic flow components: A comparison of areas MT and MST, *J. Neurophysiol.* **71:**1597–1626.

LaMantia, A.-S., and Rakic, P., 1990, Cytological and quantitative characteristics of four cerebral commissures in the rhesus monkey, *J. Comp. Neurol.* **291:**520–537.

Lamme, V. A. F., 1995, The neurophysiology of figure ground segregation in primary visual cortex, *J. Neurosci.* **15:**1605–1615.

Lee, B. B., Cleland, B. G., and Creutzfeldt, O. D., 1977, The retinal input to cells in area 17 of the cat's cortex, *Exp. Brain Res.* **30:**527–538.

Levick, W. R., 1973, Variation in the response latency of cat retinal ganglion cells, *Vision Res.* **13:**837–853.

Levitt, J. B., Lund, J. S., and Yoshioka, T., 1996, Anatomical substrates for early stages in cortical processing of visual information in the macaque monkey, *Behav. Brain Res.* **76:**5–19.

Livingstone, M. S., and Hubel, D. H., 1984, Anatomy and physiology of a color system in the primate visual cortex, *J. Neurosci.* **4:**309–356.

Livingstone, M. S., and Hubel, D. H., 1987, Connections between layer 4b of area 17 and the thick cytochrome oxydase stripes of area 18 in the squirrel monkey, *J. Neurosci.* **7:**3371–3377.

Lohmann, H., and Rörig, B., 1994, Long-range horizontal connections between supragranular pyramidal cells in the extrastriate visual cortex of the rat, *J. Comp. Neurol.* **344:**543–558.

Löwel, S., and Singer, W., 1992, Selection of intrinsic horizontal connections in the visual cortex by correlated neuronal activity, *Science* **255:**209–212.

Malpeli, J. G., Schiller, P. H., and Colby, C. L., 1981, Response properties of single cells in monkey striate cortex during reversible inactivation of individual geniculate laminae, *J. Neurophysiol.* **46:**1102–1119.

Marrocco, R. T., 1976, Sustained and transient cells in monkey lateral geniculate nucleus: Conduction velocities and response properties, *J. Neurophysiol.* **39:**340–353.

Martin, K. A. C., and Whitteridge, D., 1984, Form function and intracortical projections of spiny neurons in the striate visual cortex of the cat, *J. Physiol.* **356:**463–504.

Mason, A., Nicoll, A., and Stratford, M., 1991, Synaptic transmission between individual pyramidal neurons of the rat visual cortex *in vitro*, *J. Neurosci.* **11:**72–84.

Mastronarde, D. N., 1987, Two classes of single-input X-cells in cat lateral geniculate nucleus. II. Retinal inputs and the generation of receptive-field properties, *J. Neurophysiol.* **576:**381–413.

Mastronarde, D. N., 1992, Nonlagged relay cells and interneurons in the cat lateral geniculate Nucleus—Receptive-field properties and retinal inputs, *Visual Neurosci.* **8:**407–441.

Matsumura, M., 1979, Intracellular synaptic potentials of primate motor cortex neurons during voluntary movement, *Brain Res.* **163:**33–48.

Maunsell, J. H. R., 1987, Physiological evidence for two visual subsystems, in: *Matters of intelligence* (L. M. Vaina, ed.), Reidel, Dordrecht, Holland, pp. 59–87.

Maunsell, J. H. R., and Gibson, J. R., 1992, Visual response latencies in striate cortex of the macaque monkey, *J. Neurophysiol.* **4:**1332–1334.

McClurkin, J. W., and Optican, L. M., 1996, Primate striate and prestriate cortical neurons during discrimination. I. Simultaneous temporal encoding of information about color and pattern, *J. Neurophysiol.* **75:**481–495.

McCormick, D. A., and Van Krosigk, M., 1992, Corticothalamic activation modulates thalamic firing through glutamate "metabotropic" receptors, *Proc. Natl. Acad. Sci. USA* **89:**2774–2778.

McCourt, M. E., Thalluri, J., and Henry, G. H., 1990, Properties of area 17/18 border neurons contributing to the visual transcallosal pathway in the cat, *Visual Neurosci.* **5:**83–98.

Michalski, A., Gestin, G. L., Czarowska, J., and Tarneki, R., 1983, Interactions between cat striate cortex neurons, *Exp. Brain Res.* **51:**97–107.

Mitchison, G., 1992, Axonal trees and cortical architecture, *Trends Neurosci.* **15:**122–126.

Mitzdorf, U., and Singer, W., 1979, Excitatory synaptic ensemble properties in the visual cortex of the macaque monkey: A current source density analysis of electrically evoked potentials, *J. Comp. Neurol.* **187:**71–84.

Movshon, J. A., and Newsome, W. T., 1996, Visual response properties of striate cortical neurons projecting to area MT in macaque monkeys, *J. Neurosci.* **16:**7733–7741.

Mukerjee, P., and Kaplan, E., 1995, Dynamics of neurons in the cat lateral geniculate nucleus: *In vivo* electrophysiology and computational modeling, *J. Neurophysiol.* **74:**1222–1243.

Munk, M. H. J., Nowak, L. G., Girard, P., Chounlamountri, N., and Bullier, J., 1995, Visual latencies in cytochrome oxidase bands of macaque area V2, *Proc. Natl. Acad. Sci. USA* **92:**988–992.

Murre, J. M. J., and Sturdy, D. P. F., 1995, The connectivity of the brain: Multi-level quantitative analysis, *Biol. Cybernet.* **73:**529–545.

Naito, H., Miyakawa, F., and Ito, N., 1971, Diameters of callosal fibers interconnecting cat sensorimotor cortex, *Brain Res.* **27:**369–372.

Nakamura, H., Gattass, R., Desimone, R., and Ungerleider, L. G., 1993, The modular organization of projections from areas V1 and V2 to areas V4 and TEO in macaques, *J. Neursci.* **13:**3681–3691.

Nealey, T. A., and Maunsell, J. H. R., 1994, Magnocellular and parvocellular contributions to the responses of neurons in macaque striate cortex, *J. Neurosci.* **14:**2069–2079.

Nelson, J. I., Salin, P. A., Munk, M. H. J., Arzi, M., and Bullier, J., 1992, Spatial and temporal coherence in corticocortical connections: A cross-correlation study in areas 17 and 18 in the cat, *Visual Neurosci.* **9:**21–38.

Nelson, M. E., and Bower, J. M., 1990, Brain maps and parallel computers, *Trends Neurosci.* **13:**403–408.

Nicoll, A., and Blakemore, C., 1993, Single-fibre EPSPs in layer 5 of rat visual cortex *in vitro, NeuroReport* **4:**167–170.

Nowak, L. G., and Bullier, J., 1997a, Axons, but not cell bodies, are activated by electrical stimulation in cortical gray matter: I. Evidence from chronaxie measurements, *Exp. Brain Res.*, in press.

Nowak, L. G., and Bullier, J., 1997b, Axons, but not cell bodies, are activated by electrical stimulation in cortical gray matter: II. Evidence from selective inactivation of cell bodies and axon initial segments, *Exp. Brain Res.*, in press.

Nowak, L. G., Munk, M. H. J., Chounlamountri, N., and Bullier, J., 1994, Temporal aspects of information processing in areas V1 and V2 of the macaque monkey, in: *Oscillatory Event Related Brain Dynamics* (C. Pantev, T. Elbert, and B. Lutkenhöner, eds.), Plenum Press, New York, pp. 85–98.

Nowak, L. G., Munk, M. H. J., Girard, P., and Bullier, J., 1995, Visual latencies in areas V1 and V2 of the macaque monkey, *Visual Neurosci.* **12:**371–384.

Nowak, L. G., James, A. C., and Bullier, J., 1997, Corticocortical connections between visual areas 17 and 18a of the rat studied *in vitro:* Spatial and temporal organisation of functional synaptic responses, *Exp. Brain Res.* **117:**283–305.

Oram, M. W., and Perrett, D. I., 1992, Time course of neural responses discriminating different views of the face and head, *J. Neurophysiol.* **68:**70–84.

Orbach, H. S., and Van Essen, D. C., 1993, *In vivo* tracing of pathways and spatio-temporal activity patterns in rat visual cortex using voltage sensitive dyes, *Exp. Brain Res.* **94:**371–392.

Orban, G. A., Hoffmann, K.-P., and Duyssens, J., 1985, Velocity selectivity in the cat visual system. I. Responses of LGN cells to moving bar stimuli: A comparison with cortical areas 17 and 18, *J. Neurophysiol.* **54:**1026–1049.

Palmer, L. A., and Rosenquist, A. C., 1974, Visual receptive fields of single striate cortical units projecting to the superior colliculus in the cat, *Brain Res.* **67:**27–42.

Payne, B. R., 1993, Evidence for visual cortical area homologs in cat and macaque monkey, *Cerebral Cortex* **3:**1–25.

Pei, X., Vidyasagar, T. R., Volgushev, M., and Creutzfeldt, O. D., 1994, Receptive field analysis and orientation selectivity of postsynaptic potentials of simple cells in cat visual cortex, *J. Neurosci.* **14:**7130–7140.

Perkel, D. H., Gerstein, G. L., and Moore, G. P., 1967, Neuronal spike trains and stochastic point processes. II: Simultaneous spike trains, *Biophys. J.* **7:**419–440.

Perrett, D. L., Rolls, E. T., and Caan, W., 1982, Visual neurons responsive to faces in the monkey temporal cortex, *Exp. Brain Res.* **47:**329–342.

Peters, A., and Sethares, C., 1996, Myelinated axons and the pyramidal cell modules in monkey primary visual cortex, *J. Comp. Neurol.* **365:**232–255.

Peters, A., Payne, B. R., and Budd, J., 1994, A numerical analysis of the geniculocortical input to striate cortex in the monkey, *Cerebral Cortex* **4:**215–229.

Raiguel, S. E., Lagae, L., Gulyas, B., and Orban, G. A., 1989, Response latencies of visual cells in macaque areas V1, V2 and V5, *Brain Res.* **493:**155–159.

Ranck, J. B., 1975, Which elements are excited in electrical stimulation of mammalian central nervous system: A review, *Brain Res.* **98:**417–440.

Reese, B. E., and Guillery, R. W., 1987, Distribution of axons according to diameter in the monkey's optic tract, *J. Comp. Neurol.* **260:**453–460.

Reese, B. E., and Ho, K.-Y., 1988, Axon diameter distributions across the monkey's optic nerve, *Neuroscience* **27:**205–214.

Reyes, A. D., and Fetz, E. E., 1993a, Two modes of interspike interval shortening by brief transient depolarizations in cat neocortical neurons, *J. Neurophysiol.* **69:**1661–1672.

Reyes, A. D., and Fetz, E. E., 1993b, Effects of transient depolarizing potentials on the firing rate of cat neocortical neurons, *J. Neurophysiol.* **69:**1673–1683.

Richmond, B. L., and Optican, L. M., 1990, Temporal encoding of two dimensional patterns by single units in primate primary visual cortex. II. Information transmission, *J. Neurophysiol.* **64:**307–380.

Rockland, K. S., 1989, Bistratified distribution of terminal arbors of individual axons projecting from area V1 to middle temporal area (MT) in the macaque monkey, *Visual Neurosci.* **3:**155–170.
Rockland, K. S., 1992, Configuration, in serial reconstruction, of individual axons projecting from area V2 to V4 in the macaque monkey, *Cerebral Cortex* **2:**353–374.
Rockland, K. S., 1995, Morphology of individual axons projecting from area V2 to Mt in the macaque, *J. Comp. Neurol.* **355:**15–26.
Rockland, K. S., and Virga, A., 1989, Terminal arbors of individual "feedback" axons projecting from area V2 to V1 in the macaque monkey: A study using immunohistochemistry of anterogradely transported *Phaseolus vulgaris*-leucoagglutinin, *J. Comp. Neurol.* **285:**54–72.
Rockland, K. S., and Virga, A., 1990, Organization of individual cortical axons projecting from area V1 (area 17) to V2 (area 18) in the macaque monkey, *Visual Neurosci.* **4:**11–28.
Rockland, K. S., Saleem, K. S., and Tanaka, K., 1994, Divergent feedback connections from areas V4 and TEO, *Visual Neurosci.* **11:**579–600.
Rolls, E. T., and Tovee, M. J., 1994, Processing speed in the cerebral-cortex and the neurophysiology of visual masking, *Proc. R. Soc. Lond. B* **257:**9–15.
Rossetti, Y., 1997, Implicit perception in action: Short-lived motor representations of space, in: *Advances in Consciousness Research* (P. G. Grosenbacher, ed.), Benjamins Publishers, Amsterdam.
Rushton, W. A., 1951, A theory of the effects of fibre size in medullated nerve, *J. Physiol.* **115:**101–122.
Salin, P.-A., and Bullier, J., 1995, Corticocortical connections in the visual system: Structure and function, *Physiol. Rev.* **75:**107–154.
Sato, H., Katsuyama, N., Tamura, H., Hata, Y., and Tsumoto, T., 1994, Broad-tuned chromatic inputs to color-selective neurons in the monkey visual cortex, *J. Neurophysiol.* **72:**163–168.
Saul, A. B., and Humphrey, A. L., 1990, Spatial and temporal response properties of lagged and non-lagged cells in cat lateral geniculate nucleus, *J. Neurophysiol.* **64:**206–224.
Schall, J. D., Morel, A., King, D. J., and Bullier, J., 1995, Topography of visual-cortex connections with frontal eye field in macaque—Convergence and segregation of processing streams, *J. Neurosci.* **15:**4464–4487.
Schiller, P. H., and Malpeli, J. G., 1977a, The effect of striate cortex cooling on area 18 cells in the monkey, *Brain Res.* **126:**366–369.
Schiller, P. H., and Malpeli, J. G., 1977b, Properties and tectal projections of monkey retinal ganglion cells, *J. Neurophysiol.* **40:**428–445.
Schiller, P. H., and Malpeli, J. G., 1978, Functional specificity of lateral geniculate nucleus laminae of the rhesus monkey, *J. Neurophysiol.* **41:**788–797.
Segal, M., and Barker, J. L., 1984, Rat hippocampal neurons in culture: Potassium conductances, *J. Neurophysiol.* **51:**1409–1433.
Sestokas, A. K., and Lehmkuhle, S., 1986, Visual latency of X- and Y-cells in the dorsal lateral geniculate nucleus of the cat, *Vision Res.* **26:**1041–1054.
Shadlen, M. N., and Newsome, W. T., 1994, Noise, neural codes and cortical organization, *Curr. Opin. Neurobiol.* **4:**569–579.
Shipp, S., and Grant, S., 1991, Organization of reciprocal connections between area 17 and the lateral suprasylvian area of cat visual cortex, *Visual Neurosci.* **6:**339–355.
Shipp, S., and Zeki, S. M., 1985, Segregation of pathways leading from area V2 to areas V4 and V5 of macaque monkey visual cortex, *Nature* **315:**322–325.
Sillito, A. M., Kemp, J. A., Milson, J. A., and Berardi, N., 1980, A re-evaluation of the mechanisms underlying simple cell orientation selectivity, Brain Res. **194:**517–520.
Singer, W., Tretter, F., and Cynader, M., 1975, Organization of cat striate cortex: A correlation of receptive-field properties with afferent and efferent connections, *J. Neurophysiol.* **38:**1080–1098.
Softky, W. R., 1995, Simple codes versus efficient codes, *Curr. Opin. Neurobiol.* **5:**239–247.
Stafstrom, C. E., Schwindt, P. C., Chubb, M. C., and Crill, W. E., 1985, Properties of persistent sodium conductance and calcium conductance of layer V neurons from cat sensorimotor cortex *in vitro*, *J. Neurophysiol.* **53:**153–170.
Stanford, L. R., 1987, Conduction velocity variations minimize conduction time differences among retinal ganglion cell axons, *Science* **238:**358–360.
Stone, J., 1983, *Parallel Processing in the Visual System. The Classification of Retinal Ganglion Cells and Its Impact on the Neurobiology of Vision,* Plenum Press, New York.

Sutor, B., and Hablitz, J. J., 1989, EPSPs in rat neocortical neurons *in vitro*. II. Involvement of N-methyl-D-aspartate receptors in the generation of EPSPs, *J. Neurophysiol.* **61**:621–634.

Swadlow, H. A., 1974, Systematic variations in the conduction velocity of slowly conducting axons in the rabbit corpus callosum, *Exp. Neurol.* **43**:445–451.

Swadlow, H. A., 1992, Monitoring the excitability of neocortical efferent neurons to direct activation by extracellular current pulses, *J. Neurophysiol.* **68**:605–619.

Swadlow, H. A., 1994, Efferent neurons and suspected interneurons in motor cortex of the awake rabbit: Axonal properties, sensory receptive fields, and subthreshold synaptic inputs, *J. Neurophysiol.* **71**:437–453.

Swadlow, H. A., and Weyand, T. G., 1981, Efferent systems of the rabbit visual cortex; Laminar distribution of the cells of origin, axonal conduction velocities and identification of axonal branches, *J. Comp. Neurol.* **203**:799–822.

Swadlow, H. A., Rosene, D. L., and Waxman, S. G., 1978, Characteristics of interhemispheric impulse conduction between prelunate gyri of the rhesus monkey, *Exp. Brain Res.* **33**:455–467.

Tanaka, K., 1983, Cross-correlation analysis of geniculostriate neuronal relationships in cats, *J. Neurophysiol.* **49**:1303–1318.

Tanaka, M., Weber, H., and Creutzfeldt, O. D., 1986, Visual properties and spatial distribution of neurons in the visual association area of the prelunate gyrus of the awake monkey, *Exp. Brain Res.* **65**:11–37.

Thalluri, J., and Henry, G. H., 1989, Neurons of the striate cortex driven trans-synaptically by electrical stimulation of the superior colliculus, *Vision Res.* **10**:1319–1323.

Thompson, K. G., Hanes, D. P., Bichot, N. P., and Schall, J. D., 1996, Perceptual and motor processing stages identified in the activity of macaque frontal eye field neurons during visual search, *J. Neurophysiol.*, in press.

Thomson, A. M., and West, D. C., 1993, Fluctuations in pyramid-pyramid excitatory postsynaptic potentials modified by presynaptic firing pattern and postsynaptic membrane potential using paired intracellular recordings in rat neocortex, *Neuroscience* **54**:329–346.

Thomson, A. M., Girdlestone, D., and West, D. C., 1988, Voltage-dependent currents prolong single-axon postsynaptic potentials in layer III pyramidal neurons in rat neocortical slices, *J. Neurophysiol.* **60**:1896–1907.

Thomson, A. M., Deuchars, J., and West, D. C., 1993, Single axon excitatory postynaptic potentials in neocortical interneurons exhibit pronounced paired pulse facilitation, *Neuroscience* **54**:347–360.

Thorpe, S. J., and Imbert, M., 1989, Biological constraints on connectionist models, in: *Connectionism in Perspective* (R. Pfeiffer, Z. Schreter, F. Fogelamn-Soulié, and L. Steels, eds.), Elsevier, Amsterdam, pp. 63–92.

Thorpe, S. J., Rolls, E. T., and Maddison, S., 1983, The orbitofrontal cortex: Neuronal activity in the behaving monkey, *Exp. Brain Res.* **49**:93–115.

Thorpe, S., Fize, D., and Marlot, C., 1996, Speed of processing in the human visual system, *Nature* **381**:520–522.

Tovee, M. J., Rolls, E. T., Treves, A., and Bellis, R. P., 1993, Information encoding and the responses of single neurons in the primate temporal visual cortex, *J. Neurophysiol.* **70**:640–654.

Toyama, K., Matsunami, K., Ohno, T., and Tokashiki, S., 1974, An intracellular study of neuronal organisation in the visual cortex, *Exp. Brain Res.* **21**:45–66.

Toyama, K., Kimura, M., and Tanaka, K., 1981, Cross-correlation analysis of interneuronal connectivity in cat visual cortex, *J. Neurophysiol.* **2**:191–201.

Tsumoto, T., Creutzfeldt, O. D., and Legendy, C. R., 1978, Functional organization of the corticofugal system from visual cortex to lateral geniculate nucleus in the cat, *Exp. Brain Res.* **32**:345–364.

Ullman, S., 1995, Sequence seeking and counter streams—A computational model for bidirectional information-flow in the visual-cortex, *Cerebral Cortex* **5**:1–11.

Vogels, R., and Orban, G. A., 1990, How well do response changes of striate neurons signal differences in orientation: A study in the discriminating monkey, *J. Neurosci.* **10**:3543–3558.

Vogels, R., and Orban, G. A., 1994, Activity of inferior temporal neurons during orientation discrimination with successively presented gratings, *J. Neurophysiol.* **71**:1428–1451.

Volgushev, M., Vidyasagar, T. R., and Pei, X., 1995, Dynamics of orientation tuning of postsynaptic potentials in the cat visual cortex, *Visual Neurosci.* **12**:621–628.

Wässle, H., Levick, W. R., Kirk, D. L., and Cleland, B. G., 1975, Axonal conduction velocity and perikaryal size, *Exp. Neurol.* **49**:246–251.

Waxman, S. G., and Bennett, M. V. L., 1972, Relative conduction velocities of small myelinated and non-myelinated fibres in the central nervous system, *Nature* **238:**217–219.
Waxman, S. G., and Swadlow, H. A., 1976, Ultrastructure of visual callosal axons in the rabbit, *Exp. Neurol.* **536:**115–127.
Weiss, G., 1901, Sur la possibilité de rendre comparables entre eux les appareils servant à l'excitation électrique, *Arch. Ital. Biol.* **35:**431–446.
Welker, E., Armstrong-James, M., Van der Loos, H., and Kraftsik, R., 1993, The mode of activation of a barrel column: Response properties of single units in somatosensory cortex of the mouse upon whisker deflection, *Eur. J. Neurosci.* **5:**691–712.
Wiesel, T. N., and Hubel, D. H., 1966, Spatial and chromatic interactions in the lateral geniculate body of the rhesus monkey, *J. Neurophysiol.* **29:**1115–1156.
Wilson, P. D., Rowe, M. H., and Stone, J., 1976, Properties of relay cells in the cat's lateral geniculate nucleus: A comparison of W-cells with X- and Y-cells, *J. Neurophysiol.* **39:**1193–1209.
Yoshioka, T., and Dow, B. M., 1996, Color, orientation and cytochrome oxidase reactivity in areas V1, V2 and V4 of macaque monkey visual cortex, *Behav. Brain Res.* **76:**71–88.
Yoshioka, T., Levitt, J. B., and Lund, J. S., 1994, Independence and merger of thalamocortical channels within macaque monkey primary visual cortex: Anatomy of interlaminar projections, *Visual Neurosci.* **11:**467–489.
Young, M. P., 1992, Objective analysis of the topological organization of the primate cortical visual system, *Nature* **358:**152–155.

6

Elements of Cortical Architecture

Hierarchy Revisited

KATHLEEN S. ROCKLAND

1. Introduction

Extrastriate visual cortex consists of multiple areas. As reviewed elsewhere (Kaas, 1989; Colby and Duhamel, 1991; and several chapters in this volume), there are still many questions concerning specific boundaries and subdivisions, and the criteria for area identification themselves remain under discussion. How areas interact is even less well known and is very much a topic of active research. Is there an overall architecture? Are there patterns of sequential or synchronous coactivation?

One approach to these questions is the anatomical analysis of connectivity patterns among cortical areas. Ipsilateral corticocortical connections have traditionally been subdivided into several major systems on the basis of the layers containing their neurons of origin and their terminal fields. One major system, "feedforward," originates primarily from neurons in layer 3 and terminates primarily in layer 4 of the target area. "Feedback" systems originate primarily from neurons in layer 6 and terminate primarily in layer 1.

That connections are distinguishable on the basis of terminal distribution in layer 1 or 4 was recognized relatively early, even with silver degeneration techniques (Kuypers *et al.*, 1965; Tigges *et al.*, 1974, 1977; in auditory cortex, Pandya

KATHLEEN S. ROCKLAND • Department of Neurology, 2007 RCP, College of Medicine, University of Iowa, Iowa City, Iowa 52242-1053.

Cerebral Cortex, Volume 12, edited by Rockland *et al.* Plenum Press, New York, 1997.

and Sanides, 1973). A little later, the horseradish peroxidase technique demonstrated distinct patterns of projection neurons (Spatz 1977; Wong-Riley, 1978; Tigges *et al.*, 1981; Weller and Kaas, 1983). With the combination of anterograde and retrograde tracers, the laminar complementarity of feedforward and feedback systems was suggested as a general principle, at least for early visual areas (Rockland and Pandya, 1979), although further work has revealed significant variations in the basic laminar patterns. For example, feedforward connections from V1 to V2, when labeled with fluorescent tracers, are reported to originate from neurons in layer 5 as well as layer 3 (Kennedy and Bullier, 1985), and feedback connections from MT/V5 to V1 have a significant contribution from neurons in layer 3 (Weller and Kaas, 1983; Kennedy and Bullier, 1985; Krubitzer and Kaas, 1990). Further work has also distinguished an additional broad category, defined as "lateral connections" (Maunsell and Van Essen, 1983). These tend to be between higher order areas, originate from neurons in both layers 3 and 5, and terminate in a "columnar" fashion through several layers.

Analysis of cortical connections at a microcircuitry level is lacking, however, with the arguable exception of area V1 (among recent studies, see Casagrande and Kaas, 1994; Lund *et al.*, 1994, 1995; and Wiser and Callaway, 1996; for V2, see Valverde, 1978, and Lund *et al.*, 1981). For extrastriate areas, there is very little electron microscopy, and technical difficulties have largely precluded the kind of combined intracellular staining and physiology that has been achieved for geniculocortical connections (Blasdel and Lund, 1983) or for intrinsic cortical connections (in cat, Gilbert and Wiesel, 1983,Martin and Whitteridge, 1984, and Ojima *et al.*, 1991, 1992; in slice, Wiser and Callaway, 1996; in primate area V1, Anderson *et al.*, 1993).

Microanatomical investigations of cortical connections have been significantly advanced by the introduction of high-resolution anterograde tracers (*Phaseolus vulgaris* leucoagglutinin, PHA-L: Gerfen and Sawchenko, 1984; biocytin: Horikawa and Armstrong, 1988; King *et al.*, 1989; and biotinylated dextran amine: Veenman *et al.*, 1992). These tracers can be injected extracellularly and yet result in Golgi-like labeling of individual axons. Serial section reconstruction of labeled profiles then allows morphometric analysis and identification of three-dimensional geometric configurations. Significantly, these tracers easily distinguish between actual terminations and preterminal fibers, and allow more accurate evaluations of terminal density. Several studies using these techniques have now been carried out for visual cortical connections (Rockland and Virga, 1989, 1990; Rockland, 1989, 1992, 1994, 1995; Rockland *et al.*, 1994; see also, for related studies of the cat callosal system, Houzel *et al.*, 1994, and Innocenti *et al.*, 1994) and for inferotemporal connections (Saleem *et al.*, 1993; Saleem and Tanaka, 1996).

The present chapter is in part intended to review new data on cortical connectivity, especially taking into account single-axon analysis. Discussion is geared as much as possible to a general level, with less consideration of connections between specific areas (these are treated in several other chapters in this volume). The implicit emphasis, however, is on connections of the early visual areas (V1, V2, V4, and MT/V5), largely because of the availability of relevant data. Callosal, pulvinocortical, and other important systems are only briefly treated. Lateral connections (between higher order areas), and interneuronal inhibition are also treated only in passing. Functional issues are not discussed, such as localization of function (see others chapters in this volume) or specific connectional interactions (see Ullman, 1995, and references therein).

Another goal of this chapter is to assess microcircuitry data in relation to present models of cortical organization. One of the most successful of these has emphasized a hierarchical framework (for example, Maunsell and Van Essen, 1983; Andersen *et al.,* 1990; Boussaoud *et al.,* 1990; Felleman and Van Essen, 1991, and references therein). In this framework, area V1 is commonly considered in a "low" position (nearer peripheral inputs), and extrastriate areas are assigned progressively higher levels, in part on the basis of laminar patterns of connectivity. Other frameworks, however, have been suggested (Goldman-Rakic, 1988; Zeki and Shipp, 1988; Zeki, 1990; Tononi *et al.,* 1992; Van Essen and DeYoe, 1995), and several recent results also suggest other aspects to be considered. For example, area MT/V5, although densely connected with areas V1 and V2, is still visually responsive after inactivation of V1 (reviewed in Bullier *et al.,* 1994; and in the chapter by Orban in this volume). Presumably, this is because there are other connections (colliculopulvinocortical) that are effective for conveying visual information. Imaging studies in humans suggest that area V1 is involved in nonretinal processes during mental imagery (Ishai and Sagai, 1995; Kosslyn *et al.,* 1995; Miyashita, 1995). Experiments on the latency of neuronal responses to visual stimuli indicate a large temporal scatter within any single cortical area. This implies that several cortical areas, or portions of areas, may be active nearly simultaneously (see chapters by Nowak and Bullier and by Orban in this volume). Finally, there is growing physiological evidence of the importance of temporal synchronization and grouping of spatially separated neural assemblages (J. I. Nelson *et al.,* 1992; Singer, 1994, 1995; Bressler, 1995, 1996); and the importance of temporal interactions has also been stressed from the viewpoint of experimental neuropsychology (Damasio, 1989; Damasio and Damasio, 1994).

2. Elements of a "Computational" Architecture

2.1. Laminar Terminations of Single Axons

Feedforward connections, as established by numerous studies based on autoradiography or wheat germ agglutinin–horseradish peroxidase (WGA–HRP), terminate heavily in the middle cortical layers, and mainly in layer 4. Analysis of individual axons has largely confirmed this distribution (Fig. 1). Exceptions have, however, been observed. Some of these are projection-specific; namely, projections from V1 to MT/V5 have a bistratified distribution in layers 4 and 6 (Figs. 2, 3) This pattern has not been observed in connections from V1 to V2 or from V2 to MT/V5, and may be explained by the specialized projection neurons from which this projection originates. Other results are less consistent and harder to interpret without more data. An axon which terminates mainly in layer 4 can have collaterals in layer 5 (V1 to V2: Fig. 4 in Rockland and Virga, 1990; V2 to V4: Figs. 11 and 13 in Rockland, 1992). Some axons may more directly target layer 3 than layer 4 (Fig. 12 in Rockland, 1992). Some axons, more surprisingly,

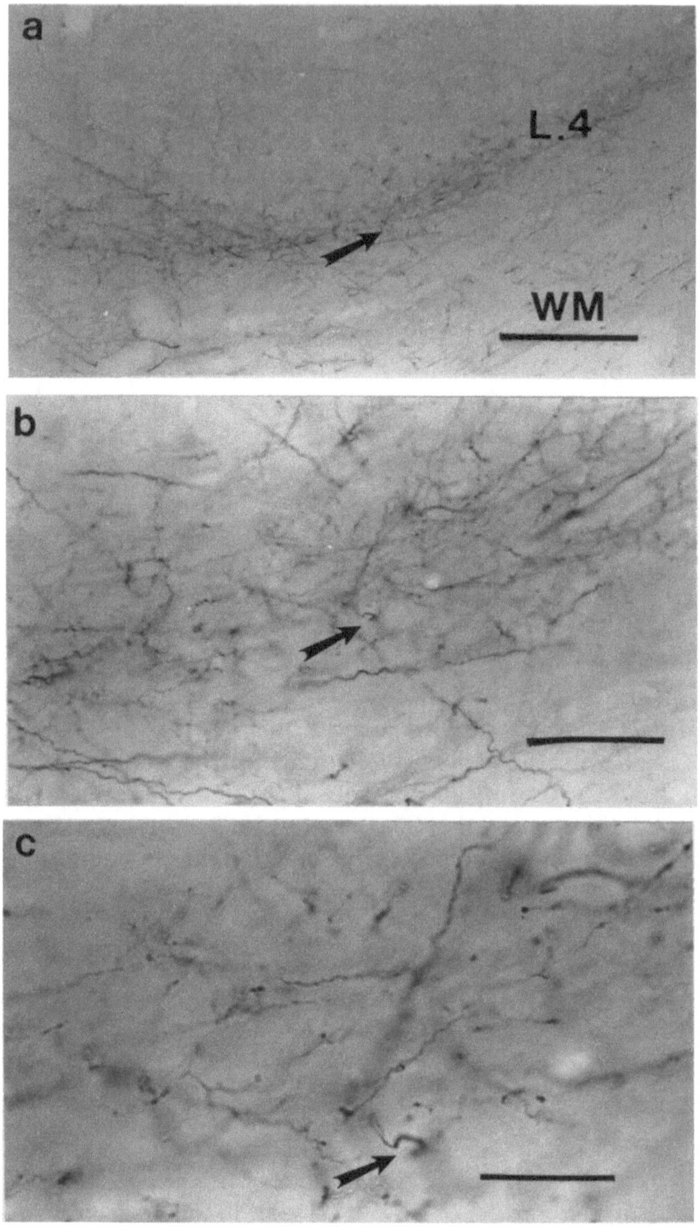

Figure 1. Photomicrographs of feedforward axons anterogradely labeled by an injection of biotinylated dextran amine (BDA) in area MT (*Saimiri*). Terminations are shown anteriorly in the superior temporal sulcus, mainly in layer 4. Arrows point to corresponding feature at higher magnification. Bars: (a) 200 μm, (b) 50 μm; (c) 20 μm.

preferentially terminate in layers 1 and 2 (V2 to V4: Figs. 15 and 16 in Rockland, 1992; TEO to TE: Fig. 7 in Saleem *et al.,* 1993).

Another important observation from single-axon reconstruction is the clarification that terminations are distributed variably within adjoining portions of layers 3 and 4. Some axons have terminations confined to layer 4 (V1 to V2: Fig.

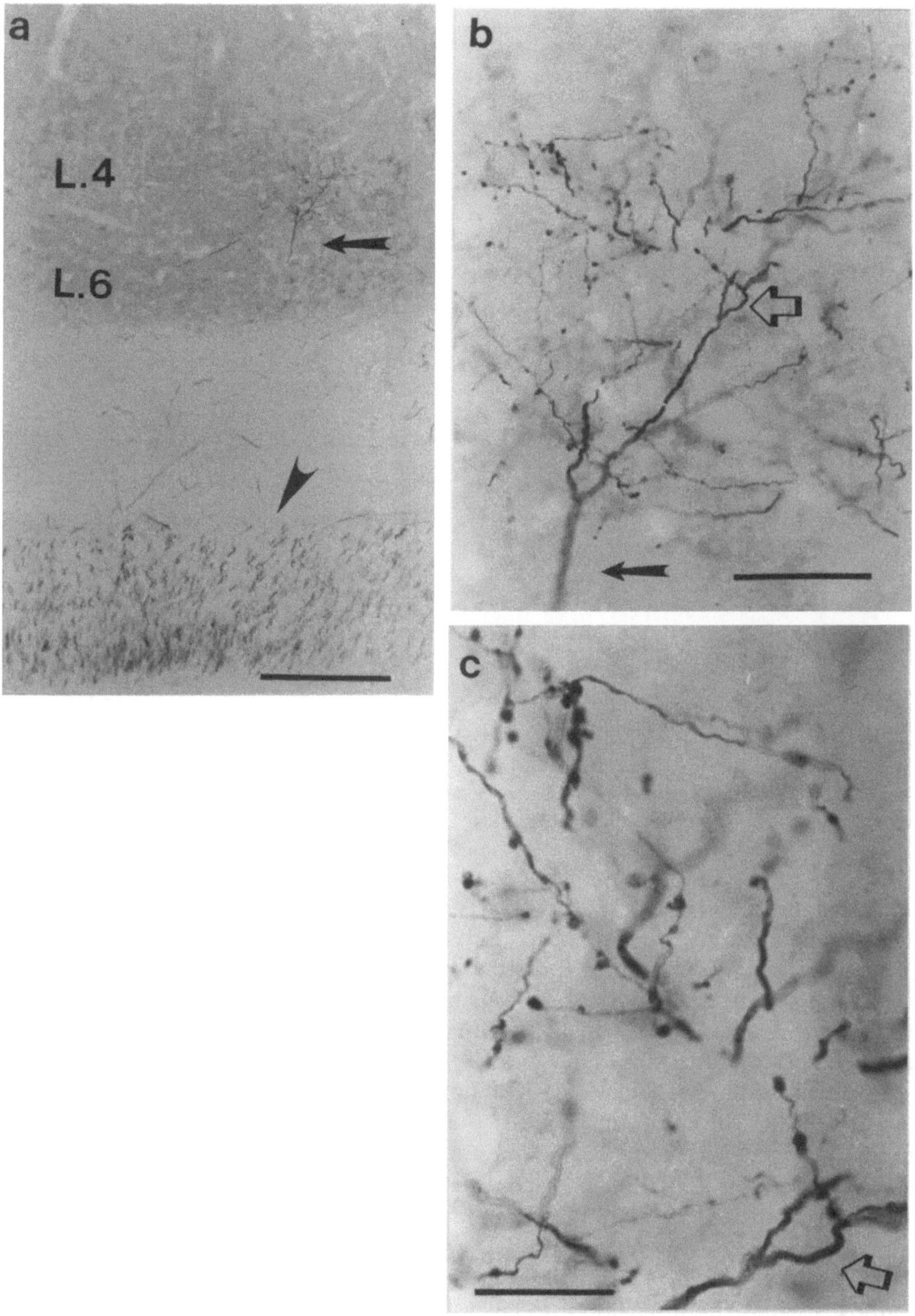

Figure 2. Photomicrographs of feedforward axons anterogradely labeled by an injection of BDA in area V1 (*Saimiri*). (a) An exceptionally well-isolated arbor (arrow) is illustrated in layer 4 of area MT. Segments of single axons (arrowhead) course toward the termination focus in MT from a medially situated fiber bundle. (b,c) Higher magnification. Solid arrows point to corresponding features in panels a and b; hollow arrows point to corresponding features in panels b and c. Terminal specializations are large and beaded (compare Figs. 3c and 1c). Bars: (a) 300 μm; (b) 200 μm; (c) 20 μm.

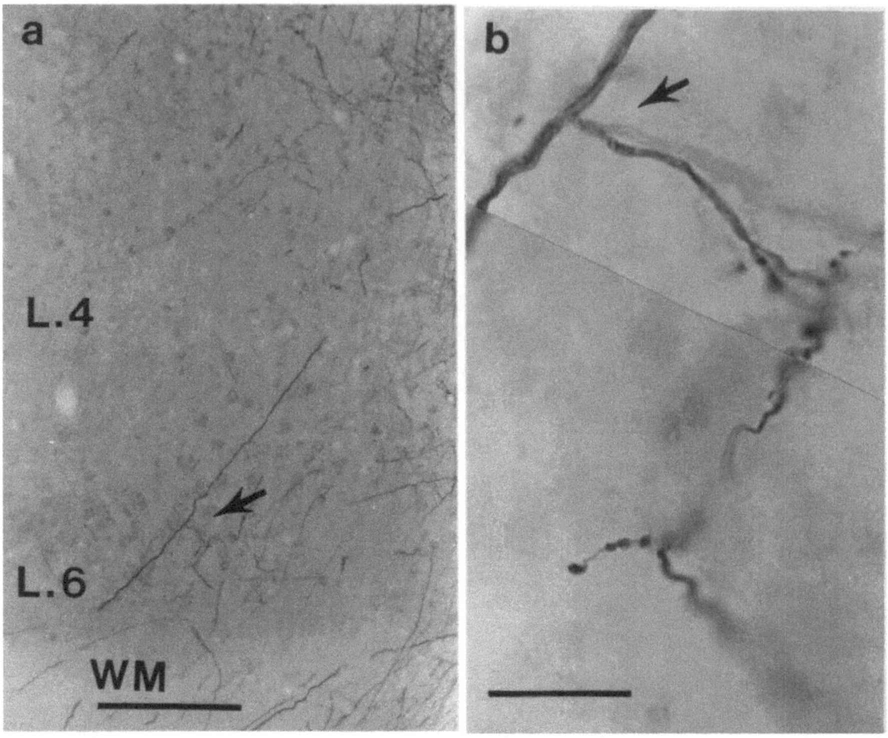

Figure 3. Photomicrographs of feedforward axons anterogradely labeled by an injection of BDA in area V1 (*Saimiri*). Illustrated is a terminal arbor in layer 6 of MT (arrows point to corresponding features at different magnifications in panels a and b). Terminal specializations are large and beaded, but the arbor is smaller than those in layer 4 (see Fig. 2). Bars: (a) 200 μm; (b) 20 μm.

5 in Rockland and Virga, 1990; V2 to V4: Fig. 3 in Rockland, 1992), but frequently, terminations are distributed between the upper half of layer 4 and the deeper part of layer 3, almost indiscriminately. Particularly striking are differences among multiple arbors of a single axon. These typically terminate differentially within layers 3 and 4 in all feedforward systems investigated so far (V1 to V2: Rockland and Virga, 1989; V2 to V4: Rockland, 1992; and both V1 and V2 to MT/V5: Rockland, 1989, 1994). The implication is that multiple arbors of one axon may have a nonidentical mix of postsynaptic populations, depending on whether contacts are made with elements mainly in layer 4 or mainly in layer 3 (see below, Section 2.3). This pattern is distinct from the elaborate substratification of geniculocortical terminations in layer 4 in area V1. In fact, this may be a major difference in the organization of the middle input layers (layers 4A and 3C) beyond area V1. This possibility has previously been suggested on other grounds, such as the intrusion into layer 4 of V2, but not V1, of basal dendrites belonging to overlying pyramidal neurons (Valverde, 1978; Lund *et al.*, 1981).

Feedback connections characteristically terminate mainly in layer 1, with possible involvement of other layers, but not layer 4. Further studies at the single-axon level have confirmed that feedback axons terminate mainly in layers 1 and 5 of V1, or 1 and 6 of other areas (Fig. 4). As described above for feedforward

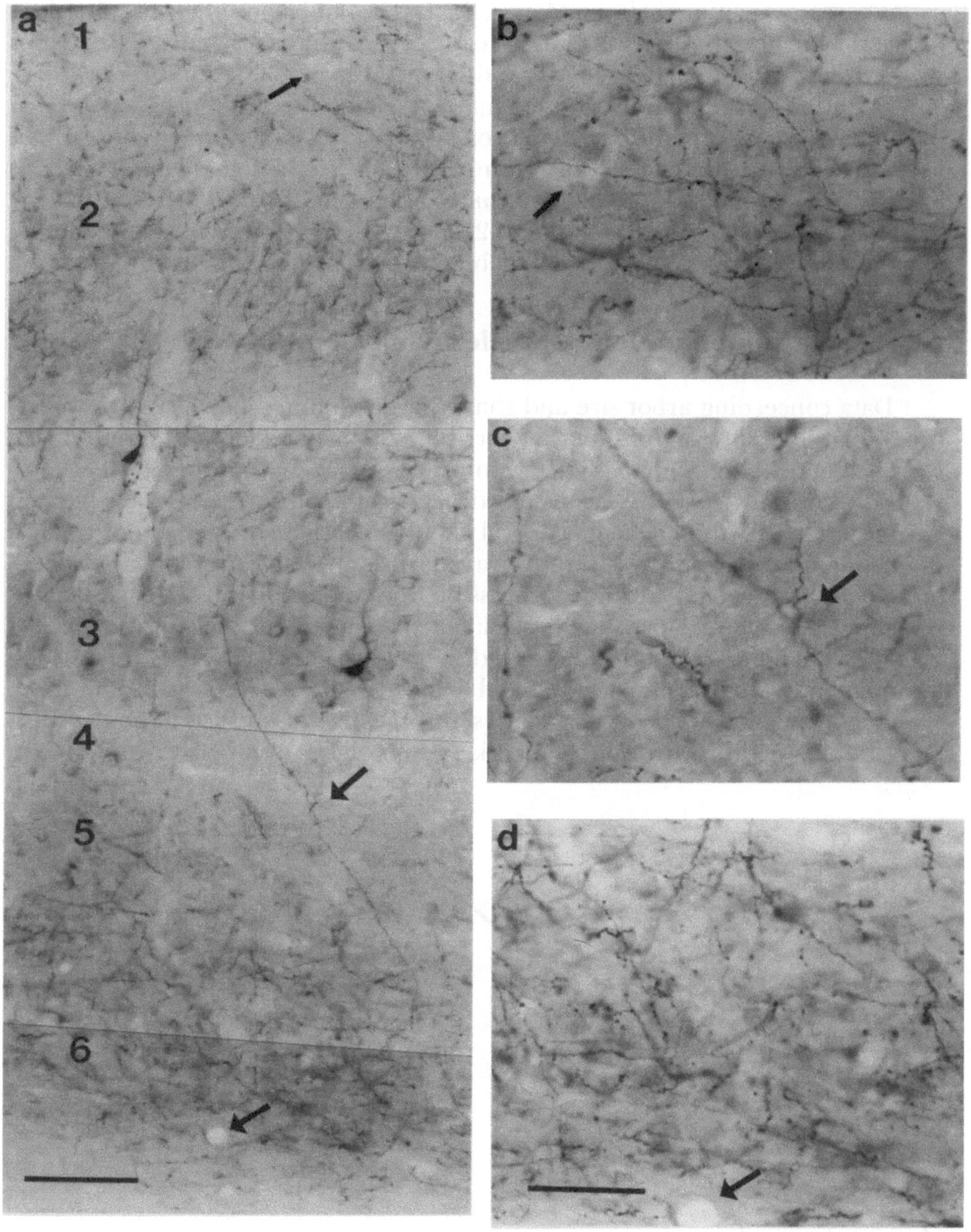

Figure 4. Photomicrographs of feedback axons anterogradely labeled by an injection of PHA-L in area TEO. Terminations are shown in area V4 of the occipitotemporal sulcus, mainly in layers 1 and 6. Two feedforward neurons in layer 3 have been retrogradely labeled by the injection (as sometimes occurs, although only sporadically and unreliably). Three fields in panel a are illustrated at higher magnification in panels b–d (arrows point to corresponding features at the different magnifications). Bars: (a) 100 μm; (b–d) 50 μm.

connections, however, there are considerable variations, even for connections from V2 to V1. Some axons have infragranular collaterals in layer 6 instead of 5 (Fig. 3 in Rockland, 1994), or there may be no infragranular collaterals at all. Some axons have terminations in layer 3A as well as layer 1, and many of the terminations in layer 1 intrude into layer 2. These features also characterize feedback axons projecting from V4 or TEO to V1, and from V4 to V2 (Rockland *et al.*, 1994).

Variations in laminar distribution may correlate with differences—of type or of laminar position—in the parent neuron population. As mentioned above, this is likely to be the case for connections from V1 to MT/V5. In general this is perhaps less likely, since multiple arbors of a *single* axon terminate in different layers. Alternately, certain layers (layers 1 and 2, layers 3C and 4, or even layers 5 and 6) may be more or less similar. Laminar boundaries are to some extent artificial.

2.2. Spatial Configuration of Single Axons

Data concerning arbor size and shape have prompted key insights into the functional anatomy of the geniculocortical system (Blasdel and Lund, 1983; Florence and Casagrande, 1987; Freund *et al.*, 1989; Lachica *et al.*, 1992). For several technically related reasons (corticocortical axons are less accessible and harder to identify than geniculocortical axons; intracellular injection of cortical projection neurons is difficult; tracer transport is unreliable over long distances), these data have been practically inaccessible for corticocortical systems (but see DeFelipe *et al.*, 1986). Data from intracellular experiments are still not available for primates, but high-resolution anterograde labels have now facilitated direct visualization of individual terminal arbors and detailed analysis of arbor configuration.

Arbor analysis of corticocortical systems has produced several new results. One is that feedforward and feedback systems have terminal fields that are

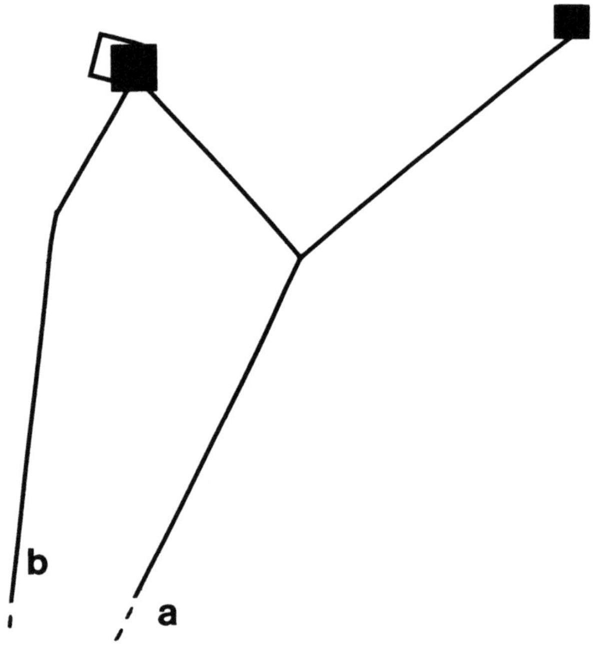

Figure 5. Feedforward axons (except for those projecting from V1 to V2) commonly have two or more spatially separate arbors (a). One, but not necessarily both of these, might converge with arbors of other axons (b) labeled by the same injection site.

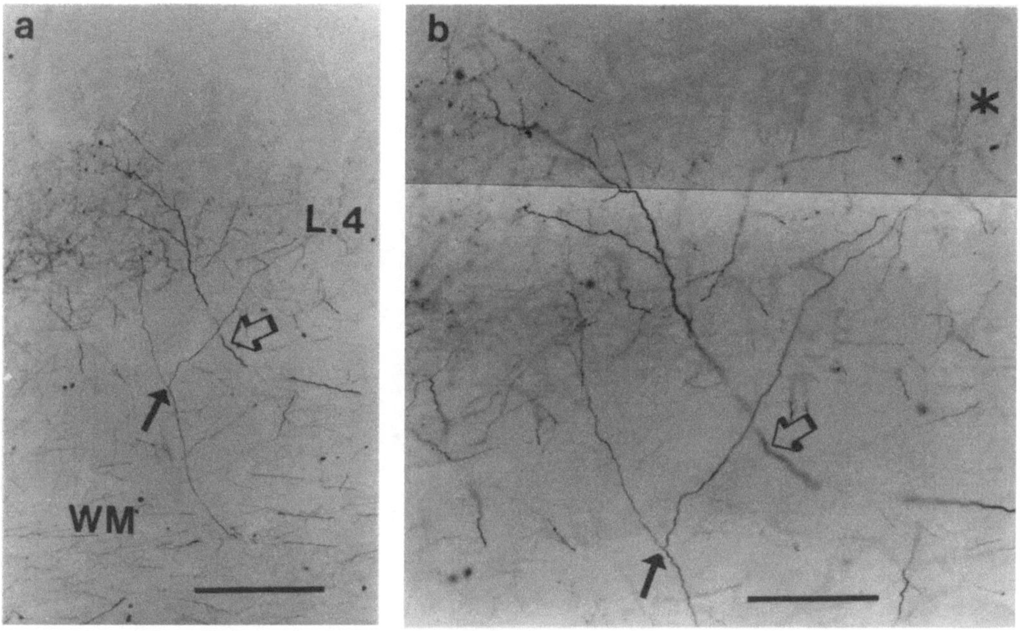

Figure 6. Photomicrographs of feedforward axons anterogradely labeled by an injection of biocytin in area V1. Terminations are shown in area V2, mainly in layer 4. Two axons (hollow and solid arrows) can be seen to enter the terminal focus, where they converge, but only partially. One arbor of the second axon (*) remains separate. Panel a is lower magnification of panel b. Bars: (a) 200 μm; (b) 100 μm.

spatially distinctive. Feedforward arbors are delimited and typically measure about 200 μm in diameter. There can, however, be multiple arbors (although this is not common for connections from V1 to V2; Figs. 5 and 6). Feedback fields are elongated and typically diverge over 1.0–2.0 mm (Fig. 7). The divergence of feedback terminations, which had been inferred from overall projection paterns (Krubitzer and Kaas, 1989; Shipp and Zeki, 1989a), means that these connections may cross over functionally defined domains, such as ocular dominance columns or the cytochrome oxidase (CO) compartments in area V1.

Another result, unexpected from earlier reports, is that collaterals in the infragranular layers (Figs. 2, 3, 8, and 9), when they occur, are consistently smaller than terminal fields in layers 1 or 4 (of feedback and feedforward axons, respectively). A similar asymmetry characterizes geniculocortical axons (Blasdel and Lund, 1983; Florence and Casagrande, 1987; Freund *et al.,* 1989). In both cases, size differences may be related to laminar-specific differences in spatial mapping (Blasdel and Fitzpatrick, 1984).

A third result is that the arbor size of feedforward axons may be relatively constant across areas (200–250 μm in diameter). The *number* of arbors, however, seems to increase for connections progressively beyond V1. That is, most axons

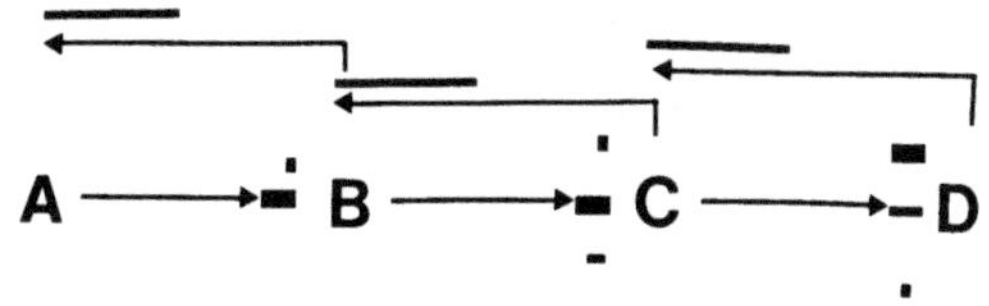

Figure 7. Schematic diagram to summarize the different configurations of feedforward and feedback arbors. These are, respectively, delimited or divergent.

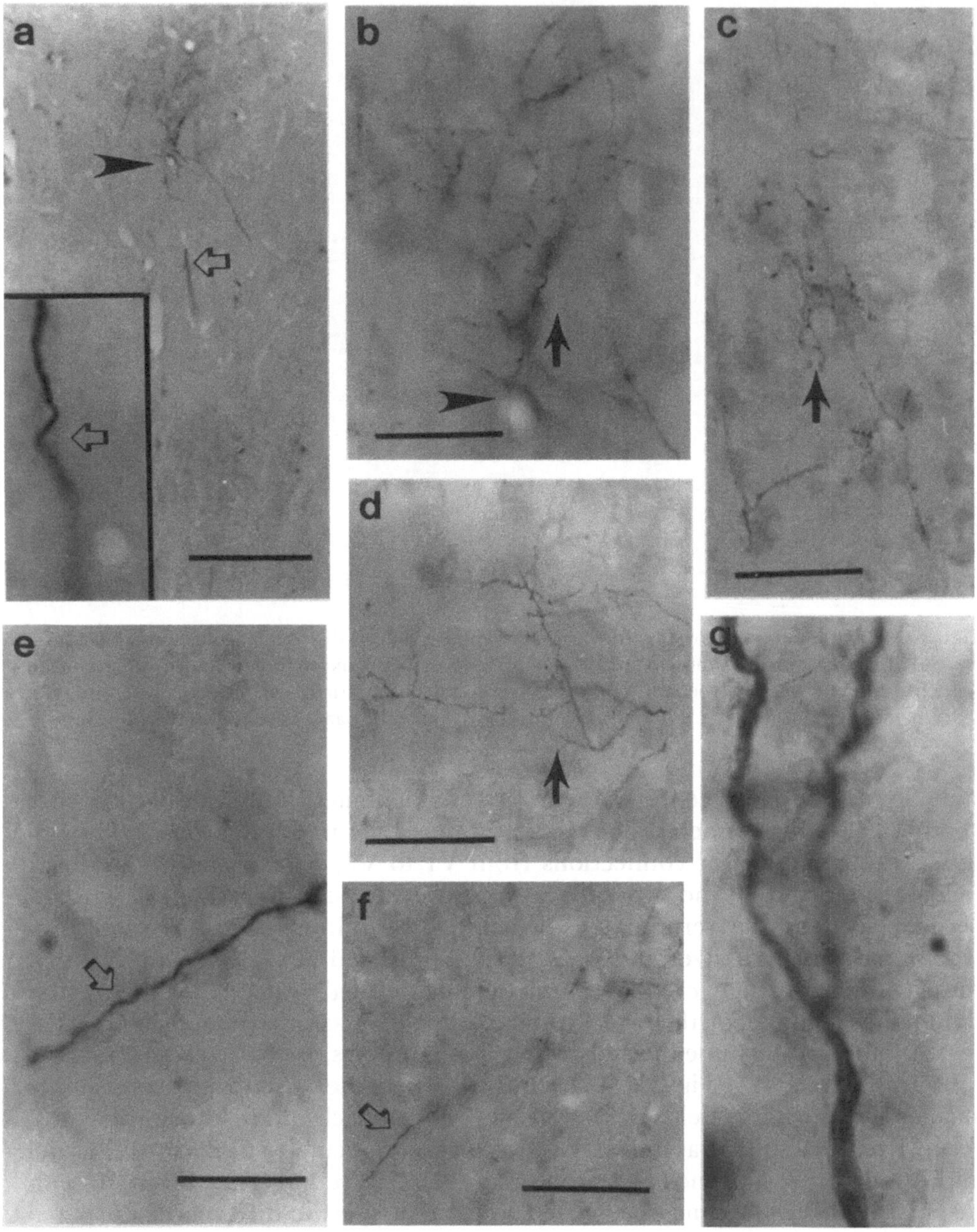

Figure 8. Photomicrographs of feedforward arbors and axons anterogradely labeled by injections of PHA-L. (a) Arbor terminating in layer 4 of V2, originating from V1. A portion of preterminal trunk (hollow arrow) is shown at higher magnification in inset. (b) Portion from arrowhead shown at higher magnification. (c) Collateral in layer 6 of area MT (V5) of a feedforward axon originating from V1. Note beaded shape of terminations, and small size of the arbor (vertical arrows in b–d designate approximate center point of three separate arbors). (d) Arbor terminating in layer 4 of V2, originating from V1. (f) Arbor terminating in layer 4 of V4, originating from V2. (e) Portion of preterminal trunk (hollow arrow) shown at higher magnification. (g) High magnification of large-caliber axon in the white matter below MT (V5), originating from V1. For comparison, see panel e and inset of panel a (respectively, V2 to V4, and V1 to V2 axons). Bars: (a) 200 μm; (b–d) 50 μm, (f) 100 μm; (e, g, inset of a) 20 μm.

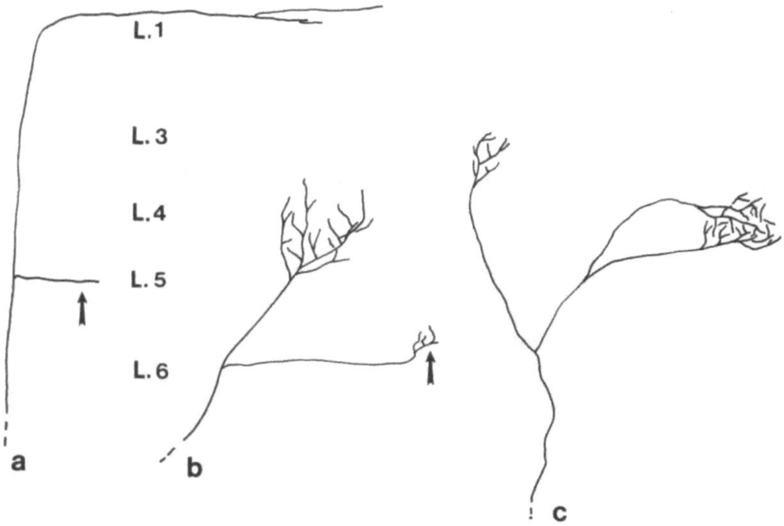

Figure 9. Terminal fields of a single axon differ in size. For feedback axons (a), collaterals in the deeper layers are smaller (arrow) than the terminal field in layer 1. For feedforward axons (b), collaterals (arrow) in layers 5 or 6, if they occur, are smaller than the main arbor. For feedforward axons with multiple arbors (c), these will differ in size.

projecting from V1 to V2 terminate in one or two distinct arbors, but those projecting from V2 to V4 commonly have two or three spatially separate arbors (Rockland, 1992), as do those projecting from area TEO to TE (Saleem *et al.*, 1993) and from anterior TE (Saleem and Tanaka, 1996).

Fourth, despite the tendency toward size constancy, there is a range of arbor size and shape in both feedforward and feedback systems. Minor variability in shape, as also exhibited by geniculocortical axons, can probably be discounted as quasirandom or "biological." The size range is harder to interpret. Most axons projecting from V1 to V2, for example, have arbors in layer 4 that are 200–250 μm in diameter, but the range is 100–400 μm and, in a few instances, even up to 1.0 mm (Rockland and Virga, 1990). Size differences may reflect subtypes of parent neurons, and/or correlate with distinctive functional domains within the target area. Similar differences have been noted in the callosal pathway of the cat. Size differences between multiple arbors of a single callosal axon have been interpreted in a temporal context, as a possible mechanism for achieving degrees of synchronous or serial coactivation (Houzel *et al.*, 1994; Innocenti *et al.*, 1994).

Questions of termination size and uniformity have been raised previously for systems of extrinsic (Bugbee and Goldman-Rakic, 1983) and intrinsic connections. Horizontal intrinsic connections, labeled by extracellular tracer injections, have a patchy distribution, and several studies have suggested a progressive increase in "patch" size as assessed in V1, V2, V4, and other areas (Yoshioka *et al.*, 1992; Amir *et al.* 1993; Lund *et al.*, 1993; Fujita and Fujita, 1996). Given the results on single-arbor analysis described above, some caution may be advisable until more intracellular data and larger samples are available. Differences in patch size, especially in the case of extracellular labels, may be attributable to differences in the overlap or number of axons contributing to a patch (convergence factor), rather than to particular arbor size.

2.3. Postsynaptic Population

No recent electron microscopic (EM) data are available for feedforward connections in primates. From general principles of pyramidal cell connectivity and other systems or species (McGuire *et al.*, 1984; Peters, 1987; White, 1989; Beaulieu *et al.*, 1992), a reasonable prediction is that feedforward contacts will be

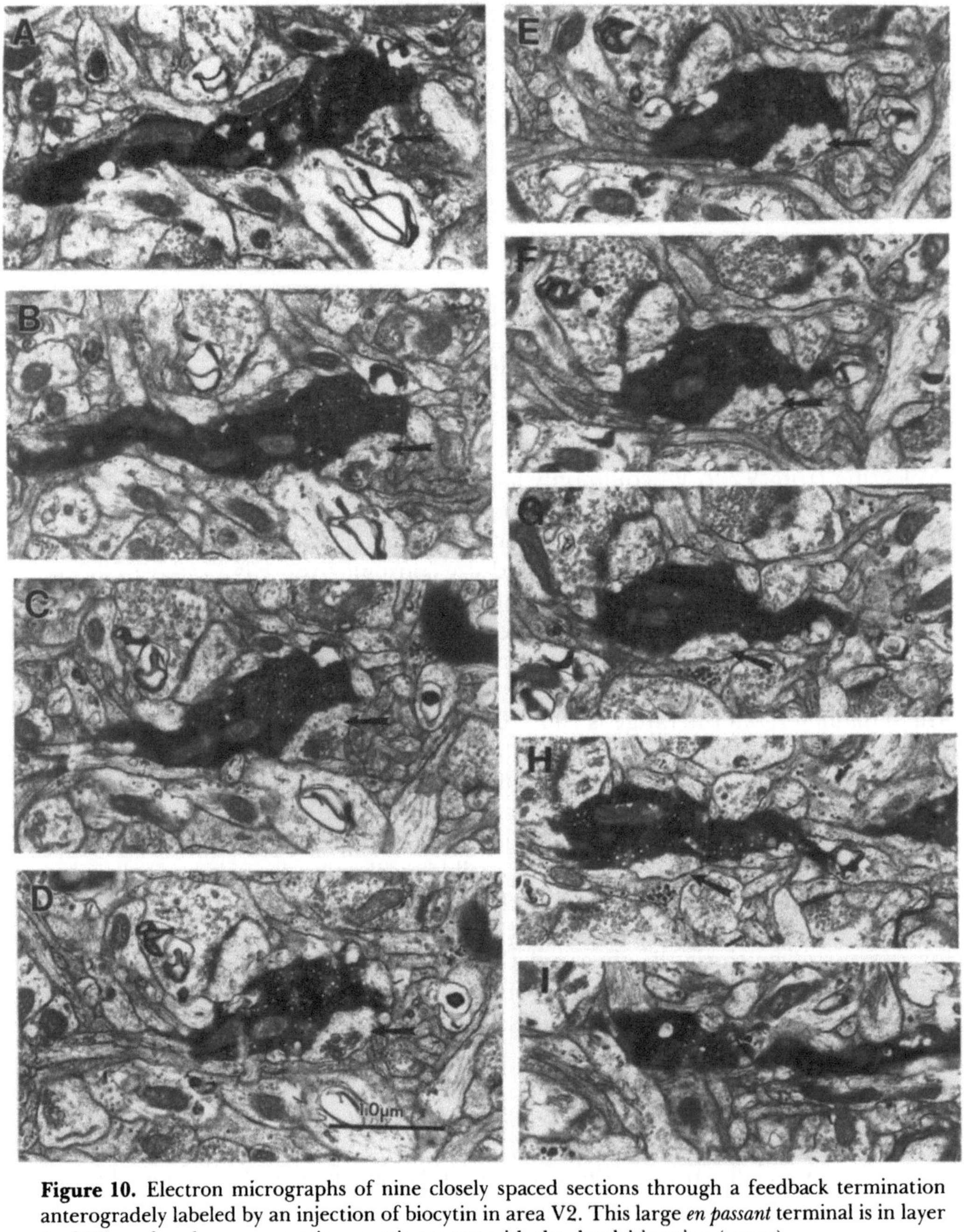

Figure 10. Electron micrographs of nine closely spaced sections through a feedback termination anterogradely labeled by an injection of biocytin in area V2. This large *en passant* terminal is in layer 1 of V1, and makes asymmetric synaptic contact with the dendritic spine (arrow).

excitatory in nature, and that about 80% of the contacts will be with other pyramidal cells and 20% with inhibitory interneurons. On the basis solely of proximity, synapses can be inferred with basal dendrites of pyramids in layer 3, with apical dendrites of pyramids in the infragranular layers, and with small pyramids and interneurons located in layer 4.

For feedback connections, there are slightly more experimental data. In a

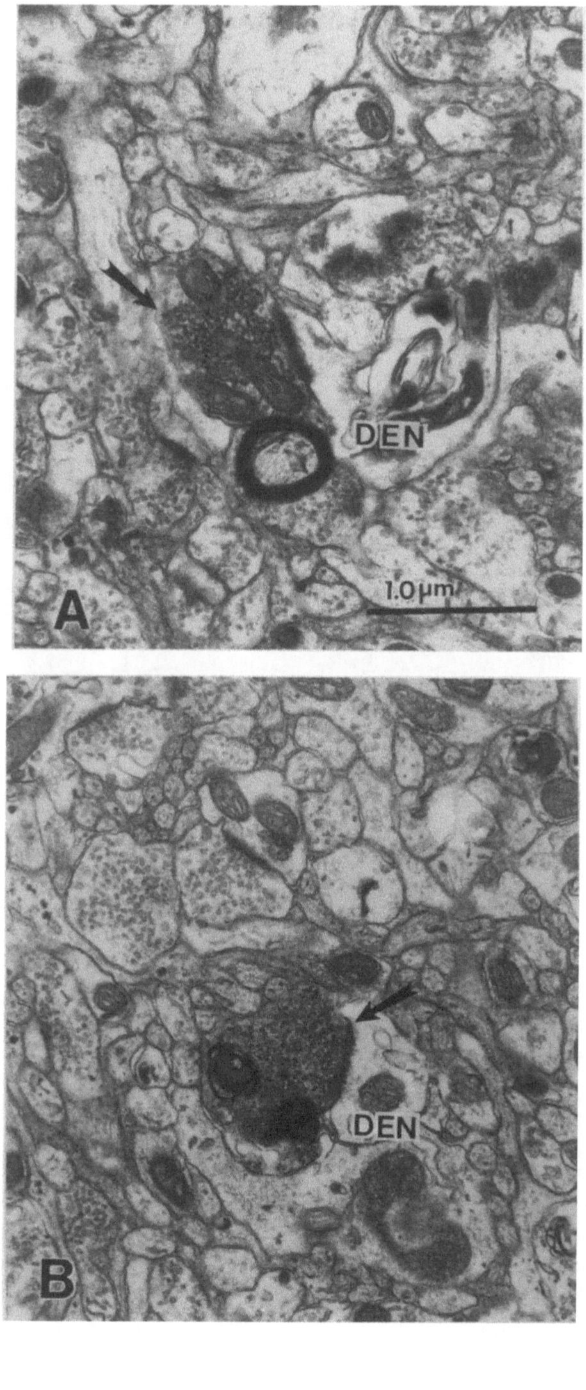

Figure 11. Electron micrographs of feedback terminations anterogradely labeled by a biocytin injection in area V2. Terminals (arrows), in layer 1 of area V1, are making asymmetric contacts with profiles that, on the basis of size and organelle content, can be identified as dendrites (DEN).

study of connections from V2 to V1 that combined anterograde labels with electron microscopy, Rockland and Douglas (1993) found that of 44 well-defined synapses (as followed in short series of 11 thin sections), 36 involved dendritic spines and 8 involved dendritic shafts (82% and 18%, respectively; Figs. 10–12). Of these 36, 4 synapses were onto spines that were clearly identifiable as belonging to pyramidal cell apical dendrites (Fig. 13). These may be apical dendrites of feedforward-projecting neurons in layer 3 of V1 (and thus form part of a monosynaptic feedforward–feedback loop between areas V1 and V2). Just as possibly, they may belong to other pyramidal neurons in layer 3 (without projections to area V2), or to pyramidal cells whose somas are in the infragranular layers.

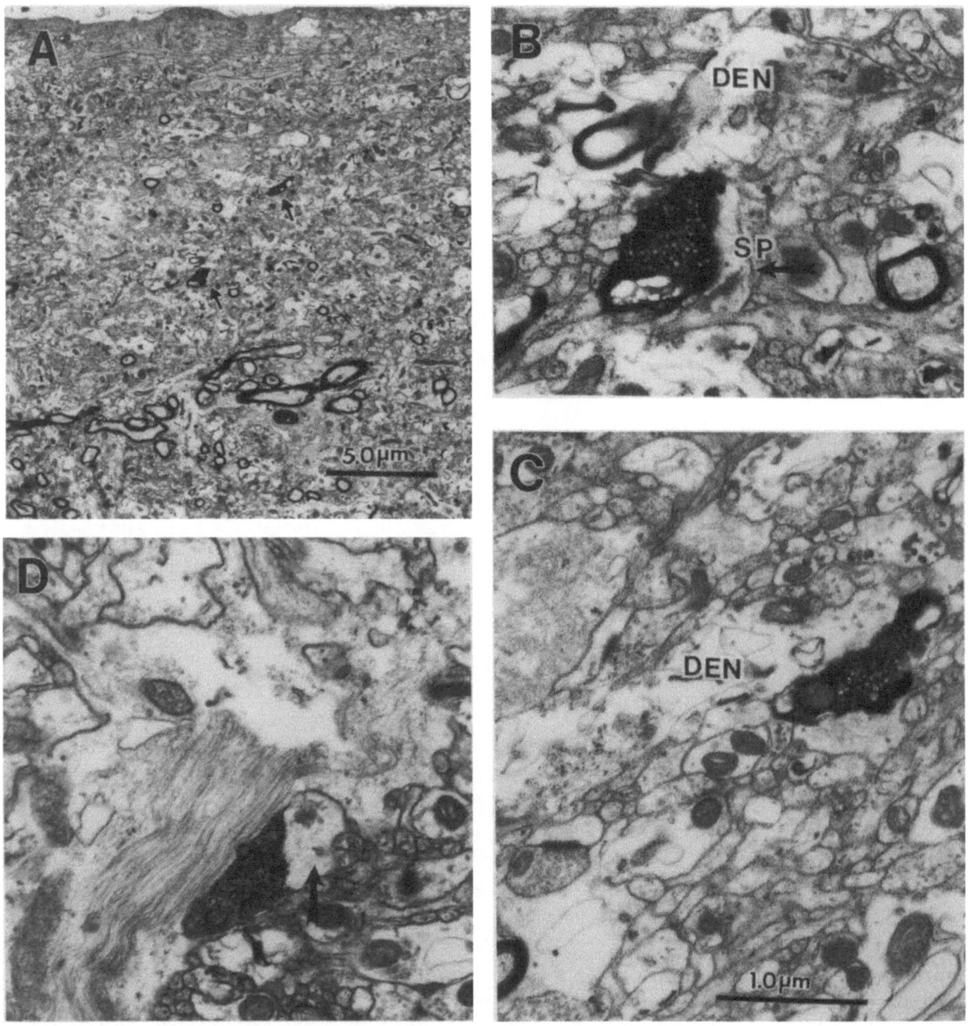

Figure 12. (A) Low-power electron micrograph through layer 1 of area V1. Arrows point to two profiles anterogradely labeled by a biocytin injection in V2. These are shown at higher magnification in panels B and C, where they can be seen to make asymmetric contacts with a dendritic spine (B) and shaft (C). (D) Another example of a feedback termination synapsing with what is probably a dendritic spine (arrow).

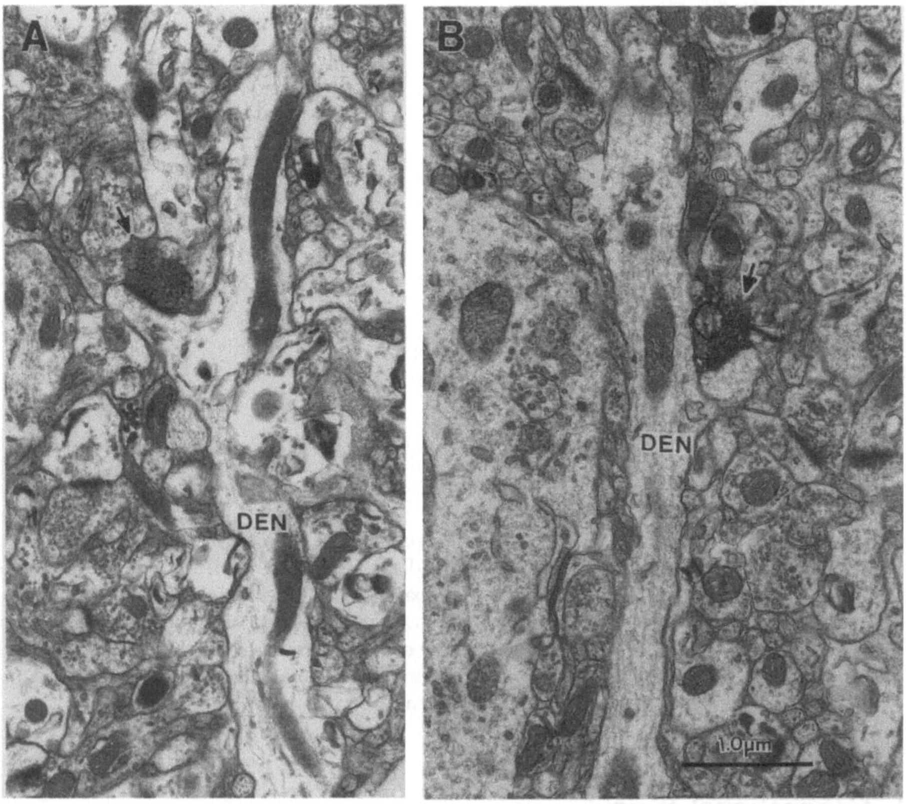

Figure 13. Electron micrographs of feedback terminations anterogradely labeled by a biocytin injection in area V2. Terminals (arrows), in layer 1 of V1, are making asymmetric contacts with spines belonging to distal apical dendrites. These can be identified on the basis of caliber and organelle content (elongated mitochondria).

In a recent study on rat visual cortex, Johnson and Buckhalter (1996) used GABA immunocytochemistry and short serial sections to analyze the postsynaptic contacts of feedforward or feedback axons labeled by PHA-L. According to their findings, feedforward connections from area 17 to LM (the secondary extrastriate lateromedial area) are distributed mainly to pyramidal neurons, although also to some GABA-labeled interneurons (in proportions of 90% and 10%, respectively). They find that feedback connections from LM to area 17, in contrast, are almost exclusively (98%) to pyramidal cells. Further studies are necessary to determine whether this may be a species difference between rats and primates.

Feedback terminations in layer 1 are likely to heavily target distal apical dendrites. Feedback collaterals to layers 5 or 6 are likely to terminate on basal dendrites or proximal apical dendrites of infragranular neurons. Synapses from feedforward terminations in the middle layers might in some cases be in a distal location on the same apical dendrite (Fig. 14). Clearly, there are multiple potential circuits (see discussion in Jones, 1984), and further work is necessary to establish the actual microanatomy.

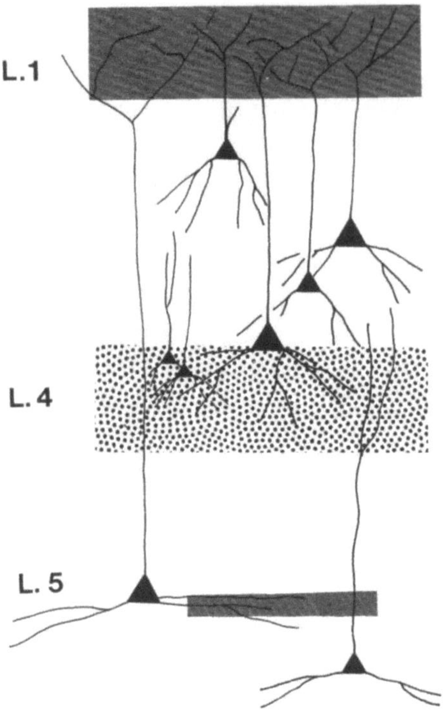

Figure 14. Schematic representation of the laminar distribution of feedforward (stipple) and feedback (shading) terminations. A subpopulation of pyramidal neurons is also shown to suggest how these connections might contact different dendritic segments. Actual postsynaptic contacts are not known.

2.4. Connectional Weight

An approach for evaluating the relative importance of different connections is to assess the number of neurons contributing to each. Kennedy and Bullier (1985), after a quantitative analysis of neurons retrogradely labeled by double injections in V1 and V2, concluded that cortical inputs are stronger to V2 than to V1, but that the superior temporal sulcus projected equally heavily to both areas. There is also the interesting result that, in the cat at least, there are more corticogeniculate than geniculocortical neurons (Sherman and Koch, 1986).

As a step toward determining the density of cortical connections, we estimated the number of projection neurons in several cortical systems (Tables I–III). It must be emphasized that these are only estimates, which need to be confirmed, in particular by optical dissector techniques. (Samples were taken form both fluorescent-labeled and HRP material. All labeled cells were counted within a 200- or 250-μm-wide stripe superimposed horizontally within layers 3 or 6, and subtotals were then multiplied by 100 or 80 (respectively) to determine the number of labeled neurons below 1 mm^2 of cortex. This procedure does not correct for shrinkage or for neurons split between two histological sections, and probably results in overestimates.) These numbers are also subject to variability related to technical factors (size and placement of injections; parameters of tracer uptake, transport, and sensitivity; interanimal variability) and mechanical factors, notably regional differences in laminar thickness, related to sulcal and gyral deformations (compare Figs. 17 and 18; see Kennedy and Bullier, 1985, for further discussion of these issues).

Table I. Numbers of Neurons in Area V1

	Number of neurons beneath 1 mm^2 of cortical surface	Layer 2–3	Layer 4B	Layer 6
O'Kusky and Colonnier (1982)	202,000	56,000	—	23,400 (layer 6A) 6,700 (layer 6B) 30,100
Beaulieu *et al.* (1992)	174,400	52,100	—	19,100 (layer 6A) 7,600 (layer 6B) 26,700
Fitzpatrick *et al.* (1994)	—	—	—	12,604

With these disclaimers, several observations are suggested. First, the number of feedforward-projecting neurons from V2 to V4 is less than that of neurons projecting from V1 to V2 (Figs. 15, 16), although this might simply follow from the larger number of neurons in V1 (Peters, 1987). Second, the number of feedback-projecting neurons from V4 to V2 is less than that of neurons projecting from V2 to V1 (Figs. 17, 18). Third, for this particular set of areas, the number of feedback-projecting neurons is overall less than that of feedforward-projecting neurons, but only slightly less (70% for V1 and V2; 75% for V2 and V4; see also the suggestion in Rockland, 1994, that feedforward neurons in V2 comprise a slightly higher proportion of the total neuronal population in their home layer than do feedback-projecting neurons, 59–84% versus 41–68%).

Other factors affecting connectional strength are the number and efficacy of synapses, and degrees of convergence and divergence. Here again there is relatively little data on corticocortical connections. There are not even estimates of convergence, as are available for several types of interneurons (Kisvardy and Eysel, 1992), because of the greater difficulty in identifying extrinsic connections in EM material. In principle, some indirect gauge of convergence can be derived by comparing the size of an injection site (of retrograde tracer) with the resultant size of projection focus, although this approach has its own limitations. As a simple alternative, we might start with the total number of synapses per given neuron, and "guess" the proportion claimed by particular projection systems. For example, we might allow 6000 synapses as the total number of synapses for a cortical pyramidal neuron (Peters, 1987; DeFelipe and Farinas, 1992). Data suggest that 80% of these contacts will be of intrinsic origin from pyramidal neurons in the same area. This would leave, as a ballpark figure, 1200 synapses to be distributed among extrinsic connections, including feedforward, feedback, pulvinocortical, callosal, and monoaminergic.

Table II. Approximate Numbers of Neurons in Areas V2 and V4

	Number of neurons beneath 1 mm^2	Layer 3	Layer 4	Layer 6
Area V2	92,600	31,200	37,200	10,600
Area V4	83,960	16,480	30,200	18,600

Table III. Approximate Numbers of Projection Neurons Beneath 1 mm² of Cortex

	V1 → V2	V2 → V1	V2 → V4	V4 → V2
Case 23	8,800 (DY)[a]		5,120	
	10,560		4,400	
	11,040		7,200	
	21,600			
Case 24	12,700 (DY)		5,360	
	13,300 L 4B:		6,000	
	17,700 2,000		6,080	
	18,200 2,800		7,446	
Case 19		L 6 6,800		
		10,000		
		10,400		
		10,800		
		10,640		
		L 3A 1,200		
		1,600		
		2,000		
Case 20			4,800	
			6,000	
			8,160	
Case 21	14,160			3,280
	18,160			3,200
F5 (HRP)			6,500	6,500
			4,500	6,800
			6,800	
			10,000	
BMQ (HRP)		5,200		
		7,500		
		7,500		
		L 3A 600		
		4,320[b]		
Zeki and Shipp (1989)			2,240	

[a]DY = Diamidine yellow.
[b]Tangential.

From light microscopy alone, estimates of divergence are more feasible, if we are willing to assume that boutons correspond to synapses and that—in accordance with the postulate of "sparse connectivity"—pyramidal neurons make only one or a few synapses with any postsynaptic neuron. We know from single-arbor analysis that there are 200–400 boutons per arbor for feedforward axons, or 600–1000 boutons per axon if there are multiple arbors (Rockland and Virga, 1990; Rockland, 1992). About the same numbers apply to feedback axons, at least from V2 to V1 (Rockland and Virga, 1989). Thus, a single neuron may have divergent connections to 200–1000 other neurons (or 100–500, if we allow for two synapses with a given postsynaptic neuron). About the same number may apply for feedforward and feedback axons, but as noted earlier (Section 2.2), the two systems have terminal field with very different geometries. Feedforward arbors are more focused than those of feedback axons, and contact spatially delimited groups of neurons, whereas those of feedback axons influence a series of neurons, more or less in a row.

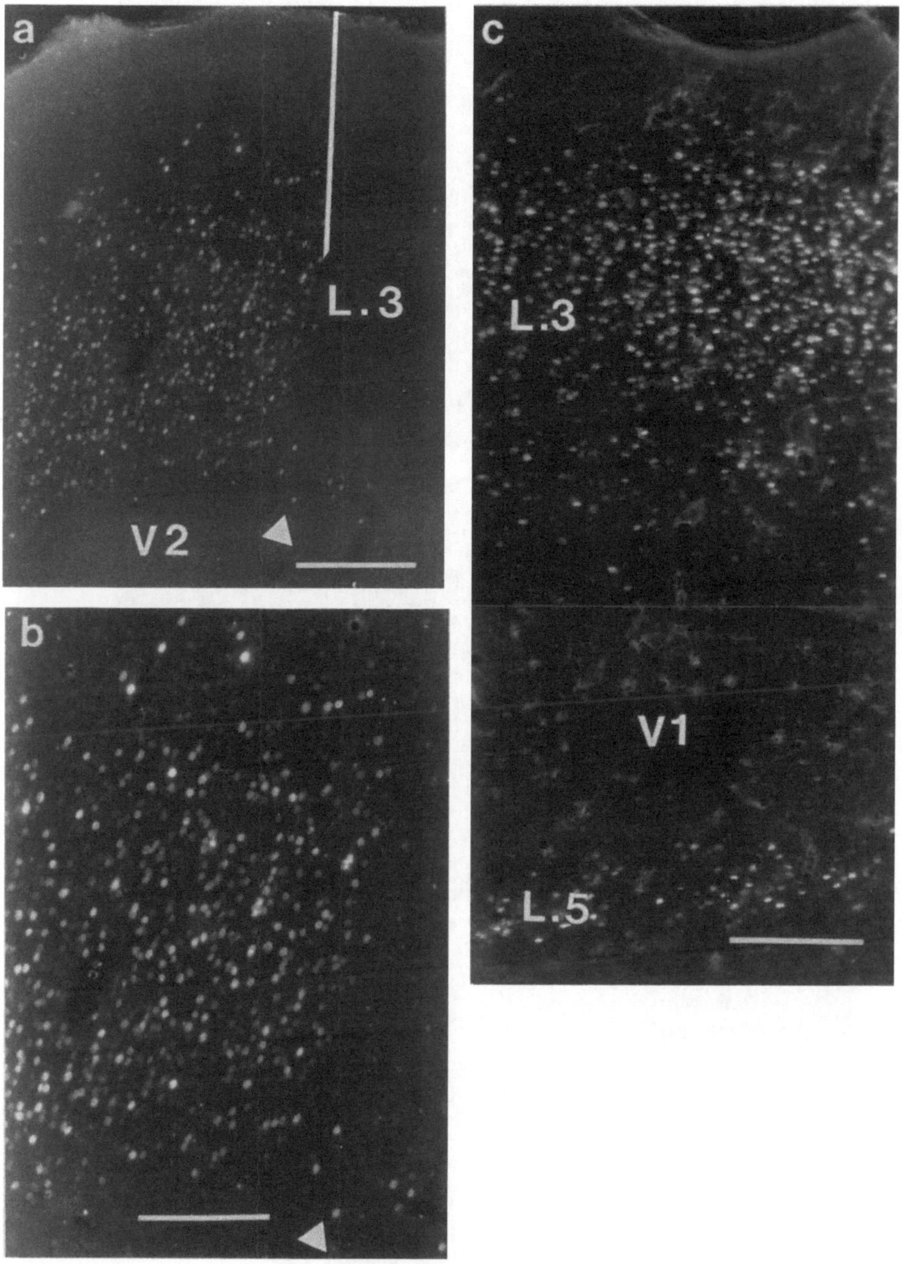

Figure 15. Photomicrographs of feedforward-projecting neurons retrogradely labeled by diamidino yellow. (a and [higher magnification from arrowhead] b) Neurons in layer 3 of area V2 labeled by an injection in area DL (*Saimiri*). Short line demarcates the border between areas V1 and V2. (c) Neurons in layers 3A and, in smaller numbers, 5 of area V1 labeled by an injection in V2. From case 24. Bars: (a) 300 μm; (b, c) 200 μm.

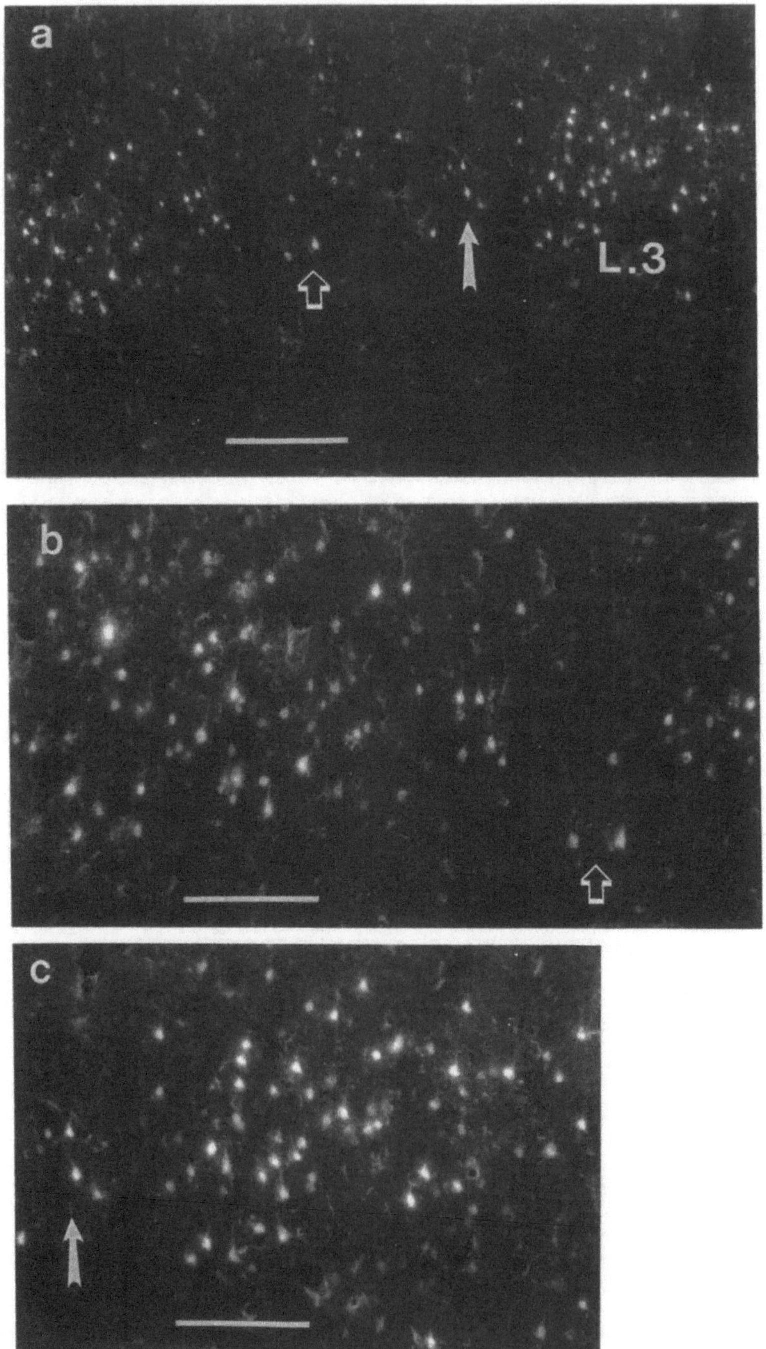

Figure 16. Photomicrographs of feedforward-projecting neurons retrogradely labeled by True Blue. (a) Two clusters of neurons in layer 3 of area V2 labeled by an injection in V4. (b, c) Portions from arrows shown at higher magnification (hollow and solid arrows, respectively). From case 23. Bars: (a) 300 μm; (b, c) 200 μm.

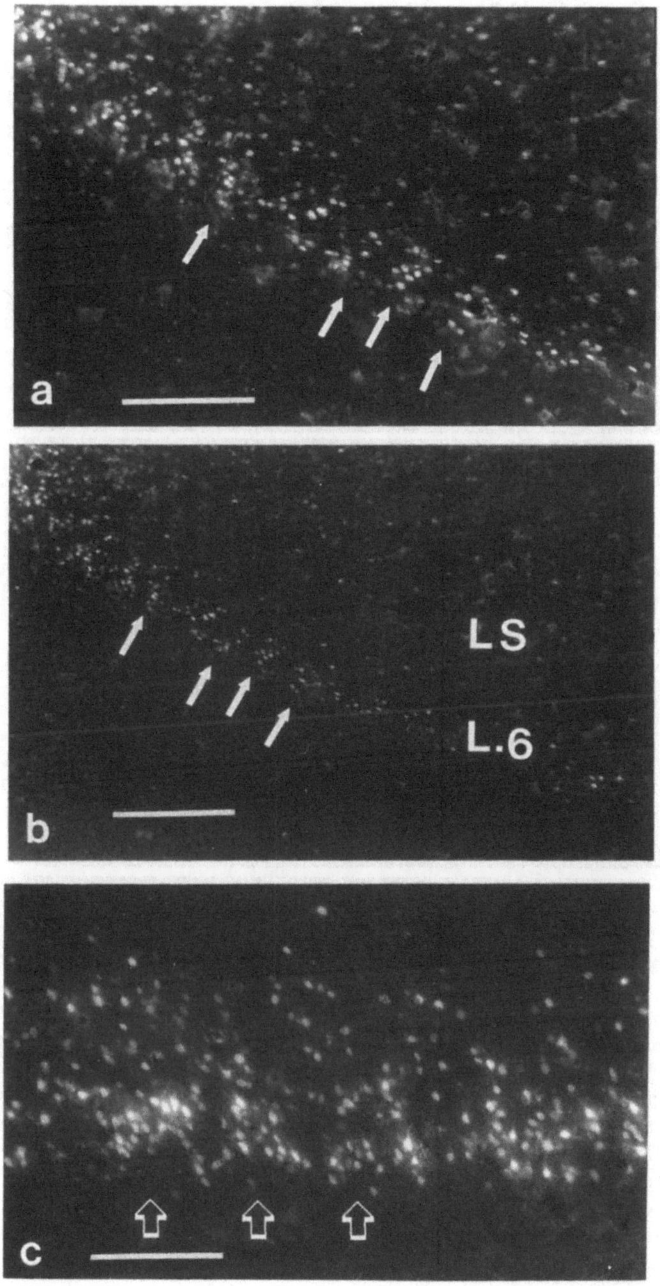

Figure 17. Photomicrographs of feedback-projecting neurons in layer 6 retrogradely labeled by injections of diamidino yellow (a, b) or True Blue (c). (a and [lower magnification] b) Neurons in area V3 labeled by an injection in V2. Note the tendency for neurons to aggregate in small cords or clusters (four solid arrows). From case 24. (c) For comparison, neurons in area V2 (of the inferior occipital sulcus) labeled by an injection in V1. Again, note the tendency for neurons to aggregate in clusters (three hollow arrows). From case 23. LS, lunate sulcus. Bars: (a, c) 200 μm; (b) 300 μm.

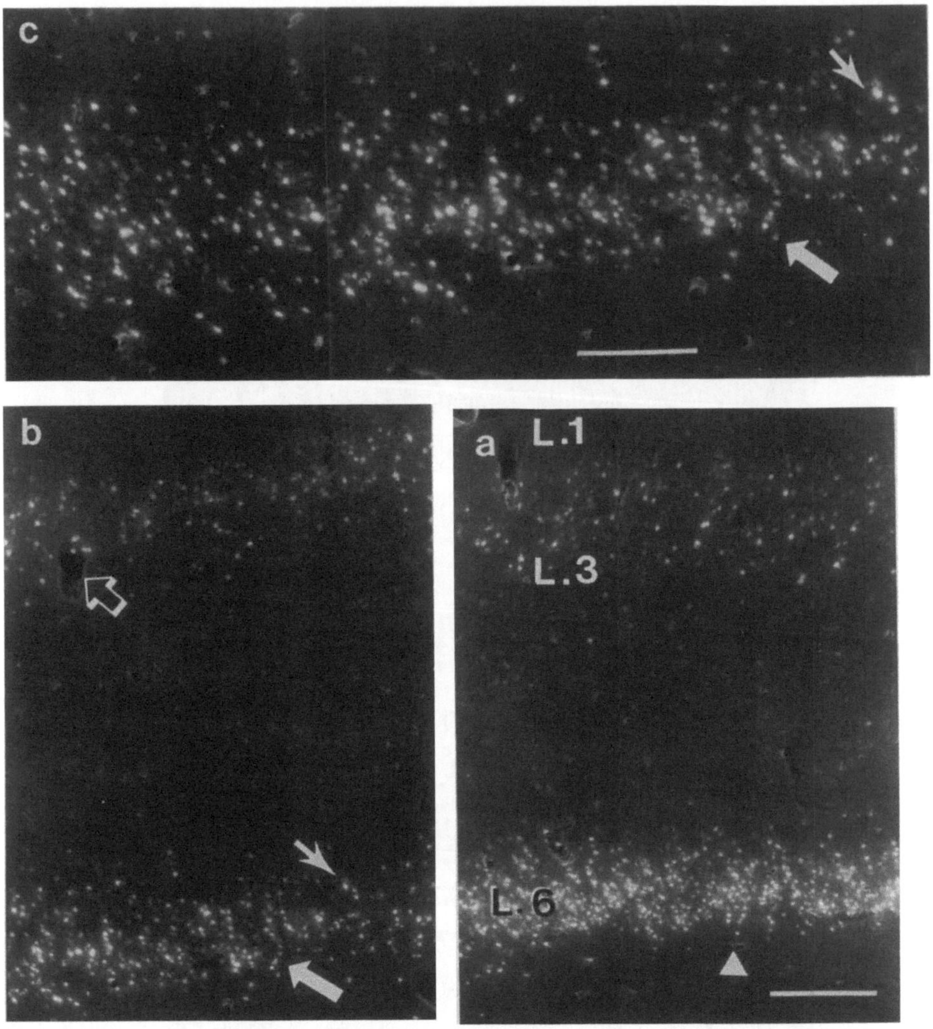

Figure 18. Photomicrographs of feedback-projecting neurons in area V2 retrogradely labeled by injections of diamidino yellow in area V1 (from case 19). (a, b) Two fields in the inferior occipital sulcus (IOS: panel a is 1.6 mm anterior to panel b), with dense labeling in layer 6 and secondary labeling in 3A. The depth of the sulcus is toward the right-hand margin. (c, d) Higher magnification of label in layer 6, from panel b (arrows point to corresponding features). (e) Higher magnification of label in layer 3A (hollow arrow points to corresponding features at panel b). (f) Higher magnification from the second field shown in a (arrowheads point to corresponding features). Note the variability in density of labeled neurons in the two fields. Note also the tendency for neurons to aggregate in small clusters (vicinity of lower arrow in panel c and arrowhead in panel f; see also Fig. 17). Bars: (a, b) 300 μm; (c–f) 200 μm.

2.5. Some Quantitative Parameters

Single-axon analysis is a rich source of functionally relevant parameters, what has been called "computational structure" (Innocenti *et al.*, 1994). This type of analysis has been applied to advantage in the geniculocortical system (Blasdel and Lund, 1983; Freund *et al.*, 1989; Florence and Casagrande, 1990), where physiologically characterized differences between the magno- and parvocellular-

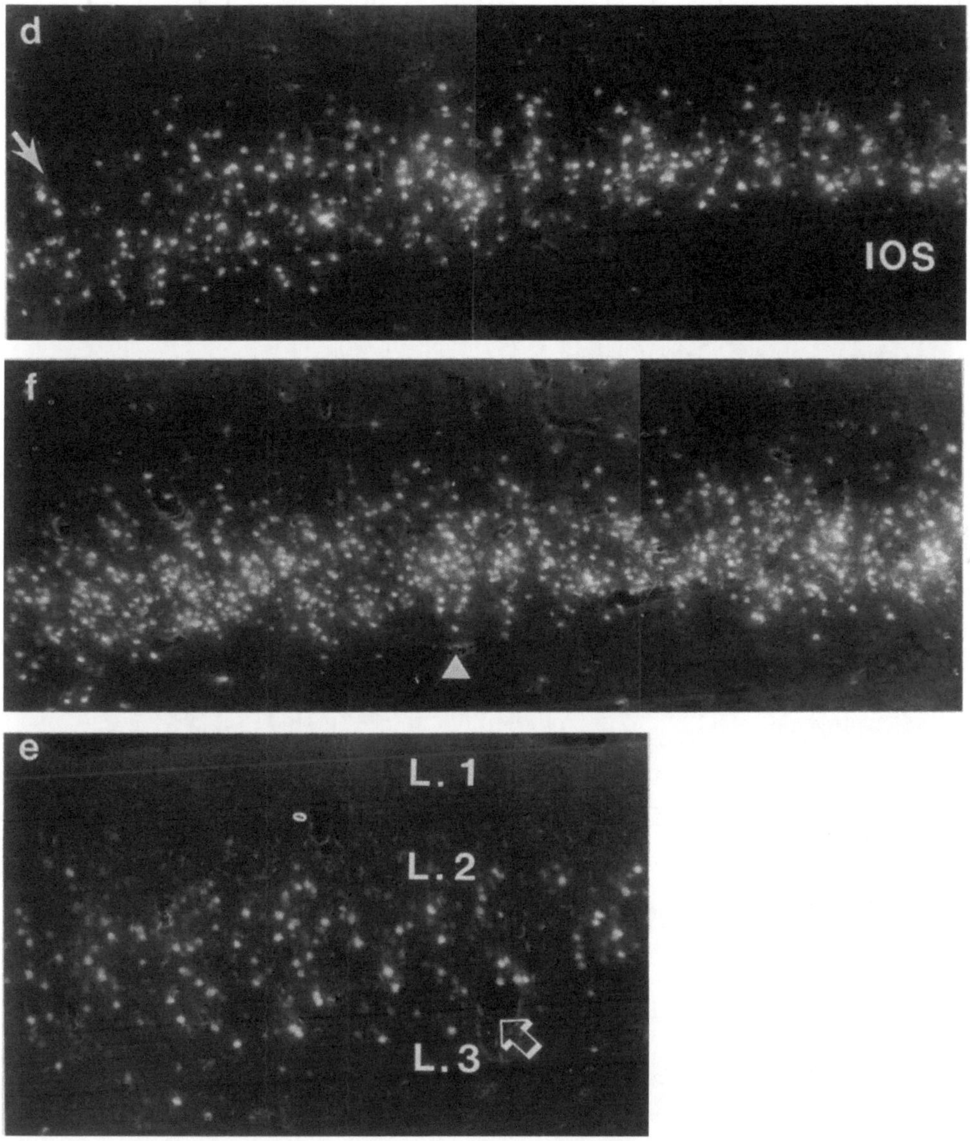

Figure 18. (*Continued*).

derived axons have been correlated with distinctive morphological features. Magnocellular axons, in accord with their faster conduction velocity and lower resolution, are slightly larger in caliber than parvocellular ones (3–4 μm vs. 1–3 μm, respectively), have larger arbors (0.3–0.5 mm × 0.6–1.2 mm vs. 0.2 mm, respectively), and have more terminal boutons (>3200 vs. >1500, respectively).

Single-axon analysis of callosal connections in the cat has been combined with "computer-injected current" to simulate the invasion and progression of an orthodromic action potential (Houzel *et al.*, 1994; Innocenti *et al.*, 1994). The underlying rationale is that the architecture of axonal arbors (axon caliber, arbor size, branching pattern, bouton distribution) has significance not only in terms of spatial mapping, but also for operations in the temporal domain. Among other interesting results, Innocenti *et al.* propose, on the basis of modeling, that activation latencies across the multiple arbors of a single axon vary from 0.3 to 2.7 msec. Arbors separated in space can have temporally synchronous, partially synchronous, or serial activation patterns, depending on factors such as axon diameter and pathway length, as influenced by trajectory and branching pattern. These authors further suggest, from the results of "fudging" real arbors, that different geometries can lead to similar patterns of spatiotemporal activa-

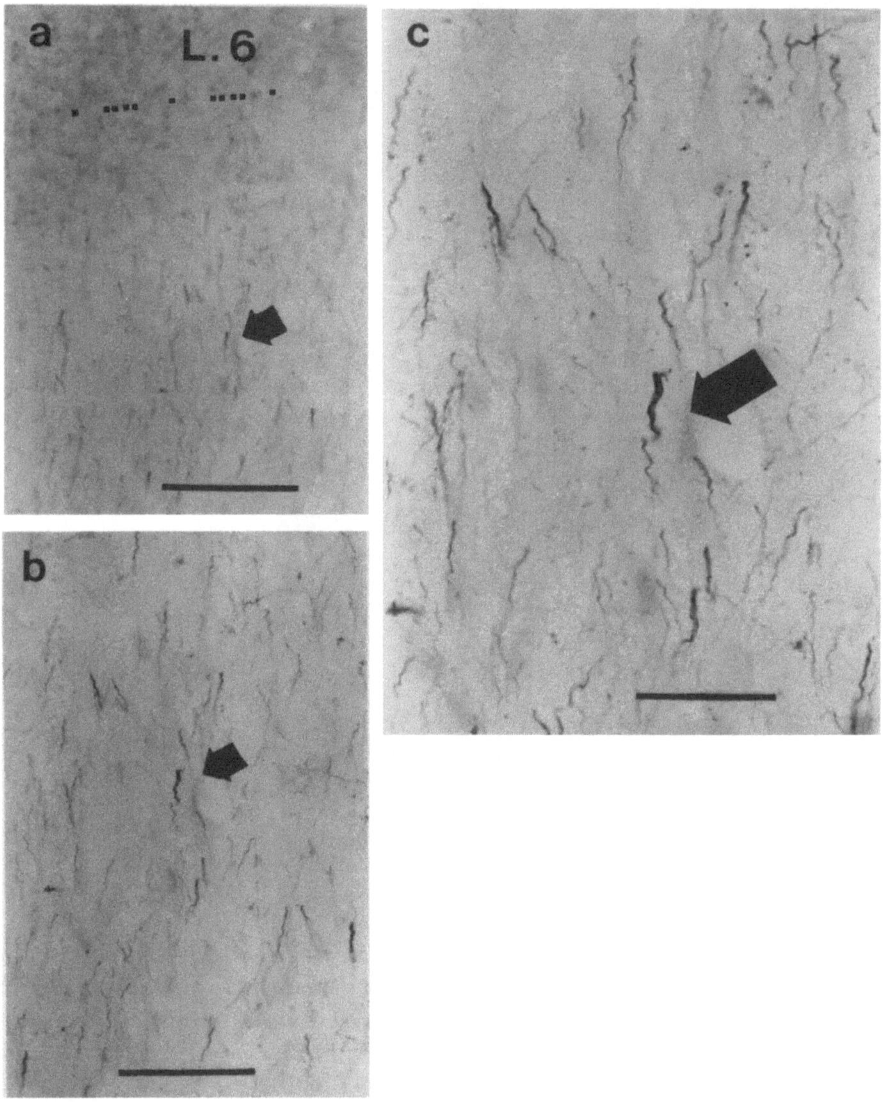

Figure 19. Photomicrographs of axons in the subgriseal white matter below a projection focus in area V4. These are feedforward-projecting axons, which have been anterogradely labeled by an injection of PHA-L in area V2. Arrows point to corresponding features at progressively higher magnification in panels a, b, and c. Note range of axon calibers, including a small number of large axons (2.0 μm, at arrow). Bars: (a) 100 μm; (b) 50 μm; (c) 20 μm.

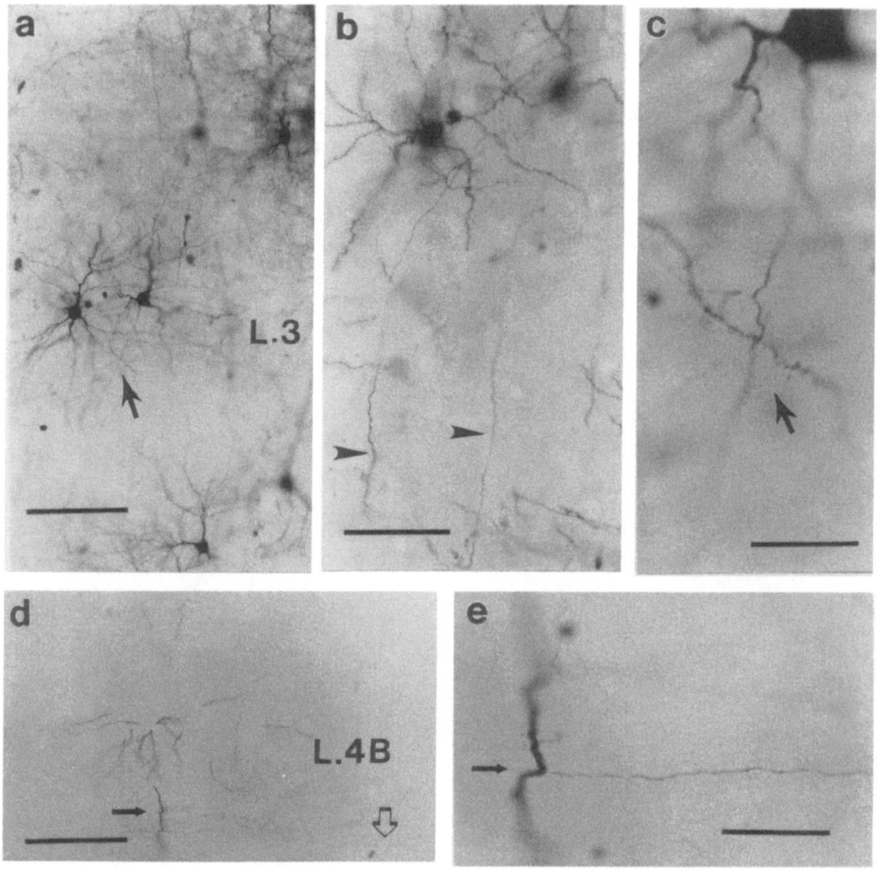

Figure 20. Photomicrographs of feedforward-projecting neurons in V1, retrogradely labeled by injections of PHA-L (a–c) or biocytin (d, e). (a) Field of several neurons in layer 3 of V1, which send feedforward terminations to V2. (b) Higher magnification (from arrow) of panel a. Arrowheads point to the axons descending from these neurons. (c) Higher magnification of a portion of descending axon (arrow). (d and [higher magnification from arrow] e) Stellate cell in layer 4B of V1, which sends feedforward terminations to V2. Note large caliber of descending main axon (compare with descending axons of layer 3 pyramidal cells, in panel c and in Fig. 21). Thin collateral branch at right (hollow arrow in panel d) contributes to the system of intrinsic connections within layer 4B. Bars: (a) 200 μm; (b) 500 μm; (c, e) 20 μm; (d) 50 μm.

tion, and that, therefore, there may be no *unique* correspondence between morphology and computational properties of an axon (Innocenti *et al.*, 1994, p. 934). These conclusions may also apply to corticocortical connections (e.g., Manor *et al.*, 1991). More data, including ultrastructural data, are clearly necessary, but several relevant findings can be adduced.

First, most axons, feedback as well as feedforward, are about 1.0 μm in diameter (Figs. 19–22). The single clear exceptions so far are axons projecting from area V1 to MT/V5, which are commonly 2–3.0 μm in diameter (Fig. 8; and Rockland, 1989). This difference probably correlates with the distinctive cell types in V1 (spiny stellates in layer 4B and giant Meynert cells in layers 5/6) which give rise to this projection (Lund *et al.*, 1975; Shipp and Zeki, 1989a). Perhaps, as is true for the large-caliber magnocellular-derived geniculocortical axons, the larger caliber of these axons is associated with fast conduction velocity. Such large-caliber axons have not been reported in the connections from V2 to MT/V5 (Rockland, 1994).

Second, both feedback and feedforward axons exhibit a range in diameter. Feedback axons projecting from V2 to V1 range from about 0.4 to 1.2 μm (outer

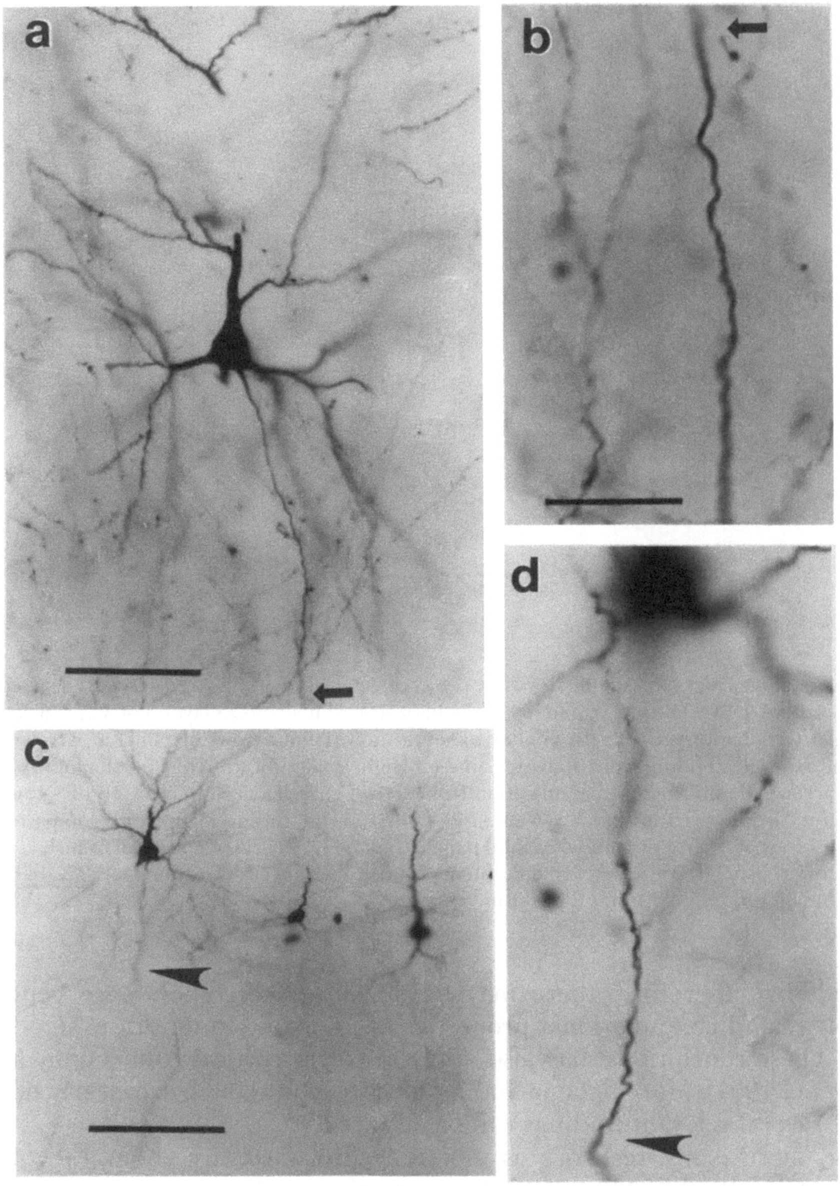

Figure 21. Photomicrographs of projection neurons, retrogradely filled by injections of biocytin. (a) Neuron in layer 3A of V2 sending feedback terminations to V1. (b) Portion of descending axon (from arrow) shown at higher magnification. (c) Three neurons in layer 3 of V1 sending feedforward axons to V2. (d) Portions of descending axon of one neuron (from arrowhead) shown at higher magnification. Comparison of panels b and d shows no obvious difference in the caliber of feedback- and feedforward-projecting axons. Bars: (a) 50 μm; (b, d) 20 μm; (c) 100 μm.

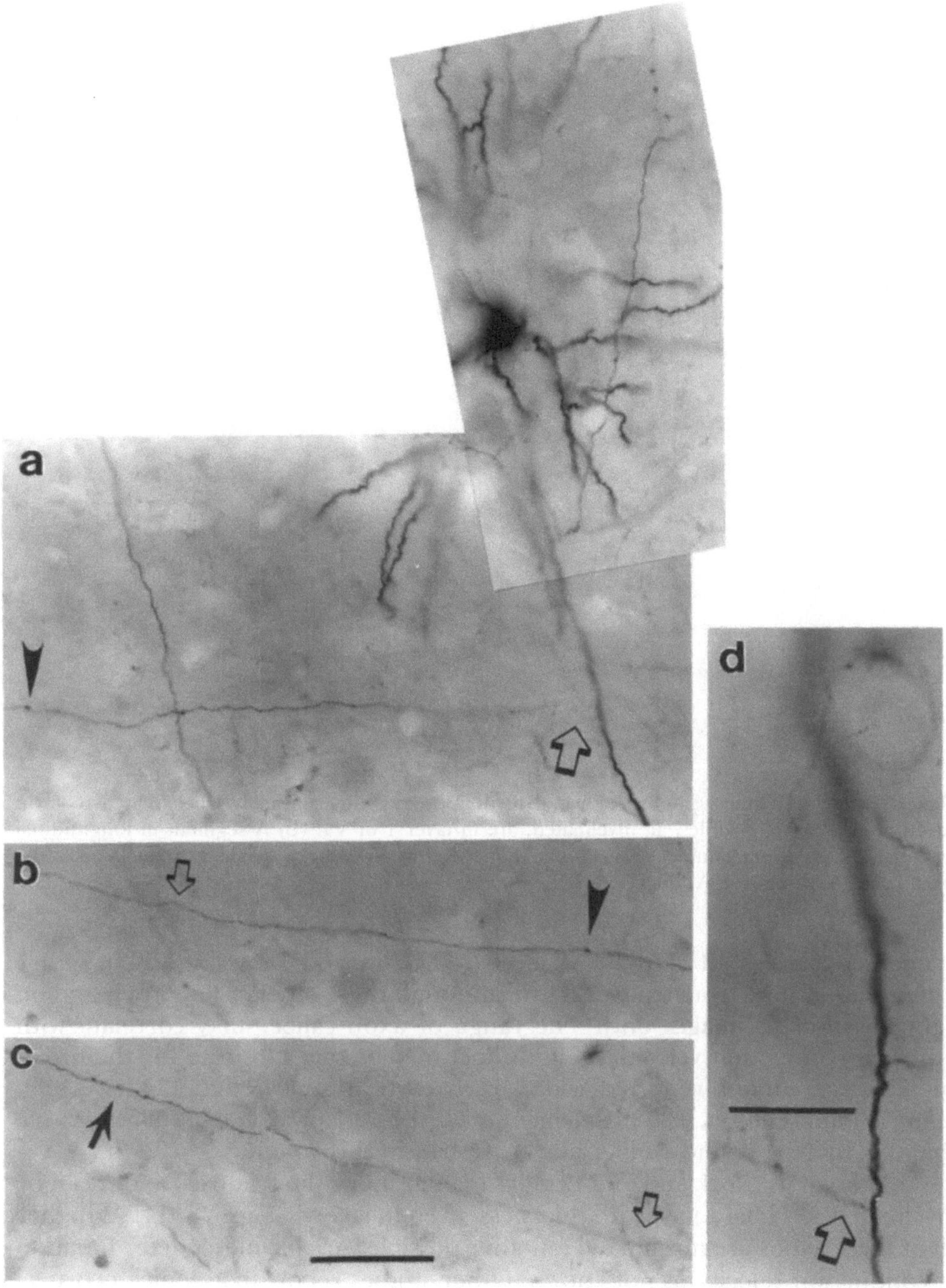

Figure 22. Photomicrograph of a neuron in layer 3A of area V4, retrogradely labeled by an injection of PHA-L in area V2. (a) Portion of the cell body, descending main axon (hollow arrow), and intrinsically projecting collateral branch (arrowhead). (b, c) Continuing portions of the intrinsic collateral (small hollow arrows point to duplicated portion). Solid arrow in panel c points to a group of small terminations, about 600 μm from the original branch point. (d) Higher magnification of descending axon (hollow arrows point to corresponding features in panels a and d). A second intrinsic collateral branch is visible, but not illustrated further. Axon caliber is about the same (1.0 μm) as for feedback-projecting neurons in V2 and for many feedforward-projecting neurons in V1. Bars: (a–c) 50 μm; (d) 20 μm.

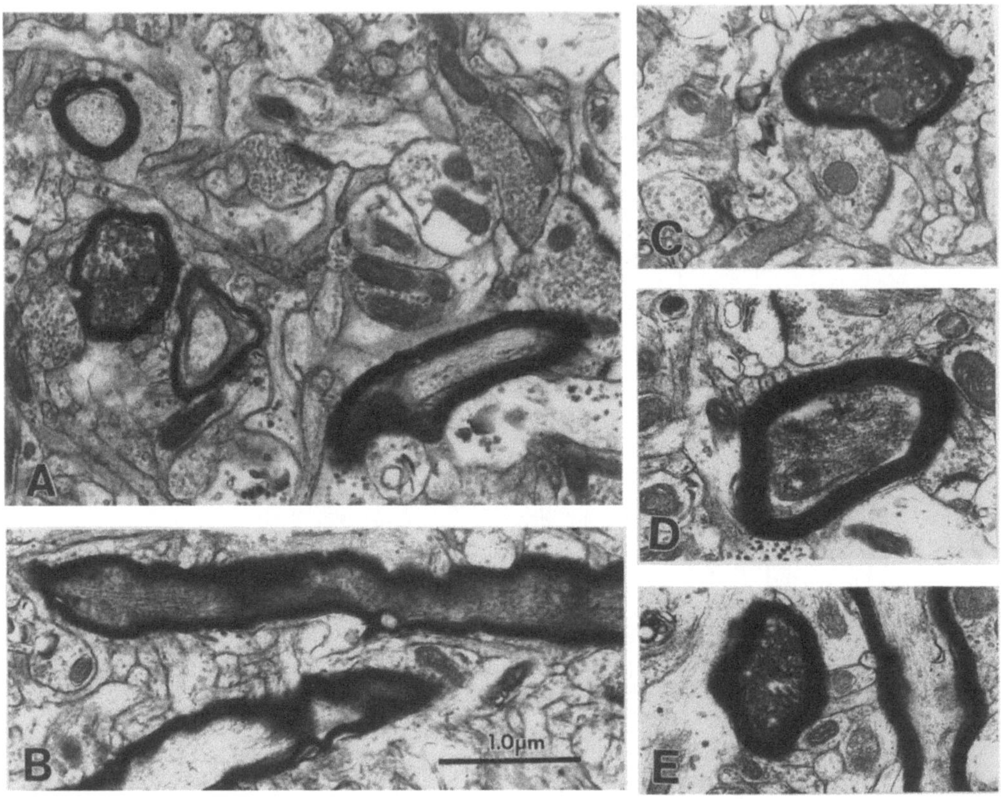

Figure 23. Electron micrographs of several feedback axons anterogradely labeled by a biocytin injection in area V2. Profiles, all myelinated, are in layer 1 of V1. (B) A segment in longitudinal section; (A, C–E) four examples cut in cross/oblique section. Note slight variability in thickness of the myelin sheath (thinner in A, thicker in D).

diameter), as verified at the EM level (Rockland and Douglas, 1993). Other systems also range from 0.5 to 2.0 μm, although these observations are based on light microscopic images only. The range of axon caliber may correlate with different conduction velocities, and in fact may be one substrate for the broad latency distributions reported within each cortical area (Raiguel *et al.*, 1989; Maunsell and Gibson, 1992; Celebrini *et al.*, 1993; Nowak *et al.*, 1995; and see chapters by Nowak and Bullier and by Orban in this volume).

EM analysis (Rockland and Douglas, 1993) shows the feedback axons can be myelinated (0.1–0.4 μm sheath thickness) even in layer 1 (Figs. 23, 24). Whether all of these axons are myelinated (another possible variable in the determination of conduction velocity) is not known. Other studies report that geniculocortical (Freund *et al.*, 1989) and callosal axons (Houzel *et al.*, 1994) are myelinated at their point of entry into and within the gray matter; large basket cell axons (Jones, 1984; Kisvardy and Eysel, 1992) are also myelinated.

A third observation is that axons can change caliber along their trajectory, even in the gray matter (Rockland and Drash, 1996). Such diameter modulations, as remarked by Innocenti *et al.* (1994), could be a means of achieving temporal synchronicity between spatially separated arbors.

Fourth, terminal specializations of both feedback and feedforward connections appear to be similar. Endings consist of a mix of stalked and beaded profiles (*boutons terminaux* and *en passant*), and range from 0.3 to 1.5 μm in size (Figs. 25–27). Interbouton intervals typically range from 3–15 μm. There can be an appearance of regular grouping or periodicity (Fig. 25), but no systematic pattern or trend has been determined. The single exception is again the projection from V1 to MT/V5, where endings are predominantly beaded and large, up to 2–3 μm in size (Fig. 2).

In contrast with the morphological uniformity of terminal specializations of feedforward and feedback connections, corticopulvinar connections, originating from neurons in the infragranular layers, do have recognizably different endings (Rockland, 1996). One type (classical type 1) is spiny, almost like thin dendritic spines, and the second (classical type 2) is beaded and large (2–3 μm) (Ogren and Hendrickson, 1977; Guillery, 1995). These two types are associated with a different geometry of terminal field (spinous, type 1 endings are arrayed along thin, elongated segments; beaded, type 2 endings are concentrated in small, round arbors), and probably derive from different neurons of origin in layer 5 and/or 6 (Lund *et al.*, 1981; Levitt *et al.*, 1995). Both these types of corticopulvinar neurons might be expected to have local intracortical collaterals;

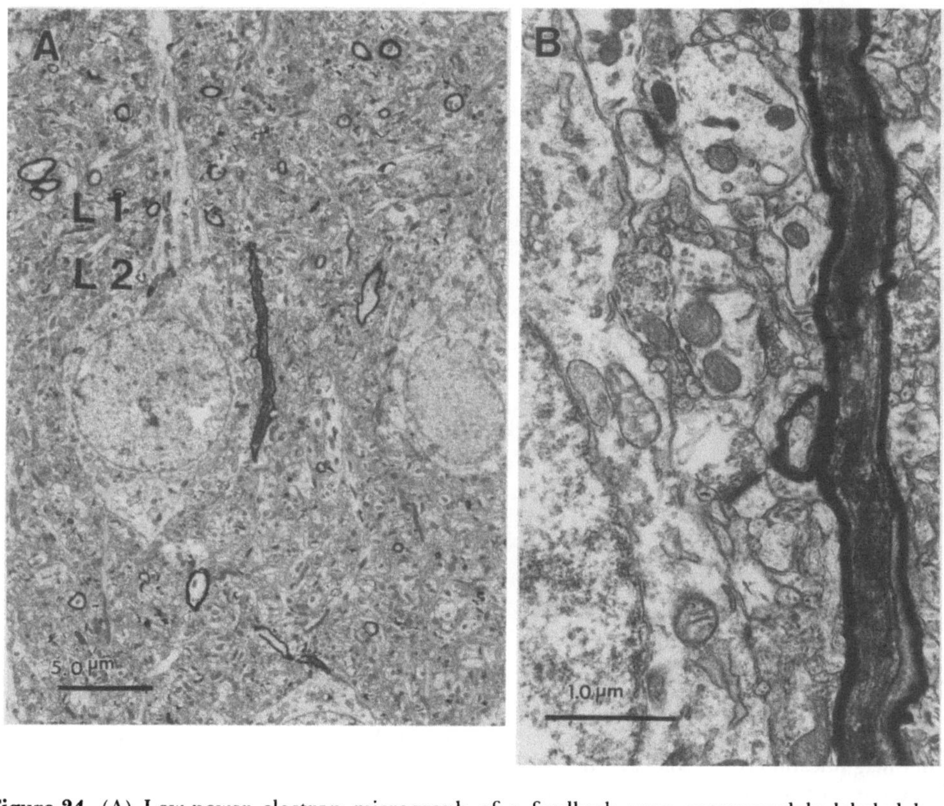

Figure 24. (A) Low-power electron micrograph of a feedback axon anterogradely labeled by a biocytin injection in area V2. Axon ascends in layer 2, perpendicular to the pia. (B) Higher magnification to show myelin sheath in greater detail.

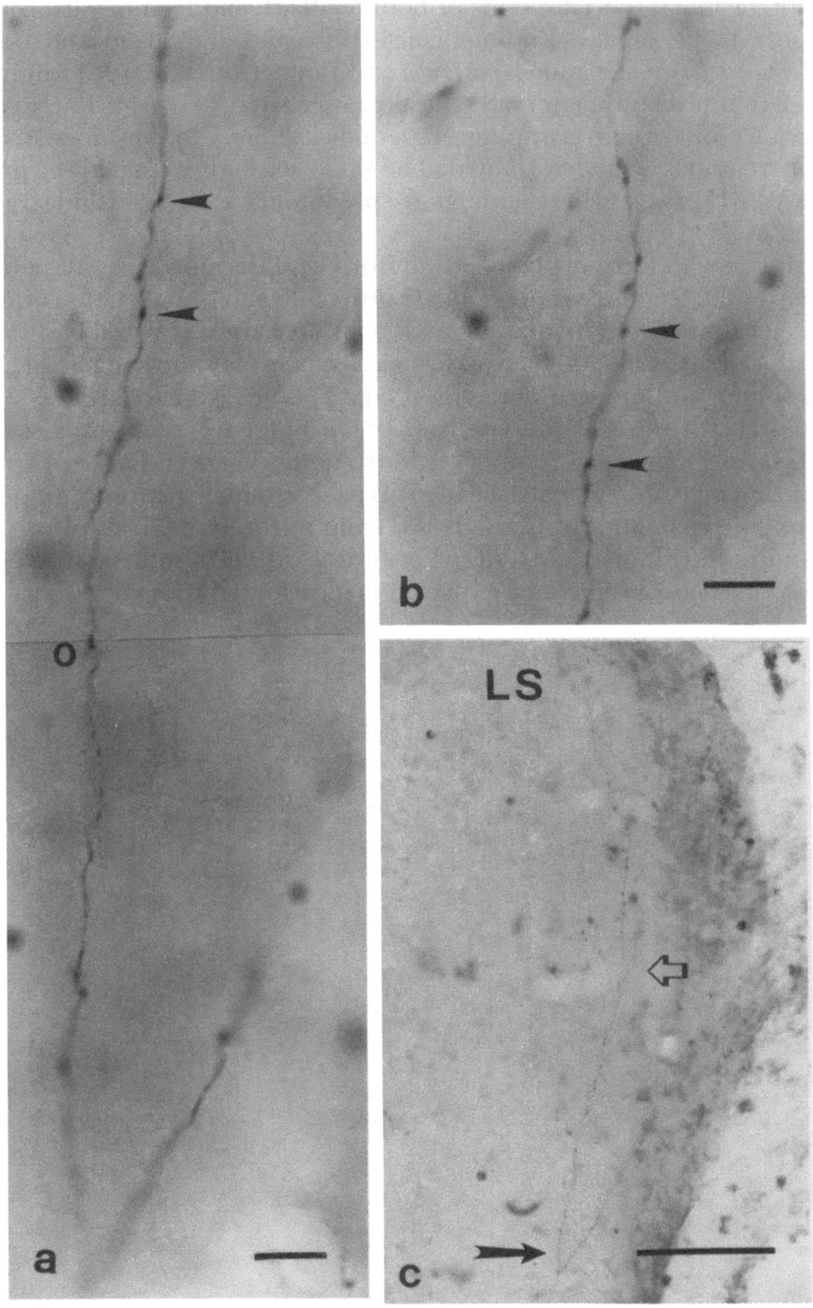

Figure 25. Photomicrographs of a feedback axon in layer 1 of area V2 (LS, lunate sulcus; depth of the sulcus would be off the top margin) anterogradely labeled by a PHA-L injection in ventromedial area TE. (c) Low-magnification view of a branched terminal segment near its distal portion. (a, b) Higher magnification (from solid and hollow arrows in panel c, respectively). Note density of terminal boutons, and variations in their size and shape. Arrowheads point to representative beaded specializations; O = one complex consisting of a beaded and stalked profile together. Bars: (a, b) 10 μm; (c) 100 μm.

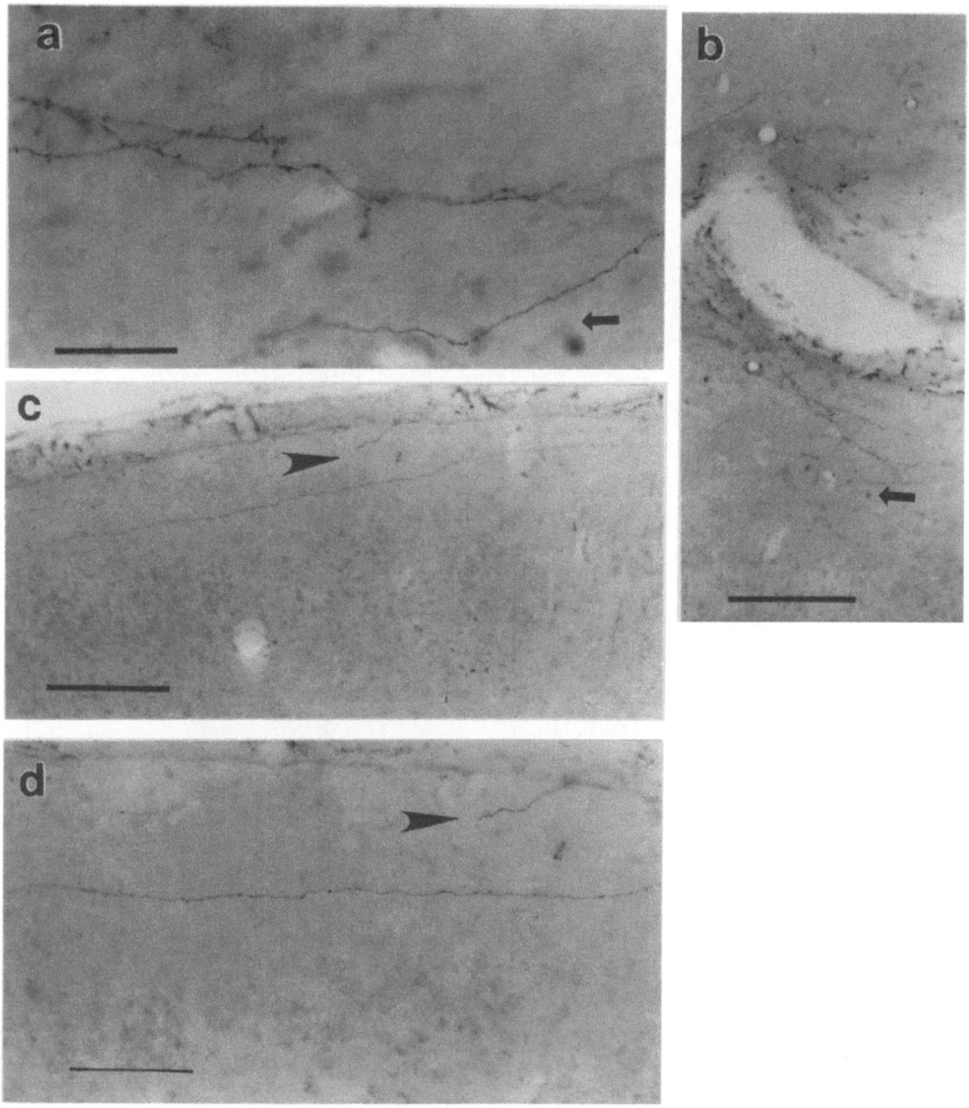

Figure 26. Photomicrographs of feedback axons in layer 1 of area V2, anterogradely labeled by a PHA-L injection in ventromedial TEO. (a and [lower magnification from arrow] b) A field of several axons in the annectent gyrus. (c and [higher magnification from arrowhead] d) A second, sparser field, 0.6 mm anterior. Bars: (a) 50 μm; (b, c) 200 μm; (d) 100 μm.

occasionally, both conspicuously spiny and conspicuously beaded intracortical terminations can in fact be identified by local transport in the vicinity of an injection site (Figs. 28, 29).

It is interesting to note that systematic variations in "bouton architecture" have been reported along parallel fibers in the cerebellum. When these are visualized by extracellular injections of biocytin, mean intervaricosity interval and intervaricosity size vary systematically as a function of proximal-to-distal location. Mean intervaricosity interval is reported as 3.7 μm within 250 μm of

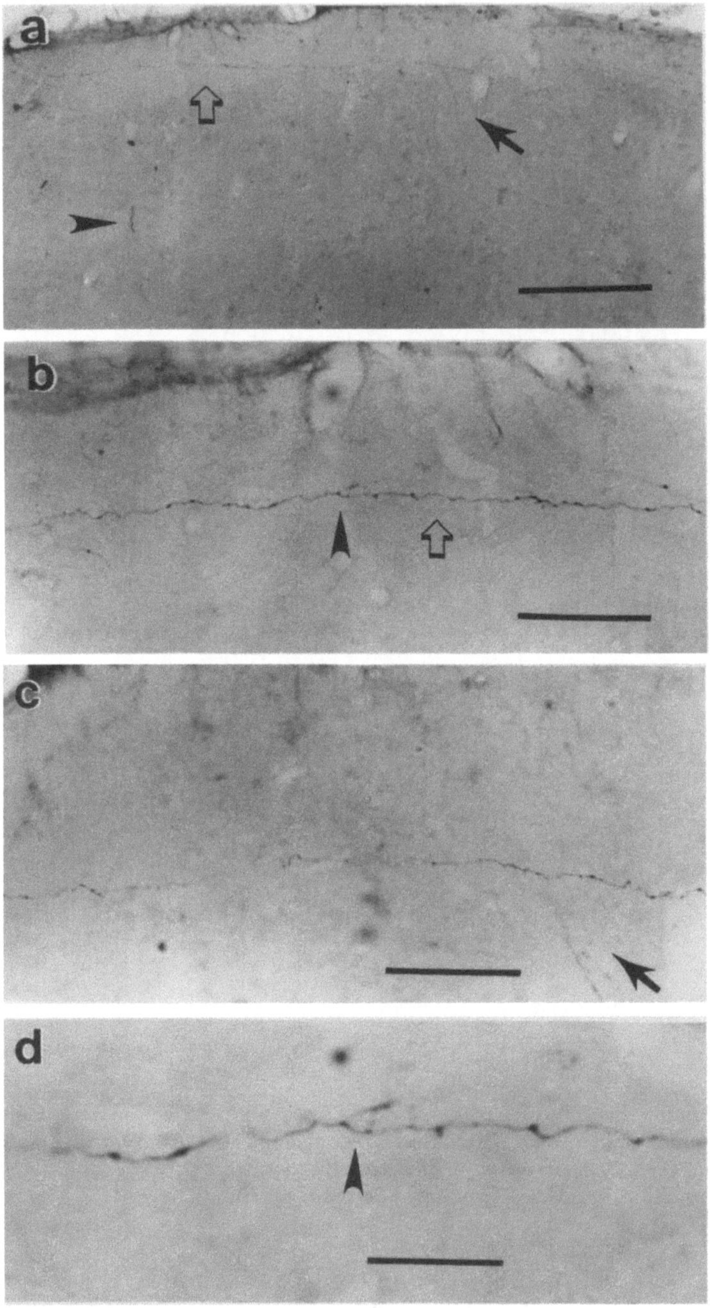

Figure 27. Photomicrographs of a feedback axon in layer 1 of V1, anterogradely labeled by a PHA-L injection in area V2. (a) Low-magnification view. Portions from hollow and solid arrows are shown at higher magnification in panels b and c, respectively. Arrowhead indicates portion of ascending trunk of another axon. (d) Higher magnification (from arrowhead in panel b), indicating typical axon caliber, and density, size, and shape of terminal specializations. These do not notably differ from nonreciprocal feedback axons (see Figs. 25 and 26, which illustrate terminations from ventromedial TE to V2). Bars: (a) 200 μm; (b, c) 50 μm; (d) 20 μm.

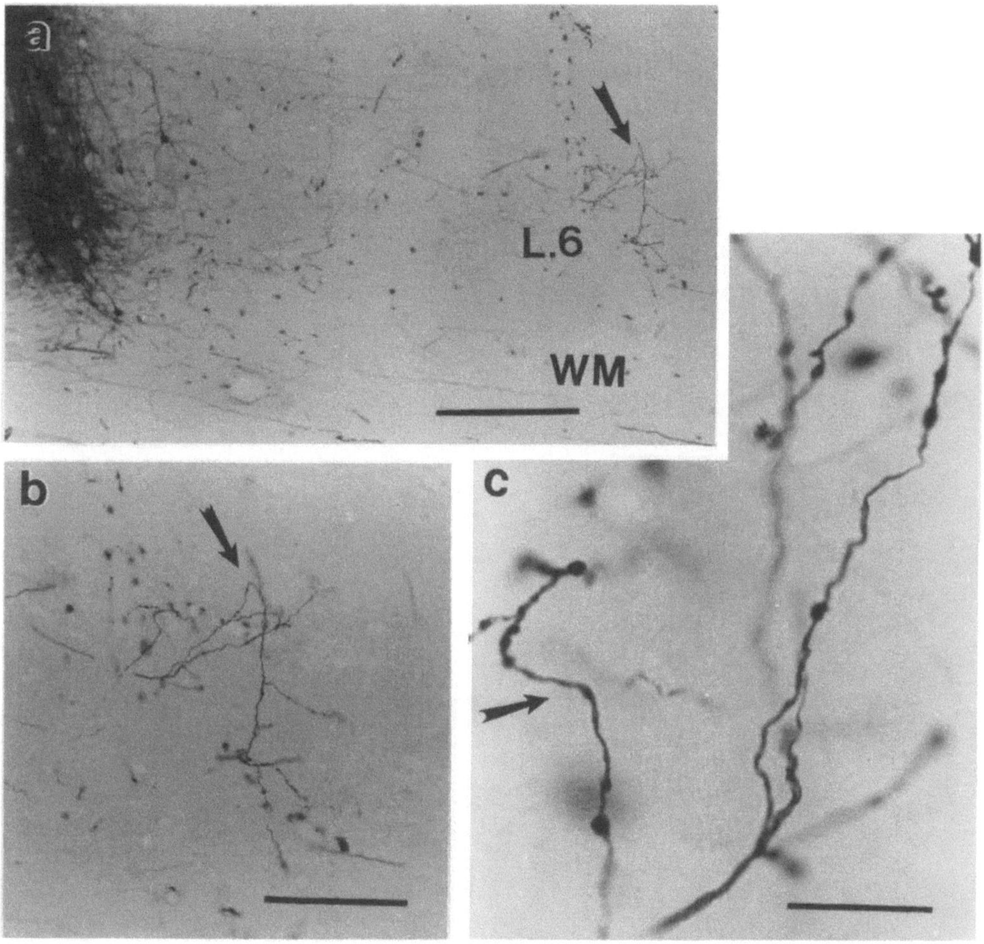

Figure 28. Photomicrographs of profiles in V1 labeled by local transport of biocytin from an injection in area V1. Collaterals with large, beaded terminations sometimes occur in layer 6 (a–c). Arrows point to corresponding feature at different magnifications in panels a–c. Bars: (a) 200 μm; (b) 100 μm; (c) 20 μm.

the fiber bifurcation, but 7.4 μm distally. Mean varicosity size is 0.82 μm^2 in proximal regions, but 0.47 μm^2 about 1.2 mm distally (Pitchitpornchai *et al.*, 1994). No comparably systematic differences have been identified for corticocortical connections that might correlate with connections, layers, or areas.

2.6. Subtypes of Connections?

As reviewed in the preceding sections, corticocortical axons exhibit a range in axon caliber, number of arbors, and spatial extent. The question naturally arises of whether these features indicate functionally significant subtypes. There is some evidence that parent cell populations are not homogeneous. Neurons

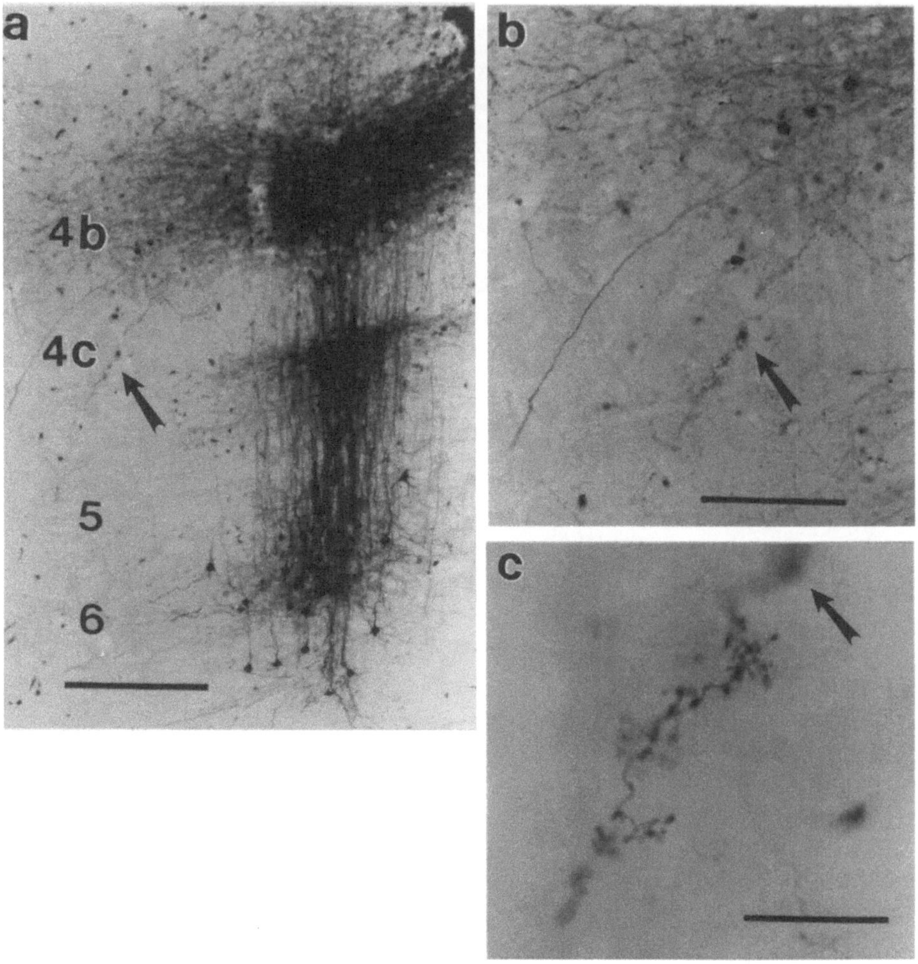

Figure 29. Photomicrographs of profiles in V1 labeled by local transport of biocytin from an injection of area V1. Collaterals with extremely spinous terminations sometimes occur, as illustrated here in layer 4C (a–c). Arrows point to corresponding features at different magnifications in panels a–c. Bars: (a) 200 μm; (b) 100 μm; (c) 20 μm.

within layer 6 differ in soma size and dendritic arborization (reviewed in Tombol, 1984; see also Valverde, 1978; Lund *et al.*, 1981; deLima *et al.*, 1990), and probably also in the pattern of local axon collaterals (specific data on this point are available only for V1: Anderson *et al.*, 1993; Wiser and Callaway, 1996). Certainly, neurons in layer 3A are morphologically distinct from those in layer 6, although both groups contribute to feedback connections. Pyramidal neurons in layer 3, the primary source of feedforward connections, differ in soma size and dendritic patterns among themselves and even more so in relation to neurons in layer 5, which are a secondary but consistent source of feedforward connections. Combined experiments to colocalize neurofilament protein and specific populations of retrogradely labeled projection neurons suggest that corticocortical systems have distinguishable proportions of neurofilament-containing neurons

(Hof *et al.*, 1996). Feedback projections from areas MT/V5, V4, and V3 have a relatively consistent proportion of neurofilament-containing neurons (70–80%), regardless of whether the target is V1 or V2. Feedforward projections from V1, V2, V3, and V3A to MT/V5 originate from a population 57–100% of which is positive for neurofilament protein. Feedforward neurons projecting to area V4 are less frequently positive for neurofilament protein (25–36%). More data are clearly necessary, however, concerning physiological and other features, for further assessment of possible pyramidal cell subcategories.

2.7. Intrinsic Connections

An important aspect of areal interactions is how feedforward and feedback connections combine with other cortical inputs—functionally or anatomically (see also Section 3.4). This is largely still unknown. For other extrinsic connections to extrastriate areas (see Felleman and Van Essen, 1991, and, for area V1, Tigges and Tigges, 1984), there are in fact no EM or even single-axon analyses.

Cortical connections interact with intrinsic connections, deriving from the horizontal collaterals of pyramidal neurons. This system, which seems common to all extrastriate areas and probably to all cortical areas, has been extensively investigated, including some intracellular studies (in monkey: McGuire *et al.*, 1991; and in cat: Gilbert and Wiesel, 1983; Martin and Whitteridge, 1984; Ojima *et al.*, 1991, 1992). It is particularly interesting to compare this system with extrinsic feedforward connections, since they are frequently different parts of the axonal arborization of a single neuron.

Horizontal collaterals of layer 3 pyramidal neurons consistently have multiple, spatially discrete arbors (Fig. 30). As summarized above for feedforward arbors (Section 2.2), these vary in size, shape, and exact laminar disposition. There is characteristically a dense concentration of terminations in the immediate vicinity of the soma, with a progressive falloff in more distal patches. Representative numbers, available for five cells in layer 3 of cat auditory cortex are (Dr. H. Ojima, personal communication)

	Dorsal patch	Soma patch	Ventral patch
Cell 1	106	881	126
Cell 2	—	1467	198
Cell 3	706	945	535, 41
Cell 4	—	1334	863, 86
Cell 5	206, 957	2063	—

The exact relationship, numerical and spatial, between the intrinsic and extrinsic arbors of individual neurons is not known. For neurons in area V1 the intrinsic projection field may be larger than the extrinsic one (Fig. 31). That is, neurons in layer 3 have multiple local arbors (McGuire *et al.*, 1991), but typically only one or two extrinsic arbors in area V2 (Rockland and Virga, 1990). For neurons in extrastriate areas, which more consistently have multiple extrinsic arbors, the match between intrinsic and extrinsic fields may be closer.

Another question still to be investigated is what horizontal collaterals do at

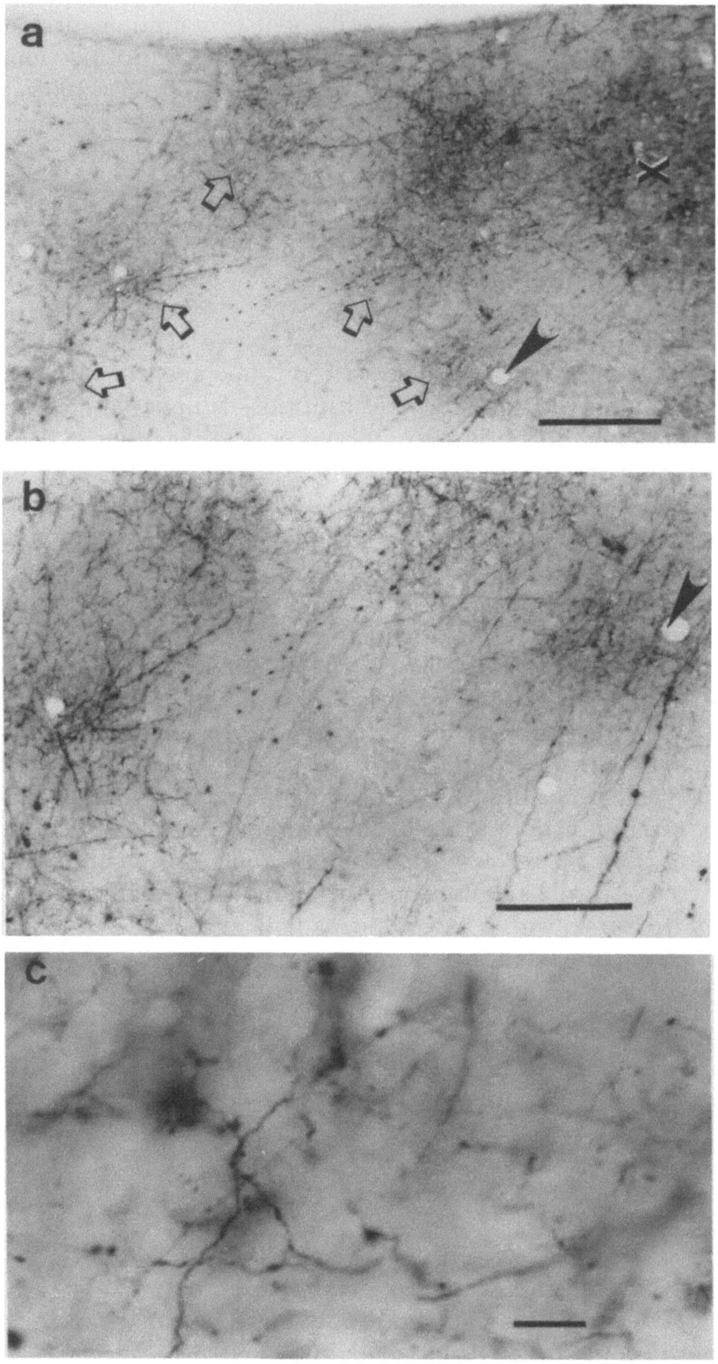

Figure 30. Photomicrographs of patchy intrinsic connections in area V2 anterogradely labeled by an injection of PHA-L (X). Hollow arrows (in panel a) point to five patches. Portion from arrowhead is shown at higher magnification in panel b. Note the marked falloff in terminations between the patches. (c) Higher magnification within patch region to show the relative density of terminations. Bars: (a) 300 μm; (b) 200 μm; (c) 20 μm.

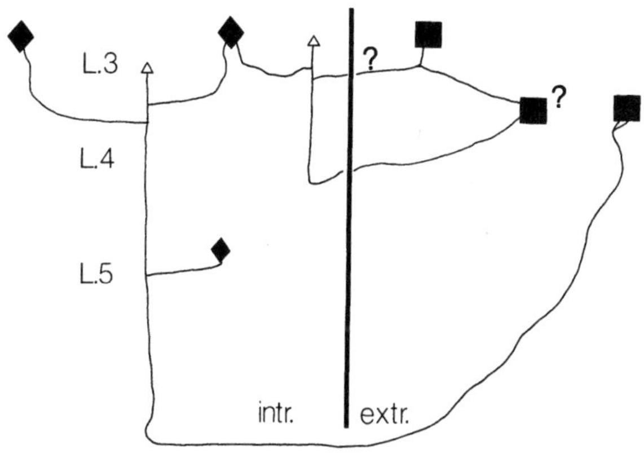

Figure 31. Pyramidal neurons typically have both intrinsic (diamonds) and extrinsic (squares) terminations. The relationship between these two components is unknown. It is also unknown whether neurons near a border region send "intrinsic" connections across the border.

the border between two areas. Do they continue across the border, as if this were invisible; and if so, what dictates the laminar disposition of the terminations? Intrinsic connections in general avoid layer 4; but extrinsic connections terminate mainly in layer 4. (Laminar shifts of a single axon over distance have been reported in the piriform cortex of the rat. Associational axons terminate, over short distances, onto basal dendrites of layer 3 pyramids, but over long distances, they shift to apical dendrites in layer 1B; Haberly and Presto, 1986.)

3. Hierarchy Revisited

The tendency for laminar complementarity and reciprocity of feedforward and feedback connections has been interpreted in a hierarchical framework or serial progression of areas. In the visual system, area V1 has commonly been considered as "low" in the hierarchy, and extrastriate areas are placed at different higher "levels." Areas connected by "lateral" connections (as defined by laminar patterns) are often assigned to the same level of the hierarchy (Felleman and Van Essen, 1991; Young, 1993). The concept of hierarchy is not without controversy (Bullier *et al,* 1994; Bullier and Nowak, 1995; Salin and Bullier, 1995; Hilgetag *et al.,* 1996; Van Essen and Felleman, 1966) and as currently used has several meanings. All versions share an emphasis on "levels." In a strict hierarchy, these are construed as relatively fixed and in a serial arrangement. In a "distributed" (Felleman and Van Essen, 1991) or "quasi"-hierarchy (Van Essen and Felleman, 1996), some degree of intermixing and irregularity is recognized. ("It is worth noting in general terms that information flow in a hierarchical system 1) can go in both directions . . . , 2) can skip over intermediate levels to go directly from a low to a high level, and 3) can travel in parallel through multiple, functionally distinct channels"; Felleman and Van Essen, 1991, p. 14). Strict hierarchical interpretations have been criticized as incompatible with several recent results (Bullier *et al.,* 1994; Salin and Bullier, 1995). The next sections of this chapter will reexamine some of the basic anatomical criteria for hierarchical organization, and, in conclusion, discuss possible modifications or alternatives.

3.1. Laminar Patterns

For several reasons, laminar patterns may be of limited usefulness as an index of relationship and hierarchical rank. First, as noted elsewhere (Boussaoud *et al.*, 1990; Felleman and Van Essen, 1991; Salin and Bullier, 1995) there are frequent variations and exceptions to any "rule." Connections from V1 to MT/V5 are not "strictly" feedforward in terms of laminar patterns. They originate from giant Meynert cells at the border of layers 5 and 6, as well as from neurons in layer 4B (Lund *et al.*, 1975; Tigges *et al.*, 1981; Shipp and Zeki, 1989a; Vogt Weisenhorn *et al.*, 1995); and single axons from V1 have been found to have collaterals in layer 6 in addition to their dense terminations in layer 4 (Rockland, 1989). Connections from V1 to V2 originate from some neurons in layer 5 as well as from the supragranular layers (Kennedy and Bullier, 1985).

Some variations can be systematically related to visual field topography. Connections from the central field representation of MT/V5 terminate in layers 4B and 6 of V1, but those from the peripheral representation terminate in layer 1 as well (Ungerleider and Desimone, 1986; Krubitzer and Kaas, 1989, 1990; Shipp and Zeki, 1989a). Curiously, a comparable asymmetry does not seem to occur in the pathway from MT/V5 to V2; Shipp and Zeki, 1986a; Stewart Shipp, personal communication).

Other variations in the laminar patterns conform to a "distance rule" (Kennedy and Bullier, 1985; Salin and Bullier, 1995). That is, connections between areas that are physically adjacent or comparatively "close" in the brain tend to be denser and to involve more layers. This observation applies (as a tendency, if not a rule) both to projection neurons and to their terminations. Feedback connections from area MT/V5 and parahippocampal areas to V1 originate from neurons in layer 6 only, but those from V2 to V1, or V4 or V2, also involve neurons in layer 3A (Rockland and Van Hoesen, 1994). Feedforward connections from V1 to MT/V5 are concentrated in layer 4, but those from V2 to MT/V5 extend into layer 1 (Rockland, 1995). Feedforward connections from V4 to TE target layer 4, but those from TEO to TE are more widely distributed (Saleem *et al.*, 1993). The significance of the distance rule is not known, but may be related to the composition of postsynaptic pools (see above, Section 2.3), or possibly to the temporal relationships among sets of connections. Corticocollicular terminations are another instance of a depthwise gradient in the target area (Kaas and Huerta, 1988; Harting *et al.*, 1992).

Second, the functional significance of laminar distributions depends on both the postsynaptic pool and on interactions with other connectional systems. To the extent that these components both differ from area to area, they will alter the effect of cortical connections. "Layer 4" in area MT/V5, for example, may be a distinctive connectional environment from layer 4 of V2 or other areas. For that matter, within a single area, the patchiness of many connectional systems (callosal, pulvinocortical, or intrinsic) will result in internal discontinuities across layer 4 within the same area (this is dramatically shown by the stripelike CO compartments in V2; see chapter by Roe and Ts'o in this volume).

Third, lamination is not a particularly successful criterion of hierarchical levels when applied more broadly to other connectional systems (Selemon and Goldman-Rakic, 1988). It does not make sense to rank the pulvinar "above" the amygdala, on the basis of their terminations, respectively, in layers 3/4 and 1 of extrastriate areas (Benevento and Rezak, 1976; Ogren and Hendrickson, 1977; Curcio and Harting, 1978; Iwai and Yukie, 1987; Levitt *et al.*, 1995).

3.2. Reciprocity

Cortical connections are commonly reciprocal between areas. As just discussed for laminar patterns, however, reciprocity is a complex phenomena.

First, even where there are dense connectional links between two areas, reciprocal connections may not be reciprocal at a neuron-to-neuron level. Feedforward terminations to layer 4 cannot directly contact feedback neurons in layer 3A and would only contact feedback neurons of layer 6 along the distal extent of their dendrites (Fig. 32). Thus, for any two interconnected areas, feedforward terminations probably contact feedback neurons mainly via multiple, short- and long-chain intra- and interlaminar routes (Fig. 33). Feedback connections may, to some extent, form a monosynaptic feedback–feedforward loop, that is, it seems likely that these target distal dendrites of projection neurons; but in this case also there are likely to be multiple circuits (see Section 2.3).

A second observation is that reciprocal connections seem not to be point-to-point, but rather are spatially asymmetrical. Terminal arbors of feedforward connections are relatively focused, whereas those of feedback connections are typically divergent. The divergent character of feedback connections was observed for projection foci (Krubitzer and Kaas, 1989; Shipp and Zeki, 1989a,b) and has now been confirmed at the single-axon level (Rockland and Virga, 1989; Rockland *et al.*, 1994; Rockland and Drash, 1996). Corticogeniculate axons are also reported to be widely divergent (Murphy and Sillito, 1996).

Third, there are many examples of connections that are one-way, frequently in the feedback direction; and these connections as a class may be more divergent than the feedforward. Area TEO receives input from area 36 and possibly area 35, but does not itself project to these perirhinal regions (Felleman and Van Essen, 1991; Webster *et al.*, 1994; Distler *et al.*, 1993). Areas V1 and V2 receive widespread connections from the ventromedial inferotemporal and parahippocampal regions, but do not themselves project directly to these regions (Doty, 1983; Kennedy and Bullier, 1985; Rockland and Van Hoesen, 1994; Rockland and Drash, 1996).

3.3. Collateralized Connections

In a strictly serial hierarchy, connections are stepwise and reciprocal. This is not immediately compatible with reports of collateralized connections, which seem to be a common, if secondary component of many connection systems. Nakamura *et al.* (1993) have reported secondary, "bypass" projections from V1 to V4, and from V2 to TEO. These investigators estimate that 6–20% of these are collaterals of the denser primary projections from V1 to V2, and from V2 to V4. Similarly, some feedback connections from areas beyond V2 project to several visual areas (Rockland and Van Hoesen, 1994; Rockland *et al.*, 1994; Rockland and Drash, 1996). Ten percent of the feedback projections from temporal

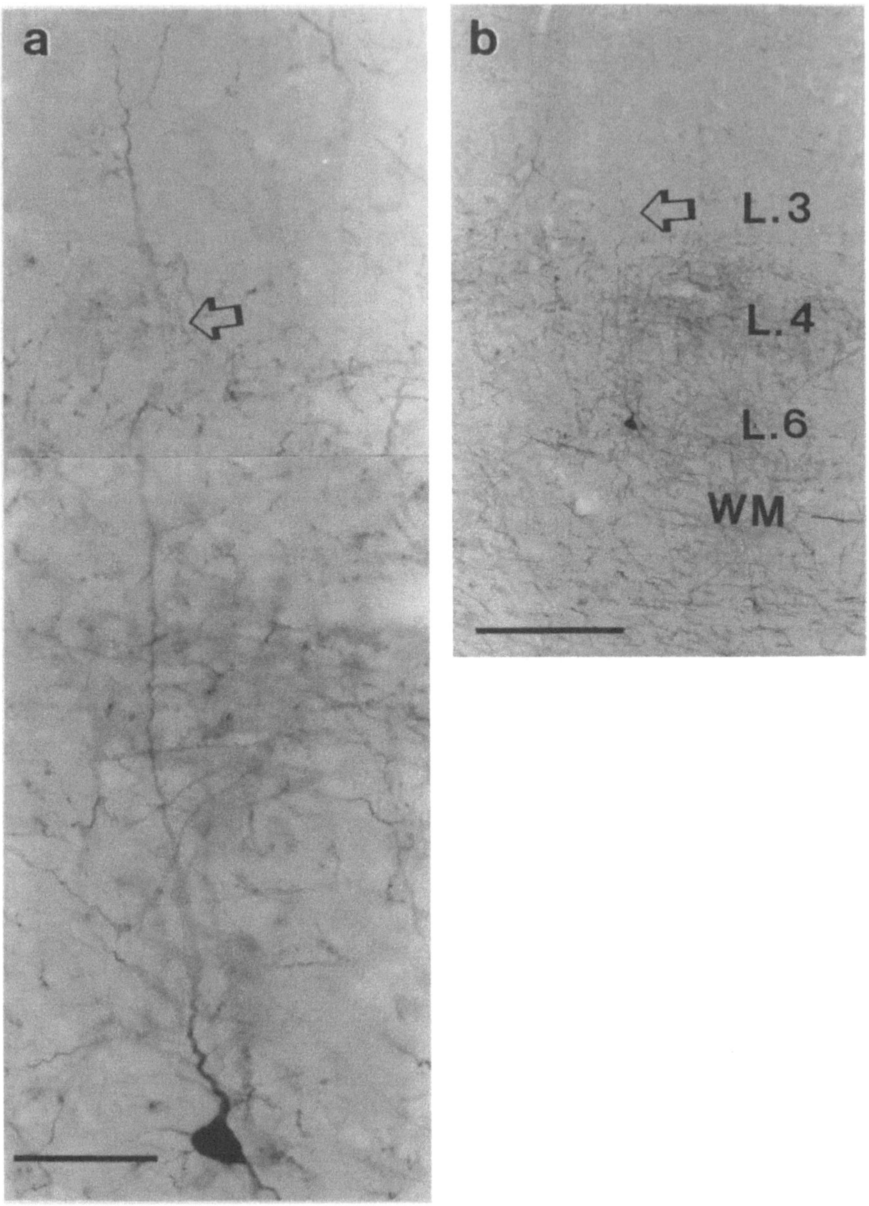

Figure 32. Photomicrographs of feedforward axons anterogradely labeled by an injection of BDA in area V1 (*Saimiri*). Terminations are in layer 4 of area V2. There is also a retrogradely filled neuron in layer 6. Its distal apical dendrite (hollow arrows in panels a and b) passes through the terminal field into layer 3. Potential contacts would occur in layer 4. Bars: (a) 50 μm; (b) 200 μm.

areas to V1 and V2 have been estimated as having collaterals to both areas (Rockland and Van Hoesen, 1994).

Collateralized connections have been reported from MT/V5 to V1 and V2 (Kennedy and Bullier, 1985), and from several subcortical centers—such as the amygdala, the lateral geniculate nucleus, and the inferior and lateral pulvinar—to areas V1 and V2 (Kennedy and Bullier, 1985). Some large Meynert cells of

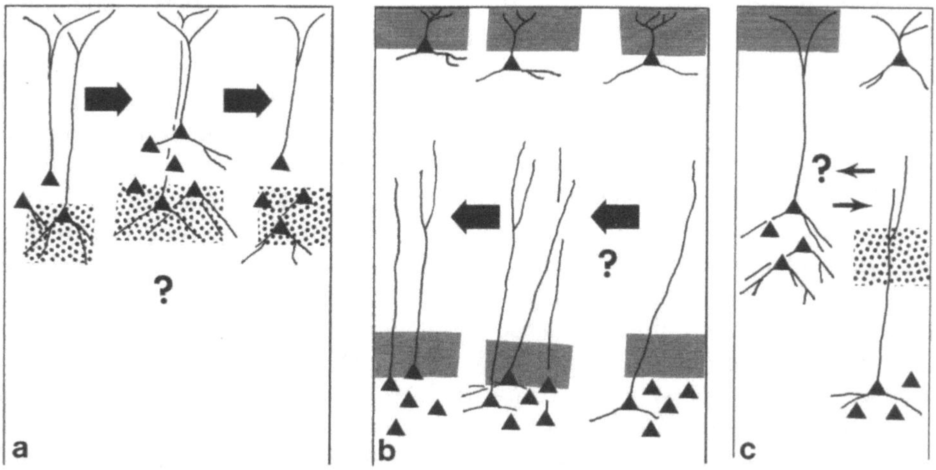

Figure 33. Schematic representation to suggest some possible interareal circuits. (a) Feedforward chain, (b) feedback chain, (c) feedforward/feedback loops. These would be short-synaptic pathways; multisynaptic pathways are not shown.

area V1 project via collaterals to MT/V5, DM, and the superior colliculus (Fries and Distel, 1983; Fries *et al.*, 1985; Vogt Weisenhorn *et al.*, 1995). In other regions outside the visual cortex, collateralized projections have been reported to the primary motor, supplementary motor, and prefrontal cortex of owl monkeys from single dopaminergic neurons (Gaspar *et al.*, 1992). In the somatosensory system, single neurons (filled by white matter injection of HRP) have collaterals directed to several architectonic fields (DeFelipe *et al.*, 1986). Thalamocortical connections from the ventroposterior thalamus project to several somatosensory areas via collaterals (Cusick *et al.*, 1985), and those from the medial geniculate nucleus (magnocellular division) project via collaterals to several auditory fields (Hashikawa *et al.*, 1995).

Collaterialization thus occurs in feedforward, feedback, and thalamocortical pathways, although the rules of collateral connectivity are not particularly clear. If, however, a single neuron in V1 projects to both V2 and V4, these areas for some purposes may function as an ensemble rather than as serial modules. There is a sense in which, because most projection neurons in layer 3 also have extensive intrinsic collaterals, *all* feedforward connections can be viewed as collateralized projections (Fig. 31). Might intrinsic collaterals in fact extend over area borders (see Bonhoeffer *et al.*, 1995, for physiological evidence)? From this perspective, V2 is an ensemble with V1 rather than a "higher level" in a ranked hierarchy (and similarly for any two interconnected areas).

3.4. Other Connections

Hierarchical frameworks have been based on patterns of corticocortical connections. Too often this translates into a "corticocentric" view which tends to deemphasize other connectional systems. Other major structures that have dense connections with visual cortical areas include the amygdala and the pulvinar complex. The amygdala projects to layer 1 throughout occipitotemporal cortex (Iwai and Yukie, 1987), and the pulvinar complex projects to layers 3 and 4, or to layer 1 in area V1 (Benevento and Rezak, 1976; Ogren and Hendrickson, 1977; Levitt *et al.,* 1995). These connections can be inferred to interact closely with cortical connections, feedback to layer 1, and feedforward to layers 3 and 4, and to participate with these in multiple inter- and intralaminar processing routes. The resulting scenario is of massively parallel interactive pathways.

The role of corticothalamocortical connections, although still unclear, has consistently been advocated as integral and crucial to cortical function and organization: "Thalamocortical and cortico-thalamo-cortical circuits are extensive in their distribution . . . and these transthalamic pathways appear to be systematically interlocked with lines of transcortical conduction. To try to appreciate the function of the cerebral cortex by studying it alone would, therefore, surely be a mistake; for much of what we ordinarily think of as processing by the neocortex is actually the result of reciprocal and nonreciprocal circuiting of information between thalamus and neocortex" (Graybiel and Berson, 1981). And "complex maps and analysis of such corticocortical pathways have been published, and have led to some interesting speculations about distinct pathways in the corticocortical system of connections. However, it should be clear that possibly a major system of corticocortical intercommunication has been overlooked . . ." (Guillery, 1995).

4. Summary: Cortical Architectures

Many recent studies of cortical organization and architecture have stressed some kind of hierarchical framework. Common components of these hierarchies are the construction of more or less distinct levels and the incorporation of a "low-to-high" progression, where "lower" is typically "closer" to peripheral sensory input. As surveyed above, however, the hierarchical concept has limitations. Perhaps foremost, although this has been only an incidental point in the present chapter, hierarchy models are essentially corticocentric, with little or no consideration of other connectional systems. *A priori,* nothing precludes an "amygdalocentric" or "thalamocentric" framework of cortical organization, although these would each look quite different. Analysis of corticostriatal projections has in fact suggested alternative ordering of cortical sectors, if not individual areas (Yeterian and Van Hoesen, 1978; Selemon and Goldman-Rakic, 1985; Ragsdale and Graybiel, 1990). Ultimately, models of cortical organization need to account for all these systems and their network interactions.

Even within the context of corticocortical connections, hierarchical models tend not to incorporate features such as convergence and divergence, connectional subtypes, or combinations of direct and indirect ("bypass") pathways. These features, as this chapter has tried to review, point to a rich multiplicity of interareal interactions. They are not immediately compatible with a strictly serial arrangement (as noted also in Felleman and Van Essen, 1991). A similar conclusion results from recent investigations of visual response latencies. These form broad temporal ranges, which are partially overlapping even for areas as distant as V1 and inferotemporal cortex (see chapters by Nowak and Bullier and by

Orban in this volume). Hierarchy frameworks also tend to neglect temporal interactions, concurrency and synchronization of neural activity, waveforms of global activation, and functional interdependency (for recent reviews, see Singer and Gray, 1993; Salin and Bullier, 1994; Bressler, 1996; Grinvald *et al.*, 1994). Nor do they take into account (although this could be incorporated) the possibility of dynamic shifts in coactivation patterns as a function of the internal state of the organism and/or the external stimuli.

Despite the limitations of the hierarchical framework, there are relatively few alternative models. The term "complex" has been used, for example, to group MT, MST, and FST, areas closely associated with motion. "Family," in a somewhat broader sense, has been suggested to designate a set of areas in a systematic relationship to particular subdivisions of the thalamus (Graybiel and Berson, 1981). There is of course "stream," as popularly used in differentiating parietal and temporal areas (Ungerleider and Mishkin, 1982). All these terms have a connotation of some kind of supra-areal organization. The phenomenon of cortical connections that are collateralized across areas also points to a supra-areal organization.

Evidence of a gradient-like organization along the occipitotemporal axis has been reported (reviewed in Fujita and Fujita, 1996): connectional gradients, with denser callosal (Van Essen *et al.*, 1982) and amygdalocortical connections (Turner *et al.*, 1980; Iwai and Yukie, 1987); and neurochemical gradients, with progressive increase (along the same axis) in the number of calbindin-immunoreactive pyramidal cells (Kondo *et al.*, 1994) and in phosphorylation levels of protein kinase C substrates (R. B. Nelson *et al.*, 1987). Preferential concentrations of nonadrenergic and serotonergic terminations have been localized, respectively, to parietal or inferotemporal sectors (Morrison and Foote, 1986).

In this summary, it is appropriate to point out that other cortical regions have not conformed particularly well to a strict hierarchical framework. In the somatosensory system, for example, the ventroposterior thalamus projects to areas 3, 1, and 2, so that the serial ordering of even these early thalamorecipient areas is not straightforward (Jones, 1986). In the motor system, the existence of multiple corticospinal pathways is now generally accepted, and in part on this basis a recent article concludes, "A serial, hierarchical model of cortical motor control is no longer tenable" (Galea and Darian-Smith, 1994). This same article suggests five factors that must be incorporated into any alternative model. With some modification (reading corticothalamic for corticospinal), these are also applicable to visual cortex, and are listed here in abbreviated form: (1) the "territorial" complexities of thalamic input to each corticospinal population, (2) the many cortical and spinal connections between the component corticospinal neuron populations, (3) the greater access through multiple corticospinal projections of the forebrain to spinal motoneuron populations, direct and indirect, (4) the potentially greater role of spinal circuitry in the control of movement, and (5) the inevitable distribution of function within a complex system, and the importance of temporal interactions across spatially separate populations.

Finally, there is the strong possibility that entirely new mapping approaches will be taken to the problem of area organization and interactions. For example, Friston *et al.* (1996) describe a variant of multidimensional scaling that maps anatomy into a functional space. Proximity relationships in this space represent functional connectivity among brain regions, defined as the correlation between remote neurophysiological events. McIntosh *et al.* (1994) apply network analysis to pathways mapped with positron emission tomography.

5. Concluding Remarks

This chapter has reviewed several facets relevant to continued discussions of cortical architectures. What emerges is the formidable number of uncertainties and unkonwns that characterize the current state of our understanding of cortical connectivity, particularly at the microcircuitry level. In many cases there are still no answers to even simple questions. We would like to know the relative numbers of neurons contributing to specific connections, the synaptic distribution of different connections in different areas, the degree of variability, and modes of connectional interactions. Not only are there no definite answers, but there is often such a lack of data that is difficult to formulate even an educated guess. Inescapably, as frequently repeated in these pages, more experimental data are urgently required.

Despite the critical lack of data, it is possible at this stage to frame some statements about what the cortex is *not.* For example, cortex is *not* homogeneous. Areas are identifiable and they are distinct, as distinct as subcortical nuclei. This is conspicuously true of heterotypical areas (as classically distinguished; see, e.g., von Bonin and Bailey, 1947), such as the primary sensory, primary motor, and entorhinal areas; but differentiating features can also be defined for more "homotypic" areas. Differentiation of areas, as noted in other chapters in this volume, has become less ambiguous now that connectional and other results can be combined with probes for chemical or pharmacological characteristics (e.g., Hof and Morrison, 1995). This does not necessarily mean, however, that areas are rigid entities, either in their functional interactions or perhaps even their anatomical borders. As noted above, it seems likely (although not proven) that pyramidal cell collaterals, as horizontal intrinsic connections, cross over area boundaries. Cortex is also *not* a flat sheet. Primate extrastriate cortex is about 2.0 mm in thickness. This, by way of comparison, is 10 times the thickness of the retina, and the increase in connectional complexity is bound to be considerably greater.

It might be useful, in thinking about a systems level of organization, to turn briefly to current research directed at the level of single neurons. A recent review of neuronal time constants, for example, emphasizes that neuronal properties are to be viewed in the context of a tightly interwoven network of other nerve cells, that a property as basic as the membrane time constant is not a fixed, but rather a dynamic variable which can be rapidly modulated by overall network activity, and that neural functions are likely to be linked to multiple mechanisms (Koch *et al.*, 1996, and references listed therein). The same qualities may apply to systems level performance. Cross correlation analyses suggest that neurons rapidly and dynamically associate into functional groups (e.g., Vaadia *et al.*, 1995). Similar conclusions have been derived from considerations of thalamocortical connections alone: "These facts force us to think about intracortical circuitry not in terms of a single elaborate pathway but rather in terms of multiple potential circuits . . . Some elements may, under one set of conditions, form part of one circuit and, under another, part of a second . . . If to these considerations we add temporal factors and increased synaptic weighting due to increased numbers of a particular synaptic type or to increased efficacy at single synapses . . . , we obtain some glimmer of the dynamism that must pervade cortical activity . . ." (Jones, 1984, p. 22).

The past few years have brought exciting technical advances, such as the development of multielectrode arrays and various imaging procedures, which are well suited to address interareal interactions and ensemble organization. It is safe to predict a continuation if not explosion of new findings and new ideas.

ACKNOWLEDGMENTS. I would like to thank Benjamin Romick for his contributions to quantitative data analysis (Section 2.4), Kristopher Kinsey for his assistance with photographic preparation, and Diane Topinka for her patient help with manuscript preparation. EM technical support was expertly provided by Hong Yi and the Central Microscopy Facility at the University of Iowa. Dr. H. Ojima kindly provided morphometric data on bouton density (Section 2.7). I am also grateful to Drs. Stephen Bressler, Sherre Florence, Philippe Grandguillaume, Jon Kaas, and Gary Van Hoesen for critical comments on the manuscript. Work described from the author's laboratory was supported by federal funding through IBN 9421970, MH53598, and NS 19632.

6. References

Amir, Y., Harel, M., and Malach, R., 1993, Cortical hierarchy reflected in the organization of intrinsic connections in macaque monkey visual cortex, *J. Comp. Neurol.* **334:**19–64.

Andersen, R. A., Asanuma, C., Essick, G., and Siegel, R. M., 1990, Corticocortical connections of anatomically and physiologically defined subdivisions within the inferior parietal lobule, *J. Comp. Neurol.* **296:**65–113.

Anderson, J. C., Martin, K. A. C., and Whitteridge, D., 1993, Form, function, and intracortical projections of neurons in the striate cortex of the monkey *Macacus nemestrinus, Cerebral Cortex* **3:**412–420.

Beaulieu, C., Kisvardy, Z., Somogyi, P., Cynader, M., and Cowey, A., 1992, Quantitative distribution of GABA-immunopositive and immunonegative neurons and synapses in the monkey striate cortex (area 17), *Cerebral Cortex* **2:**295–309.

Benevento, L. A., Rezak, M., 1976, The cortical projections of the inferior pulvinar and adjacent lateral pulvinar in the rhesus monkey (*Macaca mulatta*): An autoradiographic study, *Brain Res.* **108:**1–24.

Blasdel, G. G., and Fitzpatrick, D., 1984, Physiological organization of layer 4 in macaque striate cortex, *J. Neurosci.* **4:**880–895.

Blasdel, G. G., and Lund, J. S., 1983, Termination of afferent axons in macaque striate cortex, *J. Neurosci.* **3:**1389–1413.

Bonhoeffer, T., Kim, D.-S., Malonek, D., Shoham, D., and Grinvald, A., 1995, Optical imaging of the layout of functional domains in area 17 and across the area 17/18 border in cat visual cortex, *Eur. J. Neurosci.* **7:**1973–1988.

Boussaoud, D., Ungerleider, L. G., and Desimone, R., 1990, Pathways for motion analysis: Cortical connections of the medial superior temporal and fundus of the superior temporal visual areas in the macaque, *J. Comp. Neurol.* **296:**462–495.

Bressler, S. L., 1995, Large-scale cortical networks and cognition, *Brain Res. Rev.* **20:**288–304.

Bressler, S. L., 1996, Interareal synchronization in the visual cortex, *Behav. Brain Res.* **76:**37–49.

Bugbee, N. M., and Goldman-Rakic, P. S., 1983, Columnar organization of corticocortical projections in squirrel and rhesus monkeys: Similarity of column width in species differing in cortical volume, *J. Comp. Neurol.* **220:**355–364.

Bullier, J., and Nowak, L. G., 1995, Parallel versus serial processing: New vistas on the distributed organization of the visual system, *Curr. Opin. Neurobiol.* **5:**497–503.

Bullier, J., Girard, P., and Salin, P. A., 1994, The role of area 17 in the transfer of information to extrastriate visual cortex, in: *Cerebral Cortex,* Volume 10, *Primary Visual Cortex in Primates,* (A. Peters and K. S. Rockland, eds.), Plenum Press, New York, pp. 301–330.

Casagrande, V. A., and Kaas, J. H., 1994, The afferent, intrinsic, and efferent connections of primary cortex in primates, in: *Cerebral Cortex,* Volume 10, *Primary Visual Cortex in Primates* (A. Peters and K. S. Rockland, eds.), Plenum Press, New York, pp. 201–259.

Celebrini, S., Thorpe, S., Trotter, Y., and Imbert, M., 1993, Dynamics of orientation coding in area V1 of the awake primate, *Visual Neurosci.* **10:**811–826.

Colby, C. L., and Duhamel, J. R., 1991, Heterogeneity of extrastriate visual areas and multiple parietal areas in the macaque monkey, *Neuropsychologia* **29:**517–537.

Curcio, C. A., and Harting, J. K., 1978, Organization of pulvinar afferents to area 18 in the squirrel monkey: Evidence for stripes, *Brain Res.* **143:**155–161.

Cusick, C. G., Steindler, D. A., and Kaas, J. H., 1985, Corticocortical and collateral thalamocortical connections of postcentral somatosensory cortical areas in squirrel monkeys: A double-labeling study with radiolabeled wheatgerm agglutinin conjugated to horseradish peroxidase, *Somatosens. Res.* **3:**1–31.

Damasio, A., 1989, The brain binds entities and events by multiregional activation from convergence zones, *Neural Comput.* **1:**123–132.

Damasio, A., and Damasio, H., 1994, Cortical systems for retrieval of concrete knowledge: The convergence zone framework, in: *Large-Scale Neuronal Theories of the Brain* (C. Koch, and J. L. Davis, eds.), MIT Press, Cambridge, MA, pp. 61–74.

DeFelipe, J., and Farinas, I., 1992, The pyramidal neuron of the cerebral cortex: Morphological and chemical characteristics of the synaptic inputs, *Prog. Neurobiol.* **39:**563–607.

DeFelipe, J., Conley, M., and Jones, E. G., 1986, Long-range focal collateralization of axons arising from cortico-cortical cells in monkey sensory-motor cortex, *J. Neurosci.* **6:**3749–3766.

deLima, A. D., Voigt, T., and Morrison, J. H., 1990, Morphology of the cells within the inferior temporal gyrus that project to the prefrontal cortex in the macaque monkey, *J. Comp. Neurol.* **296:**159–172.

Distler, C., Boussaoud, D., Desimone, R., and Ungerleider, L. G., 1993, Cortical connections of inferior temporal area TEO in macaque monkeys, *J. Comp. Neurol.* **334:**125–150.

Doty, R. W., 1983, Nongeniculate afferents to striate cortex in macaques, *J. Comp. Neurol.* **218:**159–173.

Felleman, D. J., and Van Essen, D. C., 1991, Distributed hierarchical processing in the primate cerebral cortex, *Cerebral Cortex* **1:**1–47.

Fitzpatrick, D., Usrey, W. M., Shofield, B. R., and Einstein, G., 1994, The sublaminar organization of corticogeniculate neurons in layer 6 of macaque striate cortex, *Visual Neurosci.* **11:**307–316.

Florence, S. L., and Casagrande, V. A., 1987, The organization of individual afferent axons in layer IV of striate cortex of a primate (*Galago senegalensis*), *J. Neurosci.* **7:**3850–3868.

Florence, S. L., and Casagrande, V. A., 1990, The development of geniculocortical axon arbors in a primate, *Visual Neurosci.* **5:**291–311.

Freund, T. F., Martin, K. A. C., Soltesz, I., Somogyi, P., and Whitteridge, D., 1989, Arborisation pattern and postsynaptic targets of physiologically identified thalamocortical afferents in striate cortex of the macaque monkey, *J. Comp. Neurol.* **289:**315–336.

Fries, W., and Distel, H., 1983, Large layer VI neurons of monkey striate cortex (Meynert cells) project to the superior colliculus, *Proc. R. Soc. Lond. (Biol.)* **219:**53–59.

Fries, W., Keizer, K., and Kuypers, H. G. J. M., 1985, Large layer VI cells in macaque striate cortex (Meynert cells) project to both superior colliculus and prestriate area V5, *Exp. Brain Res.* **58:**613–616.

Friston, K. J., Frith, C. D., Fletcher, P., Liddle, P. F., and Frackowiak, R. S. J., 1996, Functional topography: Multidimensional scaling and functional connectivity in the brain, *Cerebral Cortex* **6:**156–164.

Fujita, I., and Fujita, T., 1996, Intrinsic connections in the macaque inferior temporal cortex, *J. Comp. Neurol.* **368:**467–486.

Galea, M. P., and Darian-Smith, I., 1994, Multiple corticospinal neuron populations in the macaque monkey are specified by their unique cortical origins, spinal terminations, and connections, *Cerebral Cortex* **4:**166–194.

Gaspar, P., Stepniewska, I., and Kaas, J. H., 1992, Topography and collateralization of the dopaminergic projections to motor and lateral prefrontal cortex in owl monkeys, *J. Comp. Neurol.* **325:**1–21.

Gerfen, C. R., and Sawchenko, P. E., 1984, An anterograde neuroanatomical tracing method that shows the detailed morphology of neurons, their axons and terminals: Immunohistochemical localization of an axonally transported plant lectin. *Phaseolus vulgaris* leucoagglutinin (PHA-L), *Brain Res.* **290:**219–238.

Gilbert, C. D., and Wiesel, T. N., 1983, Clustered intrinsic connections in cat visual cortex, *J. Neurosci.* **3:**1116–1133.

Goldman-Rakic, P., 1988, Topography of cognition: Parallel distributed networks in primate association cortex, *Annu. Rev. Neurosci.* **11:**137–156.

Graybiel, A. M., and Berson, D. M., 1981, On the relation between transthalamic and transcortical pathways in the visual system, in: *The Organization of the Cerebral Cortex* (F. O. Schmitt, F. G. Worden, and F. Dennis, eds.), MIT Press, Cambridge, MA, pp. 285–319.

Grinvald, A., Lieke, E. E., Frostig, R. D., and Hildesheim, R., 1994, Cortical point-spread function and long-range lateral interactions revealed by real-time optical imaging of macaque monkey primary visual cortex, *J. Neurosci.* **14:**2545–2568.

Guillery, R. W., 1995, Anatomical evidence concerning the role of the thalamus in corticocortical communication: A brief review, *J. Anat.* **185:**583–592.

Haberly, L. B., and Presto, S., 1986, Ultrastructural analysis of synaptic relationships of intracellularly stained pyramidal cell axons in piriform cortex, *J. Comp. Neurol.* **284:**464–474.

Harting, J. K., Updyke, B. V., and Lieshout, D. P. V., 1992, Corticotectal projections in the cat: Anterograde transport studies of twenty-five cortical areas, *J. Comp. Neurol.* **324:**379–414.

Hashikawa, T., Molinari, M., Rausell, E., and Jones, E. G., 1995, Patchy and laminar terminations of medial geniculate axons in monkey auditory cortex, *J. Comp. Neurol.* **362:**195–208.

Hilgetag, C.-C., O'Neill, M. A., and Young, M. P., 1996, Indeterminate organization of the visual system, *Science* **271:**776–777.

Hof, P. R., and Morrison, J. H., 1995, Neurofilament protein defines regional patterns of cortical organization in the macaque monkey visual system: A quantitative immunohistochemical analysis, *J. Comp. Neurol.* **352:**161–186.

Hof, P., Ungerleider, L., Webster, M., Gattass, R., Adams, M., Sailstad, C., and Morrison, J., 1996, Neurofilament protein is differentially distributed in subpopulations of corticocortical projection neurons in the macaque monkey visual pathways, *J. Comp. Neurol.* **376:**112–127.

Horikawa, K., and Armstrong, W. E., 1988, A versatile means of intracellular labeling: Injection of biocytin and its detection with avidin conjugates, *Neurosci. Meth.* **25:**1–11.

Houzel, J.-C., Milleret, C., and Innocenti, G., 1994, Morphology of callosal axons interconnecting areas 17 and 18 of the cat, *Eur. J. Neurosci.* **6:**898–917.

Innocenti, G. M., Lehmann, P., and Houzel, J.-C., 1994, Computational structure of visual callosal axons, *Eur. J. Neurosci.* **6:**918–935.

Ishai, A., and Sagi, D., 1995, Common mechanisms of visual imagery and perception, *Science* **268:**1722–1774.

Iwai, E., and Yukie, M., 1987, Amygdalofugal and amygdalopetal connections with modality-specific visual cortical areas in macaques (*Macaca fuscata, M. mulatta,* and *M. fascicularis*), *J. Comp. Neurol.* **261:**362–387.

Johnson, R. R., and Burkhalter, A., 1996, Microcircuitry of forward and feedback connections within rat visual cortex, *J. Comp. Neurol.* **368:**383–398.

Jones, E. G., 1984, Identification and classification of intrinsic circuit elements in the neocortex, in: *Dynamic Aspects of Neocortical Organization* (G. M. Edelman, W. M. Cowan, and W. E. Gall, eds.), Wiley, New York, pp. 7–40.

Jones, E. G., 1986, Connectivity of the primate sensory-motor cortex, in: *Cerebral Cortex,* Volume 5, *Sensory-Motor Areas and Aspects of Cortical Connectivity* (E. G. Jones and A. Peters, eds.), Plenum Press, New York, pp. 113–183.

Kaas, J. H., 1989, Why does the brain have so many visual areas? *J. Cognitive Neurosci.* **1:**121–135.

Kaas, J. H., and Huerta, M. F., 1988, Subcortical visual system of primates, in: *Comparative Primate Biology,* Volume 4 (H. P. Steklis, ed.), Liss, New York, pp. 327–391.

Kennedy, H., and Bullier, J., 1985, A double-labeling investigation of the afferent connectivity to cortical areas V1 and V2 of the macaque monkey, *J. Neurosci.* **5:**2815–2830.

King, M. A., Lewis, P. M., Hunter, B. E., and Walker, D. W., 1989, Biocytin: A versatile anterograde neuroanatomical tract-tracing alternative, *Brain Res.* **497:**361–367.

Kisvardy, Z. F., and Eysel, Y. T., 1992, Cellular organization of reciprocal patchy networks in layer III of cat visual cortex (area 17), *Neuroscience* **46:**275–286.

Koch, C., Rapp, M., and Segev, I., 1996, A brief history of time (constants), *Cerebral Cortex* **6:**93–101.

Kondo, H., Hashikawa, T., Tanaka, K., and Jones, E. G., 1994, Neurochemical gradient along the monkey occipito-temporal cortical pathway, *NeuroReport* **5:**613–616.

Kosslyn, S. M., Thompson, W. L., Kim, I. J., and Alpert, N. M., 1995, Topographic representations of mental images in primary visual cortex, *Nature* **378:**496–498.

Krubitzer, L. A., and Kaas, J. H., 1989, Cortical integration of parallel pathways in the visual system of primates, *Brain Res.* **478:**161–165.

Krubitzer, L. A., and Kaas, J. H., 1990, Cortical connections of MT in four species of primates: Areal, modular, and retinotopic patterns, *Visual Neurosci.* **5:**165–204.

Kuypers, H. G. J. M., Szwarcbart, M. K., Mishkin, M., and Rosvold, H. E., 1965, Occipitotemporal corticocortical connections in the rhesus monkey, *Exp. Neurol.* **11:**245–262.

Lachicha, E. A., Beck, P. D., and Casagrande, V. A., 1992, Parallel pathways in macaque monkey striate cortex: Anatomically defined columns in layer III, *Proc. Natl. Acad. Sci. USA* **89:**3566–3570.

Levitt, J. B., Yoshioka, T., and Lund, J. S., 1995, Connections between the pulvinar complex and cytochrome oxidase-defined compartments in visual area V2 of macaque monkey, *Exp. Brain Res.* **104:**419–430.

Lund, J. S., Lund, R. D., Hendrickson, A. E., and Fuchs, A. F., 1975, The origin of efferent pathways from the primary visual cortex, area 17, of the macaque monkey as shown by the retrograde transport of horseradish peroxidase, *J. Comp. Neurol.* **164:**287–304.

Lund, J. S., Hendrickson, A. E., Ogren, M. P., and Tobin, E. A., 1981, Anatomical organization of primate visual cortex area VII, *J. Comp. Neurol.* **202:**19–45.

Lund, J. S., Yoshioka, T., and Levitt, J. B., 1993, Comparison of intrinsic connectivity in different areas of macaque monkey cerebral cortex, *Cerebral Cortex* **3:**148–162.

Lund, J. S., Yoshioka, T., and Levitt, J. B., 1994, Substrates for interlaminar connections in area V1 of macaque monkey cerebral cortex, in: *Cerebral Cortex,* Volume 10, *Primary Visual Cortex in Primates* (A. Peters and K. S. Rockland, eds.), Plenum Press, New York, pp. 37–60.

Lund, J. S., Wu, Q., Hadingham, P. T., and Levitt, J. B., 1995, Cells and circuits contributing to functional properties in area V1 of macaque monkey cerebral cortex: Bases for neuroanatomically realistic models, *J. Anat.* **185:**563–581.

Manor, Y., Koch, C., and Segev, I., 1991, Effect of geometrical irregularities on propagation delay in axonal trees, *Biophys. J.* **60,** 1424–1437.

Martin, K. A. C., and Whitteridge, D., 1984, Form, function, and intracortical projections of spiny neurons in the striate visual cortex of the cat, *J. Physiol.* **353:**463–504.

Maunsell, J. H. R., and Gibson, J. R., 1992, Visual response latencies in striate cortex of the macaque monkey, *J. Neurophysiol.* **68:**1332–1344.

Maunsell, J. H. R., and Van Essen, D. C., 1983, The connections of the middle temporal visual area (MT) and their relationship to a cortical hierarchy in the macaque monkey, *J. Neurosci.* **3:**2563–2586.

McGuire, B. A., Hornung, J.-P., Gilbert, C. D., and Wiesel, T. N., 1984, Patterns of synaptic input to layer 4 of cat striate cortex, *J. Neurosci.* **4:**3021–3033.

McGuire, B. A., Gilbert, C. D., Rivlin, P. K., and Wiesel, T. N., 1991, Targets of horizontal connections in macaque primary visual cortex, *J. Comp. Neurol.* **305:**370–392.

McIntosh, A. R., Grady, C. L., Ungerleider, L. G., Haxby, J. V., Rapoport, S. L., and Horwitz, B., 1994, Network analysis of cortical visual pathways mapped with PET, *J. Neurosci.* **14:**655–666.

Miyashita, Y., 1995, How the brain creates imagery: Projection to primary visual cortex, *Science* **268:**1719–1720.

Morrison, J. H., and Foote, S. L., 1986, Noradrenergic and serotoninergic innervation of cortical, thalamic, and tectal visual structures in Old and New World monkeys, *J. Comp. Neurol.* **243:**117–138.

Murphy, P. C., and Sillito, A. M., 1996, Functional morphology of the feedback pathway from area 17 of the cat visual cortex to the lateral geniculate nucleus, *J. Neurosci.* **16:**1180–1192.

Nakamura, H., Gattass, R., Desimone, R., and Ungerleider, L., 1993, The modular organization of projections from areas V1 and V2 to areas V4 and TEO in macaques, *J. Neurosci.* **13:**3681–3691.
Nelson, J. I., Salin, P. A., Munk, M., Arzi, M., and Bullier, J., 1992, Spatial and temporal coherence in cortico-cortical connections: A cross-correlation study in areas 17 and 18 in the cat, *Visual Neurosci.* **9:**21–37.
Nelson, R. B., Friedman, D. P., O'Neill, J. B.,, Mishkin, M., and Routtenberg, A., 1987, Gradients of protein kinase C substrate phosphorylation in primate visual system peak in visual memory storage areas, *Brain Res.* **416:**387–392.
Nowak, L. G., Munk, M. H. J., Girard, P., and Bullier, J., 1995, Visual latencies in areas V1 and V2 of the macaque monkey, *Visual Neurosci.* **12:**371–384.
Ogren, M. P., and Hendrickson, A. E., 1977, The distribution of pulvinar terminals in areas 17 and 18 of the monkey, *Brain Res.* **137:**343–350.
Ojima, H., Honda, C. N., and Jones, E. G., 1991, Patterns of axon collateralization of identified supragranular pyramidal neurons in the cat auditory cortex, *Cerebral Cortex* **1:**80–94.
Ojima, H., Honda, C. N., and Jones, E. G., 1992, Characteristics of intracellularly injected infragranular pyramidal neurons in cat primary auditory cortex, *Cerebral Cortex* **2:**197–216.
O'Kusky, J., and Colonnier, M., 1982, A laminar analysis of the number of neurons, glia and synapses in the visual cortex (area 17) of adult macaque monkeys, *J. Comp. Neurol.* **210:**291–306.
Pandya, D. N., and Sanides, F., 1973, Architectonic parcellation of the temporal operculum in rhesus monkey and its projection pattern, *Z. Anat. Entwickl.-Gesch.* **13:**127–161.
Peters, A., 1987, Number of neurons and synapses in primary visual cortex, in: *Cerebral Cortex,* Volume 6, *Further Aspects of Cortical Function, Including Hippocampus* (E. G., Jones and A. Peters, eds.), Plenum Press, New York, pp. 267–294.
Pitchitpornchai, C., Rawson, J. A., and Rees, S., 1994, Morphology of parallel fibers in the cerebellar cortex of the rat: An experimental light and electron microscopic study with biocytin, *J. Comp. Neurol.* **342:**206–220.
Ragsdale, C. W., and Graybiel, A. M., 1990, A simple ordering of neocortical areas established by the compartmental organization of their striatal projections, *Proc. Natl. Acad. Sci. USA* **87:**6196–6199.
Raiguel, S. E., Lagae, L., Gulyas, B., and Orban, G. A., 1989, Response latencies in macaque areas V1, V2, and V5, *Brain Res.* **493:**155–159.
Rockland, K. S., 1989, Bistratified distribution of terminal arbors of individual axons projecting from area V1 to middle temporal area (MT) in the macaque monkey, *Visual Neurosci.* **3:**155–170.
Rockland, K. S., 1992, Configuration, in serial reconstruction, of individual axons projecting from area V2 to V4 in the macaque monkey, *Cerebral Cortex* **2:**353–374.
Rockland, K. S., 1994, The organization of feedback connections from area V2 (18) to area V1 (17), in *Cerebral Cortex,* Volume 10, *Primary Visual Cortex in Primates* (A. Peters and K. S. Rockland, eds.), Plenum Press, New York, pp. 261–299.
Rockland, K. S., 1995, Morphology of individual axons projecting from area V2 to MT in the macaque, *J. Comp. Neurol.* **355:**15–26.
Rockland, K. S., 1996, Two types of corticopulvinar terminations: round (type 2) and elongate (type 1), *J. Comp. Neurol.,* **368:**57–87.
Rockland, K. S., and Douglas, K. L., 1993, Excitatory contacts of feedback connections in layer 1 of area V1: An EM-biocytin study in the macaque, *Neurosci. Abstr.* **19:**424.
Rockland, K. S., and Drash, G. W., 1996, Collateralized divergent feedback connections that target multiple cortical areas, *J. Comp. Neurol.,* **373:**529–548.
Rockland, K. S., and Pandya, D. N., 1979, Laminar origins and terminations of cortical connections of the occipital lobe in the rhesus monkey, *Brain Res.* **179:**3–20.
Rockland, K. S., and Van Hoesen, G. W., 1994, Direct temporal-occipital feedback connections to striate cortex (V1) in the macaque monkey, *Cerebral Cortex* **4:**300–313.
Rockland, K. S., and Virga, A., 1989, Terminal arbors of individual "feedback" axons projecting from area V2 to V1 in the macaque monkey: A study using immunohistochemistry of anterogradely transported *Phaseolus vulgaris* leucoagglutinin, *J. Comp. Neurol.* **285:**54–72.
Rockland, K. S., and Virga, A., 1990, Organization of individual cortical axons projecting from area V1 (area 17) to V2 (area 18) in the macaque monkey, *Visual Neurosci.* **4:**11–28.
Rockland, K. S., Saleem, K. S., and Tanaka, K., 1994, Divergent feedback connections from areas V4 and TEO in the macaque, *Visual Neurosci.* **11:**579–600.
Saleem, K. S., and Tanaka, K., 1996, Divergent projections for the anterior inferotemporal area TE to the perirhinal and entorhinal cortices in the macaque monkey, *J. Neurosci.* **4:**4757–4775.

Saleem, K. S., Tanaka, K., and Rockland, K. S., 1993, Specific and columnar projections from area TEO to TE in the macaque inferotemporal cortex, *Cerebral Cortex* **3:**454–464.

Salin, P. A., and Bullier, J., 1995, Corticocortical connections in the visual system: Structure and function, *Physiol. Rev.* **75:**107–154.

Selemon, L. D., and Goldman-Rakic, P. S., 1985, Longitudinal topography and interdigitation of corticostriatal projections in the rhesus monkey, *J. Neurosci.* **5:**776–794.

Selemon, L. D., and Goldman-Rakic, P. S., 1988, Common cortical and subcortical targets of the dorsolateral prefrontal and posterior parietal cortices in the rhesus monkey: Evidence for a distributed neural network subserving spatially guided behavior, *J. Neurosci.* **8:**4049–4068.

Sherman, S. M., and Koch, C., 1986, The control of retinogeniculate transmission in the mammalian lateral geniculate necleus, *Exp. Brain Res.* **63:**1–20.

Shipp, S., and Zeki, S., 1989a, The organization of connections between areas V5 and V1 in macaque monkey visual cortex, *Eur. J. Neurosci.* **1:**309–332.

Shipp, S., and Zeki, S., 1989b, The organization of connections between areas V5 and V2 in macaque monkey visual cortex, *Eur. J. Neurosci.* **1:**333–354.

Singer, W., 1994, Putative functions of the temporal correlations in neocortical processing, in: *Large-Scale Neuronal Theories of the Brain* (C. Koch and J. L. Davis, eds.), MIT Press, Cambridge, MA, pp. 201–237.

Singer, W., 1995, Development and plasticity of cortical processing architectures, *Science* **270:**758–764.

Singer, W., and Gray, C. M., 1995, Visual feature integration and the temporal correlation hypothesis, *Annu. Rev. Neurosci.* **18:**555–586.

Spatz, W. B., 1977, Topographically organized reciprocal connections between areas 17 and MT (visual area of superior temporal sulcus) in marmoset (*Callithrix jacchus*), *Exp. Brain Res.* **27:**91–108.

Tigges, J., and Tigges, M., 1984, Subcortical sources of direct projections to visual cortex, in: *Cerebral Cortex,* Volume 3, *Visual Cortex* (A. Peters, and E. G. Jones, eds.), Plenum Press, New York, pp. 351–378.

Tigges, J., Spatz, W. B., and Tigges, M., 1974, Efferent corticocortical fiber connections of area 18 in the squirrel monkey (*Saimiri*), *J. Comp. Neurol.* **158:**219–236.

Tigges, J., Tigges, M., and Perachio, A. A., 1977, Complementary laminar terminations of afferents to area 17 originating in area 18 and in the lateral geniculate nucleus in squirrel monkey, *J. Comp. Neurol.* **176:**87–100.

Tigges, J., Tigges, M., Anschel, S., Cross, N. A., Letbetter, W. D., and McBride, R. L., 1981, Areal and laminar distribution of neurons interconnecting the central visual cortical areas 17, 18, 19, and MT in squirrel monkey (*Saimiri*), *J. Comp. Neurol.* **202:**539–560.

Tombol, T., 1984, Layer VI cells, in: *Cerebral Cortex,* Volume 4, *Association and Auditory Cortices* (A. Peters and E. G. Jones, eds.), Plenum Press, New York, pp. 479–519.

Tononi, G., Sporns, O., and Edelman, G. M., 1992, Reentry and the problem of integrating multiple cortical areas: Simulation of dynamic integration in the visual system, *Cerebral Cortex* **2:**310–335.

Turner, B. H., Mishkin, M., and Knapp, M., 1980, Organization of the amygdalopetal projections from modality-specific cortical association areas in the monkey, *J. Comp. Neurol.* **191:**515–543.

Ullman, S., 1995, Sequence seeking and counter streams: A computational model for bidirectional information flow in the visual cortex, *Cerebral Cortex* **5:**1–11.

Ungerleider, L. G., and Desimone, R., 1986, Cortical connections of visual area MT in the macaque, *J. Comp. Neurol.* **248:**190–222.

Ungerleider, L. G., and Mishkin, M., 1982, Two cortical visual systems, in: *Analysis of Visual Behavior* (D. J., Ingle, M. A., Goodale, and R. J. W. Mansfield, eds.), MIT Press, Cambridge, MA, pp. 549–586.

Vaadia, E., Haalman, I., Abeles, M., Bergman, H., Prut, Y., Slovin, H., and Aertsen, A., 1995, Dynamics of neuronal interactions in monkey cortex in relation to behavioral events, *Nature* **373:**515–518.

Valverde, F., 1978, The organization of area 18 in the monkey: A Golgi study, *Anat. Embryol.* (Berl.) **154:**305–334.

Van Essen, D. C., and DeYoe, E. A., 1995, Concurrent processing in the primate visual cortex, in: *The Cognitive Neurosciences* (M. S. Gazzaniga, ed.), MIT Press, Cambridge, MA, pp. 383–401.

Van Essen, D. C., and Felleman, D. J., 1996, On hierarchies, *Science* **271:**777.

Van Essen, D. C., Newsome, W. T., and Bixby, J. L., 1982, The pattern of interhemispheric connections and its relationship to extrastriate visual areas in the macaque monkey, *J. Neurosci.* **2:**265–283.

Veenman, C. L., Reiner, A., and Honig, M. G., 1992, Biotinylated dextran amine as an anterograde tracer for single- and double-labeling studies, *J. Neurosci. Meth.* **41:**239–254.

Vogt Weisenhorn, D. M., Illing, R. B., and Spatz, W. B., 1995, Morphology and connections of neurons in area 17 projecting to the extrastriate areas MT and 19 DM and to the superior colliculus in the monkey, *Callithrix jacchus, J. Comp. Neurol.* **362:**233–255.

Von Bonin, G., and Bailey, P., 1947, *The Neocortex of Macaca mulatta,* University of Illinois Press, Urbana, IL.

Webster, M. R., Bachevalier, J., and Ungerleider, L. G., 1994, Connections of inferior temporal areas TEO and TE with parietal and frontal cortex in macaque monkeys, *Cerebral Cortex* **4:**470–483.

Weller, R. E., and Kaas, J. H., 1983, Retinotopoic patterns of connections of area 17 with visual areas V-II and MT in macaque monkeys, *J. Comp. Neurol.* **220:**253–279.

White, E. L., 1989, *Cortical Circuits,* Birkhauser, Boston.

Wiser, A. K., and Callaway, E. M., 1996, Contributions of individual layer 6 pyramidal neurons to local circuitry in macaque primary visual cortex, *J. Neurosci.* **16:**2724–2739.

Wong-Riley, M. T. T., 1978, Reciprocal connections between striate and prestriate cortex in squirrel monkey as demonstrated by combined peroxidase histochemistry and autoradiography, *Brain Res.* **147:**159–164.

Yeterian, E. H., and Van Hoesen, G. W., 1978, Cortico-striate projections in the rhesus monkey: The organization of certain cortico-caudate connections, *Brain Res.* **139:**43–63.

Yoshioka, T., Levitt, J. B., and Lund, J. S., 1992, Intrinsic lattice connections of macaque monkey visual cortical area V4, *J. Neurosci.* **12:**2785–2802.

Young, M. P., 1993, The organization of neural systems in the primate cerebral cortex, *Proc. R. Soc. Lond. B* **256:**1327–1331.

Zeki, S., 1990, Parallelism and functional specialization in human visual cortex, *Cold Spring Harbor Symp. Quant. Biol.* **55:**651–661.

Zeki, S., and Shipp, S., 1988, The functional logic of corticocortical connections, *Nature* **335:**311–317.

Zeki, S., and Shipp, S., 1989, Modular connections between areas V2 and V4 of macaque monkey visual cortex, *Eur. J. Neurosci.* **1:**494–506.

7

The Functional Architecture of Area V2 in the Macaque Monkey

Physiology, Topography, and Connectivity

ANNA WANG ROE and DANIEL Y. TS'O

1. Introduction

One of the most striking aspects of the primate visual cortex is the remarkable extent of internal structure and organization within V1, and the second visual area, V2. Architecturally, V2 is composed of an interleaved series of bands specialized for the processing of the visual submodalities of form, color, and depth, known as the V2 stripes. The unique position of V2 in the visual hierarchy at the juncture of the higher visual pathways suggests that its segregated functional architecture facilitates the channeling of visual input into these pathways for handling "what" and "where" visual information. Thus the organization and connectivity of V2 perhaps most clearly embodies the notion of parallel and integrated processing of form, color, motion, and stereopsis, as observed psychophysically in studies of human visual performance. In comparison to primary visual cortex V1, the architecture and functional role of V2 has only recently been studied in earnest. This chapter reviews the current understand-

ANNA WANG ROE • Section in Neurobiology, Yale University School of Medicine, New Haven, Connecticut 06510. DANIEL Y. TS'O • Department of Neurobiology, Rockefeller University, New York, New York 10021.

Cerebral Cortex, Volume 12, edited by Rockland *et al.* Plenum Press, New York, 1997.

ing of the modular organization within stripes, the functional properties of V2 stripes, the visual map with respect to the stripes, and the functional connectivity of V2 stripes with other cortical areas.

Early physiological exploration of the second visual area in the primate by Cowey (1964) showed that focal electrical stimulation of striate cortex resulted in responses in lateral prestriate cortex, the organization of which was topographically mirrored across the anterior border of primary visual cortex (V1). Subsequent studies, both anatomical and physiological (Zeki, 1969; Allman and Kaas, 1974; Gattass *et al.*, 1981), established V2 as a separate, topographically organized area. V2 occupies a long, narrow strip of cortex surrounding and anterior to V1, and in the macaque monkey, is located on the lip and in the depths of the lunate sulcus (see inset, Fig. 8). V2 is quite a large cortical area, second only to V1, and occupies roughly 10% of all neocortex in the macaque. Historically this prestriate area has been referred to as area 18. However, since area 18 as defined by Brodmann (1909) has been shown to include parts of several other visual areas, the term V2 should be used when referring to the topographically defined second visual area in the macaque monkey. In some other species (such as the owl monkey; Allman and Kaas, 1974), the terms V2 and area 18 are indeed equivalent.

Cytochrome Oxidase

A significant milestone in the study of V2 came with the development of cytochrome oxidase staining (Wong-Riley, 1979a; Horton and Hubel, 1981) as well as 2-deoxyglucose methods (Tootell *et al.*, 1983; Humphrey and Hendrickson, 1983). These metabolic labeling methods revealed exquisite and distinctive structures within V1 and V2. As seen in Fig. 1, cytochrome oxidase histology reveals a lattice of darkly stained patches in V1, termed blobs, separated by

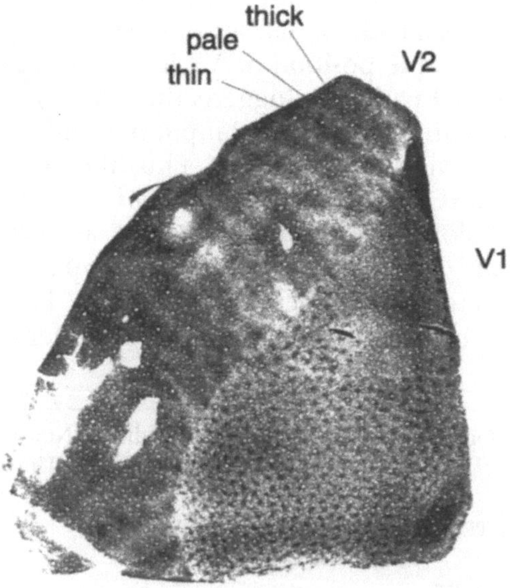

Figure 1. Cytochrome oxidase staining of V1 and V2. This tangential section of visual cortex in squirrel monkey reveals the lattice of darkly staining cytochrome oxidase blobs separated by lightly staining interblobs in V1 and the alternating cycle of thin/pale/thick/pale stripes in V2. The boundary between these two staining patterns is the V1/V2 border.

regions of lighter stain, termed interblobs (Livingston and Hubel, 1984). In contrast, V2 is characterized by a series of stripes extending orthogonally from the V1/V2 border. These stripes, which are alternatively dark *thin* and dark *thick* stripes separated by lightly staining *pale* stripes, cycle across V2 in thin/pale/thick/pale sequence. Relative to the cytochrome oxidase blobs in V1, V2 stripes are large (roughly 1–1.5 mm in width) and few in number (only 10–12 stripe cycles in dorsal V2). An accumulation of anatomical and physiological studies has lent support to the notion that the thin, pale, and thick stripes comprise functionally distinct compartments and are specialized for color, form, and disparity processing, respectively (DeYoe and Van Essen, 1985; Shipp and Zeki, 1985; Hubel and Livingstone, 1987; Ts'o *et al.*, 1990b).

Thin (color) stripes are not always thin; thick (disparity) stripes are not always thick. A word should be said about the equivalence of the stripes seen with cytochrome oxidase stain and the functionally defined stripes in V2. The nomenclature 'thin,' 'pale,' and 'thick' derives from descriptions of stripe patterns revealed by cytochrome oxidase staining. However, there remains a significant degree of variability in both the widths and the regularity of their pattern across V2, particularly in the macaque (Hubel and Livingstone, 1987; DeYoe *et al.*, 1990; Ts'o *et al.*, 1990b; Roe and Ts'o, 1995). There is substantial individual variation in the pattern of stripes and deviations from the strict cycle of thin/pale/thick/pale stripe structure. Furthermore, physiologically defined 'thin' stripes are not consistently narrower than 'thick' stripes as assessed by length of tangentially recorded sequences of receptive field types across stripes in V2 (Hubel and Livingstone, 1987; Zeki and Shipp, 1987; Roe and Ts'o, 1995). This observation calls into question the correlation between stripe width as seen in cytochrome oxidase staining and stripe width determined functionally. That is, the width of the physiologically defined color stripe often exceeds the width of the cytochrome oxidase thin stripe, and the width of the physiologically defined pale stripe is often narrower than the apparent width of the cytochrome oxidase pale stripe. Thus, it may be misleading to assume that the thick, pale, and thin stripes as defined histologically bear an absolute correspondence to the distribution of functional properties.

2. Receptive Field Properties with Respect to V2 Stripes

The distinctive striped functional organization in V2 is complemented by the emergence of new functional properties. The earliest studies describing the functional specialization of each of the three cytochrome oxidase stripe types, thick, thin, and pale, did not characterize the receptive field properties of cells found in each stripe with much detail. The descriptions were largely limited to noting the extent of orientation tuning, color selectivity, and binocular interaction (e.g., DeYoe and Van Essen, 1985). However, studies prior to the knowledge of the cytochrome oxidase stripes had already reported several receptive field types in V2 that indicated a further elaboration of receptive field properties beyond those observed in V1 (Hubel and Wiesel, 1970; Baizer *et al.*, 1977; Poggio and Fischer, 1977).

One of the first prominent differences noted between V1 and V2 is the lack of ocular segregation in V2, where most cells are binocularly driven (e.g., Hubel and Livingstone, 1987). In fact, for a population of the cells in V2 ocular information is combined in highly specific ways. As first reported by Hubel and Wiesel (1970), in their V2 recordings there was a high proportion of cells exquisitely sensitive to retinal disparity, which they called "depth cells" (Fig. 2). These retinal disparity cells (now also called tuned-excitatory obligatory binocular) only responded to simultaneous stimulation of both eyes with a very narrow range of disparities (e.g., 0.125–0.25 deg). Subsequent studies have localized these and other disparity cells primarily to the V2 thick stripes (DeYoe and Van Essen, 1985; Hubel and Livingstone, 1987; Ts'o *et al.*, 1990b; Roe and Ts'o, 1995). One of the important transformations in the representation of disparity information from V1 to V2 is the establishment of a definitive functional organization for disparity in V2. Not only are the disparity cells concentrated within the thick stripes, but within the thick stripes there is a clustering of disparity cell types, with tuned excitatory, tuned inhibitory, and near and far (crossed and uncrossed) disparity cells (Poggio and Fischer, 1977; LeVay and Voigt, 1988) each segregated into their own set of patches within a single thick stripe (Ts'o *et al.*, 1990b). Although the thick stripes have also been cited as the compartment for motion processing in V2, most studies have not reported a particularly high concentration of directionally selective cells anywhere within V2 (Livingstone and Hubel, 1987; Ts'o *et al.*, 1990b; Peterhans and von der Heydt, 1993; Levitt *et al.*, 1994a). Thus the mainstream motion pathway may bypass V2 and enter V5 (MT) directly from V1.

In the color domain, the receptive field types that have been reported include several that have also been described in V1: type I (single color opponent center surround), type II (center-only color opponent), modified type II (center color opponent with broadband suppressive surround), and color oriented (Ts'o and Gilbert, 1988). One marked distinction of these V2 color cells in comparison

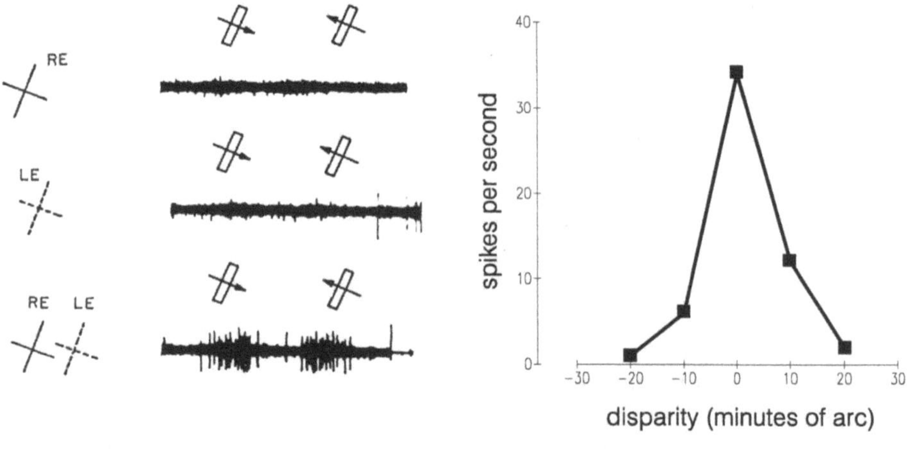

Figure 2. Disparity cell. An obligatory binocular cell recorded in a thick stripe in V2. It is not responsive to monocular stimulation, neither left eye nor right eye alone, but responds robustly to binocular stimulation (left bottom) within a very narrow range of retinal disparities (right). After Hubel and Wiesel (1970).

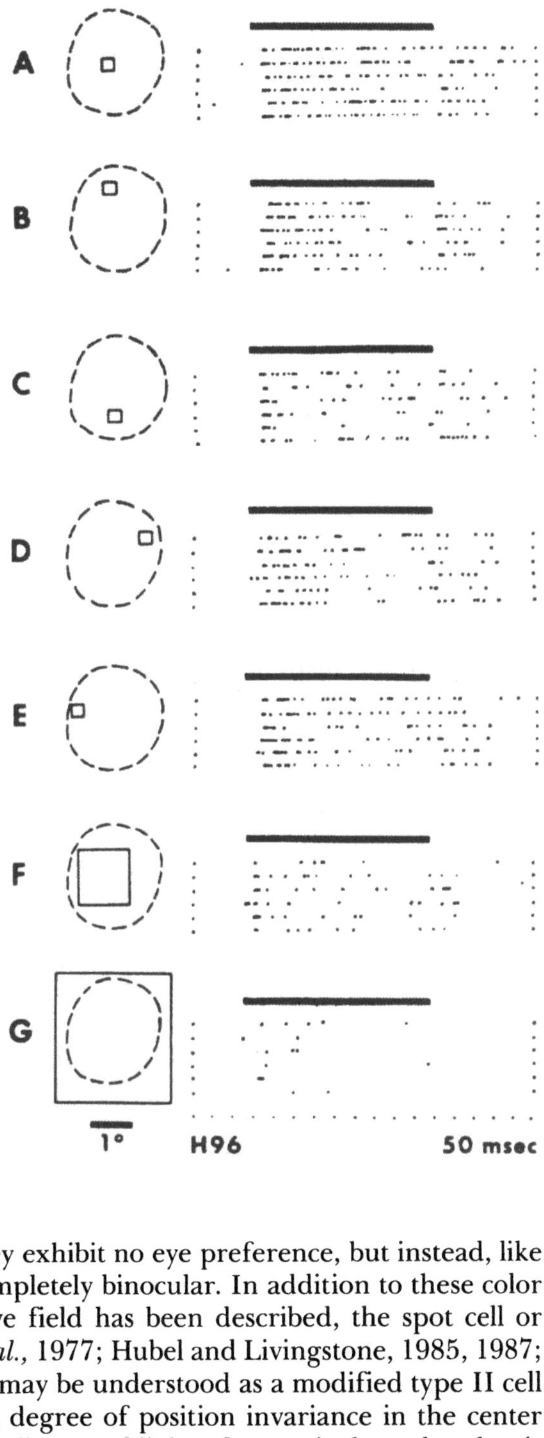

Figure 3. Spot cell. Responses of a spot cell to stimuli at different positions and of different sizes. Responses (raster plots on right) are robust to small (much smaller than receptive field size) spots of light within the receptive field (A–E). Larger spots tend to reduce the response (F) and illumination of the entire receptive field virtually silences the cell (G). From Baizer *et al.* (1977).

to their V1 counterparts is that they exhibit no eye preference, but instead, like most V2 nondisparity cells, are completely binocular. In addition to these color cells, a new type of color receptive field has been described, the spot cell or complex unoriented cell (Bazier *et al.,* 1977; Hubel and Livingstone, 1985, 1987; Roe and Ts'o, 1995). The spot cell may be understood as a modified type II cell that has "complexified" and has a degree of position invariance in the center response. It responds well to a small spot of light of a particular color that is positioned somewhere within its receptive field, which may be 10 times or more larger than the optimal spot size (Fig. 3). Another transformation in color responses that is apparent when comparing V2 with V1 color cells is that the range of color preferences is much greater in V2, suggesting that there is considerable mixing of the two color opponent systems, red/green and blue/yellow, in or prior to V2 (Ts'o *et al.*, 1990b).

In V2 there is a greater proportion of color oriented cells (Ts'o and Gilbert, 1988; Ts'o *et al.*, 1990b) in comparison to V1, and they are often found at the borders between the thin and pale stripes. Among the population of color oriented cells in V2, a particularly interesting type, the "color border" cell, has been described (Ts'o *et al.*, 1990b). This new type of color oriented cell is somewhat analogous to the double color opponent cell in that its optimal stimulus is an oriented color edge with one color (e.g., red) on one side and a different color (e.g., green) on the other. It is then an appropriate cell type for detecting borders or contours defined by color contrast. Another novel class of color oriented cell found in V2 is the color disparity cell (Ts'o *et al.*, 1991, 1995). These cells are found at the borders between color and disparity regions, and exhibit both color and disparity tuning. The mixing of the submodalities of color and orientation and of color and disparity is also indicated in the pattern of anatomical connections (Ts'o *et al.*, 1991; Levitt *et al.*, 1994b; Malach *et al.*, 1994), and represents interactions between the otherwise segregated processing of form, color, and disparity information. Such cross-modal interactions may aid in the maintenance of the registration of the separate maps and computations of the form, color, and disparity systems and thus contribute to a unified visual percept.

In the form domain, it has been reported that a substantial portion of V2 oriented cells, particularly in the pale stripes, exhibit end-inhibition. A further elaboration of this property has been described by von der Heydt and Peterhans (1989; Peterhans and von der Heydt, 1989) in which a class of oriented cells, in addition to responding to true contours, also respond to illusory contours (also termed subjective contours or anomalous contours, see Fig. 4). These investigators suggested a model for illusory contour cells built from input provided by end-inhibited cells. Motion coherence cells have also been described which respond to a coherently moving group of dots. While their responses to single dots are unoriented, a colinear array of dots yields an oriented response which matches the orientation selectivity of the cell to a luminance contrast border. These illusory contour cells have been found primarily in the thick and pale stripes (Peterhans and von der Heydt, 1993).

A controversy that began with several physiological studies exploring the degree of functional segregation between the blobs and interblobs of V1 has raised similar questions in V2. In V1 it has been reported that color and orientation information is largely segregated, with the blobs primarily containing unoriented monocular color cells and the interblobs primarily oriented noncolor-selective cells (Livingstone and Hubel, 1984; Tootell *et al.*, 1988b; Ts'o and Gilbert, 1988). However, some investigators, using somewhat different methods, report little or nor segregation or specialization for color in the blobs (Lennie *et al.*, 1990; Leventhal *et al.*, 1995). A similar, but lesser discrepancy exists for the issue of whether the V2 thin stripes are specialized for color processing, and whether there is substantial functional segregation in the stripes overall (Levitt *et al.*, 1994a). The approaches the various studies have taken in examining the

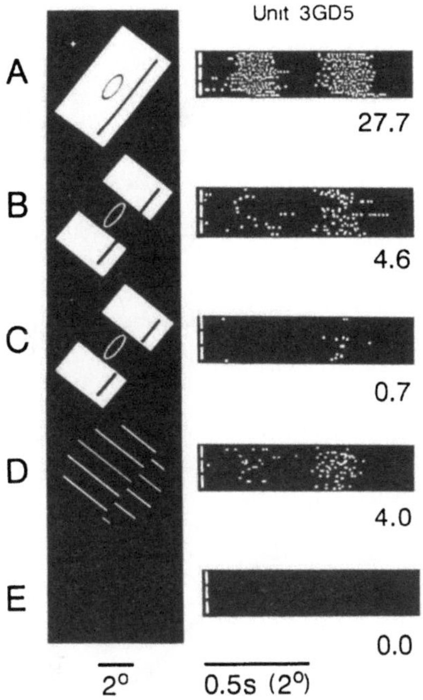

Figure 4. Illusory contour cell recorded in V2 of an alert monkey. This cell (response field indicated by ellipse) responds robustly (raster plots on right) to an oriented dark bar moving on a light background (A). Its responses to similarly oriented illusory contours are also quite strong (B, D). However, with the closing lines the illusory-bar response is virtually abolished (C). From Peterhans and von der Heydt (1989).

color properties of blob and interblob cells differ greatly, particularly in the means for generating color stimuli, but also in localizing the recording site. It seems clear that the extent of functional segregation in V2 is much more apparent when considering the distribution of some functional properties, such as color and disparity, as opposed to others, such as spatial and temporal frequency tuning. Results from functional imaging and 2-deoxyglucose studies do indicate that the blobs of V1 and the thin stripes of V2 are selectively activated with color stimuli and in general confirm the notion of functional segregation within the stripes of V2 (Tootell and Hamilton, 1989; Ts'o *et al.*, 1990b, Malach *et al.*, 1994). Nevertheless more detailed corroboration between functional anatomical methods and electrophysiology will be necessary to fully resolve these issues.

In summary, although our understanding of the range of receptive field properties of V2 cells must be regarded as preliminary, it is clear that in each of the color, form, and disparity domains significant functional transformations are performed. While V1 is characterized by cells concerned with ocularity, orientation of contrast borders, and unoriented color properties, one observes in V2 the emergence of a distinctive organization for binocular disparity, and of cells signalling perceptual contours, complex color properties, and interactions between color/form and color/disparity information. These properties are summarized in Fig. 5. It is apparent that one common theme in V2 and beyond (e.g., V4, MT, MST) is the convergence of inputs to generate higher order receptive field properties (such as spot cells or illusory contour cells) that integrate information across increasing spatial extents (Hubel and Livingstone, 1985; Peterhans and von der Heydt, 1993). However, much of the relationship between specific cells types and higher order visual perceptions needs to be further investigated (Merigan *et al.*, 1993).

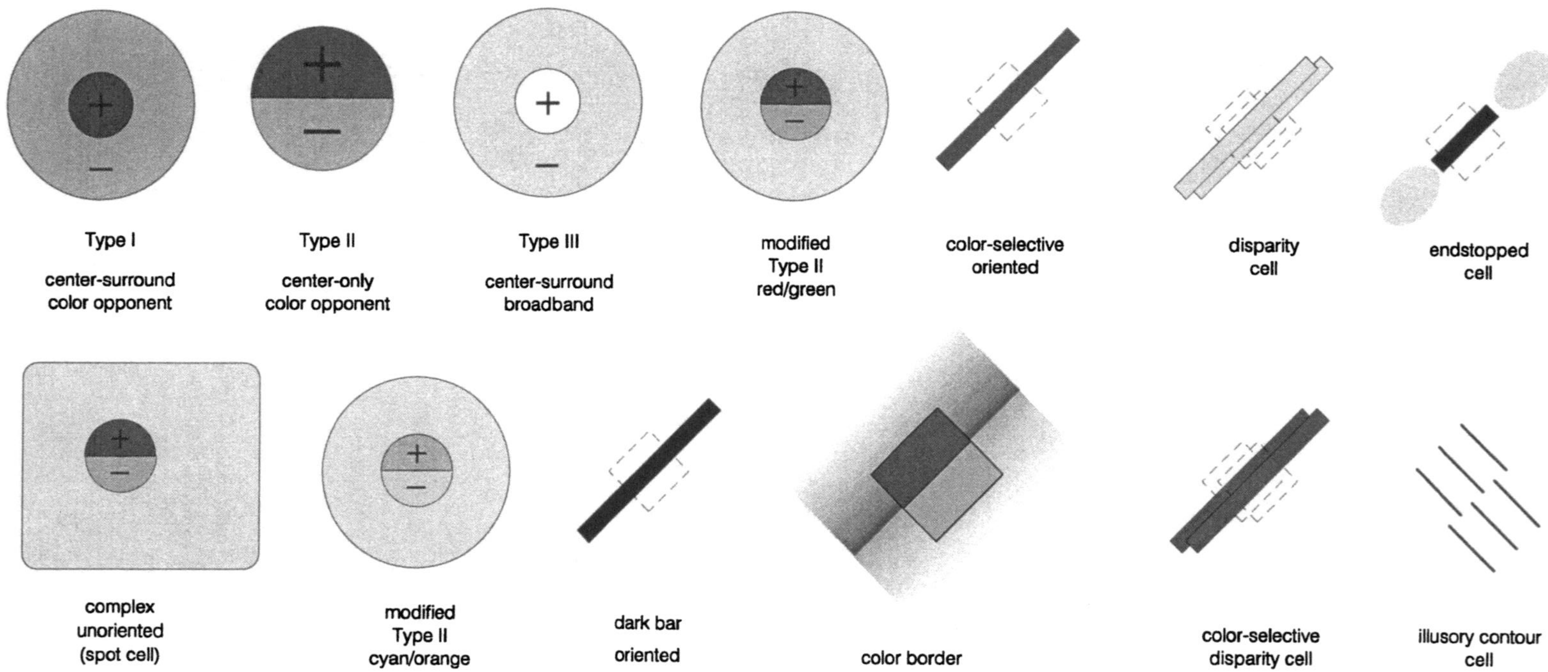

Figure 5. Summary of schematics of receptive field types found in V2. The top row depicts cell types that are also found in V1, while the bottom row shows cell types that are either particularly notable in V2 or described only in V2. Certain other cell types, such as directional cells, have been omitted for the sake of brevity.

A further examination of the architecture of the V2 stripes has revealed an additional level of substructure and modularity within single stripes. Histologically, cytochrome oxidase stains of the V2 stripes often have an uneven, mottly appearance. The earliest evidence of subcompartments within V2 stripes came from anatomical studies examining the pattern of corticocortical connections between V1 and V2 (Rockland and Pandya, 1979; Wong-Riley, 1979b; Weller and Kaas, 1983). In these experiments, injections of anterograde tracers into V1 resulted in clusters of terminal label in V2 measuring roughly 400 μm in diameter (Wong-Riley, 1979b). These clustered connections from V1 are likely to arise from single axons terminating in V2 in clusters 200 μm in size and separated by 200–500 μm in layers 4 and lower 3 (Rockland and Virga, 1990). Patchy patterns of connectivity have also been observed with other areas, such as V4 (DeYoe *et al.*, 1994), MT (Shipp and Zeki, 1989; Krubitzer and Kaas, 1989; DeYoe *et al.*, 1990), and the pulvinar (Ogren and Hendrickson, 1977; Curcio and

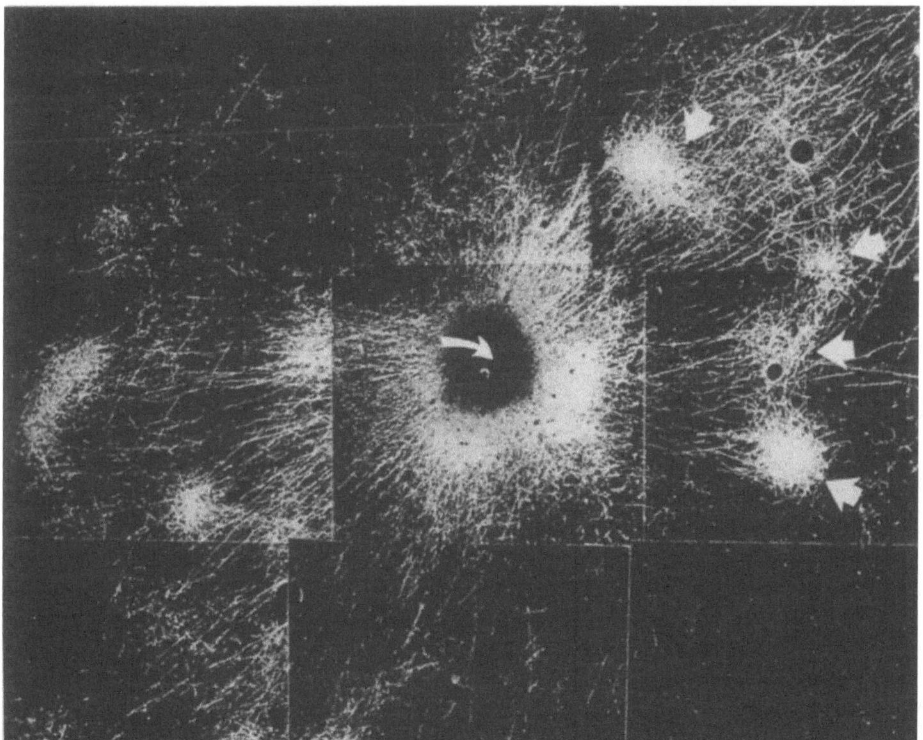

Figure 6. Intrinsic horizontal connections in area V2. This is a dark-field, low-power montage of a tangential section from V2 that has been flattened. Transported label from a biocytin injection (curved arrow) reveals a network of radiating fibers and clustering into patches (large arrows) farther away. Scale bar = 1 mm. From Malach *et al.* (1994).

Harting, 1978; Horton, 1984; Levitt *et al.*, 1995), as well as from intrinsic connections in V2 (Rockland, 1985; Amir *et al.*, 1993; Levitt *et al.*, 1994b; Malach *et al.*, 1994) (Fig. 6). These findings strongly suggest anatomical and therefore physiological subcompartments within single V2 stripes.

2-Deoxyglucose (Tootell and Hamilton, 1989), optical imaging (Malach *et al.*, 1994; Ts'o *et al.*, 1990b, 1991, 1995), and electrophysiological (Ts'o *et al.*, 1990b, 1991; Roe and Ts'o, 1995) methods have gone on to demonstrate a functional modularity within single V2 stripes that is likely to correspond to the stripe subcompartments revealed anatomically. Optical imaging reveals functionally uniform domains, about 0.4–0.5 mm in diameter, in each of the three stripe types. These domains contain clusters of cells with similar receptive field properties. For example, in thin stripes, there are subcompartments dedicated to the representation of color versus luminance response properties (Fig. 7). Disparity stripes have been shown to be characterized by clusters of near, far, tuned-excitatory, or tuned-inhibitory cells (Ts'o *et al.*, 1991). Orientation subcompartments in V2, prominent in pale stripes and disparity stripes, contain cells sharing

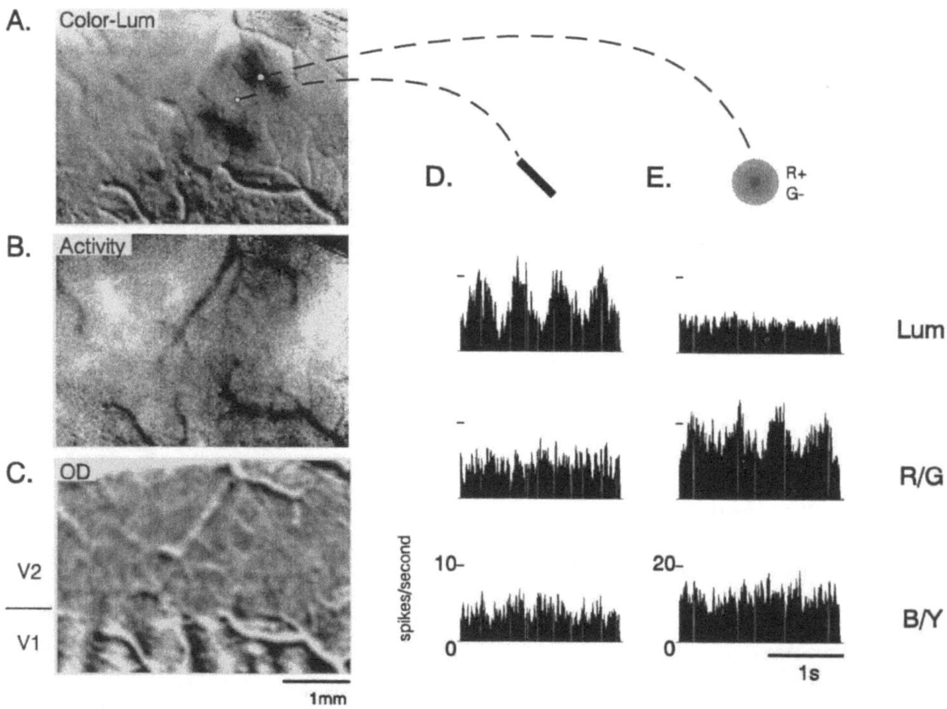

Figure 7. Optical imaging and single-unit electrophysiology in subcompartments of a single V2 color stripe. On the left, three images taken of the same region of V2, under three different imaging paradigms, designed to reveal the ocular dominance columns of V1 (C), the stripes of V2 (B), and regions that prefer color over luminance stimuli (A). When imaging for color versus luminance (A), the single V2 thin stripe can be seen to be actually composed of several distinct modules or subcompartments. On the right, single-unit recording confirms that light subcompartments prefer low-spatial-frequency (1 cyc/deg) luminance gratings (D). Dark subcompartments prefer color isoluminant red/green (R/G) gratings (E); this particular cell was qualitatively described as red-ON, green-OFF type II and did not respond to modulations in the blue/yellow (B/Y) direction in color space. From Ts'o *et al.* (1997).

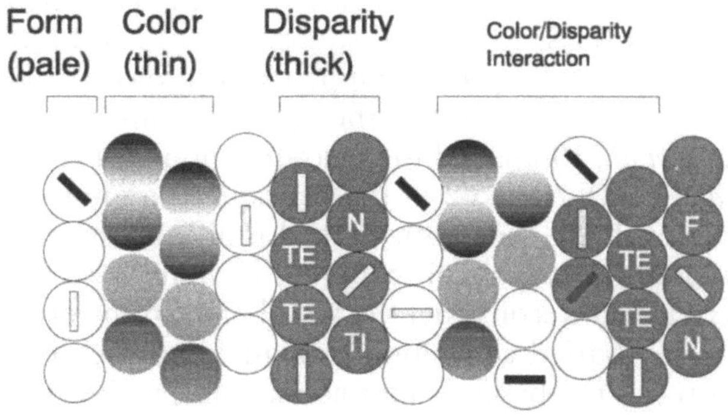

Figure 8. Schematic model of modular organization within V2. Within thin stripes, different color subcompartments process information along different axes in color space. Pale stripes contain orientation-selective subcompartments, some of which are also color-selective. Within disparity stripes are subcompartments containing either near cells, far cells, tuned excitatory cells, or tuned inhibitory cells. At mergings between thin and thick stripes, clusters of color disparity cells are observed.

a single preferred orientation and are adjacent to other orientation subcompartments of quite different orientation selectivity, unlike the smoothly changing orientation representation within V1 (Ts'o *et al.*, 1990b). Such anatomical, functional, and topographic (see Section 4) evidence suggests that these modules may constitute the fundamental units of representation in V2 (Ts'o *et al.*, 1997). Thus, each stripe within V2 may be seen as a collection of ½-mm-sized modules, each of which is dedicated to processing a specific submodality within either the color, form, or disparity domain (Fig. 8).

In describing the V2 stripe subcompartments, roughly conserved in size regardless of whether they subserve color, orientation, or disparity processing, it is interesting to note the parallel to the blob/interblob compartments of V1. The blobs of V1, though they apparently subserve color and brightness processing, are roughly the same size as a single orientation column in the interblobs of V1. These observations lead to the idea that in each cortical area, there is a characteristic functional module size, roughly independent of the particular functional specificity. As the blobs and orientation columns of V1 occupy about one-quarter the cortical surface area of a V2 stripe subcompartment, we note that the characteristic module size increases from V1 to V2. Anatomical and physiological studies of V4 (Amir *et al.*, 1993; DeYoe *et al.*, 1994; Ghose *et al.*, 1997; Ts'o *et al.*, 1997) suggest that similar, but even larger, stereotyped functional modules also exist in V4. Thus, although when one considers the V2 stripes as a whole, it appears that the functional organization of V2 is quite different from that of V1, greater similarity may be found by comparing the blob/orientation modules of V1 to the subcompartments of the V2 stripes.

4. Topography

The traditional physiological method of sensory cortical mapping has been to make a gridlike map with multiple vertical penetrations throughout a cortical area. This method is well suited to sampling extensively over an entire cortical area, including regions which are buried in cortical convolutions. This approach has been commonly used in mapping the visual field representation in V1 of macaque monkeys (Talbot and Marshall, 1941; Daniel and Whitteridge, 1961; Dow *et al.*, 1981), vervet monkeys (Guld and Bertulis, 1976), owl monkeys (Allman and Kaas, 1971), and squirrel monkeys (Cowey, 1964). With similar methods, area V2 has been mapped in the owl monkey (Allman and Kaas, 1974; Sereno *et al.*, 1994, 1995), the squirrel monkey (Cowey, 1964), the cebus monkey (Rosa *et al.*, 1988), and the macaque monkey (Van Essen and Zeki, 1978; Gattass *et al.*, 1981). Area V2 in the macaque is a quite convoluted structure, covering roughly 1000^2 μm, with both dorsal and ventral territories, little of which is exposed for easy areal mapping. Despite these difficulties, Gattass *et al.* (1981) were able to reconstruct the visual map from vertical penetrations which passed through buried convolutions. Figure 9 (which is based on Gattass *et al.*, 1981) illustrates the basic visuotopic mapping in V2. Inferior visual fields are repre-

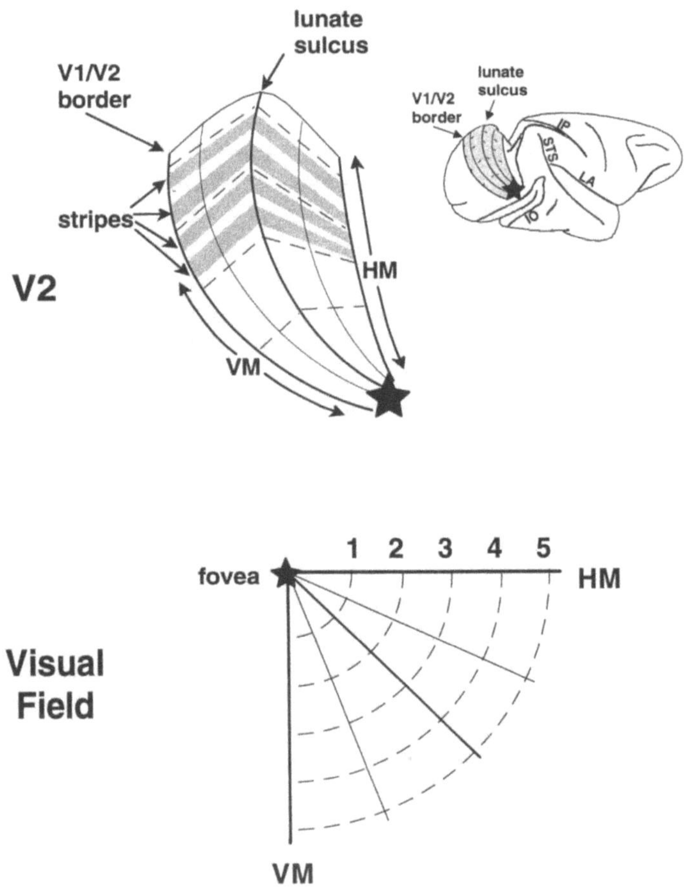

Figure 9. Schematic of visual representation in dorsal V2. Area V2 is located on the caudal bank of the lunate sulcus and extends onto the operculum in the macaque brain (shaded region in upper left inset). An expanded view of this region (above) illustrates the representation of the visual field (below) in V2. Representation of the vertical meridian (VM) lies on the V1/V2 border and that of the horizontal meridian (HM) lies in the depths of the lunate sulcus. Lines of isoeccentricity (dashed lines) run perpendicular to the V1/V2 border. Lines of isopolarity (solid lines) run roughly parallel to the V1/V2 border and converge at the foveal representation (star), which is located laterally on the brain surface. The orientation of stripes in V2 is indicated (gray stripes).

sented dorsally and superior fields ventrally. Area V2 is bounded posteriorly by the vertical meridian representation (VM), located at the V1/V2 border, and anteriorly by the horizontal meridian representation (HM), located in the depths of the lunate sulcus. Thus, isopolar representations (solid lines) run roughly parallel to the VM and HM, while isoeccentricity is represented in the anterior–posterior direction (dashed lines). Since this type of mapping results in a split horizontal meridian representation (one in dorsal V2 and one in ventral V2), it has been termed a "second-order transformation" (Allman and Kaas, 1974). In fact, there appears to be an overrepresentation of a few degrees above and below the horizontal meridian; receptive fields at the anterior border of dorsal V2 extend above the HM and those at the anterior border in ventral V2 extend below the HM (Gattass *et al.*, 1981).

One of the most definitive criteria for the delineation of cortical areas has been that of a single topographic representation. Within an area, representations are usually continuous, as evidenced by gradually shifting receptive field locations. Boundaries between areas are often marked by reversals or discontinuities in topography, changes in receptive field properties, changes in receptive field scatter (a measure of the orderliness of receptive field progression), and changes in cortical magnification. However, the functional specializations in visual cortex, particularly in V2 with its dozen or so millimeter-sized stripes, have prompted a reexamination of these topographic principles.

4.1. Multiple Maps

The division of V2 into functionally distinct stripes raised new questions with respect to topography. In V2, since stripes are oriented anteroposteriorly (Fig. 9, gray stripes), isoeccentricity lines map parallel to the stripes, and lines of isopolarity map across the stripes. How do each of the three stripe types provide sufficient coverage of the visual world within their respective modalities?

Recent studies (Zeki and Shipp, 1987; Zeki, 1990; Roe and Ts'o, 1995a) have shown that the visual map in V2 actually consists of three distinct, interleaved maps. By using optical imaging (Grinvald *et al.*, 1986, 1988; Frostig *et al.*, 1990; Ts'o *et al.*, 1990a) to determine the location of stripes and stripe borders *in vivo,* Roe and Ts'o investigated the topography of V2 with respect to these functionally identified stripes with long, tangential electrode penetrations which densely sampled each stripe and which crossed all stripes within a thin/pale/thick/pale stripe cycle (Fig. 10). These studies found that each region of visual space is represented three times, once by color nonoriented cells in the thin stripes, once by broadband oriented cells in pale stripes, and once by disparity-selective cells in the thick stripes. Receptive field progressions within stripes were accompanied by discontinuities or "jumps back" in progression at stripe borders. For example, thin/pale stripe sequences contain progressions of nonoriented color cells followed by a jump back in receptive field position at the thin/pale border followed by a progression of oriented broadband cells (Fig. 11). Each progression within a stripe is slightly shifted relative to that in the previous stripe. In addition, to avoid any holes or gaps in topographic representation within any single functional submodality, continuity in coverage is achieved from one stripe to the next like stripe (e.g., from thin stripe to thin stripe). Thus,

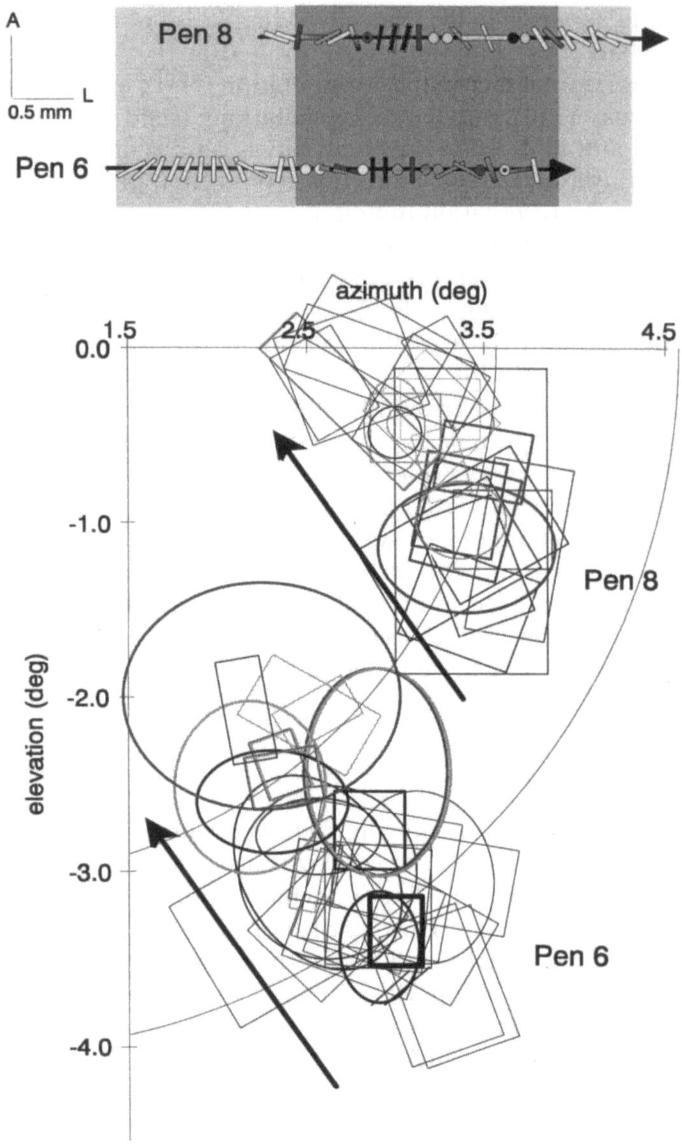

Figure 10. Parallel tangential penetrations oriented across stripes record parallel receptive field progressions. These progressions cross lines of isoeccentricity (two arcs shown). Above: Illustrated at the top are the receptive field types recorded along two penetrations (Pen 6 and Pen 8) and the positions of these recording sites with respect to the color stripe (dark gray) and pale (light gray) stripes (A, anterior; L, lateral), as determined by optical imaging. Broadband oriented cells are indicated by white bars; color nonoriented and oriented cells by light and dark gray circles and bars, respectively; and disparity cells by black bars (color-selective disparity cells are indicated with outlines). Symbols with mixed grays indicate those color cells responsive to either red and green or blue and yellow wavelengths. Spot cells are indicated by symbols with black dots. Below: Receptive field plots of cells shown above. Receptive fields of broadband cells are drawn in black outline and color cells in dotted outline. Rectangles and circles/ovals indicate oriented and nonoriented receptive fields, respectively. Disparity cells are coded with bold lines. Large arrows indicate direction of overall receptive field progression recorded in each penetration.

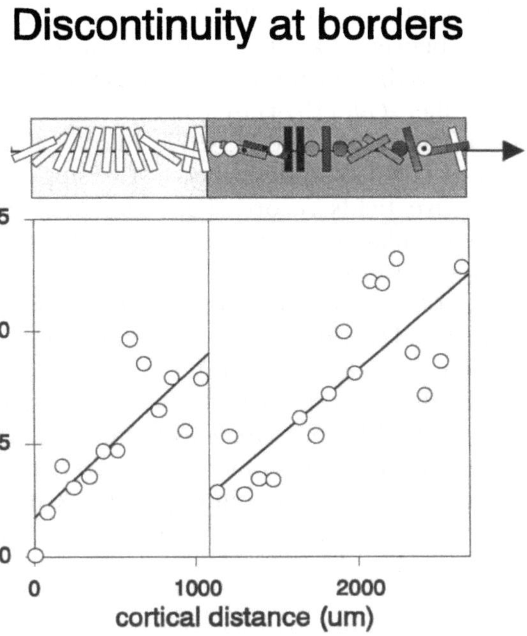

Figure 11. Discontinuity or jumps back in receptive field progression at stripe borders. Top: Same as Fig. 10, Pen 6. Bottom: abscissa: position along tangential electrode penetration that receptive fields were recorded; ordinate: size of receptive field shift from first receptive field recorded in penetration. Points are fit by linear regression. The data suggest the presence of two progressions separated by a discontinuity at the pale/color stripe border.

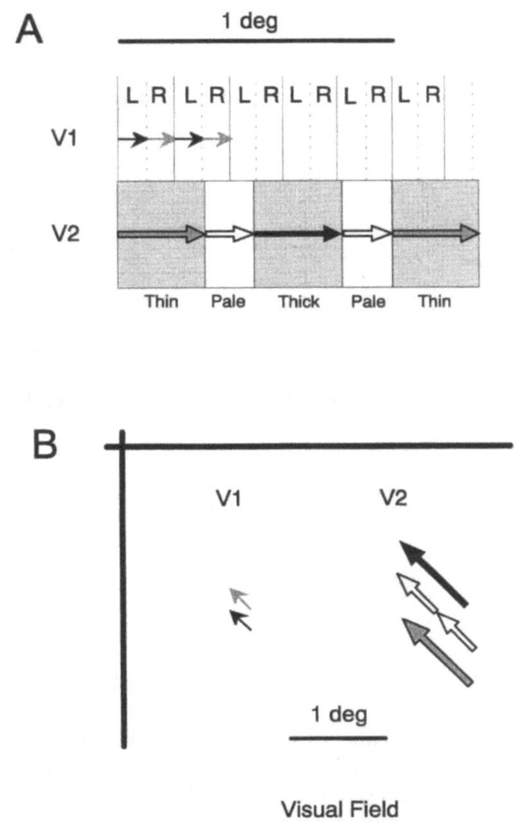

Figure 12. Summary of interleaved "one step forward, two steps back" mapping schemes in both V1 and V2. In V1, ocular dominance is doubly represented. In V2, color, form, and disparity are separately represented, resulting in three interleaved maps.

representation in multiple domains, discontinuity at stripe borders, and continuity between like stripes are characteristic of the visual map in V2.

This discontinuous and interleaved strategy is very similar to that described by Hubel *et al.* (1974) in V1 with respect to representation by the left and right eyes (Hubel and Wiesel, 1977; Blasdel and Fitzpatrick, 1984). In V1, Hubel and Wiesel (1977) had described a "two steps forward, one step back" progression of receptive fields recorded in tangential penetrations through layer 4C of V1 across ocular dominance columns. That is, each locus in visual space is represented twice in succession, once by the right eye and once by the left, accompanied by a slight shift in overall progression between the two eyes (Fig. 12A). The visual world is thus represented twice in V1, once by each eye. In a manner similar to that taken in V1, receptive fields in V2 progress in a "two steps forward, one step back" fashion from stripe to stripe, albeit with much larger steps (roughly four times as large) (Fig. 12B). These data suggest that V2 sees three different versions of the world, one in color (thin stripes), one in form (pale stripes), and one in depth (thick stripes). In a sense, V2 takes these three views, slices each one into stripes, and them interleaves them. This idea is illustrated in Fig. 13.

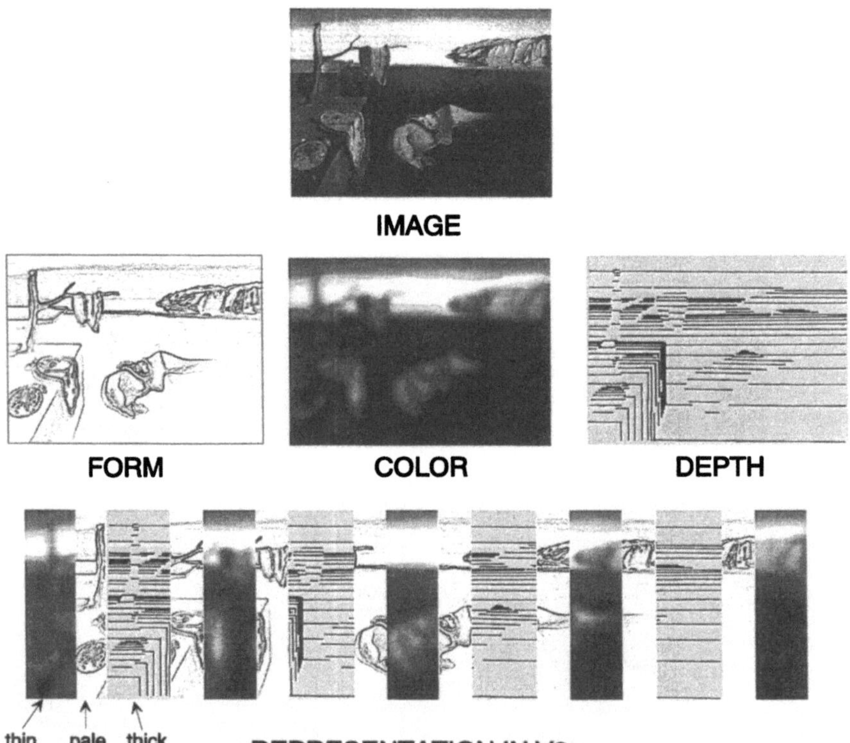

Figure 13. How V2 sees the world. Top: Original image: *Persistence of Memory* by Salvadore Dali. Middle: Decomposition into three images, each defined by either color, form, or depth information. Bottom: Interleaving of these three views in V2. Note the resulting relationships of objects when multiply represented in the different stripe compartments.

4.2. Cortical Magnification, Anisotropy, Modular Maps, and Receptive Field Size

The term cortical magnification was originally coined by Daniel and Whitteridge (1961) to refer to the amount of cortex (in linear distance) devoted to representing each degree of visual space. In V1, this quantity varies (roughly) inversely as a function of eccentricity; greater cortical area is occupied by central fields and less by peripheral visual field representation. While linear magnification is related to a particular axis in cortex, areal magnification is calculated from total area of cortex representing a particular locus in space. Thus, it is possible for two different cortical areas to have similar areal magnification factors for a particular eccentricity, but different component linear magnification factors.

4.2.1. Anisotropy

In comparing cortical magnifications between V1 and V2, it is important to take into account the two axes of representation. V1 contains an inherent anisotropy due to the rerepresentation between the two eyes; the cortical magnification in the axis across ocular dominance columns is roughly twice that along ocular dominance columns (Sakitt, 1982; Tootell *et al.*, 1983, 1988a; Grinvald *et al.*, 1994; Roe and Ts'o, 1994; cf. Malach *et al.*, 1993). As would be predicted by the reflection of the visual field across the V1/V2 border, in the axis parallel to the border (that is, *across* stripes in V2), overall cortical magnification is similar to that in V1. However, in the axis parallel to the stripes in V2, the magnification is much lower, leading to significant anisotropy in the map (Roe and Ts'o, 1995a; cf. Rosa *et al.*, 1988). The cortical dimensions of area V2 (including both dorsal and ventral V2) are at least 5–7 times greater in extent across the stripes than along them (personal observations; cf. Tootell *et al.*, 1983; Van Essen and Maunsell, 1983). If anisotropy in V2 were due solely to the triple representation of the visual map, then the cortical magnification *across* the stripes should simply be three times that *along the stripes*. Each of the three maps taken individually would then be isotropic. However, recent evidence suggests that there is a 2:1 (across:along) anisotropy *within single stripes*, resulting in an overall 6:1 anistropy in V2, a ratio consistent with the anatomical dimensions of V2 (Roe and Ts'o, 1995).

4.2.2. Cortical Magnification in Stripes

At visual eccentricities of 2–5 deg, the cortical magnification factors *across* stripes are roughly 1.3 mm/deg *for each of the three stripe systems* (Roe and Ts'o, 1995). That is, at this eccentricity a full cycle of representation would cover on average roughly 4 mm (= 1.3 + 1.3 + 1.3 mm) of cortex per degree of visual representation. This overall magnification factor in V2 approximates those reported in the 2- to 5-deg eccentricity of V1 (approximately 2–5 mm/deg) (Daniel and Whitteridge, 1961; Hubel and Wiesel, 1974; Dow *et al.*, 1981; Van Essen *et al.*, 1984). Since there are twice as many pale as color or disparity stripes, one might have expected cortical magnification in pale stripes to be somewhat higher. However, that is not the case: cortical magnification in pale stripes is not significantly different from that in color stripes. Coupled with the fact that color stripes are on average twice the width of pale stripes (as indicated by the length of recording sequences in each stripe), this argues that the amount of cortex devoted to broadband orientation representation is similar to that of color representation. In other words, on average, color and disparity stripes are twice as wide as pale stripes.

4.2.3. Modular Maps

Anisotropy within V2 stripes may be directly related to the modular organization within stripes. As we have seen, anisotropy can arise from duplicating the map in one axis of visual field representation. Such remapping also occurs within single stripes, resulting in a cortical magnification factor across the stripe which is twice that along the stripe. Two remapping strategies within stripes have been observed. The first is quite similar to that of the "two steps forward, one step back" described in V1. The second way in which anisotropy within stripes may be created results from larger receptive field scatter in one axis relative to another. Some color stripes resort to large receptive field size and scatter to achieve coverage by each modular domain. Another indication that color, pale, and disparity stripes in V2 may have different coverage strategies derives from the fact that receptive field sizes and scatter in color stripes are generally greater than those in either pale or disparity stripes. Thus, a single cortical area may take advantage of a combination of different mapping strategies to achieve coverage. At a global level coverage is achieved by rerepresentation across the stripes; at a more local level both rerepresentation and large receptive field size and scatter are called upon to provide coverage.

4.2.4. Relative Receptive Field Size and Scatter in Different Stripes

There is greater receptive field size and scatter for the color stripes than for the disparity and pale stripes (Roe and Ts'o, 1995). Although the exact relationship between these findings to psychophysical measures of color, form, and depth perception has not been explored, it is reasonable that the color system would require a less precise mapping than either the form or depth system. The visual system's ability to localize accurately isoluminant color borders is far less precise than its ability to localize luminance edges or disparity boundaries (cf. Morgan and Aiba, 1985). Unoriented color information may serve to fill in the regions bounded by luminance or disparity contours and therefore would require less positional precision.

4.3. Point Image versus Point Set

The extent of the receptive field of a cell reflects the region of visual space which, when appropriately stimulated, causes the cell to respond. The size of an aggregate receptive field, defined as the size of visual space covered by the receptive fields of all cells within a column of cortex, reflects the minimal amount of visual space that a single column of cortex "sees." The term *point image* refers to the amount of cortex activated by a point stimulus in space and is calculated as the product of cortical magnification and local scatter.

Receptive field size in V2 (defined as the square root of receptive field area) is roughly 2–3 times that in V1 at corresponding eccentricities (see Gattass *et al.*, 1981, Fig. 13). Figure 13 illustrates the typical scatter recorded in single vertical penetrations in a pale stripe and a color stripe at any eccentricity of 2–5 deg. While there is some suggestion of anisotropy within the scatter, it is not sufficient to account for the 2:1 anisotropy within single V2 stripes.

The point image in V2 is somewhat different than that in V1. We find it useful to introduce a term distinct from point image, which we refer to as *point set*. Whereas point image refers to the amount of cortex which represents a point in space "by a comfortable margin" (Hubel and Wiesel, 1974), point set refers to the smallest fundamental unit which contains all relevant stimulus dimensions. In V1, the point image is roughly a 2 mm × 2 mm block of cortex which contains two sets of ocular dominance columns. Point set then refers to the 1 mm × 1 mm unit which contains a single ocular dominance hypercolumn, orientation hypercolumn, and possibly a single color hypercolumn (cf. Landisman *et al.*, 1994). In V2, it is likely that the color hypercolumn, the orientation hypercolumn, and the disparity hypercolumn is each contained within a single thin, pale, and thick stripe, respectively. Thus, a point set in V2 would be equivalent to the territory within one stripe cycle. However, the hypercolumn in each visual modality in V2 resides in nonoverlapping cortical territory, unlike the orientation and ocular dominance hypercolumns of V1. The size of this minimal unit of representation is roughly four times greater in extent in V2 than in V1 (4 mm vs. 1 mm in extent). Optical imaging experiments using small point stimuli (0.1–0.5 deg in size) have directly shown a full stripe cycle of activation and two or three stripe cycles of secondary activation, likely due to lateral or divergent activation. Thus, point image in V2 is a significantly larger construct and spans at least two stripe cycles (Roe and Ts'o, 1994). This suggests a V1/V2 divergence ratio of at least two and perhaps as much as six in the cortical axis across stripes. Furthermore, in contrast to the point image in V1, in V2 the point images in the color domain, the orientation domain, and the disparity domain are represented separately in different pieces of cortical tissue.

4.4. What Makes V2 a Single Cortical Area?

The criterion of a single topographic map has traditionally been a strong basis for defining a cortical area. However, the finding of multiple visual representations within V2 reintroduces the issue of what it is that defines a cortical area. The anisotropic cortical magnification within V2 is consistent with the presence of at least three, and perhaps six, different representations, interleaved among and perhaps within the V2 stripes. Each stripe also contains its own characteristic receptive field sizes. The point images in each of the color, form, and disparity domains span distinct pieces of cortical tissue within V2. Given this architecture, one could view each stripe system as a complete and independent representation. What then qualifies V2 as a single cortical area, rather than three areas?

Although these multiple, discontinuous representations exist in V2, since they are roughly in register when considering an entire stripe set, a more macroscopic description of V2 would find a single topographic map, albeit with local distortions and discontinuities. Several other considerations help justify the view that V2 is indeed a single visual area. Although V2 contains segregated multiple functional representations, we would suggest that each submodality within V2 performs a similar and characteristic functional transformation on its particular inputs. These stereotyped transformations, we speculate, are implemented with similar circuits residing within or associated with the fundamental V2 module and construct higher order receptive field properties, regardless of the type of visual input. Thus, these common transformations shared by modules of different modalities within V2 are another justification for characterizing V2 as a single cortical area. A second important component in unifying the three stripes in V2 is the network of intrinsic connections known to link stripes of different modalities. These intracortical connections may form the mechanism by which close coordination of different submodalities is achieved, perhaps for the purpose of binding multiple views into one. Thick, pale, and thin stripes would then behave not as completely independent domains, but rather as a coordinated whole. Thus we suggest that the overall topography of V2, its stereotyped functional transformations, and the interactive relationships between the otherwise segregated stripe compartments are all justifications for characterizing V2 as a single cortical area.

5. Connectivity

We now turn to the patterns and dynamics of V2's connections with other cortical areas, particularly to its major source of ascending input, area V1. As we have seen, the organization of information in area V1 and V2 is quite different. While in V1 color and orientation are largely segregated into the blob and interblob compartments, in V2 these properties are organized into the V2 stripes. The organization for properties such as ocular dominance, so prominent in V1, are completely absent in V2, wherein, instead, other properties such as disparity become well organized. These marked changes in organization lead to questions regarding how different types of information are redistributed and recombined as they pass from V1 to V2. One claim that has been made with respect to V1–V2 connections is that they maintain a segregation of color, form, and motion information. Another issue concerns the notion that the relationship between visual areas such as V1 and V2 is hierarchical and the flow of information is serial in nature. Below we review the evidence surrounding both of these questions and offer some further discussion on the issue of interareal processing.

5.1. Color, Form, and Motion: How Segregated Are They?

Anatomical studies have indicated a significant degree of segregation between the so-called color, form, motion, and depth pathways. As shown by several studies, injections of anatomical tracers into the thin stripes in V2 preferentially label the blobs in V1, while injections into pale stripes in V2 preferentially label interblob regions in V1 (Livingstone and Hubel, 1984; Horton, 1984; DeYoe and Van Essen, 1985; Shipp and Zeki, 1985). Thick stripes in V2 receive inputs primarily from the magno-dominated layer 4B of V1 (Livingstone and Hubel, 1987a). Each of these stripes in turn has a different pattern of feedfoward connections with higher cortical areas. For example, the thin and pale stripes in V2 have interconnections with area V4 and thick stripes with area MT (Shipp and Zeki, 1985; DeYoe and Van Essen, 1985; Krubitzer and Kaas, 1990a,b; Nakamura *et al.*, 1993; DeYoe *et al.*, 1994). Within V4 there are segregated color and orientation patches (Ghose and Ts'o, 1997), consistent with the segregated connectivity of V4 with the V2 thin and pale stripes (DeYoe *et al.*, 1994).

Though the pattern of the V2–V1 feedback connections has not been well studied, the segregation between the blob–thin stripe connections and the interblob–pale stripe connections has been reported to be reciprocal in nature and patchy as well (Wong-Riley, 1979b; Livingstone and Hubel, 1984). Within the V1 intrinsic connections, the pattern of functional segregation is apparently maintained, since injections into blobs label neighboring blobs and avoid interblob regions, while those in interblobs label neighboring interblobs preferentially (Livingstone and Hubel, 1984). This anatomical pattern is also mirrored in the pattern of functional interactions as measured by single-unit physiology coupled with cross-correlation analysis (Ts'o and Gilbert, 1988). These findings as well as a number of psychophysical (e.g., Cavanagh, 1989; Livingstone and Hubel, 1987b) and clinical and human (Vaina *et al.*, 1990; Baker *et al.*, 1991; Haxby *et al.*, 1991; Zeki *et al.*, 1991; Tootell *et al.*, 1995) studies served as a basis for the idea that color, form, and disparity/motion information are processed in separate parallel visual pathways (Livingstone and Hubel, 1987b; DeYoe and Van Essen, 1988; cf. Merigan and Maunsell, 1993; Cavanagh, 1989):

	V1	V2	
Color	Blob	Thin stripe	V4 color patch
Orientation	Interblob	Pale stripe	V4 orientation patch
Motion/disparity	layer 4B	Thick stripe	MT

5.1.1. Segregation of M and P Inputs

The question of color, form, and motion segregation has been related to the functional and anatomical segregation of the magnocellular (M) and parvocellular (P) "pathways," so named for their laminar localization in the lateral geniculate nucleus (LGN). The P system has been proposed as the primary contributor of visual input to the ventral ("what") pathway, and is though to be responsible or form and color perception. The M system has been suggested to be the primary source to the dorsal pathway, which handles primarily motion and disparity information (for reviews see Livingstone and Hubel, 1988; Merigan and Maunsell, 1993). There are certain anatomical and functional grounds for these claims. Anatomical studies suggest parallel and largely separate retinocortical pathways: the P pathway arises in the midget retinal ganglion cell and ascends through the parvocellular LGN to layer 4cβ in V1, both blobs and interblobs in V1, thin stripes and pale stripes of V2, V4, and the temporal lobe; the M pathway beings with the parasol retinal ganglion cell and continues to magnocellular LGN, layer 4Cα in V1, layer 4B in V1 and blobs in V1, thick stripes in V2, V3d, V5 (MT), and beyond, into parietal cortex. Further circumstantial evidence has been provided by the staining pattern for the antibody CAT 301 (Hockfield and McKay, 1983), which seems to selectively stain components of the M pathway, from the retina, through the magnocellular layers of the LGN, to the thick stripes of V2 and onward to V5 (MT) (DeYoe *et al.*, 1990; Krubitzer and Kaas, 1990a, b). From a functional standpoint, the most distinctive properties of the cells of the early P system, that of color opponency (Wiesel and Hubel, 1966), are indeed likely to be the source of the color input into the ventral stream (Merigan, 1989). Similarly, the specialization of the M system for the highest temporal properties (for review, see Shapley, 1990) are likely to be responsible for at least the fastest motion processing abilities of the dorsal pathway.

However, a number of studies now suggest that there is considerable mixing of M and P inputs in both the dorsal and ventral streams. In V1 it appears that blobs receive both M and P inputs (Lachica *et al.*, 1992; Edwards *et al.*, 1995) and, contrary to the initial anatomically based notions, M inputs contribute to both blobs and interblobs (Nealey *et al.*, 1991). In V2, evidence for further mixing occurs at places where thin and thick stripes merge, where patches of color disparity cells are found (Ts'o *et al.*, 1991). In V4, the inactivation of either M or P layer of the LGN significantly reduces responses of V4 cells (Ferrera *et al.*, 1992), thus indicating that this component of the ventral stream receives significant input from both M and P sources. However, responses of cells in V5 (MT) are strongly affected by M, but not P, inactivation (Maunsell *et al.*, 1991), suggesting that at least this component of the dorsal stream is M dominated. It has been shown that responses of single cells in the M pathway fall to a minimum with isoluminant chromatic stimuli, though residual frequency doubled responses may remain (Lee *et al.*, 1988; Gegenfurtner *et al.*, 1994; Dobkins and Albright, 1993). These observations along with psychophysical demonstrations of a reduction in motion perception near isoluminance suggest an absence of strong P input to MT. However, lesions to the P layers of the primate LGN are reported to severely affect high-spatial-frequency stereopsis (Schiller *et al.*, 1990), suggesting that for stereopsis, a property often associated with the dorsal pathway, there may be significant P input. Thus, the weight of evidence suggests that for most aspects of normal vision, both the P and M systems are likely to contribute in both the dorsal and ventral streams.

Similarly, several lines of evidence suggest that the notion of a strict segregation of form, color, motion, and depth processing is too simplistic a view. While it is clear that in V1 and V2 color, form, and motion are segregated, there is also clearly some degree of interaction between these pathways. Anatomically, within V2, for example, stripes are highly interconnected by horizontal connections (Ts'o *et al.*, 1991; Levitt *et al.*, 1994b; Malach *et al.*, 1994). Injections of tracer into any stripe result in patches of label in each of the other stripe types. Furthermore, both in V1 and V2, cells of mixed functional properties are often found. For example, color oriented cells are prominently found at the borders between thin and pale stripes in V2 (Ts'o *et al.*, 1991; Roe and Ts'o, 1995) and also in periblob regions in V1 (Ts'o and Gilbert, 1988); color disparity cells have also been described at regions of merging between thin and thick stripes (Ts'o *et al.*, 1997). Other investigators have also emphasized that the spatial and temporal tuning properties of the cells found in each type of V2 stripe are quite similar (Levitt *et al.*, 1994a). Thus, although at the level of V2 the extent of functional and anatomical segregation is quite striking, there is also ample evidence that this segregation is not strict and the pathways for form, color, motion, and depth processing have substantial opportunities for interaction.

5.1.2. Specificity of Functional Interactions

Recent cross-correlation studies between V1 and V2 have added further insights into the interactions between V1 and V2 and the issue of segregation versus interaction. The strategy of Roe and Ts'o (1992, 1997) involved the use of multiple microelectrodes to target specific structures (such as blobs or interblobs in V1 or specific stripe compartments in V2) localized by optical imaging, followed by the functional characterization of single cells in each electrode and collection of spike trains in response to particular visual stimuli. Cross-correlation analysis was then applied to these spike trains to characterize the functional interactions between these pairs of identified cells. By inference, these interactions are also indicative of the connectivity between specific structures in V1 and V2, thus providing a functional correlate to known anatomical connections between V1 and V2.

By studying hundreds of cell pairs using this approach, certain patterns or rules of V1–V2 interactions have emerged (Fig. 14). Similar to the pattern of interactions found between V1–V1 cell pairs (Ts'o *et al.*, 1986; Ts'o and Gilbert, 1988), among oriented cells, peak correlograms were found primarily among V1–V2 cell pairs with matching orientation preference (Fig. 14A). However,

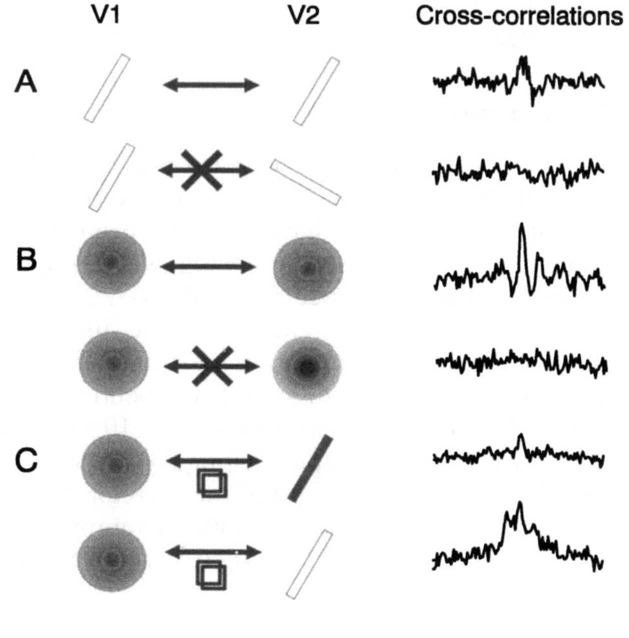

Figure 14. Summary of functionally specific V1–V2 interactions. Cell pairs are schematized in the left column and representative correlograms are shown in the right column. "Like to like" interactions are commonly seen: (A) Orientation-selective broadband cell pairs in V1 and V2 show peaked correlograms when orientation selectivities match, but flat correlograms otherwise. (B) Color-selective cell pairs interact only when color selectivities match. (C) While interactions in A and B do not require receptive field overlap, those between nonoriented color cells in V1 and oriented cells in V2 have a strong dependence on receptive field overlap.

such interactions were not restricted to those between interblobs and pale stripes. Interactions between broadband oriented cells in V1 interblobs and color oriented cells at thin/pale borders as well as disparity cells in thick stripes were also observed. In the color domain, peaked correlations occur most often between nonoriented color cells located in V1 blobs and cells in V2 thin stripes with similar color specificities (Fig. 14B). Interactions between nonoriented color cells in V1 blobs and oriented broadband cells in V2 pale stripes were also found (Fig. 14C). Cell pairs with dissimilar properties most often exhibited flat correlograms.

The presence of interaction apparently depended on receptive field overlap for certain functional pairings (Fig. 14C) but not for others. In some examples (such as color nonoriented V1–V2 cell pairs), interactions were found between V1–V1 pairs and V1–V2 pairs which had nonoverlapping receptive fields separated by several receptive field diameters. For other functional cell pairings, such as nonoriented V1 color cell paired with an oriented V2 color cell, interactions were only observed when there was significant receptive field overlap. This differing dependence on receptive field overlap suggests that different functional compartments within the stripes may have differing degrees of spatial divergence and convergence.

While these results may be viewed as supportive of the general notion of functional segregation, a more comprehensive examination of the data reveals that different receptive field types within a single stripe type may exhibit different patterns of connectivity, some of which may conform to the notion of segregation and others which may not. The possibility that there are more than three pathways (e.g., a color oriented pathway?) should also be considered. And finally, even cells which appear functionally similar may in fact be part of different pathways. For example, it is possible that one class of broadband oriented cell in V2 may simply have its properties conferred on it by other broadband oriented cells in V1, and another broadband oriented cell class be constructed from converging nonoriented broadband (or color) inputs. Thus there is a great deal of complexity in the details of functional connectivity that make "segregated or not segregated" an overly simplistic question.

5.1.3. Dynamics

Yet another aspect of cortical connectivity that greatly increases the complexity of the issue is the dynamic nature of these connections. For example, cross-correlation studies have reported dramatic changes in V1–V2 effective connectivity, including strengthening, weakening, as well changes in the signature and oscillatory characteristics of V1–V2 correlograms (Nelson *et al.*, 1992; Frien *et al.*, 1994; Ts'o *et al.*, 1993; Roe and Ts'o, 1997). These changes are thought to reflect a modulation or shift in synaptic efficacy or coupling. The changes may be context or stimulus dependent, or may relate to attentional or behavioral state (cf. Gilbert and Wiesel, 1990; Knierem and Van Essen, 1992; Lamme, 1995). For example, among color cells, the type of color stimuli influenced the shape and strength of the correlogram (Roe and Ts'o, 1997; Fig. 15). Continuity of contour may be another example of context dependence, and has been linked to perceptual binding (Gray and Singer, 1989; Nowak *et al.*, 1995b;

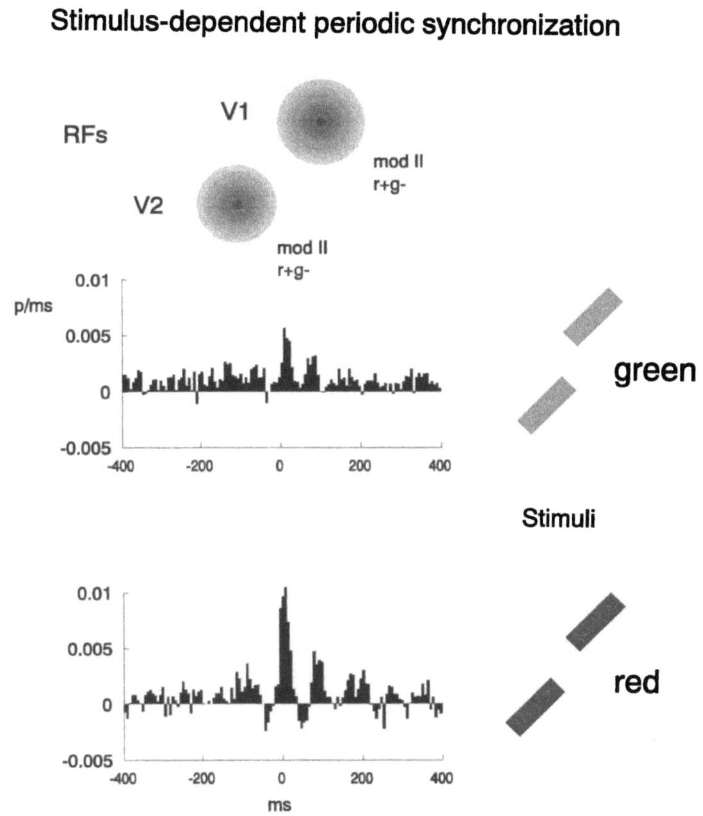

Figure 15. Stimulus dependence in the color domain. Two nonoverlapping red on-center/green off-center color-opponent cells, one in V1 and one in V2. The correlograms were significantly stronger and exhibited more periodicity when stimulated with red moving bars (below) than with green (above), even though both stimuli drove the cell pair nearly equally well, with similar resulting firing rates.

Ts'o *et al.*, 1993; Roe and Ts'o, 1997). Importantly, it appears that the dynamics may be functionally specific, being more prevalent for some cell types than others. Interactions between orientation-selective cells can be influenced by the orientation of presented stimulus. Interactions between color-selective cells are changed by stimulus color. It is also conceivable that, for example, an oriented V2 cell is signalled by orientation information in V1 under one stimulus condition and convergent color cells under another. Thus the finding that these large V1–V2 networks might change or reconfigure themselves in very specific ways depending on state or context suggests ongoing and frequent shifts in the degree of functional segregation of the form, color, motion, and depth pathways during normal vision and visually guided behavior.

In considering the overall task of visual perception, the question of whether the color, form, and motion/disparity pathways are segregated must be considered too simplistic. For example, there is physiological and anatomical evidence that columns of cells tuned for a particular orientation are interconnected not only within a single area (e.g., V1), but across corticocortical boundaries (e.g.,

V1–V2). In one sense then, the degree of functional segregation, in view of this general rule of like-to-like connectivity, is even higher than that suggested by the generally more macroscopic anatomical studies. On the other hand, when one considers, for example, that a color border might appear as part of the foreground in one context and part of the background in another, it becomes clear that dynamic interactions between color and form are necessary to define and interpret that border, and that strict functional segregation is not workable. Thus, while at early stages there may be a significant extent of segregation between color, form, and motion/disparity, interactions between these pathways must become an increasingly crucial aspect of higher order visual processing.

5.2. Serial or Parallel?

5.2.1. Traditional Hierarchical View

The hierarchical view of the architecture of visual cortical processing is in part based on the observation that outputs from upper cortical layers of "lower" cortical areas terminate in layer 4 of "higher" cortical areas, and feedback from "higher" areas originate and project to layers other than layer 4 (Rockland and Pandya, 1979; Maunsell and Van Essen, 1983; Felleman and Van Essen, 1991). In considering the interconnectivity of V1 and V2, projections from V1, arising primarily from layers 2, 3, 4A, and 4B in V1, terminate primarily in layers 3B and 4 in V2 both in the macaque (Lund *et al.*, 1981; Livingstone and Hubel, 1984, 1987a; Kennedy and Bullier, 1985; Van Essen, *et al.*, 1986; Rockland, 1992) and in the squirrel monkey (Spatz *et al.*, 1970; Wong-Riley, 1979b; Tigges *et al.*, 1981; Fitzpatrick *et al.*, 1983; Rockland, 1992). Feedback projections from V2 arise primarily in layer 6, though also in layer 3A in V2, and terminate in layers 1, 5, and 6 in V1 (Rockland and Virga, 1989; Weller and Kaas, 1983).

The hierarchical view of the visual cortical pathways leads to the expectation that V2 is higher in the cortical hierarchy than V1 and that it receives its primary driving input from V1. This is consistent with the fact that virtually all ascending input, with the exception of some relatively sparse input from the pulvinar (Curcio and Harting, 1978; Horton, 1984), to V2 is provided by V1. Inactivation of V1 clearly silences cells in V2, including those in all three compartments (Schiller and Malpeli, 1977; Girard and Bullier, 1989), thus demonstrating that V1 input to V2 is necessary for the activation of V2 neurons. It is likely that other inputs to V2, such as those from the pulvinar, V4, or MT, may provide only a modulatory role which alone are not sufficient to drive V2 cells.

5.2.2. Parallel Processing in V1 and V2

5.2.2a. Inactivation Studies. Further physiological examination of this issue, however, has put this strict hierarchical scheme into question. While inactivation of V1 silences V2, the effect of V2 inactivation on V1 can also be profound. Both supragranular and infragranular layer responses in V1 are also vastly reduced by V2 inactivation with either cooling or cobalt chloride lesion methods (Sandell and Schiller, 1982; Mignard and Malpeli, 1991). In addition, Mignard and Malpeli (1991) report that following inactivation of LGN with ibotenic acid lesions, receptive field properties, such as orientation selectivity, of

V1 cells outside of layer 4 remain intact; additional cobalt chloride lesions of V2 lead to profound shutdown of V1 responsiveness, suggesting that V2 can be a major driving source for V1 cells.

5.2.2b. Latency Studies. Studies of visual latencies of cells in V1 and V2 suggest that much neuronal activation occurs simultaneously in V1 and V2 (Raiguel *et al.,* 1989; Nowak *et al.,* 1995a). Neuronal activation latencies (as measured by onset of response to flashing spots or bars) of V1 cells are shortest in layer 4Cα and layer 4B (Maunsell and Gibson, 1992) and precede the shortest latencies in V2 (which surprisingly occur in the infragranular layers) by up to 10 msec. While it is true that as a population, V2 cells lag those of V1 cells, neuronal latency distributions in V1 and V2 overlap greatly such that by the time half of the sampled V2 population has responded, roughly half the cells in V1 have yet to respond (Nowak *et al.,* 1995a). This is true even within each of the color and motion channels. Layer 4B latencies in V1 have great overlap with those of thick stripes in V2, as do supragranular layer cells in V1 (which would include those of both blob and interblob cells) and those in thin and pale stripes in V2 (compare Nowak *et al.,* 1995a, Fig. 5 to Munk *et al.,* 1995, Fig. 3A). Cross-correlation studies of V1–V2 cell pairs also indicate that the predominant contributor to V1–V2 interaction is common input (Roe and Ts'o, 1992, 1997; see also Nelson *et al.,* 1992).

5.2.2c. Sources of Common Input. Possible sources of common input to V1 and V2 are numerous, both cortically and subcortically. The most prevalent source of common input to V1 and V2 may well be other projection neurons within V1 and V2. These are typically pyramidal cells in the supra- or infragranular layers, which are known to have dense local axonal plexuses (e.g., Valverde, 1985). Other possible sources are much less prominent and unlikely to form major sources of the common input observed. In double-labeling studies with fluorescent tracer injections into V1 and V2, Kennedy and Bullier (1985) report that on the order of 10% of all labeled neurons in the lateral and inferior pulvinar and the claustrum were double-labeled. Inputs from the LGN, however, are segregated with inputs to V1 arising primarily from the magno and parvo layers and those to V2 from the interlaminar S layers (Bullier and Kennedy, 1983). In other cortical areas, neurons which bifurcate to both V1 and V2 reside primarily in the infragranular layers of the anterior lunate, the prelunate gyrus, and the superior temporal sulcus. These include areas V3, V4, MT, TEO, as well as others (Van Essen, 1985; Rockland *et al.,* 1994).

5.2.2d. Network Size. Another striking observation is that correlograms between V1–V2 cell pairs are often quite broad in comparison to the V1–V1 cell pairs (cf. Ts'o *et al.,* 1986), indicating a greater temporal dispersion in V1–V2 interactions. Figure 16 shows a comparison of typical correlogram peak widths between V1–V1, V1–V2, and V2–V2 cell pairs. The V1–V1 correlogram peak widths typically fall in the range of 2–10 msec (Ts'o *et al.,* 1986; Ts'o and Gilbert, 1988). The V1–V2 correlogram peak widths, in contrast, are typically 30–50 msec and range quite broadly up to 100 msec or more. The V2–V2 correlogram peaks as a whole also tend to be sharper than V1–V2 interactions, and range

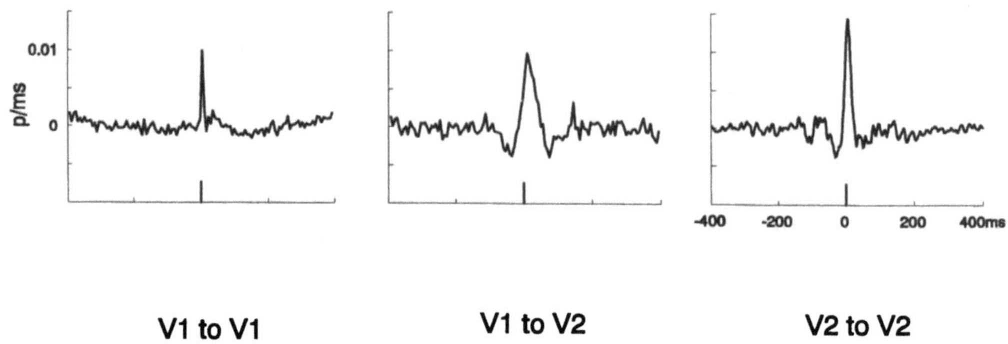

Figure 16. Comparison of V1–V1, V1–V2, and V2–V2 correlogram peak widths. V1–V2 correlogram peaks are much broader than those of either V1–V1 or V2–V2 peaks, indicating a greater temporal dispersion in corticocortical connections. Peak widths of V1–V1 (left), V1–V2 (middle), and V2–V2 correlograms shown are 10, 50, and 20 msec, respectively.

from 5 to 20 msec. These results indicate that V1–V2 interactions are much more temporally dispersed than either V1–V1 or V2–V2 interactions. This greater temporal dispersion suggests that V1–V2 interactions, and perhaps corticocortical interactions in general, are part of larger, more complex cortical networks than the cortical circuitry intrinsic to a single area.

5.2.3. Revised View of Interareal Interaction

The overwhelming prevalence of common input interactions (Nelson *et al.*, 1992; Roe and Ts'o, 1992, 1997) raises questions concerning the relative roles of V1 and V2 in the transfer and transformation of incoming visual inputs. One traditional view of cortical processing has held that, since visual information to higher cortical areas passes through V1, information passes more or less serially from V1 to V2. This view has been strengthened as well by anatomical studies suggesting segregated channeling of visual information from V1 to V2. However, findings that V1 and V2 cells are often simultaneously activated lead to a slightly different view of corticocortical connectivity. Perhaps cortical areas should not be considered separate entities linked by a few functional streams, but rather they should be regarded as highly coupled and interdependent coprocessors. Instead of passively transmitting or replicating the outputs of V1 in the input layer of V2, the interareal connections themselves could participate significantly in computational transformations, bringing about the generation of new receptive field properties. These observations suggest that cortical areas process at least a portion of visual information concurrently, in parallel, and serve to deemphasize a serial or hierarchical view.

6. Relationship of Structure and Circuitry to Perception

6.1. The Role of V2 in Perceptual Binding across Modalities

In the past several sections, we have noted that while the extent of segregation of the form, color, motion, and depth pathways is remarkable, there is also substantial evidence for interaction and cooperativity between these pathways. These themes continue in higher areas such as V4 and MT. Consistent with physiological characteristics of cells in V4 and MT, V4 receives anatomical inputs preferentially from the thin and pale stripes of V2, while MT receives projections from thick stripes in V2 (DeYoe and Van Essen, 1985; Shipp and Zeki, 1985, 1989; DeYoe *et al.,* 1994). The functional segregation that occurs in V2 allows the distribution of visual information to either the ventral ("what") or dorsal ("where") pathways. Thus, V2 sits at an important juncture in the information flow to other cortical areas. Its position in the cortical hierarchy would be well suited to the context-dependent gating of ascending information. Given its position, V2 conceivably could play a crucial role, via feedback connections, in the binding of what and where percepts.

The multiple, interleaved mapping architecture of V2 may be ideally suited for stripe–stripe interactions that may contribute to perceptual binding. Several lines of evidence suggest that such functional binding could take place via horizontal connections within V2. Injections of anatomical tracers into V2 stripes revealed patches of label contained primarily within the injected stripe cycle (i.e., within 4 mm of the injection site; Amir *et al.,* 1993; Levitt *et al.,* 1994b). Furthermore, these patches occurred both within the stripe injected and within each of the other two stripe types (Ts'o *et al.,* 1991; Levitt *et al.,* 1994b; Malach *et al.,* 1994). Supplementing these anatomical findings are functional studies using cross-correlation analysis that also demonstrate interactions between different stripe compartments in V2 up to a distance of several stripes (Roe and Ts'o, 1992, 1997). Therefore V2 seems to possess the functional and topographic architecture to contribute to the perceptual binding of the submodalities of visual information.

6.2. Role of V2 in Contour Perception: An Implementation

In this section, as a way of summarizing, we give an example of how the organization and circuitry discussed in this chapter might be used to help construct a coherent view of object contours. These suggestions are not meant to be a comprehensive summary of all that V2 does. That is, contour identification is by no means the only or even the primary role of V2, as V2 is also likely to be involved in other aspects of visual processing.

One of the primary functions of the visual system is the identification of objects. Inherent in object identification is the delineation of object contour for the purposes of figure–ground segregation. It has been proposed (Cavanagh, 1989) from psychophysical studies that the visual system initially encodes two-dimensional shape in several different domains (including those of luminance, color, texture, motion, and binocular disparity) which are subsequently abstracted to generate a generalized shape representation. Such a "segregate and generalize" strategy has also been proposed for the representation of three-dimensional surfaces (Carman and Welch, 1992). However, the cortical locus of these events is unknown. We propose that area V2 may play a key role in the extraction of generalized shape information from the visual scene. Specifically, the evidence reviewed in this chapter suggests that V2 is involved in the extraction of higher order contour information using inputs from several different

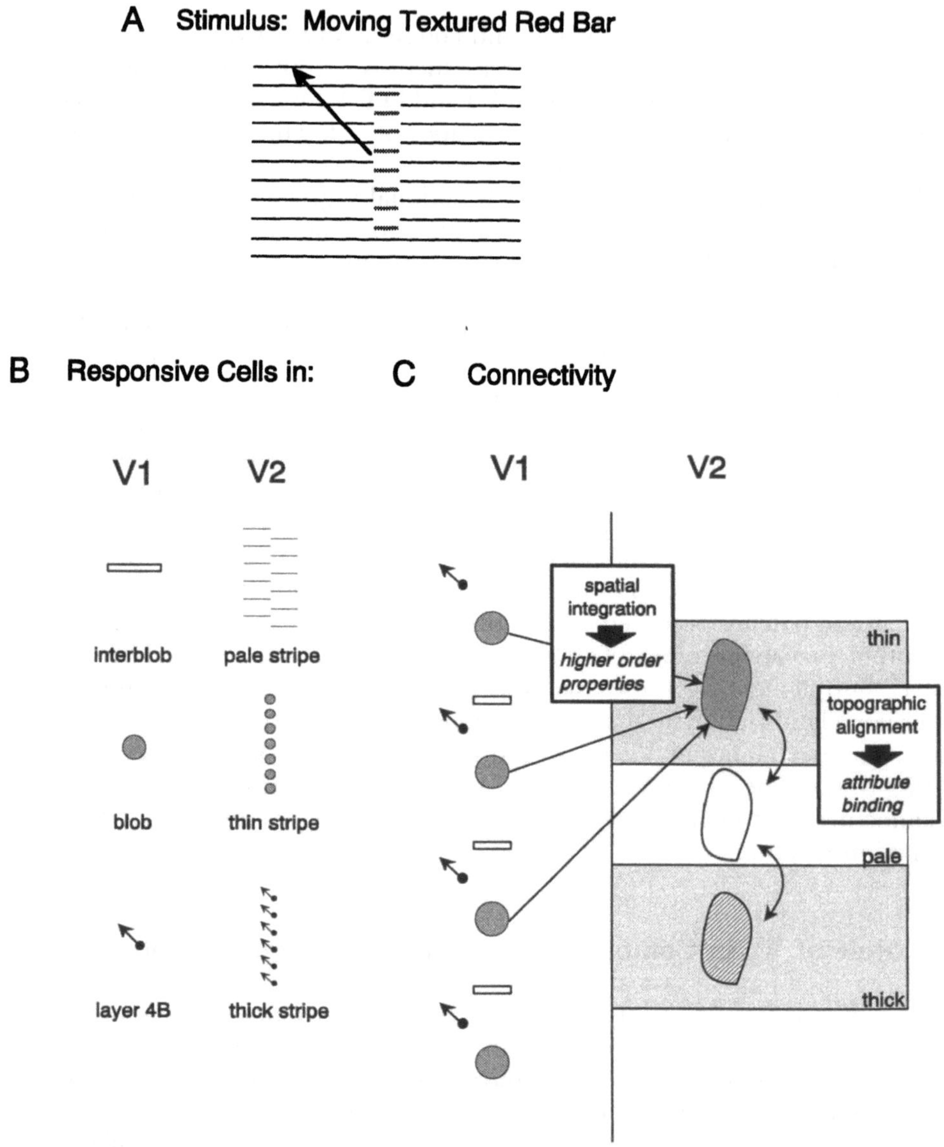

Figure 17. Building of object contour representation by V2 organization and circuitry. How V2 sees a red bar moving in space. (A) The stimulus is a red bar moving in space whose borders are defined by color, by illusory contours, and by motion and depth cues. (B) V1 cells and V2 cells see different aspects of this stimulus. In V1, the following cells are responsive to this stimulus: interblob cells with horizontal orientation selectivity, blob cells with red selectivity, motion cells with upward and leftward direction, and perhaps disparity cells of the appropriate disparity selectivity (not shown). In V2, there are more-specialized cells which see higher order features of this stimulus: illusory contour cells in pale (and perhaps thick) stripes with vertical orientation selectivity, color oriented cells (at pale/thin stripe border zones) with vertical orientation selectivity, disparity cells in thick stripes with appropriate disparity selectivity (not shown). Response to the vertical motion contrast border may be encoded by either motion coherence cells in V2 (Peterhans and von der Heydt, 1993) or by MT cells. (C) Possible circuitries between V1 and V2 cell populations. Convergent connections (large arrows) from functionally specific V1 sites serve, either directly or indirectly, to generate higher order receptive field properties in V2 which require spatial integration. Interstripe connections in V2 serve to ensure close visuotopic alignment of multiple views and contribute to perceptual unity.

submodalities, the segregation of the visual scene into multiple views, and the alignment of these multiple views. This example is illustrated in Fig. 17.

6.2.1. Feature (Contour) Extraction within Separate Functional Domains

In Fig. 17A we depict a red bar moving in space whose borders are defined by color, by illusory contours, and by motion and depth cues. In V1, several different cell types (Fig. 17B) are responsive to this stimulus: interblob cells with horizontal orientation selectivity (possibly end-stopped), nonoriented blob cells with red selectivity, motion cells selective for upward and leftward direction, and disparity cells of the appropriate disparity selectivity. In V2, there are more specialized cells which see higher order features of this stimulus: illusory contour cells in pale (and perhaps thick) stripes with vertical orientation selectivity, color oriented cells (at pale/thin stripe border zones) with vertical orientation selectivity, and disparity and motion contrast cells which also signal vertical orientation. The presence of this contour is thus signalled in several different ways within V2.

Connections between these several cell populations are illustrated in Fig. 17C (arrows indicate either direct or indirect connections). Red-selective blob cells in V1 may converge, either directly or indirectly, onto a single red orientation-selective cell either in a V2 thin stripe or at a color orientation-selective cluster at a thin/pale border. End-stopped V1 cells located in interblobs which are broadband and selective for horizontal orientations may converge to generate vertically oriented illusory contour cells in V2 pale stripes (cf. Peterhans and von der Heydt, 1989). Motion-selective cells in layer 4B of V1 which prefer motion in the upward and leftward directions may converge onto cells in V2 thick stripes or in MT which signal the presence of a vertical motion border. Similarly, perhaps via additional circuitry within V2, the convergence of disparity information from the two eyes may signal the presence of a disparity contour (not shown). As we have seen from both anatomical and functional connectivity studies, each of the three stripe systems in V2 serves to collect, in a convergent fashion, functionally specific information from V1. The purpose of collecting this information across an expanse of V1 into a single large stripe may be to construct higher order receptive fields that integrate across spatially distributed local detectors, such as the elements forming an illusory contour percept.

6.2.2. Topographic Alignment of Multiple Views

Given this decomposition into multiple views, by what mechanism does the visual system know that these three representations refer to the same contour in space? A likely candidate for bringing these three representations together is the extensive network of clustered interconnections between the three functional stripe types. These horizontal connections, which extend primarily to within a single stripe cycle, could serve to bind the several attributes together within a stripe cycle and thereby provide close visuotopic alignment of the different visual maps (Fig. 17C). The thin/pale/thick/pale stripe set is thus a functional unit which shares and processes contour information in concert. The output of such processing in V2 may then serve to produce an abstracted contour representation of a vertical bar at a higher cortical level.

7. Why Another Stage of Cortical Processing Apart from V1?

Thus far, we have skirted around the question as to the raison d'être for V2, and more generally, the principles for the existence of multiple visual areas. This question is perhaps particularly pertinent in the case of V2, given that the types of visual inputs and outputs for V1 and V2 are quite similar and both subserve both the dorsal and ventral visual pathways. Several notions of the functional role of V2 and the necessity for multiple visual areas in general revolve around the idea that a single cortical area is limited in the amount and extent of processing and transformations it can perform, either spatially or computationally or both. A complex computational task such as those required for vision may be too difficult or impossible for one area to handle, or may be biologically/developmentally "easier" to distribute among several cortical areas. A few lines of circumstantial evidence can be offered:

1. The continued elaboration from area to area of higher order receptive field properties, particularly those that integrate information across an increasing extent of visual space. This trend may be viewed as multiple representations in scale. This observation might suggest that the computational demands of the task require more machinery than a single cortical area possesses or can easily mobilize.

2. The establishment of distinctive cortical organizations for a different mix of functional properties from area to area. Functional properties that are organized in one area may not be organized in a "higher" area even though that functional property may be present. New types of functional organization often emerge, instead, in the higher area.

3. The apparent conservation from area to area in the extent and character of the intrinsic anatomical connections, and similarly the stereotyped nature of cortical circuitry. These presumably biological/developmental limitations must set an upper bound as to how much a single cortical area can accomplish.

4. Despite the hierarchical evidence that implies a serial flow of information from V1 to V2, other physiological results suggests that visual input arrives in V2 at nearly the same time as in V1 and therefore should permit processing in V2 to occur largely in parallel to that in V1.

5. The anatomical and physiological nature of corticocortical connections is quite different from that of connections and intrinsic circuitry within a cortical area. Therefore the computational capabilities and the nature of the interactions between different functional units within a single cortical area versus units distributed in separate cortical areas are likely to be quite different.

Given these observations, we further suggest that:

1. There are limits for a single cortical area in the number of different representations of visual information possible in the cortical architecture.

2. There is a computational payoff in using multiple intermediate representations for the ease of computing different transformations on the visual information.

3. There are biophysical/energetic/developmental pressures that favor cortical representations that are implemented as columns or clusters of neighboring neurons rather than as spatially dispersed or disorganized ensembles.

4. The nature of corticocortical connections may be more suited to a restricted or different class of computations or information transfer, perhaps including those employed by computational units that are loosely coupled and operate in parallel.

5. Certain aspects of the whole computational task are more easily separable and parallelized, and together with the considerations of differences between corticocortical and intrinsic circuitry, may dictate which aspects or submodalities are explicitly represented, parallelized, and delegated to different cortical areas. In other words, some computations are so inherently interactive that the implementing circuits must reside within the same cortical area, while others may be more easily broken down into intermediate computations that are more easily distributed across separate cortical areas.

These ideas and observations prompt us to suggest that the biological pressure to create a V2 stems from the computational limitations of a single cortical area and the need for a number of different intermediate representations. The advantages of using well-chosen intermediates toward tackling a complex problem can be found in many disciplines, such as engineering, chemistry, and of course mathematics. Such intermediate forms often transform a difficult task into a much easier procedure. Similarly, cortical areas such as V1 and V2 may be viewed as platforms for staging the computational problem in terms of useful intermediates so as to facilitate parallel processing and ultimately the completion of vision further down the visual pathways. For example, the extraction of perceptual contours in a visual scene might be staged by computing the distribution of the orientation of edges (defined by luminance, color, etc.), one useful intermediate, and then from orientation proceeding with endstopping and curvature, perhaps another useful intermediate, and then onto the incorporation of illusory contours and the binding of local features to extract a more global contour. Several of these stages of the overall computation and their intermediate results are sufficiently independent so as to facilitate their implementation in separate cortical areas, and separate functional modules within these cortical areas.

The available visual information to V1 and V2 is similar if not essentially identical, yet the mix of functional properties that are organized in each area are different. Both V1 and V2 have disparity cells and end-stopped cells, but only in V2 does a clear organization for these properties emerge. Both V1 and V2 contain directionally selective cells, yet only in V5 (MT) does a clear columnar organization for directionality emerge. On the other hand, V1 contains the most precise retinotopy, as well as the only organization for ocular dominance. These observations suggest not only that no single area can provide all the explicit representations necessary to facilitate visual processing, but that there is a natural sequence or succession of computations and their resultant intermediate forms in visual processing. If some of these intermediates can be built and used concurrently (in parallel), so much the better.

In this context, we restate some observations from a previous section by noting that nowhere beyond V2 in the visual hierarchy does the early and separate computations on form, color, disparity, and motion ever again benefit from the possibilities for close interaction as in the neighboring stripes of V2. Thus an attractive view as to the special role of V2 is as an area for the staging of the intermediates of visual computation, promoting the parallelization of the processing of certain visual submodalities, and facilitating interactions and perhaps binding among these segregated submodalities, in preparation for their distribution to the even more separate dorsal and ventral pathways.

8. References

Allman, J. M., and Kaas, J. H., 1971, Representation of the visual field in striate and adjoining cortex of the owl monkey (*Aotus trivirgatus*), *Brain Res.* **35:**89–106.

Allman, J. M., and Kaas, J. H., 1974, The organization of the second visual area (V-II) in the owl monkey: A second order transformation of the visual hemifield, *Brain Res.* **76:**247–265.

Amir, Y., Harel, M., and Malach, R., 1993, Cortical hierarchy reflected in the organization of intrinsic connections in macaque monkey visual cortex, *J. Comp. Neurol.* **334:**19–46.

Baizer, J. S., Robinson, D. L., and Dow, B. M., 1977, Visual responses of area 18 neurons in awake, behaving monkey, *J. Neurophysiol.* **40:**1024–1037.

Baker, C. L., Hess, R. F., and Zihl, J., 1991, Residual motion perception in a "motion-blind" patient, assessed with limited-lifetime random dot stimuli, *J. Neurosci.* **11:**454–461.

Blasdel, G. G., and Fitzpatrick, D., 1984, Physiological organization of layer 4 in macaque striate cortex, *J. Neurosci.* **4:**880–895.

Brodmann, K., 1909, *Vergleichende Lokalisationslehr der Grosshirnrinde,* Barth, Leipzig.

Bullier, J., and Kennedy, H., 1983, Projection of the lateral geniculate nucleus onto cortical area V2 in the macaque monkey, *Exp. Brain Res.* **53:**168–172.

Carman, G. J., and Welch, L., 1992, Three-dimensional illusory contours and surfaces, *Nature* **360:**585–587.

Cavanagh, P., 1989, Multiple analyses of orientation in the visual system, in: *Neural Mechanisms of Visual Perception* (D. M. Lam and C. D. Gilbert, eds.), Portfolio, Woodlands, TX, pp. 261–279.

Cowey, A., 1964, Projection of the retina on to striate and prestriate cortex in the squirrel monkey, *Saimiri sciureus, J. Neurophysiol.* **27:**366–393.

Curcio, C. A., and Hartin, J. K., 1978, Organization of pulvinar afferents to area 18 in the squirrel monkey: Evidence for stripes, *Brain Res.* **143:**155–161.

Daniel, P. M., and Whitteridge, D., 1961, The representation of the visual field in the cerebral cortex in monkeys, *J. Physiol.* (Lond.) **159:**302–321.

DeYoe, E. A., and Van Essen, D. C., 1985, Segregation of efferent connections and receptive field properties in visual area V2 of the macaque, *Nature* **317:**58–61.

DeYoe, E. A., and Van Essen, D. C., 1988, Concurrent processing streams in monkey visual cortex, *Trends Neurosci.* **11:**219–226.

DeYoe, E. A., Hockfield, S., Garren, H., and Van Essen, D. C., 1990, Antibody labeling of functional subdivisions in visual cortex: Cat-301 immunoreactivity in striate and extrastriate cortex of the macaque monkey, *Visual Neurosci.* **5:**67–81.

DeYoe, E. A., Felleman, D. J., Van Essen, D. C., and McClendon, E., 1994, Multiple processing streams in occipitotemporal visual cortex, *Nature* **371:**151–154.

Dobkins, K. R., and Albright, T. D., 1993, What happens if it changes color when it moves? The nature of chromatic input to macaque visual area MT, *J. Neurosci.* **14:**4854–4870.

Dow, B. M., Snyder, A. Z., Vautin, R. G., and Bauer, R., 1981, Magnification factor and receptive field size in foveal striate cortex of the monkey, *Exp. Brain Res.* **44:**213–228.

Edwards, D. P., Purpura, K. P., and Kaplan, E., (1995), Contrast sensitivity an spatial frequency response of primate cortical neurons in and around the cytochrome blobs, *Vision Res.* **35:**1501–1523.

Felleman, D. J., and Van Essen, D. C., 1991, Distributed hierarchical processing in the primate cerebral cortex, *Cerebral Cortex* **1:**1–47.

Ferrera, V. P., Nealey, T. A., and Maunsell, J. H. R., 1992, Mixed parvocellular and magnocellular geniculate signals in visual area V4, *Nature* **358:**756–758.

Fitzpatrick, D., Itoh, K., and Diamond, I. T., 1983, The laminar organization of the lateral geniculate body and the striate cortex in the squirrel monkey (Saimiri sciureus), *J. Neurosci.* **3**:673–702.

Frien, A., Eckhorn, R., Bauer, R., Woelber, T., and Kehr, H., 1994, Stimulus-specific fast oscillations at zero phase between visual areas V1 and V2 of awake monkey, *NeuroReport,* in press.

Frostig, R. D., Lieke, E. E., Ts'o, D. Y., and Grinvald, A. M., 1990, Cortical functional architecture and local coupling between neuronal activity and the microcirculation revealed by *in vivo* high-resolution optical imaging of intrinsic signals, *Proc. Natl. Acad. Sci. USA* **87**:6082–6086.

Gattass, R., Gross, C. G., and Sandell, J. H., 1981, Visual topography of V2 in the macaque, *J. Comp. Neurol.* **201**:519–539.

Gegenfurtner, K. R., Kiper, D. C., Beusmans, J. M. H., Caradini, M., Zaidi, Q., and Movshon, J. A., 1994, Chromatic properties of neurons in macaque MT, *Visual Neurosci.* **11**:455–466.

Ghose, G. M., and Ts'o, D. Y., 1997, Form processing modules in primate area V4, *J. Neurophysiol.* **77**:2191–2196.

Gilbert, C. D., and Wiesel, T. N., 1990, The influences of contextual stimuli on the orientation selectivity of cells in primary visual cortex of the cat, *Vision Res.* **11**:1689–1701.

Girard, P., and Bullier, J., 1989, Visual activity in area V2 during reversible inactivation of area 17 in the macaque monkey, *J. Neurophysiol.* **62**:1287–1302.

Gray, C. M., and Singer, W., 1989, Stimulus-specific neuronal oscillations in orientation columns of cat visual cortex, *Proc. Natl. Acad. Sci. USA* **86**:1698–1702.

Grinvald, A., Lieke, E., Frostig, R. D., Gilbert, C. D., and Wiesel, T. N., 1986, Functional architecture of cortex revealed by optical imaging of intrinsic signals, *Nature* **324**:361–364.

Grinvald, A., Frostig, R. D., Lieke, E., and Hildesheim, R., 1988, Optical imaging of neuronal activity, *Physiol. Rev.* **68**:1285–1365.

Grinvald, A., Lieke, E. E., Frostig, R. D., and Hildesheim, R., 1994, Cortical point-spread function and long-range lateral interactions revealed by real-time optical imaging of macaque monkey primary visual cortex, *J. Neurosci.* **14**:2545–2568.

Guld, C., and Bertulis, A., 1976, Representation of fovea in the striate cortex of vervet monkey, cercopithecus aethiops pygerythrus, *Vision Res.* **16**:629–631.

Haxby, J. V., Grady, C. L., Horwitz, B., Ungerleider, L. G., Mishkin, M., Carson, R. E., Herscovitch, P., Schapiro, M. B., and Rapoport, S. I., 1991, Dissociation of object and spatial visual processing pathways in human extrastriate cortex, *Proc. Natl. Acad. Sci. USA* **88**:1621–1625.

Hockfield, S., and McKay, R. D. G., 1983, A surface antigen expressed by a subset of neurons in the vertebrate central nervous system, *Proc. Natl. Acad. Sci. USA* **88**:1621–1625.

Horton, J., 1984, Cytochrome oxidase patches: A new cytoarchitectonic feature of monkey visual cortex, *Phil. Trans. R. Soc. Lond.* **304**:199–253.

Horton, J., and Hubel, D. H., 1981, A regular patchy distribution of cytochrome-oxidase staining in primary visual cortex of the macaque monkey, *Nature* **292**:762–764.

Hubel, D. H., and Livingstone, M. S., 1984, Complex-unoriented cells in a subregion of primate area 18, *Nature* **315**:325–327.

Hubel, D. H., and Livingstone, M. S., 1984, Segregation of form, color, and stereopsis in primate area 18, *J. Neurosci.* **7**:3378–3415.

Hubel, D. H., and Wiesel, T. N., 1970, Cells sensitive to binocular depth in area 18 of the macaque monkey cortex, *Nature* **225**:41–42.

Hubel, D. H., and Wiesel, T. N., 1974, Uniformity of monkey satiate cortex: A parallel relationship between field size, scatter, and magnification factor, *J. Comp. Neurol.* **158**:295–305.

Hubel, D. H., and Wiesel, T. N., 1977, Functional architecture of macaque monkey visual cortex, *Proc. R. Soc. Lond. B* **198**:1–59.

Hubel, D. H., Wiesel, T. N., and LeVay, S., 1974, Visual field representation in layer IVC of monkey striate cortex, in: *Society for Neuroscience, 4th Annual Meeting, St. Louis,* Abstract 264.

Humphrey, A. L., and Hendrickson, A. E., 1983, Background and stimulus-induced patterns of high metabolic activity in the visual cortex (area 17) of the squirrel and macaque monkey, *J. Neurosci.* **3**:345–358.

Kennedy, H., Bullier, J., 1985, A double-labeling investigation of the afferent connectivity to cortical areas V1 and V2 of the macaque monkey, *J. Neurosci.* **5**:2815–2830.

Knierem, J. J., and Van Essen, C. D., 1992, Neuronal responses to static texture patterns in area V1 of the alert macaque monkey, *J. Neurophysiol.* **67**:961–980.

Krubitzer, L. A., and Kaas, J. H., 1989, Cortical integration of parallel pathways in the visual system of primates, *Brain Res.* **478:**161–165.

Krubitzer, L. A., and Kaas, J. H., 1990a, Cortical connections of MT in four species of primates: Areal, modular, and retinotopic patterns, *Visual Neurosci.* **5:**165–204.

Krubitzer, L., and Kaas, J. H., 1990b, Convergence of processing channels in the extrastriate cortex of monkeys, *Visual Neurosci.* **5:**609–613.

Lachica, E. A., Beck, P. D., and Casagrande, V. A., 1992, Parallel pathways in macaque monkey striate cortex: Anatomically defined columns in layer III, *Proc. Natl. Acad. Sci. USA* **89:**3566–3570.

Lamme, V. A., 1995, The neurophysiology of figure–ground segregation in primary visual cortex, *J. Neurosci.* **15:**1605–1615.

Landisman, C. E., Roe, A. W., and Ts'o, D. Y., 1994, The relationship of receptive field coverage to functional modules in primate V1, *Soc. Neurosci. Abstr.* **20:**1477.

Lee, B. B., Martin, P. R., and Valberg, A., 1988, The physiological basis of heterochromatic flicker photometry demonstrated in the ganglion cells of the macaque retina, *J. Physiol.* **404:**323–347.

Lennie, P., Krauskopf, J., and Sclar, G., 1990, Chromatic mechanisms in striate cortex of macaque, *J. Neurosci.* **10:**649–669.

LeVay, S., and Voigt, T., 1988, Ocular dominance and disparity coding in cat visual cortex, *Visual Neurosci.* **1:**395–414.

Leventhal, A. G., Thompson, K. G., Liu, D., Zhou, Y., and Ault, S. J., 1995, Concomitant sensitivity to orientation, direction, and color of cells in layers 2, 3, and 4 of monkey striate cortex, *J. Neurosci.* **15:**1808–1818.

Levitt, J. B., Kiper, D. C., and Movshon, J. A., 1994a, Receptive fields and functional architecture of macaque V2, *J. Neurophysiol.* **71:**2517–2542.

Levitt, J. B., Yoshioka, T., and Lund, J. S., 1994b, Intrinsic cortical connections in macaque visual area V2: Evidence for interaction between different functional streams, *J. Comp. Neurol.* **342:**551–570.

Levitt, J. B., Yoshioka, T., and Lund, J. S., 1995, Connections between the pulvinar complex and cytochrome oxidase-defined compartments in visual area V2 of macaque monkey, *Exp. Brain Res.* **104:**419–430.

Livingstone, M. S., and D. H., 1984, Anatomy and physiology of a color system in the primate visual cortex, *J. Neurosci.* **4:**309–356.

Livingstone, M. S., and Hubel, D. H., 1987a, Connections between layer 4B of area 17 and the thick cytochrome oxidase stripes of area 18 in the squirrel monkey, *J. Neurosci.* **7:**3371–3377.

Livingstone, M. S., and Hubel, D. H., 1987b, Psychophysical evidence for separate channels for the perception of form, color, and movement, and depth, *J. Neurosci.* **7:**3416–3468.

Livingstone, M., and Hubel, D., 1988, Segregation of form, color, movement, and depth: Anatomy, physiology, and perception, *Science* **240:**740–749.

Lund, J. S., Hendrikson, A. E., Ogren, M. P., and Tobin, E. A., 1981, Anatomical organization of primate visual cortical area V2, *J. Comp. Neurol.* **202:**19–45.

Malach, R., Amir, Y., Harel, M., and Grinvald, A., 1993, Relationship between intrinsic connections and functional architecture revealed by optical imaging and *in-vitro* targeted biocytin injections in primate striate cortex, *Proc. Natl. Acad. Sci. USA* **90:**10469–10473.

Malach, R., Tootell, R. B. H., and Malonek, D., 1994, Relationship between orientation domains, cytochrome oxidase stripes, and intrinsic horizontal connections in squirrel monkey area V2, *Cerebral Cortex* **4:**151–165.

Maunsell, J. H. R., and Gibson, J. R., 1992, Visual response latencies in striate cortex of the macaque monkey, *J. Neurophysiol.* **68:**1332–1344.

Maunsell, J. H. R., and Van Essen, D. C., 1983, The connections of the middle temporal visual area (MT) in the macaque monkey, *J. Neurosci.* **3:**2563–2586.

Maunsell, J. H. R., Nealy, T. A., and Depriest, D. D., 1991, Magnocellular and parvocellular contributions to responses in the middle temporal visual area (MT) of the macaque monkey, *J. Neurosci.* **10:**3323–3334.

Merigan, W. H., 1989, Chromatic and achromatic vision of macaques: Role of the P pathway, *J. Neurosci.* **9:**776–783.

Merigan, W. H., Maunsell, J. H. R., 1993, How parallel are the visual pathways? *Annu. Rev. Neurosci.* **16:**369–402.

Merigan, W. H., Nealey, T. A., and Maunsell, J. H. R., 1993, The effects of lesions of cortical area V2 in macaques, *J. Neurosci.* **13:**3180–3191.

Mignard, M., and Malpeli, J. G., 1991, Paths of information flow through visual cortex, *Science* **251:**1249–1251.

Morgan, M. J., and Aiba, T. S., 1985, Positional acuity with chromatic stimuli, *Vision Res.* **25:**689–695.

Munk, M. H. J., Nowak, L. G., Girard, P., Chounlamountri, N., and Bullier, J., 1995, Visual latencies in cytochrome oxidase bands of macaque area V2, *Proc. Natl. Acad. Sci. USA* **92:**988–992.

Nakamura, H., Gattass, R., Desimone, R., and Ungerleider, L. G., 1993, The modular organization of projections from areas V1 and V2 to areas V4 and TEO in macaques, *J. Neurosci.* **13:**3681–3691.

Nealey, T. A., Ferrera, V. P., Maunsell, J. H. R., 1991, Magnocellular and parvocellular contributions to the ventral extrastriate cortical processing stream, *Soc. Neurosci. Abstr.* **17:**525.

Nelson, J. I., Salin, P. A., Munk, H. J., Arzi, M., and Bullier, J., 1992, Spatial and temporal coherence in cortico-cortical connections: A cross correlation study in areas 17 and 18 in the cat, *Visual Neurosci.* **9:**21–37.

Nowak, L. G., Munk, M. H. J., Girard, P., and Bullier, J., 1995a, Visual latencies in areas V1 and V2 of the macaque monkey, *Visual Neurosci.* **12:**371–384.

Nowak, L. G., Munk, M. H. J., Nelson, J. I., James, A. C., and Buller, J., 1995b, The structural basis of cortical synchronization. I. Three types of interhemispheric coupling, *J. Neurophysiol.* **74:**2379–2400.

Ogren, M. P., and Hendrickson, A. E., 1977, The distribution of pulvinar terminals in visual areas 17 and 18 of the monkey, *Brain Res.* **137:**343–350.

Peterhans, E., and von der Heydt, R., 1989, Mechanisms of contour perception in monkey visual cortex. II. Contours bridging gaps, *J. Neurosci.* **9:**1749–1763.

Peterhans, E., and von der Heydt, R., 1993, Functional organization of area V2 in the alert macaque, *Eur. J. Neurosci.* **5:**509–524.

Poggio, G. F., and Fischer, B., 1977, Binocular interaction and depth sensitivity in striate and prestriate cortex of behaving rhesus monkey, *J. Neurophysiol.* **40:**1392–1405.

Raiguel, S. E., Lagae, L., Bulyas, B., and Orban, G. A., 1989, Response latencies of visual cells in macaque areas V1, V2, and V5, *Brain Res.* **493:**155–159.

Rockland, K. S., 1985, A reticular pattern of intrinsic connections in primate area V2 (area 18), *J. Comp. Neurol.* **235:**467–478.

Rockland, K. S., 1992, Laminar distribution of neurons projecting from area V1 to V2 in macaque and squirrel monkeys, *Cerebral Cortex* **2:**38–47.

Rockland, K. S., and Pandya, D. N., 1979, Laminar origins and terminations of cortical connections to the occipital lobe in the rhesus monkey, *Brain Res.* **179:**3–20.

Rockland, K. S., and Virga, A., 1989, Terminal arbors of individual "feedback" axons projecting from area V2 to V1 in the macaque monkey: A study using immunohistochemistry of anterogradely transported *Phaseolus vulgaris*-leucoagglutinin, *J. Comp. Neurol.* **285:**54–72.

Rockland, K. S., and Virga, A., 1990, Organization of individual cortical axons projecting from area V1 (area 17) to V2 (area 18) in the macaque monkey, *Visual Neurosci.* **4:**11–28.

Rockland, K. S., Saleem, K. S., and Tanaka, K., 1994, Divergent feedback connections from areas V4 and TEO in the macaque, *Visual Neurosci.* **11:**579–600.

Roe, A. W., and Ts'o, D. Y., 1992, Functional connectivity between V1 and V2 in the primate, *Soc. Neurosci. Abstr.* **18:**11.

Roe, A. W., and Ts'o, D. Y., 1994, Relationships between topographic maps in V1 and V2 revealed by optical imaging with spot stimuli, *Soc. Neurosci. Abstr.* **20:**840.

Roe, A. W., and Ts'o, D. Y., 1995, Visual topography in primate V2: Multiple representation across functional stripes, *J. Neurosci.* **15:**3689–3715.

Roe, A. W., and Ts'o, D. Y., 1997, Functional connectivity between primate V1 and V2, submitted.

Rosa, M. G. P., Sousa, A. P. B., and Gatass, R., 1988, Representation of the visual field in the second visual area in the *Cebus* monkey, *J. Comp. Neurol.* **275:**326–345.

Sakitt, B., 1982, Why the cortical magnification factor in rhesus cannot be isotropic, *Vision Res.* **22:**417–421.

Sandell, J. H., and Schiller, P. H., 1982, Effect of cooling area 18 on striate cortex cells in the squirrel monkey, *J. Neurophysiol.* **48:**38–48.

Schiller, P. H., and Malpeli, G. J., 1977, The effect of cooling area 18 on striate cortex cells in the squirrel monkey, *Brain Res.* **126:**366–369.

Schiller, P. H., Logothetis, N. K., and Charles, E. R., 1990, Role of the color-opponent and broad-band channels in vision, *Visual Neurosci.* **5**:321–346.

Sereno, M. I., McDonald, C. T., and Allman, J. M., 1994, Analysis of retinotopic maps in extrastriate cortex, *Cerebral Cortex* **4**:601–620.

Sereno, M. I., Dale, A. M., Repas, J. B., Lwong, K. K., Belliveau, J. W., Brady, T. J., Rosen, B. R., and Tootell, R. B. H., 1995, Borders of multiple visual areas in humans revealed by functional magnetic resonance imaging, *Science* **268**:889–893.

Shapley, R., 1990, Visual sensitivity and parallel retinocortical channels, *Annu. Rev. Psychol.* **41**:635–658.

Shipp, S., and Zeki, S., 1985, Segregation of pathways leading from area V2 to areas V4 and V5 of macaque monkey, *Nature* **315**:322–325.

Shipp, S., and Zeki, S., 1989, The organization of connections between areas V5 and V2 in macaque monkey visual cortex, *Eur. J. Neurosci.* **1**:333–354.

Spatz, W. B., Tigges, J., and Tigges, M., 1970, Subcortical projections, cortical associations, and some intrinsic laminar connections of the striate cortex in the squirrel monkey (Saimiri), *J. Comp. Neurol.* **140**:155–174.

Talbot, S. A., and Marshall, W. H., 1941, Physiological studies on neural mechanisms of visual localization and discrimination, *Am. J. Opthalmol.* **24**:1255–1263.

Tigges, J., Tigges, M., Anschel, S., Cross, N. A., Letbetter, W. D., and McBride, R. L., 1981, Areal and laminar distribution of neurons interconnecting the central visual cortical areas 17, 18, 19, and MT in the squirrel monkey (Saimiri), *J. Comp. Neurol.* **202**:539–560.

Tootell, R. B. H., Silverman, M. S., De Valois, R. L., and Jacobs, G. H., 1983, Functional organization of the second cortical visual area in primates, *Science* **220**:737–739.

Tootell, R. B. H., and Hamilton, S. L., 1989, Functional anatomy of the second visual area (V2) in the macaque, *J. Neurosci.* **9**:2620–2644.

Tootell, R. B. H., Switkes, E., Silverman, M. S., and Hamilton, S. L., 1988a, Functional anatomy of macaque striate cortex. II. Retinotopic organization, *J. Neurosci.* **8**:1531–1568.

Tootell, R. B. H., Silverman, M. S., Hamilton, S. L., DeValois, R. L., and Switkes, E., 1988b, Functional anatomy of macaque striate cortex. III. Color, *J. Neurosci.* **8**:1569–1593.

Tootell, R. B. H., Reppas, J. B., Kwong, K. K., Malach, R., Born, R. T., Brady, T. J., Rosen, B. R., and Belliveau, J. W., 1995, Functional analysis of human MT and related visual cortical areas using magnetic resonance imaging, *J. Neurosci.* **15**:3215–3230.

Ts'o, D. Y., and Gilbert, C. D., 1988, The organization of chromatic and spatial interactions in the primate striate cortex, *J. Neurosci.* **8**:1712–1727.

Ts'o, D. Y., Gilbert, C. D., and Wiesel, T. N., 1986, Relationships between horizontal interactions and functional architecture in the cat striate cortex as revealed by cross-correlation analysis, *J. Neurosci.* **6**:1160–1170.

Ts'o, D. Y., Frostig, R. D., Lieke, E. E., and Grinvald, A., 1990a, Functional organization of primate visual cortex revealed by high resolution optical imaging, *Science* **249**:417–420.

Ts'o, D. Y., Gilbert, C. D., and Wiesel, T. N., 1990b, Functional architecture of color and disparity in visual area 2 of macaque monkey, *Soc. Neurosci. Abstr.* **16**:293.

Ts'o, D. Y., Gilbert, C. D., and Wiesel, T. N., 1991, Orientation selectivity of and interactions between color and disparity subcompartments in area V2 of macaque monkey, *Soc. Neurosci. Abstr.* **17**:1089.

Ts'o, D. Y., Roe, A. W., and Shey, J., 1993, Functional connectivity within V1 and V2: Patterns and dynamics, *Soc. Neurosci. Abstr.* **19**:1499.

Ts'o, D. Y., Gilbert, C. D., and Roe, A. W., 1997, Functional architecture of color and disparity processing in primate visual area 2, submitted.

Vaina, L. M., Lemay, M., Bienfang, D. C., Choi, A. Y., and Nakayama, K., 1990, Intact "biological motion" and "structure from motion" perception in a patient with impaired motion mechanisms: A case study, *Visual Neurosci.* **5**:353–369.

Valverde, F., 1985, The organizing principles of the primary visual cortex in the monkey, in: *Cerebral Cortex,* Volume 3, *Visual Cortex* (E. G., Jones, and A. A. Peters, eds.), Plenum Press, New York, pp. 207–257.

Van Essen, D. C., 1985, Functional organization of primate visual cortex, in: *Cerebral Cortex,* Volume 3, *Visual Cortex* (E. G., Jones, and A. A. Peters, eds.), Plenum Press, New York, pp. 259–329.

Van Essen, D. C., and Zeki, S. M., 1978, The topographic organization of rhesus monkey prestriate cortex, *J. Physiol.* (Lond.) **277:**193–226.

Van Essen, D. C., and Maunsell, J. H. R., 1983, Hierarchical organization and functional streams in the visual cortex, *Trends Neurosci.* **6:**370–375.

Van Essen, D. C., Newsome, W. T., and Maunsell, J. H. R., 1984, The visual field representation in striate cortex of the macaque monkey: Asymmetries, anisotropies, and individual variability, *Vision Res.* **24:**429–448.

Van Essen, D. C., Newsome, W. T., Maunsell, J. H. R., and Bixby, J. L., 1986, The projections from striate cortex (V1) to areas V2 and V3 in the macaque monkey: Asymmetries, areal boundaries, and patchy connections, *J. Comp. Neurol.* **244:**451–480.

Von der Heydt, R., and Peterhans, E., 1989, Mechanisms of contour perception in monkey visual cortex. I. Lines of pattern discontinuity, *J. Neurosci.* **9:**1731–1748.

Weller, R. E., and Kaas, J. H., 1983, Retinotopic patterns of connections of area 17 with visual areas V-II and MT in macaque monkeys, *J. Comp. Neurol.* **220:**253–279.

Wiesel, T. N., and Hubel, D. H., 1966, Spatial and chromatic interactions in the lateral geniculate body of the rhesus monkey, *J. Neurophys.* **29:**1115–1156.

Wong-Riley, M. T. T., 1979a, Changes in the visual system of monocularly sutured or enucleated cats demonstrable with cytochrome oxidase histochemistry, *Brain Res.* **171:**11–28.

Wong-Riley, M. T. T., 1979b, Columnar cortico-cortical interconnections within the visual system of the squirrel and macaque monkeys, *Brain Res.* **162:**201–217.

Wong-Riley, M. T. T., Hevner, R. F., Cutlan, R., Earnest, M., Egan, R., Frost, J., and Nguyen, T., 1993, Cytochrome oxidase in the human visual cortex: Distribution in the developing and the adult brain, *Visual Neurosci.* **10:**41–58.

Zeki, S., 1969, The secondary visual cortex of the monkey, *Brain Res.* **13:**197–226.

Zeki, S., 1990, Functional specialization in the visual cortex: The generation of separate constructs and their multistage integration, in: *Signal and Sense: Local and Global Order in Perceptual Maps* (G. M. Edelman, W. E. Gall, and W. M. Cowan, eds.), Wiley-Liss, New York, pp. 85–130.

Zeki, S., and Shipp, S., 1987, Functional segregation within area V2 of macaque monkey visual cortex, in: *Seeing Contour and Color* (J. J. Kulikowski, C. M. Dickinson, and I. J. Murray, eds.), Pergamon Press, Oxford, pp. 120–124.

Zeki, S., Watson, J. D. G., Lueck, C. J., Friston, K. J., Kennard, C., and Frackowiak, R. S. J., 1991, A direct demonstration of functional specialization in human visual cortex, *J. Neurosci.* **11:**641–649.

8

Functional Organization of Area V2 in the Awake Monkey

ESTHER PETERHANS

1. Introduction

The second area of primate visual cortex, area V2, is the first and one of the largest areas of extrastriate cortex, yet relatively little is known about its function in vision. This chapter provides insight into the functional organization of area V2 by discussing the correlation between anatomy (cytochrome oxidase pattern) and single-cell physiology.

Area V2 has an internal structure that can be demonstrated anatomically by staining for the enzyme cytochrome oxidase (Wong-Riley and Carroll, 1984; Tootell *et al.*, 1983). This structure consists of alternating thin and thick dark stripes with high concentrations of the enzyme, separated by pale stripes which contain less of the enzyme. This pattern embraces a complex network of intrinsic and extrinsic connections which has been demonstrated anatomically in tracer studies (Rockland, 1995; Levitt *et al.*, 1994b; Malach *et al.*, 1994; Rockland, 1992; Rockland and Virga, 1990; Zeki and Shipp, 1988; Livingstone and Hubel, 1987; DeYoe and van Essen, 1985; Rockland, 1985; Shipp and Zeki, 1985). On the one hand, these studies suggest that each of the three compartments of area V2 (thin, thick, and pale stripes) receives its major input from a different source of area V1; the thin stripes from the cytochrome oxidase-rich blobs of laminae

---

ESTHER PETERHANS • Department of Neurology, University Hospital Zurich, CH-8091 Zurich, Switzerland.

Cerebral Cortex, Volume 12, edited by Rockland *et al.* Plenum Press, New York, 1997.

2+3, the thick stripes from lamina 4B (magnocellular pathway), and the pale stripes from the interblob regions (parvocellular pathway). The major projections from these compartments to other cortical areas also seem to segregate: from the thin stripes to area V4 (compartments concerned with color), from the thick stripes to areas V3 and MT, and from the pale stripes to area V4 (compartments concerned with form). On the other hand, these studies also demonstrated a wide distribution of the back projections from these higher cortical areas to area V2, which suggests an overlap between functional streams. Similarly, the high density of intrinsic connections suggests a high level of neuronal interaction both within and between these compartments.

Studies which correlated the cytochrome oxidase pattern of area V2 with physiology also suggest a segregation of function. In most of these studies, the physiological results have been recorded in anesthetized and paralyzed animals (Gegenfurter *et al.*, 1996; Tamura *et al.*, 1996; Roe and Ts'o, 1995; Levitt *et al.*, 1994a; Ts'o *et al.*, 1990; Tootell and Hamilton, 1989; Hubel and Livingstone, 1987; DeYoe and Van Essen, 1985; Shipp and Zeki, 1985). These authors investigated classical neuronal properties such as selectivity for orientation and length of contrast borders, sensitivity to spatial frequency, chromaticity, direction of movement, and binocular disparity. The correlation with anatomy revealed that some functions (or combinations of functions) concentrated in specific cytochrome oxidase compartments, while others were distributed throughout. Overall, these results support the hypothesis of a complex modular structure of area V2, both at the compartmental and subcompartmental level. Nevertheless, it seems justified to conclude from these results that the three compartments of area V2 concentrate on different aspects of vision—color being dominant in the thin stripes, form and color in the pale stripes, and motion and stereoscopic depth in the thick stripes (see also Roe and Ts'o, this volume).

Comparatively little is known about this functional organization in the awake monkey. Classical neuronal properties such as orientation and direction selectivity for contrast borders and sensitivity to binocular disparity have been studied, though not in relation to the cytochrome oxidase pattern (Poggio *et al.*, 1985; Baizer *et al.*, 1977; Poggio and Fischer, 1977). In our laboratory we studied neural mechanisms of contour perception and figure–ground segregation in the awake, behaving monkey (Baumann *et al.*, 1997; Peterhans and von der Heydt, 1991a; von der Heydt *et al.*, 1984). In this chapter the correlation of these results with the cytochrome oxidase pattern will be discussed and compared with results obtained from anesthetized, paralyzed animals (for an overview of this comparison see Table I).

2. Methodological Considerations

The interpretation of the functional organization of the visual cortex depends critically on the physiological approach and on the precision of the correlation between anatomy and physiology. Therefore, I will discuss briefly the principle of our physiological experiments and of our method of reconstructing the recording sites of cortical neurons (for details see Peterhans and von der Heydt, 1993).

Table I. Functional Segregation of Monkey Area V2[a]

Cytochrome oxidase stripes	Thin		Pale		Thick	
	Anesth.	Alert	Anesth.	Alert	Anesth.	Alert
Orientation						
DeYoe and Van Essen, (1985, Table 1)	20		17		51	
Hubel and Livingstone, (1987)[b]	~25					
Levitt *et al.* (1994a, Fig. 16)	48		78		87	
Roe and Ts'o (1995, Table 2)	34		84		72	
Gegenfurter *et al.* (1996, Fig. 4B)	73		96		84	
Peterhans and von der Heydt (1993, Fig. 5)		64		82		87
Illusory contours						
Peterhans and von der Heydt (1993, Fig. 8)		3		48		50
End-stopping						
Levitt *et al.* (1994a, Fig. 16)	30		18		21	
Roe and Ts'o (1995, Table 2)	16		37		42	
Gegenfurter *et al.* (1996, p. 166)	19		41		13	
Peterhans and von der Heydt (1993, Fig. 16)		22		30		20
Color						
DeYoe and Van Essen (1995, Table 1)	86		64		16	
Levitt *et al.* (1994a, Fig. 16)	27		27		10	
Roe and Ts'o (1995, Table 2)	75		12		16	
Gegenfurter *et al.* (1996, Fig. 4B)	65		33		39	
Direction						
DeYoe and Van Essen (1985, Table 1)	7		0		19	
Levitt *et al.* (1994a, Fig. 16)	4		13		50	
Gegenfurter *et al.* (1996, Fig. 4B)	9		25		28	
Peterhans and von der Heydt (1993, Fig. 14)		21		34		30
Lines defined by motion						
Peterhans and von der Heydt (1993, Fig. 12)		3		25		52
Binocular disparity						
DeYoe and Van Essen (1985, Table 1)	33		22		68	
Roe and Ts'o (1995, Table 2)	10		1		77	
Peterhans and von der Heydt (1993, Fig. 19)		21		15		38

[a]DeYoe and Van Essen (1985) recorded mulitunit activity. For the other authors, numbers indicate the percentages of neurons selective for a particular stimulus parameter of the total number of neurons studied in a particular cytochrome oxidase strip. Animals were either anesthetized (anesth.) or alert.
[b]From Figs. 18, 19, and 22; for further results see text.

2.1. Animal Training

Rhesus monkeys were trained on a visual fixation task that reinforced foveal viewing. They were trained on this task up to a performance rate of 85% or better, and were then surgically prepared for recording. Immediately before the experiment they were retrained with visual stimuli presented besides the fixation target. By this mean they learned to concentrate on the fixation task and to ignore the experimental stimuli as much as possible. The aim of the training was to minimize 'top-down' influences on the responses of cortical neurons in order to study the more-hard-wired 'bottom-up' processes of visual perception.

2.2. Visual Stimulation

For an initial characterization of receptive fields we used classical stimuli such as light and dark bars, edges, and square-wave gratings. The stimuli and the fixation target were generated on a flat-faced, high-resolution oscilloscope, two images side by side (one for each eye), presented to the animal via a stereoscope. Binocular disparity was introduced by varying for each eye the relative (horizontal) position of the stimulus with respect to the fixation target. More complex aspects of contour processing and figure–ground segregation were studied with illusory-contour stimuli as shown in Figs. 5 and 7, and lines defined by coherent motion as shown in Fig. 8. To correlate neuronal function and perception, we compared the responses elicited in stimulus conditions that did or did not induce such percepts.

2.3. Relating Anatomy and Physiology

The recording chambers were implanted over the operculum, with its center about 12–15 mm lateral to the midsagittal plane and 4–6 mm occipital to the lunate sulcus (Fig. 1A). The microelectrodes were inserted *perpendicular* to the cortical surface (Fig. 1B), which allowed access in sequence to areas V1 and V2 and to the area V3/V3A complex. The animals were studied over periods of up to 2 years. Thus, only the last 5–7 microelectrode tracks could be marked by electrolytic lesions. Immediately before perfusion of the animal, tungsten pins were inserted as additional markers (see also Poggio *et al.*, 1977). They were inserted at the limits of the X/Y stage that had been used during the experiment for positioning of the microelectrode (Fig. 1A, open circles). These pins marked the block of brain studied and defined two orthogonal planes which ran approximately parallel (X) and orthogonal (Y) to the lunate sulcus. After perfusion, the blocks of brain were removed, sectioned, and stained alternatingly for Nissl substance and cytochrome oxidase. For reconstruction of the cytochrome oxidase pattern of area V2, the sections were cut parallel to the X plane, photographed, and aligned using the Y plane for reference. This technique allowed a reconstruction of the microelectrode tracks both with regard to cortical lamination and cytochrome oxidase pattern with a precision that was comparable to reconstructions made in acute experiments with anesthetized animals. An example of such a reconstruction is shown in Fig. 2. The outline of this cytochrome oxidase pattern and the distribution of the microelectrode penetrations made in this area are shown in Fig. 3B. The mean position and the scatter of the receptive fields mapped in each penetration are shown in Fig. 3A. They represented the central 1–6 deg of visual field.

3. Visual Contours

Pyschophysical evidence suggests that contour perception involves several mechanisms which are activated either alone or in conjunction. On the retinal

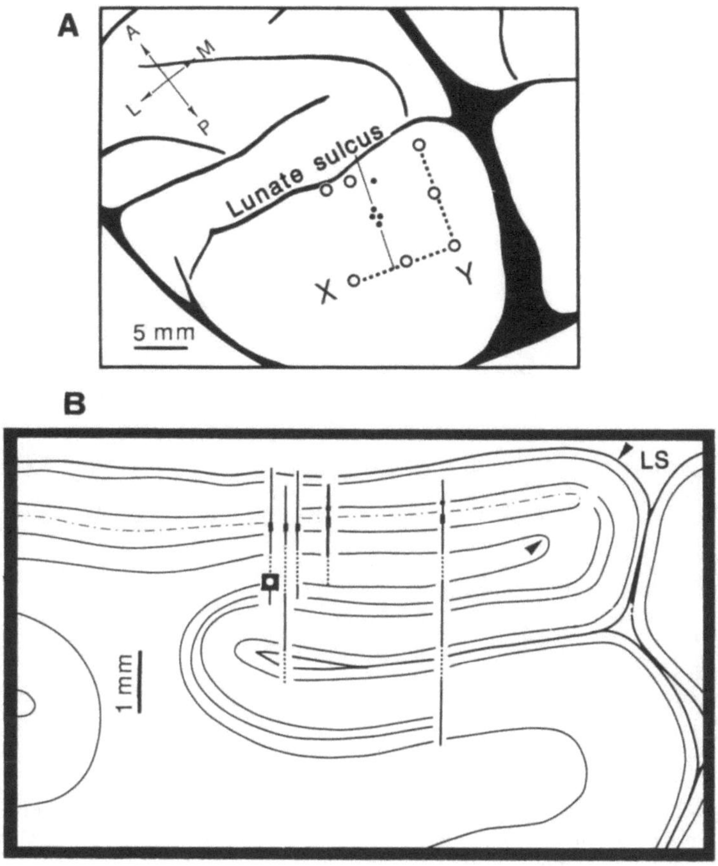

Figure 1. Access to area V2 as studied in the alert monkey. (A) Occipital view of the surface of the left hemisphere of a rhesus monkey. At the end of the experiment the block of brain studied was marked by inserting tungsten pins perpendicular to the cortical surface (open circles); the two planes defined by the marker pins (X and Y) served as references for sectioning and reconstruction. (B) Section taken at the position of the straight line in A. It shows the reconstruction of five microelectrode tracks marked by dots in A. The filled rectangles indicate high level of background activity; the white dot on a dark square marks a neuron sensitive to illusory contours. The arrows indicate the V1/V2 border. Reproduced with permission from Peterhans and von der Heydt (1993).

level, contours can be defined by discontinuities of luminance, chromaticity, motion, texture, or binocular disparity. Each mechanism alone is sufficient to induce the perception of a contour or an object boundary. In natural vision these mechanisms are often activated together, a conjunction which enhances perceptual saliency and stability. This section is devoted to contours defined by luminance contrast or chromaticity, Section 4 to boundaries produced by more complex image configurations which require the detection of occlusion cues, coherent motion, or binocular disparity. By definition, we called a neuron sensitive to a particular type of contour if it was activated by a stimulus which induced that contour in perception, and if it was selective for the orientation of this contour. (For a comparison with results obtained in anesthetized, paralyzed animals, see Table I.)

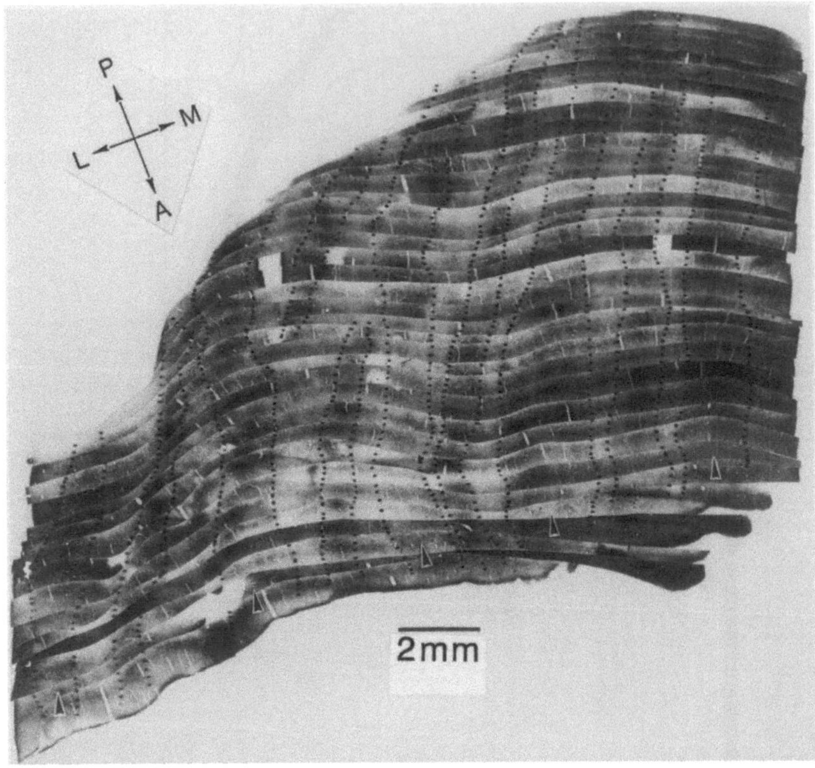

Figure 2. Cytochrome oxidase pattern of area V2 (rhesus monkey, right hemisphere). Reconstruction from serial sections made parallel to the X plane of Fig. 1A, stained for cytochrome oxidase, enlarged photographically, and mounted using the Y plane of Fig. 1A for reference. The dotted lines show the stripe borders as determined by eye, and subsequently verified by densitometry (for details see Peterhans and von der Heydt, 1993). Reproduced with permission from Peterhans and von der Heydt (1993).

3.1. Contrast Borders

3.1.1. Lines, Edges

In the awake monkey, orientation-selective neurons as determined with contrast borders (lines, edges) were found in all compartments of the cytochrome oxidase pattern, though somewhat less frequently in the thin stripes (82% and 87% in the pale and in the thick stripes, respectively, 64% in the thin stripes; Peterhans and von der Heydt, 1993). Some studies of anesthetized monkeys suggested similar or even higher proportions of orientation-selective cells (Levitt *et al.*, 1994a; Gegenfurter *et al.*, 1996). Others suggested lower percentages of oriented cells in the thin stripes (34%, Roe and Ts'o, 1995; about 25%, Hubel and Livingstone, 1987; and 20%, DeYoe and Van Essen, 1985). The reason for these different proportions is not clear. Considering the findings of Levitt *et al.* (1994a) and Gegenfurter *et al.* (1996), we see that the state of the animal cannot be the only explanation. More likely, these differences may be due to differences in classification, to a sampling bias induced by the subcompart-

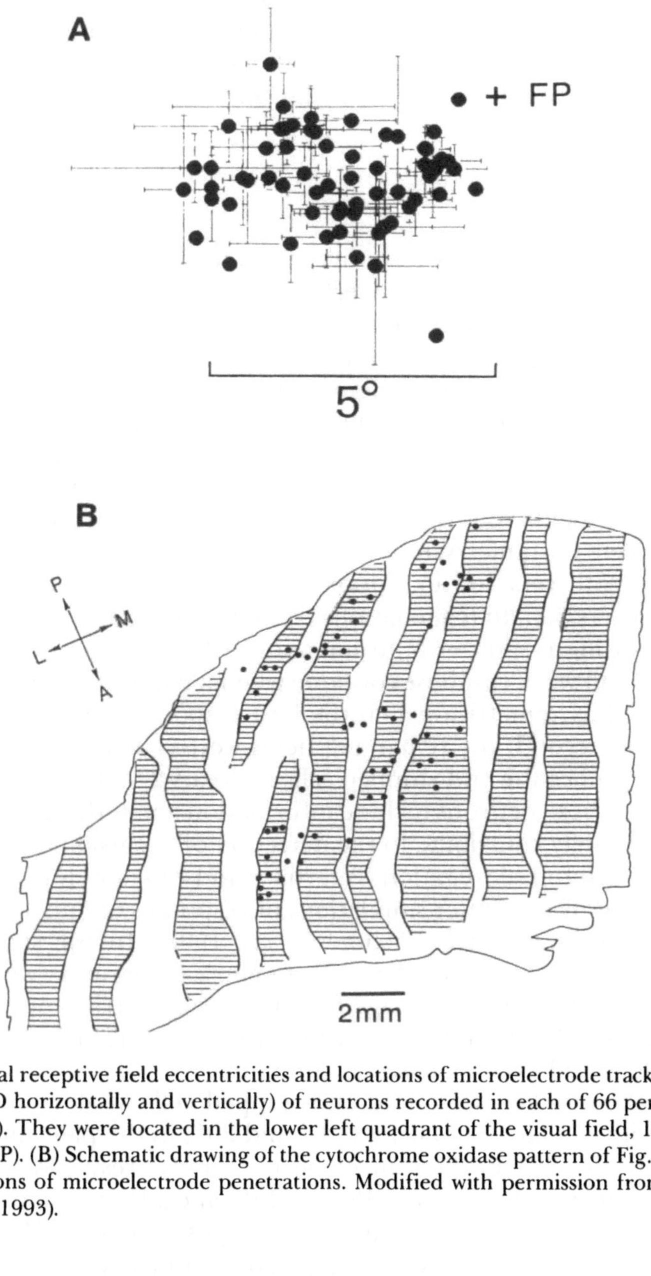

Figure 3. Typical receptive field eccentricities and locations of microelectrode tracks. (A) Mean field positions (±1 SD horizontally and vertically) of neurons recorded in each of 66 penetrations shown in panel B (dots). They were located in the lower left quadrant of the visual field, 1–6 deg from the fixation point (FP). (B) Schematic drawing of the cytochrome oxidase pattern of Fig. 2; the filled dots mark the positions of microelectrode penetrations. Modified with permission from Peterhans and von der Heydt (1993).

mental structure of the stripe pattern (Roe and Ts'o, 1995; Tootell and Hamilton, 1989), and to complex intrinsic connections which suggest interactions between stripes (Levitt *et al.,* 1994b; Rockland, 1985). Nevertheless, it seems justified to conclude that orientation selectivity is a prominent neuronal property in the pale and in the thick stripes, but less so in the thin stripes. Selectivity for the direction of movement of contrast borders was not a prominent property of neurons in the awake monkey (Peterhans and von der Heydt, 1993). Direction-selective neurons were found in all types of stripes and with similar frequencies

(34% and 30% in the pale and in the thick stripes, respectively, 21% in the thin stripes). In the anesthetized monkey, the overall incidence of direction-selective neurons was also low (Zeki, 1978). Some authors found that these neurons clustered in the thick stripes (Levitt *et al.*, 1994a; DeYoe and Van Essen, 1985; Shipp and Zeki, 1985).

3.1.2. Line-Ends, Corners

Neurons with end-stopped receptive fields were found in area V2 of both anesthetized and awake monkeys (Peterhans and von der Heydt, 1991b; Hubel and Livingstone, 1987; Burkhalter and Van Essen, 1986). These neurons preferred short bars or edges and gave only weak responses or none to long stimuli. Hubel and Wiesel (1968), in their original study of end-stopped cells in monkey visual cortex (area V1), found two types of cell: (a) 'single-stopped cells' with a single inhibitory zone at one end of the receptive field, and (b) 'double-stopped cells' with two inhibitory zones, one at either end of the field. In the awake monkey, these two field types represented the extremities of a continuum ranging from complete asymmetry to complete symmetry (Peterhans and von der Heydt, 1991b). With very few exceptions, end-stopped cells also responded to line-ends or corners, i.e., to terminations of contours. Neurons with asymmetrical fields gave strong responses to terminations when the stimulus covered one end of the field and weak responses or none when it covered the other. Receptive fields were called asymmetrical when the first response was twice as strong or stronger than the second; the remaining fields were called symmetrical. By this criterion, about half of the end-stopped cells of area V2 had asymmetrical fields. The responses of these neurons include information about the *direction* of terminations, and thus may contribute to the detection of occlusion cues and mechanisms of figure–ground segregation (Peterhans and von der Heydt, 1991b). This result is supported by computer simulations of responses of end-stopped cells to binary and gray-valued images (Heitger *et al.*, 1992). In the simulation only completely symmetrical and completely asymmetrical operators were used. The two types reproduced different image features. Asymmetrical operators reproduced the position and direction of terminations. Symmetrical operators were activated by blobs and points of strong curvature.

In the awake monkey, we found end-stopped cells in all compartments of area V2, and with similar frequencies (30% and 20% in the pale and in the thick stripes, respectively, 22% in the thin stripes). A similar result was found in the anesthetized monkey by Levitt *et al.* (1994a) (18% and 21% in the pale and in the thick stripes, respectively, 30% in the thin stripes), whereas Hubel and Livingstone (1987) and Gegenfurter *et al.* (1996) found end-stopped cells most frequently in the pale stripes.

3.2. Color Borders

Less is known about representations of color borders. In area V2 of the anesthetized monkey, neurons have been reported to be selective for the orientation of chromaticity borders (Gegenfurter *et al.*, 1996; Tamura *et al.*, 1996;

Burkhalter and Van Essen, 1986). Several reports suggested that these neurons cluster in the thin and in the pale cytochrome oxidase stripes (Levitt *et al.* 1994a), or at the borders between thin and pale stripes (Roe and Ts'o, 1995; Ts'o *et al.*, 1990). Most of these neurons responded to both contrast and chromaticity borders, but some clearly preferred chromaticity borders and responded in equiluminant conditions. Overall, neurons preferring chromaticity borders seemed to be less common than neurons sensitive to contrast borders. Recent studies specifically tested for a combined selectivity for orientation and chromaticity in neurons of area V2 and found that about 21–43% of the neurons were of this type (Gegenfurter *et al.*, 1996; Tamura *et al.*, 1996). In our study in the awake monkey we had no color stimuli available. However, all neurons that were unresponsive to our monochromatic stimulation concentrated in the thin stripes (Peterhans and von der Heydt, 1993). It is tempting to conclude that these neurons were of the unoriented type with a strong preference for chromatic stimuli, neurons which seem to concentrate in the thin stripes (Hubel and Livingstone, 1987).

4. Mechanisms of Figure–Ground Segregation

Contrast and chromaticity borders are insufficient for defining object borders. In cluttered visual scenes borders fade when objects that occlude one another have similar texture, luminance, or color. Nevertheless, the visual system is efficient at segmenting such scenes. Physiological evidence suggests that area V2 includes three mechanisms which contribute to this task, the first using occlusion cues, the second motion cues, and the third stereoscopic cues.

4.1. Occlusion Cues

In the retinal image, cluttered visual scenes produce occlusion cues such as line-ends, corners, and different types of junctions (particularly *T*-junctions) indicating spatial interposition (Nakayama *et al.*, 1989). Such cues indicate the position and direction of occluding objects, and the visual system can use them to infer occluding contours and the direction of figure (occluding surface) and ground at such contours. Recently it has been shown that similar mechanisms also initiate the depth segregation of moving surfaces (Stoner and Albright, 1996).

4.1.1. Illusory Contours

In situations of spatial occlusion, the visual system uses occlusion cues to produce illusory contours at sites of fading contrast (Petry and Meyer, 1987). This mechanism enhances the perception of occluding contours and thus stabilizes the segregation of figure and ground (Peterhans and von der Heydt, 1991a). Two examples of artificial figures illustrating this point are shown in Fig. 4. Both Figs. 4A and 4B induce the perception of an opaque surface bounded by illusory contours which appears to overlay a set of striped objects in the back-

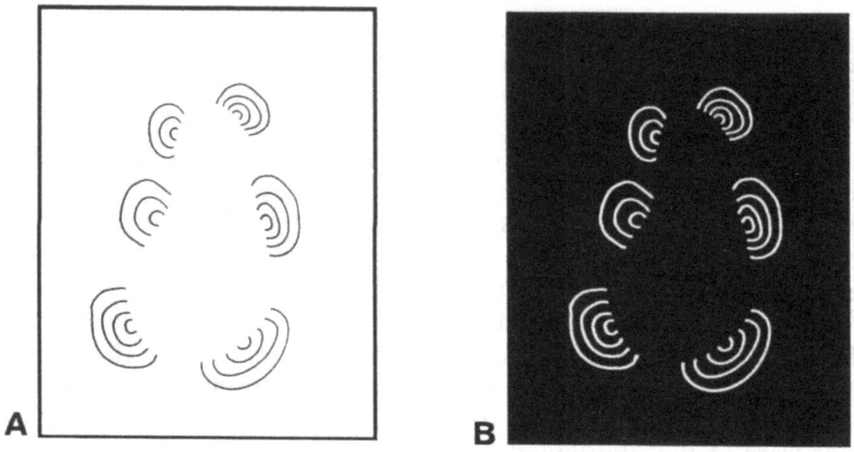

Figure 4. Illusory contour figures which mimic situations of spatial occlusion. (A) A bright surface bounded by illusory contours appears to overlay a set of striped objects in the background. (B) The same figure, but opposite contrast polarity. Modified with permission from Kanizsa (1979).

ground. Depending on the contrast polarity of the occlusion cues (line-ends defining the illusory contour), the occluding surface appears brighter (Fig. 4A) or darker (Fig. 4B) than the background. The figures show that it is the direction of the occlusion cues which determines the depth order at the contour, and that this mechanism is independent of contrast polarity. Both dark (Fig. 4A) and light (Fig. 4B) line-ends produce the same figure–ground direction as long as they point in the same direction.

We studied whether neurons that were selective for the orientation of contrast borders showed a similar sensitivity for illusory borders (Peterhans and von der Heydt, 1989; von der Heydt and Peterhans, 1989). The results suggest that about one-third of the neurons of area V2 show this property. An example of the responses of such a neuron is shown in Fig. 5. The ellipses indicate the neuron's response field for a dark bar. The stimuli were moved back and forth over this field; the dot displays show the responses to a solid bar (A), to an illusory bar (B), and to an illusory contour between abutting line gratings (C). The neuron was not spontaneously active, as one can see from the bottom display (D). While these records were made, the monkey concentrated on its fixation task, and thereby ignored the experimental stimuli as much as possible. (Note that the fixation target is indicated by a cross to the left of stimulus A.) The neuron preferred similar orientations for all three types of contour. The responses to the two illusory-contour stimuli (B, C) were similar, but weaker than those to the solid bar (A).

In neurons of area V2, the response strength to illusory-contour stimuli was on the average about 60% of that to bars or edges. This result parallels perception—illusory contours appear more frail than comparable contrast borders. In a series of control experiments we correlated the responses to illusory-contour stimuli with perception. The results suggest parallels, particularly with regard to the number, the spatial position, and the orientation of the inducing elements. For example, at least two stimulus elements are needed to perceive an illusory contour (Schumann, 1900), and similarly at least two corners (stimulus B in Fig.

5) or line ends (stimulus C in Fig. 5) were needed for a neuronal response. A single stimulus element usually failed to evoke a response, or it produced similar activity at all orientations, a result not related to the contour. Line-ends oriented orthogonal to the contour evoked the strongest response in cortical neurons, and similarly produce the strongest contours in perception (Westheimer and Li, 1996; Soriano *et al.*, 1996; Kennedy, 1978). The responses of cortical neurons weakened when the spacing of the inducing elements was made wider than about 1 deg, a condition which also reduced the saliency of illusory contours in perception (Soriano *et al.*, 1996).

Apart from one exception, we failed to record these responses in area V1. However, Grosof *et al.* (1993), using a different set of stimuli, reported that about half of the 25 neurons they studied in area V1 of the anesthetized monkey did signal illusory contours. Their stimuli consisted of either abutting sine-wave gratings displaced by half a cycle (spatial frequencies 0.7–1.3 cycles/deg), or edges of single sine-wave gratings which had the same mean luminance as the abutting uniform surface. The authors tested the correlation between neuronal responses and perception by analyzing the sensitivity of neurons to the spatial phase of the inducing gratings. Only neurons not sensitive to spatial phase were called sensitive to illusory contours. However, responses invariant with respect to spatial phase can also be produced by the alternating light and dark stimulus elements, since each of these element was bounded by a local edge at the tip, either light/dark or reverse. Further, the authors did not report how many stimulus elements these neurons actually needed for response. Also, they found line width to be a critical stimulus parameter, a parameter which determines the length of the edge at the tip. None of the neurons they studied responded to

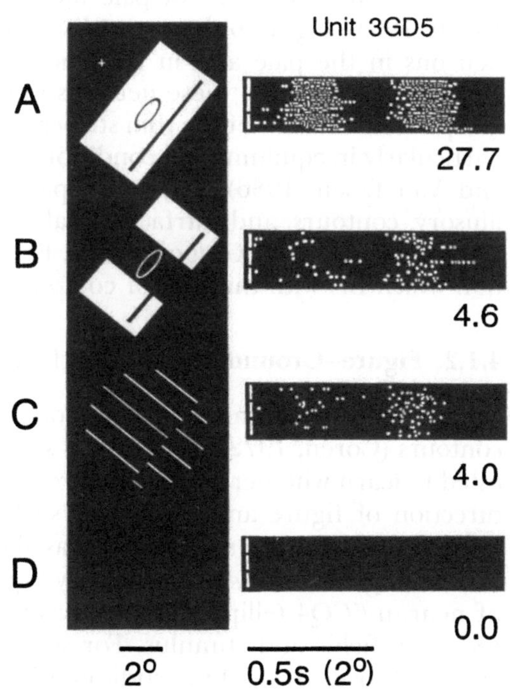

Figure 5. Responses of a 'contour neuron' of area V2 (thick stripes, lamina 4). It responded to a dark bar of optimal orientation (A), to an illusory bar (B), and to a contour between abutting line gratings (C). The stimuli were moved over the response field as mapped with a dark bar (ellipse). Each dot display shows responses for 24 cycles of stimulus movement (forth and back sweep on the left and right halves, respectively). Numbers underneath represent mean numbers of spikes per motion cycle. The bottom display (D) shows spontaneous activity for an equivalent time interval. Reproduced with permission from Peterhans and von der Heydt (1991b).

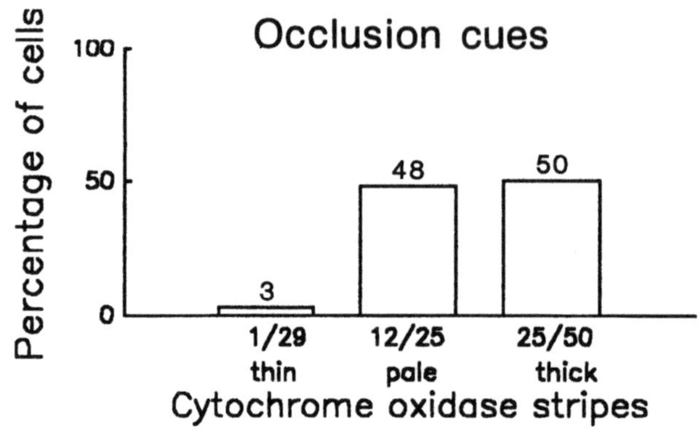

Figure 6. Distribution of contour neurons across the cytochrome oxidase pattern. About half of the neurons in the pale and the thick stripes were of this type; only one neuron was recorded in the thin stripes. Modified with permission from Peterhans and von der Heydt (1993).

abutting line-gratings of the type we used in the awake monkey (line width, 1 min arc).

In the visual cortex of the anesthetized cat, Redies *et al.* (1986) and Leventhal and Zhou (1994) found that sensitivity to illusory-contour stimuli were found more frequently in extrastriate cortex (areas 18 and LS) than in striate cortex, where simple cells predominate. These results were confirmed by a recent study that combined two techniques (optical and single cell recording) for showing representations of illusory contours in cat visual cortex (Sheth *et al.*, 1996). This study further showed that this representation of illusory contours is regular, forming orientation maps just as had been shown for solid contrast borders (Bonhoeffer and Grinvald, 1993).

In our study, we have correlated sensitivity to illusory contours with the cytochrome oxidase pattern of area V2 and found that neurons showing this property concentrated in the pale and in the thick stripes (Peterhans and von der Heydt, 1993). Figure 6 shows the distribution; one can see that about half of the neurons in the pale and in the thick stripes showed this property. With the exception of one cell, these neurons were not found in the thin stripes. If one accepts that neurons of the thin stripes are concerned with chromaticity borders, particularly in equiluminant conditions (Roe and Ts'o, 1995; see also Burkhalter and Van Essen, 1986), this result parallels perception. It is well known that illusory contours and surfaces weaken or fade in equiluminant conditions (Brussell *et al.*, 1977; Gregory, 1977; Brigner and Gallagher, 1974), a phenomenon which fits with this lack of contour neurons in the thin stripes.

4.1.2. Figure–Ground Segregation from Occlusion Cues

Since it is well known that illusory contours often coincide with occluding contours (Coren, 1972; for a review see Petry and Meyer, 1987), we were interested to learn whether contour neurons of area V2 also provide a signal for the direction of figure and ground at such contours (Baumann *et al.*, 1997). We studied this question using stimuli as shown in Fig. 7 (A–D), stimuli that were akin to single elements of the illusory figures shown in Fig. 4. The response field of neuron 6CQ4 (ellipse) is shown as an example of the relationship between receptive fields and stimulus. For stimulation, the contours were moved back and forth over this field as indicated by arrows, or were kept stationary in the center of the field. In human observers, the stimuli of Fig. 7 induced the perception of a light (A, C) or dark (B, D) tongue sliding over a line-grating of opposite contrast. These stimuli were designed to reduce the figure–ground problem to a single, straight contour, a stimulus known to be effective in cortical neurons. However, such a simplified stimulus has the disadvantage that it can be ambiguous in human perception. Indeed, when one looks at the stationary stimuli shown in Fig. 7 the perception alternates between an occluding rectangle and a notch cut out of the grating pattern. This ambiguity is reduced in the moving condition, where the first percept dominates. Nevertheless, the signals conveyed at the neuronal level are unambiguous. They indicate the spatial arrangement and the direction of line terminations (occlusion cues). Such local signals can represent basic elements of a mechanism which, at later stages of processing, defines more global aspects of figure–ground segregation.

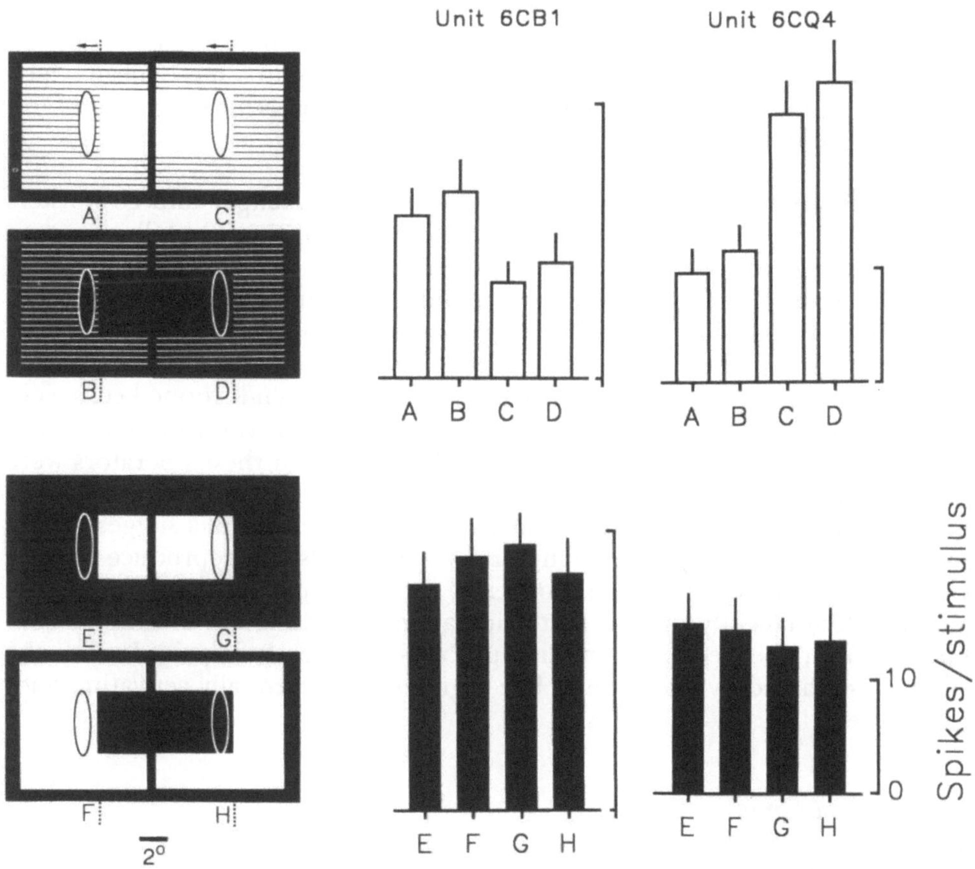

Figure 7. Sensitivity to figure–ground direction of two neurons of area V2. Neuron (6CB1) preferred contours with terminations pointing to the right (A, B), neuron (6CQ4) contours with terminations pointing to the left (C, D). Note that this result was independent of contrast polarity: First, the contours of each pair evoked similar responses although they had opposite contrast polarity. Second, all solid edges (E–H) evoked similar responses. Bars show mean responses to 24 cycles of stimulus movement (frequency: 1.5 Hz, unit 6CB1; 1 Hz, unit 6CQ4); vertical lines indicate standard deviations. The ellipses mark the response field of unit 6CQ4 as defined with a bar stimulus. Reproduced with permission from Baumann *et al.* (1996).

At each orientation there are two possible directions of figure and ground which are represented by different neurons, one for each direction. An example of such a neuronal pair is shown in Fig. 7. Neuron 6CB1 gave stronger responses to the contours A and B than to the contours C and D; and the reverse was true for neuron 6CQ4. Thus, neuron 6CB1 preferred contours with the occluding surface to the right, and neuron 6CQ4 contours with the occluding surface to the left. These responses were independent of contrast polarity—note that the contours of each stimulus pair (A, B, and C, D) had opposite contrast polarity. This result was confirmed by the responses to the corresponding solid edges. Both neurons gave similar responses to all four types of edges (Fig. 7E–7H). We have quantified neuronal sensitivity to figure-ground direction by the following index (see also Baumann *et al.*, 1997):

$$I_S = |(R_A + R_B) - (R_C + R_D)/(R_A + R_B + R_C + R_D)|$$

In the 66 neurons studied in area V1 we never found an index greater than 0.2, whereas 22% of the neurons of area V2 (25/116) had an index greater than 0.2. This result suggests that neurons of area V2 contribute to a mechanism which uses the position and direction of occlusion cues to infer occluding contours and the direction of figure and ground at such contours.

Computational Study. In a computational study Heitger and co-workers (Heitger *et al.*, 1997; Heitger and von der Heydt, 1993) simulated illusory contours and the perceived depth order at such contours using a simple, feedforward model based on our physiological results (Peterhans *et al.*, 1986; Baumann *et al.*, 1997). In the simulation, the image was convolved with different sets of operators which mimicked the function of cortical neurons. In succession, operators simulating the properties of simple, complex, and end-stopped cells were used. The activity of end-stopped operators with asymmetrical fields was used to reproduce the direction of occlusion cues. The signals of these operators were then used in a nonlinear grouping process to produce illusory contours and the direction of figure and ground at such contours. These results suggest that a simple feedforward model involving end-stopped cells can reproduce illusory contours and the perceived depth order at such contours. Indeed, Westheimer and Li (1996) recently showed that orientation discrimination of illusory contours (contours between abutting line gratings) is severely impaired when the stimulus is masked by a texture of line segments preferentially activating end-stopped cells.

4.2. Motion Cues

4.2.1. Lines Defined by Coherent Motion

The results reviewed so far suggest that neurons of area V2 contribute to figure–ground segregation from occlusion cues. Since motion is another effective cue for defining figure and ground in perception, we studied whether area V2 also includes neurons which might be involved in this task. In preliminary experiments, we found neurons that were selective for the orientation of contrast borders and rows of dots moving on a uniformly illuminated background (Peterhans and von der Heydt, 1991b). Rows of dots (size 1 min arc, spacing 12–60 min arc) and contrast lines evoked similar responses as long as the two types of stimulus had the same overall size and the dots were moved coherently. When the dots were moved out of phase, these responses of these neurons were reduced or abolished. The neurons tolerated large misalignments of the dots (sometimes up to three times the width of their response field as mapped with a solid bar), as long as they were moved coherently. These neurons were orientation–selective for such stimuli as long as the stimuli induced elongated objects in perception. Circular objects evoked similar responses at all orientations.

More recently, we found that these neurons also responded when the dots were moved relative to a background of stationary dots of similar density, a stimulus condition in which motion was the only cue for response (Peterhans and Baumann, 1994). To study the correlation between neuronal response and perception, we have tested these neurons in four stimulus conditions, as shown in Fig. 8 (note that the row of dots was enlarged for ease of discrimination; during stimulation all dots had the same size). In conditions A and B the row of dots was moved relative to a background of dots that was either stationary (A) or moving in antiphase (B). In condition C all dots were moved in phase, and in condition D all dots were kept stationary. In conditions A and B a human observer perceived a rigid, rodlike object that appeared to oscillate back and forth relative to a dotted background of approximately square shape. Conditions C and D were control conditions which failed to induce such percepts. Rather, they were perceived as dotted surfaces that either moved as a whole C or were stationary D. Figure 8 shows the responses of three neurons of area V2 to the four stimulus conditions. One can see that all neurons gave stronger responses to conditions A and B than to the control conditions C and D. All neurons of this type were as selective for the orientation of motion-defined lines as they were for lines defined by luminance contrast. Interestingly, these neurons were often not selective for the direction of movement of the line itself. Thus, rather than being involved in local aspects of motion analysis, these neurons seem to contribute to more complex aspects of dynamic form analysis. The results suggest that these neurons detect orientation from motion cues and thus may contribute to mechanisms segregating figure and ground based on motion cues.

Neurons sensitive to lines defined by coherent motion concentrated in the thick cytochrome oxidase stripes, but some were also found in the pale stripes (Peterhans and von der Heydt, 1993). Figure 9 shows the distribution. With one exception, these neurons were not found in the thin stripes. This result parallels perception, since form perception from motion cues is much reduced or lost in equiluminant conditions. For example, in random-dot kinematograms, object boundaries disappear, and figure–ground segregation is lost in these conditions, even though the movement of individual dots is still perceived (Cavanagh *et al.*, 1985; Ramachandran and Gregory, 1978). In the thick stripes of area V2 about half of the neurons showed this property, and this proportion was even higher (75%) in the area V3/V3A complex (Peterhans and Baumann, 1994). This fits with the anatomical connection of the two regions, and with the fact that both receive input from lamina 4B of area V1 (see Zeki and Shipp, 1988, for review).

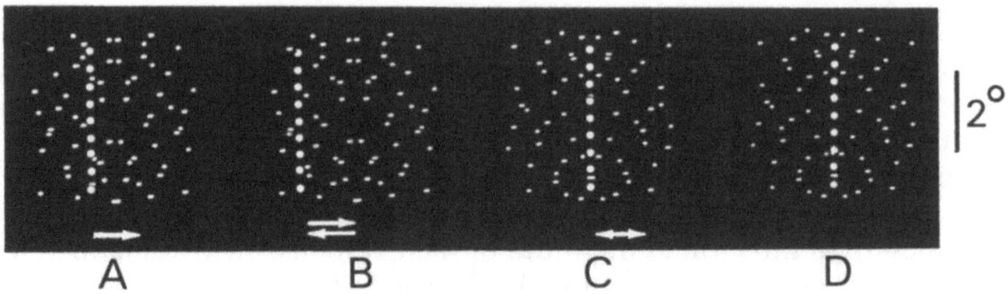

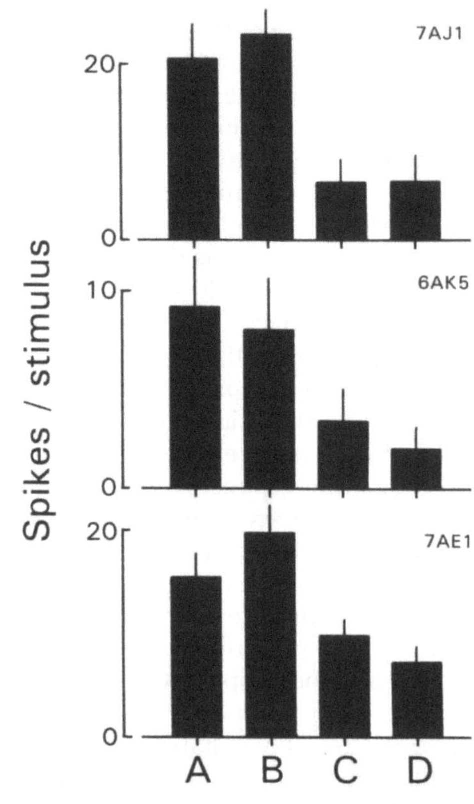

Figure 8. Sensitivity to lines defined by coherent motion of three neurons of area V2. A row of dots produced a response when the dots were moved coherently and relative to a dotted background that was either kept stationary (A) or moved in antiphase (B). These responses were reduced when the dots and the background were moved in phase (C) or were kept stationary (D). This result parallels perception—conditions A and B evoked the percept of an oscillating, rodlike object that was not perceived in conditions C and D. All dots were 1 min arc in size (the row of dots has been enlarged for illustration only). Bars show mean responses to 24 cycles of stimulus movement (or a corresponding time interval for stationary stimuli); vertical lines indicate standard deviations. (E. Peterhans and R. Baumann, unpublished results.)

4.2.2. Figure–Ground Segregation from Motion Cues

Neurons sensitive to lines defined by coherent motion, as described above, have not been recorded in striate cortex. A similar result was found for neurons sensitive to kinetic boundaries (Orban, 1994). Further, neurons of area V1 of the

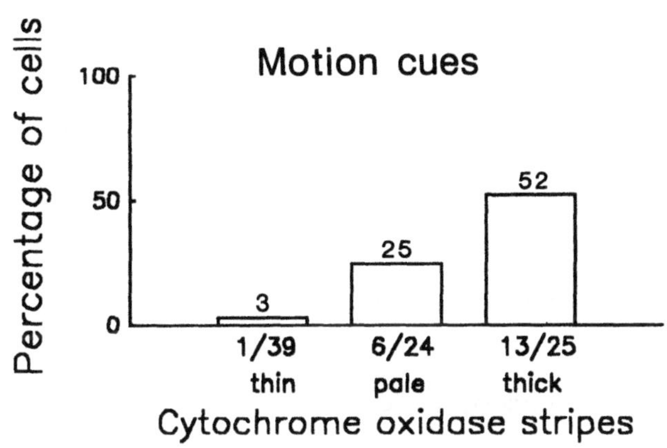

Figure 9. Distribution of neurons sensitive to coherence of motion across the cytochrome oxidase pattern. These neurons were found most frequently in the thick stripes, less frequently in the pale stripes, and generally not in the thin stripes. Modified with permission from Peterhans and von der Heydt (1993).

anesthetized owl monkey were generally poorly activated by patches of randomly spaced dots that were moved relative to a background of dots of similar density (Allman *et al.,* 1990), and V1 neurons of the anesthetized rhesus monkey were almost unaffected by random-dot stimuli that induced transparent surfaces in human perception (Snowden *et al.,* 1991). With plaid patterns, neurons of area V1 of anesthetized rhesus monkeys were sensitive to the direction of movement of each component, but not to the perceived motion direction of the plaid (Movshon *et al.,* 1985). This suggests that area V1 deals with local aspects of motion analysis involving the local direction and speed of contrast borders (see also Orban, 1994).

More complex mechanisms of motion processing, though not directly related to form vision, have been found in area MT (or V5). Allman *et al.* (1985) found neurons in area MT of the anesthetized owl monkey that were sensitive to patches of randomly spaced dots that moved *relative* to a background of dots of similar density. These neurons, although selective for the direction of motion of the patch, were not sensitive to its form. Marcar *et al.* (1995) studied representations of kinetic boundaries in the anesthetized rhesus monkey. They found that neurons of area MT were selective for the direction of motion of the stimulus texture, but not for the orientation of the boundaries. Snowden *et al.* (1991) recorded responses in area MT that were related to the perception of transparent surfaces, and Movshon *et al.* (1985) found neurons that were sensitive to the direction of motion of plaid patterns. These results suggest that neuronal responses in area MT are related to surface qualities, and not to local contours or object forms (see Stoner and Albright, 1993, for a review). These neurons may contribute to more global aspects of scene segmentation which relate to the depth order of surfaces. Recent studies in the awake monkey challenge the view that such representations can be found in area MT but not in area V1 (Lamme *et al.,* 1995; Zipser *et al.,* 1996). These studies showed that neurons of area V1 can be sensitive to patches of texture which differ from a surrounding texture in luminance, color, orientation, direction of motion, or stereoscopic depth.

The results of studies in the awake monkey (Peterhans and von der Heydt, 1993, Peterhans and Baumann, 1994) suggest that neurons of area V2 and the area V3/V3A complex contribute to form vision based on motion cues. Other types of neurons also involved in dynamic form analysis have been found in areas V2 and V4 and in the inferotemporal cortex. For example, Marcar *et al.* (1992)

recorded from neurons in area V2 of the anesthetized monkey that were selective for the orientation of kinetic boundaries, and similar neurons were found in area V4 (Logothetis and Charles, 1990). Sáry *et al.* (1993) reported neurons in the inferotemporal cortex of the awake monkey that were sensitive to two-dimensional forms (triangles, stars) defined by kinetic boundaries. Although the precise anatomical connections of these areas is still under investigation (see Felleman and Van Essen, 1991, for a review), the physiology suggests that they are functionally related in that they all contribute to form perception and figure–ground segregation from motion cues.

4.3. Stereoscopic Cues

4.3.1. Contrast Borders

Poggio and Fischer (1977) found neurons that were sensitive to the binocular disparity of contrast borders already in area V1. These neurons were also recorded in the adjacent cortical region, in area V2 of both anesthetized and awake monkeys (Burkhalter and Van Essen, 1986; Poggio and Fischer, 1977; Hubel and Wiesel, 1970). More recent studies correlated this neuronal property with the cytochrome oxidase pattern of area V2 (Roe and Ts'o, 1995; Peterhans and von der Heydt, 1993; Hubel and Livingstone, 1987; DeYoe and Van Essen, 1985). The results showed that disparity-sensitive neurons occurred in clusters, and that these clusters concentrated in the thick, dark cytochrome oxidase stripes. Figure 10 shows the distribution for the awake monkey. These recordings also revealed that neurons which failed to respond to single-eye stimulation (binocular neurons) also concentrated in the thick stripes. This suggests that it is mainly the thick stripes which contribute to mechanisms of stereoscopic depth perception. The neurons of these stripes may provide basic information for defining the depth order of contrast borders from stereoscopic cues. This fits with the anatomical connections of these stripes with the area V3/V3A complex and with area MT, both regions which include neurons sensitive to binocular disparity (Felleman and Van Essen, 1987; Maunsell and Van Essen, 1983).

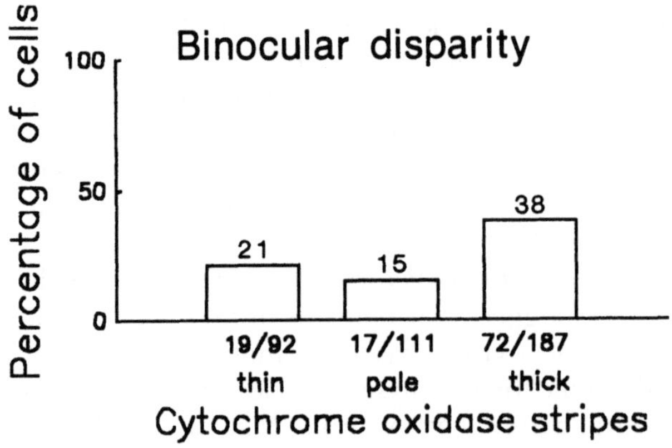

Figure 10. Distribution of neurons sensitive to binocular disparity across the cytochrome oxidase stripes. These neurons were found in all types of stripes, but most frequently in the thick stripes. Modified with permission from Peterhans and von der Heydt (1993).

4.3.2. Figure–Ground Segregation from Stereoscopic Cues

Sensitivity to stereoscopic depth, as described so far, implied sensitivity to the binocular disparity of contrast borders. Poggio *et al.* (1985) studied the disparity sensitivity of neurons in areas V1 and V2 using dynamic random-dot stereograms (Julesz, 1971). As previously with contrast borders, they found neurons which preferred zero or close to zero disparities ('tuned excitatory cells'), neurons that were inhibited by zero and close to zero disparities ('tuned inhibitory cells'), and others which preferred either crossed or uncrossed disparities ('near or far cells'). While these neurons preferred a certain orientation and size for stimuli defined by luminance contrast, the responses to random-dot stereograms were not related to the two-dimensional form that these stimuli induced in perception. Often, large squares at any orientation produced maximum response. This suggests that these responses were not related to the local definition of contours. Rather, they seemed to contribute to more global aspects of scene segmentation, for example, to the definition of surfaces and their depth order from stereoscopic cues. Neurons sensitive to borders defined by random dot stereograms have been described only recently (von der Heydt *et al.*, 1995).

5. Conclusions

The correlation of anatomy and physiology, as discussed in the present review, revealed parallels between different studies of anesthetized and awake animals, although the exact proportions of neurons exhibiting a particular physiological function varied from study to study (Table I). Orientation-selective neurons have been found in all compartments of area V2, but least frequently in the thin stripes. Neurons with end-stopped receptive fields have been found in all types of stripes, as have neurons selective for the direction of motion. Some functions have been found to concentrate in specific compartments: selectivity for chromatic stimuli in the thin and pale stripes; sensitivity to contours produced by occlusion cues (illusory contours) in the pale and thick stripes; and sensitivity to lines defined by coherent motion, and to binocular disparity, in the thick stripes. Overall the results suggest that the different compartments of area V2, though highly interconnected, contribute to different aspects of vision, especially in view of the more complex aspects of form processing and figure–ground segregation.

The results recorded in the awake monkey also support evidence from psychophysical studies which suggests that object boundaries and the depth order at such boundaries are defined early in visual processing. For example, Nakayama and Shimojo (1990) showed in a study of the da Vinci stereopsis that unpaired monocular stimulus elements can induce illusory contours, and that the figure–ground direction at such contours is determined by the unpaired stimulus elements. They concluded that figure–ground segregation at contours occurs early in visual processing, possibly already in area V1, where the signals of the two eyes converge. Ramachandran and Anstis (1986) showed that the perception of apparent motion of a scene segment can be reinforced for figures and suppressed for ground. They concluded that figure and ground have to be defined *before* additional attributes are assigned to the segments of a visual scene. They also showed that solid lines, illusory contours, and clusters of dots can be equally effective in defining figures and selecting them from ground, a finding which fits with the physiology discussed in this review. The results suggest that cue integration leading to a more generalized representation of contours begins in area V2, and that mechanisms segregating figure and ground at such contours are also implemented at this early level of processing.

ACKNOWLEDGMENTS. I thank F. Heitger and L. Spillmann for valuable comments on the manuscript. This research was supported by ESPRIT-Insight II 6019 (SNF #31.31970.91) and HFSP RG-31/93.

6. References

Allman, J., Miezin, F., and McGuinness, E., 1985, Direction- and velocity-specific responses from beyond the classical receptive field in the middle temporal visual area (MT), *Perception* **14:**105–126.

Allman, J., Miezin, F., and McGuinness, E., 1990, The effects of background motion on the responses of neurons in the first and second cortical visual areas (V-I and V-II), in: *Signal and Sense: Local and Global Order in Perceptual Maps* (G. M. Edelman, W. E. Gall, and W. M. Cowan, eds.), Wiley-Liss, New York, pp. 131–141.

Baizer, J. S., Robinson, D. L., and Dow, B. M., 1977, Visual responses of area 18 neurons in awake, behaving monkey, *J. Neurophysiol.* **40:**1024–1037.

Baumann, R., Sauvan, X. M., and Peterhans, E., 1996, Neural mechanism of figure–ground segregation at occluding contours in monkey prestriate cortex, in: *Brain Theory: Biological Basis and Computational Theory of Vision* (A. Aertsen and V. Braitenberg, eds.), Elsevier, Amsterdam, pp. 53–72.

Baumann, R., van der Zwan, R., and Peterhans, E., 1997, Figure–ground segregation at contours: A neural mechanism in the visual cortex of the alert monkey, *Eur. J. Neurosci.* **9:**1129–1303.

Bonhoeffer, T., and Grinvald, A., 1993, The layout of iso-orientation domains in area 18 of cat visual cortex-optical imaging reveals a pinwheel-like organization, *J. Neurosci.* **13:**4157–4180.

Brigner, W. L., and Gallagher, M. B., 1974, Subjective contour: Apparent depth or simultaneous brightness contrast? *Percept. Mot. Skills* **38:**1047–1053.

Brussell, E. M., Stober, S. R., and Bodinger, D. M., 1977, Sensory information and subjective contour, *Am. J. Psychol.* **90:**145–156.

Burkhalter, A., and Van Esssen, D. C., 1986, Processing color, form and disparity information in visual areas VP and V2 of ventral extrastriate cortex in the macaque monkey, *J. Neurosci.* **6:**2327–2351.

Cavanagh, P., Boeglin, J., and Favreau, O. E., 1985, Perception of motion in equiluminous kinematograms, *Perception* **14:**151–162.

Coren, S., 1972, Subjective contours and apparent depth, *Psychol. Rev.* **79:**359–367.

DeYoe, E. A., and Van Essen, D. C., 1985, Segregation of efferent connections and receptive field properties in visual area V2 of the macaque, *Nature* **317:**58–61.

Felleman, D. J., and Van Essen, D. C., 1987, Receptive field properties of neurons in area V3 of macaque monkey extrastriate cortex, *J. Neurophysiol.* **57:**889–920.

Felleman, D. J., and Van Essen, D. C., 1991, Distributed hierarchical processing in the primate cerebral cortex, *Cerebral Cortex* **1:**1–47.

Gegenfurter, K. R., Kiper, D. C., and Fenstemaker, S. B., 1996, Processing of color, form, and motion in macaque area V2, *Visual Neurosci.* **13:**161–172.

Gregory, R. L., 1977, Vision with isoluminant color contrast: 1. A projection technique and observations, *Perception* **6:**113–119.

Grosof, D. H., Shapley, R. M., and Hawken, M. J., 1993, Macaque V1 neurons can signal 'illusory' contours, *Nature* **365:**550–552.

Heitger, F., and von der Heydt, R., 1993, A computational model of neural contour processing: Figure–ground segregation and illusory contours, in: *Proceedings Fourth International Conference on Computer Vision, Berlin, Germany,* IEEE Computer Society Press, Los Alamitos, CA, pp. 32–40.

Heitger, F., Rosenthaler, L., von der Heydt, R., Peterhans, E., and Kübler, O., 1992, Simulation of neural contour mechanisms: From simple to end-stopped cells, *Vision Res.* **32:**963–981.

Heitger, F., von der Heydt, R., Peterhans, E., Rosenthaler, L., and Kübler, O., 1997, Simulation of neural contour mechanisms: Representing anomalous contours, *Image and Vision Computing,* in press.

Hubel, D. H., and Livingstone, M. S., 1987, Segregation of form, color, and stereopsis in primate area 18, *J. Neurosci.* **7:**3378–3415.

Hubel, D. H., and Wiesel, T. N., 1968, Receptive fields and functional architecture of monkey striate cortex, *J. Physiol. (Lond.)* **195:**215–243.

Hubel, D. H., and Wiesel, T. N., 1970, Cells sensitive to binocular depth in area 18 of the macaque monkey cortex, *Nature* **225:**41–42.

Julesz, B., 1971, *Foundations of Cyclopean Perception,* University of Chicago Press, Chicago.

Kanizsa, G., 1979, *Organization in Vision. Essays on Gestalt Perception,* Praeger, New York.

Kennedy, J. M., 1978, Illusory contours and the ends of lines, *Perception* **7:**605–607.

Lamme, V. A. F., 1995, The neurophysiology of figure–ground segregation in primary visual cortex, *J. Neurosci.* **15:**1605–1615.

Leventhal, A. G., and Zhou, Y., 1994, Cat visual cortical cells are sensitive to the orientation and direction of 'illusory' contours, *Soc. Neurosci. Abstr.* **20:**1053.

Levitt, J. B., Kiper, D. C., and Movshon, J. A., 1994a, Receptive fields and functional architecture of macaque V2, *J. Neurophysiol.* **71:**2517–2542.

Levitt, J. B., Yoshioka, T., and Lund, J. S., 1994b, Intrinsic cortical connections in macaque visual area V2: Evidence for interaction between different functional streams, *J. Comp. Neurol.* **342:**551–570.

Livingstone, M. S., and Hubel, D. H., 1987, Connections between layer 4B of area 17 and the thick cytochrome oxidase stripes of area 18 in the squirrel monkey, *J. Neurosci.* **7:**3371–3377.

Logothetis, N. K., and Charles, E. R., 1990, V4 Responses to gratings defined by random dot motion, *Invest. Ophthalmol. Vis. Sci.* **31:**90.

Malach, R., Tootell, R. B. H., and Malonek, D., 1994, Relationship between orientation domains, cytochrome oxidase stripes, and intrinsic horizontal connections in squirrel monkey area V2, *Cerebral Cortex* **4:**151–165.

Marcar, V. L., Raiguel, S. E., Xiao, D., Maes, H., and Orban, G. A., 1992, Do cells in area V2 respond to the orientation of kinetic boundaries? *Soc. Neurosci. Abstr.* **18:**1275.

Marcar, V. L., Xiao, D.-K., Raiguel, S. E., Maes, H., and Orban, G. A., 1995, Processing of kinetically defined boundaries in the cortical motion area MT of the macaque monkey, *J. Neurophysiol.* **74:**1258–1270.

Maunsell, J. H. R., and Van Essen, D. C., 1983, Functional properties of neurons in middle temporal visual area of the macaque monkey. II. Binocular interactions and sensitivity to binocular disparity, *J. Neurophysiol.* **49:**1148–1167.

Movshon, J. A., Adelson, E. H., Gizzi, M. S., and Newsome, W. T., 1985, The analysis of moving visual patterns, in: *Pattern Recognition Mechanisms* (C. Chagas, R. Gattass, and C. Gross, eds.), Pontifical Academy of Sciences, Vatican City, pp. 117–151.

Nakayama, K., and Shimojo, S., 1990, Da Vinci stereoposis: Depth and subjective occluding contours from unpaired image points, *Vision Res.* **30:**1811–1825.

Nakayama, K., Shimojo, S., and Silverman, G. H., 1989, Stereoscopic depth: its relation to image segmentation, grouping, and the recognition of occluded objects, *Perception* **18:**55–68.

Orban, G. A., 1994, Motion processing in monkey striate cortex, in: *Cerebral Cortex,* Volume 10, *Primary Visual Cortex in Primates* (A. Peters and K. S. Rockland, eds.), Plenum Press; New York, pp. 413–441.

Peterhans, E., and Baumann, R., 1994, Elements of form processing from motion in monkey prestriate cortex, *Soc. Neurosci. Abstr.* **20:**1053.

Peterhans, E., and von der Heydt, R., 1989, Mechanisms of contour perception in monkey visual cortex. II. Contours bridging gaps, *J. Neurosci.* **9:**1749–1763.

Peterhans, E., and von der Heydt, R., 1991a, Subjective contours—bridging the gap between psychophysics and physiology, *Trends Neurosci.* **14:**112–119.

Peterhans, E., and von der Heydt, R., 1991b, Elements of form perception in monkey prestriate cortex, in: *Representations of Vision: Trends and Tacit Assumptions* (A. Gorea, Y. Frégnac, Z. Kapoulis, and J. Findlay, eds.), Cambridge University Press. Cambridge, pp. 111–124.

Peterhans, E., and von der Heydt, R., 1993, Functional organization of area V2 in the alert macaque, *Eur. J. Neurosci.* **5:**509–524.

Peterhans, E., von der Heydt, R., and Baumgartner, G., 1986, Neuronal responses to illusory contour stimuli reveal stages of visual cortical processing, in: *Visual Neuroscience* (J. D. Pettigrew, K. J. Sanderson, and W. R. Levick, eds.), Cambridge University Press, Cambridge, pp. 343–351.

Petry, S., and Meyer, G. L., 1987, *The Perception of Illusory Contours,* Springer, New York.

Poggio, G. F., and Fischer, B., 1977, Binocular interaction and depth sensitivity in striate and prestriate cortex of behaving rhesus monkey, *J. Neurophysiol.* **40:**1392–1405.

Poggio, G. F., Doty, Jr., R. W., and Talbot, W. H., 1977, Foveal striate cortex of behaving monkey: Single-neuron responses to square-wave gratings during fixation of gaze, *J. Neurophysiol.* **40:**1369–1391.

Poggio, G. F., Motter, B. C., Squatrito, S., and Trotter, Y., 1985, Responses of neurons in visual cortex (V1 and V2) of the alert macaque to dynamic random-dot stereograms, *Vision Res.* **25:**397–406.

Ramachandran, V. S., and Anstis, S., 1986, Figure–ground segregation modulates apparent motion, *Vision Res.* **26:**1969–1975.

Ramachandran, V. S., and Gregory, R. L., 1978, Does colour provide an input to human motion perception? *Nature* **275:**55–56.

Redies, C., Crook, J. M., and Creutzfeldt, O. D., 1986, Neuronal responses to borders with and without luminance gradients in cat visual cortex and dorsal lateral geniculate nucleus, *Exp. Brain Res.* **61:**469–481.

Rockland, K. S., 1985, A reticular pattern of intrinsic connections in primate area V2 (area 18), *J. Comp. Neurol.* **235:**467–478.

Rockland, K. S., 1992, Configuration, in serial reconstruction, of individual axons projecting from area V2 to V4 in the macaque monkey, *Cerebral Cortex* **2:**353–374.

Rockland, K. S., 1995, Morphology of individual axons projecting from area V2 to MT in the macaque, *J. Comp. Neurol.* **355:**15–26.

Rockland, K. S., and Virga, A., 1990, Organization of individual cortical axons projecting from area V1 (area 17) to V2 (area 18) in the macaque monkey, *Visual Neurosci.* **4:**11–28.

Roe, A. W., and Ts'o, D. Y., 1995, Visual topography in primate V2: Multiple representation across functional stripes, *J. Neurosci.* **15:**3689–3715.

Sáry, G., Vogels, R., and Orban, G. A., 1993, Cue-invariant shape selectivity of macaque inferior temporal neurons, *Science* **260:**995–997.

Schumann, F., 1990, Beiträge zur Analyse der Gesichtswahrnehmungen. Erste Abhandlung. Einige Beobachtungen über die Zusammenfassung von Gesichtseindrücken zu Einheiten, *Z. Psychol.* **23:**1–32.

Sheth, B. R., Sharma, J., Rao, S. C., and Sur, M., 1996, Orientation maps of subjective contours in visual cortex, *Science* **274:**2110–2115.

Shipp, S., and Zeki, S., 1985, Segregation of pathways leading from area V2 to areas V4 and V5 of macaque monkey visual cortex, *Nature* **315:**322–325.

Snowden, R. J., Treue, S., Erickson, R. G., and Andersen, R. A., 1991, The response of area MT and V1 neurons to transparent motion, *J. Neurosci.* **11:**2768–2785.

Soriano, M., Spillman, L., and Bach, M., 1996, The abutting grating illusion, *Vision Res.* **36:**109–116.

Stoner, G. R., and Albright, T. D., 1993, Image segmentation cues in motion processing: Implications for modularity in vision, *J. Cognitive Neurosci.* **5:**129–149.

Stoner, G. R., and Albright, T. D., 1996, The interpretation of visual motion: Evidence for surface segmentation mechanisms, *Vision Res.* **36:**1291–1310.

Tamura, H., Sato, H., Katsuyama, N., Hata, Y., and Tsumoto, T., 1996, Less segregated processing of visual information in V2 than in V1 of the monkey visual cortex, *Eur. J. Neurosci.* **8:**300–309.

Tootell, R. B. H., and Hamilton, S. L., 1989, Functional anatomy of the second visual area (V2) in the macaque, *J. Neurosci.* **9:**2620–2644.

Tootell, R. B. H., Silverman, M. S., De Valois, R. L., and Jacobs, G. H., 1983, Functional organization of the second cortical visual area in primates, *Science* **220:**737–739.

Ts'o, D. Y., Gilbert, C. D., and Wiesel, T. N., 1990, Functional architecture of color and disparity in visual area 2 of macaque monkey, *Soc. Neurosci. Abstr.* **16:**293.

von der Heydt, R., and Peterhans, E., 1989, Mechanisms of contour perception in monkey visual cortex. I. Lines of pattern discontinuity, *J. Neurosci.* **9:**1731–1748.

von der Heydt, R., Peterhans, E., and Baumgartner, G., 1984, Illusory contours and cortical neuron responses, *Science* **224:**1260–1262.

von der Heydt, R., Zhou, H., Friedman, H., and Poggio, G. F., 1995, Neurons of area V2 of visual cortex detect edges in random-dot stereograms, *Soc. Neurosci. Abstr.* **21:**18.

Westheimer, G., and Li, W., 1996, Classifying illusory contours by means of orientation discrimination, *J. Neurophysiol.* **75:**523–528.

Wong-Riley, M. T. T., and Carroll, E. W., 1984, Quantitative light and electron microscopic analysis of cytochrome-oxidase rich zones in VII prestriate cortex of the squirrel monkey, *J. Comp. Neurol.* **222:**18–37.

Zeki, S. M., 1978, Uniformity and diversity of structure and function in rhesus monkey prestriate visual cortex, *J. Physiol. (Lond.)* **277:**273–290.

Zeki, S., and Shipp, S., 1988, The functional logic of cortical connections, *Nature* **335:**311–317.

Zipser, K., Lamme, V. A. F., and Schiller, P. H., 1996, Contextual modulation in primary visual cortex, *J. Neurosci.* **16:**7376–7389.

9

Visual Processing in Macaque Area MT/V5 and Its Satellites (MSTd and MSTv)

GUY A. ORBAN

1. Introduction

It is now well established that monkey extrastriate cortex, the visual cortex beyond primary or striate cortex, contains many different areas (for review, see Van Essen *et al.*, 1992; Felleman and Van Essen, 1991). Of these 30 or so extrastriate areas, a small group in the caudal superior temporal sulcus (STS) stands out because their neurons share the property of direction selectivity, suggesting that these areas might be involved in the analysis of retinal motion and in motion perception. A large number of studies have been devoted to substantiating and clarifying the role of the middle temporal (MT) area, also referred to as V5, and that of its satellites, the dorsal and ventral parts of the medial superior temporal (MST) visual area. Although area MT/V5 was discovered almost simultaneously in macaque monkeys (Zeki, 1969, 1971) and in owl monkeys (Allman and Kaas, 1971) and there are a number of similarities between these areas of the two species, this review will be restricted to the macaque monkey. Indeed, with the passage of time, differences between MT of the two species have become appar-

GUY A. ORBAN • Laboratorium voor Neuro- en Psychofysiologie, KU Leuven, Medical School, Campus Gasthuisberg, B-3000 Leuven, Belgium.

Cerebral Cortex, Volume 12, edited by Rockland *et al.* Plenum Press, New York, 1997.

ent (Sereno and Allman, 1991; Zeki, 1980) and the macaque as a species is closer to the human (Ciochon and Chiarelli, 1980). A further restriction will be that for the physiological studies preference will be given to the more recent, quantitative data. Since the physiology of macaque visual cortex, and particularly of area MT/V5 and its satellites, has been a major source of inspiration for recent human functional imaging work, the homologues of MT/V5 and MST in humans will be briefly discussed.

1.1. Definition of Areas

Area MT/V5 is located in the posterior bank of the caudal STS and is defined by the association of four features: a direct input from striate cortex, a large proportion of direction-selective cells, a complete representation of the contralateral visual hemifield, and a heavy myelination. The first three features were discovered by Zeki (Zeki, 1974; Dubner and Zeki, 1971; Zeki, 1969), although others had found indications of a projection from striate cortex to the STS (Ungerleider and Mishkin, 1979; Cragg, 1969). Zeki (1980) proposed the term V5 to refer to the area which he had initially labeled the "visual area in the posterior STS" (Zeki, 1974) or the "motion area in posterior STS" (Zeki, 1978). The combination of these three features with the myelination was demonstrated by Van Essen *et al.* (1981), who coined the term middle temporal (MT), in keeping with the term used to describe the homologous area in New World monkeys (Spatz and Tigges, 1972; Allman and Kaas, 1971). It therefore seems logical to refer to this area in the macaque as area MT/V5. Of the four criteria defining MT/V5, the myelination and high incidence of direction-selective neurons have proven the most useful in subsequent studies and most definitions of MT/V5 are based on these two criteria.

There has been discussion as to whether the area defined by heavy myelination is completely coextensive with the area receiving projections from V1 (Ungerleider and Mishkin, 1979) and with the region representing the complete contralateral hemifield. In particular, Ungerleider and Desimone (1986a) have proposed that a part of the peripheral field representation, especially that in the upper quadrant, is located outside the heavily myelinated region and refer to this part of posterior STS as MTp. This issue was resolved by Maunsell and Van Essen (1987), who concluded that the representation of the periphery in the upper quadrant extended at most 1–2 mm beyond the medial border of MT/V5, as defined by the myelination. Our own results (Raiguel *et al.*, 1995) support their view. Thus, one may conclude that the regions defined by the four criteria are indeed congruent, at least within the likely imprecision (1–2 mm) of the techniques.

MT/V5 is a relatively small area, much smaller than its afferent V1. Originally Van Essen *et al.* (1981) estimated the average surface of area MT/V5 to be 33 mm^2, 3% of the striate cortex, compared to 68 mm^2 (7% of striate cortex) reported subsequently by Weller and Kaas (1983). In a later report, Maunsell and Van Essen (1987) resolved this apparent discrepancy: the surface area of MT/V5 depends on body weight and its surface area in mm^2 equals 14 times the animal's body weight in kg.

Van Essen *et al.* (1981) were the first to notice that direction selectivity in posterior STS was not restricted to area MT/V5: they noted that the lateral or outer border of MT (i.e., the border with V4t) corresponded rather well to a sharp change in direction selectivity, but that the medial or inner border (i.e., toward the fundus of the sulcus) did not. Maunsell and Van Essen (1983c) then showed that area MT/V5 projects with a forward pattern (see below, Fig. 2) to an area located medially to MT, where cells are direction-selective and have large receptive fields (RFs). This area they deemed the medial superior temporal (MST) visual area. Desimone and Ungerleider (1986) described two satellite areas neighboring MT on the medial side: MST and the visual area in the fundus of the superior temporal sulcus, abbreviated FST (Fig. 1A). Desimone and Ungerleider (1986) distinguished FST from MST based on topographic criteria, a lower incidence of direction-selective cells in FST (32%) compared to MST (almost 100%) and the presence of separate projections from MT (Ungerleider and Desimone, 1986b). Komatsu and Wurtz (1988a) reported that visual pursuit cells—cells which are direction-selective and respond during visual pursuit—occur in three regions within caudal STS: the foveal representation of MT, a dorsomedial part of MST in the anterior bank referred to as MSTd, and a lateral anterior part of MST in the fundus and the posterior bank of STS, designated MSTl (Fig. 1). They argued that MSTl represented the most posterior part of FST and should be considered distinct from the rest of FST. There is now general agreement that MSTd is a separate area, characterized by direction-selective cells with large RFs (Lagae *et al.*, 1994; Komatsu and Wurtz, 1988a; Desimone and Ungerleider, 1986; Tanaka *et al.*, 1986), many of which are selective for rotation and "expansion/contraction" (Graziano *et al.*, 1994; Lagae *et al.*, 1994; Duffy and Wurtz, 1991; Saito *et al.*, 1986), and many of which require large field stimuli (Lagae *et al.*, 1994; Komatsu and Wurtz, 1988b; Tanaka *et al.*, 1986). MSTd includes, but is not restricted to, the densely myelinated region (DMZ) of the anterior bank of STS (Lagae *et al.*, 1994; Tanaka *et al.*, 1993; Desimone and Ungerleider, 1986). There is now also agreement among three groups (Lagae *et al.*, 1994; Tanaka *et al.*, 1993; Komatsu and Wurtz, 1988a) that most cells in the posterior part of what was originally described as FST are more frequently direction-selective than described by Desimone and Ungerleider (1986) and must be distinguished from FST proper. The most complete description of this area has been provided by Tanaka *et al.* (1993), who labeled the area MSTv (ventral MST) in contrast to MSTd. Both areas share the features of relatively large RFs with direction-selective responses and input from MT. The ventral region of MST is distinguished from the dorsal part by: (1) a near complete absence of cells selective for rotation and "expansion/contraction," (2) a clear preference for small, moving stimuli, and (3) an intervening strip of cells with small RF sizes with respect to their eccentricity, located in the dorsal part of MSTv and delineating the border between the two areas (Fig. 1D). Whether the remainder of FST, exclusive of MSTv, receives input from area MT and hence could be considered a third satellite must await further study, although the distribution of labeled terminals described by Ungerleider and Desimone (1986b) seems to support such a view. The two subdivisions of MST occupy roughly the fundus and the inner third of the anterior bank of caudal STS. Initially Desimone and Ungerleider (1986) described the more outer region of the anterior bank as posterior parietal (PP), which contains direction-selective cells, but receives no projection from area MT/V5. This region was explored in greater detail by Hikosaka *et al.* (1988). They subdivided the anterior bank into three parts (Fig. 1C): MST, the "mostly unresponsive visual region," and caudal superior temporal polysensory region (CSTP). They agreed with Desimone and

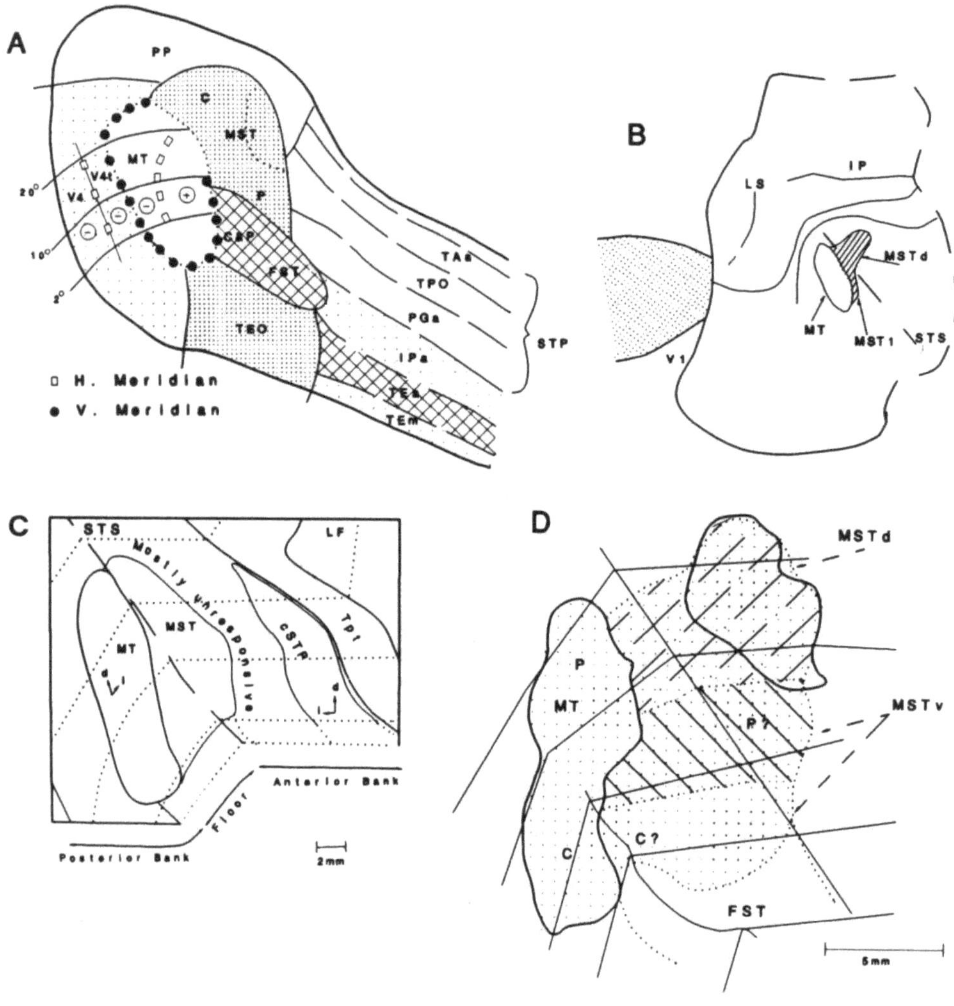

Figure 1. Definition of area MT/V5 and satellites: (A) Original proposal by Desimone and Ungerleider (1986) (from Boussaoud *et al.*, 1990); (B) modification proposed by Komatsu and Wurtz (1988a) (from Wurtz *et al.*, 1990); (C) diagram of areas outside of MST by Hikosaka *et al.* (1988); (D) definition of areas MSTd and MSTv by Tanaka *et al.* (1993). In parts A and D, the letters C and P indicate representations of central and peripheral visual fields, respectively. In part A, the dotted line within MST indicates the border of the densely myelinated zone (DMZ): minus and plus signs indicate parts devoted to the lower and upper quadrants, respectively. In part D, the stippled area including all of MT, MSTd, and MSTv indicates large proportions of direction-selective neurons; /// hatching indicates presence of rotation- and "expansion/contraction"-selective neurons; \\\hatching indicates small RFs compared to eccentricity. In part D, the two regions of heavy myelinations (MT and DMZ) are outlined in heavy lines. Abbreviations: (A) PP, posterior parietal sulcal zone; STP, superior temporal polysensory area; TEO, TEa, TEm, and IPa, parts of inferotemporal cortex. (B) LS, lunate sulcus; IP, intraparietal sulcus. (C) d, dorsal; I, inner; cSTP, caudal STP; Tpt, part of auditory cortex.

Ungerleider (1986) that the outer border of the DMZ was a reliable landmark for the outer border of MSTd. Thus, the picture which has emerged over the last 25 years is that caudal STS contains three distinct, adjoining areas all characterized by a high incidence of direction-selective neurons: MT/V5, receiving from V1, and MSTd and MSTv, which both receive from MT/V5 and will be referred to as satellites of MT/V5. I will refer to these areas collectively as the MT/V5 complex.

1.2. Connections

In addition to input from V1, MT/V5 also receives afferents from V2 and V3 and V4. These connections are reciprocal and topographically organized (Maunsell and Van Essen, 1983c). The same authors noted that the laminar patterns of efferent connections from area MT/V5 fall into three groups (Fig. 2): forward projections to layer IV extending into the supragranular layers of MST and the ventral intraparietal (VIP) area; feedback projections sparing layer IV, to areas V2, V3, and VP, and intermediate projections to all layers or to all layers except IV of area V4. Conversely, the afferent projections to area MT/V5 display a complementary dissociation: forward afferents (from V2 and V3) arise from supragranular layers (II and III), feedback projections (from MST and VIP) mainly from infragranular layers (VI and V) and to a lesser extent from layer III, and intermediate-type afferents (from V4) from supragranular layers and layer V. The laminar pattern of the V1–MT connection was unique, since the projections to MT originate in layer IVB and VI of striate cortex and the projections returning from MT terminate in the same layer (see below). These authors (Maunsell and Van Essen, 1983c), building on earlier work, notably that of Rockland and Pandya (1979), proposed using the laminar pattern of corticocortical connections to derive a hierarchy of visual cortical areas, a proposal which has now won general acceptance. Ungerleider and Desimone (1986b) extended the description of MT/V5 (reciprocal) connections as follows: afferent connections from V1, V2, and V3, intermediate connections with V4, V4t, V3A, and parietooccipital (PO) area, and forward connection to MST, FST, and VIP. The latter connections imply the existence of two satellite areas neighboring MT, which we here refer to as MSTd and MSTv.

The afferents from V1 to area MT/V5 end largely in layers IV and III, and to a lesser extent in layer VI (Rockland, 1989; Weller and Kaas, 1983). These afferents differ from those originating in V2, which are thinner, have finer terminal specializations, and do not involve layer VI (Rockland, 1995). As noted earlier, the V1–MT projection arises from layers IVb and VI, which contain a high incidence of direction-selective cells (Hawken *et al.*, 1988; Orban *et al.*, 1986) with a preference for faster speeds (Orban *et al.*, 1986). The input from V2 was shown (Shipp and Zeki, 1985; DeYoe and Van Essen, 1985) to arise from the thick cytochrome oxidase stripes, which also contain many direction-selective cells. Thus direct connections link compartments of V1 and V2, MT/V5, MSTd, and MSTv, and VIP, which all contain very high proportions of direction-selective cells and may be viewed as forming a motion analysis pathway.

A distinctive feature of the connections between V1 or V2 and MT/V5 is the reciprocal asymmetry (Shipp and Zeki, 1989a). Feedback connections terminate in a more widespread region than that containing the cells of origin of the feedforward connections. In area V1 the MT efferent neurons occupy the full thickness of layer IVb and tend to cluster together, while those in layer VI occupy only the most superficial portion of the layer (Shipp and Zeki, 1989a; Maunsell and Van Essen, 1983c; Lund *et al.*, 1975). In contrast, the feedback terminals diffusely label the whole of layer VI and wide patches of layer IVB centered around the MT efferent cells (Shipp and Zeki, 1989a). In addition, feedback projections terminate in layer 1 of V1, at least in the representation of the visual field beyond 10° eccentricity (Shipp and Zeki, 1989a; Ungerleider and Desimone, 1986b; Maunsell and Van Essen, 1983c). In area V2, MT efferent cells are located mainly in deep layer III, and to a lesser degree in layers II, superficial layer III, and layer V (Shipp and Zeki, 1989b; Ungerleider and Desimone, 1986b; Maunsell and Van Essen, 1983c; Lund *et al.*, 1981), while

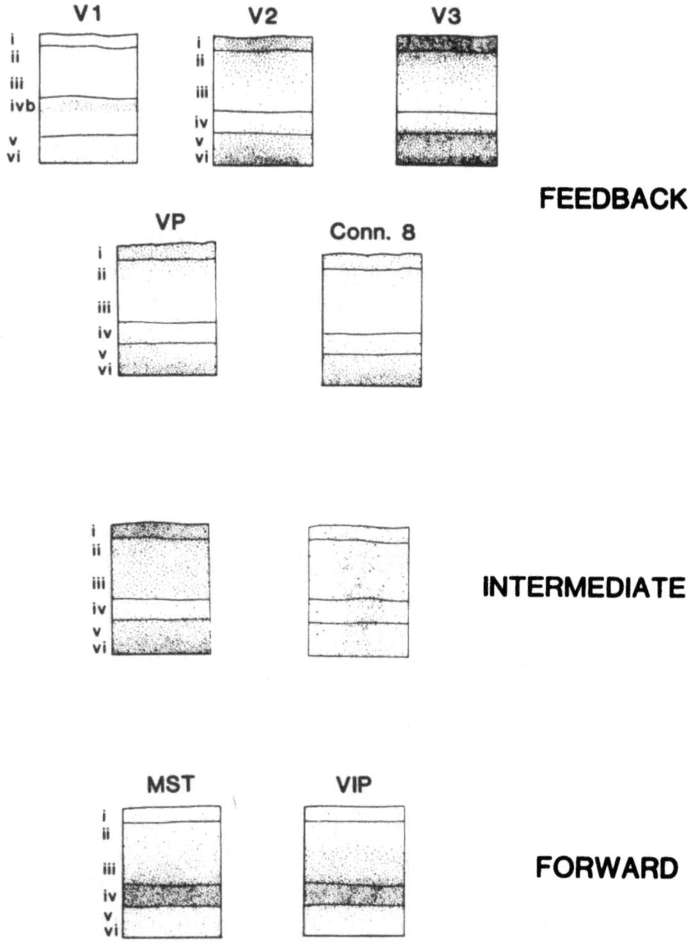

Figure 2. The laminar distribution of anterogradely transported label after MT injections. The laminar distribution of label is shown for each cortical target of MT. Feedback projections from MT are characterized by terminating largely in layers I and VI and sparing layer IV. Forward projections terminate mainly in layers IV and III. Two different laminar distributions were found in V4, one sparing layer IV and the other labeling all layers about equally (see the text). From Maunsell and Van Essen (1983c).

feedback projections from MT terminate in layers I and VI and less so in layers II, III, and V (Shipp and Zeki, 1989b; Ungerleider and Desimone, 1986b; Maunsell and Van Essen, 1983c). The V2 cells projecting to areas MT/V5 are located in the thick cytochrome oxidase stripes (Shipp and Zeki, 1989b; Shipp and Zeki, 1985; DeYoe and Van Essen, 1985), but the feedback projections terminate in all three cytochrome compartments, with the heaviest termination in the thick stripes.

The cortical connections of the MT/V5 satellites have been investigated by only a single group, who at that time referred to MST and FST as satellites of MT/V5 (Boussaoud *et al.*, 1990), and thus it is difficult to ascertain the connections of MSTd and MSTv precisely. From inspection of the injection sites, it seems that MSTp and MSTc in that study (Boussaoud *et al.*, 1990) together correspond to MSTd, and FST to MSTv. These connections are summarized in Fig. 3. Only the projections of MSTd and MSTv into the anterior STS and to FEF (not shown) are of the forward type; those with other parietal areas are of the intermediate type and those to the prestriate areas of the feedback type. Projections from MT/V5 and its satellites to FEF have been reported by Schall *et al.* (1995).

Area MT/V5 also has a substantial number of efferent projections to subcortical regions: the superior colliculus, inferior and lateral pulvinar, reticular nucleus of the thalamus, caudate, putamen and claustrum, and pons (Ungerleider *et al.*, 1984). MT/V5 and possibly MSTv also project to the pretectum (Hoffmann *et al.*, 1991). Areas MST and FST have been shown to have reciprocal connections with the pulvinar and claustrum and efferent projections to the reticular thalamic nucleus, the striatum (caudate, putamen), and the pons (Boussaoud *et al.*, 1992).

In conclusion, MT and its satellites constitute an important part of the dorsal stream (Ungerleider and Mishkin, 1982) leading from V1 into parietal cortex. This stream has been implicated in spatial vision and guidance of movements (Ungerleider and Mishkin, 1982; Milner *et al.*, 1991. It is important to stress that MT/V5 is not the only gateway to parietal cortex, but that early areas such as PO and PIP provide alternative routes (Van Essen *et al.*, 1992). After all, motion information is an important component of the visual information used to guide movements, but not the only one, since three-dimensional position and shape are also important for guidance of saccades, pointing, and grasping.

1.3. Nature of the Input to Area MT/V5

From the description of the connections of area MT/V5 it is clear that its cortical afferents originate directly or indirectly from striate cortex. The question then arises whether activity in area MT/V5 depends entirely on striate input as is the case for V2 (Schiller and Malpeli, 1977) and inferotemporal cortex (Rocha-Miranda *et al.*, 1975). Converging lines of evidence from ablation (Rodman *et al.*, 1989) and cooling (Girard *et al.*, 1992) of striate cortex show that area MT/V5 does not depend entirely on striate input. A substantial proportion of cells (60–80%) remain visually responsive after striate destruction or deactivation, although responses are much weaker than in normal animals. Interestingly, in both studies it was clearly demonstrated that the remaining responses were generally direction-selective. In a further study, Rodman *et al.* (1990) showed that while removal of the superior colliculus alone had little influence on the

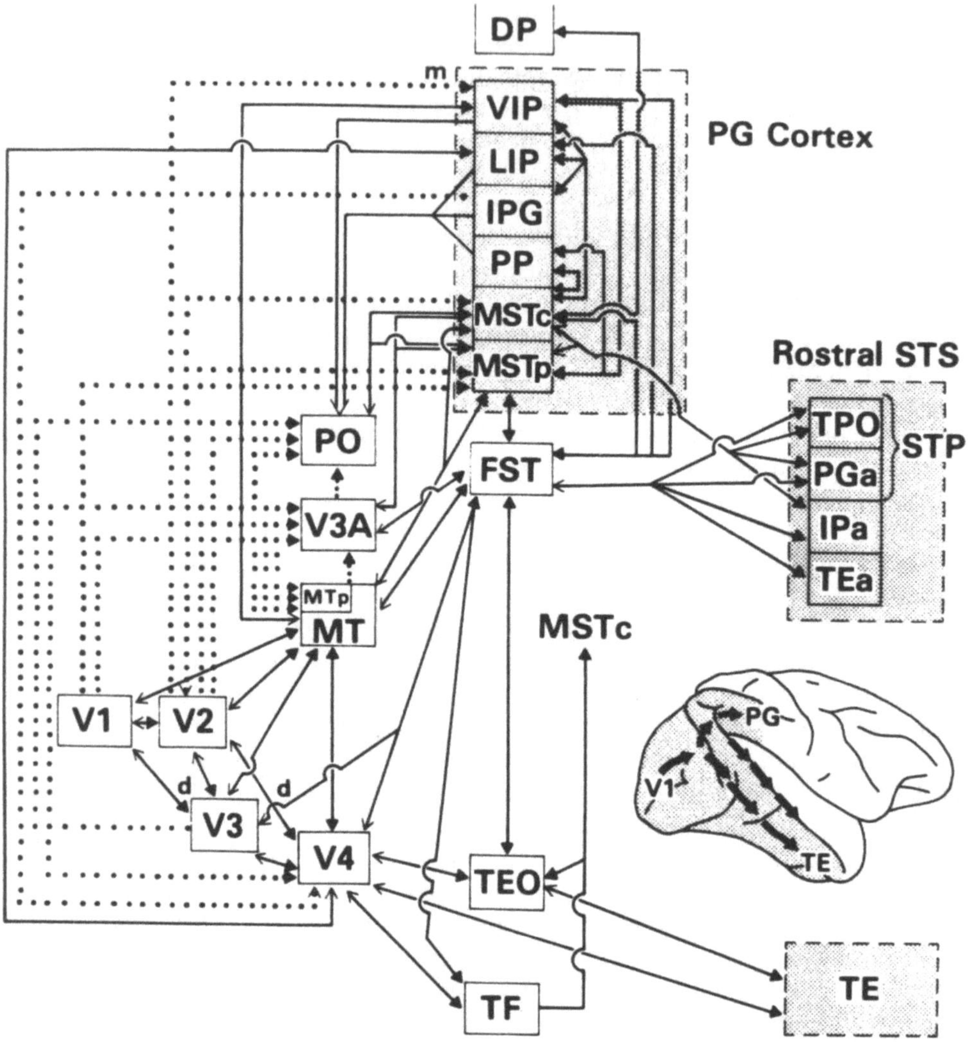

Figure 3. Summary of visual cortical areas in STS and their connections. MSTc and MSTp both correspond to MSTd in the present terminology, FST stands for both MSTv and FST proper, thus their connections need further investigation. Solid lines indicate projections that originate from both central and peripheral visual field representations, and dotted lines indicate projections that arise exclusively from peripheral field representations. The letter d indicates that the connections of V3 with both V1 and FST are limited to the dorsal portion of V3; m indicates that the projection from V2 to VIP is limited to VIP's medial portion. The complete connections of MTp and V4t are unknown and are therefore not illustrated separately from those of MT and V4, respectively. The connection between V4 and FST includes V4t. The diagram suggests that, in addition to the occipitotemporal (ventral) stream for object vision and the occipitoparietal (dorsal) stream for spatial vision, there may be a third processing stream which relays motion information from MST and FST forward into the cortex of the rostral superior temporal sulcus. Heavy arrowheads indicate forward projections, light arrowheads feedback projections. Intermediate-type projections are indicated by two reciprocal heavy arrowheads. DP, Dorsal prelunate area; IPG, inferior parietal gyrus; LIP, lateral intraparietal area; PO, parieto-occipital area; PP, posterior parietal sulcal zone; STP, superior temporal polysensory zone; VIP, ventral intraparietal area. From Boussaoud *et al.* (1990).

responses and properties of MT neurons, removal of the superior colliculus following striate ablation abolished all responsiveness to visual stimuli in area MT/V5. Thus there is an alternative subcortical route by which retinal information may reach area MT/V5. The most likely pathway according to the anatomical studies passes through the pulvinar, which receives input from the superior colliculus and projects to area MT (Standage and Benevento, 1983), although other routes through the pretectum or lateral genicular nucleus (LGN) cannot be excluded. Evidence for a direct geniculate projection to area MT/V5 is inconclusive since the injection into the posterior bank of STS made by Fries (1981) probably invaded V4t as well as MT/V5. Overall, area MT/V5 behaves very differently from the ventral stream areas, which depend entirely on V1 input. Indeed lesion or inactivation of V1 has been shown to abolish visual responses in ventral stream areas such as area V2 (Girard and Bullier, 1989; Schiller and Malpeli, 1977), area V4 (Girard *et al.,* 1991), and inferotermporal cortex (Rocha-Miranda *et al.,* 1975).

Within the retinostriate projection, two parallel streams have been identified, the magno- and parvocellular streams (Lennie, 1980). The question then arises as to which of these pathways provides input to area MT/V5. This issue has been settled by Maunsell *et al.* (1990), who showed that blocking the magnocellular geniculate layers decreased MT responsiveness to the point that generally responses were abolished, while blocking the parvocellular layers had an effect only in a few multiunit sites. This is again in striking contrast to ventral areas such as area V4, which depend on both parvo- and magnocellular input (Ferrera *et al.,* 1994a). Of further interest in the Maunsell *et al.* (1990) study was the attempt to inactivate all LGN input by blocking both magno- and parvocellular layers, which eliminated MT responses completely. However, this procedure was carried out in only three MT recording sites and only responses to moving spots and bars were tested. It is possible that the responses of MT neurons to moving random dot patterns depend to a much greater extent upon subcortical input than do the responses to moving bars and spots. Hence the observations made by Maunsell *et al.* (1990) do not necessarily contradict the existence of an alternative subcortical route to MT/V5 via superior colliculus and pulvinar.

In conclusion, area MT/V5 does not absolutely require V1 integrity and operates mainly on magnocellular input. This is in contrast to the ventral stream areas, such as area V4, which require V1 input and operate both on parvo- and magnocellular input. Yet the difference between MT and the ventral areas may not be as great as it would appear. While neurons in ventral areas are driven by magno- and parvocellular striate input, MT/V5 neurons are also driven by two inputs: magnocellular striate input and subcortical input.

2. Physiological Exploration of Macaque Area MT/V5 and Its Satellites

2.1. General Response Properties

2.1.1. Nature of the Moving Stimulus

Most physiological studies of area MT/V5 and satellites are performed either in awake, fixating or in paralyzed, anesthetized monkeys. The stimuli used in these experiments move on a screen and it is then impossible to determine whether neurons simply react to any retinal image motion, which can result from eye movements or from relative motion between the subject and scene, or whether they react exclusively to image movement induced by relative motion between the observer's head and the outside world. This question has been addressed by Erickson and Thier (1991). MT neurons were found to respond to retinal image motion since they respond equally well in both active and passive conditions (Fig. 4). MSTd and MSTv neurons respond almost exclusively to the retinal image motion induced by relative motion between the observer's head and the outside world, since they react chiefly in passive conditions (Fig. 4). According to Erickson and Thier (1991), by this behavior MSTd and MSTv neurons could contribute to the stability of the visual world during voluntary eye movement, while stability during head movements is ensured by vestigular ocular reflex (VOR) and optokinetic nystagmus (OKN).

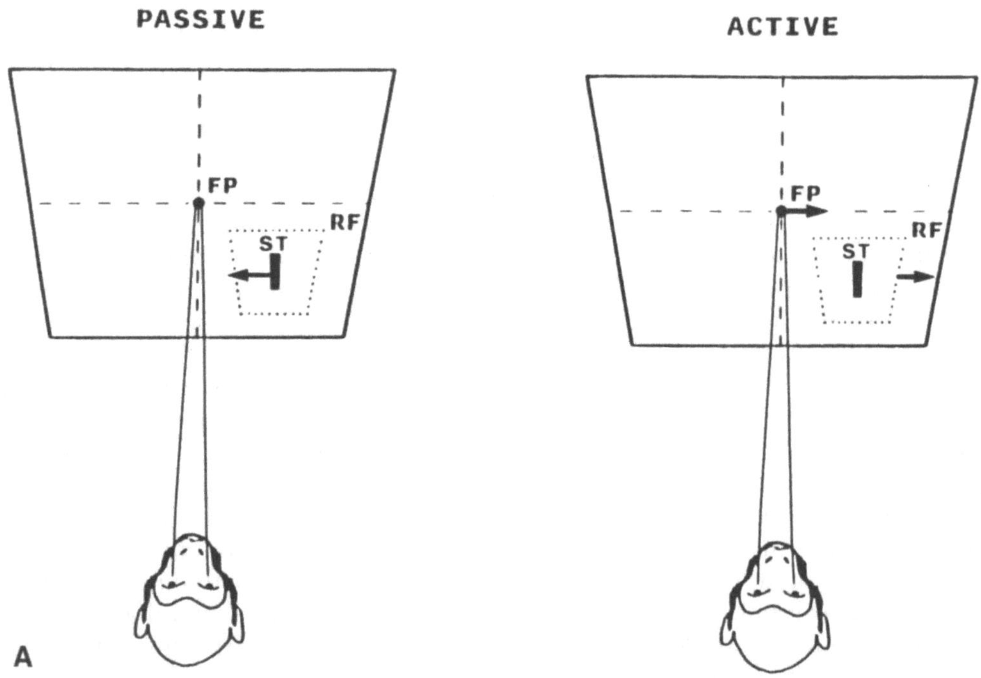

Figure 4. (A) Testing method of Erickson and Thier (1991). Passive: visual testing during periods of stable fixation (eyes stationary, stimulus ST moves through stationary receptive field RF). Active: visual testing during visual pursuit (with eyes moving, the receptive field moves over stationary stimulus). FP, Fixation point. (B) Distribution of passive/active comparison in area MT/V5 and satellites. Abscissa, passive/active index (1-Da/Dp, with Da and Dp Direction indexes [DI = $R_P - R_{NP}/R_P + R_{NP}$] in active and passive conditions), bin width = 0.1 for all three plots. Ordinate, Number of cells. Top: Cells from visual area MT (N = 24, two values less than −10 not shown). Most indices are clustered near zero, indicating inability to discriminate passively versus actively induced retinal slip. Middle: Cells from visual area MSTv (N = 43, one value less than −10 and four greater than 2.0 not shown). The range of MSTv (indicated as MSTi) indices spans those of MT and MSTd. The majority of indices are between 0.5 and 1.0, indicating greater directional modulations during passive visual stimulation (see text). Bottom: Cells from visual area MSTd (N = 32, two values greater than 2.0 not shown). Cells with large index values correspond to visual tracking or visual pursuit cells.

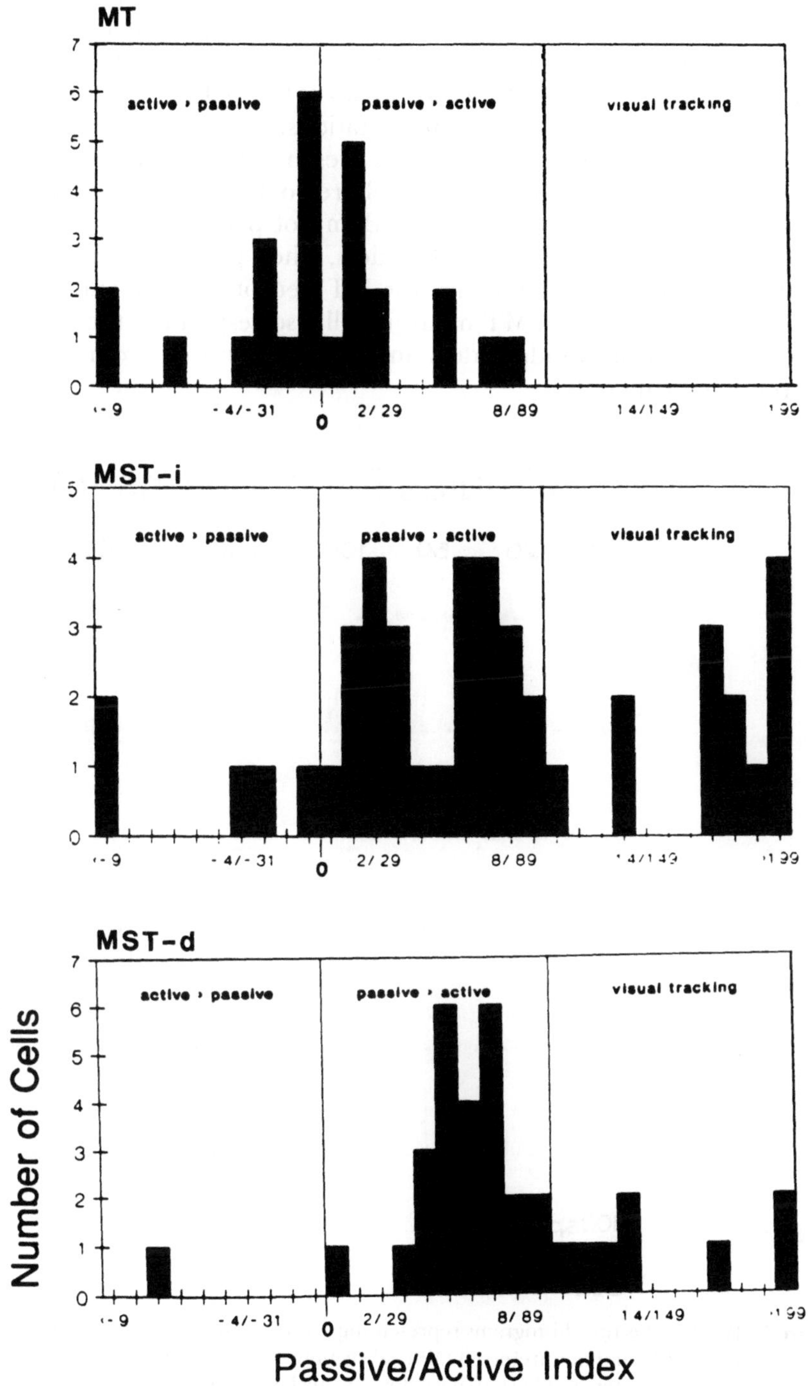

Figure 4. (*Continued*).

Two types of moving stimuli have been used in the physiological studies: restricted stimuli such as moving bars or slits which move through the receptive field, and large stimuli such as gratings or random dot patterns which cover the receptive field at all times. In the latter case the receptive field is stimulated by the motion of the elements in the stimulus, whereas the edges of the stimulus remain stationary. The question then arises as to what the stimulus should do in the interval between the motion presentations. Generally the same stimulus as that used for the motion presentation is present but stationary, but this need not be the case. In a few exceptional cases where no dots were present in the interval between motion presentations the random dot pattern was switched on when motion started (see Section 3.2). The dots, when present between motion presentations, need not be stationary for MT neurons to respond to the motion onset. As shown in Fig. 5, MT neurons will also respond to the transition from flickering to moving random dots and to the transition from incoherently to coherently moving dots.

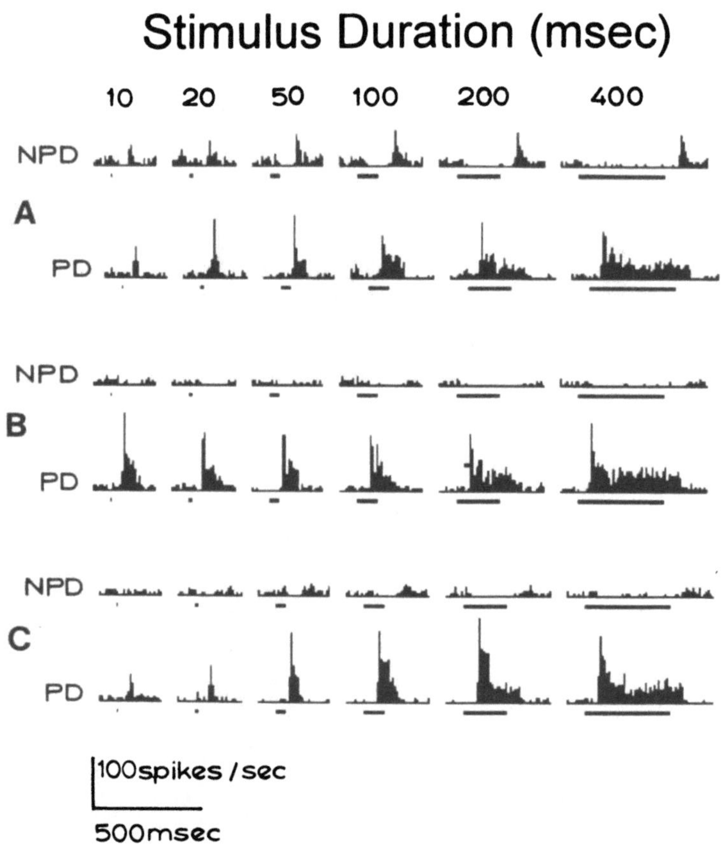

Figure 5. Peristimulus time histograms representing average responses of an MT neuron to motion in the preferred (PD) and nonpreferred (NPD) directions for different motion durations and three types of intermotion interval: (A) stationary dots, (B) flickering dots, (C) incoherently moving dots. Speed of motion was 10°/sec. Notice the suppression of spontaneous activity in the nonpreferred direction in all three conditions, which, however, was followed by a rebound only in A. Cell recorded in layers 2–3, eccentricity of receptive field 11°.

2.1.2. Receptive Field

Neurons in the MT/V5 complex will respond to visual stimuli only in a restricted part of the visual field, referred to as their receptive field (RF). Before describing this RF in greater detail, one has to address the question of whether responses depend only on visual stimulus position or on both stimulus and eye position, as has been shown for parietal cortex (Andersen *et al.,* 1985) and for some lower order visual cortical areas (Galletti and Battaglini, 1989). In fact, the responses to visual stimuli are influenced by gaze position in 60% of the MT cells and 80% of the MST cells (most were probably MSTd cells) in both fixation and pursuit tasks (Bremmer *et al.,* 1996). The modulation of visual responsiveness by eye position is distinct from the extraretinal signals related to eye movements. The latter are confined to MST neurons (see Section 3.2), the former predominates in the MT/V5 complex.

There is general agreement that RF size increases with eccentricity in area MT/V5 (Mikami *et al.,* 1986b; Gattass and Gross, 1981; Van Essen *et al.,* 1981; Zeki, 1974) and that neurons in MST have larger RFs than those in area MT (Tanaka *et al.,* 1993; Desimone and Ungerleider, 1986; Van Essen *et al.,* 1981). Several authors (Tanaka *et al.,* 1993; Rodman *et al.,* 1989; Albright and Desimone, 1987; Maunsell and Van Essen, 1987; Desimone and Ungerleider, 1986; Tanaka *et al.,* 1986) have estimated the dependence of RF size on eccentricity in area MT/V5: RF size, generally defined as the square root of the area, increases linearly with eccentricity, with the values reported for the intercept ranging from 0.3° to 3.2° and those for the slope from 0.47° to 0.72°. Unfortunately, the RF was determined with hand plotting in these studies, leading to an underestimation of RF size. Using quantitative measurements (Fig. 6), we have recently estimated the RF size of MT neurons (Raiguel *et al.,* 1995). Defining RF size as the square root of the area at 50% of maximum response yielded a slope of 0.67 and an intercept of 4.1° (Fig.7). Thus RFs in MT are uniformly larger than estimated by earlier studies. Eccentricity explained only one-third of the variance in RF size and at all eccentricities there is a tenfold range in RF size (Fig. 7). Raiguel *et al.* (1995) were also the first to report a relationship between RF size and layer: RFs were much larger in lamina V than in other layers or sublaminae (Fig. 21A).

Desimone and Ungerleider (1986) reported that RF size increases with eccentricity in both MSTd and MSTv (their MST and FST), while Tanaka *et al.* (1986, 1993) observed an increase in RF size only in MSTv, but not in MSTd. Celebrini and Newsome (1994) even reported a decrease in RF size with increasing eccentricity in MSTd. S. Raiguel, L. Lagae, D.-K. Xiao, and G. A. Orban (unpublished) were also unable to detect a change in RF size with eccentricity in MSTd and quantitatively estimated the average RF size at 39°, compared to 41° reported by Tanaka *et al.* (1986) and 43° by Tanaka *et al.* (1993). It should be pointed out that the quantitative measure was made at 50% of the peak value, while the measurements reported by Tanaka and co-workers were based on the edges of the RF determined with hand plotting.

While RF size has received a great deal of attention, the RF structure has received much less, probably because MT cells invariably display overlapping discharge regions for moving light and dark bars and thus all qualify as complex cells (Hubel and Wiesel, 1968). This does not mean that it is not important to map the RF; in fact, a number of errors could have been avoided had this been done systematically, even in awake animals. Interest in the RF structure was reawakened by the finding of a silent antagonistic surround in MT neurons of the owl monkey by Allman *et al.* (1985). These authors refer to the RF mapped using excitatory responses to motion in the preferred direction as the classical RF (CRF). The silent region surrounding the CRF and suppressing its responses was designated the antagonistic surround. Antagonistic surrounds were described in MT and in MSTd and MSTv of the macaque by Tanaka *et al.* (1986, 1993) and Raiguel *et al.* (1995). Most other studies, however, have simply ignored the surround and studied the MT and MST neurons with the surround masked.

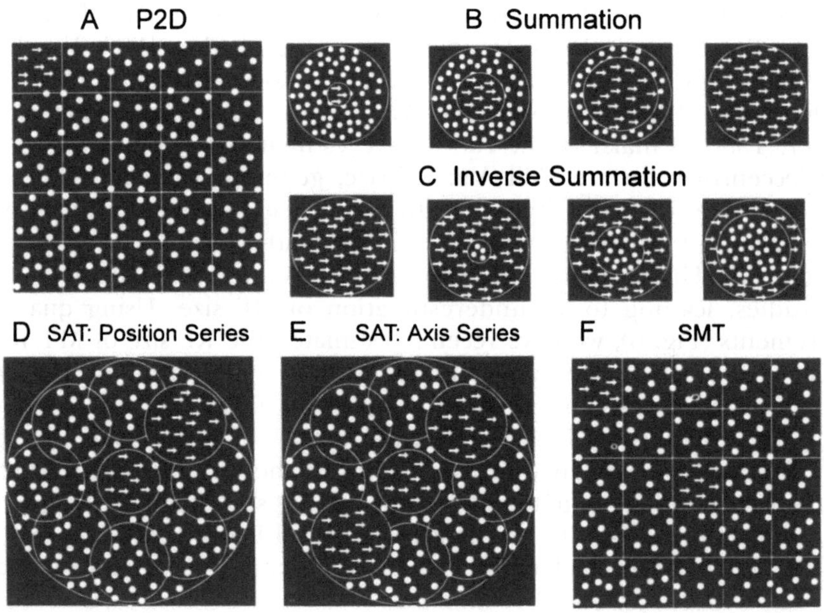

Figure 6. Tests with moving random dot patterns used to investigate the RFs of MT and MST neurons in our laboratory. (A) 2D position (P2D) test, (B) summation series, (C) inverse summation series, (D, E) surround asymmetry test (SAT: position [D] and axis series [E]), (F) surround mapping test (SMT). In the 2D position, test squares of a moving random dot pattern test 25 different positions; size of the square could be 2.5° × 2.5°, 5° × 5°, or 10° × 10°. In the summation test, circular patches of a moving random dot pattern of increasing diameter are tested. This test is performed with the centers of the patches aligned with the RF center derived from panel A. For optic flow components which contain a speed gradient, this test was run at two speeds to disambiguate areal and speed effects. The inverse summation test starts with a maximum-diameter circular patch of a moving random dot pattern and increases the diameter of patches containing stationary dots presented in the center. In the surround asymmetry test a center stimulus is combined with a single surround patch (position series) or a pair of patches at opposite sides of the center (axis series). In the surround map test the same grid of squares is used as in the 2D position test, but with a random dot pattern moving in two squares: the center square and another square appearing in one of the 24 remaining positions.

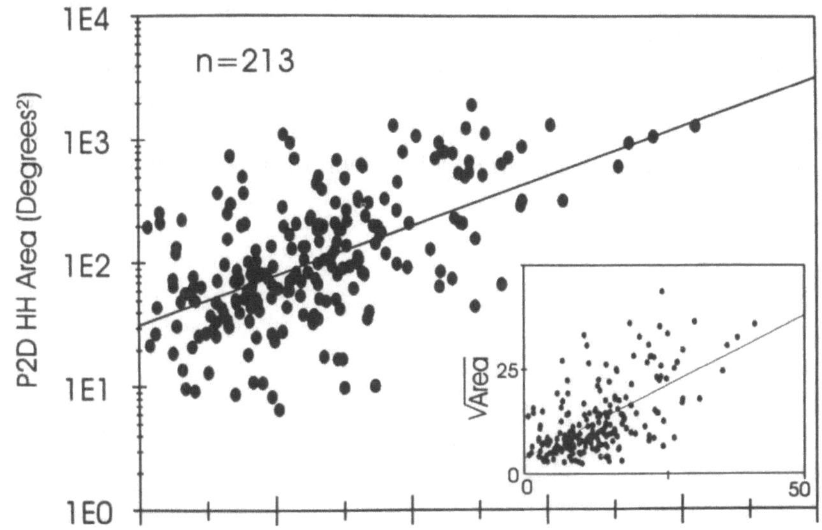

Figure 7. Relationship between 2D position test (P2D) half-height (HH) area (RF area at 50% of peak value derived from 2D position test; see Fig. 6A) and eccentricity. A log-linear plot is used, since measurements of RF areas, whether by hand plotting or curve fitting, otherwise tend to be heteroscedastic, i.e., show greater scatter at the high end of the abscissa. The relationship appears to be log-linear, with a slope of 0.044, an intercept (RF area at the fovea) of 32 deg^2, and a correlation coefficient R^2 of 0.33. Scatter at any given eccentricity is ~2 log units. Inset: Same data, with square root of area as the ordinate, for comparison with previously published data. Slope 0.67; intercept 4.08; R^2 = 0.36.

2.1.3. Topographical Organization

There is general agreement that area MT is retinotopically organized (Gattass and Gross, 1981; Van Essen *et al.*, 1981; Zeki, 1974), but the retinotopy is less regular than in V1 (Albright and Desimone, 1987; Maunsell and Van Essen, 1987) and often biased to the lower quadrant (Maunsell and Van Essen, 1987): central vision is represented anteroventrally (i.e., at the tip of MT), large eccentricities posterodorsally, the lower field laterally, and the upper field (which is often compressed) medially (Fig. 8). There is less agreement on the retinotopical organization of the satellites. Desimone and Ungerleider (1986) originally described some retinotopic organization in MSTd (their MST), but found very little evidence for a retinotopic order in MSTv (their FST). However, both Tanaka's group (Tanaka *et al.*, 1986, 1993) and our group have been unable to find any retinotopic order in MSTd. A possible explanation is that the most dorsal-medial part of MSTd, where the fovea would project according to Desimone and Ungerleider (1986), was little explored by the other two groups and that the foveal representation in MSTd may have been missed. On the other hand, Komatsu and Wurtz (1988a), who explored MSTd quite extensively, found visual pursuit cells, whose RFs include the fovea, to be scattered throughout MSTd, suggesting that no particular part of MSTd is involved in the processing of foveal visual information. A weak retinotopic organization seems to be present in MSTv according to the results of Tanaka *et al.* (1993), since small eccentricities are represented anterolaterally in this area (Fig. 1D), in agreement with the report by Komatsu and Wurtz (1988a).

In addition to representations of the contralateral hemifield, area MT/V5 and its satellites contain a representation of the portion of the ipsilateral field close to the vertical meridian. Since qualitative determinations of RFs underestimate their size, the ipsilateral overlap has been underestimated: Desimone and

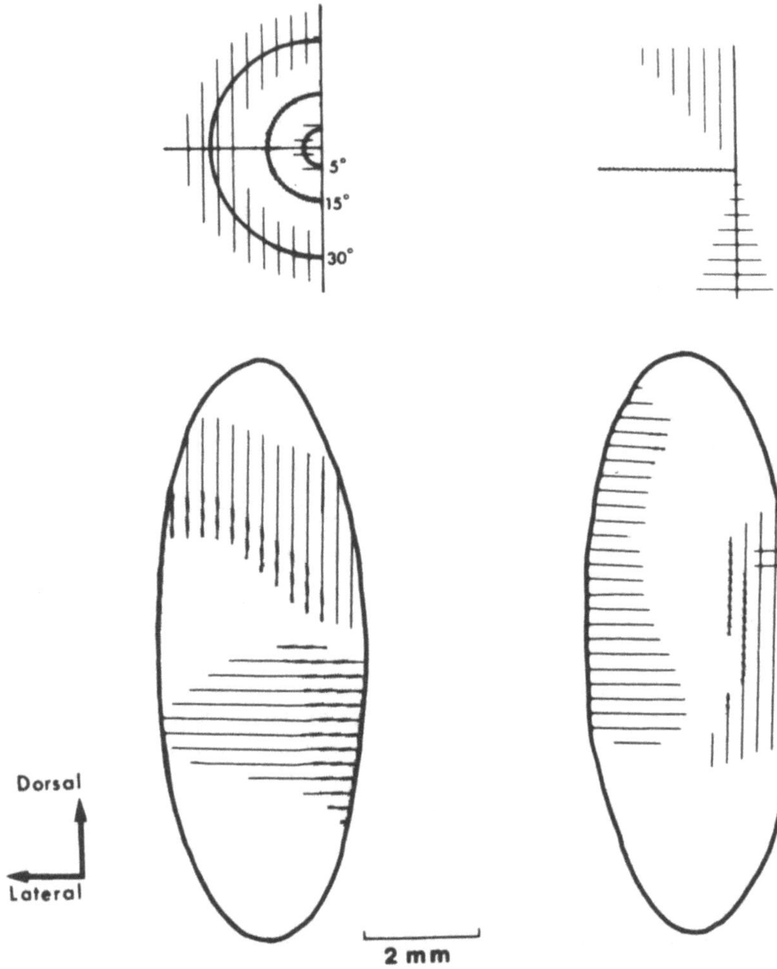

Figure 8. Summary diagram of visual topography in area MT/V5. Left: Representation of eccentricity on the standardized map of MT. Three types of shading are used to indicate the portions of MT in which particular ranges of eccentricity (5° ± 3°, 15° ± 4°, 30° ± 7°) were represented. Right: Representation of vertical and horizontal meridians on the standardized map. The hatched regions indicate the portions of MT in which receptive field centers were encountered within 15° polar angle of the inferior vertical meridian, 10° polar angle of the horizontal meridian, or 40° polar angle of the superior vertical meridian. The magnitude of the window around each meridian was adjusted to be large enough to include a substantial number of points while still retaining adequate resolution; hence, the window is large for the superior vertical meridian because of the underrepresentation of superior fields. From Van Essen *et al.* (1981). Subsequent work (Maunsell and Van Essen, 1987) has stressed the variability of the topography between individuals and the overrepresentation of inferior oblique diagonal of the visual hemifield.

Ungerleider (1986) observed an overlap reaching 4–8° in area MT and up to 30° in MST and especially FST. The recent quantitative study of Raiguel *et al.* (1995) yielded an ipsilateral overlap in area MT of 15–20°. Similar techniques yield estimates of 35° overlap for MSTd (S. Raiguel, L. Lagae, D.-K. Xiao, and G. A. Orban, unpublished). The large degree of ipsilateral overlap in area MT/V5 might be related to the pattern of callosal inputs. According to Maunsell and Van Essen (1987), the callosal input is not restricted to cells with RFs close to the vertical meridian as is the case in V1 or V2: cells recorded in callosal recipient regions of MT could have their RF borders up to 15–20° from the VM.

Since the RF scatter is similar in areas MT/V5 and V1, a logical consequence of the increased RF size in area MT compared to V1 is the decrease of the magnification factor (MF) (Albright and Desimone, 1987; Maunsell and Van Essen, 1987; Gattass and Gross, 1981). Albright and Desimone (1987) reported that in area MT the MF is equal to 1.14 E-0.76, compared to an MF of 1.95E-0.9 estimated by Maunsell and Van Essen (1987). Roughly, the magnification in MT is one-fifth that in V1, although at eccentricities of 20° the factor might be as little as one-third (Albright and Desimone, 1987). Conversely, if the magnification is lower in MT than in V1, then the point image is larger in MT than in V1 (Albright and Desimone, 1987; Maunsell and Van Essen, 1987; Gattass and Gross, 1981). In addition, the point image depends on the polar angle in the visual field, and is largest along the lower diagonal, a consequence of the representational bias favoring this part of the visual field (Maunsell and Van Essen, 1987).

2.1.4. Variability of Neuronal Responses

Response variability is a key element of neuronal reactivity. Without some estimation of response variability, it is impossible to determine how well cortical neurons signal outside events, particularly changes in visual stimuli (Britten *et al.*, 1992; Hawken and Parker, 1990; Vogels and Orban, 1990; Bradley *et al.*, 1987). One would expect variability to be influenced by behavioral state. Yet the first estimates obtained from striate cortex of the awake monkey (Vogels *et al.*, 1989) were remarkably similar to those obtained earlier in area 17 of the anesthetized and paralyzed cat and monkey (Tolhurst *et al.*, 1983; Dean, 1981). Vogels *et al.* (1989) observed that in striate neurons as a population the relationship between response variance and average response followed a power law. Thus this relationship can be fitted by a linear regression on a log–log plot. They estimated the average slope to be 1.11 and average intercept to be 1.9 for striate neurons (N = 183). They also obtained similar average values for individual striate neurons with a mean slope of 1.10 (N = 10) and mean intercept of 2.12. Subsequently Snowden *et al.* (1992) reported a mean slope of 1.21 and a mean intercept of 1.08 for individual striate neurons (N = 41). Similar estimations were made by Snowden *et al.* (1992) and Britten *et al.* (1993) for individual MT neurons: these two groups obtained average slopes of 1.10 (N = 39) and 1.20 (N = 190), respectively, and average intercepts of 1.37 and 1, respectively. Britten *et al.* (1993) showed that the average relationship for the individual cells was very similar to the relationship across the population of cells. Hence Celebrini and Newsome (1994) estimated only the population function for MSTd neurons and obtained a slope of 1.12 and an intercept of 1.09. Thus response variability is similar in area MT/V5, MSTd, and striate cortex. In all three areas the variance remains closely proportional to the mean. Hence as a first approximation, the variance can be considered to equal 1.5–2 times the mean (Shadlen *et al.*, 1996).

2.1.5. Temporal Aspects of Responses

Bair *et al.* (1994) have examined the interspike intervals of 212 MT neurons recorded during near-threshold, opposed-direction discrimination (Britten *et al.*, 1992; see Section 3.1) and found two groups of cells: nonbursting (one-third) and bursting neurons (two-thirds). The bursting neurons exhibit a peak in their power spectrum between 20 and 60 Hz, but the presence of this peak does not correlate with the neuronal threshold (see Section 3.1), nor does the strength of this peak correlate with the decision made by the monkey. Thus the functional role of stimulus-induced semisynchronous oscillations of MT neurons as reported in the awake monkey (Kreiter and Singer, 1992), remains unclear.

Relatively little attention has been given to the latency of neuronal responses in caudal STS. One reason may be that the traditional way of measuring latencies, using stationary flashed stimuli, seemed ill-suited for areas apparently involved in motion analysis. Response latencies to moving stimuli can be assessed by using bars at different speeds. Using this procedure, Raiguel *et al.* (1989) reported that area MT neurons, like V2 neurons, responded on average 10 msec after V1 neurons. Maunsell (1987) had made a similar observation on a somewhat smaller MT sample. An alternative technique for measuring the latency of motion responses is to measure response latencies to motion onset of random dot patterns which cover the RF at all times (Movshon *et al.*, 1990). Using this technique, Lagae *et al.* (1994) showed that the MSTd responses were on average delayed by 10 msec compared to MT responses. This was true over the entire range of speeds tested, an important observation, as motion onset latencies depend on speed (Kawano *et al.*, 1994; Movshon *et al.*, 1990). The estimated 10-msec delay introduced by processing in each cortical area fits with other observations made in this laboratory comparing latencies of V1 and inferotemporal neurons (Vogels and Orban, 1994). This 10-msec delay stands in sharp contrast to the 2 msec required for the activation of a subcortical target by a cortical area (Kawano *et al.*, 1994). Another interesting aspect of these latencies is their high degree of scatter within a single cortical area: the median latency of MT neurons reported by Raiguel *et al.* (1989) for moving bars was 95 msec, but values as short as 40 msec and as long as 200 msec were also observed. An important consequence of this larger scatter is that most cortical areas are nearly simultaneously active, reducing the functional impact of the anatomically defined hierarchical organization of the macaque visual system.

2.2. Exploration of the Classical RF: Simple Parameter Selectivities

2.2.1. Direction of Motion

For obvious reasons, direction selectivity (i.e., the preference for one of the opposite directions along the optimal axis) and directional tuning (i.e., the direction–response curve of MT neurons) have been explored quantitatively by a large number of groups (Marcar *et al.*, 1995; Lagae *et al.*, 1993, 1994; Snowden *et al.*, 1992; Mikami *et al.*, 1986a; Tanaka *et al.*, 1986; Albright, 1984; Maunsell and Van Essen, 1983a). Similar results have been obtained with either awake or paralyzed and anesthetized preparations. These studies have used various indices as measures of selectivity. The index for direction selectivity is the direction index (DI), defined as $(R_{PD}-R_{NPD}/R_{PD}$ (Baker *et al.*, 1981; Orban *et al.*, 1981b), where R_{NPD} and R_{PD} are net responses in nonpreferred direction (NPD) and in preferred direction (PD), respectively. The traditional measure for selectivity in the directional tuning is the bandwidth, defined as the width of tuning at half-maximum response. We (Vogels and Orban, 1994) and others have introduced a selectivity index (SI) based on circular statistics and taking all data points into account. This selectivity index is different from that used by Ferrera *et al.* (1994b) and Maunsell *et al.* (1991), which is based only on the extreme responses. Our selectivity index, which ranges from 0 to 100, has the additional benefit that it can be applied to any degree of tuning, avoiding arbitrary distinctions between biased and selective neurons. Marcar *et al.* (1995) have used such a direction selectivity index for MT neurons, where a single number incorporates both the width of tuning and the difference in response to opposite directions. Another way to visualize the degree of selectivity is to plot the average direction–response curve of all neurons (Maunsell and Van Essen, 1983a). Examples of directional tunings obtained for two MT neurons over a 16-fold range of speeds are shown in Fig. 9. The values of the DI, bandwidth, and direction selectivity index are indicated in the legend.

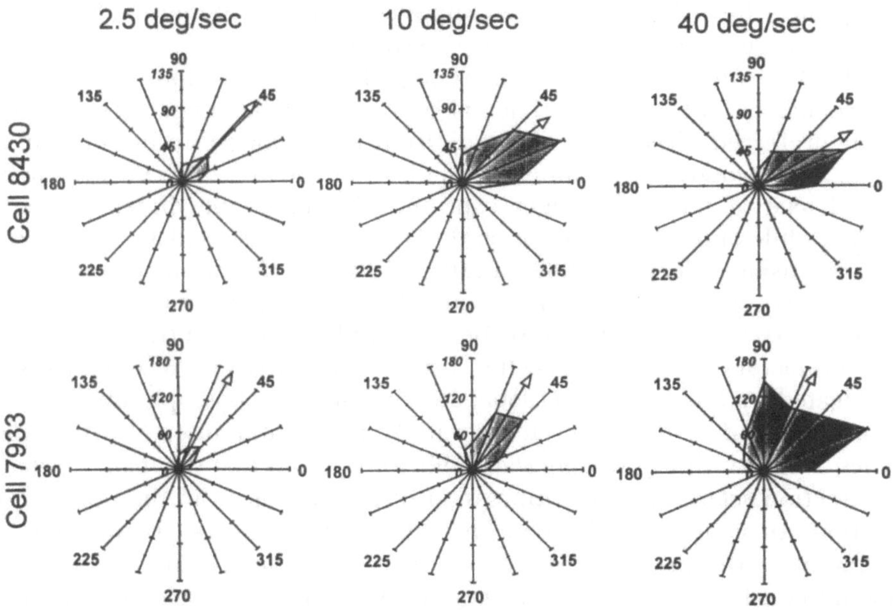

Figure 9. Directional tuning of MT neurons at three speeds. Response is plotted as a function of the direction of random dot pattern motion (16 directions tested). The arrows indicate preferred directions for each speed. Cell 8430 recorded in layer 3 had an RF at 7.6° from the fixation point (FP). Direction SI was 0.86, 0.84, and 0.83 at 2.5, 10, and 40°/sec, respectively, DI was 100 for all three speeds, and bandwidth was 90°, 56°, and 51°, respectively. Cell 7933 was recorded in layer V and had an RF at 0.9° from FP. Direction SI was 0.84, 0.82, and 0.72 at 2.5, 10, and 40°/sec, respectively, DI was 100 at all three speeds, and bandwidth was 67°, 52°, and 96°, respectively.

All studies agree that MT neurons (range of eccentricities tested up to 40°) are very direction-selective. Maunsell and Van Essen (1983a) reported that on average MT neurons responded 10.9 times more strongly in the preferred than in the opposite direction and that the average response fell to 50% of the peak value for directions 30° from the optimum. Albright (1984) compared direction tunings for moving slits, spots, and random dot patterns. The average DI for MT neurons was 1, 0.98, and 1.05 for these three types of stimuli respectively, while the average bandwidths were 91°, 105°, and 83° respectively. Studies using moving bars (Lagae *et al.,* 1993; Mikami *et al.,* 1986a) as well as those using random dot patterns (Lagae *et al.,* 1994; Snowden *et al.,* 1992; Britten *et al.,* 1992) confirmed that the average DI in area MT/V5 is close to 1. Using a DI of 0.9 as the cutoff value, Tanaka *et al.* (1986) reported that 76% of the MT neurons were direction-selective, with 10% unresponsive and the remaining non-direction-selective. Finally, Marcar *et al.* (1995) reported a median value of 76 for the direction selectivity index of MT/V5 neurons. Related to the issue of selectivity is the smallest difference in direction of motion MT neurons can signal. Snowden *et al.* (1992) measured the average response and variance for a range of directions and compiled the neurometric curves to calculate a neuronal just-noticeable difference (JND). The average neuronal JND was 5.5°, and most neuronal JNDs exceeded those of human observers. Monkeys have JNDs in this type of task very similar to that of humans (2°) (Lauwers *et al.,* 1995). Thus in signaling small differences in direction most MT cells performed more poorly than the subject, a finding very similar to that reported for V1 neurons signaling small differences in orientation (Vogels and Orban, 1990). It is noteworthy that for signaling the difference between opposite directions of motion, MT neurons are on average as good as the monkey itself (Newsome *et al.,* 1989).

Maunsell and Van Essen (1983a) reported that all directions were uniformly represented by MT neurons. They also noted that neighboring neurons had similar preferred directions. This indication was followed up by Albright *et al.* (1984), who proposed a columnar organization for area MT/V5 in which the optimal axis of motion changes continuously along one dimension in area MT/V5, while opposite directions on this axis are represented along the orthogonal dimension. Albright (1989) confirmed that all directions of motion were uniformly covered for MT neurons with RFS within 12° from the fixation point, but reported a centrifugal bias for neurons with more eccentric fields.

The factors contributing to the direction selectivity of MT neurons were investigated by Mikami *et al.* (1986a, 1986b) using stroboscopically illuminated moving slits and manipulating temporal (Δt) and spatial intervals (Δx) between flashes. These authors (Mikami *et al.,* 1986a) reported that the main mechanism for direction selectivity was suppression of the responses in the nonpreferred direction. In about half the neurons this mechanism was complemented by facilitation in the preferred direction. In a subsequent study, Mikami (1992) showed that while the suppression generally required only two successive flashes, the facilitation required several. Mikami *et al.* (1986b) also investigated the maximum spatial interval still permitting direction-selective interactions and found that this maximum was larger in area MT/V5 than in V1. This observation related to the fact that MT neurons maintain their direction selectivity for higher speeds than V1 cells. These findings also seemed in line with other studies indicating that MT cells are more direction-selective than V1 cells (Albright, 1984) and that they are tuned to faster speeds than V1 (Maunsell and Van Essen, 1983a). These statements, however, must be qualified since both direction selectivity and speed sensitivity are strongly dependent on laminar position in V1 (Hawken *et al.*, 1988; Orban *et al.*, 1986). Hence if one considers only the V1 neurons in laminae projecting to MT, the difference between these two processing stages is in fact very small (Orban, 1994). It has been argued that the additional processing within MT is used not to accommodate faster speeds, but rather to build more invariant responses (Orban, 1994).

In fact, a substantial amount of evidence has been amassed for the invariance of direction selectivity of MT neurons. Before reviewing this evidence it is worthwhile to state clearly what I mean by invariance. From our work in inferotemporal cortex (Kovács *et al.*, 1995; Sáry *et al.*, 1993, 1995) we have arrived at the following concept of invariance: invariance does not mean that the response remains constant, nor does it imply that the degree of selectivity remains completely unaltered, but that the preference of the neuron remains the same (Kovács *et al.*, 1995). This notion is similar to the relative invariance of Duffy and Wurtz (1995a). Applied to direction selectivity and direction tuning, this means that the response to the preferred direction can change when another parameter is changed, and that the degree of selectivity (as captured by the DI, bandwidth, or SI) can also change somewhat, but that the preferred direction remains fairly constant. Substantial evidence indicates that this is indeed the case for a number of stimulus parameters such as contrast polarity, speed, duration, position, size, and type of pattern. Figure 9 demonstrates the invariance of the direction tuning for changes in speed using a random dot pattern as stimulus. Obviously the response at optimal direction changes with the speed, and depending on the neuron, the degree of selectivity may change at slower speeds (Fig. 9A) or faster speeds (Fig. 9B), but the preferred direction is largely unchanged. This is a general finding for MT neurons explored over a 16-fold range of speeds bracketing the optimal speed. Rodman and Albright (1987) also reported that direction tuning changed little with the speed of bar motion. They found that the average bandwidth and preferred direction were both unaltered. Lagae *et al.* (1993) explored a much wider range of speeds, but measured only direction selectivity. Over a 35-fold range (2–70°/sec) the average DI remained above 75, but decreased at more extreme speeds. Because of the clear speed dependence of direction selectivity—if a wide enough range is explored—we have advocated the use of the MDI, the weighted mean of the DIs at the speeds to which the neuron is responsive (Orban *et al.*, 1981b). Over the range of speeds to which the MT neurons respond, there is rarely a reversal in preferred direction (Lagae *et al.*, 1993), indicating that the preferred stimulus is indeed the most invariant characteristic. The same study also demonstrated that direction selectivity was invariant for the contrast polarity of the bar. The MDI almost always had the same sign for both polarities, but the values were not always exactly equal (see their Fig. 4A). We also have evidence that the direction selectivity is invariant with motion duration: reducing the duration of random dot pattern motion to 10 msec affects the DI of MT neurons very little (Xiao *et al.*, 1993).

Tests using random dot pattern stimuli have shown that the direction selectivity is also invariant with respect to stimulus position and stimulus size. Raiguel *et al.* (1995) investigated direction selectivity at different positions in the RF and concluded that direction selectivity is maintained over the RF, but the DI averaged over the different positions (SADI) is generally lower than the peak DI (PDI) measured at the RF center. This implies that the DI (the degree of selectivity) must decrease at the edges of the RF. Figure 10 shows that this is indeed the case. This figure was prepared as follows: we selected 41 MT cells with similar RF sizes and plotted the relative response and DI as a function of distance from the center. At 5° from the center the average response is reduced to 30% while the average DI decreases by less than 10% (change significant at $p <$ .0001 level). At 7.5° the response drops to 10% and the DI falls to 40. This illustrates perfectly our concept of invariance: the response level and the degree of selectivity change, but the preferred stimulus does not, since the DI keeps the same sign.

Albright (1992) examined the invariance of direction tuning for the type of moving pattern. He compared the direction tuning for a first-order stimulus (a classical light or dark bar) and for second-order motion stimuli (Lu and Sperling, 1995; Cavanagh and Mathers, 1989; Chubb and Sperling, 1988). The latter were either a moving bar defined by flickering (temporal frequency noise) on a static two-dimensional random dot pattern background or a moving texture bar defined by a spatial frequency difference (larger dots in the bar than in the background). Albright (1992) observed that the preferred direction was gener-

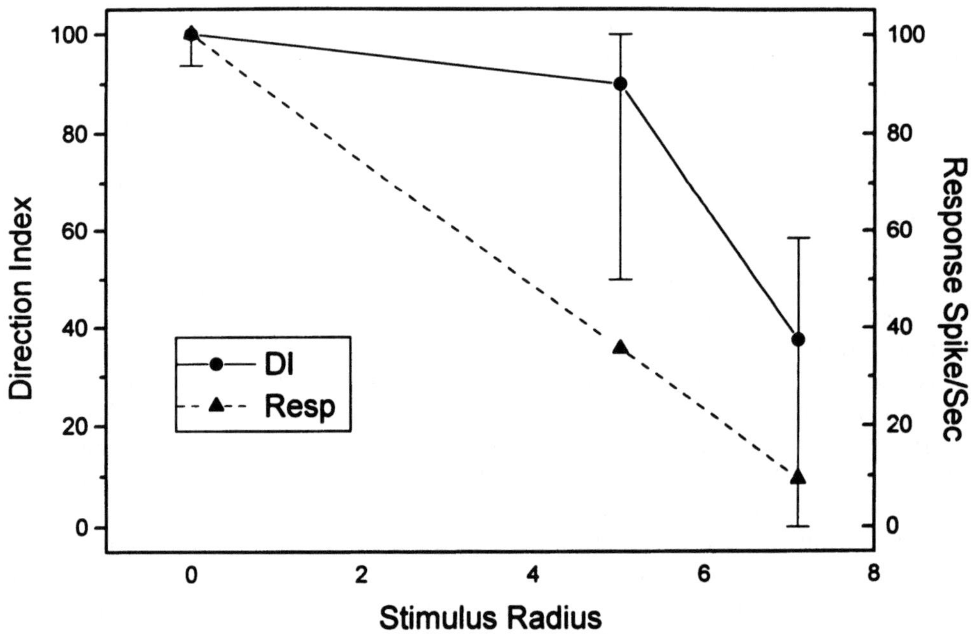

Figure 10. Change in direction selectivity with stimulus position in the RF. Median direction index and response are plotted as a function of distance of the stimulus square (5° × 5°) from RF center. Average data for 41 MT neurons with similar RF size. For DI the quartiles are indicated by vertical lines. The difference in DI between RF center and 5° radius was significant (Wilcoxon paired sample test).

ally constant: taking the average difference in PD over the population yielded zero difference. On the other hand, the response level was weaker for second-order motion, as was the degree of selectivity: the bandwidth remained unaltered, but the DI was reduced for second-order motion. Similar observations were made by Olavarria *et al.* (1992), who compared direction selectivities for a luminance-defined bar and a texture bar, defined by a difference in the orientation of the texture elements. Again they observed a change in response strength and degree of selectivity, although the preferred direction was generally unaltered despite claims made by the authors. Olavarria *et al.* (1992) stressed that in a number of cells the preferred direction reversed, since the DI changed sign. However these reversals were only minor, since the DI for the texture bar never exceeded a value of 50; hence these cells should be better considered as non-direction-selective for the texture bar. In fact, the degree of invariance for the cue defining the moving bar is strikingly similar to that for the contrast polarity of the bar, as attested by the similarity between Fig. 11 of Olavarria *et al.* (1992) and Fig. 4A of Lagae *et al.* (1993).

Directional tuning of MST neurons has been less well documented, since most quantitative studies (Graziano *et al.*, 1994; Tanaka *et al.*, 1986, 1993; Duffy and Wurtz, 1991a) have investigated only four cardinal directions (i.e., directions on vertical and horizontal axes). Tanaka and co-workers, again using the criterion of a DI of 0.9 (10/1 ratio), reported that 56% of the MSTd and 65% of the MSTv neurons are direction-selective (compared to 75% in MT; see above), while the remaining cells were selective for rotation or "expansion/contraction" (28% in MSTd, 3% in MSTv), non-direction-selective (5% in MSTd and 16% in MSTv), or unresponsive (11% in MSTd and 16% in MSTv). Figure 11 shows an example of a complete direction tuning from Lagae *et al.* (1994). These authors compared the DI of MT cells and MST cells and found both to average about 1. Celebrini and Newsome (1994) also report an average DI value of 1 for MSTd neurons ($N = 120$). Kawano *et al.* (1994) estimated the average direction tuning curves of MST neurons (MSTd and MSTv) and reported that average responses decrease to 50% of maximum at 60° from optimal direction. This figure might overestimate the tuning width because only neurons sensitive to fast speeds were tested. Figure 11 also illustrates that direction selectivity is generally invariant with stimulus position and with stimulus size. A number of exceptions have been noted by Komatsu and Wurtz (1988b) with respect to the size invariance (see Section 3.2). According to Tanaka *et al.* (1993), reversals in preferred direction with increasing stimulus size are rare (10/271 cells) and are somewhat more common in MSTd than in MSTv.

Tanaka *et al.* (1993) described a peculiar behavior in MSTv cells. The reader will remember that these neurons generally respond only to small random dot patterns (typically 10 or 20 deg in diameter). These neurons could be made responsive to a large random dot pattern when the central 10° was blanked out by a static foreground disk (Fig. 12), but the preferred direction reversed. Tanaka *et al.* (1993) took this as evidence that the MSTv cell was signaling the relative direction of motion of the foreground stimulus: indeed, when the disk moved with respect to a large, static random dot pattern, the PD was the same as for the small random dot pattern (Fig. 12). Sugita and Tanaka (1991) have shown that the responses of MSTv cells to background dots in the direction *opposite* to the

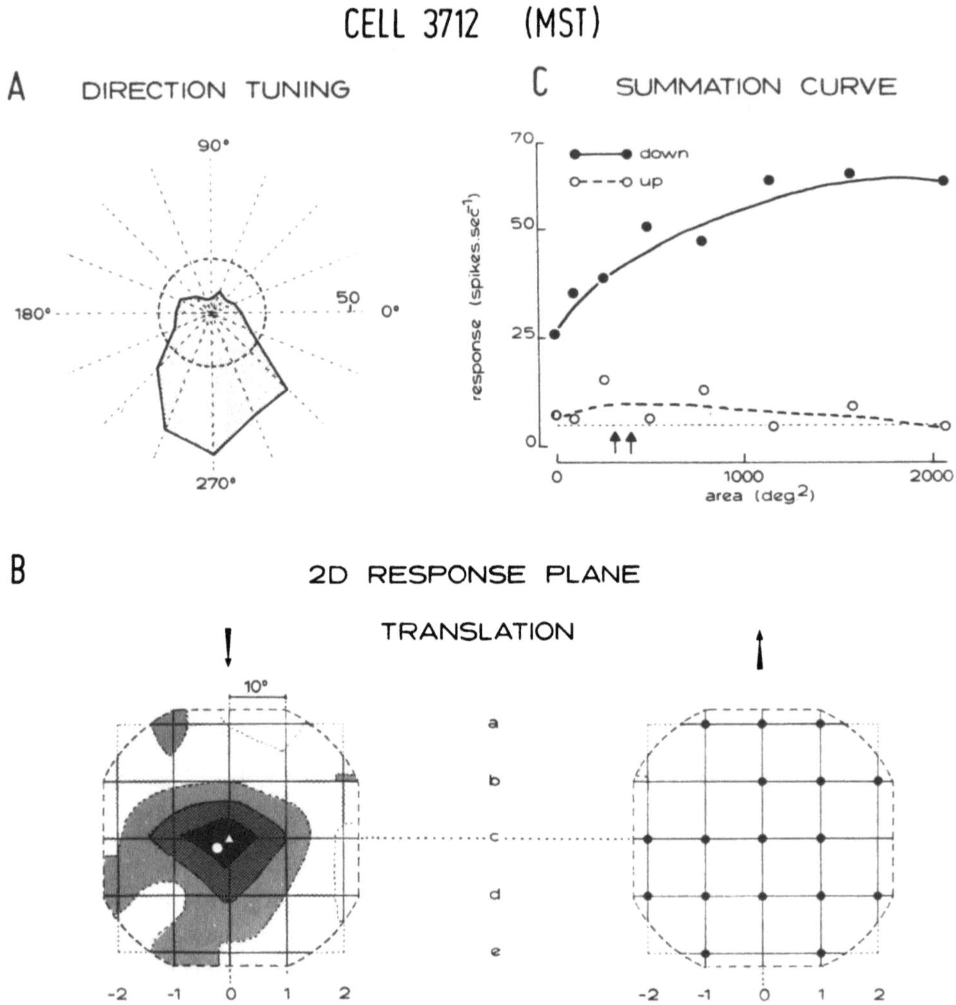

Figure 11. Preliminary testing of an MSTd neuron responsive to translation. (A) Directional tuning, (B) two-dimensional position response plane in preferred and opposite directions, and (C) summation curve. In all three tests the speed of translation was 10°/sec, although the directional tuning was also tested at 5°/sec and the summation also at 20°/sec. (A) Zero degrees corresponds to rightward motion and 90° to upward motion. Calibration: 50 spikes/sec. (B) Response levels are plotted as a function of stimulus position for the preferred (left) and opposite (right) directions of translation. The positive responses are represented in five levels obtained by interpolation and indicated by darkness of shading. Black: 80–100%. White: 0–20%. Intermediate shading: the three intermediate levels (60–80%, 40–60%, and 20–40%). One hundred percent corresponds to 61 spikes/sec. Filled dots: negative responses. The responses obtained at the four corners were not used in the calculations because the stimulus size (20° × 20°) exceeded the step size (10°) and are not drawn. Open dot: center of mass of the response, indicating the weighted average. The spatial scale indicates the step size. The eccentricity of RF center was 37° from the fixation point (15° azimuth, 34° elevation). (C) Response is plotted as a function of the stimulus area for preferred and opposite directions of translation. Arrows: areas of 312 and 400 deg^2, corresponding to truncated and untruncated (outside positions) stimuli in part B, respectively. These tests are routinely used in all MT and MST experiments carried out in our laboratory. For most neurons the P2D test was tested with step size equal to stimulus size, so that no truncation occurred in the corners. For stimuli for the P2D test producing the two-dimensional response plane, and the summation test yielding the summation curve, see Figs. 6A and 6B, respectively. From Lagae *et al.* (1994).

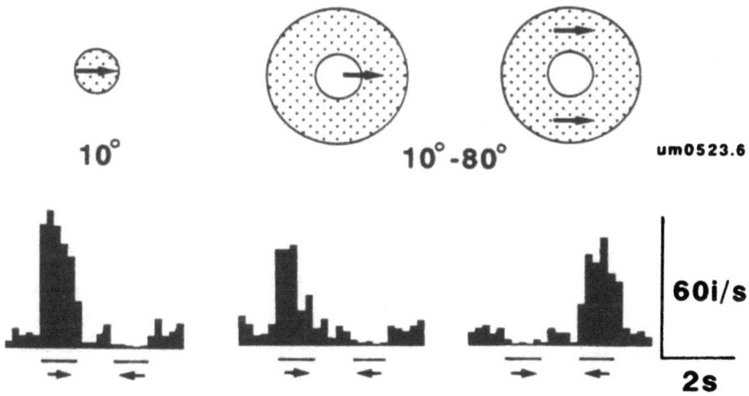

Figure 12. Behavior typical of MSTv neurons. These neurons respond only to small-diameter, moving random dot patterns (as on the left) and not to large-diameter random dot patterns (not shown). However, when a stationary window is present in the middle of the large random dot pattern the cell responds to motion in the opposite direction (right). This is in fact a response signaling motion of the stationary window relative to the background random dot pattern, as is demonstrated by a response of the neuron to a small disk moving in front of the large random dot pattern kept stationary (middle). Thus the neurons' response in the three configurations is consistent with motion of a small stimulus to the left, relative to the background. From Tanaka *et al.* (1993).

preferred direction (for a small random dot pattern) depends on the sharpness of the disk edge. They interpreted this as evidence that MSTv neurons are able to use dynamic occlusion cues. Quite different behavior is displayed by MSTd cells. Lagae and co-workers observed that when the central part of a large random dot pattern is blanked out in the inverse summation test (Fig. 6), the cell retains its response and direction selectivity (Fig. 13). In fact, motion confined to the four corners of the display was sufficient to maintain a direction-selective response (Lagae, 1991). Tanaka and Saito (1989) reported a few similar observations (their Fig. 3). Such a behavior seems to indicate that MSTd cells signal direction of motion in the background independently of the presence of objects in the foreground. Thus the opposite behavior of MSTv and MSTd cells fits nicely with the proposal of Tanaka *et al.* (1993) that MSTd subserves analysis of egomotion signals, while MSTv subserves analysis of object motion signals.

2.2.2. Speed of Motion

A number of studies have been devoted to the speed selectivity of MT neurons. Most studies (Cheng *et al.*, 1994; Mikami *et al.*, 1986a; Maunsell and Van Essen, 1983a) have used moving slits as stimuli and average firing rate as a response measure. However, the use of this measure with slit responses presumes that the shape of the response peak remains constant with speed. This is unlikely, since the shape will reflect the spatial profile of the RF at slow speeds and its derivative at fast speeds. Therefore we have advocated the use of maximum firing rate as a measure of responses to slit motion. Unfortunately, these two measures, average and maximum firing rate, yield different speed–response curves, with those prepared using average firing rate being much more narrowly tuned (Orban, 1991; Orban *et al.*, 1986). Indeed all authors using moving slits and average firing rate (Cheng *et al.*, 1994; Mikami *et al.*, 1986a; Maunsell and

Van Essen, 1983a) report that most MT neurons are speed tuned and that the average optimal speed is close to 32°/sec in both awake and paralyzed, anesthetized monkeys. Furthermore, there is little change in the optimal speed with eccentricity within area MT/V5 (Cheng *et al.*, 1994; Maunsell and Van Essen, 1983a). Finally, two areas as different as MT and V4 yield very similar optimal speed distributions, the median value being 32°/sec in both areas when average firing rate is used as response metric (Cheng *et al.*, 1994). Quite a different view emerges when maximum firing rate is used as the measure for moving slit responses. Three different types of speed–responses curves can be observed: low-pass curves, tuned, and broadband curves (Lagae *et al.*, 1993). In the monkey, high-pass curves (Orban *et al.*, 1981a) have never been observed in response to moving slits. The proportions of these three speed–response types change with eccentricity in MT: the low-pass curves predominate in the representation

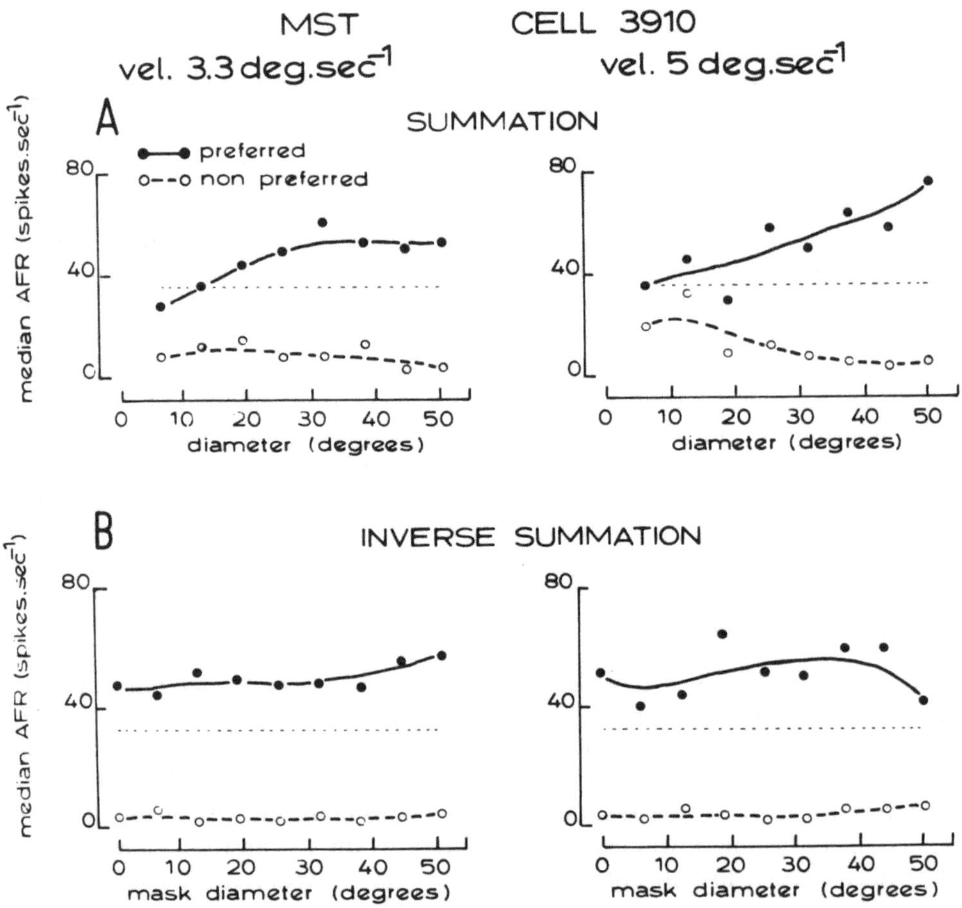

Figure 13. Summation curves and inverse summation curves of an MSTd neuron taken at two translation speeds. AFR is firing rate, averaged over stimulus duration. The dotted line indicates the significance level. Stimuli are shown in Figs. 6B and 6C. Notice that in contrast to the MSTv neuron of Fig. 12, the preferred direction was not changed by introduction of central stationary windows in the inverse summation test. From Lagae (1991).

of the central 2° of vision, tuned in the representation of pericentral vision, and broadband in the representation of peripheral vision (Lagae *et al.*, 1993). The observation that cells with RFs close to the fovea generally have low-pass speed–response curves fits with a report of visual pursuit cells located in the foveal part of MT (Komatsu and Wurtz, 1988a). Indeed, during visual tracking the retinal image is very close to the fovea and the retinal slip speed is small. Lagae *et al.* (1993) have proposed that low-pass and speed-tuned cells play a role in the maintenance and initiation of pursuit, respectively. Using moving random dot patterns as stimuli eliminates the ambiguity in the response measurement: since the stimulus moves over the RF continuously, the effect of the RF profile is removed and either average or maximum firing rate will yield similar results. Using a moving random dot pattern, we observed speed–response curves similar to those described with moving bars, except that in this case a few MT cells displayed a high-pass behavior (Fig. 14). For these tests using random dot patterns it did not matter whether a constant duration or constant amplitude of motion was used.

In their study with stroboscopically illuminated slits Mikami *et al.* (1986a) also investigated the mechanism producing speed tuning in area MT/V5. They reported that one-third of the neurons tested showed a suppression acting at the high-speed cutoff, which seems to complement the restrictions placed on the upper speed range by energy limitations. This fits with the observation that spontaneous activity can be inhibited in some cells by fast motion in the preferred direction. These authors (Mikami *et al.*, 1986a) also claimed that suppres-

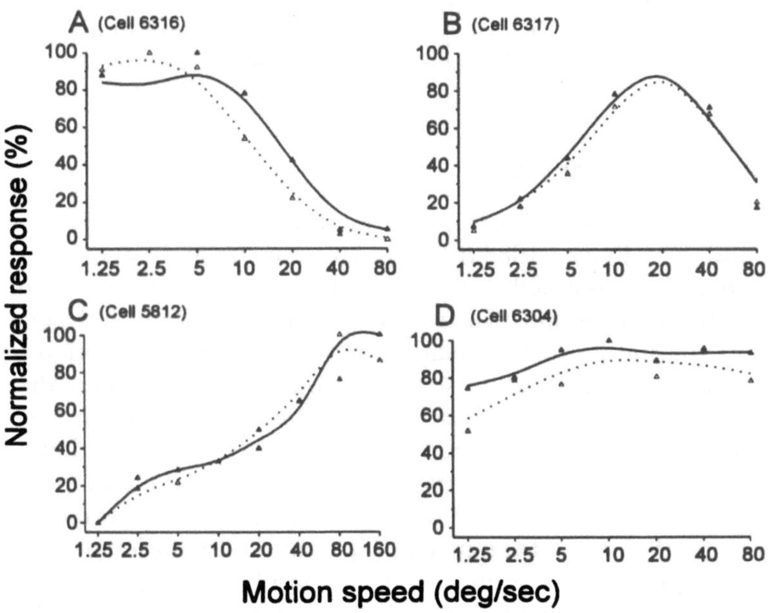

Figure 14. Speed–response curves obtained with moving random dot pattern in MT neurons: (A) low-pass curve, (B) tuned curve, (C) high-pass curve, (D) broadband curve. Solid lines: test with constant duration of motion (320 msec); dotted lines: test with constant amplitude (8°). Only responses in the preferred direction are shown. All neurons had RFs within 13–16° from the fixation point; laminar position: cell 6316, layer 3b; cell 6317, layer 3c; cell 5812, layer 3a; cell 6304, layer 5.

sion produces the low-speed cutoff in the majority of MT neurons. Indeed, many neurons responded to slow speeds at large Δt, but failed to respond to these speeds at very small temporal intervals. An alternative explanation for the latter observation is that this effect was due to the change in steepness of the illumination onset over an RF locus. In the isolated flashes the illumination transient is very steep, while in a smoothly moving stimulus moving at low speed the illumination change will be very gradual. At least for cat cortical neurons, it has been shown using moving neutral-density wedges that neurons are very sensitive to alterations in the slope of illumination changes (Orban, 1986). Thus no cortical mechanism may be needed for the low-speed cutoff, since the change in stimulus alone might explain the difference in response of MT neurons.

Little work has been done with respect to the invariance of speed tuning in MT neurons. One exception is the report of Lagae *et al.* (1993), which compared the speed–response curves of MT neurons for moving light and dark bars. Most (77%) MT neurons had the same speed–response curve for both contrast polarities. For the speed-tuned cells (tuned for both polarities) the optimal speed was very similar for the two polarities, but the bandwidth was much more dependent on polarity, again illustrating that the preferred parameter value is more invariant than the degree of selectivity. We also have evidence that speed tuning is invariant for changes in motion duration (Xiao *et al.*, 1993).

Little has been reported about the speed characteristics of MST cells. We have preliminary evidence that moving random dot patterns yield speed–response curves similar to those in MT (Orban *et al.*, 1995a; Lagae, 1991). Kawano *et al.* (1994) reported that the responses of MST cells (both MSTd and MSTv) to random dot patterns increase with increasing speed and level off between 40 and 160°/sec. This preference for fast speeds might be the result of a screening bias, since the authors searched for cells using a random dot pattern moving at 80°/sec.

2.2.3. Orientation and Disparity

Three studies have investigated orientation selectivity in MT neurons. Maunsell and Van Essen (1983a) tested 70 neurons and reported that these neurons generally responded to the flashed bar with a brief transient. Three-quarters of the neurons were clearly tuned for orientation and generally the optimum orientation was orthogonal to the best direction of motion of the slit. The orientation average tuning of all neurons tested indicated that an MT neuron's response to the optimal orientation is on average eight times the response to the orthogonal orientation. Albright (1984) observed that 74/89 MT cells were selective for orientation (flashed slit), and that the average bandwidth was 64°, a value slightly larger than for V1 neurons. Furthermore, Albright (1984) reported that 61% of the neurons had an optimal orientation within 30° of a line orthogonal to the preferred direction of slit motion, while 29% had their optimal orientation within 30% of the preferred direction. Marcar *et al.* (1995) tested MT neurons with grating patterns flashed on and off and obtained an average orientation SI of 32, confirming that most of the MT neurons are tuned for orientation (i.e., have an orientation SI larger than 15). This study also confirmed that the optimal orientation was nearly parallel to the preferred direction of random dot pattern motion in about one-third of the neurons, while for the remaining two-thirds the optimal orientation was nearly orthogonal to the PD.

The influence of bar length on the responses of MT neurons to moving bars was investigated by Maunsell and Van Essen (1983a), who studied 18 neurons, and by Cheng *et al.* (1994), who reported on 75 MT neurons. According to Maunsell and Van Essen (1983a), only 3/18 neurons showed substantial endstopping (sharp decrease in response at long bar lengths). According to Cheng *et al.* (1994), about half the neurons showed a reduction in response to less than 50% of the peak value at longer lengths. It is probably more meaningful to discuss how invariant the direction and speed tuning are for stimulus length changes. In fact, Cheng *et al.* (1994) tested 10 neurons for changes in direction tuning with varying stimulus length and observed that in most neurons the preferred direction was unaffected by the change in length. Maunsell and Van Essen (1983a) also tested 18 MT neurons for influence of stimulus width and reported no significant changes.

Maunsell and Van Essen (1983b) studied the disparity sensitivity of MT neurons using moving slits. About two-thirds (52/76) were found to be profoundly sensitive to horizontal disparity. About equal proportions of these disparity-sensitive cells were near or far (54%) and tuned (46%). These authors also searched for selectivity to direction of motion in depth, but failed to find evidence for such selectivity in MT. Zeki (1974) has reported qualitative observations of MT neurons with opposite preferred directions in the two eyes. Such a neuron is quantitatively documented in Orban (1991) for a wide range of speeds. These neurons, which could effectively signal direction of motion in depth, are exceedingly rare in MT (about 1% of the neurons).

Roy *et al.* (1992) studied the disparity selectivity of MSTd and MSTv neurons using a moving random dot pattern: 149 neurons were studied in MSTd and 103 in what appears to be the dorsal part of MSTv. In both regions the vast majority of neurons (90%) were disparity-selective and of these the preponderance (over 90%) were either near or far neurons. In three-quarters of the 20 cells tested the disparity–response curves was unaffected by changes in vergence angle, suggesting that these neurons signal distance with respect to the fixation point. In 40% of the cells tested (26/65) with the PD and NPD, there was a reversal of direction selectivity with disparity (DDD cells). Roy *et al.* (1992) correctly pointed out that such neurons would be extremely useful for extracting direction of egomotion from the direction of retinal image motion, but unfortunately did not specify the relative proportions of these interesting neurons in MSTd and MSTv. Further support for the use of MSTd signals in analysis of egomotion was provided by the observation that while changes in speed did not alter the DDD characteristic, the optimum disparity shifted to nearer values with increasing speed. The fact that most MSTv cells were also far or near neurons is consistent with the view that these neurons contribute to control of vergence.

2.2.4. Contrast and Wavelength

The contrast response curves of MT neurons were studied by Sclar *et al.* (1990) and Cheng *et al.* (1994). Sclar *et al.* (1990) compared the MT curves with those of V1 neurons and of magno- and parvocellular geniculate neurons. These authors observed that semisaturation contrast is markedly lower in MT and magnocellular neurons than in parvocellular geniculate neurons, in accordance with the predominantly magnocellular input to MT (Maunsell *et al.*, 1990). Compared to V1 neurons, MT neurons were more sensitive. However, some overlap exists, in agreement with the combined magno- and parvocellular input to V1 (Nealey and Maunsell, 1994). Cheng *et al.* (1994) observed somewhat more overlap in the contrast thresholds of MT and V4 neurons, again consistent with the combined magno- and parvocellular input to V4 (Ferrera *et al.*, 1994a).

There is now clear evidence (Gegenfurtner *et al.*, 1994; Saito *et al.*, 1989) that a number of MT neurons remain direction-selective at isoluminance and thus can, when necessary, rely on color cues. These responses are much weaker than those to luminance-modulated stimuli, however. Gegenfurtner *et al.* (1994) commented that these MT responses are unlikely to sustain the perceptual responses to isoluminant stimuli. Using phase-reversing heterochromatic gratings, Dobkins and Albright (1994) tried to identify the color cues underlying the direction selectivity in MT neurons under isoluminant conditions. They obtained evidence for both an unsigned color cue (based on chromatic borders), compatible with magnocellular input to area MT/V5, and a signed color input (based on the color per se), which is less compatible with magnocellular input. They speculate that the proportion of magno- and parvocellular input one observes in MT might depend on the nature of the stimuli used, since Maunsell *et al.* (1990) used light bars as stimuli in their experiments. Dobkins and Albright (1995) showed that the weak direction selectivity of MT neurons at isoluminance might have a behavioral correlate with the oculomotor response of monkeys to moving stimuli, suggesting an interesting dissociation between motion perception and control of eye movements elicited by isoluminant stimuli.

2.3. Exploration of the Classical RF: Responses to Complex Stimuli

2.3.1. Responses to Moving Plaids

Motion of one-dimensional patterns such as moving slits and gratings is inherently ambiguous for a restricted RF, since it can correspond to a large variety of translation vectors: this is referred to as the aperture problem. This problem vanishes as soon as two contours are present on an object. In order to investigate whether cortical cells are able to combine information from two contours to solve the aperture problem and thus to signal the true direction of motion (translation), Movshon *et al.* (1985) devised the following test. They compared direction tuning for a moving grating and a moving plaid made of two identical gratings differing in orientation by 90° or 135°. By correlating the plaid directional tuning with the predictions for pattern- and component-direction selectivity derived from the grating directional tuning (Fig. 15), they were able to show that MT cells could be either pattern- or component-direction selective, while monkey and cat striate neurons generally fell into the component-direction-selective category. Unfortunately, because of the breadth of the direction tuning curves of MT neurons, one-third of neurons remained unclassified. One way to view the pattern-direction-selective neurons is that their direction selectivity is invariant for the type of stimulus pattern (grating versus plaid). Movshon *et al.* (1985) reported preliminary evidence that pattern-direction-selective neurons occurred mainly in layers 2–3 and 5, while component cells occurred mainly in layers 4 and 6, but no explicit laminar data were presented. These authors went on to explicitly study V1 neurons with demonstrated projections to MT and (although the numbers were small) they confirmed that these cells were component-direction-selective (Movshon and Newsome, 1984).

Rodman and Albright (1989) confirmed the existence of pattern- and com-

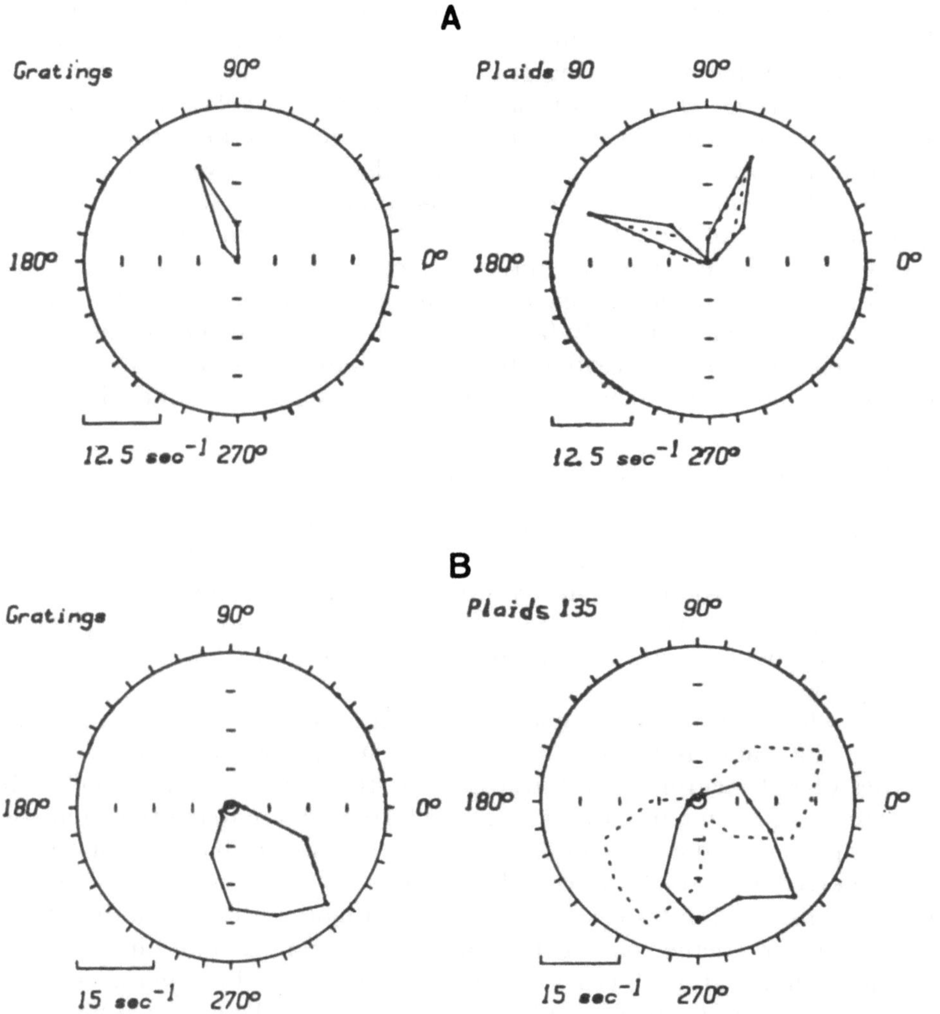

Figure 15. Directional tuning of two MT neurons. (A) Component-direction-selective neuron, (B) pattern-direction-selective neuron. On the left the neuron's tuning is shown for moving gratings, on the right for moving plaids. The dashed curve on the right shows the predicted response of a component-direction-selective neuron. The inner circles in each plot show the maintained discharge. (A) Spatial frequency 3.6 cycles/deg, drift rate 4 Hz, component correlation 0.991, pattern correlation −0.092 (N = 16). (B) Spatial frequency 2.7 cycles/deg, drift rate 4 Hz, component correlation 0.349, pattern correlation 0.940 (N = 16). From Movshon *et al.* (1985).

ponent-direction-selective MT neurons and showed that neurons with their optimal orientation close to the parallel of the preferred motion direction are generally pattern-direction-selective, while those with their preferred orientation close to the orthogonal of the PD generally were component-direction-selective.

In psychophysical experiments, it has been shown that two moving gratings do not always produce the percept of a coherently moving plaid. The spatial frequency of the two gratings has to be within 1.5–2 octaves (Adelson and Movshon, 1982). Another requirement is that the luminance of the intersections has to be compatible with coherence. When using gratings made of narrow, dark bars separated by wide, light bars, the intersection must be slightly darker than the dark bars. Indeed, markedly darker or brighter intersections produce the percept of transparently moving gratings rather than that of coherent motion (Stoner *et al.*, 1990). However, Kim and Wilson (1993) have shown that both these requirements are subordinate to the difference in direction between the two gratings. Small differences in direction (typically 20–25°) always produce coherence. For differences larger than 45° the percept depends on spatial frequency or on luminance of the intersections. Stoner and Albright (1992) showed that MT neurons behave in the same way as perception: pattern-direction selectivity will disappear when the luminance of intersection is not compatible with the percept of coherent motion. In fact all MT neurons shift to a more component-dominated direction selectivity in these conditions. Thus there is some "intelligence" in the way MT neurons combine motion signals over their RF.

2.3.2. Responses to Transparent Stimuli

The stimuli used by Andersen and co-workers discussed in this section differ in a fundamental way from those used by Stoner and Albright (1992) and described above: Andersen and co-workers used stimuli moving in opposite (rather than orthogonal) directions. In this case the alternative to transparency is flicker, and indeed two gratings moving in opposite directions yield a phase-reversing grating.

Snowden *et al.* (1991) compared the responses of MT and V1 neurons to random dot patterns moving transparently in opposite directions. The responses of MT neurons to the transparent stimuli were clearly smaller than those to a single stimulus moving in the preferred direction, but the attenuation was less pronounced in V1. Snowden *et al.* (1991) showed that the motion in the nonpreferred direction was equally effective at suppressing MT responses whether presented in transparent stimuli or in motion discontinuities. Thus it did not matter whether or not the stimuli moving in opposite directions were spatially segregated. Both types of stimuli were restricted to the classical RF in these experiments. Qian and Andersen (1995) used locally paired and unpaired random dot patterns moving in opposite directions as examples of stimuli with locally balanced and unbalanced motion signals. The former produced no transparency but flicker, whereas the latter produced transparency. They observed that in general, V1 cells could not distinguish between paired and unpaired stimuli, but that MT cells as a population responded better to the unpaired than to the paired stimuli. These authors suggested that the fundamental role of the inhibition in the nonpreferred direction is noise reduction, since MT neurons are much less sensitive to flicker than V1 cells. In this respect it is worth mentioning that Lagae *et al.* (1994) have reported that MSTd neurons respond even less to flicker than MT neurons. Thus there seems to be a continual reduction of noise along the motion processing pathway rather than just the two-step process suggested by Qian and Andersen (1995). This still leaves the question of how we can perceive two separate coherently moving surfaces when confronted with the transparent stimuli. This percept is greatly aided by the addition of binocular disparity. Bradley *et al.* (1995) showed that the reduction in the MT response to transparent stimuli or motion discontinuities is largely reversed when a disparity is introduced between the two sets of moving random dot patterns (Fig. 16). This experiment and that of Stoner and Albright (1992) have interesting implications: both show that MT neurons behave 'intelligently' by integrating other visual information (stereo, luminance) into the motion processing, and second, both lend support to the idea that area MT/V5 is involved in the processing of surfaces specified by motion.

2.3.3. Responses to Optic Flow Stimuli

Although Sakata *et al.* (1985) had described parietal neurons (area 7a) that were selective for the direction of rotation or "expansion/contraction" (described as sensitive to movement in depth), Saito *et al.* (1986) were the first to describe such cells in area MST and to show that their occurrence was restricted to a region they had initially labeled "DSR" (i.e., containing displacement, size change, and rotation cells), but later (Tanaka and Saito, 1989) renamed "MSTd." Saito *et al.* (1986) used two criteria to define cells as selective for "expansion/contraction" (E cells) or rotation (R cells): (1) a direction-selective response for "expansion/contraction" or rotational movement and (2) the absence of any response to translation (Fig. 17). The same group also reported that very few MT cells met these criteria (Tanaka *et al.*, 1986), a finding confirmed by the quantitative study of Lagae *et al.* (1994). Saito *et al.* (1986) further reported that very few cells in the ventral part of MST met the criteria for E and R cells, a finding later amplified by Tanaka *et al.* (1993). Indeed, the presence or absence of cells selective for rotation and for "expansion/contraction" is one of the criteria distinguishing MSTd from MSTv (see Section 1.1). Most subsequent studies on responses to optic flow stimuli have thus concentrated their efforts on MSTd.

Duffy and Wurtz (1991a) criticized the approach of Tanaka and co-workers because it *a priori* excluded combined selectivities (e.g., for rotation and translation). Hence Duffy and Wurtz (1991a) required only that cells be responsive to rotation, "expansion/contraction," or translation when testing all three motion types with large random dot patterns (100° × 100°). They reported a response continuum in MSTd, in which some cells responded to a single type of motion (23%), some to two types (34%), others to three types (29%), with the remaining being unselective (14%). Interestingly, among the 74 double-component cells, none were responsive to both rotation and "expansion/contraction." Hence all multiple-component cells responded to translation, which raises the question of whether the responses to rotation and expansion are due to an accidental match between the translation vector distribution and a translation-selective RF, or whether they reflect a receptive field with true selectivity for rotation or "expan-

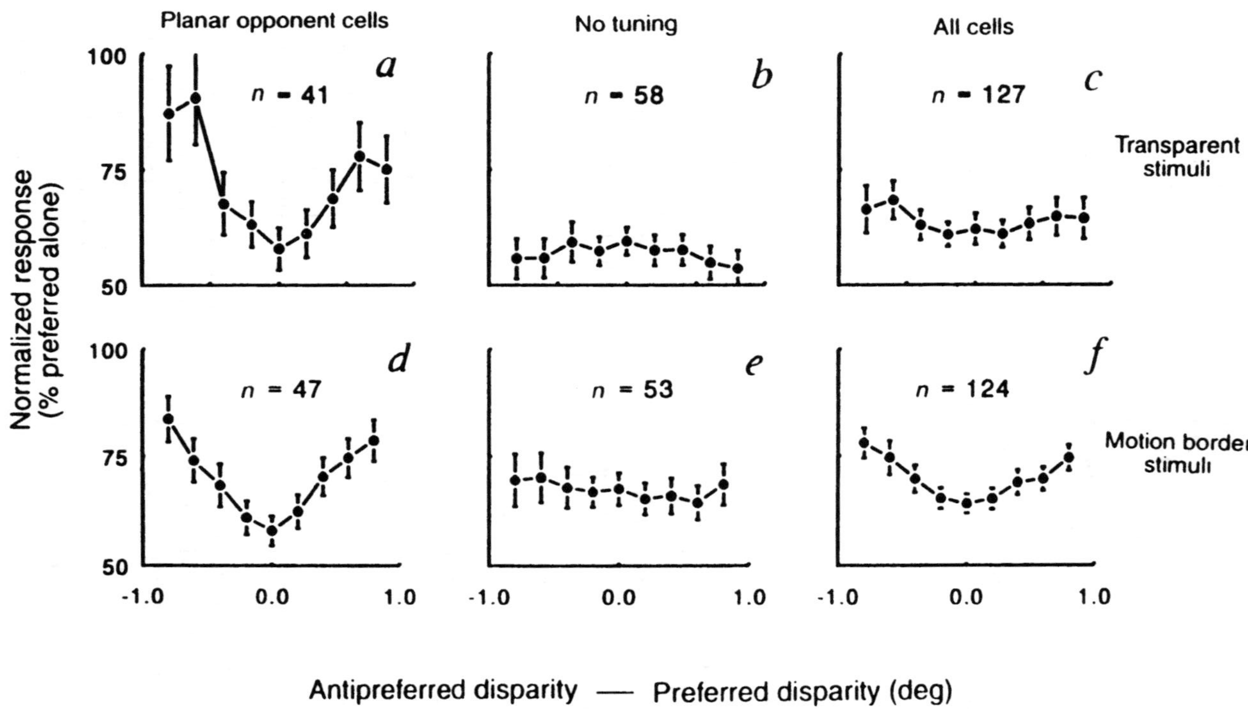

Figure 16. Average responses of MT neurons to transparent and motion border stimuli (or antiphase kinetic boundaries) depend on the disparity between the two moving random dot patterns. Firing rates (normalized to responses to preferred motion alone) as a function of the difference between preferred and antipreferred disparities (reflecting the apparent depth between the opponent surfaces). Top row: data for transparent stimuli. Bottom row: data for motion border stimuli. Left: Average responses for cells classified as 'planar opponent' (i.e., cells for which the disparity causing strongest excitation in the PD also caused strongest inhibition in the NPD). Middle: Average responses for cells without apparent effect of disparity on suppression. Right: Average responses for all cells. Values are means ± standard error. The planar opponency trend seen in the population data for motion border stimuli (f) was highly significant ($p < .001$: linear regression of normalized firing rate versus absolute antipreferred − preferred disparity difference). For the planar opponent cells (left column), the average tuning width for half-maximal effect (estimated graphically) was 1.0° for transparent stimuli and 0.9° for motion border stimuli. From Bradley *et al.* (1995).

sion/contraction." Thus there is a clear need for a criterion with which to distinguish between these two alternatives. Although they explored position invariance, Duffy and Wurtz (1991b) did not fully appreciate this necessity, perhaps because they were using large stimuli (100° × 100°).

All studies mentioned so far have restricted their investigations to two optic flow components (OFC): rotation and "expansion/contraction." In fact, there is a third component: deformation (Koenderink and van Doorn, 1975), and our group (Lagae *et al.*, 1994) has systematically compared responses to translation and to all three optic flow components. As did Duffy and Wurtz (1991a), Lagae *et al.* (1994) observed a response continuum in MSTd (Fig. 18A), but also found one in MT, underscoring the need for a criterion with which to distinguish RFs specialized for optic flow components. The criterion Lagae *et al.* (1994) proposed is the position invariance of the direction selectivity for flow components. The reader will notice that this is exactly the sort of invariance that has been demonstrated for translational-direction-selective responses in MT: only the preferred direction (e.g., anticlockwise rotation in Fig. 19B) is required to remain constant

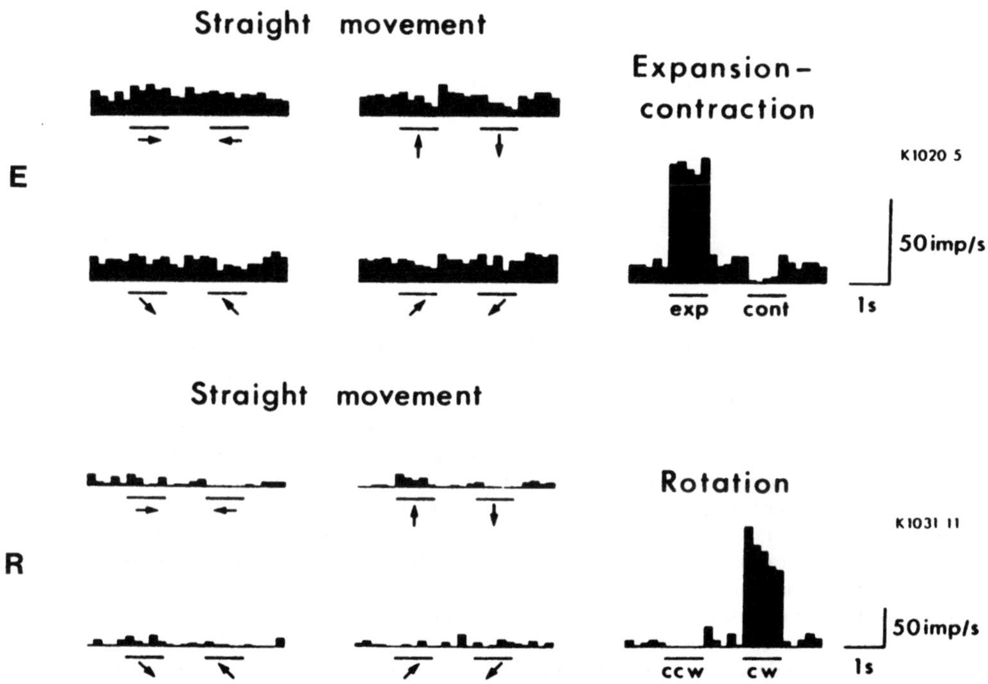

Figure 17. Responses of an expansion (E) MSTD cell (top) and a rotation (R) MSTd cell (bottom) to straight movements (i.e., translations) compared to their responses to an "expansion/contraction" or rotation. Left and middle: A dot pattern was moved straight on the screen in one of eight different directions (13°/sec). Right: The same dot pattern was either expanded and contracted (doubled or halved in 1 sec) or rotated in the counterclockwise and clockwise directions (0.11 revolutions in 1 sec). Lines beneath the poststimulus time histograms indicate the periods during which the pattern moved; arrows indicate the direction of straight movements; and exp, cont, ccw, and cw indicate expansion, contraction, counterclockwise rotation, and clockwise rotation, respectively. Cells were tested with this set of modes of motion, and a cell that responded only to expansions or contractions was classified as an "expansion/contraction" cell. A cell that responded only to clockwise or counterclockwise rotations was classified as a rotation cell. From Tanaka and Saito (1989).

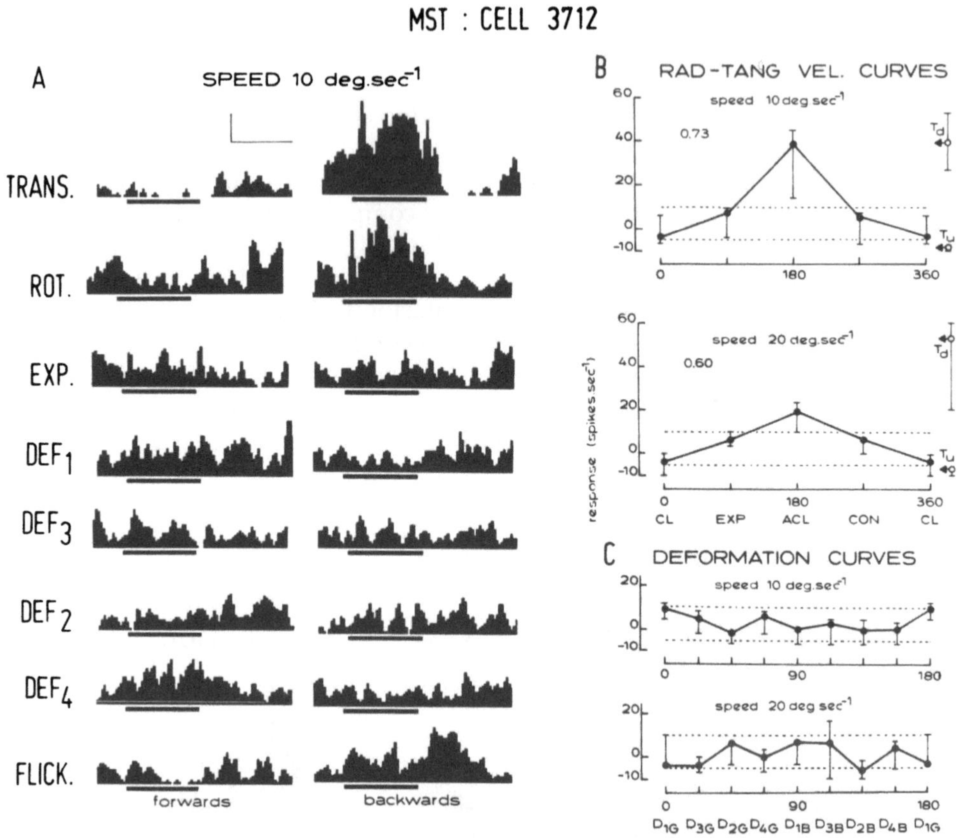

Figure 18. Responses of a rotation-selective MSTd cell to optimal (vertical) translation, rotation, "expansion/contraction," four types of deformation, and flicker. (A) Peristimulus time histograms indicating the average response of a layer 4 MST cell, cell 3712, to the different motion types tested. Black bars below histograms: motion and stimulus duration. Calibration bars: 25 spikes/sec and 500 msec. For this cell the PDs of translation and rotation were downward and anticlockwise. Notice the spontaneous bursts in the cell's activity (before and after rotation forward, before translation backward, and after flicker backward) that were not eliminated by averaging more than seven runs. The influence of these bursts was minimized by using median responses rather than averages that are shown in the histograms. Notice also that although the flicker responses differed in forward (−7 spikes/sec) and in backward (+3 spikes/sec) directions, they were small compared to the spontaneous activity: median 10 spikes/sec and third quartile 20 spikes/sec. (B and C) Radial–tangential (R–T) velocity response curves (B) and deformation response curves (C) of cell 3712 obtained at two speeds, 10 and 20°/sec. The data points corresponding to clockwise rotation and deformation$_1$ forward are plotted twice because both variables are circular. The curves at 10°/sec correspond to the histograms shown in part A. In parts B and C, the responses to the two opposite directions of translation are also indicated by horizontal arrows. Vertical bars: upper and lower quartiles of the responses, which are median net firing rates. Horizontal stippled lines: the quartiles of the spontaneous firing rate. The numbers indicate the value of the selectivity index (SI) associated with the curves. No SI was calculated for deformation response curves. CL, Clockwise rotation; EXP, expansion; ACL, anticlockwise rotation; CON, contraction; T_u, translation upward; T_d, translation downward; D_{1G}, forward direction of deformation$_1$; D_{1B}, backward direction of deformation$_1$. Similar conventions were used for the other three deformations. Translation was optimized by the preliminary testing shown in Fig. 11 (same cell). From Lagae *et al.* (1994).

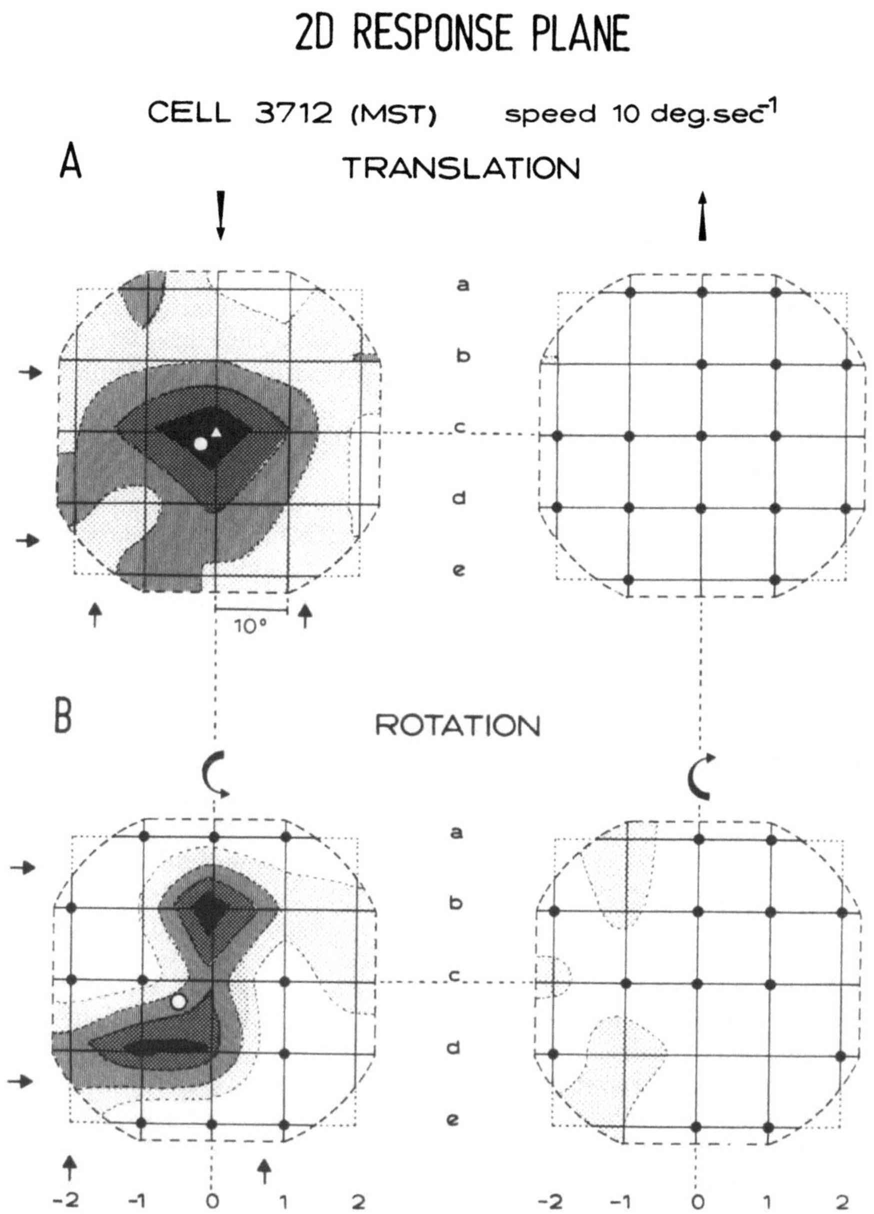

Figure 19. Two-dimensional position response planes for opposite directions of (A) translation and (B) rotation for MSTd cell 3712 (see Figs. 11 and 18). The two motions were tested at the same nominal speed. Same conventions as in Fig. 11B. The data in part A are the same as in Fig. 11B. One hundred percent response corresponds to 61 spikes/sec in part A and 25 spikes/sec in part B. Notice that the cell is direction-selective for translation as well as for rotation in all locations. Notice also how close the center of mass for rotation (B, ○) is to the center of mass for translation (A, ○). Finally, despite a similarity in average location captured by the center of mass, the rotation RF contains two peaks, whereas the translation RF has only one. The SADI (DI averaged over different positions in RF) of this cell was 117 for translation and 52 for rotation. Arrows: vertical and horizontal dimensions of the RF for translation and rotation estimated from the 50% contour. From Lagae *et al.* (1994).

over the RF; the response level and even the degree of selectivity (captured by a DI for the flow component) may change. It should be stressed that in order to demonstrate position invariance of direction selectivity for flow components one needs to determine the translation RF and to explore the visual field over a sufficiently wide enough region on all sides of the RF. With this criterion, about 20% of the MSTd cells were selective for rotation, 14% for "expansion/contraction" and only 6% for deformation (Lagae *et al.*, 1994). These proportions are somewhat higher than those of Saito *et al.* (1986) (16% "expansion/contraction"- and 14% rotation-selective MSTd cells) or of Duffy and Wurtz (1991a) (4% rotation and 9% "expansion/contraction"). It is worth noting that among the cells selective for "expansion/contraction" there is a clear bias favoring expansion: 7:1 ratio in Tanaka and Saito's sample (1989) and 6:1 in the report of Lagae *et al.* (1994).

Lagae *et al.* (1994) observed that the combined selectivity usually involved translation as the additional component, and that no cell showed a genuine selectivity for both rotation and "expansion/contraction," in agreement with Duffy and Wurtz (1991a). Yet Graziano *et al.* (1994) reported that a number of MSTd cells are selective for spiral motion, i.e., a combination of rotation and "expansion/contraction." Given the contradiction with the other work, it is likely that these responses do not reflect genuine RF selectively, but accidental matches with translation RFs, since the stimuli used were small (20° × 20°), and the region over which position invariance was tested was also small (20° × 20°) and of unknown position with respect to the translation RF. With such a stimulus setup a substantial proportion of unselective cells would appear selective for flow components (Orban *et al.*, 1995a). Perhaps a distinct class of spiral-selective neurons is not required by the visual system at all. In this regard Orban *et al.* (1992) have shown that MSTd cells selective for a flow component and unresponsive to another component stop responding to their preferred component when a sufficient proportion of the second component is introduced. Thus MSTd cells do not extract flow components from a complex motion in a mathematical sense, but rather match the pattern of directions present over their RF to that of their selectivity. However, in this study, rotation-selective cells were still responsive when "expansion/contraction" was added in equal measure, and conversely for "expansion/contraction"-selective cells. Hence spiral motion is covered well enough by rotation- and "expansion/contraction"-selective cells.

The two other combinations of MSTd neuron selectivity observed by Duffy and Wurtz (1991a) and by Lagae *et al.* (1994) might be behaviorally more meaningful. Since translation in fact represents rotation around axes orthogonal to the line of sight, the combination of rotation and translation corresponds to selectivity for rotation around an axis making an oblique angle with the frontoparallel plane. Combinations of "expansion/contraction" with translation correspond to displacement of the focus of expansion away from the center of the visual field (Orban, 1992). Duffy and Wurtz (1995a) have examined the effect of center-of-expansion displacement using large (100° × 100°) random dot patterns centered on the visual field center. Duffy and Wurtz (1995a) reported that MSTd neurons are selective for different positions of the focus of expansion. Unfortunately, all tests were carried out at a single speed, creating a confound between the pattern of directions and the speed at a given position. For example, the cell in their Fig. 9B was found to be selective for expansion with the center of expansion at the center of the screen, yet its RF was in the right hemifield. Thus, different positions of the center of expansion create different speeds of translation over the RF, faster when the focus is displaced to the right and slower when displaced to the left. A neuron tuned to a medium speed of translation to the right would therefore produce the sort of responses documented. Unfortunately, Duffy and Wurtz (1959a) did not mention the preferred direction of translation, which would allow one to preclude the alternative explanation. It thus seems that the selectivity of MSTd neurons for combined types of motion requires further study.

In a very careful study Tanaka *et al.* (1989) investigated all the cues included in rotation and "expansion/contraction" and demonstrated that the real determinant of the selectivity was the pattern of translation directions present in the stimulus, with four different directions being the minimum needed for selectivity and eight giving responses equal to the complete stimulus. In particular, they showed that the speed gradient which is normally present in a flow component can be removed without affecting the direction selectivity. This was confirmed by Orban *et al.* (1995a), who showed in addition that the speed tuning, which depends on the average speed, is also unaffected by removal of the speed gradient. Since MSTd cells require at least four directions of translations to mimic responses to optic flow components, it is not surprising that they respond to isotropic expansion less than to expansion, and to shear less than to rotation, since each of these pairs of motion patterns share only two directions of translation (Tanaka and Saito, 1989). These authors observed a few exceptional cells responding better to shear or isotropic expansion, which might correspond to some of the deformation-selective cells of Lagae *et al.* (1994). Indeed, shear and isotropic contraction can be seen as composites of deformation and another flow component, and the deformation-selective cells of Lagae *et al.* (1994) generally had combined selectivities. Tanaka and Saito (1989) further reported that MSTd cells selective for rotation or "expansion/contraction" clearly exhibit spatial summation, a finding fully confirmed by Lagae *et al.* (1994). In this latter study only 3 of the 20 OFC-selective cells tested with the summation test (Fig. 6), performed at two speeds to disambiguate area and speed effects, had an antagonistic surround.

Tanaka and Saito (1989) also showed that the flow component selectivity was little affected by the shape of the pattern (windmills and rings in place of random dot patterns) or by the size and the spacing of the dots in the random dot pattern. Lagae *et al.* (1994) added invariance for speed to the list of invariances displayed by direction selectivity for flow components. Finally, Tanaka and Saito (1989) reported that responses to flow components increased with increasing speed, and suggested that MST neurons were not tuned for flow component speed. Extending the range of speeds beyond those used by Tanaka and Saito (1989), we (Orban *et al.*, 1995a) were able to show that in all cases responses to flow components decrease at faster speeds, and that most speed–response curves for flow components were tuned or broadband. Finally, it is worth mentioning that some evidence for a clustering of rotation- and "expansion/contraction"-selective cells has been reported (Duffy and Wurtz, 1991a; Lagae *et al.*, 1994).

Thus there is considerable agreement among the various studies that MSTd is involved in the analysis of optic flow, supporting the general role of MSTd in the control of egomotion. The expansion-selective cells (and those with combined expansion–translation selectivity) can provide signals very useful for heading, as has been pointed out by many authors, not the least by Duffy and Wurtz (1995a). The use of the rotation signals seems somewhat less clear. Lagae *et al.* (1994) have proposed that these can play a very important role in visual stabilization of posture during leaping, an important asset for an arboreal species such as monkeys.

2.4. The Antagonistic Surround

In their seminal paper, Allman *et al.* (1985) described the antagonistic surround as a region from which no response could be evoked, but by which responses elicited from the excitatory RF could be suppressed, hence the term "silent antagonistic surround." They added two elements to this description: first, that the surround was generally direction-selective with the same preferred direction as the classical receptive field (CRF), and second, that the surround was an extensive region, up to 7–10 times the diameter of the CRF, encircling the classical RF. From that description many deduced that the surround, although modulating the responses of MT neurons, did not contribute to their selectivity. Hence most studies, in fact all those reviewed above, used an 'optimal' stimulus, which means that the surround was masked and excluded from stimulation. Increasing evidence suggests that this picture of the surround is incomplete: the surround region can be fairly restricted and confined on one side of the CRF and can possess definite parameter selectivities. Finally, the laminar distribution suggests that these surrounds arise from processing within MT. Hence, study of the surround should reveal a great deal about the function of MT and its satellites.

2.4.1. Incidence

In the macaque, most information about the antagonistic surround of MT neurons comes from the study by Tanaka *et al.* (1986) and from this laboratory. Tanaka *et al.* (1986) used a bar as the center stimulus and a random dot pattern as the surround stimulus, whereas Raiguel *et al.* (1995) used mainly random dot patterns to stimulate both center and surround. Tanaka *et al.* (1986) estimated that 40% of the MT neurons have an antagonistic surround as witnessed by a decrease in the center response of at least 50%. Raiguel *et al.* (1995) used summation curves to estimate surround strength (Fig. 20). They found a proportion of 60% surround cells using a similar 50% criterion. But in general, bars are a somewhat stronger stimulus for MT cells than random dot patterns (Tanaka *et al.,* 1986) and hence it is possible that more suppression can be observed when using a weaker stimulus to drive the CRF. Within area MT there is now firm evidence for a laminar distribution of surrounds. Tanaka *et al.* (1986) observed more surrounds in the superficial layers and Lagae *et al.* (1989) provided the initial evidence that surrounds occurred primarily outside layer 4. Raiguel *et al.*

(1995), in a study of 237 MT neurons, showed that surround strength is weakest in layers 4 and 6 and strongest in layer 2 (Fig. 21B). These authors also provided some evidence for a horizontal grouping of surround strengths, reminiscent of the band and interbands described by Born and Tootell (1992) in area MT of the owl monkey.

In area MSTd, surrounds occur almost as frequently as in MT, at least for cells selective for translation (Lagae *et al.*, 1994; Raiguel *et al.*, 1993). As men-

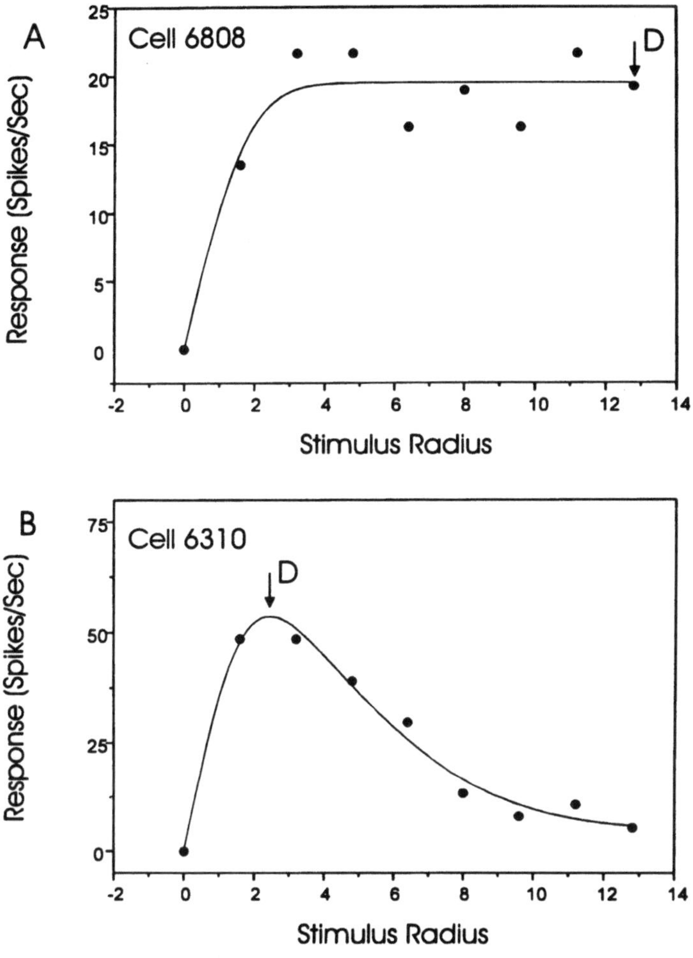

Figure 20. The summation curves of two example MT neurons 6808 and 6310, fitted with the ERF and DOE (difference of ERFs) functions, respectively. Arrows at D show optimum radii. (A) Cell 6808 had no surround and thus the optimum stimulus radius (arrow) was the maximum size tested, 12.8°. The response in such a nonsurround cell rises to a plateau, then remains at this level or increases very slightly. However, the ERF function gives a half-height radius of 2.8°, which is closer to the 'true' size of the field. (B) Cell 6310, fitted with the DOE function, in view of its strong surround. The response in this surround cell rises to a maximum at the optimum radius (arrow), then declines as the stimulus impinges on the surround, finally approaching a constant low level at the largest radii. In this case the fitted half-height radius (3.7°) is close to the optimum radius (2.9°). From Raiguel *et al.* (1995).

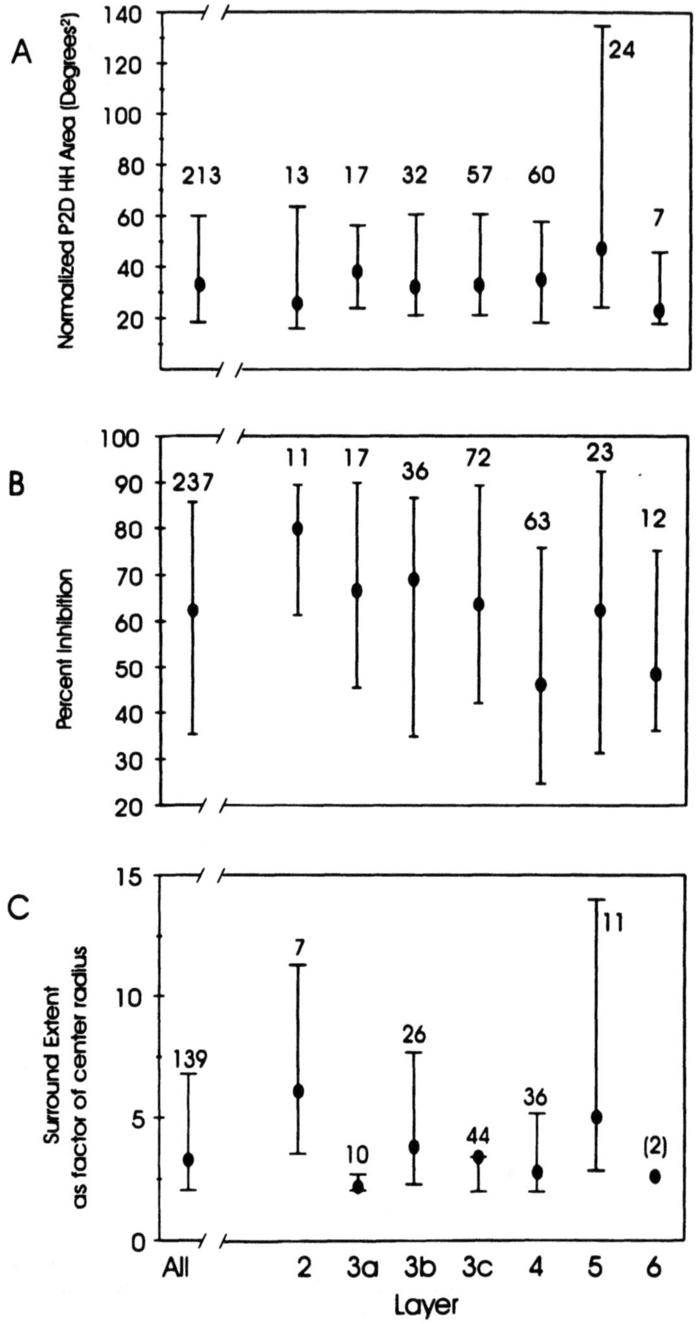

Figure 21. Changes in RF characteristics with laminar position of MT neurons: (A) RF size captured by the normalized P2D half-height area, (B) surround strength captured by the inhibition at largest diameter tested, and (C) surround extent captured by the outer radius of surround at half-height. Numbers above the quartile bars show the number of neurons represented in each subsample. (A) Medians and quartiles of RF areas as determined by the half-height of the P2D generalized Gaussian fit, by lamina. All areas have been normalized to 0° eccentricity based on the regression line in Fig. 7.

tioned earlier, MSTd cells selective for flow components generally have no antagonistic surround (Lagae *et al.*, 1994; Tanaka and Saito, 1989). This is very different from area MSTv neurons, which, according to Tanaka *et al.* (1993), respond only to small-diameter random dot patterns and not to large patterns, suggesting that indeed many have an antagonistic surround.

2.4.2. Parameter Selectivities

There is clear evidence that the surround is direction-selective in most MT neurons. Tanaka *et al.* (1986) found all 20 cells tested to have direction-selective surrounds. While this is also the case in the majority of MT cells we tested, there are exceptions; see Fig. 22. This figure also illustrates that surrounds in MSTd are direction-selective. The speed of motion in the surround also influences surround inhibition. Tanaka *et al.* (1986) reported different types of MT neurons classified according to whether the surround suppression was released at slow speed, fast speed, both slow and fast speeds, or not at all. We have had difficulty replicating this finding with random dot patterns. Using the same range of speeds, from four times slower to four times faster than the center, we observed that all speeds are nearly equally suppressive and that if release of suppression occurs, it does so only at slow speeds (Xiao *et al.*, 1995a). One can also show that the strength of the surround, while depending on the relative speed between center and surround, depends little on the absolute center speed. Figure 23 shows that surround suppression is similar for two speeds in both MT and MST cells (mainly MSTd cells). We have begun to explore the time course of the surround suppression in area MT/V5. Preliminary findings suggest that the time course is very similar to that of the CRF response (Fig. 24).

2.4.3. Areal Extent of the Surround

Tanaka *et al.* (1986) estimated the surround diameter to be about four to five times the CRF size. They also reported that the surround was circularly uniform—i.e., equally effective all around the CRF—because stimulation with

---

Figure 21 *(Continued).* Layer 5 appears to contain a population of cells which had large areas compared with other laminae, shown by the high median and upper quartile, and was statistically distinct from all other layers combined ($p = .03$, Mann-Whitney U-test) and from sublaminae 3c ($p = .018$). (B) Surround inhibition by layer. Strongest surround inhibitions are to be found in the superficial layers, especially layer 2, where inhibition was significantly higher than in layers 4 or 6 ($p = .008$ and .02; Mann-Whitney U-test), but not all others combined ($p = .06$). Surround inhibition is especially weak in the input layer 4 (inhibition in layer 4 compared with 2, 3c, 3b, or all others combined: $p =$ .008, .01, .05, and .005, respectively). (C) Surround extents by layer. Extent of the surround, measured from the center to the outer half-height boundary, expressed as a factor of the half-height radius of the excitatory center. Surrounds extend the farthest out from the center in layers 2 and 5 (2 and 5 combined versus all others, $p = .02$; Mann-Whitney U-test) and cover the smallest area in layer 3a, but not significantly so ($p = .1$, 3a against all others combined). Lamina 3a, however, was significantly different from layer 2 ($p = .01$) and layer 5 ($p = .05$), but the only other significant comparison was between 2 and 4 ($p = .05$). From Raiguel *et al.* (1995).

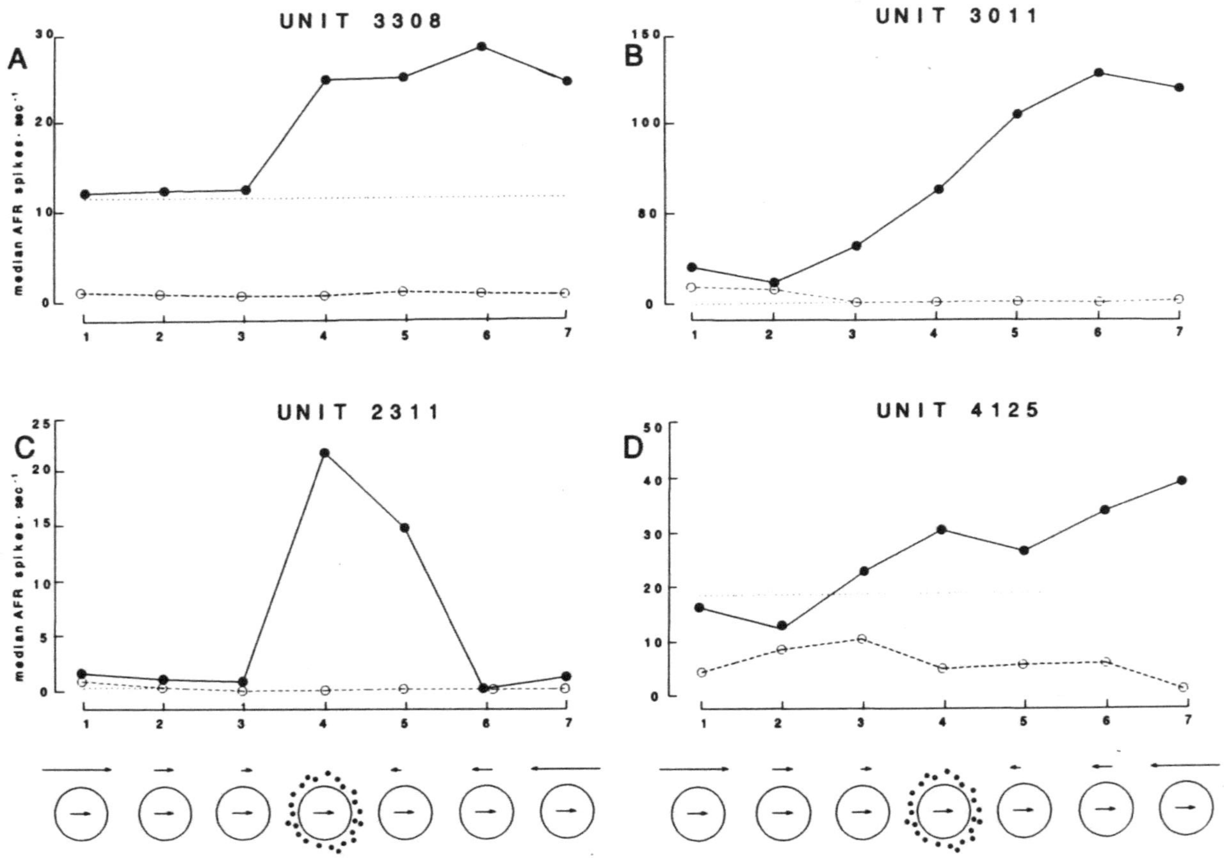

Figure 22. Direction selectivity of antagonistic surround in area MT and MSTd. Center–surround interactions for different directions and speed of motion in the surround: in (A, B) two MT neurons with direction-selective surround, (C) an MT neuron with non-direction-selective surround, and (D) an MSTd cell with direction-selective surround. The surround conditions are indicated below the abscissa; length of arrows indicates speed: equal to center speed, four times faster, four times slower. Stimuli in center and surround were segregated random dot patterns. Stippled line: significance level. From Lagae (1991).

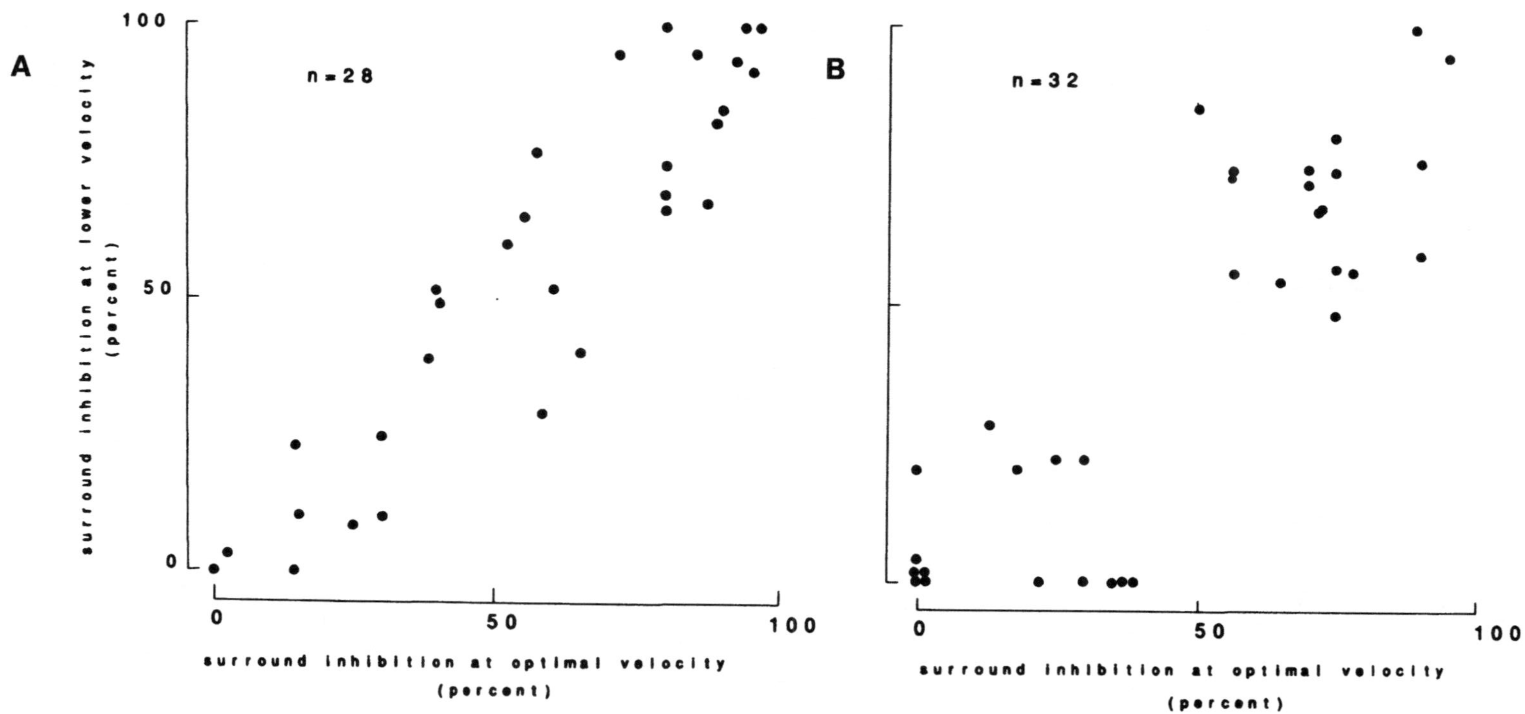

Figure 23. Invariance of surround inhibition for center speed: (A) MT neurons (N = 28) and (B) MSTd neurons (N = 32). Surround inhibition obtained from the summation curve at lower than optimal speed is plotted as a function of surround inhibition obtained at optimal speed. From Lagae (1991).

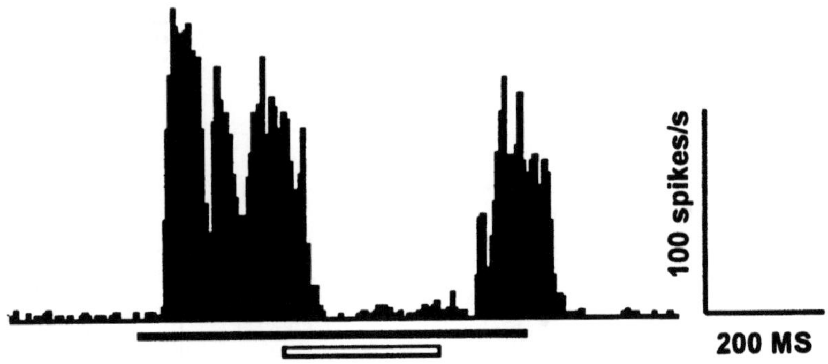

Figure 24. Time course of surround inhibition in an MT neuron. The CRF stimulus was on for 640 msec, the surround stimulus for 250 msec. The latency of the excitatory response (measured with cusum method; see Lagae *et al.*, 1994) was 40 msec, as was the latency of the inhibition. Also the end of both responses was very similar: they stopped 60 msec after stimulus end. Average of 201 presentations. From D.-K. Xiao, S. Raiguel, V. Marcar, and G. A. Orban (unpublished).

pairs of wedges positioned on opposite sides of the surround produced equal suppression. By modeling the summation curves, Raiguel *et al.* (1995) observed that the surround extent (50% of maximum suppression) was on average three times the CRF diameter, in good agreement with the results of Tanaka *et al.* (1986). Raiguel *et al.* (1995) further reported that surrounds were broadest in layers 2 and 5 (Fig. 21C), the latter not completely surprising, since the CRF was also largest in this layer. Xiao *et al.* (1995a, b) probed the surround with eight relatively small stimuli (Fig. 6) and observed that in most neurons the surround was not circularly uniform at all. In fact, in a substantial proportion of cells the surround was completely asymmetric and flanked the CRF on only one or two sides. Thus MT neurons are endowed with the capacity to take directional derivatives of the velocity field, just as V1 simple cells take derivatives of the luminance field, suggesting that the surround is intrinsic to the RF of the MT neurons and must contribute directly to the selectivity of MT neurons. Experiments to test this proposition explicitly are under way and thus far confirm the prediction.

3. Behavioral Exploration of Macaque Area MT/V5 and Its Satellites

Motion signals can be used for many different behavioral purposes (Nakayama, 1985). Obviously they can be used for motion perception, although motion can be perceived without much actual retinal image motion (i.e., during pursuit). In addition, motion signals can be used to guide different types of body movements. Finally motion can be used to recover the shape of objects (structure from motion) in either two or three dimensions. The role of MT/V5 and satellites in motion perception and in guidance of eye movements has been intensively explored. In the type of studies to be reported in this section, the monkey is not a passive subject, but must react behaviorally to the stimulus: he either has to make a movement under the guidance of the stimulus, which is thus used in a direct sense, or to signal symbolically his decision concerning the motion signals contained in the stimulus. Three strategies have been used to explore the involvement of the MT/V5 complex in such behaviors: recording of single cells during the behavioral reaction, observing the impairments caused by lesions of area MT/V5 and satellites, and finally studying the effects of electrical microstimulation of area MT/V5 and satellites on the monkey's behavior.

3.1. Motion Perception

There is increasing evidence that lesions restricted to area MT itself have only mild effects on motion perception tasks. Schiller (1993) reported that MT lesions moderately impair motion detection and speed discrimination as well as flicker detection. From our own work we also have evidence that just-noticeable differences (JNDs) in direction are permanently but modestly elevated after restricted MT lesions (Lauwers *et al.*, 1995). Newsome and Paré (1988) reported that MT lesions cause a strong transient impairment of opposed-direction discrimination in which the signal strength is manipulated by changing the percentage of correlated dots moving in a single direction and at a given speed (Fig. 25). The transient effect was large and selective, as contrast sensitivity was not influenced. The degree of recovery was very substantial and the permanent deficit was an increase in threshold of only a few percent (Fig. 26). This finding had

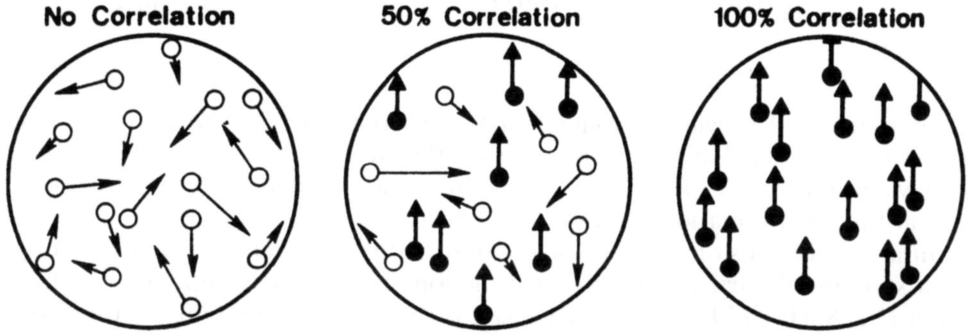

Figure 25. Schematic representation of the motion stimuli used by Newsome and co-workers in their studies of opposed-direction discrimination. A sequence of dots is plotted under computer control at random positions on a cathode ray tube screen. Each dot survives for a brief interval (20–30 μsec), after which it disappears and is replaced by another randomly placed dot. The operator may specify that a certain percentage of the dots be replotted with a fixed spatial and temporal offset from their partner dots. This subset of dots constitutes a net motion signal that is embedded within a mask of random motion events. Left: The condition in which all dots are plotted randomly (no, or 0%, correlation). There is no net motion in such a display, although there are many local motion events due to fortuitous pairings among the stream of randomly positioned dots (arrows). Right: The other extreme, in which all dots are replotted with a fixed spatiotemporal offset so that the motion of each dot is identical to that of the entire pattern (100% correlation). Center: An intermediate display in which 50% of the dots comprise the 'correlated' motion signal, while the other 50% are randomly plotted in a masking motion noise (50% correlation). The task of the monkey is to signal the direction of motion (up or down) by a saccade to one of two response LEDs. From Newsome and Paré (1988).

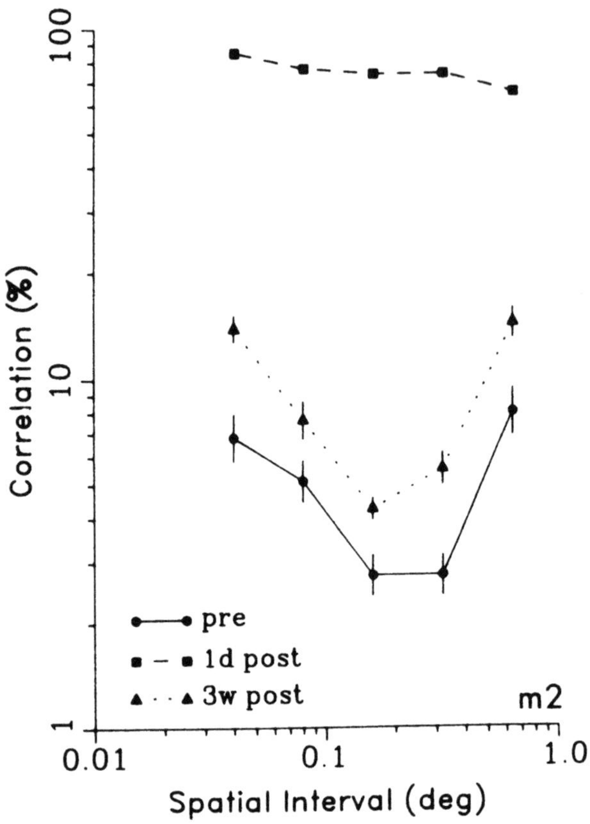

Figure 26. Transient and permanent impairment of opposed-direction discrimination after complete unilateral MT lesion. Experiment *m2*. Solid line: prelesion thresholds for this experiment; dashed line: postlesion thresholds obtained 24 hr after the second round of injections in this hemisphere; dotted line: postlesion thresholds 2–3 weeks postlesion. Symbols for the prelesion and the 3-week postlesion data represent the mean value of at least 10 threshold measurements for each condition; error bars, standard error of the mean. We tested this animal until 5 months postlesion, but observed little recovery beyond that evident at 3 weeks. The residual deficit thus appeared to be permanent. From Newsome and Paré (1988).

been interpreted as evidence that normally (before the lesion) the motion signals travel via MT, but that after the lesion another area such as, e.g., MST, which receives input from V1 and V3A (see Fig. 3), takes over. An alternative possibility is that this discrimination is based on a network of those areas containing substantial proportions of direction-selective neurons and that the whole network, or a substantial part, transiently malfunctions because it is deprived of input from MT. Such an idea is not novel, as it is related to the diaschisis—the functional standstill of an area due to deprivation of input from another area—first formulated by von Monakow (1914) and revived by Yamasaki and Wurtz (1991) in their interpretation of remote effects resulting from MT lesions. Here, however, we consider a malfunction of the entire network, or at least the subgroup which receives input from MT, and not the dysfunction of any single remote area. In this alternative view, the recovery is due to a return of the network to normality rather than to a shift of the function from one area to another.

On the other hand, combined lesions of area MT/V5 and its satellites result in impairment, which seems to vary with the type of motion discrimination task. JNDs in direction were elevated only by a factor of 2–3 in the study of Pasternak and Merigan (1994), who also reported a mild deficit in the opposite-direction discrimination task developed by Newsome *et al.* (1988). These authors also reported a 2- to 3-fold increase in JNDs in speed, but these were measured with moving gratings, for which temporal frequency can be a confounding factor. Cowey and Marcar (1992) observed impairments in the perception of dot motion after combined MT/MST lesions, but only after removal of the flicker cue present in the distinction the monkey was required to make between motion and nonmotion conditions. In a similar vein, we observed very large deficits in speed discrimination for moving random dot patterns after combined MT/MST lesions (Fig. 27), but only modest deficits when moving bars were used (Orban *et al.*, 1995c). The most likely explanation for this range of effects is that depending on the stimulus pattern, the motion attribute to be used, and the type of task, different ensembles of areas can provide signals useful for the decision process and that the MT/V5 complex is only one element of that larger network active in motion perception. Within that overall view, several factors can explain the comparatively larger impairment observed in the experiment of Orban *et al.* (1995c) compared to that of Pasternak and Merigan (1994). The network involved in discriminating the speed of a random dot pattern might be much more restricted than that involved in direction discrimination. Also the MT/V5 complex may assume a much more critical position in the speed-discriminating network, such as might be the case if MT lesions prevent output to the decision-making process.

In their attempt to further link activity in the MT/V5 complex to motion perception, Newsome and colleagues (Murasugi *et al.*, 1993; Salzman *et al.*, 1992) showed that the decision made by the monkey in the opposed-direction discrimination can be biased by electrical stimulation of a direction column in area MT (Fig. 28). Despite the fact that the authors could demonstrate that the effects were not due to the spread of current within MT, they were careful in their conclusion to state that although these experiments establish a link between direction-selective cells and perception of direction, these direction-selective cells need not all be located in area MT/V5. Indeed, as Salzman *et al.* (1992) acknowledged in their discussion, the electrical stimulation might well spread to other areas connected to MT, in such a way that the activity of the entire network is modified rather than just that in area MT. Microstimulation of area MST (i.e., mainly area MSTd) produced behavioral effects similar to stimulation of MT, except on a coarser scale (Celebrini and Newsome, 1995), leading to similar interpretations: either areas MT and MSTd belong to a restricted pathway leading to decisions concerning motion direction, or they are part of a much wider network to which activation spreads, at least in part. The latter view is the counterpart of the idea discussed above, that MT lesions lead to a dysfunction of the network. Given the close temporal synchrony of activation in cortical areas (a latency difference which is small compared to the 2-sec stimulation period used in the opposed-direction discrimination), the "spread of activation to the network" view does not need to imply a parallel connection scheme between the elements of the network and the decision center as presented in Fig. 11B of Celebrini and Newsome (1995).

The single-cell studies do not contradict the view that relatively large networks of areas featuring direction-selective neurons, including the MT/V5 complex, contribute to opposite-direction discrimination. The neuronal threshold and the behavioral thresholds (both in % correlation) for direction discrimination are similar (average ratio close to 1) in both area MT/V5 (Britten *et al.*, 1992)

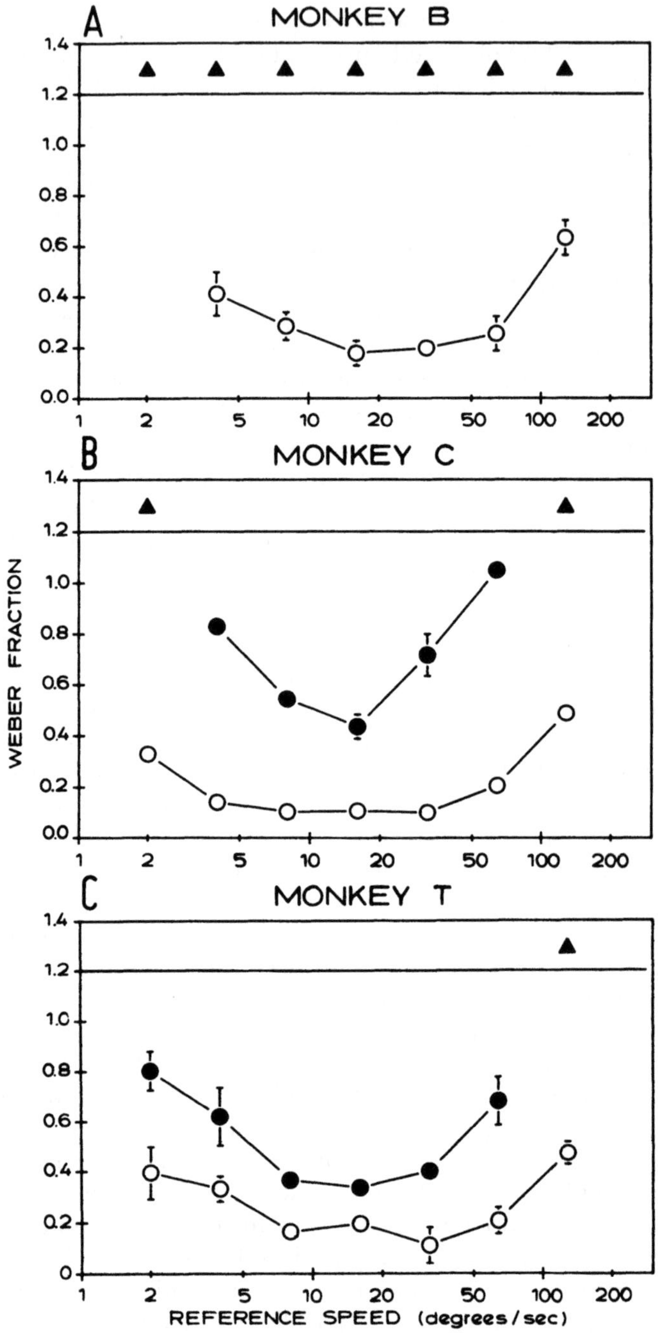

Figure 27. Permanent impairment of speed discrimination after MT/MST lesion. Pre- and postoperative speed discrimination curve of monkeys (A) B, (B) C, (C) T, tested with random textured fields (RTF). Just-noticeable differences (JNDs) in speed expressed as Weber fraction are plotted as a function of reference speed. Filled symbols indicate postoperative JNDs. Triangles indicate absence of stimulus control. Error bars are SEM of 12 JNDs obtained in two or three sessions. Each data point corresponds to 2000–3000 trials. Postoperative data obtained 7–8 months after the lesion. From Orban *et al.* (1995c).

and both subdivisions of area MST (Celebrini and Newsome, 1994), provided that the stimulus is optimized for the neuron under study. In area MT/V5 the close relationship between neuronal and psychophysical threshold is maintained even when both vary during the course of a session (Zohary *et al.*, 1994a). Newsome and colleagues stressed the need for optimizing the stimulus for the neu-

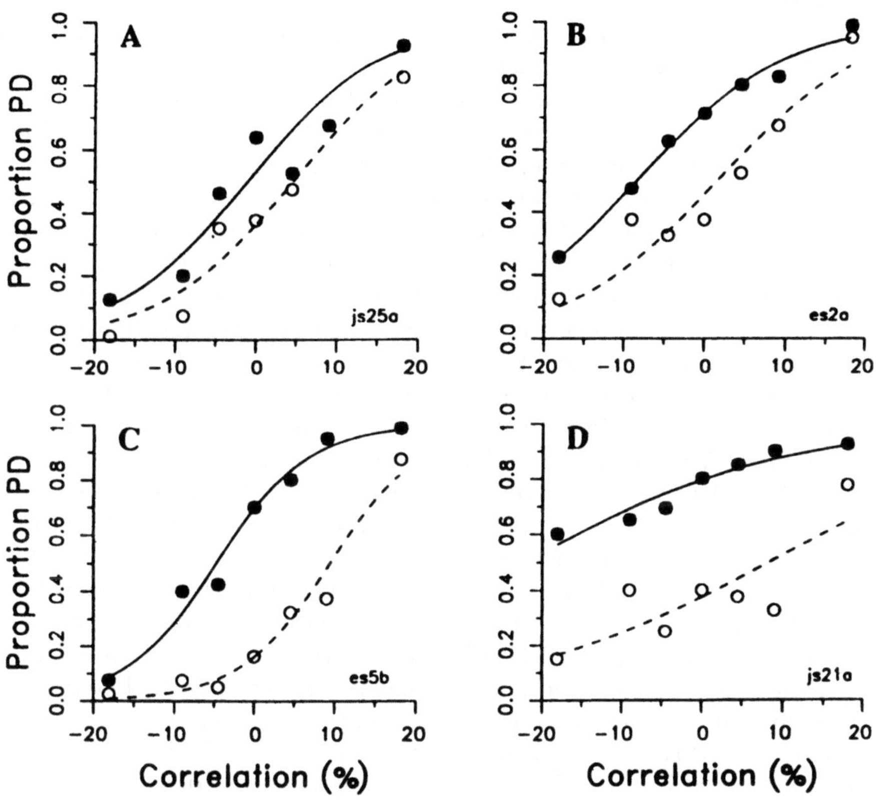

Figure 28. Effect of microstimulation on psychophysical performance in opposed-direction discrimination task for four stimulation sites in area MT. The proportion of decisions favoring the neurons' preferred direction (preferred decisions, PD) is plotted as a function of strength of the motion signal expressed in percentage of correlated dots (see Fig. 25). The abscissa has been opened to indicate separately motion in the preferred direction (positive values) and motion in the opposite or 'null' direction (negative values). Trials that did not contain microstimulation are shown with open symbols, trials that contained microstimulation are represented with solid symbols. Each data point is based on 40 trials, except for 0% correlation, where 80 trials were conducted. The psychometric functions depicted were fitted with a logistic regression model in which the shapes of the stimulated and nonstimulated psychometric functions were constrained to be identical. (A) A small but significant ($p \leq .0001$, logistic regression analysis) effect of microstimulation in monkey J (equivalent visual stimulus = 5.6% correlated dots). (B) A medium-sized effect in monkey E (equivalent visual stimulus = 10.0% correlated dots). (C) A larger effect of microstimulation in monkey E (equivalent visual stimulus = 14.2% correlated dots). (D) A large effect of microstimulation in monkey J (equivalent visual stimulus = 30.6% correlated dots). The difficulty in each experiment varied depending upon several variables, including the eccentricity of the motion display, the size of the visual stimulus aperture, and the speed of the motion stimulus. The variability in difficulty is reflected by the different slopes in each experiment. More difficult discriminations imply that the monkey made fewer correct choices, and the slope of the psychometric function therefore appears flatter (e.g., part D). From Salzman *et al.* (1992).

ron under study, something others (Vogels and Orban, 1990) found unnecessary. Indeed, any psychophysical stimulus will be optimal for some neurons and less than optimal for others, and it seems difficult to understand how the visual system could ignore the multitude of signals from the less than optimally stimulated neurons. Initially the Stanford group in fact favored the view that the system might use only a very small set of very direction-selective cells, although they stressed that additional information was required to confirm this conclusion (Newsome *et al.*, 1989). In their search for such additional evidence this group established two other important facts: (1) the activity in neighboring MT cells is weakly correlated (Zohary *et al.*, 1994b) and (2) the activity of MT and MST neurons is weakly correlated with the decision made by the monkey (Britten *et al.*, 1996; Celebrini and Newsome, 1994). This correlation, to which Britten *et al.* (1996) refer as the "choice behavior" of MT and MST neurons, is reminiscent of the correlation between some MT neurons and the monkey's decision about the direction of motion under binocular rivalry conditions (Logothetis and Schall, 1989). Bringing these three elements (similarity of neuronal and psychophysical thresholds, weak correlation between neighboring neurons, and choice behavior of individual neurons) together in a more formal attempt to model the link between the activity of direction-selective cells and psychophysical decisions led this group (Shadlen *et al.*, 1996) to reach a conclusion somewhat contradictory to their initial view. According to Shadlen *et al.* (1996), the decisions are based on pools of at least 100 weakly correlated neurons, and the neurons composing these pools must have a substantially broader range of sensitivities than the MT cells recorded in the combined psychophysical–electrophysiological experiments of Britten *et al.* (1992). Thus the network need not consist of areas featuring solely strongly direction-selective neurons such as components of V1 and V2, the MT/V5 complex, and VIP (the motion pathway, see Section 1.2), but other areas such as V3, V3A, and even V4 may participate as well. The participation of widely scattered cortical regions in reaching decisions regarding motion direction has been directly demonstrated by Ferrera *et al.* (1949b). These authors recorded in areas MT, MST, 7a, and V4 during a motion direction match-to-sample task. The key finding is that while the activity in areas MT and MST had relatively little to do with the direction of the cue, that in areas 7a and V4 (Fig. 29) did, suggesting that motion signals can also be analyzed in ventral stream areas, as is also evident from our own work (Sáry *et al.*, 1993, 1995).

In conclusion, the behavioral experiments set out to show that area MT/V5 plays a key role in the analysis of retinal image motion signals used in motion perception. This hypothesis has not been borne out, given the relatively mild deficits observed in several motion perception tasks after MT lesions. An alternative view, still within the realm of strong cortical specialization, holds that area MT is part of a motion pathway comprising the cortical areas endowed with strong direction selectivity: i.e., MT/V5 and satellites, their inputs V1, V2, and V3, and VIP, and that motion perception is critically dependent on the integrity of this pathway. In this view, the mild effect of lesions in area MT is easily explained by the connectivity, since higher order areas such as MSTd or MSTv can receive inputs from lower order areas such as V3A, thereby bypassing area MT/V5. The much stronger effect of combined MT/MST lesions supports this view. However, all available evidence is compatible with a much less specialized view of the cortex stating that each motion task involves a different network of areas containing more or less direction-selective neurons, including the motion pathway mentioned above, but also other areas such as V4, etc. The exact configuration of a given network depends on the stimulus (bar or random dot pattern), the nature of the attribute (speed or direction), and the nature of the task. Opposed-direction discrimination (Newsome and Paré, 1988), fine direction

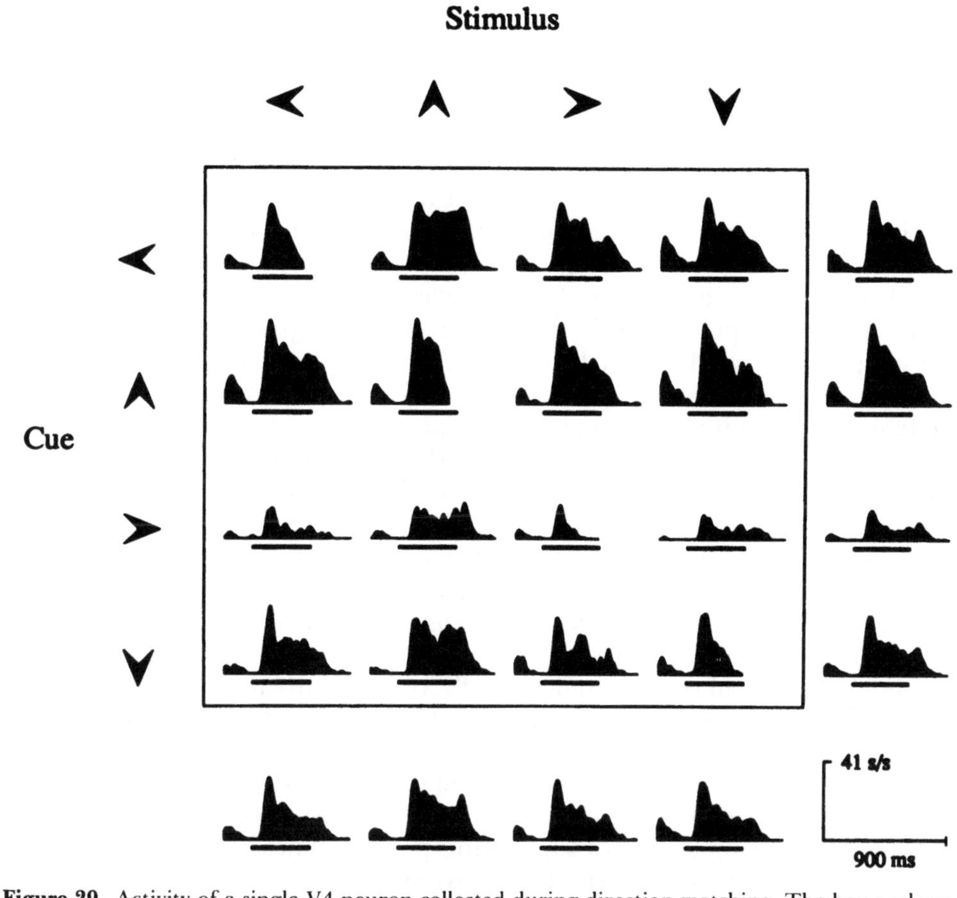

Figure 29. Activity of a single V4 neuron collected during direction matching. The box encloses a four-by-four array of response histograms that represent the 16 different combinations of cue and test stimulus directions that occurred during the matching task. Each column represents responses to a particular test stimulus, sorted into rows according to the direction that the animal was seeking to match at the time the stimulus appeared. Responses to the presentation of the cue itself are not shown. Each plot is the average of 16 stimulus presentations, and the bars below each show the period during which the stimulus was on. The response histograms were smoothed by convolving with a unit Gaussian. Responses to matching conditions (main diagonal) were truncated at the time the monkey made its response, which varied from trial to trial. Therefore, plots on the principal diagonal are responses to matching conditions, and are truncated at the time the animal made its behavioral response. As the monkey's reaction time varied from trial to trial, the average neuronal response appears to taper off when in fact each individual response ended abruptly (with some latency) at the time the monkey responded and the stimulus was turned off. Plots below and to the right of the box show the column and row averages. The response of this V4 neuron showed very little selectivity for stimulus direction, but appeared to be strongly modulated by the direction of the cue, even though it was no longer present at the time the response occurred. From Ferrera *et al.* (1994b).

discrimination (Lauwers *et al.*, 1995; Pasternak and Merigan, 1994), identification of one of eight directions (Salzman and Newsome, 1994), detection of an outlier direction (Schiller, 1993), and direction match to sample (Ferrera *et al.*, 1994b) are all examples of different tasks using the same attribute (direction). The effect of combined MT/MST lesions will then depend on how widespread a given network is and how critical the position of the MT/V5 complex is in that network.

3.2. Guidance of Eye Movements

It has been proposed that the study of eye movements, pursuit in particular, enables one to study the cortical processing of motion signals (Newsome *et al.*, 1985). Although it is likely that the networks involved in motion perception and those involved in control of pursuit overlap to a certain degree, a number of dissociations between pursuit and perceptual reports have been noted (see above, isoluminant stimuli; and also Milner *et al.*, 1991). Thus, it seems more fruitful to let the experiments themselves decide the degree of overlap between the two types of networks. Visual motion information is used in the guidance of three types of eye movements: pursuit of a small target, ocular following (Miles *et al.*, 1986), and optokinetic nystagmus. The latter two are evoked by large, moving stimuli such as random dot patterns. Motion processing is irrelevant for saccades toward a stationary target, and it has been repeatedly shown that saccades are not affected by MT lesions (Schiller and Lee, 1994; Newsome *et al.*, 1985) nor by MST lesions (Dürsteler and Wurtz, 1988).

The physiological studies of Komatsu, Wurtz, and Newsome have clarified which parts of the MT/V5 complex might contribute to visual pursuit. Komatsu and Wurtz (1988a) described three populations of visual pursuit cells: those occurring in the foveal representation of MT, others in the foveal representation of MSTv (what Komatsu and Wurtz labeled MSTl), and a third group scattered through MSTd. The main features of these cells were direction selectivity and inclusion of the fovea within their RFs. From the study of Lagae *et al.* (1993), we know that at least the MT pursuit cells have a low-pass speed–response curve, appropriate to the small retinal slip during pursuit. These visual pursuit cells described by Komatsu and Wurtz (1988a) are synonymous with the visual tracking cells of Erickson and Dow (1989) and of Erickson and Thier (1991). It is noteworthy that Erickson and Dow (1989) also observed a few tracking cells, all in the portion of SST probably corresponding to MSTv, with no visual response. Such neurons may exclusively carry extraretinal signals related to pursuit (see below). Wurtz and colleagues (Newsome *et al.*, 1988) have further characterized the visual pursuit cells and determined whether they were purely visual or also carried an extraretinal signal. By blinking the target during the pursuit or by stabilizing the target on the retina, they showed that pursuit cells could either be purely visual or carry an additional extraretinal signal, which generally started at least 50 msec after pursuit onset. The purely visual cells were observed in foveal MT and MSTv, whereas the cells with an additional extraretinal signal occurred in MSTv and MSTd. Thier and Erickson (1992) went on to show that many MSTv tracking or pursuit cells have an extraretinal signal related not only to eye

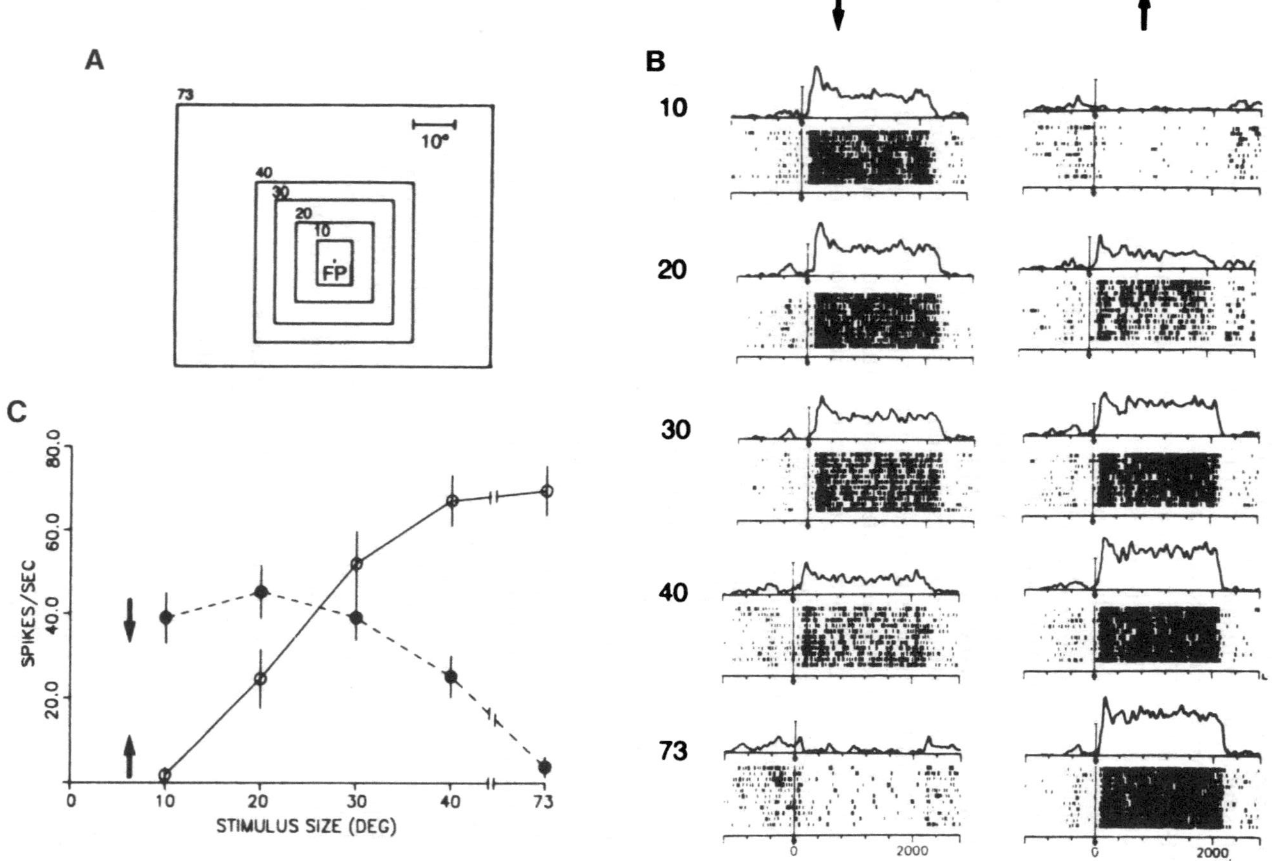

Figure 30. Reversal of preferred direction of stimulus motion with increase in stimulus size for an MSTv neuron. The receptive field of this cell in the lateral–anterior region (see text) of the medial superior temporal area (MSTl) extended over the entire screen (80° × 80°) with the strongest response in the center and progressively weaker responses toward the periphery. (A) Schematic illustration of the five sizes of random-dot fields used. The number above each rectangle indicates the square root of the area of the random-dot field (the 10° size was actually 9° × 11°; 73° was 80° × 66°). All sizes of fields were centered on the fixation point (FP) except for the largest, which stimulated an area 40° above but only 25° below the FP. (B) Responses to random-dot fields whose size is indicated on the left side of each row. The speed of the stimulus motion was 11°/sec and the direction was downward in the left column and upward in the right column. Responses are aligned on stimulus onset. (C) Graph of the relationship between size of dot field and magnitude of response. Each point indicates the mean and standard deviation of the number of spikes per second calculated during a 1- to 1.5-sec period after stimulus onset. The same time period was used to calculate the response magnitude for bar graphs. Directions of stimulus motion are indicated by arrows on the left side of the graph. This cell showed weak responses to small spots moving downward. From Komatsu and Wurtz (1988b).

position change, but also to head position change. Thus these cells carry a signal which represents velocity in space, potentially useful for head tracking as well as driving pursuit eye movements. Kawano *et al.* (1994) also reported that MST cells (both in MSTd and MSTv) carry an extraretinal signal related to pursuit. They showed that the average MST response decreased during blinking of the target, and that this was paralleled by a slowing down of the pursuit. Komatsu and Wurtz (1988b), in a third step, characterized the visual response of pursuit cells. Again two type of cells were observed: those which responded best to small spots and those which preferred large random dot patterns. The latter occurred in MSTd and MSTv, the former in foveal MT and MSTv. Those cells which preferred large random dot patterns reversed their preferred direction with changes in size (Fig. 30) and their pursuit direction was opposite to the preferred visual direction when tested with a large random dot pattern, but the same when tested with a small visual target. Those responsive to a small spot did not, on the other hand, reverse preferred direction with increasing size, and the preferred directions of pursuit and of visual stimulus always matched (Table I). These properties have led to the proposal that foveal MT and MSTv neurons can contribute to pursuit maintenance, while MSTd neurons and nonfoveal MSTv neurons are useful for perceptual purposes during pursuit. With respect to pursuit initiation, Movshon *et al.* (1990) noted that a number of aspects of pursuit, notably the change in latency with speed and the acceleration effects, might reflect properties of MT cells rather than motor processes.

Kawano *et al.* (1994) studied MST cells (both MSTd and MSTv) during ocular following responses. Responses of MST neurons increased with stimulus speed, while their latency and that of ocular following responses decreased. Changing the stimulus by blurring or replacing random dot patterns with checkerboards had similar effects on MST cells and ocular following responses. These authors compared the two types of responses in 35 MST cells which responded in the same direction during pursuit and ocular following responses: response latencies of the neurons were longer for the moving spot than the moving ran-

Table I. Summary of Pursuit Cells[a]

		MSTv			
	MTf	MTf-like		MSTd-like	MSTd
Visual directional	Yes		Yes		Yes
RF including fovea	Yes		Yes		Yes
RF size	Small	Small		Large	Large
Preferred stimulus	Spot	Spot		Field	Field
Pursuit directional	Yes		Yes		Yes
Pursuit in dark	Yes		Yes		Yes
Extraretinal input	No	No/yes		No/yes	Yes
Direction: spot and pursuit	Same	Same		Same	Same
Direction: field and pursuit	—	Same		Opposite[b]	Opposite[b]
Pursuit light/pursuit dark	Lt ≤ Dk	Lt ≤ Dk		Lt > Dk	Lt >> Dk

[a]Mtf, foveal region of middle temporal area; MSTv, lateral–anterior region of medial superior temporal area; MSTd, dorsal–medial region of medial superior temporal area; RF, receptive field; Lt, light; Dk, dark.
[b]Most but not all respond in this way.

dom dot pattern (16.5 msec difference on average) and pursuit latency also exceeded the latency of ocular following responses. The main difference between ocular following responses and OKN is that the random dot pattern is visible but stationary before ocular following responses, while the pattern is switched on as it starts moving in the OKN. Kawano *et al.* (1994) reported that in response to an OKN stimulation with random dot patterns, MST neurons show first a non-direction-selective response of short latency, followed by a direction-selective response at 90–100 msec. For 32 MST cells tested both with stimuli eliciting ocular following responses and OKN, the OKN direction-selective response lagged the response during ocular following by 37 msec. Similarly, the OKN latency is longer than the latency of the ocular following response.

Just as the direction selectivity of MT and MST cells is an indication that these cells may be involved in motion perception, the occurrence of visual pursuit cells in these areas indicates only that they might be involved in pursuit. Wurtz and colleagues used lesion and microstimulation techniques to clarify the role of these areas in pursuit. They basically observed two types of deficits in horizontal pursuit after ibotenic acid lesions in areas MT and satellites: a retinotopic deficit and a directional deficit (Fig. 31). The retinotopic deficit occurred for motion localized in contralateral visual space and affected the initiation of pursuit (decreased eye speed during the first 100-msec pursuit and a decreased-amplitude initial saccade to the target in step ramp paradigm). This deficit was observed after lesions of extrafoveal parts of area MT (Newsome *et al.*, 1985) and after lesion of any part of MST (Dürsteler and Wurtz, 1988). The only difference is that the retinotopic deficit was more restricted within the visual field in the case of MT lesions. On the other hand, the directional deficit affected pursuit maintenance, as eye speed was reduced 200–400 msec after stimulus onset. It was specific for the direction of motion to the side of the lesion and occurred throughout the visual field (Fig. 31). Such deficits were observed following lesions of foveal MT and of MSTv (Dürsteler and Wurtz, 1988; Dürsteler *et al.*, 1987). By stabilization and superimposition of additional retinal slip Dürsteler and Wurtz (1988) showed that the directional deficit was due mainly to a reduced sensitivity to retinal slip, i.e., a deficit in the visual motion signal. Chemical lesions of MST also affected OKN, particularly the slow buildup and the initial rapid rise (Dürsteler and Wurtz, 1988). All these lesion effects were very short-lived, typically lasting a few days. The mechanisms of this recovery were studied by Yamaskai and Wurtz (1991). These authors showed that the duration of the retinotopic effect of MT lesions correlated with lesion size. They also showed that a combined MT/MST lesion had permanent effects, more clearly so for the directional deficit than for the retinotopic deficit.

Finally, Komatsu and Wurtz (1989) studied the effects of microstimulation of area MT and satellites on pursuit. They observed that during the maintenance of pursuit the stimulation effect depended on the direction of pursuit: pursuit in all directions was slowed, with the exception of pursuit along the horizontal axis toward the side of stimulation, which increased in speed. The latter effect is consistent with the decrease in speed typical of the directional deficit following MST lesions. Komatsu and Wurtz (1989) provided evidence that the electrical stimulation substituted for visual motion input, since the present system was insensitive to the slip of the target over the retina during stimulation. This is again consistent with the reduced slip sensitivity underlying the directional deficit after MST lesions. Electrical stimulation had a clear effect in the foveal representations of MT and of MSTv. Together with the lesion effects, these data strongly suggest that only these two regions, and not MSTd, are involved in pursuit maintenance. It is noteworthy that the effects of microstimulation are more in accordance with the lesion effects on pursuit maintenance than those on the opposed-direction discrimination. In the latter case, lesions increased the threshold (Newsome and Paré, 1988), while stimulation had no effect on the threshold (Salzman *et al.,* 1992). Indeed, these authors insisted that the psychophysical curves did not change in slope. This occurred only when stimulation involved cells with different preferred directions due to increased stimulation strength or mispositioning of the electrode (Murasugi *et al.,* 1993). Electrical stimulation not only influenced pursuit maintenance, it also had a deleterious effect on pursuit initiation. This effect might be due to the fact that Komatsu and Wurtz (1989) made no attempts to restrict stimulation to a single direction column and were in fact injecting noise into the system. A similar explanation might also apply to the speed reduction observed during pursuit maintenance in all directions, mentioned earlier. It remains unclear, however, why MT and MSTv stimulation and lesions should have a directional

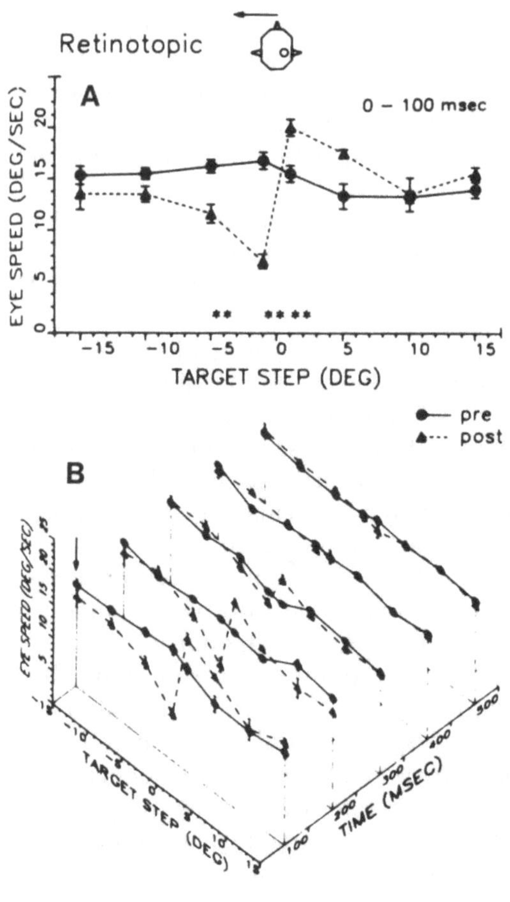

Figure 31. Visual field specificity of retinotopic and directional pursuit deficits following MT and MST chemical lesions. (A) Deficit for pursuit away from the side of the lesion. Target motion was to the left at 16°/sec. Abscissa: Step size from the primary position with rightward steps positive, leftward steps negative. Ordinate: Horizontal eye speed average over the first 100 msec following identification of the end of initial saccade. The means and standard errors for each point were derived from 10 pursuit responses for each step location before (pre-) and 24 hr after (post-) the injection. Along the abscissa: **significance level of $p \leq .01$ as determined by a Student's test; *level of $p \leq .01$. The retinotopic deficit was small, 1° and 5° on the left, and was accompanied by a significantly increased pursuit speed at 1° on the right. (B) Pursuit deficit throughout the visual field at successive time periods after the initial saccade to the pursuit target. Pursuit was always away from the side of the lesion as in part A. Axes: Target step, the mean and standard error of eye speed, and the time after the end of the saccade at which eye speed was computed. The 0- to 100-msec range

effect on the maintenance of horizontal pursuit: Wurtz and colleagues (Komatsu and Wurtz, 1989; Dürsteler and Wurtz, 1988) proposed that this is due to an overrepresentation of ipsilateral directions in the preferred directions of MT and MSTv pursuit cells, for which there is unfortunately little evidence.

Thus these experiments show that, as for motion perception, the area MT/V5 complex is part of the cortical network involved in control of pursuit and probably those controlling OKN and ocular following response. How many other areas may be included in these networks is at present unresolved, although many authors (Albright, 1993, for review; Yamasaki and Wurtz, 1991) favor the view that area MT/V5 and satellites are key components of these networks. The networks for eye movement control are likely to be somewhat more restricted than those involved in motion perception. Indeed the projections to the pons, the apparent subcortical relay for these signals, arise mainly from caudal STS, intraparietal sulcus, and parietooccipital sulcus (Glickstein *et al.*, 1980), restricting the networks for eye movement control to regions described as belonging to the dorsal stream. However, within the MT/V5 complex, distinct restricted populations are involved in the maintenance of pursuit, namely the foveal parts of MT and MSTv, or in the perceptual use during pursuit, namely MSTd and MSTv.

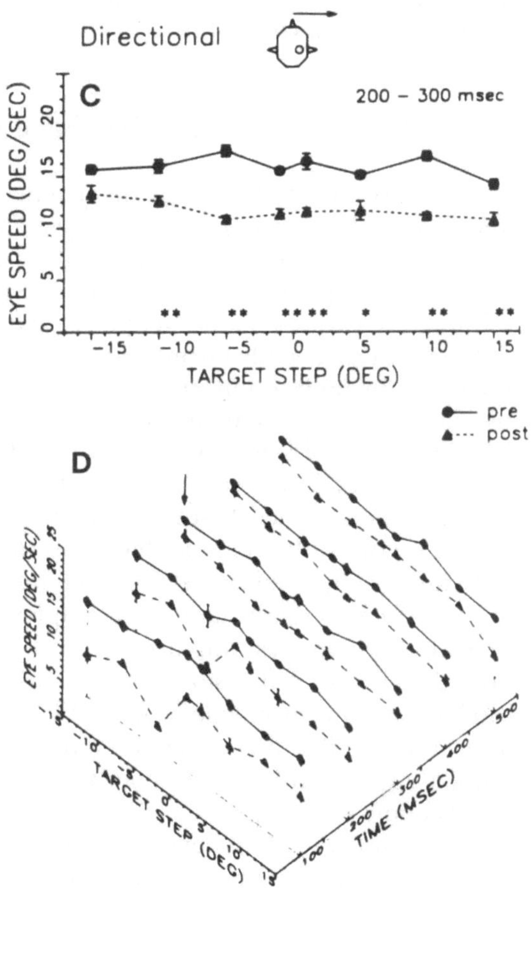

Figure 31. (*Continued*) shown (indicated by the arrow) is the same as that in part A. The mean of any time range on the graph is plotted at the middle point so that the 0- to 100-msec mean is plotted at 50 msec. The retinotopic deficit (in the left visual field contralateral to the side of the lesion) disappeared by the 200- to 300-msec period. (C) Pursuit deficit for target motion toward the side of the lesion. Pursuit speed is for the 200- to 300-msec period after the monkey acquired the target moving to the right. The decreased pursuit speed was clear for target steps extending throughout the visual field, was significant for all points but one, and illustrates the directional pursuit deficit. (D) Eye speed for different target steps at different times after the monkey made saccades to a target moving toward the side of the lesion. The 200- to 300-msec period on this graph (indicated by the arrow) is the same as the record in C. At all times the eye speed for this direction was reduced, but the amount of decrease clearly declined as time after the initial saccade increased. From Dürsteler and Wurtz (1988).

3.3. Guidance of Body Movements

For obvious technical reasons, very little work has been carried out in this arena. The only exception is the recent study by Duffy and Wurtz (1995b) of optic flow effects on posture and the modification of these effects produced by MST lesions. These authors could induce postural sway forward and backward by contracting and expanding optic flow stimuli, effective in driving MST cells. However, the postural effects were highly variable, because of the intrusion of other factors, probably vestibular and proprioceptive signals. Bilateral reversible lesions, including all of MSTd and most of MT and MSTv, produced postural instability, as the monkeys were unable to stay on the platform during a 24-sec trial. Monkeys recovered after each injection.

While at least an attempt has been made to clarify the role of area MT/V5 and satellites in the control of whole-body movement, no work has to our knowledge been done on the control of arm and hand movements, although Maunsell and Van Essen (1987) speculated that the bias in topographical organization of area MT, favoring the lower quadrant, might be related to extensive use of the hand. In this respect it may be noted that activation of the human homologue of area MT/V5 (see Section 4) has been noted when human subjects observed the moment of an 'alien' arm (Decety *et al.*, 1994). This activation, however, might simply reflect the activation by motion in general, since this aspect was absent from the control condition.

3.4. Extraction of 2D and 3D Shape from Motion

With respect to 2D shape determined by motion, or kinetic shape, Marcar and Cowey (1992) reported that large STS lesions, removing area MT and most of MSTd and MSTv, impaired the ability of monkeys to use kinetic boundaries as a basis for shape discrimination. The lesioned monkeys were incapable of discriminating two test shapes (e.g., triangle versus square) when no uncorrelated dots were added, while normal monkeys and monkeys with other types of lesions tolerated the addition of up to 60–65% of random movement to the coherently oscillating displays. Marcar and Cowey (1992) indicated that one of the animals in which area MT was mostly spared displayed little deficit, suggesting that either area MT or the combination of lesioned areas is critical for the extraction of kinetic boundaries. It seems unlikely that area MT as such is critical. Schiller (1993) reported that kinetic shape discrimination (square versus rectangle) was only mildly impaired by an area MT lesion, but again alternative cues (differences in flicker) might have been employed by the monkeys. We also observed, however, only modest impairments in the orientation discrimination of kinetic boundaries after MT lesions (Lauwers *et al.*, 1995). While it is possible that the satellites of MT supplement area MT/V5 for the extraction of kinetic boundaries, it is possible that other cortical areas could also participate. Such a possibility is supported by the human positron emission tomography (PET) studies (see Section 4).

Marcar *et al.* (1995) have tested whether MT neurons can signal the orientation and position of kinetic boundaries created by motion in opposite directions.

This type of boundary was employed rather than those created by stationary versus moving dots, since the latter contain cues other than motion differences, notably differences in temporal frequency and in blur. Marcar *et al.* (1995) disambiguated direction of motion (of the dots) versus orientation (of the boundary) by comparing tuning for orthogonal and parallel boundaries, in which the axis of motion of the dots was either orthogonal or parallel to the boundary, respectively. Without exception, MT neurons coded the direction of the moving dots and not the orientation of the boundary (Fig. 32). Furthermore, they observed that responses to the kinetic stimuli were weaker than those to uniform motion, in agreement with the results of Snowden *et al.* (1991). One could argue that at least those MT cells with an antagonistic surround should be able to code for the position of kinetic boundaries. Marcar *et al.* (1995) found that this was not the case for the antiphase type of kinetic boundaries used (Fig. 33), although it may be that boundaries created by motion versus stationary dots yield different results. Since then we have gone on to show that neurons in area V2 are able to code the orientation of antiphase kinetic boundaries (Marcar *et al.*, 1992, 1994). This finding might explain why such selectivity is also observed further along the ventral pathway (Sáry *et al.*, 1993, 1995; Logothetis and Charles, 1990).

With respect to extraction of 3D structure from motion by area MT and satellites, little information is available. Siegel and Andersen (1986) reported in abstract form that lesions of area MT/V5 impaired the perception of 3D structure, more precisely the transition from incoherent motion to coherent motion corresponding to dots painted on a rotating cylinder. No further details about these experiments have appeared in the literature. Treue and Andersen (1993) failed to obtain a clear tuning for the direction of tilt of planar surfaces specified by motion in MT neurons. However, these authors masked the surround of the MT cells tested. We also failed to find selectivity for 3D shapes defined by motion using a similar strategy (D.-K. Xiao, V. L. Marcar, S. Raiguel, and G. A. Orban, unpublished). Theoretical studies (Buracas and Albright, 1994, 1996; Droulez and Cornilleau-Pérès, 1990) suggest that the surround must play a critical role in the depth from motion extraction. This proposal receives strong support from the experiments of Xiao *et al.* (1995b) describing marked asymmetries in the surround of MT neurons. One would therefore expect that when the surround is left uncovered, MT cells show selectivity for the direction of tilt of planar moving surfaces. Recent results (Orban *et al.*, 1996; Xiao *et al.*, 1997) support this view.

4. Functional Imaging of Area MT/V5 and Satellites in Humans

The area considered to be the putative homologue of area MT/V5 in humans was first localized by Zeki *et al.* (1991) by virtue of its differential activation by moving dots as compared to stationary dots. Watson *et al.* (1993) replicated this finding in both single subjects and in averaged data (Table II). These authors stressed the anatomical variability in the position of the human homologue of MT/V5. The stimulus they used is relatively weak, owing to its low dot density,

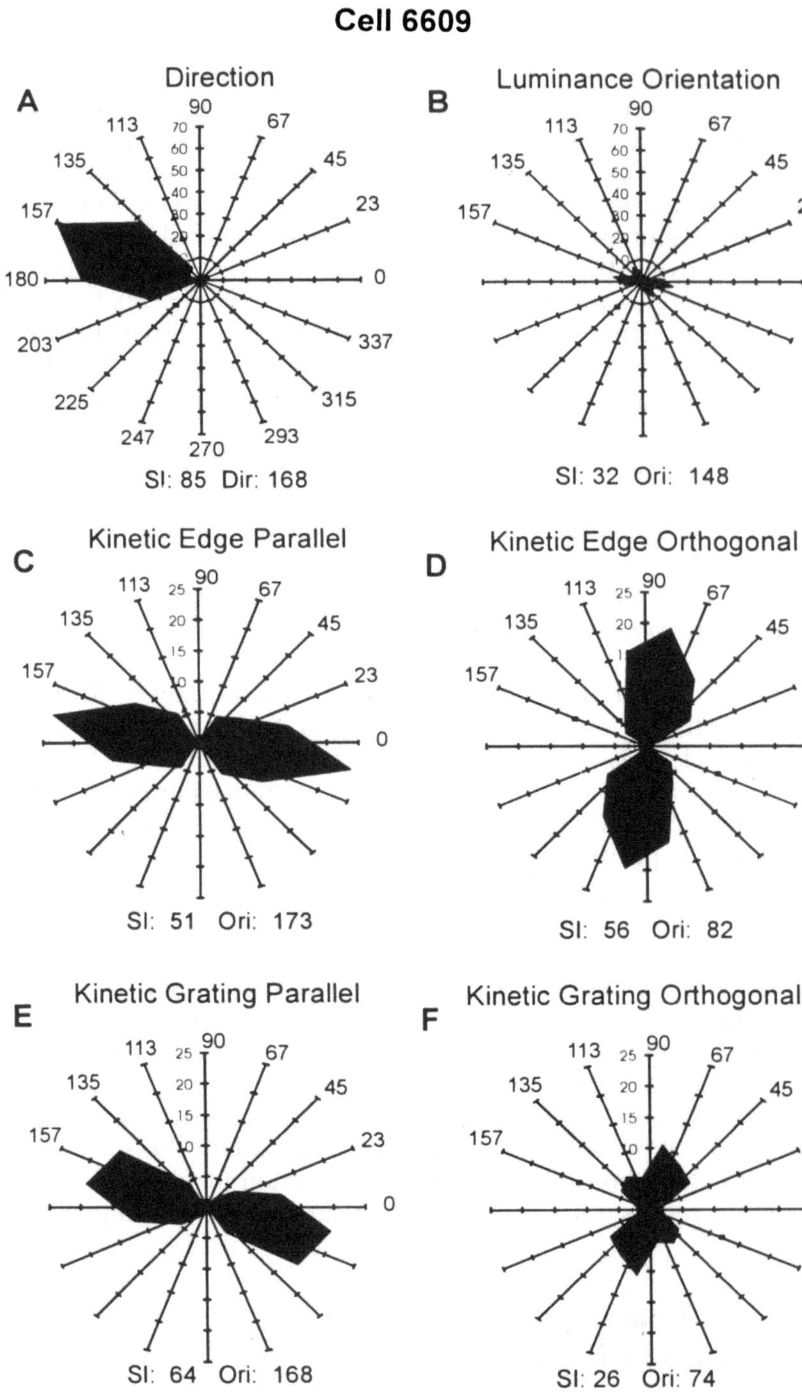

Figure 32. Example of an MT neuron showing 90° shift in preferred orientation for parallel and orthogonal kinetic boundaries: behavior characteristic of a neuron encoding motion direction, but not kinetic boundary orientation. Responses of cell 6609 to (A) uniform motion in 16 directions, (B) a stationary luminance contrast stimulus, and (C, D) our standard kinetic edges and (E, F) kinetic grating stimuli, in the form of polar plots. The cell shows a preferred orientation (parallel to its preferred direction of motion, thus making it a type II cell; Albright, 1984). The data for the orientation tests (stationary luminance contrast and kinetic boundaries) were plotted twice, following the convention of Maunsell and Van Essen (1983a), to complete the circle. The direction or orientation selectivity index (SI) value for each test and the calculated preferred direction (Dir) or preferred orientation (Ori) are shown below the appropriate polar plot. From Marcar *et al.* (1995).

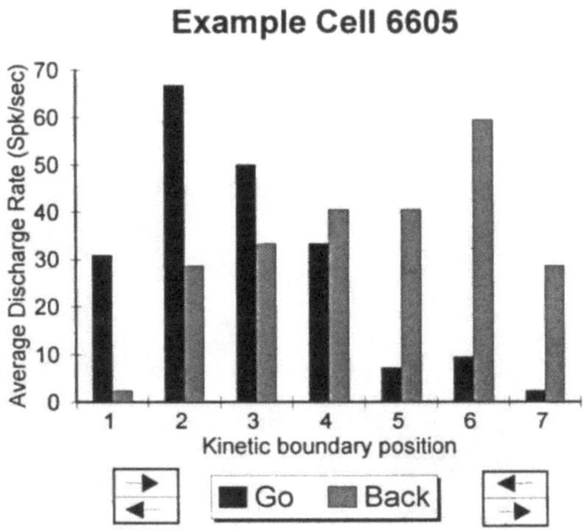

Figure 33. Response levels of MT neuron (cell 6605) at different locations across its RF where an antiphase kinetic boundary was presented: positions 2 and 6 corresponded to the edges between the CRF and surround. Black bars: responses with the local motion vectors in their normal direction. Hatched bars: responses with the local motion vectors in the reversed direction. The two boxes on either side of the legend indicate the direction of motion of the motion vectors in the "forward" and "backward" direction. From Marcar *et al.* (1995).

which reduces the level of responses in area MT/V5 of the monkey (Snowden *et al.,* 1991), raising the possibility that some of the variability might be due to the stimulus rather than to the subjects. Watson *et al.* (1993) noted a relatively close relationship between the area differentially activated by motion and a densely myelinated area described by Tootell and Taylor (1995) and by Clarke and Miklossy (1990). Thus at least two characteristics of monkey MT/V5 have been demonstrated for the area identified as its human homologue. Watson *et al.* (1993) also reported the activation of other areas in addition to MT/V5 and V1/V2, notably an area in the cuneus and another one in parietal cortex. This result was expanded upon by Dupont *et al.* (1994), who reported differential activation by motion versus stationary dots (3°-diameter random dot pattern in central vision) in many cortical regions outside the homologue of MT/V5 and of V1/V2 (Fig. 34): in the cuneus, in lower area 19, in parieto-occipital sulcus, and also in two apparently vestibular areas, a retroinsular area and area 2v. The activation in the cuneus might correspond to area V3A in the monkey, a view supported by later functional magnetic resonance imaging (fMRI) work by Tootell and co-workers (Tootell *et al.,* 1995c). Dupont *et al.* (1994) suggested that the retroinsular area was the homologue of the vestibular PIVC area described by Grüsser *et al.* (1990) in the monkey. O.-J. Grüsser (personal communication) has suggested that it might instead correspond to a visual area T3, located next to VIPC in the New World monkey and receiving input from MST. This anterior

Table II. Talairach Coordinates of Human MT/V5[a]

Zeki *et al.* (1991)[b]	$N = 3$	lat 44	post 64	sup 2
Watson *et al.* (1993)[b]	$N = 12$	lat 42	post 69	sup 0
Dupont *et al.* (1994)[b]	$N = 14$	lat 39	post 73	sup 4
Tootell *et al.* (1995b)[c]	$N = 6$	lat 45	post 76	sup 3
Sunaert *et al.*[c,d]	$N = 4$	lat 42	post 77	sup 4

[a]Relative to the anterior commissure (Talairach and Tournoux, 1988).
[b]PET data.
[c]fMRI data.
[d]S. Sunaert, S. Van Oostende, P. Van Hecke, G. Marchal, and G. A. Orbian, unpublished.

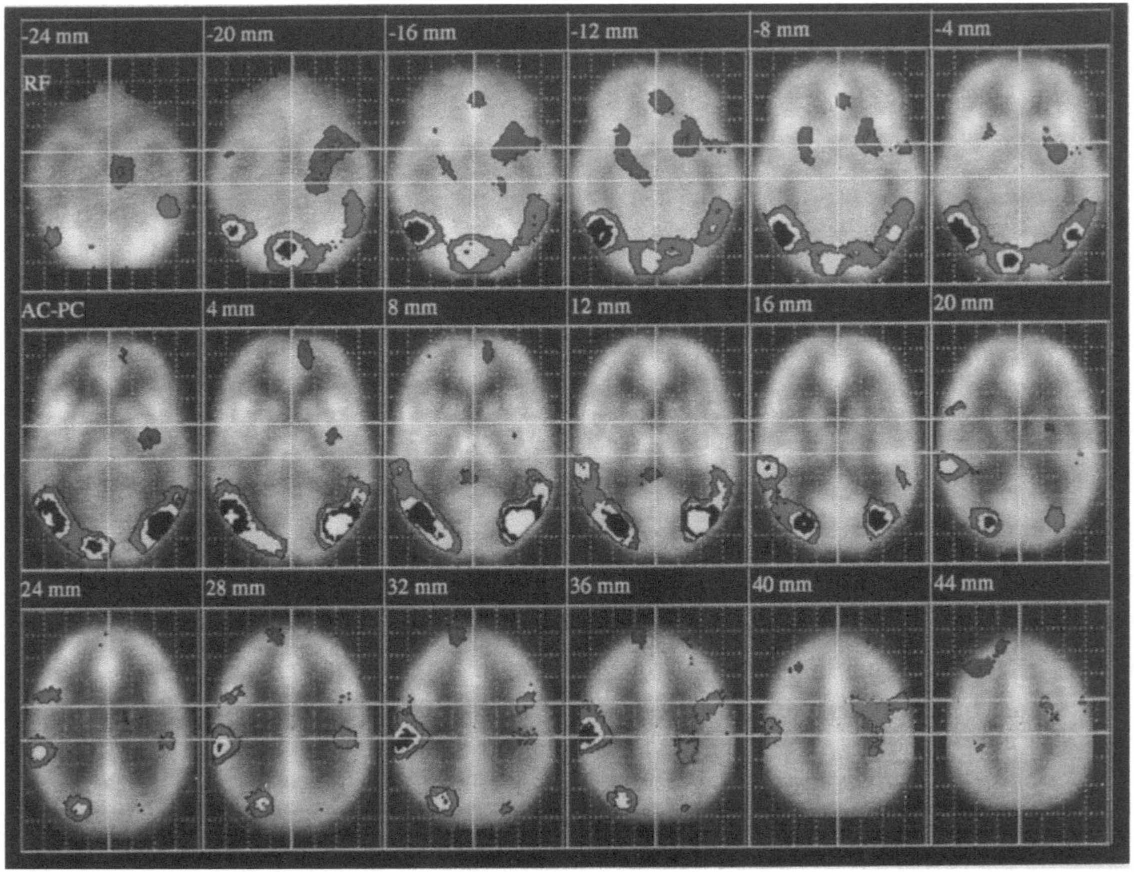

Figure 34. Many regions in human cortex are differentially activated by a moving random dot pattern compared to a stationary random dot pattern. Regions differing significantly in regional cerebral blood flow (rCBF) between the experimental condition (motion) and the control condition (no motion) are indicated by gray, black, and white pixels. Decreased rCBF is indicated by light and dark gray pixels corresponding to z scores of 2.33 ($p < .01$) and 3.09 ($p < .001$) respectively. Increased rCBF is indicated by alternating black and white pixels corresponding to z scores of 2.33 (black), 3.09 (white), 3.75 (black), 4.5 (white), the lowest significance being on the edges of the activation sites. Notice that higher significance levels are reached for activated than for deactivated regions. These labels are superimposed on horizontal sections, from -24 below to 44 mm above the anterior commissure (AC)–posterior commissure (PC) line, through the average rCBF scans from all conditions and subjects, to show the anatomic brain features. RF: right frontal, indicating that in these sections right and left are reversed. From Dupont *et al.* (1994).

motion area has since been observed with fMRI also (R. B. H. Tootell, unpublished). It is likely that Dupont *et al.* (1994) underestimated the number of motion-related areas, as the scans were restricted to *z* planes below 40 and that dorsal parietal areas might have been missed. Tootell and co-workers have identified a number of visual areas with the motion/no-motion paradigm in fMRI. In addition to the homologue of MT/V5, they have identified the two satellites MSTd and MSTv, and also V3A and the homologue of VIP (Tootell *et al.*, 1995b, c; Dale *et al.*, 1995; Kennedy *et al.*, 1995). While the homologue of VIP has a high contrast threshold, area V3A is the only area, outside MT/V5, activated by low contrast motion (Tootell *et al.*, 1995a). In the monkey, V3A has reciprocal connections with area MT and both satellites, but it is not traditionally considered as part of the motion pathway, which is generally restricted to the MT/V5 complex, the areas afferent to MT (V1, V2, and V3), and VIP.

Thus in humans a large number of cortical areas are differentially activated by uniform dot motion whether compared to stationary dots or to incoherently moving dots (Cheng *et al.*, 1995). Yet an additional motion area has been discovered in humans which seems specialized in the processing of kinetic boundaries (Dupont *et al.*, 1995; Orban *et al.*, 1995b). This area is located more posteriorly than area MT/V5, almost exactly halfway between MT/V5 and V1/V2. Comparison of this location with the location of areas V1, V2, V3 obtained by Sereno *et al.* (1995) and with area V3A (R. B. H. Tootell, personal communication) suggests that this additional kinetic occipital (KO) area lies outside these known areas.

Using fMRI, Tootell and co-workers (Tootell *et al.*, 1995b) studied the functional properties of the human homologue of area MT/V5. They showed that fMRI responses over MT/V5 are sensitive to very low contrast (Fig. 35), as are monkey MT neurons (Sclar *et al.*, 1990). They also showed that these signals are strongly reduced for isoluminant gratings, again as MT neuronal responses are (Gegenfurtner *et al.*, 1994; Saito *et al.*, 1989). Furthermore, they reported substantial ipsilateral field input to the human homologue of area MT/V5, as is the case in monkey MT/V5 (Raiguel *et al.*, 1995). Finally, the motion versus nonmotion paradigm in fact contains a confound: a difference in temporal frequency or flicker. Indeed, Tootell *et al.* (1995b) have shown that MRI signals in the homologue of area MT/V5 increase in response to flicker as much as to motion. Also MT neurons respond quite well to flicker (Lagae *et al.*, 1994). The issue left unresolved is why retinotopic organization is so diffuse in human MT/V5 while it is quite evident in the monkey (Maunsell and Van Essen, 1987; Van Essen *et al.*, 1981). Tootell *et al.* (1995a) have also reported a prolonged activation of area MT/V5 after unidirectional movement correlating with the perceptual motion aftereffect. This might be related to the rebound activation which can be observed in MT (see Fig. 5) and MSTd neurons after the end of stimulation in the nonpreferred direction of motion (H. Maes, L. Lagae, S. Raiguel, and G. A. Orban, unpublished).

Thus functional imaging studies in humans have, on one hand, led to the identification of the human area equivalent to area MT/V5 in the monkey. The similarities between the macaque area MT/V5 and its human homologue are so striking that one can conclude that there is a human area MT/V5 just as there is in macaque and owl monkey (Tootell *et al.*, 1995b). On the other hand, the functional studies have also called attention to a large set of areas which could be involved in motion processing in addition to area MT/V5 and its satellites.

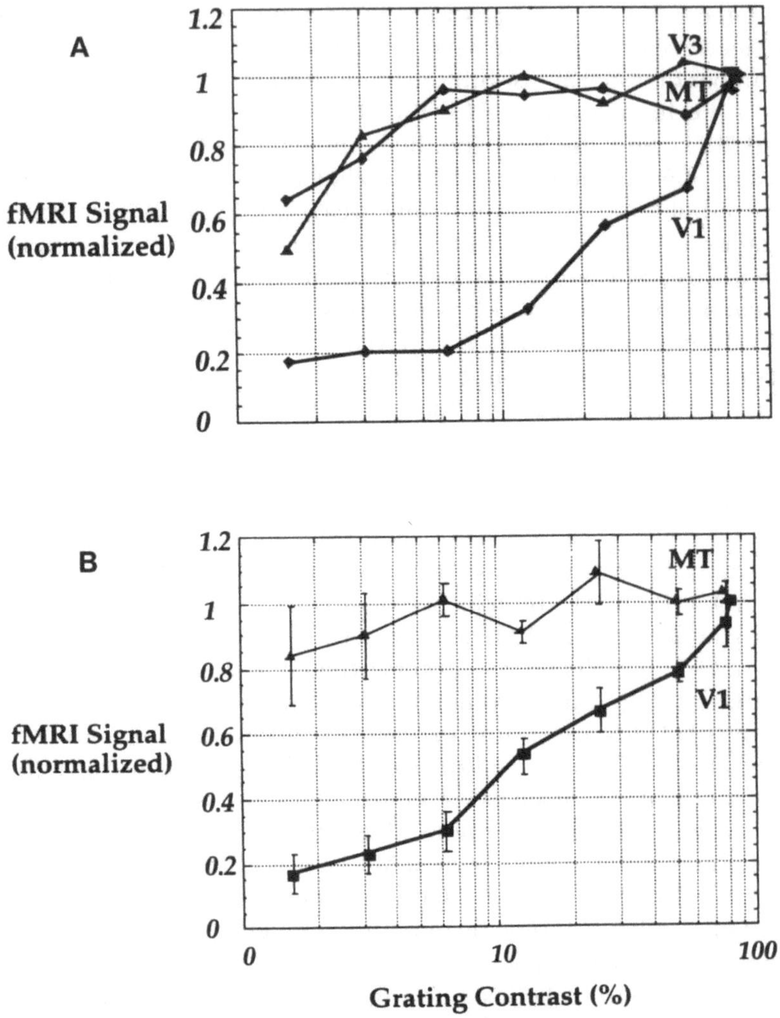

Figure 35. Luminance contrast gain curves derived from MRI measurements. (A) Averaged data from two complete contrast gain functions in a single subject, from areas MT, V1, and V3. The amplitudes in each area were normalized relative to the amplitude produced by the highest contrast presented (82%). (B) Averaged data from all subjects tested with these contrasts, amounting to eight complete stimulus presentations in four subjects. SE bars are indicated for each data point. The amplitudes from each area are normalized as in part A, since absolute levels of activation are not meaningful in this context. In both parts A and B, the contrast sensitivity in area MT is much higher than that in V1 over most of the contrast range. The half-amplitude of the contrast gain function in MT is near 1%, varying slightly in different subjects. At most contrasts above that, MRI amplitudes in MT are saturated. In V1, one needs approximately 15% luminance contrast to produce a half-amplitude response, and the response continues to increase up to the highest contrast tested. The high contrast sensitivity in human MT strongly suggest that it receives input predominantly from magnocellular stream inputs, as in macaque MT. Subsequent work has shown that contrast gain curves are similar in area V3A and V3, but only V3A, and not V3, is selective for motion. From Tootell *et al.* (1995b).

5. Conclusions

There is marked contrast between the detailed anatomical and physiological knowledge which has accumulated about area MT and satellites and the relative uncertainty about the behavioral role of these areas. On one hand, the experiments which have been performed to clarify the role of areas MT and satellites in motion perception and control of eye movements are consistent with a key contribution by these areas to those behaviors, a view defended by most workers in the field (Albright, 1993; Britten *et al.,* 1992; Yamasaki and Wurtz, 1991). On the other hand, these same experiments, as discussed above, are vulnerable to less enthusiastic interpretations (Merigan and Maunsell, 1993; Schiller, 1993). A possible explanation for this somewhat unsatisfactory state of affairs lies in the fact that the behavioral role of cortical areas has been approached with techniques emphasizing the single-cell or areal level. What is needed are techniques which allow us to capture brain function at the network level (Mesulam, 1990) and the human work suggests that mapping techniques might provide this sort of information. Recently such techniques have begun to be used in monkeys: 2-deoxyglucose (Friedman and Goldman-Rakic, 1988; Vanduffel *et al.,* 1993, 1995) and PET applied to awake monkeys (Takechi *et al.,* 1994). It seems that use of these mapping techniques in combination with the more traditional techniques holds great future in understanding the functional role of cortical areas, not only area MT/V5 and satellites, but extrastriate areas in general.

6. References

Adelson, E. H., and Movshon, J. A., 1982, Phenomenal coherence of moving visual patterns, *Nature* **300:**523–525.

Albright, T. D., 1984, Direction and orientation selectivity of neurons in visual area MT of the macaque, *J. Neurophysiol.* **52:**1106–1130.

Albright, T. D., 1989, Centrifugal directional bias in the middle temporal visual area (MT) of the macaque, *Visual Neurosci.* **2:**177–188.

Albright, T. D., 1992, Form-cue invariant motion processing in primate visual cortex, *Science* **255:**1141–1143.

Albright, T. D., 1993, Cortical processing of visual motion, in: *Visual Motion and its Role in the Stabilization of Gaze* (F. A. Miles, and J. Wallman, eds.), Elsevier, Amsterdam, pp. 177–201.

Albright, T. D., and Desimone, R., 1987, Local precision of visuotopic organization in the middle temporal area (MT) of the macaque, *Exp. Brain Res.* **65:**582–592.

Albright, T. D., Desimone, R., and Gross, C. G., 1984, Columnar organization of directionally selective cells in visual area MT of the macaque, *J. Neurophysiol.* **51:**16–31.

Allman, J. M., and Kaas, J. H., 1971, A representation of the visual field in the caudal third of the middle temporal gyrus of the owl monkey (*Aotus trivirgatus*), *Brain Res.* **31:**85–105.

Allman, J., Miezin, F., and McGuinness, E. L., 1985, Direction- and velocity-specific responses from beyond the classical receptive field in the middle temporal visual area (MT), *Perception* **14:**105–126.

Andersen, R. A., Essick, G. K., and Siegel, R. M., 1985, Encoding of spatial location by posterior parietal neurons, *Science* **230:**456–458.

Bair, W., Koch, C., Newsome, W., and Britten, K., 1994, Power spectrum analysis of bursting cells in area MT in the behaving monkey, *J. Neurosci.* **14:**2870–2892.

Baker, J. F., Petersen, S. E., Newsome, W. T., and Allman, J. M., 1981, Visual response properties of neurons in four extrastriate visual areas of the owl monkey (*Aotus trivirgatus*), *J. Neurophysiol.* **45:**397–416.

Born, R. T., and Tootell, R. B. H., 1992, Segregation of global and local motion processing in primate middle temporal visual area, *Nature* **357**:497–499.

Boussaoud, D., Ungerleider, L. G., and Desimone, R., 1990, Pathways for motion analysis: Cortical connections of the medial superior temporal and fundus of the superior temporal visual areas in the macaque, *J. Comp. Neurol.* **296**:462–495.

Boussaoud, D., Desimone, R., and Ungerleider, L. G., 1992, Subcortical connections of visual areas MST and FST in macaques, *Visual Neurosci.* **9**:291–302.

Bradley, A., Skottun, B. C., Ohzawa, I., Sclar, G., and Freeman, R. D., 1987, Visual orientation and spatial frequency discrimination: A comparison of single cells and behavior, *J. Neurophysiol.* **57**:755–772.

Bradley, D. C., Qian, N., and Andersen, R. A., 1995, Integration of motion and stereopsis in middle temporal cortical area of macaques, *Nature* **373**:609–611.

Bremmer, F., Ilg, U. J., Thiele, A., and Hoffmann, K.-P., 1996, Eye position effects in monkey cortex, I: Visual and pursuit related activity in extrastriate areas MT and MST, *J. Neurophysiol.*, in press.

Britten, K. H., Shadlen, M. N., Newsome, W. T., and Movshon, J. A., 1992, The analysis of visual motion: A comparison of neuronal and psychophysical performance, *J. Neurosci.* **12**:4745–4765.

Britten, K. H., Shadlen, M. N., Newsome, W. T., and Movshon, J. A., 1993, Responses of neurons in macaque MT to stochastic motion signals, *Visual Neurosci.* **10**:1157–1169.

Britten, K. H., Newsome, W. T., Shadlen, M. N., Celebrini, S., and Movshon, J. A., 1996, A relationship between behavioral choice and the visual responses of neurons in macaque MT, *Visual Neurosci.* **13**:87–100.

Buracas, G. T., and Albright, T. D., 1994, The role of MT neuron receptive field surrounds in computing object shape from velocity fields, *Adv. Neural Information Processing Syst.* **6**:969–976.

Buracas, G. T., and Albright, T. D., 1996, Contribution of area MT to perception of three-dimensional shape: A computational study, *Vision Res.* **36**:869–888.

Cavanagh, P., and Mathers, G., 1989, Motion: The long and short of it, *Spatial Vis.* **4**:103–129.

Celebrini, S., and Newsome, W. T., 1994, Neuronal and psychophysical sensitivity to motion signals in extrastriate area MST of the macaque monkey, *J. Neurosci.* **14**:4109–4124.

Celebrini, S., and Newsome, W. T., 1995, Microstimulation of extrastriate area MST influences performance on a direction discrimination task, *J. Neurophysiol.* **73**:437–448.

Cheng, K., Hasegawa, T., Saleem, K. S., and Tanaka, K., 1994, Comparison of neuronal selectivity for stimulus speed, length, and contrast in the prestriate visual cortical areas V4 and MT of the macaque monkey, *J. Neurophysiol.* **71**:2269–2280.

Cheng, K., Fujita, H., Kanno, I., Miura, S., and Tanaka, K., 1995, Human cortical regions activated by wide-field visual motion: An $H_2{}^{15}O$ PET study, *J. Neurophysiol.* **74**:413–427.

Chubb, C., and Sperling, G., 1988, Drift-balanced random stimuli: A general basis for studying non-Fourier motion perception, *J. Opt. Soc. Am. A* **5**:1986–2006.

Ciochon, R. L., and Chiarelli, A. B., 1980, *Evolutionary Biology of the New World Monkeys and Continental Drift*, Plenum Press, New York.

Clarke, S., and Miklossy, J., 1990, Occipital cortex in man: Organization of callosal connections, related myelo- and cytoarchitecture, and putative boundaries of functional visual areas, *J. Comp. Neurol.* **298**:188–214.

Cowey, A., and Marcar, V. L., 1992, The effect of removing superior temporal cortical motion areas in the macaque monkey: I. Motion discrimination using simple dots, *Eur. J. Neurosci.* **4**:1219–1227.

Cragg, B. G., 1969, The topography of the afferent projections in circumstriate visual cortex of the monkey studied by the Nauta method, *Vision Res.*, **9**:733–747.

Dale, A. M., Ahlfors, S. P., Aronen, H. J., Belliveau, J. W., Huotilainen, M., Ilmoniemi, R. J., Kennedy, W. A., Korvenoja, A., Liu, A. K., Reppas, J. B., Rosen, B. R., Sereno, M. I., Simpson, G. V., Standertskjöld-Nordenstam, C.-G., Virtanen, J., and Tootell, R. B. H., 1995, Spatiotemporal imaging of coherent motion selective areas in human cortex, *Soc. Neurosci. Abstr.* **21**:1275.

Dean, A. F., 1981, The variability of discharge of simple cells in cat striate cortex, *Exp. Brain Res.* **44**:437–440.

Decety, J., Perani, D., Jeannerod, M., Bettinardi, V., Tadary, B., Woods, R., Mazziotta, J. C., and Fazio, F., 1994, Mapping motor representations with positron emission tomography, *Nature* **371**:600–602.

Desimone, R., and Ungerleider, L. G., 1986, Multiple visual areas in the caudal superior temporal sulcus of the macaque, *J. Comp. Neurol.* **248:**164–189.

DeYoe, E. G., and Van Essen, D. C., 1985, Segregation of efferent connections and receptive field properties in visual area V2 of the macaque, *Nature* **317:**58–61.

Dobkins, K. R., and Albright, T. D., 1994, What happens if it changes color when it moves? The nature of chromatic input to macaque visual area MT, *J. Neurosci.* **14:**4854–4870.

Dobkins, K. R., and Albright, T. D., 1995, Behavioral and neural effects of chromatic isoluminance in the primate visual motion system, *Visual Neurosci.* **12:**321–332.

Droulez, J., and Cornilleau-Pérès, V., 1990, Visual perception of surface curvature. The species variation and its physiological implications, *Biol. Cybernet.* **62:**211–224.

Dubner, R., and Zeki, S. M., 1971, Response properties and receptive fields of cells in an anatomically defined region of the superior temporal sulcus in the monkey, *Brain Res.* **35:**528–532.

Duffy, C. J., and Wurtz, R. H., 1991a, Sensitivity of MST neurons to optic flow stimuli. I. A continuum of response selectivity to large-field stimuli, *J. Neurophysiol* **65:**1329–1345.

Duffy, C. J., and Wurtz, R. H., 1991b, Sensitivity of MST neurons to optic flow stimuli. II. Mechanisms of response selectivity revealed by small-field stimuli, *J. Neurophysiol.* **65:**1346–1359.

Duffy, C. J., and Wurtz, R. H., 1995a, Response of monkey MST neurons to optic flow stimuli with shifted centers of motion, *J. Neurosci.* **15:**5192–5208.

Duffy, C. J., and Wurtz, R. H., 1995b, Optic flow, posture, and the dorsal visual pathway, in: *Perception, Memory, and Emotion: Frontier in Neuroscience* (T. Ono, B. L. McNaughton, S. Molotchnikoff, E. T. Rolls, and H. Nishijo, eds.), Pergamon Press, Oxford.

Dupont, P., Orban, G. A., De Bruyn, B., Verbruggen, A., and Mortelmans, L., 1994, Many areas in the human brain respond to visual motion, *J. Neurophysiol.* **72:**1420–1424.

Dupont, P., Rosier, A., Vandenberghe, R., De Bruyn, B., Bormans, G., Mortelmans, L., and Orban, G. A., 1995, Regions in the human brain involved in the processing of motion discontinuities and of uniform motion: A PET study, *Soc. Neurosci. Abstr.* **21:**663.

Dürsteler, M. R., and Wurtz, R. H., 1988, Pursuit and optokinetic deficits following chemical lesions of cortical areas MT and MST, *J. Neurophysiol.* **60:**940–965.

Dürsteler, M. R., Wurtz, R. H., and Newsome, W. T., 1987, Directional pursuit deficits following lesions of the foveal representation within the superior temporal sulcus of the macaque monkey, *J. Neurophysiol.* **57:**1262–1287.

Erickson, R. G., and Dow, B. M., 1989, Foveal tracking cells in the superior temporal sulcus of the macaque monkey, *Exp. Brain Res.* **78:**113–131.

Erickson, R. G., and Thier, P., 1991, A neuronal correlate of spatial stability during periods of self-induced visual motion, *Exp. Brain Res.* **86:**608–616.

Felleman, D. J., and Van Essen, D. C., 1991, Distributed hierarchical processing in the primate cerebral cortex, *Cerebral Cortex* **1:**1–47.

Ferrera, V. P., Nealey, T. A., and Maunsell, J. H. R., 1994a, Responses in macaque visual area V4 following inactivation of the parvocellular and magnocellular LGN pathways, *J. Neurosci.* **14:**2080–2088.

Ferrera, V. P., Rudolph, K. K., and Maunsell, J. H. R., 1994b, Responses of neurons in the parietal and temporal visual pathways during a motion task, *J. Neurosci.* **14:**6171–6186.

Friedman, H. R., and Goldman-Rakic, P. S., 1988, Activation of the hippocampus and dentate gyrus by working-memory: A 2-deoxyglucose study of behaving rhesus monkeys, *J. Neurosci.* **8:**4693–4706.

Fries, W., 1981, The projections from the lateral geniculate nucleus to the prestriate cortex of the macaque monkey, *Proc. R. Soc. Lond. B* **213:**73–80.

Galletti, C., and Battaglini, P. P., 1989, Gaze-dependent visual neurons in area V3A of monkey prestriate cortex, *J. Neurosci.* **9:**1112–1125.

Gattass, R., and Gross, C. G., 1981, Visual topography of striate projection zone (MT) in posterior superior temporal sulcus of the macaque, *J. Neurophysiol.* **46:**621–638.

Gegenfurtner, K. R., Kiper, D. C., Beusmans, J. M. H., Carandini, M., Zaidi, Q., and Movshon, J. A., 1994, Chromatic properties of neurons in macaque MT, *Visual Neurosci.* **11:**455–466.

Girard, P., and Bullier, J., 1989, Visual activity in area V2 during reversible inactivation of area 17 in the macaque monkey, *J. Neurophysiol.* **62:**1287–1302.

Girard, P., Salin, P.-A., and Bullier, J., 1991, Visual activity in macaque area V4 depends on area 17 input, *Neuroreport* **2:**81–84.

Girard, P., Salin, P. A., and Bullier, J., 1992, Response selectivity of neurons in area MT of the macaque monkey during reversible inactivation of area V1, *J. Neurophysiol.* **67:**1437–1446.

Glickstein, M., Cohen, J. L., Dixon, B., Gibson, A., Hollins, M., Labossiere, E., and Robinson, F., 1980, Corticopontine visual projections in macaque monkeys, *J. Comp. Neurol.* **190:**209–229.

Graziano, M. S. A., Andersen, R. A., and Snowden, R. J., 1994, Tuning of MST neurons to spiral motions, *J. Neurosci.* **14:**54–67.

Grüsser, O.-J., Pause, M., and Schreiter, U., 1990, Vestibular neurones in the parieto-insular cortex of monkeys (*Macaca fascicularis*): Visual and neck receptor responses, *J. Physiol.* (Lond.) **430:**559-583.

Hawken, M. J., and Parker, A. J., 1990, Detection and discrimination mechanisms in the striate cortex of the old-world monkey, in: *Vision: Coding and Efficiency* (C. Blakemore, ed.), Cambridge University Press, Cambridge, pp. 103–116.

Hawken, M. J., Parker, A. J., and Lund, J. S., 1988, Laminar organization and contrast sensitivity of direction-selective cells in the striate cortex of the old world monkey, *J. Neurosci.* **8:**3541–3548.

Hikosaka, K., Iwai, E., Saito, H., and Tanaka, K., 1988, Polysensory properties of neurons in the anterior bank of the caudal superior temporal sulcus of the macaque monkey, *J. Neurophysiol.* **60:**1615–1637.

Hoffmann, K. P., Distler, C., and Erickson, R., 1991, Functional projections from striate cortex and superior temporal sulcus to the nucleus of the optic tract (NOT) and dorsal terminal nucleus of the accessory optic tract (DTN) of macaque monkeys, *J. Comp. Neurol.* **313:**707–724.

Hubel, D. H., and Wiesel, T. N., 1968, Receptive fields and functional architecture of monkey striate cortex, *J. Physiol.* (Lond.) **195:**215–243.

Kawano, K., Shidara, M., Wanatabe, Y., and Yamane, S., 1994, Neural activity in cortical area MST of alert monkey during ocular following responses, *J. Neurophysiol.* **71:**2305–2324.

Kennedy, W. A., Dale, A. M., Reppas, J. B., Liu, A. K., Belliveau, J. W., Rosen, B. R., and Tootell, R. B. H., 1995, fMRI responses to 180 degree reversals of 3D optical flow fields reveal direction selectivity in human area MT, *Soc. Neurosci. Abstr.* **21:**663.

Kim, J., and Wilson, H. R., 1993, Dependence of plaid motion coherence on component grating directions, *Vision Res.* **33:**2479–2489.

Koenderink, J. J., and van Doorn, A. J., 1975, Invariant properties of the motion parallax field due to the movement of rigid bodies relative to an observer, *Optica Acta* **22:**773–791.

Komatsu, H., and Wurtz, R. H., 1988a, Relation of cortical areas MT and MST to pursuit eye movements. I. Localization and visual properties of neurons. *J. Neurophysiol.* **60:**580–603.

Komatsu, H., and Wurtz, R. H., 1988b, Relation of cortical areas MT and MST to pursuit eye movements. III. Interaction with full-field visual stimulation, *J. Neurophysiol.* **60:**621–644.

Komatsu, H., and Wurtz, R. H., 1989, Modulation of pursuit eye movements by stimulation of cortical areas MT and MST, *J. Neurophysiol.* **62:**31–47.

Kovács, Gy., Vogels, R., and Orban, G. A., 1995, Selectivity of macaque inferior temporal neurons for partially occluded shapes, *J. Neurosci.* **15:**1984–1997.

Kreiter, A. K., and Singer, W., 1992, Oscillatory neuronal responses in the visual cortex of the awake macaque monkey, *Eur. J. Neurosci.* **4:**369–375.

Lagae, L., 1991, A neurophysiological study of optic flow analysis in the monkey brain, Ph.D. Thesis, Faculty of Medicine, KU Leuven.

Lagae, L., Gulyás, B., Raiguel, S., and Orban, G. A., 1989, Laminar analysis of motion information processing in macaque V5, *Brain Res.* **496:**361–367.

Lagae, L., Raiguel, S., and Orban, G. A., 1993, Speed and direction selectivity of macaque middle temporal neurons, *J. Neurophysiol.* **69:**19–39.

Lagae, L., Maes, H., Raiguel, S., Xiao, D.-K., and Orban, G. A., 1994, Responses of macaque STS neurons to optic flow components: a comparison of areas MT and MST, *J. Neurophysiol.* **71:**1597–1626.

Lauwers, K., Saunders, R. C., De Bruyn, B., Vogels, R., Vandenbussche, E., and Orban, G. A., 1995, The effect of MT lesions on direction discrimination and on orientation discrimination of kinetic gratings in the macaque, *Soc. Neurosci. Abstr.* **21:**280.

Lennie, P., 1980, Parallel visual pathways: A review, *Vision Res.* **20:**561–594.

Logothetis, N. K., and Charles, E. R., 1990, V4 responses to gratings defined by random dot motion, *Invest. Ophthalmol. Vis. Sci. Suppl.* **31:**444.

Logothetis, N. K., and Schall, J. D., 1989, Neuronal correlates of subjective visual perception, *Science* **245:**761–763.

Lu, Z.-L., and Sperling, G., 1995, The functional architecture of human visual motion perception, *Vision Res.* **35:**2697–2722.

Lund, J. S., Lund, R. D., Hendrickson, A. E., Bunt, A. M., and Fuchs, A. F., 1975, The origin of efferent pathways from the primary visual cortex (area 17) of the macaque monkey as shown by retrograde transport of horseradish peroxidase, *J. Comp. Neurol.* **164:**287–304.

Lund, J. S., Hendrickson, A. E., Ogren, M. P., and Tobin, E. A., 1981, Anatomical organization of primate visual cortex area VII, *J. Comp. Neurol.* **202:**19–45.

Marcar, V. L., and Cowey, A., 1992, The effect of removing superior temporal cortical motion areas in the macaque monkey: II. Motion discrimination using random dot displays, *Eur. J. Neurosci.* **4:**1228–1238.

Marcar, V. L., Raiguel, S. E., Xiao, D., Maes, H., and Orban, G. A., 1992, Do cells in area V2 respond to the orientation of kinetic boundaries? *Soc. Neurosci. Abstr.* **18:**1275.

Marcar, V. L., Xiao, D.-K., Raiguel, S.E., and Orban, G. A., 1994, Selectivity of area V2 of the macaque to kinetic- and other types of boundaries, *Soc. Neurosci. Abstr.* **20:**1740.

Marcar, V. L., Xiao, D.-K., Raiguel, S. E., Maes, H., and Orban, G. A., 1995, Processing of kinetically defined boundaries in the cortical motion area MT of the macaque monkey, *J. Neurophysiol.* **74:**1258–1270.

Maunsell, J. H. R., 1987, Physiological evidence for two visual subsystems, in: *Matters of Intelligence* (L. Vaina, ed.), Reidel, Dordrecht, Holland, pp. 59–87.

Maunsell, J. H. R., and Van Essen, D. C., 1983a, Functional properties of neurons in middle temporal visual area of the macaque monkey. I. Selectivity of stimulus direction, speed, and orientation, *J. Neurophysiol.* **49:**1127–1147.

Maunsell, J. H. R., and Van Essen, D. C., 1983b, Functional properties of neurons in middle temporal visual area of the macaque monkey. II. Binocular interactions and sensitivity to binocular disparity, *J. Neurophysiol.* **49:**1148–1167.

Maunsell, J. H. R., and Van Essen, D. C., 1983c, The connections of the middle temporal visual area (MT) and their relationship to a cortical hierarchy in the macaque monkey, *J. Neurosci.* **3:**2563–2586.

Maunsell, J. H. R., and Van Essen, D. C., 1987, Topographic organization of the middle temporal visual area in the macaque monkey: Representational biases and the relationship to callosal connections and myeloarchitectonic boundaries, *J. Comp. Neurol.* **266:**535–555.

Maunsell, J. H. R., Nealey, T. A., and DePriest, D. D., 1990, Magnocellular and parvocellular contributions to responses in the middle temporal visual area (MT) of the macaque monkey, *J. Neurosci.* **10:**3323–3334.

Maunsell, J. H. R., Sclar, G., Nealey, T. A., and DePriest, D. D., 1991, Extraretinal representations in area V4 in the macaque monkey, *Visual Neurosci.* **7:**561–573.

Merigan, W. H., and Maunsell, J. H. R., 1993, How parallel are the primate visual pathways? *Annu. Rev. Neurosci.* **16:**369–402.

Mesulam, M.-M., 1990, Large-scale neurocognitive networks and distributed processing for attention, language, and memory, *Ann. Neurol.* **28:**597–613.

Mikami, A., 1992, Spatiotemporal characteristics of direction-selective neurons in the middle temporal visual area of the macaque monkey, *Exp. Brain Res.* **90:**40–46.

Mikami, A., Newsome, W. T., and Wurtz, R. H., 1986a, Motion selectivity in macaque visual cortex. I. Mechanisms of direction and speed selectivity in extrastriate area MT, *J. Neurophysiol.* **55:**1308–1327.

Mikami, A., Newsome, W. T., and Wurtz, R. H., 1986b, Motion selectivity in macaque visual cortex. II. Spatiotemporal range of directional interactions in MT and V1, *J. Neurophysiol.* **55:**1328–1339.

Miles, F. A., Kawano, K., and Optican, L. M., 1986, Short-latency ocular following responses of monkey. I. Dependent on temporospatial properties of visual input, *J. Neurophysiol.* **56:**1321–1354.

Milner, A. D., Perrett, D. I., Johnston, R. S., Benson, P. J., Jordan, T. R., Heeley, D. W., Bettucci, D., Mortara, F., Mutani, R., Terazzi, E., and Davidson, D. L. W., 1991, Perception and action in "visual form agnosia," *Brain* **114:**405–428.

Movshon, J. A., and Newsome, W. T., 1984, Functional characteristics of striate cortical neurons projecting to MT in the macaque, *Soc. Neurosci. Abstr.* **10:**933.

Movshon, J. A., Adelson, E. H., Gizzi, M. S., and Newsome, W. T., 1985, The analysis of moving visual patterns, in: *Pattern Recognition Mechanisms* (R. Gattass, and C. Gross, eds.), Pontificia Academia Scientiarium, Vatican City, pp. 117–151.

Movshon, J. A., Lisberger, S. G., and Krauzlis, R. J., 1990, Visual cortical signals supporting smooth pursuit eye movements, in: *Cold Spring Harbor Symposia on Quantitative Biology,* Volume LV, *The Brain,* Cold Spring Harbor Laboratory Press, Cold Spring Harbor, NY, pp. 707–716.

Murasugi, C. M., Salzman, C. D., and Newsome, W. T., 1993, Microstimulation in visual area MT: Effects of varying pulse amplitude and frequency, *J. Neurosci.* **13:**1719–1729.

Nakayama, K., 1985, Biological image motion processing: A review, *Vision Res.* **25:**625–660.

Nealey, T. A., and Maunsell, J. H. R., 1994, Magnocellular and parvocellular contributions to the responses of neurons in macaque striate cortex, *J. Neurosci.* **14:**2069–2079.

Newsome, W. T., and Paré, E. B., 1988, A selective impairment of motion perception following lesions of the middle temporal visual area (MT), *J. Neurosci.* **8:**2201–2211.

Newsome, W. T., Wurtz, R. H., Dürsteler, M. R., and Mikami, A., 1985, Deficits in visual motion processing following ibonetic acid lesions of the middle temporal visual area of the macaque monkey, *J. Neurosci.* **5:**825–840.

Newsome, W. T., Wurtz, R. H., and Komatsu, H., 1988, Relation of cortical areas MT and MST to pursuit eye movements. II. Differentiation of retinal from extraretinal inputs, *J. Neurophysiol.* **60:**604–620.

Newsome, W. T., Britten, K. H., and Movshon, J. A., 1989, Neuronal correlates of a perceptual decision, *Nature* **341:**52–54.

Olavarria, J. F., DeYoe, E. A., Knierim, J. J., Fox, J. M., and Van Essen, D. C., 1992, Neural responses to visual texture patterns in middle temporal area of the macaque monkey, *J. Neurophysiol.* **68:**164–181.

Orban, G. A., 1986, Processing of moving images in the geniculocortical pathway, in: *Visual Neuroscience* (J. D. Pettigrew, K. J. Sandersen, and W. R. Levick, eds.), Cambridge University Press, Cambridge, pp. 121–141.

Orban, G. A., 1991, Quantitative electrophysiology of visual cortical neurones, in: *Vision and Visual Dysfunction,* Volume 4, *The Neural Basis of Visual Function* (J. Cronly-Dillon, gen. ed., and A. G. Leventhal, ed.), Macmillan, London, pp. 173–222.

Orban, G. A., 1992, The analysis of motion signals and the nature of processing in the primate visual system, in: *Artificial and Biological Vision Systems* (G. A. Orban and H. H. Nagel, eds.), Springer-Verlag, Berlin, pp. 24–56.

Orban, G. A., 1994, Motion processing in monkey striate cortex, in: *Cerebral Cortex,* Volume 10, *Primary Visual Cortex in Primates* (A. Peters and K. S. Rockland, eds.), Plenum Press, New York, pp. 413–441.

Orban, G. A., Kennedy, H., and Maes, H., 1981a, Response to movement of neurons in areas 17 and 18 of the cat: Velocity sensitivity, *J. Neurophysiol.* **45:**1043–1058.

Orban, G. A., Kennedy, H., and Maes, H., 1981b, Response to movement of neurons in areas 17 and 18: Direction selectivity, *J. Neurophysiol.* **45:**1059–1073.

Orban, G. A., Kennedy, H., and Bullier, J., 1986, Velocity sensitivity and direction selectivity of neurons in areas V1 and V2 of the monkey: Influence of eccentricity, *J. Neurophysiol.* **56:**462–480.

Orban, G. A., Lagae, L., Verri, A., Raiguel, S., Xiao, D., Maes, H., and Torre, V., 1992, First-order analysis of optical flow in monkey brain, *Proc. Natl. Acad. Sci. USA* **89:**2595–2599.

Orban, G. A., Lagae, L., Raiguel, S., Xiao D., and Maes, H., 1995a, The speed tuning of medial superior temporal (MST) cell responses to optic-flow components, *Perception* **24:**269–285.

Orban, G. A., Dupont, P., De Bruyn, B., Vogels, R., Vandenberghe, R., and Mortelmans, L., 1995b, A motion area in human visual cortex, *Proc. Natl. Acad. Sci. USA* **92:**993–997.

Orban, G. A., Saunders, R.C., and Vandenbussche, E., 1995c, Lesions of the superior temporal cortical motion areas impair speed discrimination in the macaque monkey, *Eur. J. Neurosci.* **7:**2261–2276.

Orban, G. A., Xiao, D.-K., Marcar, V., and Raiguel, S., 1996, Selectivity of macaque MT neurons for direction of tilt in depth, *Invest. Ophthalmol. Vis. Sci.* **37:**485.

Pasternak, T., and Merigan, W. H., 1994, Motion perception following lesions of the superior temporal sulcus in the monkey, *Cerebral Cortex* **4:**247–259.

Qian, N., and Andersen, R.A., 1995, V1 responses to transparent and nontransparent motions, *Exp. Brain Res.* **103:**41–50.

Raiguel, S. E., Lagae, L., Gulyás, B., and Orban, G. A., 1989, Response latencies of visual cells in macaque areas V1, V2 and V5, *Brain Res.* **493:**155–159.

Raiguel, S., Marcar, V., Xiao, D.-K., Maes, H., and Orban, G. A., 1993, Summation properties of macaque MT and MST neurons, *Soc. Neurosci. Abstr.* **19:**1283.

Raiguel, S., Van Hulle, M., Xiao, D.-K., Marcar, V. L., and Orban, G. A., 1995, Shape and spatial distribution of receptive fields and antagonistic motion surrounds in the middle temporal area (V5) of the macaque, *Eur. J. Neurosci.* **7:**2064–2082.

Rocha-Miranda, C., Bender, D., Gross, C. G., and Mishkin, M., 1975, Visual *activation* of neurons in inferotemporal cortex depends on striate cortex and the forebrain commissures, *J. Neurophysiol.* **38:**475–491.

Rockland, K. S., 1989, Bistratified distribution of terminal arbors of individual axons projecting from area V1 to middle temporal area (MT) in the macaque monkey, *Visual Neurosci.* **3:**155–170.

Rockland, K. S., 1995, Morphology of individual axons projecting from area V2 to MT in the macaque, *J. Comp. Neurol.* **355:**15–26.

Rockland, K. S., and Pandya, D. N., 1979, Laminar origins and terminations of cortical connections of the occipital lobe in the rhesus monkey, *Brain Res.* **179:**3–20.

Rodman, H. R., and Albright, T. D., 1987, Coding of visual stimulus velocity in area MT of the macaque, *Vision Res.* **27:**2035–2048.

Rodman, H. R., and Albright, T. D., 1989, Single-unit analysis of pattern-motion selective properties in the middle temporal visual area (MT), *Exp. Brain Res.* **75:**53–64.

Rodman, H. R., Gross, C. G., and Albright, T. D., 1989, Afferent basis of visual response properties in area MT of the macaque. I. Effects of striate cortex removal, *J. Neurosci.* **9:**2033–2050.

Rodman, H. R., Gross, C. G., and Albright, T. D., 1990, Afferent basis of visual response properties in area MT of the macaque. II. Effects of superior colliculus removal, *J. Neurosci.* **10:**1154–1164.

Roy, J.-P, Komatsu, H., and Wurtz, R. H., 1992, Disparity sensitivity of neurons in monkey extrastriate area MST, *J. Neurosci.* **12:**2478–2492.

Saito, H., Yukie, M., Tanaka, K., Hikosaka, K., Fukada, Y., and Iwai, E., 1986, Integration of direction signals of image motion in the superior temporal sulcus of the macaque monkey, *J. Neurosci.* **6:**145–157.

Saito, H., Tanaka, K., Isono, H., Yasuda, M., and Mikami, A., 1989, Directionally selective response of cells in the middle temporal area (MT) of the macaque monkey to the movement of equiluminous opponent color stimuli, *Exp. Brain Res.* **75:**1–14.

Sakata, H., Shibutani, H., Kawano, K., and Harrington, T. L., 1985, Neural mechanisms of space vision in the parietal association cortex of the monkey, *Vision Res.* **25:**453–463.

Salzman, C. D., and Newsome, W. T., 1994, Neural mechanisms for forming a perceptual decision, *Science* **264:**231–237.

Salzman, C. D., Murasugi, C. M., Britten, K. H., and Newsome, W. T., 1992, Microstimulation in visual area MT: Effects on direction discrimination performance, *J. Neurosci.* **12:**2331–2355.

Sáry, Gy., Vogels, R., and Orban, G. A., 1993, Cue-invariant shape selectivity of macaque inferior temporal neurons, *Science* **260:**995–997.

Sáry, Gy., Vogels, R., Kovács, Gy., and Orban, G. A., 1995, Responses of monkey inferior temporal neurons to luminance-, motion-, and texture-defined gratings, *J. Neurophysiol.* **73:**1341–1354.

Schall, J. D., Morel, A., King, D. J., and Bullier, J., 1995, Topography of visual cortex connections with frontal eye field in macaque: Convergence and segregation of processing streams, *J. Neurosci.* **15:**4464–4487.

Schiller, P. H., 1993, The effects of V4 and middle temporal (MT) area lesions on visual performance in the rhesus monkey, *Visual Neurosci.* **10:**717–746.

Schiller, P. H., and Lee, K., 1994, The effects of lateral geniculate nucleus area V4, and middle temporal (MT) lesions on visually guided eye movements, *Visual Neurosci.* **11:**229–241.

Schiller, P. H., and Malpeli, J. G., 1977, The effect of striate cortex coding on area 18 cells in the monkey, *Brain Res.* **126:**366–369.

Sclar, G., Maunsell, J. H. R., and Lennie, P., 1990, Coding of image contrast in central visual pathways of the macaque monkey, *Vision Res.* **30:**1–10.

Sereno, M. I., and Allman, J. M., 1991, Cortical visual areas in mammals, in: *The Neural Basis of Visual Function* (A. G. Leventhal, ed.), Macmillan, London, pp. 160–172.

Sereno, M. I., Dale, A. M., Reppas, J. B., Kwong, K. K., Belliveau, J. W., Brady, T. J., Rosen, B. R., and

Tootel, R. B. H., 1995, Borders of multiple visual areas in humans revealed by functional magnetic resonance imaging, *Science* **268:**889–893.

Shadlen, M. N., Britten, K. H., Newsome, W. T., and Movshon, J. A., 1996, A computational analysis of the relationship between neuronal and behavioral responses to visual motion, *J. Neurosci.* **16:**1486–1510.

Shipp, S., and Zeki, S. M., 1985, Segregation of pathways leading from area V2 to areas V4 and V5 of macaque monkey visual cortex, *Nature* **315:**322–325.

Shipp, S., and Zeki, S., 1989a, The organization of connections between areas V5 and V1 in macaque monkey visual cortex, *Eur. J. Neurosci.* **1:**309–332.

Shipp, S., and Zeki, S., 1989b, The organization of connections between areas V5 and V2 in macaque monkey visual cortex, *Eur. J. Neurosci.* **1:**333–354.

Siegle, R. M., and Andersen, R. A., 1986, Motion perceptual deficits following ibotenic acid lesions of the middle temporal area (MT) in the behaving rhesus monkey, *Soc. Neurosci. Abstr.* **12:**1183.

Snowden, R. J., Treue, S., Erickson, R. G., and Andersen, R. A., 1991, The response of area MT and V1 neurons to transparent motion, *J. Neurosci.* **11:**2768–2785.

Snowden, R. J., Treue, S., and Andersen, R. A., 1992, The response of neurons in areas V1 and MT of the alert rhesus monkey to moving random dot patterns, *Exp. Brain Res.* **88:**389–400.

Spatz, W. B., and Tigges, J., 1972, Experimental anatomical studies on the "middle temporal visual area (MT)" in primates. I. Efferent cortico-cortical connections in marmoset *Callithrix jacchus*, *J. Comp. Neurol.* **146:**451–464.

Standage, G. P., and Benevento, L. A., 1983, The organization of connections between the pulvinar and visual area MT in the macaque monkey, *Brain Res.* **262:**288–294.

Stoner, G. R., and Albright, T. D., 1992, Neural correlates of perceptual motion coherence, *Nature* **358:**412–414.

Stoner, G. R., Albright, T. D., and Ramachandran, V. S., 1990, Transparency and coherence in human motion perception, *Nature* **344:**153–155.

Sugita, Y., and Tanaka, K., 1991, Occlusion-related cue used for analysis of motion in the primate visual cortex, *NeuroReport* **2:**751–754.

Takechi, H., Onoe, H., Imamura, K., Onoe, K., Kakiuchi, T., Nishiyama, S., Yoshikawa, E., Mori, S., Kosugi, T., Okada, H., Tsukada, H., and Watanabe, Y., 1994, Brain activation study by use of positron emission tomography in unanesthetized monkeys, *Neurosci. Lett.* **182:**279–282.

Talairach, J., and Tournoux, P., 1988, *Co-Planar Stereotaxic Atlas of the Human Brain*, Thieme, New York, p. 122.

Tanaka, K., and Saito, H., 1989, Analysis of motion of the visual field by direction, expansion/contraction, and rotation cells clustered in the dorsal part of the medial superior temporal area of the macaque monkey, *J. Neurophysiol.* **62:**626–641.

Tanaka, K., Hikosaka, K., Saito, H., Yukie, M., Fukada, Y., and Iwai, E., 1986, Analysis of local and wide-field movements in the superior temporal visual areas of the macaque monkey, *J. Neurosci.* **6:**134–144.

Tanaka, K., Fukada, Y., and Saito, H., 1989, Underlying mechanisms of the response specificity of expansion/contraction and rotation cells in the dorsal part of the medial superior temporal area of the macaque monkey, *J. Neurophysiol.* **62:**642–656.

Tanaka, K., Sugita, Y., Moriya, M., and Saito, H., 1993, Analysis of object motion in the ventral part of the medial superior temporal area of the macaque visual cortex, *J. Neurophysiol.* **69:**128–142.

Thier, P., and Erickson, R. G., 1992, Responses of visual-tracking neurons from cortical area MST-L to visual, eye and head motion, *Eur. J. Neurosci.* **4:**539–553.

Tolhurst, D. J., Movshon, J. A., and Dean, A. F., 1983, The statistical reliability of signals in single neurons in cat and monkey visual cortex, *Vision Res.* **23:**775–785.

Tootell, R. B. H., and Taylor, J. B., 1995, Anatomical evidence for MT/V5 and other cortical visual areas in man, *Cerebral Cortex* **5:**39–55.

Tootell, R. B. H., Reppas, J. B., Dale, A. M., Look, R. B., Sereno, M. I., Malach, R., Brady, T. J., and Rosen, B. R., 1995a, Visual motion aftereffect in human cortical area MT revealed by functional magnetic resonance imaging, *Nature* **375:**139–141.

Tootell, R. B. H., Reppas, J. B., Kwong, K. K., Malach, R., Born, R. T., Brady, T. J., Rosen, B. R., and Belliveau, J. W., 1995b, Functional analysis of human MT and related visual cortical areas using magnetic resonance imaging, *J. Neurosci.* **15:**3215–3230.

Tootell, R. B. H., Reppas, J. R., and Rosen, B. R., 1995c, Functional analysis of human visual cortical areas V3, VP and V3A using magnetic resonance imaging, *Human Brain Mapping Suppl.* **1:**62.

Treue, S., and Andersen, R. A., 1993, Tuning of MT cells to velocity gradients, *Invest. Ophthalmol. Vis. Sci. Suppl.* **34:**813.

Ungerleider, L. G., and Desimone, R., 1986a, Projections to the superior temporal sulcus from the central and peripheral field representations of V1 and V2, *J. Comp. Neurol.* **248:**147–163.

Ungerleider, L. G., and Desimone, R., 1986b, Cortical connections of visual area MT in the macaque, *J. Comp. Neurol.* **248:**190–222.

Ungerleider, L. G., and Mishkin, M., 1979, The striate projection zone in the superior temporal sulcus of *Macaca mulatta:* Location and topographic organization, *J. Comp. Neurol.* **188:**347–366.

Ungerleider, L. G., and Mishkin, M., 1982, Two cortical visual systems, in: *The Analysis of Visual Behavior* (D. J. Ingle, R. J. W. Mansfield, and M. S. Goodale, eds.), MIT Press, Cambridge, MA, pp. 549–586.

Ungerleider, L. G., Desimone, R., Galkin, T. W., and Mishkin, M., 1984, Subcortical projections of area MT in the macaque, *J. Comp. Neurol.* **223:**368–386.

Vanduffel, W., Vandenbussche, E., Singer, W., and Orban, G. A., 1993, Bar orientation discrimination in the cat: A 2-deoxyglucose study, *Soc. Neurosci. Abstr.* **19:**772.

Vanduffel, W., Tootell, R. B. H., and Orban, G. A., 1995, Metabolic mapping of an orientation discrimination task in the macaque using the double-label deoxyglucose technique, *Soc. Neurosci. Abstr.* **21:**771.

Van Essen, D. C., Maunsell, J. H. R., and Bixby, J. L., 1981, The middle temporal visual area in the macaque: Myeloarchitecture, connections, functional properties and topographic organization, *J. Comp. Neurol.* **199:**293–326.

Van Essen, D. C., Anderson, C. H., and Felleman, D. J., 1992, Information processing in the primate visual system: An integrated systems perspective, *Science* **255:**419–423.

Vogels, R., and Orban, G. A., 1990, How well do response changes of striate neurons signal differences in orientation: A study in the discriminating monkey, *J. Neurosci.* **10:**3543–3558.

Vogels, R., and Orban, G. A., 1994, Activity of inferior temporal neurons during orientation discrimination with successively presented gratings, *J. Neurophysiol.* **71:**1428–1451.

Vogels, R., Spileers, W., and Orban, G. A., 1989, The response variability of striate cortical neurons in the behaving monkey, *Exp. Brain Res.* **77:**432–436.

Von Monakow, C., 1914, *Die Lokalisation im Grosshirn und der Abbau der Funktion Durch Kortikale Herde,* Bergmann, Wiesbaden, Germany.

Watson, J. D. G., Myers, R., Frackowiak, R. S. J., Hajnal, J. V., Woods, R. P., Mazziotta, J. C., Shipp, S., and Zeki, S., 1993, Area V5 of the human brain: Evidence from a combined study using positron emission tomography and magnetic resonance imaging, *Cerebral Cortex* **3:**79–94.

Weller, R. E., and Kaas, J. H., 1983, Retinotopic patterns of connections of area 17 with visual areas V-II and MT in macaque monkeys, *J. Comp. Neurol.* **220:**253–279.

Wurtz, R. H., Yamasaki, D. S., Duffy, D. J., and Roy, J.-P., 1990, Functional specialization for visual motion processing in primate cerebral cortex, in: *Cold Spring Harbor Symposia on Quantitative Biology,* Volume LV, *The Brain,* Cold Spring Harbor Laboratory Press, Cold Spring Harbor, NY, pp. 717–727.

Xiao, D.-K., Marcar, V., Raiguel, S., Maes, H., and Orban, G. A., 1993, Influence of stimulus duration on speed tuning of MT neurons, *Soc. Neurosci. Abstr.* **19:**769.

Xiao, D.-K., Marcar, V. L., Raiguel, S. E., and Orban, G. A., 1995a, Spatial heterogeneity in the surround of area MT for motion vectors differing in direction or speed from the CRF, *Soc. Neurosci. Abstr.* **21:**663.

Xiao, D.-K., Raiguel, S., Marcar, V., Koenderink, J., and Orban, G. A., 1995b, Spatial heterogeneity of inhibitory surrounds in visual area MT, *Proc. Natl. Acad. Sci. USA* **92:**11303–11306.

Xiao, D.-K., Marcar, V. L., Raiguel, S. E., and Orban, G. A., 1997, Selectivity of macaque MT/V5 neurons for surface orientation in depth, specified by motion, *Eur. J. Neurosci.* **9:**956–964.

Yamasaki, D. S., and Wurtz, R. H., 1991, Recovery of function after lesions in the superior temporal sulcus in the monkey, *J. Neurophysiol.* **66:**651–673.

Zeki, S. M., 1969, Representation of central visual fields in prestriate cortex of monkey, *Brain Res.* **14:**271–291.

Zeki, S. M., 1971, Cortical projections from two prestriate areas in the monkey, *Brain Res.* **34:**19–35.

Zeki, S. M., 1974, Functional organization of a visual area in the posterior bank of the superior temporal sulcus of the rhesus monkey, *J. Physiol.* (Lond.) **236**:549–573.

Zeki, S. M., 1978, Uniformity and diversity of structure and function in rhesus monkey prestriate visual cortex, *J. Physiol.* (Lond.) **277**:273–290.

Zeki, S. M., 1980, The response properties of cells in the middle temporal area (area MT) of owl monkey visual cortex, *Proc. R. Soc. Lond. B* **207**:239–248.

Zeki, S., Watson, J. D. G., Lueck, C. J., Friston, K. J., Kennard, C., and Frackowiak, R. S. J., 1991, A direct demonstration of functional specialization in human visual cortex, *J. Neurosci.* **11**:641–649.

Zohary, E., Celebrini, S., Britten, K. H., and Newsome, W. T., 1994a, Neuronal plasticity that underlies improvement in perceptual performance, *Science* **263**:1289–1292.

Zohary, E., Shadlen, M. N., and Newsome, W. T., 1994b, Correlated neuronal discharge rate and its implications for psychophysical performance, *Nature* **370**:140–143.

10

The Superior Temporal Polysensory Region in Monkeys

CATHERINE G. CUSICK

This chapter will review the progressive refinement of the concept of superior temporal polysensory (STP) cortex in macaque monkeys, describe the different organizational schemes in current use, and highlight a number of unresolved issues. Studies of the STP region have been limited by the difficulty of reconciling disparate physiological findings, suggesting, on the one hand, a role in visuospatial functions or eye movement control, and on the other, contributions to complex visual recognition functions. It is perhaps premature to assign STP cortex to either the dorsal or ventral stream of visual processing, thought to be crucial for visuospatial functions and object recognition, respectively, and the potential integrative functions of this region should not be ignored.

1. Definition, Location, and Extent of Superior Temporal Polysensory Cortex in Monkeys

1.1. Early Views of the Organization of Cerebral Cortex

Initial studies of functional localization in the mammalian cerebral cortex stressed a fundamental division of sensory cortex into modality-specific areas and

---

CATHERINE G. CUSICK • Department of Anatomy SL49, Tulane School of Medicine, New Orleans, Louisiana 70112.

Cerebral Cortex, Volume 12, edited by Rockland *et al.* Plenum Press, New York, 1997.

multimodal cortex. Electrophysiological mapping from the cortical surface revealed modality-specific, topographically organized units termed areas. For each sensory modality, a primary area (A-I, S-I, or V-I) was flanked by secondary areas that were also topographically organized and responded to a single sensory modality (e.g., Woolsey, 1964). Large intervening regions of cortex were thought to be true "association" areas, responsible for intersensory integration of modality-specific information. Early connectional evidence, based on comparatively large cortical lesions and degeneration methods for tracing pathways, supported the existence of three broad regions of polysensory cortex located in the posterior parietal, superior temporal, and parahippocampal regions (Jones and Powell, 1970).

Since the advent of microelectrode mapping methods and axon transport-based methods for tracing neuronal pathways, physiological and connectional evidence has accumulated that the cerebral cortex of most mammals is dominated by single sensory modalities and relatively little is association cortex in the original sense.* In macaque monkeys, it has been estimated that visual cortex comprises fully half of the surface area of the cerebrum and is composed of at least 24 separate areas. The region of cortex devoted to somatosensory cortex has grown to encompass at least a dozen areas, and the auditory cortex probably contains a comparable number (see Felleman and Van Essen, 1991, for review).

Along with the expansion of the cortical territories devoted to single sensory modalities, however, the existence of cortex with putative polysensory functions has been confirmed (Colby *et al.,* 1993; Oram *et al.,* 1993; Stein *et al.,* 1993; Desimone and Gross, 1979; see also Suzuki and Amaral, 1994). Thus, a current view would recognize polysensory areas as restricted portions of the parietal and temporal lobes that respond in two or three modalities. The validity of the concept of polysensory areas may still be called into question by obvious dominance of "polysensory cortex" by one of the modalities it represents. The posterior parietal, superior temporal, and parahippocampal areas, originally designated as bi- or multimodal, are dominated by the visual modality and may alternatively be considered high-order visual areas. The dichotomy of uni- and polysensory cortex may not be a fundamental, functional way of classifying cortex. For example, polysensory information may be used within different visual processing streams for global orientation or recognition of objects. Thus, polysensory areas may function separately or together, depending on the cortical processing strategy, and strict cross-modal integration of sensory information may not be their only or their main function.

The superior temporal polysensory (STP) area, with which this chapter deals, was originally described by Desimone and Gross (1979) as responding to visual, auditory, and somatic stimulation. The STP is a complex region of the brain, interesting not only as a possible site for integrating information from three sensory modalities, but also as a candidate area for integration of dorsal and ventral cortical visual streams. STP cortex occupies a large portion of the upper bank of the superior temporal sulcus (STS), but precise concepts of the size and location of STP have been modified in different ways by different investigators.

*It should be noted that while studies of the areal organization of human visual cortex are progressing rapidly (e.g., Kaas, 1995; Sereno *et al.,* 1995; Tootell and Taylor, 1995; Haxby *et al.,* 1991, 1994; Zeki *et al.,* 1991), it is uncertain how the proportion of polysensory association cortex in humans compares to the best-studied animal models.

The function of the STP region is still largely unknown, although ablation studies suggest participation in a number of visuospatial and complex discrimination functions (see section 2.4.). There are many fundamental issues of its organization that remain unresolved, including the extent of STP cortex, whether it contains separate cortical areas or modules, its position within hierarchical schemes of cortical processing, and its relations to the dorsal and ventral streams within the cortical visual system.

1.2. Early Studies of the Location and Extent of the Superior Temporal Polysensory Area of Monkeys

The STP region, as originally described (Desimone and Gross, 1979), encompassed most of the upper bank of the STS (Fig. 1). As so defined, the STP

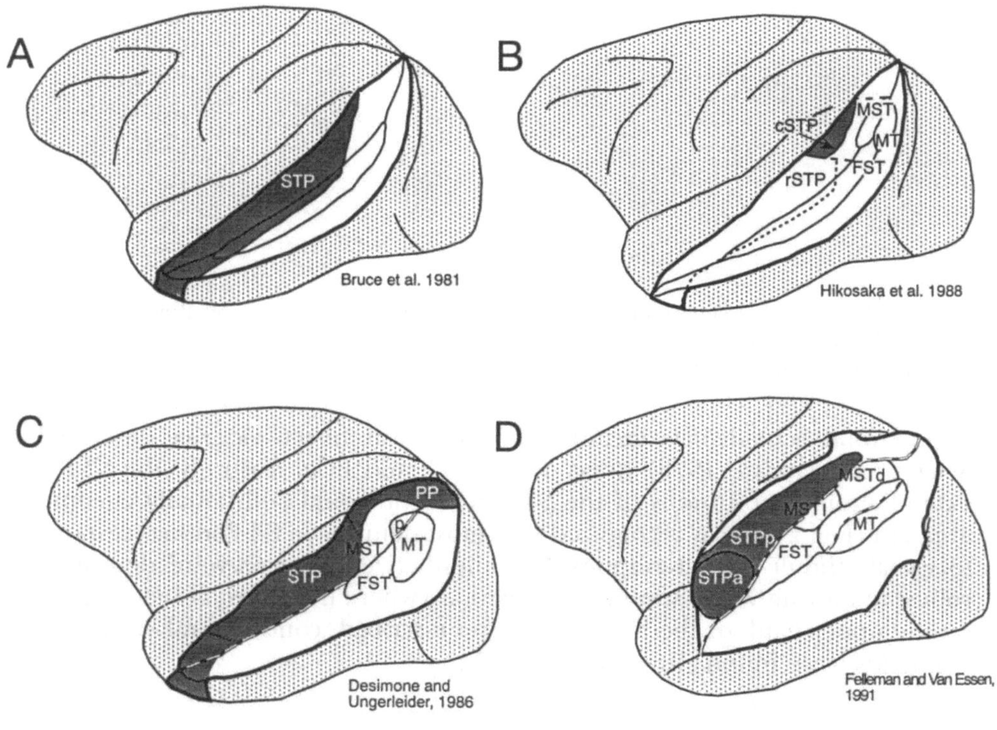

Figure 1. Different views of the organization of the superior temporal polysensory (STP) region in macaques. (A) Location of STP from Bruce *et al.* (1981). (B) Location of caudal superior temporal region from Hikosaka *et al.* (1988). (C) Areas of the upper bank and fundus of the STS according to Desimone and Ungerleider (1986), modified from Boussaoud *et al.* (1990). (D) Areas of the upper bank and fundus of the STS proposed by Felleman and Van Essen (1991). In each part, the cortex in the depths of the STS is shown in white and includes polysensory cortex, shown in dark gray. In parts A and B, the fundus of the STS is shown schematically in thinner lines; in parts C and D, the fundus is represented by a thicker dashed line. cSTP, Caudal superior temporal polysensory area; FST, fundal superior temporal area; MST, medial superior temporal area; MSTd, dorsal division of MST; MSTl, lateral division of MST; MT, middle temporal area; p, region thought to contain the peripheral field representation of area MT; rSTP, rostral superior temporal polysensory area; STPa, anterior division of STP; STPp, posterior division of STP.

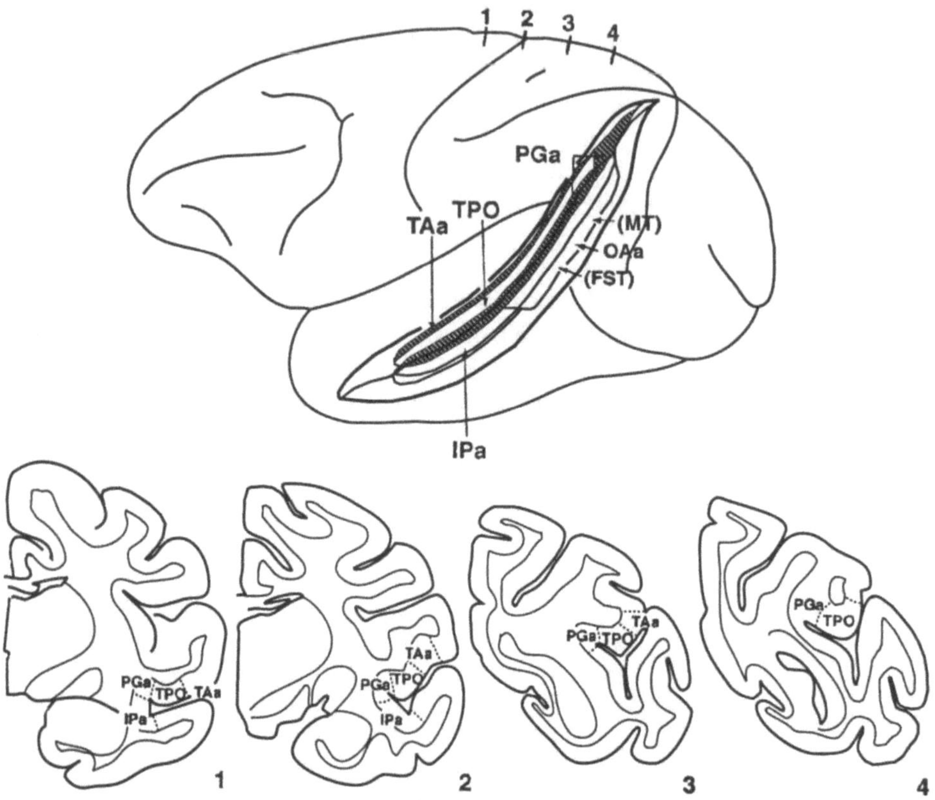

Figure 2. Location of architectonic areas of the STS defined by Seltzer and Pandya (1978). For clarity, only the areas of the upper bank and fundus of the STS are shown. The rostrocaudal levels of sections 1–4 are indicated at the top of the brain drawing.

region would actually have included three of the architectonic areas described by Seltzer and Pandya (1978), areas TAa, TPO, and PGa. Rostrally, the STP also extended onto the fundus of the STS, where it would additionally encompass area IPa (Fig. 2). It was recognized that most neurons in STP in fact responded to visual stimuli. Visual responses in anesthetized animals were described as robust to moving visual stimuli presented anywhere in the visual field. Caudally on the upper bank of the STS, more circumscribed, contralateral visual receptive fields were found (Bruce *et al.,* 1981).

1.3. Initial Revision of the Extent of the STP Region: Visual Areas of the Upper Bank of the STS

Soon after the discovery of STP cortex, it became apparent that the caudal portion of the upper bank of the STS also contained areas that would more accurately be described as visual than polysensory. Area MST, in the approximate location of caudal architectonic area PGa (Seltzer and Pandya, 1978), was first identified by Maunsell and Van Essen (1983) as a cortical region that received a fourth-layer projection from area MT. Areas MST and a small and variable region interpreted as the representation of the visual periphery within MT, termed MTp, were identified as visuotopically organized zones (Desimone and Ungerleider, 1986; Ungerleider and Desimone, 1986a; Fig. 1C). A densely myelinated zone extended to the lateral border of MST. By contrast, Desimone and Ungerleider (1986) found both visual and polysensory responses in the most lateral cortex of the upper bank of the STS, which they considered to be a part of posterior parietal cortex and termed PP.

Uncertainties remain regarding the precise border between MST and polysensory cortex, in part because unfolded summary views of cortex sometimes seem to imply that most of the upper bank of the STS constitutes area MST (e.g., Fig. 1C). In contrast, several physiological studies suggest that MST is restricted to the medial one-third to one-half of the STS on cross sections of the caudal STS (Saito *et al.*, 1986; Tanaka *et al.*, 1986). If MST is defined as the projection field of MT, MST would appear to occupy a limited, medial region extending from the upper bend of the STS to the lateral edge of the densely myelinated zone. If we assume that projections of the MTp zone can be used to identify the lateral border of MST (Ungerleider and Desimone, 1986b), then more than half of the upper bank of the STS could still be considered candidate STP cortex.

Interestingly, in three species of New World monkeys and in the prosimian galago, an MT projection zone has been identified in a similar medial relation to area MT and termed MST (Krubitzer and Kaas, 1990a). However, in squirrel and owl monkeys, the MST region appears to extend across the upper bank of the STS and well onto the superior temporal gyrus, leaving very little cortex that could occupy a position similar to the caudal STP region in those species. Thus, in several primate species, including macaques, the extent of caudal STS polysensory region remains unclear.

Distinguishing the STP region from MST is potentially complicated by the existence of other cortical areas adjoining the border of area MT (Kaas and Morel, 1993; Allman and Kaas, 1974) and of different subdivisions of MST itself. Dorsal (MSTd) and lateral (MSTl) subdivisions of MST have been proposed (Fig. 1D), which have separate representations of the fovea and may subserve different aspects of motion processing (Orban *et al.*, 1995; Graziano *et al.*, 1994; Tanaka *et al.*, 1993; Duffy and Wurtz, 1991a, b; Tanaka and Saito, 1989; Komatsu and Wurtz 1988a–c). The mixture of unimodal visual and polysensory responses of neurons in the adjacent portion of the STP region, by definition, contrasts with the purely visual responses of area MST (Hikosaka *et al.*, 1988; Desimone and Ungerleider, 1986). The unimodal visual cells in caudal STP comprise about 51% of the population and are not selective for visual form. Somewhat surprisingly, only about half of the visually responsive cells within polysensory cortex are reported to show directional selectivity for moving visual stimuli (Hikosaka *et al.*, 1988; Desimone and Gross, 1986), although 36% of visual responses were selective for movement in depth (Desimone and Ungerleider, 1986). A lower percentage of directionally selective neurons in caudal STP may help distinguish it from dorsal MST (MSTd) and might suggest that caudal STP is less directly involved in visual motion processing than MSTd. However, further studies are warranted, since area MSTd projects directly to adjacent polysensory cortex (area PP; Boussaoud *et al.*, 1990; Desimone and

Ungerleider, 1986) and such connections may convey information on motion processing.

2. Alliances of Superior Temporal "Polysensory" Cortex with Dorsal and Ventral Visual Processing Streams

One approach to understanding the functional organization of STP cortex is to view the STP region as a high-order component of the cortical visual system. Segregation of visual function within the parietal and temporal lobes of monkeys has been extensively reviewed elsewhere (Milner, 1995; Van Essen and Gallant, 1994; Merigan and Maunsell, 1993; Milner and Goodale, 1993; Felleman and Van Essen, 1991; Ungerleider and Mishkin, 1982; Trevarthen, 1968). Posterior parietal cortex appears to participate in several complex visual functions, including object localization (Ungerleider and Mishkin, 1982), oculomotor control (Colby and Duhamel, 1991; Andersen *et al.,* 1990), control of visually guided movements (Milner and Goodale, 1993; Hyvärinen, 1982; Mountcastle *et al.,* 1975), and processing of visual motion (Merigan and Maunsell, 1993). Inferotemporal visual cortex also subserves complex visual functions, but primarily those concerned with object recognition or visual memory (Milner and Goodale, 1993; Ungerleider and Mishkin, 1982). Visual information is transmitted to parietal and temporal cortex by numerous corticocortical relays arising within area 17. "Early" visual areas sometimes can be related to one or the other processing streams, but it is clear that interaction between streams occurs at early stages of the system and probably to some extent at all levels (e.g., Ferrera *et al.,* 1994; Leavitt *et al.,* 1994; Webster *et al.,* 1994; Krubitzer and Kaas, 1989, 1990b; see Merigan and Maunsell, 1993, and chapter by Schiller in this volume, for review).

At first glance it would seem that STP cortex is ideally situated anatomically to be a potential site of integration of the parietal and temporal processing streams. However, studies that have tested this idea directly, with injections of different tracers into posterior parietal and inferotemporal cortex of the same hemispheres, have identified overlapping label not within STP cortex, but rather near the fundus of the STS (Baizer *et al.,* 1991; Morel and Bullier, 1990). Based on the pattern of cortical connections, it seems plausible that the caudal and middle portions of the STP cortex are most closely associated with the functions of the dorsal visual pathway, and that integration of the parietal and temporal processing streams may occur within the rostral fundus and the most rostral portion of the upper bank of the STS (see Seltzer and Pandya, 1994, and section 2.1.).

2.1. Dorsal Stream Cortical Connections of STP Cortex

The major visual cortical inputs to the area TPO/STP region arise from areas associated with the dorsal pathway. There are presently no clear examples of fourth-layer inputs from the ventral pathway. In fact, the original definition of area TPO (as temporal-parietal-occipital association cortex; Seltzer and Pandya, 1978) recognized its strong feedforward inputs from posterior parietal cortex, especially the caudal inferior parietal lobule (area PGa, Seltzer and Pandya, 1978, 1980, 1994; area 7a, Andersen *et al.*, 1990; see also Cusick *et al.*, 1995; Neal *et al.*, 1990; Cavada and Goldman-Rakic, 1989a; Cipolloni and Pandya, 1989). The lateral intraparietal area (LIP), whose neurons can carry a large number of different visual and eye movement-related signals (Colby and Duhamel, 1991), also sends forward-going connections to caudal TPO (Blatt *et al.*, 1990; Seltzer and Pandya, 1994), as do STS areas associated with visual motion processing, MST and FST (Boussaoud *et al.*, 1990).

2.2. Pulvinar Connections

The thalamic projections to area TPO, as a whole, are predominantly with the medial pulvinar, mediodorsal, limitans-suprageniculate, and intralaminar nuclei (Yeterian and Pandya, 1989, 1991). Recent studies show that thalamic connections with dorsal and ventral stream areas are almost completely segregated within the pulvinar (Baizer *et al.*, 1993; Baleydier and Morel, 1992). Ventral stream, inferotemporal connections from areas TEO and TE apparently overlap the caudal pole of the visuotopically organized, striate sector of the pulvinar (termed inferior pulvinar, PI; Gutierrez and Cusick, 1997; Gutierrez *et al.*, 1995; Cusick *et al.*, 1993) or are immediately caudal to it. In contrast, the dorsal stream affiliated posterior parietal cortex lacks connections with PI, having connections mainly with the medial pulvinar, PM, and adjacent dorsal portions of the lateral pulvinar, PL (Baizer *et al.*, 1991; Baleydier and Morel, 1992; Hardy and Lynch, 1992). Since pulvinar connections of area TPO resemble most closely those of posterior parietal cortex and area TPO has almost no connections with the location of PI, these data provide further evidence for the association of area TPO with the dorsal stream of visual processing.

2.3. Cortical versus Subcortical Influences on STP Cortex

Early concepts of the organization of the mammalian visual system emphasized a role for the superior colliculus in visual orienting behavior (Schneider, 1967), a function which may also have contributions from the dorsal stream cortical areas in primates (e.g., Lynch *et al.*, 1977). Gross (1991) has reviewed the effects of two-stage lesions, of striate cortex followed by the superior colliculus, on response properties of STP neurons. Visual receptive fields in STP are normally large and nearly all bilateral. Following unilateral striate lesions, many cells in STP continue to respond to the damaged, contralateral hemifield. In fact, the percentages of bilateral and ipsilateral (from the intact hemifield) responses were nearly equal, although responses to moving stimuli in the damaged hemifield were notably weaker and less directionally selective. After removing the superior colliculus on the side of the striate lesion, however, the ipsilateral STP cortex loses all contralateral visual responses (Bruce *et al.*, 1986).

As Gross (1991) has emphasized, these results imply a dependence of STP cortex on *both* striate cortex *and* the superior colliculus for visual activation. These results are generally consistent with an association of STP cortex with the dorsal processing stream. However, it is important to consider the following apparent paradox. Primate superior colliculus neurons cannot provide STP cortex with information about direction of motion because they show relatively little directional selectivity (Goldberg and Wurtz, 1972). Thus, striate, not colliculus lesions must be mediating the impairment of directional selectivity in STP. However, striate cortex inactivation reportedly has only minimal effects on selectivity and sharpness of directional tuning area MT (Girard *et al.*, 1992; Rodman *et al.*, 1989). Apparently, in striate-lesioned animals the information about direction of stimulus motion, preserved within area MT, is not available to STP neurons. This is puzzling in view of known dorsal stream connections, from MT to MST and FST (Ungerleider and Desimone, 1986b), and thence to STP (Boussaoud *et al.*, 1990).

2.4. Behavioral Studies of Visuospatial and Object Recognition Functions after Ablation of STP Cortex

Following bilateral lesions of the STS, inattention to visual, auditory, and somatosensory stimuli has been reported (Luh *et al.*, 1986; Petrides and Iversen, 1979). However, it was not possible to interpret these studies as supporting a role of STS cortex in trimodal attention, because the very large lesions possibly extended into posterior parietal cortex. More recently, transient "trimodal neglect" has been reported in monkeys following lesions restricted to the STS, but not following inferior parietal lobule lesions (Watson *et al.*, 1994). Unfortunately, the STS lesions in that study were variable; the two lesions associated with the most severe postoperative neglect had damaged mainly the superior temporal gyrus and very little of the upper bank of the STS. Such global contralateral neglect, sometimes ascribed to lesions in STS polysensory cortex of monkeys, is a syndrome that cannot be conceptualized as a deficit in either dorsal or ventral stream processing. As pointed out by Milner (1995), an individual or an animal showing hemispatial neglect can identify neither "what" nor "where" an object is.

Control of eye movements is among the hypothesized functions of posterior parietal cortex (e.g., Lynch, 1992; Colby and Duhamel, 1991; Andersen *et al.*, 1990; Blatt *et al.*, 1990; Cavada and Goldman-Rakic, 1989b; Lynch and McLaren, 1989; Mountcastle *et al.*, 1975) and generally associated with the dorsal stream of visual processing. A recent report has shown latencies of saccadic eye movements are transiently increased to contralateral targets, especially peripheral ones, following large lesions of the upper bank of the STS (Scalaidhe *et al.*, 1995). Saccadic latencies along the vertical meridian or in the ipsilateral visual hemifield are unaffected. The effects are restricted to saccadic latency and no impairment is found in saccadic accuracy or the ability to fixate. Monkeys with STP lesions also show decreased speed of smooth pursuit eye movements. The impairments of saccadic eye movements consequent to STP lesions resemble those reported following ablations of posterior parietal cortex (Lynch and McLaren, 1989) as well as the frontal eye fields (FEF; Lynch, 1992). The recovery from transient deficits in saccadic eye movements (median, within 2 weeks) after STP lesions (Scalaidhe *et al.*, 1995) may reflect participation of several different cortical areas in visuomotor control. After sequential lesions of STP cortex followed by unilateral or bilateral lesions of FEF cortex, impairment of saccadic eye movements is greater than after either type of lesion alone (Skelly *et al.*, 1991). Thus, the STP region, among its other possible "visuospatial" functions associated with the dorsal stream of cortical processing, may normally participate with the FEF and posterior parietal cortex in control of eye movements.

Consistent with the view that STP cortex is not involved with ventral stream functions, several studies have found that large ablations of both banks of the STS do not impair visual discrimination or performance on a cross-modal (visual–tactile) recognition task (Streicher and Ettlinger, 1987; Ettlinger and Garcha, 1980; Petrides and Iversen, 1979). By contrast, other studies have reported discrimination deficits. For example, large lesions of the STS cortex produced a long-lasting (1 year postoperative) deficit in *acquisition* of a visual–auditory association task (Petrides and Iversen, 1978). Moreover, deficits in visual discrimination learning after STS lesions have been found and are described as about equal for stimulus shape and orientation (Eacott *et al.*, 1993). Large STS lesions are also reported to impair discriminations of the direction of eye gaze in photographs of faces (Eacott *et al.*, 1993; Campbell *et al.*, 1990), a deficit interpreted by some as a nonspecific decrement in visual discrimination ability (Eacott *et al.*, 1993). On the other hand, others have emphasized that these results may represent an impairment in recognizing pattern orientation rather than pattern per se (Scalaidhe *et al.*, 1995).

Conclusions on the role of STP cortex in visual pattern discrimination based on the above-mentioned studies remain controversial because lesions included the lower bank or rostral fundus of the STS (Eacott *et al.*, 1993; Campbell *et al.*, 1990; Petrides and Iversen, 1978), regarded as a component of inferotemporal (IT) cortex (see Felleman and Van Essen, 1991). Nevertheless, connectional evidence suggests the possibility that the most rostral portion of the upper bank of the STS, as the area with strongest links to IT cortex (Seltzer and Pandya, 1994), may function within the ventral stream of visual processing or be an area of potential convergence of the two visual streams. Overall, the lesion studies support a role for STP cortex in visuospatial functions, confirm an association with the dorsal stream of visual processing, and provide further evidence of a network of areas subserving visuospatial functions with centers in the frontal eye fields, posterior parietal cortex, and STP region (Morecraft *et al.*, 1993; Seltzer, 1992).

2.5. Neurophysiological Evidence for Functional Heterogeneity within Polysensory Cortex of the Upper Bank of the STS

Initial studies in anesthetized animals described visually responsive neurons in rostral and caudal portions of STP to be insensitive to stimulus form (Hikosaka *et al.*, 1988; Bruce *et al.*, 1981). In contrast, studies in awake animals indicated neurons in rostral portions of the STS to be concerned with form identification, particularly of faces (Baylis *et al.*, 1987; Perrett *et al.*, 1982). These apparently contradictory observations have led to two major hypotheses on the role of the STP region in visual behavior.

The first hypothesis, as proposed by Bruce *et al.* (1981), is that the STP cortex is a multisensory integration area subserving orienting behavior to novel stimuli. The visual functions of STP would then be considered as primarily concerned with visuospatial analysis rather than integration of aspects of object form. Indeed, somatosensory responses of neurons in the rostral portion of the STS of awake animals appear to depend on stimulus novelty, supporting the view that STP is involved in orienting behavior (Hietanen and Perrett, 1993; Mistlin and Perrett, 1990). A majority of these somatosensory cells respond in a direction-selective manner to visual stimuli and are insensitive to visual form. Furthermore, in a preliminary report, Colby and Miller (1986) describe many cells within STP of awake monkeys as showing elevated discharge rates before and during saccadic eye movements. They identify visual, motor, and visuomotor neurons by time-locked firing patterns, as well as neurons whose activity is modulated by smooth pursuit eye movements. These findings, together with known connections of the STP region with prefrontal and posterior parietal cortex (Seltzer and Pandya, 1994; Cusick *et al.,* 1995), suggest that STP neurons may participate in directed attention and process information related to eye movements.

The second school of thought is that the upper bank of the STS, especially rostrally, is concerned with complex visual recognition functions. Many neurons in that vicinity are reported to be strongly activated by faces (Perrett *et al.,* 1988; Baylis *et al.,* 1987; Perrett *et al.,* 1982), specific views of the head (Perrett *et al.,* 1985b), and moving heads or body parts (Perrett *et al.,* 1985a). For the majority of face-selective cells within the fundus of the STS, however, rotational movements produce diminished responses or abolish them altogether, and the great majority of such cells are not responsive to either auditory or somatic stimuli (Perrett *et al.,* 1982; 1985b; 1988). Such purely visual responses suggest these cells may be located outside the STP area (Bruce *et al.,* 1981; Desimone and Gross, 1979). The role of STP cortex in object, or moving object, recognition is complicated by the difficulty of precisely localizing recording sites in awake, behaving animals.

Different views of the role of STP cortex may be partly reconciled by the recent confirmation, using awake animals, that an anterior STP region exists in which neurons are highly sensitive to visual motion and insensitive to visual form (Oram *et al.,* 1993). Since sensory responses within the upper bank of the STS tend to be highly clustered (Oram *et al.,* 1993; Perrett *et al.,* 1992; Mistlin and Perrett, 1990), it remains possible that the region of the STS in which face recognition cells are found is largely rostral and medial to the STP area, which contains visual motion-sensitive cells that are not selective for stimulus form.

While functional heterogeneity in STP cortex and segregation of response types may explain some of the different physiological results, it is probably too simplistic to predict exclusive selectivity for visual form or visual motion. Neurons responsive to different angular views of faces or to moving faces have been reported in a relatively caudal location (Perrett *et al.,* 1985a, b) that may have overlapped the "classical" STP region. Such selectivity could be involved in recognizing the same object in motion or in different orientations. Form-and-movement-selective cells, unresponsive to static stimuli, have also been reported in the upper bank, fundus, and medial lower bank of the anterior STS (Perrett *et al.,* 1985a). Such characteristics might be expected within an area in which dorsal and ventral processing streams converge. It is tempting to speculate that this physiological convergence zone lies outside of STP/TPO cortex, in the rostral fundus of the STS, where connections from posterior parietal and inferotemporal cortex converge (Baizer *et al.*, 1991; Morel and Bullier, 1990). Clearly, there is a large number of issues to be addressed on the functional architecture of STP cortex.

3. Subparcellation of the STP Region: Alternative Views

Along with the refinement of concepts of the location of STP cortex, evidence has accumulated for further morphological/functional subdivisions of STP cortex. Some of these subdivisions, which may constitute separate cortical areas, are on the order of size of areas MT and MST (Fig. 1B–D). In order to illustrate the discrepancies that motivated our recent efforts to subdivide STP cortex on connectional/chemoarchitectonic grounds, previous concepts of the areal organization of the STP region are discussed in detail below.

3.1. Caudal and Rostral STP

As reviewed above, the original descriptions of the STP region noted that neurons in caudal STP had smaller receptive fields, encompassing less than the entire visual field, while rostral STP contained neurons that were robustly activated from moving stimuli anywhere in the visual field (Bruce *et al.*, 1981; Desimone and Gross, 1979). Hikosaka *et al.* (1988) termed this zone a separate cortical area, caudal STP (cSTP), and studied its neuronal responses in anesthetized animals. They interpreted this region as concerned with global alerting of the animal to its environment, as opposed to focal attention to particular stimuli.

Hikosaka *et al.* (1988) described the upper bank of the STS as having three approximately equal mediolateral zones (Fig. 1B). The inner (medial) third corresponds to the strictly visual MST, the middle third is mostly unresponsive to sensory stimuli, although some visually driven responses are observed there, and the outer (lateral) third corresponds to cSTP. The majority of cSTP neurons exhibit unimodal visual, somatosensory, or auditory responses. All three possible types of bimodal cells are observed, as well as a small number of trimodal cells. Although visual receptive fields in cSTP are quite large, with a mean diameter of 59°, a surprising two-thirds are restricted to the contralateral hemifield. Auditory spatial receptive fields are large, mostly contralateral, and respond to a variety of complex stimuli. Somatosensory receptive fields in cSTP are often less than the entire body surface, located mainly on the trunk, and sensitive to light cutaneous stimuli. Thus, two STP areas can be defined physiologically, since the more restricted sensory receptive fields in cSTP allow this area to be distinguished from the large rostral STP area, rSTP.

3.2. PP and STP Proper

The PP area (Fig. 1C) described by Desimone and Ungerleider (1986) is a polysensory area considered to be an extension of posterior parietal cortex that folds into the caudal upper bank of the STS. PP matches closely the size and location of cSTP of Hikosaka *et al.* (1988). In the organizational scheme proposed by Desimone and Ungerleider (1986), the STP area lies entirely rostral to area MT and occupies about the anterior two-thirds of the upper bank of the STS.

3.3. Connectional/Architectonic Subdivisions of "Area TPO" (TPO-1–TPO-4)

Following their original analysis of the cyto- and myeloarchitecture of STS cortex, Seltzer and Pandya (1989a b, 1991, 1994) embarked on a series of combined architectonic–connectional studies. As a result of these studies, they described differential connections for four rostrocaudal sectors of area TPO. From rostral to caudal, these divisions were termed areas TPO-1, TPO-2, TPO-3, and TPO-4, and were schematically shown as approximately equal in size. The exact boundaries between individual putative divisions and their locations relative to better known visual areas such as MT and MST have not been closely specified. One of the most intriguing observations in these studies is that restricted deposits of tracers within area TPO reveal, at several different rostrocaudal levels, patterns of fourth-layer connections rostral to the injection site. Concomitantly, connections to layers I and VI are found caudal to the injection site (Seltzer and Pandya, 1989b). These different laminar patterns of connections within area TPO raise the possibility that the original area TPO (Seltzer and Pandya, 1978; Fig. 2) contains separate cortical areas.

One way to understand possible roles of STP cortex in behavior is to consider what better known areas interconnect with it. The connections of area TPO as a whole are extremely diverse (see Seltzer and Pandya, 1994, for review), and the purpose here is not to present a complete catalog, but to highlight connections that may reveal functional differences and support the concept of separate areas within STP cortex. Along its rostrocaudal extent, area TPO has connections with visual areas of the caudal STS, MST, and fundal superior temporal area (FST) (Boussaoud *et al.*, 1990); the superior temporal gyrus; cortex of the lower bank of the lateral sulcus from the supratemporal plane to the insula; portions of the posterior parietal cortex including the exposed cortex of the inferior parietal lobule (area 7a or caudal PG/area OPt), medial parietal lobe (area PGm), and lateral bank of the intraparietal sulcus (area POa or LIP); medial "paralimbic" cortex (areas 23, 24, and retrosplenial cortex); the parahippocampal region; and extensive portions of prefrontal cortex (Seltzer and Pandya, 1989a, 1994).

Clear connectional differences exist among the caudal, middle, and rostral TPO divisions (Fig. 3). For example, only TPO-4 is reported to have connections with area POa (area LIP; Anderson *et al.*, 1990) and the medial surface of the parietal lobe (PGm; (Seltzer and Pandya, 1994). Major differences also exist in connections of the rostralmost division TPO-1 compared to the remainder of area TPO. The caudal zones each receive strong feedforward connections from area 7a, and thus may be associated with the dorsal visual pathway. In contrast, TPO-1 is reported to lack input from posterior parietal cortex (Seltzer and Pandya, 1994) and instead receives an indirect relay of ventral stream informa-

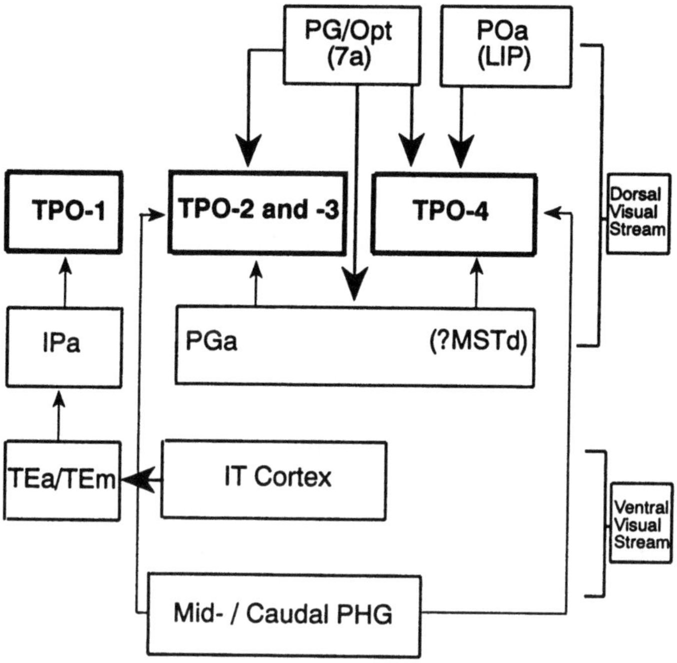

Figure 3. Differential connection patterns of rostrocaudal subdivisions of architectonic area TPO from Seltzer and Pandya (1994). Fourth-layer projections are shown as large and "lateral" projections as small arrows.

tion from inferotemporal areas TEa/TEm (Fig. 3). This occurs via a relay through area IPa, in the fundus of the STS (Seltzer and Pandya, 1994). A further difference from the other subdivisions is that TPO-1 connects with rostral portions of the superior temporal gyrus, while caudal parts of area TPO connect with superior temporal gyrus (STG) auditory association areas.

3.4. Other Proposals: STPp, STPa

As described above, the parcellation of Ungerleider and colleagues (e.g., Boussaoud *et al.,* 1990) contains two polysensory subdivisions, PP in the caudal third and STP in the rostral two-thirds of the upper bank of the STS. Focusing on the different connections along the upper bank of the STS, Felleman and Van Essen (1991) also proposed two rostrocaudal subdivisions, but with their relative sizes reversed (Fig. 1D). They distinguished posterior and anterior STP areas, with the larger, STPp, adjacent to MSTd, MSTl, and FST, while STPa occupied about the rostral one-third of the upper bank. This parcellation appears to be based mainly on the connections of the dorsal division of anterior inferotemporal cortex (area IPa of Seltzer and Pandya, 1994; or AITd of Felleman and Van Essen, 1991; Fenstemaker, 1986) with STPa and not STPp. In addition, MSTd has connections throughout Felleman and Van Essen's (1991) STPp region, including PP (Boussaoud *et al.,* 1990).

These differing schemes for subdividing the STP cortex underscore some of the persistent difficulties in identifying valid subdivisions of this region. As a common theme, each of the different ways of dividing the STP region emphasizes rostrocaudal differences in cortical connections. In order to reconcile the various descriptions, reexamination of the structural and connectional organization of the upper bank of the STS is needed.

Our recent studies (Cusick *et al.*, 1995; Seltzer *et al.*, 1996) have focused on evidence for separate areas within presumptive polysensory cortex. We provisionally term this polysensory cortex area TPO, retaining the original name for the largest and best-distinguished architectonic zone of the upper bank of the STS. As discussed above, differential patterns of connections support the view that the larger, caudal part of area TPO receives few connections from the ventral visual stream and major connections from the dorsal. We have provisionally removed the rostralmost part of the upper bank of the STS from area TPO based on its distinct architecture and connectional relations.

4. A Revision of the STP Region Based on Chemoarchitecture and Connections with Other High-Order Cortical Areas

A major obstacle to understanding the organization of the upper bank of the STS has been the difficulty of reliably identifying the same cortical area in different hemispheres. Analysis of Nissl- and myelin-stained material suggests the architecture of the upper bank of the STS is heterogeneous, but in places the distinctions are subtle (Seltzer and Pandya, 1989b). Therefore, we undertook a different approach to the architectonics of the STP region. Previous work had shown that neurochemical markers, including cytochrome oxidase histochemistry and Cat-301 immunoreactivity, had the potential to reveal different functional streams and even novel cortical subdivisions (DeYoe *et al.*, 1990; McGuire *et al.*, 1989; Tootell *et al.*, 1985; Livingstone and Hubel, 1984). Our objective was to provide a consistent neurochemical map of the STS visual and polysensory areas, using markers that might indicate the distribution of certain classes of pyramidal cells (SMI-32 staining; Hof and Morrison, 1995; Campbell and Morrison, 1989) and the density patterns of thalamocortical axons (parvalbumin immunocytochemistry; Blümcke *et al.*, 1991; DeFelipe and Jones, 1991; Jones and Hendry, 1989). In order to focus on regional staining differences that would presumably have functional significance, the chemoarchitectonic analysis was correlated with studies of connections that would potentially distinguish the locations of polysensory versus unimodal visual areas of the STS (Seltzer *et al.*, 1996; Cusick *et al.*, 1995).

4.1. Revision of Architectonic Subdivisions of the Upper Bank and Fundus of the STS

The present chemoarchitectonic/connectional parcellation of the upper bank of the STS is shown in Fig. 4, and the appearance of these three divisions in SMI-32 and parvalbumin stains is shown in Figs. 5–7. The caudal, intermediate, and rostral subdivisions of area TPO (TPOc, TPOi, and TPOr) are of unequal sizes, TPOr being the largest and TPOi the smallest subdivision (Fig. 4). The term PGa has not been used because caudally it appears to correspond at least in part to area MST (Seltzer and Pandya, 1994). Rostrally, area PGa and area IPa have been combined, on the grounds that a single distinct architectonic area appears to occupy the fundus of the STS and adjacent portions of the upper and

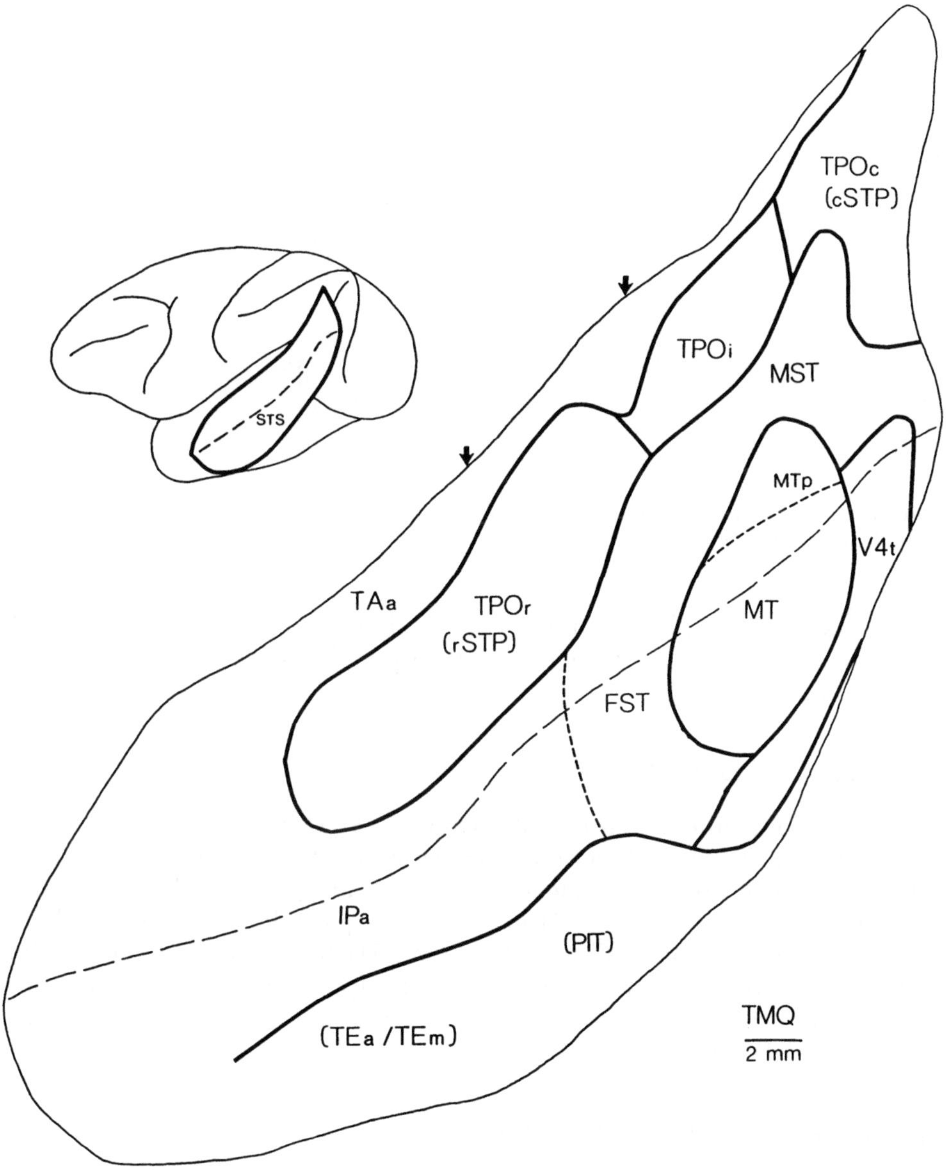

Figure 4. Architectonic areas of the STS identified by parvalbumin and SMI-32 immunostaining and patterns of corticocortical connections. The arrows denote the location of a small accessory gyrus on the upper bank of the STS. cSTP, Hypothesized location of the caudal superior temporal polysensory area; PIT, posterior inferotemporal area; rSTP, hypothesized location of the rostral superior temporal area. From Cusick *et al.* (1995).

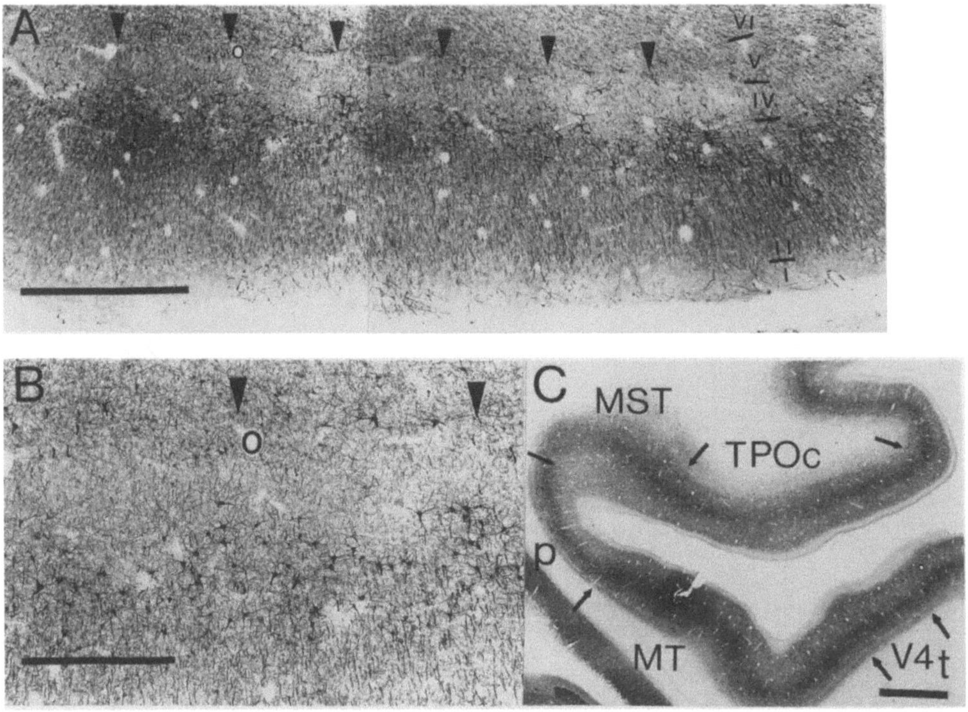

Figure 5. Chemoarchitecture of TPOc shown by SMI-32 (A, B) and parvalbumin (C) immunostaining. Scale: (A) 1 mm; (B) 0.5 mm; (C) 2 mm.

lower banks (Fig. 8). The extent of area TAa has also been modified, since this zone is most distinct along the middle of the upper bank of the STS and may not form a continuous striplike area.

4.1.1. Caudal Area TPO (TPOc)

This region, immediately lateral to caudal MST (MSTd), extends to near the rostral edge of area MT. A periodic pattern is clearly visible in SMI-32-stained sections, in which unstained patches in layer IV are partly enclosed by narrow gaps containing invading layer III neurons and apical dendrites of layer V cells (Fig. 5A). Thus, SMI-32-positive cells in layer III form a scalloped edge along layer IV (Fig. 5B). TPOc also shows a periodic appearance in parvalbumin immunostaining, corresponding to variations in the density of the fiber plexus in layers III and IV (Fig. 5C).

4.1.2. Intermediate Area TPO (TPOi)

With both markers, the intermediate subdivision of area TPO begins near the rostral border of area MT with an abrupt loss of the obvious periodicity in TPOc. TPOi occupies a short rostrocaudal distance (about 4 mm) along the STS. In SMI-32 staining, TPOi shows a distinct band of cells in layer III, a medium number of layer V cells, and a moderately thick layer IV (Fig. 6A). In parvalbumin stains, TPOi is lighter than adjacent cortex of the superior temporal gyrus and area MST (MSTl; Fig. 6B).

4.1.3. Rostral TPO (TPOr)

The pattern of SMI-32-stained cells in TPOr is bilaminar, in layers III and V, but compared to adjacent areas TPOi, TAa, and IPa there is a marked decrease in the number of immunopositive cells in layer V (Fig. 7A). SMI-32 staining in layer III is not completely homogeneous, although a patchy appearance, obvious in TPOc, is less pronounced in TPOr. In parvalbumin stains, TPOr has a thicker layer IV than TPOi (Fig. 7B). Rostral to TPOr, there is a transition to a distinct region of cortex with a narrow, less-well-developed parvalbumin fiber plexus in layer IV. In SMI-32 stains, the thin layer IV contains numerous immunoreactive pyramidal cells. This rostral border of TPOr also corresponds to that estimated from connections with prefrontal cortex.

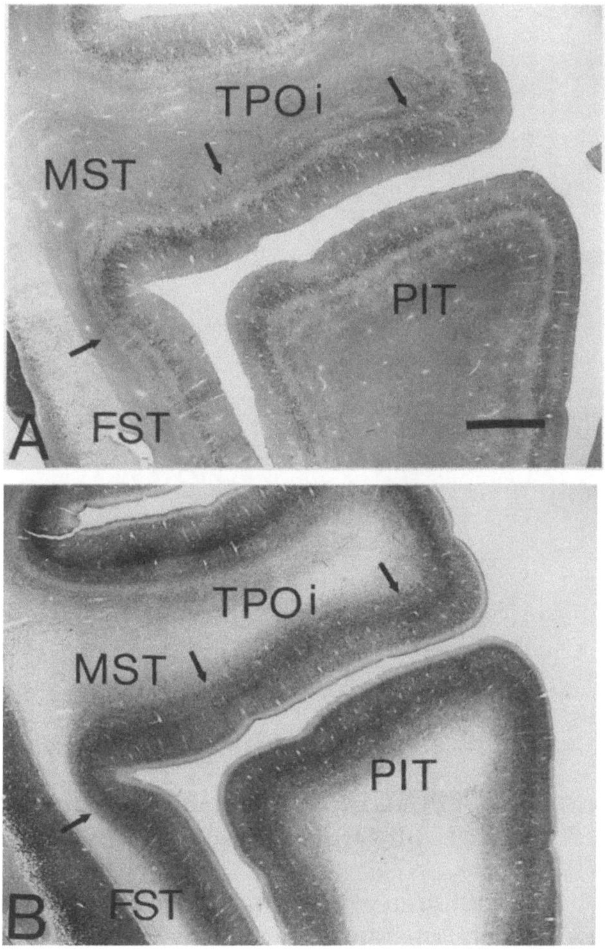

Figure 6. (A) SMI-32 and (B) parvalbumin immunostaining of adjacent sections at the level of area TPOi. The scale is 2 mm.

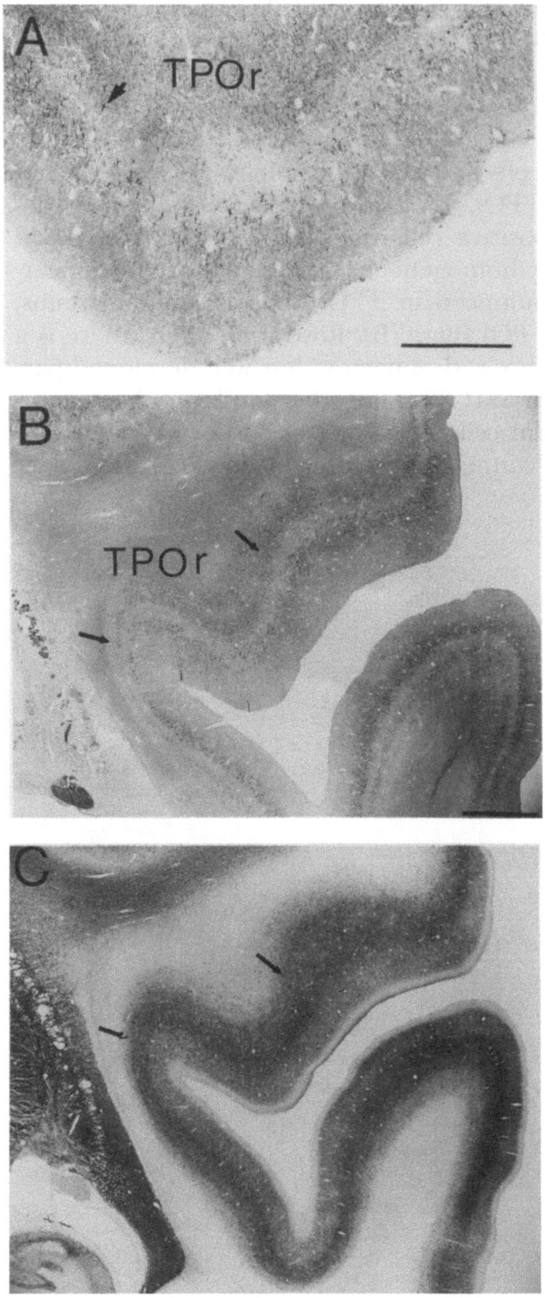

Figure 7. (A, B) SMI-32 and (C) parvalbumin immunostaining of area TPOr. Scale for A is 1 mm; for the adjacent sections in B and C, 2 mm.

4.2. Area TPO Can Be Identified by Distinctive Connections with "High-Order" Association Cortex

A useful strategy for initially identifying separate cortical areas has been to trace widely separated connections from a better known cortical subdivision (e.g., Steele *et al.*, 1991; Cusick and Kaas, 1988; Ungerleider and Desimone, 1986a, b; Andersen *et al.*, 1985; Maunsell and Van Essen, 1983). A second important strategy for distinguishing different areas relies on different laminar patterns of connections. Three laminar patterns of axonal terminations have been described: a "feedforward" connection type focused on layer IV, a "feedback" connection avoiding layer IV and concentrated in layers I and VI, and a "columnar" or "lateral" pattern that terminates with similar density throughout all cortical layers (Felleman and Van Essen, 1991; Seltzer and Pandya, 1989b; Maunsell and Van Essen, 1983; Rockland and Pandya, 1979). For visual cortex, termination patterns between large numbers of different areas have been examined, and support the conclusion that projections to one cortical area from another exhibit only one type of laminar pattern (Felleman and Van Essen, 1991). As a cautionary note, this conclusion applies to bulk labeling of pathways and may not accurately describe the distinguishable termination patterns of individual axons (see chapter by Rockland in this volume). With this caveat, two

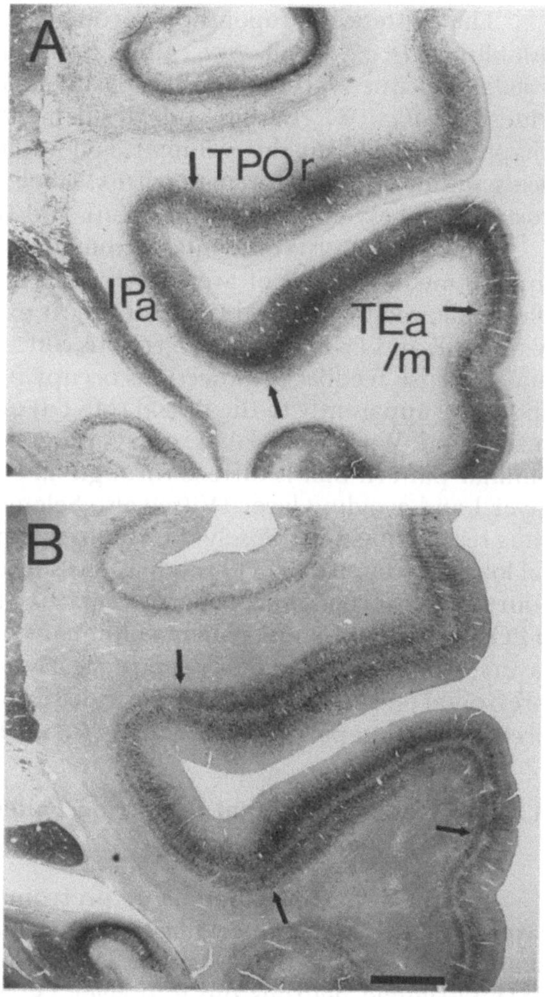

Figure 8. Architectonic appearance of area IPa in the rostral fundus of the STS. (A) Parvalbumin, (B) SMI-32 immunostaining. The scale is 2 mm. From Cusick *et al.* (1995).

projection foci with different laminar patterns from one cortical area can reasonably be interpreted as occupying separate cortical areas.

Based on laminar differences in connections, hierarchical schemes have been developed to describe interconnections of visual areas (Young, 1992, 1993; Felleman and Van Essen, 1991; Maunsell and Van Essen, 1983). If such schemes can be extended beyond the early stages of visual processing, laminar projections from high-order areas may help resolve the longstanding problem of distinguishing STP cortex from adjacent visual areas. In the case of dorsal prearcuate cortex, for example, areas MST and MT are predicted to receive "feedback" connections from prefrontal cortex, presumed to be at a "higher" cortical level. By contrast, STS polysensory areas, thought to be nearer the same level as prefrontal cortex, might be predicted to exhibit a columnar or other pattern of terminations.

4.2.1. Connections of Prefrontal Cortex Discriminate Area TPO from MST and Other STS Visual Areas

The existence of topographic frontal connections to the STP region is well established (e.g., Schall *et al.,* 1995; Stanton *et al.,* 1995; Seltzer and Pandya, 1989a; Leichnetz, 1989; Barbas, 1988; Selemon and Goldman-Rakic, 1988; Huerta *et al.,* 1987; Barbas and Mesulam, 1981; Künzle and Akert, 1977; see chapter by Schall in this volume). Following injections of radiolabeled amino acids within dorsal prearcuate cortex, labeled terminations occur over a broad rostrocaudal extent of the upper bank and fundus of the STS (Fig. 9).

Laminar patterns of frontal terminations are potentially useful for identifying the medial border of area TPO along its length, beginning with MST caudally, followed by FST and IPa rostrally. As expected, in the STS visual areas MT, V4t, MST, and FST, frontal projections concentrate outside layer IV, in layers VI and I. Dense feedback connections occupy two distinct locations in MST, corresponding apparently to the dorsal (MSTd) and lateral (MSTl) subdivisions (Komatsu and Wurtz, 1988a). Frontal projections to area IPa show an unusual laminar pattern that is diffuse through all cortical layers, but clearly densest in layer I and avoiding layer IV (see also Selemon and Goldman-Rakic, 1988), and thus resembling most closely the feedback type of connection. In contrast, frontal lobe projections to TPO subdivisions show roughly alternating columnar and fourth-layer termination patterns (Fig. 9). The caudal- and rostralmost foci, in TPOc and TPOr, show patchy columnar connections throughout the cortical layers, while label in TPOi is more clearly focused on layer IV. The different prefrontal termination patterns provide support for the view that subdivisions of area TPO are separate cortical areas (see below).

4.2.2. Different Prefrontal Projection Patterns to Areas TPOc, TPOi, and TPOr

In TPOc, narrow columns of prefrontal projections traverse the cortical layers in TPOc, with a periodicity that resembles the patchy SMI-32 immunostaining. The projections in layer IV often coincide with locations where the SMI-32-stained neurons dip into layer IV and associate with the edges of the

SMI-32-poor patches in layer IV (Fig. 10). However, the frontal columns and the immunostaining pattern do not correspond completely, possibly due to a nonoptimal plane of section. Within TPOi, prefrontal connections focus markedly on layer IV and are reduced in layer V, as is typical of a forwardgoing connection (Fig. 11). Prefrontal projections to TPOr form semiregular patches in a columnar laminar pattern, with increased density in layer IV in places. Anterior to TPOr, label is greatly reduced or absent. The semiregular patches of label seen in areas TPOc and TPOr suggest a modular organization of these

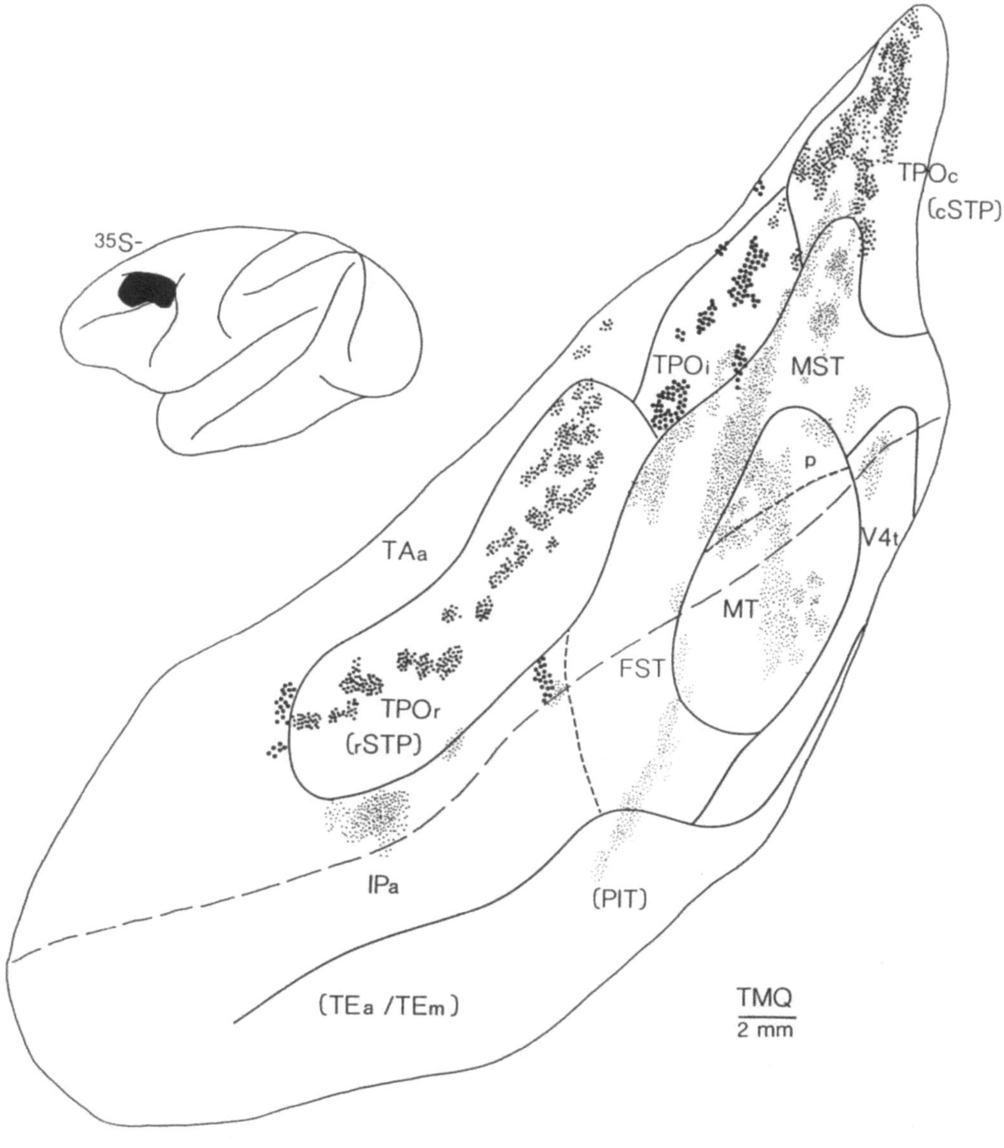

Figure 9. Distribution of prefrontal terminals shown on an unfolded reconstruction of the banks of the STS. Feedback connections are shown as fine dots, lateral connections as medium dots, and fourth-layer projections as large dots. From Cusick *et al.* (1995).

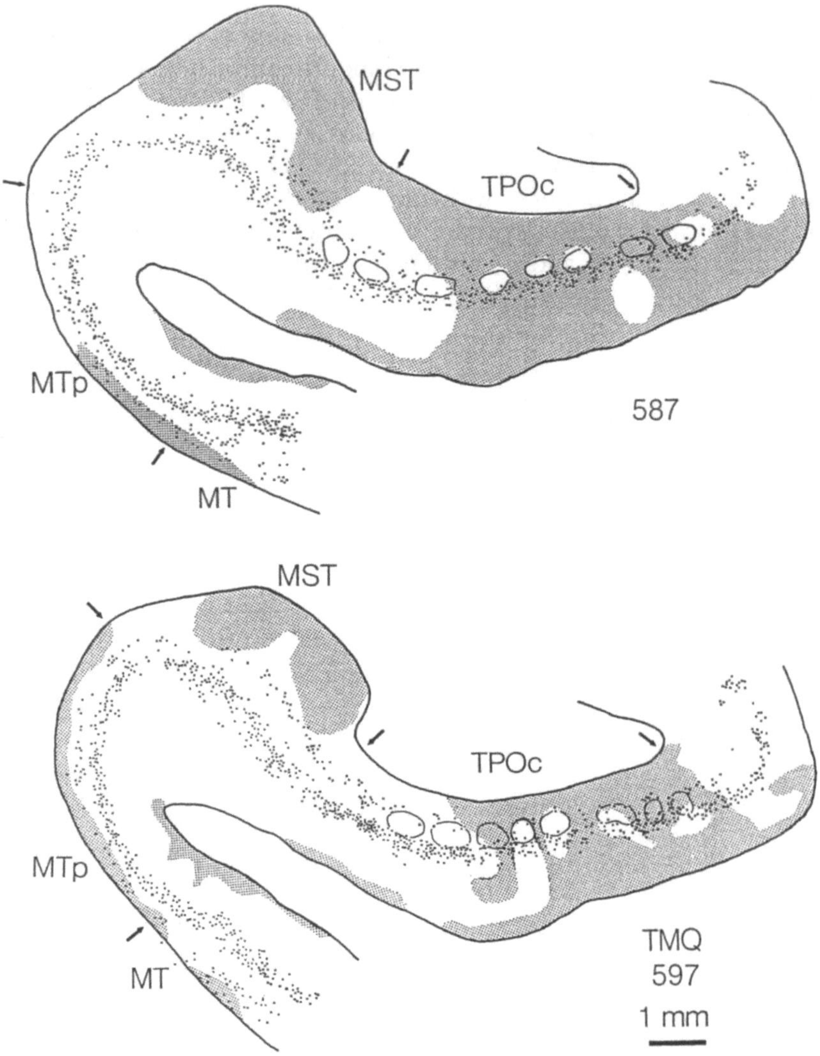

Figure 10. The relation of prefrontal terminations (shading) to the pattern of SMI-32-stained cells (dots) and more lightly stained neuropil (ovals) in TPOc. From Cusick *et al.* (1995).

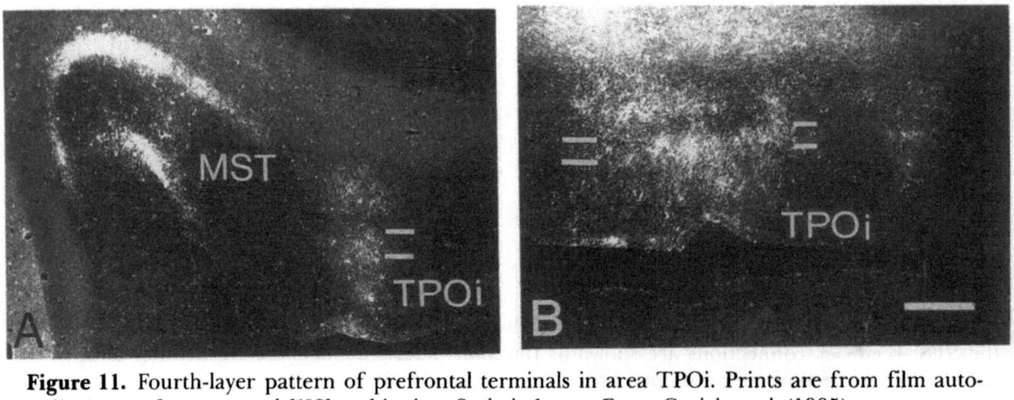

Figure 11. Fourth-layer pattern of prefrontal terminals in area TPOi. Prints are from film autoradiograms of transported [^{35}S]methionine. Scale is 1 mm. From Cusick *et al.* (1995).

zones, and the patch size of about 250 μm may be a fundamental feature of corticocortical connections (Amir *et al.*, 1993; Lund *et al.*, 1993).

The overall pattern of columnar and fourth-layer connections to the upper bank of the STS appears to identify the polysensory cortex. The question remains whether the different architectonic–connectional zones within "area" TPO are separate areas or constitute anatomical modules within a single area. In the present concept of area TPO, the entire region receives strong connections from the more dorsal prefrontal cortex. Area TPO-1 of Seltzer and Pandya (1989a, b) has not been included in the present scheme (Seltzer *et al.*, 1996; Cusick *et al.*, 1995), since it appears to be rostral to the dorsal prefrontal projection zone and receives input from ventral visual pathways. The exact sources of the prefrontal projections to TPO subdivisions are uncertain, because tracer injections involved most of the dorsal wing of area 46, area 8a (Walker, 1940), and a minor extent of prearcuate cortex lateral to the principal sulcus. The injections thus encompassed several distinct functional areas, including both the large and small saccade portions of the FEF (Schall *et al.*, 1995; Stanton *et al.*, 1995) as well as the more rostral prearcuate cortex. More dorsal FEF, from which large saccades are elicited, has been shown to be connected with caudal portions of STP cortex (chapter by Schall in this volume; Schall *et al.*, 1995; Stanton *et al.*, 1995; Huerta *et al.*, 1987). The more ventral FEF, representing short saccades, appears to have fewer connections with the upper bank of the STS (Schall *et al.*, 1995; Stanton *et al.*, 1995). Thus, the TPO subdivisions may have differential connections, yet to be examined, with separate functional zones of the frontal lobe.

4.2.3. Projections to the STP Region from the Inferior Parietal Lobule and Superior Temporal Gyrus Also Identify Area TPO and Its Subdivisions

Many previous studies (reviewed above) have identified projections from posterior parietal cortex to the upper bank of the STS. As presently defined, the entire area TPO receives inputs from the inferior parietal lobule (IPL). Injections of tracers that infiltrate most of the exposed surface of the IPL and adjacent parts of the lower bank of the IPS produce columnar terminal label in TPOc and label focused on layer IV in TPOi and TPOr (Fig. 12). These results are consistent with the view that area TPO/STP receives visual information from dorsal stream cortex.

Connections from the superior temporal gyrus (STG) are particularly useful for defining the extent of STS polysensory cortex. Based on connectional relays from auditory cortex within the lateral sulcus to the superior temporal gyrus (Morel *et al.*, 1993; Galaburda and Pandya, 1983), the STG is presumed to represent unimodal auditory cortex. Projections from the superior temporal gyrus have previously been shown with area TAa on the most lateral portion of the upper bank of the STS and area TPO (Seltzer and Pandya, 1994; Barnes and Pandya, 1992; Cipolloni and Pandya, 1989; Seltzer and Pandya, 1978). The connections from the exposed surface of the middle and caudal STG to the upper bank of the STS clearly avoid the visual areas MT, MST, and FST, as well as area IPa (Fig. 13). Instead, STG projections are predominantly to area TAa and all three divisions of area TPO. Projections from the STG form patches with feedforward (TAa, TPOc), columnar (TPOr), or a mixture of both (TPOi) laminar patterns.

Interestingly, in experiments with paired injections of tracers in prearcuate and STG cortex in the same hemisphere (Seltzer *et al.*, 1996), the densest STG inputs target the STS region with densest inputs from prearcuate cortex. This suggests that either the STG or prearcuate termination field can usefully identify the location and extent of polysensory cortex. Furthermore, paired injections

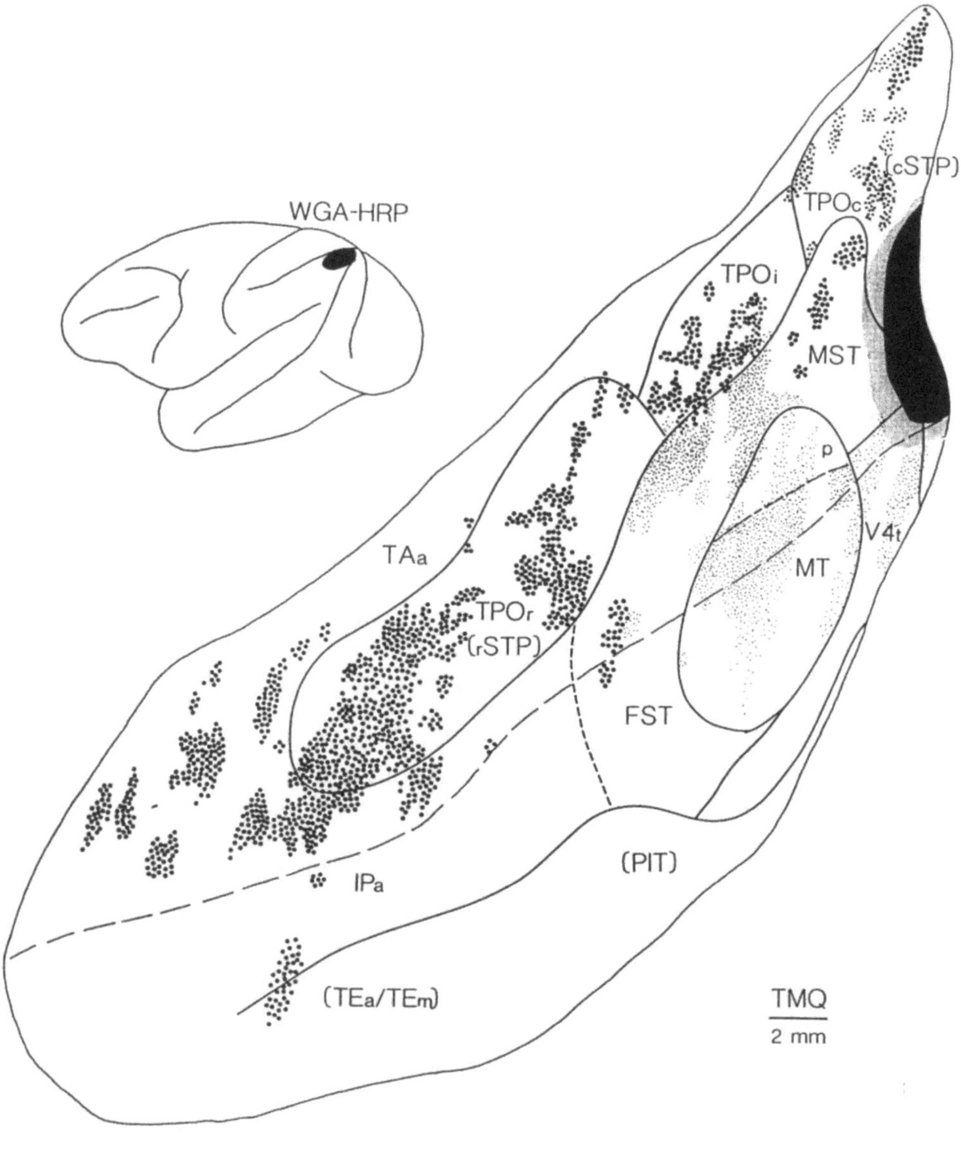

Figure 12. Posterior parietal projections to subdivisions of area TPO. Conventions as in Fig. 9. From Cusick *et al.* (1995).

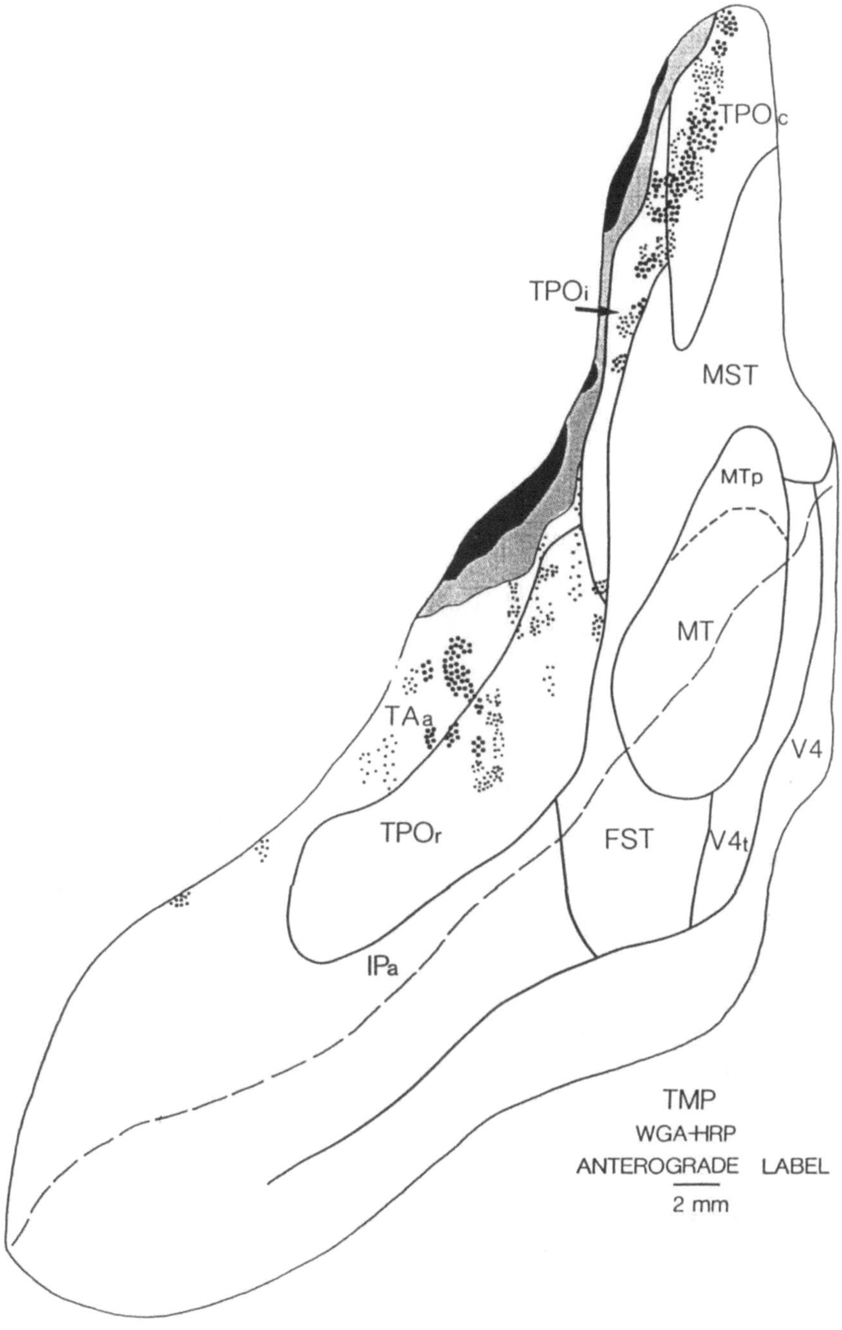

Figure 13. Superior temporal gyrus projections to subdivisions of area TPO. Conventions as in Fig. 9. From Cusick *et al.* (1995).

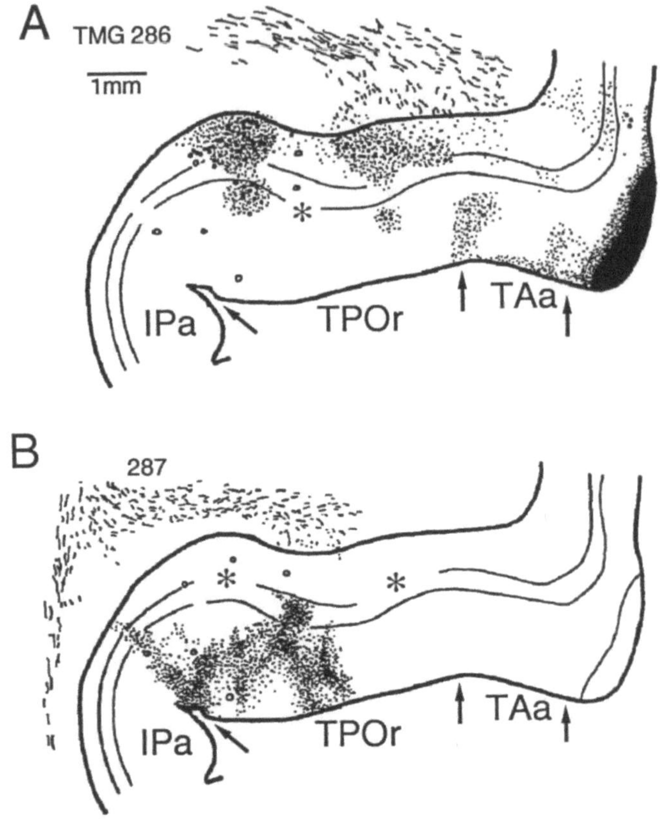

Figure 14. Comparison of (A) superior temporal gyrus and (B) posterior parietal projections in the same hemisphere. The superior temporal gyrus received multiple, closely spaced injections of wheat germ agglutinin–horseradish peroxidase, and the inferior parietal lobule was infiltrated with [^{3}H]proline. Although both tracers are transported to area TPOr, the labeled zones are segregated. The plane of cut is semitangential medially near a small accessory gyrus, and different radial columns are labeled by the different methods. Asterisks on each panel indicate locations where the other type of label crosses layer IV. Modified from Seltzer *et al.* (1996).

within IPL (visual) and STG (auditory) cortex connect to the same regions of area TPO. While these connections are also presumed to identify polysensory cortex, they do not directly overlap, but tend to interdigitate in adjacent cortical columns (Fig. 14; Seltzer *et al.*, 1996).

4.3. Proposed Physiological Correspondences of Subdivisions of Area TPO

There are obviously many questions regarding functional correlates of TPO subdivisions that cannot be answered at this time, but some general observations can be made. As reviewed above, physiological and connectional studies support the concept of rostrocaudal subdivisions within STP cortex. Considering the proposed organization of Felleman and Van Essen (1991), their large caudal STPp may correspond to the entire area TPO as defined here, whereas their

STPa may be either rostral to the present area TPO or encompass TPOr. Due to uncertainties about boundaries and differences in terminology, further studies are clearly needed.

Focusing on the smaller zones, the cSTP region of Hikosaka *et al.* (1988) and the PP area defined by Desimone and Ungerleider (1986) correspond closely in size and location to TPOc. However, those physiological zones appear to be restricted to a lateral part of the upper bank of the STS. Moreover, Hikosaka *et al.* (1988) observed a "mostly unresponsive" zone interposed between cSTP and MST. Thus, it is premature to equate area TPOc with the caudal physiological polysensory cortex. While a number of previous studies appear to include TPOc within area MST (e.g., Hof and Morrison, 1995), the distinct feedback versus columnar laminar patterns of projections from the frontal lobe can help resolve the transition between MST and TPOc. Functional correspondences of TPOi and TPOr are also uncertain. These two zones together may comprise the STP area of Ungerleider and colleagues (Boussaoud *et al.*, 1990). In broad terms, the preponderance of the physiological evidence does support a major distinction between rostral and caudal STP, and it is tempting to speculate that these functional zones correspond largely to TPOr and TPOc, respectively.

4.4. Are Subdivisions within the STP Region (Area TPO) Separate Cortical Areas?

Multiple foci of labeled terminals were found in area TPO following injections in frontal, parietal, and STG cortex (Seltzer *et al.*, 1996; Cusick *et al.*, 1995). Since a single cortical area typically has multiple patches of connections with another single cortical area (e.g., Krubitzer, 1995; Distler *et al.*, 1993; Steele *et al.*, 1991; Krubitzer and Kaas, 1990a; Colby *et al.*, 1988; Cusick and Kaas, 1988; Selemon and Goldman-Rakic, 1988; DeYoe and Van Essen, 1985; Shipp and Zeki, 1985; Livingstone and Hubel, 1984), the existence of patchy connections does not establish the existence of different cortical areas within TPO cortex. The subdivisions TPOc, TPOi, and TPOr may either be cortical *areas,* i.e., basic cortical units with internally coherent architecture, separate patterns of topographic connections, and possibly distinct physiological roles, or they may be anatomical *modules* within a single, heterogeneous cortical area. The CO modules within V2 are conspicuous examples of functionally different modules within a single area (DeYoe and Van Essen, 1985), although CO banding patterns are smaller (narrower) units than the subdivisions of area TPO.

Physiological studies summarized above broadly support functional differences between rostral and caudal portions of the STP region. If these functional subdivisions correspond to TPOr and TPOc, respectively, one possible interpretation is that these two divisions constitute separate *areas.* The intervening region, TPOi, could be either a separate area or a subcomponent of either TPOr or TPOc. Although small (4 mm rostrocaudally vs. 6–8 mm mediolaterally), TPOi is on the order of the size of other distinct areas such as MT and MST. Moreover, evidence that area TPOi has a prominent feedforward input from prefrontal cortex, compared to TPOc and TPOr, favors TPOi being distinct. Therefore, we provisionally interpret the subdivisions of polysensory cortex on the upper bank of the STS to be three separate *areas* TPOc, TPOi, and TPOr. As a recipient of "feedforward" connections from prefrontal cortex, TPOi may be a specialized zone for integrating high-order information for control of eye movements or spatially guided behavior. The finding of fourth-layer projections from prefrontal to superior temporal cortex presents an interesting anomaly for constructing hierarchical relationships among cortical areas (Young, 1992, 1993; Felleman and Van Essen, 1991), and suggests caution in interpreting laminar connection patterns between higher-order association areas.

5. Concluding Remarks

Disparate physiological findings described for the STP region probably reflect the fact that this region, as originally described, is an amalgam of separate cortical areas. Our combined connectional–chemoarchitectonic studies offer a new approach toward understanding the complex morphology of this region. These studies were initiated in the hope of anchoring future work on functional specialization and connections of physiologically identified STP cortex in an independent, reproducible structural frame of reference. We have retained the term "area TPO" for the cortex of the STP region because physiological correlates are necessarily tentative now. A further, perhaps stronger, reason for using the term TPO in lieu of STP is conceptual. Physiological recordings throughout the extent of "STP" support the view that the region is visual, and that the "polysensory" aspect reflects the complexity of visual functions performed within higher cortical areas.

There are many unanswered questions on the functional contributions of TPO cortex. Its possible roles in dorsal or ventral stream processing, or integration of the two streams, remain to be resolved. As concepts of the functional organization of TPO cortex in macaque monkeys become further developed, they will potentially contribute to an understanding the organization of high-order cortex in the human brain. Heilman and his co-workers have argued that portions of the monkey superior temporal region comprise the homologue to the human inferior parietal lobule that includes the angular and supramarginal gyri (Heilman *et al.*, 1983; Watson *et al.*, 1994). Milner (1995) has stressed that the human inferior parietal lobule is not strictly a component of either the dorsal or ventral visual stream, but requires information from both to organize spatial representations or visual scenes or complex patterns. Object identities may be necessary to represent spatial relationships among objects.

The human angular and supramarginal gyri are located dorsorostral to the parieto-temporo-occipital junction, the area of the human brain most sensitive to visual motion and the presumed homologue of area MT (Sereno *et al.*, 1995; Zeki *et al.*, 1991), and thus appear to occupy a relative location similar to the STP region in monkeys. But, with the scarcity of data regarding the discrete functions of STP cortex, any proposal of homologies with human cortical regions must be tentative. Despite remaining unresolved issues for establishing valid subdivisions of visual cortex (see chapter by Kaas in this volume), evidence is accumulating that similar patterns of "lower" visual areas do exist in a range of primate species, including prosimians, New and Old World monkeys, and humans (Sereno *et al.*, 1995; Krubitzer and Kaas, 1990a; Cusick *et al.*, 1984; Van Essen and Zeki, 1978). A comparative approach, emphasizing areal and functional organization in monkeys and the contributions of the human posterior temporal lobe to behavior, promises new insights into the functional organization of high-order association cortex.

ACKNOWLEDGMENTS. I would like to thank Drs. K. S. Rockland and B. Seltzer for helpful comments on an earlier version of the manuscript and Carolina Gutierrez for assistance with the figures. Supported by NIH Grant EY08906.

6. References

Allman, J. M., and Kaas, J. H., 1974, A crescent-shaped cortical visual area surrounding the middle temporal area (MT) in the owl monkey (*Aotus trivirgatus*), *Brain Res.* **81:**199–213.

Amir, Y., Harel, M., and Malach, R., 1993, Cortical hierarchy reflected in the organization of intrinsic connections in macaque monkey visual cortex, *J. Comp. Neurol.* **334:**19–46.

Andersen, R. A., Asanuma, C., and Cowan, W. M., 1985, Callosal and prefrontal associational projecting cell populations in area 7A of the macaque monkey: A study using retrogradely transported fluorescent dyes, *J. Comp. Neurol.* **232:**443–455.

Andersen, R. A., Asanuma, C., Essick, G., and Siegel, R. M., 1990, Cortico-cortical connections of anatomically and physiologically defined subdivisions within the inferior parietal lobule, *J. Comp. Neurol.* **296:**65–113.

Baizer, J. S., Ungerleider, L. G., and Desimone, R., 1991, Organization of visual inputs to the inferior temporal and posterior parietal cortex in macaques, *J. Neurosci.* **11:**168–190.

Baizer, J. S., Desimone, R., and Ungerleider, L. G., 1993, Comparison of subcortical connections of inferior temporal and posterior parietal cortex in monkeys, *Visual Neurosci.* **10:**59–72.

Baleydier, C., and Morel, A., 1992, Segregated thalamocortical pathways to inferior parietal and inferotemporal cortex in macaque monkey, *Visual Neurosci.* **8:**391–405.

Barbas, H., 1988, Anatomic organization of basoventral and mediodorsal visual recipient prefrontal regions in the rhesus monkey, *J. Comp. Neurol.* **276:**313–342.

Barbas, H., and Mesulam, M.-M., 1981, Organization of afferent input to subdivisions of area 8 in the rhesus monkey, *J. Comp. Neurol.* **200:**407–431.

Barnes, C. L., and Pandya, D. N., 1992, Efferent cortical connections of multimodal cortex of the superior temporal sulcus in the rhesus monkey, *J. Comp. Neurol.* **318:**222–244.

Baylis, G. C., Rolls, E. T., and Leonard, C. M., 1987, Functional subdivisions of the temporal lobe neocortex, *J. Neurosci.* **7:**330–342.

Blatt, G. J., Andersen, R. A., and Stoner, G. R., 1990, Visual receptive field organization and cortico-cortical connections of the lateral intraparietal area, area LIP, in the macaque, *J. Comp. Neurol.* **299:**421–445.

Blümcke, I., Hof, P. R., Morrison, J. H., and Celio, M. R., 1991, Parvalbumin in the monkey striate cortex: A quantitative immunoelectron-microscopy study, *Brain Res.* **554:**237–243.

Boussaoud, D., Ungerleider, L. G., and Desimone, R., 1990, Pathways for motion analysis: Cortical connections of the medial superior temporal and fundus of the superior temporal visual areas in the macaque, *J. Comp. Neurol.* **296:**462–495.

Bruce, C., Desimone, R., and Gross, C. G., 1981, Visual properties of neurons in a polysensory area in superior temporal sulcus of the macaque, *J. Neurophysiol.* **46:**369–384.

Bruce, C., Desimone, R., and Gross, C. G., 1986, Both striate cortex and superior colliculus contribute to visual properties of neurons in superior temporal polysensory area of macaque monkey, *J. Neurophysiol.* **55:**1057–1075.

Campbell, M. J., and Morrison, J. H., 1989, Monoclonal antibody to neurofilament protein, SMI-32, labels a subpopulation of pyramidal neurons in the human and monkey neocortex, *J. Comp. Neurol.* **282:**191–205.

Campbell, R., Heywood, C. A., Cowey, A., Regard, M., and Landis, T., 1990, Sensitivity to eye gaze in prosopagnosic patients and monkeys with superior temporal sulcus ablation, *Neuropsychologia* **28:**1123–1142.

Cavada, C., and Goldman-Rakic, P. S., 1989a, Posterior parietal cortex in rhesus monkey: I. Parcellation of areas based on distinctive limbic and sensory corticocortical connections, *J. Comp. Neurol.* **287:**393–421.

Cavada, C., and Goldman-Rakic, P. S., 1989b, Posterior parietal cortex in rhesus monkey: II. Evidence for segregated corticocortical networks linking sensory and limbic areas with the frontal lobe, *J. Comp. Neurol.* **287:**422–445.

Cipolloni, P. B., and Pandya, D. N., 1989, Connectional analysis of the ipsilateral and contralateral afferent neurons of the superior temporal region in the rhesus monkey, *J. Comp. Neurol.* **281:**567–585.

Colby, C. L., and Duhamel, J. R., 1991, Heterogeneity of extrastriate visual areas and multiple parietal areas in the macaque monkey, *Neuropsychologia* **29:**517–537.

Colby, C. L., and Miller, E. K., 1986, Eye movement related responses of neurons in superior temporal polysensory area of the macaque, *Soc. Neurosci. Abstr.* **12:**1184.

Colby, C. L., Gattass, R., Olson, C. R., and Gross, C. G., 1988, Topographical organization of cortical afferents to extrastriate visual area PO in the macaque: A dual tracer study, *J. Comp. Neurol.* **269:**392–413.

Colby, C. L., Duhamel, J.-R., and Goldberg, M. E., 1993, Ventral intraparietal area of the macaque: Anatomic location and visual response properties, *J. Neurophysiol.* **69:**902–914.

Cusick, C. G., and Kaas, J. H., 1988, Cortical connections of area 18 and dorsolateral visual cortex in squirrel monkeys, *Visual Neurosci.* **1:**211–237.

Cusick, C. G., Gould, H. J., III, and Kaas, J. H., 1984, Interhemispheric connections of visual cortex of owl monkeys (*Aotus trivirgatus*), marmosets (*Callithrix jacchus*), and galagos (*Galago crassicaudatus*), *J. Comp. Neurol.* **230:**311–336.

Cusick, C. G., Scripter, J. L., Darensbourg, J. G., and Weber, J. T., 1993, Chemoarchitectonic subdivisions of the visual pulvinar in monkeys and their connectional relations with the middle temporal and rostral dorsolateral visual areas, MT and DLr, *J. Comp. Neurol.* **336:**1–30.

Cusick, C. G., Seltzer, B., Cola, M., and Griggs, E., 1995, Chemoarchitectonics and cortico-cortical terminations within the superior temporal sulcus of the rhesus monkey: Evidence for subdivisions of superior temporal polysensory cortex, *J. Comp. Neurol.* **360:**513–535.

DeFelipe, J., and Jones, E. G., 1991, Parvalbumin immunoreactivity reveals layer IV of monkey cerebral cortex as a mosaic of microzones of thalamic afferent terminations, *Brain Res.* **562:**39–47.

Desimone, R., and Gross, C. G., 1969, Visual areas in the temporal cortex of the macaque, *Brain Res.* **178:**363–380.

Desimone, R., and Ungerleider, L. G., 1986, Multiple visual areas in the caudal superior temporal sulcus of the macaque, *J. Comp. Neurol.* **248:**164–189.

DeYoe, E. A., and Van Essen, D. C., 1985, Segregation of efferent connections and receptive field properties in visual area V2 of the macaque, *Nature* **317:**58–61.

DeYoe, E. A., Hockfield, S., Garren, H., and Van Essen, D. C., 1990, Antibody labeling of functional subdivisions in visual cortex: Cat-301 immunoreactivity in striate and extrastriate cortex of the macaque monkey, *Visual Neurosci.* **5:**67–81.

Distler, C., Boussaoud, D., Desimone, R., and Ungerleider, L. G., 1993, Cortical connections of inferior temporal area TEO in macaque monkeys, *J. Comp. Neurol.* **334:**125–150.

Duffy, C. J., and Wurtz, R. H., 1991a, Sensitivity of MST neurons to optic flow stimuli. I. A continuum of response selectivity to large-field stimuli, *J. Neurophysiol.* **65:**1329–1345.

Duffy, C. J., and Wurtz, R. H., 1991b, Sensitivity of MST neurons to optic flow stimuli. II. Mechanisms of response selectivity revealed by small-field stimuli, *J. Neurophysiol.* **65:**1346–1359.

Eacott, M. J., Heywood, C. A., Gross, C. G., and Cowey, A., 1993, Visual discrimination impairments following lesions of the superior temporal sulcus are not specific for facial stimuli, *Neuropsychologia* **31:**609–619.

Ettlinger, G., and Garcha, H. S., 1980, Cross-modal recognition by the monkey: The effects of cortical removals, *Neuropsychologia* **18:**685–692.

Felleman, D. J., and Van Essen, D. C., 1991, Distributed hierarchical processing in the primate cerebral cortex, *Cerebral Cortex* **1:**1–47.

Fenstemaker, S. B., 1986, The organization and connections of visual cortical area TEO in the macaque, PhD thesis, Princeton University, Princeton, NJ.
Ferrera, V. P., Rudolph, K. K., and Maunsell, J. H. R., 1994, Responses of neurons in the parietal and temporal visual pathways during a motion task, *J. Neurosci.* **14**:6171–6186.
Galaburda, A. M., and Pandya, D. N., 1983, The intrinsic architectonic and connectional organization of the superior temporal region in the rhesus monkey, *J. Comp. Neurol.* **221**:169–184.
Girard, P., Salin, P. A., and Bullier, J., 1992, Response selectivity of neurons in area MT of the macaque monkey during reversible inactivation of area V1, *J. Neurophysiol.* **67**:1437–1446.
Goldberg, M. E., and Wurtz, R. H., 1972, Activity of superior colliculus in behaving monkey. I. Visual receptive fields of single neurons, *J. Neurophysiol.* **35**:542–559.
Graziano, M. S. A., Andersen, R. A., and Snowden, R. J., 1994, Tuning of MST neurons to spiral motions, *J. Neurosci.* **14**:54–67.
Gross, C. G., 1991, Contribution of striate cortex and the superior colliculus to visual function in area MT, the superior temporal polysensory area and the inferior temporal cortex, *Neuropsychologia* **29**:497–515.
Gutierrez, C., and Cusick, C. G., 1997, Area V1 in macaques projects to multiple histochemically defined subdivisions of the inferior pulvinar complex, *Brain Res.* **765**:349–356.
Gutierrez, C., Yaun, A., and Cusick, C. G., 1995, Neurochemical subdivisions of the inferior pulvinar in macaque monkeys, *J. Comp. Neurol.* **363**:545–562.
Hardy, S. G., and Lynch, J. C., 1992, The spatial distribution of pulvinar neurons that project to two subregions of the inferior parietal lobule in the macaque, *Cerebral Cortex* **2**:217–230.
Haxby, J. V., Grady, C. L., Ungerleider, L. G., Horwitz, B., 1991, Mapping the functional neuroanatomy of the intact human brain with brain work imaging, *Neuropsychologia* **29**:539–555.
Haxby, J. V., Horwitz, B., Ungerleider, L. G., Maisog, J. M., Pietrini, P., and Grady, C. L., 1994, The functional organization of human extrastriate cortex: A PET-rCBF study of selective attention to faces and locations, *J. Neurosci.* **14**:6336–6353.
Heilman, K. M., Watson, R. T., Valenstein, E., and Damasio, A. R., 1983, Localization of lesions in neglect, in: *Localization in Neuropsychology* (A. Kertesz, ed.), Academic Press, Orlando, FL.
Hietanen, J. K., and Perrett, D. I., 1993, Motion sensitive cells in the macaque superior temporal polysensory area. I. Lack of response to the sight of the animal's own limb movement, *Exp. Brain Res.* **93**:117–128.
Hikosaka, K., Iwai, E., Saito, H.-A., and Tanaka, K., 1988, Polysensory properties of neurons in the anterior bank of the caudal superior temporal sulcus of the macaque monkey, *J. Neurophysiol.* **60**:1615–1637.
Hof, P. R., and Morrison, J. H., 1995, Neurofilament protein defines regional patterns of cortical organization in the macaque monkey visual system: A quantitative immunohistochemical analysis, *J. Comp. Neurol.* **352**:161–186.
Huerta, M. F., Krubitzer, L. A., and Kaas, J. H., 1987, Frontal eye field as defined by intracortical microstimulation in squirrel monkeys, owl monkeys, and macaques. II. Cortical connections, *J. Comp. Neurol.* **265**:332–361.
Hyvärinen, J., 1982, *The Parietal Cortex of Monkey and Man,* Springer, New York.
Jones, E. G., and Hendry, S. H. C., 1989, Differential calcium binding protein immunoreactivity distinguishes classes of relay neurons in monkey thalamic nuclei, *Eur. J. Neurosci.* **1**:222–246.
Jones, E. G., and Powell, T. P. S., 1970, An anatomical study of converging sensory pathways within the cerebral cortex of the monkey, *Brain* **93**:793–820.
Kaas, J. H., 1995, Human visual cortex. Progress and puzzles, *Curr. Biol.* **1**:1126–1128.
Kaas, J. H., and Morel, A., 1993, Connections of visual areas of the upper temporal lobe of owl monkeys: The MT crescent and dorsal and ventral subdivisions of FST, *J. Neurosci.* **13**:534–546.
Komatsu, H., and Wurtz, R. H., 1988a, Relation of cortical areas MT and MST to pursuit eye movements. I. Localization and visual properties of neurons, *J. Neurophysiol.* **60**:580–603.
Komatsu, H., and Wurtz, R. H., 1988b, Relation of cortical areas MT and MST to pursuit eye movements. II. Differentiation of retinal from extraretinal inputs, *J. Neurophysiol.* **60**:580–603.
Komatsu, H., and Wurtz, R. H., 1988c, Relation of cortical areas MT and MST to pursuit eye movements. III. Interaction with full field visual stimulation, *J. Neurophysiol.* **60**:621–644.
Künzle, H., and Akert, K., 1977, Efferent connections of cortical area 8, frontal eye field, in *Macaca fascicularis.* A reinvestigation using the autoradiographic technique, *J. Comp. Neurol.* **173**:147–164.

Krubitzer, L. A., 1995, The organization of neocortex in mammals: Are species differences really so different? *Trends Neurosci.* **18:**408–417.
Krubitzer, L. A., and Kaas, J. H., 1989, Cortical integration of parallel pathways in the visual system of primates, *Brain Res.* **478:**161–165.
Krubitzer, L. A., and Kaas, J. H., 1990a, Cortical connections of MT in four species of primates: Areal, modular, and retinotopic patterns, *Visual Neurosci.* **5:**165–204.
Krubitzer, L., and Kaas, J., 1990b, Convergence of processing channels in the extrastriate cortex of monkeys, *Visual Neurosci.* **5:**609–613.
Leavitt, J. B., Yoshioka, T., and Lund, J. S., 1994, Intrinsic cortical connections in macaque visual area V2: Evidence for interaction between different functional streams, *J. Comp. Neurol.* **342:**551–570.
Leichnetz, G. R., 1989, Inferior frontal eye field projections to the pursuit-related dorsolateral pontine nucleus and middle temporal area (MT) in the monkey, *Visual Neurosci.* **3:**171–180.
Livingstone, M. S., and Hubel, D. H., 1984, Anatomy and physiology of a color system in the primate visual cortex, *J. Neurosci.* **4:**309–356.
Luh, K. E., Butter, C. M., and Buchtel, H. A., 1986, Impairments in orienting to visual stimuli in monkeys following unilateral lesions of the superior temporal sulcal polysensory cortex, *Neuropsychologia* **24:**461–470.
Lund, J. S., Yoshioka, T., and Levitt, J. B., 1993, Comparison of intrinsic connectivity in different areas of macaque monkey cerebral cortex, *Cerebral Cortex* **3:**148–162.
Lynch, J. C., 1992, Saccade initiation and latency deficits after combined lesions of the frontal and posterior eye fields in monkeys, *J. Neurophysiol.* **68:**1913–1916.
Lynch, J. C., and McLaren, J. W., 1989, Deficits of visual attention and saccadic eye movements after lesions of parieto-occipital cortex in monkeys, *J. Neurophysiol.* **61:**74–89.
Lynch, J. C., Mountcastle, V. B., Talbot, W. H., and Yin, T. C. T., 1977, Parietal lobe mechanisms of directed visual attention, *J. Neurophysiol.* **40:**362–389.
Maunsell, J. H. R., and Van Essen, D. C., 1983, The connections of the middle temporal visual area (MT) and their relationship to a cortical hierarchy in the macaque monkey, *J. Neurosci.* **3:**2563–2586.
McGuire, P. K., Hockfield, S., and Goldman-Rakic, P. S., 1989, Distribution of Cat-301 immunoreactivity in the frontal and parietal lobes of the macaque monkey, *J. Comp. Neurol.* **288:**280–296.
Merigan, W. H., and Maunsell, J. H. R., 1993, How parallel are the primate visual pathways? *Annu. Rev. Neurosci.* **16:**369–402.
Milner, D. A., 1995, Cerebral correlates of visual awareness, *Neuropsychologia* **33:**1117–1130.
Milner, D. A., and Goodale, M. A., 1993, Visual pathways to perception and action, *Prog. Brain Res.* **95:**317–337.
Mistlin, A. J., and Perrett, D. I., 1990, Visual and somatosensory processing in the macaque temporal cortex: The role of 'expectation,' *Exp. Brain Res.* **82:**437–450.
Morecraft, R. J., Geula, C., and Mesulam, M.-M., 1993, Architecture of connectivity within a cingulo-fronto-parietal neurocognitive network for directed attention, *Arch. Neurol.* **50:**279–284.
Morel, A., and Bullier, J., 1990, Anatomical segregation of two cortical visual pathways in the macaque monkey, *Visual Neurosci.* **4:**555–578.
Morel, A., Garraghty, P. E., and Kaas, J. H., 1993, Tonotopic organization, architectonic fields, and connections of auditory cortex in macaque monkeys, *J. Comp. Neurol.* **335:**437–459.
Mountcastle, V. B., Lynch, J. C., Georgopoulos, A., Sakata, H., and Acuña, C., 1975, Posterior parietal association cortex of the monkey: Command functions for operations within extrapersonal space, *J. Neurophysiol.* **38:**871–908.
Neal, J. W., Pearson, R. C., and Powell, T. P., 1990, The connections of area PG, 7a, with cortex in the parietal, occipital and temporal lobes of the monkey, *Brain Res.* **532:**249–264.
Oram, M. W., Perrett, D. I., and Hietanen, J. K., 1993, Directional tuning of motion-sensitive cells in the anterior superior temporal polysensory area of the macaque, *Exp. Brain Res.* **97:**274–294.
Orban, G. S., Lagae, L., Raiguel, S., Xiao, D., and Maes, H., 1995, The speed tuning of medial superior temporal (MST) cell responses to optic-flow components, *Perception* **24:**269–285.
Perrett, D. I., Rolls, E. T., and Caan, W., 1982, Visual neurones responsive to faces in the monkey temporal cortex, *Exp. Brain Res.* **47:**329–342.
Perrett, D. I., Smith, P. A. J., Mistlin, A. J., Chitty, A. J., Head, A. S., Potter, D. D., Broennimann, R.,

Milner, A. D., and Jeeves, M. A., 1985a, Visual analysis of body movements by neurones in the temporal cortex of the macaque monkey: A preliminary report, *Behav. Brain Res.* **16:**153–170.

Perrett, D. I., Smith, P. A. J., Potter, D. D., Mistlin, A. J., Head, A. S., Milner, A. D., and Jeeves, M. A., 1985b, Visual cells in the temporal cortex sensitive to face view and gaze direction, *Proc. R. Soc. Lond. B* **223:**293–317.

Perrett, D. I., Mistlin, A. J., Chitty, A. J., Smith, P. A. J., Potter, D. D., Broennimann, R., and Harries, M. H., 1988, Specialized face processing and hemispheric asymmetry in man and monkey: Evidence from single unit and reaction time studies, *Behav. Brain Res.* **29:**245–258.

Perrett, D. I., Hietanen, J. K., Oram, M. W., and Benson, P. J., 1992, Organization and functions of cells responsive to faces in the temporal cortex, *Phil. Trans. R. Soc. Lond. B* **335:**23–30.

Petrides, M., and Iversen, S. D., 1978, The effect of selective anterior and posterior association cortex lesions in the monkey on performance of a visual–auditory compound discrimination test, *Neuropsychologia* **16:**527–537.

Petrides, M., and Iversen, S. D., 1979, Restricted posterior parietal lesions in the rhesus monkey and performance on visuospatial tasks, *Brain Res.* **161:**63–77.

Rockland, K. S., and Pandya, D. N., 1979, Laminar origins and terminations of cortical connections of the occipital lobe in the rhesus monkey, *Brain Res.* **179:**3–20.

Rodman, H. R., Gross, C. G., and Albright, T. D., 1989, Afferent basis of visual response properties in area MT of the macaque. I. Effects of striate cortex removal, *J. Neurosci.* **9:**2033–2050.

Saito, H. A., Yukie, M., Tanaka, K., Hikosaka, K., Fukada, Y., and Iwai, E., 1986, Integration of direction signals of image motion in the superior temporal sulcus of the macaque monkey, *J. Neurosci.* **6:**145–157.

Scalaidhe, S. P., Albright, T. D., Rodman, H. R., and Gross, C. G., 1995, Effects of superior temporal polysensory area lesions on eye movements in the macaque monkey, *J. Neurophysiol.* **73:**1–19.

Schall, J. D., Morel, A., King, D. J., and Bullier, J., 1995, Topography of visual cortex connections with frontal eye field in macaque: Convergence and segregation of processing streams, *J. Neurosci.* **15:**4464–4488.

Schneider, G. E., 1967, Contrasting visuomotor functions of tectum and cortex in the golden hamster, *Psychol. Forsch.* **31:**52–62.

Selemon, L. D., and Goldman-Rakic, P. S., 1988, Common cortical and subcortical targets of the dorsolateral prefrontal and posterior parietal cortices in the rhesus monkey: Evidence for a distributed neural network subserving spatially guided behavior, *J. Neurosci.* **8:**4049–4068.

Seltzer, B., 1992, An anatomy of parallel distributed processing, *Brain Behav. Sci.* **15:**796.

Seltzer, B., and Pandya, D. N., 1978, Afferent cortical connections and architectonics of the superior temporal sulcus and surrounding cortex in the rhesus monkey, *Brain Res.* **149:**1–24.

Seltzer, B., and Pandya, D. N., 1980, Further observation on parieto-temporal connections in the rhesus monkey, *Exp. Brain Res.* **55:**301–312.

Seltzer, B., and Pandya, D. N., 1989a, Frontal lobe connections of the superior temporal sulcus in the rhesus monkey, *J. Comp. Neurol.* **281:**97–113.

Seltzer, B., and Pandya, D. N., 1989b, Intrinsic connections and architectonics of the superior temporal sulcus in the rhesus monkey, *J. Comp. Neurol.* **290:**451–471.

Seltzer, B., and Pandya, D. N., 1991, Post-rolandic cortical projections of the superior temporal sulcus in the rhesus monkey, *J. Comp. Neurol.* **312:**625–640.

Seltzer, B., and Pandya, D. N., 1994, Parietal, temporal, and occipital projections to cortex of the superior temporal sulcus in the rhesus monkey: A retrograde tracer study, *J. Comp. Neurol.* **343:**445–463.

Seltzer, B., Cola, M., Gutierrez, C., Massee, M., Weldon, C., and Cusick, C. G., 1996, Overlapping and non-overlapping cortical projections to cortex of the superior temporal sulcus in the rhesus monkey: Double anterograde tracer studies, *J. Comp. Neurol.* **370:**173–190.

Sereno, M. I., Dale, A. M., Reppas, J. B., Kwong, K. K., Belliveau, J. W., Brady, T. J., Rosen, B. R., and Tootell, R. B., 1995, Borders of multiple visual areas in humans revealed by functional magnetic resonance imaging, *Science* **268:**889–893.

Shipp, S., and Zeki, S., 1985, Segregation of pathways leading from area V2 to areas V4 and V5 of macaque monkey visual cortex, *Nature* **315:**322–325.

Skelly, J. P., Albright, T. D., Rodman, H. R., and Gross, C. G., 1991, The effects of combined superior temporal polysensory area and frontal eye fields lesions on eye movements in the macaque, *Soc. Neurosci. Abstr.* **17:**863.

Stanton, G. B., Bruce, C. J., and Goldberg, M. E., 1995, Topography of projections to posterior cortical areas from the macaque frontal eye fields, *J. Comp. Neurol.* **353:**291–305.

Steele, G. E., Weller, R. E., and Cusick, C. G., 1991, Cortical connections of the caudal subdivision of the dorsolateral area, V4, in monkeys, *J. Comp. Neurol.* **306:**495–520.

Stein, B. E., Meredith, M. A., and Wallace, M. T., 1993, The visually responsive neuron and beyond: Multisensory integration in cat and monkey, *Prog. Brain Res.* **95:**79–90.

Streicher, M., and Ettlinger, G., 1987, Cross-modal recognition of familiar and unfamiliar objects by the monkey: The effects of ablations of polysensory neocortex or of the amygdaloid complex, *Behav. Brain Res.* **23:**95–107.

Suzuki, W. A., and Amaral, D. G., 1994, Perirhinal and parahippocampal cortices of the macaque monkey: Cortical afferents, *J. Comp. Neurol.* **350:**497–533.

Tanaka, K., and Saito, H.-A., 1989, Analysis of motion of the visual field by direction, expansion/contraction, and rotation cells clustered in the dorsal part of the medial superior temporal area of the macaque monkey, *J. Neurophysiol.* **62:**626–641.

Tanaka, K., Hikosaka, K., Saito, H., Yukie, M., Fukada, Y., and Iwai, E., 1986, Analysis of local and wide-field movements in the superior temporal visual areas of the macaque monkey, *J. Neurosci.* **6:**134–144.

Tanaka, K., Sugita, Y., Moriya, M., and Saito, H., 1993, Analysis of object motion in the ventral part of the medial superior temporal area of the macaque visual cortex, *J. Neurophysiol.* **69:**128–142.

Tootell, R. B., and Taylor, J. B., 1995, Anatomical evidence for MT and additional cortical visual areas in humans, *Cerebral Cortex* **5:**39–55.

Tootell, R. B. H., Hamilton, S. L., and Silverman, M. S., 1985, Topography of cytochrome oxidase activity in owl monkey cortex, *J. Neurosci.* **5:**2786–2800.

Trevarthen, C., 1968, Two mechanisms of vision in primates, *Psychol. Forsch.* **31:**299–337.

Ungerleider, L. G., and Desimone, R., 1986a, Projections to the superior temporal sulcus from the central and peripheral field representations of V1 and V2, *J. Comp. Neurol.* **248:**147–163.

Ungerleider, L. G., and Desimone, R., 1986b, Cortical connections of visual area MT in the macaque, *J. Comp. Neurol.* **248:**190–222.

Ungerleider, L. G., and Mishkin, M., 1982, Two cortical visual systems, in: *Analysis of Visual Behavior* (D. G. Ingle, M. A. Goodale, and R. J. Q. Mansfield, eds.), MIT Press, Cambridge, MA, pp. 549–586.

Van Essen, D. C. and Gallant, J. L., 1994, Neural mechanisms of form and motion processing in the primate visual system, *Neuron* **13:**1–10.

Van Essen, D. C., and Zeki, S., 1978, The topographic organization of rhesus monkey prestriate cortex, *J. Physiol.* (Lond.) **277:**193–226.

Walker, A. E., 1940, A cytoarchitectural study of the prefrontal areas of the macaque monkey, *J. Comp. Neurol.* **73:**59–86.

Watson, R. T., Valenstein, E., Day, A., and Heilman, K. M., 1994, Posterior neocortical systems subserving awareness and neglect: Neglect associated with superior temporal sulcus but not area 7 lesions, *Arch. Neurol.* **51:**1014–1021.

Webster, M. J., Bachevalier, J., and Ungerleider, L. G., 1994, Connections of inferior temporal areas TEO and TE with parietal and frontal cortex in macaque monkeys, *Cerebral Cortex* **4:**470–483.

Woolsey, C. N., 1964, Organization of somatic sensory and motor areas of the cerebral cortex, in: *Biological and Biochemical Bases of Behavior* (H. F. Harlow and C. N. Woolsey, eds.), University of Wisconsin Press, Madison, WI, pp. 63–81.

Yeterian, E. H., and Pandya, D. N., 1989, Thalamic connections of the cortex of the superior temporal sulcus in the rhesus monkey, *J. Comp. Neurol.* **282:**80–97.

Yeterian, E. H., and Pandya, D. N., 1991, Cortiocothalamic connections of the superior temporal sulcus in rhesus monkeys, *Exp. Brain Res.* **83:**268–284.

Young, M. P., 1992, Objective analysis of the topological organization of the primate cortical visual system, *Nature* **358:**152–155.

Young, M. P., 1993, The organization of neural systems in the primate cerebral cortex, *Proc. R. Soc. Lond. B* **252:**13–18.

Zeki, S., Watson, J. D. G., Lueck, C. J., Friston, K. J., Kennard, C., and Frackowiak, R. S. J., 1991, A direct demonstration of functional specialization in human visual cortex, *J. Neurosci.* **11:**641–649.

11

Columnar Organization in the Inferotemporal Cortex

KEIJI TANAKA

1. Introduction

A key function of the primate brain is recognition of objects from their visual images. The recognition process is flexible, tolerating marked changes in images due to changes in illumination, viewing angle, and pose of the object. Moreover, the primate visual system processes images of novel objects based on previous visual experience of similar objects. Generalization may be an intrinsic property of the primate visual system. In this review the neural organization essential for these flexible aspects of visual object recognition in TE of the inferotemporal cortex is discussed.

Area TE represents the final stage of the ventral visual cortical pathway, which is thought to be essential for visual object recognition. The ventral visual cortical pathway starts at the primary visual cortex (V1) and leads to TE after relays at V2, V4, and TEO (V1-V2-V4-TEO-TE). Although there are also skipping projections, such as those from V2 to TEO (Nakamura *et al.*, 1993) and those from V4 to the posterior part of TE (Saleem *et al.*, 1992; Desimone *et al.*, 1980), the step-by-step projections are more numerous. TE projects to various polymodal brain sites outside the visual cortex, including the perirhinal cortex (areas 35 and 36), the prefrontal cortex, the amygdala, and the striatum of the basal ganglia. The projections to these targets are more numerous from TE, especially from the anterior part of TE, than from the areas at earlier stages

---

KEIJI TANAKA • Brain Science Institute, RIKEN, 2-1 Hirosawa, Wako-shi, Saitama, 351-01, Japan.
Cerebral Cortex, Volume 12, edited by Rockland *et al.* Plenum Press, New York, 1997.

(Suzuki and Amaral, 1995; Cheng *et al.*, 1993; Ungerleider *et al.*, 1989; Iwai and Yukie, 1987). Therefore, there is a sequential cortical pathway from V1 to TE, and outputs from the pathway originate mainly in TE.

Monkeys that underwent bilateral TE ablation showed severe and selective deficits in learning tasks that required the visual recognition of objects (Dean, 1976; Gross, 1973). Based on these behavioral results together with the above-described important anatomical position of TE, TE is assumed as the site of neural organization essential for the flexible properties of visual object recognition in the primate.

I first summarize the data from single-cell recording experiments to show that cells in TE respond to moderately complex object features and that those responding to similar features cluster in a columnar region. The optical imaging data are next introduced to discuss the function of the TE columns. I expand the discussion to consider how the concept of the object emerges in the brain. Our recordings that are introduced in this chapter were all conducted in anesthetized monkeys (*Macaca fuscata*), and they were from the dorsolateral part of TE, lateral to the lateral lip of the anterior middle temporal sulcus. This part is often referred to as TEd (dorsal part of TE). Because the ventromedial part of TE (TEv, ventral part of TE), medial to the lateral lip of AMTS, has anatomical connections slightly different from those of TEd (Saleem *et al.*, 1994, 1995; Cheng *et al.*, 1993; Martin-Elkins and Horel, 1992; Yukie *et al.*, 1992; Iwai *et al.*, 1987), there may be differences between the functional architecture of TEd and TEv.

2. Stimulus Selectivity of Cells in TE

An obstacle in the study of neuronal mechanisms of object vision has been the difficulty in determining the stimulus selectivity of individual cells. There is a great variety of object features in the natural world, and we do not know how the brain scales down the variety.

We have developed a systematic reduction method which involves the use of a specially designed image-processing computer system (Ito *et al.*, 1994, 1995; Kobatake and Tanaka, 1994; Fujita *et al.*, 1992; Tanaka *et al.*, 1991). After spike activities from a single cell were isolated, many three-dimensional animal and plant models were first presented to find the effective stimuli. Different aspects of the objects were presented in different orientations. Second, the images of the effective stimuli were taken with a video camera and presented on a TV monitor by the computer to determine the most effective stimulus. Finally, the image of the most effective stimulus was simplified step by step to determine which feature or combination of features contained in the image was essential for the maximal activation. The minimal requirements for the maximal activation were determined as the critical feature for the cell. Figure 1 exemplifies the process for a cell for which the effective stimulus was reduced from the view of a water bottle to a combination of a vertical ellipse and a downward projection from the ellipse.

After the reduction was completed, the image was modified to examine the

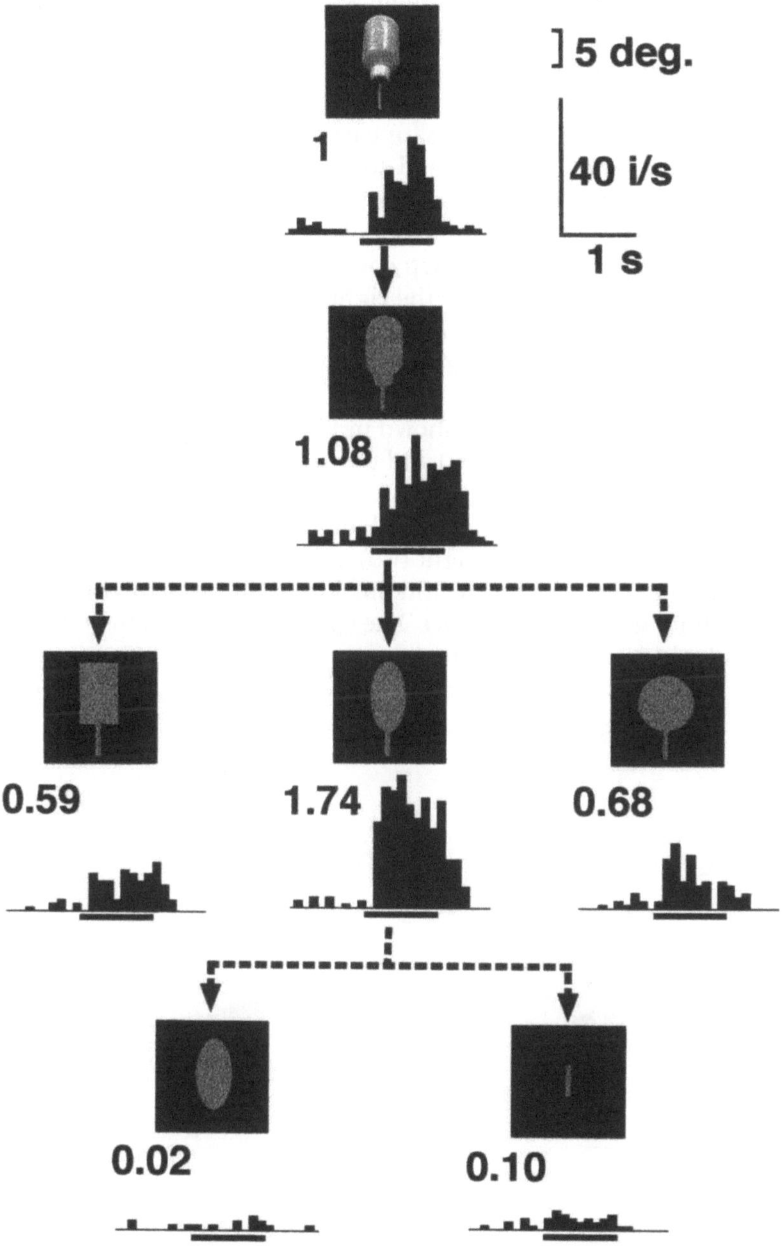

Figure 1. An example of the reduction process to determine the feature critical for the activation of cells in the ventral visual pathway. The responses were averaged over 10 repetitions of the stimuli. The underlines indicate the period of stimulus presentation, and the numbers above the histograms indicate the magnitude of the responses normalized by the response to the image of a water bottle. This cell was recorded from TE.

selectivity further. Figure 2 exemplifies this latter process for a second TE cell, which is one of the cases in which the domain of selectivity was most clearly determined. The cell responded maximally to a pear model within the routine set of object stimuli, and the critical feature was determined as "a rounded protrusion from a rounded body with a concave smooth neck." The body or the head by itself did not evoke any responses (first line in Fig. 2). The head had to be rounded because the response disappeared when the rounded head was replaced by a square, and the body had to be rounded because the response decreased by 51% when the body was cut in half (second line, left). The neck had to be smooth and concave because the response decreased by 78% or 85% when the neck was replaced by one with sharp corners or a straight one (second line, right). The critical feature was neither the right upper contour alone nor the left lower contour alone, because either half of the stimulus did not evoke responses (third line, left). The width and length of the projection were not very critical (third line, right).

Responses of a third TE cell are shown in Fig. 3. The cell responded strongly to the face of a toy monkey (top left) and the critical feature was determined as a configuration in which two black spots and one horizontal black bar were arranged in a gray disk (top line, second from the left). Both the bar and the spots were indispensable (top line, right two), as was the circular outline (bottom left). The contrast between the inside and outside of the circular contour was not critical (bottom line, second from right). The spots and bar had to be darker than the background within the outline (bottom right).

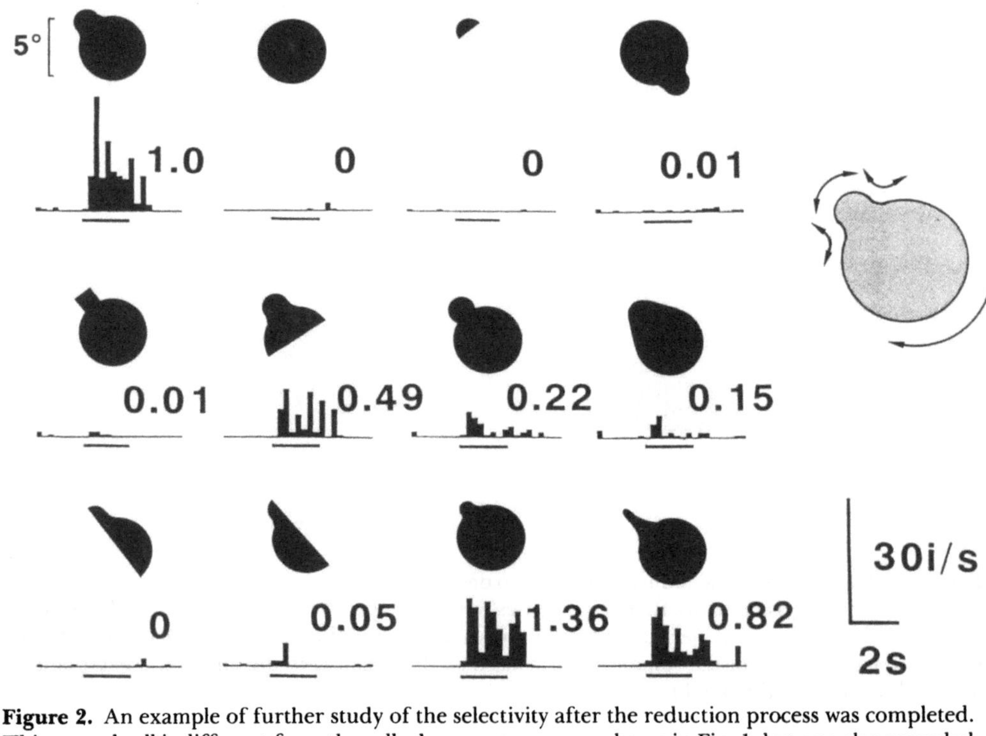

Figure 2. An example of further study of the selectivity after the reduction process was completed. This second cell is different from the cell whose responses are shown in Fig. 1, but was also recorded from TE.

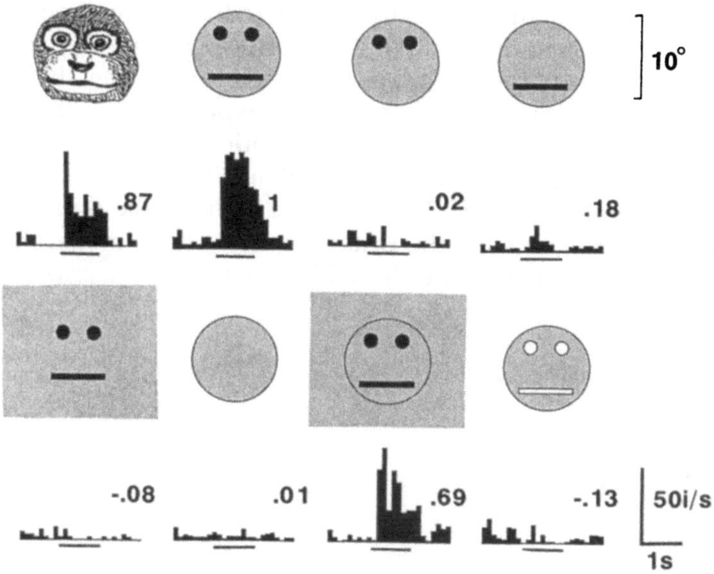

Figure 3. Responses of a third TE cell.

After determining the critical features for hundreds of cells in TE, we reached the conclusion that most cells in TE required moderately complex features for their activation, like the examples shown in Figs. 1–3. The critical features were more complex than orientation, size, color, and simple textures, which are known to be extracted and represented by cells in V1, but at the same time not complex enough to represent the image of a natural object through the activity of single cells. Combined activation of multiple cells, each of which represents a partial (either holistic or local) feature of the object image, is necessary. Some of the features were shapes that were moderately complex, while others were combinations of such shapes with color or texture.

Although the critical features for the activation of TE cells are only moderately complex in general, there are cells which respond to faces and require nearly all the essential features of the face. Such cells were originally found deep in the superior temporal sulcus (Perrett *et al.*, 1992; Bruce *et al.*, 1981), but they have also been found in TE. Thus, there is more convergence of information onto single cells for faces than nonface objects. This difference may be because discrimination of faces from other objects is not the final goal and much more processing of face images is needed to discriminate individuals and expressions, while discrimination of a nonface object from other objects is close to the final goal.

3. Position, Orientation, Size, and Attributes Defining the Contours

Gross *et al.* (1972) in their pioneering experiments found that cells in TE had large receptive fields across which a stimulus continuously evoked responses. By using a set of shape stimuli composed of the critical feature independently determined for individual cells by the reduction method and several other shape stimuli made by modifying the critical feature, we have demonstrated that the selectivity for shape was mostly preserved over the large receptive fields (Ito *et al.*, 1995). An example of such experiments is shown in Fig. 4.

In contrast, responses of TE cells to their critical features were more selective for the orientation in the frontoparallel plane (Tanaka *et al.*, 1991) and for the size of the stimuli (Ito *et al.*, 1995). Figure 5 shows the data of eight cells for the tuning of responses for the orientation in the frontoparallel plane to indicate the general properties of cells in TE. Rotation of the critical feature by

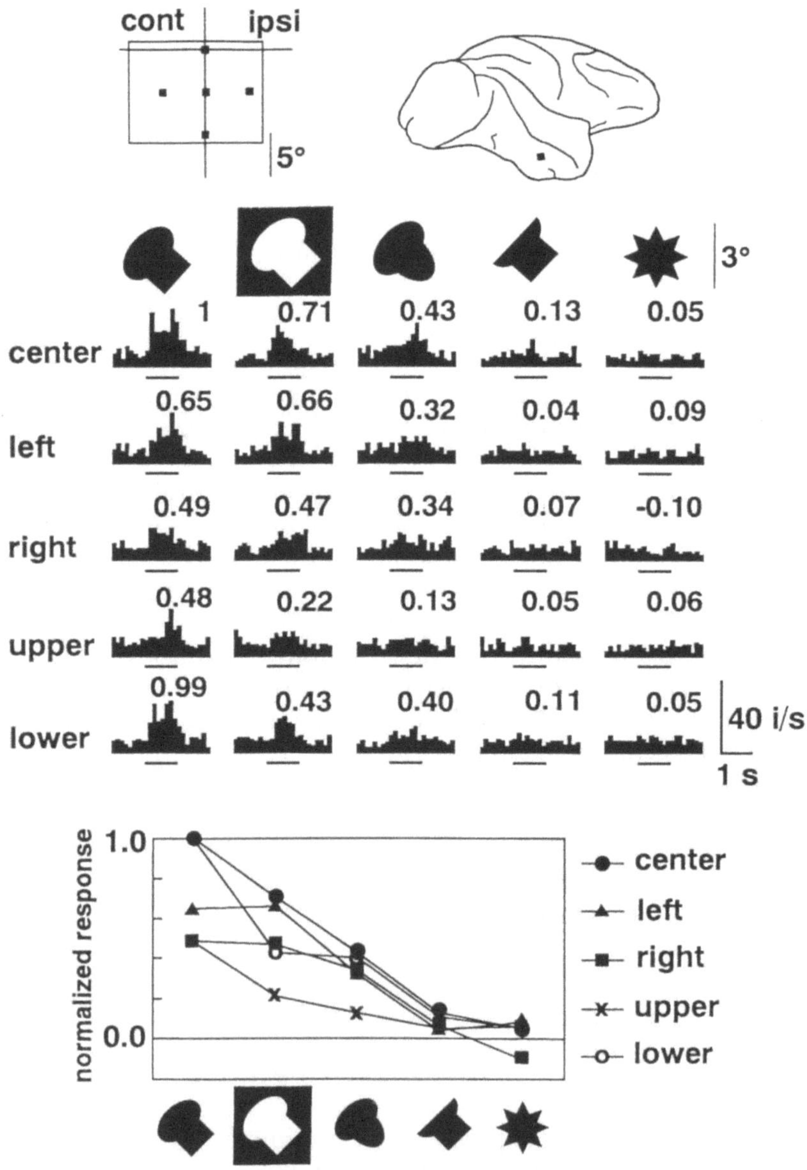

Figure 4. Stimulus selectivity of a TE cell invariant for the stimulus position.

90 deg decreased the response by >50% for most cells (Fig. 5A–F). The tuning of the remaining cells was broader: the response was reduced by >50% only by a rotation of 180 deg (Fig. 5G) or the cell showed only <50% change (Fig. 5H).

Effects of changes in stimulus size varied more among cells. Twenty-one percent of TE cells tested responded to ranges of size of >4 octaves of the critical features, whereas 43% responded to size ranges of <2 octaves. Tuning curves of four TE cells, two from each group, are shown in Fig. 6. The tuned cells may be immature and those with various optimal sizes may converge to make the size-invariant responses. Alternatively, both size-dependent and size-independent processing of images may occur in TE.

Responses of TE cells were also selective for the contrast polarity for the shapes: the contrast reversal of the critical feature reduced the response by >50% in 60% of tested cells and replacement of the solid critical features by line drawings of the contour reduced the response by >50% in 70% of tested cells (Ito *et al.*, 1994). However, a minor group of cells continued to respond to particular shapes tolerating a contrast reversal and a solid-outline conversion. Sary *et al.* (1993) found that some TE cells responded commonly to shapes defined by difference in luminosity, direction of motion of texture components, and the coarseness of texture, while maintaining selectivity for shapes.

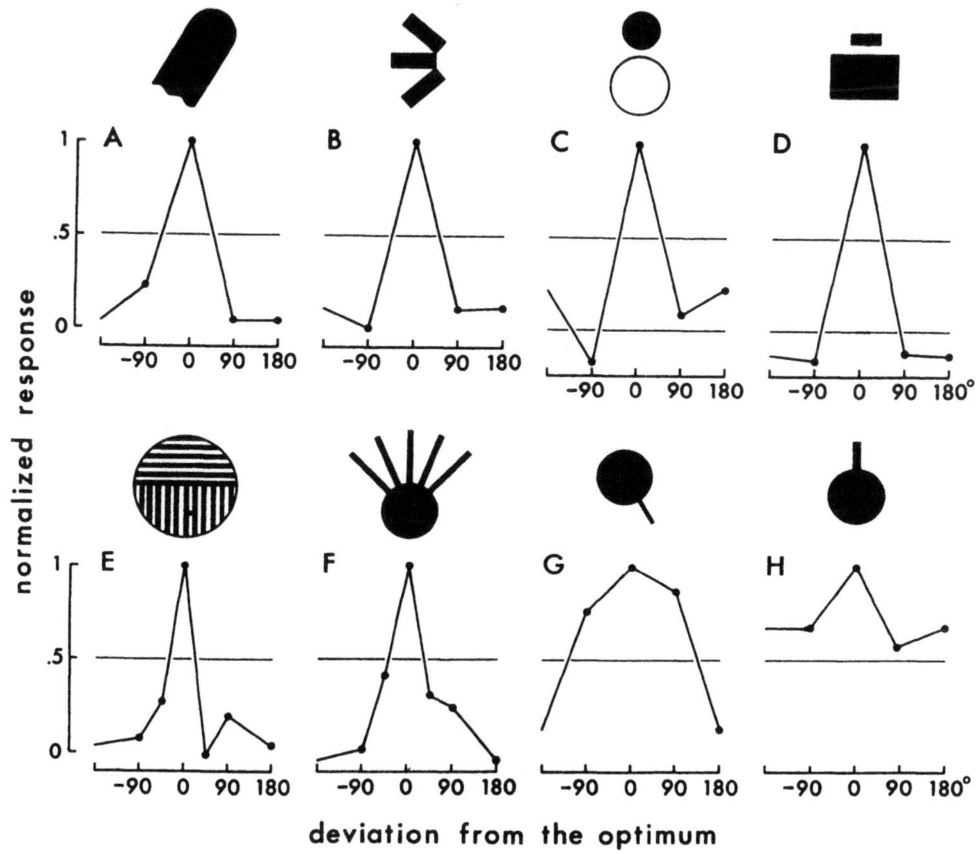

Figure 5. Tuning of responses of eight TE cells for the stimulus orientation.

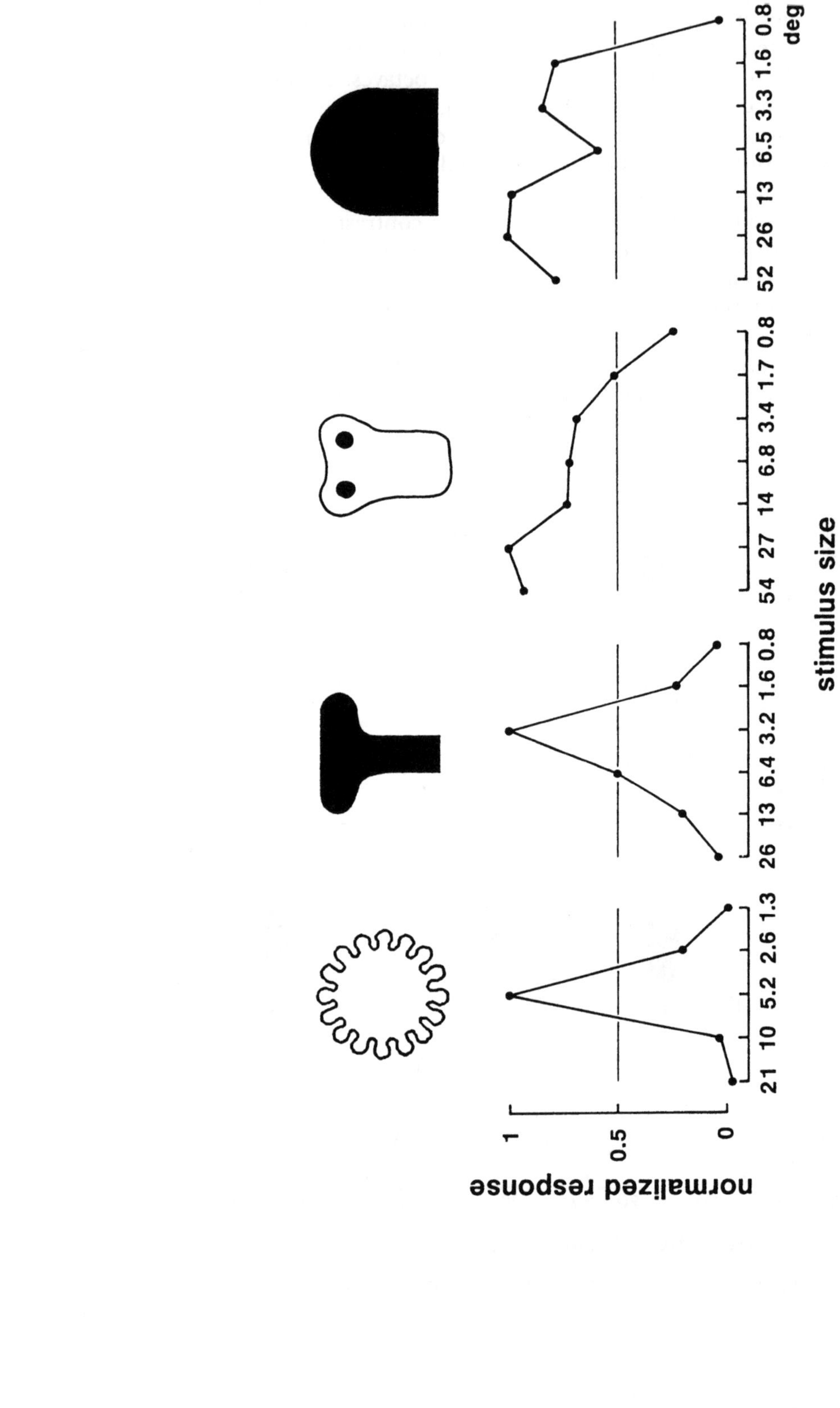

Figure 6. Tuning of responses of four TE cells for the stimulus size.

4. Columnar Organization in TE

We examined the spatial distribution of cells with various critical features in TE. By recording from two TE cells simultaneously with a single electrode, we found that cells located close together in the cortex have similar stimulus selectivities (Fujita *et al.,* 1992). The critical feature of one isolated cell was determined by using the same procedure as described above, while responses of another isolated cell, or nonisolated multiunits, were simultaneously recorded. In most cases, the second cell responded to the optimal and suboptimal stimuli of the first cell. The selectivities of the two cells differed slightly, however, in that the maximal response was evoked by slightly different stimuli, or the mode of the decrease in response was different when the stimulus was changed from the optimal stimulus. Figure 7 shows an example of the latter cases.

To determine the spatial extent of the clustering of cells with similar selectivities, we examined responses of cells recorded successively along long penetrations made vertical or oblique to the cortical surface (Fujita *et al.,* 1992). The critical feature of a cell located at the middle of the penetration was first determined. A set of stimuli, including the optimal feature for the first cell, its rotated versions, and ineffective control stimuli, was constructed, and cells recorded at different positions along the penetration were tested only with the fixed set of stimuli. As in the example shown in Fig. 8, cells recorded along the vertical penetrations commonly responded to the critical feature of the first cell or some related stimuli. The span of the commonly responsive cells covered nearly the entire thickness from layers 2 to 6.

The situation was different in the penetrations made oblique to the cortical surface. The cells that were commonly responsive to the critical feature of the first cell or related versions were limited to a short span around the first cell. The horizontal extent of the span was on the average 400 μm. The cells outside the span were not responsive to any of the stimuli included in the set, or responded to some stimuli that were not effective for the first cell and included in the set as ineffective control stimuli.

The TE region is thus composed of columnar modules in which cells respond to similar features (Fig. 9). Cells within a columnar module respond to similar but not identical features. The width of a columnar module across the cortical surface may be slightly greater than 400 μm. The span of a columnar module along an oblique penetration should be smaller than the actual size of the column if the penetration crosses its periphery. The number of modules, which was estimated by dividing the entire surface area of TEd into 500 by 500 μm squares, was 1300. Although the columns are drawn as discrete units in Fig. 9, the overlap and continuity between neighboring columns will be considered based on the optical imaging data.

5. Organization of Afferents to TE

The selective responses to complex features, which were first found in TE cells, have been traced to earlier stages in the afferent pathway to TE. We found

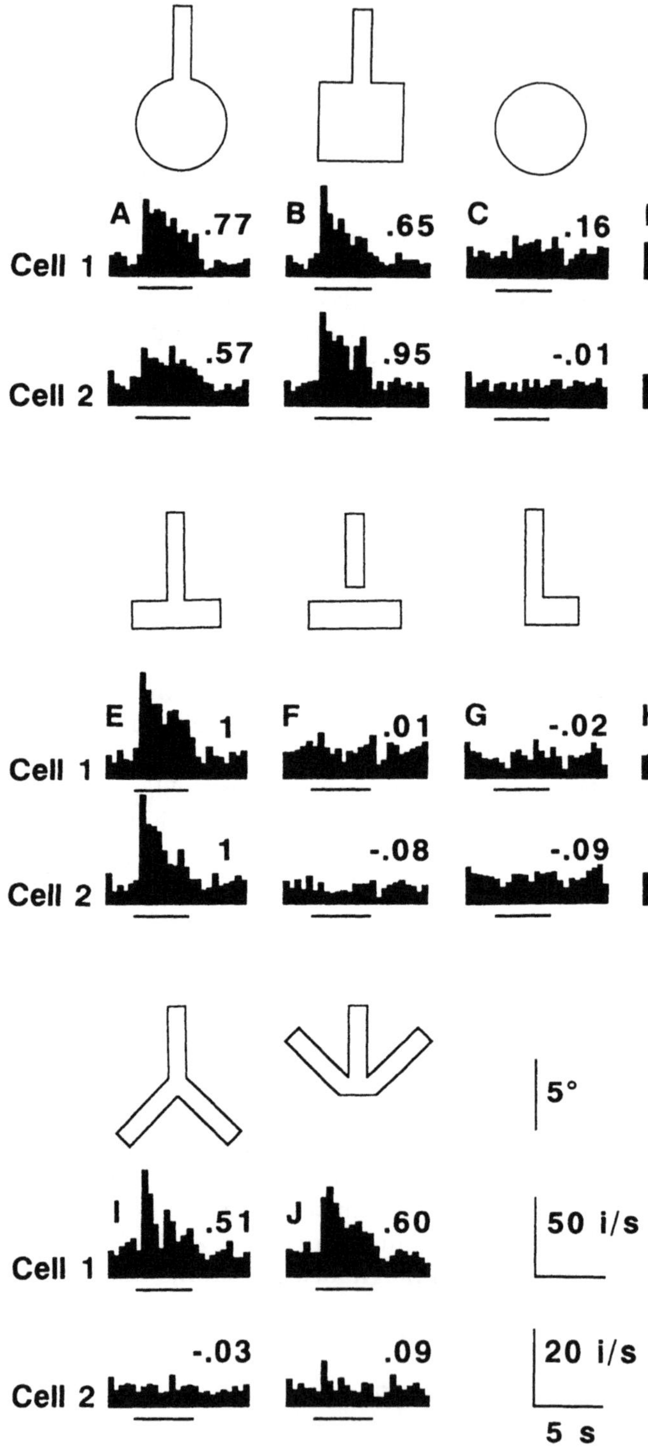

Figure 7. An example of simultaneous recording from two nearby neurons in TE.

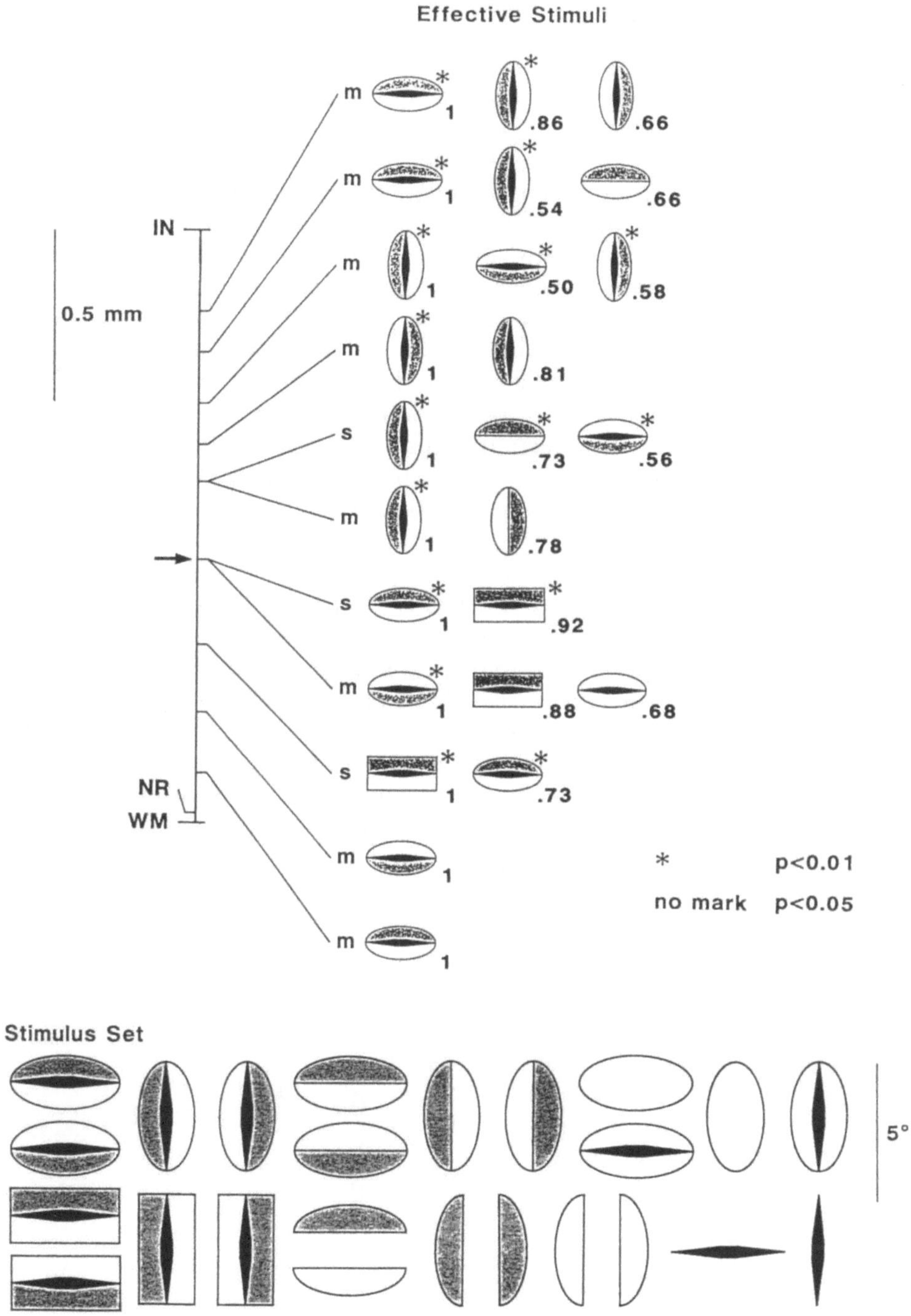

Figure 8. Responses of cells recorded along a vertical penetration in TE. The responsiveness of the cells was tested with the set of stimuli shown at the bottom, which were constructed in reference to the critical feature of the first cell indicated by the arrow. Effective stimuli are listed separately for individual recording sites, in the order of effectiveness. The letter m indicates recording from multiunits, and s that from a single unit.

that cells requiring such complex features for the maximal activation were already present in TEO and V4 (Kobatake and Tanaka, 1994), although their proportion was smaller than that in TE.

To compare such cells in TEO and V4 with cells in TE, we tested responses of individual cells to a fixed set of stimuli of simple features in comparison to their responses to individually determined critical features (Kobatake and Tanaka, 1994). The set of stimuli is composed of 16 bars of four different orientations with 45 deg intervals and two different sizes and 16 colored squares of four different colors and two different sizes (Fig. 10). Stimuli both darker than and lighter than the background were included. This set was composed so that some of the stimuli in this set would evoke some strong, though not maximal, responses in cells in V2 and V4 that showed selectivity only in the domain of orientation or color, and size. Most cells in TE did not respond to the simple stimuli included in the set, or the responses were negligible compared with their responses to the individually determined complex critical features (like the cell whose responses are shown in Fig. 11, left), whereas cells in TEO and V4 showed divergent properties in terms of responses to the simple stimuli. Some TEO and V4 cells showed no or negligible responses to any of the simple stimuli, similar to cells in TE, others showed moderately strong responses to some of the simple stimuli in addition to the maximum response to the complex critical features (like the cell whose responses are shown in Fig. 11, right), and the remaining cells even maximally responded to some of the simple stimuli. Figure 12 shows the proportion of three groups of cells that were classified by setting arbitrary borders at 0.25 and 0.75 for the magnitude of the maximum response to the simple stimuli normalized by the overall maximum response of the cell.

TEO and V4 were thus characterized by a mixture of cells with various levels of selectivity. We may take this mixture of various cells as evidence that selectivity

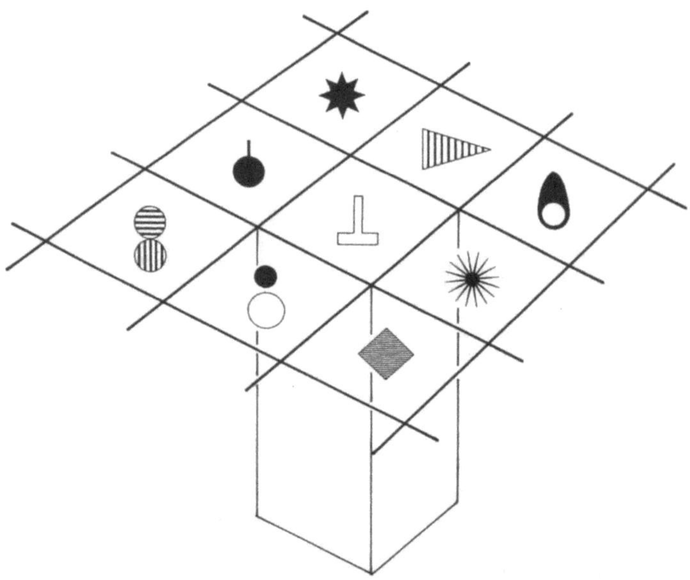

Figure 9. Schematic drawing of columnar organization in TE.

is constructed through local networks in these regions. If we randomly sample cells from a local network in which the selective responses to complex features are constructed by integrating simple features, the sample should include cells with various levels of selectivity. Cells located close to the input end should be maximally activated by simple features, cells close to the output end should respond only to the complex features, and cells at intermediate stages should show some intermediate properties. The areas satisfying these conditions were TEO and V4. It may be proposed that the selectivity to features of medium complexity is mainly constructed in local networks in TEO and V4.

Although we did not find clear evidence of selectivity to complex features in V2, there is a suggestion that slightly stronger responses to a complex pattern than the best simple stimulus (bar or grating) are not unusual among cells in V1 (Lehky *et al.*, 1992). It is possible that selectivity to moderately complex features gradually develops throughout and becomes apparent in V4 and TEO.

The anatomical organization of the forward projection from TEO to TE is consistent with the idea that the selectivity to moderately complex features is

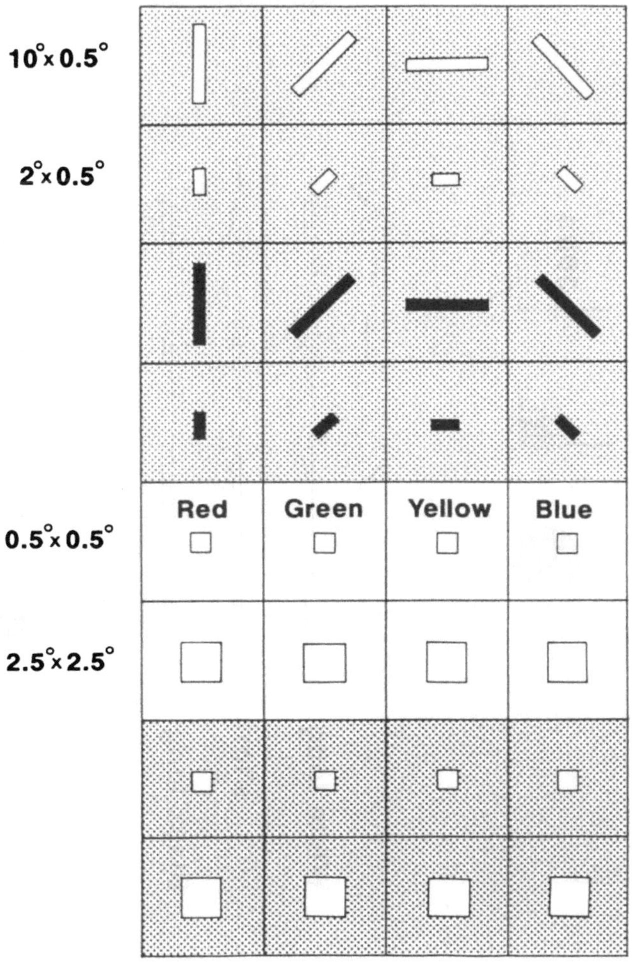

Figure 10. The set of simple stimuli used to quantify the distinctiveness of the stimulus selectivity to particular complex features.

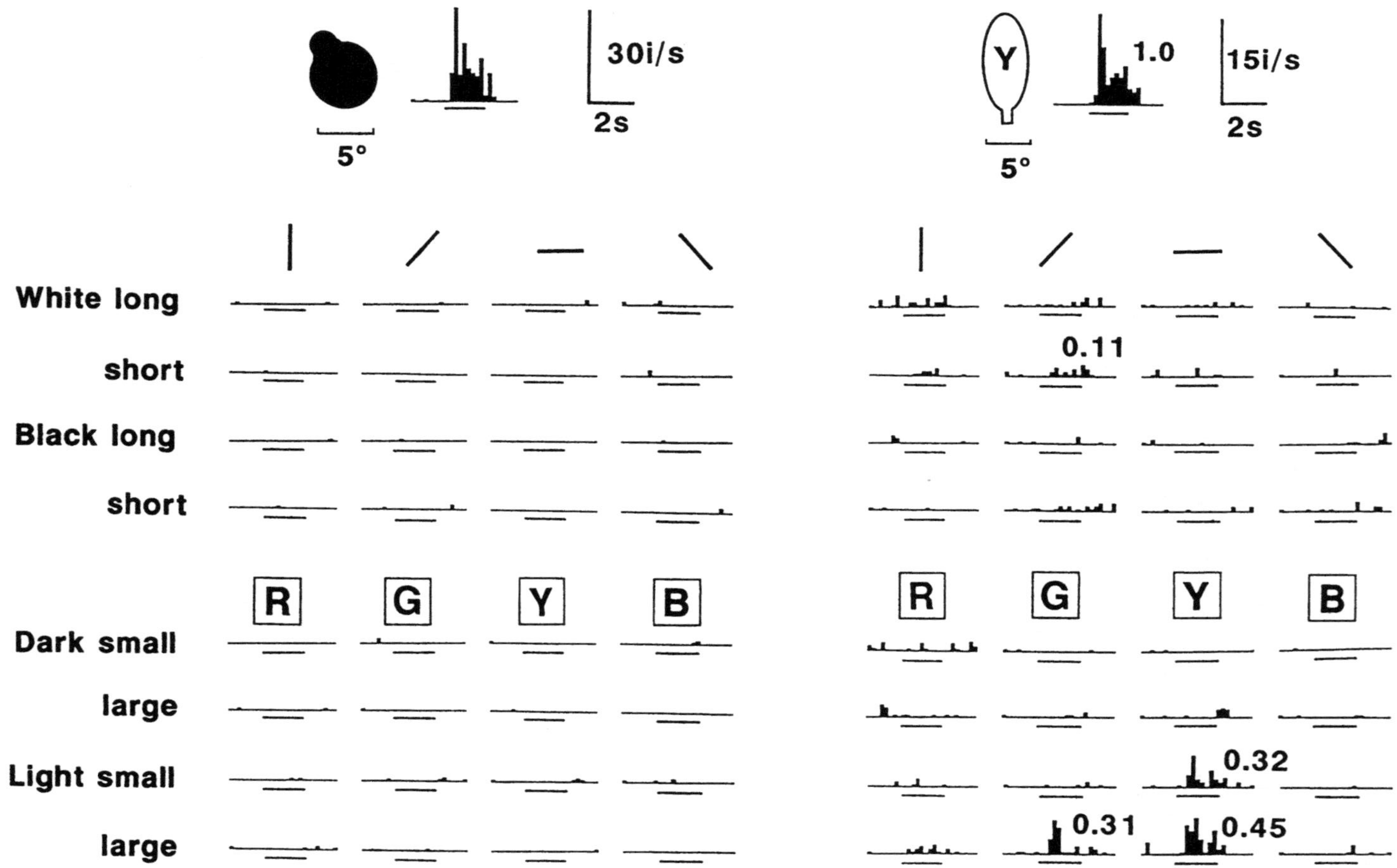

Figure 11. Responses of a TE cell (left) and TEO cell (right) to the set of simple stimuli. Their responses to individually determined critical features are shown at the top.

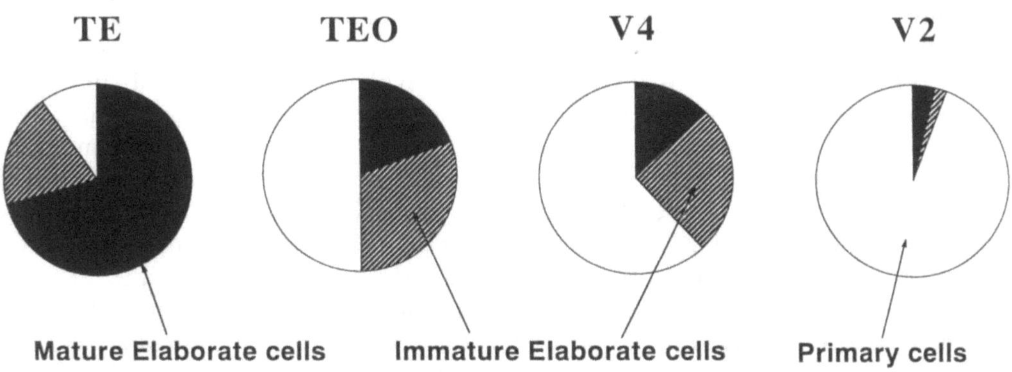

Figure 12. Proportion of cells with different levels of selectivity to complex features. The maximum response to the simple stimuli was <0.25 of the response to the complex critical feature for mature elaborate cells, between 0.25 and 0.75 for immature elaborate cells, and >0.75 for primary cells.

already developed in the circuit up to TEO. We injected an anterograde tracer, *Phaseolus vulgaris* leucoagglutinin (PHA-L), into a single small region (the horizontal width of the injection sites was 330–600 μm) of the part of TEO representing the central visual field, and observed labeled axon terminals in TE (Saleem *et al.*, 1993). Labeled terminals were nearly limited to three to five focal regions in TE (Fig. 13). In each of the projection foci, the labeled terminals were

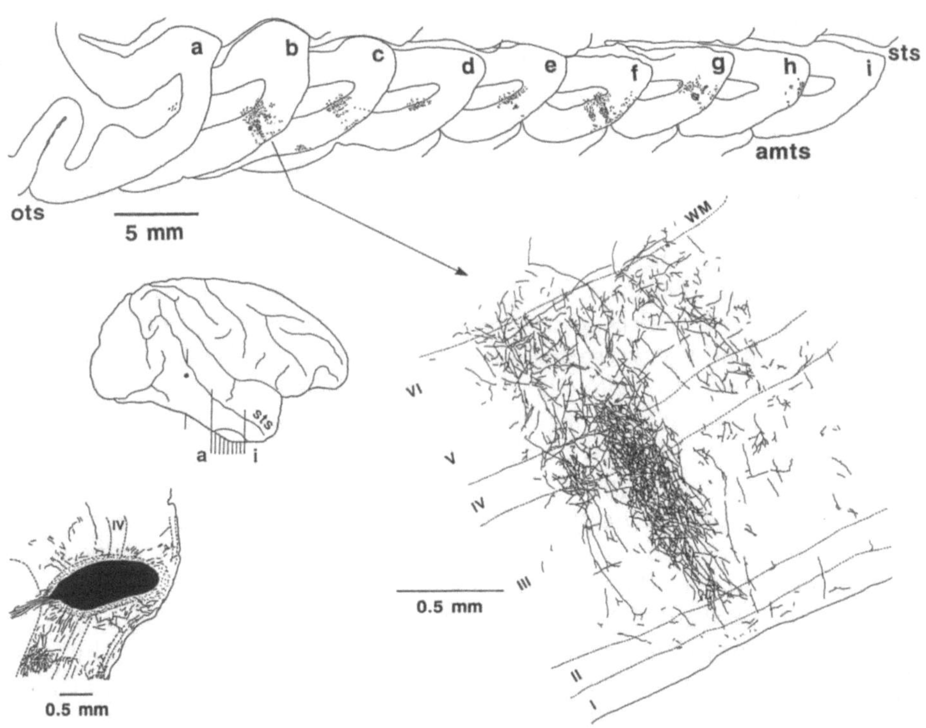

Figure 13. Distribution of labeled axon terminals in TE after a single focal injection of PHA-L into TEO.

not limited to the middle layers, but were distributed to form a columnar region encompassing layers 1–6. The horizontal width of the columnar foci was 200–380 μm, slightly smaller than the physiologically determined width of columns in TE. As noted above, the receptive fields of cells in TE are large, usually including the fovea, and no retinotopical organization has been found in TE. Thus, the specificity of the TEO-to-TE connections should be defined in the feature space, and not in the retinotopical space. It is suggested that outputs from a single site in TEO may carry information about a particular complex feature, and be sent, therefore, to only a limited number of foci in TE.

Although the overall distribution of labeled terminals was elongated to form a columnar region, this does not necessarily mean that individual axons form arbors in columnar regions. Indeed, single axons reconstructed from serial sections were heterogeneous in shape. Some axons terminated exclusively in layer 4 and the bottom of layer 3, like the one shown in Fig. 14, right. Other axons terminated only in layers 1 and 2, like the one shown in Fig. 14, left. There were also single axons that terminated in elongated regions including both the middle and superficial layers. It is possible that a single site in TEO sends multiple different kinds of information about a complex feature to a column in TE, and they interact with each other through the local network within the TE column.

Thus, there are two things first achieved in TE: one is the columnar organization, namely, arrangement of cells with overlapping and slightly different selectivities in local columnar regions; and the other is the invariance of responses for the stimulus position. The receptive fields of cells in TE are large, usually including the fovea, and the selectivity of responses is essentially constant throughout the receptive fields. A significant proportion of cells in TEO and V4 respond to moderately complex stimuli as in TE. However, the receptive fields of the cells in TEO and V4 are much smaller than those of cells in TE and are retinotopically organized (Boussaoud *et al.* 1991; Kobatake and Tanaka, 1994). This means that there are two steps in the construction of cells responding to integrated features with invariance to changes in stimulus position. First, the selectivity is constructed for stimuli at a particular retinal position in TEO and V4, and then the invariance is achieved in TE by obtaining inputs of the same selectivity but with the receptive fields at different retinal positions.

A problem in this two-step structure is that individual cells or columns in TE each require a set of input cells with receptive fields distributed over the large receptive fields of the TE cells. Because the central visual field is magnified in TEO (Boussaoud *et al.*, 1991), cells in the peripheral TEO may not be sufficient in number to extract a great number of integrated features. I would raise the possibility that the inputs from the peripheral TEO convey information on primitive features and that the selectivity in the periphery is constructed at synapses of TE cells. The selectivity can be generalized in TE cells from the central to the peripheral visual field by modifying the synapses of the peripheral inputs according to the generalized Hebbian rule (Foldiak, 1991) whenever the object containing the critical feature moves from the center to the periphery of the visual field. The selective inputs from the central TEO are used as seeds. This hypothesis can be tested by examining properties of responses of cells in the part of TEO representing the peripheral visual field.

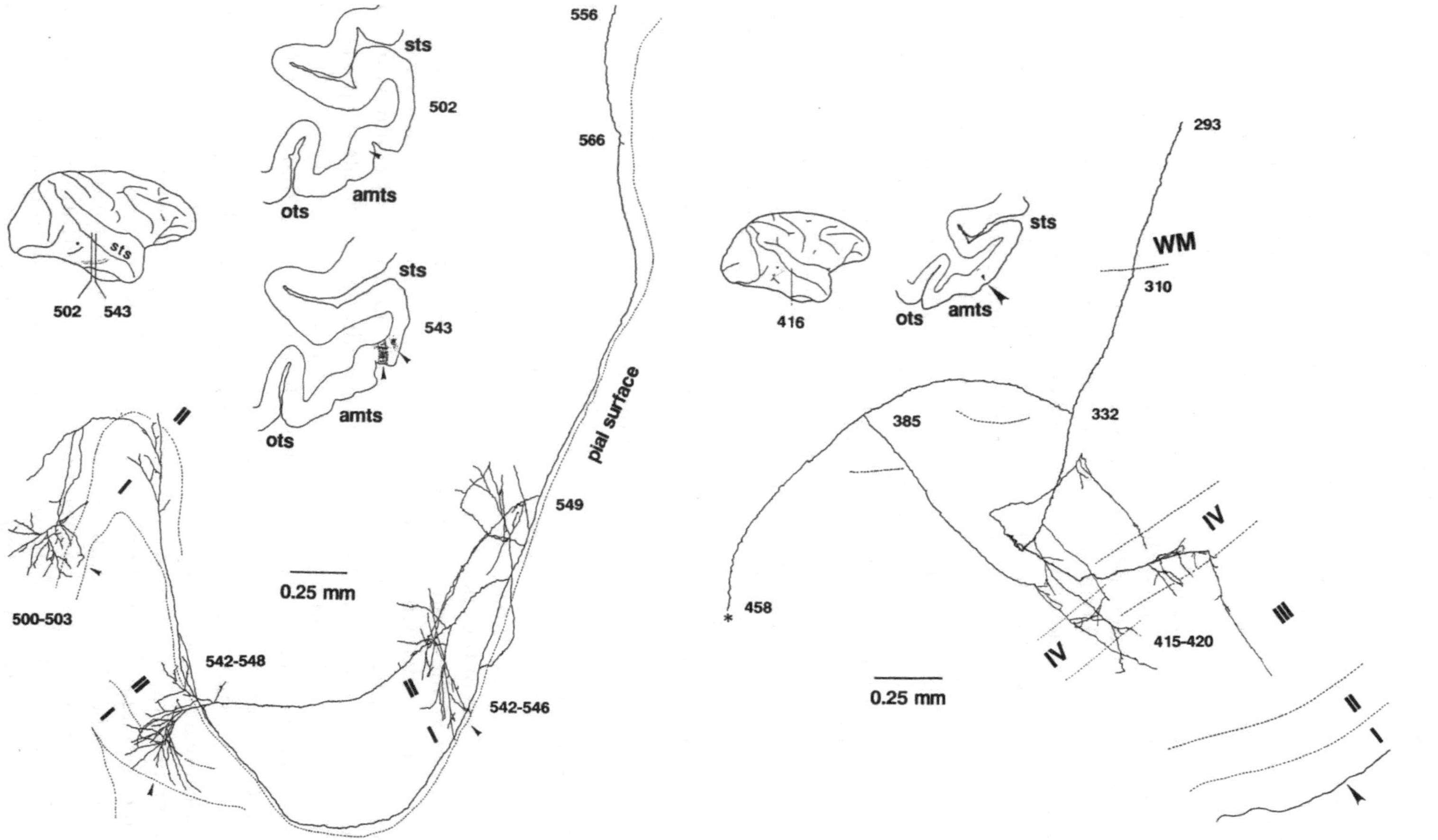

Figure 14. Two single axons projecting from TEO to TE. The axon shown on the left side formed terminal arbors only in layers 1 and 2, whereas the axon on the right formed arbors only around layer 4.

6. Optical Imaging of the Columnar Organization

To further study the spatial properties of columnar organization in TE, we introduced the technique of optical imaging (Wang *et al.*, 1996). The intrinsic signals, which are thought to originate in the increase in the concentration of deoxygenated hemoglobin in capillaries around activated neuronal activities (Frostig *et al.*, 1990), were measured. The cortical surface was exposed and illuminated with red light tuned to 605 nm. Oxygen is released from hemoglobin in activated neuronal tissue, and so the concentration of deoxygenated hemoglobin in nearby capillaries increases. Because deoxygenated hemoglobin absorbs much more light than oxidized hemoglobin at that wavelength, the region of the cortex with elevated neuronal activity becomes darker in the reflected image.

To find visual stimuli that elicit neuronal responses in the part of TE exposed for optical imaging and to establish the relationship between the optical changes and elevated neuronal activity in TE, we combined electrophysiological recordings from single cells with the optical imaging. In contrast to the previous studies performed in V1 and V2, the single-cell recordings were conducted in separate sessions prior to the optical imaging session. Activities of single cells were recorded with a microelectrode, and the critical features were determined for 15–25 cells recorded in six to eight penetrations at different positions. The optical imaging was performed with the critical features and other, control stimuli. Five to 25 different stimuli were combined in a recording session and each of them was presented 24 or 40 times each for 4 sec. The images for individual stimuli were divided by the image without stimuli (blank reference) to remove the basic level unrelated to the visual stimuli, and the images were further subtracted by the image averaged over all the different stimuli combined in the recording session (cocktail reference) to remove nonselective activation of the cortex by visual stimuli.

Each of the critical features determined in the preceding single-cell-recording sessions evoked 2–12 dark spots within the imaged region of TE (3.3 by 6.1 mm). The locations of the spots were different for different features, and one of the spots covered the position of the electrode penetration from which the critical feature was determined (Fig. 15). The average diameter of individual spots was 490 μm, which roughly coincided with the width of columns in TE inferred from the results of the single-cell recording experiments (Fujita *et al.*, 1992). The features should have activated a large proportion of cells within the region to evoke the metabolic change. Thus, clustering of cells that responded to a moderately complex feature was confirmed.

The set of visual stimuli used in one session of optical imaging included three critical features, which were determined for three different cells recorded in the same penetration. Two of them were combinations of two colors of different luminosities, and the third one included a gradation of color from light blue to dark blue. The three stimuli all evoked dark spots around the electrode penetration. All of the spots covered the position of penetration, but they only partially overlapped. Each of the spots was about 500 μm in diameter, and the length of the long axis of the overall region was 1100 μm. The three stimuli were similar to each other in that they all contained a change in luminosity.

Partially overlapping activation elicited by similar stimuli was also observed in a case in which a series of faces of different viewing angles was used. All of the five cells recorded in an electrode penetration selectively responded to the sight of a face. The image of a face could be simplified to a combination of the eyes and nose for one of them with no change in activation level, but we failed to find simpler stimuli eliciting the same level of activation for the remaining four cells. Three of the five cells maximally responded to front faces, and the other two maximally responded to profiles. Thus, we included five different views of one face in the stimulus set used for the optical imaging. In the optical imaging experiment, all of the face stimuli evoked activation spots around the penetration (Fig. 16). Each of the activation spots partially overlapped with the other four spots, and the center position of the spots systematically moved in one direction as the face was turned from the left profile to the right profile through the front and 45 deg faces. Individual spots were 300–400 μm in diameter and the overall region covered by them was 800 μm in the long axis along which the center of the spots moved. These cortical regions were not activated by 10 nonface stimuli included in the stimulus set.

Similar results, i.e., selective activation by faces and systematic shift of activation with the rotation of the face, were observed in three other monkeys. In one of three cases, we presented two different faces of the same series of viewing angles. The two different faces at the same angle activated the same cortical regions. The small possibility that the shift of activation was caused by the displacement of some part of the face which accompanied the rotation was excluded, because the front face presented to horizontally shifted positions activated the same center region.

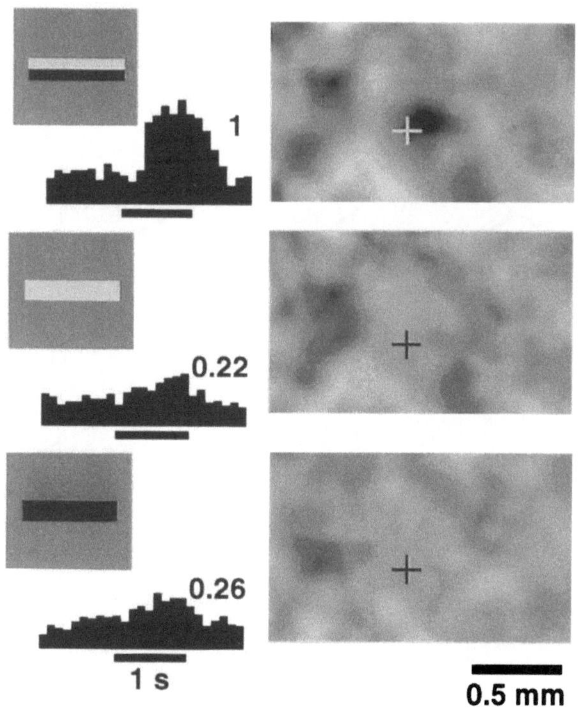

Figure 15. Correspondence of optical signals with neuronal activity. The left histograms show responses of a cell recorded at the site indicated by crosses in the right optical images. The cell selectively responded to a combination of a white bar and black bar. Either a white bar alone or a black bar alone evoked much smaller responses. Correspondingly, the optical image showed a black spot covering the recording site while the monkey was seeing the combination shape, whereas there were no dark spots with the control stimuli.

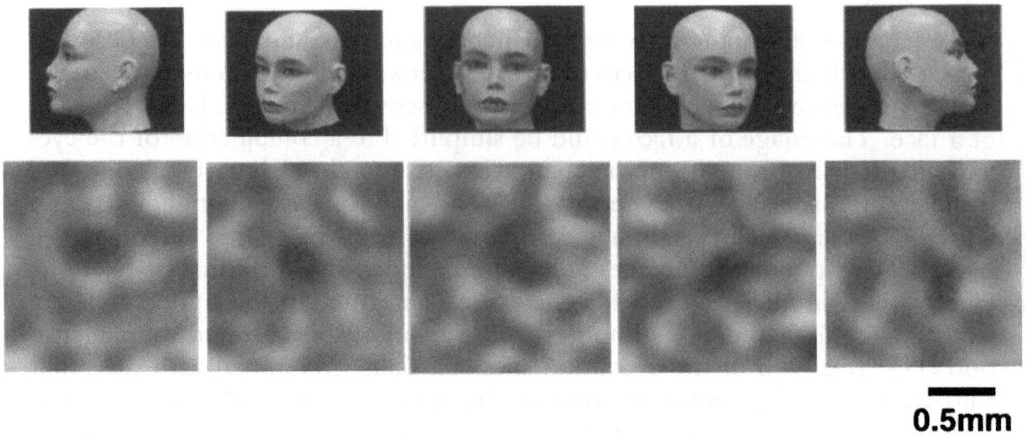

Figure 16. Shift of activation spot with rotation of face. Optical images while the monkey was seeing five different views of the same doll face.

These findings may suggest that multiple columns that represent different but related features overlap with one another and as a whole compose a larger scale unit in TE. The face case further suggests that the arrangement of features within the larger scale unit, at least in some cases, obeys some rule, that is, some parameter of complex features is continuously mapped (Fig. 17). Whether the

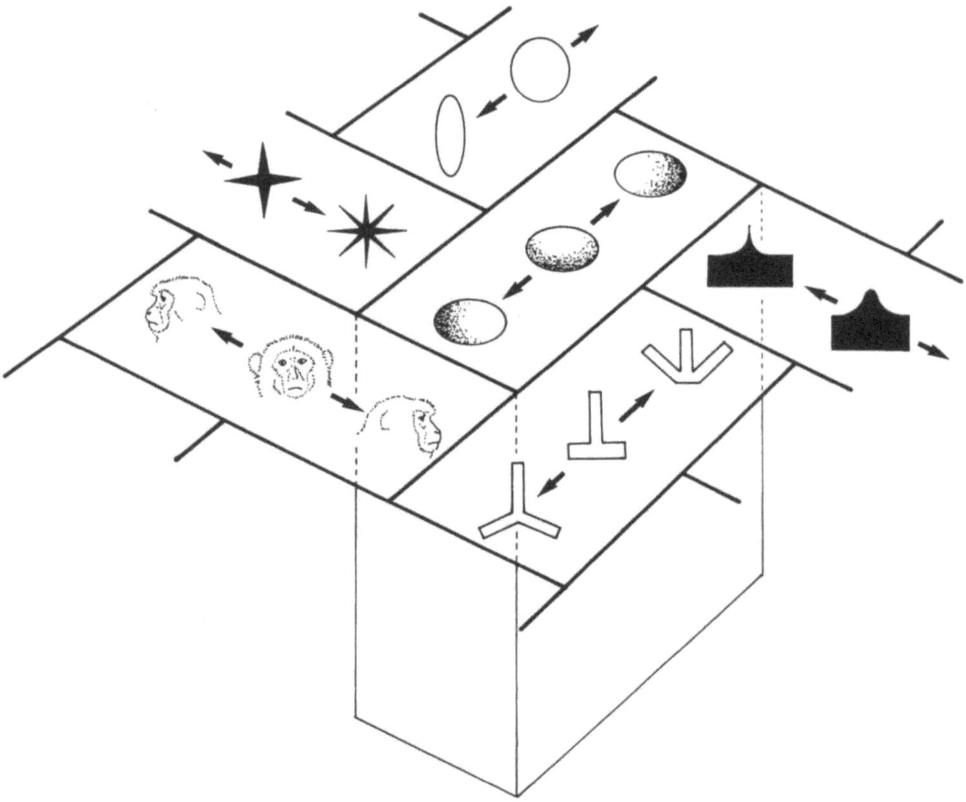

Figure 17. Revised schematic diagram of columnar organization in TE.

mapping is continuous throughout a large part of TE, or there are discontinuities between the units that are probably around 1 mm in length, remains to be determined. Considering that the dimension of the feature space that TE should represent is so high, the latter is more likely.

The continuous mapping of different views of faces cannot be generalized to nonface objects. Because the critical features of TE cells are only moderately complex except faces, the image of a nonface object has to be represented by a combination of activations at multiple cortical sites. Rotation of a nonface object causes shifts of activation at multiple cortical sites, each of which corresponds to change of a partial feature. The parameters along which the activation moves in nonface columns should be examined to uncover further the functional architecture in TE.

7. Changeability of the Selectivity in the Adult

The selectivity to complex critical features is subject to changes according to changes in the visual environment in the adult. We trained two adult monkeys to discriminate the 28 moderately complex shapes shown in Fig. 18 with a standalone apparatus including a display, computer, and touch screen, using a task similar to the delayed-matching-to-sample task (Kobatake *et al.*, 1993, 1992). One stimulus, randomly selected from among the 28 stimuli, appeared on the display (sample stimulus) and disappeared when the monkey touched it. After a 16-sec blank-screen delay, five shapes, including the sample, appeared on the display. The monkey had to select the sample and touch it to obtain a drop of juice. After 1 year of training, the monkeys were prepared for repeated recordings. The

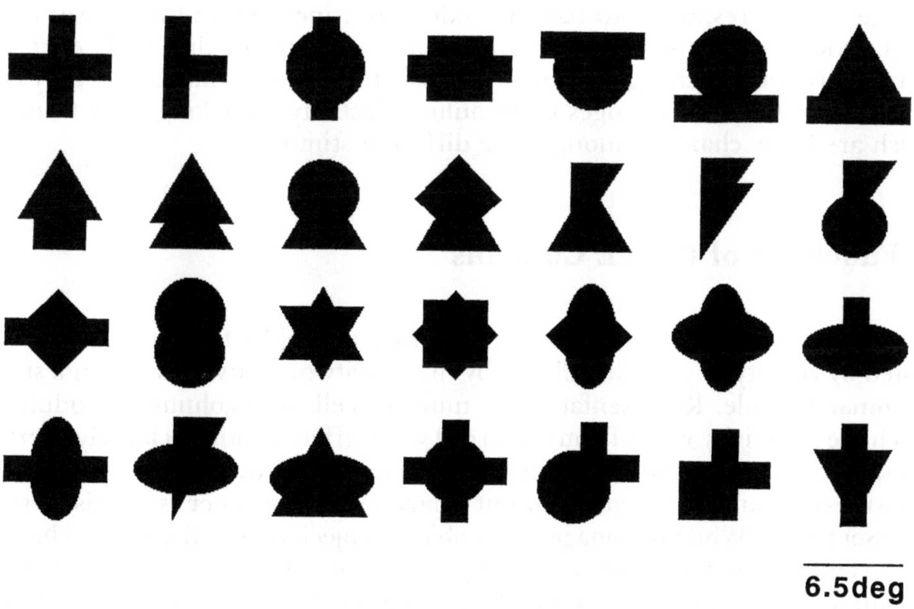

Figure 18. Twenty-eight shape stimuli used in the discrimination training of adult monkeys.

recordings were performed from cells in TE under anesthesia. We determined for individual cells the best stimulus from the set of animal and plant models that we had previously prepared to investigate the critical features in naive monkeys. The response to the best object stimulus was then compared with responses of the same cell to the shape stimuli used in the training. We did not perform the reduction process in this experiment, for the sake of time.

In TE of the trained monkeys, about 25% of the cells gave a maximum response to some of the stimuli used in the training. Conversely, 5% of the TE cells in the untrained animals responded maximally to these stimuli. These results indicate that the number of cells that responded to training stimuli increased due to the 1-year-long discrimination training. However, the spatial organization of the modified cells in the cortex is yet to be studied. We do not know whether new columns were formed for the discrimination of training stimuli, or some of the cells included in the columns present since before the training were tuned to the training stimuli. Whether or not similar changes occurred in TEO and V4, and the time course of the change, also remain to be investigated.

The results of two previous studies showed that responses of cells in TE and surrounding regions changed during the course of recording from the same single cell. Miller *et al.* (1991; Li *et al.*, 1993) found that, as the newly introduced stimulus became familiar, responses of cells at the border region between TE and the perirhinal cortex to the stimuli gradually decreased. This effect is different from the habituation of responses to successive presentation of the same stimulus, because the decrease occurred even after several intervening presentations of different stimuli. However, because the changes that Miller *et al.* observed were opposite in direction to those we observed after the long training, the two phenomena are not likely to be related. Rolls *et al.* (1989) found that responses of cells in TE and the ventral bank of the superior temporal sulcus to faces changed rather rapidly after introduction of a new set of faces. The changes of the responses to faces included both increase and decrease of the magnitude of responses. Because the immediate changes that Rolls *et al.* observed were relative changes among responses to similar stimuli (faces), they may be different from the changes of stimulus selectivity after long-term training, which are likely changes among more different stimuli.

8. Functions of the TE Columns

The columnar organization suggests that an object feature is not represented by activity of a single cell, but by the activity of many cells within a single columnar module. Representation by multiple cells in a columnar module, in which the selectivity varies from cell to cell while effective stimuli largely overlap, can satisfy two apparently conflicting requirements in visual recognition: one is robustness to subtle changes in input images, and the other is preciseness of representation. While the image of an object projected onto the retina changes due to changes in illumination, viewing angle, and pose of the object, the global organization of outputs from TE should change little. The clustering of cells with overlapping and slightly different selectivities works as a buffer to absorb the changes. Although we found that most TE cells are selective for the size, orientation, and the contrast polarity of the stimuli, cells preferring different sizes, orientations, and contrast polarities of the same shape may be packed in a column to provide invariant outputs. Whether signals from these selective cells converge to a group of single cells which show invariant responses is a matter of discussion and for further investigation. An alternative possibility is that outputs of the cells preferring different sizes, orientations, and contrast polarities of the same shape overlap in the target structure to evoke the same effects.

The representation by multiple cells with overlapping selectivities can be more precise than a mere summation of representations by individual cells. A similar argument has been made for retinal cells and the hyperacuity (Snippe and Koenderink, 1992; Erickson, 1968). The position of the receptive fields changes gradually on the retina with a large overlap among nearby cells. By taking the difference between the activities of nearby cells, an acuity much smaller than the size of the receptive fields is produced. A similar mechanism to that in retinal space may work in feature space with largely overlapping and gradually changing selectivities. A subtle change in a particular feature, which does not markedly change the activity of individual cells, can be coded by the differences in the activities of cells with overlapping and slightly different selectivities.

The function of the columnar organization in TE may go beyond the discrimination of input images. The results of the optical imaging experiments suggested that there is a continuous mapping of features within cortical units about 1 mm in length across the cortical surface. There may be twofold functional significance of this continuous mapping. One is to form an evenly distributed variety of cell properties along the feature axis. The continuous mapping may be a tool to achieve the full divergence without omission (Malach, 1994; Purves *et al.,* 1992). The other one is to conduct computations involving the mapped features based on the local neuronal connections between the cells representing the various features. The computations may serve to transfer the image of an object for three-dimensional rotations, or production of the image under different illumination conditions or at different articulation poses.

The results of a series of studies recently performed in slices of the rat motor cortex suggest that there are two kinds of connections between pyramidal neurons through their axon collaterals (Thomson and Deuchars, 1994). Pyramidal cells located within a narrow columnar region 50–100 μm in width are tightly connected by synapses on the basal dendrites or the proximal part of the apical dendrites. They tend to fire together with sharp-rising large excitatory postsynaptic potentials (EPSPs) propagated through these connections. There is another anatomical structure which provides similar response properties to pyramidal cells within a narrow column. The shafts of their apical dendrites converge to compose a bundle, on which input axons may form synapses without discrimination (Peters and Yilmaz, 1993). This narrow column corresponds to the "minicolumn" described by Mountcastle (1978). In contrast, pyramidal cells which are further apart are connected by synapses at the distal part of the apical dendrites. The EPSPs are small and slowly rising, while they are long-lasting: the synapses may contain *N*-methyl-D-aspartate glutamate receptors.

Considering all of the above, we may draw a schema of TE in which cells within the minicolumn compose a unit by receiving common inputs and exciting each other, and nearby minicolumns send long-lasting but weak excitatory inputs to each other. After a minicolumn is activated by the retinal visual input, subthreshold activation propagates from it to nearby minicolumns forming a pattern of activation with a focus. The focus of activation may move from one minicolumn to another, through interaction with distant activation foci in TE or interaction with other brain sites (as will be described in Section 11). This mechanism may be used for various kinds of computation that the visual system must perform to achieve flexibility in visual recognition, such as transfer of the image of an object for three-dimensional rotations, production of the image under different illumination conditions, and, in the case of a face, production of the image with different expressions.

Thus, the columnar organization of TE may enable overlapping and continuous representation of object features, upon which various kinds of calculations can be performed.

9. Binding Activities in Distant Columns

Because object features to which individual TE cells responded were only moderately complex and cells within a single column responded to similar features, the calculation performed within a column can provide only information on partial (either holistic or local) features of object images. To represent the entire image of an object, calculation in several or several tens of different columns must be combined. This evokes the problem of "binding," that is, how to discriminate different sets of activity when there is more than one object in the nearby retinal positions. The receptive fields of TE cells are too large to discriminate different objects according to their retinal positions. The problem of "binding" exists regardless of the presence or absence of the "concept units" representing the concept of objects through their activity alone in brain sites outside TE. The concept units, if present, have to discriminate different sets of TE activity elicited by images of different objects.

One possible mechanism to overcome this problem is the synchronization of firings (Singer, 1993; Engel *et al.*, 1992). If firing of cells that is elicited by the image of the same object is synchronized and firing that is elicited by images of different objects is desynchronized, the different sets of firings will be discriminated from each other. Firing synchronized with oscillations has been found between cells in the cat visual cortex, and some context dependence of the synchronization has also been reported. Although oscillating firing has not been found in TE (Young *et al.*, 1992; Tovee and Rolls, 1992), nonperiodic synchronization may be present in TE.

Another possible mechanism of overcoming the problem of binding in TE is selection by attention (Crick, 1984). We can pay attention to only one, or at most a few, objects at a time. If the representation of features of an attended object is enhanced and that of other objects is suppressed, the binding problem will disappear. This mechanism is likely to exist, because strong effects of attention have been found on responses of TE cells (Chelazzi *et al.*, 1993; Spitzer *et al.*, 1988; Moran and Desimone, 1985; Richmond and Sato, 1987).

A third possibility is that the set of distant activities in TE elicited from the image of a single object is combined by the formation of loops of activity with activity in retinotopically organized areas at former stages in the ventral pathway. TE projects back to TEO, V4, V2, and even V1 (Rockland *et al.*, 1994; Rockland and Van Hoesen, 1994). There are also step-by-step feedback projections. Kawato *et al.* (1991, 1993) have indicated the importance of feedback projections from a similar viewpoint.

10. "Face Neurons" in the Anterior Part of the Superior Temporal Sulcus

Cells that selectively responded to the sight of a face were found in STPa in the early 1980s (Perrett *et al.*, 1982; Bruce *et al.*, 1981), and such cells, so-called "face neurons," have been extensively studied. There are reports that such cells are also present in TE itself (Tanaka *et al.*, 1991; Baylis *et al.*, 1987), area TG (Nakamura *et al.*, 1994), and the amygdala (Nakamura *et al.*, 1992; Leonard *et al.*, 1985). The meaning of "selective" varied among these studies: in some of them only several nonface stimuli were presented, and in most of them partial features of the image of a face were not tested. However, in a few of them a scrambled face, which consisted of scrambled patches of the picture of a face, was used and found not to be effective for the neuronal activation (Bruce *et al.*, 1981). In other studies it was found that there were cells that were not activated by a face missing the eyes or by the eyes only (Rolls *et al.*, 1985; Perrett *et al.*, 1982). We found systematically arranged columns in TE that responded to different views of faces. These data suggest that there are cells that require all the essential features that compose the image of a face. The image of a face is more complex than the other features represented by cells in TE.

The presence of face neurons cannot be generalized to representation of other objects. Faces of monkeys are special to monkeys, and those of human beings to laboratory monkeys, in that faces are important media for social communication between individuals. There are reports that activities of the face neurons in STPa represent individual differences and expressions by population coding (Rolls and Tovee, 1995; Young and Yamane, 1992; Baylis *et al.*, 1985). Discrimination of faces from other objects is only a preliminary stage in the representation of information of expressions or features of individual faces. The units prepared for population coding of individual differences and expressions of faces may appear as "concept units" of faces. This view is supported by the fact that there are cells that respond to views of body poses, body movements, or hand actions in STPa (Oram and Perrett, 1994; Perrett *et al.*, 1985, 1989, 1992). Body movements and hand actions often express important information about the relation between individuals in a scene or between an individual in a scene and the observer. These groups of cells in STPa may be specially developed for social communication.

Sakata and his colleagues recently found shape-selective cell activities in areas of the intraparietal sulcus (Sakata and Kusunoki, 1992; Taira *et al.*, 1990). Many cells in the lateral bank of the sulcus selectively responded to visual images of switches that the monkey was trained to manipulate. The monkey grasped the switches and pulled them. The switches varied in shape, and the monkey shaped its hand differently before it reached for different switches. The discharges started when the monkey saw the switches, and because the discharges decreased in frequency when the task was performed in a dark room, the discharges might be caused, at least partially, by the visual inputs.

Sakata and his colleagues further found that some cells in a more posterior part of the sulcus responded to more primitive features of stimuli, including the 3D orientation of a pole or 3D tilt of a plane. They presented the stimuli on a computer graphic system with binocular disparity (Tanaka *et al.*, 1992, 1994; Kusunoki *et al.*, 1993). The responses were reduced in magnitude when the binocular disparity was removed, which indicates that the binocular disparity gave an essential cue for the responses.

Considering all of these findings, there seems to exist a flow of information of the 3D shape of objects that the monkey may manipulate. These findings are consistent with the proposition by Goodale *et al.* (1991, 1992), based on human clinical data, that the dorsal pathway leading to the parietal cortex is responsible for "visuomotor control," but not for "spatial vision" (Mishkin *et al.*, 1983). To manipulate an object, its 3D structure must be perceived. There may be a representation of the shape of objects in the dorsal pathway, independent of the representation of objects in the ventral pathway, and shapes are probably only roughly represented. This dorsal representation of objects may influence the representation of objects in the ventral pathway through the indirect connection via the parahippocampal structures (Suzuki and Amaral, 1995; Van Hoesen, 1982) or via regions in the superior temporal sulcus (Seltzer and Pandya, 1978, 1984, 1989, 1994).

The accumulated findings support the idea that there are no cognition units representing the concept of objects, but that the concept of objects is only represented by the activities distributed over various regions of the brain. The visual image of an object is processed in the ventral visual pathway, and the representation showing similarity to other representations and association with images of the same object under different conditions is reconstructed there. This representation in the ventral visual pathway utilizes two levels of population coding. First, the image of the object is represented by a combination of multiple partial (either holistic or local) features represented by different columns in TE. Because the presence of the partial features is represented in an analog manner, the combinatorial representation can be understood as a combination of similarity to different prototypes of object images (Edelman, 1995). The second level of population coding is in the representation of partial features. The features are represented by multiple cells within a TE column that have overlapping selectivities. This level of population coding was extensively discussed in Section 8. Triggered by inputs from the inferotemporal cortex, emotional information of the object is read out in the amygdala, association with other objects is read out through the perirhinal cortex, and behavioral significance emerges in the prefrontal cortex. The visual image of the object is also processed in the dorsal pathway, and information necessary for the monkey to manipulate the object is read out in the parietal cortex. All this recovered information of the object, distributed over various regions of the brain, may represent the concept of the object.

12. References

Baylis, G. C., Rolls, E. T., and Leonard, C. M., 1985, Selectivity between faces in the responses of a population of neurons in the cortex in the superior temporal sulcus of the monkey, *Brain Res.* **342**:91–102.

Baylis, G. C., Rolls, E. T., and Leonard, C. M., 1987, Functional subdivisions of the temporal lobe neocortex, *J. Neurosci.* **7**:330–342.

Boussaoud, D., Desimone, R., and Ungerleider, L. G., 1991, Visual topography of area TEO in the macaque, *J. Comp. Neurol.* **306**:554–575.

Bruce, C., Desimone, R., and Gross, C. G., 1981, Visual properties of neurons in a polysensory area in superior temporal sulcus of the macaque, *J. Neurophysiol.* **46**:369–384.

Chelazzi, L., Miller, E. K., Duncan, J., and Desimone, R., 1993, A neural basis for visual search in inferior temporal cortex, *Nature* **363**:345–347.

Cheng, K., Saleem, K. S., and Tanaka, K., 1993, PHA-L study of the subcortical projections of the macaque inferotemporal cortex, *Soc. Neurosci. Abstr.* **19**:971.

Crick, F., 1984, Function of the thalamic reticular complex: The searchlight hypothesis, *Proc. Natl. Acad. Sci. USA* **81**:4586–4590.

Dean, P., 1976, Effects of inferotemporal lesions on the behavior of monkeys, *Psychol. Bull.* **83**:41–71.

Desimone, R., Fleming, J., and Gross, C. D., 1980, Prestriate afferents to inferior temporal cortex: An HRP study, *Brain Res.* **184**:41–55.

Edelman, S., 1995, Representation, similarity and the chorus of prototypes, *Minds Machines* **5**:45–68.

Engel, A. K., Konig, P., Kreiter, A. K., Schillen, T. B., and Singer, W., 1992, Temporal coding in the visual cortex: New vistas on integration in the nervous system, *Trends Neurosci.* **15**:218–226.

Erickson, R. P., 1968, Stimulus coding in topographic and nontopographic afferent modalities: On the significance of the activity of individual neurons, *Psychol. Rev.* **75**:447–465.

Foldiak, P., 1991, Learning invariance from transformation sequences, *Neural Comput.* **3**:194–200.

Frostig, R. D., Lieke, D. E., Ts'o, D. Y., and Grinvald, A., 1990, Cortical functional architecture and local coupling between neuronal activity and the microstimulation revealed in *in vivo* high-resolution optical imaging of intrinsic signals, *Proc. Natl. Acad. Sci. USA* **87**:6082–6086.

Fujita, I., Tanaka, K., Ito, M., and Cheng, K., 1992, Columns for visual features of objects in monkey inferotemporal cortex, *Nature* **360**:343–346.

Goodale, M. A., and Milner, A. D., 1992, Separate visual pathways for perception and action, *Trends Neurosci.* **15**:20–25.

Goodale, M. A., Milner, A. D., Jakobson, L. S., and Carey, D. P., 1991, A neurological dissociation between perceiving objects and grasping them, *Nature* **349**:154–156.

Gross, C. G., 1973, Visual functions of inferotemporal cortex, in: *Handbook of Sensory Physiology*, Vol. 7, Part 3B (R. Jung, ed.), Springer-Verlag, Berlin, pp. 451–482.

Gross, C. G., Rocha-Miranda, C. E., and Bender, D. B., 1972, Visual properties of neurons in inferotemporal cortex of the macaque, *J. Neurophysiol.* **35**:96–111.

Ito, M., Fujita, I., Tamura, H., and Tanaka, K., 1994, Processing of contrast polarity of visual images in inferotemporal cortex of the macaque monkey, *Cerebral Cortex* **5**:499–508.

Ito, M., Tamura, H., Fujita, I., and Tanaka, K., 1995, Size and position invariance of neuronal responses in monkey inferotemporal cortex, *J. Neurophysiol.* **73**:218–226.

Iwai, E., and Yukie, M., 1987, Amygdalofugal and amygdalopetal connections with modality-specific visual cortical areas in macaques (*Macaca fuscata, M. mulatta,* and *M. fascicularis*), *J. Comp. Neurol.* **261:**362–387.

Iwai, E., Yukie, M., Suyama, H., and Shirakawa, S., 1987, Amygdalar connections with middle and inferior temporal gyri of the monkey, *Neurosci. Lett.* **83:**25–29.

Kawato, M., Inui, T., Hongo, S., and Hayakawa, H., 1991, Computational theory and neural network models of interaction between visual cortical areas, ATR Technical Report, TR-A-0105 (Kyoto, ATR).

Kawato, M., Hayakawa, H., and Inui, T., 1993, A forward-inverse optics model of reciprocal connections between visual cortical areas, *Network* **4:**415–422.

Kobatake, E., and Tanaka, K., 1994, Neuronal selectivities to complex object features in the ventral visual pathway of the macaque cerebral cortex, *J. Neurophysiol.* **71:**856–867.

Kobatake, E., Tanaka, K., and Tamori, Y., 1992, Long-term learning changes the stimulus selectivity of cells in the inferotemporal cortex of adult monkeys, *Neurosci. Res.* **S17:**S237.

Kobatake, E., Tanaka, K., Wang, G., and Tamori, Y., 1993, Effects of adult learning on the stimulus selectivity of cells in the inferotemporal cortex, *Soc. Neurosci. Abstr.* **19:**975.

Kusunoki, M., Tanaka, Y., Ohtsuka, H., Ishiyama, K., and Sakata, H., 1993, Selectivity of the parietal visual neurons in the axis orientation of objects in space, *Soc. Neurosci. Abstr.* **19:**770.

Lehky, S. R., Sejnowski, T. J., and Desimone, R., 1992, Predicting responses of nonlinear neurons in monkey striate cortex to complex patterns, *J. Neurosci.* **12:**3568–3581.

Leonard, C. M., Rolls, E. T., Wilson, F. A. W., and Baylis, G. C., 1985, Neurons in the amygdala of the monkey with responses selective for faces, *Behav. Brain Res.* **15:**159–176.

Li, L., Miller, E. K., and Desimone, R., 1993, The representation of stimulus familiarity in anterior inferior temporal cortex, *J. Neurophysiol.* **69:**1918–1929.

Malach, R., 1994, Cortical columns as devices for maximizing neuronal diversity, *Trends Neurosci.* **17:**101–104.

Martin-Elkins, C. L., and Horel, J. A., 1992, Cortical afferents to behaviorally defined regions of the inferior temporal and parahippocampal gyri as demonstrated by WGA-HRP, *J. Comp. Neurol.* **321:**177–192.

Miller, E. K., Li, L., and Desimone, R., 1991, A neural mechanism for working and recognition memory in inferior temporal cortex, *Science* **254:**1377–1379.

Mishkin, M., Ungerleider, L. G., and Macko, K. A., 1983, Object vision and spatial vision: Two cortical pathways, *Trends Neurosci.* **6:**414–417.

Moran, J., and Desimone, R., 1985, Selective attention gates visual processing in the extrastriate cortex, *Science* **229:**782–784.

Mountcastle, V. B., 1978, An organizing principle for cerebral function: The unit module and the distributed system, in: *The Mindful Brain* (V. B. Mountcastle and G. M. Edelman, eds.), MIT Press, Cambridge, MA, pp. 7–50.

Nakamura, H., Gattass, R., Desimone, R., and Ungerleider, L. G., 1993, The modular organization of projections from areas V1 and V2 to areas V4 and TEO in macaques, *J. Neurosci.* **13:**3681–3691.

Nakamura, K., Matsumoto, K., Mikami, A., and Kubota, K., 1994, Visual response properties of single neurons in the temporal pole of behaving monkeys, *J. Neurophysiol.* **71:**1206–1221.

Nakamura, K., Mikami, A., and Kubota, K., 1992, Activity of single neurons in the monkey amygdala during performance of a visual discrimination task, *J. Neurophysiol.* **67:**1447–1463.

Oram, M. W., and Perrett, D. I., 1994, Responses of anterior superior temporal polysensory (STPa) neurons to "biological motion" stimuli, *J. Cognitive Neurosci.* **6:**99–116.

Perrett, D. I., Rolls, E. T., and Caan, W., 1982, Visual neurones responsive to faces in the monkey temporal cortex, *Exp. Brain Res.* **47:**329–342.

Perrett, D. I., Smith, P. A. J., Mistlin, A. J., Chitty, A. J., Head, A. S., Potter, D. D., Broennimann, R., Milner, A. D., and Jeeves, M. A., 1985, Visual analysis of body movements by neurons in the temporal cortex of the macaque monkey: A preliminary report, *Behav. Brain Res.* **16:**153–170.

Perrett, D. I., Harries, M. H., Bevan, R., Thomas, S., Benson, P. J., Mistlin, A. J., Chitty, J. K., Hietanen, J. K., and Ortega, J. E., 1989, Frameworks of analysis for the neural representation of animate objects and actions, *J. Exp. Biol.* **146:**87–113.

Perrett, D. I., Hietanen, J. K., Oram, M. W., and Benson, P. J., 1992, Organization and functions of cells responsive to faces in the temporal cortex, *Phil. Trans. R. Soc. Lond. B* **335:**23–30.

Peters, A., and Yilmaz, E., 1993, Neuronal organization in area 17 of cat visual cortex, *Cerebral Cortex* **3:**49–68.

Purves, D., Riddle, D. R., and LaMantia, A.-S., 1992, Iterated patterns of brain circuitry (or how the cortex gets its spots), *Trends Neurosci.* **15:**362–368.

Richmond, B. J., and Sato, T., 1987, Enhancement of inferior temporal neurons during visual discrimination, *J. Neurophysiol.* **58:**1292–1306.

Rockland, K. S., and Van Hoesen, G. W., 1994, Direct temporal-occipital feedback connections to striate cortex (V1) in the macaque monkey, *Cerebral Cortex* **4:**300–313.

Rockland, K. S., Saleem, K. S., and Tanaka, K., 1994, Divergent feedback connections from areas V4 and TEO in the macaque, *Visual Neurosci.* **11:**579–600.

Rolls, E. T., and Tovee, M. J., 1995, Sparseness of the neuronal representation of stimuli in the primate temporal visual cortex, *J. Neurophysiol.* **73:**713–726.

Rolls, E. T., Baylis, G. C., and Leonard, C. M., 1985, Role of low and high spatial frequencies in the face-selective responses of neurons in the cortex in the superior temporal sulcus in the monkey, *Vision Res.* **25:**1021–1035.

Rolls, E. T., Baylis, G. C., Hasselmo, M. E., and Nalwa, V., 1989, The effect of learning on the face-selective responses of neurons in the cortex in the superior temporal sulcus of the monkey, *Exp. Brain Res.* **76:**153–164.

Sakata, H., and Kusunoki, M., 1992, Organization of space perception: Neural representation of three-dimensional space in the posterior parietal cortex. *Curr. Opin. Neurobiol.* **2:**170–174.

Saleem, K. S., Tanaka, K., and Rockland, K. S., 1992, PHA-L study of connections from TEO and V4 to TE in the monkey visual cortex, *Soc. Neurosci. Abstr.* **18:**294.

Saleem, K. S., Tanaka, K., and Rockland, K. S., 1993, Specific and columnar projection from area TEO to TE in the macaque inferotemporal cortex, *Cerebral Cortex* **3:**454–464.

Saleem, K. S., Cheng, K., and Tanaka, K., 1994, Differential projection from ventral and dorsal parts of the anterior TE to perirhinal cortex in the macaque monkey, *Neurosci. Res.* **19:**S201.

Saleem, K. S., Cheng, K., and Tanaka, K., 1995, Differential cortical projection of dorsal and ventral sub-regions of the area TE in the macaque inferotemporal cortex, *IBRO Abstr.* **4:**284.

Sary, G., Vogels, R., and Orban, G. A., 1993, Cue-invariant shape selectivity of macaque inferior temporal neurons, *Science* **260:**995–997.

Seltzer, B., and Pandya, D. N., 1978, Afferent cortical connections and architectonics of the superior temporal sulcus and surrounding cortex in the rhesus monkey, *Brain Res.* **149:**1–24.

Seltzer, B., and Pandya, D. N., 1984, Further observations on parieto-temporal connections in the rhesus monkey, *Exp. Brain Res.* **55:**301–312.

Seltzer, B., and Pandya, D. N., 1989, Intrinsic connections and architectonics of the superior temporal sulcus in the rhesus monkey, *J. Comp. Neurol.* **290:**451–471.

Seltzer, B., and Pandya, D. N., 1994, Parietal, temporal, and occipital projections to cortex of the superior temporal sulcus in the rhesus monkey: A retrograde tracer study, *J. Comp. Neurol.* **343:**445–463.

Singer, W., 1993, Synchronization of cortical activity and its putative role in information processing and learning, *Annu. Rev. Physiol.* **55:**349–374.

Snippe, H. P., and Koenderink, J. J., 1992, Discrimination thresholds for channel-coded systems, *Biol. Cybernet.* **66:**543–551.

Spitzer, H., Desimone, R., and Moran, J., 1988, Increased attention enhances both behavioral and neuronal performance, *Science* **240:**338–340.

Suzuki, W. A., and Amaral, D. G., 1995, Perirhinal and parahippocampal cortices of the macaque monkey: Cortical afferents, *J. Comp. Neurol.* **349:**1–36.

Taira, M., Mine, S., Georgopoulos, A. P., Murata, A., and Sakata, H., 1990, Parietal cortex neurons of the monkey related to the visual guidance of hand movement, *Exp. Brain Res.* **83:**29–36.

Tanaka, K., Saito, H., Fukada, Y., and Moriya, M., 1991, Coding visual images of objects in the inferotemporal cortex of the macaque monkey, *J. Neurophysiol.* **66:**170–189.

Tanaka, Y., Kusunoki, M., Ohtsuka, H., Takiura, K., and Sakata, H., 1992, Analysis of three dimensional directional selectivity of the monkey parietal depth-movement-sensitive neurons using a stereoscopic computer display system, *Neurosci. Res.* **S17:**S238.

Tanaka, Y., Murata, A., Taira, M., Shikata, E., and Sakata, H., 1994, Responses of the parietal visual neurons to stereoscopic stimuli on the computer graphic display in alert monkeys, *Neurosci. Res.* **S19:**S200.

Thomson, A. M., and Deuchars, J., 1994, Temporal and spatial properties of local circuits in neocortex, *Trends Neurosci.* **17:**119–126.

Tovee, M. J., and Rolls, E. T., 1992, Oscillatory activity is not evident in the primate temporal visual cortex with static stimuli, *NeuroReport* **3:**369–372.

Ungerleider, L. G., Gaffan, D., and Pelak, V. S., 1989, Projections from inferior temporal cortex to prefrontal cortex via the uncinate fascicle in rhesus monkeys, *Exp. Brain Res.* **76:**473–484.

Van Hoesen, G. W., 1982, The parahippocampal gyrus: New observations regarding its cortical connections in the monkey, *Trends Neurosci.* **5:**345–350.

Wang, G., Tanaka, K., and Tanifuji, M., 1996, Optical imaging of functional organization in the monkey inferotemporal cortex, *Science* **272:**1665–1668.

Young, M. P., and Yamane, S., 1992, Sparse population coding of faces in the inferotemporal cortex, *Science* **29:**1327–1331.

Young, M. P., Tanaka, K., and Yamane, S., 1992, On oscillating neuronal responses in the visual cortex of the monkey, *J. Neurophysiol.* **67:**1464–1474.

Yukie, M., Hikosaka, K., and Iwai, E., 1992, Organization of cortical visual projections to the dorsal and ventral parts of area TE of the inferotemporal cortex in macaques, *Soc. Neurosci. Abstr.* **18:**294.

12

Construction and Representation of Visual Space in the Inferior Parietal Lobule

RALPH M. SIEGEL and HEATHER L. READ

1. Introduction

In human subjects, neurological case histories have provided details of striking deficits of cognitive functions. These findings have been supplemented by comparative studies in nonhuman primates that permit invasive manipulations and measurements of brain tissue, as well as theories that provide a structure for the synthesis and distillation of such clinical and experimental results into general principles. The formation of visuospatial representations of the surrounding world has proven to be particularly amenable to this multifaceted approach. Although far from being completely understood, substantial progress has been made in the past 20 years in the analysis of this cognitive process. One patient, suffering from profound hemi-inattention has:

> lost her idea of 'left,' both with regard to the world and her own body. Sometimes (when eating) she complains that her portions are too small, but this is because she eats only from the right half of the plate—it does not occur to her that it has a left half as well. Sometimes, she will put on lipstick, and make up the right half of her face, leaving the

RALPH M. SIEGEL and HEATHER L. READ • Center for Molecular and Behavioral Neuroscience, Rutgers University, Newark, New Jersey 07102.

Cerebral Cortex, Volume 12, edited by Rockland *et al.* Plenum Press, New York, 1997.

> left half completely neglected: It is almost impossible to treat these things, because her attention cannot be drawn to them, and she has no conception that they are wrong. She knows it intellectually, and can understand, and laugh; but it is impossible for her to know it directly. (Sacks, 1985)

This particular type of spatial and attentional deficit arises from damage to associational cortical regions which were initially defined on the basis of late myelination during development (Flechsig, 1876). This patient does not have a hemianopia (half-vision field blindness), for her vision is normal. But her cognitive disassociations are in a sense more profound than blunt dissections of primary sensory or motor function. Removal of primary sensory or motor cortex leads to devastating blindness, deafness, or paralysis. But such loss can be overcome, at least in the sensory modality, by reliance on spared sensory capabilities. However, damage to the association cortex strikes at the very core of being. The patient can see but cannot *attend* to half of her visual world. She cannot intellectually compensate, because her *spatial* intellect itself is damaged. The challenge to neuroscience is to generate mechanistic explanations, perhaps reductionist, perhaps synthetic, to account for normal cognitive deficits on this scale. These explanations might explain cognitive function in the normal individual.

In this chapter, the spectrum of spatial deficits found in humans is discussed first. Then a nonhuman primate model is described, pinpointing regions of cortex most likely to be involved in one type of visuospatial impairment. Particular attention is given to the properties of single neurons in the monkey cortex with respect to the different behavioral components of spatial deficits. Theoretical constraints are extracted from the experimental data, and finally a computational model that constructs a representation of space is presented.

2. Neurological Findings

Lesions of the inferior parietal cortex in humans produce a consistent and readily recognized set of clinical signs and symptoms, which include deficits in visuospatial perception, spatial memories, route following, and oculomotor control (Critchley, 1953; Lynch, 1980). This set of deficits, called "parietal neglect syndrome" or simply "hemispatial neglect," can be associated with the disruption of spatial perception. The syndrome includes disturbances of the awareness of body, of the relation of body to the environment, and of the perception of the topographic organization of the environment. Patients report that the environment appears "confused" or "jumbled" (Critchley, 1953) and cannot judge the location of two objects relative to one another (Stenvers, 1951). Deficits in visual localization following inferior parietal lesions are indicated by mistakes in pointing to visual targets (Stenvers, 1951; Ratcliff and Davies-Jones, 1972). Other visuospatial defects occur in motion perception (Riddoch, 1917; Regan *et al.*, 1992), distance estimation, appreciation of the relative lengths and sizes of objects, and the occasional loss of stereopsis (Faust, 1947; Holmes, 1927; Paterson and Zangwill, 1944). Recently, a dissociation between hemifield neglect and visuospatial defects has been made. The former is said to arise from difficulties or "restrictions" on attentional capacities in the spatial domain, while the latter may appear to arise from an inability to represent information about spatial relationships (Newcombe and Ratcliff, 1989; Stark *et al.*, 1996). This chapter is primarily concerned with deficits of representation rather than of attention.

How do these spatial deficits arise from lesions in neuronal tissue? Or, posing this question differently, how does intact neuronal tissue produce a representation of space? Until the recent advent of positron emission tomography and magnetic resonance imaging, the primary tools for monitoring the neural activity of the brain were electrical. Electroencephalography provided gross measures of cortical function, but suffered from an inability to locate precisely the sources of the observed activity. The only way to measure the activity of individual localized neurons under behavioral control was, and is, through the invasive placement of recording electrodes directly into the brain. But to do this, it is necessary to develop a nonhuman primate model for deficits in visuospatial perception.

3. The Macaque Model

The macaque monkey (genus *Macaca*) homologue of the human inferior parietal lobe is a cortical region that lies lateral to the intraparietal sulcus and rostral to the lunate sulcus. Von Bonin and Bailey (1947) performed a series of detailed cytoarchitectonic studies on the brain of *Macaca mulatta* and named portions of the parietal lobule PF and PG based on comparison to the human parietal lobe's cytoarchitecture (Economo, 1929). Surprisingly, these cytoarchitectonic divisions have retained their validity over the past half-century, as confirmed by numerous anatomical investigations using very different histological methods (Pandya and Seltzer, 1982a; Selemon and Goldman-Rakic, 1988; Cavada and Goldman-Rakic, 1989; Andersen *et al.*, 1990a; Blumcke *et al.*, 1990; Hof and Morrison, 1990). There has been a proliferation of nomenclature, however, and it is now common practice to refer to PG and PF as areas 7a and 7b, respectively. In addition to these two areas, the inferior parietal lobule in the macaque monkey includes the ventral and lateral intraparietal areas, LIP_v and LIP_d respectively, also called area 7_{ip} (Selemon and Goldman-Rakic, 1988; Cavada and Goldman-Rakic, 1989) and the dorsal prelunate area (DP) (Pandya and Seltzer, 1982a; Andersen *et al.*, 1990a). Some of the subdivisions are subtle. For example, the subdivisions of LIP can only be differentiated using myelin stains, which reveal the deeper ventral cortex as being more heavily myelinated (Ungerleider and Desimone, 1986; Andersen *et al.*, 1990a; Colby *et al.*, 1993).

Parietal-like deficits were established in monkey by a seminal series of studies in which either the parietal or the temporal lobe was ablated by suction (Ungerleider and Mishkin, 1982; Mishkin *et al.*, 1983). Monkeys were trained to perform tasks in which they identified either the location of an object or its shape. Spatial deficits in tasks, such as remembering the right or left location of a cube placed over a food well, were found to be specifically associated with the parietal lesions. In contrast, deficits in shape perceptions, such as the ability to remember the particular triangular or square shape of an object placed over a food well, were found to be associated with temporal lesions (Klüver and Bucy,

1939; Mishkin and Pribram, 1954; Mishkin *et al.*, 1983). These results paralleled thc human clinical literature on prosopagnosias and other deficits in shaped-based knowledge (Tranel and Damasio, 1985; Damasio *et al.*, 1990; Warrington and Shallice, 1984; Warrington and McCarthy, 1987; Young *et al.*, 1989; Gross, 1994). Mishkin *et al.* (1983) concluded that the monkey parietal cortex subsumed spatial processing, while the temporal cortex subsumed form processing. This provided a nonhuman primate model for cortical lesions in humans. Moreover, they proposed that not only were there two regions for processing visual input, but that these two regions are at the apex of two *pathways* originating in the striate cortex, referred to as the "what" and "where" pathways (Mishkin *et al.*, 1983)*. Although it is impossible to ascertain whether the deficits in monkeys caused by parietal lesions entail the same kind of loss of the highest cognitive spatial functions as observed in human patients, sufficient similarity at the behavioral level exists between the two species to justify further use of the monkey as a model for the human (Ettlinger, 1990; Gross, 1994; Vaina, 1994).

4. The Connections of the Macaque Inferior Parietal Lobule

The neuronal connections of the inferior parietal lobule have been worked out in some detail through injections of anterograde- or retrograde-transported substances, such as horseradish peroxidase (Mesulam *et al.*, 1980), tritiated amino acids (Cowan *et al.*, 1972), and fluorescent substances (Cavada and Goldman-Rakic, 1989). A hierarchical projection pattern was inferred from the finding that there are primarily two types of anterograde and retrograde corticocortical projections (Pandya and Seltzer, 1982b; Maunsell and Van Essen, 1983; Barbas, 1986; Ungerleider and Desimone, 1986; Felleman and Van Essen, 1991). Correlations between these projection types and postulated hierarchical arrangements in the primary visual areas have led to an acceptance of the definition of a hierarchical arrangement based upon the anatomy. Thus, regions that project upward in the hierarchy have afferents from neurons of layer II that terminate in layer IV of the target cortex; regions that project downward in the hierarchy have afferents that originate in the neurons in the infragranular and supragranular layers and terminate in layers I and VI of the target cortex. An intermediate type that appears to represent projections at the same level of the hierarchy received afferent projections evenly across all layers.

Such tracing studies have placed area 7a at the apex of the "where" pathway and the center of a network of regions involved in spatial processing (Fig. 1). Area 7a receives strong ascending inputs from the cortical areas LIP_v and LIP_d, the medial superior temporal area (MST), from the dorsal prelunate sulcus (DP), and from the medial area PO (Selemon and Goldman-Rakic, 1988; Cavada and Goldman-Rakic, 1989; Andersen *et al.*, 1990a). It sends ascending projections to the prefrontal cortex, notably the frontal eye fields, Walker's (1940) areas 46 and

*The question of integration of these two perceptual pathways is presently under active consideration and will not be further considered here (Gray and Singer, 1989; Edelman, 1990; Mesulam, 1990; Friedman and Goldman-Rakic, 1994; Knowlton and Squire, 1993; Wilson *et al.*, 1993).

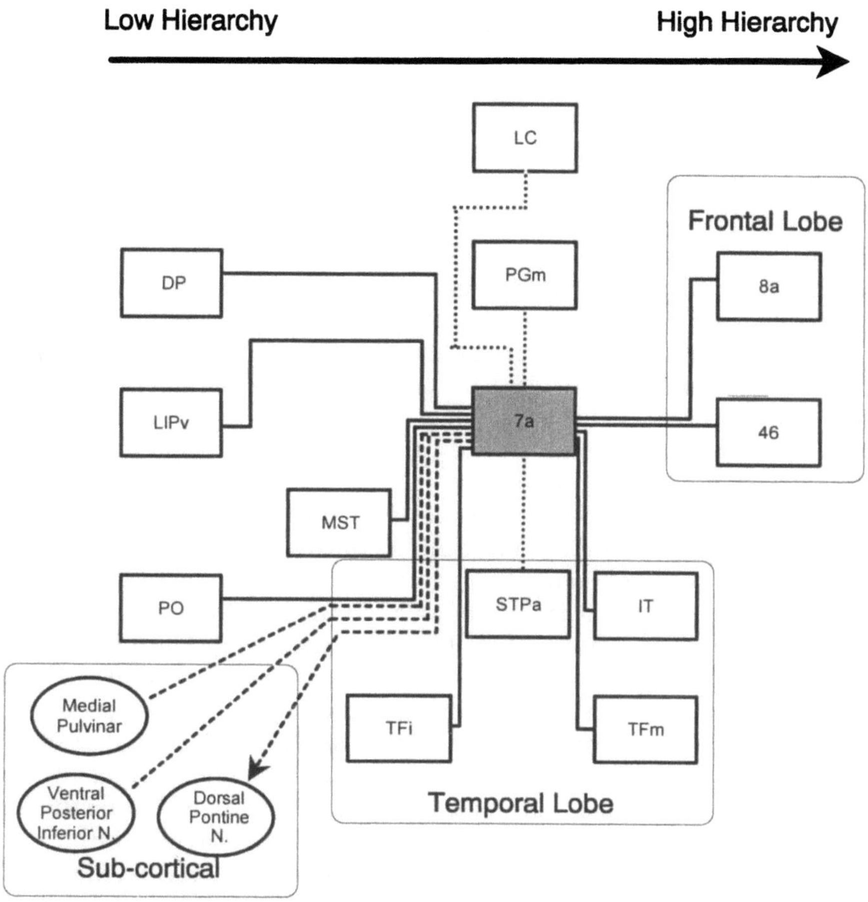

Figure 1. Cortical and subcortical connections of area 7a. Cortical regions are marked by rectangles, while subcortical nuclei are marked by ovals. Solid lines indicate both feedback and feedforward connections. Area 7a (shaded) sends its neural activity to parts of the temporal lobe and frontal lobe. It receives visual and putative oculomotor input from other parietal and temporal regions and oculomotor signals from subcortical structures. Attentional signals may be gleaned from the pulvinar as well as from cortical structures. There are also cortical regions which cannot be assigned a position above or below area 7a (dotted lines). These regions include LC in the posterior cingulate and PGm (medial area PG) on the medial surface of the brain. Dashed lines designate the subcortical nuclei connections to area 7a to indicate that there is no hierarchy associated with them. Only the pontine nuclei receive projections from cortex. Approximately 20 visual areas are *not* shown. (See Felleman and van Essen, 1991, for complete description of the cortical hierarchy.) DP, Dorsal prelunate sulcus; LIP_v, ventral lateral intraparietal area; MST, medial superior temporal area; STP_a, polysensory area of the superior temporal sulcus; IT, inferior temporal region. A number of cortical areas from the nomenclature of von Bonin and Bailey reflect the parcellation of the cortex into frontal (F), parietal (P), occipital (O), and temporal (T) regions with subclassifications generally taken alphabetically (von Bonin and Bailey, 1947) and additional designators added at later dates. Thus TF_i and TF_m refer to inferior and medial TF, respectively, with PG_m referring to medial PG.

8a (Selemon and Goldman-Rakic, 1988; Cavada and Goldman-Rakic, 1989; Andersen *et al.*, 1990a). It has projections to and from memory and limbic structures via the cingulate (LC in von Bonin and Bailey's, 1947, terminology) and the parahippocampal gyrus (TF_i) and sparse, although consistent, projections to the temporal lobule, in particular to the polysensory area (STPa) and inferotemporal cortex (Mesulam *et al.*, 1977; Pandya and Seltzer, 1982b; Boussaoud *et al.*, 1990; Baizer *et al.*, 1991; Andersen *et al.*, 1990a; Ungerleider and Desimone, 1991). The type of neural signal coming from some of these areas is also known. For example, DP (Andersen *et al.*, 1990a), MST (Saito *et al.*, 1986; Tanaka *et al.*, 1986; Duffy and Wurtz, 1991a), and STPa (Bruce *et al.*, 1981) are sources of visual motion signals, while the activity of neurons of LIP are correlated with oculomotor behavior (Andersen *et al.*, 1987, 1990b; Gnadt and Andersen, 1988; Colby *et al.*, 1993). The inferior parietal lobule also receives eye movement information from the frontal eye fields (FEF), area 8a, and area 46 in the frontal lobe (Andersen *et al.*, 1990b). This wide range of connections provides area 7a with an abundance of neural signals of the current state of the internal visual and motor representations in the brain.

Area 7a, as well as many parts of the inferior parietal lobule, also has subcortical connections. These subcortical connections serve as the source of nonvisual inputs. There are connections between area 7a and the medial pulvinar (Asanuma *et al.*, 1985), which could play a role in attentional modulation (Robinson *et al.*, 1986), as well as connections to the limbic system (Mesulam *et al.*, 1977). The inferior parietal lobe is reciprocally connected with a number of subcortical structures related to vestibular and vestibulo-ocular function, including the ventral posterior inferior nucleus, the magnocellular part of the medial geniculate nucleus, and a number of intralaminar nuclei (Faugier-Grimaud and Ventre, 1989). There are also a number of efferents from inferior parietal lobule to vestibular nuclei, including the medial and superior vestibular nuclei, the "y" group, and the dorsolateral pontine nucleus (Ventre and Faugier-Grimaud, 1988; Faugier-Grimaud and Ventre, 1989). These connections seem to be largely dedicated to control of the vestibulo-ocular response and smooth pursuit eye movements. Lastly, the inferior parietal lobule is well situated to incorporate postural position, as it receives such input from the ventral posterior inferior nucleus, which receives proprioceptive signals from the suboccipital muscles (Deecke *et al.*, 1977). Thus, with numerous cortical and subcortical connections, area 7a is well placed to integrate multiple sources of information needed to represent space and send the result of its constructions on to other cortical areas (Fig. 1).

5. Behavioral Methods

These structural studies provide evidence beyond the primate lesion data (Klüver and Bucy, 1939; Mishkin *et al.*, 1983; Warrington and Shallice, 1984; Tranel and Damasio, 1985; Young *et al.*, 1989) about the role of area 7a as a collection and dissemination point for neural signals related to spatial representation. However, to form hypotheses about the functions performed by area 7a and the inferior parietal lobe, it is necessary to determine what the neural signals are actually representing. Most, if not all, of the electrophysiological studies have been performed in the awake, behaving macaque, since anesthesia would preclude testing the involvement of voluntary eye movements and spatial attention. The use of the behaving monkey, pioneered by Evarts (1966), allows experimental control of the performance in a number of different tasks. The simplest of these tasks is a fixation dimming task. The monkey is trained to pull a lever at the onset of a small illuminated dot and to release it when the dot dims slightly. The time at which the dot dims is randomized, so that the animal must pay attention to the fixation point and not utilize timing cues. A device to monitor eye position is commonly used to ensure fixation when other stimuli are added. Initially this was done with electro-oculograms, but more recently the eye coil technique (Judge *et al.,* 1980) and infrared eye tracking devices (Bach *et al.,* 1983; Siegel and Read, 1997) have been used to measure eye position with a precision of at least 1 deg. The fixation dimming task can be much more elaborate, with the animal performing multiple saccades (Colby *et al.,* 1993; Gnadt and Andersen, 1988), varying its locus of attention (Lynch *et al.,* 1977; Yin and Mountcastle, 1978; Mountcastle *et al.,* 1975, 1981, 1987; Mountcastle, 1976; Moran and Desimone, 1985), or responding behaviorally to psychophysical motion stimuli (Newsome and Pare, 1988; Siegel and Andersen, 1988). Once an animal can perform these tasks, a recording chamber is affixed to the skull and repeated electrode penetrations into the cortical tissue are made over a period of weeks to months to observe the action potential patterns of single cortical neurons during behavior.

6. Electrophysiological Studies

Most single-unit electrophysiological studies in the behaving monkey are performed while the head is fixed to a rigid post to eliminate head movements that can jeopardize the long-term stability of the electrical recordings. In studies of primary visual tissue (Dow *et al.,* 1985), the head immobilization has no conceptual impact on the experimental design. However, in studies of parietal cortex, head immobilization drastically alters the framework of the problem. This is because there are many cues related to the visual input and the location of body parts (e.g., head location on shoulders, torso orientation) which might alter the activity of inferior parietal neurons. When the monkey's head is fixed, there is no need for the cortex (or monkey) to take into account variations in head and body movements while performing a visuospatial task. In terms of experimental design, head immobilization changes and reduces the problem of spatial location from a body-centered coordinate system to a problem of localization relative to the head. There are only two free variables in the physiological optics that control the representation of space: the position of the eye in the orbit and the location of the target or object in the visual environment.

6.1. Visual Properties

Two visual cues important for extracting spatial information are the location of an object's image on the retina and its motion. The influence of both of these stimuli on the response of area 7a neurons will be described.

6.1.1. Stationary Stimuli as a Spatial Cue

The responses of neurons in area 7a utilizing stationary bar stimuli as stand-ins for real objects have been extensively studied (Lynch *et al.*, 1977; Robinson *et al.*, 1978; Motter and Mountcastle, 1981b; Andersen *et al.*, 1985b, 1987, 1990a; Read *et al.*, 1994). The majority of these studies have used stimuli at a fixed viewing depth in the frontoparallel plane, although some studies have examined the effect of disparity and other depth cues (Sakata *et al.*, 1984; Gnadt and Mays, 1989). The firing rate of the neurons is measured in response to the onset of a square subtending 5–10° of visual angle while the animal is performing the fixation task. The location of the test square is varied systematically in the frontoparallel plane. Neurons in area 7a have large receptive fields that are often bilateral. The size of these receptive fields can be as much as 60° in width (Fig. 2). Even though the receptive field is large, there are distinct variations of the response within it. A number of different properties have been attributed to the

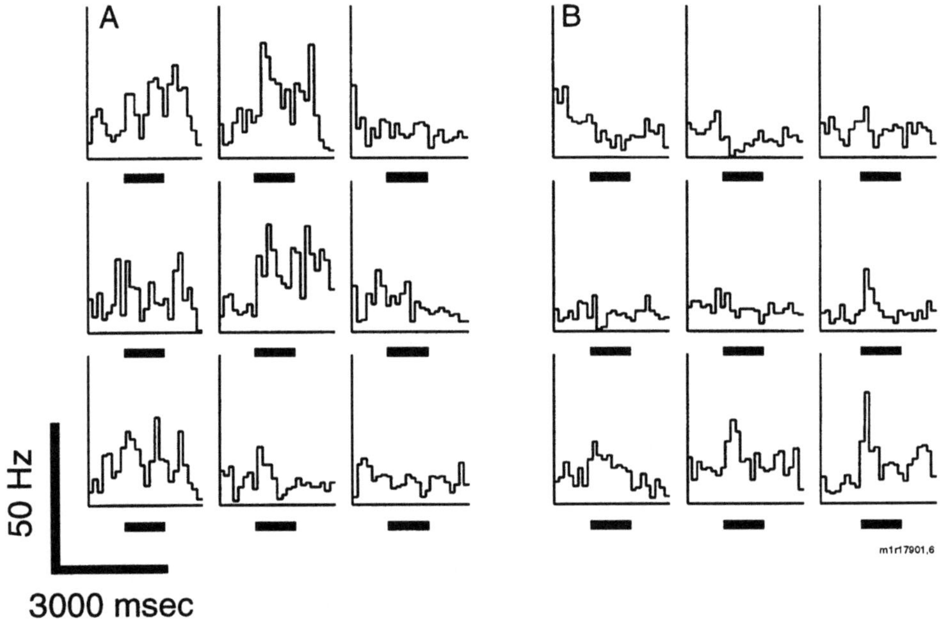

Figure 2. Receptive field map of a single typical area 7a neuron. (A) The neuron has a visual receptive field in the upper, left visual field. A 10 deg square stimulus is placed in one of nine locations relative to the fixation point at (0°, 0°). The stimuli are located on a three-by-three grid, with one side being 40°. The square stimulus, which is indicated by the dark bars, is turned on 2000 msec into the trial and turned off at 3000 msec. The animal releases the key when the fixation point dims, which always occurs at random times longer than 3500 msec. The location of each peristimulus time histogram in this panel indicates the location of the stimulus. (B) Modulation of the visual response by the position of the eye in the orbit. In this experiment, the animal fixates at one of nine locations on a three-by-three grid, with one side being 20°. The visual stimulus is always at the same retinotopic location. This neuron's activity is modulated initially by the fixation position. The activity for the upper left fixation is increased as compared to the lower, right fixation prior to the visual stimulus onset. This recording was made in the right hemisphere. All histograms begin 1000 msec into the trial, so that any changes in neuron activity due to the initial fixation or pulling of the key are not seen. The calibration bars apply to all the histograms. After Read and Siegel (1997).

receptive field shape (Motter and Mountcastle, 1981a; Andersen *et al.*, 1985a, 1990a; Siegel and Read, 1994). All studies agree, however, that the receptive fields are found throughout both hemispheres.

A conceptual problem arises from the large size of the receptive fields. Reaching and pointing movements can be very accurate, corresponding to visual errors of 1 deg or less (Georgopoulos, 1986; Hocherman, 1993; Walker *et al.*, 1993). How can visually guided movements be so precise when the receptive fields in parietal cortex are so large? Computational studies of the general problem of representations have suggested two means by which a particular property can be coded: by small, finely tuned units (Barlow, 1972) or by large, coarsely tuned units (Ballard *et al.*, 1983; Georgopoulos *et al.*, 1986; Churchland and Sejnowski, 1992). Both means have advantages and disadvantages.

With the small, finely tuned units, also called "grandmother cells," identification of an event can be made simply by locating the finely tuned unit which is active. But this has the disadvantage that one neuron is dedicated to subserving only one very specific event. Hence with the environment being very rich and the number of possible combinations of visual events, even in the world of spatial vision, being very large, their proper representation would require more cerebral neurons than are available. (For a strong counterargument, see Barlow, 1972.)

With large, coarsely tuned units, identification of an event requires a complex algorithm by which multiple rough measurements are summed or compared. One well-known example of coarse tuning is the identification of an enormous number of spectral hues by the three cone types in the retina (von Helmholtz, 1924). Coarsely tuned units solve the problem of many combinatorial possibilities by using a limited number of representations and mixing them in appropriate proportions. What is lost in precision of measurement is gained by a mixture of representations.

The large size of the receptive fields in area 7a by itself suggests a coarse rather than fine tuning of spatial location. One corollary of a coarse coding tuning scheme is the requirement for distributed representation. Quantitative analysis of the receptive fields of area 7a neurons suggests that the receptive field centers are evenly scattered across the retinotopic representation (Andersen *et al.*, 1990a; Siegel and Read, 1994). Thus all of space is sampled, albeit without great detail. The question remains as to how coarse tuning of spatial location might be implemented in area 7a. One possible solution is to use other sources of visual information.

6.1.2. Motion as a Spatial Cue

Motion is an important cue for spatial perception (Gibson, 1966; Beverley and Regan, 1975; Regan and Beverley, 1979; Regan, 1986; Koenderink, 1986). Motion of a visual image can be generated by the movement of the observer relative to a stationary object or by the movement of an object relative to a stationary observer. There are multiple types of motion information that can be utilized for spatial perception of an object. Here we will concentrate on three essential characteristics that are well defined both mathematically and ecologically (upper panels of Fig. 3). They will be defined by movement of the observer

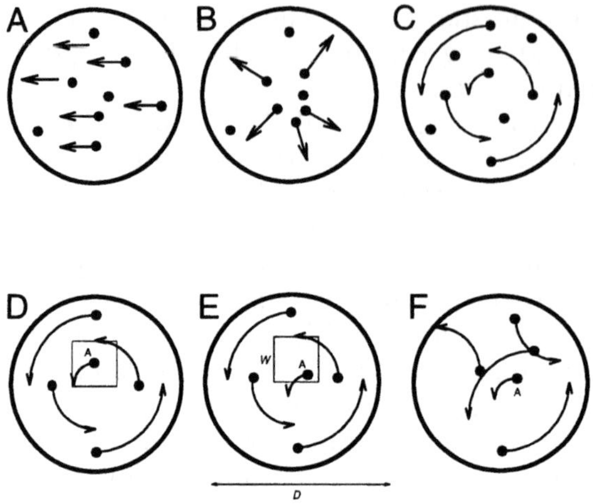

Figure 3. Different optic flow stimuli used. (A) translation motion to the left, (B) radial expansion motion, (C) counterclockwise rotation. (D–F) The process by which a control unstructured motion display is generated. (D) A counterclockwise motion trajectory with a FOS of 1 with one trajectory labeled A. (E) The entire motion trajectory A is displaced to a random location with a window of width *W*. Note that A has been displaced relative to its initial location in panel D. (F) This process is repeated for all the motion trajectories with a window size equal to the diameter *D* of the display. The motion display, with a FOS of 0, is now jumbled and has no coherent percept.

relative to a stationary environment. Consider yourself as simply looking straight ahead. Your *line of sight* is defined by the vector originating at your nose and ending at the point you are fixating in space. Translation motion occurs when you move along a line perpendicular to the line of sight. Planar rotation motion occurs when you rotate around your line of sight. Lastly, radial motion occurs when you move along your line of sight. The radial motion will be *expansive* if you move forward and *compressive* if you move backward along your line of sight. Translation, planar rotation, and radial motion are mathematically defined as the *first derivative,* the *curl,* and the *divergence* functions (Werkhoven and Koenderink, 1991) and are illustrated schematically in Figs. 3A–3C, respectively.

As you drive an automobile down the road, looking in the direction of your forward movement, the lampposts will move *radially* away (i.e., expansively) from your fixation point. If you turn your head fully to the left to look at the view, you will then see *translation* of the scenery along the side of the road. If you turn the steering wheel and fixate its center rather than the road or the scenery, you will see *planar rotation of* the steering wheel.

Combinations are often seen. You see radial combined with translation motion if you look down the road, slightly off the direction of motion of your automobile (Duffy and Wurtz, 1991a). Spiral motion is seen by aircraft pilots engaged in corkscrew maneuvers. There are other classes of motions that provide spatial cues, e.g., shearing motion (Golomb *et al.,* 1985) and occlusion motion, which will not be considered here.

6.1.2a. Radial Motion and Opponent Vector Neurons. The first attempts to study cells selectively responsive to complex motion in the inferior parietal lobe of the monkey were made by Sakata *et al.* (1984). They found that there are parietal neurons whose activity varied in response to translation motion along the line of sight, with differential responses seen depending on whether an illuminated square moves toward or away from the fixation point. This stimulus was delivered with the square object moving along a motorized track. This stimulus has three components that could give rise to directional selectivity: its actual motion, its change in size, and its change in luminosity.

A different experimental design was used by Mountcastle and colleagues (Motter and Mountcastle, 1981a; Motter *et al.*, 1987; Steinmetz *et al.*, 1987) to determine whether parietal neurons are selectively responsive also to radial motion. Rather than have a single stimulus change size, they used a square stimulus whose motion trajectory would mimic single motion components of radial motion. A square that was 5° on a side was moved across the visual field in a manner akin to a standard directional selectivity test. The stimulus begins its movement at a peripheral location relative to the fixation point, then moves toward the fixation point, and then past it, with its total sweep generally being 40–60°. During the first half of its sweep, the square would replicate compressive radial motion, while during the latter half it would replicate expansive motion. Neurons of the inferior parietal lobe were found that responded only during the compressive phase of the motion and were inhibited during the expansive phase. These were termed *opponent-vector motion neurons* and were proposed to encode radial optic flow that would arise from forward body movement. It was also suggested that these neurons would be useful for representing views of reaching movement by the limbs (Motter *et al.*, 1987). Subsequent studies explored the nonlinear spatial and temporal interactions within the receptive field that gave rise to the opponent-vector motion neurons (Steinmetz *et al.*, 1987). These studies provided support for Sakata *et al.*'s (1984) contention that area 7a neurons have signals related to radial optic flow. However, these studies do not eliminate the possibility that some of the directional signals of the area 7a neurons were in fact caused by form cues of the motion stimulus (e.g., the contours of the square).

6.1.2b. Selectivity to Motion Flow. To circumvent any residual form cues in the motion stimulus, a new class of motion stimuli was devised to stimulate parietal neurons. These stimuli were taken from psychophysical studies (Ullman, 1979; Regan and Beverley, 1979; Williams and Sekuler, 1984; Nakayama and Silverman, 1984; Chang and Julesz, 1985; Regan *et al.*, 1992). With this new class of stimuli, an overall motion is presented by the computer-generated movement of independently flickering white dots on a black background, eliminating all form cues (Morgan and Ward, 1980) and presenting a pure, formless motion stimulus. These stimuli resemble points of light flickering on a river, whose overall translational motion is imposed upon them by the flow of the river.

A further modification of the stimuli (lower panels of Fig. 3) was made to provide a control condition that could be used to "ask" monkey subjects behaviorally about the stimuli (Siegel and Andersen, 1988, 1990). This control condition is called "unstructuring" of the motion stimuli, and is generated by moving *entire* motion trajectories randomly within a prescribed area. The process is illustrated for an expansive radial motion field of diameter D. One of the motion trajectories, labeled A, is randomly displaced within a box of size W (Fig. 3E). This process is repeated for all the motion trajectories (Fig. 3F), destroying the global percept of motion. The fraction of structure (FOS) is quantified as $OS = 1 - W/D$. A FOS of 0 would indicate an unstructured display, while a FOS of 1 would indicate a completely structured display.

Psychophysical studies have been performed with these stimuli, establishing that the animals were able to extract complex motion information in a manner similar to that of human subjects (Siegel and Andersen, 1988; 1990). Monkey and human subjects were trained to pull a lever at the onset of unstructured motion and release it when the display changed to a structured motion. Both species had no trouble doing this basic task. Measurements of the subject's ability to perform this task were obtained as a function of various parameters of the display (e.g., the FOS, the number of points, and the temporal rate of flickering). Comparison of the thresholds between monkey and human subjects indicated that performance was statistically the same for both species over many variations of stimulus parameters. The performance of monkeys and humans has been compared for rotation and radial motion (Siegel and Andersen, 1990). Some differences have been observed, particularly with respect to reaction time. Monkeys release the key faster than human subjects. More recent work has suggested that these intraspecies differences are due to the extensive overtraining of the monkeys (H. L. Read, K. C. Anderson, and R. M. Siegel, unpublished results).

(i.) Optic Flow Selectivity. The response of neurons in the inferior parietal lobule of monkeys was tested with these flickering, random dot stimuli (Siegel and Read, 1997). Since parietal neurons respond selectively to eye position, it was necessary to ensure that the monkey was fixating during the psychophysical task. The sequence of the behavioral task was somewhat different from that used in the above psychophysical studies (Fig. 4). The trial began with the onset of a fixation point. After a delay, the motion stimulus came on, changing some way at a random later time. The monkey has to release the key within a reaction time window to receive a reward. If the animal breaks fixation at any time during the trial, the stimulus is immediately shut off and no reward is given. Use of this sequence permitted precise control of the attentional state of the animal, his eye position, and the retinotopic location of the stimulus. Moreover, this experimental design

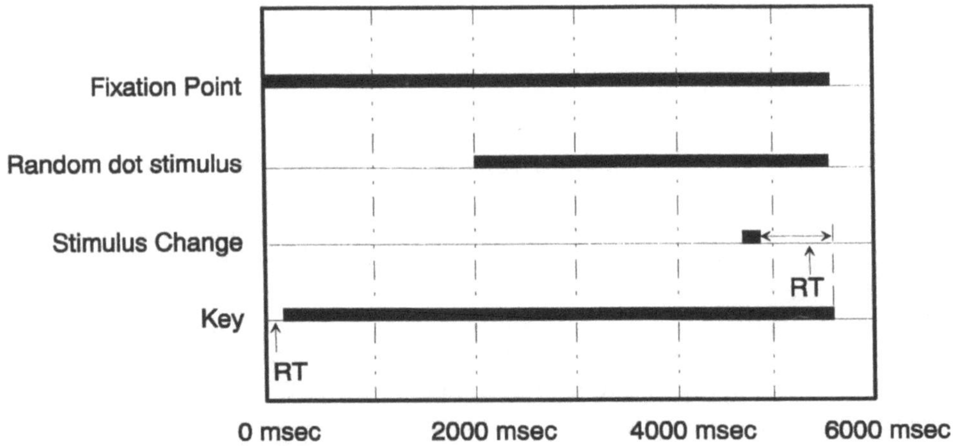

Figure 4. Sequence of events that occur during a behavioral task with optic flow stimuli. At 0 msec, the 0.1° fixation point is illuminated at the center of the screen. The subject must pull the key back within a reaction time (RT) window of 800 msec. The test optic flow stimulus comes on 2000 msec after the fixation point appears. At some time occurring randomly in the time interval between 3500 and 6000 msec, the optic flow stimulus changes in some way (e.g., the FOS changes; the dots stop moving). The subject releases the key within a reaction time window for a liquid reward.

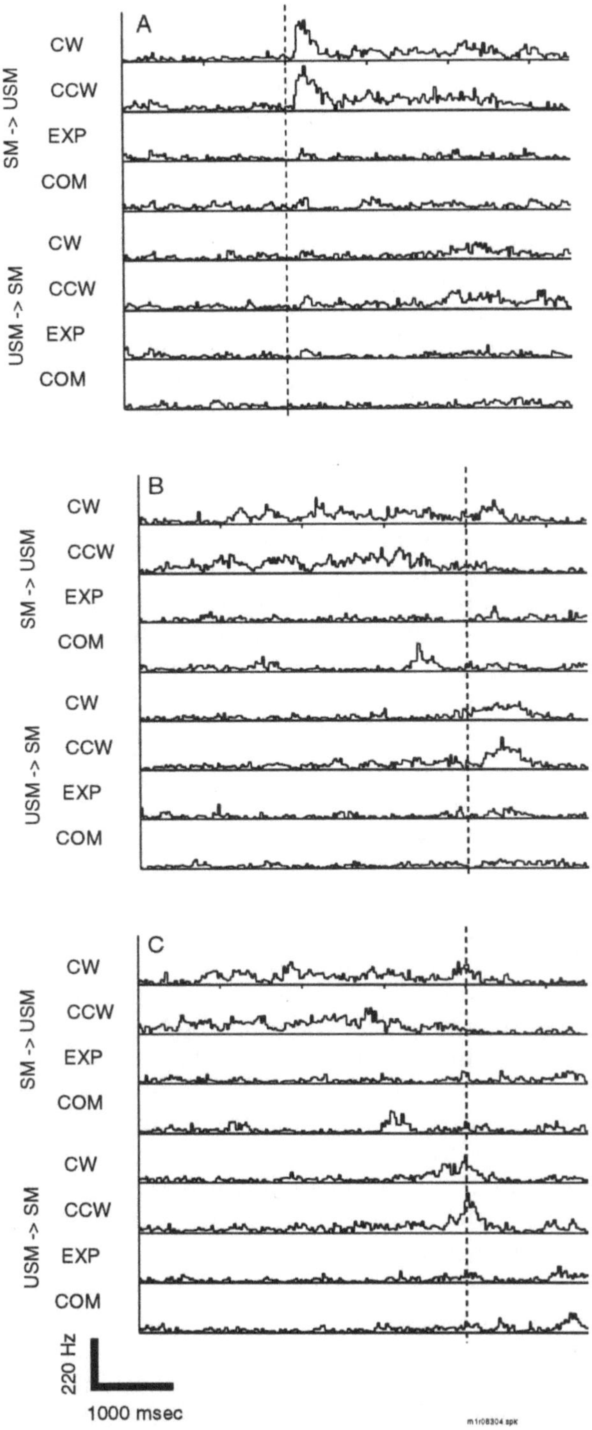

Figure 5. Example of a neuron that responds selectively to rotation optic flow. The monkey was tested with four different types of motions: clockwise rotation (CW), counterclockwise rotation (CCW), radial expansive (EXP), and radial compressive (COM). Two different sequences were presented to the animal for each type of motion. In one sequence the initial stimulus was structured motion (SM), which was followed by unstructured motion (USM). In the other sequence the initial stimulus was USM, which was followed by SM. (A) The peristimulus time histograms are synchronized to the stimulus onset at 2000 msec following fixation point onset (indicated by vertical dashed line). When the initial motion is CW or CCW, the cell responds with a burst of activity. Under all other conditions, there is a weak or no response. (B) The histograms are synchronized to the randomized times when the stimulus begins to change (indicated by vertical dashed line). Here a strong response is seen at the time the stimulus changes from unstructured rotation to structured rotation. (C) The histograms are synchronized to the time of key release (indicated by vertical dashed line). The neuronal response increases prior to key release only when the appropriate motion is presented. The upper histogram of each panel has small ticks indicating 1000 msec; the same calibration bars apply to all histograms. From Siegel and Read (1997).

permitted the correlation of the firing activity of the extracellularly recorded neuron with different events in the task (e.g., stimulus onset, time of stimulus change, time of key release). These events need to be separable, because there is the possibility that neuronal activity could be stimulated by any of them.

Some area 7a neurons indeed respond selectively to structured motion (Fig. 5). Neurons were generally tested with four types of structured motion with two different sequences in one block of trials. In order to control for any slow changes in neuronal excitability, the eight conditions were presented in a random order. This fixed block of eight stimuli was then repeated in a different random order until the desired number of repetitions was obtained. In all three panels of Fig. 5, in the upper four traces, the initial stimuli are clockwise rotation, counterclockwise rotation, radial expansion, and compression with FOS = 1. In the lower four traces, the initial stimuli are unstructured versions of each of these. The monkey subject was trained to release the key when he perceived a change from unstructured motion to structured motion or a change from structured motion to unstructured motion. Figure 5A shows peristimulus histograms synchronized to the stimulus onset. It can be seen that substantial responses occur at the onset of the clockwise and counterclockwise *structured* motion, while almost no response occurs to expansive or compressive radial motion or to the *unstructured* motion. This is one indication that neurons in this region of area 7a neurons respond selectively to optic flow.

Figure 5B shows a reanalysis of the same data with the spike trains synchronized to the change in stimulus. It can be seen that the neuron responds selectively for the structured motion, since its activity increases at the time of stimulus change for the rotation motions. Figure 5C shows the same data with the spike trains synchronized to the time of key release. It can be seen that the increase in activity of the neuron occurs prior to the key release, indicating that the modulation is not motor in origin. Thus this neuron responds to the change from unstructured motion to structured motion as well as to the onset of structured motion. The selectivity of the neuron is to clockwise and counterclockwise rotations. The response pattern of this neuron is therefore *prima facie* evidence for a representation of optic flow in area 7a.

Neurons in area 7a have also been found that respond to other types of optic flow, such as radial expansion and/or compression, while others are simply selective to translation (Read *et al.*, 1994; Siegel and Read, 1997). Thus, these area 7a neurons are selective to a variety of optic flows that may be used to determine position in the surrounding environment. If there is indeed a range of selective responses to optic flow, then area 7a can be said to have a *distributed representation* of optic flow. This possibility was examined by computing selectivity indices for each of the neurons, and then looking at the distribution of the indices.

Neurons were chosen for further analysis if they had a significant response, computed using a two-way analysis of variance, to either the onset of optic flow or to the change from structured to unstructured optic flow (Read *et al.*, 1994; Siegel and Read, 1997). Selectivity indices (SI) were computed separately for the radial and for the rotation motions and were defined as the probability that the analysis of variance indicated that the neuronal responses were taken from different groups. A value of 0 for SI indicates there is powerful optic flow selectivity, while SI greater than 0.05 indicates weak or no optic flow selectivity. The

SI was computed for the rotational and radial stimuli independently. A graph of all the selectivity indices indicates that very few neurons are selective to both rotation and radial stimuli at stimulus onset. Most of the neurons were selective to either rotation or radial motion. There was also a large number of neurons that were unable to differentiate clockwise versus counterclockwise rotation, nor expansion versus compression radial motion.

It is possible that this last large group of neurons might be differentially selective to rotation versus radial motion regardless of the direction of motion. In the neuron of Fig. 5, responses are found to both clockwise and counterclockwise rotation, indicating that it was not selective to the direction of rotation. However, the area 7a neuron clearly provided differential signals for rotation and radial movement. To reveal such neurons, this was determined by computing the selectivity index for all the rotational responses versus all the radial responses on a neuron by neuron basis. There is a substantial population of neurons that were able to differentiate between radial and rotational motion while being unable to tell the direction of motion of either (Siegel and Read, 1997). Thus it appears that area 7a has a novel representation of optic flow that, in principle, could underlie spatial localization from motion regardless of the direction of motion.

(ii) Source of Optic Flow Selectivity. Do the translation, radial, and rotation motion selectivities arise *de novo* in area 7a neurons? Or do other areas lower in the visual hierarchy (e.g., MT/V5 or MST) extract optic flow information and send it to area 7a via afferents? Translation selectivity to both moving bars and dots is found in numerous cortical areas. For instance, striate cortex has a class of neurons, termed "special hypercomplex" neurons, that are selective to the movement of a drifting random dot field (Hammond and MacKay, 1975). Neurons in area V5/MT are also selective to moving dots (Allman *et al.*, 1973; Lagae *et al.*, 1989; Orban *et al.*, 1986), as are neurons in the medial superior temporal area (MST) (Duffy and Wurtz, 1991a, b; Graziano *et al.*, 1994; Lagae *et al.*, 1989). MST neurons project directly to area 7a and thus could serve as a source of the translation signal in area 7a (Fig. 1). Moreover, MST neurons are selective to radial, rotation, and spiral optic flow (Saito *et al.*, 1986; Tanaka *et al.*, 1986; Duffy and Wurtz, 1991a, b; Graziano *et al.*, 1994; Orban *et al.*, 1992; Lagae *et al.*, 1994). Lastly, neurons in the polysensory area of the superior temporal sulcus (STP_a), which has projections at the same hierarchical level as area 7a (Felleman and Van Essen, 1991), also have rotational and radial optic flow selectivities (Bruce *et al.*, 1981). The several cortical regions with motion- and direction-selective neurons could thus indeed serve as the source of the motion selectivity in area 7a via afferents. Another possibility is that area 7a works in concert with these areas to extract optic flow.

There are some differences in the optic flow selectivity of MST and area 7a neurons, however, which would support the concept of a hierarchical processing. MST has a preponderance of neurons that are selective to only one of the different types of optic flow (Duffy and Wurtz, 1991a, b; Graziano *et al.*, 1994). Duffy and Wurtz have hypothesized that this collection of "pure" optic flow neurons indicates that MST is the locale for the emergence of representations of radial and rotation optic flow. Like MST, area 7a has neurons tuned for specific directions of optic flow. Unlike MST, area 7a has neurons tuned to a *class* of motion. The broad tuning of area 7a neurons could thus be built from summated outputs of MST neurons.

Thus it seems that there are a number of likely candidates prior to area 7a in the "where" processing stream for the source of the motion floww selectivity in area 7a. If this is indeed the case, then why does the brain generate an additional representation of the same motion information in area 7a?

6.2. Extraretinal Signals

It was noted earlier that in the head-immobilized monkey, only two measurements are needed to locate a location in space—the retinal locus of the visual stimulus and the position of the eye in the orbit. The retinal signals are represented in area 7a, with about 40% of the neurons being visually responsive. The details of the representation of extraretinal (e.g., eye position) signals present in area 7a are considered next.

6.2.1. Fixation Neurons

Neurons in area 7a have been found to respond to the position of the eye in the orbit. A typical experiment to demonstrate this would be to have the monkey fixate one of nine locations on a tangent screen arranged in a three-by-three square grid. The firing activity of the area 7a neurons is modulated according to the horizontal and vertical position of the eye in the orbit (Andersen *et al.*, 1985b; Siegel and Read, 1994; Read and Siegel, 1997). Eye position signals can be seen by inspection of the neural signal prior to the stimulus onset for neurons that respond both to stationary visual stimuli (Fig. 2B) and to optic flow (Fig. 6B) for the 1000 msec prior to the stimulus onset. Statistical analysis of the response activity patterns indicates that this modulation often varies linearly with eye position. The dependence of the neural activity on eye position is termed the "gaze field." The role of visual background in creating the gaze field can be ruled out by showing that these neurons have the same gaze fields in complete darkness as in light (Lynch *et al.*, 1977; Motter and Mountcastle, 1981b; Yin and Mountcastle, 1977; Sakata *et al.*, 1980; Andersen *et al.*, 1985b, 1987, 1990a; Mountcastle *et al.*, 1987). It is likely that these eye position signals arise from brainstem projections (May and Andersen, 1986; Baizer *et al.*, 1993) or at least are extraretinal in origin.

6.2.2. Modulation of Visual Responses by Eye Position

The presence of the eye position and visual signal in area 7a is sufficient information to determine the spatial location of an object. These two signals could coexist in one population of neurons or could be found in two separate populations of neurons. During the fixation studies, it was found that many cells that had eye-position-related activity were also responsive to visual input, with a smaller subpopulation only responsive to one or the other input (Andersen and Mountcastle, 1983; Andersen *et al.*, 1985b, 1990a). For some of these neurons with a dual sensitivity, the modulation was simply additive, i.e., the response was a linear sum of the eye position signal and the visual signal. However, for the vast majority of these neurons, the visual response was attenuated or amplified by eye position.

6.2.2a. Modulation of Stationary Visual Responses by Eye Position. In order to explore the nature of the eye position modulation of visual responses, a monkey was trained to fixate different locations on a screen. During the fixation of this location, say a point directly ahead of his nose, a stationary stimulus was turned on at various locations relative to the fixation point allowing the determination of the visual receptive field of the area 7a neuron. This receptive field determination was then repeated with the monkey fixating a different point, say a point in the upper right visual field. It was found that the receptive field varied with eye position. By contrast, if the recordings were taken from a neuron in V1, one would find that the visual receptive field would be exactly the same for both fixation points.*

After the subtraction of the fixation signal from the total response, it was found that the visual response was still modulated by the position of the eye in the orbit. Quantitative analysis of this modulation suggested that the location of the receptive field remained fixed in retinotopic coordinates; however, the intensity of the response at all retinal locations was increased. This multiplicative modulation of the visual receptive field signal by eye position was termed a "gain field" and was often linear with eye position (Andersen *et al.*, 1985b; Siegel and Read, 1994; Read and Siegel, 1997). The combined eye position and retinotopic signal could be used to construct a map of space. Similar "gain field" effects were seen for opponent vector neurons (Steinmetz *et al.*, 1987) in area 7a and for saccade neurons (Andersen *et al.*, 1990b) in LIP. As optic flow selectivity is also found in area 7a, it is possible that this signal could be modulated by eye position.

6.2.2b. Modulation of Optic Flow Responses by Eye Position. Recently, the effect of eye position on the response to optic flow has been explored in area 7a (Siegel and Read, 1994; Read and Siegel, 1997). In these experiments, a three-step procedure was performed (Fig. 6). First, the optimal type of stimulus motion is determined for a particular neuron, by means of experiments similar to that of Fig. 5. Then the receptive field is mapped with smaller optic flow stimuli using the optimal type of optic flow stimulus in various retinotopic locations. Unlike neurons in MST (Duffy and Wurtz, 1991a, b), the retinal location of the motion stimulus appears to modulate the response to optic flow. These motion fields appear to be large, covering up to 40° of the visual field, with the selectivity of the neuron maintained at all locations (Siegel and Read, 1994; Read and Siegel, 1997). For about two-thirds of the neurons responding to motion, a modulation of the visual response is seen with eye position (Siegel and Read, 1994; Read and Siegel, 1997). As with stationary stimuli, part of this modulation can simply be due to the additive eye position modulation of the neuron. Subtraction of the eye position signal leads to the uncovering of a directional modulation of the visual response, i.e., a "gain field." Therefore, the signals allowing determination of the direction of movement relative to the head are present in area 7a.

In summary, the electrophysiological studies of area 7a neurons demonstrate that both visual and oculomotor signals are present in the inferior parietal

*A small percentage (15%) of V1 neurons have actually been shown to have visual movement signals that are modulated by tracking eye movements in the monkey (Battaglini *et al.*, 1993; Galletti *et al.*, 1984). There is a possibility that these signals could form part of the origin of the eye position modulation in area 7a.

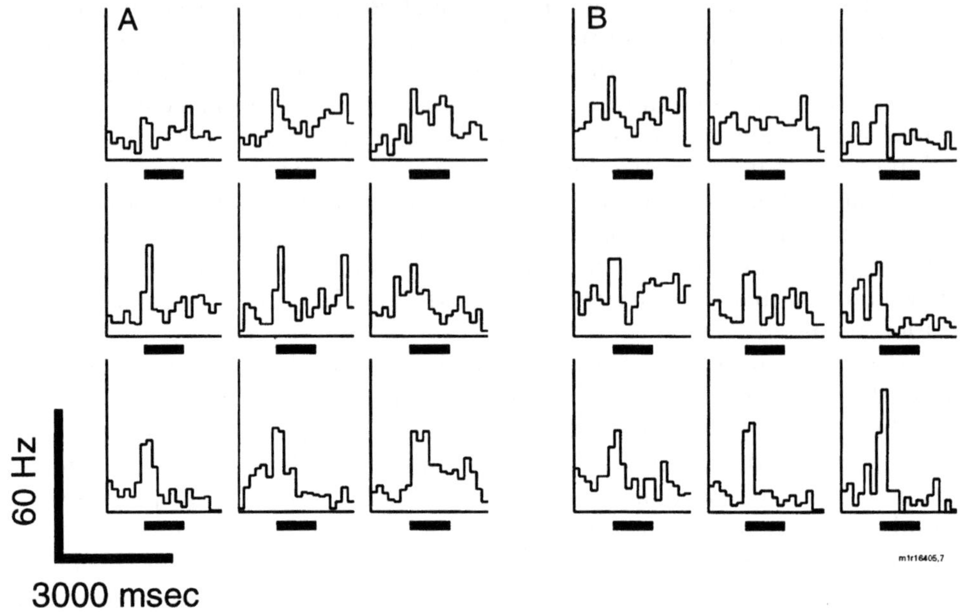

Figure 6. Retinotopic receptive and gain field for an area 7a neuron tested with optic flow. When tested with the various types of optical flow patterns, this neuron had the greatest firing rate to a counterclockwise compressing (CCC) optic flow. A 10° diameter CCC spiral was then used to map the neuron's retinotopic receptive and gain fields. (A) Receptive field. The neuron was visually responsive at almost all locations in the 20° by 20° retinotopic receptive field tested with the strongest responses in the lower right retinal sector. (B) Gain field. The fixation point was moved to the nine locations in a 20° by 20° grid. The CCC stimulus was always placed at (−10°, 10°) in retinal coordinates. This cell had a fixation-dependent response signal that was strongest in the upper left retinal sector. The visual response, however, was maximal in the lower right sector. All other conditions are as in Fig. 2. The same calibration bars apply to all histograms. This recording was made in the right hemisphere. After Read and Siegel (1997).

lobule. These signals appear to be a modulation of the visual response to either stationary or optic flow stimuli by the position of the eye in the orbit. The visual signals appear to be broadly tuned in regard to both retinotopic domains and selectivity to stimulus properties. These results are consistent with the hypothesis generated by the anatomical studies that area 7a is a central hub in a network collecting diverse sources of information about visual and extraretinal signals needed to construct a spatial representation of the surround. Moreover, the physiological properties of area 7a neurons are consistent with the neurological findings of defects in spatial perception following inferior parietal damage in both human and nonhuman primates.

7. A Mathematical Model for Spatial Representation

How might the cortex utilize the activity of parietal neurons to construct a representation of space? Three experimental results are relevant for this question. First, the response of some neurons to either stationary or moving visual stimuli is modulated by eye position in the orbit. Second, the receptive fields of parietal neurons are large. Third, the receptive field centers for either stationary or moving visual stimuli appear to be evenly distributed across the visual field, as are the optimal selectivities for eye position signals. These latter two attributes seem appropriate for coarsely tuned neurons whose concurrent activity is combined for the extraction of a precise representation. But how *is* their activity combined?

It is possible to estimate how the intensity of a visual response varies as a function of eye position and visual stimulus location (i.e., in the case of a moving stimulus, the center of the optic flow pattern), by deriving a simple set of equations that describes the neural activity by considering only eye position and retinotopic locations of the visual stimulus along the vertical meridian (Siegel, 1998). For the sake of simplicity, the other variables that can also affect neuronal activity are ignored in this derivation, including the specific tuning to the visual stimulus and attention.

The eye position contribution E to the response is a linear function of the vertical eye position p:

$$E(e) = ap + b \tag{1}$$

where a and b are constants, with b representing the eye position contribution while the monkey is looking straight ahead ($p = 0$).

The visual contribution R to the response as a function of retinotopic stimulus location r is a humped function, where $r = 0$ at the fovea. A simple choice for that function, which matches much of the data, is a Gaussian curve, although this is not a critical assumption:

$$R(r) = c \exp\left[\frac{-(r - r_0)^2}{\lambda^2} \right] \tag{2}$$

where r_0 is the retinal location of the receptive field center, c is the maximal rate of firing when the stimulus is over the receptive field center, and λ governs the vertical width of the receptive field. Based on the experimental results, the response of the neuron would be given by the product of the eye position contribution and the visual contribution; thus the overall response A would be given by

$$A(p, r) = (ap + b)\, c \exp\left[\frac{-(r - r_0)^2}{\lambda^2} \right] \tag{3}$$

as shown graphically in Fig. 7B. Changes in eye position linearly scale the peak response of the neuron to the visual signal.

To describe the dependence of the neuronal response found in area 7a on the location of the stimulus in head-centered coordinates, a simple geometric relation between the location of the stimulus in head-centered coordinates and the retinal locus and eye position is used (Fig. 7A). The location l in head-centered coordinates is simply given as the sum of the location of the stimulus on the retina relative to the fovea, r, and the position of the eye in the orbit, p:

$$l = r + p \tag{4}$$

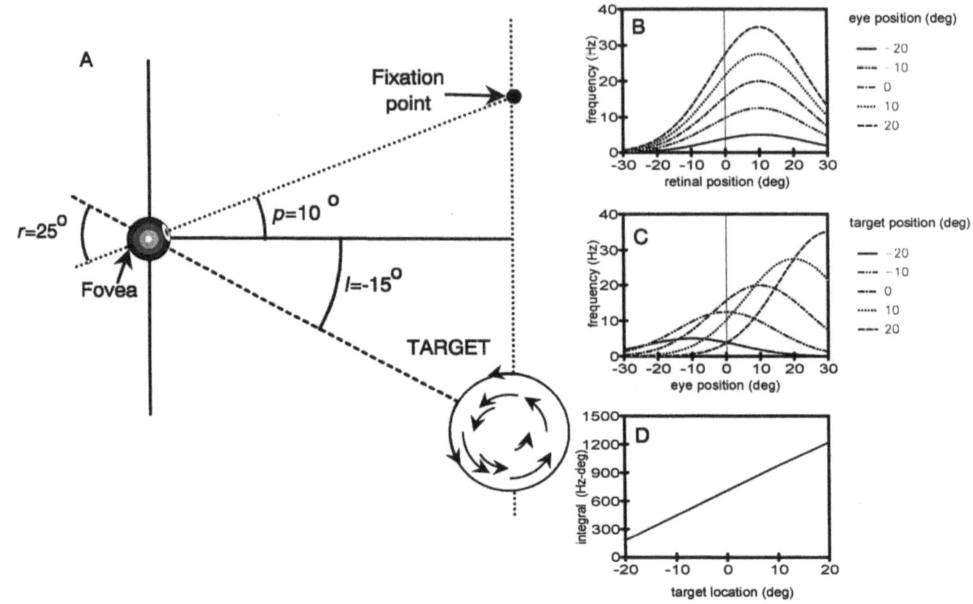

Figure 7. Simple model for constructing a spatial representation. (A) Physiological optics defines how the eye position and the location of the target in space interact to yield the location of the target on the retina. A fixation point is located up 10° on the vertical meridian ($p = +10°$). A stimulus is located 15° below the animal. The retinal image of the target is thus 25° from the fovea. (B) The firing rate of the modeled neuron is plotted as a function of the retinotopic location of a stimulus for different eye positions. Note that the receptive field remains centered on the retina, while the amplitude is linearly modulated by eye position, i.e., as the eye position is moved up, the peak of the retinotopic response scales linearly. This is the basis of the gain field. (C) The firing rate is plotted as a function of the eye position for different targets in head-centered coordinates. Note that as the target's location is varied, the center of the response in eye coordinates changes, as does the strength of the response. (D) Plot of the integrated activity of a single area 7a neuron. The response of this neuron varies linearly with the location of the target in head-centered coordinates. The constants for Eq. (5) and panels B and C are $ac = 0.75$ Hz/deg, $bc = 20$ Hz, $r_0 = 10°$, $\lambda = 20°$. The constants for Eq. (6) and panel D were determined by trapezoidal integration of Eq. (5): $k_1 = 26$ Hz, $k_2 = 709$ Hz-deg (after Siegel, 1998).

Solving Eq. (4) for r and substituting it in Eq. (3) yields a description of how the firing rate of the area 7a neuron scales as a function of the head-centered location of the stimulus and the position of the eye in the orbit:

$$A(p, l) = (ap + b)\, c \exp \left[\frac{-(l - p - r_0)^2}{\lambda^2} \right] \tag{5}$$

This description is plotted in Fig. 7C. Integration of Eq. (6) over eye position yields

$$\int A(p, l)\, dp = k_1 l + k_2 \tag{6}$$

where k_1 and k_2 are constants (see Feynman *et al.*, 1963, for the integral of the formally equivalent center-of-mass equation of mechanics). In biological terms, this means that if a neuron integrates the activity of a single area 7a neuron while the animal scans along the vertical meridian, the head-centered coordinate of the location of the target can be inferred by virtue of the integral's linear dependence on it. Equation (6) is illustrated graphically in Fig. 7D, where the integrated firing activity of an area 7a neuron is plotted as a function of various target locations in space. It can be seen that as the target is moved to different locations the response is *linearly* scaled up or down (Siegel, 1998).

Of course one would not expect the cortex to use the activity of a single neuron to determine spatial location of a target, and indeed the physiological studies point toward distributed large, coarsely tuned assemblages. With a distributed representation of area 7a neurons mapped onto the cortical surface, the cortex can perform its computations *in parallel*. This entails a simultaneous mapping of area 7a neuron properties into a new distributed, second-tier representation, according to which the neuronal response changes linearly with changes in eye position or object location, since linear signals can be incorporated into feedback neural circuits (Waitzman *et al.*, 1988; Kapoula *et al.*, 1990; Miles *et al.*, 1986; Tomlinson and Robinson, 1984) with the output of area 7a neurons being converted to signals for motor neurons driving the eyes (Pellionisz and Ramos, 1993).

The covariance of eye position signals and retinotopic signals has some physiological and anatomical implications. First, the afferents to area 7a that provide the eye position signals (presumably from subcortical nuclei, or cortex) must be in register with the retinotopic field center. This suggests that there ought to be topographic concordance of signal input across area 7a. Second, the efferents from area 7a neurons need to converge in a manner that allows the integration of their output. Some evidence exists for an anatomical convergence of afferents to, and of efferents from, area 7a (Asanuma *et al.*, 1985; Andersen *et al.*, 1985a, 1990a; May and Andersen, 1986; Selemon and Goldman-Rakic, 1988). Both the afferents and efferents have elongated interdigitated, as well as overlapping, patchy distributions. These patches may represent regions with covaried tuning between the eye position signals and the retinotopic signals.

8. Conclusion

Neurons in area 7a of the inferior parietal lobule appear to construct representations of visual space. They have visual and oculomotor inputs, as well as outputs to structures involved in guiding eye movements. The distribution of neural events, both internally and externally generated, that alter the activity of these neurons are quite broad (i.e., neurons have large receptive fields). Hence, a coarse tuning mechanism has been suggested for computing the location of a target or movement field. A simple mathematical relation can be derived to describe how these neurons might effect the transformation of mixed sensorimotor coordinates, by which area 7a neurons represent their visual and oculomotor inputs, into head-centered coordinates of visual space.

There are neurons in other areas of cortex that are tuned to more than one sensory and/or motor signal, such as neurons in the motor system. For instance, some neurons in the motor cortex of a monkey trained to move his hand to a target location for a reward respond prior to the actual execution of the movement (Georgopoulos *et al.*, 1986; Caminiti *et al.*, 1990; Kalaska *et al.*, 1989). Some of these neurons in the motor cortex also take account of mechanical load on the representation of movement (Kalaska *et al.*, 1989). A motor neuron that is broadly tuned to the direction of movement in a two-dimensional plane in the absence of a load either magnifies or attenuates its directional response when a load is added to the limb. Yet other motor neurons in a monkey trained to make reaching movements in different regions of his workspace change the preferred direction of movement with the starting location of the hand (Caminiti *et al.*, 1990). This modulation of neural activity guiding a given hand path takes into account different arm postures relative to the torso (Scott and Kalaska, 1994).

These studies in motor cortex illustrate a modulation of a motor response that has a preferred direction in space by a second signal (i.e., load, location of movement in space, or arm posture). This modulation of a motor signal is formally similar to the modulation in inferior parietal cortex of the visual response by eye position. Hence the multisensory input to the motor cortex neurons might modulate their activity by the same mathematical summation as that put forward in this chapter for area 7a visual neurons.

It is this integration or association of multiple signals that is critical in constructing new representations. In the motor cortex, transformations are made which combine motor intent and sensory function, while in inferior parietal cortex, visual and oculomotor signals are mixed for spatial representation. The generation of these new internal representations forms the basis of cognitive functions, which are so profoundly disrupted with damage to the cerebral cortex. In inferior parietal lobule damage, many phenomena are left unexplained by the present state of knowledge, suggesting that more than eye position and vision are associated there. It remains an open question and a challenge to define not only the other types of inputs, but the modes of their interactions, if we are to fully understand the mechanisms of spatial representation in the human brain.

ACKNOWLEDGMENTS. We thank Gunther Stent for extensive and thoughtful comments, and Kathy Anderson, Kate Edgar, Dr. Howard Poizner, and Dr. James Tepper for careful reading of the manuscript. This work was funded by National Institutes of Health 1R01 EY09223 and Office of Naval Research N000014-93-1-0334.

9. References

Allman, J. M., Kaas, J. H., and Lane, R. H., 1973, The middle temporal visual area (MT) in the bushbaby, *Galago senegalensis, Brain Res.* **57:**197–202.

Andersen, R. A., and Mountcastle, V. B., 1983, The influence of the angle of gaze upon the excitability of the light-sensitive neurons of the posterior parietal cortex, *J. Neurosci.* **3:**532–548.

Andersen, R. A., Asanuma, C., and Cowan, W. M., 1985a, Callosal and prefrontal associational projecting cell populations in area 7A of the macaque monkey: A study using retrogradely transported fluorescent dyes, *J. Comp. Neurol.* **232:**443–455.

Andersen, R. A., Essick, G. K., and Siegel, R. M., 1985b, The encoding of spatial location by posterior parietal neurons, *Science* **230:**456–458.

Andersen, R. A., Essick, G. K., and Siegel, R. M., 1987, Neurons of area 7a activated by both visual stimuli and oculomotor behavior, *Exp. Brain Res.* **67:**316–322.

Andersen, R. A., Asanuma, C., Essick, G. K., and Siegel, R. M., 1990a, Cortico-cortical connections of anatomically and physiologically defined subdivisions within inferior parietal lobule, *J. Comp. Neurol.* **232:**443–455.

Andersen, R. A., Bracewell, R. M., Barash, S., Gnadt, J. W., and Fogassi, L., 1990b, Eye position effects on visual, memory, and saccade-related activity in areas LIP and 7a of macaque, *J. Neurosci.* **10:**1176–1196.

Asanuma, C., Andersen, R. A., and Cowan, W. M., 1985, The thalamic relations of the caudal inferior parietal lobule and the lateral prefrontal cortex in monkeys: Divergent cortical projections from cell clusters in the medial pulvinar nucleus, *J. Comp. Neurol.* **241:**357–381.

Bach, M., Bouis, D., and Fischer, B., 1983, An accurate and linear infrared oculometer, *J. Neurosci. Meth.* **9:**9–14.

Baizer, J. S., Ungerleider, L. G., and Desimone, R., 1991, Organization of visual inputs to the inferior temporal and posterior parietal cortex in macaques, *J. Neurosci.* **11:**168–190.

Baizer, J. S., Desimone, R., and Ungerleider, L. G., 1993, Comparison of subcortical connections of inferior temporal and posterior parietal cortex in monkeys, *Visual Neurosci.* **10:**59–72.

Ballard, D. H., Hinton, G. E., and Sejnowski, T. J., 1983, Parallel visual computation, *Nature* **306:**21–26.

Barbas, H., 1986, Pattern in the laminar origin of corticocortical connections, *J. Comp. Neurol.* **252:**415–422.

Barlow, H. B., 1972, Single units and sensation: A neuron doctrine for perceptual psychology? *Perception* **1:**371–394.

Battaglini, P. P., Galletti, C., and Fattori, P., 1993, Functional properties of neurons in area V1 of awake macaque monkeys: Peripheral versus central visual field representation, *Arch. Ital. Biol.* **131:**303–315.

Beverley, K. I., and Regan, D., 1975, The relation between discrimination and sensitivity in the perception of motion in depth, *J. Physiol.* **249:**387–398.

Blumcke, I., Hof, P. R., Morrison, J. H., and Celio, M. R., 1990, Distribution of parvalbumin immunoreactivity in the visual cortex of Old World monkeys and humans, *J. Comp. Neurol.* **301:**417–432.

Boussaoud, D., Ungerleider, L. G., and Desimone, R., 1990, Pathways for motion analysis: Cortical connections of the medial superior temporal and fundus of the superior temporal visual areas in the macaque, *J. Comp. Neurol.* **296:**462–495.

Bruce, C., Desimone, R., and Gross, C. G., 1981, Visual properties of neurons in a polysensory area in superior temporal sulcus of the macaque, *J. Neurophysiol.* **46:**369–384.

Caminiti, R., Johnson, P. B., and Urbana, A., 1990, Making arm movements within different parts of space: Dynamic aspects in the primate motor cortex, *J. Neurosci.* **10:**2039–2058.

Cavada, C., and Goldman-Rakic, P. S., 1989, Posterior parietal cortex in rhesus monkey: I. Parcellation of areas based on distinctive limbic and sensory corticocortical connections, *J. Comp. Neurol.* **287:**393–421.

Chang, J. J., and Julesz, B., 1985, Cooperative and non-cooperative processes of apparent movement of random-dot cinematograms, *Spat. Vis.* **1:**39–45.

Churchland, P. M., and Sejnowski, T. J., 1992, *The Computational Brain,* MIT Press, Cambridge, MA.

Colby, C. L., Duhamel, J. R., and Goldberg, M. E., 1993, Ventral intraparietal area of the macaque: Anatomic location and visual response properties, *J. Neurophysiol.* **69:**902–914.

Cowan, W. M., Gottlieb, D. I., Hendrickson, A. E., Price, J. L., and Woolsey, T. A., 1972, The autoradiographic demonstration of axonal connections in the central nervous system, *Brain Res.* **37:**21–51.

Critchley, M., 1953, *The Parietal Lobes,* Hafner Press, New York.

Damasio, A. R., Damasio, H., Tranel, D., and Brandt, J. P., 1990, Neural regionalization of knowledge access: Preliminary evidence, *Cold Spring Harbor Symp. Quant. Biol.* **55:**1039–1047.

Deecke, L., Schwartz, D. W., and Fredrickson, J. M., 1977, Vestibular responses in the rhesus monkey ventroposterio thalamus. II. Vestibulo-proprioceptive convergence at thalamic neurons, *Exp. Brain Res.* **30:**219–232.

Dow, B. M., Vautin, R. G., and Bauer, R., 1985, The mapping of visual space onto foveal striate cortex in the macaque monkey, *J. Neurosci.* **5:**890–902.

Duffy, C. J., and Wurtz, R. H., 1991a, Sensitivity of MST neurons to optic flow stimuli. I. A continuum of response selectivity to large-field stimuli, *J. Neurophysiol.* **65:**1329–1345.

Duffy, C. J., and Wurtz, R. H., 1991b, Sensitivity of MST neurons to optic flow stimuli. II. Mechanisms of response selectivity revealed by small-field stimuli, *J. Neurophys.* **65:**1346–1359.

Economo, C., 1929, *The Cytoarchitectonics of the Human Cerebral cortex,* Oxford University Press, Oxford.

Edelman, G. M., 1990, *The Remembered Present,* Basic Books, New York.

Ettlinger, G., 1990, Object vision and spatial vision: The neuropsychological evidence for the distinction, *Cortex* **26:**319–341.

Evarts, E. V., 1966, Methods for recording activity of individual neurons in moving animals, in: *Methods in Medical Research,* Year Book, Chicago, pp. 241–250.

Faugier-Grimaud, S., and Ventre, J., 1989, Anatomic connections of inferior parietal cortex (area 7) with subcortical structures related to vestibulo-ocular function in a monkey (*Macaca fascicularis*), *J. Comp. Neurol.* **280:**1–14.

Faust, C., 1947, Ubergestaltzerfall als symptom des parieto-ocipitalen ubergangs gebietes bei doppelseitige, *Nervenarzt* **18:**103–115.

Felleman, D. J., and Van Essen, D. C., 1991, Distributed hierarchial processing in primate cerebral cortex, *Cerebral Cortex* **1:**1–47.

Feynman, R. P., Leighton, R. B., and Sands, M., 1963, *The Feynman Lectures on Physics,* Addison-Wesley, Reading, MA.

Flechsig, P., 1876, *Die Leitungsbahnen im Gehirn und Ruckenmark des Menschen auf Grund Entwickelungsgeschichtlicher Untersuchungen,* Engelmann, Leipzig.

Friedman, H. R., and Goldman-Rakic, P. S., 1994, Coactivation of prefrontal cortex and inferior parietal cortex in working memory tasks revealed by 2DG functional mapping in the rhesus monkey, *J. Neurosci.* **14:**2775–2788.

Galletti, C., Squatrito, S., Battaglini, P. P., and Maioli, M. G., 1984, Real-motion cells in the primary visual cortex of macaque monkeys, *Brain Res.* **301:**95–110.

Georgopoulos, A. P., 1986, On reaching, *Annu. Rev. Neurosci.* **9:**147–170.

Georgopoulos, A. P., Schwartz, A. B., and Kettner, R. E., 1986, Neuronal population coding of movement direction, *Science* **233:**1416–1419.

Gibson, J. J., 1966, *The Senses Considered as Perceptual Systems,* Houghton Mifflin, Boston.

Gnadt, J. W., and Andersen, R. A., 1988, Memory related motor planning activity in posterior parietal cortex of macaque, *Exp. Brain Res.* **70:**216–220.

Gnadt, J. W., and Mays, L. E., 1989, Posterior parietal cortex, the oculomotor near response and spatial coding in 3-D space, *Soc. Neurosci. Abstr.* **19:**786.

Golomb, B., Andersen, R. A., Nakayama, K., Macleod, D. I. A., Wong, A., 1985, Visual thresholds for shearing motion in monkey and man, *Vision Res.* **25:**813–820.

Gray, C. M., and Singer, W., 1989, Stimulus-specific neuronal oscillations in orientation columns of cat visual cortex, *Proc. Natl. Acad. Sci. USA* **86:**1698–1702.

Graziano, M., Andersen, R. A., and Snowden, R. J., 1994, Tuning of MST to spiral motions, *J. Neurosci.* **14:**56–67.

Gross, C. G., 1994, How inferior temporal cortex became a visual area, *Cerebral Cortex* **4:**455–69.

Hammond, P., and MacKay, D. M., 1975, Differential responses of cat visual cortical cells to textured stimuli, *Exp. Brain Res.* **22:**427–430.

Hocherman, S., 1993, Proprioceptive guidance and motor planning of reaching movements to unseen targets, *Exp. Brain Res.* **95:**349–58.

Hof, P. R., and Morrison, J. H., 1990, Quantitative analysis of a vulnerable subset of pyramidal neurons in Alzheimer's disease: II. Primary and secondary visual cortex, *J. Comp. Neurol.* **301:**55–64.

Holmes, G., 1927, Disorders of sensation produced by cortical lesions, *Brain* **50:**413–427.

Judge, S. J., Richmond, B. J., and Chu, F. C., 1980, Implantation of magnetic search coils for measurement of eye position: An improved method, *Vision Res.* **20:**535–538.

Kalaska, J. F., Cohen, D. A., Hyde, M. L., and Prud'homme, M., 1989, A comparison of movement direction-related versus load direction-related activity in primate motor cortex, using a two-dimensional reaching task, *J. Neurosci.* **9:**2080–2102.

Kapoula, Z., Optican, L. M., and Robinson, D. A., 1990, Retinal image motion alone does not control disconjugate postsaccadic eye drift, *J. Neurophysiol.* **63:**999–1009.

Klüver, H., and Bucy, P. C., 1939, Preliminary analysis of functions of the temporal lobes in monkeys, *Arch. Neurol. Psychiatry* **42:**979–1000.

Knowlton, B. J., and Squire, L. R., 1993, The learning of categories: Parallel brain systems for item memory and category knowledge, *Science* **262:**1747–1749.

Koenderink, J. J., 1986, Optic flow, *Vision Res.* **25:**161–180.

Lagae, L., Gulyas, B., Raiguel, S., and Orban, G. A., 1989, Laminar analysis of motion information processing in macaque V5, *Brain Res.* **496:**361–367.

Lagae, L., Maes, H., Raiguel, S., Xiao, D. K., and Orban, G. A., 1994, Responses of macaque STS neurons to optic flow components: A comparison of areas MT and MST, *J. Neurophys.* **71:**1597–1626.

Lynch, J. C., 1980, The functional organization of the posterior parietal cortex, *Behav. Brain Sci.* **3:**485–534.

Lynch, J. C., Mountcastle, V. B., Talbot, W. H., and Yin, T. C., 1977, Parietal lobe mechanisms for directed visual attention, *J. Neurophysiol.* **40:**362–389.

Maunsell, J. H., and Van Essen, D. C., 1983, The connections of the middle temporal visual area (MT) and their relationship to a cortical hierarchy in the macaque monkey, *J. Neurosci.* **3:**2563–2586.

May, J. G., and Andersen, R. A., 1986, Different patterns of corticopontine projections from separate cortical fields within the inferior parietal lobule and dorsal prelunate gyrus of the macaque, *Exp. Brain Res.* **63:**265–278.

Mesulam, M.-M., 1990, Large scale neurocognitive networks and distributed processing for attention, language, and memory, *Ann. Neurol.* **28:**597–612.

Mesulam, M.-M., Van Hoesen, G. W., Pandya, D. N., and Geschwind, N., 1977, Limbic and sensory connections of the inferior parietal lobule (area PG) in the rhesus monkey, *Brain Res.* **136:**393–414.

Mesulam, M., Hegarty, E., Barbas, H., Carson, K. A., Gower, E. C., Knapp, A. G., Moss, M. B., and Mufson, E. J., 1980, Additional factors influencing sensitivity in the tetramethyl benzidine method for horseradish peroxidase neurohistochemistry, *J. Histochem. Cytochem.* **28:**1255–1259.

Miles, F. A., Kawano, K., and Optican, L. M., 1986, Short-latency ocular following responses of monkey. I. Dependence on temporospatial properties of visual input, *J. Neurophysiol.* **56:**1321–1354.

Mishkin, M., and Pribram, K. H., 1954, Visual discrimination performance following partial ablations of the temporal lobe: I. Ventral vs. lateral, *J. Comp. Physiol. Psychol.* **47:**14–20.

Mishkin, M., Ungerleider, L. G., and Macko, K. A., 1983, Object vision and spatial vision: Two cortical pathways, *TINS* **6:**414–417.

Moran, J., and Desimone, R., 1985, Selective attention gates visual attention processing in extrastriate cortex, *Science* **229:**782–784.

Morgan, M. J., and Ward, R., 1980, Conditions for motion flow in dynamic visual noise, *Vision Res.* **20:**431–435.

Motter, B. C., and Mountcastle, V. B., 1981a, The functional properties of the light-sensitive neurons of the posterior parietal cortex studied in waking monkeys: Foveal sparing and opponent vector organization, *J. Neurosci.* **1:**3–26.

Motter, B. C., and Mountcastle, V. B., 1981b, The functional properties of the light-sensitive neurons of the posterior parietal cortex, *J. Neurosci.* **1:**3–26.

Motter, B., Steinmetz, M. A., Duffy, C. J., and Mountcastle, V. B., 1987, Functional properties of parietal visual neurons: Mechanisms of directional selectivity, *J. Neurosci.* **7:**154–176.

Mountcastle, V. B., 1976, The world around us: Neural command function for selective attention, *Neurosci. Res. Program Bull.* **14:**1–47.

Mountcastle, V. B., Lynch, J. C., Georgopoulos, A., Sakata, H., and Acuna, C., 1975, Posterior parietal association cortex of the monkey: Command functions for operations within extrapersonal space, *J. Neurophys.* **78:**871–908.

Mountcastle, V. B., Andersen, R. A., and Motter, B. C., 1981, The influence of attentive fixation upon the excitability of the light sensitive neurons of the posterior parietal cortex, *J. Neurosci.* **1:**1218–1235.

Mountcastle, V. B., Motter, B. C., Steinmetz, M. A., and Sestokas, A. K., 1987, Common and differential effects of attentive fixation on the excitability of parietal and prestriate (V4) cortical visual neurons in the macaque monkey, *J. Neurosci.* **7:**2239–2255.

Nakayama, K., and Silverman, G. H., 1984, Temporal and spatial characteristics of the upper displacement limit for motion in random dots, *Vision Res.* **24:**293–299.

Newcombe, F., and Ratcliff, G., 1989, Disorders of visuospatial analysis, in: *Handbook of Neuropsychology,* Elsevier, New York, pp. 333–356.

Newsome, W. T., and Pare, E. B., 1988, A selective impairment of motion perception following lesions of the middle temporal visual area (MT), *J. Neurosci.* **8:**2201–2211.

Orban, G. A., Kennedy, H., and Bullier, J., 1986, Velocity sensitivity and direction selectivity of neurons in areas V1 and V2 of the monkey, *Exp. Brain Res.* **56:**462–480.

Orban, G. A., Lagae, L., and Verri, A., 1992, First order analysis of optical flow in monkey brain, *Proc. Natl. Acad. Sci. USA* **89:**2595–2599.

Pandya, D. N., and Seltzer, B., 1982a, Intrinsic connections and architectonics of posterior parietal cortex in the rhesus monkey, *J. Comp. Neurol.* **204:**196–210.

Pandya, D. N., and Seltzer, B., 1982b, Intrinsic connections and architectonics of posterior parietal cortex in the rhesus monkey, *J. Comp. Neurol.* **204:**196–210.

Paterson, A., and Zangwill, O. L., 1944, Disorders of visual space perception associated with lesions of right cerebral hemisphere, *Brain* **67:**331–358.

Pellionisz, A. J., and Ramos, C. F., 1993, Geometrical approach to neural net control of movements and posture [Review], *Prog. Brain Res.* **97:**245–256.

Ratcliff, G., and Davies-Jones, G. A., 1972, Defective localization in focal brain wounds, *Brain* **95:**49.

Read, H. L., Anderson, K. C., Noguiera, C. A. M., and Siegel, R. M., 1994, Space cells in the dorsal stream: Optic flow sensitivity in area 7a, *Abstr. Soc. Neurosci.* **20:**1278.

Read, H. L., and Siegel, R. M., 1997, Modulation of responses to optic flow in area 7a by retinotopic and oculomotor cues in monkey, *Cerebral Cortex* **7:**1–16.

Regan, D., 1986, Visual processing of four kinds of relative motion, *Vision Res.* **26:**127–145.

Regan, D., and Beverley, K. I., 1979, Visually guided locomotion: Psychophysical evidence for a neural mechanism sensitive to flow patterns, *Science* **205:**311–313.

Regan, D., Giaschi, D., Sharpe, J. A., and Hong, X. H., 1992, Visual processing of motion-defined form: Selective failure in patients with parietotemporal lesions, *J. Neurosci.* **12:**2198–2210.

Riddoch, G., 1917, On the relative perception of movement of a stationary object in certain visual disturbance, *Proc. R. Soc. Med.* **10:**13–34.

Robinson, D. L., Goldberg, M. E., and Goldberg, G. B., 1978, Parietal association cortex in the primate: Sensory mechanisms and behavioral modulations, *J. Neurophysiol.* **41:**910–932.

Robinson, D. L., Petersen, S. E., and Keys, W., 1986, Saccade-related and visual activities in the pulvinar nuclei of the behaving rhesus monkey, *Exp. Brain Res.* **62:**625–634.

Sacks, O. W., 1985, Eyes right! in: *The Man Who Mistook His Wife for a Hat and Other Clinical Stories,* Summit, New York, pp. 73–75.

Saito, H., Yukie, M., Tanaka, K., Hikosaka, K., Fukada, Y., and Iwai, E., 1986, Integration of direction signals of image motion in the superior temporal sulcus of the macaque monkey, *J. Neurosci.* **6:**145–157.

Sakata, H., Shibutani, H., and Kawano, K., 1980, Spatial properties of visual fixation neurons in posterior parietal association cortex of the monkey, *J. Neurophysiol.* **43:**1654–1672.

Sakata, H., Shubutani, H., Ito, Y., and Tsurugai, K., 1984, Parietal visual neurons respond to rotary movements of the visual stimulus, *J. Physiol. Soc. Jpn.* **46:**445.

Scott, S. H., and Kalaska, J. F., 1994, Cell activity in monkey motor cortex is altered by changes in arm posture for movements with identical hand trajectories, *Abstr. Soc. Neurosci.* **20:**982.

Selemon, L. D., and Goldman-Rakic, P. S., 1988, Common cortical and subcortical targets of the dorsolateral prefrontal and posterior parietal cortices in the rhesus monkey: Evidence for a distributed neural network subserving spatially guided behavior, *J. Neurosci.* **8:**4049–4068.

Siegel, R. M., 1998, Representation of visual space in area 7a neurons: A theoretical approach using the center of mass, *J. Comp. Neurosci.*

Siegel, R. M., and Andersen, R. A., 1988, Perception of three-dimensional structure from two-dimensional motion in monkey and man, *Nature* **3319:**259–261.

Siegel, R. M., and Andersen, R. A., 1990, The perception of structure from motion in monkey and man, *J. Cognitive Neurosci.* **2:**306–319.

Siegel, R. M., and Read, H. L., 1994, Egocentric motion from optic flow and eye position in area 7a of the behaving macaque, *Abstr. Soc. Neurosci.* **20:**1278.

Siegel, R. M., and Read, H. L., 1997, Analysis of optic flow in the monkey brain, *Cerebral Cortex* **7**:327–346.

Stark, M., Coslett, H. B., and Saffran, E. M., 1996, Impairment of an egocentric map of locations: implications for perception and action, *Cog. Neuropsych.* **13**:481–523.

Steinmetz, M. A., Motter, B., Duffy, C. J., and Mountcastle, V. B., 1987, Functional properties of parietal visual neurons: Radial organization of directionalities, *J. Neurosci.* **7**:177–191.

Stenvers, H. W., 1951, Localization direct et localization par image dans les champs visuals periphe-riques, *Rev. Oto-neuro-opthalmol.* **23**:6.

Tanaka, K., Hikosaka, K., Saito, H., Yukie, M., Fukada, Y., and Iwai, E., 1986, Analysis of local and wide-field movements in the superior temporal visual areas of the macaque monkey, *J. Neurosci.* **6**:134–144.

Tomlinson, R. D., and Robinson, D. A., 1984, Signals in vestibular nucleus mediating vertical eye movements in the monkey, *J. Neurophysiol.* **51**:1121–1136.

Tranel, D., and Damasio, A. R., 1985, Knowledge without awareness: An autonomic index of facial recognition by prosopagnosics, *Science* **228**:1453–1454.

Ullman, S., 1979, *The Interpretation of Visual Motion,* MIT Press, Cambridge, MA.

Ungerleider, L. G., and Desimone, R., 1986, Cortical connections of visual area MT in the macaque, *J. Comp. Neurol.* **248**:190–222.

Ungerleider, L. G., and Desimone, R., 1991, Organization of visual inputs to the inferior temporal and posterior parietal cortex in macaques, *J. Neurosci.* **11**:168–190.

Ungerleider, L. G., and Mishkin, M., 1982, Two cortical systems, in: *Analysis of Visual Behavior,* MIT Press, Cambridge, MA, pp. 549–586.

Vaina, L. M., 1994, Functional segregation of color and motion processing in the human visual cortex: Clinical evidence, *Cerebral Cortex* **5**:555–572.

Ventre, J., and Faugier-Grimaud, S., 1988, Projections of the temporo-parietal cortex on the vestibu-lar complex in the macaque monkey (*Macaca fascicularis*), *Exp. Brain Res.* **72**:653–658.

Von Bonin, G., and Bailey, P., 1947, *The Neocortex of Macaca Mulatta,* University of Illinois Press, Urbana, IL.

Von Helmholtz, L. F., 1924, *Helmholtz's Treatise on Physiological Optics* [translated from 3rd German ed.], Dover, New York.

Waitzman, D. M., Ma, T. P., Optican, L. M., and Wurtz, R. H., 1988, Superior colliculus neurons provide the saccadic motor error signal, *Exp. Brain Res.* **72**:649–652.

Walker, A. E., 1940, A cytoarchitectural study of the prefrontal area of the macaque monkey, *J. Comp. Neurol.* **73**:86.

Walker, N., Meyer, D. E., and Smelcher, J. B., 1993, Spatial and temporal characteristics of rapid cursor-positioning movements with electromechanical mice in human–computer interaction, *Human Factors* **35**:431–458.

Warrington, E. K., and McCarthy, R. A., 1987, Categories of knowledge: Further fractionations and an attempted integration, *Brain* **110**:1273–1296.

Warrington, E. K., and Shallice, T., 1984, Category specific semantic enhancements, *Brain* **107**:829–854.

Werkhoven, P., and Koenderink, J. J., 1991, Visual processing of rotary motion, *Percept. Psychophys.* **49**:73–82.

Williams, D. W., and Sekuler, R., 1984, Coherent global motions from stochastic local motions, *Vision Res.* **24**:55–62.

Wilson, F. A., Scalaidhe, S. P., and Goldman-Rakic, P. S., 1993, Dissociation of object and spatial processing domains in primate prefrontal cortex, *Science* **260**:1955–1958.

Yin, T. C., and Mountcastle, V. B., 1977, Visual input to the visuomotor mechanisms of the monkey's parietal lobe, *Science* **197**:1381–1383.

Yin, T. C., and Mountcastle, V. B., 1978, Mechanisms of neural integration in the parietal lobe for visual attention, *Fed. Proc.* **37**:2251–2257.

Young, A. W., Newcombe, D. H., and de Haan, E., 1989, Implicit access to semantic information, *Brain Cognition* **11**:186.

13

Visuomotor Areas of the Frontal Lobe

JEFFREY D. SCHALL

1. Introduction

That frontal cortex is involved in visually guided behavior has been known for over a century. Since the pioneering work of Ferrier, several areas in frontal cortex have been identified as having visual responses and playing some role in producing movements of the eyes, head, and limbs. Comprehensive reviews of frontal lobe organization and function have appeared (Fuster, 1989; Goldman-Rakic, 1987, 1988; Passingham, 1993; Levin *et al.*, 1991; Perecman, 1987; Stuss and Benson, 1986; Petrides and Pandya, 1994). This chapter will survey recent findings regarding the possible roles of the different areas of frontal cortex in the production of visually guided movements. Areas of disagreement in the literature will be examined. Although some neurons in primary motor cortex are visually responsive (e.g., Kwan *et al.*, 1985), such signals seem to be fairly nonspecific activations. Most emphasis will be on eye movements and the function of the frontal and supplementary eye fields. Recent experiments will also be reviewed that examine the function of the agranular premotor cortex and the granular dorsolateral and ventrolateral prefrontal cortex. The state of knowledge and the author's competence wane toward the rostral pole. Nevertheless, the orbitofrontal and cingulate cortex are integral if poorly understood parts of frontal cortex that relate to affect and personality.

As reviewed in this volume, the visual system in primates seems to be orga-

JEFFREY D. SCHALL • Vanderbilt Vision Research Center, Department of Psychology, Vanderbilt University, Nashville, Tennessee 37240.

Cerebral Cortex, Volume 12, edited by Rockland *et al.* Plenum Press, New York, 1997.

nized into at least two pathways (reviewed by Colby and Duhamel, 1991; Felleman and Van Essen, 1991; Merigan and Maunsell, 1993; see also the chapters by Schiller, Kaas, and Rockland in this volume). One pathway leads into inferior temporal cortex and subserves object vision. The other pathway leads into posterior parietal cortex and subserves spatial vision and the guidance of movements. It is important to realize, though, that neural processing is not concluded in the far reaches of the temporal and parietal lobes. Both of these "end stages" of visual processing as well as most other extrastriate visual areas are reciprocally connected with various parts of the frontal lobe (e.g., Baizer *et al.*, 1991; Schall *et al.*, 1995b; Stanton *et al.*, 1995). One of the topics of current interest and anticipated future progress is the nature of the visual processing conveyed to and carried out in the frontal lobe. One might even expect that understanding the function of extrastriate visual cortex will entail knowing what frontal cortex does with its visual afferents and what signals are fed back to the visual areas. For example, current hierarchical schemes developed from anatomical data (e.g., Felleman and Van Essen, 1991) could be construed to imply that visual processing proceeds in a sequential fashion. However, visually responsive cells in some parts of frontal cortex are activated earlier than or as early as many cells in striate and prestriate cortex. Thus, temporally if not also anatomically, concurrent processing seems to be the rule.

Physiological studies of extrastriate visual cortex have shown that visual response properties become more elaborate and specialized the further from striate cortex one records. Neuronal recordings in frontal cortex have demonstrated that the concepts of localized receptive fields and topographic maps are still useful descriptors for several areas. However, the nature of the mapping within an area or the relationship of the maps between two areas may be somewhat different in frontal cortex as compared to the visual cortical areas. For instance, evidence will be reviewed below indicating that the pattern of connectivity between the frontal eye field and the supplementary eye field is difficult to reconcile with the hypothesis that both areas have the same kind of mapping of saccade direction and amplitude. Also, the "trigger features" of neurons in frontal cortex are typically somewhat different from what is observed in occipital, temporal, and parietal visual cortical areas. We will review data from numerous studies showing that the visual responses of neurons in frontal cortex generally do not represent visual stimulus features. Instead, the visual responsiveness of frontal neurons is a function of the instructional significance or reinforcement value of stimuli and the motivational state of the monkey. In other words, certain neurons in frontal cortex seem to register not just the properties, but the meaning or value of a visual stimulus. The properties of premotor cortex, for instance, may provide a useful perspective for readers of this volume on extrastriate visual cortex. Visual perception generally entails recognizing the meaning of a stimulus. Meaning is defined in terms of the action made in response to the perceived stimulus. Thus, the movement responses called forth by stimuli represents a fundamental stage of visual perception. Premotor cortex seems to be necessary for associating the appropriate movement with an arbitrary stimulus (reviewed by Passingham, 1993). In addition, neurons in inferior orbital frontal cortex, an area linked with the limbic and reward systems of the brain, respond to visual stimuli according to how palatable an associated reward will be (Thorpe *et al.*, 1983).

The foregoing observations have stimulated quite provocative proposals regarding frontal cortex. For example, Crick and Koch (1995) have recently proposed that to reach the state of conscious awareness, visual signals must be registered in prefrontal cortex. Whether or not this proposal proves true, investigations of frontal cortex are necessary to understand how perception guides action and how action influences perception. Whereas our understanding of striate and prestriate visual cortex has been anchored in the retina, our understanding of frontal cortex has been anchored in the motor neurons. Investigations of the visuomotor areas in frontal cortex benefit from both perspectives.

Organization of Frontal Cortex

Progress on the functional organization of frontal cortex has proceeded in parallel with progress in distinguishing the anatomical areas. Figure 1 illustrates a parcellation of frontal cortex that represents most currently accepted subdivisions (e.g., Matelli *et al.*, 1991; Preuss and Goldman-Rakic, 1991; see also Walker, 1940; von Bonin and Bailey, 1947; Barbas and Pandya, 1987). Unfortunately, a

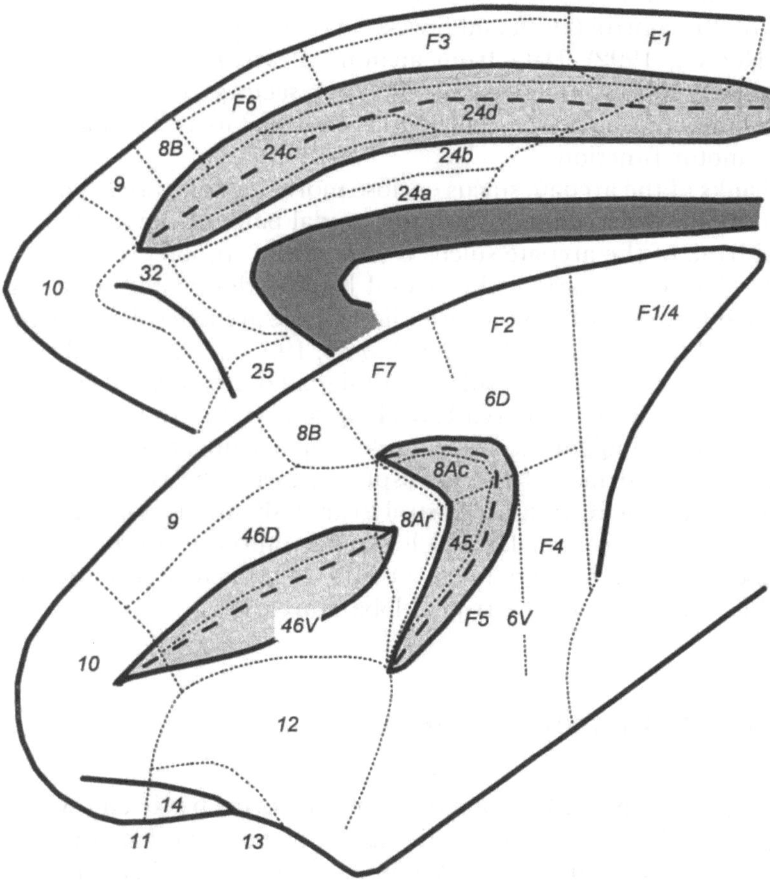

Figure 1. Anatomical subdivisions of macaque frontal cortex. Thicker lines represent the lips of opened sulci, showing the fundus as a dashed line. Boundaries between labeled areas are indicated by dotted lines.

uniform nomenclature has not been developed, nor will one be proposed in this chapter. Rizzolatti and co-workers have introduced a nomenclature scheme for somatic motor cortex that parallels that used for the visual cortical areas. Area F1 is primary motor cortex (area 4). Between area F1 and the arcuate sulcus lies premotor cortex in area 6. Anatomical and physiological evidence supports the distinction of dorsal and ventral subdivisions of area 6 (Barbas and Pandya, 1987; Matelli *et al.*, 1985, 1991; Kurata, 1991; Dum and Strick, 1991; Preuss and Goldman-Rakic, 1991; reviewed by Wiesendanger and Wise, 1992; Kurata, 1994). The ventral division, referred to functionally as PMv and anatomically as area 6V, may be further subdivided into rostral (area F5) and caudal (area F4) areas. The dorsal premotor area (PMd or area 6D) may also have rostral–caudal differences, but in this scheme, it is referred to wholly as area F2. On the dorsal cortex rostral to primary motor cortex, area F1, lies the supplementary motor area. This area has been subdivided into dorsal area F2 and mesial area F3. Rostral to area F2 on the convexity is a region designated area F7; this is the supplementary eye field (SEF) (Luppino *et al.*, 1991; Schall *et al.*, 1993). On the mesial surface adjacent to area F7 is area F6, which evidently corresponds to the presupplementary motor area (Tanji, 1994).

Recent work has revealed areas in anterior cingulate cortex that are also related to motor control (Dum and Strick, 1991; Luppino *et al.*, 1991; Morecraft and van Hoesen, 1992). Data from anatomical architecture and connectivity studies support subdividing area 24 into four sectors. The evidence indicates that the subdivisions buried in the cingulate sulcus (areas 24c and 24d) at least subserve a motor function.

The banks of the arcuate sulcus enclose more than one area. The dorsal and ventral premotor areas extend down the caudal bank of the arcuate sulcus. In the rostral bank of the arcuate sulcus is located the frontal eye field (Stanton *et al.*, 1989). The dorsal part of the rostral bank is designated area 8Ac and is distinguished from area 45 in the lower limb of the arcuate sulcus (Walker, 1940; Preuss and Goldman-Rakic, 1991; Schall *et al.*, 1995b). The lip and convexity of the rostral bank of the arcuate sulcus is designated area 8Ar (Preuss and Goldman-Rakic, 1991). Rostral to area 8Ar along and within the principal sulcus is area 46. The upper limb of the arcuate sulcus opens into area 8B, which, passing over to the mesial surface of the hemisphere, forms the rostral border of the supplementary eye field, area F7. Rostral to area 8B and dorsal to area 46 is area 9. Ventral to area 46 is area 12, which extends onto the ventral surface of the frontal lobe. Area 12 borders areas 11 and 13, which occupy orbital frontal cortex. The rostral pole of the frontal lobe is area 10.

2. Postarcuate Premotor Cortex

The organization and function of premotor cortex has been reviewed previously (e.g., Wise, 1985; Gentilucci and Rizzolatti, 1989; Wiesendanger and Wise, 1992; Kurata, 1994). This cortical region has traditionally been associated with the skeletal motor system. Some neurons in premotor cortex discharge in response to sensory stimuli that guide movements, and other neurons discharge before limb movements (Wise *et al.*, 1986; Godschalk *et al.*, 1985; Kurata, 1989).

Other neurons exhibit what has been referred to as preparatory set-related activity, discharging during the interval when a monkey can program, but not yet execute a movement (Kurata and Wise, 1988; Weinrich and Wise, 1982; Wise and Kurata, 1989; Wise and Mauritz, 1985). This review will selectively emphasize recent findings involving the visual properties of cells in the premotor cortex and also possible relations to gaze behavior. Premotor cortex has been hypothesized to function by retrieving appropriate motor actions in response to sensory stimuli and environmental contingencies (Passingham, 1993).

2.1. Ventral Premotor Cortex

A consensus is emerging that the ventral portion of premotor cortex (PMv) may be specialized for the visual guidance of forelimb movements. Visually responsive cells are found more commonly in ventral as compared to dorsal premotor cortex (Boussaoud *et al.*, 1993a; Fogassi *et al.*, 1992). Compared to the dorsal premotor area, PMv is more heavily innervated by prefrontal cortex (Matelli *et al.*, 1986; Preuss and Goldman-Rakic, 1989; Lu *et al.*, 1994) and by the rostral inferior parietal lobule (Godschalk *et al.*, 1984; Kurata, 1991; Cavada and Goldman-Rakic, 1989; Deacon, 1992; Petrides and Pandya, 1984). The face, mouth, and forelimb are represented in this part of cortex (Kurata *et al.*, 1985; Kurata, 1989; Gentilucci *et al.*, 1988; Stepniewska *et al.*, 1993; Preuss *et al.*, 1995). Thus, a key function of this cortical area may be related to visually guided grasping and manipulating objects, such as, for example, in feeding.

Based on functional and structural grounds the ventral premotor area has been further subdivided into caudal (area F4) and rostral (area F5) zones each with distinctive characteristics. The caudal zone is more closely associated with trunk or proximal forelimb movements, and the rostral zone is more related to distal limb and orofacial movements (Gentilucci *et al.*, 1988). Neurons in area F4 have somatosensory responses, and some cells are also visually responsive; approaching objects are the most effective stimuli (Gentilucci *et al.*, 1988). The "trigger features" of the visual cells in ventral premotor cortex are particular, requiring both proximity and incentive (Rizzolatti *et al.*, 1988). Neurons in area F5 are active in relation to various goal-directed movements to reach and grasp food and bring it to the mouth (Rizzolatti *et al.*, 1988). Neurons in area F4 are more likely to respond to visual stimuli than are cells in area F5, which required motivationally significant stimuli. A recent study has reported that visually responsive neurons in area F5 that discharge in relation to particular reaching or grasping movements also respond when a monkey simply observes another individual making the same movements of the hand (di Pellegrino *et al.*, 1992).

Another intriguing possible set of functions for PMv involves vocalization and speech, because the rostral part of PMv in the macaque may be homologous with Broca's area in humans (Petrides and Pandya, 1994). Positron emission tomography (PET) studies reveal distal forelimb and orofacial representation in ventral area 6 and area 44 of human (Colebatch *et al.*, 1991; Petersen *et al.*, 1988; Seitz and Roland, 1992). These loci correspond to areas that are activated in association with various language tasks (Démonet *et al.*, 1992; Paulescu *et al.*, 1993; Zatorre *et al.*, 1992) as well as during working memory tasks (Jonides *et al.*, 1993).

Coordinates of PMv Visual Responses

A key problem the brain solves in producing reaching and grasping movements is transforming the location of a visual object from retinal coordinates to limb coordinates (Soechting and Flanders, 1992; Flanders *et al.*, 1992). The location of an object in space could be converted from retinal coordinates to craniocentric coordinates by making use of eye position information. Head position information could then be used to derive spatial location. In seeking to understand this transformation, several investigators have explored whether neurons exist that encode the location of stimuli in nonretinotopic, craniotopic, spatial, or body-centered coordinates. Studies from different laboratories have yielded somewhat different conclusions about the extent to which visual neuronal responses in PMv are modulated by the angle of gaze or the location of the limbs.

Visually responsive neurons have been recorded in ventral premotor cortex by more than one laboratory. Many neurons recorded caudal to the arcuate sulcus are in fact responsive to both visual and tactile stimuli (Rizzolatti *et al.*, 1981; Gentilucci *et al.*, 1983; Fogassi *et al.*, 1992; Graziano *et al.*, 1994). Such neurons seem to be more or exclusively responsive to visual stimuli approaching close enough to the monkey to permit grasping (Rizzolatti *et al.*, 1981). The visual and somatosensory receptive fields exhibit spatial correspondence; that is, the visual receptive fields cover the space from which objects arrive to stimulate the tactile response field.

Some investigators have reported that the visual receptive fields of PMv cells do not shift with gaze, but rather remain stable in space or anchored to the tactile receptive field on the forelimb (Gentilucci *et al.*, 1983; Fogassi *et al.*, 1992; Graziano *et al.*, 1994). Based on these findings, Fogassi *et al.* (1992) and Graziano *et al.*, (1994) conclude that PMv encodes the location of visual stimuli in body-centered rather than retinal coordinates. An illustration from the study of Graziano *et al.* (1994) is shown in Fig. 2. In this study a monkey was trained to fix gaze either directly ahead or 20° to the left or right. During fixation a 10-cm white ball was moved toward the monkey along one of four trajectories, bringing it close enough that the monkey could grasp it. The cell illustrated had a tactile receptive field on the arm and was visually responsive to the moving ball. The visual response to the approaching ball was tested when the arm was fixed on the right and on the left. When the arm was fixed in a rightward posture, the cell illustrated in Fig. 2 responded to the approaching ball only when it followed the rightwardmost trajectory. Moreover, the magnitude of response was the same when the monkey's gaze was fixed in the center or 20° to the left or right. When the monkey's arm was placed toward the left, the response to movement of the ball along the rightmost trajectory was diminished somewhat, and the response to the ball moving along the trajectory closer to the midline increased. Other cells with tactile receptive fields on the face responded to the approaching stimulus along the same trajectory regardless of gaze angle. In an awake monkey Graziano and co-workers reported that essentially all of the cells tested had visual receptive fields that represented a constant spatial location regardless of gaze angle, but the magnitude of the visual response of most of these cells was significantly modulated by gaze angle. Also, most of the bimodal cells with receptive fields on the arm showed shifts of visual receptive field location as arm position changed. All of the subgroup of this cell population that was tested with varying gaze angles exhibited an apparent constancy of receptive field position regardless of gaze fixation location.

In contrast to these findings, Boussaoud *et al.* (1993), using different methods and more quantitative analyses, concluded that the visual responses of most PMv cells do not explicitly encode spatial location, but instead vary significantly with gaze angle. Their essential finding is illustrated in Fig. 3. Stimuli were presented on a video monitor in front of the monkey. When a stimulus was presented at the same retinal locus, the visual responsiveness of PMv cells varied with the direction of gaze. When a stimulus was presented at the same position on the screen and gaze shifted around that location, then the response of the cell varied greatly, indicating that due to shifting gaze, the stimulus was no longer in the receptive field. These investigators measured the degree of modulation of

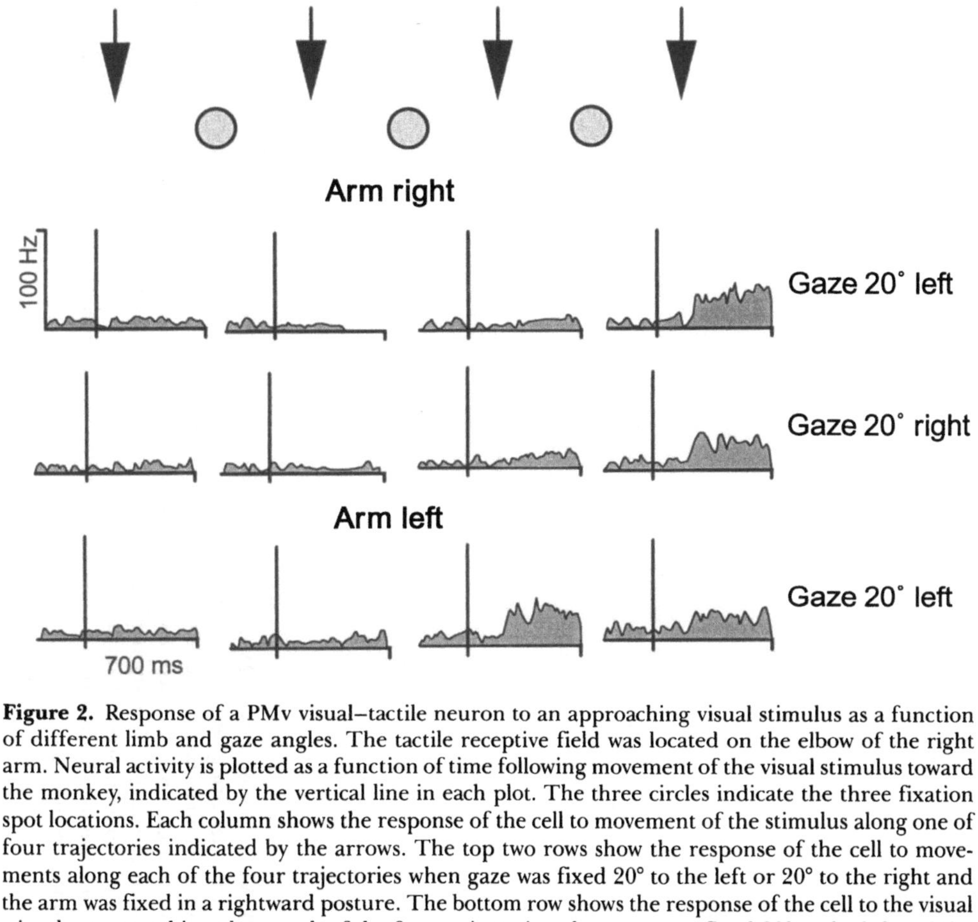

Figure 2. Response of a PMv visual–tactile neuron to an approaching visual stimulus as a function of different limb and gaze angles. The tactile receptive field was located on the elbow of the right arm. Neural activity is plotted as a function of time following movement of the visual stimulus toward the monkey, indicated by the vertical line in each plot. The three circles indicate the three fixation spot locations. Each column shows the response of the cell to movement of the stimulus along one of four trajectories indicated by the arrows. The top two rows show the response of the cell to movements along each of the four trajectories when gaze was fixed 20° to the left or 20° to the right and the arm was fixed in a rightward posture. The bottom row shows the response of the cell to the visual stimulus approaching along each of the four trajectories when gaze was fixed 20° to the left and the arm was fixed in a leftward posture. Apparently regardless of gaze angle, this neuron responded best when the stimulus moved along the rightmost trajectory when the arm was fixed in a rightward posture. When the arm was placed in the leftward posture, the cell responded less to movement along the rightmost trajectory and best to movement along the right of center trajectory. Redrawn from Graziano *et al.* (1994).

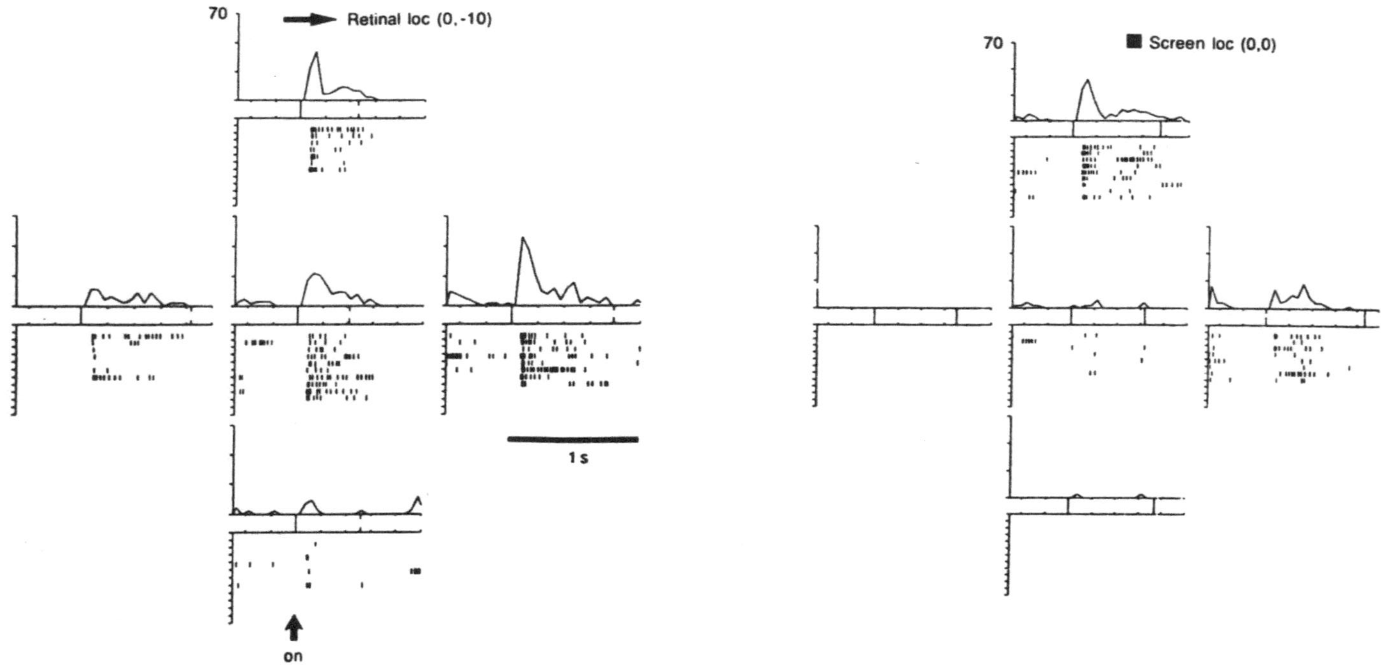

Figure 3. Visual response of a PMv neuron as a function of retinal locus, spatial locus, and gaze angle. (Left) Neural activity represented in raster and average firing rate plots aligned on the time of visual stimulus presentation, indicated by the arrow labeled 'on' and the first vertical line between the time axes of the raster and discharge rate plots. The stimulus was presented 10° below the fovea on the vertical meridian (0, −10) when gaze was directed toward the screen center or 8° left, right, up, or down from screen center. The displays are arranged according to gaze angle. Note the modulation of the visual response with gaze angle. (Right) Response of the same cell to a visual stimulus presented at a constant screen location as a function of gaze angle. The stimulus was always presented at the center of the screen (0, 0) when gaze was directed toward the screen center or 8° left, right, up, or down. Note the modulation of response with gaze angle. From Boussaoud *et al.* (1993) with permission.

visual responses as a function of gaze angle for a sample of PMv cells. Their data indicate that the response of nearly all of the PMv cells tested varied significantly with gaze angle. Boussaoud *et al.* (1993) concluded that the representation of visual stimulus location by PMv cells resembles what has been found in posterior parietal cortex (e.g., Andersen *et al.*, 1990; see also Galletti and Battaglini, 1989) and therefore does not encode absolute stimulus position in space. Boussaoud (1995) has also reported similar gaze angle dependence of activity of cells in PMd.

There are a number of differences in the experimental design of the Boussaoud *et al.* (1993) study as compared to those of Fogassi *et al.* (1992) and Graziano *et al.* (1994) which may account for their different conclusions. On the one hand the studies that claim to have found spatial constancy in the visual responses were recording from cells that had tactile receptive fields. Also, these studies presented stimuli that were physical manipulanda moving to within grasping distance of the monkey. Boussaoud *et al.* (1993) did not test for tactile responses, and they presented static stimuli on a video monitor that was beyond the reach of the monkey. So, because different stimulus tests and conditions were used, we cannot presently rule out the possibility that different populations of cells were studied by the different investigators. This possibility is suggested by the report that PMv cells with tactile receptive fields respond preferentially to visual objects moving toward the monkey that are close enough to grasp; such cells reportedly do not respond reliably to visual stimuli presented on a distant screen (Rizzolatti *et al.*, 1981). The relevance of the context of stimulus presentation as a factor explaining the different conclusions can be contested, though. Graziano *et al.* (1994) reported finding visual receptive fields that shifted with arm position in anesthetized monkeys. We should also not overlook the possibility that the various studies may have been sampling neurons in slightly different cortical zones. Another factor to consider in the interpretation of these experiments is the quantitative degree of gaze angle independence. Based on qualitative testing, Fogassi *et al.* and Graziano *et al.* emphasize the lack of variation of response with gaze angle. Based on quantitative testing, Boussaoud *et al.* emphasize the degree of variation of response with gaze angle. It may be that the different reports in this literature reflect different emphases or perspectives on the same phenomenon.

2.2. Dorsal Premotor Cortex

This cortical region has been investigated most often in relation to reaching movements produced in conditional motor tasks. In a conditional motor task a subject is cued by a stimulus (tactile, acoustic, or visual) to make one of two or more different movements, e.g., different directions to reach. A number of variations of this kind of task have been employed to investigate different aspects of movement programming and production. For instance, the meaning of the cue stimulus, i.e., which movement will be rewarded in response to which stimulus, can be constant, or the mapping of stimuli onto responses can be variable across trials or sessions. When the mapping of stimuli to responses is variable, an explicit, arbitrary stimulus–response mapping can be instructed by another stimulus presented before the movement cue. For example, a movement cue (e.g., green or red light) can signal either a rightward or leftward movement, depending on the property of a prior instruction stimulus (e.g., high-frequency tone or low-frequency tone). In another version of the task an implicit stimulus–response mapping can be discovered by subjects through the reward contingencies. Many studies have now shown that premotor cortex cells respond in such a way as to suggest that they play a role in the selection and guidance of reaching movements in particular directions (e.g., Weinrich and Wise, 1982; Riehle and Requin, 1989; Caminiti *et al.*, 1991; Kalaska and Crammond, 1992; reviewed by Kurata, 1994). PMd consists of a significant fraction of cells that begin to discharge when informative stimuli are presented that instruct monkeys about upcoming movements (Kurata, 1989; Kurata and Wise, 1988; Weinrich *et al.*, 1984). Visually responsive cells in premotor cortex have large receptive fields that commonly occupy a quadrant if not a hemifield, and are likely to represent the ipsilateral hemifield as well as the contralateral one (Boussaoud and Wise, 1993a).

2.2.1. Contingency of PMd Visual Responses

During natural behavior not every visual stimulus evokes a movement or conscious perception. This fact suggests that certain neurons in the brain ought to produce an apparent visual response contingent on whether the stimulus evokes a movement. In other words, one ought to find neurons that do not respond to a visual stimulus if that stimulus does not result in a movement. Boussaoud and Wise (1993a, b) designed a task for macaque monkeys in which the same stimulus, presented more than once, could either designate a location to attend and remember or could specify a particular movement to make. Upon its first presentation the stimulus instructed monkeys where an upcoming stimulus would appear and so guided a covert shift of attention (designated SAM in Fig. 4). When it appeared again at the cued location, the stimulus instructed a direction to move (designated MIC in Fig. 4). This experimental dissociation of stimulus from response has been done before in previous studies of frontal cortex, although with no control over gaze (e.g., Alexander and Crutcher, 1990; Watanabe, 1990, 1992). Boussaoud and Wise recorded neurons in postarcuate premotor cortex as well as in ventrolateral prefrontal cortex. They identified neurons which tended to respond differently to the same stimulus when it guided an attention shift versus when it instructed a movement direction. In particular, over half of the neurons in PMd respond preferentially or exclusively to a stimulus when it signals movement direction. The same retinal stimulus evokes little or no response when it signals a location to attend and remember (Fig. 4). This property was significantly less prevalent in PMv and ventral prefrontal cortex, in which a higher fraction of cells responded to the first cue stimulus.

The increased discharge rate of a PMd cell following a visual stimulus that instructs a movement may be interpreted not so much as a visual response as rather a signal to prepare a particular movement. This finding is consistent with the hypothesis that PMd plays a key role in the selection of movements based on the sensory array (Passingham, 1993). In contrast, most cells in ventral prefrontal cortex and in PMv cortex responded more to the stimulus that cued the location of the upcoming movement instruction cue. These results demonstrate that the responses of neurons in different areas of frontal cortex register different aspects of the visual, cognitive, and motor processing giving rise to behavior.

Another way to dissociate sensory or attention-related activation from motor planning-related activation is to manipulate stimulus–response compatibility. Almost always eye or limb movements are directed toward the location of stimuli in the world. That is, the direction of the stimulus and of the evoked movement are compatible or congruent. It is possible, however, for primates to generate arbitrary movements in a direction that is different from that of a stimulus. For example, humans and monkeys can be instructed to generate an antisaccade, that is, a gaze shift in the direction opposite a visual stimulus (Hallett and Adams, 1980). In conditions that require movements that are incompatible with stimulus locations, neurons in prefrontal cortex tend to discharge more in relation to the location of the stimulus (di Pellegrino and Wise, 1993a, b;

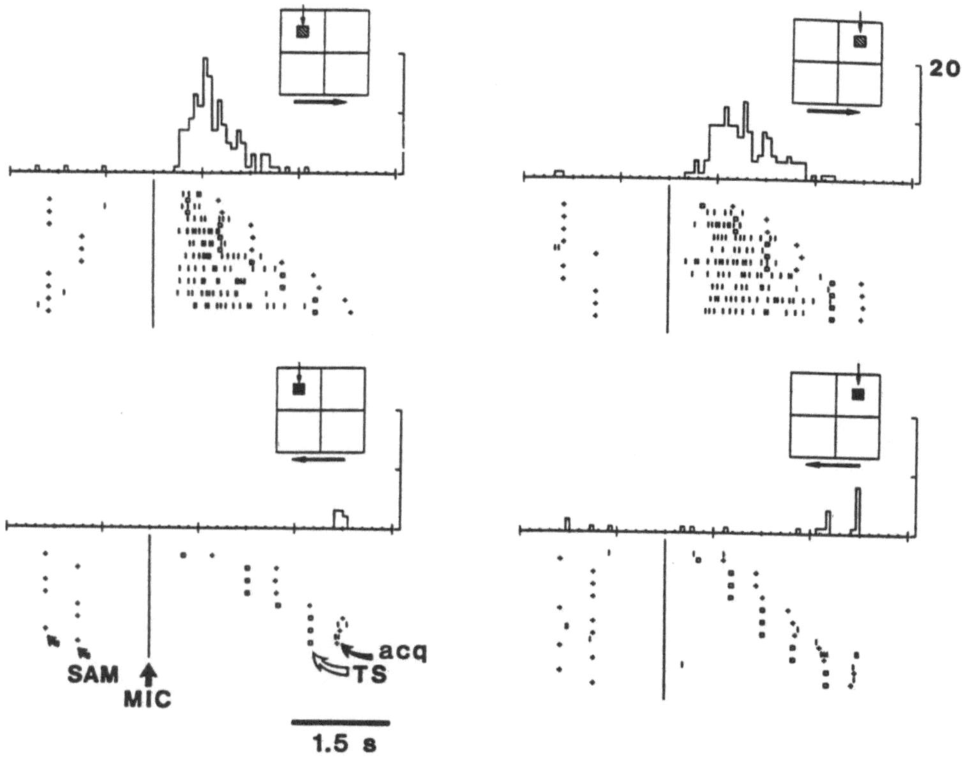

Figure 4. Responses of a PMd neuron to visual stimuli with different meanings. The rasters and histograms are aligned on the time of presentation of a motor instructional/conditional cue (MIC). The plus sign to the left indicates the time of presentation of the spatial-attentional/mnemonic stimulus (SAM). The open square to the right indicates the time of the stimulus triggering the movement (TS), and the next plus sign to the right is the time of the movement. The scale for all histograms is 20 Hz. The top panels illustrate activity recorded when the motor instruction cue instructed a rightward movement; the bottom panels illustrate trials when leftward movements were instructed. The left panels show trials when the motor instruction cue appeared on the left, and the right panels show trials when the motor instruction stimulus appeared on the right. This neuron responded only to visual stimuli that instructed a rightward movement and did not respond to identical stimuli that appeared otherwise. From Boussaoud and Wise (1993a) with permission.

Funahashi *et al.*, 1993b). However, in cortical areas linked to movement production, cells are more commonly found that are active in relation to the movement direction as opposed to the stimulus direction (di Pellegrino and Wise, 1993a, b; Alexander and Crutcher, 1990; see also Crammond and Kalaska, 1994).

Such an observation would not be surprising for cells that discharge in relation to movement generation. However, Fig. 5 shows that even the visual

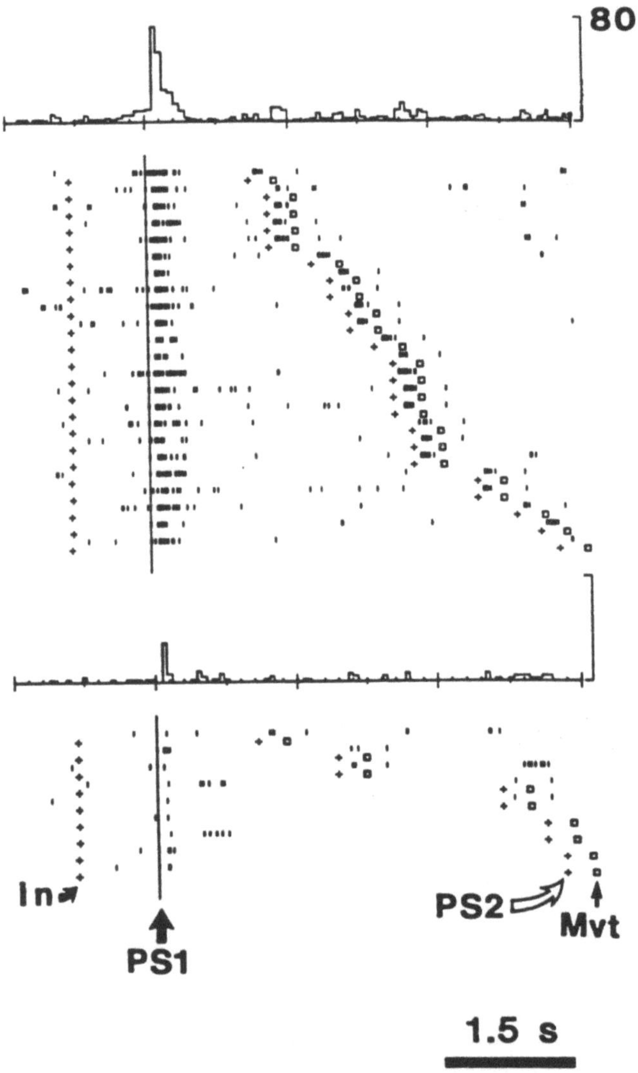

Figure 5. Visual response of a premotor cortex neuron to a stimulus that directs attention to a location that will (top) or will not (bottom) be the target for a movement. The rasters and histograms are aligned on the time of presentation of the original prime stimulus (PS1). The plus sign to the left represents the initiation of each trial (In). The plus sign to the right shows the time of presentation of the second prime stimulus (PS2) appearing at the location of PS1. The open square shows the time of the movement (Mvt). The top panel shows the response to the PS1 when the movement was directed to the subsequent target at that location. Also note the vigorous response to PS2. The bottom panel shows the attenuated response to PS1 when the movement would not be directed to the stimulus when it appeared at that location. The scale of the histograms was 80 Hz. Modified from di Pelligrino and Wise (1993a) with permission.

responses of PMd cells reflect whether or not the stimulus and response are compatible. In this study by di Pellegrino and Wise (1993a, b) monkeys were trained to make forelimb movements in one of eight directions guided by a circular array of visual stimuli. After fixating a central spot, the trial began with a flashed presentation of a stimulus at one of the eight locations (designated the first prime stimulus, PS1). Then the other stimuli flashed in random sequence. When the stimulus at the original location flashed again (designated the second prime stimulus, PS2), the monkey was rewarded for making a limb movement. Thus, the monkey had to attend to and/or remember the first stimulus location until that stimulus appeared again to trigger the movement. The investigators challenged the monkey further, though, by including blocks of trials in which reward was contingent on limb movements directly to the location of the second flashed target (stimulus–response compatible) and other blocks of trials in which monkeys were rewarded for moving toward the top stimulus regardless of where the cued stimulus was located (stimulus–response incompatible). The response of the visually responsive cell in Fig. 5 was significantly modulated by stimulus–response compatibility. As shown in the top panel, the cell responded well to the stimulus flashed in its response field when it cued the primed location and also when it triggered the movement. In contrast, when the same stimulus was flashed at the same location except that the monkey would not ultimately make a movement to its location, the cell response was markedly attenuated following a very brief initial response.

For those interested in visual processing, the time course of these modulatory effects are worth noting. In the extreme instances, the effect of stimulus meaning on visual responses occurs within the response latency of the cell. The latencies of cells in premotor cortex average 138 msec with minimum values of 60 msec (Weinrich *et al.*, 1984). If the response of such cells is to be modulated, then within this latent interval several events must occur. The stimulus must be discriminated based on its properties and evaluated based on the instructions and reward contingencies. In addition, the appropriate modulatory control must be exerted over the afferents to premotor visual cells. It is important to note that this extraretinal modulation has been observed under blocked trial conditions; that is, monkeys could adopt a particular strategy throughout a series of trials until the reward contingency changed. Thus, the modulation of the visual responses could be established over a period of several minutes. How visual responses are modulated or gated is not known, so future work should focus on possible mechanisms of such control.

2.2.2. Role of PMd in Conditional Motor Learning

Ablations of premotor cortex cause impaired learning of conditional motor tasks (Halsband and Freund, 1990; Halsband and Passingham, 1985; Petrides, 1987; Crowne *et al.*, 1989; Passingham, 1988, 1989; see also Kurata and Hoffman, 1994). Motivated by this observation, Mitz *et al.* (1991) recorded in dorsal premotor cortex of monkeys while they learned novel, arbitrary stimulus–response associations. Mitz and co-workers trained monkeys to move a handle in one of three directions or not at all depending on which visual stimulus was presented. Arbitrary stimulus–response associations were established through reward contingency. Once a monkey responded consistently to a particular set of stimuli, a new set was introduced for the monkey to learn. Neural activity was recorded from PMd neurons when performance was perfect and also throughout the learning phase. The investigators identified PMd cells that were active when a particular stimulus resulted in a particular movement. They then found that such neurons became much less active when a new stimulus set was introduced calling for a different stimulus–response mapping. However, as a monkey's performance improved, the PMd neural activity grew. Notably, though, the growth in neural activity lagged a few trials behind the improvement in performance. The authors interpret this lag as indicating that PMd cells may not necessarily function in the learning of the new stimulus–response association, but may instead subserve a movement retrieval or preparation process. This learning-related modulation of PMd neuronal activity was observed for preparatory set- and movement-related activity. But Mitz and co-workers also observed this phenomenon in the visually responsive cells. Furthermore, there was a significant relation between the magnitude of visual and set-related activity in response to given stimuli and whether the subsequent movement was correct or incorrect. This is further evidence that the visual response of premotor cells is contingent on behavioral relevance.

3. Frontal Eye Field

Investigations of frontal eye field (FEF) have been motivated by its association with the oculomotor system. The very well developed understanding of the oculomotor system has provided a basis for developing and evaluating relatively sophisticated and mature hypotheses as compared to studies of other areas of frontal cortex. The organization and function of frontal eye field has been reviewed previously (Goldberg and Segraves, 1989; Bruce, 1990; Schall, 1991c). This section will summarize the anatomical inputs and outputs of FEF and will review recent findings regarding the role of FEF in eye movement production and target selection. New information from neuroimaging and lesion studies in humans will also be presented.

3.1. Overview of the Oculomotor System

FEF is one node in a complex network of structures mediating gaze control. A basic understanding of the oculomotor system is necessary to understand and evaluate recent experiments investigating FEF. Figure 6 is a simplified schematic diagram of key structures and connections in the visuomotor system responsible for saccade production. A network located in the brainstem is responsible for generating saccadic eye movements (reviewed by Hepp *et al.*, 1989; Keller, 1991). In general terms, a horizontal saccade is initiated when burst neurons in the paramedian pontine reticular formation activate motor neurons innervating the medial and lateral recti muscles. Burst neurons are gated by pause neurons which are located in the nucleus raphe interpositus. The rapid saccadic eye movement is driven by an error signal generated by a local feedback loop. The error is the difference between the desired and the current eye position. The current eye position is represented by a neural integrator localized mainly in the nucleus prepositus hypoglossi. When the motor error signal is reduced to zero, the drive on the burst cells is removed, and the pause cells reinstate their inhibition on the network for the next period of fixation.

The saccade generation network requires two inputs, one signaling the desired direction and amplitude of the movement ("where") and the other triggering the initiation of the movement ("when"). One main source of these signals is the superior colliculus (reviewed by Sparks and Hartwich-Young, 1989; Guitton, 1991). The superior colliculus receives visual afferents directly from the retina as well as descending inputs from many cortical areas, but particularly for the present discussion the lateral intraparietal area (LIP) in posterior parietal cortex, FEF, and the supplementary eye field (SEF). FEF and SEF each also project directly to the brainstem.

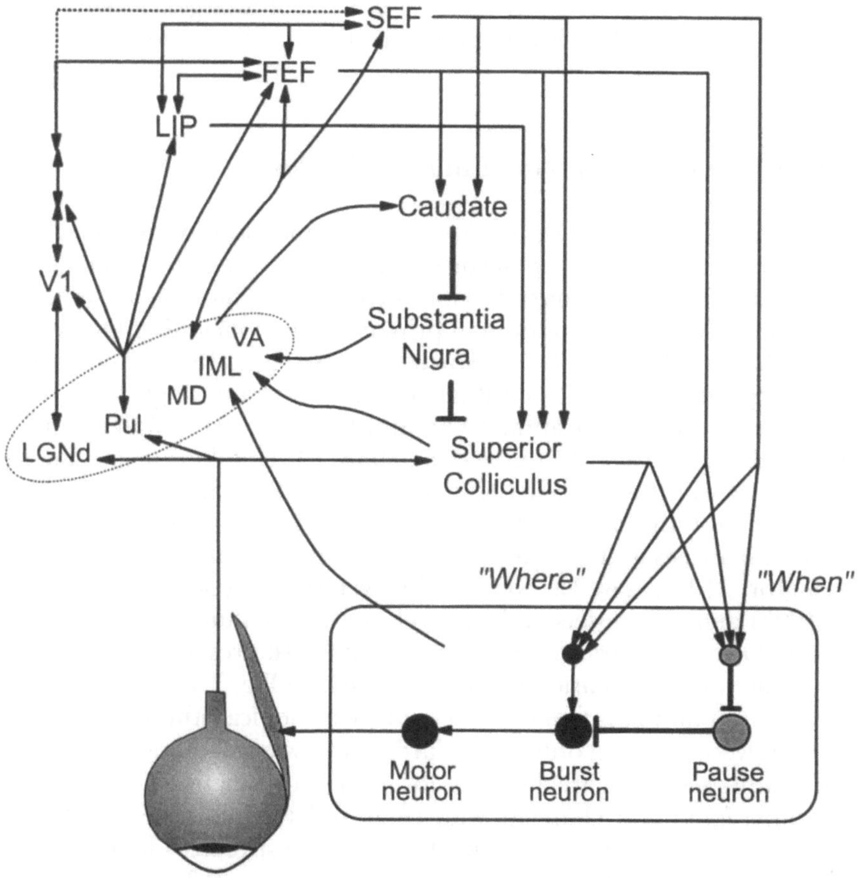

Figure 6. Simplified diagram of the oculomotor system. Excitatory connections are represented by black arrows. Inhibitory connections are indicated by solid bars. The units enclosed in the rectangle represent elements of the brainstem saccade-generating network. Thalamic nuclei are enclosed by the ellipse. Details in text. IML, Internal medullary lamina thalamus; LGNd, dorsal lateral geniculate nucleus; LIP, lateral intraparietal area; MD, mediodorsal nucleus; Pul, pulvinar; VA, ventroanterior nucleus; V1, primary visual cortex.

An oculomotor circuit also passes through the basal ganglia, through which FEF and SEF may regulate the activity of neurons in the superior colliculus. Movement cells in the superior colliculus receive tonic inhibition from the substantia nigra pars reticulata (Hikosaka and Wurtz, 1983b, 1985a, b). Neurons in the substantia nigra pars reticulata are themselves largely inhibited in relation to visually guided saccades by an oculomotor region of the caudate nucleus (Hikosaka *et al.*, 1989a–c, 1993). The oculomotor region of the caudate nucleus is innervated by both FEF and SEF (Parthasarathy *et al.*, 1992; Shook *et al.*, 1991; Stanton *et al.*, 1988a).

Visual signals reach FEF and SEF via a number of extrastriate visual areas and also possibly thalamic afferents. While FEF is innervated to an extent by pulvinar, FEF and SEF receive denser input from the segments of the ventroanterior and mediodorsal thalamic nuclei adjacent to the internal medullary lamina. These thalamic nuclei are themselves mainly innervated by afferents from the superior colliculus and the substantia nigra as well as cerebellar nuclei (e.g., Lynch *et al.*, 1994). Clearly, FEF is a component of a larger, integrated system, functioning in parallel with a number of other cortical and subcortical structures.

3.2. Location of FEF in Nonhuman Primates

Following Ferrier's (1874, 1875) original demonstration that electrical stimulation of certain parts of frontal cortex elicits eye movements, many subsequent workers mapped the regions of frontal cortex from which eye movements could be elicited in a variety of nonhuman primate species including Old World monkeys (Beevor and Horsley, 1888; Horsley and Schäfer, 1888; Mott and Schäfer, 1890; Sherrington, 1893; Risien Russell, 1894; Vogt and Vogt, 1907, 1919; Jolly and Simpson, 1907; Levinsohn, 1909; Smith, 1936, 1940; Walker and Weaver, 1940; Crosby *et al.*, 1952; Wagman *et al.*, 1958, 1961; Robinson and Fuchs, 1969; Marrocco, 1978; Schiller *et al.*, 1979; Bruce *et al.*, 1985; Schall, 1991b; Russo and Bruce, 1993), orangutan (Beevor and Horsley, 1890; Leyton and Sherrington, 1917), chimpanzees and gorilla (Grunbaum and Sherrington, 1901; Leyton and Sherrington, 1917), as well as New World owl monkeys (Gould *et al.*, 1986; Preuss *et al.*, 1995) and marmosets and squirrel monkeys (Blum *et al.*, 1982; Huerta *et al.*, 1986). Over the years a number of authors have presented summaries of the eye movement representation in frontal cortex. In Fig. 7, the current addition to this compendium, maps from a variety of physiological stimulation and recording studies were redrawn onto a standardized view of macaque frontal cortex. As can be readily seen, the areal extent, location, and functional mapping of areas and subdivisions have changed with time and technique. Nevertheless, FEF, located rostral to the arcuate sulcus, has been consistently identified.

Ferrier (1874) evoked contraversive eye movements by surface stimulation of a region extending from the dorsal part of the prearcuate cortex to the dorsomedial midline (Fig. 7A). This region encompasses cortex that is currently defined as FEF and SEF. Mott and Schäfer (1890) evoked eye movements over the same general area of cortex (Fig. 7B). Besides extending the eye field ventrolaterally to the lower limb of the arcuate sulcus, they reported a systematic localization of different directions of gaze deviations. Levinsohn (1909) delimited the ventral caudal boundary of the FEF to the bank of the arcuate sulcus (Fig. 7C). He noted a lower threshold region at the genu of the arcuate sulcus, surrounded rostrally by a higher threshold zone. Head movements were evoked caudal and medial to the arcuate sulcus with ear movements localized in a dorsomedial sector. This map presaged many features of the current view. Smith (1936, 1940) identified different kinds of eye movements evoked by surface stimulation. Smith found extensive intermingling of the kinds of eye movements evoked by stimulation, but he summarized the general tendencies indicated in Fig. 7D. Sites evoking ocular rotations were localized mainly rostral to the arcuate sulcus. Sites evoking eyeblink and pupil dilation were located ventral and dorsal, respectively, to the arcuate sulcus. Crosby *et al.* (1952) evoked eye movements only from the rostral bank of the arcuate sulcus in lightly anesthetized monkeys (Fig. 1E). They noted systematic variation in the direction of contraversive deviation as indicated in the figure. They also noted that the nature of the eye movement evoked could vary with the level of anesthesia. Wagman *et al.* (1961) evoked eye movements over a more extensive part of frontal cortex (Fig. 7F). These investigators stimulated the surface of the cortex as well as subcortically using large electrodes in nonanesthetized, spinal transected monkeys and took pains to detect any eye movement evoked by the stimulation. The figure represents their summary of the data. They evoked eye movements over an area extending from the rostral end of the principal sulcus back to the superior central sulcus caudal to the arcuate sulcus, from the ventral limb of the arcuate sulcus up to the dorsomedial convexity. The highest density of excitable points, though, was immediately rostral to the arcuate sulcus.

Robinson and Fuchs (1969) performed the first study of the effects of frontal lobe stimulation using modern methods of intracortical microstimulation and the scleral search coil to monitor eye position in alert monkeys. The use of electrical stimulation through microelectrodes permitted the investigators to deliver smaller effective currents, so localization was improved. The use of the scleral search coil in the alert monkey provided a sensitive and reliable measure of eye position over time. These methods continue to yield useful information, which will be reviewed below. The zone from which Robinson and Fuchs evoked eye movements with microampere currents was restricted to the rostral bank of the arcuate sulcus (Fig. 7G). Robinson and Fuchs were the first to show that the eye movements evoked by FEF stimulation are saccades as characterized by the quantitative relationship between eye velocity and movement amplitude. They noted that the direction and amplitude of the evoked saccade did not vary too much with initial gaze angle, but did vary with location of the stimulating electrode in the cortex. They also discovered that the amplitude of the evoked saccade varied systematically in FEF. Ventrolateral sites evoked smaller amplitude saccades, and saccade amplitude increased gradually as more dorsomedial sites were stimulated along the arcuate sulcus. The lowest thresholds for stimulation Robinson and Fuchs found were in the genu of the arcuate sulcus.

Suzuki and Azuma (1983) recorded single-cell responses to visual stimuli. They found a map of receptive field eccentricity in the cortex immediately rostral to the arcuate sulcus, extending midway along the principal sulcus (Fig. 7H). Suzuki and Azuma also found that receptive field size (Fig. 7I) increased

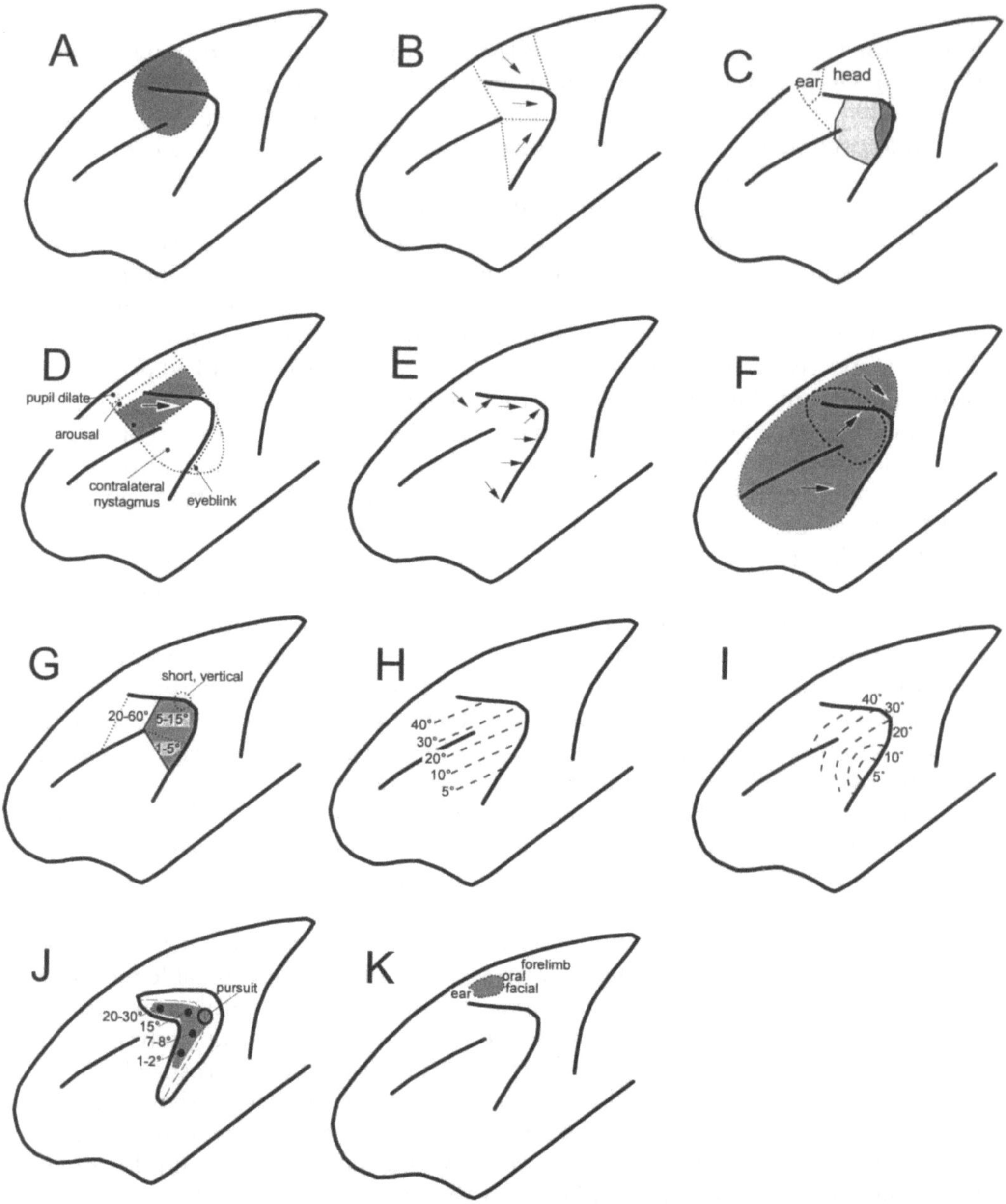

Figure 7. Overview of areas in frontal cortex related to eye movement generation. (A) Original map from Ferrier (1875). Eye movement were evoked from the shaded region. (B) Map from Mott and Schäfer (1890). Zones from which different directions of eye movements were evoked are indicated. (C) Map from Levinsohn (1909). Eye movements were evoked from the shaded region; the darker shading indicates the lower threshold region. (D) Map from Smith (1940). The regions from which the different types of responses were observed are indicated. (E) Map from Crosby *et al.* (1952). The different directions of eye movements evoked are indicated. (F) Map from Wagman *et al.* (1961). Regions from which different directions of eye movements were evoked are indicated by different stipples and line widths. (G) Map from Robinson and Fuchs (1969). Zones are shown from which different amplitude saccades were evoked. (H) Map of visual response field eccentricity from Suzuki and Azuma (1983). (I) Map of visual response field size from Suzuki and Azuma (1983). (J) Map of

from lateral to medial, and from caudal to rostral. Bruce *et al.* (1985) used currents less than 50 μA to localize FEF in the rostral bank and fundus of the arcuate sulcus (Fig. 7J). They replicated and refined the map of saccade amplitude as indicated in the figure. Shorter (<2°) saccades are represented ventrolaterally, and progressively longer saccades (15–20°) are represented dorsomedially. Stimulation in the cortex of the upper limb of the arcuate sulcus can also elicit pinna movements in macaques (Parthasarathy *et al.*, 1992; Schall *et al.*, 1993; Bon and Lucchetti, 1994). A region related to slow tracking eye movements was localized by MacAvoy *et al.* (1991) and Gottlieb *et al.* (1994) at the fundus of the arcuate sulcus immediately caudal to the principal sulcus (Fig. 7J). We should note that the map of visual receptive field size and eccentricity in the rostral bank of the arcuate sulcus corresponds roughly to the map of saccade amplitude.

The FEF, as defined by the lowest current thresholds for eliciting an eye movement, is now localized to the cortex on the rostral bank of the arcuate sulcus (Bruce *et al.*, 1985). This area of cortex has a distinctive cyto- and myeloarchitecture (Walker, 1940; von Bonin and Bailey, 1947; Stanton *et al.*, 1989; Preuss and Goldman-Rakic, 1991). The region in the rostral bank of the arcuate sulcus containing the highest concentration of large pyramidal cells in layer V corresponds to the region in which saccades are elicited with the lowest thresholds (Stanton *et al.*, 1989). This cortex is also characterized by having a thinner granular layer 4 as compared to the more rostral cortex. The ventrolateral portion of this zone, designated area 45, is distinguished by the presence of large pyramidal cells in layer 3 as well as layer 5 (Walker, 1940; Preuss and Goldman-Rakic, 1991). The dorsomedial zone, referred to as area 8A, can be further parcellated based on cyto- and myeloarchitectonic differences (Preuss and Goldman-Rakic, 1991). The cortex within the medial portion of the rostral bank of the arcuate sulcus contains fewer large pyramidal cells in layer 3 and a loosely organized granular layer; it is referred to as area 8Ac. In addition to this subdivision, there is a transitional zone, designated area 8Ar, with fewer large pyramidal cells and a thicker, more clearly defined granular layer. In myelin stains area 8Ar lacks the thick fascicles that are observed in area 8Ac. This transition zone forms the rostral boundary of area 8Ac and area 45a and joins the caudal boundary of area 46. Area 8Ar should probably be considered as functionally distinct from FEF. It may correspond to the area FV that has been distinguished from the heavily myelinated part of FEF in owl monkeys based on patterns of connectivity with prestriate visual areas (Weller and Kaas, 1987; Krubitzer and Kaas, 1990).

3.3. Location of FEF in Humans

Much current interest is focused on determining whether cortical areas defined in the human are homologous to areas in the brains of nonhuman primates (e.g., Preuss, 1995; Kaas, 1995; Kaas, Chapter 3, this volume). Consid-

Figure 7. *(Continued)* frontal eye field from Bruce *et al.* (1985) and Gottlieb *et al.* (1994). The arcuate sulcus is represented as opened; the thick line is the lip, and the thin dashed line is the fundus. The low-threshold zone is indicated by the shaded region. (K) Location of supplementary eye field from Schlag and Schlag-Rey (1987), Schall (1991a), and Parthasarathy *et al.* (1992).

eration of FEF provides an interesting test case. The location of FEF in humans has been identified by the location of lesions affecting gaze control, the effects of electrical stimulation, and also through modern neuroimaging techniques. By mapping the region from which eye movement could be evoked by surface electrical stimulation, Foerster (1931, 1936) located FEF at the caudal end of the middle frontal gyrus. In contrast, Penfield and co-workers (Penfield and Boldrey, 1937; Rasmussen and Penfield, 1948) evoked eye movements over a wider area of frontal cortex. In this view, FEF in humans extends more caudally onto the precentral gyrus. A recent study that delivered electrical stimulation through implanted subdural electrode arrays (Godoy *et al.*, 1990) located FEF in a zone that was typically 2 cm in diameter, rostrally contiguous with the motor cortex representation of head and forelimb.

Human FEF has also been localized in many PET studies (Melamed and Larsen, 1979; Fox *et al.*, 1985; Paus *et al.*, 1993, 1995; Petit *et al.*, 1993, 1995, 1996; Anderson *et al.*, 1994; Lang *et al.*, 1994; Nakashima *et al.*, 1994; O'Driscoll *et al.*, 1995; O'Sullivan *et al.*, 1995; Sweeney *et al.*, 1995) and a recent functional MRI study (Darby *et al.*, 1996). Paus (1996) reviewed eight PET blood flow studies and found reasonably good agreement in the localization of human FEF—on the precentral gyrus and in the precentral sulcus, rostral to the primary motor cortex hand representation (Fig. 8). Paus noted variation across the PET studies of the mediolateral locus of elevated blood flow that could be related to the amplitude of saccades used in the various studies. The locus of elevated blood flow associated with larger amplitude saccades was displaced laterally and caudally from the locus associated with smaller amplitude saccades. This relative positioning of larger as compared to smaller saccade representations is inverted relative to what has been observed in macaques (see above). One possible reason for an apparent lateral–caudal FEF locus is associated with large-amplitude saccades is that such saccades are typically associated with neck contractions to rotate the head (e.g., Zangemeister and Stark, 1982; Tomlinson and Bahra, 1986; Guitton and Volle, 1987). In addition, eye blinks are commonly associated with saccades (e.g., Evinger *et al.*, 1994). Thus, blood flow may change in cortical areas neighboring FEF that are related to facial and neck skeletal movements when subjects are instructed to make saccades. The lateral–caudal locus of elevated blood flow may be attributed in part to the neck and face representation in dorsal premotor and primary motor cortex (Preuss *et al.*, 1995). Further experiments are needed to resolve this issue.*

The location of the frontal cortex lesions that selectively disrupt gaze behavior coincides with the location determined by the functional methods just described (reviewed by Pierrot-Deseilligny *et al.*, 1995). Early studies noted gaze

*In more recent work, preliminary findings by Sweeney and colleagues have localized subregions of the frontal eye field in humans using fMRI at high field strength (Berman *et al.*, 1996). During performance of saccadic eye movements, activation was localized specifically to the rostral bank of the precentral sulcus. Further, during performance of smooth pursuit eye movements, they have now observed activation localized specifically to the fundus and caudal bank of the precentral sulcus. As reviewed above, this localization of a pursuit region relative to the saccade region corresponds precisely to what has been observed in macaques. Thus, higher resolution functional MRI has demonstrated that the organization of the frontal eye field in humans corresponds to what has been observed in macaques.

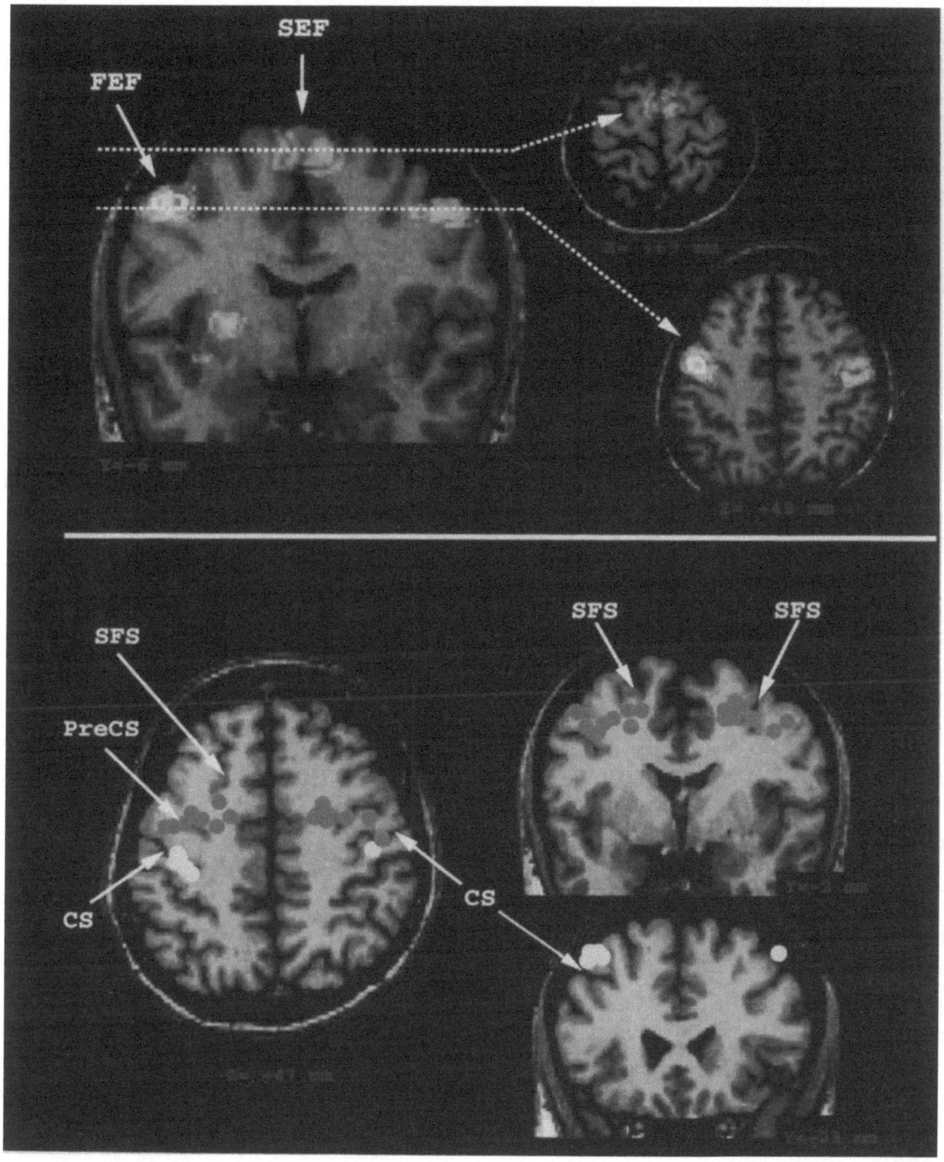

Figure 8. Location of frontal and supplementary eye movement fields in human frontal cortex. The images are statistical maps of the difference between the blood flow measured when subjects made self-paced large-amplitude (40°) saccade in complete darkness and the blood flow measured when subjects were in a resting condition making no eye movements. The top panel shows a coronal section and two associated horizontal sections. Color intensity reflects statistical reliability of the difference in blood between the saccade and fixation conditions. The locations of FEF and SEF are indicated. The bottom panel shows a horizontal and two coronal sections illustrating the location of the central focus of FEF identified in eight PET studies (reviewed in Paus, 1996) of oculomotor and manual control. CS, Central Sulcus; PreCs, precentral sulcus; SFS, superior frontal sulcus. A four-color reproduction of this figure appears following page xxiii.

deficits with damage involving large parts of frontal cortex. Only recently have data been collected with lesions more restricted to FEF (Rivaud *et al.*, 1994).

We should note an interesting and hopefully ultimately informative discrepancy that exists between the apparent location of FEF in humans and nonhuman primates. In the human brain FEF appears to be situated in Brodmann's area 6. This assignment contrasts with the localization of FEF in macaques to the granular cortical areas 8 and 45, rostral to area 6 (Petrides and Pandya, 1994), although in his original work Foerster (1936) identified FEF with Vogt's cytoarchitectural area 8αβδ in the rostral bank of the precentral sulcus. Thus, as the localization of functional areas in human cortex proceeds, concurrent architectural work will be required. Furthermore, the view that eye-movement-related cortex is solely located in prearcuate FEF of macaques may be too restrictive. Surrounding cortex may be related to gaze behavior more broadly defined. Recent intracortical microstimulation studies (Mitz and Godschalk, 1989; Preuss *et al.*, 1995) have evoked eye movements over a wider range of cortex than just the prearcuate FEF. Also, as described above, the activity of neurons in postarcuate premotor and granular prefrontal cortex is modulated in association with gaze behavior. Hence, the localization of the field of frontal cortex associated with gaze behavior may, in the end, be a matter of criteria.

3.4. FEF Connectivity

3.4.1. Subcortical

FEF influences saccade production through three pathways (Fig. 6). One pathway is a major projection to the ipsilateral superior colliculus concentrated in the intermediate layers but extending to superficial and deep layers (e.g., Leichnetz *et al.*, 1981; Fries, 1984; Komatsu and Suzuki, 1985; Huerta *et al.*, 1986; Stanton *et al.*, 1988b; Shook *et al.*, 1991). Another major pathway is through the basal ganglia via the ipsilateral striatum and subthalamic nucleus (Selemon and Goldman-Rakic, 1985; Stanton *et al.*, 1988a; Shook *et al.*, 1991; Parthasarathy *et al.*, 1992). FEF efferents terminate in the region of caudate where neural activity related to saccade production is recorded (Hikosaka *et al.*, 1989a–c). The terminations in the striatum are topographically organized; the medial aspect of FEF projects to the central part of the head and body of the caudate and dorsomedial putamen, while the lateral portion of FEF terminates ventrolaterally in the caudate and ventromedial in the putamen (Stanton *et al.*, 1988a).

The third pathway is a projection to mesencephalic and pontine nuclei (e.g., Leichnetz *et al.*, 1984a, b; Schnyder *et al.*, 1985; Huerta *et al.*, 1986; Stanton *et al.*, 1988b; Shook *et al.*, 1990). FEF projects weakly and inconsistently to the ipsilateral nucleus of Darkschewitsch, interstitial nucleus of Cajal, and rostral interstitial nucleus of the medial longitudinal fasciculus. FEF also projects weakly to the paramedian pontine reticular formation and nucleus prepositus hypoglossi and slightly more strongly to the nucleus raphe interpositus. These projections tend to be mainly ipsilateral, but some studies report some contralateral fibers as well. The FEF projection is stronger and clearly bilateral to the nucleus reticularis tegmenti pontis.

Many studies have shown that FEF is reciprocally connected in a topographic manner with a longitudinal zone of thalamic nuclei bordering the internal medullary lamina extending from the ventroanterior nucleus to the medial pulvinar (most recently see Huerta *et al.,* 1986; Stanton *et al.,* 1988a; Shook *et al.,* 1991). The general organization of these thalamocortical connection is shown in Fig. 25. The densest connections of FEF are with the lateral part of the mediodorsal nucleus (mainly the multiform and parvicellular sectors) and the medial part of the ventroanterior nucleus (mainly the magnocellular division). FEF is more weakly connected with the more medial and caudal parts of the mediodorsal nucleus, with area X of the ventrolateral nucleus and with the caudal ventrolateral nucleus and medial pulvinar. Some but not all studies have reported weak FEF connections with the paracentral, centrolateral, and central superior lateral intralaminar nuclei. The FEF connections with the paralaminar nuclei are topographically organized, with the dorsomedial part of FEF projecting dorsally and the ventrolateral part of FEF projecting ventrally. The thalamic zones most heavily connected with FEF are themselves innervated by oculomotor afferents from the intermediate and deep layers of the superior colliculus, the substantia nigra pars reticulata, and the dentate nucleus of the cerebellum (Ilinsky *et al.,* 1985; Lynch *et al.,* 1994).

3.4.2. Intracortical

Because so much other evidence indicates that FEF is a key site of sensorimotor integration, it is not surprising that FEF is connected with a wide variety of cortical areas. Within the frontal lobe FEF is interconnected with SEF, with prefrontal areas 46 and 12, with anterior cingulate area 24, and with postarcuate premotor cortex (Barbas and Mesulam, 1981; Huerta *et al.,* 1987; Stanton *et al.,* 1993). The pattern of connectivity of FEF with SEF will be discussed below. Afferents to FEF from prefrontal areas 12 and 46 and anterior cingulate cortex may mediate regulatory control over gaze, although the nature of the signals that may be generated in prefrontal cortex is an active area of research. Connections of FEF with postarcuate cortex probably subserve the coordination of eye, head, and possibly forelimb movements (e.g., Gielen *et al.,* 1984; Fisk and Goodale, 1985; Bock, 1987; Tomlinson and Bahra, 1986; Baedeker and Wolf, 1987; Guitton and Volle, 1987; Vercher and Gauthier, 1992; Bekkering *et al.,* 1994; Rossetti *et al.,* 1994).

FEF is interconnected with nearly all of the extrastriate visual areas (Schall *et al.,* 1995b; Stanton *et al.,* 1995). Thus, FEF receives a rich constellation of visual afferents that represent various stages and streams of visual processing. The areas most heavily connected with FEF are generally several anatomical steps removed from V1. The primate visual system has been viewed as consisting of at least two processing streams, one passing ventrally into temporal cortex, responsible for object vision, and the other running dorsally into parietal cortex, responsible for spatial vision and the guidance of movement (Ungerleider and Mishkin, 1982; Merigan and Maunsell, 1993; Goodale and Milner, 1992). Tracer injections placed in posterior parietal cortex and inferior temporal cortex result in very few regions of overlapping distributions of labeled neurons (Morel and Bullier, 1990; Baizer *et al.,* 1991). Nevertheless, in evaluating the evidence for a segregation and independence of processing streams, it is well to remember that action and perception are usually unified. That is, an eye movement, for example, can only go to one target at a time. This notion is consistent with the finding that FEF is distinguished from most other cortical areas by being connected reciprocally with areas assigned to both the dorsal and the ventral visual processing streams. Lateral FEF, which is responsible for generating short saccades, receives visual afferents from the foveal representation in retinotopically organized areas, from areas that represent central vision in inferotemporal cortex, and from other areas having no retinotopic order. In contrast, medial FEF, which is responsible for generating longer saccades, is innervated by the peripheral representation of retinotopically organized areas, from areas that emphasize peripheral vision or are multimodal, and from other areas that have no retinotopic order or are auditory. These data demonstrate clearly that ventrolateral FEF is a site of convergence of signals associated with the two processing streams.

Some of the evidence for these conclusions is illustrated in Figs. 9 and 10 and is summarized in Fig. 26. The distribution of neurons labeled by injections of different tracers into medial and lateral FEF is illustrated in Fig. 9. Note that neurons in both inferior temporal and posterior parietal cortex project to the lateral part of FEF. Ventrolateral FEF is also innervated by cells near the foveal representation in areas MT and V4. Dorsomedial but not ventrolateral FEF is innervated by area PO, posterior cingulate area 23, the peripheral visual field representation of areas MT and V4, and the densely myelinated zone in MST. Dense collections of neurons projecting to both ventral and dorsal FEF are found in LIP and in areas MST, FST, IPa, and PGa in the fundus of the superior temporal sulcus. It should be noted that areas IPa and PGa were distinguished in earlier anatomical studies by being unique points of convergence of connections with inferior temporal and posterior parietal cortex (Morel and Bullier, 1990; Bazier *et al.,* 1991). Neurons in areas IPa and PGa are multimodal, responding to acoustic, tactile, and visual stimuli, but they exhibit little stimulus specificity (Baylis *et al.,* 1987). Further work is clearly needed to provide more information about the role of these cortical areas in visuomotor behavior and their relationship to frontal cortex.

The prestriate afferents from inferior temporal and posterior parietal cortex to FEF are reciprocated (Distler *et al.,* 1993; Webster *et al.,* 1994; Schall *et al.,* 1995b). Figure 10 shows the distribution of neurons in arcuate cortex labeled by tracer injections placed in areas TEO plus TE and in area LIP. A large part of the cortex in and around the arcuate sulcus and caudal principal sulcus project to LIP. Smaller regions in the ventrolateral arcuate sulcus and in area 12 project to LIP as well as to inferior temporal cortex.

These anatomical data highlight a number of questions about FEF function. Do the afferents from the various prestriate areas converge on individual neurons in FEF? What use is made of the sometimes elaborate stimulus specificity of visual afferents arriving in FEF? Can FEF cells be selective for visual features? Visual cells in FEF have been regarded as largely not selective for features such as color, form, or motion (Mohler *et al.,* 1973). However, the activity of FEF neurons does reflect the properties of stimuli in their receptive field under particular conditions (Schall *et al.,* 1995a; see also Bichot *et al.,* 1996). These conditions will be reviewed in Section 3.6.

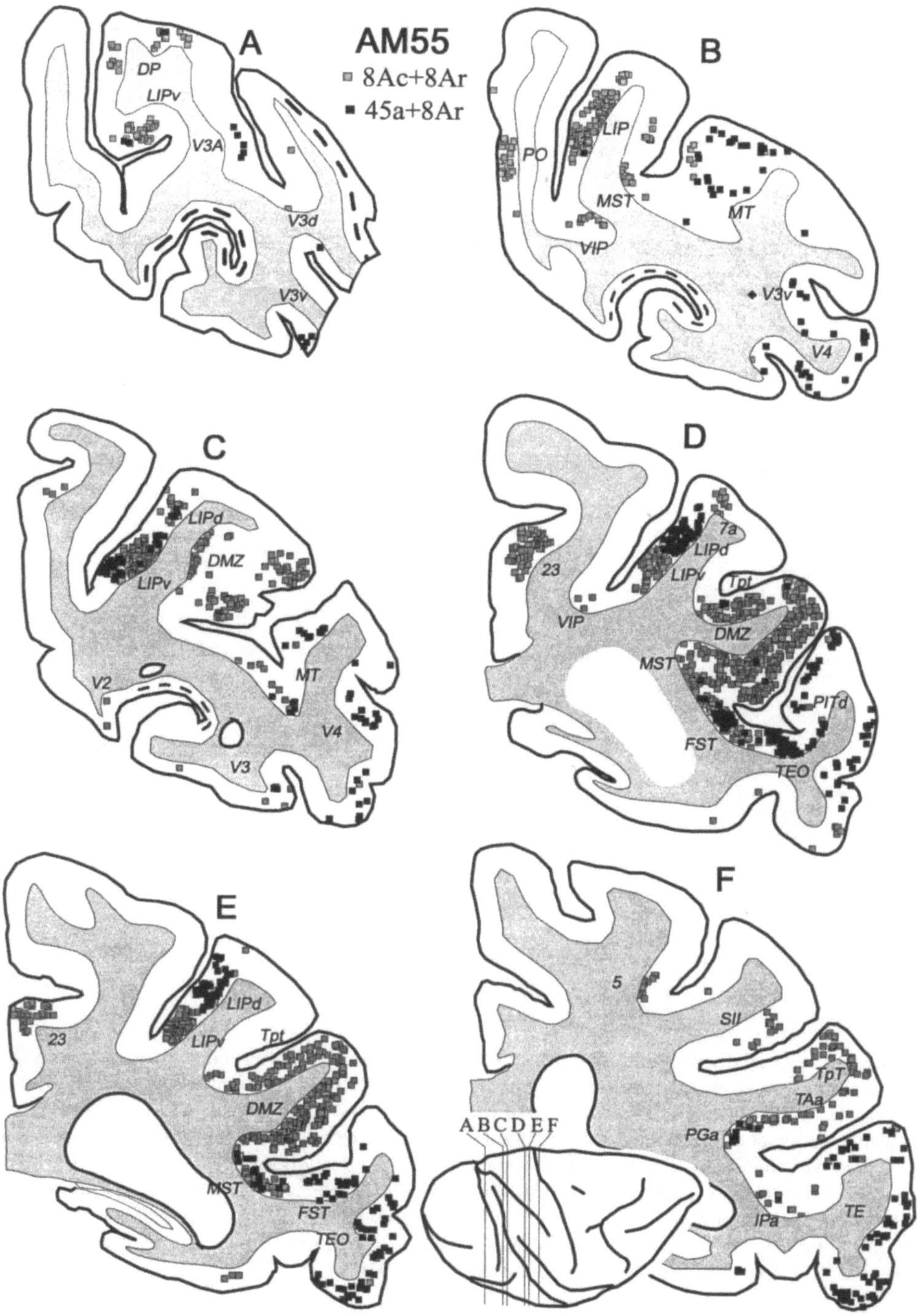

Figure 9. Distribution of neurons in prestriate cortex retrogradely labeled by injections of tracers into ventrolateral (black squares) and dorsomedial (gray squares) FEF. The identity of cortical areas is indicated in the coronal sections. The levels of the sections are indicated on the dorsolateral view of the brain. From Schall *et al.* (1995b) with permission.

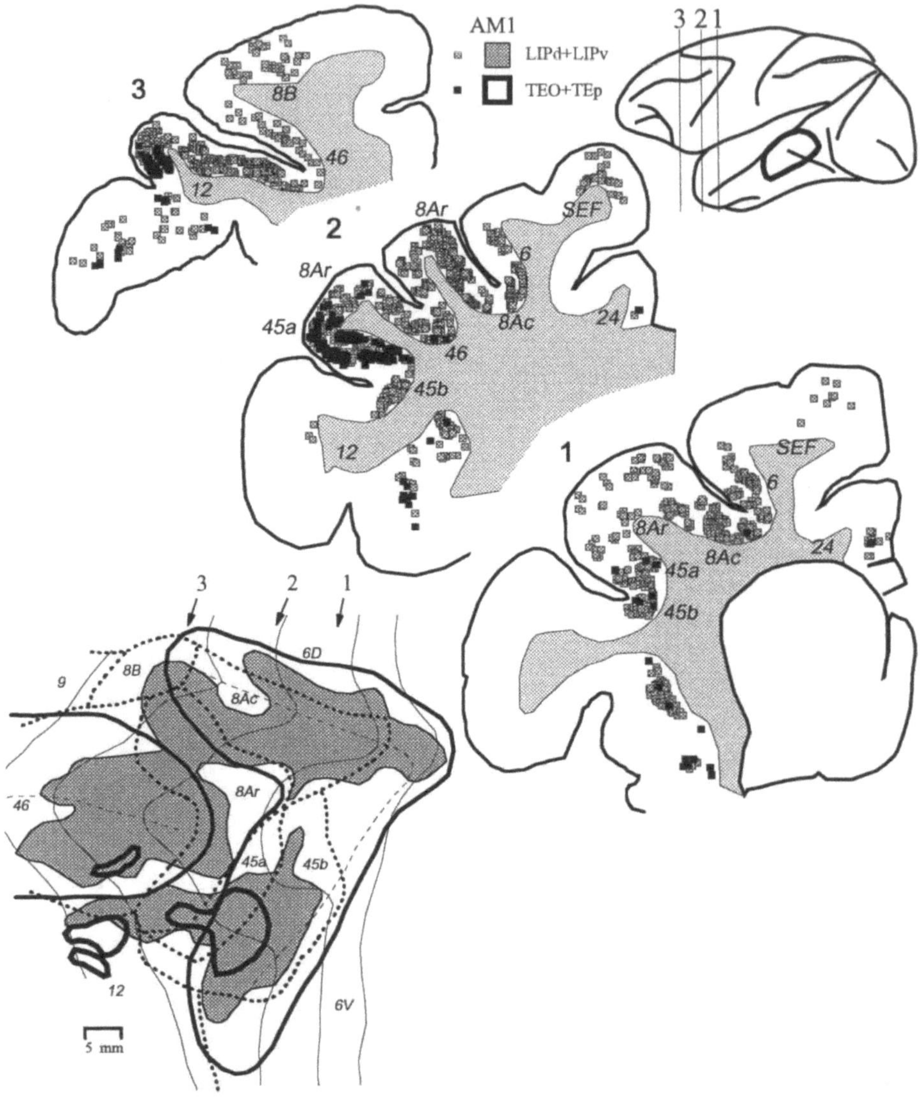

Figure 10. Spatial distributions of neurons in frontal cortex retrogradely labeled by tracer injections into inferior temporal and posterior parietal cortex. The location of the injections is indicated on the lateral view of the brain. The distribution of retrogradely labeled neurons from the two injections is shown in representative coronal sections and in a flattened reconstruction of the arcuate sulcus and caudal end of the principal sulcus. Neurons labeled by the injection in LIP are indicated by the gray squares in the coronal sections and by gray shaded regions in the flattened reconstruction. Neurons labeled by the injection in TEO and caudal TE are represented by black squares in the sections and by the open area enclosed by the thick line in the two-dimensional view. Architectural areas are labeled. Modified with permission from Schall *et al.* (1995b).

3.5. FEF Role in Saccade Production

The original neural recordings in the FEF of awake monkeys making spontaneous, unrewarded eye movements found less than 10% of neurons were modulated in relation to saccadic eye movements (Bizzi, 1968; Bizzi and Schiller,

1970). However, when an operant reward is made contingent on a monkey's saccade or other manual response to visual stimuli, nearly half of the cells in FEF exhibit visual responses (Mohler *et al.*, 1973; Wurtz and Mohler, 1976; Pigarev *et al.*, 1979; Kubota *et al.*, 1980; Suzuki and Azuma, 1977, 1983; Goldberg and Bushnell, 1981; Bruce and Goldberg, 1985; Schall, 1991b; Schall and Hanes, 1993; Burman and Segraves, 1994; Schall *et al.*, 1995a). The responses of these neurons can be phasic, signaling the presentation and removal of a stimulus, or tonic, lasting throughout a delay period, in some cases even if the stimulus has been removed. Neurons in FEF also discharge in relation to the production of saccades, the generation of pursuit eye movements, or the maintenance of fixation (Bruce and Goldberg, 1985; Schall, 1991b; Segraves and Park, 1993; Hanes *et al.*, 1995). Certain saccade-related neurons are found in FEF that discharge specifically before and during the eye movement, and other neurons fire specifically postsaccadically. The following sections will review the evidence for the role that FEF plays in saccade programming, saccade execution, and saccade target selection.

3.5.1. Maintenance of Spatial Constancy

Each movement of the eyes causes a shift of the retinal image. A longstanding problem has been understanding how the brain registers whether a movement in the retinal image is due to a movement in the world or a movement of the eyes.* It was recognized early on that the brain must have some representation of eye position (or the change in eye position) that was updated with each eye movement (Helmholtz, 1866/1962; Sperry, 1950; Von Holst and Mittelstaedt, 1950; but see Steinbach, 1987; Bridgemen *et al.*, 1994). The nature of this internal representation has been the focus of an extensive line of research over the last several years. The central issues involve whether the oculomotor system represents absolute or relative eye position and how the visual system relates to the oculomotor system.

Viewing the oculomotor system as a feedback control system has proven effective in guiding theoretical and experimental work (Robinson, 1986, 1991). Successful models of motor control make use of error signals to drive feedback loops. An exemplar of a feedback control system is represented by a thermostat. The difference between the actual temperature and the desired temperature is the error that drives the circuit to activate either a heating or a cooling system. This conceptualization and the mathematical approach it renders has proven quite effective at simulating the oculomotor system and generating useful hypotheses (Robinson, 1986, 1991; but see Steinman, 1986). Early models of saccade programming drove saccadic eye movements with a retinal error signal, i.e., the location of a visual target on the retina relative to the fovea (Young and Stark, 1963). This retinal error signal, registered, for example, by cells in the upper layers of the superior colliculus, was then converted into a motor error signal, represented by cells in the deeper layers of the superior colliculus (Schiller and Koerner, 1971). The initial motor error signal was supposed to be

*The reader can explore this phenomenon by gently moving one eye manually by gently pressing on the side of the orbit with a finger. The phenomenal experience of the world moving is believed to be due to the fact that the eye was moved without the oculomotor system updating its internal representation of eye position.

"latched" by the brainstem circuitry that drove the eyes in a saccade until a dynamic motor error signal was eliminated when gaze reached the target. For what follows it may be useful to think of the retinal error and motor error signals as vectors with their tails at the fovea aligned on the current line of gaze and their heads at the target of the gaze shift.

Normally the retinal error and motor error leading to a gaze shift coincide. But clever experiments have demonstrated conditions under which retinal error and motor error can be dissociated, and their results have cast doubt on the generality of the simple scheme described above of converting retinal error directly into motor error. Hallett and Lightstone (1976) reported that human subjects can make an accurate saccade to fixate a spot that was flashed briefly in the dark while another saccade was in flight. In other words, subjects were able to make a saccade based on a motor error vector that was different from the retinal error vector. This finding by Hallett and Lightstone stood in interesting contrast to well-documented findings of systematic errors of perceptual localization of stimuli flashed around the time of saccadic eye movements (reviewed by Matin, 1985). Subsequent work by a number of investigators has further examined this phenomenon. Several laboratories have shown that the oculomotor system actually is unable to accurately guide gaze to a second target if it is flashed around the time of an initial saccade (Honda, 1990, 1991; Dassonville *et al.*, 1992a, 1995; Gellman and Fletcher, 1992; Schlag and Schlag-Rey, 1995; see also Becker and Jürgens, 1979). This general experimental paradigm has been referred to as the *double-step saccade task.* It is important to understand that the double-step task is designed to dissociate a motor error from a retinal error. If the second target is flashed less than 100–200 msec before the first saccade, then the second target is mislocalized as if it were further in the direction of the upcoming first saccade. The saccade response errors grow as the second target is flashed closer to the time that the first saccade will begin; thereafter the errors of localization decrease.

This error between where the second light flashed and where the second saccade landed the focus of gaze provides information about the internal representation of eye position used by the oculomotor system. Subtracting the actual endpoint of the saccade to the second target from the actual retinal location of the second target provides an estimate of the eye position signal used to generate the movement. When this calculation is done, it is found that the estimated eye position signal does not correspond to the actual change of eye position of the second saccade. The estimated eye position begins to change around 100 msec before the actual shift of gaze, and the estimated internal representation of eye position does not match the actual eye position until around 50 msec after the saccade.

How the oculomotor system updates its representation of eye position can be investigated neurophysiologically. For instance, what happens if, during the interval before a saccade is made to a flashed target, the eyes are displaced by electrical stimulation of an oculomotor structure? Is an error introduced, or does the brain compensate for the experimental perturbation? Sparks and Mays (1983) demonstrated that monkeys could make a reasonably accurate saccade to a flashed target even if stimulation of the superior colliculus drove gaze to a new position in the dark before the saccade was generated. Other work showed that monkeys can compensate for deflections of the eyes caused by stimulation of FEF prior to saccade initiation, and this compensation is not impaired by lesions of the superior colliculus (Schiller and Sandell, 1983).

These findings prompted the search for cells carrying signals that are sufficient to mediate the computations needed to accurately guide the second saccade based on a motor error that had no corresponding retinal error. To generate an accurate second saccade in the double-step task, the retinal locus of the second target had to be combined with the change in eye position produced by the saccade made to the first target. Originally, it was proposed that the double-step saccade task is performed by converting the retinal coordinates of the second target to craniotopic coordinates that could then be used to accurately guide the second saccade (Robinson, 1973; Mays and Sparks, 1980). The conversion from retinal coordinates to craniotopic coordinates was achieved by utilizing a signal representing the position of the eyes in the orbit. Subsequent experimental and modeling work has prompted a revision of this view. More recent models posit that the motor error is based on a representation of eye displacement instead of absolute eye position using a second neural integrator that is reset after each saccade (Jürgens *et al.*, 1981; Scudder, 1988).

A number of experiments have been performed to provide data to demonstrate whether specific stations in the oculomotor system play a role in maintaining spatial constancy. The conclusions drawn by different laboratories have been at odds. With regard to FEF, one neuronal recording experiment using the double-step task concluded that the output of FEF is a motor error signal appropriate for the upcoming saccade (Goldberg and Bruce, 1990). On the other hand, a set of experiments using electrical stimulation of FEF and other structures concluded that the output of FEF is a retinotopic goal for the next saccade (Dassonville *et al.*, 1992b). However, the observed effects of electrical stimulation may have an alternative interpretation (Nicholls and Sparks, 1995; Kustov and Robinson, 1995). These experiments will now be reviewed.

Goldberg and Bruce (1990) examined FEF cell activity recorded when monkeys performed a double-step saccade task. By rapidly flashing two targets in succession before the saccade to the first target was begun, the second saccade must be based on a motor error that was different from the initial retinal error. They found cells in FEF that appeared to register the motor error needed to make the second saccade in the double-step paradigm. It may not be surprising that all movement cells did so. It is more interesting, though, that visually responsive cells did so as well. As illustrated in Fig. 11, a cell that ordinarily responded to the appearance of a visual target that guided a single saccade began to fire in the double-step condition after the first saccade that brought the location of the flashed target for the second saccade into its receptive field. Recall that at no time in the double-step trials did a visual stimulus actually fall on the retinal locus of the cell's receptive field. Thus, the apparent visual activity of these cells in FEF depends not necessarily on the retinal stimulation, but rather on the saccade that will be produced. This is analogous to what was described above in premotor cortex, and represents another example showing that the visual responses of cells in frontal cortex are not necessarily stimulus-bound, but instead depend on behavioral context. We should also note that other visual cells in FEF did require direct retinal stimulation to be activated. Thus,

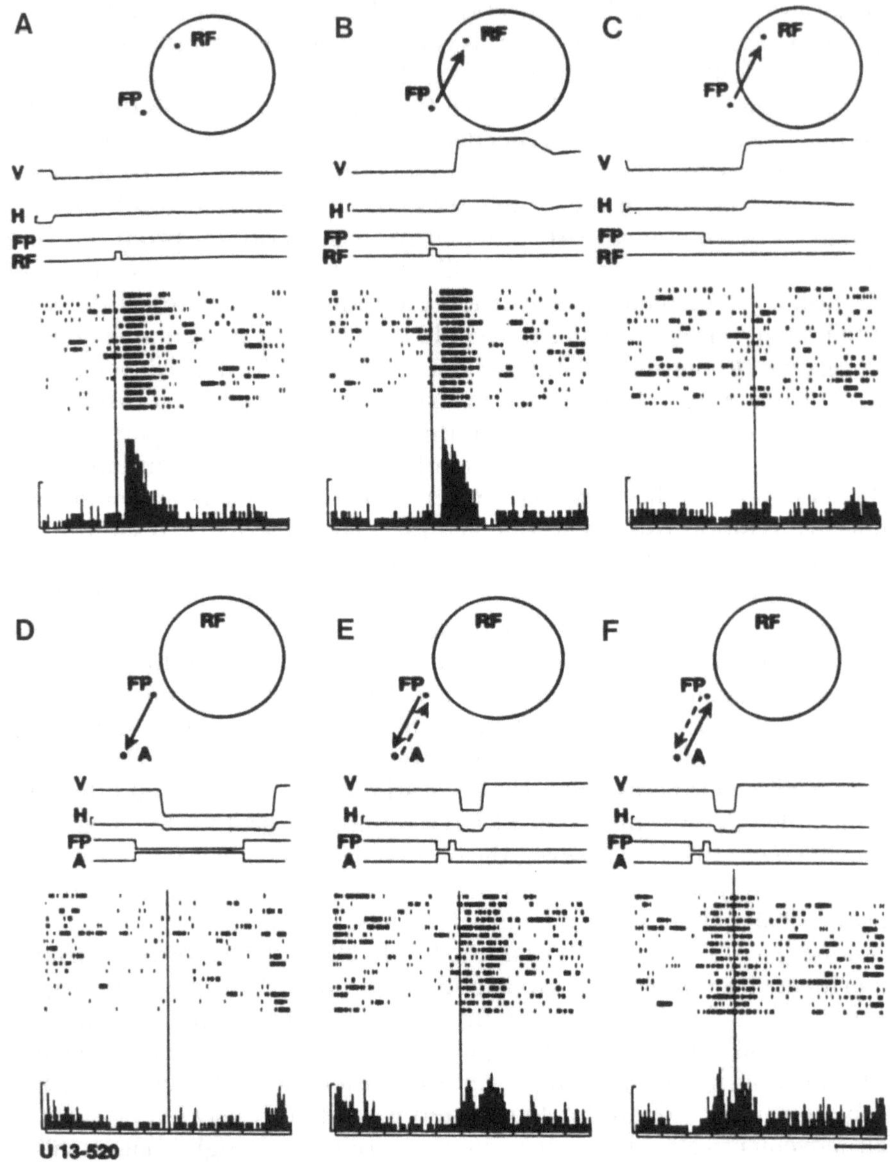

Figure 11. Activity of an FEF visual cell tested in no-saccade, single-step, and double-step trials. Each panel shows horizontal (H) and vertical (V) eye position and fixation point (FP) and stimulus (RF, A) state traces for a sample trial. Histogram scales represent 100 Hz. The time scale (lower right) represents 400 msec. The scale on the horizontal eye position trace represents 5°. The solid circle indicates the location of the cell's receptive field. (A) Visual response of neuron to a stimulus flashed for 50 msec in its receptive field (RF). The raster and histogram are aligned on the time of stimulus presentation. (B) Response of the neuron when a saccade was made to the stimulus flashed for 50 msec in its receptive field. The raster and histogram are aligned on the time of stimulus presentation. (C) Absence of response when a saccade is made to the same location in the cell's receptive field when no stimulus is presented. The raster and histogram are aligned on the time of saccade initiation. (D) Absence of activity associated with the first saccade of a double-step trial to a stimulus presented outside the receptive field. The raster and histogram are aligned on the end of the saccade. (E) Activation following the first saccade in a double-step task when the second saccade has a

both retinocentric and "transformed" representations are present in FEF. This latter finding is further indication that different stages of processing are represented within a single visuomotor cortical area.

Goldberg and Bruce (1990) also observed that cells in FEF with postsaccadic activity exhibited some tuning for saccade dimensions, discharging preferentially in relation to a limited range of saccade vectors. They also noted that the activity of postsaccadic cells was effectively suppressed at the initiation of subsequent saccades. This suppression, they thought, may indicate that whatever the postsaccadic cell was signaling was no longer needed as the next gaze shift got underway. Goldberg and Bruce also found that many of the visual cells that registered the second motor error also showed a postsaccadic discharge after saccades in the direction opposite their receptive field.

Based on these data, Goldberg and Bruce suggest that FEF neurons carry all of the signals needed to perform successfully the double-step saccade task by using a vector subtraction mechanism. The motor error vector of the second saccade is computed by vector subtraction of the dimension of the first saccade from the retinal error vector of the retinal location of the target. The activity of postsaccadic and visual cells in FEF can be the basis for this computation. Therefore, based on this reasoning they suggest that no explicit craniotopic eye position signal is needed. The output signal of FEF is, on this view, the motor error necessary for the next saccade.

In a series of studies using a different research strategy, Schlag, Schlag-Rey, and Dassonville also investigated how the visuomotor system maintains spatial constancy. They examined how the saccade evoked by electrical stimulation of visuomotor structures changed as the stimulation was delivered at different times relative to the initiation of a visually triggered saccade. Coining the phrase *colliding saccade* for this paradigm, they supposed that the electrical stimulation would be considered as an artificial injection of the location of the second target in a double-step task (reviewed by Schlag and Schlag-Rey, 1990). They found that the vector of the saccade evoked by microstimulation of the deep layers of the superior colliculus did not change if the stimulation occurred around the time of another saccade (Schlag-Rey *et al.*, 1989). The fact that no compensation for the intervening saccade happened indicates that the deep layers of the superior colliculus represent a stage of processing that occurs after the visuomotor transformation needed to ensure direction constancy.

Two other laboratories have recently investigated the saccades evoked by microstimulation of the superior colliculus shortly before or after another visually guided saccade (Nicholls and Sparks, 1995; Kustov and Robinson, 1995). Both of these studies did find an interaction between the visually guided saccade and the stimulation-evoked saccade. The magnitude of the effect, though, decreased with time after the visually guided saccade ended. Both of these studies interpret their findings in terms of the resettable integrator, the theoretical

---

Figure 11. *(Continued)* motor error corresponding to the location of the cell's receptive field. The raster and histogram are aligned on the beginning of the first saccade. (F) Activation preceding the second saccade made to fixate the location of the second flashed target. The raster and histogram are aligned on the beginning of the second saccade. From Goldberg and Bruce (1990) with permission.

network that registers eye displacement associated with each saccade. The authors of these studies argued that as the visually guided saccade is executed, the resettable integrator registers the actual change in eye position. Electrical stimulation of the superior colliculus instates a new representation of the desired change in eye position, different from the original one that drove the visually guided saccade. The eye movement evoked by the electrical stimulation is driven by the new motor error, which is just the difference between the new desired change in eye position and the old actual change in eye position. Thus, the variation of the saccade evoked by microstimulation of the superior colliculus shortly after a visually guided saccade had begun reflects the state of the resettable integrator as it decayed and reset for the next movement. The two studies agree on an estimate of the time constant of the integrator at close to 50 msec.

Schlag-Rey and co-workers found that the vector of the saccade evoked by stimulation of FEF as well as of the superficial, visual layers of the superior colliculus changed dramatically if the stimulation was delivered around the time of another saccade. They reported further that the nature of the compensation was different in FEF than it was in the superficial superior colliculus. The saccades evoked by electrical stimulation of the superficial superior colliculus compensated for the change of eye position that occurred just between the moment of electrical stimulation and the beginning of the evoked movement (Schlag-Rey *et al.,* 1989). In contrast, the saccades evoked by microstimulation of FEF compensated for the change in eye position that occurred from a point in time *before* the stimulation was turned on until the beginning of the evoked movement (Dassonville *et al.,* 1992b). The additional compensation associated with stimulation of FEF, the authors reasoned, occurred because the brain interpreted the electrical stimulation as if it were activation due to a visual stimulus. Moreover, Dassonville *et al.* suggested that this representation was referenced to the time that an actual visual stimulus would have appeared on the retina to result in this FEF activation; in other words, the representation incorporated an afferent delay of around 50 msec. Dassonville *et al.* observed complete compensation following stimulation of all sites in FEF, even those at which movement cells were recorded. These data led the investigators to conclude that the signal output by FEF, unlike that of the motor layers of the superior colliculus, was not a motor error signal, but instead a retinotopic goal for the next saccade. They reasoned that FEF must represent a stage of processing that occurs prior to the stage that compensates for the intervening saccade, and that the compensation is accomplished by subcortical efferent structures. In other words, according to Dassonville *et al.,* the FEF appears to represent a retinotopic goal referenced to an earlier eye position. This goal is subsequently and elsewhere converted into a motor error. This hypothesis should be contrasted with the alternative hypothesis that explains these effects in terms of the resettable neural integrator mentioned above.

To summarize, on the one hand, Goldberg and Bruce (1990) concluded, based on single-unit recordings in the double-step paradigm, that all movement cells, most visuomovement cells, and some visual cells in FEF encode motor error and not retinal error. In contrast, Dassonville *et al.* (1992b) concluded, based on the colliding saccade paradigm, that FEF represents a retinal goal, but not a motor error. What is the basis for this apparent disagreement? One question is whether differences in the experimental conditions could account for the different conclusions. Were there differences in stimulus timing, location, and visibility that could result in the monkeys in the two studies adopting different behavior strategies? Were there (as is likely) fundamental differences in the brain state evoked by microstimulation and that inferred by single-unit recordings? More experimental work will be needed to reconcile these two findings. One possible inroad to resolving the conclusions of the various studies may be the realization that one's emphasis in interpreting the physiological findings may depend on whether one is trying to account for perfect performance as indicated by Hallet and Lightstone (1976) or to account for the systematic errors found in subsequent studies (Honda, 1990, 1991; Dassonville *et al.*, 1992a, 1995; Gellman and Fletcher, 1992; Schlag and Schlag-Rey, 1995). The conditions used by Hallet and Lightstone permitted subjects to predict with some certainty where the second target would be, so their performance was better than perhaps it ought to have been (Dassonville *et al.*, 1995). Thus, the most recent behavioral data indicate that the brain registers the position that the eye is going to be before the saccade actually happens. This sluggish change in the representation of eye position before a saccade actually may be based on a phenomenon identified by Duhamel *et al.* (1992) in LIP and subsequently in FEF (Umeno and Goldberg, 1994). Certain visually responsive cells when recorded in a double-step task begin to fire in anticipation of the saccade that will bring a visual stimulus into its receptive field. This activation of what otherwise appear to be visual cells may contribute to the pattern of errors observed in the double-step paradigm.

3.5.2. FEF Motor Signal

FEF generates signals sufficient but not necessary to generate saccadic eye movements (e.g., Schiller *et al.*, 1980). Electrical stimulation of FEF evokes saccades, and there is very good agreement between the various modern studies of the effects of electrical stimulation of FEF (Robinson and Fuchs, 1969; Marrocco, 1978; Bruce *et al.*, 1985; Goldberg *et al.*, 1986; Schall, 1991b; Russo and Bruce, 1993). Electrical microstimulation of FEF results in saccades with direction and amplitude determined by the location in FEF and not influenced much, if at all, by gaze angle. Prolonged stimulation (over 200 msec or so) results in multiple saccades, all of the same direction and amplitude (until the eyes rotate to the limit of the orbit). The latency of the saccade elicited by suprathreshold stimulation is 30–45 msec. The threshold for eliciting saccades from FEF depends on the state of the monkey. The lowest thresholds are obtained when the monkey is alert but not fixating intently, and thresholds are elevated if the monkey is actively fixating or pursuing a target (Marrocco, 1978; Goldberg *et al.*, 1986) or if the monkey is drowsy or pharmacologically depressed (Robinson and Fuchs, 1969).

As described above, FEF generates saccades via efferents to the superior colliculus and the brainstem. Using antidromic stimulation while recording in FEF, Segraves determined the functional properties of neurons projecting specifically to the superior colliculus (Segraves and Goldberg, 1987) and to the pons (Segraves, 1992). These projection neurons in FEF can be assigned to three particular subpopulations. Half of the neurons antidromically activated by stimulation of either the superior colliculus or the pons are movement neurons of the type illustrated in Fig. 12; one-quarter of the corticotectal and corticopontine cells have foveal receptive fields and discharge when monkeys fixate stimuli; an example of such a cell is shown in Fig. 13; the remaining one-quarter of corticotectal and corticopontine cells in FEF are heterogeneous. Equally notable is the finding that visually responsive cells with receptive fields excluding the fovea do not innervate the oculomotor efferent structures.

Segraves measured the latency of the antidromic activation of the different cell classes in FEF. Figure 14 summarizes his results incorporating a speculative morphological correlate for the two main efferent functional cell classes. Fries (1984) illustrated the diverse morphology of corticotectal neurons partially filled by horseradish peroxidase (HRP) injections placed in the superior colliculus. From area 8 he showed one neuron with a large soma and extensively branching dendrites. As described above, such large pyramidal cells in layer 5 are a defining cytoarchitectural feature of FEF (Walker, 1940; Stanton *et al.*, 1989). The other corticotectal cell shown from area 8 was smaller with more sparsely branching dendrites. The latency of antidromic activation from superior colliculus stimulation was significantly shorter for movement cells than for cells with foveal receptive fields. Given the correlation between axon size and conduction velocity, it seems plausible that the large pyramidal cells are movement cells and the smaller pyramidal cells are the cells with foveal receptive fields and the other functional types. Determining the morphological correlates of functional cell types is a very important issue that is basic to understanding the organization of processing in FEF. This question may be addressed by quantitatively determining the frac-

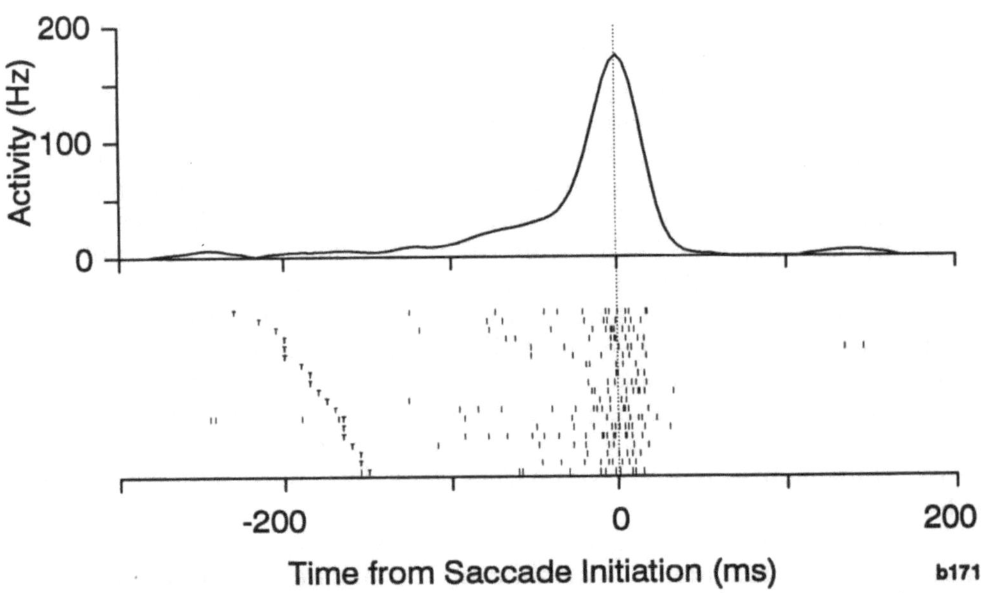

Figure 12. FEF presaccadic movement cell. Negative values on the abscissa indicate times before saccade initiation. The top panel shows the spike density function averaged across trials. In the lower raster display the T signifies the time of presentation of the target. Trials were sorted by saccade latency. Modified from Hanes *et al.* (1995) with permission.

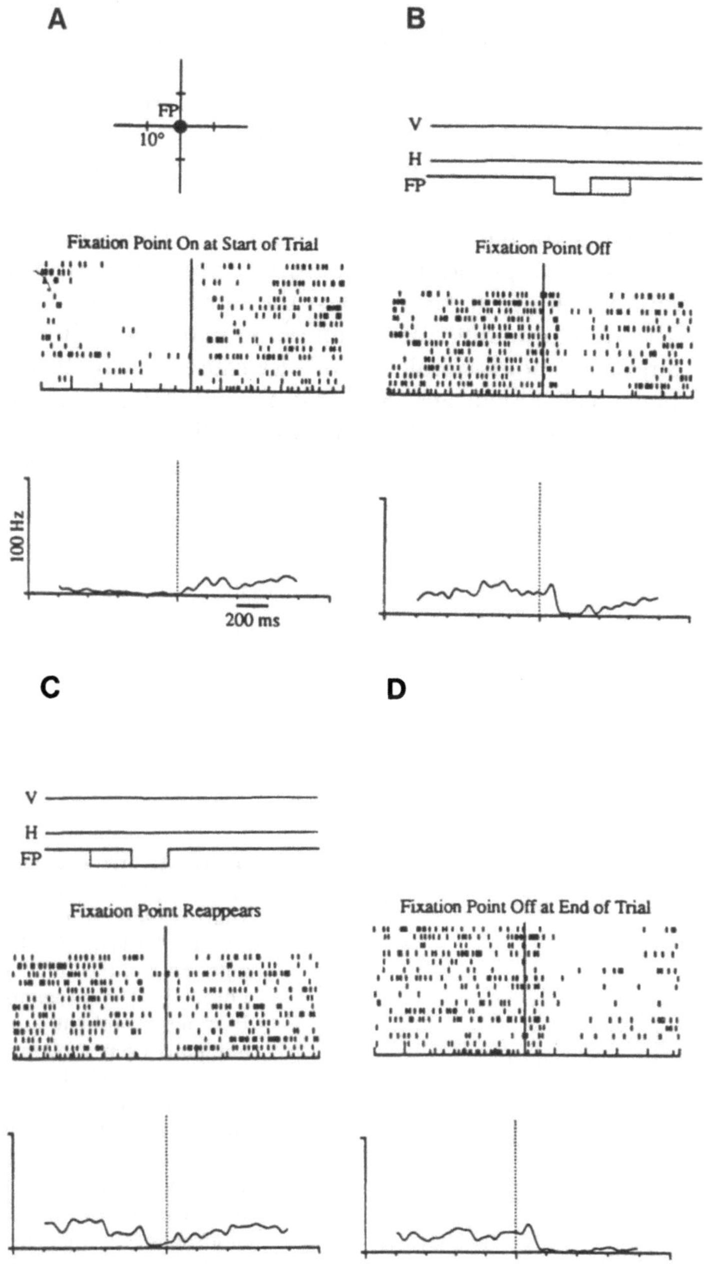

Figure 13. Corticopontine FEF neuron activity during fixation task. The monkey was required to fixate the location of a fixation light (FP) for the duration of the trial. During the trial, the fixation point was turned off for a period of 250–500 msec. The presence of the fixation light is indicated by the FP trace directly below the vertical (V) and horizontal (H) eye position traces. Downward deflections correspond to the disappearance of the fixation light, and upward deflections correspond to its reappearance. Solid and dashed lines outline the shortest and longest times for the fixation light off-period. Rasters and spike-density profiles are aligned on the appearance of the fixation point at the beginning of the trial (A), the turning off (B) and on (C) of the fixation point during the trial, and its disappearance at the end of the trial (D). From Segraves (1992) with permission.

tions of corticotectal cells with different morphologies. A strict structure–function correlation would predict that half of the cells would be large enough to support rapidly conducting axons, and the remainder would be smaller cells. This research strategy was used to elucidate structure–function correlations of retinal ganglion cells (reviewed by Stone, 1983).

Segraves (1992) also provided information about the cellular targets of FEF efferents in the pons. He found that only long-lead burst neurons in the paramedian pontine reticular formation are activated by FEF stimulation, these with a latency of 4–10 msec. Long-lead burst neurons begin to discharge 100 msec or so before saccade initiation, at more or less the same time that FEF movement cells begin to fire. FEF also innervates omnipause neurons in the nucleus raphe interpositus. Suprathreshold stimulation of FEF causes a virtually instantaneous cessation in pause cell discharges. Stimulation with lower electrical currents produces a reduction in pause cell activity with a 25- to 30-msec latency. Normally the cessation of pause cell activity precedes saccade initiation by 10–15

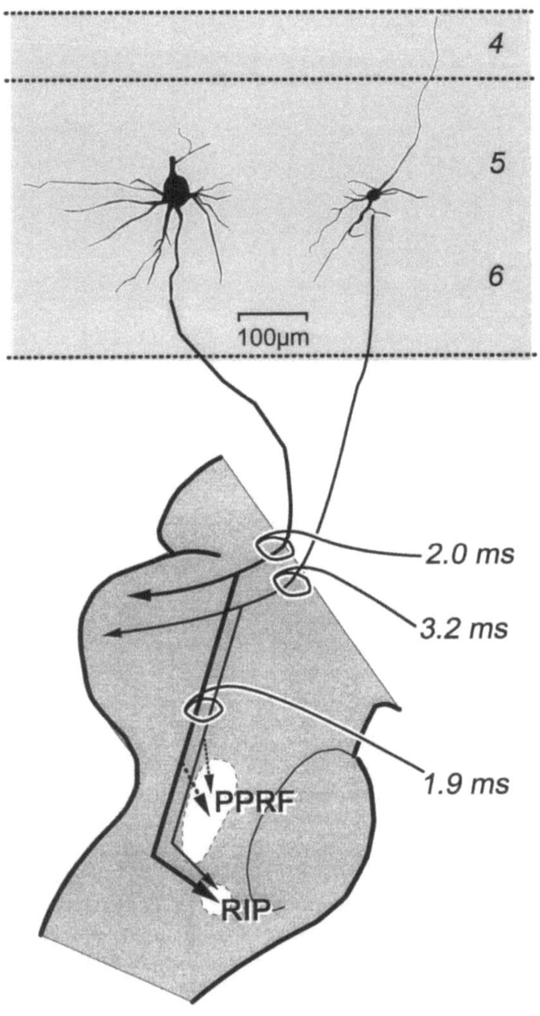

Figure 14. FEF oculomotor efferents. The morphology of corticotectal cells from area 8 is illustrated at the top (redrawn from Fries, 1984). The approximate width of the lower cortical layers is indicated. Efferent axons are drawn as diverging to both the superior colliculus and pons even though no direct evidence is available for this. Mean antidromic activation latencies are indicated for superior colliculus stimulation (from Segraves and Goldberg, 1987) and pontine stimulation (Segraves, 1992). Movement neurons had significantly shorter antidromic activation latencies following tectal stimulation than did foveal neurons. The figure is drawn based on the speculation that the larger layer 5 pyramidal cell is a movement cell by virtue of the fact that larger axons conduct faster. The fusiform pyramidal cell, being smaller and thereby conducting slower, may correspond to the cells with foveal receptive fields. PPRF, Paramedian pontine reticular formation; RIP, nucleus raphe interpositus.

msec. In all probability the FEF influence on the pause cells is mediated by an inhibitory interneuron.

3.5.3. FEF Trigger Signal

By innervating brainstem omnipause neurons, FEF is in a position to directly trigger saccades. Still to be characterized, though, is the specific signal generated by FEF (as well as other structures) that reliably predicts when a saccade will be initiated, although this is obviously basic to understanding the neural regulation of saccade production. In developing hypotheses about particular trigger signals it will be important to account for the distribution of saccade latencies observed under different circumstances (e.g., Fisher and Weber, 1993; Carpenter and Williams, 1995).

Some progress has been made on this question. Three recent studies have examined the timecourse of activation of movement cells in FEF. Schall (1991b) defined the beginning of activity of FEF presaccadic movement neurons based on the change in the slope of the cumulative sum of the spikes averaged across all trials, determining the mean onset time of FEF presaccadic movement neurons to be 130 msec before saccade initiation. Segraves and Park (1993) analyzed the FEF presaccadic movement neurons identified by antidromic stimulation of the nucleus raphe interpositus and superior colliculus. Using Gaussian spike density functions averaged across trials, they defined the onset of activity as the moment the activity rose above the baseline firing rate. Based on this analysis, Segraves and Park found that the FEF corticopontine and corticotectal neurons begin to discharge 150 msec before the saccade. Segraves and Park also related the time of the peak of activity to saccade initiation; they found that corticotectal cells had a mean time of peak activity 34 msec before saccade initiation, and corticopontine cells had a mean time of peak activity 2 msec before saccade initiation. These authors proposed that the peak of discharge may represent the trigger signal to initiate a gaze shift. Using a different analytical technique, Hanes *et al.* (1995) also examined the timecourse of growth of activity of FEF movement cells. The analysis they used determined periods of activation by analyzing individual spike trains based on Poisson statistics. Two periods of activation were distinguished, an early prelude followed by a burst of activity. The beginning of activation determined by Schall (1991b) and Segraves and Parks (1993) defined by increases from a baseline level were closer to the time Hanes *et al.* found for the beginning of the prelude. Hanes *et al.* showed that the beginning of the burst of FEF presaccadic movement neurons occurred, on average, 28 msec before saccade initiation, with a mode around 8 msec before saccade initiation. Segraves and Park (1993) point out that signals from FEF can influence pontine cells directly within 1.4–4.0 msec and indirectly, via the superior colliculus, within 2.3–4.9 msec. Collectively, these findings seem to indicate that FEF movement cells must reach a fairly high rate of discharge to trigger initiation of a saccade.

The growing prelude of activity preceding saccade initiation may be a preparatory signal that readies motor circuitry for action, specifying where possible targets may be (Glimcher and Sparks, 1992). A number of findings indicate that the neural representation of the direction and amplitude of an eye movement take some time (~100 msec) to build up (Sparks *et al.*, 1987; Stanford *et al.*, 1990; see also Ghez *et al.*, 1990). Some dissociation of the movement specification and

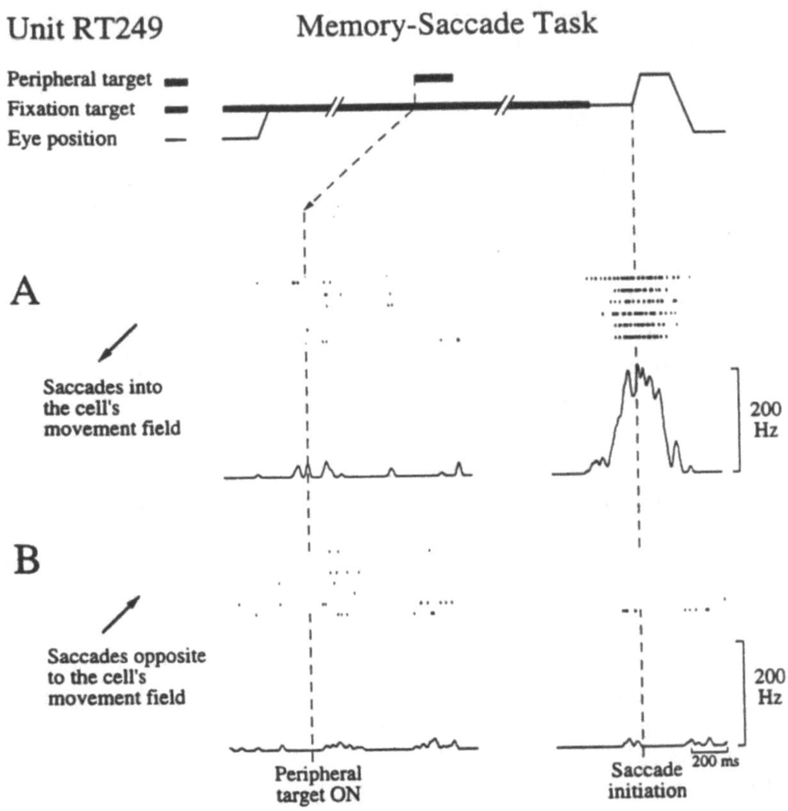

Figure 15. FEF movement cell activity during the gap task. The top rasters and spike density functions are from trials when the target fell in the cell's movement field. The bottom plots are from trials when the target fell outside the movement field. Activity of a frontal eye field neuron with presaccadic movement activity on the memory-saccade task and a fixation-disengagement discharge on the gap tasks. (Left) Memory-saccade task. As diagrammed at top, the peripheral target is briefly (200 msec) flashed while the monkey fixates the centrally located fixation target. When the fixation target is extinguished (1 sec later), the monkey is rewarded for making a saccade to the remembered location of the peripheral target. Two target locations were randomly interspersed, one in the center of the cell's response field (A) and one in the opposite direction (B). Raster plots and spike density histograms on the left are aligned to the presentation of the target; there was no response of this neuron to visual stimulation. The rasters and spike density functions on the right are aligned on the beginning of the saccade; there is a strong presaccadic response on trials calling for saccades into the cell's movement field. (Right) Gap task. As diagrammed at top, there was a 200-msec gap between the disappearance of the fixation spot and the appearance of the peripheral saccadic target. Trials into (A) and opposite (B) the cell's movement field were randomly interspersed. The plots of activity on the left are aligned in relation to the disappearance of the fixation target and appearance of the peripheral target. The plots of activity on the right are aligned on saccade initiation. The median saccade latency in gap trials was 166 msec when the target fell in the cell's movement field. For comparison, the median saccade latency in a conventional no-gap task (not shown) was 245 msec for the movement field case. In the memory-saccade trials the median saccade latency (relative to when the fixation light went off) was 318 msec when the target was in the movement field. In the gap task the cell discharged following the disappearance of the fixation target in all trials, but this preactivation leads to a presaccadic burst only when the target resulted in a saccade into the cell's movement field. The discharge observed in the gap trials was not due simply to the removal of the fixation spot; otherwise the discharge would have been observed in the memory-saccade trials when the target was not in the cell's movement field (B, left).

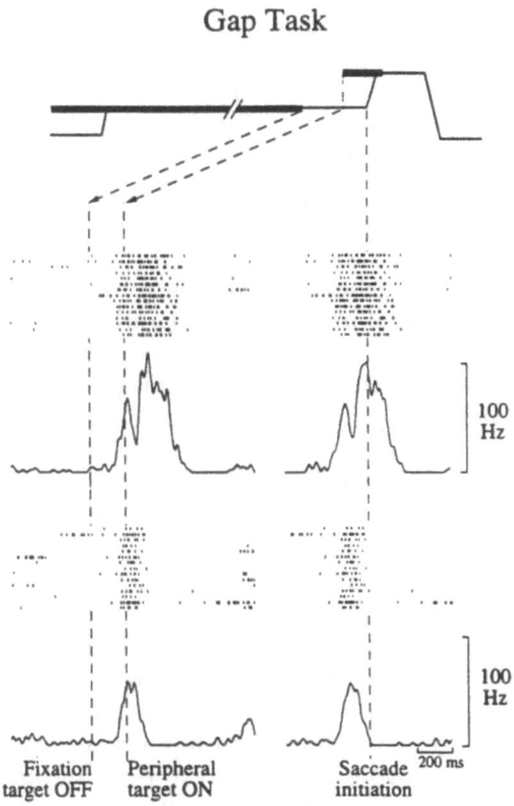

Figure 15. *(Continued).*

movement trigger signals must exist to permit speed–accuracy tradeoffs. In fact, evidence indicates that prematurely triggered saccades have limited metric specification (Sparks *et al.,* 1987).

The role of FEF cell activity in initiating saccades can be investigated by exploring how FEF cell activity changes as saccade latency changes. A variety of experimental conditions exist through which saccade latency can be systematically manipulated. In one condition, referred to as *gap,* the fixation spot disappears before a target is presented. If 200 msec or so intervenes between fixation spot removal and target presentation, then saccade latencies can become very short. Such saccades have been referred to as *express saccades,* and the general reduction in saccade latency under this condition is known as the *gap effect* (reviewed by Fischer and Weber, 1993). A plausible substrate for the gap effect is that during the gap interval presaccadic cells begin to discharge in anticipation of the presentation of the expected target. The growth of activity would produce a condition in which less additional growth would be needed to reach the threshold level to trigger the saccade. Preliminary data indicate that cells in FEF do become active during the gap interval (Dias and Bruce, 1994). After the fixation spot is removed and the monkey can reliably anticipate the presentation of the target, a prelude of activity begins in FEF movement cells (Fig. 15). If the target does not fall in the cell's movement field, then the anticipatory activity decays before the saccade is made elsewhere. If the target falls in the cell's movement field, then less change in discharge rate is needed to bring the cell to the burst threshold that triggers the saccade. This preactivation coupled with analogous changes in superior colliculus cell activity (Dorris and Munoz, 1995) could account for the reduction in saccade latencies in the gap condition. However, nothing is ever so simple. It is clear in Fig. 15 that following presentation of the target in the FEF cell's movement field, the discharge rate was actually reduced transiently before increasing to generate the saccade to the target. The temporal dynamics of neural modulation in the frontal cortex as elsewhere is complex and will need further work to be understood.

Our laboratory has developed a different behavioral paradigm with which to investigate the neural regulation of saccade initiation. This paradigm is referred to as the *countermanding paradigm* (Hanes and Schall, 1995). Monkeys are trained to make a saccade to a peripheral target unless a *stop signal* is presented, in which case they must withhold the movement. By systematically varying the interval between presentation of the saccade target and the stop signal, this task manipulates monkeys' ability to inhibit a planned movement. Performance on this task can be accounted for with a model of a race between a process that generates the movement and a process that inhibits the movement. The race model provides an estimate of how long it takes to countermand a planned movement.

With this critical information provided by the countermanding paradigm we can evaluate whether neurons are directly involved in controlling the production of voluntary movements (Schall and Hanes, 1996), and if so, what signals they generate to either initiate or prevent movements (Hanes and Schall 1996). We have discovered a surprisingly straightforward relationship between the growth of activation of movement-related neurons in frontal eye field and the time of initiation of eye movements. We have discovered that a voluntary gaze shift begins when the movement activity of FEF cells reaches a critical threshold level; the threshold trigger level remains constant for all saccade latencies (Hanes and Schall, 1996). A significant advantage of the countermanding paradigm is that we can compare the neural activity on trials when a movement was inhibited with the activity on trials when a movement was produced but would have been inhibited if the stop signal had been given. We found that on trials when the monkeys inhibited the saccade because of the stop signal the neural activity did not reach the critical threshold level. If the variability of behavioral response times is not due to variability in the trigger level of neural activity, then the range of reaction times arises from stochastic variability in the rate of growth of the neural activity toward that threshold. Lower growth rates result in longer reaction times and vice versa. To test further the reliability of this relationship, we implemented a linear accumulator model architecture to simulate the distribution of saccade latencies (Carpenter and Williams, 1995). Using parameters obtained from single FEF neurons, this simple model could reconstruct the distribution of saccade latencies made by the monkey while the neuron was recorded.

This evidence for a precise relationship between presaccadic movement activation levels and saccade initiation was obtained from single cells recorded in FEF. An fundamental question this finding raises is whether the same relationship between the growth of movement-related neural activity and the time of movement initiation holds across neurons within and across cortical areas as well as in subcortical structures. Two extreme possibilities present themselves. On the one hand, each neuron may have an idiosyncratic threshold, but the rate of growth of activity to reach that threshold is correlated across areas, so that there is temporal agreement on when to move. That is, all movement-related neurons contributing to a given movement reach their trigger threshold at essentially the same time. On the other hand, the activation of different neurons may grow in a variable fashion, and movements are produced when a critical number of neurons reach their respective thresholds. To test the hypothesis that the growth of movement-related activity in multiple neurons is correlated, future experiments are needed to record simultaneously from multiple movement cells in multiple saccade-related structures such as FEF, SEF, superior colliculus, the oculomotor thalamus, and the basal ganglia.

3.5.4. FEF Projection to the Superior Colliculus

To understand the role of FEF in saccade production, it is necessary to consider the functional relation with the superior colliculus. Simultaneous stimulation of FEF and the superior colliculus elicits a saccade with amplitude and direction that are the vector sum of the individual saccades represented at the stimulation sites in either structure weighted by the stimulus current (Schiller *et al.*, 1979). This result indicates a high degree of organization in the mapping of these two structures relative to one another. Anatomical evidence also indicates topographic connectivity (Komatsu and Suzuki, 1985; Stanton *et al.*, 1988b). Further evidence for a topographic projection from the FEF to the superior colliculus is provided by the fact that the lowest thresholds for antidromic activation were found at sites which represented the same saccade vector as that recorded in FEF (Segraves and Goldberg, 1987).

A recent physiological experiment has revealed more details about the functional mapping of FEF onto the superior colliculus (Schlag-Rey *et al.*, 1992). Data from this clever but complex experiment are shown in Fig. 16. Saccades were evoked by stimulation of FEF while neural activity was recorded from superior colliculus saccade-related burst cells. In this experiment, preferred saccade vectors for pairs of movement cells isolated in FEF and superior colliculus were first determined, then the activity of the superior colliculus cell was tested under several conditions. Two superior colliculus cells are contrasted in the left and right columns. As indicated in left panels 1 and 2, the saccade represented by superior colliculus cell E and the particular FEF stimulation site were both upward to the right. When a saccade was evoked by FEF stimulation, the superior colliculus cell discharged briskly (left panel 3). To test whether this excitation was anatomically hard-wired, Schlag-Rey and co-workers utilized the colliding saccade paradigm (see above) to manipulate the vector of the saccade evoked by FEF stimulation. The superior colliculus cell E discharged even when stimulation of the FEF site evoked a compensatory saccade that was leftward (left panel 4). The superior colliculus cell I was preferentially active for right, downward saccades (right panel 2), which did not coincide with the right, upward saccade evoked by stimulation of another FEF site (right panel 1). Cell I exhibited no activity associated with the saccade evoked by stimulation of the FEF site (data not shown). To observe the effect of FEF stimulation on cell I, it was necessary to make it fire beforehand (right panel 3). This was achieved by visually inducing a saccade in the cell's preferred direction and then stimulating FEF at the beginning of that saccade; this explains the U-shape of the saccade trajectory. Under this condition, stimulation of FEF caused suppression of the superior colliculus (SC) activity. By the same procedure the direction of the FEF-evoked saccade was made to correspond to the movement field of cell I (right panel 4). Cell I never became excited when the trajectory of the FEF-evoked saccade was made in this cell's preferred direction. These results suggest that whether SC cells are activated or inhibited depends on the actual FEF input to SC, not on the direction of the electrically induced saccade. Indeed, when stimulation parameters were below the threshold for evoking saccades, weaker excitation or inhibition of SC by FEF could still be detected. These data indicate that the functional relations of common-vector excitation and dissimilar-vector inhibition are hardwired.

The mapping of FEF into the motor layers of the superior colliculus is

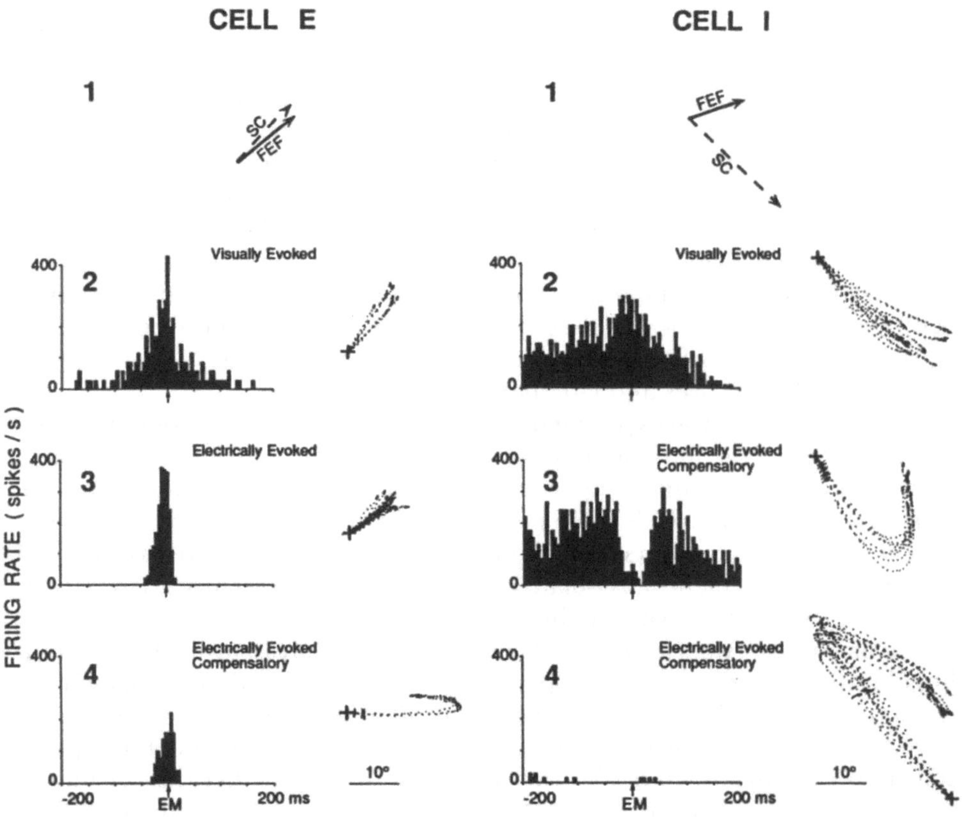

Figure 16. Effects of FEF stimulation on superior colliculus (SC) movement cell activity. The effects of FEF stimulation are shown for two SC cells. (1) The relation between the saccade evoked by FEF stimulation and the saccade represented by each SC cell. The movement field of cell E (left) coincided with the vector of saccade evoked by FEF stimulation. The movement field of cell I (right) differed from the vector of saccade evoked by FEF stimulation. (2) The saccade-related activity of each SC cell associated with a visually guided saccade to a target in the respective movement field. (3) The modulation of SC cell activity evoked by FEF stimulation (7 μA for cell E, 10 μA for cell I). (4) The modulation of SC cell activity evoked by FEF stimulation when the direction of the evoked saccade was manipulated in the colliding saccade paradigm. FEF stimulation always excites superior colliculus movement cells if the saccade vector represented at the SC recording site corresponds to that represented at the FEF stimulation site. Conversely, the same FEF stimulation always inhibits SC cells if the FEF and SC saccade vectors differ. These relationships holds regardless of the actual vector of the saccade produced. Details in text. From Schlag-Ray *et al.* (1992) with permission.

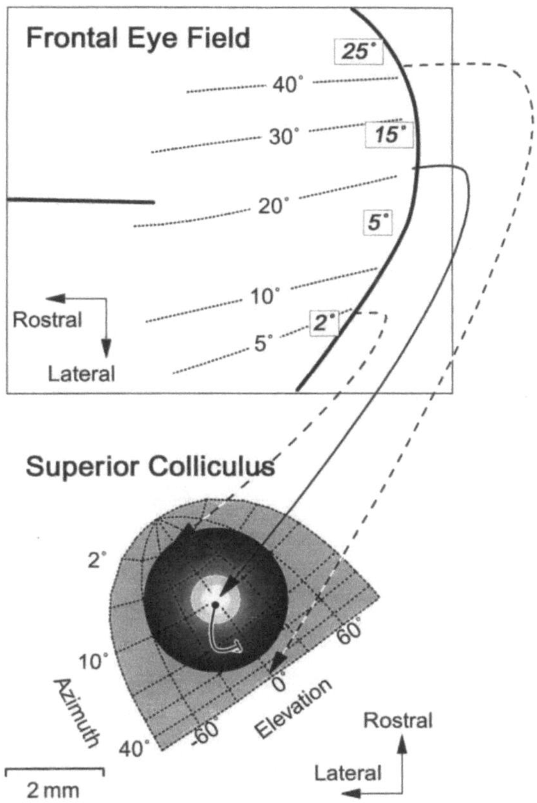

Figure 17. Mapping from FEF to the motor layers of superior colliculus. The prearcuate cortex is illustrated on top. Dashed lines indicate visual receptive field eccentricity. Amplitudes of saccades evoked by microstimulation are written in the boxes. The superior colliculus motor map is illustrated on the bottom. The scales of FEF and superior colliculus are equated approximately. The projection from FEF to superior colliculus for a 10° saccade is drawn. The lighter stipple indicates the extent of the zone of excitatory influence. The darker stipple indicates the zone of surrounding inhibition. Further details in the text.

summarized in Fig. 17. The visual receptive field eccentricity (Suzuki and Azuma, 1983) and amplitudes of saccades evoked by microstimulation (Bruce *et al.,* 1985) characterize the map of FEF. The motor map of the superior colliculus is summarized from data derived from electrical stimulation studies (Robinson, 1972; see also van Gisbergen *et al.,* 1987). The projection for a 10° saccade is indicated in the figure. The approximate extent of the zone of excitatory influence (indicated by the lighter stipple) corresponds to the area of superior colliculus activated before a 10° saccade (Ottes *et al.,* 1986). The zone of surrounding inhibition is also indicated. The inhibition seems most likely mediated by intracollicular inhibition (e.g., van Opstal and van Gisbergen, 1989).

3.6. FEF Sensory Function

3.6.1. Auditory Responses

By all accounts FEF subserves the sensory guidance of gaze. Most studies of FEF have concentrated on visually guided saccades. Saccadic eye movements, although usually guided by visual stimuli, may also direct gaze toward the source of acoustic stimuli (e.g., Zambarbieri *et al.,* 1982). Indeed, neurons in FEF and surrounding prearcuate and postarcuate cortex, besides being activated by visual stimuli, are also responsive to acoustic stimuli (Azuma and Suzuki, 1984; Suzuki, 1985; Vaadia *et al.,* 1986; Vaadia, 1989; Watanabe, 1992). Figure 18

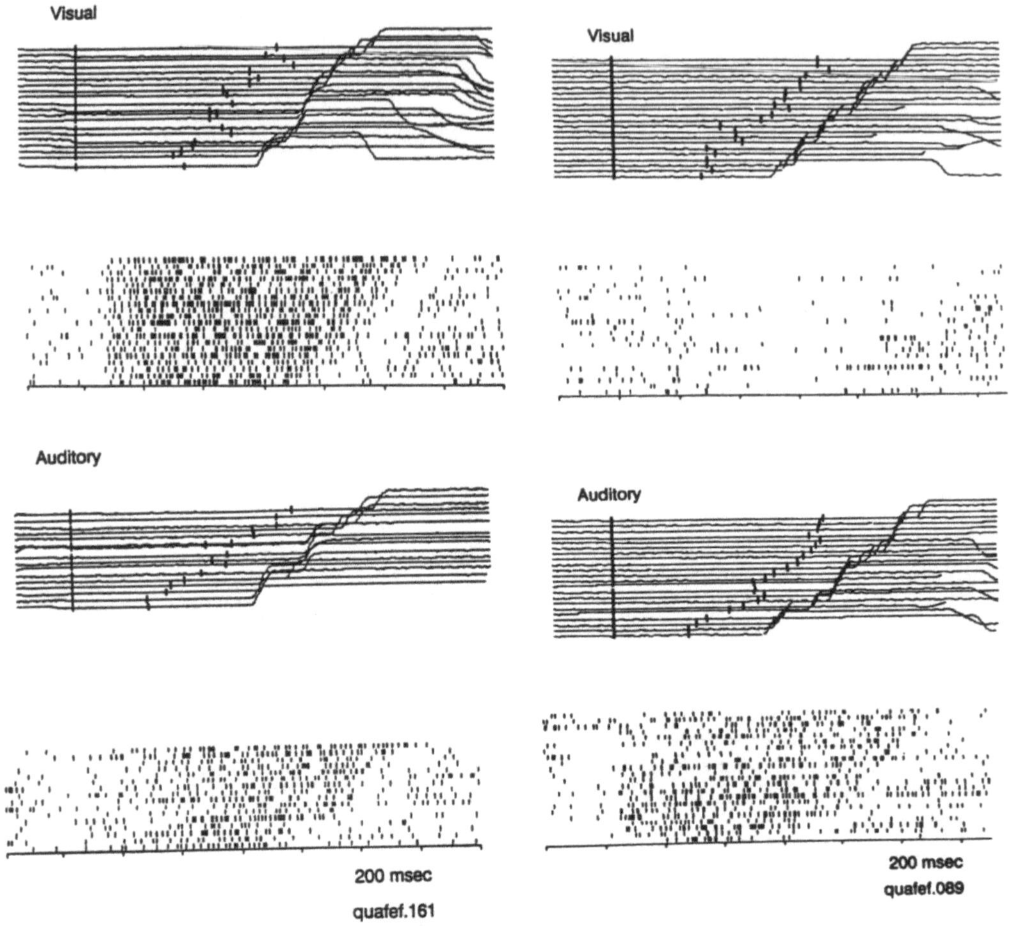

Figure 18. (Left) Visual-movement FEF cell. The eye position traces and rasters are aligned on the time of presentation of the visual (top) and auditory (bottom) targets. Following the visual target, this cell showed a sustained elevation of activity which decayed after the saccade was in flight. The acoustic target evoked no sensory response even though a rise in activity preceded the saccade. (Right) Auditory-movement FEF cell. Eye position traces and rasters are aligned on the visual (top) or auditory (bottom) target. This cell did not respond to the visual target, but was active following the auditory stimulus. Modified from Schall (1991b) with permission.

shows two examples of the differential responsiveness of FEF neurons to visual and acoustic targets (Schall, 1991b). Vaadia *et al.* (1986) determined that both prearcuate and postarcuate cells located mainly around the upper limb of the arcuate sulcus responded to acoustic stimuli either exclusively or along with visual stimuli. This location of neurons responsive to auditory stimuli is concordant with the distribution of afferents from auditory cortex to the dorsal arcuate region (e.g., Barbas and Mesulam, 1981).

Electrical stimulation studies have shown that ear movements can be evoked by microstimulation around the upper limb of the arcuate sulcus (Bon and Luchetti, 1994; Schall *et al.,* 1993). The responsiveness of the cells is contingent, however, on whether monkeys must localize the stimuli to guide movements (Vaadia *et al.,* 1986; Russo and Bruce, 1994). Cells responding to auditory stimuli had delimited response fields in the contralateral hemifield extending to the midline. Bimodal cells had visual receptive fields overlapping the acoustic response field in space. Vaadia *et al.* (1986) noted that postarcuate cells were more likely to be bimodal, and a higher fraction of prearcuate cells responded exclusively to visual or acoustic stimuli.

Investigating the nature of the tuning of FEF neurons to visual and acoustic stimuli provides more information about the coordinate transformations underlying the guidance of gaze. Whereas acoustic stimuli are referred to head-centered coordinates, visual stimuli are referred at least initially to retinal coordinates, which can change relative to head-centered coordinates as gaze changes. Physiological recordings have shown that the auditory response fields and presaccadic movement fields of cells in FEF shift with the angle of gaze (Russo and Bruce, 1989, 1994). In other words, FEF appears to represent the location of acoustic stimuli in terms of gaze motor error, i.e., the saccade needed to fixate the auditory stimulus. This finding is analogous to what has been observed for acoustic responsive cells in the superior colliculus (Jay and Sparks, 1987a, b).

3.6.2. Saccade Target Selection

Typically saccades direct gaze to conspicuous points in the image (Yarbus, 1967; Keating and Keating, 1993; Viviani, 1990). How the brain selects the targets for eye movements is not understood, but this research problem has begun to attract more attention (reviewed by Schall, 1995). Our laboratory has performed single-unit recordings in FEF of monkeys selecting the target for a saccade during visual search (Schall and Hanes, 1993; Schall *et al.*, 1995a; Bichot *et al.*, 1996; Thompson *et al.*, 1996). In monkeys trained to locate the salient oddball stimulus in a visual search array, we find that the initial visual response evoked by the search array does not discriminate whether the target (e.g., a red square) or a distractor (e.g., a green square) is in an FEF cell's response field. Subsequently, though, the activity evolves to signal whether the target is in the response field before the saccade is made. We also find that, besides failing to discriminate the target from a distractor, the initial visual response of FEF neurons is modulated by the presence of stimuli outside the excitatory response field and by whether a saccade is going to be produced. We have also determined that target selection occurs through suppression of distractor-evoked activity contingent on the location of the target relative to the receptive field. The cell shown in Fig. 19 illustrates most of these findings. When the target was presented alone in the response field of the cell, it elicited a brisk, transient response with a 50 msec latency. In contrast, when the target was presented in the hemifield opposite the receptive field, the cell became suppressed until the saccade was made, even if the saccade was delayed by instruction. In catch detection trials in which no target was presented and the monkey was required to maintain fixation of the central spot, the neuron continued to fire at its resting rate.

During visual search this cell exhibited a very different pattern of activity. Presentation of the search display evoked an early, transient burst of activity that was attenuated relative to the response to the target presented alone. We observed an attenuated response to the search stimulus array for over one-third of the visually responsive cells in FEF. The visual response to the search array may have been reduced due to the suppressive influence invoked by the presence of stimuli in the ipsilateral visual field. Alternatively, earlier work has shown

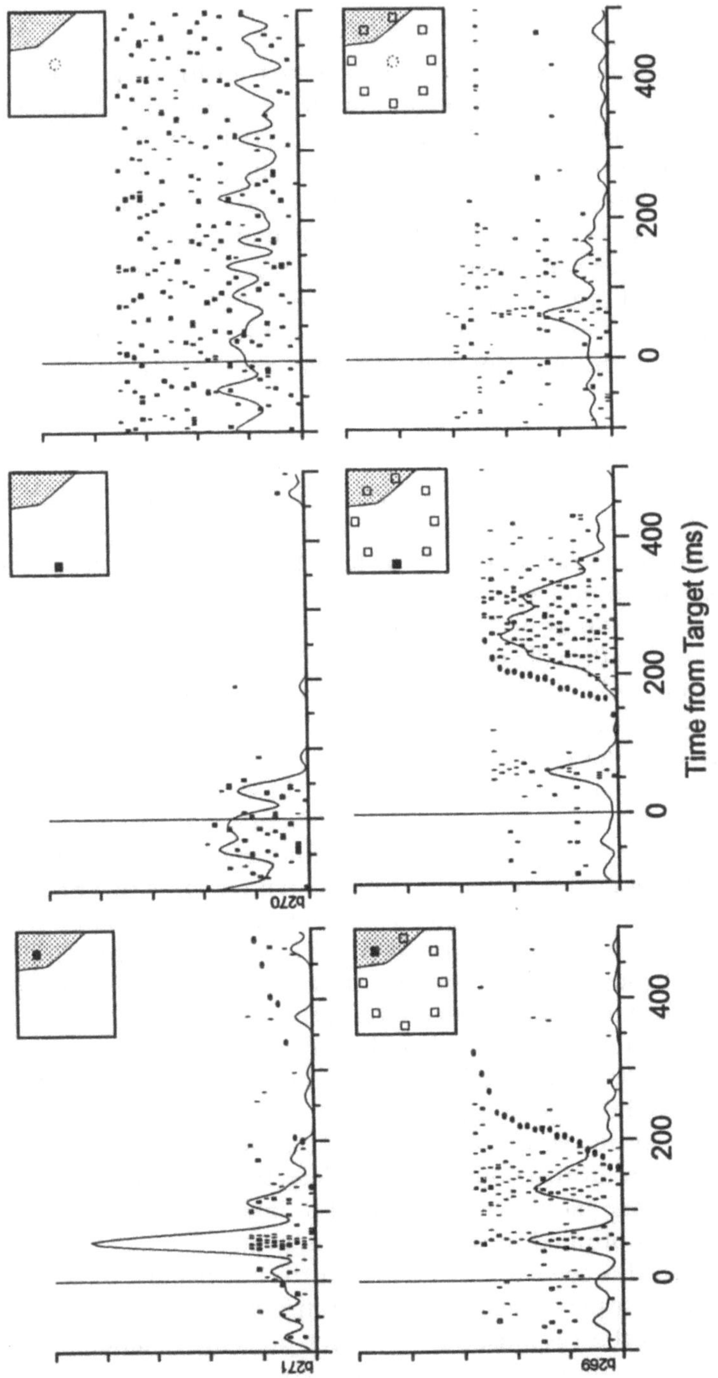

Figure 19. Responses of a visual neuron to the target presented alone, with distractors, and in catch trials. The rasters and superimposed spike density functions are aligned on target presentation and sorted according to the interval between target presentation and saccade initiation. The time of occurrence of saccade initiation is indicated by the solid circle in each raster line. The top panels show the activity collected when the target was presented alone in the receptive field (left), alone outside the receptive field (middle), or not at all in catch trials (right). The bottom panels show the activity collected in search trials with the target in the most sensitive part of the receptive field (left), with distractors in the receptive field and the target in the opposite hemifield (middle), or with only distractors in catch trials (right). The ordinate scale represents 250 Hz. From Schall *et al.* (1995a) with permission.

that the activation of many visually responsive cells in FEF is enhanced if the stimulus is used as the target for a saccade (Goldberg and Bushnell, 1981). It may be that the initial activation of the FEF cells in response to visual search displays is reduced because the target for the saccade has not been selected. The evidence for this is that the magnitude of the initial visual response did not distinguish whether the target or a distractor fell in the receptive field. We observed this lack of differential initial response in virtually all of the visually responsive cells in FEF. However, the neuron shown in Fig. 19 exhibited a subsequent reactivation specifically when the stimulus in the receptive field was the target of the search array. When a distractor stimulus was in the receptive field, suppression ensued after the initial burst until the saccade was made to the target located elsewhere. In catch trials of visual search in which only the distractors were presented and the monkey was rewarded for maintaining fixation, the activity after the initial burst decayed more gradually. Evidently the suppression observed when the target was in the hemifield opposite the receptive field was an active process contingent on preparation of a saccade to the target outside the receptive field. Overall, essentially all of the visually responsive cells we have recorded in FEF that have sustained activity (visuomovement cells) discriminate whether the target or the distractor of the search array is in their response field. The cells respond initially indiscriminantly to the target or distractor stimuli in their receptive field. Subsequently, before the eyes move, the activation of frontal eye field cells evolves to signal whether the target or a distractor is in the receptive field. This target selection is achieved by a process that suppresses responses evoked by the distractor, leaving only cells with the target in their receptive field with maximal activity. These data indicate that one way to conceive of FEF function is as a saliency map, registering the location of stimuli that can become the targets for eye movements.

Another property of FEF neurons was revealed in the visual search condition. Certain FEF neurons appear to have a center–surround movement field organization when tested with the search arrays. In other words, the magnitude of the distractor suppression reflects the proximity of the target to the receptive field, often being greater if the target flanks the receptive field. This is illustrated in the activity of the FEF neuron shown in Fig. 20. This neuron had a relatively long visual response latency of 100 msec, but the fact that the onset of activation was synchronized on the time of target presentation indicates that the early activity was not related exclusively to movement generation. When the target of the search array appeared in the center of the response field, the neuron discharged in the same manner as when the target appeared alone in the response field. When the target of the search array appeared at a location distant from the response field, the cell responded to the distractor in the response field at an initial rate comparable to that observed when the target was in the response field. The level of activity then, however, never grew to reach the level it did when the target was in the response field. When the target of the search array appeared at a location adjacent to the response field, this cell exhibited a brief period of stimulus-evoked activation followed by a sustained period of suppression leading up to the saccade. Thus, during visual search trials the initial activation of this neuron did not predict saccade direction because it did not distinguish whether the target or a distractor was in the receptive field.

During the latent period before the saccade, the activity of this cell changed according to where the target was located relative to the receptive field. Ultimately, in the period immediately preceding saccade initiation the discharge rate varied significantly with target direction. Notably, the magnitude of presaccadic activation decreased from the center of the response field in a nonmonotonic fashion. The presence of suppression mechanisms beside the response field of this cell was tested by comparing how well a single Gaussian function and a difference of Gaussian function fit the data. The variation of activity in the 50 msec preceding initiation of the saccade to the search target was not as well fit by a single Gaussian function as it was by a difference-of-Gaussians function (Fig. 20). This indicates that a specific suppressive mechanism modulated the activation of this neuron before saccade generation.

The findings just described were made in monkeys that were trained to search for the oddball stimulus regardless of the particular visual feature that

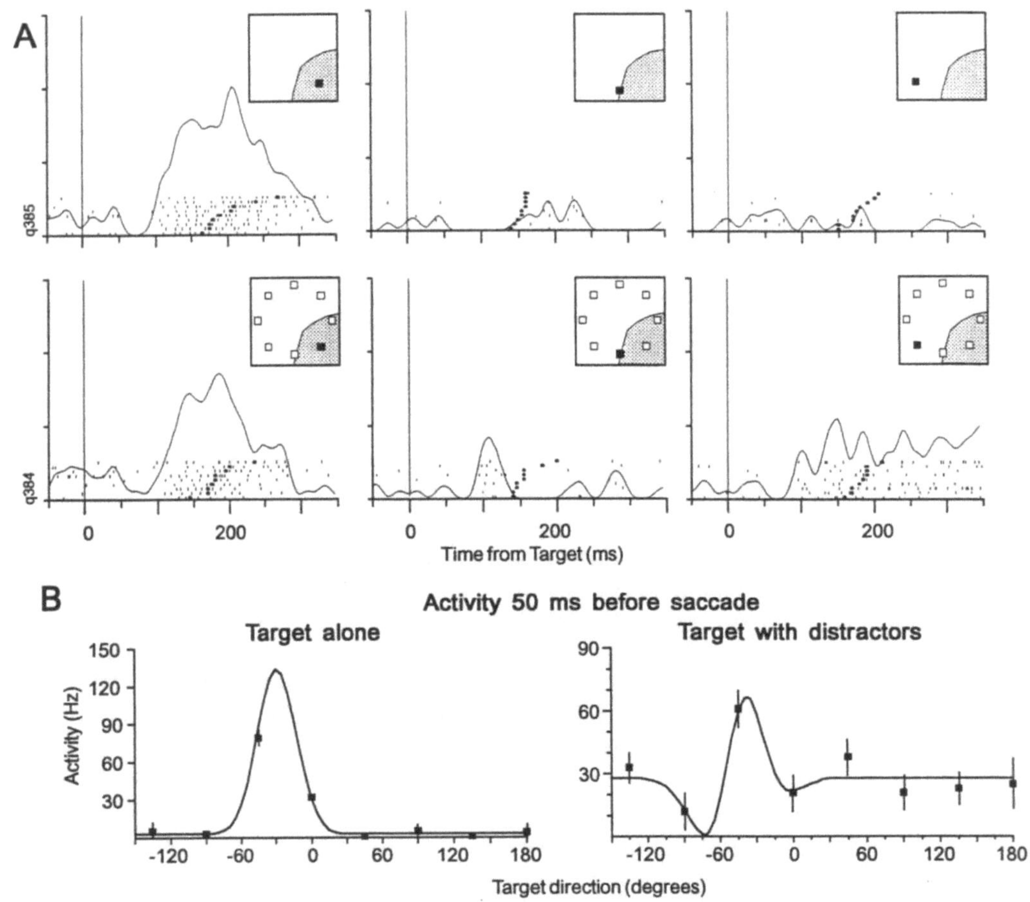

Figure 20. Activity of FEF cell with visual and premovement activity during detection and search trials. (A) Raster displays of neural activity evoked by the target presented alone (top) or with distractors (bottom). The ordinate represents 150 Hz. (B) Discharge rate measured in the 50-msec interval before saccade initiation as a function of target direction for detection and search trials. Solid lines are the Gaussian function (left) or difference-of-gaussians function (right) best fit to the activity preceding the saccade. From Schall *et al.* (1995a) with permission.

defined it (e.g., both red among green and green among red); the monkeys were employing a generalized strategy of searching for the salient oddball stimulus. But daily experience and controlled experiments show quite clearly that cognitive considerations can override stimulus attributes; that is, gaze can be directed to nonsalient stimuli. In the course of this work on saccade target selection, we have made a surprising discovery that the initial visual responses of frontal eye field neurons to flashed stimuli can be modulated by a learned cognitive set. We have found that when monkeys are given exclusive experience with one visual search array (e.g., only red among green), they adopt the strategy of ignoring stimuli that are distractors, even when those same stimuli become the target in another context (Bichot *et al.,* 1996). In monkeys employing this behavioral strategy we discovered that nearly half of the visually responsive cells in FEF (Bichot *et al.,* 1996) exhibited feature selectivity unlike what has been observed before. In other words, the response to the learned target was greater than the suppressed response to the learned distractor as soon as the neurons responded. Furthermore, the visual response latencies at which this modulation occurred were comparable to the visual response latencies of color-selective neurons in primary visual cortex (see Fig. 22, and the chapter by Nowak and Bullier in this volume), permitting insufficient time for conventional attentional modulation (reviewed by Desimone and Duncan, 1995). Therefore, we surmise that the induced visual selectivity of FEF neurons for the learned target may be a result of reduced efficacy of synapses conveying the original response to the distractor stimuli. If the experience-induced visual feature selectivity in frontal eye field is based on functional plasticity, then it represents a novel manifestation of this phenomenon. Previous demonstrations of experience-dependent plasticity have shown expansions of representations in topographic maps consequent to extensive selective stimulation of a part of the sensory surface or particular motor effectors (reviewed by Weinberger, 1995). The visual feature selectivity we observed, which was induced throughout FEF representing all parts of the visual field, rather seems to reflect learning a particular stimulus–response mapping. This phenomenon may correspond to the changes in visual responses associated with learning that have been observed in other parts of frontal cortex.

To the extent that saccades are normally produced in a cluttered visual environment, studying saccade target selection with the visual search tasks just described will provide necessary but ultimately limited information. In this context, a recent study by Burman and Segraves (1994) is important because they investigated the activity of neurons in FEF of monkeys scanning natural images. Burman and Segraves found that the activity of movement cells was the same for saccades produced in response to light spots as for saccades produced to scan a natural image. The activity of visually responsive cells during scanning behavior, however, varied according to whether the subsequent saccade was made to a feature in the cell's receptive field (Fig. 21). On average, visual cells were activated if the next saccade was to a feature located in their receptive field, and were suppressed before saccades to features outside the receptive field. This finding indicates that the visual cells in FEF on average signal whether the upcoming saccade will be into the cell's receptive field. Close examination of the data, though, reveals that this relationship can be quite variable. For example, the top panel of Fig. 21 shows a raster display of the activity of a visual cell before

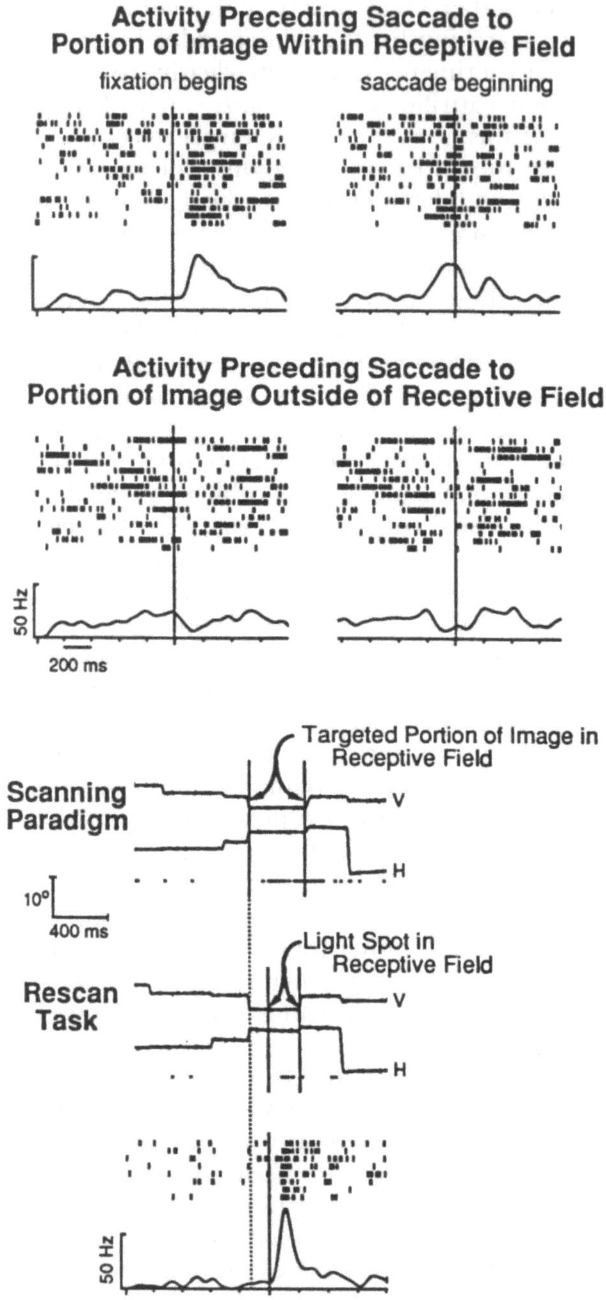

Figure 21. FEF visual neuron during scanning eye movements. Rasters and spike density functions in the top four panels are aligned on the beginning of a fixation (left) and the beginning of the subsequent saccade (right). The top two panels show the activity preceding saccades to portions of the image falling within the receptive field. The middle two panels show the activity preceding saccades to portions of the image falling outside the receptive field. The bottom panels compare the activity observed during the natural scanning condition and a rescan condition. Two representative trials are shown from the scanning paradigm and the rescan task. Vertical (V) and horizontal (H) eye positions are illustrated with the associated neural activity. The bottom panels show the response evoked before several rescan gaze shifts. Modified from Burman and Segraves (1994) with permission.

saccades made to parts of the image that were in the cell's receptive field. Sometimes the neuron was activated after a specific, short latency once gaze landed in the current fixation. This tended to occur near the bottom of the raster, which represents chronologically later saccades in the scanning (D. D. Burman and M. A. Segraves, personal communication). Other times the same neuron was not activated or not modulated before the saccade. This tended to occur near the top of the raster, which represents earlier saccades in the scanning. Thus, the visual cell activity reliably predicted where the upcoming saccade would go only following certain fixations. The fact that the visual response, when it was observed, occurred with a consistent short latency following fixation seems to indicate that the decision about the next saccade target was made either very rapidly following fixation or even before the fixation in question.

But how could the decision be made so soon? One possibility is based on the observation that the pattern of eye movements made to inspect a complex image often exhibits a repetitive scanpath (Yarbus, 1967; Norton and Stark, 1971). According to this idea, perhaps in the early period of scanning the image the monkey had not established a scanpath, so the choice of saccade target was being made only during each fixation. However, as the scanpath may have become established, the target for each saccade following a given fixation became more regular and predictable. In this later period the visual neuron could discharge with a constant and short latency following fixation because target selection was essentially already completed. Interestingly, there appeared to be a corresponding change in fixation duration as scanning progressed during collection of the data from the cell shown in Fig. 21. Early on, the fixation durations tended to be long and variable, whereas later in the scanning the fixation durations became shorter and more regular. Curiously, however, this evolution of fixation durations is actually opposite the general tendencies that have been observed for humans viewing pictures (Antes, 1974).

Burman and Segraves made another very interesting finding. When the monkey retraced the original scanpath over the image by following a projected spot of light, the visual cell responses were delayed until the presentation of the tracking spot. This happened even though the visual image and subsequent saccade were identical to what had previously elicited activation earlier during the fixation (Fig. 21, bottom). This finding shows that the visual responses of FEF neurons are sensitive to the context in which the saccade target identification occurs. Also, these data provide further evidence that the visual responses of FEF cells are gated by rapid and potent mechanisms. These mechanisms involve not only visual, but also mnemonic and motivational elements.

In summary, the findings from our laboratory and from the work of Burman and Segraves show quite clearly that FEF is active during the process of saccade target selection during visual search and natural scanning eye movements. We should note that following ablation of FEF, monkeys can still make accurate saccades in a visual search task (Schiller *et al.*, 1987), but they do show impairments in more complex scanning conditions (Latto, 1978a, b; Collin *et al.*, 1982; see also Teuber *et al.*, 1949; Luria *et al.*, 1966; Karpov *et al.*, 1968; Chedru *et al.*, 1973; Moser and Kömpf, 1990). Even if the neural processing in FEF is not necessary for selecting the targets for individual saccades, it is apparent that the activity in FEF does reflect the evolving state of a distributed system. This work frames many questions for future research. Where and how are the visual responses of FEF neurons gated? Are intrinsic interactions in FEF responsible for the evolution of activity in visual and visuomovement cells? What do the visual afferents to FEF signal about stimulus properties and target location? How is the processing in FEF linked to what is happening in other visuomotor centers such as SEF and LIP as well as subcortical structures?

At least two alternative hypotheses can be framed about the role of FEF and other brain areas in target selection. One hypothesis is that the visual afferents to

FEF discriminate the target from surrounding distractors and convey that information to FEF. Evidence for this hypothesis would be that the afferents to FEF signal target location before cells in FEF do. The alternative hypothesis is that the neural processing responsible for saccade target selection occurs concurrently in many brain regions. Evidence for this hypothesis would be that neural signals of target location arise concomitantly throughout the visuomotor system.

Our laboratory has developed methods with which to estimate the time required for FEF neurons to discriminate whether the target or a distractor is in their response field. We applied a Poisson spike-train analysis (Hanes *et al.*, 1995; Thompson *et al.*, 1996) to determine the visual latency of FEF neurons in response to the target presented alone, to the target presented with distractors, and to the distractors in the response field. Figure 22 shows the distributions of visual response latencies of FEF neurons under several conditions. The latencies of the visual response to the target presented alone in the receptive field, the latencies of the response to the target presented with distractors, and the latencies of the response to the distractors in the receptive field are not significantly different from one another. All of the latencies averaged 67 msec. Thus, even though the average magnitude of response to the search array is attenuated relative to the response to the target alone (Schall *et al.*, 1995b), the response latencies are not different. Superimposed on the FEF visual latencies in Fig. 22 is the distribution of V1 visual latencies measured by Nowak *et al.* (1995). The overlap is virtually complete except for the lowest time bin in V1, which represent the short-latency responses of cells in layers 4cα and 4B (Maunsell and Gibson, 1992; Nowak *et al.*, 1995). Thus, according to these data, FEF becomes active at essentially the same time as does V1 (and presumably the other visual areas) in response to visual stimuli. This fact indicates that anatomical hierarchies do not necessarily imply temporal sequences (see the chapter by Nowak and Bullier in this volume).

Given that FEF neurons in monkeys trained to search for oddball stimuli discharge at the same initial rate with the same response latency to either the target or distractors in their response field, we wanted to determine when the target and distractor-evoked activity become different (Thompson *et al.*, 1996).

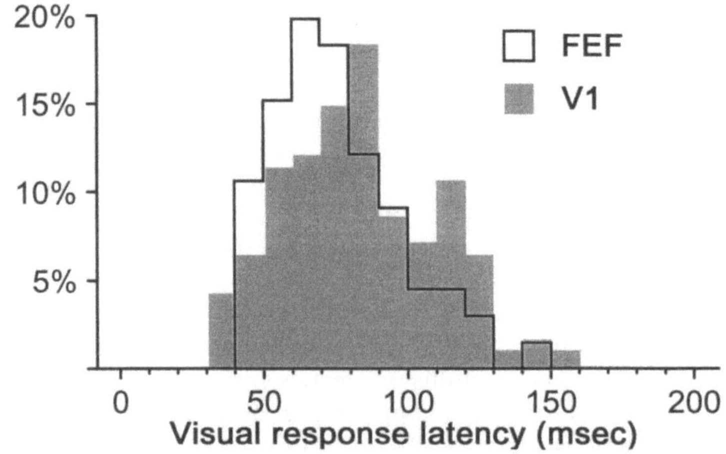

Figure 22. Distributions of visual response latencies of FEF cells (open histogram). V1 visual response latencies are shown in the shaded region. Data from Nowak *et al.* (1995).

To do this, we employed a method derived from signal detection theory motivated by previous studies relating neural activity to behavior (Cohn *et al.*, 1975; Bradley *et al.*, 1987; Vogels and Orban, 1991; Britten *et al.*, 1992). The difference in firing rate when the target versus distractors fell in the response field was determined as a function of time by calculating the receiver operating characteristic (ROC) curves based on the two distributions of activity measured from spike density functions. The area under the ROC curve estimates the probability that an ideal observer could correctly discriminate the signal (target in the response field) from the noise (distractor in the response field). An example of this analysis is shown in Fig. 23. For this cell the area under the ROC curves increased monotonically as a function of time following search array presentation from an initial value of 0.5 to reach nearly 1.0. To estimate the timecourse of FEF cell target discrimination, we fit the plot of ROC area as a function of time with a cumulative Weibull function. We defined the target discrimination time as the instant when the best-fit function reached a value of 0.75. Essentially all of the visuomovement cells we have recorded had activity that met this criterion before saccade initiation and thus could contribute to target selection. The average target discrimination time was 140 msec after presentation of the search array.

It has been thought for many years that behavioral reaction times such as saccade latencies may be comprised of different stages of processing that may or may not be discrete and separable (e.g., Sternberg, 1969; Taylor, 1976; J. Miller,

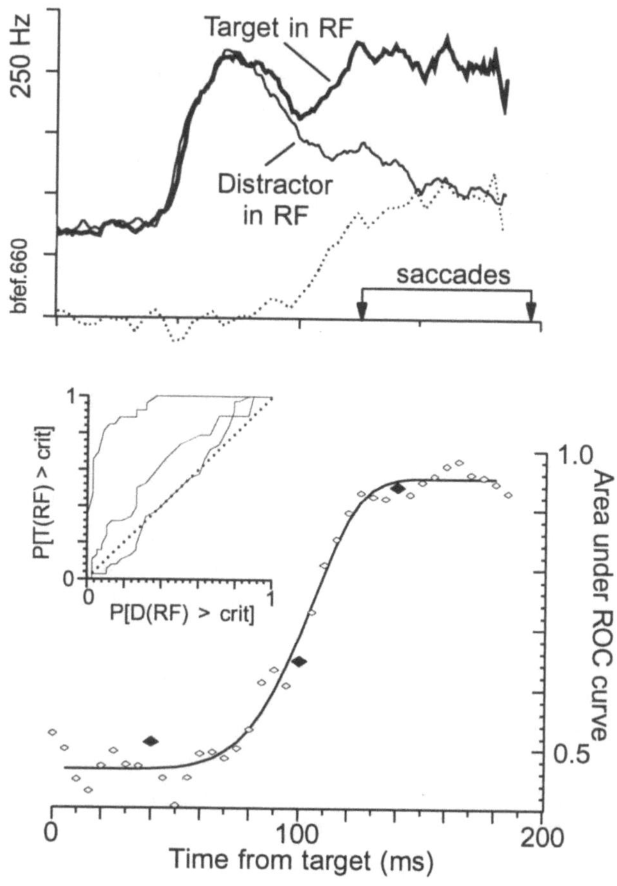

Figure 23. Timecourse of target discrimination by an FEF neuron. (Top) Average discharge rate when the target (thick line) and when distractors (thin line) were in the response field. Dotted line is their difference. The range of saccade latencies is indicated on the abscissa. (Bottom) Area under ROC curves as a function of time. The filled symbols are from the three ROC curves shown in the inset. The ROC curves plot the probability that the activity when the target is in the response field is greater than a criterion, as a function of the probability that the activity when a distractor is in the response field is greater than that criterion. With time the ROC curves bow away from the diagonal. This is quantified by plotting the area under the ROC curve as a function of time; the values increase from around 0.5 to less than 1.0. The solid line is the best-fit cumulative Weibull function.

1988). The process of discriminating the location of the target of a search array represents a key stage of processing. It has been suggested that much of the time taken in saccade latencies is occupied by the decision of where to shift gaze (e.g., Carpenter and Williams, 1995). If this is so, then we might expect that the latencies of saccades made to fixate the target of a search display ought to be predicted by the time taken by individual FEF neurons to discriminate whether the target is in their response field. We approached this question by performing the ROC analysis on groups of trials separated into early, middle, and late saccade latency groups. We found that the time at which the activity of individual FEF neurons discriminated whether the target was in the response field was more closely synchronized with the time of stimulus presentation than with the time of saccade initiation. In other words, trials with longer saccade latencies did not occur because more time was taken to discriminate the target. Thus, what was true for the FEF cell illustrated in Fig. 19 was the general rule for all visually responsive cells in FEF. These data demonstrate at the single-cell level a dissociation between perceptual processing and response programming in the population of FEF cells. Further work is needed to determine what the timecourse of discrimination under these conditions is in other visual and visuomotor areas. This finding also highlights the need to understand better how saccade initiation is regulated.

3.7. FEF Role in Slow Pursuit Eye Movements

Once fixated, a feature in the image that moves will be tracked with pursuit eye movements (Pola and Wyatt, 1991). This behavior requires voluntary control and directed attention (Kowler, 1990). Recent work has discovered a region near the fundus of the arcuate sulcus that is involved in producing pursuit eye movements (MacAvoy *et al.*, 1991; Gottlieb *et al.*, 1993, 1994). Ablations of the prearcuate cortex that extend to the fundus of the arcuate sulcus cause reduced gain of ipsilateral pursuit (Lynch, 1987; Keating, 1991; MacAvoy *et al.*, 1991). Electrical microstimulation of this zone evokes slow eye movements (usually around 11°/sec) with a latency averaging 39 ± 2.5 msec (Gottlieb *et al.*, 1993). Although the dynamics of the slow eye movement vary with stimulation intensity, the direction of the movement is constant for a given site, tending to be mainly ipsilateral. If no visual stimulus is visible, then the smooth eye movement evoked by microstimulation accelerates more than if a visual stimulus is fixated. The difference in eye velocity under open-loop and closed-loop conditions occurs after 75–100 msec, corresponding to the visual afferent latency of FEF. Based on these data, Gottlieb *et al.* (1993) suggest that FEF generates an eye acceleration signal. Smooth eye movements were evoked only within a small region ($\leq$6 mm^2) at the fundus of the arcuate sulcus located posterior to the tip of the principal sulcus. The smooth eye movement region was surrounded rostrally by saccade-related cortex and caudally by skeletomotor premotor cortex. FEF can contribute to slow eye movement generation via efferents to the dorsolateral pontine nuclei and the nucleus reticularis tegmenti pontis (reviewed by Keller, 1991).

Neurons recorded in the arcuate fundus region typically discharge before and during smooth pursuit, but not saccadic eye movements (MacAvoy *et al.*, 1991; Gottlieb *et al.*, 1994). Figure 24 shows the activity of an FEF cell that was

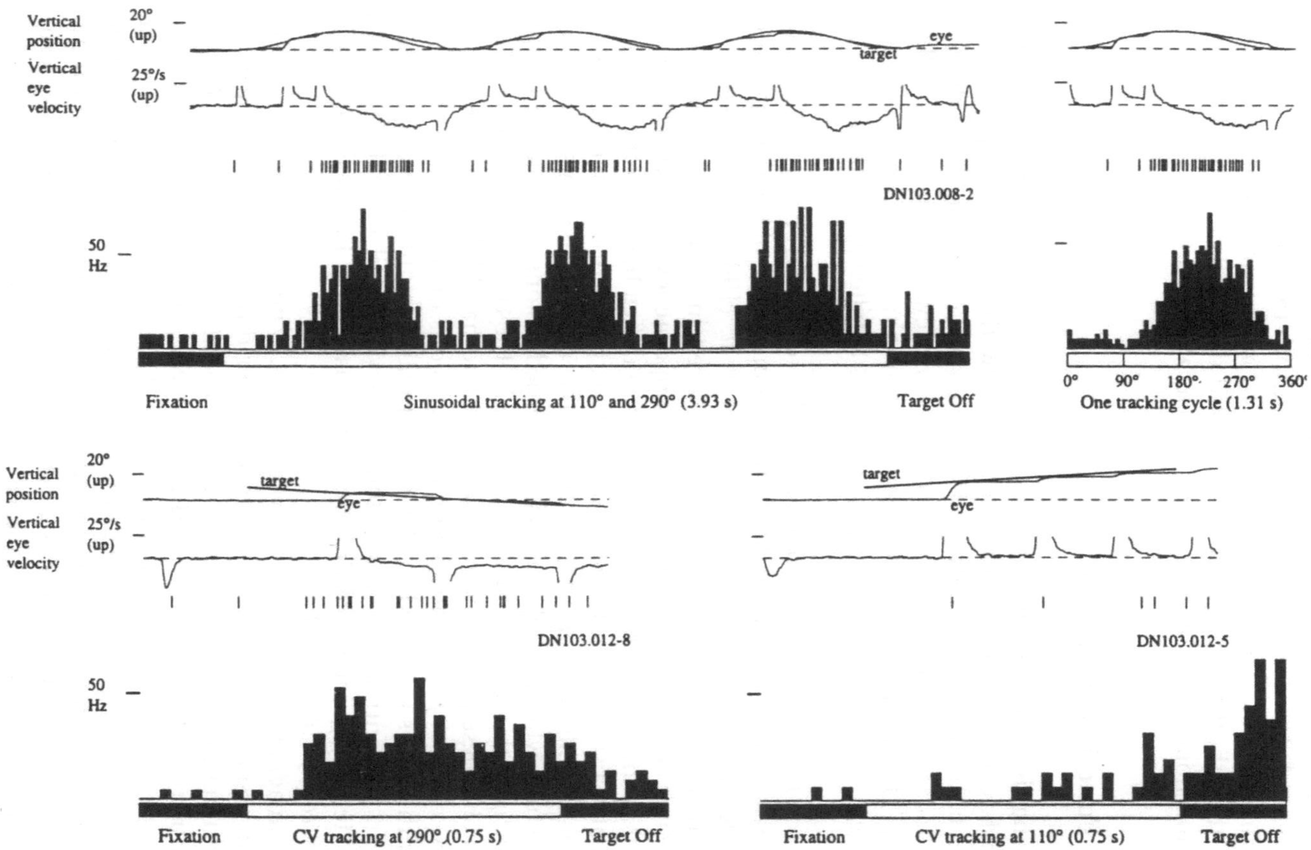

Figure 24. Pursuit responses of an FEF neuron in the right hemisphere. (Top) Responses during tracking sinusoidal motion (0.762 Hz, 15° peak-to-peak amplitude, directed along the 110°–290° axis). Vertical eye position and velocity traces are shown with discharges from a single trial. The top left histogram is averaged from six trials and aligned on the beginning of target motion in each trial. The top right histogram shows the same responses aligned by phase on each movement cycle (average of 18 cycles). Notice that the neuron's response profile is approximately sinusoidal, slightly leading the 180° phase (corresponding with the onset of motion in its preferred direction) and peaking before the 270° phase (corresponding with the peak target velocity in its preferred direction). (Bottom) The same neuron's responses during constant-velocity (CV) tracking directed at 290° and 110°. Target speed was 20°/sec, and the 10° steps were opposite the cell's preferred pursuit direction in both conditions. The time scales for each histogram and corresponding eye records are given by open bars indicating the tracking epoch. Bin widths for these histograms are 20 msec. Note the lack of correlation, in the single records, between the neural activity and the saccades made during tracking. Also note the lack of response during stationary fixation.

active in relation to pursuit eye movements. The activity begins from up to 100 msec before to 50 msec after initiation of the eye movement. On average, neural activity precedes the initiation of the slow eye movement by 19 msec. Eye velocity was positively correlated with discharge rate. Neurons respond differentially for smooth eye movements in different directions, but exhibit one best direction. The direction of slow eye movement associated with the most activity coincided with the direction of eye movement evoked by microstimulation of that site. Curiously, though, the distribution of preferred directions measured to date is uniform, and notably not biased toward ipsilateral directions.

3.8. FEF Function in Human

Studying the effects of ablation of FEF and surrounding cortex in humans on gaze behavior has produced the general view that FEF is responsible for preventing undesirable gaze shifts and for generating purposive saccades (e.g., Holmes, 1938; Sharpe, 1986; Henik *et al.*, 1994; reviewed by Pierrot-Deseilligny *et al.*, 1995). The results from experiments with nonhuman primates are generally consistent with this view (Deng *et al.*, 1986; Schiller *et al.*, 1987; Keating, 1991).

However, the interpretation of many studies relating frontal cortex damage to gaze impairments is complicated by variations in the size of the damaged zone. For example, an influential study by Guitton *et al.* (1985) demonstrated that patients with frontal lobe ablations were impaired in producing antisaccades, that is, eye movements made in the direction opposite a visual target. However, the patients in this study who showed the greatest deficit had dorsomedial cortex damage that may have included SEF and more extensive prefrontal damage (see also Pierrot-Deseilligny *et al.*, 1991a). In one recent study patients were identified successfully who had relatively circumscribed lesions centered on the precentral sulcus at the level of FEF (Rivaud *et al.*, 1994). The contralesional saccades made by these patients were hypometric and had higher latencies (except in the gap condition). In contrast to the conclusions of the earlier studies, the patients with lesions restricted to FEF did not make an unusually high number of errors in the antisaccade task, although they did fail to make predictive saccades. Memory-guided saccades are also impaired by FEF lesions (Pierrot-Deseilligny *et al.*, 1991b, 1993).

Neuroimaging studies offer a new approach to investigate whether FEF is activated in association with saccades made under different circumstances. Initial PET studies (Melamed and Larsen, 1979; Fox *et al.*, 1985; Petit *et al.*, 1993) measured increased blood flow in FEF in subjects performing self-generated saccadic eye movements with or without visual feedback. Subsequent studies have explored the magnitude of activation associated with saccades made under more challenging conditions, e.g., saccades to visual targets based on arbitrary instructions, antisaccades, saccades to the remembered location of a flashed stimulus, and sequences of remembered saccades (Paus *et al.*, 1993; Andersen *et al.*, 1994; O'Sullivan *et al.*, 1995; O'Driscoll *et al.*, 1995; Sweeney *et al.*, 1995; Petit *et al.*, 1996). The results of these various studies are not in full agreement. Whereas O'Driscoll *et al.* (1995) and Sweeney *et al.* (1995) reported a relative increase in blood flow through FEF associated with antisaccades as compared to prosaccades, Paus *et al.* (1993) did not. The basis for this difference may be in the design of the task or the training of the subjects. O'Driscoll *et al.* (1995) began trials with fixation times of 800, 1000, or 1200 msec; the fixation spot went off simultaneously with target (1° square) presentation for 100 msec, 15° left or right. Prosaccade latencies averaged 225 msec, and antisaccade latencies averaged 242 msec. Sweeney *et al.* (1995) began trials with a fixation time of 2000–3000 msec followed by target (0.45° spot) presentation for 2000–3000 msec at one of four random target positions (4.8° and 9.6° left or right). No saccade latencies were reported. The conditions used by Paus *et al.* (1993) were somewhat different. They began trials with a shorter 300-msec fixation time, followed by a 200-msec gap before the target (0.3° square) was presented for 200 msec, 5° left or right of the fixation spot. Prosaccade latencies averaged 6 msec; this remarkably low value was due to use of the gap interval and the fact that subjects could predict target location. Antisaccade latencies averaged a little over 200 msec in part because target location was randomized. The lack of blood flow in FEF associated with antisaccades reported by Paus *et al.* (1993) may therefore be a result of the different nature of performance called for by their task.

Another significant difference is that Sweeney *et al.* (1995) reported a relative increase in FEF blood flow associated with memory-guided as compared to visually guided saccades, but Anderson *et al.* (1994) did not. Petit *et al.* (1996) also did not report differences in blood flow through human FEF associated with voluntary saccades made under different conditions. Once again the conditions of the task are the probable basis for the discrepancy. Sweeney *et al.* (1995) flashed the target for 100 msec followed by a delay period ranging from 1500 to 2500 msec before the saccade could be initiated, but Anderson *et al.* (1994) flashed the target for 200 msec followed by a delay period ranging from 400 to 600 msec. It may be that the short delay period used by Andersen *et al.* did not permit or require a sufficient increase in blood flow through FEF. We should note that discrepancies between studies at this very early stage of research should not necessarily challenge the validity of any single study. The fact that the various studies employed slight variations in task design and obtained apparently different results only means that a larger parameter space is being represented in the literature.

One of the advantages of PET neuroimaging is the ability to observe coordinated modulation of blood flow throughout the entire brain. This permits analyses of how activation in different brain regions may be linked. A recent study has made the interesting observation that blood flow in FEF increases as a function of the frequency of saccades made by subjects in absolute darkness (Paus *et al.,* 1995). Moreover, the change in FEF blood flow was paralleled by decreased blood flow in areas of visual cortex including V1. The fact that the eye movements were made in complete darkness means that the modulation of blood flow was of extraretinal origin. The evidence is interpreted as indicative of mechanisms that may be responsible for saccadic suppression, i.e., the reduction of vision associated with saccadic eye movements (e.g., Volkman *et al.,* 1968, 1978). Also, in an analysis of blood flow changes associated with memory-guided saccades relative to visually guided saccades, Sweeney *et al.* (1995) noted significant positive correlations between blood flow changes in FEF and those in dorsolateral prefrontal cortex, SEF, and posterior parietal cortex. Significant negative correlations were observed between blood flow in FEF and blood flow in anterior cingulate cortex and medial orbital cortex.

4. Supplementary Eye Field

The eye field defined in Ferrier's original map of frontal cortex extended medially to the dorsal surface of the brain (Fig. 7A). Based on the experimental efforts of the last century reviewed above, the frontal eye field proper shrunk into the rostral back of the arcuate sulcus. However, several lines of evidence indicated that the dorsomedial part of the frontal lobe does play some role in the generation of eye movements. For example, in his pioneering electrical stimulation studies in humans Penfield noted gaze shifts evoked by stimulation of the rostral part of the supplementary motor area (SMA) (Penfield and Welch, 1949, 1951). This finding has been confirmed recently using subdural electrode arrays to stimulate (Fried *et al.,* 1991; Lim *et al.,* 1994) and record in humans (Ikeda *et al.,* 1992). An early neurophysiological recording study in behaving monkeys also noted neurons in rostral SMA that were active in relation to gaze shifts (Brinkman and Porter, 1979). Cerebral blood flow studies during saccades also distinguished two distinct loci of elevated cerebral blood flow (Melamed and Larsen, 1979; Orgogozo and Larsen, 1979; Fox *et al.,* 1985). The lateral locus corresponds to the FEF, and the dorsomedial locus corresponds to the rostral end of SMA. In an electrical stimulation mapping study Gould *et al.* (1986) demonstrated in anesthetized owl monkeys that conjugate eye movements were elicited by intracortical microstimulation of rostral SMA. Ultimately, Schlag and Schlag-Rey (1985, 1987) defined the supplementary eye field (SEF) as the region in which low currents ($<50\ \mu A$) evoked saccades, and unit activity was related to saccadic eye movements (Fig. 7K). Based on unit recording and microstimulation it now seems clear that SEF is contiguous caudally with the part of SMA cortex that represents orofacial, pinna, and forelimb movements (Schlag and Schlag-Rey, 1987; Schall, 1991a; Parthasarathy *et al.,* 1992). SEF is located in area 6aβ of Vogt and Vogt (1919), area 6DR of Barbas and Pandya (1987), and corresponds to area F7 of Matelli *et al.* (1991). Figure 8 shows the location of the SEF in the human brain.

We currently do not understand what functions are subserved by SEF relative to those mediated by FEF. However, several studies have demonstrated particular differences between these two eye fields in the frontal cortex. Recent reviews on SEF organization and function have appeared (Schlag and Schlag-Rey, 1992; Tehovnik, 1995).

4.1. Supplementary Motor Area

At this relatively early stage of investigation it seems useful to be guided in our thinking about SEF by information we have about the supplementary motor area (SMA) (reviewed by Goldberg, 1985; Tanji, 1994). Single-unit recordings from SMA on the medial surface of the hemispheres have identified a number of different functional cell classes. Many neurons in SMA begin to discharge following the presentation of visual, auditory, and tactile stimuli (Brinkman and Porter, 1979; Wise and Tanji, 1981; Tanji and Kurata, 1982; Okano and Tanji, 1987; Romo and Schultz, 1987; Hummelsheim *et al.,* 1988; Alexander and Crutcher 1990; Chen *et al.,* 1991; Schall, 1991a; Okano, 1992). Some of these neurons discharge transiently at the presentation of the stimulus, but others continue to discharge until another signal triggers the movement. The tonic activation has been interpreted as contributing to the preparation of the movement (referred to as preparatory set). Subsequent work has subdivided SMA into caudal (area F3) and rostral (area F6) areas. Besides the physiological differences that will be described, these two regions have different anatomical relations. Area F3 is more closely related to motor cortex, and area F6 is more closely related to prefrontal cortex (Luppino *et al.*, 1993; Bates and Goldman-Rakic, 1993). Recent neuroimaging studies in human have distinguished other functional differences between rostral and caudal SMA (Matelli *et al.*, 1993; Stephan *et al.*, 1995). Visually responsive cells are found more commonly in area F6 (also referred to as pre-SMA) than in area F3 (SMA proper) (Matsuzaka *et al.*, 1992; Halsband *et al.*, 1994). Pre-SMA also exhibits more neural activity during preparatory periods and less movement-related activity. Many of these neurons discharge in response to stimuli according to their instructional significance for guiding movements. For example, certain neurons respond to a sensory stimulus only if it signals the monkey to make (or to withhold) a movement (Sakai, 1978; Tanji *et al.*, 1980; Kurata and Tanji, 1985; Tanji and Kurata, 1985; Romo and Schultz, 1992). If the same stimulus signals withholding (or making) the movement, SMA cells do not respond. Thus, the sensory responses of SMA cells represent a stage of processing that evaluates stimuli according to their relevance for movement production.

Other neurons in SMA discharge in relation to orofacial or limb movements (e.g., Okano and Tanji, 1987; Tanji and Kurata, 1979, 1981, 1982; Brinkman and Porter, 1979; Romo and Schultz, 1987, 1992; Tanji *et al.*, 1988; Matsuzaka *et al.*, 1992). Movement-related neural activity is found predominantly in the caudal SMA proper (Matsuzaka *et al.*, 1992). Neural activity in pre-SMA has been observed in relation to more coordinated reaching and grasping movements (Rizzolatti *et al.*, 1990). The activity of SMA cells is related to movement production in complex contingencies. For example, in monkeys trained to make sequential movements, certain cells in SMA discharge specifically in relation to a particular movement only when it occurs at a particular point in the sequence (Mushiake *et al.*, 1991). Other SMA neurons fire specifically before a particular movement sequence will be performed (Tanji and Shima, 1994). Pre-SMA neurons, in contrast, tend to be activated in relation to the stimuli instructing the sequence of movements (Halsband *et al.*, 1994).

4.2. Connectivity of SEF

The connectivity of SEF is distinct from that of the skeletal motor cortex surrounding it, and SEF shares many but not all efferent targets and afferent sources with FEF (Huerta and Kaas, 1990; Luppino *et al.*, 1990; Shook *et al.*, 1990, 1991).

4.2.1. Subcortical

Figure 25 compares the striatal and thalamic connectivity of SEF with that of FEF. The figure is intended to summarize the densest connections observed consistently across recent studies. The rostral part of SMA including SEF projects to a central longitudinal zone in the caudate nucleus (Selemon and Goldman-Rakic, 1985; Arikuni and Kubota, 1986; Shook *et al.*, 1991). The SEF efferents to striatum largely overlap those from FEF, except that the SEF efferents extend further rostrally than do those from FEF (Parthasarathy *et al.*, 1992). SEF is reciprocally connected with the medial part of the ventroanterior nucleus, part of nucleus X of the ventrolateral complex, the multiformis and parvicellular parts of the mediodorsal nucleus, and mainly the paracentral, centrolateral, and central superior lateral of the intralaminar thalamic nuclei (Huerta and Kaas, 1990; Shook *et al.*, 1991). In comparison to FEF, which entertains its heaviest thalamic connectivity with the lateral part of the mediodorsal nucleus, SEF is most heavily connected with the ventroanterior nucleus (Shook *et al.*, 1991). The medial zone of the ventroanterior nucleus and of area X that innervates SEF is distinct from the more lateral parts of these nuclei that are associated with the skeletal motor system. Overall it appears that SEF is interconnected with a spatially more distributed set of thalamic nuclei. Further work is needed, however, to determine whether the apparently broader thalamic connectivity of SEF as compared to FEF is actually the case or is an artifact of injections aimed at SEF that were actually placed in surrounding skeletomotor cortex.

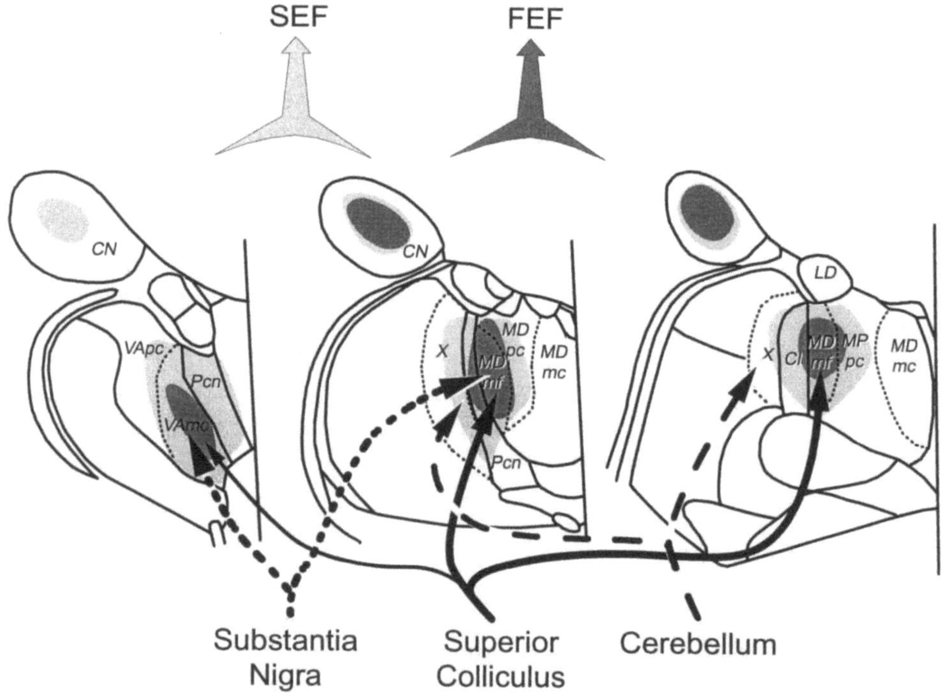

Figure 25. Thalamic innervation of FEF and SEF. The lighter shading indicates connections with SEF, and the darker shading, connections with FEF. The general organization of thalamic afferents from substantia nigra pars reticulata, superior colliculus, and cerebellar nuclei are shown by arrows. The sections are arranged rostral to caudal, left to right. CI, central lateral nucleus; CN, caudate nucleus; LD, lateral dorsal nucleus; MDmc, medial dorsal nucleus magnocellular; MDmf, medial dorsal nucleus multiformis; MDpc, medial dorsal nucleus parvicellular; Pcn, paracentral nucleus; VAmc, ventral anterior nucleus magnocellular; VApc, ventral anterior nucleus parvicellular; X, nucleus X.

The thalamic nuclei primarily connected with SEF are themselves innervated by other subcortical visuomotor structures (Fig. 25). The medial part of the ventroanterior nucleus is innervated by the visuomotor zone of the substantia nigra pars reticulata (Ilinksy *et al.*, 1985) and the intermediate layers of the superior colliculus (Harting *et al.*, 1980). The medial part of area X is innervated by the face representation of the deep cerebellar nuclei (Asanuma *et al.*, 1983). The lateral segments of the mediodorsal nucleus also receive afferents from the superior colliculus and substantia nigra pars reticulata (Harting *et al.*, 1980; Ilinsky *et al.*, 1985).

The superficial as well as the intermediate layers of the superior colliculus receive afferents from SEF (Leichnetz *et al.*, 1981; Fries, 1984; Huerta and Kaas, 1990; Shook *et al.*, 1990). The terminals from SEF appear to be more widespread than are those from FEF. Bilateral innervation of the superior colliculus from SEF has been observed by Shook *et al.* (1990), but not by Huerta and Kaas (1990). Projections have also been reported from SEF to specific brainstem oculomotor regions such as the nucleus raphe interpositus, the nucleus reticularis tegmenti pontis, the nucleus prepositus hypoglossi, the mesencephalic reticular formation, the interstitial nucleus of Cajal, and the pontine reticular formation as well as the dorsomedial pontine nuclei (Huerta and Kaas, 1990; Shook *et al.*, 1990). Not every terminal zone is found in each study; but such differences may be attributed to the precise location of the tracer injections. In general, though, SEF innervates the same oculomotor structures that FEF does, although there appear to be quantitative differences in the density and topography of the projections.

4.2.2. Intracortical

SEF seems to be innervated by fewer cortical areas than is FEF (Huerta and Kaas, 1990) (Fig. 26). In contrast to the extensive array of prestriate visual areas connected with FEF, SEF receives afferents only from area MST, the superior temporal polysensory area, and LIP (Huerta and Kaas, 1990). The prestriate visual areas projecting to SEF also project to the dorsomedial, but not the ventrolateral FEF (Schall *et al.*, 1995b). The visual areas associated with the ventral visual processing stream do not project to SEF (Huerta and Kaas, 1990; Schall *et al.*, 1995b). The limited visual input to SEF as compared to FEF is consistent with the different respective roles of SEF and FEF.

SEF is reciprocally connected with anterior and posterior cingulate cortex (Huerta and Kaas, 1990). SEF is connected with the postarcuate premotor areas as well as SMA. The connections with premotor cortex and SMA may be an artifact of tracer spread or, if not, may mediate links of SEF function with the skeletal motor system. SEF is reciprocally connected with prefrontal cortex in areas 12 and 46 (Huerta and Kaas, 1990).

4.2.3. Connectivity between FEF and SEF

As reviewed above, saccade amplitude is represented in a roughly mediolaterally oriented map within FEF. If saccade amplitude is mapped in the same way in SEF, then one ought to observe systematically displaced loci of neurons retrogradely labeled by injections of different tracers into different parts of FEF.

The pattern of connectivity between SEF and FEF is shown by the distribution of retrogradely labeled cells in SEF following an injection of two different tracers into FEF (Fig. 27) (Schall *et al.*, 1993; Stanton *et al.*, 1993). The injection into dorsomedial FEF labeled cells throughout the region defined by cytoarchitecture as area F7. In contrast, the injection into ventrolateral FEF only labeled cells in a more restricted portion of SEF that was always medial and often caudal. Thus, a limited sector within SEF projects to the short-saccade and the long-saccade representation in FEF. But the remainder of SEF appears to project preferentially to the longer saccade representation in FEF.

The literature indicates that the pattern of connectivity between most visual cortical areas conforms more or less to a point-to-point mapping between two retinotopic representations. The organization of the connectivity from SEF to FEF is qualitatively different from what has been reported for visual cortical areas. The input to FEF from SEF does not conform to the uniform, one-to-one mapping one might expect between two cortical areas with common representations of the visual field and saccade vectors. This observation indicates that saccade vector may be represented differently in SEF as compared to the mapping observed in FEF. Data from a number of electrical stimulation and neural recording studies also lead to this conclusion. If this conclusion is valid, then the anatomical relation between FEF and SEF would represent a first demonstration of an organized pattern of connectivity between cortical fields instantiating different representations of visual or motor space.

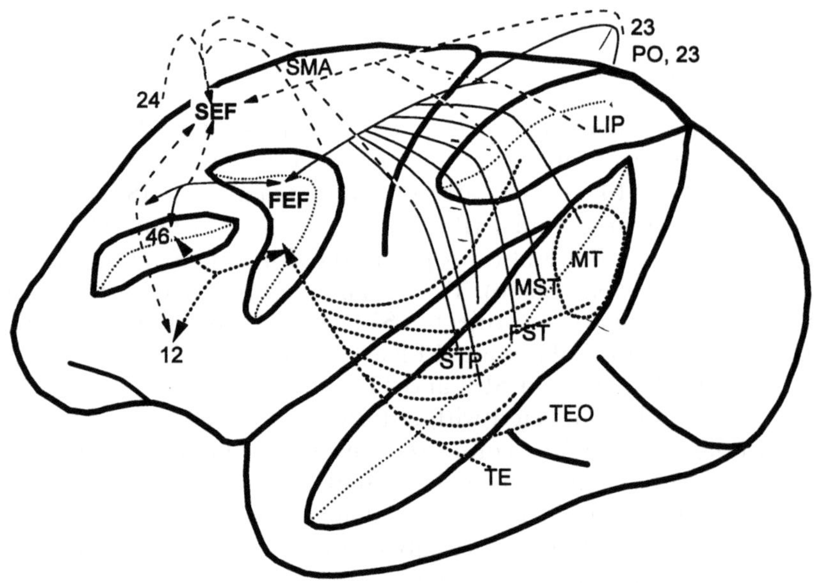

Figure 26. Summary of cortical connections of FEF and SEF. The dotted lines indicated connections of ventrolateral FEF. Solid lines show afferents of dorsomedial FEF. Dashed lines show connections of SEF. Reciprocal connectivity between FEF and SEF is not illustrated. FST, Fundus of superior temporal sulcus area; LIP, lateral intraparietal area; MST, medial superior temporal area; MT, middle temporal area; PO, parietooccipital area; STP, superior temporal polysensory area; TE, area TE; TEO, area TEO; SMA, supplementary motor area; 12, area 12; 23, cingulate cortex area 23; 24, cingulate cortex area 24; 46, area 46.

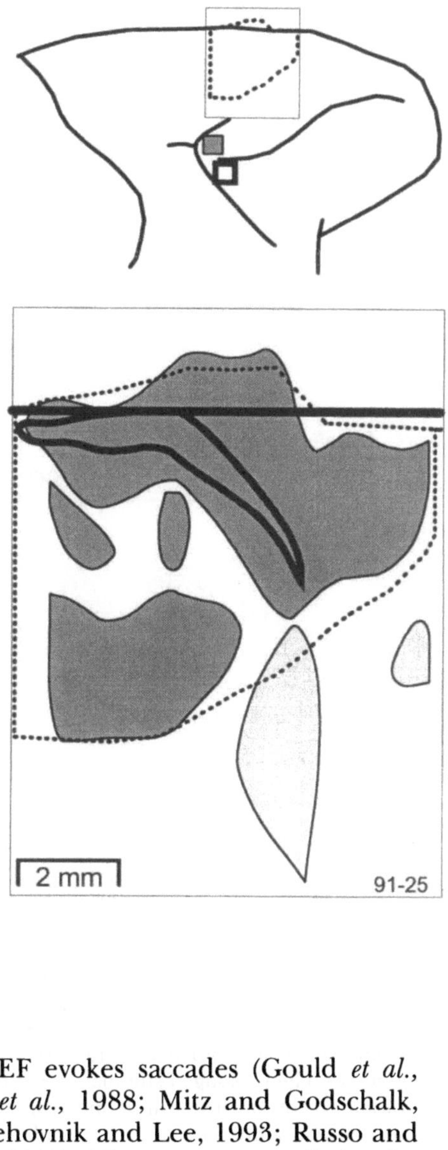

Figure 27. Spatial distributions of neurons in SEF projecting to medial and lateral FEF. The upper diagram shows a dorsolateral view of the frontal lobe. The relative locations of the different tracer injections as well as of the cytoarchitectural borders of SEF are indicated. Distributions of labeled neurons in SEF are portrayed in a flattened reconstruction of dorsomedial cortex. Shaded areas encompass cells projecting to medial FEF; the lighter shading signifies a less dense projection. The area enclosed by the thick line encompasses the region projecting to ventrolateral FEF. The midline convexity is indicated by the thick line at the top of the reconstruction. Adapted from Schall *et al.* (1993) with permission.

4.2.4. Effects of Electrical Stimulation

Intracortical microstimulation of the SEF evokes saccades (Gould *et al.*, 1986; Schlag and Schlag-Rey, 1987; Mann *et al.*, 1988; Mitz and Godschalk, 1989; Luppino *et al.*, 1991; Schall, 1991b; Tehovnik and Lee, 1993; Russo and Bruce, 1993; Tehovnik *et al.*, 1994; Fujii *et al.*, 1995). The parameters of stimulation necessary to evoke saccades from SEF are comparable to those that are effective in FEF with thresholds less than 50 μA (and as low as 3 μA), and latencies of 30–50 msec. One of the most notable findings made about SEF in several laboratories is that the vector of the saccades evoked by stimulation of a given site in SEF seems to vary with the initial angle of gaze. Stimulation of the superior colliculus or of FEF evokes saccades with the same direction and amplitude regardless of the initial position of the eye in the orbit (referred to as *fixed vector* saccades). In contrast, stimulation of many sites in SEF tends to bring the eye to particular positions in the orbit (referred to as *convergent* saccades) (Schlag and Schlag-Rey, 1987; Mann *et al.*, 1988; Mitz and Godschalk, 1989; Luppino *et al.*, 1991; Schall, 1991b; Tehovnik and Lee, 1993). This effect is shown in Fig. 28.

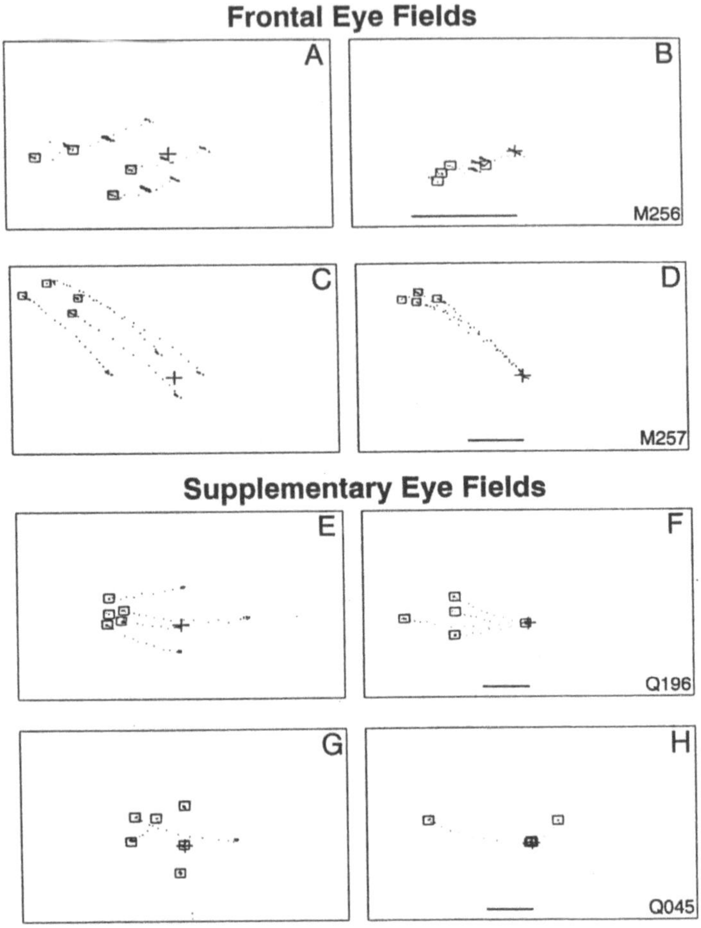

Figure 28. Eye movements elicited by intracortical microstimulation of (A–D) FEF and (E–H) SEF (A, C, E, G) absolute eye position; (B, D, F, H) eye position relative to a common starting point. The cross represents the central fixation spot. The horizontal bar in panels B, D, F, and H represents 10°. A square is drawn around the final eye position. The initial positions correspond to the locations of the one central and four peripheral stimuli. (A, B) Saccades evoked from one site in FEF (40 μA for 500 msec). The prolonged stimulation elicited a sequence of two leftward saccades from each initial position except for the leftmost one, from which a single saccade resulted. As shown in panel B, the direction and amplitude of each saccade were essentially the same. (C, D) Saccades elicited from stimulation of another site in FEF (40 μA for 500 msec). Multiple saccades were not evoked, presumably because the saccades were of such large amplitude. While there is a degree of curvature in the trajectories, panel D shows that the saccades were essentially all of the same amplitude and direction. (E, F) Saccades evoked from a site in SEF (90 μA for 1000 msec). The prolonged stimulation elicited a single saccade from each initial position except for the leftward one. The direction and amplitude of the saccades varied in such a manner that the eye tended to be moved to one general orbital position. As shown in panel F, the amplitude and direction of the saccades evoked from different initial positions were not the same. (G, H) Saccades evoked from another site in SEF (20 μA for 800 msec). The prolonged stimulation elicited a leftward saccade from the right initial position. In contrast, no saccades were evoked when the eye was at the top, central, or bottom initial positions, as indicated by the square around a constant fixation eye position in panel G and by the number of squares around the central point in panel H. Two trials are illustrated with the eye at the left initial position. In the typical trial illustrated, no saccade was elicited. However, in a single trial a reversal, right-upward saccade was evoked from the left initial position. From Schall (1991b) with permission.

The pattern of eye movements evoked by long stimulus trains also distinguishes SEF from FEF. Prolonged stimulation of FEF as well as of superior colliculus elicits successive saccades all of the same direction and amplitude (Robinson, 1972; Schiller and Stryker, 1972; Schiller *et al.*, 1979). In contrast to this, prolonged stimulation of the SEF does not result in multiple saccades, but rather the eye is moved to and then stays at a specific position (Schall, 1991b). If SEF is stimulated when gaze is already in the "termination zone," then the eyes remain fixed for the duration of the stimulation train. Inhibiting movements by SEF stimulation is consistent with observations that SMA stimulation can interrupt or prevent movements (Penfield and Welch, 1951; Fried *et al.*, 1991). Tehovnik *et al.* (1994) have found that this saccade inhibition evoked by SEF stimulation survives unilateral ablation of FEF, but not of the superior colliculus.

Recall that saccade amplitude is mapped mediolaterally in FEF. Recent work has provided evidence for a different map organization in SEF (Tehovnik and Lee, 1993; Lee and Tehovnik, 1995). The nature of the map is illustrated in Fig. 29. Stimulation of rostral SEF evokes saccades that bring gaze to a location in the peripheral contralateral hemifield. Stimulation of a more caudal site in SEF evokes saccades only when gaze is directed into the ipsilateral hemifield, as if the termination zone is less eccentric in the contralateral hemifield. Stimulation at more caudal sites evokes saccades that bring gaze toward the central resting gaze position. Stimulating the most caudal sites, Lee and Tehovnik report evoking ipsiversive saccades when gaze was directed in the contralateral hemifield (see also Fig. 28G, H). The anatomical substrate for these ipsiversive saccades may be the projection from SEF to the contralateral superior colliculus (Shook *et al.*, 1990). Lee and Tehovnik also report that stimulation of medial SEF brings gaze to a termination zone in the lower visual field, and stimulation of lateral FEF brings gaze to a termination zone in the upper visual field. A conclusive interpretation of these mapping data is premature, though, because these investigators did not perform a direct anatomical reconstruction of the stimulation sites.

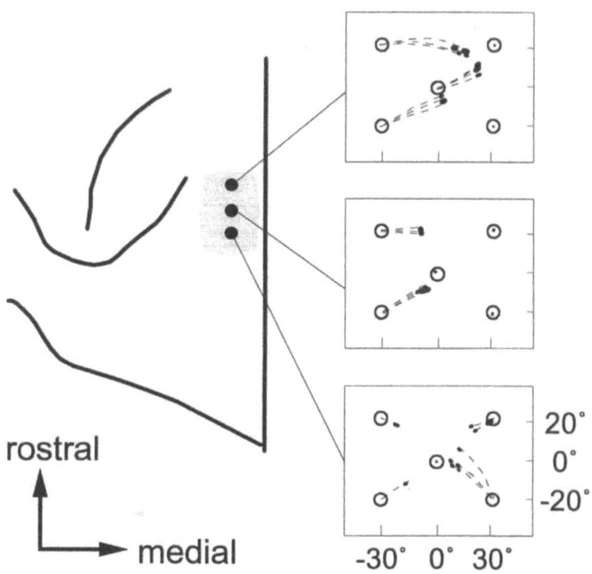

Figure 29. Gaze map in SEF. A dorsal view of left frontal cortex is shown with SEF indicated by the shaded region; three sites of microstimulation are shown. The eye movements evoked by stimulation of each of the three sites are shown in panels to the right. The open circles indicate the five initial gaze angles. Stimulation of the most rostral site evoked contraversive saccades except when the gaze was already directed into the contralateral visual field. Stimulation at the middle site evoked shorter saccades only when the gaze was in the ipsilateral visual field. Stimulation of the caudal site evoked short contraversive saccades when the gaze was in the ipsilateral visual field and ipsiversive saccades when the gaze was in the contralateral visual field. Redrawn from Lee and Tehovnik (1995).

The pattern of anatomical projections from SEF to FEF described above provides further information about the nature of the saccade representation in the two areas. If both FEF and SEF represented saccades retinotopically and had isomorphic maps, then the pattern of projection from SEF to FEF should exhibit a point-to-point topography with uniform convergence and divergence across the two areas. Instead, current data show that whereas virtually all of SEF projects to the medial FEF, which generates longer amplitude saccades, only a limited, medial part of SEF projects to the lateral FEF, which generates shorter amplitude saccades. This pattern of connectivity seems to imply either that SEF does not represent saccades retinotopically or that the mapping of saccade amplitude in SEF is markedly different from that in FEF. Additional evidence that the mapping of saccade vector is different in SEF as compared to FEF is the topographically broad projection from SEF to the intermediate layers of the superior colliculus (see above).

Russo and Bruce (1993) reexamined the question of distinct saccade representations in SEF as opposed to FEF. They quantitatively compared the vectors of saccades evoked by microstimulation of FEF and SEF as a function of different initial eye positions. These investigators observed fixed vector and convergent saccades evoked from sites in both FEF and SEF. However, in their data, sites identified as fixed vector far outnumbered sites from which convergent saccades were evoked in both FEF and SEF. Russo and Bruce did find, however, a significant correlation between saccade amplitude and the degree of convergence. Sites evoking larger amplitude saccades tended to exhibit more convergent saccades. Based on this finding, they suggest that the convergence represents an abnormal lack of compensation by saccade-related circuits in the cerebellum because of the unnatural nature of electrical stimulation of the brain. Overall, Russo and Bruce could find no significant difference in the nature of the saccades evoked by stimulation of FEF or SEF.

What is the basis for the discrepancy between Russo and Bruce (1993) on the one hand and, for instance, Tehovnik and Lee (1993) on the other? One possibility is that different parameters for the electrical stimulation were used by different investigators. Tehovnik and Lee (1993) used 0.1-msec pulses at 150 Hz and 400 μA for 400-msec pulse trains. These parameters are different from typical values used in other studies. Most other investigators have used 0.2-msec pulse durations, at 250–350 Hz, with <100 μA for 10- to 70-msec train durations to evoke saccades from SEF (Schlag and Schlag-Rey, 1987; Mitz and Godschalk, 1989; Schall, 1991b; Russo and Bruce, 1993). Perhaps the high currents and long train duration used by Tehovnik and Lee resulted in a higher incidence of convergent saccades. However, other investigators have observed convergent saccades in SEF using more standard, low currents and shorter stimulus trains (Schlag and Schlag-Rey, 1987; Mitz and Godschalk, 1989; Schall, 1991b). Another peculiarity of Tehovnik and Lee's (1993) data is the very long latencies of the evoked saccades they report. Whereas most saccades evoked by microstimulation of the superior colliculus or FEF have latencies of less than 50 msec, more than half of the data reported by Tehovnik and Lee included saccades evoked with latencies exceeding 50 msec, with some ranging to over 250 msec. Tehovnik and Lee did note that the latency of evoked saccades from SEF increased the closer the initial gaze was to the "termination zone."

Other questions are prompted by the discrepancy under consideration. Is it possible that Russo and Bruce did not observe convergent saccades because they were stimulating a different location from that explored by other investigators? Comparing their maps to those of other investigations of SEF, this does not seem likely. Tehovnik and Lee (1993), on the other hand, appear to be obtaining eye movements from stimulation of sites over a much wider area of cortex than other investigators. They emphasize this difference by referring to the area of cortex in their investigation as *dorsomedial frontal cortex*. Specifically, Tehovnik and Lee report evoking saccades from sites caudal to the area that most investigators refer to as SEF. To evoke saccades from the caudal sites, Tehovnik and Lee note that longer stimulation durations and eccentric initial gaze must be used. The caudal sites of Tehovnik and Lee (1993) probably coincide with the cortex in SMA (area F2) associated with forelimb movements.* Was the statistical analysis of Russo and Bruce not sensitive enough to identify convergent saccade sites? One could certainly imagine alternative tests, but it seems clear that the analysis used by Russo and Bruce did identify convergent saccades when they were observed. Did Russo and Bruce sample a higher proportion of shorter amplitude saccades than other investigators? Whereas stimulation of approximately 90% of the sites in SEF and FEF studied by Russo and Bruce (1993) evoked saccades with amplitudes less than 20°, over 90% of the sites stimulated by Tehovnik and Lee (1993) evoked saccades with amplitudes over 20°. It appears that Schlag and Schlag-Rey (1987) also found many sites in SEF evoking saccades longer than 20°. If the degree of convergence of evoked saccades increases with saccade amplitude, then perhaps the real issue is why Tehovnik and Lee as compared to Russo and Bruce encountered so many sites representing very long saccades. Thus, the apparent discrepancy in the literature on this point may represent a matter of emphasis in describing the same phenomenon. Further work and other kinds of data are therefore needed to resolve whether SEF has a robust or an incidental eye position representation including convergent saccades with a systematic map of endpoints. This issue is important to resolve in order to decide whether it is legitimate to conclude that SEF has a craniotopic rather than a retinotopic saccade representation.

*Tehovnik and Lee (1993) have not reported histological reconstructions of the stimulation and recording sites in dorsomedial frontal cortex in their monkeys. One of the animals they used, however, has also been investigated in two other studies. Monkey Q reported by Tehovnik and Lee (1993) (see their Fig. 2) was also mapped by Schall (1991a) (see his Fig. 24). It appears that the caudal sites in Tehovnik and Lee's map of this monkey correspond to sites at which Schall (1991a) reported finding most often cells related to oral and forelimb movements as well as infrequent cells related to eye movements. Monkey Q was subsequently used in a study of the projections from SEF to striatum (designated M7 by Parthasarathy *et al.*, 1992). Tracer injections concentrated in the SEF cortex from which only eye movements were evoked in other monkeys in that study resulted in terminal label concentrated in the caudate nucleus. In contrast, the terminal label from the injection made throughout the dorsomedial frontal cortex defined by Tehovnik and Lee in monkey Q resulted in dense terminal label in the putamen as well as caudate nucleus. This indicates that the cortex caudal to the SEF proper is linked to skeletomotor function. Finally, we have recently examined the cytoarchitecture of the medial frontal gyrus in sections from monkey Q (unpublished observations). The cortex conforming to the criteria of area F7 was located in the rostral region in which eye-movement-related activity was concentrated in Schall (1991a) and contraversive saccades were evoked by Tehovnik and Lee (1993). The cortex in which forelimb-movement-related cell activity was recorded by Schall (1991a) and the least eccentric termination zones were found by Tehovnik and Lee (1993) conformed to the description of area F2.

Another line of evidence for an eye position signal in SEF is the finding that neurons with activity modulated according to the position of the eyes in the orbit have been identified (Schall, 1991a; Bon and Lucchetti, 1992; Schlag *et al.*, 1992; Lee and Tehovnik 1995; but see Russo and Bruce, 1996). These SEF units commonly discharge before saccades directed to a particular range of endpoints; they may also discharge during tracking eye movements having the same endpoints and also as well during attentive fixation in the appropriate direction. This set of properties is reminiscent of, but somewhat distinct from, the fixation and tracking units of posterior parietal cortex (Lynch *et al.*, 1977; Robinson *et al.*, 1978; Sakata *et al.*, 1980, 1983; Komatsu and Wurtz, 1988; Erickson and Dow, 1989), prefrontal fixation units (Suzuki and Azuma, 1977), and the eye position units originally described in FEF (Bizzi, 1968; Bizzi and Schiller, 1971; Bruce and Goldberg, 1985). Supporting correlated anatomical evidence is that SEF but not FEF receives input from the central superior lateral thalamic nucleus (Huerta and Kaas, 1990). Schlag-Rey and Schlag (1984) reported a high incidence of eye position cells in this nucleus. Finally, a comparative study of individuals with lesions of SEF, FEF, and dorsolateral prefrontal cortex found that patients with SEF damage were impaired in generating a memory-guided saccade that compensated for a whole-body (with fixed head) rotation (Pierrot-Deseilligny *et al.*, 1993). This finding seems to indicate that SEF utilizes an extraretinal, vestibular signal in programming saccades, consistent with other evidence for a craniotopic representation in SEF.

4.3. Neuronal Properties

4.3.1. Saccade-Related Activity

A variety of cell types and functional properties have been observed in single-unit recording studies of SEF (Schlag and Schlag-Rey, 1987; Mann *et al.*, 1988; Schall, 1991a; Bon and Lucchetti, 1991, 1992; Hanes *et al.*, 1995; Chen and Wise, 1995a; Lee and Tehovnik, 1995; Schlag *et al.*, 1992; Olson and Gettner, 1995; Heinen, 1995; Russo and Bruce, 1996). In summary, some neurons have been recorded in SEF that discharge in response to visual stimuli and/or before and during saccadic eye movements, while other neurons discharge during attentive fixation, both during smooth pursuit eye movements and during instructed delay periods.

We have seen that the sensory responses of many cells in frontal cortex are contingent on the behavioral state of the monkey. Evidence from SEF also indicates that the same contingency can be observed in movement-related neural activity. The original reports by Schlag and Schlag-Rey (1985, 1987) described SEF cells as discharging in relation to spontaneous saccades. Subsequently, Schall (1991a) reported that SEF neurons fired in relation to saccades that were the operant response of a behavioral paradigm, but did not fire in relation to spontaneous saccades made outside the task. Figure 30 illustrates modulation of saccade-related activity of an SEF neuron by behavioral state. Neurons in SEF discharge in relation to saccades that are goal-directed to receive a reward, but they fire less, if at all, in relation to saccades that are self-generated, made spontaneously, and are unrewarded. This resembles findings for cells in other areas, including the substantia nigra pars reticulata (Hikosaka and Wurtz, 1983a), the caudate nucleus (Hikosaka *et al.*, 1989a), and FEF (Bruce and Goldberg, 1985; Schall 1991b).

The apparent discrepancy between the observations of Schlag and Schlag-Rey (1987) and those of Schall (1991a) most likely has to do with specific differences in the behavioral states of the monkeys related to the task design in the two studies. Schall trained monkeys to make saccades to stimuli in a structured trial

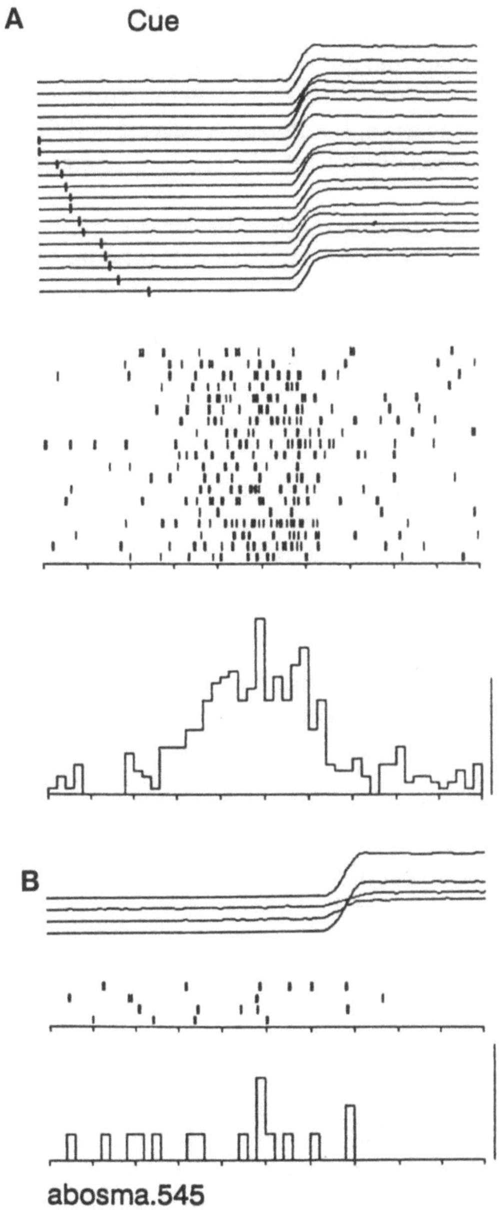

Figure 30. Comparison of presaccadic eye movement activity associated with goal-directed (A) and spontaneous, self-generated (B) saccades. The timescale is 50 msec and the histogram scale is 100 Hz. The activity is significantly modulated in relation to goal-directed, rewarded but not spontaneous, unrewarded saccades. From Schall (1991a) with permission.

with clear intertrial intervals in which no juice was ever delivered. In contrast, Schlag and Schlag-Rey rewarded monkeys on a variable schedule for making saccades to stimuli presented at unpredictable intervals and locations. The drive state of monkeys in the Schlag and Schlag-Rey study would accordingly be different from that of monkeys in the Schall study (e.g., Bolles, 1975). By delivering rewards on an unpredictable schedule, Schlag and Schlag-Rey encouraged a sustained state of motivation. In contrast, Schall's (1991a) monkeys never received a reward during the intertrial interval or when the task was not running, so their drive state would be higher during the trial and attenuated or absent otherwise.

Further work is required to investigate the mechanisms underlying this potent modulation. Recent work has documented reward-related modulation of dopaminergic cells in the ventral tegmentum in a conditional motor learning task (Ljungberg *et al.*, 1992; Schultz *et al.*, 1993). Thus, the modulatory effects of motivational state on SEF cell activity may be based on catecholaminergic afferents (Lidow *et al.*, 1991; Gaspar *et al.*, 1992; Williams and Goldman-Rakic, 1993).

The distinction between "motivated" and "unmotivated" saccades, so to speak, is different from the distinction between saccades made in response to visual stimuli and those that are self-generated. SEF cells do discharge before purposive self-generated saccades made in the dark. Moreover, the timecourse of activity in SEF cells relative to visually triggered and self-generated saccades can vary. As shown in Fig. 31, SEF cells begin to fire as much as 500 msec before self-generated saccades. In contrast, the response of the same SEF cell begins more briskly following a short latent period after a visual target is presented. What triggered the onset of activation before the visually guided saccade is clear, but it is presently unclear what signal was responsible for generating the slow increment of activity before the self-generated saccade.

This observation at the single-cell level may correspond to the well-known *readiness potential* (e.g., Deecke *et al.*, 1969; Deecke and Kornhuber, 1978; reviewed by Hallett, 1994). The readiness potential (also referred to as bereitschaftspotential) is a negative-going electrical potential that is recorded over SMA and surrounding cortex before movements are made. Studies to determine whether the readiness potential is actually generated in SMA or in other cortical areas have led to different conclusions (e.g., Lang *et al.*, 1991; Ikeda *et al.*, 1992; Bötzel *et al.*, 1993). Before self-generated movements the readiness potential can begin as much as 1000 msec or more before movement initiation. This evidence for a specific modulation of brain state so long before a movement is made has led to a number of provocative experiments attempting to relate the timecourse of the readiness potential to the phenomenal state associated with movement generation (reviewed by Libet, 1985). For example, a number of studies have measured the readiness potential preceding movements that can be categorized as voluntary (being accompanied by conscious awareness) versus involuntary (occurring without the subjects later reporting awareness of the act). Readiness potentials are observed before unconscious, involuntary movements, but are of lower amplitude and different cortical distributions than are the readiness potentials observed before conscious, voluntary movements (reviewed by Libet, 1985; see also Keller and Heckhausen, 1990; Praamstra *et al.*, 1995). We should note that the readiness potential occurs before saccadic eye movements and appears to be no different in character from that observed before other movements (Kurtzberg and Vaughan, 1982; Moster and Goldberg, 1990; Thickbroom and Mastaglia, 1990; Evdokimidis *et al.,* 1992; Klostermann *et al.,* 1994).

Data like these have led to the hypothesis that SMA (and by inference SEF) is specialized by virtue of its inputs from cingulate cortex for the production of self-generated movements (Eccles, 1982; Goldberg, 1985). A corollary of this hypothesis is that the postarcuate premotor cortex (and by inference FEF) is specialized for generating movements guided and triggered by sensory stimuli. Four independent experiments have been performed to test this hypothesis in SMA and postarcuate premotor cortex (Okano and Tanji, 1987; Romo and Schultz, 1987; Kurata and Wise, 1988; Thaler *et al.,* 1988). Each study reported neurons in both areas that were active before both self-generated and externally triggered movements. However, there was a bias in the distribution of responses seen in the two areas that was consistent with what was predicted by the hypothesis. Thus, physiological data seem to be consistent with a weaker, nonexclusive version of the hypothesis.

Other physiological data indicate that SEF plays a role in the generation of eye movements that are not triggered by external events. For instance, Heinen (1995) has recorded from SEF neurons that are active in relation to the production of pursuit eye movements. When the stimulus that the monkey is tracking undergoes a predictable change of direction, the SEF neurons modulate their activity before the anticipated change in the direction of pursuit. Also, Schall (1991a) identified a subpopulation of the neurons in SEF that was referred to as preparatory set cells because their timecourse of activation resembled that of cells by the same name identified in postarcuate premotor cortex. In an in-

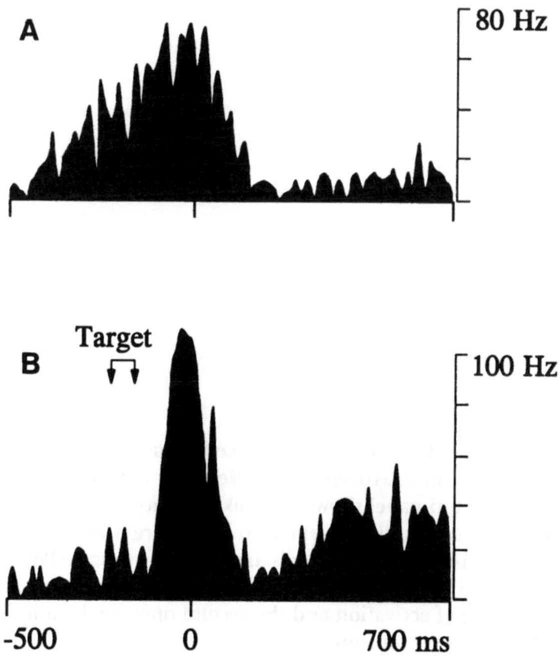

Figure 31. Comparison of presaccadic activity accompanying self-initiated (A) and visually evoked (B) saccades. The histograms are aligned on the time of saccade initiation. Visually guided and self-generated saccades of 10–20° amplitude in the optimum direction are compared. As compared to the activation associated with the visually guided saccade, the activation associated with the self-generated saccade grew over a longer time and reached a lower peak level. Redrawn from Schlag and Schlag-Rey (1987).

structured delay paradigm, these neurons discharge following the presentation of the target stimulus until a consistent interval before saccade initiation, which usually follows a signal triggering the movement (Fig. 32). This temporal pattern of modulation has been observed in primary motor cortex (Tanji and Evarts, 1976) and in prefrontal cortex (Kubota and Funahashi, 1982) as well as in other studies of SMA (Tanji *et al.*, 1980; Tanji and Kurata, 1985). Hanes *et al.* (1995) have shown, using a statistical spike train analysis, that the time of termination of activity of the preparatory set cells occurs consistently before saccade initiation. The activity ceased even if the saccade was self-initiated and began before the trigger signal was given.

Using a statistical analysis of the time of termination relative to the trigger signal and to saccade initiation, the time of termination was found to be more related to saccade initiation across the population of cells (Fig. 33). The cessation of discharge of preparatory set cells before saccadic eye movements suggests the possibility that these neurons may serve a permissive gating function within SEF,

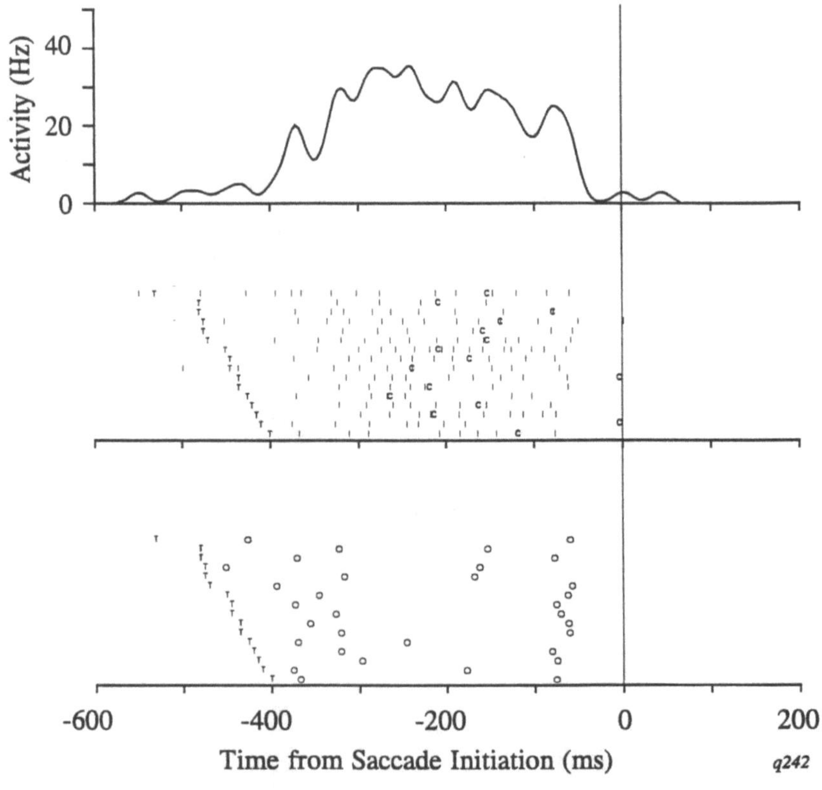

Figure 32. SEF preparatory set cell. This activity was collected during a saccade task to a single target with an instructed delay. Negative values on the abscissa indicate time before saccade initiation. The top panel shows the spike density function averaged across trials. In the lower two panels the T represents the presentation of the target and the C, the trigger signal. The middle panel shows a raster display in which trials are sorted by saccade latency. The bottom panel shows the results of a spike train analysis from each trial in the middle panel. The first open circle in each row indicates the beginning of activation and the second open circle indicates the end of activation. From Hanes *et al.* (1995) with permission.

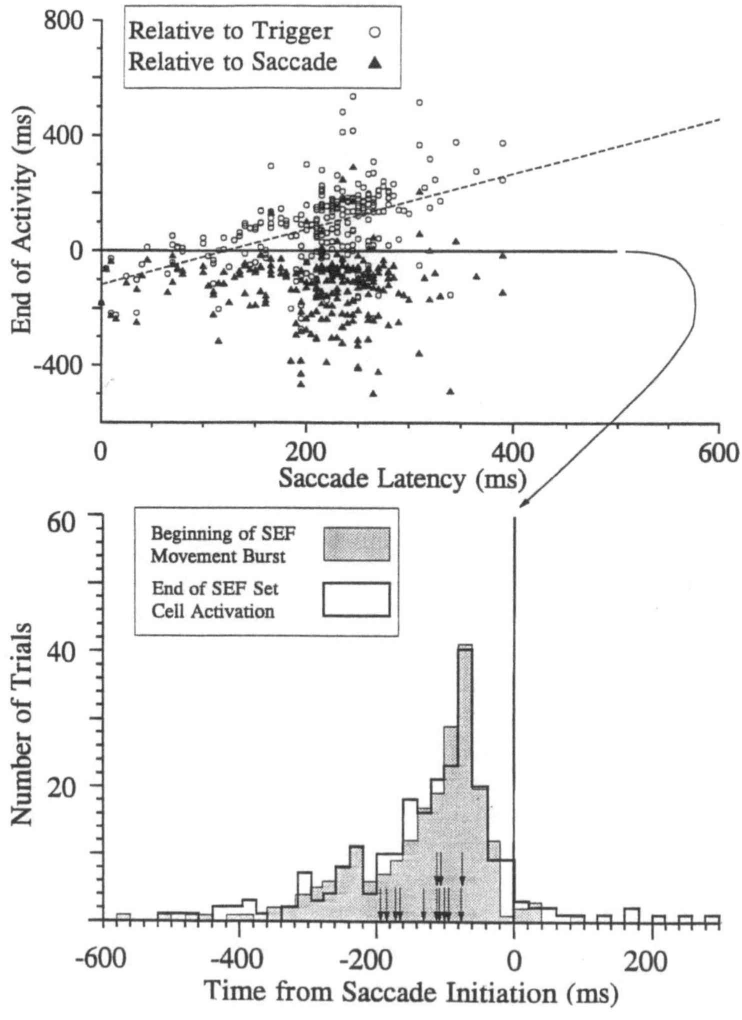

Figure 33. Temporal relation of the end of activation relative to the time of the movement trigger signal and the time of saccade initiation for all rewarded trials from a sample of SEF preparatory set cells. The scatterplot shows the end of activation for each trial for every cell plotted twice, once relative to the trigger signal (open circles) and once relative to saccade initiation (solid triangles). The time of neuronal modulation is negative if it occurred before the event relative to which it is measured. Note that the termination of activation of preparatory set cells occurring before the shortest latency saccades (<100 msec) actually occurred before the trigger signal was given. The dashed diagonal line indicates a significant linear regression of the end of activation measured relative to the trigger signal as a function of saccade latency. The histogram plots the end of activation relative to saccade initiation (open histogram). Overlayed on the histogram of the end of activation times for the SEF preparatory set cells is the histogram of the beginning of burst times for SEF presaccadic movement cells (shaded area). The beginning of the burst of SEF presaccadic movement cells coincided with the end of activation of SEF preparatory set cells. From Hanes *et al.* (1995) with permission.

analogous to the influence of cells in the substantia nigra pars reticulata (Hikosaka and Wurtz, 1983b), fixation cells in the rostral superior colliculus (Munoz and Wurtz, 1993), or the omnipause neurons in the brainstem (Keller, 1991). To test this possibility, the distribution of the time of termination of preparatory set cell activity was compared to the distribution of time of onset of activity of movement cells in SEF. As shown in Fig. 33, the distributions of neural modulation times for the two types of neurons coincided. This concurrence between the cessation of set cell activity and the onset of movement cell activity is consistent with the possibility that the onset of movement cell activity in SEF is gated by the activity of preparatory set cells (Hanes *et al.,* 1995). Of course, the next question is what controls the preparatory set cells?

Further evidence for a role of SEF in regulating the initiation of gaze changes is the finding that some cells in SEF are specifically or differentially activated when monkeys withhold saccades following a no-go instruction (Mann *et al.,* 1988; Schall, 1991a). No-go-related activity has also been observed in SMA (Tanji and Kurata, 1985) as well as in prefrontal cortex. Additional work using tasks designed to probe the control of voluntary movements (e.g., Hanes and Schall, 1995) is needed to clarify whether and in what way cells in SEF as well as in other visuomotor structures regulate saccade initiation.

A different and surprising property of the saccade-related activity of SEF cells was recently reported by Olson and Gettner (1995). Monkeys were trained to shift gaze to one or the other end of a horizontal bar. A visual cue, presented a specific interval before the horizontal bar, instructed the monkey which end to look at. The bar was presented in three different locations in such a fashion that a saccade of a particular direction and amplitude could be made to shift gaze to either end of the bar on different trials. Olson and Gettner found that many cells in SEF were active according to the end of the bar to which gaze would shift and not according to the metrics of the saccade. This specific preference was observed not only in the neural activity recorded before saccades, but also during a delay period after the cue had been presented instructing the monkey which end of the bar to fixate, but before the bar was presented. The authors interpret their data as the first neurophysiological evidence for an object-centered representation in the brain.

4.3.2. Stimulus-Evoked Activity

SEF neurons seem to respond less consistently or vigorously to visual or auditory stimuli than do their counterparts in FEF (Schall, 1991a, b; see also Russo and Bruce, 1996). The visual response latencies in SEF are significantly greater than the latencies measured in FEF, too. Evidence about the nature of the visual activation in SEF and FEF was obtained by comparing the patterns of responses in go trials requiring execution of a saccade with the patterns of activation in no-go trials requiring withholding of a planned saccade (Schall, 1991a). The visually responsive tonic neurons in SEF (preparatory set cells and sensory-movement cells) cease firing after a no-go cue is given even though the visual target stimulus is still present. In contrast, the tonic neurons in FEF continue to discharge in such no-go trials until the stimuli are turned off at the conclusion of the trial (Schall, 1991a, b). One interpretation of these results is that the activity of the SEF cells may signal target location only when a saccade is being programmed. Thus, their response could represent not only stimulus location, but also movement intention. Other evidence also shows that at least some of the neurons in SEF which appear to be visually responsive are not actually stimulus-bound, that is, that their response was not necessarily linked to the actual physical presentation of the target. First, a higher proportion of the cells responding to the target in SEF than in FEF exhibited anticipatory activity. Second, whereas most of the preparatory set and sensory-movement cells in SEF respond equally to visual and auditory targets, most of the cells in FEF respond preferentially for visual or auditory stimuli.

The absence of short-latency, strong visual responses and the incidence of extraretinal modulation of SEF cells is consistent with the constellation of visual afferents to SEF. As described above, SEF is not innervated by areas in inferior temporal cortex associated with object recognition, but is innervated by LIP and MST as well as visually responsive areas in prefrontal cortex (Huerta and Kaas, 1990; Schall *et al.,* 1995b). Thus, SEF visual responses reflect less about the properties of the image and more about the stimulus in the context of the ongoing behavior.

Chen and Wise (1995a) have investigated how SEF neural activity is related to motor learning processes. The design of the task and analyses was similar to that of Mitz *et al.* (1991) described above. Monkeys were trained to make a saccade to one of four targets based on an arbitrary visual instruction given at the fixation spot. Neural activity was recorded when the monkeys performed the task with familiar cues and also while monkeys learned the correct saccades associated with novel cues. Figure 34 shows the result for one SEF sensory cell. During trials with familiar cues the cell responded reasonably well, although with some variability, to the stimulus instructing a downward gaze shift. When a novel stimulus was introduced that instructed a downward saccade, the monkey initially failed to make the correct eye movement. After just three trials, though, the monkey performed correctly. This SEF cell did not respond to the novel cue in the early trials, but after the monkey's performance improved, the cell responded as well as ever to the new cue instructing a downward saccade. Chen and Wise found other cells that were exclusively active while monkeys learned new stimulus–response associations and did not discharge once monkeys had mastered the new stimulus–response association. The changes in activity associated with motor learning were found in delay period activity and in pre- and postsaccadic activity as well. Also, while some cells performed like the one illustrated in Fig. 34, other cells also showed systematic reductions in activity levels as learning progressed. Overall, around half of the neurons sampled in SEF exhibited some form of modulation related to the motor learning. In contrast, of neurons recorded in FEF under the same conditions, fewer than one in six were modulated in relation to the motor learning (Chen and Wise, 1995b).

4.4. Role of SEF in Saccade Production

Lesions studies have shown that the FEF and superior colliculus are necessary for saccade production. Combined bilateral ablation of both the FEF and

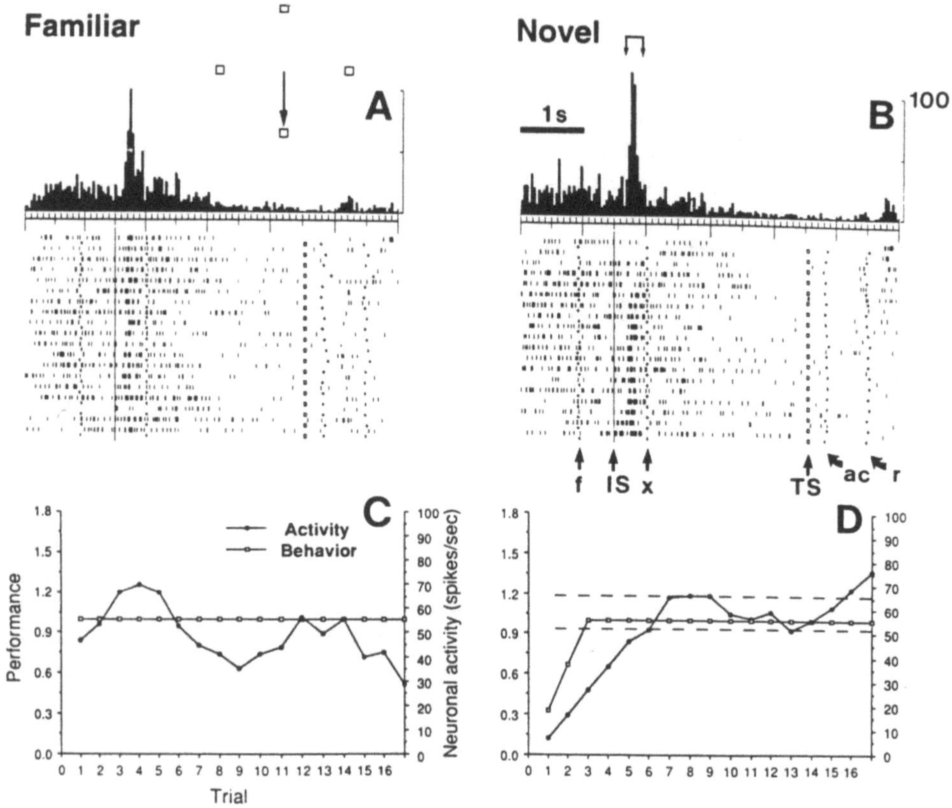

Figure 34. A learning-dependent SEF neuron during the presentation of a familiar stimulus (A, C) and a novel stimulus (B, D) instructing a saccade to the downward target. Each line of the raster shows the cell's activity during a trial and each of the marks beneath the raster line indicates the time of an event: f, Central fixation required (although the monkey usually fixated earlier); IS, instruction stimulus onset; x, instruction stimulus offset; TS, movement trigger stimulus; ac, target acquisition; r, reward. Above the raster is a perievent spike histogram aligned on the presentation of the instruction stimulus. The ordinate represents 100 Hz for both histograms. The first correctly performed trial is at top, the last at the bottom. Above the histogram, the horizontal bar and downward-pointing arrows mark the boundary of the time window for which data are plotted in panels C and D. Note that trials with the novel instruction stimulus were intermixed with other trials having different novel or familiar instruction stimuli. (C) Performance and neural activity during the instruction period from trials with a familiar instruction stimulus (from panel A). The three-point moving average of performance (proportion of correct responses/three trials) is shown (□), as is the centered, three-point moving average of activity (●). (D) Performance and activity during the instruction period from trials with a novel instruction stimulus (from panel B). Note that the performance (□) rapidly improves. The response evoked by the novel instruction stimulus (●) also increases as performance improves. These three-point averages contain data from correctly executed trials only. Thus, point 1 in panel D shows the average activity from the first, second, and third correct trials (centered on the second trial in panel B). The dashed line in panel D indicates ±1 standard deviation of the mean activity for the familiar instruction stimulus directing downward saccades. The monkey made errors on trials 1, 2, and 4 in response to the novel instruction stimulus. From Chen and Wise (1994a) with permission.

superior colliculus results in a complete loss of ability to shift gaze (Schiller *et al.*, 1980). In this study SEF (which had not been identified at that time) was left intact. Thus, SEF alone does not appear to be sufficient to generate saccades. This observation is consistent with the relationship of SMA to primary motor cortex (e.g., Hummelsheim *et al.*, 1986). However, such a conclusion seems difficult to reconcile with the anatomical and physiological observations that were cited above to establish that SEF does play a role in the production of saccades. Based on the observation of Schiller *et al.* (1980), we must surmise that the role of SEF in saccade generation is secondary or higher order, although specifying precisely what that means is difficult. For example, does microstimulation of SEF evoke saccades after ablation of FEF and superior colliculus? It is now known that the effectiveness of SEF microstimulation survives ablation of either superior colliculus or FEF (Tehovnik *et al.*, 1994). We do not yet know whether saccades can be evoked by SEF stimulation following ablation of both FEF and superior colliculus.

The effects of ablation also indicate that SEF is not necessary for the production of individual saccades. Damage to SEF in humans causes no evident problem with making single visually guided or memory-guided saccades (Pierrot-Deseilligny *et al.*, 1991a, b, 1993), but a clear deficit is observed in the production of a sequence of memory-guided saccades (Gaymard *et al.*, 1990). The nature of the deficit, more precisely, was shifting gaze in the wrong order to stimuli that had appeared in a particular sequence. Notably, this deficit is observed exclusively following left-hemisphere, but not right-hemisphere lesions; moreover, patients with left-hemisphere SMA lesions were impaired in generating a sequence of memory-guided saccades that began in either the contralesional or the ipsilesional direction (Gaymard *et al.*, 1993). These observations are consistent with the nature of deficits observed following damage to SMA which involve a deficit in the production of coordinated sequences of multijointed movements of the limbs in both humans (Dick *et al.*, 1986; Watson *et al.*, 1986; Massion *et al.*, 1989; Halsband *et al.*, 1993) and monkeys (Brinkman, 1984; Chen *et al.*, 1995). We also note that SMA lesions cause greater deficits in the production of self-generated as compared to externally triggered movements (Passingham *et al.*, 1989; Thaler *et al.*, 1995). Other evidence indicates a role for SMA in the generation of complex movement sequences. For example, regional blood flow increases are observed in SMA in subjects performing complex movement sequences (Orgogozo and Larsen, 1979; Roland *et al.*, 1980; Grafton *et al.*, 1992; Rao *et al.*, 1993; Shibasaki *et al.*, 1993; Remy *et al.*, 1994; Wessel *et al.*, 1995). Also, several laboratories have shown that the readiness potential recorded over SMA is larger before movement sequences as compared to single movements (Benecke *et al.*, 1985; Deecke *et al.*, 1985; Lang *et al.*, 1989, 1990; Kitamura *et al.*, 1993).

Based on this evidence, it has been suggested that SEF may play a role in programming sequences of saccades. This idea has merit on two fronts. First, normal visual perception is accomplished with incessant scanning eye movements that often follow idiosyncratic but somewhat regular paths over the image (Yarbus, 1967; Noton and Stark, 1971). Such scanpaths may represent a programmed sequence of saccades rather than a succession of individually programmed movements. Second, as mentioned above, neurons have been recorded in SMA that discharge specifically in association with movements that are part of a sequence. Although no laboratory has recorded such cells in SEF, it seems likely that they will be found because the generation of a sequence of saccades appears to be based on the same sort of motor program as the generation of a sequence of limb, finger, or vocal movements (Zingale and Kowler, 1987). Sequences of gaze shifts are naturally produced concurrently with reaching and orienting movements. Evidence indicates that neurons in SEF may be specifically associated with eye-hand coordination (Mann *et al.*, 1988; Mushiake *et al.*, 1996).

Experiments using transcranial magnetic stimulation to inactivate transiently a local cortical area provide further support for a fundamental role of SEF in generating sequences of remembered saccades. Production of a memorized sequence of eye movements is impaired if SEF is inactivated by transcranial stimulation when the sequence of stimuli to be remembered is presented or when the movement sequence is initiated (Müri *et al.*, 1994, 1995). If SEF is inactivated by the transcranial stimulation once the movement sequence begins, no deficit is observed. A corollary of the hypothesis that SEF is responsible for the production of sequences of saccades is that a given site in SEF would be connected with multiple sites in FEF or superior colliculus representing a range of saccade vectors. This prediction seems consistent with the pattern of anatomical projections from SEF described above. The hypothesis that SEF plays a role in generating sequences of saccades may also help explain why the effects of electrical microstimulation of SEF may be different from those observed from microstimulation of FEF or superior colliculus.

5. Prefrontal Cortex

This portion of the chapter will survey what is known about the function of prefrontal cortex in visuomotor behaviors. Recent neurophysiological studies will be highlighted. Prefrontal cortex is distinguished from motor and premotor cortex by the presence of a pronounced granular cell layer and by connections with the mediodorsal nucleus of the thalamus. The lateral prefrontal cortex of the macaque has been partitioned into three main areas: area 9 on the convexity medial to the principal sulcus, area 46, which occupies the banks and lips of the principal sulcus, and area 12, on the convexity ventrolateral to the principal sulcus. These regions have somewhat different constellations of connections. For example, areas 9 and 46 are heavily connected with posterior parietal cortex (e.g., Andersen *et al.*, 1985; Schall *et al.*, 1995b; Cavada and Goldman-Rakic, 1989; Petrides and Pandya, 1984). Area 12, on the other hand, is more heavily interconnected with inferior temporal cortex (Kawamura and Naito, 1984; Barbas, 1988; Ungerleider *et al.*, 1989; Baizer *et al.*, 1991; Webster *et al.*, 1994).

Various views about the role of prefrontal cortex in behavior have been reviewed previously (e.g., Stuss and Benson, 1986; Goldman-Rakic, 1987, 1988, 1990; Fuster, 1989; Petrides, 1994a). Much current interest in prefrontal cortex is motivated by evidence indicating that certain symptoms of disorders like schizophrenia, Alzheimer's disease, Parkinson's disease, and senile dementia may be attributable to prefrontal dysfunction (Levin *et al.*, 1991; Weinberger and

Berman, 1996). For example, the gaze control deficits of schizophrenic patients have been attributed to a dysfunction of prefrontal cortex (Levin, 1984; Sereno and Holzman, 1993; Fukushima *et al.*, 1994; MacAvoy and Bruce, 1995). This hypothesis is supported by other evidence for frontal lobe dysfunction in schizophrenics: reduced prefrontal blood flow and glucose metabolism (Weinberger *et al.*, 1986; Buchsbaum and Haier, 1987; Wolkin *et al.*, 1992; Siegel *et al.*, 1993), frontal cortex structural abnormalities (Andreasen *et al.*, 1986), and impaired performance on other tasks diagnostic for frontal lobe dysfunction (e.g., Goldberg *et al.*, 1987; Franke *et al.*, 1993). In fact, the elevated blood flow through FEF normally associated with eye movements is absent in schizophrenics (Nakashima *et al.*, 1994).

5.1. Dorsolateral Prefrontal Cortex—Areas 46 and 9

The delayed-response task has been the central paradigm with which dorsolateral prefrontal cortex has been studied. In this task a stimulus is presented (e.g., a light flashes on a screen or a piece of food is placed in a well), but the movement afforded by the stimulus (e.g., making a saccade or reaching for the food) must be delayed for an arbitrary period of time while the target for the movement is removed from view. During the delay period the position or properties of the stimulus must be remembered to accurately guide subsequent action. Lesion data demonstrate that dorsolateral prefrontal cortex is necessary for performing this task (e.g., Jacobsen, 1936; reviewed by Fuster, 1989). The processing subserved by dorsolateral prefrontal cortex has been characterized as spatial working memory (e.g., Goldman-Rakic, 1987, 1988; Petrides, 1994a, 1994b) or alternatively as temporally organizing behavior, mediating the linkage across time of stimulus and response (e.g., Milner *et al.*, 1985; Fuster, 1989).

5.1.1. Functional Properties: Stimulus-Evoked Activity

Visually responsive neurons have been recorded in and around the principal sulcus, mainly concentrated in the caudal half (Fuster, 1973; Kubota *et al.*, 1974, 1980; Suzuki and Azuma, 1977, 1983; Suzuki *et al.*, 1979; Kojima, 1980; Mikami *et al.*, 1982; Ito, 1982; Sawaguchi, 1987a; Quintana *et al.*, 1988; Yajeya *et al.*, 1988; Boch and Goldberg, 1989; Funahashi *et al.*, 1989, 1990, 1993b; Watanabe, 1990, 1992). The response latencies can be under 100 msec, but commonly range to over 200 msec. The receptive fields of the cells are large and commonly emphasize the contralateral hemifield. Many cells in the ventral aspect of the dorsolateral prefrontal cortex have receptive fields including the fovea, and discharge in relation to fixation of visual stimuli (Suzuki and Azuma, 1977; Mikami *et al.*, 1982). Receptive fields that include the fovea are smaller than receptive fields representing the peripheral visual field.

Most if not all cells in dorsolateral prefrontal cortex receive potent extraretinal inputs. A fair number of cells in dorsolateral prefrontal cortex exhibit anticipatory activity, beginning to discharge before a predictable visual stimulus appears (Boch and Goldberg, 1989; Fuster, 1973). Anticipatory visual activity has also been observed in FEF (Goldberg and Bruce, 1985; Schall, 1991b) and even more commonly in SEF (Schall, 1991a) as well as in other parts of prefrontal cortex (Niki, 1974a; Niki and Watanabe, 1979; Sakagami and Niki, 1994a). Some but not all visually responsive neurons in the principal sulcus give a spatially specific enhanced response if a visual stimulus is the target for a saccade (Boch and Goldberg, 1989). Some evidence also indicates that the visual responses are enhanced when the monkey directs attention without making a saccade to the peripheral target (Mikami *et al.*, 1982). In a task in which a saccade to a target is delayed by instruction, certain units exhibit two bursts of activity, one following the presentation of the target and the second immediately prior to and during the saccade (Boch and Goldberg, 1989). The second burst does not occur, however, if the same saccade is made to the remembered location of the target. Accordingly, the second discharge was interpreted as a reactivation of the original visual response. Its functional purpose, however, is unclear.

The visual responses of few prefrontal neurons reflect stimulus properties such as color, shape, or direction of motion, but the responses of many prefrontal neurons reflect the behavioral meaning of the stimuli (Kojima, 1980; Quintana *et al.*, 1988; Yajeya *et al.*, 1988). The timecourse of the selection in the visual responses of prefrontal cells has not been carefully investigated though. The relatively long visual response latencies of the prefrontal neurons would permit stimulus evaluation by striate and prestriate visual areas to occur well before the response of most prefrontal neurons begins.

5.1.2. Functional Properties: Activity Related to Movement

Neurons in principal cortex have also been recorded that discharge in relation to eye or limb movements (Kojima, 1980; Funahashi *et al.*, 1989, 1990, 1993b; Boch and Goldberg, 1989; Kubota and Niki, 1971; Watanabe, 1986b; Kubota and Funahashi, 1982). We should note that area 46 projects to the superior colliculus (Fries, 1984; Leichnetz *et al.*, 1981), as well as to FEF, SEF, and premotor cortex. But while these results may indicate some role of prefrontal cortex in actually generating movements such as saccades, the role must be indirect, because low-intensity microstimulation in the cortex of the principal sulcus does not elicit saccades (Boch and Goldberg, 1989).

5.1.3. Functional Properties: Delay Period Activity

Most unit recordings that have been carried out in and around the principal sulcus of monkeys have used some version of delayed-response, delayed-alternation, or delayed-discrimination tasks and found neurons firing in a sustained fashion during the delay period (Fuster, 1973; Kubota and Niki, 1971; Kubota *et al.*, 1974; Niki, 1974a–c; Batuev *et al.*, 1985; Fuster *et al.*, 1982; Joseph and Barone, 1987; Kojima and Goldman-Rakic, 1982; Funahashi *et al.*, 1989, 1990). This delay period activity has been thought to be registering a mnemonic signal necessary to perform the delayed-response task (e.g., Niki, 1974a; Funahashi *et al.*, 1989). However, only recently have investigators designed experiments to examine precisely what may be remembered. Every study of dorsolateral prefrontal cortex has found some cells responding to visual stimuli as well others that maintain activation during the delay periods intervening between presentation of a stimulus that will guide the movement and a trigger signal permitting the initiation of the movement. Similar delay period activity has been recorded in area 9 (Fuster, 1973; Joseph and Barone, 1987; Quintana *et al.*, 1988), frontal eye field (Funahashi *et al.*, 1989), posterior parietal cortex (Gnadt and Andersen, 1988), inferior temporal cortex (Fuster and Jervey, 1982), anterior cingulate cortex (Fuster, 1973; Niki and Watanabe, 1976), and orbital frontal cortex (Rosenkilde *et al.*, 1981). It was commonly found that single prefrontal cells exhibited preferential delay period activity for stimuli on one side of the midline. However, the early work did not typically control for eye movements.

This work has been elaborated by using an oculomotor delayed-response task in which monkeys must make a saccade to the remembered location of a

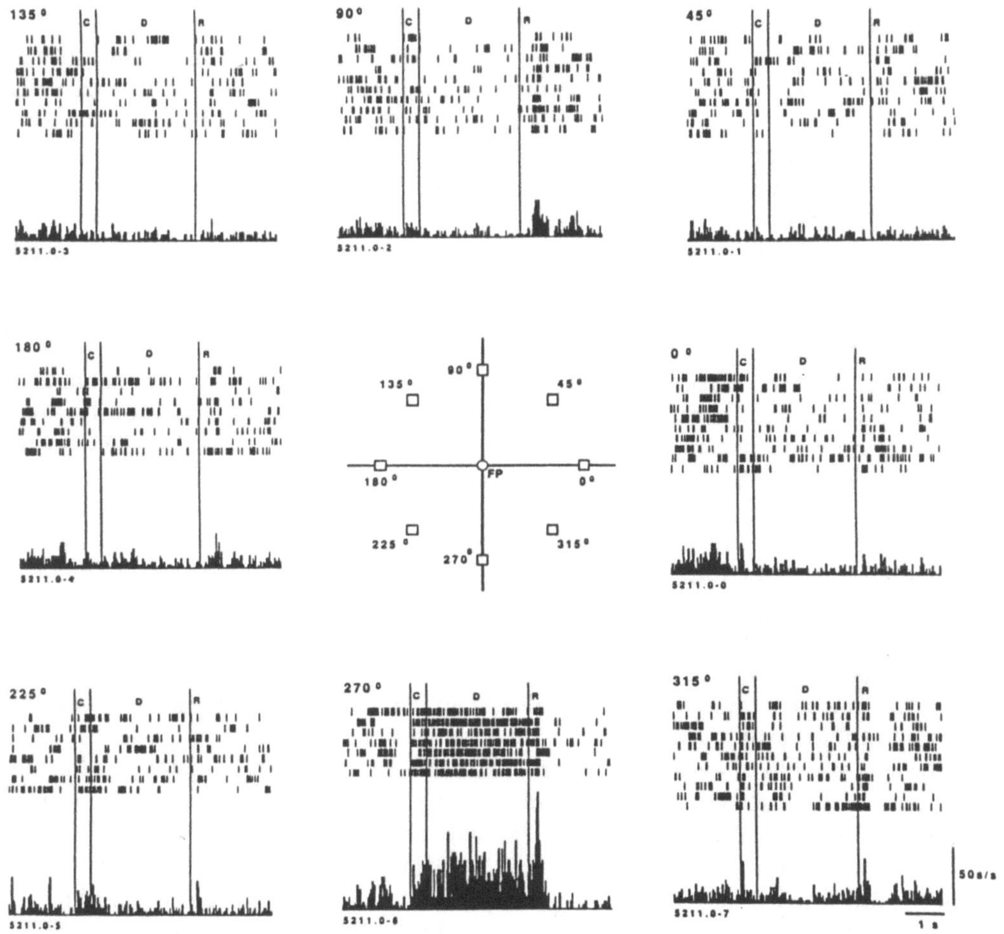

Figure 35. Activity of a principal sulcus neuron during an oculomotor delayed-response task. Neural activity is represented as rasters and histograms aligned on the time of presentation of the visual cue (C), which was followed by the delay period (D) and the saccade response (R). Each panel shows the activity recorded when the visual cue was presented at each of eight locations. The cell was activated when the cue target flashed at the lower location and remained active throughout the 3-sec delay period until the saccade. Suppression of ongoing activity may be evident when the target cue was presented at the top location. From Funahashi *et al.* (1989) with permission.

target that was flashed 1–2 sec and up to 6 sec earlier (Fig. 35). During such a delayed-response task, some neurons in area 46 respond only to the visual stimulus. But another group of cells exhibit sustained activation or suppression following the brief presentation of the target until the signal to move is given (Funahashi *et al.,* 1989; 1990). An important finding coming out of this work with the oculomotor task is that many cells in area 46 show delay period activity in association with remembered stimuli at specific locations in the visual field. Most neurons are preferentially activated when stimuli in the contralateral visual field are to be remembered. This observation extends earlier work showing exclusive hemifield specificity. The observation of reasonably delimited single-cell "memory fields" analogous to visual receptive fields may represent evidence for a more refined representational system for working memory than had hitherto been suspected.

Further evidence that neural activity during the delay period is necessary for performing delayed-response tasks is provided by error trials. When monkeys generate errors in the oculomotor delayed-response task, prefrontal cells discharge less during the delay period (Funahashi *et al.,* 1989). Similar correlations of level of activity with accurate performance have been made in other studies (Suzuki and Azuma, 1977; Batuev *et al.,* 1985; Niki and Watanabe, 1976; Watanabe, 1986b).

Additional evidence for the discrete representation of the remembered location of visual stimuli by principal sulcus is provided by a recent ablation study. Unilateral localized lesions in principal sulcus area 46 cause a specific deficit in monkeys performing the oculomotor delayed-response task (Funahashi *et al.,* 1993a). Monkeys exhibited specific errors in generating memory-guided saccades to stimuli flashed in restricted regions of contralateral visual space. The propensity to make an error increased with longer delays. However, monkeys could make accurate visually guided saccades to stimuli presented in the "mnemonic scotomas." Thus, the absence of a specific oculomotor impairment indicates that the basis of the deficit could be mnemonic. A similar deficit in making memory-guided saccades has been observed following ablation of FEF (Deng *et al.,* 1986). Also, dorsolateral prefrontal cortex damage in humans results in an impaired ability to generate memory-guided saccades (Pierrot-Deseilligny *et al.,* 1991b, 1993).

Pharmacological methods have also been used to investigate the function of dorsolateral prefrontal cortex in working memory (e.g., Sawaguchi, 1987b). This ongoing work has shown quite clearly that the synaptic mechanisms mediating the delay period activity are very complex. Performance is impaired in working memory tasks when antagonists to the D1 dopamine receptor are delivered locally or systemically (Sawaguchi and Goldman-Rakic, 1991; Arnsten *et al.,* 1994). One study (Sawaguchi *et al.,* 1990) found that the delay period activity of area 46 cells was suppressed by application of nonspecific dopamine antagonists. However, a subsequent study has shown that the delay period activity of area 46 cells is actually increased in a dose-dependent manner by iontophoresis of selective dopamine D1 receptor antagonists (Williams and Goldman-Rakic, 1995). Application of a D1 agonist reverses the effect. The elevated activity was interpreted as enhancing the memory of stimulus location represented by the cell. The neural activity evoked by the stimuli or generated in association with the oculomotor responses was not modulated by the D1 antagonist. The pharmacology of this modulatory effect appears to be quite specific; a D2 receptor antagonist only suppressed neural activity nonspecifically.

The interpretation of the sustained activity of prefrontal neurons as representing a spatial working memory to guide action leaves open the question of whether the activity has to do with representing the location of the visual stimulus or representing the goal of the upcoming movement. We reviewed above a series of experiments that addressed this issue. Boussaoud and Wise (1993a, b) designed tasks for macaque monkeys in which the same stimulus, presented more than once, had different meanings for the monkey because of the reward contingencies. The same stimulus could mark a location to be remembered and guide a covert shift of attention or it could instruct the direction of a movement. Among other findings, these investigators reported that whereas most neurons in the dorsal postarcuate premotor cortex (PMd) respond more to the stimulus when it signals movement direction, most cells in prefrontal cortex (area 12 mainly) respond more when the stimulus cues the location of the upcoming movement stimulus. This finding seems to indicate that the maintained activity of prefrontal cells during the instructed delay period may have more to do with registering visual locations to attend or remember than with representing the direction of the movement.

Another way to dissociate sensory or attentional activation from motor planning-related activation is to manipulate stimulus–response compatibility. In conditions that require movements that are incompatible with stimulus locations, i.e., directed to a location different from the location of the stimulus, most neurons in prefrontal cortex tend to discharge more in relation to the location of the stimulus (di Pellegrino and Wise, 1993a, b; Funahashi *et al.*, 1993b). However, in cortical areas linked to movement production, cells are more commonly found that are active in relation to the movement direction as opposed to the stimulus direction (Alexander and Crutcher, 1990; di Pellegrino and Wise, 1993a, b). It is important to note that these distinctions represent trends across populations of neurons, with clear counterexamples in each area.

The interpretation of the delay period activity as representing spatial working memory has been queried as incomplete. Fuster (1989), for example, has proposed a more inclusive hypothesis which holds that dorsolateral prefrontal cortex subserves the temporal organization of behavior by mediating links between stimuli and responses that are separated in time and sustaining the appropriate behavioral strategy. This view is motivated by studies that have reported delay period activity in principal sulcus cortex occurring in non-spatial-task conditions (e.g., Watanabe, 1981; Quintana *et al.*, 1988). Lesion and inactivation data also indicate some role for principal sulcus cortex in nonspatial behavior, specifically in the performance of tasks that require generating a response based on a remembered or arbitrary relation to a stimulus (Bauer and Fuster, 1976; Quintana and Fuster, 1993).

Other data are relevant for Fuster's hypothesis; for instance, studies have shown that the cortex medial to the principal sulcus plays a role in the production of sequential behavior. Ablations of the cortex in dorsal area 46 and lateral area 9 result in profound deficits in monkeys' ability to perform nonspatial self-ordered tasks (Petrides 1991, 1995). In these tasks monkeys must monitor their behavior so that subsequent actions can be accurately based on prior performance. The deficit is apparently not due to an impaired ability to generate sequential movements (Passingham, 1985; Colombo *et al.*, 1993). Notably, neural activity recorded in the cortex dorsal to the principal sulcus in the arcuate sulcus has been shown to represent the temporal order of stimuli when monkeys must generate a remembered sequence of movements (Barone and Joseph, 1989). Monkeys with lesions located further posteriorly, in arcuate cortex, are not impaired in the nonspatial self-ordered task (Petrides 1991, 1995), but are impaired in learning and performing conditional association tasks (Halsband and Passingham, 1982; Petrides, 1985). Evidence for analogous differential function of dorsal area 46 and area 8 in human has been obtained using PET (Petrides *et al.*, 1993).

5.2. Ventrolateral Prefrontal Cortex—Area 12

Whereas ablation of area 46 has most commonly been reported to cause deficits mainly in tasks requiring spatial memory, lesions of area 12 have been shown most often to cause deficits in tasks that require working memory of a nonspatial, visual recognition character (e.g., Mishkin and Manning, 1978; reviewed by Fuster, 1989; Petrides, 1994b). Numerous investigators have recorded visually responsive cells in the ventrolateral convexity of prefrontal cortex (Sakagami and Niki, 1994a, b; Rosenkilde *et al.*, 1981; Suzuki and Azuma, 1977; Wilson *et al.*, 1993; Watanabe, 1990, 1992; Boussaoud and Wise, 1993a; de Pelligrino and Wise, 1993a). Area 12 cells respond better to stimuli presented foveally (Mikami *et al.*, 1982; Suzuki and Azuma, 1983). Most cells respond with long (100–300 msec) latencies. Although some cells respond nonspecifically to visual stimuli that are presented as instructional cues in a delayed-response task, a consistent fraction of cells respond differentially according either to the stimulus color or to the behavioral context of the stimulus. Most studies that have reported some stimulus feature specificity, e.g., color, shape, or complex form, in prefrontal cortex have located the neurons in the ventrolateral convexity as opposed to the principal sulcus (Rosenkilde *et al.*, 1981; Watanabe, 1990; Wilson *et al.*, 1993; Sakagami and Niki, 1994a).

A number of investigators have used go/no-go visual discrimination tasks to investigate the signals carried by prefrontal neurons (Kubota and Komatsu, 1985; Sasaki and Gemba, 1986; Watanabe 1986a, b; Sakagami and Niki, 1994a, b). For example, Sakagami and Niki (1994a, b) used a go/no-go task in which the discriminative stimulus was defined by either color, shape, or position. The color of the fixation spot at the beginning of the trial specified whether color, shape, or position would be the relevant cue. After training on a standard stimulus set, monkeys were tested with unpredictable, irrelevant probe stimuli. Monkeys generally ignored these probes, which indicates that they had generalized the behavioral strategy and learned to use the stimulus combinations appropriately. Many cells recorded in this study responded to the visual cue stimuli. The visual response latencies ranged from 80 to 140 msec, and most cells responded best to stimuli presented in the contralateral hemifield. Although a few neurons responded to the cues according to their visual properties, most of the cells responded to the go/no-go cues according to their meaning in the context of the task (Fig. 36). In other words, certain neurons responded to the visual cue only if the stimulus signaled a go trial, not responding at all if the same visual stimulus signaled a no-go trial. Neurons were found that responded preferentially to stimuli specifying no-go as well as go trials. More than two-thirds of the cells tested showed changes in responses when color, shape, or position became the relevant cue for guiding the behavior.

As in the principal sulcus, cells on the ventral convexity discharge during the delay period of memory tasks. A recent neurophysiological study has highlighted a functional difference between the processing occurring in principal sulcus and that occurring on the ventrolateral convexity (Wilson *et al.*, 1993). Neurons were recorded in area 12 of ventrolateral prefrontal cortex in monkeys

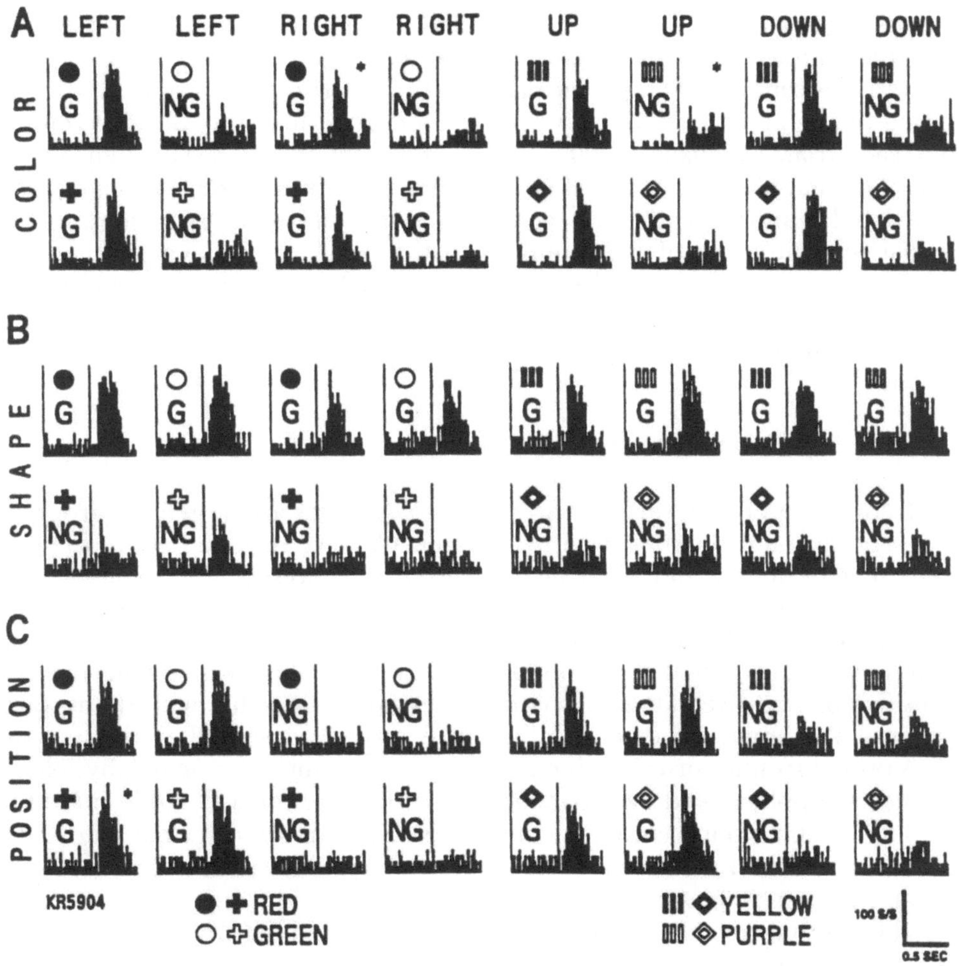

Figure 36. Ventrolateral prefrontal cell activity in go/no-go visual discrimination task. Poststimulus time histograms of responses to visual cues presented for 160 msec during color (A), shape (B), and position (C) discrimination trials. The vertical line indicates the time of stimulus presentation. The color of stimuli is indicated by the legend. The position of stimuli is indicated by the labels at the top of each column. Histograms are labeled according to whether the stimulus called for a go (G) or a no-go (NG) response. This neuron responded to stimuli if and only if the stimulus cued a go trial. If the stimulus cued a no-go trial, then the cell did not respond. From Sakagami and Niki (1994a) with permission.

performing two versions of a delayed-response task. One version was the standard spatial oculomotor delayed-response paradigm described above. The second version required monkeys to shift gaze based on the memory of complex visual objects such as faces. Specific objects cued different directions of movement. Neurons in area 12 were not active during the delay period when monkeys were to remember a spatial location, but they were active during delay periods when monkeys had to remember a complex object. These visual properties may be conferred on the area 12 cells by their afferents from inferior temporal cortex (Barbas, 1988; Ungerleider *et al.*, 1989; Baizer *et al.*, 1991; Webster *et al.*, 1994) where neurons also discharge during delay periods when monkeys must remember the identity of a visual stimulus (E. K. Miller *et al.*, 1993; Fuster and Jervey, 1982; Miyashita and Chang, 1988; Colombo and Gross, 1994).

The activity of other area 12 neurons appears to register receipt of reward (Rosenkilde *et al.*, 1981; Watanabe, 1990, 1992). Most such cells discharge specifically when a juice reward is given. This response occurs regardless of whether the reward is contingent on performance of the task (Rosenkilde *et al.*, 1981). By pairing different cue stimuli with presentation or withholding of juice reward after each trial, Watanabe (1990, 1992) demonstrated that certain neurons in area 12 discharge according to whether or not a stimulus predicts receipt of the reward following the trial. This activity was interpreted as representing the associative significance of the stimulus. Further work should explore the relationship of these neurons to the dopaminergic cells of the ventral tegmentum that discharge contingent on reward contingency and associations (Ljungberg *et al.*, 1992; Schultz *et al.*, 1993).

5.3. Inferior Orbital Prefrontal Cortex—Area 13

In oversimplified terms, the prestriate visual areas are largely regarded as representing the properties of visual stimuli. We have reviewed above the evidence that premotor and prefrontal cortex may represent the meaning of a stimulus relative to ongoing behavior. In this section we will review the evidence that inferior orbital frontal cortex may represent the value associated with a visual stimulus, i.e., whether a stimulus is reliably associated with a reinforcer, be it positive or negative. Thus, this cortical region is associated with the reward system of the brain. An extensive literature concerned with clinical issues involves orbital frontal cortex. This chapter will only summarize neurophysiological studies supplemented by the minimal necessary anatomical and lesion data.

The orbital frontal cortex is comprised of a number of cytoarchitectural zones (Amaral and Price, 1984; Preuss and Goldman-Rakic, 1991; Barbas, 1993; Ray and Price, 1993). Although typically associated with the olfactory system, orbital frontal cortex does receive visual afferents from rostral inferior temporal cortex (Moran *et al.*, 1987; Ungerleider *et al.*, 1989; Barbas, 1993; Webster *et al.*, 1994) and posterior parietal cortex (Cavada and Goldman-Rakic, 1989; Preuss and Goldman-Rakic, 1991). Orbital frontal cortex is connected mainly with the medial sector of the mediodorsal nucleus (Giguere and Goldman-Rakic, 1988; Goldman-Rakic and Porrino, 1985; Russchen *et al.*, 1987; Ray and Price, 1993), which is innervated by the amygdala, hippocampus, and lateral hypothalamus (Porrino *et al.*, 1981; Russchen *et al.*, 1987; Gower, 1989). Orbital frontal cortex is also innervated directly by the amygdala and hippocampus (Amaral and Price, 1984). Orbital frontal cortex projects to the ventral striatum (Selemon and Goldman-Rakic, 1985; Haber *et al.*, 1995), in which neural activity has been shown to signal the reward significance of stimuli (Apicella *et al.*, 1991).

Monkeys with orbital frontal cortex ablations show deficits in a variety of tasks that require associating stimuli with rewards or switching the reward contingencies of stimuli, e.g., delayed nonmatch to sample, go/no-go visual discriminations with reversals, object or spatial reversals or alternations (e.g., Iversen and Mishkin, 1970; Mishkin and Manning, 1978; Bachevalier and Mishkin, 1986). In general, monkeys with orbital frontal damage persevered, continuing to respond according to an outdated and maladaptive strategy. The precise anatomical focus of this effect is unknown, however. Such lesioned monkeys are also more resistant to extinction of performance following removal of reward (Butter, 1969). Other evidence indicates that inferior orbital prefrontal cortex is involved in associating the rewarding significance of visual stimuli. Monkeys with bilateral lesions of ventromedial prefrontal cortex are impaired in learning to associate arbitrary visual stimuli with reward (D. Gaffan and Murray, 1990). Similar impairments follow lesions of amygdala or the mediodorsal nucleus (E. A. Gaffan *et al.*, 1989; D. Gaffan and Murray, 1990).

Only a few physiological studies have recorded unit activity in orbital frontal cortex of behaving monkeys (Niki *et al.*, 1972; Rosenkilde *et al.*, 1981; Thorpe *et al.*, 1983; Rolls *et al.*, 1996a, b; see also Benevento *et al.*, 1977). Neurons that respond to visual stimuli and during the delay period of a delayed-response task are found. The visual responses of many such cells are nonselective, but other neurons respond preferentially for stimuli that are associated with specific gustatory salience, e.g., a syringe associated with either positive or aversive taste or food objects. Other cells are preferentially activated according to the complex visual features of the stimuli (Thorpe *et al.*, 1983). The contingencies of the visual responses of different neurons in orbital frontal cortex can be quite complex; straightforward sensory or motor explanations must be supplemented with reinforcement history. The selective response of certain orbital frontal neurons to a visual stimulus can change if the associated gustatory valence or reinforcement value is switched (e.g., Rolls *et al.*, 1996a). Neural activity associated with food items or visual stimuli associated with gustatory stimuli has been recorded also in the lateral hypothalamus and in the amygdala (Rolls *et al.*, 1979; Sanghera *et al.*, 1979), but in these structures the responses are less visually discriminating. The visual response latencies of orbitofrontal cells are typically 100–200 msec. The response latency of the visual nonselective cells appears to be slightly shorter than the latency of the selective cells.

Neurons in orbital cortex also exhibit gustatory responsiveness, being preferentially activated for different food items (e.g., Rolls *et al.*, 1996b). A fraction of cells respond bimodally to both visual and gustatory stimuli. A subset of these cells exhibit the same preference for gustatory and visual stimuli, e.g., discharging most in response to the image or the taste of a banana. The gustatory responses may arise from olfactory inputs to orbital cortex (Tanabe *et al.*, 1975; Yarita *et al.*, 1980). In a go/no-go discrimination task many neurons responded to the visual or auditory cue. Of note, too, were neurons that responded differentially to 'go' cues that signaled that the juice reward could be obtained as opposed to 'no-go' cues that signaled that no juice could be obtained by the operant response. The latency of response is long and quite variable, ranging from 100 to more than 300 msec. When the reward and no-reward cues were switched, most of the cells tested began to respond to the new reward-related cue stimulus as the monkey's performance switched. Thus, in orbital frontal cortex there are neurons that represent the reinforcement value of visual stimuli such as food items, but also of arbitrary associated stimuli.

Another class of neuron in orbital cortex seems to register the reinforcement event or the lack thereof. The most common of these discharge preferentially after trials when the reward is withheld or also when an unexpected reward is delivered (Niki *et al.*, 1972; Rosenkilde *et al.*, 1981; Thorpe *et al.*, 1983). One possible interpretation of these neurons is that they register whether a reward ought to be received by signaling deviations from expectancy. Certain other neurons in orbital cortex discharge exclusively when monkeys make errors in a visual discrimination reversal (Thorpe *et al.*, 1983). The functional properties of these neurons seems to clarify why perseveration follows their loss.

6. Cingulate Cortex

Traditionally, cingulate cortex has been considered part of the limbic system (reviewed by Vogt, 1985; Vogt *et al.*, 1992). Several recent developments have required a reevaluation of the role of cingulate cortex in behavior. For example, neuroimaging studies in humans have demonstrated elevated blood flow in anterior cingulate cortex during semantic processing of words (Petersen *et al.*, 1988). The magnitude of blood flow through anterior cingulate cortex increases when subjects must detect more targets (Posner *et al.*, 1988), when subjects must counteract the effect of interfering stimuli (Pardo *et al.*, 1990), when subjects must perform a divided attention task (Corbetta *et al.*, 1991), or when subjects generate voluntary movements (Frith *et al.*, 1991). Based on these and related data, Posner and Petersen (1990) argued that anterior cingulate cortex plays a role in directing or maintaining attention. In particular, it is thought that anterior cingulate cortex is involved in attention to response selection as opposed to internal response generation.

Recent anatomical, physiological, and neuroimaging experiments have also now demonstrated a motor-related function for anterior cingulate cortex. Tract-tracing studies have demonstrated a zone of anterior cingulate cortex that projects somatotypically to the spinal cord and primary motor cortex (Hutchins *et al.*, 1988; Baleydier and Maguire, 1980; Muakkasa and Strick, 1979; Vogt *et al.*, 1987; Morecraft and van Hoesen, 1992; Stepniewska *et al.*, 1993; He *et al.*, 1995). It may be via these pathways that electrical stimulation of parts of anterior cingulate cortex evoke body movements (Luppino *et al.*, 1991).

Only a few studies have recorded single-unit activity in the cingulate cortex of monkeys (Niki and Watanabe, 1976, 1979; Shima *et al.*, 1990). In the standard delayed-response task with stimulus–response-compatible and -incompatible conditions Niki and Watanabe found that cells in anterior cingulate cortex (both the upper and lower banks of the cingulate sulcus) discharged during the 4-sec delay period. Nearly all of the cells were active preferentially according to the location of the target stimuli (up, down, left, right), but a few cells were active in relation to movement direction as revealed in a stimulus–response-incompatible condition. These results generally resemble those of di Pelligrino and Wise reported in Section 2.2.1. The authors also reported that an absence of activity of the cingulate cells predicted trials on which the monkey would make errors. The neural registration of an error in performance has also been found in anterior cingulate cortex surface potentials (Gemba *et al.*, 1986). The similarity of anterior cingulate and dorsolateral prefrontal cortex activity is not unexpected, because the two cortical areas are reciprocally connected (Vogt and Pandya, 1987; Selemon and Goldman-Rakic, 1988). The neuroimaging studies cited above also noted that dorsolateral prefrontal cortex shows elevated blood flow during periods when anterior cingulate cortex does.

Another study recorded unit activity in cingulate cortex of monkeys making self-paced or signal-triggered movements (Shima *et al.*, 1990). Many units were encountered in anterior cingulate cortex that discharged before forelimb movements. If the movement was triggered by a visual, acoustic, or tactile stimulus, the neural activity began usually within 200 msec of the movement. In contrast, if the movement was self-generated, the activity of many anterior cingulate cells began more than 500 msec and as much as 2000 msec before the movement. Such very long-lead activation was observed much less frequently in posterior cingulate cortex (area 23) and not at all in primary motor cortex. Certain cells in anterior cingulate discharged preferentially or exclusively in relation to either self-generated or signal-triggered movements. Another subpopulation of cells in anterior cingulate cortex were activated by sensory stimuli. Although more data are needed on this point, the evidence indicated that anterior cingulate cells were multimodal in response. The cells in anterior cingulate cortex that were active before the forelimb movements were found in the upper and lower banks of the cingulate sulcus, in a region that appears to be connected with primary motor cortex.

The very long-lead activity of the anterior cingulate cells is likely related to the same pattern of activity observed in SMA and SEF (see Fig. 32), in which many cells begin to fire long before self-generated movements are initiated (see above). SMA is innervated by anterior cingulate cortex. By virtue of its afferents from limbic structures such as the amygdala (Vogt and Pandya, 1987), the anterior cingulate cortex may be involved in converting motivation and drive into action (Goldberg, 1985; Passingham *et al.*, 1989).

To further clarify the relation of anterior cingulate cortex to the motor system, a recent study by Paus *et al.* (1993) measured regional cerebral blood flow (rCBF) in the anterior cingulate cortex of humans during tasks using eye movements, speaking, and manual movements. We will review the eye movement findings. Two conditional response tasks were used to manipulate the planning and preparation of the movements. In the first task subjects made a saccade to one of three peripheral targets, specified by an arbitrary association with one of three stimuli appearing at the fixation spot. Subjects were overtrained with one stimulus–response association, and rCBF measurements were made. Then rCBF measurements were taken when the subjects had to learn a new association of stimuli to responses (this is like the conditional motor learning tasks used by Mitz *et al.*, 1991, and Chen and Wise, 1995a). In the second task subjects made prosaccades or antisaccades in response to visual stimuli. The rCBF values measured in the different task conditions were compared to a specific baseline condition when no eye movements were made. When subjects made saccades while learning a new stimulus–response association or made antisaccades, elevated blood flow was observed in the anterior cingulate cortex. The location in anterior cingulate cortex that exhibited elevated blood flow in association with the eye movement tasks was situated rostral to regions of elevated blood flow associated with forelimb tasks. Paus *et al.* suggest that the anterior cingulate cortex subserves the selection of action.

7. Conclusions

The research reviewed in this chapter shows that frontal cortex plays many important roles in the production of visually guided behavior. A large fraction of neurons in frontal cortex respond to visual stimuli. However, the visual responses do not signal as much about the properties of stimuli per se as they do the meaning and value of the stimulus. The visual responses of many neurons in frontal cortex are modulated by motor planning, memory, attention, and reward valence. The mapping of visual responses onto motor outputs is flexible and adaptive.

Despite steady progress that has been made, though, our understanding of how frontal cortex functions in the preparation and generation of visually guided movements is very incomplete. Many key questions remain unclear, awaiting future empirical work to resolve current conflicts in the literature. Nevertheless, a sense of progress is justified. The relationship of perception and action has been one of the central problems since the founding of modern neuroscience and psychology 150 years ago (e.g., Neumann and Prinz, 1990). Further research on the visuomotor areas of the frontal lobe promises to yield significant and satisfying insights into how visual signals from the constellation of extrastriate visual areas are combined, coordinated, and selected to guide movements and how motor signals from the various areas of frontal cortex influence the processing in visual areas.

ACKNOWLEDGMENTS. I thank D. Hanes and K. Thompson for reading the manuscript, B. Hendricks for typing the manuscript, T. Paus for providing Fig. 8, M. Schlag-Rey for providing Fig. 16, and M. Segraves for providing Fig. 21. The author's research has been supported by R01 EY08890, R01 MH55806, P30 EY08126, P30 HD15052, the Alfred P. Sloan Foundation, and the McDonnell–Pew Program in Cognitive Neuroscience.

8. References

Alexander, G. E., and Crutcher, M. D., 1990, Preparation for movement: Neural representations of intended direction in three motor areas of the monkey, *J. Neurophysiol.* **64:**133–150.

Amaral, D. G., and Price, J. L., 1984, Amygdalo-cortical projections in the monkey (*Macaca fascicularis*), *J. Comp. Neurol.* **230:**465–496.

Andersen, R. A., Asanuma, C., and Cowan, W. M., 1985, Callosal and prefrontal associational projecting cell populations in area 7a of the macaque monkey: A study using retrogradely transported fluorescent dyes, *J. Comp. Neurol.* **232**:443–455.

Andersen, R. A., Bracewell, R. M., Barash, S., Gnadt, J. W., and Fogassi, L., 1990, Eye position effects on visual memory, and saccade-related activity in areas LIP and 7a of macaque, *J. Neurosci.* **10**:1176–1196.

Anderson, T. J., Jenkins, I. H., Brooks, D. J., Hawken, M. B., Rackowiak, R. S. J., and Kennard, C., 1994, Cortical control of saccades and fixation in man. A PET study, *Brain* **117**:1073–1084.

Andreasen, N., Nasrallah, H. A., Dunn, V., Olson, S. C., Grove, W. M., Ehrhardt, J. C., Coffman, J. A., and Crossett, J. H., 1986, Structural abnormalities in the frontal system in schizophrenia. A magnetic resonance imaging study, *Arch. Gen. Psychiatry* **43**:136–144.

Antes, J. R., 1974, The time course of picture viewing, *J. Exp. Psychol.* **103**:62–70.

Apicella, P., Ljungberg, T., Scarnati, E., and Schultz, W., 1991, Responses to reward in monkey dorsal and ventral striatum, *Exp. Brain Res.* **85**:491–500.

Arikuni, T., and Kubota, K., 1986, The organization of prefrontocaudate projections and their laminar origin in the macaque monkey: A retrograde study using HRP-gel, *J. Comp. Neurol.* **244**:492–510.

Arnsten, A. F. T., Cai, J. X., Murphy, B. L., and Goldman-Rakic, P. S., 1994, Dopamine D1 receptor mechanisms in the cognitive performance of young adult and aged monkeys, *Psychopharmacology* **116**:143–151.

Asanuma, C., Thach, W. T., and Jones, E. G., 1983, Distribution of cerebellar terminations and their relation to other afferent terminations in the thalamic ventral lateral region of the monkey, *Brain Res. Rev.* **5**:237–265.

Azuma, M., and Suzuki, H., 1984, Properties and distribution of auditory neurons in the dorsolateral prefrontal cortex of the alert monkey, *Brain Res.* **298**:343–346.

Bachevalier, J., and Mishkin, M., 1986, Visual recognition impairment follows ventromedial but not dorsolateral prefrontal lesions in monkeys, *Behav. Brain Res.* **20**:249–261.

Baedeker, C., and Wolf, W., 1987, Influence of saccades on manual reactions—A reaction time and VEP study, *Vision Res.* **27**:609–619.

Baizer, J. S., Ungerleider, L. G., and Desimone, R., 1991, Organization of visual inputs to the inferior temporal and posterior parietal cortex in macaques, *J. Neurosci.* **11**:168–190.

Baleydier, C., and Maguire, F., 1980, The duality of the cingulate gyrus in monkey. Neuroanatomical study and functional hypothesis, *Brain* **103**:525–554.

Barbas, H., 1988, Anatomic organization of basoventral and mediodorsal visual recipient prefrontal regions in the rhesus monkey, *J. Comp. Neurol.* **276**:313–342.

Barbas, H., 1993, Organization of cortical afferent input to orbitofrontal areas in the rhesus monkey, *Neuroscience* **56**:841–864.

Barbas, H., and Mesulam, M.-M., 1981, Organization of afferent input to subdivisions of area 8 in the rhesus monkey, *J. Comp. Neurol.* **200**:407–431.

Barbas, H., and Pandya, D. N., 1987, Architecture and frontal cortical connections of the premotor cortex (area 6) in the rhesus monkey, *J. Comp. Neurol.* **256**:211–228.

Barone, P., and Joseph, J.-P., 1989, Prefrontal cortex and spatial sequencing in macaque monkey, *Exp. Brain Res.* **78**:447–464.

Bates, J. F., and Goldman-Rakic, P. S., 1993, Prefrontal connections of medial motor areas in the rhesus monkey, *J. Comp. Neurol.* **336**:211–228.

Batuev, A. S., Shaefer, V. I., and Orlov, A. A., 1985, Comparative characteristics of unit activity in the prefrontal and parietal areas during delayed performance in monkeys, *Behav. Brain Res.* **16**:57–70.

Bauer, R. H., and Fuster, J., 1976, Delayed-matching and delayed-response deficit from cooling dorsolateral prefrontal cortex in monkeys, *J. Comp. Physiol. Psychol.* **90**:293–302.

Baylis, G. C., Rolls, E. T., and Leonard, C. M., 1987, Functional subdivisions of the temporal lobe neocortex, *J. Neurosci.* **7**:330–342.

Becker, W., and Jürgens, R., 1979, An analysis of the saccadic system by means of double step stimuli, *Vision Res.* **19**:976–983.

Beevor, C. E., and Horsley, V., 1888, A further minute analysis by electric stimulation of the so-called motor region of the cortex cerebri in the monkey (*Macacus sinicus*), *Phil. Trans. R. Soc. Lond.* **179**:205–256.

Beevor, C. E., and Horsley, V., 1890, A record of the results obtained by electrical excitation of the so-called motor cortex and internal capsule in an orang-outang, *Phil. Trans. R. Soc. Lond.* **181:**129–158.

Bekkering, H., Adam, J. J., Kingma, H., Huson, A., and Whiting, H. T. A., 1994, Reaction time latencies of eye and hand movements in single- and dual-task conditions, *Exp. Brain Res.* **97:**471–476.

Benecke, R., Dick, J. P. R., Rothwell, J. C., Day, B. L., and Marsden, C. D., 1985, Increase of the Bereitschaftspotential in simultaneous and sequential movements, *Neurosci. Lett.* **62:**347–352.

Benevento, L. A., Fallon, J. H., Davis, B. J., and Rezak, M., 1977, Auditory-visual interaction in single cells of the superior temporal sulcus and orbito-frontal cortex of the macaque monkey, *Exp. Neurol.* **57:**849–872.

Berman, R. A., Luna, B., McCurtain, B. J., Strojwas, M. H., Voyvodic, J. T., Thulborn, K. R., and Sweeney, J. A., 1996, FMRI studies of human frontal eye fields, *Soc. Neurosci. Abstr.* **22:**1687.

Bichot, N. P., Schall, J. D., and Thompson, K. G., 1996, Visual feature selectivity in frontal eye fields induced by experience in mature macaques, *Nature* **381:**697–699.

Bizzi, E., 1968, Discharge of frontal eye field neurons during saccadic and following eye movements in unanesthetized monkeys, *Exp. Brain Res.* **6:**69–80.

Bizzi, E., and Schiller, P. H., 1970, Single unit activity in the frontal eye fields of unanesthetized monkeys during eye and head movement, *Exp. Brain Res.* **10:**151–158.

Blum, B., Kulikowski, J. J., Carden, D., and Harwood, D., 1982, Eye movements induced by electrical stimulation of the frontal eye fields of marmosets and squirrel monkeys, *Brain Behav. Evol.* **21:**34–41.

Boch, R., and Goldberg, M. E., 1989, Participation of prefrontal neurons in the preparation of visually guided eye movements in the rhesus monkey, *J. Neurophysiol.* **61:**1064–1084.

Bock, O., 1987, Coordination of arm and eye movements in tracking of sinusoidally moving targets, *Behav. Brain Res.* **24:**93–100.

Bolles, R. C., 1975, *Theory of Motivation,* Harper & Row, New York.

Bon, L., and Lucchetti, C., 1991, Behavioral and motor mechanisms of dorsomedial frontal cortex of macaca monkey, *Int. J. Neurosci.* **60:**187–193.

Bon, L., and Lucchetti, C., 1992, The dorsomedial frontal cortex of the macaca monkey: Fixation and saccade-related activity, *Exp. Brain Res.* **89:**571–580.

Bon, L., and Lucchetti, C., 1994, Ear and eye representation in the frontal cortex, area 8b, of the macaque monkey: An electrophysiological study, *Exp. Brain Res.* **102:**259–271.

Bötzel, K., Plendl, H., Paulus, W., and Scherg, M., 1993, Bereitschaftspotential: Is there a contribution of the supplementary motor area: *Electroenceph. Clin. Neurophysiol.* **89:**187–196.

Boussaoud, D., and Wise, S. P., 1993a, Primate frontal cortex: Neuronal activity following attentional versus intentional cues, *Exp. Brain Res.* **95:**15–27.

Boussaoud, D., and Wise, S. P., 1993b, Primate frontal cortex: Effects of stimulus movement, *Exp. Brain Res.* **95:**28–40.

Boussaoud, D., 1995, Primate premotor cortex: Modulation of preparatory neuronal activity by gaze angle, *J. Neurophysiol.* **73:**886–890.

Boussaoud, D., Barth, T. M., and Wise, S. P., 1993, Effects of gaze on apparent visual responses of frontal cortex neurons, *Exp. Brain Res.* **93:**423–434.

Bradley, A., Skottun, B. C., Ohzawa, I., Sclar, G., and Freeman, R., 1987, Visual orientation and spatial frequency discrimination: A comparison of single neurons and behavior, *J. Neurophysiol.* **57:**755–772.

Bridgeman, B., van der Heijden, A. H. C., and Velichkovsky, B. M., 1994, A theory of visual stability across saccadic eye movements, *Behav. Brain Sci.* **17:**247–292.

Brinkman, C., 1984, Supplementary motor area of the monkey's cerebral cortex: Short- and long-term deficits after unilateral ablation and the effects of subsequent callosal section, *J. Neurosci.* **4:**918–929.

Brinkman, C., and Porter, R., 1979, Supplementary motor area in the monkey: Activity of neurons during performance of a learned motor task, *J. Neurophysiol.* **42:**681–709.

Britten, K. H., Shadlen, M. N., Newsome, W. T., and Movshon, J. A., 1992, The analysis of visual motion: A comparison of neuronal and psychophysical performance, *J. Neurosci.* **12:**4745–4765.

Bruce, C. J., 1990, Integration of sensory and motor signals for saccadic eye movements in the primate frontal eye fields, in: *Signals and Sense, Local and Global Order in Perceptual Maps* (G. M. Edelman, W. E. Gall, and W. M. Cowan, eds.), Wiley, New York, pp. 261–314.

Bruce, C. J., and Goldberg, M. E., 1985, Primate frontal eye fields. I. Single neurons discharging before saccades, *J. Neurophysiol.* **53:**603–635.

Bruce, C. J., Goldberg, M. E., Bushnell, C., and Stanton, G. B., 1985, Primate frontal eye fields. II. Physiological and anatomical correlates of electrically evoked eye movements, *J. Neurophysiol.* **54:**714–734.

Buchsbaum, M. S., and Haier, R. J., 1987, Functional and anatomical brain imaging: Impact on schizophrenia research, *Schizophrenia Bull.* **13:**115–132.

Burman, D. D., and Segraves, M. A., 1994, Primate frontal eye field activity during natural scanning eye movements, *J. Neurophysiol.* **71:**1266–1271.

Butter, C. M., 1969, Perseveration in extinction and in discrimination reversal tasks following selective frontal ablations in *Macaca mulatta, Physiol. Behav.* **4:**163–171.

Caminiti, R., Johnson, P. B., and Urbano, A., 1991, Making arm movements within different parts of space: The premotor and motor cortical representation of a coordinate system for reaching to visual targets, *J. Neurosci.* **11:**1182–1197.

Carpenter, R. H. S., and Williams, M. L. L., 1995, Neural computation of log likelihood in control of saccadic eye movements, *Nature* **377:**59–62.

Cavada, C., and Goldman-Rakic, P. S., 1989, Posterior parietal cortex in rhesus monkey: II. Evidence for segregated corticocortical networks linking sensory and limbic areas with the frontal lobe, *J. Comp. Neurol.* **287:**422–445.

Chedru, F., Leblanc, M., and Lhermitte, F., 1973, Visual searching in normal and brain-damaged subjects (contribution to the study of unilateral inattention), *Cortex* **9:**94–111.

Chen, D. F., Hyland, B., Maier, V., Palmeri, A., and Wiesendanger, M., 1991, Comparison of neural activity in the supplementary motor area and in the primary motor cortex in monkeys, *Somatosensory Motor Res.* **8:**27–44.

Chen, L. L., and Wise, S. P., 1995a, Neuronal activity in the supplementary eye field during acquisition of conditional oculomotor associations, *J. Neurophysiol.* **73:**1101–1121.

Chen, L. L., and Wise, S. P., 1995b, Supplementary eye field contrasted with the frontal eye field during acquisition of conditional oculomotor associations, *J. Neurophysiol.* **73:**1122–1134.

Chen, Y.-C., Thaler, D., Nixon, P. D., Stern, C. E., and Passingham, R. E., 1995, The functions of the medial premotor cortex. II. The timing and selection of learned movements, *Exp. Brain Res.* **102:**461–473.

Cohn, T. E., Green, D. G., and Tanner, W. P., 1975, Receiver operating characteristic analysis: Application to the study of quantum fluctuation effects in optic nerve of *Rana pipiens, J. Gen. Physiol.* **66:**583–616.

Colby, C. L., and Duhamel, J.-R., 1991, Heterogeneity of extrastriate visual areas and multiple parietal areas in the macaque monkey, *Neuropsychologia* **29:**517–537.

Colebatch, J. G., Deiber, M. P., Passingham, R. E., Friston, K. J., and Frackowiak, R. S. J., 1991, Regional cerebral blood flow during voluntary arm and hand movements in human subjects, *J. Neurophysiol.* **65:**1392–1401.

Collin, N. G., Cowey, A., Latto, R., and Marzi, C., 1982, The role of frontal eye-fields and superior colliculi in visual search and non-visual search in rhesus monkeys, *Behav. Brain Res.* **4:**177–193.

Colombo, M., and Gross, C. G., 1994, Responses of inferior temporal cortex and hippocampal neurons during delayed matching to sample in monkeys (*Macaca fascicularis*), *Behav. Neurosci.* **108:**443–455.

Colombo, M., Eickhoff, A. E., and Gross, C. G., 1993, The effects of inferior temporal and dorsolateral frontal lesions on serial-order behavior and visual imagery in monkeys, *Cognitive Brain Res.* **1:**211–217.

Corbetta, M., Miezin, F. M., Dobmeyer, S., Shulman, G. L., and Petersen, S. E., 1991, Selective and divided attention during visual discriminations of shape, color and speed: Functional anatomy by positron emission tomography, *J. Neurosci.* **11:**2383–2402.

Crammond, D. J., and Kalaska, J. F., 1994, Modulation of preparatory neuronal activity in dorsal premotor cortex due to stimulus–response compatibility, *J. Neurophysiol.* **71:**1281–1284.

Crick, F., and Koch, C., 1995, Are we aware of neural activity in primary visual cortex? *Nature* **375:**121–123.

Crosby, E. C., Yoss, R. E., and Henderson, J. W., 1952, The mammalian midbrain and isthmus regions. Part II. The fiber connections. D. The pattern for eye movements in the frontal eye field and the discharge of specific portions of this field to and through midbrain levels, *J. Comp. Neurol.* **97:**357–381.

Crowne, D. P., Dawson, K. A., and Richardson, C. M., 1989, Unilateral periarcuate and posterior parietal lesions impair conditional position discrimination learning in the monkey, *Neuropsychologia* **27:**1119–1127.

Darby, D. G., Nobre, A. C., Thangaraj, V., Edelman, R., Mesulam, M. M., and Warach, S., 1996, Cortical activation in the human brain during lateral saccades using EPISTAR functional magnetic resonance imaging, *Neuroimage* **3:**53–62.

Dassonville, P., Schlag, J., and Schlag-Rey, M., 1992a, Oculomotor localization relies on a damped representation of saccadic eye displacement in human and nonhuman primates, *Visual Neurosci.* **9:**261–269.

Dassonville, P., Schlag, J., and Schlag-Rey, M., 1992b, The frontal eye field provides the goal of saccadic eye movement, *Exp. Brain Res.* **89:**300–310.

Dassonville, P., Schlag, J., and Schlag-Rey, M., 1995, The use of egocentric and exocentric location cues in saccadic programming, *Vision Res.* **35:**2191–2199.

Deacon, T. W., 1992, Cortical connections of the inferior arcuate sulcus cortex in the macaque brain, *Brain Res.* **573:**8–26.

Deecke, L., and Kornhuber, H. H., 1978, An electrical sign of participation of the mesial "supplementary" motor cortex in human voluntary finger movements, *Brain Res.* **159:**473–476.

Deecke, L., Scheid, P., and Kornhuber, H. H., 1969, Distribution of readiness potential, pre-motion positivity and motor potential of the human cerebral cortex preceding voluntary finger movements, *Exp. Brain Res.* **7:**158–168.

Deecke, L., Kornhuber, H. H., Lang, W., Lang, M., and Schreiber, H., 1985, Timing function of the frontal cortex in sequential motor and learning tasks, *Human Neurobiol.* **4:**143–154.

Démonet, J. F., Chollet, F., Ramsey, S., Cardebat, D., Nespoulous, J. L., Wise, R., Rascol, A., and Frackowiak, R., 1992, The anatomy of phonological and semantic processing in normal subjects, *Brain* **115:**1753–1768.

Deng, S.-Y., Goldberg, M. E., Segraves, M. A., Ungerleider, L. G., and Mishkin, M., 1986, The effect of unilateral ablation of the frontal eye fields on saccadic performance in the monkey, in *Adaptive Processes in Visual and Oculomotor Systems,* (E. L. Keller and D. S. Zee, eds.), Pergamon Press, Oxford, pp. 201–208.

Desimone, R., and Duncan, J., 1995, Neural mechanisms of selective visual attention, *Ann. Rev. Neurosci.* **18:**193–222.

Dias, E. C., and Bruce, C. J., 1994, Physiological correlate of fixation disengagement in the primate's frontal eye field, *J. Neurophysiol.* **72:**2532–2537.

Dick, J. P. R., Benecke, R., Rothwell, J. C., Day, B. L., and Marsden, C. D., 1986, Simple and complex movements in a patient with infarction of the right supplementary motor area, *Movement Disord.* **1:**255–266.

Di Pelligrino, G., and Wise, S. P., 1993a, Visuospatial versus visuomotor activity in the premotor and prefrontal cortex of a primate, *J. Neurosci.* **13:**1227–1243.

Di Pelligrino, G., and Wise, S. P., 1993b, Effects of attention on visuomotor activity in the premotor and prefrontal cortex of a primate, *Somotosensory Motor Res.* **10:**245–262.

Di Pelligrino, G., Fadiga, L., Fogassi, L., Gallese, V., and Rizzolatti, G., 1992, Understanding motor events: A neurophysiological study, *Exp. Brain Res.* **91:**176–180.

Distler, C., Boussaoud, D., Desimone, R., and Ungerleider, L. G., 1993, Cortical connections of inferior temporal area TEO in macaque monkeys, *J. Comp. Neurol.* **334:**125–150.

Dorris, M. C., and Munoz, D. P., 1995, A neural correlate for the gap effect on saccadic reaction times in monkey, *J. Neurophysiol.* **73:**2558–2562.

Duhamel, J.-R., Colby, C. L., and Goldberg, M. E., 1992, The updating of the representation of visual space in parietal cortex by intended eye movements, *Science* **255:**90–92.

Dum, R. P., and Strick, P. L., 1991, The origin of corticospinal projections from the premotor areas in the frontal lobe, *J. Neurosci.* **11:**667–689.

Eccles, J. C., 1982, The initiation of voluntary movements by the supplementary motor area, *Arch. Psych. Nervenkr.* **231:**423–441.

Erickson, R. G., and Dow, B. M., 1989, Foveal tracking cells in the superior temporal sulcus of the macaque monkey, *Exp. Brain Res.* **78:**113–131.

Evdokimidis, I., Mergner, T., and Lücking, C. H., 1992, Dependence of presaccadic cortical potentials on the type of saccadic eye movement, *Electroenceph. Clin. Neurophysiol.* **83:**179–191.
Evinger, C., Manning, K. A., Pellegrini, J. J., Basso, M. A., Powers, A. S., and Sibony, P. A., 1994, Not looking while leaping: The linkage of blinking and saccadic gaze shifts, *Exp. Brain Res.* **100:**337–344.
Felleman, D. J., and Van Essen, D. C., 1991, Distributed hierarchical processing in the primate cerebral cortex, *Cerebral Cortex* **1:**1–47.
Ferrier, D., 1874, The localization of function in brain, *Proc. R. Soc. Lond.* **22:**229–232.
Ferrier, D., 1875, Experiments on the brain of monkeys, *Proc. R. Soc. Lond.* **23:**409–432.
Fischer, B., and Weber, H., 1993, Express saccades and visual attention, *Behav. Brain Sci.* **16:**553–610.
Fisk, J. D., and Goodale, M. A., 1985, The organization of eye and limb movements during unrestricted reaching to targets in contralateral and ipsilateral visual space, *Exp. Brain Res.* **60:**159–178.
Flanders, M., Tillery, S. I. H., and Soechting, J. F., 1992, Early stages in a sensorimotor transformation, *Behav. Brain Sci.* **15:**309–362.
Foerster, O., 1931, The cerebral cortex in man, *Lancet* **2:**309–312.
Foerster, O., 1936, Motorische Felder und Bahnen. Sensible corticale Felder, in: *Handbuch der Neurologie,* Volume 6 (H. Bumke and O. Foerster, eds.), Springer, Berlin, pp. 1–448.
Fogassi, L., Gallese, V., di Pellegrino, G., Fadiga, L., Gentilucci, M., Luppino, G., Matelli, M., Pedotti, A., and Rizzolatti, G., 1992, Space coding by premotor cortex, *Exp. Brain Res.* **89:**686–690.
Fox, P. T., Fox, J. M., Raichle, M. E., and Burde, R. M., 1985, The role of cerebral cortex in the generation of voluntary saccades: A positron emission tomographic study, *J. Neurophysiol.* **54:**348–369.
Franke, P., Maier, W., Hardt, J., Frieboes, R., Lichtermann, D., and Hain, C., 1993, Assessment of frontal lobe functioning in schizophrenia and unipolar major depression, *Psychopathology* **26:**76–84.
Fried, I., Katz, A., McCarthy, G., Sass, K. J., Williamson, P., Spencer, S. S., and Spencer, D. D., 1991, Functional organization of human supplementary motor cortex studied by electrical stimulation, *J. Neurosci.* **11:**3656–3666.
Fries, W., 1984, Cortical projections to the superior colliculus in the macaque monkey: A retrograde study using horseradish peroxidase, *J. Comp. Neurol.* **230:**55–76.
Frith, C. D., Friston, K., Liddle, P. F., and Frackowiak, R. S. J., 1991, Willed action and the prefrontal cortex in man: A study with PET, *Proc. R. Soc. Lond. B* **244:**241–246.
Fujii, N., Mushiake, H., Tamai, M., and Tanji, J., 1995, Microstimulation of the supplementary eye field during saccade preparation, *NeuroReport* **6:**2565–2568.
Fukushima, J., Fukushima, K., Miyasaka, K., and Yamashita, I., 1994, Voluntary control of saccadic eye movement in patients with frontal cortical lesions and Parkinsonian patients in comparison with that in schizophrenics, *Biol. Psychiatry* **36:**21–30.
Funahashi, S., Bruce, C. J., and Goldman-Rakic, P. S., 1989, Mnemonic coding of visual space in the monkey's dorsolateral prefrontal cortex, *J. Neurophysiol.* **61:**331–349.
Funahashi, S., Bruce, C. J., and Goldman-Rakic, P. S., 1990, Visuospatial coding in primate prefrontal neurons revealed by oculomotor paradigms, *J. Neurophysiol.* **63:**814–831.
Funahashi, S., Bruce, C. J., and Goldman-Rakic, P. S., 1993a, Dorsolateral prefrontal lesions and oculomotor delayed-response performance: Evidence for mnemonic "scotomas," *J. Neurosci.* **13:**1479–1497.
Funahashi, S., Chafee, M. V., and Goldman-Rakic, P. S., 1993b, Prefrontal neuronal activity in rhesus monkeys performing a delayed anti-saccade task, *Nature* **365:**753–756.
Fuster, J. M., 1973, Unit activity in prefrontal cortex during delayed-response performance: Neuronal correlates of transient memory, *J. Neurophysiol.* **36:**61–78.
Fuster, J. M., 1989, *The Prefrontal Cortex,* Raven Press, New York.
Fuster, J. M., and Jervey, J. P., 1982, Neuronal firing in the inferotemporal cortex of the monkey in a visual memory task, *J. Neurosci.* **2:**361–375.
Fuster, J. M., Bauer, R. H., and Jervey, J. P., 1982, Cellular discharge in the dorsolateral prefrontal cortex of the monkey in cognitive tasks, *Exp. Neurol.* **77:**679–694.
Gaffan, D., and Murray, E. A., 1990, Amygdalar interaction with the mediodorsal nucleus of the thalamus and the ventromedial prefrontal cortex in stimulus-reward associative learning in the monkey, *J. Neurosci.* **10:**3479–3493.

Gaffan, D., Gaffan, E. A., and Harrison, S., 1989, Visual-visual associative memory and reward-association learning in monkeys: The role of the amygdala, *J. Neurosci.* **9:**558–564.

Gaffan, E. A., Gaffan, D., and Harrison, S., 1988, Disconnection of the amygdala from visual association cortex impairs visual reward-association learning in monkeys, *J. Neurosci.* **8:**3144–3150.

Galletti, C., and Battaglini, P. P., 1989, Gaze-dependent visual neurons in area V3A of monkey prestriate cortex, *J. Neurosci.* **9:**1112–1125.

Gaspar, P., Stepniewska, I., and Kaas, J. H., 1992, Topography and collateralization of the dopaminergic projections to motor and lateral prefrontal cortex in owl monkeys, *J. Comp. Neurol.* **325:**1–21.

Gaymard, B., Pierrot-Deseilligny, C., and Rivaud, S., 1990, Impairment of sequences of memory guided saccades after supplementary motor area lesions, *Ann. Neurol.* **28:**622–626.

Gaymard, B., Rivaud, S., and Pierrot-Deseilligny, C., 1993, Role of the left and right supplementary motor areas in memory-saccade sequences, *Ann. Neurol.* **34:**404–406.

Gellman, R. S., and Fletcher, W. A., 1992, Eye position signals in human saccadic processing, *Exp. Brain Res.* **89:**425–434.

Gemba, H., Sasaki, K., and Brooks, V. B., 1986, 'Error' potentials in limbic cortex (anterior cingulate area 24) of monkeys during motor learning, *Neurosci. Lett.* **70:**223–227.

Gentilucci, M., and Rizzolatti, G., 1989, Cortical motor control of arm and hand movements, in: *Vision in Action: The Control of Grasping* (M. A. Goodale, ed.), Ablex, Norwood, NJ, pp. 147–162.

Gentilucci, M., Scandolara, C., Pigarev, I., and Rizzolatti, G., 1983, Visual responses in the postarcuate cortex (area 6) of the monkey that are independent of eye position, *Exp. Brain Res.* **50:**464–468.

Gentilucci, M., Fogassi, L., Luppino, G., Matelli, M., Camarda, R., and Rizzolatti, G., 1988, Functional organization of inferior area 6 in the macaque monkey I. Somatotopy and the control of proximal movements, *Exp. Brain Res.* **71:**475–490.

Ghez, C., Hening, W., and Favilla, M., 1990, Parallel interacting channels in the initiation and specification of motor response features, in: *Attention and Performance XIII: Motor Representation and Control* (M. Jeannerod, ed.), Erlbaum, Hillsdale, NJ, pp. 265–293.

Gielen, C. C. A. M., van den Heuvel, P. J. M., and van Gisbergen, J. A. M., 1984, Coordination of fast eye and arm movements in a tracking task, *Exp. Brain Res.* **56:**154–161.

Giguere, M., and Goldman-Rakic, P. S., 1988, Mediodorsal nucleus: Areal, laminar and tangential distribution of afferents and efferents in the frontal lobe of rhesus monkeys, *J. Comp. Neurol.* **277:**195–213.

Glimcher, P. W., and Sparks, D. L., 1992, Movement selection in advance of action in superior colliculus, *Nature* **355:**542–545.

Gnadt, J. W., and Andersen, R. A., 1988, Memory related motor planning activity in posterior parietal cortex of macaque, *Exp. Brain Res.* **70:**216–220.

Godoy, J., Lüders, H., Dinner, D. S., Morris, H. H., and Wyllie, E., 1990, Versive eye movements elicited by cortical stimulation of the human brain, *Neurology* **40:**296–299.

Godschalk, M., Lemon, R. N., Kuypers, H. G. J. M., and Ronday, H. K., 1984, Cortical afferents and afferents of monkey postarcuate area: An anatomical and electrophysiological study, *Exp. Brain Res.* **56:**410–424.

Godschalk, M., Lemon, R. N., Kuypers, H. G. J. M., and van der Steen, J., 1985, The involvement of monkey premotor cortex neurones in preparation of visually cued arm movements, *Behav. Brain Res.* **18:**143–157.

Goldberg, G., 1985, Supplementary motor area structure and function: Review and hypotheses, *Behav. Brain Sci.* **8:**567–616.

Goldberg, M. E., and Bruce, C. J., 1985, Cerebral cortical activity associated with the orientation of visual attention in the rhesus monkey, *Vision Res.* **25:**471–481.

Goldberg, M. E., and Bruce, C. J., 1990, Primate frontal eye fields. III. Maintenance of a spatially accurate saccade signal, *J. Neurophysiol.* **64:**489–508.

Goldberg, M. E., and Bushnell, M. C., 1981, Behavioral enhancement of visual responses in monkey cerebral cortex. II. Modulation in frontal eye fields specifically related to saccades, *J. Neurophysiol.* **46:**773–787.

Goldberg, M. E., and Segraves, M. A., 1989, Visual and frontal cortices, in: *The Neurobiology of Saccadic Eye Movements* (R. H. Wurtz and M. E. Goldberg, eds.), Elsevier, New York, pp. 283–313.

Goldberg, M. E., Bushnell, M. C., and Bruce, C. J., 1986, The effect of attentive fixation on eye movements evoked by electrical stimulation of the frontal eye fields, *Exp. Brain Res.* **61:**579–584.

Goldberg, T. E., Weinberger, D. R., Berman, K. F., Pliskin, N. H., and Podd, M. H., 1987, Further evidence for dementia of the prefrontal type in schizophrenia? A controlled study of teaching the Wisconsin Card Sorting Test, *Arch. Gen. Psychiatry* **44:**1008–1014.

Goldman-Rakic, P. S., 1987, Circuitry of primate prefrontal cortex and regulation of behavior by representational memory, in *Handbook of Physiology,* Volume 5, *The Nervous System* (V. B. Mountcastle, F. Plum, and S. Geiger, eds.), American Physiological Society, Bethesda, MD, pp. 373–417.

Goldman-Rakic, P. S., 1988, Topography of cognition: Parallel distributed networks in primate association cortex, *Annu. Rev. Neurosci.* **11:**137–156.

Goldman-Rakic, P. S., 1990, Cellular and circuit basis of working memory in prefrontal cortex of nonhuman primates, *Prog. Brain Res.* **85:**325–335.

Goldman-Rakic, P. S., and Porrino, L. J., 1985, The primate mediodorsal (MD) nucleus and its projection to the frontal lobe, *J. Comp. Neurol.* **242:**535–560.

Goodale, M. A., and Milner, A. D., 1992, Separate visual pathways for perception and action, *TINS* **15:**20–25.

Gottlieb, J. P., Bruce, C. J., and MacAvoy, M. G., 1993, Smooth eye movements elicited by microstimulation in the primate frontal eye field, *J. Neurophysiol.* **69:**786–799.

Gottlieb, J. P., MacAvoy, M. G., and Bruce, C. J., 1994, Neural responses related to smooth-pursuit eye movements and their correspondence with electrically elicited smooth eye movements in the primate frontal eye field, *J. Neurophysiol.* **72:**1634–1653.

Gould, H. J., Cusick, C. G., Pons, T. P., and Kaas, J. H., 1986, The relationship of corpus callosum connections to electrical stimulation maps of motor, supplementary motor, and the frontal eye fields in owl monkeys, *J. Comp. Neurol.* **247:**297–325.

Gower, E. C., 1989, Efferent projections from limbic cortex of the temporal pole to the magnocellular medial dorsal nucleus in the rhesus monkey, *J. Comp. Neurol.* **280:**343–358.

Grafton, S. T., Mazziotta, J. C., Woods, R. P., and Phelps, M. E., 1992, Human functional anatomy of visually guided finger movements, *Brain* **115:**565–587.

Graziano, M. S. A., Yap, G. S., and Gross, C. G., 1994, Coding of visual space by premotor neurons, *Science* **266:**1054–1057.

Grunbaum, A. S. F., and Sherrington, C. S., 1901, Observations on the physiology of the cerebral cortex of some of the higher apes, *Proc. R. Soc.* **69:**206–209.

Guitton, D., 1991, Control of saccadic eye and gaze movements by the superior colliculus and basal ganglia, in: *Eye Movements* (R. H. S. Carpenter, ed.), CRC Press, Boca Raton, FL, pp. 244–276.

Guitton, D., and Volle, M., 1987, Gaze control in humans: Eye–head coordination during orienting movements to targets within and beyond the oculomotor range, *J. Neurophysiol.* **58:**427–459.

Guitton, D., Buchtel, H. A., and Douglas, R. M., 1985, Frontal Lobe lesions in man cause difficulties in suppressing reflexive glances and in generating goal-directed saccades, *Exp. Brain Res.* **58:**455–472.

Haber, S. N., Kunishio, K., Mizobuchi, M., and Lynd-Balta, E., 1995, The orbital and medial prefrontal circuit through the primate basal ganglia, *J. Neurosci.* **15:**4851–4867.

Hallett, M., 1994, Movement-related cortical potentials, *Electromyography Clin. Neurophysiol.* **34:**5–13.

Hallett, P. E., and Adams, B. D., 1980, The predictability of saccadic latency in a novel voluntary oculomotor task, *Vision Res.* **20:**329–339.

Hallet, P. E., and Lightstone, A. D., 1976, Saccadic eye movement towards stimuli triggered by prior saccades, *Vision Res.* **16:**99–106.

Halsband, U., and Freund, H.-J., 1990, Premotor cortex and conditional motor learning in man, *Brain* **113:**107–222.

Halsband, U., and Passingham, R., 1982, The role of premotor and parietal cortex in the direction of action, *Brain Res.* **240:**368–372.

Halsband, U., and Passingham, R. E., 1985, Premotor cortex and the conditions for movements in monkeys (*Macaca fascicularis*), *Behav. Brain Res.* **18:**269–277.

Halsband, U., Ito, N., Tanji, J., and Freund, H.-J., 1993, The role of premotor cortex and the supplementary motor area in the temporal control of movement in man, *Brain* **116:**243–266.

Halsband, U., Matsuzaka, Y., and Tanji, J., 1994, Neuronal activity in the primate supplementary, presupplementary and premotor cortex during externally and internally instructed sequential movements, *Neurosci. Res.* **20:**149–155.

Hanes, D. P., and Schall, J. D., 1995, Countermanding saccades in macaque, *Visual Neurosci.* **12:**929–937.
Hanes, D. P., and Schall, J. D., 1996, Neural control of voluntary movement initiation, *Science* **274:**427–430.
Hanes, D. P., Thompson, K. G., and Schall, J. D., 1995, Relationship of presaccadic activity in frontal eye field and supplementary eye field to saccade initiation in macaque: Poisson spike train analysis, *Exp. Brain Res.* **103:**85–96.
Harting, J. K., Huerta, M. F., Frankfurter, A. L., Strominger, N. L., and Royce, G. J., 1980, Ascending pathways from monkey superior colliculus: An autoradiographic analysis, *J. Comp. Neurol.* **192:**853–882.
He, S.-Q., Dum, R. P., and Strick, P. L., 1995, Topographic organization of corticospinal projections from the frontal lobe: Motor areas on the medial surface of the hemisphere, *J. Neurosci.* **15:**3284–3306.
Heinen, S. J., 1995, Single neuron activity in the dorsomedial frontal cortex during smooth pursuit eye movements, *Exp. Brain Res.* **104:**357–361.
Helmholtz, H., 1962, *Handbook of Physiological Optics* [1866], Dover, New York.
Henik, A., Rafal, R., and Rhodes, D., 1994, Endogenously generated and visually guided saccades after lesions of the human frontal eye fields, *J. Cognitive Neurosci.* **6:**400–411.
Hepp, K., Henn, V., Vilis, T., and Cohen, B., 1989, Brainstem regions related to saccade generation, in: *The Neurobiology of Saccadic Eye Movements* (R. H. Wurtz and M. E. Goldberg, Elsevier, Amsterdam.
Hikosaka, O., and Wurtz, R. H., 1983a, Visual and oculomotor functions of monkey substantia nigra pars reticulata. III. Memory-contingent visual and saccade responses, *J. Neurophysiol.* **49:**1268–1284.
Hikosaka, O., and Wurtz, R. H., 1983b, Visual and oculomotor functions of monkey substantia nigra pars reticulata. IV. Relation of substantia nigra to superior colliculus, *J. Neurophysiol.* **49:**1285–1301.
Hikosaka, O., and Wurtz, R. H., 1985a, Modification of saccadic eye movements by GABA-related substances. I. Effect of muscimol and bicuculline in the monkey superior colliculus, *J. Neurophysiol.* **53:**266–291.
Hikosaka, O., and Wurtz, R. H., 1985b, Modification of saccadic eye movements by GABA-related substances. II. Effects of muscimol in the monkey substantia nigra pars reticulata, *J. Neurophysiol.* **53:**292–307.
Hikosaka, O., Sakamoto, M., and Usui, S., 1989a, Functional properties of monkey caudate neurons. I. Activities related to saccadic eye movements, *J. Neurophysiol.* **61:**780–798.
Hikosaka, O., Sakamoto, M., and Usui, S., 1989b, Functional properties of monkey caudate neurons. II. Visual and auditory responses, *J. Neurophysiol.* **61:**799–813.
Hikosaka, O., Sakamoto, M., and Usui, S., 1989c, Functional properties of monkey caudate neurons. III. Activities related to expectation of target and reward, *J. Neurophysiol.* **61:**814–832.
Hikosaka, O., Sakamoto, M., and Miyashita, N., 1993, Effects of caudate nucleus stimulation on substantia nigra cell activity in monkey, *Exp. Brain Res.* **95:**457–472.
Holmes, G., 1938, The cerebral integration of the ocular movements, *Br. Med. J.* **2:**107–112.
Honda, H., 1990, Eye movements to a visual stimulus flashed before, during or after a saccade, in: *Attention and Performance XIII: Motor Representation and Control* (M. Jeannerod, ed.), Erlbaum, Hillsdale, NJ, pp. 567–582.
Honda, H., 1991, The timecourse of visual mislocalization and of extraretinal eye position signals at the time of vertical saccades, *Vision Res.* **31:**1915–1921.
Horsley, V., and Schäfer, E. A., 1888, A record of experiments upon the functions of the cerebral cortex, *Phil. Trans. R. Soc. Lond. B* **179:**1–45.
Huerta, M. F., and Kaas, J. H., 1990, Supplementary eye field as defined by intracortical microstimulation: Connections in macaques, *J. Comp. Neurol.* **293:**299–330.
Huerta, M. F., Krubitzer, L. A., and Kaas, J. H., 1986, Frontal eye field as defined by intracortical microstimulation in squirrel monkeys, owl monkeys and macaque monkeys: I. Subcortical connections, *J. Comp. Neurol.* **253:**415–439.
Huerta, M. F., Krubitzer, L. A., and Kaas, J. H., 1987, Frontal eye field as defined by intracortical microstimulation in squirrel monkeys, owl monkeys and macaque monkeys: II. Cortical connections, *J. Comp. Neurol.* **265:**332–361.

Hummelsheim, H., Wiesendanger, M., Bianchetti, M., Wiesendanger, R., and Macpherson, J., 1986, Further investigations of the efferent linkage of the supplementary motor area (SMA) with the spinal cord in the monkey, *Exp. Brain Res.* **65:**75–82.

Hummelsheim, H., Bianchetti, M., Wiesendanger, M., and Wiesendanger, R., 1988, Sensory inputs to the agranular motor fields: A comparison between precentral, supplementary motor and premotor areas in the monkey, *Exp. Brain Res.* **69:**289–298.

Hutchins, K. D., Martina, A. M., and Strick, P. L., 1988, Corticospinal projections from the medial wall of the hemisphere, *Exp. Brain Res.* **71:**667–672.

Ikeda, A., Lüders, H. O., Burgess, R. C., and Shibasaki, H., 1992, Movement-related potentials recorded from supplementary motor area and primary motor area. Role of supplementary motor area in voluntary movements, *Brain* **115:**1017–1043.

Ilinsky, I. A., Jouandet, M. L., and Goldman-Rakic, P. S., 1985, Organization of the nigrothalamocortical system in the rhesus monkey, *J. Comp. Neurol.* **236:**315–330.

Ito, S.-I., 1982, Prefrontal unit activity of macaque monkeys during auditory and visual reaction time tasks, *Brain Res.* **247:**39–47.

Iversen, S. D., and Mishkin, M., 1970, Perseverative interference in monkey following selective lesions of the inferior prefrontal convexity, *Exp. Brain Res.* **11:**376–386.

Jacobsen, C. F., 1936, Studies of cerebral function in primates: I. The functions of the frontal association areas in monkeys, *Comp. Psychol. Monogr.* **13:**3–60.

Jay, M. F., and Sparks, D. L., 1987a, Sensorimotor integration in the primate superior colliculus. I. Motor convergence, *J. Neurophysiol.* **57:**22–34.

Jay, M. F., and Sparks, D. L., 1987b, Sensorimotor integration in the primate superior colliculus. II. Coordinates of auditory signals, *J. Neurophysiol.* **57:**35–55.

Jolly, W. A., and Simpson, S., 1907, The functions of the Rolandic cortex in monkeys, *Proc. R. Soc. Edinb.* **27:**64–78.

Jonides, J., Smith, E. E., Koeppe, R. A., Awh, E., Minoshima, S., and Mintun, M. A., 1993, Spatial working memory in humans as revealed by PET, *Nature* **363:**623–625.

Joseph, J. P., and Barone, P., 1987, Prefrontal unit activity during a delayed oculomotor task in the monkey, *Exp. Brain Res.* **67:**460–468.

Jürgens, R., Becker, W., and Kornhuber, H. H., 1981, Natural and drug-induced variations of velocity and duration of human saccadic eye movements: Evidence for a control of the neural pulse generator by local feedback, *Biol. Cybernet.* **39:**87–96.

Kaas, J. H., 1995, Human visual cortex: Progress and puzzles, *Curr. Biol.* **5:**1126–1128.

Kalaska, J., and Crammond, D., 1992, Cerebral cortical mechanisms of reaching movements, *Science* **266:**1517–1523.

Karpov, B. A., Luria, A. R., and Yarbus, A. L., 1968, Disturbance of the structure of active perception in lesions of the posterior and anterior regions of the brain, *Neuropsychologia* **6:**157–166.

Kawamura, K., and Naito, J., 1984, Corticocortical projections to the prefrontal cortex in the rhesus monkey investigated with horseradish peroxidase techniques, *Neurosci. Res.* **1:**89–103.

Keating, E. G., 1991, Frontal eye field lesions impair predictive and visually-guided pursuit eye movements, *Exp. Brain Res.* **86:**311–323.

Keating, C. F., and Keating, E. G., 1993, Monkeys and mug shots: Cues used by rhesus monkeys (*Macaca mulatta*) to recognize a human face, *J. Comp. Psychol.* **107:**131–139.

Keller, E. K., 1991, The brainstem, in: *Eye Movements* (R. H. S. Carpenter, ed.), CRC Press, Boca Raton, FL, pp. 200–223.

Keller, I., and Heckhausen, H., 1990, Readiness potentials preceding spontaneous motor acts: Voluntary vs. involuntary control, *Electroenceph. Clin. Neurophysiol.* **76:**351–361.

Kitamura, J., Shibasaki, H., Takagi, A., Nabeshima, H., and Yamaguchi, A., 1993, Enhanced negative slope of cortical potentials before sequential as compared with simultaneous extensions of two fingers, *Electroenceph. Clin. Neurophysiol.* **86:**176–182.

Klostermann, W., Kompf, D., Heide, W., Verleger, R., Wauschkuhn, B., and Seyfert, T., 1994, The presaccadic cortical negativity prior to self-paced saccades with and without visual guidance. *Electroenceph. Clin. Neurophysiol.* **91:**219–228.

Kojima, S., 1980, Prefrontal unit activity in the monkey: Relation to visual stimuli and movements, *Exp. Neurol.* **69:**110–123.

Kojima, S., and Goldman-Rakic, P. S., 1982, Delay-related activity of prefrontal neurons in rhesus monkeys performing delayed response, *Brain Res.* **248:**43–49.

Komatsu, H., and Suzuki, H., 1985, Projections from the functional subdivisions of the frontal eye field to the superior colliculus in the monkey, *Brain Res.* **327:**324–327.

Komatsu, H., and Wurtz, R. H., 1988, Relation of cortical areas MT and MST to pursuit eye movements. I. Localization and visual properties of neurons, *J. Neurophysiol.* **60:**580–603.

Kowler, E., 1990, The role of visual and cognitive processes in the control of eye movement, in: *Eye Movements and Their Role in Visual and Cognitive Processes* (E. Kowler, ed.), Elsevier, New York, pp. 1–70.

Krubitzer, L. A., and Kaas, J. H., 1990, Cortical connections of MT in four species of primates: Areal, modular and retinotopic patterns, *Vis. Neurosci.* **5:**165–204.

Kubota, K., and Funahashi, S., 1982, Direction-specific activities of dorsolateral prefrontal and motor cortex pyramidal tract neurons during visual tracking, *J. Neurophysiol.* **47:**362–376.

Kubota, K., and Komatsu, H., 1985, Neuron activities of monkey prefrontal cortex during the learning of visual discrimination tasks with GO/NO-GO performances, *Neurosci. Res.* **3:**106–129.

Kubota, K., and Niki, H., 1971, Prefrontal cortical unit activity and delayed alternation performance in monkeys, *J. Neurophysiol.* **34:**337–347.

Kubota, K., Iwamoto, T., and Suzuki, H., 1974, Visuokinetic activities of primate prefrontal neurons during delayed-response performance, *J. Neurophysiol.* **37:**1197–1212.

Kubota, K., Tonoike, M., and Mikami, A., 1980, Neuronal activity in the monkey dorsolateral prefrontal cortex during a discrimination task with delay, *Brain Res.* **183:**29–42.

Kurata, K., 1989, Distribution of neurons with set and movement related activity before hand and foot movements in the premotor cortex of rhesus monkeys, *Exp. Brain Res.* **77:**245–256.

Kurata, K., 1991, Corticocortical inputs to the dorsal and ventral aspects of the premotor cortex of macaque monkeys, *Neurosci. Res.* **12:**263–280.

Kurata, K., 1994, Information processing for motor control in primate premotor cortex, *Behav. Brain Res.* **61:**135–142.

Kurata, K., and Hoffman, D. S., 1994, Differential effects of muscimol microinjection into dorsal and ventral aspects of the premotor cortex of monkeys, *J. Neurophysiol.* **71:**1151–1164.

Kurata, K., and Tanji, J., 1985, Contrasting neuronal activity in supplementary and precentral motor cortex of monkeys. II. Responses to movement triggering vs. nontriggering sensory signals, *J. Neurophysiol.* **53:**142–152.

Kurata, K., and Wise, S. P., 1988, Premotor cortex of rhesus monkeys: Set-related activity during two conditional motor tasks, *Exp. Brain Res.* **69:**327–343.

Kurata, K., Okano, K., and Tanji, J., 1985, Distribution of neurons related to a hindlimb as opposed to forelimb movement in the monkey premotor cortex, *Exp. Brain Res.* **60:**188–191.

Kurtzberg, D., and Vaughan, H. G., 1982, Topographic analysis of human cortical potentials preceding self-initiated and visually triggered saccades, *Brain Res.* **243:**1–9.

Kustov, A. A., and Robinson, D. A., 1995, Modified saccades evoked by stimulation of macaque superior colliculus account for properties of resettable integrator, *J. Neurophysiol.* **73:**1724–1728.

Kwan, H. C., Mackay, W. A., Murphy, J. T., and Wong, Y. C., 1985, Properties of visual cue responses in primate precentral cortex, *Brain Res.* **343:**24–35.

Lang, W., Zilch, O., Koska, C., Lindinger, G., and Deecke, L., 1989, Negative cortical dc shifts preceding and accompanying simple and complex sequential movements, *Exp. Brain Res.* **74:**99–104.

Lang, W., Obrig, H., Lindinger, G., Cheyne, D., and Deecke, L., 1990, Supplementary motor area activation while tapping bimanually different rhythms in musicians, *Exp. Brain Res.* **79:**504–514.

Lang, W., Cheyne, D., Kristeva, R., Beisteiner, R., Lindinger, G., and Deecke, L., 1991, Three-dimensional localization of SMA activity preceding voluntary movement. A study of electric and magnetic fields in a patient with infarction of the right supplementary motor area, *Exp. Brain Res.* **87:**688–695.

Lang, W., Petit, L., Hölander, P., Pietrzyk, U., Tzourio, N., Mazoyer, B., and Berthoz, A., 1994, A positron emission tomography study of oculomotor imagery, *NeuroReport* **5:**921–924.

Latto, R., 1978a, The effects of bilateral frontal eye-field lesions on the learning of visual search task by rhesus monkeys, *Brain Res.* **147:**370–376.

Latto, R., 1978b, The effects of bilateral frontal eye-field, posterior parietal or superior collicular lesions on visual search in the rhesus monkey, *Brain Res.* **146:**35–50.

Lee, K., and Tehovnik, E., 1995, Topographic distribution of fixation-related units in the dorsomedial frontal cortex of the rhesus monkey, *Eur. J. Neurosci.* **7:**1005–1011.

Leichnetz, G. R., Spencer, R. F., Hardy, S. G. P., and Astruc, J., 1981, The prefrontal corticotectal projection in the monkey: An anterograde and retrograde horseradish peroxidase study, *Neuroscience* **6:**1023–1041.

Leichnetz, G. R., Spencer, R. F., and Smith, D. J., 1984a, Cortical projections to nuclei adjacent to the oculomotor complex in the medial dien-mesencephalic tegmentum in the monkey, *J. Comp. Neurol.* **228:**359–387.

Leichnetz, G. R., Smith, D. J., and Spencer, R. F., 1984b, Cortical projections to the paramedian tegmental and basilar pons in the monkey, *J. Comp. Neurol.* **228:**388–408.

Levin, S., 1984, Frontal lobe dysfunctions in schizophrenia—I. Eye movement impairments, *J. Psychiatr. Res.* **18:**27–55.

Levin, H. S., Eisenberg, H. M., and Benton, A. L., 1991, *Frontal Lobe Function and Dysfunction,* Oxford University Press, Oxford.

Levinsohn, G., 1909, Uber die Beziehungen der Grosshirnrinde beim Affen zu den Bewegungen des Auges, *Graefe Arch. Ophthalmol.* **71:**313–378.

Leyton, A. S. F., and Sherrington, C. S., 1917, Observations on the excitable cortex of the chimpanzee, orang-utan and gorilla, *Q. J. Exp. Physiol.* **11:**135–222.

Libet, B., 1985, Unconscious cerebral initiative and the role of conscious will in voluntary action, *Behav. Brain Sci.* **8:**529–566.

Lidow, M. S., Goldman-Rakic, P. S., Gallager, D. W., and Rakic, P., 1991, Distribution of dopaminergic receptors in the primate cerebral cortex: Quantitative autoradiographic analysis using [3H]raclopride, [3H]spiperone and [3H]SCH23390, *Neuroscience* **40:**657–71.

Lim, S. H., Dinner, D. S., Pillay, P. K., Luders, H., Morris, H. H., Klem, G., Wyllie, E., and Awad, I. A., 1994, Functional anatomy of the human supplementary sensorimotor area: Results of extraoperative electrical stimulation, *Electroenceph. Clin. Neurophysiol.* **91:**179–193.

Ljungberg, T., Apicella, P., and Schultz, W., 1992, Responses of monkey dopamine neurons during learning of behavioral reactions, *J. Neurophysiol.* **67:**145–163.

Lu, M.-T., Preston, J. B., and Strick, P. L., 1994, Interconnections between the prefrontal cortex and the premotor areas in the frontal lobe, *J. Comp. Neurol.* **341:**375–392.

Luppino, G., Matelli, M., and Rizzolatti, G., 1990, Cortico-cortical connections of two electrophysiologically identified arm representations in the mesial agranular frontal cortex, *Exp. Brain Res.* **82:**214–218.

Luppino, G., Matelli, M., Camarda, R. M., Gallese, V., and Rizzolatti, G., 1991, Multiple representations of body movements in mesial area 6 and the adjacent cingulate cortex: An intracortical microstimulation study in the macaque monkey, *J. Comp. Neurol.* **311:**463–482.

Luppino, G., Matelli, M., Camarda, R. M., and Rizzolatti, G., 1993, Corticocortical connections of area F3 (SMA-proper) and area F6 (pre-SMA) in the macaque monkey, *J. Comp. Neurol.* **338:**114–140.

Luria, A. R., Karpov, B. A., and Yarbus, A. L., 1966, Disturbances of active visual perception with lesions of frontal lobes, *Cortex* **2:**202–212.

Lynch, J. C., 1987, Frontal eye field lesions in monkeys disrupt visual pursuit, *Exp. Brain Res.* **68:**437–441.

Lynch, J. C., Mountcastle, V. B., Talbot, W. H., and Yin, T. C. T., 1977, Parietal lobe mechanisms for directed visual attention, *J. Neurophysiol.* **40:**362–389.

Lynch, J. C., Hoover, J. E., and Strick, P. L., 1994, Input to the primate frontal eye field from the substantia nigra, superior colliculus and dentate nucleus demonstrated by transneuronal transport, *Exp. Brain Res.* **100:**181–186.

MacAvoy, M. G., and Bruce, C. J., 1995, Comparison of the smooth eye tracking disorder of schizophrenics with that of nonhuman primates with specific brain lesions, *Int. J. Neurosci.* **80:**117–151.

MacAvoy, M. G., Gottlieb, J. P., and Bruce, C. J., 1991, Smooth-pursuit eye movement representation in the primate frontal eye field, *Cerebral Cortex* **1:**95–102.

Mann, S. E., Thau, R., and Schiller, P. H., 1988, Conditional task-related responses in monkey dorsomedial frontal cortex, *Exp. Brain Res.* **69:**460–468.

Marrocco, R. T., 1978, Saccades induced by stimulation of the frontal eye fields: Interaction with voluntary and reflexive eye movements, *Brain Res.* **146:**23–34.

Massion, J., Viallet, F., Massarino, R., and Khalil, R., 1989, The supplementary motor area region is involved in the coordination between movement and posture, *C. R. Acad. Paris* **308:**417–423.

Matelli, M., Luppino, G., and Rizzolatti, G., 1985, Patterns of cytochrome oxidase activity in the frontal agranular cortex of the macaque monkey, *Behav. Brain Res.* **18:**125–136.
Matelli, M., Camarda, R., Glickstein, M., and Rizzolatti, G., 1986, Afferent and efferent projections of the inferior area 6 in the macaque monkey, *J. Comp. Neurol.* **251:**281–298.
Matelli, M., Luppino, G., and Rizzolatti, G., 1991, Architecture of superior and mesial area 6 and the adjacent cingulate cortex in the macaque monkey, *J. Comp. Neurol.* **311:**445–462.
Matelli, M., Rizzolatti, G., Bettinardi, V., Gilardi, M. C., Perani, D., Rizzo, G., and Fazio, F., 1993, Activation of precentral and mesial motor areas during the execution of elementary proximal and distal arm movements: A PET study, *NeuroReport* **4:**1295–1298.
Matin, L., 1985, Visual localization and eye movements, in: *Handbook of Perception and Human Performance* (K. R. Boff, L. Kaufman, and J. P. Thomas, eds.), Wiley, New York.
Matsuzaka, Y., Aizawa, H., and Tanji, J., 1992, A motor area rostral to the supplementary motor area (presupplementary motor area) in the monkey: Neuronal activity during a learned motor task, *J. Neurophysiol.* **68:**653–662.
Maunsell, J. H. R., and Gibson, J. R., 1992, Visual response latencies in striate cortex of the macaque monkey, *J. Neurophysiol.* **68,** 1332–1344.
Mays, L. E., and Sparks, D. L., 1980, Dissociation of visual and saccade-related responses in superior colliculus neurons, *J. Neurophysiol.* **43:**207–232.
Melamed, E., and Larsen, B., 1979, Cortical activation pattern during saccadic eye movements in humans: Localization by focal cerebral blood flow increases, *Ann. Neurol.* **5:**79–88.
Merigan, W. H., and Maunsell, J. H. R., 1993, How parallel are the primate visual pathways? *Ann. Rev. Neurosci.* **16:**369–402.
Mikami, A., Ito, S., and Kubota, K., 1982, Visual response properties of dorsolateral prefrontal neurons during visual fixation task, *J. Neurophysiol.* **47:**593–605.
Miller, E. K., Li, L., and Desimone, R., 1993, Activity of neurons in anterior inferior temporal cortex during a short-term memory task, *J. Neurosci.* **13:**1460–1478.
Miller, J., 1988, Discrete and continuous models of human information processing: Theoretical distinctions and empirical results, *Acta Psychol.* **67:**191–257.
Milner, B., Petrides, M., and Smitt, M. L., 1985, Frontal lobes and the temporal organization of memory, *Hum. Neurobiol.* **4:**137–142.
Mishkin, M., and Manning, F. J., 1978, Non-spatial memory after selective prefrontal lesions in monkeys, *Brain Res.* **143:**313–324.
Mitz, A. R., and Godschalk, M., 1989, Eye-movement representation in the frontal lobe of rhesus monkeys, *Neurosci. Lett.* **106:**157–162.
Mitz, A. R., Godschalk, M., and Wise, S. P., 1991, Learning-dependent neuronal activity in the premotor cortex: Activity during the acquisition of conditional motor associations, *J. Neurosci.* **11:**1855–1872.
Miyashita, Y., and Chang, H. S., 1988, Neuronal correlate of pictorial short-term memory in the primate temporal cortex, *Nature* **331:**68–70.
Mohler, C. W., Goldberg, M. E., and Wurtz, R. H., 1973, Visual receptive fields of frontal eye field neurons, *Brain Res.* **61:**385–389.
Moran, M. A., Mufson, E. J., and Mesulam, M.-M., 1987, Neural inputs into the temporopolar cortex of the rhesus monkey, *J. Comp. Neurol.* **256:**88–103.
Morecraft, R. J., and Van Hoesen, G. W., 1992, Cingulate input to the primary and supplementary motor cortices in the rhesus monkey: Evidence for somatotopy in areas 24c and 23c, *J. Comp. Neurol.* **322:**471–489.
Morel, A., and Bullier, J., 1990, Anatomical segregation of two cortical visual pathways in the macaque monkey, *Vis. Neurosci.* **4:**555–578.
Moser, A., and Kömpf, D., 1990, Unilateral visual exploration deficit in a frontal lobe lesion, *Neuro-ophthalmology* **10:**39–44.
Moster, M. L., and Goldberg, G., 1990, Topography of scalp potentials preceding self-initiated saccades, *Neurology* **40:**644–648.
Mott, F. W., and Schäfer, E. A., 1890, On associated eye-movements produced by cortical faradization of the monkey's brain, *Brain* **13:**165–173.
Muakkasa, K. F., and Strick, P. L., 1979, Frontal lobe inputs to primate motor cortex: Evidence for four somatotypically organized "premotor" areas, *Brain Res.* **177:**176–182.

Munoz, D. P., and Wurtz, R. H., 1993, Fixation cells in monkey superior colliculus. I. Characteristics of cell discharge, *J. Neurophysiol.* **70:**559–575.

Müri, R. M., Rosler, K. M., and Hess, C. W., 1994, Influence of transcranial magnetic stimulation on the execution of memorized sequences of saccades, *Exp. Brain Res.* **101:**521–524.

Müri, R. M., Rivand, S., Vermersch, A. I., Leger, J. M., and Pierrot-Deseilligny, C., 1995, Effects of transcranial magnetic stimulation over the region of the supplementary motor area during sequences of memory-guided saccades, *Exp. Brain Res.* **104:**163–166.

Mushiake, H., Inase, M., and Tanji, J., 1991, Neuronal activity in the primate premotor supplementary and precentral motor cortex during visually guided and internally determined sequential movements, *J. Neurophysiol.* **66:**705–718.

Mushiake, H., Fujii, N., and Tanji, J., 1996, Visually guided saccade versus eye-hand reach: Contrasting neuronal activity in the cortical supplementary and frontal eye fields, *J. Neurophysiol.* **75:**2187–2191.

Nakashima, Y., Momose, T., Sano, I., Katayama, S., Nakajima, T., Niwa, S., and Matsushita, M., 1994, Cortical control of saccade in normal and schizophrenic subjects: A PET study using a task-evoked rCBF paradigm, *Schizophrenia Res.* **12:**259–64.

Neumann, O., and Prinz, W., 1990, Prologue: Historical approaches to perception and action, in *Relationships Between Perception and Action* (O. Neumann and W. Prinz, eds.), Springer-Verlag, Berlin, pp. 4–19.

Nichols, M. J., and Sparks, D. L., 1995, Nonstationary properties of the saccade system: New constraints on models of saccadic control, *J. Neurophysiol.* **73:**431–435.

Niki, H., 1974a, Differential activity of prefrontal units during right and left delayed response trials, *Brain Res.* **70:**346–349.

Niki, H., 1974b, Prefrontal unit activity during delayed alternation in the monkey. I. Relation to direction of response, *Brain Res.* **68:**185–196.

Niki, H., 1974c, Prefrontal unit activity during delayed alternation in the monkey. II. Relation to absolute versus relative direction of response, *Brain Res.* **68:**197–204.

Niki, H., and Watanabe, M., 1976, Cingulate unit activity and delayed response, *Brain Res.* **110:**381–386.

Niki, H., and Watanabe, M., 1979, Prefrontal and cingulate unit activity during timing behavior in the monkey, *Brain Res.* **171:**213–224.

Niki, H., Sakai, M., and Kubota, K., 1972, Delayed alternation performance and unit activity of the caudate head and medial orbitofrontal gyrus in the monkey, *Brain Res.* **38:**343–353.

Noton, D., and Stark, L., 1971, Scanpaths in saccadic eye movements while viewing and recognizing patterns, *Vision Res.* **11:**929–942.

Nowak, L. G., Munk, M. H. J., Girard, P., and Bullier, J., 1995, Visual latencies in areas V1 and V2 of the macaque monkey, *Visual Neurosci.* **12:**371–384.

O'Driscoll, G. A., Alpert, N. M., Matthysse, S. W., Levy, D. L., Rauch, S. L., and Holzman, P. S., 1995, Functional neuroanatomy of antisaccade eye movements investigated with positron emission tomography, *Proc. Natl. Acad. Sci. USA* **92:**925–929.

O'Sullivan, E. P., Jenkins, I. H., Henderson, L., Kennard, C., and Brooks, D. J., 1995, The functional anatomy of remembered saccades: A PET study, *NeuroReport* **6:**2141–2144.

Okano, K., 1992, Temporal priority of premotor cortex over nearby areas in receiving visual cues in primates, *NeuroReport* **3:**389–392.

Okano, K., and Tanji, J., 1987, Neuronal activities in the primate motor fields of the agranular frontal cortex preceding visually triggered and self-paced movement, *Exp. Brain Res.* **66:**155–166.

Olson, C. R., and Gettner, S. N., 1995, Object-centered direction selectivity in the macaque supplementary eye field, *Science* **269:**985–988.

Orgogozo, J. M., and Larsen, B., 1979, Activation of the supplementary motor area during voluntary movement in man suggests it works as a supramotor area, *Science* **206:**847–850.

Ottes, F. P., Van Gisbergen, J. A. M., and Eggermont, J. J., 1986, Visuomotor fields of the superior colliculus: A quantitative model, *Vision Res.* **26:**857–873.

Pardo, J. V., Pardo, P. J., Janer, K. W., and Raichle, M. E., 1990, The anterior cingulate cortex mediates processing selection in the Stroop attentional conflict paradigm, *Proc. Natl. Acad. Sci. USA* **87:**256–259.

Parthasarathy, H. B., Schall, J. D., and Graybiel, A. M., 1992, Distributed but convergent ordering of corticostriatal projections: Analysis of the frontal eye field and the supplementary eye field in the macaque monkey, *J. Neurosci.* **12:**4468–4488.

Passingham, R. E., 1985, Prefrontal cortex and the sequencing of movement in monkeys (*Macaca mulatta*), *Neuropsychologia* **23:**453–462.

Passingham, R. E., 1988, Premotor cortex and preparation for movement, *Exp. Brain. Res.* **70:**590–596.

Passingham, R. E., 1989, Premotor cortex and the retrieval of movement, *Brain Behav. Evol.* **33:**189–192.

Passingham, R. E., 1993, *The Frontal Lobes and Voluntary Action,* Oxford University Press, Oxford.

Passingham, R. E., Chen, Y. C., and Thaler, D., 1989, Supplementary motor cortex and self-initiated movement, in: *Neural Programming* (M. Ito, ed.), Karger, Basel, pp. 13–24.

Paulescu, E., Frith, C. D., and Frackowiak, R. S. J., 1993, The neural correlates of the verbal component of working memory, *Nature* **362:**342–344.

Paus, T., 1996, Location and function of the human frontal eye field: A selective review, *Neuropsychologia* **34:**475–483.

Paus, T., Petrides, M., Evans, A. C., and Meyer, E., 1993, Role of human anterior cingulate cortex in the control of oculomotor, manual and speech responses: A positron emission tomography study, *J. Neurophysiol.* **70:**453–469.

Paus, T., Marrett, S., Worsley, K. J., and Evans, A. C., 1995, Extra-retinal modulation of cerebral blood-flow in the human visual cortex: Implications for saccadic suppression, *J. Neurophysiol.* **74:**2179–2183.

Penfield, W., and Boldrey, E., 1937, Somatic motor and sensory representation in the cerebral cortex of man as studied by electrical stimulation, *Brain* **60:**389–443.

Penfield, W., and Welch, K., 1949, The supplementary motor area in the cerebral cortex of man, *Trans. Am. Neurol. Assoc.* **74:**179–184.

Penfield, W., and Welch, K., 1951, The supplementary motor area of the cerebral cortex: A clinical and experimental study, *Arch. Neurol. Psychiat.* **66:**289–317.

Perecman, E., ed., 1987, *The Frontal Lobes Revisited,* IRBN Press, New York.

Petersen, S. E., Fox, P. T., Posner, M. I., Mintun, M., and Raichle, M. E., 1988, Positron emission tomographic studies of the cortical anatomy of single-word processing, *Nature* **331:**585–589.

Petit, L., Orssaud, C., Tzourio, N., Salamon, G., Mazoyer, B., and Berthoz, A., 1993, PET study of voluntary saccadic eye movements in humans: Basal ganglia-thalamocortical system and cingulate cortex involvement, *J. Neurophysiol.* **69:**1009–1017.

Petit, L., Tzourio, N., Orssaud, C., Pietrzyk, U., Berthoz, A., and Mazoyer, B., 1995, Functional neuroanatomy of the human visual fixation system, *Eur. J. Neurosci.* **7:**169–174.

Petit, L., Orssaud, C., Tzourio, N., Crivello, F., Berthoz, A., and Mazoyer, B., 1996, Functional anatomy of a prelearned sequence of horizontal saccades in humans, *J. Neurosci.* **16:**3714–3726.

Petrides, M., 1985, Deficits in non-spatial conditional associative learning after periarcuate lesions in the monkey, *Behav. Brain Res.* **16:**95–101.

Petrides, M., 1987, Conditional learning and the primate frontal cortex, in: *The Frontal Lobes Revisited* (E. Perecman, ed.), IRBN Press, New York, pp. 91–108.

Petrides, M., 1991, Monitoring of selections of visual stimuli and the primate frontal cortex, *Proc. R. Soc. Lond. B* **246:**293–298.

Petrides, M., 1994a, Frontal lobes and behavior, *Curr. Opin. Neurobiol.* **4:**207–211.

Petrides, M., 1994b, Frontal lobes and working memory: Evidence from investigations of the effects of cortical excisions in nonhuman primates, in: *Handbook of Neuropsychology,* Volume 9. (F. Boller and J. Grafman, eds.), Elsevier, New York, pp. 59–82.

Petrides, M., 1995, Impairments on nonspatial self-ordered and externally ordered working memory tasks after lesions of the mid-dorsal part of the lateral frontal cortex in the monkey, *J. Neurosci.* **15:**359–375.

Petrides, M., and Pandya, D. N., 1984, Projections to the frontal cortex from the posterior parietal region in the rhesus monkey, *J. Comp. Neurol.* **228:**105–116.

Petrides, M., and Pandya, D. N., 1994, Comparative architectonic analysis of the human and the macaque frontal cortex, in: *Handbook of Neuropsychology,* Volume 9 (F. Boller and J. Grafman, eds.), Elsevier, New York, pp. 17–58.

Petrides, M., Alivisatos, B., Evans, A. C., and Meyer, E., 1993, Dissociation of human mid-dor solateral from posterior dorsolateral frontal cortex in memory processing, *Proc. Natl. Acad. Sci. USA* **90:**873–877.

Pierrot-Deseilligny, C., Rivaud, S., Gaymard, B., and Agid, Y., 1991a, Cortical control of reflexive visually-guided saccades, *Brain* **114:**1473–1485.

Pierrot-Deseilligny, C., Rivaud, S., Gaymard, B., and Agid, Y., 1991b, Cortical control of memory-guided saccades in man, *Exp. Brain Res.* **83:**607–617.

Pierrot-Deseilligny, C., Israël, I., Berthoz, A., Rivaud, S., and Gaymard, B., 1993, Role of different frontal lobe areas in the control of horizontal component of memory-guided saccades in man, *Exp. Brain Res.* **95:**166–171.

Pierrot-Deseilligny, C., Rivaud, S., Gaymard, B., Müri, R., Vermersch, A.-I., 1995, Cortical control of saccades, *Ann. Neurol.* **37:**557–567.

Pigarev, I. N., Rizzolatti, G., and Scandolara, C., 1979, Neurons responding to visual stimuli in the frontal lobe of macaque monkeys, *Neurosci. Lett.* **12:**207–212.

Pola, J., and Wyatt, H. J., 1991, Smooth pursuit: Response characteristics, stimuli and mechanisms, in: *Eye Movements* (R. H. S. Carpenter, ed.), CRC Press, Boca Raton, FL, pp. 138–156.

Porrino, L. J., Crane, A. M., and Goldman-Rakic, P. S., 1981, Direct and indirect pathways from the amygdala to the frontal lobe in rhesus monkey, *J. Comp. Neurol.* **198:**121–136.

Posner, M. I., and Petersen, S. E., 1990, The attention system of the human brain, *Annu. Rev. Neurosci.* **13:**25–42.

Posner, M. I., Petersen, S. I., Fox, P. T., and Raichle, M. E., 1988, Localization of cognitive operations in the human brain, *Science* **240:**1627–1631.

Praamstra, P., Stegeman, D. F., Horstink, M. W., Brunia, C. H., and Cools, A. R., 1995, Movement-related potentials preceding voluntary movement are modulated by the mode of movement selection, *Exp. Brain Res.* **103:**429–439.

Preuss, T. M., 1995, The argument from animals to humans in cognitive neuroscience, in: *The Cognitive Neurosciences* (M. S. Gazzaniga, ed.), MIT Press, Cambridge, MA, pp. 1227–1241.

Preuss, T. M., and Goldman-Rakic, P. S., 1989, Connections of the ventral granular frontal cortex of macaques with perisylvian premotor and somatosensory areas: Anatomical evidence for somatic representation in primate frontal association cortex, *J. Comp. Neurol.* **282:**293–316.

Preuss, T. M., and Goldman-Rakic, P. S., 1991, Myelo- and cytoarchitecture of the granular frontal cortex and surrounding regions of the strepsirhine primate *Galago* and the anthropoid primate *Macaca*, *J. Comp. Neurol.* **310:**429–474.

Preuss, T. M., Stepniewska, I., and Kaas, J. H., 1995, Movement representation in the dorsal and ventral premotor areas of owl monkeys: A microstimulation study, *J. Comp. Neurol.* **371:**649–676.

Quintana, J., and Fuster, F. M., 1993, Spatial and temporal factors in the role of prefrontal and parietal cortex in visuomotor integration, *Cerebral Cortex* **3:**122–132.

Quintana, J., Yajeya, J., and Fuster, J. M., 1988, Prefrontal representation of stimulus attributes during delay tasks. I. Unit activity in cross-temporal integration of sensory and sensory-motor information, *Brain Res.* **474:**211–221.

Rao, S. M., Binder, J. R., Bandettini, P. A., Hammeke, T. A., Yetkin, F. Z., Jesmanowicz, A., Lisk, L. M., Morris, G. L., Mueller, W. M., and Estkowski, L. D., 1993, Functional magnetic resonance imaging of complex human movements, *Neurology* **43:**2311–2318.

Rasmussen, T., and Penfield, W., 1948, Movement of head and eyes from stimulation of human frontal cortex, *Res. Publ. Assoc. Nerv. Ment. Dis.* **27:**346–361.

Ray, J. P., and Price, J. L., 1993, The organization of projections from the mediodorsal nucleus of the thalamus to orbital and medial prefrontal cortex in macaque monkeys, *J. Comp. Neurol.* **337:**1–31.

Remy, P., Zilbovicius, M., Leroy-Willig, A., Syrota, A., and Samson, Y., 1994, Movement- and task-related activations of motor cortical areas: A positron emission tomographic study, *Ann. Neurol.* **36:**19–26.

Riehle, A., and Requin, J., 1989, Monkey primary motor and premotor cortex: Single-cell activity related to prior information about direction and extent of an intended movement, *J. Neurophysiol.* **61:**534–549.

Risien Russell, J. S., 1894, An experimental investigation of eye movements, *J. Physiol.* (Lond.) **17:**1–26.

Rivaud, S., Müri, R. M., Gaymard, B., Vermersch, A. I., and Pierrot-Deseilligny, C., 1994, Eye movement disorders after frontal eye field lesions in humans, *Exp. Brain Res.* **102:**110–120.

Rizzolatti, G., Scandolara, C., Matelli, M., and Gentilucci, M., 1981, Afferent properties of periarcuate neurons in macaque monkeys. II. Visual responses, *Behav. Brain Res.* **2:**147–163.

Rizzolatti, G., Camarda, R., Fogassi, L., Gentilucci, M., Luppino, G., and Matelli, M., 1988, Functional organization of inferior area 6 in the macaque monkey. II. Area F5 and the control of distal movements, *Exp. Brain Res.* **71:**491–507.

Rizzolatti, G., Gentilucci, M., Camarda, R., Gallese, V., Luppino, G., Matelli, M., and Fogassi, L., 1990, Neurons related to reaching-grasping arm movements in the rostral part of area 6 (area 6aβ), *Exp. Brain Res.* **82:**337–350.

Robinson, D. A., 1972, Eye movements evoked by collicular stimulation in the alert monkey, *Vision Res.* **12:**1795–1808.

Robinson, D. A., 1973, Models of the saccadic eye movement control system, *Kybernetik* **14:**71–83.

Robinson, D. A., 1986, The systems approach to the oculomotor system, *Vision Res.* **26:**91–99.

Robinson, D. A., 1991, Overview, in: *Eye Movements* (R. H. S. Carpenter, ed.), CRC Press, Boca Raton, FL, pp. 320–331.

Robinson, D. A., and Fuchs, A. F., 1969, Eye movements evoked by stimulation of frontal eye fields, *J. Neurophysiol.* **32:**637–648.

Robinson, D. L., Goldberg, M. E., and Stanton, G. B., 1978, Parietal association cortex in the primate: Sensory mechanisms and behavioral modulations, *J. Neurophysiol.* **41:**910–932.

Roland, P. E., Larsen, B., Lassen, N. A., and Skinhoj, E., 1980, Supplementary motor area and other cortical areas in organization of voluntary movements in man, *J. Neurophysiol.* **43:**118–136.

Rolls, E. T., Sanghera, M. K., and Roper-Hall, A., 1979, The latency of activation of neurons in the lateral hypothalamus and substantia innominata during feeding in the monkey, *Brain Res.* **164:**121–135.

Rolls, E. T., Critchley, H. D., Mason, R., and Wakeman, E. A., 1996a, Orbitofrontal cortex neurons: Role in olfactory and visual association learning, *J. Neurophysiol.* **75:**1970–1981.

Rolls, E. T., Critchley, H. D., Mason, R., and Wakeman, E. A., 1996b, Representation of olfactory information in the primate orbitofrontal cortex, *J. Neurophysiol.* **75:**1982–1996.

Romo, R., and Schultz, W., 1987, Neuronal activity preceding self-initiated or externally timed arm movements in area 6 of monkey cortex, *Exp. Brain Res.* **67:**656–662.

Romo, R., and Schultz, W., 1992, Role of primate basal ganglia and frontal cortex in the internal generation of movements. III. Neuronal activity in the supplementary motor area, *Exp. Brain Res.* **91:**396–407.

Rosenkilde, C. E., Bauer, R. H., and Fuster, J. M., 1981, Single cell activity in ventral prefrontal cortex of behaving monkeys, *Brain Res.* **209:**375–394.

Rossetti, Y., Tadary, B., and Prablanc, C., 1994, Optimal contributions of head and eye positions to spatial accuracy in man tested by visually directed pointing, *Exp. Brain Res.* **97:**487–496.

Russchen, F. T., Amaral, D. G., and Price, J. L., 1987, The afferent input to the magnocellular division of the mediodorsal thalamic nucleus in the monkey, *Macaca fascicularis, J. Comp. Neurol.* **256:**175–210.

Russo, G. S., and Bruce, C. J., 1989, Auditory receptive fields of neurons in frontal cortex of rhesus monkey shift with direction of gaze, *Soc. Neurosci. Abstr.* **15:**1204.

Russo, G. S., and Bruce, C. J., 1993, Effect of eye position with the orbit on electrically elicited saccadic eye movements: A comparison of the macaque monkey's frontal and supplementary eye field, *J. Neurophysiol.* **69:**800–818.

Russo, G. S., and Bruce, C. J., 1994, Frontal eye field activity preceding aurally guided saccades, *J. Neurophysiol.* **71:**1250–1253.

Russo, G. S., and Bruce, C. J., 1996, Neurons in the supplementary eye field of rhesus monkeys code visual targets and saccadic eye movements in an oculocentric coordinate system, *J. Neurophysiol.* **76:**825–848.

Sakagami, M., and Niki, H., 1994a, Encoding of behavioral significance of visual stimuli by primate prefrontal neurons: Relation to relevant task conditions, *Exp. Brain Res.* **97:**423–436.

Sakagami, M., and Niki, H., 1994b, Spatial selectivity of go/no-go neurons in monkey prefrontal cortex, *Exp. Brain Res.* **100:**165–169.

Sakai, M., 1978, Single unit activity in a border area between the dorsal prefrontal and premotor regions in the visually conditioned motor task of monkeys, *Brain Res.* **147:**377–383.

Sakata, H., Shibutani, H., and Kawano, K., 1980, Spatial properties of visual fixation neurons in posterior parietal association cortex of the monkey, *J. Neurophysiol.* **43:**1654–1672.

Sakata, H., Shibutani, H., and Kawano, K., 1983, Functional properties of visual tracking neurons in posterior parietal association cortex of the monkey, *J. Neurophysiol.* **49:**1364–1380.

Sanghera, M. K., Rolls, E. T., and Roper-Hall, A., 1979, Visual responses of neurons in the dorsolateral amygdala of the alert monkey, *Exp. Neurol.* **63:**610–626.

Sasaki, K., and Gemba, H., 1986, Electrical activity in the prefrontal cortex specific to no-go reaction of conditioned hand movement with colour discrimination in the monkey, *Exp. Brain Res.* **64:**603–606.

Sawaguchi, T., 1987a, Properties of neuronal activity related to a visual reaction time task in the monkey prefrontal cortex, *J. Neurophysiol.* **58:**1080–1099.

Sawaguchi, T., 1987b, Catecholamine sensitivities of neurons related to a visual reaction time task in the monkey prefrontal cortex, *J. Neurophysiol.* **58:**1100–1122.

Sawaguchi, T., and Goldman-Rakic, P. S., 1991, D1 dopamine receptors in prefrontal cortex: Involvement in working memory, *Science* **251:**947–950.

Sawaguchi, T., Matsumura, M., and Kubot, K., 1990, Effects of dopamine antagonists on neuronal activity related to a delayed response task in monkey prefrontal cortex, *J. Neurophysiol.* **63:**1401–1412.

Schall, J. D., 1991a, Neuronal activity related to visually guided saccadic eye movements in the supplementary motor area of rhesus monkeys, *J. Neurophysiol.* **66:**530–558.

Schall, J. D., 1991b, Neuronal activity related to visually guided saccades in the frontal eye fields of rhesus monkeys: Comparison with supplementary eye fields, *J. Neurophysiol.* **66:**559–579.

Schall, J. D., 1991c, Neuronal basis of saccadic eye movements, in: *The Neural Basis of Visual Function* (A. G. Leventhal, ed.), Macmillan, London, pp. 388–442.

Schall, J. D., 1995, Neural basis of saccade target selection, *Rev. Neurosci.* **6:**63–85.

Schall, J. D., and Hanes, D. P., 1993, Neural basis of saccade target selection in frontal eye field during visual search, *Nature* **366:**467–469.

Schall, J. D., and Hanes, D. P., 1996, Neural control of saccade initiation studied with the countermanding paradigm: Frontal eye field, *Soc. Neurosci. Abstr.* **22:**418.

Schall, J. D., Morel, A., and Kaas, J. H., 1993, Topography of supplementary eye field afferents to frontal eye field in macaque: Implications for mapping between saccade coordinate systems, *Visual Neurosci.* **10:**385–393.

Schall, J. D., Hanes, D. P., Thompson, K. G., and King, D. J., 1995a, Saccade target selection in frontal eye field of macaque. I. Visual and premovement activation, *J. Neurosci.* **15:**6905–6918.

Schall, J. D., Morel, A., King, D. J., and Bullier, J., 1995b, Topography of visual cortical afferents to frontal eye field in macaque: Functional convergence and segregation of processing streams, *J. Neurosci.* **15:**4464–4487.

Schiller, P. H., and Koerner, F., 1971, Discharge characteristics of single units in superior colliculus of the alert rhesus monkey, *J. Neurophysiol.* **34:**920–936.

Schiller, P. H., and Sandell, J. H., 1983, Interactions between visually and electrically elicited saccades before and after superior colliculus and frontal eye field ablation in the rhesus monkey, *Exp. Brain Res.* **49:**381–392.

Schiller, P. H., and Stryker, M., 1972, Single-unit recording and stimulation in superior colliculus of the alert rhesus monkey, *J. Neurophysiol.* **35:**915–924.

Schiller, P. H., True, S. D., and Conway, J. L., 1979, Paired stimulation of the frontal eye field and the superior colliculus of the rhesus monkey, *Brain Res.* **179:**162–164.

Schiller, P. H., True, S. D., and Conway, J. D., 1980, Deficits in eye movements following frontal eye field and superior colliculus ablations, *J. Neurophysiol.* **44:**1175–1189.

Schiller, P. H., Sandell, J. H., and Maunsell, J. H. R., 1987, The effect of frontal eye field and superior colliculus lesions on saccadic latencies in the rhesus monkey, *J. Neurophysiol.* **57:**1033–1049.

Schlag, J., and Schlag-Rey, M., 1985, Unit activity related to spontaneous saccades in frontal dorsomedial cortex of monkey, *Exp. Brain Res.* **58:**208–211.

Schlag, J., and Schlag-Rey, M., 1987, Evidence for a supplementary eye field, *J. Neurophysiol.* **57:**179–200.

Schlag, J., and Schlag-Rey, M., 1990, Colliding saccades may reveal the secret of their marching orders, *TINS* **13:**410–415.

Schlag, J., and Schlag-Rey, M., 1992, Neurophysiology of eye movements, *Adv. Neurol.* **57:**135–147.

Schlag, J., and Schlag-Rey, M., 1995, Illusory localization of stimuli flashed in the dark before saccades, *Vision Res.* **35:**2347–2357.

Schlag, J., Schlag-Rey, M., and Pigarev, I., 1992, Supplementary eye field: Influence of eye position on neural signals of fixation, *Exp. Brain Res.* **90:**302–306.

Schlag-Rey, M., and Schlag, J., 1984, Visuomotor functions of central thalamus in monkey. I. Unit activity related to spontaneous eye movements, *J. Neurophysiol.* **51:**1149–1174.

Schlag-Rey, M., Schlag, J., and Shook, B., 1989, Interactions between natural and electrically evoked saccades: I. Differences between sites carrying retinal error and motor command signals in monkey superior colliculus, *Exp. Brain Res.* **76:**537–547.

Schlag-Rey, M., Schlag, J., and Dassonville, P., 1992, How the frontal eye field can impose a saccade goal on superior colliculus neurons, *J. Neurophysiol.* **67:**1003–1005.

Schnyder, H., Reisine, H., Hepp, K., and Henn, V., 1985, Frontal eye field projection to the paramedian pontine reticular formation traced with wheat germ agglutinin in the monkey, *Brain Res.* **329:**151–160.

Schultz, W., Apicella, P., and Ljungberg, T., 1993, Responses of monkey dopamine neurons to reward and conditioned stimuli during successive steps of learning a delayed response task, *J. Neurosci.* **13:**900–913.

Scudder, C. A., 1988, A new local feedback model of the saccadic burst generator, *J. Neurophysiol.* **59:**1455–1475.

Segraves, M. A., 1992, Activity of monkey frontal eye field neurons projecting to oculomotor regions of the pons, *J. Neurophysiol.* **68:**1967–1985.

Segraves, M. A., and Goldberg, M. E., 1987, Functional properties of corticotectal neurons in the monkey's frontal eye fields, *J. Neurophysiol.* **58:**1387–1419.

Segraves, M. A., and Park, K., 1993, The relationship of monkey frontal eye field activity to saccade dynamics, *J. Neurophysiol.* **69:**1880–1889.

Seitz, R. J., and Roland, P. D., 1992, Learning of sequential finger movements in man: A combined kinematic and positron emission tomography (PET) study, *Eur. J. Neurosci.* **4:**154–165.

Selemon, L. D., and Goldman-Rakic, P. S., 1985, Longitudinal topography and interdigitation of corticostriatal projections in the rhesus monkey, *J. Neurosci.* **5:**776–794.

Selemon, L. D., and Goldman-Rakic, P. S., 1988, Common cortical/subcortical targets of the dorsolateral prefrontal and posterior parietal cortices in the rhesus monkey: Evidence for a distributed neural network subserving spatially guided behavior, *J. Neurosci.* **8:**4049–4068.

Sereno, A. B., and Holzman, P. S., 1993, Express saccades and smooth pursuit eye movement function in schizophrenic, affective disorder and normal subjects, *J. Cognitive Neurosci.* **5:**303–316.

Sharpe, J. A., 1986, Adaptation to frontal lobe lesions, in: Adaptive Processes in Visual and Oculomotor Systems (E. L. Keller and D. S. Zee, eds.), Pergamon Press, New York, pp. 239–246.

Sherrington, C. S., 1893, Further experimental note on the correlation of action of antagonistic muscles, *Proc. R. Soc. Lond.* **53:**407–420.

Shibasaki, H., Sadato, N., Lyshkow, H., Yonekura, Y., Honda, M., Nagamine, T., Suwazono, S., Magata, Y., Ikeda, A., and Miyazaki, M., 1993, Both primary motor cortex and supplementary motor area play an important role in complex finger movement, *Brain* **116:**1387–1398.

Shima, K., Aya, K., Mushiake, H., Inase, M., Aisawa, H., and Tanji, J., 1990, Two movement-related foci in the primate cingulate cortex observed in signal-triggered and self-paced forelimb movements, *J. Neurophysiol.* (Paris) **65:**188–202.

Shook, B. L., Schlag-Ray, M., and Schlag, J., 1990, Primate supplementary eye field. I. Comparative aspects of mesencephalic and pontine connections, *J. Comp. Neurol.* **301:**618–642.

Shook, B. L., Schlag-Ray, M., and Schlag, J., 1991, Primate supplementary eye field. II. Comparative aspects of connections with the thalamus, corpus striatum and related forebrain nuclei, *J. Comp. Neurol.* **307:**562–583.

Siegel, B. V., Buchsbaum, M. S., Bunney, W. E., Jr., Gottschalk, L. A., Haier, R. J., Lohr, J. B., Lottenberg, S., Najafi, A., Neuchterlein, K. H., and Potkin, S. G., 1993, Cortical-striatal-thalamic circuits and brain glucose metabolic activity in 70 unmedicated male schizophrenic patients, *Am. J. Psychiatry* **150:**1325–1336.

Smith, W. K., 1936, Ocular responses elicited by electrical stimulation of the cerebral cortex, *Anat. Rec.* **64:**45.

Smith, W. K., 1940, Electrically responsive cortex within the sulci of the frontal lobe, *Anat. Rec.* **76:**75–76.

Soechting, J. F., and Flanders, M., 1992, Moving in three-dimensional space: frames of reference, vectors and coordinate systems, *Annu. Rev. Neurosci.* **15:**167–191.

Sparks, D. L., and Hartwich-Young, R., 1989, The deep layers of the superior colliculus, in: *The Neurobiology of Saccadic Eye Movements* (R. H. Wurtz and M. E. Goldberg, eds.), Elsevier, Amsterdam, pp. 213–255.

Sparks, D. L., and Mays, L. E., 1983, Spatial localization of saccade targets. I. Compensation for stimulation-induced perturbations in eye position, *J. Neurophysiol.* **49:**45–63.

Sparks, D. L., Mays, L. E., and Porter, J. D., 1987, Eye movements induced by pontine stimulation: Interaction with visually triggered saccades, *J. Neurophysiol.* **58:**300–318.

Sperry, R. W., 1950, Neural basis of the spontaneous optokinetic response produced by visual inversion, *J. Comp. Physiol. Psychol.* **43:**482–489.

Stanford, T. R., Carney, L. H., and Sparks, D. L., 1990, The amplitude of visually guided saccades is specified gradually in humans, *Soc. Neurosci. Abstr.* **16:**901.

Stanton, G. B., Goldberg, M. E., and Bruce, C. J., 1988a, Frontal eye field efferents in the macaque monkey: I. Subcortical pathways and topography of striatal and thalamic terminal fields, *J. Comp. Neurol.* **271:**473–492.

Stanton, G. B., Goldberg, M. E., and Bruce, C. J., 1988b, Frontal eye field efferents in the macaque monkey: II. Topography of terminal fields in midbrain and pons, *J. Comp. Neurol.* **271:**493–506.

Stanton, G. B., Deng, S.-Y., Goldberg, M. E., and McMullen, N. T., 1989, Cytoarchitectural characteristics of the frontal eye fields in macaque monkeys, *J. Comp. Neurol.* **282:**415–427.

Stanton, G. B., Bruce, C. J., and Goldberg, M. E., 1993, Topography of projections to the frontal lobe from the macaque frontal eye fields, *J. Comp. Neurol.* **330:**286–301.

Stanton, G. B., Bruce, C. J., and Goldberg, M. E., 1995, Topography of projections to posterior cortical areas from the macaque frontal eye fields, *J. Comp. Neurol.* **353:**291–305.

Steinbach, M. J., 1987, Proprioceptive knowledge of eye position, *Vision Res.* **27:**1737–1744.

Steinman, R. M., 1986, The need for an eclectic, rather than systems, approach to the study of the primate oculomotor system, *Vision Res.* **26:**101–112.

Stephan, K. M., Fink, G. R., Passingham, R. E., Silbersweig, D., Ceballos-Baumann, A. O., Frith, C. D., Frackowiak, R. S., 1995, Functional anatomy of the mental representation of upper extremity movements in healthy subjects, *J. Neurophysiol.* **73:**373–386.

Stepniewska, I., Preuss, T. M., and Kaas, J. H., 1993, Architectonics, somatotopic organization and ipsilateral cortical connections of the primary motor area (M1) of owl monkeys, *J. Comp. Neurol.* **330:**238–271.

Sternberg, S., 1969, The discovery of processing stages: Extensions of Donders' method, in W. G. Koster, ed., Attention and Performance II, *Acta Psychol.* **30:**276–315.

Stone, J., 1983, *Parallel Processing in the Visual System,* Plenum Press, New York.

Stuss, D. T., and Benson, D. F., 1986, *The Frontal Lobes,* Raven Press, New York.

Suzuki, H., 1985, Distribution and organization of visual and auditory neurons in the monkey prefrontal cortex, *Vision Res.* **25:**465–469.

Suzuki, H., and Azuma, M., 1977, Prefrontal neuronal activity during gazing at a light spot in the monkey, *Brain Res.* **126:**497–508.

Suzuki, H., and Azuma, M., 1983, Topographic studies on visual neurons in the dorsolateral prefrontal cortex of the monkey, *Exp. Brain Res.* **53:**47–58.

Suzuki, H., Azuma, M., and Yumiya, H., 1979, Stimulus and behavioral factors contributing to the activation of monkey prefrontal neurons during gazing, *Jpn. J. Physiol.* **29:**471–490.

Sweeney, J. A., Mintun, M. A., Kwee, S., Wiseman, M. B., Brown, D. L., Rosenberg, D. R., and Carl, J. R., 1995, A positron emission tomography study of voluntary saccadic eye movements and spatial working memory, *J. Neurophysiol.* **75:**454–468.

Tanabe, T., Yarita, H., Iino, M., Ooshima, Y., and Takagi, S. R., 1975, An olfactory projection area in the orbitofrontal cortex of the monkey, *J. Neurophysiol.* **38:**1269–1283.

Tanji, J., 1994, The supplementary motor area in the cerebral cortex, *Neurosci. Res.* **19:**251–268.

Tanji, J., and Evarts, E. V., 1976, Anticipatory activity of motor cortex neurons in relation to direction of an intended movement, *J. Neurophysiol.* **39:**1062–1068.

Tanji, J., and Kurata, K., 1979, Neuronal activity in the cortical supplementary motor area related with distal and proximal forelimb movements, *Neurosci. Lett.* **12:**201–206.

Tanji, J., and Kurata, K., 1981, Contrasting neuronal activity in the ipsilateral and contralateral supplementary motor areas in relation to a movement of monkey's distal hindlimb, *Brain Res.* **222:**155–158.

Tanji, J., and Kurata, K., 1982, Comparison of movement-related activity in two cortical motor areas of primates, *J. Neurophysiol.* **48:**633–653.

Tanji, J., and Kurata, K., 1985, Contrasting neuronal activity in supplementary and precentral motor cortex of monkeys. I. Responses to instructions determining motor responses to forthcoming signals of different modalities, *J. Neurophysiol.* **53:**129–141.

Tanji, J., and Shima, K., 1994, Role for supplementary motor area cells in planning several movements ahead, *Nature* **371:**413–416.

Tanji, J., Okano, K., and Sato, K. C., 1988, Neuronal activity in cortical motor areas related to ipsilateral, contralateral and bilateral digit movements of the monkey, *J. Neurosphysiol.* **60:**325–343.

Tanji, J., Taniguchi, K., and Saga, T., 1980, Supplementary motor area: Neuronal response to motor instructions, *J. Neurophysiol.* **43:**60–68.

Taylor, D. A., 1976, Stage analysis of reaction time, *Psychol. Bull.* **83:**161–191.

Tehovnik, E. J., 1995, The dorsomedial frontal cortex: Eye and forelimb fields, *Behav. Brain Res.* **67:**147–163.

Tehovnik, E. J., and Lee, K., 1993, The dorsomedial frontal cortex of the rhesus monkey: Topographic representation of saccades evoked by electrical stimulation, *Exp. Brain Res.* **96:**430–442.

Tehovnik, E. J., Lee, K., and Schiller, P. H., 1994, Stimulation-evoked saccades from the dorsomedial frontal cortex of the rhesus monkey following lesions of the frontal eye fields and superior colliculus, *Exp. Brain Res.* **98:**179–190.

Teuber, H.-L., Battersby, W. S., and Bender, M. B., 1949, Changes in visual searching performance following cerebral lesions, *Am. J. Physiol.* **159:**592.

Thaler, D. E., Rolls, E. T., and Passingham, R. E., 1988, Neuronal activity of the supplementary motor area (SMA) during internally and externally triggered wrist movements, *Neurosci. Lett.* **93:**264–269.

Thaler, D. E., Chen, Y.-C., Nixon, P. D., Stern, C. E., and Passingham, R. E., 1995, The functions of the medial premotor cortex I. Simple learned movements, *Exp. Brain Res.* **102:**445–460.

Thickbroom, G. W., and Mastaglia, F. L., 1990, Premotor negativity associated with saccadic eye movement and finger movement: A comparative study, *Brain Res.* **506:**223–226.

Thompson, K. G., Hanes, D. P., and Schall, J. D., 1995, Time-course of target selection in macaque frontal eye field during visual search, *Soc. Neurosci. Abstr.* **21:**1270.

Thompson, K. G., Hanes, D. P., Bichot, N. P., and Schall, J. D., 1996, Perceptual and motor processing stages identified in the activity of macaque frontal eye field neurons during visual search, *J. Neurophysiol.* **76:**4040–4055.

Thorpe, S. J., Rolls, E. T., and Maddison, S., 1983, The orbitofrontal cortex: Neuronal activity in the behaving monkey, *Exp. Brain Res.* **49:**93–115.

Tomlinson, R. D., and Bahra, P. S., 1986, Combined eye–head gaze shifts in the primate. I. Metrics, *J. Neurophysiol.* **56:**1542–1557.

Umeno, M. M., and Goldberg, M. E., 1994, Predictive visual responses in monkey frontal eye field, *Soc. Neurosci. Abstr.* **20:**144.

Ungerleider, L. G., and Mishkin, M., 1982, Two cortical visual systems, in *Analysis of Visual Behavior* (D. J. Ingle, M. A. Goodale, and R. J. W. Mansfield, eds.), MIT Press, Cambridge, pp. 549–586.

Ungerleider, L. G., Gaffan, D., and Pelak, V. S., 1989, Projections from inferior temporal cortex to prefrontal cortex via the uncinate fascicle in rhesus monkeys, *Exp. Brain Res.* **76:**473–484.

Vaadia, E., 1989, Single-unit activity related to active localization of acoustic and visual stimuli in the frontal cortex of the rhesus monkey, *Brain Behav. Evol.* **33:**127–131.

Vaadia, E., Benson, D. A., Heinz, R. D., and Goldstein, M. H., 1986, Unit activity of monkey frontal cortex: Active localization of auditory and visual cues, *J. Neurophysiol.* **56:**934–957.

Van Gisbergen, J. A. M., Van Opstal, A. J., and Tax, A. A. M., 1987, Collicular ensemble coding of saccades based on vector summation, *Neuroscience* **21:**541–555.

Van Opstal, A. J., and Van Gisbergen, J. A. M., 1989, A nonlinear model for collicular spatial interactions underlying the metrical properties of electrically elicited saccades, *Biol. Cybernet.* **60:**171–183.

Vercher, J.-L., and Gauthier, G. M., 1992, Oculo-manual coordination control: Ocular and manual tracking of visual targets with delayed visual feedback of the hand motion, *Exp. Brain Res.* **90:**599–609.

Viviani, P., 1990, Eye movements in visual search: Cognitive, perceptual and motor control aspects, in: *Eye Movements and Their Role in Visual and Cognitive Processes* (E. Kowler, ed.), Elsevier, Amsterdam, pp. 353–393.

Vogels, R., and Orban, G. A., 1990, How well do response changes of striate neurons signal differences in orientation: A study in the discriminating monkey, *J. Neurosci.* **10:**3543–3558.

Vogt, B. A., 1985, Cingulate Cortex, in: *Cerebral Cortex,* Volume 4, *Association and Auditory Cortices* (A. Peters and E. G. Jones, eds.), Plenum Press, New York, pp. 89–149.

Vogt, B. A., and Pandya, D. N., 1987, Cingulate cortex of the rhesus monkey. II. Cortical afferents, *J. Comp. Neurol.* **262:**271–289.

Vogt, B. A., Pandya, D. N., and Rosene, D. C., 1987, Cingulate cortex of the rhesus monkey: I. Cytoarchitecture and thalamic afferents, *J. Comp. Neurol.* **262:**256–270.

Vogt, B. A., Finch, D. M., and Olson, C. R., 1992, Functional heterogeneity in cingulate cortex: The anterior executive and posterior evaluative regions, *Cerebral Cortex* **2:**435–443.

Vogt, C., and Vogt, O., 1907, Zur Kenntnis der elektrisch erregbaren Hirnrinden-Gebiete bei den Säugetieren, *J. Psychol. Neurol.* (Lpz.) **8:**277–456.

Vogt, C., and Vogt, O., 1919, Allgemeine Ergebnisse unserer Hirnforschung. Vierte Mitteilung: Die physiologische Bedeutung der architektonischen Rindenfelderung auf Grund neuer Rindenreizungen, *J. Psychol. Neurol.* (Lpz.) **25:**399–462.

Volkmann, F. C., Schick, A. M. L., and Riggs, L. A., 1968, The time course of visual inhibition during voluntary saccades, *J. Opt. Soc. Am.* **58:**562–569.

Volkmann, F. C., Riggs, L. A., White, K. D., and Moore, R. K., 1978, Contrast sensitivity during saccadic eye movements, *Vision Res.* **18:**1193–1199.

von Bonin, G., and Bailey, P., 1947, *The Neocortex of Macaca Mulatta,* University of Illinois Press, Urbana, IL.

Von Holst, E., and Mittelstaedt, H., 1950, The principle of reafference: Interactions between the central nervous system and the peripheral organs, *Die Naturwissenschaften* **37:**464–474 [English translation in: Dodwell, P. C., 1980, *Perceptual Processing: Stimulus Equivalence and Pattern Recognition,* Appleton-Century-Crofts, New York, pp. 41–71].

Wagman, I. H., Krieger, H. P., and Bender, M. B., 1958, Eye movements elicited by surface and depth stimulation of the occipital lobe of *Macaque mulatta, J. Comp. Neurol.* **109:**169–193.

Wagman, I. H., Krieger, H. P., Papatheodoron, C. A., and Bender, M. B., 1961, Eye movements elicited by surface and depth stimulation of the frontal lobe, *J. Comp. Neurol.* **117:**179–188.

Walker, A. E., 1940, A cytoarchitectural study of prefrontal area of the macaque monkey, *J. Comp. Neurol.* **73:**59–86.

Walker, A. E., and Weaver, T. A., 1940, Ocular movements from the occipital lobe in the monkey, *J. Neurophysiol.* **3:**353–369.

Watanabe, M., 1981, Prefrontal unit activity during delayed conditional discriminations in the monkey, *Brain Res.* **225:**51–65.

Watanabe, M., 1986a, Prefrontal unit activity during delayed conditional Go/No-Go discrimination in the monkey. I. Relation to the stimulus, *Brain Res.* **382:**1–14.

Watanabe, M., 1986b, Prefrontal unit activity during delayed conditional Go/No-Go discrimination in the monkey. II. Relation to Go and No-Go responses, *Brain Res.* **382:**15–27.

Watanabe, M., 1990, Prefrontal unit activity during associative learning in the monkey, *Exp. Brain Res.* **80:**296–309.

Watanabe, M., 1992, Frontal units of the monkey coding the associative significance of visual and auditory stimuli, *Exp. Brain Res.* **89:**233–247.

Watson, R. T., Fleet, W. S., Gonzalez-Rothi, L., Heilman, K. M., 1986, Apraxia and the supplementary motor area, *Arch. Neurol.* **43:**787–792.

Webster, M. J., Bachevalier, J., and Ungerleider, L. G., 1994, Connections of inferior temporal areas TEO and TE with parietal and frontal cortex in macaque monkeys, *Cerebral Cortex* **4:**470–483.

Weinberger, N. M., 1995, Dynamic regulation of receptive fields and maps in the adult sensory cortex, *Ann. Rev. Neurosci.* **18:**129–158.

Weinberger, D. R., Berman, K. F., and Zec, R. F., 1986, Physiological dysfunction of dorsolateral prefrontal cortex in schizophrenia. I. Regional cerebral blood flow evidence, *Arch. Gen. Psychiatry* **43:**114–125.

Weinberger, D. R., and Berman, K. F., 1996, Prefrontal function in schizophrenia: Confounds and controversies, *Phil. Trans. R. Soc. Lond. B* **351:**1495–1503.

Weinrich, M., and Wise, S. P., 1982, The premotor cortex of the monkey, *J. Neurosci.* **2:**1329–1345.
Weinrich, M., Wise, S. P., and Mauritz, K. H., 1984, A neurophysiological study of the premotor cortex in rhesus monkey, *Brain* **107:**385–414.
Weller, R. E., and Kaas, J. H., 1987, Subdivisions and connections of inferior temporal cortex in owl monkeys, *J. Comp. Neurol.* **256:**137–172.
Wessel, K., Zeffiro, T., Lou, J. S., Toro, C., and Hallett, M., 1995, Regional cerebral blood flow during a self-paced sequential finger opposition task in patients with cerebellar degeneration, *Brain* **118:**379–393.
Wiesendanger, M., and Wise, S. P., 1992, Current issues concerning the functional organization of motor cortical areas in nonhuman primates, in: *Frontal Lobe Seizures and Epilepsy* (P. Chauvel, A. V. Delgado-Escuerta, E. Halgren, and J. Bancaud, eds.), Raven Press, New York, pp. 117–134.
Williams, S. M., and Goldman-Rakic, P. S., 1993, Characterization of the dopaminergic innervation of the primate frontal cortex using a dopamine-specific antibody, *Cerebral Cortex* **3:**199–222.
Williams, G. V., and Goldman-Rakic, P. S., 1995, Modulation of memory fields by dopamine D1 receptors in prefrontal cortex, *Nature* **376:**572–575.
Wilson, F. A. W., Ó Scalaidhe, S. P., and Goldman-Rakic, P. S., 1993, Dissociation of object and spatial processing domains in primate prefrontal cortex, *Science* **260:**1955–1958.
Wise, S. P., 1985, The primate premotor cortex: Past, present and preparatory, *Annu. Rev. Neurosci.* **8:**1–19.
Wise, S. P., and Kurata, K., 1989, Set-related activity in the premotor cortex of rhesus monkeys: Effect of triggering cues and relatively long delay intervals, *Somatosens. Motor Res.* **6:**455–476.
Wise, S. P., and Mauritz, K.-H., 1985, Set-related neuronal activity in the premotor cortex of rhesus monkeys: Effects of changes in motor set, *Proc. R. Soc. Lond. B* **223:**331–354.
Wise, S. P., and Tanji, J., 1981, Supplementary and precentral motor cortex: Contrast in responsiveness to peripheral input in the hindlimb area of the unanesthetized monkey, *J. Comp. Neurol.* **195:**433–451.
Wise, S. P., Weinrich, M., and Mauritz, K.-H., 1986, Movement-related activity in the premotor cortex of rhesus macaques, in: *Progress in Brain Research,* Volume 64 (H.-J. Freund, U. Büttner, B. Cohen, and J. Noth, eds.), Elsevier, New York, pp. 117–131.
Wolkin, A., Sanfilipo, M., Wolf, A. P., Angrist, B., Brodie, J. D., and Rotrosen, J., 1992, Negative symptoms and hypofrontality in chronic schizophrenia, *Arch. Gen. Psychiatry* **49:**959–965.
Wurtz, R. H., and Mohler, C. W., 1976, Enhancement of visual response in monkey striate cortex and frontal eye fields, *J. Neurophysiol.* **39:**766–772.
Yajeya, J., Quintana, J., and Fuster, J. M., 1988, Prefrontal representation of stimulus attributes during delay tasks. II. The role of behavioral significance, *Brain Res.* **474:**222–230.
Yarbus, A. L., 1967, *Eye Movements and Vision,* Plenum Press, New York.
Yarita, H., Iino, M., Tanabe, T., Kogure, S., and Takagi, S. F., 1980, A transthalamic olfactory pathway to orbitofrontal cortex in the monkey, *J. Neurophysiol.* **43:**69–85.
Young, L., and Stark, L., 1963, A discrete model for eye tracking movements, *IEEE Trans. Military Elect.* **MIL-7:**113–115.
Zambarbieri, D., Schmid, R., Magenes, G., and Prablanc, C., 1982, Saccadic responses evoked by presentation of visual and auditory targets, *Exp. Brain Res.* **47:**417–427.
Zangemeister, W. M., and Stark, L., 1982, Gaze latency: Variable interactions of head and eye latency, *Exp. Neurol.* **75:**389–406.
Zatorre, R. J., Evans, A. C., Meyer, E., and Gjedde, A., 1992, Lateralization of phonetic and pitch discrimination in speech processing, *Science* **256:**846–849.
Zingale, C. M., and Kowler, E., 1987, Planning sequences of saccades, *Vision Res.* **27:**1327–1341.

14

Development and Plasticity of Extrastriate Visual Cortex in Monkeys

HILLARY R. RODMAN and TIRIN MOORE

1. Introduction and Overview

The extrastriate visual cortex of the monkey has in recent decades come to include an ever-expanding portion of the neocortical domain as more and more traditional "association" or even motor territories are shown to have significant visual connections, visual responsiveness, or role in visual behavior (Felleman and van Essen, 1991; also see the chapter by Gross in this volume). In this chapter, discussion will be restricted (or broadened, depending upon one's viewpoint) to consideration of cortical zones shown to have at least some visual sensory responsiveness and direct connectivity with other, unimodal visual areas. First, we will consider the normal anatomical, physiological, and metabolic development of extrastriate visual cortex, including the prenatal specification of visual areas. Next, we will discuss the plasticity of extrastriate visual cortex in adulthood by examining the ability of extrastriate areas to function in parallel with striate cortex and the consequences of damage to extrastriate cortex in adult monkeys. In addition, we will examine evidence for learning- or experience-dependent plasticity in the response properties of neurons in extrastriate cortex.

HILLARY R. RODMAN • Department of Psychology and Yerkes Regional Primate Research Center, Emory University, Atlanta, Georgia 30322. TIRIN MOORE • Department of Brain and Cognitive Sciences, Massachusetts Institute of Technology, Cambridge, Massachusetts 02139.
Cerebral Cortex, Volume 12, edited by Rockland *et al.* Plenum Press, New York, 1997.

In the subsequent section, we will address the special plasticity associated with damage to visual cortex in developing animals. We will then briefly compare the development and plasticity of extratriate cortex in monkeys with phenomena described for other mammalian groups. In the final section, we will summarize the data presented and comment on general principles of extrastriate and cerebral cortical development.

2. Normal Development of Extrastriate Visual Cortex

Different portions of the extrastriate cortex in monkeys have received differing amounts of attention with regard to their structural, physiological, and functional development. The development of area V2 in particular has been the subject of numerous studies, as a result of several factors. First, the border of V2 with striate cortex is probably the clearest architectonic border in the cerebral cortex. Second, in adult monkeys, V2 is exquisitely compartmentalized both anatomically and physiologically and probably plays, among extrastriate areas, a particularly important role in the channeling and segregation of subtypes of visual information (see the chapter by Roe and T'so in this volume). In addition to V2, "association" areas such as inferior temporal areas TE and TEO and frontal cortex have received considerable attention. This attention is presumably a result of their documented roles in development of cognitive function and, especially recently, in light of a growing body of evidence for considerable plasticity of response properties in these regions even in adulthood (Section 3.2.1).

2.1. Prenatal Specification of Visual Cortical Areas

One of the most intriguing of all questions in developmental neurobiology has been that of the specification of the identities of the vast array of differentiable regions that make up the cortical mantle. Because of the relative precision with which the border between striate cortex and area V2 can be specified, the genesis of this boundary has been exploited as a model system for studying mechanisms of cortical parcellation by Dehay, Kennedy, and co-workers (reviewed in Kennedy and Dehay, 1993; Dehay and Kennedy, 1993). The border between striate cortex and the neighboring territory of V2 appears to be at least partially determined prenatally at the level of the ventricular zone; an increased rate of production of precursor cells in the germinal zone of striate can largely account for the twofold higher density of neurons in striate cortex versus extrastriate cortex (Dehay *et al.*, 1993). However, in other animal models, clones of neurons from a given portion of the ventricular zone have been reported to disperse widely across the surface of the developing cortex. This result suggests that areal specification is not entirely accomplished prior to the dispersal (Walsh and Cepko, 1992; Uylings, 1994).

A frequently mentioned alternative to the prespecification of cortical territories in a germinal zone "protomap" (e.g., Rakic, 1988) is the imposition of cortical identity by the ingrowth of characteristic patterns of subcortical connections, particularly from the thalamus (Kennedy and Dehay, 1993). One potential example is provided by studies which have observed a transient expression of acetylcholinesterase reactivity in the pulvinar–extrastriate cortex projection of developing primates. In adult monkeys, the pulvinar nucleus of the thalamus provides the main subcortical input to extrastriate cortex, but only a sparse input to striate, which is instead innervated heavily by the dorsal lateral geniculate nucleus (dLGN). Kostovic and Rakic (1984) traced the ingrowth of AChE-positive fibers from the pulvinar to the developing cortex in midgestation of rhesus monkeys and observed a corresponding histochemical differentiation in AChE staining within the cortex at stages preceding the appearance of, but in the correct location for, the cytoarchitectonic border of V1 and V2. One possibility is that even prior to architectonic differentation, cortical cells may differ in surface properties or other characteristics that can be recognized by the incoming fibers (e.g., Rakic and Goldman-Rakic, 1982). Alternatively, the authors speculate, the border between striate and extrastriate territories may actually be created by competition between dLGN and pulvinar axons for cells at the border of the two sets of ingrowing fibers.

Far less is known of the specification of the identities and boundaries of extrastriate areas beyond V2. Since their visible borders are far less crisp and more gradual than the common border of V1 and V2, it is possible that extrastriate territories are not specified individually. The highly distributed contributions of remaining extrastriate areas after early lesions within extrastriate cortex (see Section 4.1) is consistent with a relatively equipotential origin of areas beyond V2. Nonetheless, as far as has been examined, individual extrastriate territories in the monkey do develop their characteristic patterns of connections, metabolic activity, response properties, neurochemistry, and receptor expression at differing rates, and show differing transient patterns of connectivity or chemical markers. These phenomena, reviewed below, may contribute to the genesis of the borders or more gradual physical gradients which characterize the extrastriate matrix of the adult primate.

2.2. Development of Connectivity of Extrastriate Areas

Much of the anatomical connectivity of extrastriate cortex, in terms of basic input and output patterns, appears to be in place at or shortly after birth in primates. In particular, basic patterns of ipsilateral feedforward connections are mature early in life. Recent data indicate, however, that patterns of feedback from "higher order" and nonvisual cortices, of callosal connectivity, and of interconnections with subcortical and limbic structures all undergo a period of postnatal refinement.

2.2.1. Ipsilateral Corticocortical Connectivity

Highly specific patterns of cortical connections of extrastriate areas begin to develop long before birth. For example, Barone *et al.* (1994b, and personal communication) have used the transient expression of AChE in V2 (Section 2.1) as a marker for the portions of this area that project to area V4. Adult patterns are present at birth. At E129, neurons projecting to area V4 are clearly segregated in regions of transient high AchE activity, and a discontinuous distribution is present as early as E112. At the other end of the cortical visual pathways, in the frontal lobe, highly segregated columnar patterns of connections are present at birth in both ipsilateral and contralateral projection systems (Goldman-Rakic, 1987; Schwartz and Goldman-Rakic, 1990).

Kennedy and colleagues (Kennedy *et al.*, 1989b; Barone *et al.*, 1995) have also provided data on the development of feedback pathways of extrastriate cortex in the context of their studies of the maturation of the connectivity of striate cortex. Injections of retrograde tracers into macaque striate cortex at various pre- and postnatal stages revealed transient feedback projections from area V2, the anterior bank of the lunate sulcus (including V4), and the posterior bank of the superior temporal sulcus (including MT). Although these areas also provide feedback projections to V1 in adult monkey, in adults such projections originate predominantly in the infragranular layers (in the case of MT and V4) or about equally in infragranular and supragranular (in the case of V2). In contrast, at birth the vast majority of extrastriate cortex neurons labeled from V1 injections were located in supragranular zones. However, even in fetuses, the different extrastriate loci showed differences in the extent of supragranular labeling, indicating that the mature pattern does not emerge from a uniform distribution. Adult-like patterns were observed in all three extrastriate areas at 1 month after birth. Further analyses led Barone *et al.* to suggest that the developmental remodeling involves both a reduction of the supragranular projection zone (i.e., the extent of an extrastriate area providing a feedback projection) and an increase in the infragranular one.

Comparable results were obtained in human by Burkhalter (1993). Using deposits of the fluorescent tracer DiI in postmortem tissue from brains of fetuses and young newborns, he demonstrated that laminar termination patterns of feedback connections from V2 to V1 mature more slowly than those of the forward pathway from V1 to V2. However, local long-range connections of the superficial layers of V2 mature in advance of similar connections in V1. On the basis of these and other observations, Burkhalter concluded that cells in layers II and III of striate cortex may be unsuitable as targets for feedback connections until after at least 4 months in human. A similar situation may obtain within the first postnatal month in the monkey.

Several groups have examined the normal postnatal development of the corticocortical connectivity of inferior temporal visual areas. In adult monkeys, inferior temporal cortex (IT, or cytoarchitectonic area TE) receives visual information originating in striate cortex only indirectly, via several stags in extrastriate cortex, such that direct visual inputs arise predominantly from areas TEO, parahippocampal cortex, and to a lesser extent V4. Significant inputs also arise from perirhinal cortex, the superior temporal polysensory area (STP) and other regions of the anterior superior temporal sulcus, and ipsilateral frontal cortex (reviewed in Desimone and Ungerleider, 1989; Felleman and Van Essen, 1991; Gross *et al.*, 1993). Injections of various kinds of retrograde tracers into infant IT indicate that the overall pattern of occipitotemporal inputs—i.e., from areas that are mainly visual in function—are similar in adult monkeys and in infants, at least by 6–7 weeks (Webster *et al.*, 1991a, Rodman and Consuelos, 1994; Rodman, 1994). However, infant IT cortex also receives transient inputs from the insula and anterior cingulate cortex just above the corpus callosum (Fig. 1). In addition to these transient cortical inputs to IT in infant monkeys, IT also provides more widespread *outputs* to perirhinal and parahippocampal cortices than are found in adult monkeys (Webster *et al.*, 1991a; Rodman and Consuelos, 1994; Fig. 2). Finally, frontal projections to IT cortex originate over a more widespread region of cortex in infant monkeys within the first month of life (Rodman and Nace, 1997; Fig. 1). Visual area TEO (also called PIT; see Felleman and Van Essen, 1991) also provides transient projections to parahippocampal cortex in newborn monkeys (Webster *et al.*, 1991a). Thus, both the cortical inputs and outputs of inferior temporal areas undergo considerable refinement in a postnatal period extending to at least 6–7 weeks after birth, and possibly longer.

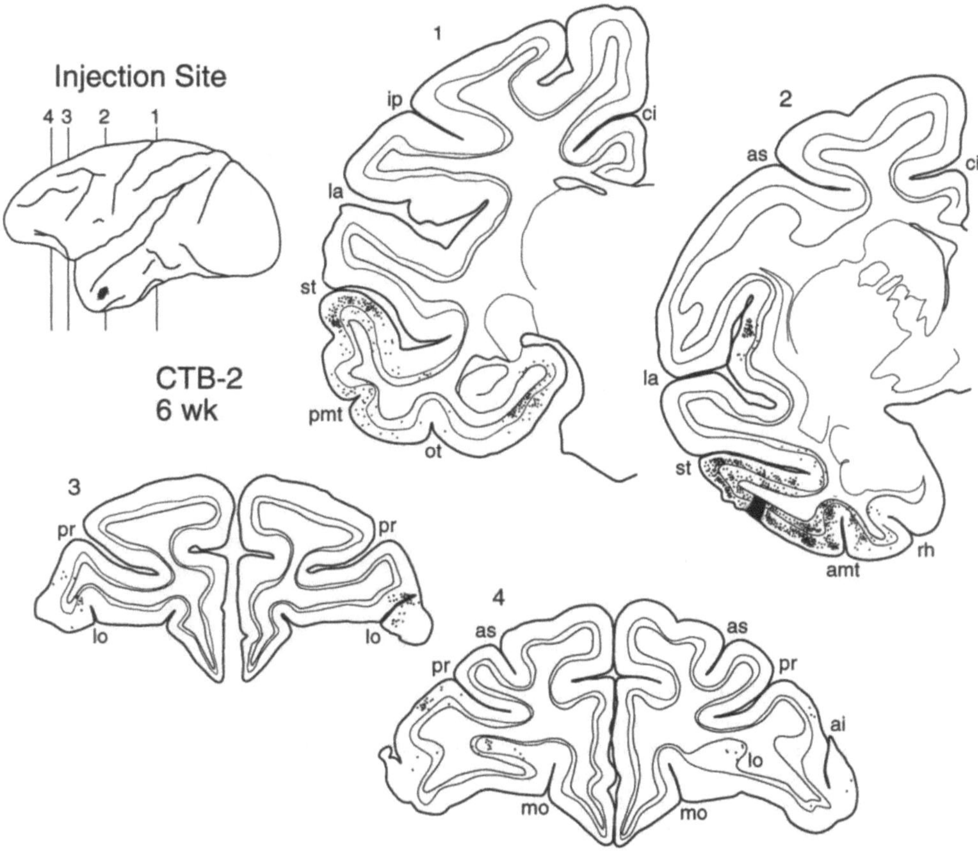

Figure 1. Cross sections illustrating locations of retrogradely labeled cells in anterior temporal lobe and frontal lobe of a 6-week-old infant monkey after an injection of cholera toxin B-subunit into anterior IT. Each dot corresponds to one labeled cell. Abbreviations: ai, inferior arcuate sulcus; amt, anterior medial temporal sulcus; as, superior arcuate sulcus; ci, cingulate sulcus; ip, intraparietal sulcus; la, lateral sulcus; lo, lateral orbital sulcus; mo, medial orbital sulcus; ot, occipitotemporal sulcus; pmt, posterior medial temporal sulcus; pr, principal sulcus; rh, rhinal sulcus; st, superior temporal sulcus.

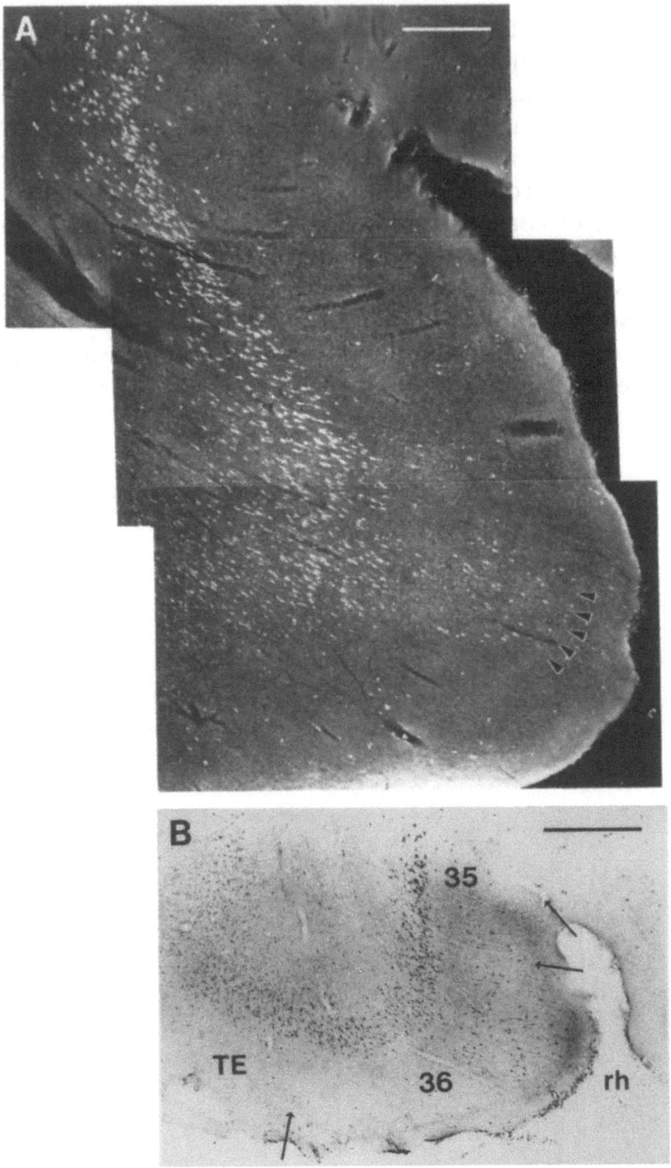

Figure 2. Labeling in ipsilateral perirhinal cortex after HRP injection into IT. (A) Dark-field photomontage. Although label was found predominantly in the deep layers, a patch of supragranular labeling is also evident at the ventral edge of the rhinal sulcus (group of arrows). Scale bar = 200 μm. (B) Bright-field photomicrograph of the same region in a neighboring section, illustrating anterograde label in layers I and IV. Scale bar = 500 μm. Abbreviations: TE, 35, 36, cytoarchitectonic areas of the temporal lobe; rh, rhinal sulcus. From Rodman and Consuelos (1994).

2.2.2. Commissural Connectivity

Because of their accessability, the cortical commissures have been utilized by many investigators as an ideal system for studying the development of corticocortical pathways in a variety of species. In monkeys, whereas some features of commissural connections are clearly laid out early, others are subject to significant postnatal refinement.

In monkeys, both the corpus callosum and the anterior commissure undergo a massive pruning of supernumerary axons during the first few months of postnatal life (LaMantia and Rakic, 1990, 1994). However, it is unclear which cortical structures are losing major proportions of connections during this period. Unlike in other mammals (see Section 5.1), in monkeys primary visual cortex never shows widespread interhemispheric connectivity, and the initially widespread origin of callosal inputs to V2 has been significantly restricted by birth (Dehay *et al.*, 1986). Some limited pruning clearly does takes place postnatally in the commissural pathways of extrastriate prior to achievement of the adult pattern. First, labeled terminals in cortex anterior to V2 show a more widespread pattern of laminar terminations after tracer injections into contralateral cortex than after comparable injections in adults (Dehay *et al.*, 1986). Second, IT cortex receives a transient projection in early postnatal life from the same portions of contralateral frontal cortex that are labeled within the ipsilateral hemisphere of both infants and adults (Rodman and Consuelos, 1994; Rodman and Nace, 1997; Figs. 1 and 3). Interestingly, in other respects, both the ipsilateral and the contralateral cortical connectivity of frontal cortex in the macaque are quite adult-like at birth (Goldman-Rakic, 1987; Webster *et al.*, 1994).

2.2.3. Connectivity with Subcortical Structures

In addition to the transient connections described above, inferior temporal areas TE and TEO also have transient connections with subcortical structures early in life (Webster *et al.*, 1991a, 1995a). In both infant and adult macaques, these regions, like other extrastriate areas, are reciprocally connected with the pulvinar nucleus, the amygdala, and the visual claustrum, and receive input from nonvisual structures including the locus coeruleus, raphe nuclei, basal forebrain, reticular formation, and several additional thalamic nuclei. The ma-

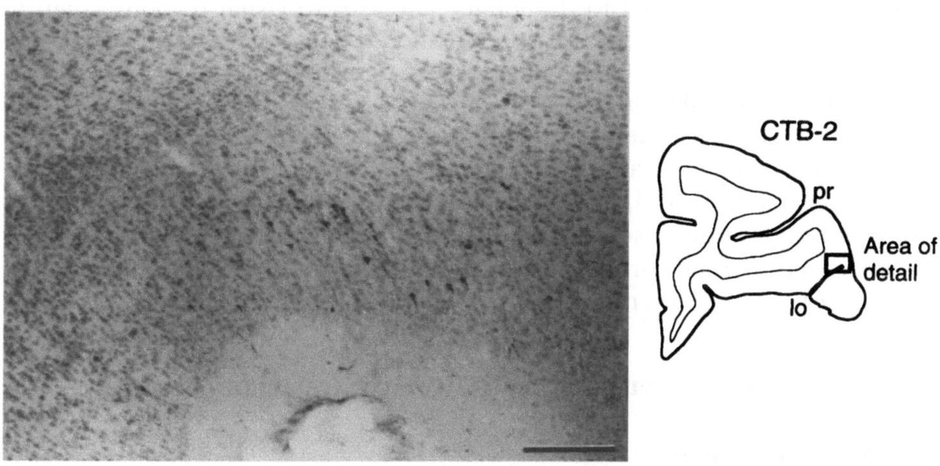

Figure 3. Bright-field photomicrograph of neurons retrogradely labeled with cholera toxin B-subunit in contralateral frontal cortex within the lateral orbital sulcus of a 6-week-old infant monkey in which tracer was placed into anterior IT. The section is counterstained with Giemsa; retrogradely filled cells appear black. Scale bar = 500 μm. See legend to Fig. 1 for abbreviations.

jor nonreciprocal subcortical output of inferior temporal visual cortex goes to the striatum. However, in infant monkeys, there is also a widespread projection from TE to the superior colliculus as well as an expanded projection to the nucleus medialis dorsalis of the thalamus and tail of the caudate nucleus. Moreover, TEO provides an expanded projection to the amygdala in infant monkeys, and receives an expanded, albeit small, projection from the dLGN. Following damage to area TE in infancy, the normally transient expanded projection from TEO to the amygdala is maintained (Webster *et al.,* 1991b).

2.2.4. Synaptic Contacts and Myelination

Within the first several months of postnatal life in the macaque, there is a massive overproduction of synaptic contacts which proceeds in synchrony across the entire cortical mantle, including striate, extrastriate, and frontal areas (Rakic *et al.,* 1986). Synaptic density peaks in all cortical areas and laminae at 2–4 months, and then only very gradually is reduced to adult levels by the age of 3 years or more. Relatively more contacts of the symmetric type (vs. the asymmetric type) are pruned off in this early period, suggesting, perhaps, a refinement in inhibitory circuitry. The synchronized time course of synaptic pruning across the neocortex in monkeys suggests that a least some aspects of the development of extrastriate circuitry may reflect very global processes of cortical maturation.

It has been known for over a century that primate "association" cortex is not fully myelinated until much later than other cortical regions, in fact not until well after the first year of life in monkeys (Flechsig, 1876; Yakovlev and Lecours, 1967). For example, at several weeks after birth, although a small number of myelinated fibers can be seen within the gray matter in "lower order" cortical areas such as V2 and MT, there are very few in "higher" areas of the parietal, anterior temporal, and frontal lobes. By 7 months, visual area MT is clearly recognizable in a myelin stain, but the dense adult pattern of myelination is just beginning to emerge in inferior temporal cortex (H. Rodman and A. M. Galioto, unpublished observations). This gradual increase in myelination in the postnatal period is likely to be related to the longer latencies to visual response seen in infant IT cortex (Section 2.3.1). However, it should be pointed out that the exact relationship between extent of myelination and neuronal activity remains uncertain, and there is no absolute correlation between dense myelination and the presence of brisk, robust neuronal responses. For example, areas that tend to have very little myelin in infancy also tend to have little myelin in adulthood, such as the principal sulcus, and, at least in adulthood, can contain cells producing strong and reliable responses. Moreover, in other mammals, some of the best responses in extrastriate cortex can be elicited from poorly myelinated areas of the lateral visual cortex (Sereno *et al.,* 1991, and unpublished observations).

2.3. Development of Neuronal Activity in Extrastriate Areas

2.3.1. Electrophysiology

Surprisingly little is known about the normal development of extrastriate visual cortex at the single-unit level in monkeys. The problem is complicated by the fact that we have an incomplete understanding of the physiological development of earlier visual structures in monkeys, including the striate cortex and especially the dLGN and pulvinar nuclei. For example, although it is generally accepted that both physiological and anatomical segregation of ocular dominance columns in V1 is complete by 8 weeks postnatal, subtle shifts are still possible many months later, and gradual improvements in contrast sensitivity and resolution take place over a very extended period reaching to beyond 1 year postnatal (Blakemore and Vital-Durand, 1981).

Thus far, virtually all published single-unit work on extrastriate development has been done in anterior temporal cortex, where Rodman and collaborators (Rodman *et al.*, 1991, 1993; Rodman, 1994) studied activity in areas IT and STP under both anesthetized and awake behaving conditions in infant macaques ranging between 5 weeks and 7 months. Under nitrous oxide anesthesia, responses were rare in both IT and STP until about 4 months of age, at which point there was a relatively sharp increase in responsive cells encountered in both areas (Fig. 4). Between 4 and 7 months, about half of the units encountered were responsive, a proportion still significantly below that found in adult monkeys studied in the same manner. Responsive neurons, however, did show evidence of stimulus selectivity (including responses selective for faces) and the large, bilateral visual receptive fields characteristic of anterior temporal cortical areas in adult monkeys, although responses tended to be weak. Responses in IT cortex were exclusively visual, whereas about half of the cells in STP were polymodal (i.e., gave some combination of visual, auditory, and/or somatosensory responses), as is the case for these regions in adults.

Recordings in awake, behaving infant monkeys provided evidence for adult-like patterns and prevalence of shape selectivity in IT cortex of the macaque within the first few months of life and also provided evidence for both a suppression effect under anesthesia at the earliest ages and for lower overall rates of

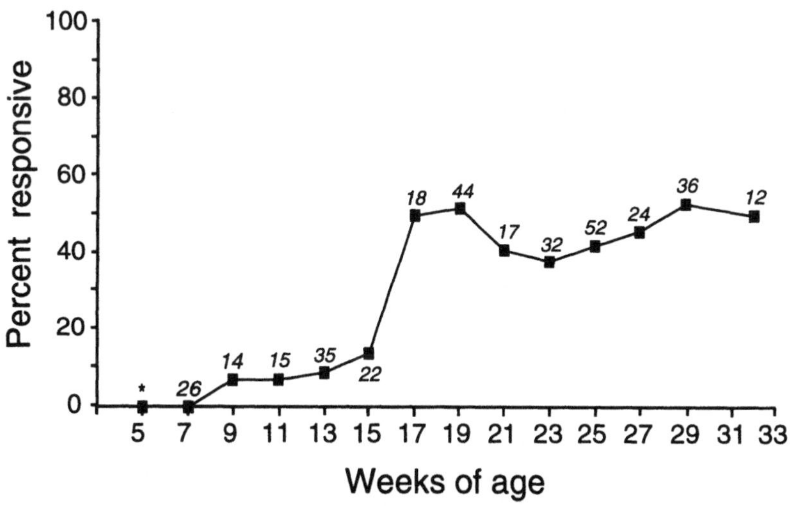

Figure 4. Proportion of responsive single units recorded in infant monkeys using the anesthetized paradigm as a function of age in weeks. Number in italics above each point indicates the number of single units tested. At 5 weeks (asterisk), under anesthesia, units were too small to be adequately isolated for study, and no responses were evident from background activity. Recordings from IT and STP have been pooled. Adapted from Rodman *et al.* (1993).

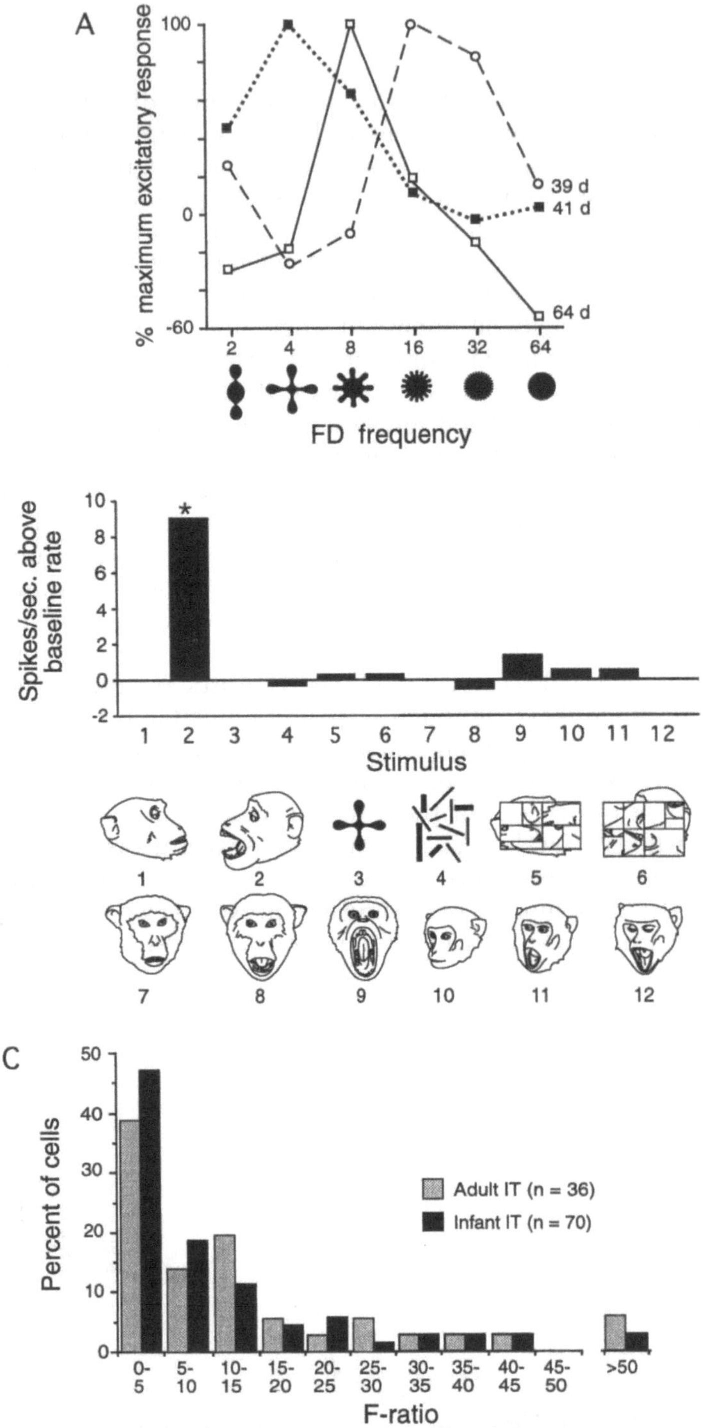

Figure 5. Selectivity of IT cells in infant monkeys. (A) Examples of responses of three cells tuned to boundary curvature (number of lobes on Fourier descriptor stimuli) in IT cortex of awake infant macaques. Age in days (d) at the time of recording is given for each tuning curve. (B) Cell in IT cortex of a 3-month-old alert infant monkey which was selective for a particular profile view. Stimuli were black and white (3, 4) or colored (remaining stimuli) images projected onto a screen and subtended 3–6 deg at the fovea. (C) Comparison of distributions of *F*-ratios from repeated-measures analysis of variance for each responsive IT cell in awake infant and adult monkeys. Adapted from Rodman *et al.* (1993).

action potential generation by IT neurons in infant monkeys. Under these conditions, comparable proportions (about two-thirds) of IT cells were visually activated in infant and adult monkeys, even below 4 months of age (Rodman *et al.*, 1991, 1993). These neurons showed tuning for shape similar to that found in adult monkeys (Fig. 5A), and, like IT cells in adults, occasionally responded specifically to faces (Fig. 5B). Overall, degree of selectivity was similarly distributed in infant and adult IT (Fig. 5C). However, responses were weaker and longer in latency than in adult monkeys, and accompanied by lower levels of spontaneous activity (Fig. 6). The generally low rates of neuronal firing in IT cortex even in the alert state may reflect processes common to developing monkey cortex (Section 2.2.4). Nonetheless, the weak signals deriving from single neurons in this tissue early in life may help explain why adult-like object recognition capacity takes several years to develop in monkeys (reviewed in Rodman, 1994). The suppression of responses in anterior temporal cortex of monkeys less than 4 months of age by anesthesia does seem to be a genuine and selective effect: individual infants lacking activity in these regions under anesthesia had visual responses in interdigitated awake recording sessions, as well as in striate cortex even when anesthetized (Rodman *et al.*, 1991, 1993).

Much less is known about the physiology of other extrastriate visual areas of monkeys in the first postnatal year. A little information exists for area MT, where, as a further control for the lack of responses in IT and STP under anesthesia, Rodman *et al.* (1991, 1993) recorded at a small number of sites (including 11 single units) in visual area MT under the same conditions in monkeys ranging from 5 to 11 weeks. Although the data are insufficient to make quantitative statements about the responsiveness of MT early in life, it was evident that cells in MT were responsive and stimulus-selective (for direction of motion, as in adult MT) on the same days when IT and STP were silent. An example of responses and direction tuning curve of an MT cell in an 11-week-old infant monkey is shown in Fig. 7. The presence of activity in MT of anesthetized infant monkeys while IT and adjacent areas are silent may reflect a general difference between rates of development between "dorsal stream" and "ventral stream" areas (see also Section 6.2). Alternatively, the difference may simply reflect the fact that cells in MT are, on average, several synapses closer to inputs from striate cortex than are cells in IT.

2.3.2. Metabolic Activity

The absence of responses in IT cortex under anesthesia and the weakness of responses early in life even in awake animals are consistent with studies on the development of energy utilization by cells of immature visual cortex. Using a 2-deoxyglucose technique, Bachevalier *et al.* (1991) showed that development of metabolic activity levels in "low-order" extrastriate areas of the ventral visual pathway such as V2 and V4 lagged behind that of striate cortex, where levels were adult-like at birth, by a few months. In inferior temporal cortex, activity levels lagged behind those of the more posterior visual areas and did not reach adult levels until about 4–6 months. However, the entire set of "dorsal stream" areas, including MT and areas within the anterior bank of the superior temporal sulcus, reach adult levels by 3 months (Distler *et al.*, 1990). This pattern suggests

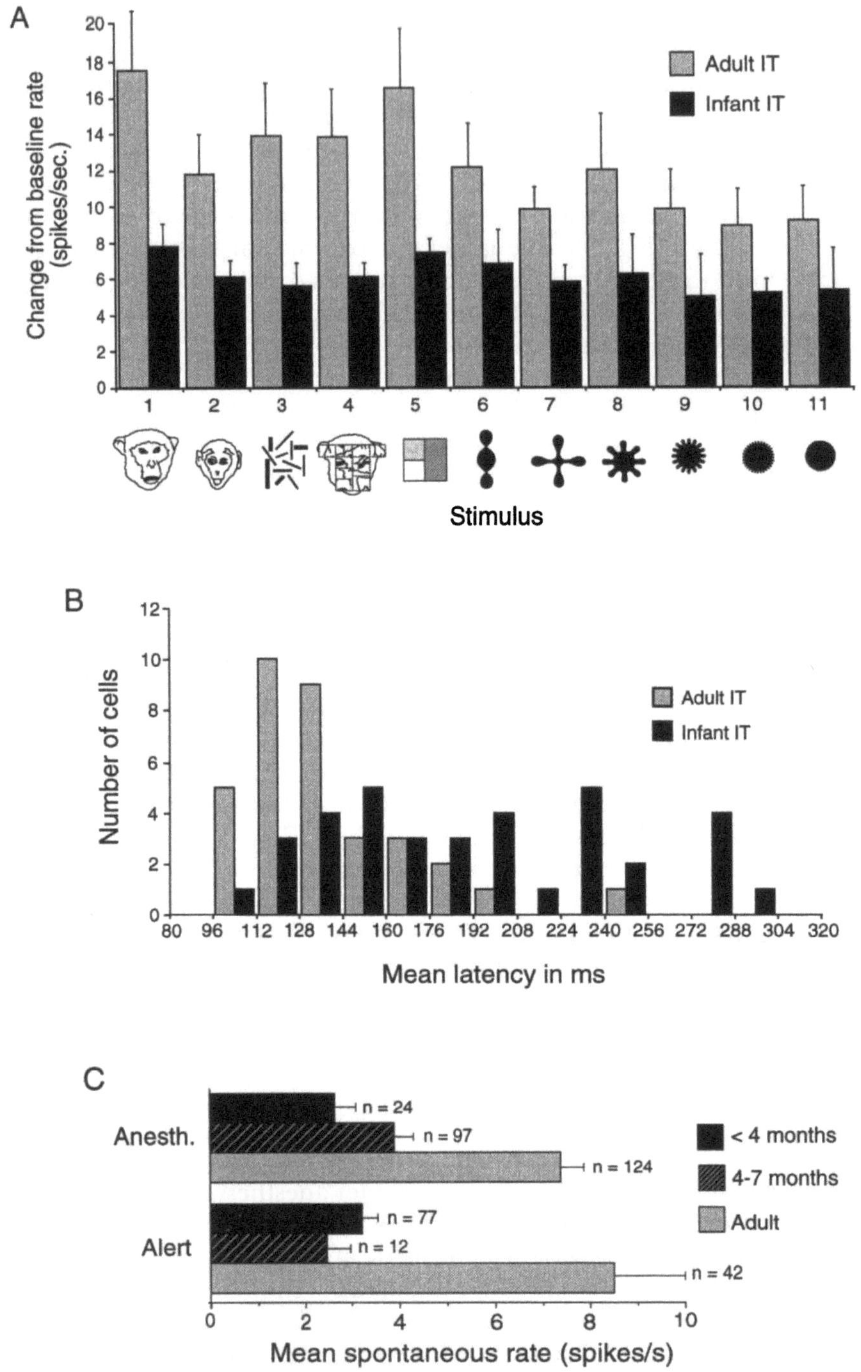

Figure 6. Signaling properties of IT cortex in infant monkeys. (A) Overall mean response magnitudes of samples of IT cells in awake infant and adult macaques to individual members of standard set of projected stimuli. (B) Overall mean latencies. (C) Mean spontaneous activity values of IT neurons in infant and adult macaques. Adapted from Rodman (1994).

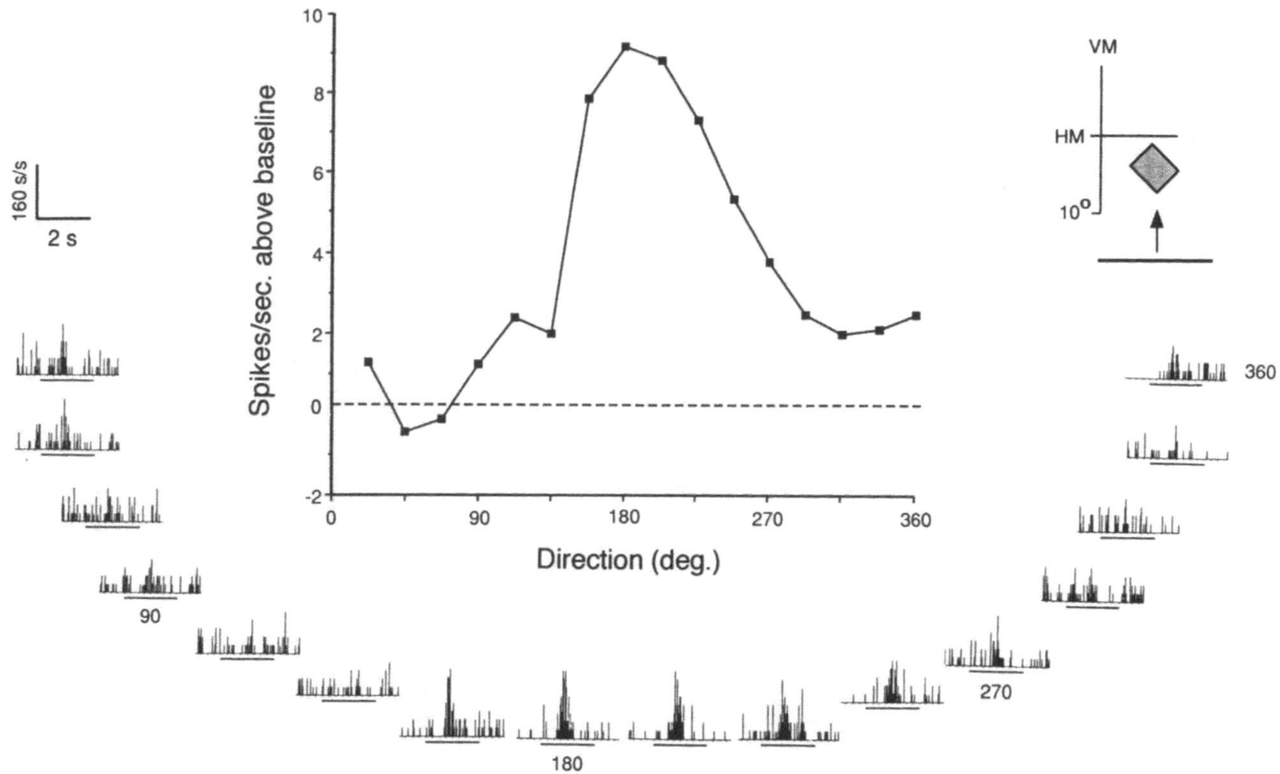

Figure 7. Examples of responses (peristimulus time histograms) and tuning curve obtained from an MT neuron in an 11-week-old animal, when IT and STP in the same animal were still unresponsive under N_2O/O_2 anesthesia. The line under each histogram indicates the time during which the stimulus was swept across the tangent screen. Dimensions of the stimulus (a bar of light moving at 16°/sec) are illustrated relative to the receptive field of the cell. Adapted from Rodman *et al.* (1993).

that while areas specialized for motion and spatial analyses are also immature at birth, they may as a group mature slightly earlier than the inferior temporal regions involved in object recognition.

Little is known about development of intrinsic regional patterns of metabolic activity in monkey extrastriate areas beyond V2. For example, patchy patterns of cytochrome oxidase (CO) staining visible in MT of adult primates (Tootell and Taylor, 1995) have apparently not been described in infant monkeys. Within V2, the pattern of CO stripes is adult-like at birth, and segregated patterns of connections related to the stripe patterns emerge during mid-gestation (Dehay and Kennedy, 1988; Sesma and Burkhalter, 1994; Barone *et al.*, 1994b).

2.4. Development of Patterns of Neurochemical Markers

A relatively large number of studies have examined the ontogeny of appearance of specific substances or their receptors in the primate cerebral cortex during both prenatal and postnatal development. Some of these substances, such as GABA and various neuropeptides, have been considered to have a direct role in synaptogenesis and formation of cortical circuitry, or to function as growth factors. Others, particularly monoamines, have been of interest because of proposed roles in maintaining cortical plasticity during a postnatal critical period (e.g., Kasamatsu, 1987) or because of involvement in arousal and orienting functions known to affect cortical processing in adult primates (Clark *et al.*, 1987; Saper, 1987). In general, a significant overproduction of neurotransmitter receptors takes place within the first 2–4 months after birth in a variety of cortical areas (Lidow *et al.*, 1991; Lidow and Rakic, 1992), in remarkable synchrony with the overproduction of synaptic contacts also seen across the cortical mantle. Concentrations and rates of synthesis of neurotransmitter molecules themselves, however, have more regionally specific developmental patterns (Goldman-Rakic and Brown, 1982).

Specifically with regard to monoamines, it has been shown than in adult monkeys dorsal extrastriate areas are relatively heavily innervated by noradrenergic fibers, whereas serotoninergic innervation predominates in the temporal cortex (Morrison and Foote, 1986). Little is known about development of detailed patterns of innervation of extrastriate cortex by monoaminergic afferents in monkeys, although it has been shown that adult laminar patterns and densities do not emerge even in striate cortex until about 2 months after birth (Foote and Morrison, 1987). Goldman-Rakic and Brown found that monoamine storage capacities and synthesis rates develop only gradually over the first several years of life. Moreover, whereas serotonin levels in the visual areas examined asymptote at about 2–5 months of age, the ability to rapidly synthesize this transmitter does not mature until about 1½ year.

Berger *et al.* (1990) studied the postnatal development of the distribution of neurons immunoreactive for DARPP-32, a phosphoprotein associated with cells bearing D1 dopamine receptors. At P3 or P42, labeled neurons were numerous throughout extrastriate visual cortex, as well as in other cortical territories, and were particularly dense in inferior temporal areas. In adults, in contrast, a far more restricted distribution was seen; within the visual cortex, densities approaching that in infants were seen only in the inferior temporal gyrus, neighboring paralimbic areas, and posterior orbitofrontal cortex, whereas other extrastriate regions (including parietal cortex and V2) were weakly stained or unstained. This retention of an infant pattern only by "high-order" stations of the ventral visual pathway is interesting in view of the predominance of "plastic" effects reported in physiological studies of this region in adult monkeys (Section 3.2).

In adult monkeys, the basal forebrain cholinergic system projects to extrastriate visual cortex, as well as to other cortical areas, and appears to play a role in the formation of visual object memories (e.g., Aigner *et al.*, 1991), possibly by influencing attention (Zola-Morgan and Squire, 1993; Voytko *et al.*, 1992). Receptor-binding studies indicate that at birth muscarinic cholinergic ligand binding is relatively low and undifferentiated in all visual areas, outside of the primary visual cortex, where the adult laminar pattern is just beginning to emerge (Bachevalier *et al.*, 1991). The immaturity of cholinergic mechanisms in temporal areas may be related to the long period of emergence of many indices of visual object memory in primates. This result also suggests that whereas "exuberant" receptor sites exist in early postnatal life for a variety of neurochemical systems, this does not necessarily translate into functional receptor activation at all or most such sites. A different type of immaturity of cholinergic circuitry is seen in V2, where the transient expression of acetylcholinesterase reactivity has been related to maturation of pulvinar inputs (Barone *et al.*, 1994a; see also Section 2). Interestingly, the transient reactivity disappears at 4–5 months, at the same time as a host of other general changes in cortical circuitry and behavior.

Yamashita and Hayashi and their colleagues (Hayashi and Oshima, 1986; Hayashi *et al.*, 1989; Yamashita *et al.*, 1989, 1990; Yamashita, 1992) have explored ontogenetic changes in neuropeptide concentrations and immunoreactivity in various neocortical regions. Within the visual cortex, for all neuropeptides examined, there is a gradual increase in immunoreactive neurons between occipital cortex through lower order extrastriate and inferior temporal cortex through the temporal pole; high concentrations and/or numbers of labeled cells are also found in other association regions of the frontal and parietal lobes. Moreover, there is a general pattern of an increase in concentration and/or cell number between about E120 and birth, whereupon there is a decrease so that adult levels are observed at about 2 months postnatal. In striate cortex, however, peptide concentrations and labeling tend to be adult-like at birth. These results further support the idea that for extrastriate cortex, as for other "association" areas, the first few postnatal months are crucial for the development of basic patterns of circuitry.

3. Plasticity of Extrastriate Visual Cortex in Adulthood

The behavior of both single neurons in visual areas and of the whole animal performing visual tasks can be shown to persist or recover after various types of insults to the cortex. In addition, the responses of neurons can alter over time as a result of learning. In both cases, a critical question becomes whether preserved or emergent properties represent adaptive reorganization of the cortical network, or the mere unmasking of inputs and signals that normally impinge upon the neurons examined.

3.1. Damage-Induced Plasticity

3.1.1. Effects of Lesions of Primary Visual Cortex

Primate extrastriate visual areas have long been thought of as performing specialized "further processing" upon specific types of information relayed to them from the primary visual cortex. In keeping with this notion were observations (reviewed below) of the generally devastating effects of damage to striate cortex in both monkey and human. As recently as 1985, it was suggested that in monkeys, unlike nonprimate species and prosimians, extrastriate areas were likely to be totally dependent on the integrity of striate cortex for their functioning (Diamond *et al.,* 1985). Contrary to these earlier notions, it is now apparent that at least some regions of monkey extrastriate cortex are capable of processing visual information in the absence of striate cortex.

The first direct evidence for preserved visual activity in the extrastriate cortex of anthropoid primates was provided by Bruce *et al.* (1986), who recorded in polysensory area STP and adjacent cortical regions of macaque monkeys after unilateral lesions of striate cortex. Because receptive fields in these portions of extrastriate cortex are bilateral, they were able to use each neuron as its own control. In so doing, they established that about half of STP cells remained responsive to visual stimuli in both half-fields of space, although responses to stimuli in the hemifield contralateral to the striate removals were weaker and devoid of selectivity for stimulus form or direction of motion. In contrast, nearly all neurons located in adjacent regions of inferior temporal cortex completely lost responsiveness in the portion of visual space contralateral to striate damage (Rocha-Miranda *et al.,* 1975; Bruce *et al.,* 1986).

The first purely visual area shown to be activated subsequent to ablation of striate cortex in the monkey was MT (Rodman *et al.,* 1989). In MT, unlike STP, most receptive fields do not extend across the midline, so responses could not be compared between half-fields with and without striate inputs as for STP. Nonetheless, after either unilateral or bilateral lesions of striate cortex, about two-thirds of MT cells remained visually responsive in the portions of visual space they coded prior to the lesion, although responses were weaker overall. Interestingly, unlike the case for STP, removal of striate inputs did not eliminate the selectivity of evoked activity in MT: the majority of MT neurons retained selectivity for direction of stimulus motion, and also retained relatively normal direction tuning width, binocularity, and receptive field size.

Because in both the Bruce *et al.* and Rodman *et al.* studies the majority of neurons tested were observed a number of weeks or months after the striate cortex ablation, the question arises as to whether the responsiveness observed was a result of normally functional inputs operating in parallel with those from striate cortex, or instead the result of a process of recovery or gradual reorganization. This question was addressed directly by Girard *et al.* (1992), who used a cooling strategy to record the responses of individual MT neurons before, during, and after the inactivation of corresponding parts of the striate representation. They confirmed that most MT neurons remained responsive without striate cortex input and that only very minor changes in directional properties could be observed, suggesting that the preserved responsiveness seen after lesions was at least largely due to the unmasking of nonstriate inputs, not to a process of reorganization. A similar conclusion was reached on the basis of a very limited sample of neurons tested with a similar cooling paradigm in the original Rodman *et al.* study. Girard *et al.* extended their cooling paradigm to examine areas V2, V3, V3A, and V4 (reviewed in Bullier *et al.*, 1994). Cooling of striate cortex wiped out responses in V2, confirming results of an earlier study by Schiller and Malpeli (1977). Responses in V3 and V4 were likewise abolished. However, approximately 30% of sites sampled in area V3A remained active during striate cortex cooling, and tended to retain the orientation or direction-of-motion selectivity they showed prior to cooling.

Examination of this pattern of results gives rise to the conjecture that parallel activation of extrastriate cortex in primates appears to be largely or entirely confined to a subset of regions of the so-called dorsal visual pathway (Rodman *et al.*, 1990; Gross, 1991; Bullier *et al.*, 1994; Fig. 8). The only areas which retain

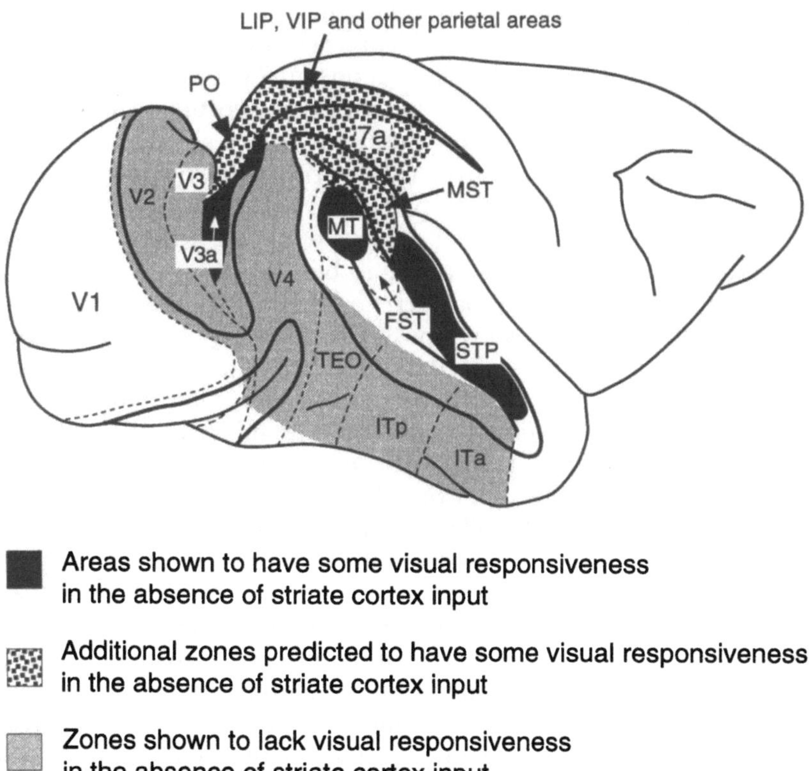

Figure 8. Lateral view of macaque monkey brain with posterior sulci "opened up" to show locations of cortical visual areas within. Black regions are components of the "dorsal" visual pathway which have been shown to have visual responses in the absence of striate cortex. Pale gray areas are regions demonstrated to be silent in the absence of striate inputs. Dotted zones are additional areas which have not yet been examined for possible responsiveness in the absence of striate cortex, and which are predicted to show at least some residual activity. Frontal visual regions are not represented in this diagram.

substantial specific activity, namely V3A and MT, are considered crucial early components of the dorsal pathway, at least that portion of it that contributes to motion processing. Area STP, which retains responsiveness but loses selectivity, receives converging projections from both inferior temporal areas and "high-order" dorsal stream areas. However, the possibility remains that at least some of the early components of the "ventral" visual pathway such as V2 and V4 might eventually recover responsiveness after permanent damage to striate cortex, since these areas have been examined thus far only in the context of a cooling paradigm. Conversely, it should also be noted that it is not certain that even "dorsal stream" areas respond without striate cortex inputs in all primate species (Kaas and Krubitzer, 1991).

As a result of observations both in humans (Holmes, 1918) and in monkeys (Kluver, 1942), it was believed until at least midcentury that primates devoid of striate cortex retained little significant visual capacity (reviewed in Weiskrantz, 1986; Bullier *et al.*, 1994). Beginning with the experiments of Weiskrantz and Cowey and the Pasiks in the 1960s and 1970s, it has been shown that monkeys with damage to striate cortex can retain a surprising amount of visually guided behavior if tested approximately, even when such lesions are made in adulthood (see Pasik and Pasik, 1982; Weiskrantz, 1986; and Cowey and Stoerig, 1991, for review). Recent experiments show that the preserved vision of destriate monkeys is similar in detail to that seen in human "blindsight" (Cowey and Stoerig, 1995; Moore *et al.*, 1995a). Preserved activity in extrastriate cortex of monkeys following damage to striate cortex may contribute to their residual visual capacities. Recent positron emission tomography (PET) studies in a human patient with striate cortex damage, moreover, show preserved activity in a proposed homolog of area MT when moving stimuli are used (Barbur *et al.*, 1993). Thus, extrastriate cortex activity may likewise play a role in blindsight.

How do visual signals reach extrastriate cortex if striate cortex is absent? Although the midbrain itself has previously been held to be sufficient for blindsight phenomena (e.g., Campion *et al.*, 1983), the dependence of residual responses in both STP (Bruce *et al.*, 1986) and MT (Rodman *et al.*, 1990) on the integrity of the superior colliculus suggests that the midbrain may be important in blindsight as a source of retinal signals to be conveyed to extrastriate. Neurons in the superficial layers of the superior colliculus remain visually responsive in the absence of V1 (Schiller *et al.*, 1974) and the proportion of cells there showing saccade-related activity increases concomitant with behavioral recovery (Mohler and Wurtz, 1977). The colliculus sends a large projection to the pulvinar complex, along with a small projection to the interlaminar zones of the dLGN (Kaas and Huerta, 1988; Harting *et al.*, 1991). In the inferior pulvinar, visual responsiveness of neurons appears to depend on descending inputs from V1 rather than the collicular inputs, but in the absence of V1, recovery of responsiveness can occur within weeks (Bender, 1983). In addition, the small population of cells that survive the massive retrograde degeneration of the dLGN following damage to V1 are cells located primarily in interlaminar zones and, in fact, project into extrastriate cortex (Cowey and Stoerig, 1989). Thus, pathways involved in recovery from damage to V1 are likely to include either or both the tectopulvinar route and a pathway from the midbrain to interlaminar layers of the dLGN and then on to extrastriate (Rodman *et al.*, 1990; Cowey and Stoerig,

1991; see also Section 4.2). In addition to routes through the midbrain, a direct retino-pulvinar-extrastriate projection (Cowey *et al.*, 1994) may play a role in the contribution of extrastriate cortex to blindsight.

3.1.2. Effects of Lesions within Extrastriate Cortex

Lesions of portions or even all of individual areas within the extrastriate cortex of monkeys can demonstrate a surprising amount of equipotentiality, depending on the exact areas lesioned and the functions investigated. For example, lesions of individual areas in the "dorsal" visual pathway which are clearly related to eye-movement control typically produce only small and transient effects on saccades or pursuit (Lynch, 1992; Ó Scalaidhe *et al.*, 1995; Yamasaki and Wurtz, 1991), and it is not clear in all cases whether the rapid recovery involves assumption of function by other areas which normally contribute to it, or reorganization of spared tissue in the lesioned cortical area.

The first mechanism must make at least a major contribution. For example, Yamasaki and Wurtz (1991) showed that extension of a lesion of MT to include much of the neighboring tissue of the superior temporal sulcus drastically increased and prolonged the small pursuit deficits seen after MT damage alone. In general, the effects of lesioning two areas involved in eye movement control (such as the frontal eye fields and posterior parietal cortex) are larger than those of lesioning one area alone (see Ó Scalaidhe *et al.*, 1997, for discussion).

Lesions within more ventral portions of extrastriate cortex tend to produce more stable and severe deficits, although the magnitude and even the existence of impairments are again intimately tied to the completeness of lesions as well as to their exact placement. For example, despite the preponderance of cells sensitive to some aspects of color in area V4, lesions of this cortex tend to produce little or no deficit on color discrimination tasks (reviewed in Merigan, 1993); preserved color discrimination capacities are apparently subserved by remaining, more ventral and anterior regions, which also contain populations of wavelength-sensitive neurons. Complete bilateral lesions of inferior temporal cortex produce severe and permanent deficits in object recognition, but unilateral lesions have little or no effect (Gross, 1973). Thus, there is rather extensive equipotentiality or redundancy between the inferior temporal cortices of the two hemispheres, a fact presumably related to the representation of both halves of visual space in the inferior temporal cortex of each hemisphere (e.g., Desimone and Gross, 1979).

3.1.3. Reorganization of Extrastriate Maps following Central or Peripheral Damage

There is ample evidence in monkeys for the ability of both somatosensory cortical maps and the visual field representation in striate cortex to reorganize following either central or peripheral damage (in the case of somatosensory cortex) or following damage to the retina (in the case of striate cortex) (see Merzenich and Sameshima, 1993, and Gilbert, 1993, and references therein). In contrast, very little attention has been paid to the possibility of realignment of extrastriate maps following visual system damage. Kaas and Krubitzer (1992)

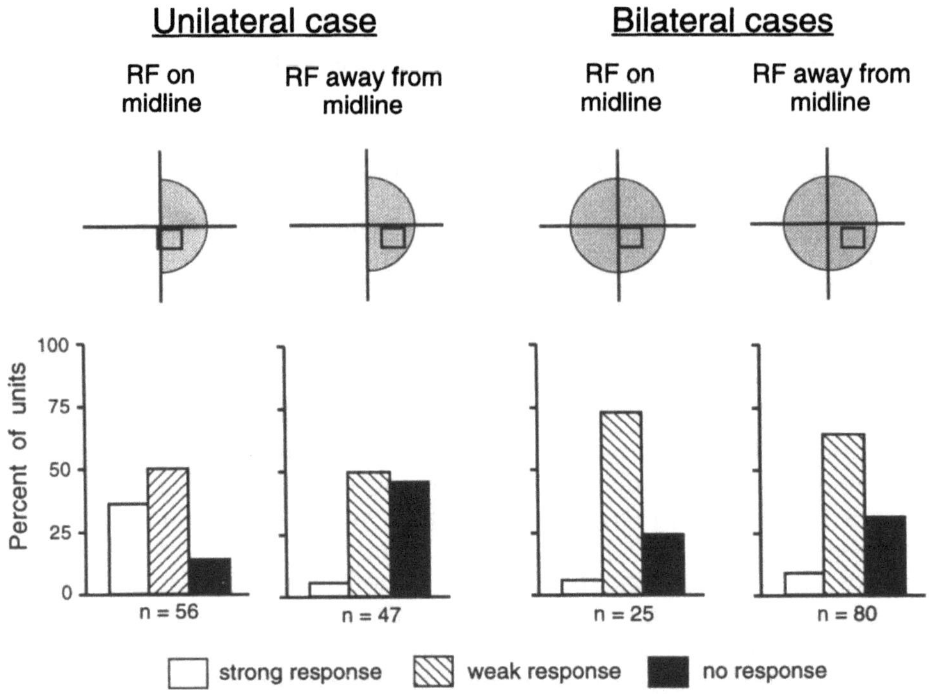

Figure 9. Distributions of recording sites within MT showing strong, weak, or no visual responses following either unilateral or bilateral lesions of striate cortex. All sites included here for which receptive fields could be plotted responded to stimulation within either the portion of the visual field represented in the excised striate cortex or, in the case of the unilateral lesion, in the visual field ipsilateral to the studied hemisphere. In normal MT, the ipsilateral visual field has only a very limited representation (e.g., Gattass and Gross, 1981). Receptive fields of neurons at unresponsive sites were estimated on the basis of those plotted for the nearest responsive units and/or from nearby control recordings made prior to the lesion. Adapted from Rodman *et al.* (1989).

noted some shifting of receptive fields in the "normal" portions of MT following acute lesions of striate cortex input in owl monkeys, and Rodman *et al.* (1989, and unpublished data) found a disproportionate emphasis on the vertical meridian and ipsilateral visual space within MT of a macaque monkey given a unilateral lesion of striate cortex (Fig. 9). In an attempt to explore mechanisms of recovery of pursuit following small lesions of MT, Yamasaki and Wurtz (1987) recorded in MT around the lesion and found a significant expansion of receptive fields of neighboring neurons. However, the receptive fields had expanded in all directions, not only in that corresponding to the destroyed part of the map, so the likelihood of a functional role for this expansion remains uncertain.

3.2. Experience or Learning-Induced Plasticity

Cells in the extrastriate cortex of monkeys have been shown to exhibit considerable plasticity of response properties as a function of experience in adulthood. Again, inferior temporal cortex, particularly its more anterior and ventral perirhinal portions, has been the focus of a large majority of these studies, as a result of the importance of these subregions for object memory as assessed from lesion studies. A few experiments, however, have focused on earlier extrastriate areas. Multiple types of physiological plasticity in extrastriate cortex of adult monkeys have been documented.

3.2.1. IT and Frontal Cortex

A large number of studies have found learning- or experience-dependent changes in the responses of IT and frontal neurons when a monkey is required to attend to visual stimuli and perform a behavioral task for reward based on some aspect of the stimuli. Such cells provide several types of signals that are candidates for memory traces or more generally indicators of stimulus familiarity or novelty. First, there can be differences in responses to match and nonmatch stimuli in delayed-match-to-sample (DMS) tasks (e.g., Gross *et al.,* 1979; Mikami and Kubota, 1980; Miller *et al.,* 1993). Second, enhanced responses can be observed when an animal performs a difficult discrimination or when a specific object is anticipated (Spitzer and Richmond, 1991; Desimone *et al.,* 1994). Third, delay-period activity in DMS or delayed-response tasks, a possible correlate of working memory or visual rehearsal when an object is not physically present, has been repeatedly demonstrated both in IT and in frontal cortex (e.g., Fuster, 1990; Wilson *et al.,* 1993; Desimone *et al.,* 1994; Colombo and Gross, 1994). Fourth, response suppression to repeated stimulation with a given object has also been observed for IT cells by a number of groups. Whereas enhancement of responses and delay activity may constitute short-term memory mechanisms, suppressive effects are also likely to be crucial for or reflect the attentional gating of visual perception (reviewed in Desimone *et al.,* 1994).

In addition, changes in patterns of stimulus selectivity or "tuning" of IT neurons with learning are observed as the animal gains experience with particular stimuli, and can be either rapid (Rolls *et al.,* 1989) or more gradual (e.g., Miyashita, 1988; Sakai and Miyashita, 1991). Of particular interest in this regard is the recent demonstration by Logothetis *et al.* (1995) that neurons in IT cortex of adult monkeys can eventually develop selectivity for novel, arbitrary geometrical stimuli when these are used as discriminanda in a difficult behavioral task.

3.2.2. MT and V4

In MT, Zohary *et al.* (1994) examined the relationship between changes in neuronal response patterns and behavioral performance in a perceptual learning paradigm involving discrimination of direction of stimulus motion. As the monkey's discrimination ability improved, there was an accompanying improvement in neuronal sensitivity which paralleled the behavioral improvement in several respects. Within V4, as for IT, changes in responsiveness (albeit not always reductions) have been observed as a direct consequence of stimulus repetition *per se* (Haenny and Schiller, 1988; Maunsell *et al.,* 1991).

3.2.3. Conditions Necessary for Learning-Dependent Plasticity in Extrastriate

There is general agreement that substantial cortical plasticity occurs in most behavioral situations where a discrimination or matching paradigm is used and in which the animal is required to attribute new or altered weighting to the significance of particular visual images. The types and extent of neuronal changes which accompany passive viewing of stimuli that do *not* have explicit significance to the animal remain less clear. Some authors have emphasized the lack of neuronal effects when stimuli are presented conjointly but without behavioral relevance (Merzenich and Sameshima, 1993; Sobotka and Ringo, 1991). Others have suggested that simple temporal contiguity of visual features is adequate for them to become associated in the responses of IT cells of adult monkeys (Miyashita, 1988; Stryker, 1991). At least one type of neuronal correlate of stimulus exposure, namely habituation to repeated presentation, is present under truly passive conditions, since it can be elicited under anesthesia in both adult and infant monkeys (Miller *et al.*, 1991; Rodman *et al.*, 1993). However, how and if this habituation is used by the visual system is not clear.

4. Plasticity of Extrastriate Visual Cortex in Development

4.1. Effects of Early Lesions of Extrastriate Cortex

Studies of early lesions of extrastriate cortex, although again clearest for "high-order" extrastriate cortex, point to a considerable plasticity and equipotentiality of extrastriate cortex and/or of its associated subcortical structures early in life. In the case of frontal cortex, lesions sustained in infancy lead to a pronounced initial sparing of delayed-responsed capacity followed by only very gradual appearance of deficits later in life (Goldman, 1972). In the inferior temporal lobe, early studies by Raisler and Harlow (1965), involving large lesions including area TEO and more posterior visual cortex as well as cytoarchitectonic area TE, indicated significant sparing of pattern recognition ability when such lesions were made as late as 1 year of age. Studies by Bachevalier and Mishkin and their colleagues have provided additional evidence for functional sparing after early lesions of temporal cortex. They found, for example, that bilateral lesions of TE made in the first month of life substantially spare rudimentary visual recognition assessed in infancy, as well as performance on a concurrent object discrimination or "habit" task (reviewed in Bachevalier, 1990). Moreover, unlike the case for early lesions of dorsolateral frontal cortex, spared abilities remain intact several years subsequent to neonatal TE damage (Malkova *et al.*, 1995). Addition of further ablations to animals sustaining TE damage in infancy indicated that the preserved capacities are subserved by a widely distributed set of "high-order" visual cortices (Webster *et al.*, 1995b).

4.2. Effects of Early Lesions of Primary Visual Cortex

Some of the most suggestive evidence for pronounced developmental plasticity of extrastriate visual cortex in monkeys derives from studies of the effects of early removal of striate cortex (Moore *et al.*, 1993, 1996). Subsequent to unilateral ablation of striate cortex at 5–6 weeks of age, monkeys show very extensive behavioral recovery, both relative to monkeys with similar lesions sustained in adulthood and relative to normal performance (Fig. 10). When mon-

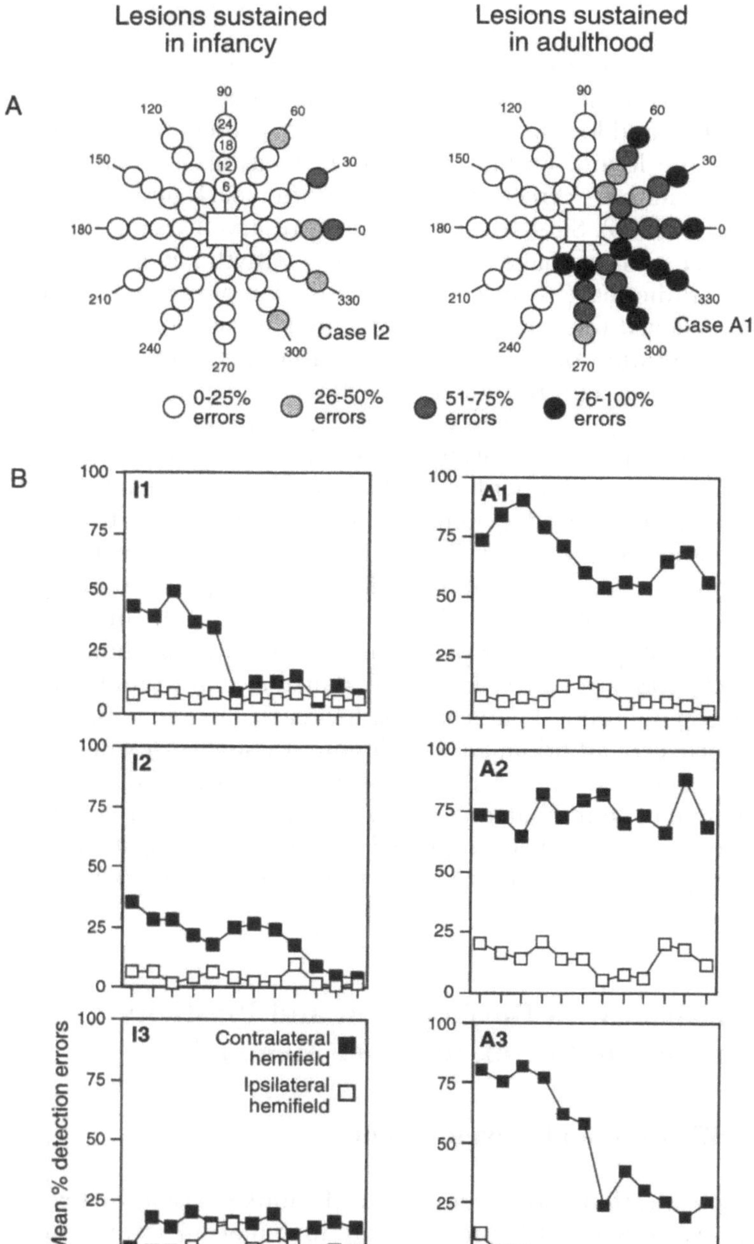

Figure 10. Comparison of detection performance after unilateral striate cortex lesions made in infancy (left side of figure, cases I1–I3) and adulthood (right side of figure, cases A1–A). (A) Graphic representation of detection errors (failures to initiate saccades to visual targets) on the grid of test points for monkeys A1 and I2. Each circle represents one test point; for each polar angle, points were tested at 6, 12, 18, and 24 deg eccentricity. The hemifield contralateral to the striate damage is represented on the right. Data are collapsed over the 12 repetitions of testing and represent a total of 48 trials/test point in the hemifield contralateral to the damage and 96 trials/test point in the hemifield ipsilateral to the lesion. The central square represents the fixation point error window (2 deg radius). (B) Overall mean detection errors in the hemifields contralateral and ipsilateral to striate lesions over the course of the 12 repetitions of testing. Data from the vertical meridian are omitted, as it is unclear in which hemifield they should be included.

keys with infant lesions were tested 2–5 years postsurgery and compared with monkeys with striate lesions sustained in adulthood, they were superior in their ability to detect and localize small visual targets presented in the hemifield contralateral to the lesion. Preliminary evidence was also obtained for preserved ability to discriminate direction of motion after early lesions of striate cortex (Moore *et al.*, 1995b). Similarly, in humans, preserved vision appears most pronounced after early damage to striate cortex (Blythe *et al.*, 1987; Payne *et al.*, 1996). PET studies have revealed strong activation of extrastriate cortex, including cortex believed to correspond to MT, in a human patient who still experiences some sensation of visual movement despite sustaining damage to striate cortex at 8 years of age (Barbur *et al.*, 1993).

Mechanisms of extrastriate involvement in the greater recovery seen after early lesions may, paradoxically, involve the retrograde degeneration which follows cortical lesions. In both infant and adult monkeys, a striate lesion is followed by extensive retrograde degeneration of the dLGN and, transneuronally, of ganglion cells in the retina (Van Buren, 1963). However, after an early lesion, cell loss in the retina takes place at a much faster rate (Dineen and Hendrickson, 1981; Weller and Kaas, 1989). While a total of about 80% of ganglion cells can ultimately be lost, populations projecting to the midbrain remain largely intact (Cowey *et al.*, 1989). As a result of early damage, more rapid ganglion cell loss may more rapidly unmask alternate pathways from retina and midbrain to extrastriate cortex. In addition, an early lesion of visual cortex might allow greater anatomical compensation or rearrangement in such pathways. Considerable rearrangement of retinothalamo–extrastriate projections after early visual cortical damage has already been reported in cats, including stabilization of early "exuberant" geniculoextrastriate projection patterns (Lomber *et al.*, 1995; Section 5.2).

5. Comparison with Development and Plasticity of Extrastriate Cortex in Other Mammals

5.1. Specification and Development

The specification and development of monkey visual cortex departs from that previously described for nonprimate mammals in several important ways. The pre- and early postnatal development of visual cortices of other species is characterized by widespread "exuberance" of both intrahemispheric and commissural pathways (e.g., Cusick and Lund, 1982; Payne *et al.*, 1988); in kittens, for example, visual cortex receives transient inputs from cortex subserving other modalities, and callosally projecting neurons interconnecting visual cortices across the two hemispheres are widely spread (Innocenti and Clarke, 1984a, b; Berman and Payne, 1988). These patterns are not found in infant monkeys, even prior to birth (Dehay *et al.*, 1988). Dehay and Kennedy (1988) have suggested that the time of callosal maturity may best be related to eye-opening, rather than birth, accounting for some of these differences.

In contrast, monkey visual cortex appears to be characterized by a fundamentally different type or types of transient circuitry (Kennedy *et al.*, 1989b).

First, there is an extensive remodeling of the precise patterns of feedback connections, possibly related to the delayed maturation of neuronal circuits in layers receiving feedback projections. Second, there is remodeling of interconnectivity between "high-order" areas of the ventral visual pathway and other "high-order," multimodal, or limbic areas. A third type of transient circuitry, which perhaps bears a closer resemblance to that seen in other species, is that seen between extrastriate areas and subcortical structures. For example, the small exuberant projection from the dLGN to area TEO of infant monkeys has a parallel in the exuberant geniculoextrastriate pathways documented in the cat (Lomber *et al.*, 1995). Other similarities and differences in transient connectivity between primate and nonprimate groups should be apparent following more study of the early connectivity of additional extrastriate areas in monkeys.

5.2. Parallel Activation and Plasticity of Extrastriate Areas

As in primates, removal or inactivation of striate cortex in rodents and cats reveals the presence of residual or compensatory activity in one or more extrastriate regions (Olavarria and Torrealba, 1978; Molotchnikoff and Hubert, 1990; see Bullier *et al.*, 1994; Payne, 1993, for review). Differences in overall patterns of residual activity in part reflect differences in thalamocortical organization among these groups. In cats, for example, strong residual activity in areas 18 and 19 is presumably a function of the strong dLGN projections to these areas. However, in cats, extrastriate areas of the lateral suprasylvian sulcus, and the more lateral area 21a, also retain significant visual responsiveness in the absence of primary visual cortex. In particular, area PMLS, proposed to be a homolog of MT (Payne, 1993), contains a number of direction-selective neurons after removal of areas 17 and 18, and shows significant anatomical reorganization (Payne and Cornwell, 1994; Lomber *et al.*, 1995) and functional compensation (Guido *et al.*, 1990) when such ablations are made early in life. Interestingly, PMLS receives projections not from the main layers of the cat dLGN, but rather from the layers believed to contain neurons homologous to the interlaminar, extrastriate-projecting cells of the primate dLGN (Payne *et al.*, 1996). Like primates, cats show increased sparing or recovery of visual capacity when primary visual cortex lesions are made in infancy (reviewed in Spear, 1983; Payne and Cornwell, 1994; Payne *et al.*, 1996). It thus appears likely that common principles characterize the means by which extrastriate cortex reorganizes, or compensates on the basis of parallel pathways, after damage involving "primary" cortical trauma in various mammals.

6. Concluding Discussion

6.1. General Principles

A number of considerations point to the first 6 months of life in monkeys as being a time of enormous maturational change for the extrastriate cortex. Relatively synchronized changes in energy utilization, overproduction of synaptic contacts, and neurotransmitter receptors point to an early exuberance of circuitry during this period. Transient projections and expression of chemical markers are present during at least some of this period, and physiological changes are present as well. Finally, this period in ontogeny includes a number of crucial behavioral milestones in macaques. For example, delayed-response capacity, delayed nonmatch-to-sample capacity, and other abilities all appear at around 4 months (reviewed in Rodman, 1994).

On the other hand, it is clear that additional maturational changes take place in extrastriate cortex at even later stages. For example, physiological response properties remain immature in inferior temporal cortex at least up to 7 months (e.g., Rodman *et al.*, 1993) and neuronal activity in frontal cortex related to performance of delayed response remains at lower than adult levels at 12 months (Alexander *et al.*, 1980). Finally, in macaques as well as humans, synaptic number does not plateau throughout neocortex for a period of years, and monoamine levels likewise take a period of years to plateau. Evidence for this extended course of extrastriate development also includes the long period of maturation of visual and cognitive abilities to adult levels seen in monkeys as well as humans. For some regions of extrastriate or more generally "association" cortex, it may be that retention of plasticity characteristic of neonatal cortex is essential for the expression of learning-dependent changes in adulthood. Alternately, adult plasticity in such areas may depend on mechanisms fundamentally different from those which shape circuitry in infants.

The presence of numerous transient pathways in the extrastriate cortex of developing primates raises anew the more general issue of the functional implications of such connections. The normal elimination of transient connections may play a role in the specification of cortical regions or their internal organization, or in achieving optimal patterns of connections based on early experience (Innocenti, 1981a, b; Berman and Payne, 1988; Dehay and Kennedy, 1993). Maintenance of transient connections may, on the other hand, provide a mechanism for plasticity and recovery after early brain damage (Innocenti, 1981a; Webster *et al.*, 1991b; Payne *et al.*, 1996). However, this latter hypothesis has yet to be supported, at least in monkey visual cortex, by direct evidence for a role for specific retained transient pathways in recovery from early trauma. Finally, loss or stabilization of transient connections may represent only a final artifact of the process of maturation, and have no organizational or adaptive functions, respectively.

6.2. Sequence of Development and Functional Streams

There is some reason to suspect that the extrastriate mantle of monkey may develop in synchrony with, or even in advance of, the primary visual cortex. There is ample evidence that many aspects of area V2 mature prior to those of V1. For example, cells destined for area V2 are generated a few days earlier in embryonic life than those for V1 (Rakic, 1976). In addition, V2 assumes its intrinsic modular organization several weeks earlier than in V1 and receives segregated patterns of thalamocortical input prior to the segregation of ocular dominance columns in V1 (Sesma and Burkhalter, 1994). As described in Section 2, the pattern of cytochrome oxidase staining as immature at birth in V1, but adult-like in V2 (Dehay and Kennedy, 1988), and long-range local connections mature earlier in V2 than in V1. However, other data on receptor development and metabolic activity point to a relative early maturity of striate cortex, while adult-like patterns take months to develop in areas farther along the visual cortical pathways (Bachevalier *et al.*, 1991).

It is tempting to speculate that the dorsal visual pathway matures at a pace somewhat ahead of that of the ventral pathway, and that this mirrors the less crucial dependence on striate cortex shown by areas of the dorsal pathway. It is generally accepted that the so-called magnocellular pathway contributes disproportionately to information processing in dorsal stream extrastriate areas, whereas the parvocellular pathway contributes relatively more to processing in ventral stream areas, although this mapping is an oversimplification, with significant magnoderived inputs also feeding into the "ventral stream" (Merigan and Maunsell, 1993). A number of lines of evidence suggest that the magnocellular pathway matures slightly in advance of the parvocellular one. For example, the magnocellular layers of the dLGN show dense staining for cytochrome oxidase at birth, whereas the parvocellular layers are only lightly reactive, compared with moderately dense staining of both sets of layers in mature monkeys (Kennedy *et al.*, 1989a). Moreover, cells in the magnocellular layers appear to mature anatomically (Garey and Saini, 1981) and physiologically (M. Hawken, personal communication, 1996) ahead of those in the parvocellular layers. In addition, subdivisions of layer IVC of striate receiving magno input mature somewhat in advance of those receiving parvo input (Lund and Harper, 1991; Lund and Holbach, 1991). Studies of perceptual development in primates also support the idea of a relatively early maturation of functions associated with "magnocellular" pathways (Mikami and Fujita, 1992; Slater *et al.*, 1985; Dobkins and Teller, 1996; K. R. Dobkins, personal communication). Taken *in toto,* these lines of evidence might be construed to indicate that dorsal stream areas primarily driven by magno inputs attain their functions somewhat in advance of ventral stream ones in both human and monkey.

On a related note, it has been shown that within human striate cortex, certain layers lag behind others in their development, and this phenomenon has been used to interpret developmental milestones in human visual attention in terms of sequential activation of different striate–extrastriate projections (reviewed in Johnson, 1990). Specifically, Johnson proposed that an early maturation of a "broadband-MT" pathway comes to underlie visual orienting in human infants before other cortical pathways are functional. The picture of inferior temporal areas TEO and TE presented here suggests a very long period of maturation, but these are areas quite far along the ventural pathway. Clearly, more work is needed to establish the relative tempo of different processing substreams within extrastriate visual cortex of the monkey.

6.3. Plasticity and Recovery of Function after Early or Late Brain Damage

In many human patients, and in most monkey studies, at least some measure of visual capacity survives trauma involving subtotal lesions of the visual cortices in adulthood, with striate and extrastriate regions taken together. In particular, spatial functions—the ability to interact with the external world through eye movements, reaching movements, postural adjustments, and the judgement of orientation—survive in some measure after lesions of extrastriate areas, and even after lesions of primary visual cortex. We have reviewed in previous sections the evidence suggesting that activity in extrastriate areas contributes to preserved capacities in "blindsight." While such a crucial contribution for extrastriate cortex remains to be confidently established, there are striking parallels between the types of capacities spared and the properties of cells in the more dorsal, motion- and spatial behavior-related areas that retain some visual activity. For example, it is noteworthy that the extrastriate areas retaining visual activity after damage to striate cortex have each been implicated in the control of eye movements (Newsome *et al.*, 1985; Galletti and Battaglini, 1989; Ó Scalaidhe *et al.*, 1995), which are largely preserved after lesions of striate cortex (Mohler and Wurtz, 1977; Segraves *et al.*, 1987; Moore *et al.*, 1995a).

Interestingly, while inferior temporal areas do not remain active following damage to striate cortex in adulthood, they do show extensive plasticity of response properties in adulthood, a long period of maturation of connectivity and response properties, and an extended postnatal period before becoming functionally committed to the roles they will play in adulthood. Thus, early in life, the more ventral cortical areas may also receive significant activation through nonstriate pathways and show more of a pluripotent character than more dorsal regions. Damage in infancy to "lower order" portions of the cortical visual pathways—at the level of striate cortex or V2—may be followed by compensatory changes in inferior temporal (and perhaps ventral frontal) regions that are not possible when similar damage is sustained in adulthood. If so, early damage of this type should be expected to preserve not only spatial abilities, but also some awareness of visual events (Barbur *et al.*, 1993; Weiskrantz *et al.*, 1995).

7. References

Aigner, T. G., Walker, D. L., and Mishkin, M., 1991, Comparison of the effects of scopolamine administered before and after acquisition in a test of visual recognition memory in monkeys, *Behav. Neural Biol.* **55:**61–67.

Alexander, G. E., Witt, E. D., and Goldman-Rakic, P. S., 1980, Neuronal activity in the prefrontal cortex, caudate nucleus and mediodorsal thalamic nucleus during delayed response performance of immature and adult rhesus monkeys, *Neurosci. Abstr.* **6:**88.

Bachevalier, J., 1990, Ontogenetic development of habit and memory formation in primates, in: *Development and Neural Bases of Higher Cognitive Functions* (A. Diamond, ed.), Academic Press, New York, pp. 457–484.

Bachevalier, J., Hagger, C., and Mishkin, M., 1991, Functional maturation of the occipitotemporal pathway in infant rhesus monkeys, in: *Brain Work and Mental Activity* (N. A. Lassen, M. E. Raichle, and L. Friberg, eds.), Munksgaard, Copenhagen, pp. 231–242.

Barbur, J. L., Watson, J. D., Frackowiak, R. S., and Zeki, S., 1993, Conscious visual perception without V1, *Brain* **116:**1293–1302.

Barone, P., Dehay, C., Berland, M., and Kennedy, H., 1994a, Developmental changes in the distribution of acetylcholinesterase in the extrastriate visual cortex of the monkey, *Dev. Brain Res.* **77:**290–294.

Barone, P., Dehay, C., Berland, M., and Kennedy, H., 1994b, Segregation of functional visual pathway in the prenatal monkey, *Neurosci. Abstr.* **20:**215.

Barone, P., Dehay, C., Berland, M., Bullier, J., and Kennedy, H., 1995, Developmental remodeling of primate visual cortical pathways, *Cerebral Cortex* **1:**22–58.

Bender, D. B., 1983, Visual activation of neurons in the primate pulvinar depends on cortex but not colliculus, *Brain Res.* **279:**258–261.

Berger, B., Febvret, A., Greengard, P., and Goldman-Rakic, P. S., 1990, DARPP-32, a phosphoprotein enriched in dopaminoceptive neurons bearing dopamine D1 receptors: Distribution in the cerebral cortex of the newborn and adult rhesus monkey, *J. Comp. Neurol.* **299:**327–348.

Berman, N. E., and Payne, B. R., 1988, Development and plasticity of visual interhemispheric connections, in: *Advances in Neural and Behavioral Development,* Volume 3 (P. Shinkman, ed.), Ablex Press, Norwood, NJ.

Blakemore, C., and Vital-Durand, F., 1981, Postnatal development of the monkey's visual system, *CIBA Symp.* **86:**152–171.

Blythe, I. M., Kennard, C., and Ruddock, K. H., 1987, Residual vision in patients with retrogeniculate lesions of the visual pathways, *Brain* **110:**887–905.

Bruce, C., Desimone, R., and Gross, C. G., 1986, Both striate cortex and the superior colliculus contribute to visual properties of neurons in the superior temporal polysensory area of the macaque, *J. Neurophysiol.* **5:**1057–1075.

Bullier, J., Girard, P., and Salin, P.-A., 1994, The role of area 17 in the transfer of information to extrastriate visual cortex, in: *Cerebral Cortex,* Volume 10, *Primary Visual Cortex in Primates* (A. Peters and K. S. Rockland, eds.), Plenum Press, New York, pp. 301–330.

Burkhalter, A., 1993, Development of forward and feedback connections between areas V1 and V2 of human visual cortex, *Cerebral Cortex* **3:**476–487.

Campion, J., Latto, R., and Smith, Y. M., 1983, Is blindsight an effect of scattered light, spared cortex, and near-threshold vision? *Behav. Brain Sci.* **6:**423–486.

Clark, C. R., G. M. Geffen, and L. B. Geffen, 1987, Catecholamines and attention. I. Animal and clinical studies, *Neurosci. Biobehav. Rev.* **11:**341–352.

Colombo, M., and Gross, C. G., 1994, Responses of inferior temporal and hippocampal neurons during delayed matching-to-sample in monkeys (macaca fascicularis), *Behav. Neurosci.* **108:**443–455.

Cowey, A., and Stoerig, P., 1989, Projection patterns of surviving neurons in the dorsal lateral geniculate nucleus following discrete lesions of striate cortex: Implications for residual vision, *Exp. Brain Res.* **75:**631–638.

Cowey, A., and Stoerig, P., 1991, The neurobiology of blindsight, *Trends Neurosci.* **14:**14–145.

Cowey, A., and Stoerig, P., 1995, Blindsight in monkeys, *Nature* **373:**247–249.

Cowey, A., Stoerig, P., and Perry, V. H., 1989, Transneuronal retrograde degeneration of retinal ganglion cells after damage to striate cortex in macaque monkeys: selective loss of Pβ cells, *Neuroscience* **29:**65–80.

Cowey, A., Stoerig, P., and Bannister, M., 1994, Retinal ganglion cells labelled from the pulvinar nucleus in macaque monkeys, *Neuroscience* **61:**691–705.

Cusick, C. G., and Lund, R. D., 1982, Modification of visual callosal projections in rats, *J. Comp. Neurol.* **212:**385–398.

Dehay, C., and Kennedy, H., 1988, The maturational status of thalamocortical and callosal connections of visual areas V1 and V2 in the newborn monkey, *Behav. Brain Res.* **29:**237–244.

Dehay, C., and Kennedy, H., 1993, Control mechanisms of primate corticogenesis, in: *Functional Organisation of the Human Cerebral Cortex* (B. Gulyas, D. Ottoson, and P. E. Roland, eds.), Pergamon Press, Oxford, pp. 13–27.

Dehay, C., Kennedy, H., and Bullier, J., 1986, Callosal connectivity of areas V1 and V2 in the newborn monkey, *J. Comp. Neurol.* **254:**20–33.

Dehay, C., Kennedy, H., Bullier, J., and Berland, M., 1988, Absence of interhemispheric connections of area 17 during development in the monkey, *Nature* **331:**348–350.

Dehay, C., Giroud, P., Berland, M., Smart, I., and Kennedy, H., 1993, Modulation of the cell cycle contributes to the parcellation of the primate visual cortex, *Nature* **366:**464–466.

Desimone, R., and Gross, C. G., 1979, Visual areas in the temporal cortex of the macaque, *Brain Res.* **178:**363–380.

Desimone, R., and Ungerleider, L. G., 1989, Neural mechanisms of visual perception in monkeys, in: *Handbook of Neuropsychology,* Volume II (R. Boller and J. Grafman, eds), Elsevier, Amsterdam, 267–299.

Desimone, R., Miller, E. K., Chelazzi, L., and Lueschow, A., 1994, Multiple memory systems in the visual cortex, in: *The Cognitive Neurosciences* (M. S. Gazzaniga, ed.), MIT Press, Cambridge, MA, 475–486.

Diamond, I. T., Fitzpatrick, D., and Sprague, J. M., 1985, The extrastriate visual cortex: A historical approach to the relation between "visuo-sensory" and "visuo-psychic" areas, in: *Cerebral Cortex,* Volume 4, *Association and Auditory Cortices* (A. Peters and E. G. Jones, eds.), Plenum Press, New York.

Dineen, J. T., and Hendrickson, A. E., 1981, Age-correlated differences in the amount of retinal degeneration after striate cortex lesions in monkeys, *Invest. Ophthalmol. Vis. Sci.* **21:**749–752.

Distler, C., Ungerleider, L. G., Bachevalier, J., and Mishkin, M., 1990, Functional development of the cortical visual pathway for motion analysis in rhesus monkeys, *Neurosci. Abstr.* **16:**5.

Dobkins, K. R., and Teller, D. Y., 1996, Infant contrast detectors are selective for direction of motion, *Vision Res.* **36:**281–294.

Felleman, D. J., and Van Essen, D. C., 1991, Distributed hierarchical processing in primate cerebral cortex, *Cerebral Cortex* **1:**1–48.

Flechsig, P., 1876, *Die Leitungsbahnen im Gehirn und Ruckenmark des Menschen auf Grund Entwicklungsgeschichtlicher Untersuchungen,* Engelmann, Liepzig.

Foote, S. L., and Morrison, J. H., 1987, Development of the noradrenergic, serotonergic, and dopaminergic innervation of neocortex, *Curr. Top. Dev. Biol.* **21:**391–423.

Fuster, J. M., 1990, Inferotemporal units in selective visual attention and short-term memory, *J. Neurophysiol.* **64:**681–697.

Galletti, C., and Battaglini, P. P., 1989, Gaze-dependent visual neurons in area V3A of monkey prestriate cortex, *J. Neurosci.* **9:**1112–1125.

Garey, L. J., and Saini, K. D., 1981, Golgi studies of the normal development of neurons in the lateral geniculate nucleus of the monkey, *Exp. Brain Res.* **44:**117–128.

Gattass, R., and Gross, C. G., 1981, Visual topography of the striate projection zone in the posterior superior temporal sulcus (MT) of the macaque, *J. Neurophysiol.* **46:**621–638.

Gilbert, C. D., 1993, Rapid dynamic changes in adult cerebral cortex, *Curr. Opin. Neurobiol.* **3:**100–103.

Girard, P., Salin, P. A., and Bullier, J., 1992, Response selectivity of neurons in area MT of the macaque monkey during reversible inactivation of area V1, *J. Neurophysiol.* **67:**1437–1446.

Goldman, P. S., 1972, Developmental determinants of cortical plasticity, *Acta Neurol. Exp.* **32:**495–511.

Goldman-Rakic, P. S., 1987, Development of cortical circuitry and cognitive function, *Child Dev.* **58:**601–622.

Goldman-Rakic, P. S., and Brown, R. M., 1982, Postnatal development of monoamine content and synthesis in the cerebral cortex of rhesus monkeys, *Dev. Brain Res.* **4:**339–349.

Gross, C. G., 1973, Inferotemporal cortex and vision, *Prog. Physiol. Psychol.* **5:**77–123.

Gross, C. G., 1991, Contributions of striate cortex and the superior colliculus to visual function in area MT, the superior temporal polysensory area and the inferior temporal cortex, *Neuropsychologia* **29:**487–515.

Gross, C. G., Bender, D. B., and Gerstein, G. L., 1979, Activity of inferior temporal neurons in behaving monkeys, *Neuropsychologia* **7:**215–229.

Gross, C. G., Rodman, H. R., Gochin, P. M., and Colombo, M. W., 1993, Inferior temporal cortex as a pattern recognition device, in: *Proceedings of the Third Annual NEC Research Symposium* (E. Baum, ed.), SIAM Press, Philadelphia, pp. 44–73.

Guido, W., Spear, P. D., and Tong, L., 1990, Functional compensation in the lateral suprasylvian visual area following bilateral visual cortex damage in kittens, *Exp. Brain Res.* **83:**219–224.

Haenny, P. E., and Schiller, P. H., 1988, State dependent activity in monkey visual cortex. I. Single cell activity in V1 and V4 on visual tasks, *Exp. Brain Res.* **69:**225–244.

Harting, J. K., Huerta, M. F., Hashikawa, T., and van Lieshout, D. P., 1991, Projection of the mammalian superior colliculus upon the dorsal lateral geniculate nucleus: Organization of tectogeniculate pathways in nineteen species, *J. Comp. Neurol.* **304:**275–306.

Hayashi, M., and Oshima, K., 1986, Neuropeptides in cerebral cortex of macaque monkey (*Macaca fuscata fuscata*): Regional distribution and ontogeny, *Brain Res.* **364:**360–368.

Hayashi, M., Yamashita, A., Shimizu, K., and Oshima, K., 1989, Ontogeny of cholecystokinin-8 and glutamic acid decarboxylase in cerebral neocortex of macaque monkey, *Exp. Brain Res.* **74:**249–255.
Holmes, G., 1918, Disturbances of visual orientation, *Br. J. Ophthalmol.* **2:**449–486.
Innocenti, G. M., 1981a, Transitory structures as substrate for developmental plasticity of the brain, in: *Functional Recovery from Brain Damage* (M. W. van Hof and G. Mohn, eds.), Elsevier, Amsterdam, pp. 305–333.
Innocenti, G. M., 1981b, Role of axon elimination in the development of visual cortex, in: *Development of Visual Pathways in Mammals,* Liss, New York, pp. 243–253.
Innocenti, G. M., and Clarke, S., 1984a, The organization of immature callosal connections, *J. Comp. Neurol.* **23:**287–309.
Innocenti, G. M., and Clarke, S., 1984b, Bilateral transitory projection to visual areas from auditory cortex in kittens, *Dev. Brain Res.* **14:**143–148.
Johnson, M. H., 1990, Cortical maturation and the development of visual attention in early infancy, *J. Cognitive Neurosci.* **26:**81–95.
Kaas, J. H., and Huerta, J. F., 1988, The subcortical visual system of primates, in: *Comparative Primate Biology,* Volume 4 (H. Steklis and J. Erwin, eds.), Liss, New York, 327–391.
Kaas, J. H., and Krubitzer, L. A., 1991, Area 17 lesions deactivate area MT in owl monkeys, *Visual Neurosci.* **9:**399–407.
Kasamatsu, T., 1987, Norepinephrine hypothesis for visual cortical plasticity: Thesis, antithesis, and recent development, *Curr. Top. Dev. Biol.* **21:**367–389.
Kennedy, H., and Dehay, C., 1993, Cortical specification of mice and men. *Cerebral Cortex* **3:**171–186.
Kennedy, H., Bullier, J., and Dehay, C., 1989a, Cytochrome oxidase activity in the striate cortex and lateral geniculate nucleus of the newborn and adult macaque monkey, *Exp. Brain Res.* **61:**204–209.
Kennedy, H., Bullier, J., and Dehay, C., 1989b, Transient projection from the superior temporal sulcus to area 17 in the newborn macaque monkey. *Proc. Natl. Acad. Sci. USA* **86:**8093–8097.
Kluver, H., 1942, Functional significance of the geniculo-striate system, *Biol. Symp.* **7:**253–299.
Kostovic, I., and Rakic, P., 1984, Development of prestriate visual projections in the monkey and human fetal cerebraum revealed by transient cholinesterase staining, *J. Neurosci.* **4:**25–42.
LaMantia, A. S., and Rakic, P., 1990, Axon overproduction and elimination in the corpus callosum of the developing rhesus monkey, *J. Neurosci.* **10:**2156–2175.
LaMantia, A. S., and Rakic, P., 1994, Axon overproduction and elimination in the anterior commissure of the developing rhesus monkey, *J. Comp. Neurol.* **340:**328–336.
Lidow, M. S., and Rakic, P., 1992, Scheduling of monoaminergic neurotransmitter receptor expression in the primate neocortex during postnatal development, *Cerebral Cortex* **2:**401–416.
Lidow, M. S., Goldman-Rakic, P. S., and Rakic, P., 1991, Synchronized overproduction of neurotransmitter receptors in diverse regions of the primate cerebral cortex, *Proc. Natl. Acad. Sci. USA* **88:**10218–10221.
Logothetis, N., Pauls, J., and Poggio, T., 1995, Shape representation in the inferior temporal cortex of monkeys, *Curr. Biol.* **5:**552–563.
Lomber, S. G., McNeil, M. A., and Payne, B. R., 1995, Amplification of thalamic projections to middle suprasylvian cortex following ablation of immature primary visual cortex in the cat, *Cerebral Cortex* **5:**166–191.
Lund, J. S., and Harper, T. R., 1991, Postnatal development of thalamic recipient neurons in the monkey striate cortex. III. Somatic inhibitory synapse acquisition by spiny stellate neurons of layer IVC, *J. Comp. Neurol.* **309:**141–149.
Lund, J. S., and Holbach, S. M., 1991, Postnatal development of thalamic recipient neurons in the monkey striate cortex. I. Comparison of spine acquisition and dendritic growth of layer 4C alpha and beta spiny stellate neurons, *J. Comp. Neurol.* **309:**115–128.
Lynch, J. C., 1992, Saccade initiation and latency deficits after combined lesions of the frontal and posterior eye fields in monkeys, *J. Neurophysiol.* **68:**1913–1916.
Malkova, L., Mishkin, M., and Bachevalier, J., 1995, Long-term effects of selective neonatal temporal lobe lesions on learning and memory in monkeys, *Behav. Neurosci.* **109:**212–226.
Maunsell, J. H. R., Sclar, G., Nealey, T. A., and DePriest, D. D., 1991, Extraretinal representations in area V4 of the macaque monkey, *Visual Neurosci.* **7:**561–573.
Merigan, W. H., 1993, Human V4? *Curr. Biol.* **3:**226–229.

Merigan, W. H., and Maunsell, J. H. R., 1993, How parallel are the primate visual pathways? *Annu. Rev. Neurosci.* **16:**369–402.

Merzenich, M. M., and Sameshima, K., 1993, Cortical plasticity in memory, *Curr. Opin. Neurobiol.* **3:**187–196.

Mikami, A., and Fujita, K., 1992, Development of the ability to detect visual motion in infant macaque monkeys, *Dev. Psychobiol.* **25:**345–354.

Mikami, A., and Kubota, K., 1980, Inferotemporal neuron activities and color discrimination with delay, *Brain Res.* **182:**65–78.

Miller, E. K., Gochin, P. M., and Gross, C. G., 1991, Habituation-like decrease in the responses of neurons in inferior temporal cortex of the macaque, *Visual Neurosci.* **7:**357–362.

Miller, E. K., Li, L., and Desimone, R., 1993, Activity of neurons in anterior inferior temporal cortex during a short-term memory task, *J. Neurosci.* **13:**1460–1478.

Miyashita, Y., 1988, Neuronal correlate of visual associative long-term memory in the primate temporal cortex, *Nature* **335:**817–820.

Mohler, C. W., and Wurtz, R. H., 1977, Role of striate cortex and superior colliculus in visual guidance of saccadic eye movements in monkeys, *J. Neurophysiol.* **40:**74–94.

Molotchnikoff, S., and Hubert, F., 1990, Susceptibility of neurons in area 18a to blockade of area 17 in rats, *Brain Res.* **510:**223–228.

Moore, T., Rodman, H. R., Repp, A. B., and Gross, C. G., 1993, Comparison of residual visual function after damage to striate cortex in infancy and adulthood, *Neurosci. Abstr.* **19:**1801.

Moore, T., Rodman, H. R., Repp, A. B., and Gross, C. G., 1995a. Localization of visual stimuli after striate cortex damage in monkeys: Parallels with human blindsight, *Proc. Natl. Acad. Sci. USA* **92:**8215–8218.

Moore, T., Repp, A. B., Rodman, H. R., and Gross, C. G., 1995b, Preserved motion discrimination in monkeys with early lesions of striate cortex, *Neurosci. Abst.* **21:**1651.

Moore, T., Rodman, H . R., Repp, A. B., Gross, C. G., and Mezrich, R., 1996, Greater residual vision after striate cortex damage in infancy, *J. Neurophysiol.* **76:**3928–3933.

Morrison, J. H., and Foote, S. L., 1986, Noradrenergic and serotoninergic innervation of cortical, thalamic, and tectal visual structures in old and new world monkeys, *J. Comp. Neurol.* **243:**117–138.

Newsome, W. T., Wurtz, R. H., Dursteler, M. R., and Mikami, A., 1985, Deficits in visual motion processing following ibotenic acid lesions of the middle temporal visual area of the macaque monkey, *J. Neurosci.* **5:**825–840.

Olavarria, J., and Torrealba, F., 1978, The effects of acute lesions of striate cortex on the retinotopic organization of lateral peristriate cortex in the rat, *Brain Res.* **151:**386–391.

Ó Scalaidhe, S. P., Albright, T. D., Rodman, H. R., and Gross, C. G., 1995, Effects of superior temporal polysensory area lesions on eye movements in the macaque monkey, *J. Neurophysiol.* **73:**1–19.

Ó Scalaidhe, S. P., Rodman, H. R., Albright, T. D., and Gross, C. G. 1997, The effects of combined superior temporal polysensory area and frontal eye field lesions on eye movements, *Behav. Brain Res.* **84:**31–46.

Pasik, P., and Pasik, T., 1982, Visual functions in monkeys after total removal of visual cerebral cortex, *Contrib. Sensory Physiol.* **7:**147–200.

Payne, B. R., 1993, Evidence for visual cortical area homologs in cat and macaque monkey, *Cerebral Cortex* **3:**1–25.

Payne, B. R., and Cornwell, P., 1994, System-wide repercussions of damage to the immature visual cortex, *Trends Neurosci.* **17:**126–130.

Payne, B. R., Pearson, H., and Cornwell, P., 1988, Development of visual and auditory cortical connections in the cat, in: *Cerebral Cortex,* Volume 7, *Development and Maturation of Cerebral Cortex* (A. Peters and E. G. Jones, eds), Plenum Press, New York, pp. 309–389.

Payne, B. R., Lomber, S. G., McNeil, M., and Cornwell, P., 1996, Evidence for greater sight in blindsight following damage to primary visual cortex early in life: A review, *Neuropsychologia* **34:**741–774.

Raisler, R. L., and Harlow, H. F., 1965, Learned behavior following lesions of posterior association cortex in infant, immature and preadolescent monkeys, *J. Comp. Physiol. Psychol.* **60:**167–174.

Rakic, P., 1976, Difference in the time of neuron origin and in eventual distribution of neurons in areas 17 and 18 of the visual cortex in rhesus monkey, *Exp. Brain Res.* **1:**244–248.

Rakic, P., 1988, Specification of cerebral cortical areas, *Science* **241:**170–176.

Rakic, P., and Goldman-Rakic, P. S., 1982, Development and modifiability of the cerebral cortex, *Neurosci. Res. Prog. Bull.* **20:**429–611.

Rakic, P., Bourgeois, J. P., Eckenhoff, M. E. F., Zecevic, N., and Goldman-Rakic, P. S., 1986, Concurrent overproduction of synapses in diverse regions of the primate cerebral cortex, *Science* **232:**232–235.

Rocha-Miranda, C., Bender, D., Gross, C. G., and Mishkin, M., 1975, Visual activation of neurons in inferotemporal cortex depends on striate cortex and the forebrain commissures, *J. Neurophysiol.* **58:**1292–1306.

Rodman, H. R., 1994, Development of inferior temporal cortex in the monkey, *Cerebral Cortex* **5:**484–498.

Rodman, H. R., and Consuelos, M. C., 1994, Cortical projections to anterior inferior temporal cortex in infant macaque monkeys, *Visual Neurosci.* **11:**119–133.

Rodman, H. R., and Nace, K. L., 1997, In: (N. A. Krasnegor, G. R. Lyon, and P. S. Goldman-Rakic, eds.), *Development of the Prefrontal Cortex: Evolution, Neurobiology and Behavior* Brookes Publishing, Baltimore, MD, pp. 167–190.

Rodman, H. R., Gross, C. G., and Albright, T. D., 1989, Afferent basis of visual response properties in area MT of the macaque: I. Effects of striate cortex removal, *J. Neurosci* **9:**203–2050.

Rodman, H. R., Gross, C. G., and Albright, T. D., 1990, Afferent basis of visual response properties in area MT of the macaque: II. Effects of superior colliculus removal, *J. Neurosci.* **10:**1154–1164.

Rodman, H. R., Skelly, J. P., and Gross, C. G., 1991, Stimulus selectivity and state dependence of activity in inferior temporal cortex of infant monkeys, *Proc. Natl. Acad. Sci. USA* **88:**7572–7575.

Rodman, H. R., Ó Scalaidhe, S. P., and Gross, C. G., 1993, Visual response properties of neurons in temporal cortical visual areas of infant monkeys, *J. Neurophysiol.* **70:**1115–1136.

Rolls, E. T., Baylis, G. C., Hasselmo, M. E., and Nalwa, V., 1989, The effect of learning on the face selective responses of neurons in the cortex in the superior temporal sulcus of the monkey, *Exp. Brain. Res.* **76:**153–164.

Sakai, K., and Miyashita, Y., 1991, Neural organization for the long-term memory of paired associates, *Nature* **354:**152–155.

Saper, C. B., 1987, Function of the locus coeruleus, *Trends Neurosci.* **10:**343–344.

Schiller, P. H., and Malpeli, J. G., 1977, The effect of striate cortex cooling on area 18 cells in the monkey, *Brain Res.* **126:**366–369.

Schiller, P. H., Stryker, M., Cynader, M., and Berman, N., 1974, Response characteristic of single cells in the monkey superior colliculus following ablation or cooling of visual cortex, *J. Neurophysiol.* **37:**181–194.

Schwartz, M. L., and Goldman-Rakic, P. S., 1990, Development and plasticity of the primate cerebral cortex, *Clin. Perinatol.* **17:**83–102.

Segraves, M. A., Goldberg, M. E., Deng, S., Bruce, C. J., Ungerleider, L. G., and Mishkin, M., 1987, The role of the striate cortex in the guidance of eye movements in the monkey, *J. Neurosci.* **7:**3040–3058.

Sereno, M. I., Rodman, H. R., and Karten, H. J., 1991, Organization of visual cortex in the California ground squirrel, *Neurosci. Abstr.* **17:**844.

Sesma, M. A., and Burkhalter, A., 1994, Development of modular organization in area V2 of monkey and human visual cortex, *Neurosci. Abstr.* **20:**215.

Slater, A., Morison, V., Town, C., and Rose, D., 1985, Movement perception and identity constancy in the new-born baby, *Br. J. Dev. Psychol.* **3:**211–220.

Sobotka, S., and Ringo, J. L., 1991, Failure of visual image pairings to produce correlated single-unit responses in inferotemporal cortex of a macaque, *Neurosci. Abstr.* **17:**660.

Spear, P. D., 1983, Neural mechanisms of compensation following neonatal visual cortex damage, in: *Synaptic Plasticity and Remodeling* (C. W. Cotman, ed.), Guilford Press, New York, 111–167.

Spitzer, H., and Richmond, B. J., 1991, Task difficulty: Ignoring, attending to, and discriminating a visual stimulus yield progressively more activity in inferior temporal neurons, *Exp. Brain Res.* **83:**340–348.

Stryker, M., 1991, Temporal associations, *Nature* **354:**108.

Tootell, R. B. H., and Taylor, J. B., 1995, Anatomical evidence for MT and additional cortical visual areas in humans, *Cerebral Cortex* **5:**39–55.

Uylings, H. B., 1994, What determines the specification of cortical areas? *Trends Neurosci.* **17:** 1–2.

Van Buren, J. M., 1963, Trans-synaptic retrograde degeneration in the visual system of primates, *J. Neurol. Neurosurg. Psychiatr,* **26:**402–409.

Voytko, M. L., Kitt, C. A., and Price, D. L., 1992, Cholinergic immunoreactive fibers in monkey anterior temporal cortex, *Cerebral Cortex* **2:**48–55.

Walsh, C., and Cepko, C. L., 1992, Widespread dispersion of neuronal clones across functional regions of the cerebral cortex, *Science* **255:**434–440.

Webster, M. J., Ungerleider, L. G., and Bachevalier, J., 1991a, Connections of inferior temporal areas TE and TEO with medial temporal-lobe structures in infant and adult monkeys, *J. Neurosci.* **11:**1095–1116.

Webster, M. J., Ungerleider, L. G., and Bachevalier, J., 1991b, Lesions of inferior temporal area TE in infant monkeys alter cortico-amygdalar projections, *NeuroReport* **2:**769–772.

Webster, M. J., Bachevalier, J., and Ungerleider, L. G., 1994, Connections of inferior temporal areas TEO and TE with parietal and frontal cortex in macaque monkeys, *Cerebral Cortex* **4:**470–483.

Webster, M. J., Bachevalier, J., and Ungerleider, L. G., 1995a, Transient subcortical connections of inferior temporal areas TEO and TEO in infant macaque monkeys, *J. Comp. Neurol.* **352:**213–226.

Webster, M. J., Bachevalier, J., and Ungerleider, L. G., 1995b, Development and plasticity of visual memory circuits, in: *Maturational Windows and Adult Cortical Plasticity. SFI Studies in the Sciences of Complexity,* Volume 14 (B. Julesz and I. Kovacs, eds.), Addison-Wesley, Reading, MA.

Weller, R. E., and Kaas, J. H., 1989, Parameters affecting the loss of ganglion cells following ablations of striate cortex in primates, *Visual Neurosci.* **3:**327–349.

Weiskrantz, L., 1986, *Blindsight: A Case Study and Implications,* Oxford University Press, Oxford.

Weiskrantz, L., Barbur, J. L., and Sahraie, A., 1995, Parameters affecting conscious versus unconscious visual discrimination with damage to V1, *Proc. Natl. Acad. Sci. USA* **92:**6122–6126.

Wilson, F. A. W., Ó Scalaidhe, S. P., and Goldman-Rakic, P. S. 1993, Dissociation of object and spatial processing domains in primate prefrontal cortex, *Science* **260:**1955–1958.

Yakovlev, P. I., and Lecours, A. R., 1967, The myelogenetic cycles of regional maturation of the brain, in: *Regional Maturation of Brain in Early Life* (A. Minkowski, ed.), Davis, Philadelphia, pp. 3–70.

Yamasaki, D. S., and Wurtz, R. H., 1987, Recovery of function following chemical lesions of cortical area MT, *Neurosci. Abstr.* **13:**625.

Yamasaki, D. S., and Wurtz, R. H., 1991, Recovery of function after lesions in the superior temporal sulcus in the monkey, *J. Neurophysiol.* **66:**651–673.

Yamashita, A., 1992, Ontogeny of cholecystokinin-immunoreactive structures in the primate cerebral neocortex, *Int. J. Neurosci.* **64:**139–151.

Yamashita, A., Hayashi, M., Shimizu, K., and Oshima, K., 1989, Ontogeny of somatostatin in cerebral cortex of macaque monkey: An immunohistochemical study, *Dev. Brain Res.* **45:**103–111.

Yamashita, A., Shimizu, K., and Hayashi, M., 1990, Ontogeny of substance P immunoreactive structures in the primate cerebral neocortex, *Dev. Brain Res.* **57:**197–207.

Zohary, E., Celebrini, S., Britten, K. H., and Newsome, W. T., 1994, Neuronal plasticity that underlies improvement in perceptual performance, *Science* **263:**1289–1292.

Zola-Morgan, S., and Squire, L. R., 1993, Neuroanatomy of memory, *Annu. Rev. Neurosci.* **16:**547–563.

15

Architecture, Connectivity, and Transmitter Receptors of Human Extrastriate Visual Cortex

Comparison with Nonhuman Primates

KARL ZILLES and STEPHANIE CLARKE

1. Historical Concepts and Early Maps of the Human Visual Cortex

The cytoarchitectonic and myelogenetic maps of the mammalian visual cortex (Brodmann, 1903, 1905, 1906, 1908a, 1908b, 1909; Flechsig, 1920; Economo and Koskinas, 1925) represented for decades widely accepted organizational concepts. Lately, they have been losing some of their importance, mainly for two reasons. First, the schematic figures published by these authors are often used for purposes of cortical localization, often by surface landmarks, without recourse to histological identification of areas and without taking into account ambiguities of definitions and interindividual variations. Second, purely architectonic maps tend to be replaced by more functionally relevant parcellations (for review see Felleman and van Essen, 1991; Kaas and Krubitzer, 1991; Kaas, 1993; Zeki, 1993). We aim at reestablishing the importance of the architectonic

KARL ZILLES • C. & O. Vogt-Institute of Brain Research, University of Düsseldorf, D-40225 Düsseldorf, Germany. STEPHANIE CLARKE • Institut de Physiologie, Université de Lausanne, CH-1005 Lausanne, Switzerland.

Cerebral Cortex, Volume 12, edited by Rockland *et al.* Plenum Press, New York, 1997.

approach to the human cortex by reviewing critically classic and modern architectonic studies and by relating them to hodological and activation studies.

The development during the last 100 years can be understood as a progression from an early bipartition of the visual cortex to the present concept of a visual cortex consisting of one primary visual area and multiple extrastriate areas arranged in a distributed hierarchical organization (for review see Felleman and van Essen, 1991; Kaas and Krubitzer, 1991; Kaas, 1993; Zeki, 1993).

1.1. Bipartition of the Visual Cortex or "Visuosensory" versus "Visuopsychic" Regions

Meynert (1872), Betz (1874), and Hammarberg (1898) proposed the first subdivision of the human visual cortex into two structurally and functionally different areas (Table I). One area corresponds approximately to the striate cortex and the second to the remaining part of the occipital cortex. Hammarberg's (1898) description identified two cytoarchitectonic features that are now recognized as typical of extrastriate cortex: poor separation of layer II from layer III and large cells present in layer III but almost absent in layer V. Bolton (1900) coined the terms "visuosensory" (striate cortex) and "visuopsychic" (extrastriate cortex) for these areas by correlating anatomical observations and clinical findings.

Table I. Architectonic Subdivisions and Nomenclature of the Human Striate and Extrastriate Visual Cortex

	Striate cortex	Extrastriate cortex	
Meynert (1872)			
Betz (1874)			
Hammarber (1898)			
Bolton (1900)	Visuosensory	Visuopsychic	
Campbell (1905)	Visuosensory	Visuopsychic	
Brodmann (1903, 1906, 1908a, b, 1909)	Area 17	Area 18	Area 19
Elliot Smith (1907)	Area Striata	Area parastriata	Area paristriata
Vogt and Vogt (1919)	Area 17	Area 18	Area 19 (19a, 19b)
Flechsig (1920)	Area 8	Part of area 23	Areas 15 and 16, parts of areas 23 and 28)
Economo and Kloskinas (1925)	OC	OB (OBγ; OBΩ)	OA (OA1, OA2, OAm)
Filimonoff (1932, 1933)	Area 17	Area 18	Area 19
Lungwitz (1937)			17 subdivisions in area 19
Bailey and von Bonin (1951)	Koniocortex	Parakoniocortex	
Heinze (1954)		Up to 39 subdivisions in area 18	
Sarkissov *et al.* (1955)	Area 17	Area 18	Area 19
Braak (1977, 1980)	Striate area	Parastriate area	Peristriate area

The cyto- and myeloarchitectonic work of Campbell (1905) confirmed this bipartition; his visuosensory area apparears coextensive with striate cortex and his visuopsychic area with extrastriate occipital cortex. Vogt (1903) described the characteristics of the "visuopsychic" area on the medial occipital gyrus in such detail that this region of the "visuopsychic" area can be easily identified as part of Brodmann's area 18 (Brodmann, 1909). According to him, the cortex of this region of the "visuopsychic" area is relatively thin compared with other isocortical areas in the parietal and frontal lobes, the packing density of cells is relatively high, layer IV is thin, but densely packed with small, round, "granular" cells, layer V contains relatively small pyramidal cells, and only rarely are large pyramids ("Solitärzellen") found in this layer.

1.2. Tripartition of the Visual Cortex and the First Hierarchical Concepts

Seminal cytoarchitectonic (Brodmann, 1903, 1905, 1906, 1908a, 1908b, 1909; von Economo and Koskinas, 1925; Filimonoff, 1932, 1933) and myeloarchitectonic (Elliot Smith, 1907) studies provided evidence for a further subdivision of Campbell's "visuopsychic" area into two fields (Table I). This tripartite organization of the occipital visual cortex is still used in much of today's anatomical and neuropsychological work; therefore, we review it here in some detail.

Brodmann's *cytoarchitectonic* studies of various mammalian brains were focused exclusively on descriptive histological and developmental (histogenetic) aspects. He defined explicitly his criteria, which he called 'tectonic' or 'architectural':

> The tectonic shifts [at the border between two areas] are based on the number and specific development of single cortical layers, or on the packing density and size of the cellular elements on a section through the whole cortical depth or within single layers, or on the differentiation of certain cell types, or finally on the total thickness of the cortex and the proportion between the thicknesses of the different layers. They [the tectonic shifts] occur in some cases as sharp borders like a line, . . . , or in other cases as blurred transitions. (Brodmann, 1909, p. 43, our translation)

Some, but not all of his cortical areas (Brodmann's areas are indicated in this chapter as BA followed by a number) were listed by him as examples of cortical regions sharply delineated by abrupt architectonic changes from the surrounding cortex, e.g., BA 4 (motor cortex), BA 17 (striate area), BA 27 (presubiculum), and BA 28 (entorhinal area).

The visual cortex is restricted in Brodmann's map to the occipital region of the human brain. This region comprises the occipital gyri on the lateral surface, the complete cuneus, and the posterior parts of the lingual and fusiform gyri on the mesial surface. He subdivided this region into BA 17 (striate area), BA 18 (occipital area), and BA 19 (praeoccipital area) (Fig. 1A, B). Area BA 18 surrounds BA 17 like a horseshoe on the mesial and lateral hemispheric surface; it is relatively wide on the lateral surface, where it extends along the lateral (superior) occipital sulcus, and it is very narrow at anterior levels of the calcarine sulcus. Area BA 19 surrounds BA 18 and invades and transgresses the interoccipital and parietooccipital sulci in the rostral direction. Apparently influenced by Brodmann's concept, a comparable map of the visual cortex with a subdivision into three areas and a topography similar to Brodmann's map was published by Sarkissov *et al.* (1995). Brodmann (1909) emphasized that neither BA 18 nor BA 19 can be delineated by using gyri and sulci as landmarks of areal borders. Additionally, Brodmann described a small area buried in the interparietal sulcus, which may be a rostral extention of BA 19. He compares this ill-defined area with Elliot Smith's (1907) "visuosensory band," which contacts rostrally the postcentral area of the parietal lobe. Extrastriate areas with a similar topology as in the human brain were also described in the brains of nonhuman primates (Brodmann, 1905, 1908b, 1909). He emphasized that the main occipital region is relatively larger in nonhuman primate brains compared with the same region in the human brain.

Despite his purely descriptive approach, Brodmann discussed his results on the background of electrophysiological and lesion studies by other authors. He concluded: "Our statements justify principally not only the assumption of a

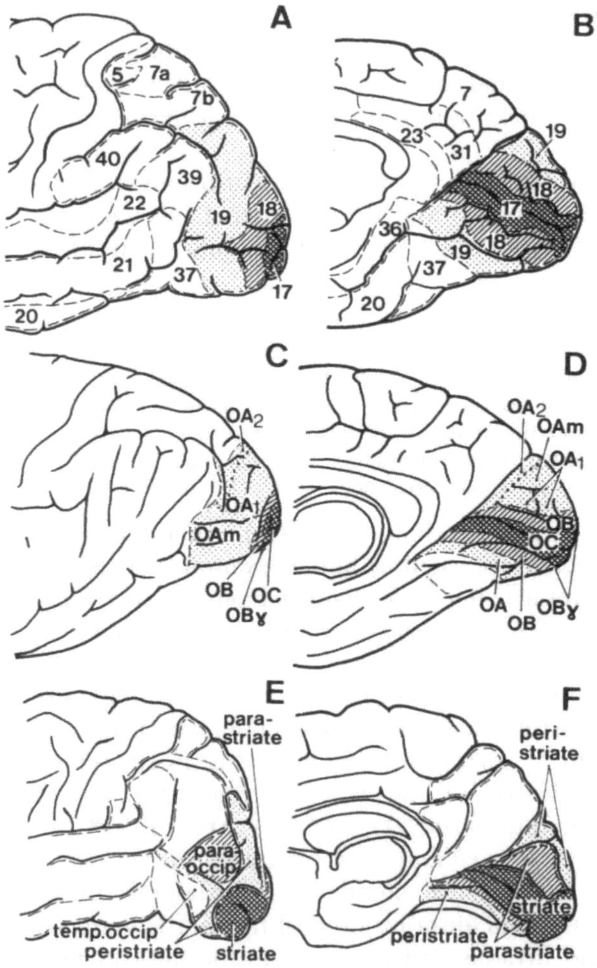

Figure 1. Cytoarchitectonic maps of the human visual cortex by (A, B) Brodmann (1909) and (C, D) von Economo and Koskinas (1925), and (E, F) a myeloarchitectonic map by Elliot Smith (1907).

localized distribution of functions within the cerebral cortex, but demonstrate also with absolute certainty the existence of a circumscribed regional localization of certain functions" (Brodmann, 1909, p. 300, our translation). Brodmann used the term "histological organ sui generis" (Brodmann, 1909, p. 301) for areas that could be sharply delineated in Nissl-stained sections, but denied strictly the existence of single "specific psychic centers of higher order" (Brodmann, 1909, p. 301) comparable to his cytoarchitectonically defined areas, in which complex mental activities such as memory, will, phantasy, intelligence, or spatial qualities such as perception of form and space (Brodmann, 1909, pp. 301–302) could be localized exclusively in one area. For example, he thought of form perception as the result of interactions of numerous cortical and subcortical areas. He did not reject cortical localization by architectonic analysis, but he promoted a concept different from the extreme localizational concept fashionable in his time. The latter culminated in a map of the human cortex (Kleist, 1934), in which precise functions were assigned to single Brodmann areas: to area 17, perception of luminance, color, shape, and motion of objects; to area 18, visual attention, perception of spatial position, and eye movements toward the upper (lower BA 18) or lower (upper BA 18) visual field; and to area 19, memory for visually identified localizations of objects, calculation, recognition of numbers, reading, visual object recognition, visual "thoughts," color vision (lateroventral part of BA 19), and recognition of colored objects (fusiform gyrus).

Brodmann emphasized the existence of basic functional units (Brodmann, 1909, p. 303), which are anatomically definable and interconnected, and which are involved in the production of complex mental and cognitive activities. Apparently, Brodmann combined the functional assumption of "visuosensory" and "visuopsychic" areas of Campbell (1905) with his histogenetic concept of heterotypical (BA 17) and homotypical (BA 18 + BA 19) cortical areas. He never used terms like "primary area" for BA 17 or "associational area" for BA 19, in order to avoid a confusion of his histological data with unproven evolutionary and functional concepts. In order to avoid such a confusion, he created his system of a "neutral" nomenclature by numbering the different cytoarchitectonic areas mainly according to the sequence (dorsal to ventral) in which he detected these areas in a horizontally sectioned human brain.

The relationship between Brodmann areas and more modern maps of the extrastriate visual cortex is difficult to establish. Brodmann never published a detailed cytoarchitectonic description of the human BA 18 and BA 19. However, he characterized extensively BA 19 of *Cercopithecus;* he described a rostral extension of BA 19 in the superior temporal sulcus which may be coextensive with the V5/MT and MST areas judging on a topographical basis. He already mentioned that BA 37, which he classified as part of the *human* temporal region (not present in his monkey maps), is not a typical temporal area, but transitional between temporal and occipital regions of the human brain (Brodmann, 1909). By that he was close to recent interpretations of this region as part of higher order visual areas within the inferior part of the temporal lobe.

Von Economo and Koskinas (1925) presented a *cytoarchitectonic* map of the human visual cortex (Fig. 1C, D), which is, by its principal tripartite concept, similar to Brodmann's map. They delineated three major fields ("*Formationen*"), OC (area striata), OB (area parastriata), and OA (area peristriata), which can be roughly compared with BA 17, 18, and 19, or with Elliot Smith's (1907) myeloarchitectonic subdivisions (striate, parastriate, and peristriate areas; Fig. 1E, F). They introduced quantitative data for supporting their delineations by objectifiable criteria (Table II). Further subdivisions ("subregions") of BA 19 were interpreted as local variation of the basic prestriate pattern. The occipital fields were distinguished from the parietal and temporal cortex by the smaller width of the occipital cortex.

Area OA is characterized by (i) a clearly visible radial arrangement of cells in layers V and VI, (ii) relatively small (only rarely are large cells found) and

Table II. The Parcellation of the Extrastriate Areas According to von Economo and Koskinas (1925)[a]

		Layer								
Area	Subarea	I	II	III	IV	V	VIa	VIb	I–VIb	Measure
OA	OA1	0.18	0.20	0.60	0.20	0.40	0.40	0.26	2.26	Depth at the free surface in mm
		0.20	0.24	0.50	0.16	0.14	0.24	0.18	1.66	Depth in the intrasulcal part in mm
		6	65	50	150	45	55	12		Neurons/0.1 mm^3
	OA2	0.18	0.26	0.56	0.24	0.38	0.40	0.46	2.50	Depth at the fee surface in mm
		0.20	0.20	0.50	0.20	0.22	0.34	0.12	1.78	Depth in the intrasulcal part in mm
		6	65	50	150	45	55	12		Neurons/0.1 mm^3
	OAm	0.20	0.20	0.52	0.18	0.26	0.22	0.10	1.70	Depth int he intrasulcal part in mm
OB		0.16	0.18	0.46	0.18	0.26	0.32	0.26	1.80	Depth at the free surface in mm
		0.20	0.18	0.48	0.18	0.26	0.18	0.10	1.58	Depth in the intrasulcal part in mm
		4	150	70	240	80	85	25		Neurons/0.1 mm^3

[a]The architectonic differences were summarized by measures of cortical thickness and cell density. Original data from von Economo and Koskinas (1925)

densely packed cells, and (iii) a clear lamination. Layer II is relatively broad (Table II) and shows a relatively low packing density of cells. Therefore, layers II and III are difficult to separate. Layer III seems to be smaller (Table II) than the respective layer in the parietal cortex. Layer IV has a higher cell packing density in OA (Table II) than the parietal and temporal cortices, but a lower density than layer IV of OB. Area OA can be further subdivided into three subregions. The OA1 subregion is separable from the OA2 subregion by the small layer V pyramids in OA1 (smaller than the neurons in layer VI) and the scattered large layer V pyramids in OA2. Subregion OA2 also shows large pyramidal cells in deeper layer III. These cells allow the delineation of a separate layer IIIc in OA2. The subregion OAm is very similar to OA2, but OAm lacks the large pyramidal cells in layer V. Subregion OA2 extends on the posterior wall of the parietooccipital sulcus on the mesial side of the hemisphere. There is no sharply delineable border between OA and OB.

Area OB consists of a very thin cortex. It is interposed between OA and OC and surrounds completely OC. The typical cytoarchitectonic features of OB are as follows: (i) An extremely high cell packing density in layer II and a relatively high density in layer IIIa (Table II). This makes a delineation between both layers difficult and gives OB the appearance of an area with a thick outer granular layer. (ii) A lightly stained layer V (low cell density) clearly delineated from the darkly stained layer VIa. (iii) A layer VI with large cells (larger than most of the cells in layer V).

Besides its general cytoarchitecture (Tables I and II), OB shows two characteristic cytoarchitectonic differentiations: a peculiar transition zone of approximately 1 mm width at the border to OC, the so-called limes parastriatus gigantocellularis (OB_{γ}), and widely distributed, patchy, 5- to 10-mm-wide conglomerates of granule cells (maculae granulosae, OBΩ) in layer III (120 cells/ 0.1 mm^3). The OBγ is easily recognizable by the giant pyramidal cells in layer IIIc (see also Vogt, 1929; Sanides and Vitzthum, 1965a, b), but the lateral border of this subarea is difficult to define, because the size of the pyramidal cells decreases progressively. The observation of the OBΩ maculae remains, however, enigmatic. OB can be delineated from OA by the larger cells in layers IIIb, IIIc, and V as well as the more cell-dense layer IV in OB.

Filimonoff (1932, 1933) subdivided the primate (human, orangutan, *Cercopithecus* sp.) visual cortex according to Brodmann's tripartite concept. His *cytoarchitectonic* studies are remarkable, because (i) they are the first systematic approach to intersubject variability in size and position of the primate BA 17–19 (Fig. 2), and (ii) he gives a detailed description of the cytoarchitectonic characteristics of BA 18 and BA 19, which had not been previously published for the human brain. Filimonoff emphasized the decreasing cell packing density in layers III, IV, and VI from BA 17 through BA 18 to BA 19 as a general cytoarchitectonic trend in the primate visual cortex. Area BA 18 does not show a light IIIb, and in that it differs from BA 19. The border between layers II and IIIa is difficult to recognize in BA 18. Although the II/IIIa border of BA 19 is less clear than in the rostrally adjoining parietal areas, layer II can be separated from layer IIIa in BA 19 of the human and nonhuman primate brains. This is the most consistent criterion for the BA 18/19 subdivision at the level of higher primates (Filimonoff, 1933). Further differences between BA 18 and BA 19 are the higher cell density in layer IV of BA 18 and the larger width of layer IIIc in BA 19. The radial arrangement of perikarya in BA 19 and to a minor degree also in BA 18 is less clearly visible in nonhuman primates. The considerable variability of the topography both in the human and in the orangutan visual areas can be seen in Fig. 2.

By observing macroscopically the regional heterogeneity of the outer and inner stripes of Baillarger in unstained sections, Elliot Smith (1907) subdivided the occipital cortex into three parts, the striate, parastriate, and peristriate areas (Fig. 1E, F). Despite a rough agreement in number and topography of the visual areas between Elliot Smith's and Brodmann's maps, many differences, especially in the topographic relations between the para- and peristriatae areas, are obvious. The schematic nature of both maps and the lack of sufficiently detailed descriptions of the topography do not allow a discussion of these differences.

The myeloarchitectonic method as a tool for mapping the cortex was developed to perfection by Cecile and Oskar Vogt (Vogt and Vogt, 1919). Although they never published a detailed and comprehensive myeloarchitectonic analysis of the extrastriate areas of human and monkey brains, Brodmann's and Economo and Koskinas' three major cytoarchitectonic fields of the occipital cortex

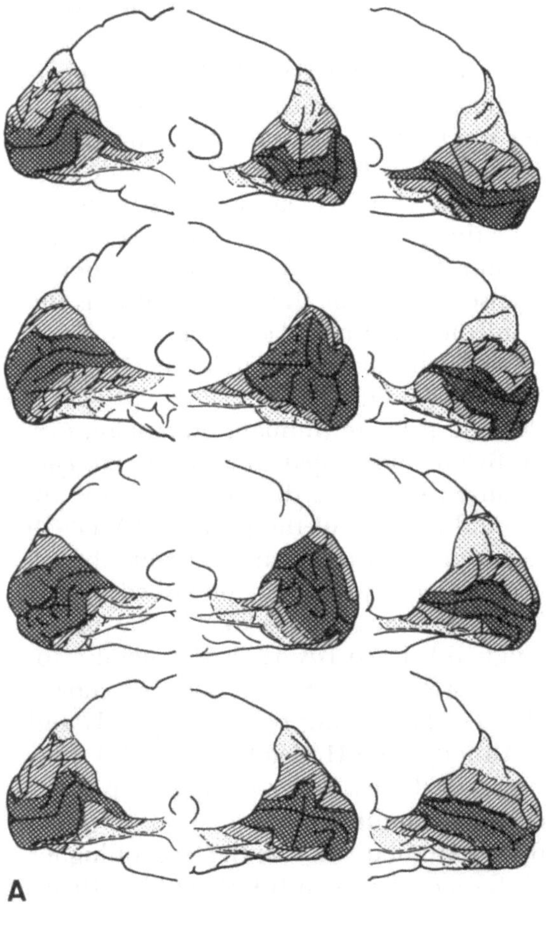
A

Figure 2. The intersubject variability of position and size of BA 17–19 in (A) human and (B) orangutan brains after Filimonoff (1932, 1933).

were found correspondingly by the Vogts in myelin-stained sections. Their "occipital area" (comparable to BA 18) shows a relatively higher density of myelinated fibers compared with the neighboring parts of their "praeoccipital area" (comparable to BA 19). The occipital area is characterized mainly by the densely myelinated inner stripe of Baillarger (Fig. 3A, B). The outer stripe of Baillarger is much less obvious or even not visible in BA 18. In contrast, both stripes of Baillarger are recognizable in many areas of BA 19 (Fig. 3C–P) by an interposed and poorly myelinated layer. A further myeloarchitectonic difference between BA 18 and BA 19 is the existence of the stripe of Kaes-Bechterew (horizontally directed, myelinated fibers) in the upper part of layer III in BA 18 and the lack of this stripe in nearly the complete BA 19 region. Based on a combination of architectonic and neurophysiological studies, a subdivision of the macaque BA 19 (but not of the human BA 19) into two areas, 19a and 19b, was reported (Vogt and Vogt, 1919). Area 19a occupies the lateral hemispheric surface from the superior temporal sulcus to the lunate sulcus. Weak electrical stimulations of this area induced adversive eye movements, whereas such movements can be elicited only with much stronger stimulations in area 19b (Vogt and Vogt, 1919). Area 19b is located below the superior temporal sulcus in the most occipital part of the temporal lobe immediately in front of the preoccipital incisure.

The tripartition of the occipital cortex was not generally accepted. In their influential book, Bailey and von Bonin (1951) believed that only the border between BA 17 and BA 18 was reliable. According to them, the occipital koniocortex (BA 17) is surrounded by a broad belt region, the occipital parakoniocortex. This belt region is, however, much larger than BA 18 and BA 19 together and is part of a large homotypical isocortical area which extends to the postcentral and central sulci and comprises also major parts of the temporal cortex on the lateral and basal hemispheric surface. Their map is clearly an oversimplification of cytoarchitectonic facts (for critical evaluation of this map, see also Braak, 1980). Hesitation to take into account small architectonic varia-

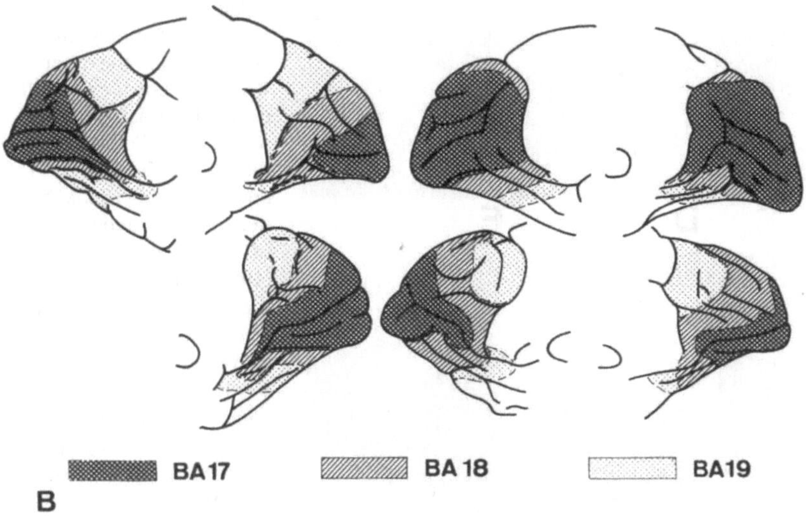

Figure 2. *(Continued).*

tions may be appropriate in some instances, but hinders the search for the structural basis of multiple functionally relevant extrastriate areas, as recognized in modern studies.

1.3. Further Cyto-, Myelo-, and Pigmentoarchitectonic Parcellations

Brodmann emphasized that his map did not give a complete list of *all* cytoarchitectonically definable cortical areas. An extremely detailed *cytoarchitectonic* subdivision of BA 18 was proposed by Heinze (1954) in a few human brains and one gorilla brain. He described 39 small cortical units in BA 18 of a single human brain (Table I). However, the functional meaning of these numerous cortical segments remains enigmatic, since most of these small units were ob-

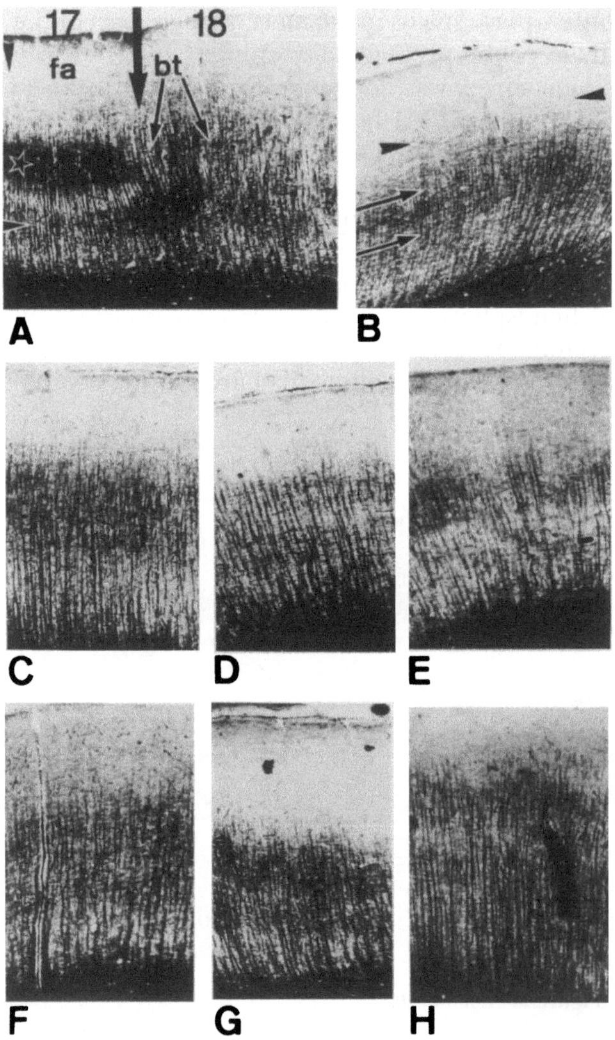

Figure 3. Myeloarchitecture of various extrastriate areas of the human brain according to Lungwitz (1937). Micrographs were taken at appropriate positions (carefully documented in the original paper by Lungwitz, 1937) from the original myelin-stained sections which Lungwitz used for his observations. The border (large arrow) between BA 17 (left of the arrow) and BA 18 (right of the arrow). The large star labels the stripe of Gennari. Near the BA 17/18 (dorsal part) border the inner stripe of Baillarger (arrowhead), which is not distinct within BA 17 proper, gradually becomes visible. This part of BA 17 was called the "fringe area" (fa; "*Randsaum*") by Sanides and Vitzthum (1965a, b). On the other side of the BA 17/18 border a bundle of myelinated radial fibers is visible, which was called the "border-tuft" (bt; "Grenzbüschel") by Sanides and Vitzthum (1965a, b). The border-tuft is located within area OBγ of von Economo and Koskinas (1925). Both the fringe area and the border-tuft area are regions where callosal fibers originate and terminate. (B) BA 18 (ventral part); (C–P) various prestriate areas of Lungwitz (1927): (C) scel, (D) pc, (E) del, (F) dpp, (G) sced, (H) scet, (I) pp, (J) scd, (K) sct, (L) eld, (M) elsc, (N) edpc, (O) eda, (P) elt. For the positions of the different areas identified by Lungwitz see Fig. 4A, B.

served only in a single brain, but not in others (Heinze, 1954). The delineations were based on tiny details registered by pure visual inspection without any attempt to test rigorously the reliability and significance of these details. In particular, the variation of cytoarchitectonic parameters within a cortical area (Vogt, 1929) and between different subjects was not sufficiently discussed.

Lungwitz (1937), a student of the Vogt school, demonstrated a remarkable structural heterogeneity of Brodmann's BA 19 by a detailed *myeloarchitectonic* study of the human occipital cortex (Figs. 3C–P and 4A, B), although number ($n = 17$) of his preoccipital (BA 19) areas did not reach the number of visual areas now shown in the macaque cortex. Since an architectonic map of the whole *human* extrastriate cortex is not available, an identification of all his fields on the basis of more recent architectonic data is not possible. We compared his descriptions and drawings with the original histological slides of his study, which are preserved in the Brain Collection of the C. & O. Vogt Brain Research Institute,

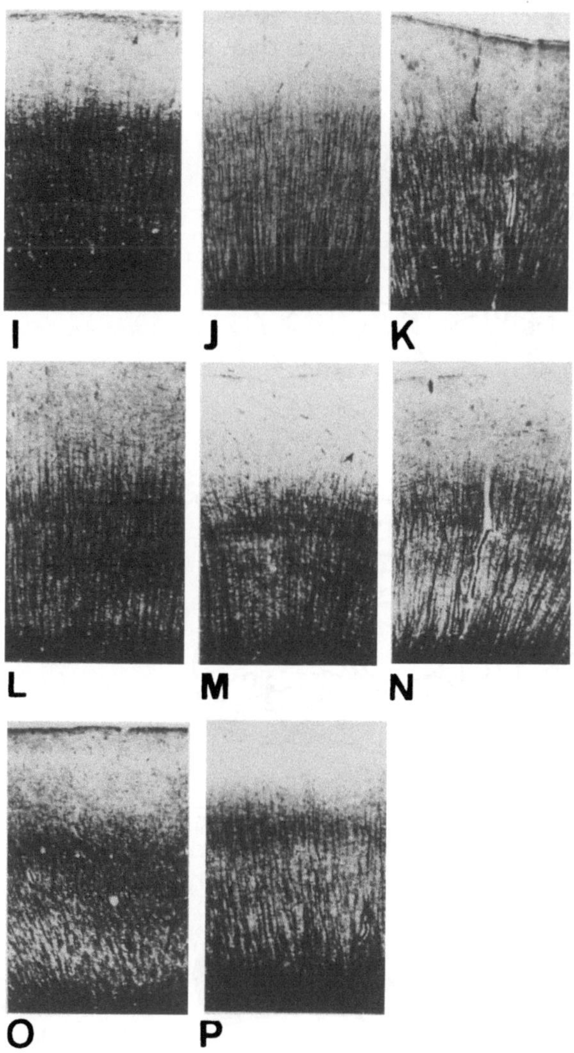

Figure 3. *(Continued).*

Düsseldorf (Fig. 3A–P). All prestriate areas of Lungwitz have the following features in common: (i) the outer stripe of Baillarger is visible, (ii) the inner stripe of Baillarger is detctable as a clearly separated structure in most, but not all of the prestriate areas, and (iii) if both stripes are expressed, the outer stripe is more densely myelinated. The overall density of myelinated fibers and differences in the structure of the radial fiber bundles (thickness and density of fibers, length of fibers) are additional structural features which make the myeloarchitectonic approach a promising method for mapping the prestriate areas. Lungwitz (1937) has grouped his myeloarchitectonic fields into several structurally related families.

Pigmentoarchitectonic studies of the human occipital cortex (Braak, 1977, 1980) make use of lipofuchsin staining in cell bodies. A variety of extrastriate fields was demonstrated, which are summarized as para- and peristriate areas in Fig. 4E–F. Whereas the parastriate area is clearly comparable to BA 18, the different peristriate areas cannot be identified reliably as regions comparable to specific architectonically or functionally defined cortical areas of the macaque or human brain. Braak (1977) emphasized that, in contrast to earlier cytoarchitec-

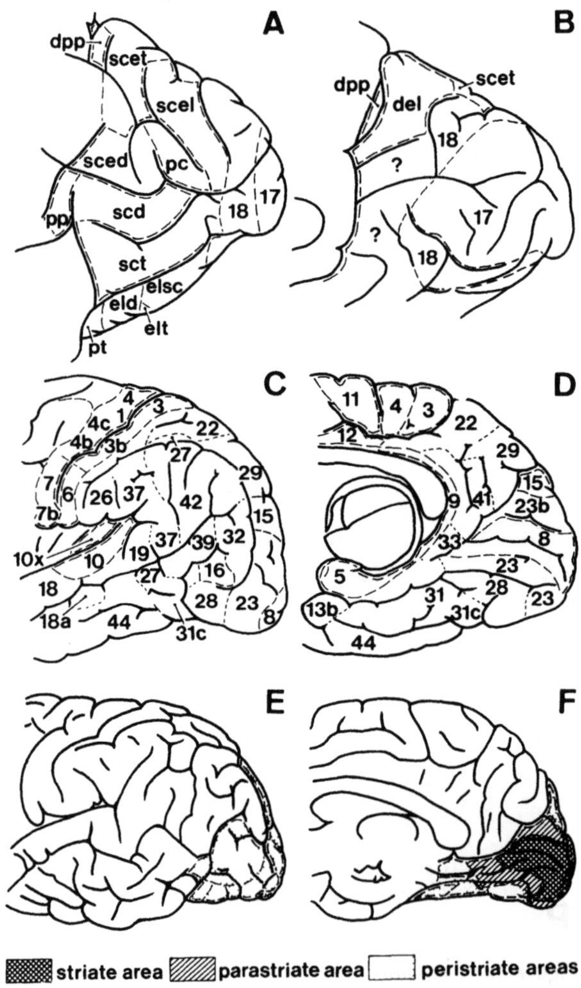

Figure 4. Architectonic maps of the human occipital cortex based on (A, B) myelarchitectonic (Lungwitz, 1937). myelogenetic (C, D) (Flechsig, 1920), and (E, F) pimentoarchitectonic (Braak, 1977) studies.

tonic observations, the border between the parastriate and peristriate areas could be precisely defined in the lipofuscin-stained sections by the disappearance of moderately pigmented and densely packed granule cells in the center of layer IV. This layer ("external taenia" in Braak's nomenclature) of the parastriate area shows a tripartition into an outer and an inner sublayer with neurons sparsely pigmented, and a centrally localized and moderately pigmented stripe. Braak (1977) stated that the parastriate area does not completely surround the striate area and as such, notably, it does not surround the oralmost tip of the striate area in the common trunk of the parietooccipital and calcarine sulci ("horseshoe" pattern of the parastriate area).

1.4. Subdivisions of the Visual Cortex Derived from Myelogenetic Studies

Flechsig's (1920) studies of the progressive myelination during maturation of the human forebrain led to a subdivision of the cortex into three fields: primordial fields ("*Primordialgebiet*"), which underwent myelination starting before birth; intermediate fields ("*Intermediärgebiet*"), between birth and the 6th postnatal week; and terminal fields ("*Terminalgebiet*"), after the second postnatal month. Flechsig speculated that only primordial fields receive sensory projection systems (e.g., optic radiation) or give origin to motor projection systems (e.g., corticospinal tract), whereas the intermediate and terminal fields are dominated by fiber tracts from cortical areas. Therefore, he called the intermediate and terminal fields association areas.

The early myelinated optic radiation terminates in Flechsig's area 8 (Fig. 4C, D), which corresponds to BA 17. Areas 15 and 16 were also classified as primordial regions because of early myelination, although, as became known later, they do not receive direct input through the optic radiation. A drawback of Flechsig's approach is that his myelogenetic maps were based mainly on the locally heterochronous *myelination of the subcortical white matter,* i.e., fiber bundles with unknown connectivity, *and not of the cortical area itself* (Vogt and Vogt, 1929). Thus, the myelogenetically determined "areas" may not match the cyto- or myeloarchitectonically defined cortical areas (Vogt and Vogt, 1929). Moreover, it is difficult to standardize the staining intensities in fetal and early postnatal human brains. This technical problem impairs the reproducibility of many myelogenetic findings, and is reflected in the numerous modifications which Flechsig provided over the years (e.g., Flechsig, 1898, Flechsig, 1920). All these arguments should, however, not impede a critical use of data from myelogenetic observations. For example, Flechsig's area 16 (his subangular gyrus) was recently proposed to be a good candidate for V5/MT (Watson *et al.*, 1993). A preliminary attempt of an overview of comparable occipital areas in the maps of different authors is given in Table I.

2. Extrastriate Visual Cortex in Prosimians and New and Old World Monkeys

Human extrastiriate cortex is largely understood in terms of work done in nonhuman primates. Therefore, we summarize here briefly evidence from work on prosimians and New and Old World monkeys which is relevant to the interpretation of human data. We only discuss those extrastriate areas (V2, V3 region, VP, V4 region, and V5/MT) for which putative human equivalents were proposed.

2.1. Maps of Visual Cortical Areas

Electrophysiological and anatomical studies of the last three decades provide a very different picture from Brodmann's tripartition of the visual cortex. The occipitotemporal cortices of the *Galago* (prosimian), marmoset, and squirrel and owl monkeys (New World monkeys) contain at least eight different visual areas with much functional and myeloarchitectonic similarities between species (Kaas and Preuss, 1993; Krubitzer and Kaas, 1990a, b, Fig. 5). These areas occupy 30% of the total cortex in nocturnal species and 40% in diurnal species. Twenty visual and visuomotor areas were identified in owl monkey (*Aotus trivirgatus,* nocturnal New World monkey; (Kaas, 1986, 1988, 1989; for review and Kaas and Krubitzer, 1991). Thirty-two visual areas were identified in the macaque cortex; 9 areas in the occipital, 10 in the parietal, 11 in the temporal, and 2 in the frontal cortex. Only two-thirds of the total macaque visual cortex are comprised within BA 17–19 (Fellemann and Van Essen, 1991).

It is evident from Filimonoff's (1933) measurements that the surface areas of human BA 18 and BA 19 together are a larger proportion of the total occipital visual cortex than the proportionate size of BA 17 in the human brain. In contrast, the orangutan shows a BA 17 relatively bigger than BA 18 or BA 19. Approximately the same relation is found in *Cercopithecus* sp. (BA 17: 41%, BA 18: 32%, BA 19: 27%; all three areas together: 100%), but a pronounced reduction of the proportionate size of BA 18 and BA 19 occurs in a prosimian species (*Galago* sp.), where BA 17 occupies approximately 95% of the surface of the occipital visual cortex (Fig. 6). The values calculated for *Cercopithecus* are similar to those given for the macaque (V2 as an area approximately comparable to BA 18: 35%; V3 + V4 + VP + VOT + V5/MT + MST as an area approximately comparable to BA 19: 32%; calculated from data of Felleman and van Essen, 1991). In both reports BA 19 is only slightly smaller than BA 18. We think that this similarity between the monkey data from two independent observations further supports the identification proposed here of areas comparable to BA 19. The relative proportion of the striate versus extrastriate cortex varies among species; macaques have a relatively smaller V1, but a larger extrastriate portion than prosimians and New World monkeys.

Although the concept of multiple visual areas is widely accepted, two issues remain open. First, some of the extrastriate areas, mainly those in the inferior temporal and posterior temporal cortex, are only roughly defined, and it is not entirely clear whether they are cortical areas to the same extent as the early-stage visual areas (e.g., V2, V3, V4, V5/MT). Second, the concept of homology of areas between species is far from clear. Nevertheless, it is largely accepted that some areas are found with similar functional specialization, architecture, connectivity, and relative position in different species, including human (see the chapter by Kaas in this volume).

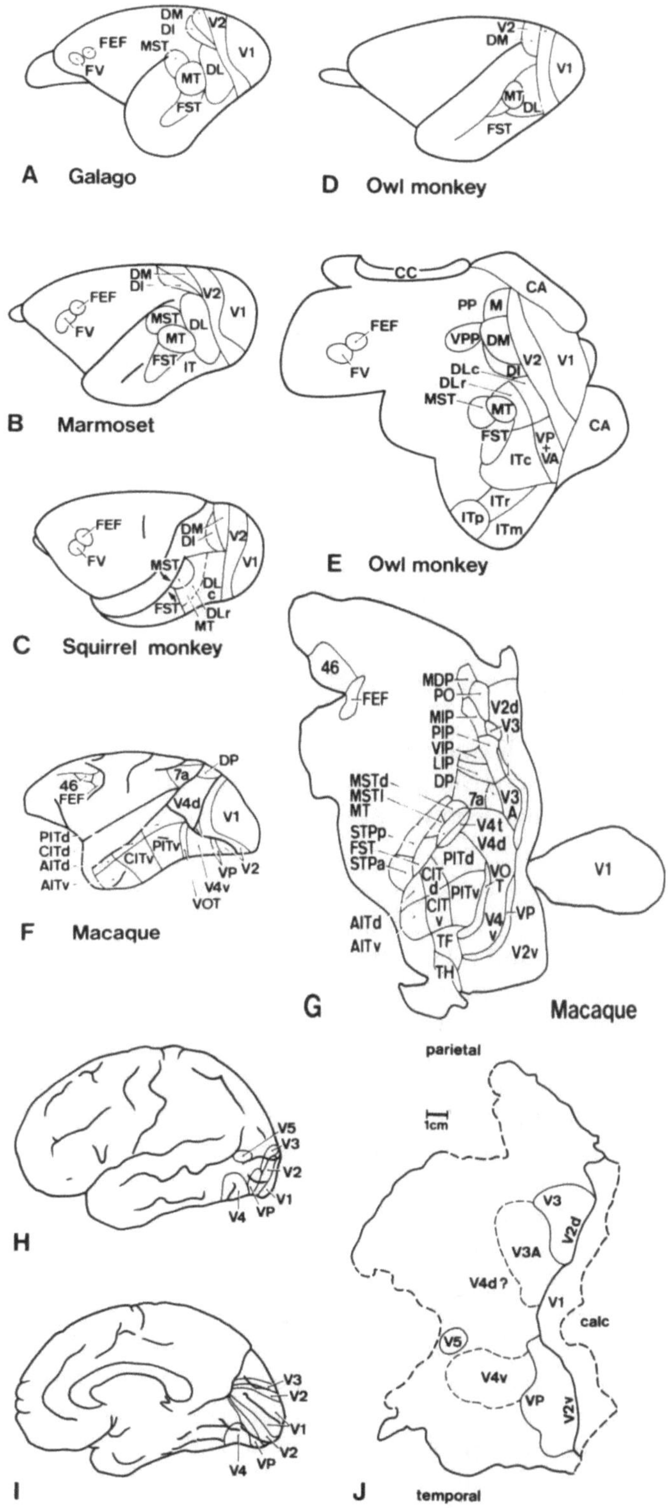

Figure 5. Location and extent of striate and extrastriate visual cortical areas in nonhuman primates and in the human brain. (A–D, F, H, I) Lateral (A–D, F, H) and medial (I) views (A) galago, (B) marmoset, (C) squirrel monkey, (D) owl monkey (all redrawn from Kaas and Krubitzer, 1991), (F) macaque (redrawn from Felleman and Van Essen, 1991), and (H, I) human brain (S. Clarke *et al.*, unpublished observations). (E, G, J) Flat maps of the left hemispheres of (E) owl monkey (redrawn from Kaas and Krubitzer, 1991), (G) macaque (redrawn from Felleman and Van Essen, 1991), and (J) human brain (S. Clarke and D. Van Essen, unpublished observations). For nomenclature see text and original publications of Kaas and Krubitzer (1991) and Felleman and Van Essen (1991).

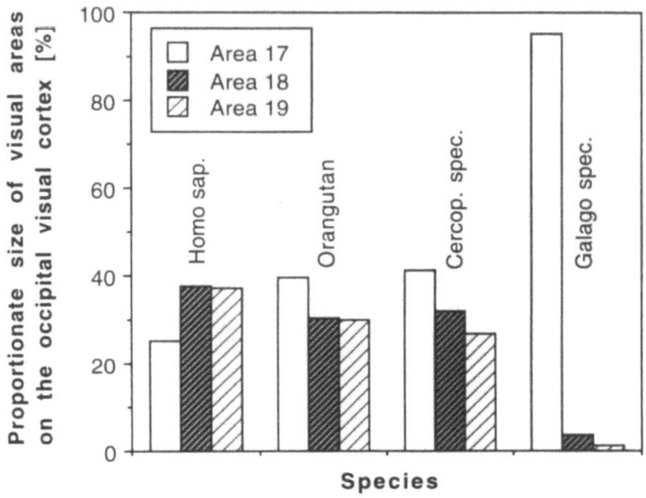

Figure 6. Proportionate sizes of the visual areas BA 17, BA 18, and BA 19 in the human, orangutan, monkey, and galago cortex. Human, orangutan, and monkey data from Filimonoff (1932, 1933) and galago data from Kaas and Krubitzer (1991).

2.2. Architecture of Extrastriate Areas

The distinctive features of architecture as seen in Nissl, myelin, histochemical, and immunohistochemical stains are summarized here for areas whose putative human equivalents are described below.

The secondary visual area, V2 (Gattass *et al.*, 1981; Van Essen *et al.*, 1986), can be characterized cytoarchitectonically (Rockland and Pandya, 1981) by the cell-dense layer III with large pyramidal cells in the lower part of layer IIIC; light layer V; and relatively cell-dense layers IV and VI. In comparison to the rostrally adjoining extrastriate areas, V2 has a broader and denser layer IV and denser layer III with large pyramidal neurons in IIIC. The areas bordering V2 rostrally show a more pronounced vertical arrangement ("columnarity") of their neurons, a darker layer VA, and a broader and less dense layer VI, which has no sharp border to the white matter. Over 70% of the cells in V2 of macaques are orientation selective, but only <15% are directionally selective and only 8% have opponent color properties (Zeki, 1978c).

Cytochrome oxidase (CO) staining of V2 reveals a modular organization with thick and thin CO-rich stripes separated by CO-sparse interstripes both in New (Livingstone and Hubel, 1987; Tootell *et al.*, 1993) and Old World monkeys (DeYoe and Van Essen, 1985; Horton, 1984; Hubel and Livingston, 1987; Livingstone and Hubel, 1983; Tootell and Hamilton, 1989; Tootell *et al.*, 1993; Wong-Riley, 1994; Wong-Riley and Carroll, 1984). In squirrel monkeys, the CO-rich stripes are parvalbumin-positive compartments (Blümcke and Celio, 1992). As in V1, a periodicity of dense and light stripes can be seen in myelin-stained flattened sections through V2 (Tootell *et al.*, 1983). Whereas the width of the densely myelinated periodic structures (Table III) is small in V1 of Old World and New World monkeys, similar structures appear on a much larger scale in V2

(Krubitzer and Kaas, 1990a; Tootell and Taylor, 1995). The densely myelinated stripes in V2 are found in a position resembling that of CO-sparse interstripes (see below) in squirrel monkeys (Krubitzer and Kaas, 1990a). Since Tootell *et al.* (1993) described a match of CO-rich and myelin-dense bands in V2 of macaques, it is not clear whether this divergence reflects methodological problems or a species difference.

In macaque monkey, the alternate stripes known to be part of the magnocellular stream are strongly immunoreactive for the CAT-301 antibody (DeYoe *et al.,* 1990); this antibody is directed against a chondroitin sulfate proteoglycan and was shown to label preferentially, but not exclusively, the magnocellular layers in the lateral geniculate body and the broadband neurons (Hendry *et al.,* 1984, 1988).

The SMI-32 antibody stained a subset of neurofilament-containing pyramidal (in layers II, III, V, and VI) and fusiform (in layer VI) neurons, which exhibit heterogeneous laminar and regional distributions throughout the extrastriate cortex. A strip of V2 directly adjacent to the V1/V2 boundary differs from the remaining V2; it has large SMI-32-positive pyramidal neurons in superficial layer V, but fewer SMI-32-positive neurons in layer III. The dorsal part of V2 has a higher density of SMI-32-immunoreactive (ir) neurons than the ventral part. Dorsal V2 differs from the adjacent area PO by its higher density of SMI-32-ir neurons in layers II–III and V–VI (Hof and Morrison, 1995).

As in V1, no calbindin-ir pyramidal cells are found in V2. In V2, 5% of the total neuronal population are calbindin-ir nonpyramidal neurons, whereas 8% are parvalbumin-ir interneurons (Kondo *et al.,* 1994). Layers II–III are most densely populated by calbindin-ir nonpyramidal neurons, whereas the parvalbumin-ir neurons reach the highest density in layer IV.

With fluorescence histochemistry, catecholamine-containing fibers were shown to be most densely packed in the deeper cortical layers of V2 (Levitt *et al.,* 1984). Serotoninergic axons are most densely packed in layer IV of V2 (Takeuchi and Sano, 1983). Cholecystokinin-immunoreactive (CCK-ir) neuronal cell bodies and fibers show a rather uniform regional distribution in the cerebral cortex

Table III. Extent of Anatomically Identified Visual Areas[a]

	Macaque	Human
Area of V1 (cm)	6.0 × 2.7 (4)	11 × 4 (3)
Width of V2 (cm)		1.5–3 (1)
		1.5–2.5 (2)
		1.8–3.4 (3)
		1.0–2.0 (5)
Width of dV3 (cm)		1.5–4.5 (5)
Width of VP		1.0–1.5 (5)
Extent of V5/MT (cm)	0.8 × 1.6 (3)	1.0 × 2.0 (3)
Distance from anterior foveal V1/V2 border (cm)		4.5–6 (3)

[a]References: (1) Hockfield *et al.* (1990)(CAT-301); (2) Wong-Riley *et al.*(1993) (CO); (3) Tootell and Taylor (1995) (CAT-301, CO; myelin); (4) Felleman and Van Essen (1991); (5) Clarke and Miklossy (1990), Clarke (1993).

of monkeys with highest numbers and densities in layers II–III and VI (Hendry *et al.*, 1983).

The third visual area, V3, corresponding to a third retinotopic representation within the occipital cortex of macaque monkeys, is often subdivided into dorsal (V3d) and ventral parts (VP). Anterior to V3d, in the lunate and parietooccipital sulci, is another representation of the lower visual field quadrant (V3A). The vertical meridian is located along the V3/V3A and the V3A/V4 borders. In V3 of the rhesus monkey, 70% of neurons are orientation-selective, fewer than 15% are directionally selective, and no color-opponent-selective cells were described (Zeki, 1978b, c). Area V3d is heavily myelinated (Van Essen, 1985; Van Essen *et al.*, 1986); the high degree of myelination overlays the inner and outer bands of Baillarger, which are difficult to distinguish. Areas V3 and V3A show a strong CAT-301 immunoreactivity (DeYoe *et al.*, 1990).

SMI-32 staining in macaque cortex was reported to provide architectonic criteria for establishing the boundaries between V2/V3, V3/V4, V3/V3A, and V3A/PO (Hof and Morrison, 1995). Area V3 contains strongly SMI-32-ir intermediate-size neurons in layers II and IIIA, a characteristic not shared with V2. The SMI-32-ir neurons are more frequent in layers IIIB–C and less frequent in layer VI of V3 than V2. The SMI-32-ir intermediate-size and small neurons in layers II–IIIA and in layers V–VI are more abundant in V4 than in V3. In contrast to V4, V3 contains large, strongly immunoreactive pyramidal cells in layer VA. Area V3A shows a decrease in SMI-32-ir neurons in layers II–III and an increase in layers V–VI compared with V3 (Hof and Morrison, 1995). Area V3A differs from the adjacent area PO by an increase of SMI-32-ir neurons in layer IIIC, their disappearance in layersII–IIIB, and a weaker labeling in layer VI in PO.

The ventroposterior area, VP, was described on the ventral cortical surface in the owl monkey (Newsome and Allman, 1980). It extends ventrolaterally and ventromedially, and lies in front of V2 and below and behind areas DL and DA. It has been proposed that the ventral part of macaque V3, which contains a representation of the upper visual field quadrant, is an area equivalent to New World monkey VP (Felleman and Van Essen, 1991). The anterior border of VP represents the vertical meridian and the posterior border represents the horizontal meridian in both New and Old World monkeys (Newsome and Allman, 1980; Newsome *et al.*, 1986). The VP of macaques (Newsome *et al.*, 1986) is interposed between ventral V2 and V4. It extends from the posterior bank of the inferior occipital sulcus laterally to the ventral lip of the calcarine sulcus medially and rostrally.

Area V4 of Old World monkeys, originally identified by Zeki (1971b) is located in the prelunate gyrus (representation of the lower visual field), to a small part in the lateral bank of the superior temporal sulcus and more ventrally on the inferior occipital gyrus (upper visual field) (Gattass *et al.*, 1988; Zeki, 1978c). Fifty-four percent of V4 neurons have been reported to have color-opponent properties (Schein and Desimone, 1990; Zeki, 1973, 1978a, c, 1980c, 1983a, b, 1985), fewer than 50% are orientation-selective, and fewer than 15% are directionally selective (Zeki, 1978c). There is a controversy as to the extent to which selective lesions of V4 impair color discrimination (Heywood *et al.*, 1995; Merigan, 1993; Ungerleider and Haxby, 1994; Walsh *et al.*, 1993; Zeki, 1993). Area V4t lies completely in the lateral part of the posterior bank of the superior temporal sulcus (Desimone and Ungerleider, 1986; Gattas *et al.*, 1988).

Areas V4 and V4t can be recognized in myelin-stained sections by distinct outer (in layer IIIC) and inner (in layer V) bands of Baillarger (Desimone and Ungerleider, 1986; Gattass *et al.*, 1988; Morel and Bullier, 1990; Rockland, 1992b; Rosa *et al.*, 1993; Schein *et al.*, 1982; Van Essen *et al.*, 1981). The overall myelination of V4t is relatively low (Ungerleider and Desimone, 1986a, b). Thus, the myelination of V4 is between the extremely high myelination of V5/MT and the relatively light myelination of V4t.

SMI-32 immunohistochemistry demonstrates a distinctive feature of V4: layer III shows a thin band of SMI-32-positive pyramidal cells with bifurcated apical dendrites in layers II–IIIA, which is not seen in adjacent V4t and V5/MT (Cusick *et al.*, 1995; Hof and Morrison, 1995). This band of upper layer III is separated from a thick band in lower layer III. Moreover, SMI-32-ir neurons are more abundant in layers V–VI of the dorsal than the ventral part of V4 (Hof and Morrison, 1995). This may indicate a cytoarchitectonic heterogeneity within V4. Area V4t differs from both V4 and V5/MT by its lower number of SMI-32-ir neurons in layers IIIB+C and V–VI. The border between V4 and VOT is difficult to establish by SMI-32 immunohistochemistry.

In contrast to V1 and V2, calbindin-ir pyramidal cells are found in V4 (less than 1% of the total neuronal population; located in layers II–III), and the frequency of calbindin-ir nonpyramidal neurons and parvalbumin-ir interneurons increases slightly in V4 as compared with V2 (Kondo *et al.*, 1994). Layers II–III are most densely populated by calbindin-ir nonpyramidal neurons, whereas the parvalbumin-ir neurons reach their highest density in layer IV.

The middle temporal or fifth visual area, *V5/MT,* of the macaque (Desimone and Ungerleider, 1986; Maunsell and Van Essen, 1987) is a relatively small cortical area (less than 3% of the size of the primary visual cortex; Van Essen *et al.*, 1981) located in the dorsal part of the superior temporal sulcus, mainly on its posterior bank. It is believed to be the equivalent area of MT or OAa in prosimians, and of MT in New World monkeys. In different primate species, direction-, speed-, and orientation-selective neurons have been described (Albright, 1984; Baker *et al.*, 1981; Born and Tootell, 1992; Felleman and Kaas, 1984; Marcar *et al.*, 1995; Maunsell and Van Essen, 1983a, b; Newsome *et al.*, 1991; Tootell and Born, 1990; Van Essen *et al.*, 1981; Zeki, 1971a, 1974, 1975, 1980a). The V5/MT of owl monkeys is bordered by the MT-crescent (=MTc; caudomedial and -ventral border; Kaas and Morel, 1993), MST (rostromedial border), and FST (rostroventral border).

Area V5/MT is very heavily myelinated in prosimians and New and Old World monkeys (Allman and Kaas, 1971; Allman *et al.*, 1973; Fiorani *et al.*, 1989; Gattass and Gross, 1981; Kaas, 1993; Krubitzer and Kaas, 1990a; Maunsell and Van Essen, 1983a; Ungerleider and Desimone, 1986a, b; Van Essen *et al.*, 1981. The high overall intensity of myelin staining makes the separation of the outer and inner bands of Baillarger difficult in some species (Rosa *et al.*, 1993).

The high cytochrome oxidase (CO) activity and the patchy distribution of this enzyme in layers II–IV is another characteristic architectonic feature of V5/MT in New World and Old World monkeys (Jen and Zeki, 1984; Kaas and Morel, 1993; Tootell *et al.*, 1985). The distance between single CO patches is about 0.85 mm in an Old World monkey (*Cercopithecus aethiops*) (Table V).

Area V5/MT shows a high parvalbumin immunoreactivity with dense fiber plexuses in lower layer III + layer IV, layer V, and layer VI (Cusick *et al.*, 1995) and a strong Cat-301 immunoreactivity (DeYoe *et al.*, 1990). The adjacent areas MST and V4t do not have a plexus in layer V, and the other two plexuses are thinner and more lightly stained.

3. Connectivity of Extrastriate Areas in Nonhuman Primates

3.1. Thalamic Connections

The macaque extrastriate visual cortex has reciprocal connections with several thalamic nuclei. The occipital and the posterior parietal and temporal cortices are connected with the pulvinar (Table IV). Different cortical areas appear to have distinct connection profiles within the pulvinar. Furthermore, parts of areas 18 and 19, the inferior temporal convexity, the superior temporal sulcus, and areas V4 and V5/MT were shown to be weakly connected with the dorsal lateral geniculate nucleus (Lysakowski *et al.*, 1988).

Nonreciprocal corticothalamic connections were described from areas V5/MT (Maunsell and Van Essen, 1983a; Standage and Benevento, 1983; Un-

Table IV. Reciprocal Thalamocortical Connections of the Macaque Extrastriate Cortex

	Reference for observation[a]			
Region/area	dLGN	Pulvinar inferior	Pulvinar medial	Pulvinar lateral
BA 18 + 19	3, 16	2		2
V2	5, 9	9		9
V5/MT		11, 12, 14		11, 12, 14
MST, FST			4	
Superior temporal sulcus	8		6	
BA 7a			6	
LIP			7	7
PO			1	1
PG			1	1
V4	10	13	13	1
TE		15	1, 15	1, 15
TEO		15	15	1, 15
Inferior temporal convexity	8			

[a]References: (1) Baleydier and Morel (1992); (2) Benevento and Davis (1977); (3) Benevento and Yoshida (1981); (4) Boussaoud *et al.* (1992); (5) Bullier and Kennedy (1983); (6) Cusick *et al.* (1995); (7) Hardy and Lynch (1992); (8) Hernandez-Gonzalez *et al.* (1994); (9) Kennedy and Bullier (1985); (10) Lysakowski *et al.* (1988); (11) Maunsell and Van Essen (1983a); (12) Standage and Benevento (1983); (13) Tanaka *et al.* (1990); (14) Ungerleider *et al.* (1984); (15) Webster *et al.* (1993); (16) Yoshida and Benevento (1981).

gerleider *et al.*, 1984) as well as MST and FST (Boussaoud *et al.*, 1992) to the reticular nucleus of the thalamus; and from areas TE and TEO to the nucleus medialis dorsalis (Webster *et al.*, 1993). Nonreciprocal thalamocortical connections were described from the nuclei paracentralis, ventralis anterior, centralis, and limitans to areas TE and TEO (Webster *et al.*, 1993).

3.2. Intrahemispheric Connections

Although far from complete, a large number of intrahemispheric connections of visual cortical areas has been found in nonhuman primates. Felleman and van Essen (1991) published a hierarchical connectivity scheme with 305 connections. This may be approximately one-third of the maximum possible connectivity. The reader is referred to Felleman and Van Essen (1991) for further details as well as for standard nomenclature as used here.

The ascending projections of V1 to V2 originate in (i) the CO-rich blobs of layers II+III, (ii) the CO-poor interblobs of layers II+III, and (iii) to a minor degree in layer IVB. Thus, V2 receives major input from the parvocellular layers of the lateral geniculate body via V1 (Casagrande and Kaas, 1994; Tootell *et al.*, 1988b). The input from the blobs to V2 terminates in the CO-rich thin stripes, from the interblobs in the interstripes, and from layer IVB in the CO-rich thick stripes of V2 (Livingston and Hubel, 1984, 1987; Rockland and Pandya, 1979). Intrahemispheric connections of V2, V3, VP, V3A, V4t, V4, and V5/MT are schematically shown in Fig. 7A–G. Major sources of input to the V4 region are, besides some input from V1, the thin CO-rich stripes and the CO-poor interstripes of V2 (DeYoe and Van Essen, 1985; Shipp and Zeki, 1985, 1995). Some regions of V4 have more thin-stripe afferents, whereas others are dominated by interstripe input, but no further divergence is found in the output channels of V4 to other extrastriate areas (Shipp and Zeki, 1995). Area V5/MT receives major input from layers IVB and VI of V1, which are involved in the magnocellular stream originating in the magnocellular layers of the lateral geniculate body (Tootell *et al.*, 1988b), and from the thick stripes of V2. Area V5/MT projects via MST and FST (Krubitzer and Kaas, 1990a) to the posterior parietal cortex (Kaas and Lin, 1977; Lin *et al.*, 1982; Weller *et al.*, 1984). In prosimians and New World monkeys, the primary visual cortex V1 projects mainly to V2 and V5/MT (about 90% of its output; Kaas and Krubitzer, 1991), to DM (V3 region) (Allman and Kaas, 1975), and to DL (V4 region).

3.3. Interhemispheric Connections

Interhemispheric connections have their origin mainly in the supragranular layers. The terminals are found in all cortical layers of V2 and are focused in layer IV of V3, V3A, and V4 (Kennedy *et al.*, 1986, 1991). Interhemispheric fibers terminate in a narrow strip along the border between V1/V2 (Kennedy *et al.*, 1986; Newsome and Allman, 1980; Spatz and Kunz, 1984; Van Essen and Zeki, 1978; Van Essen *et al.*, 1982; Zeki, 1969, 1970) and are found both in V1 and V2 (representation of the vertical meridian), along the anterior border of

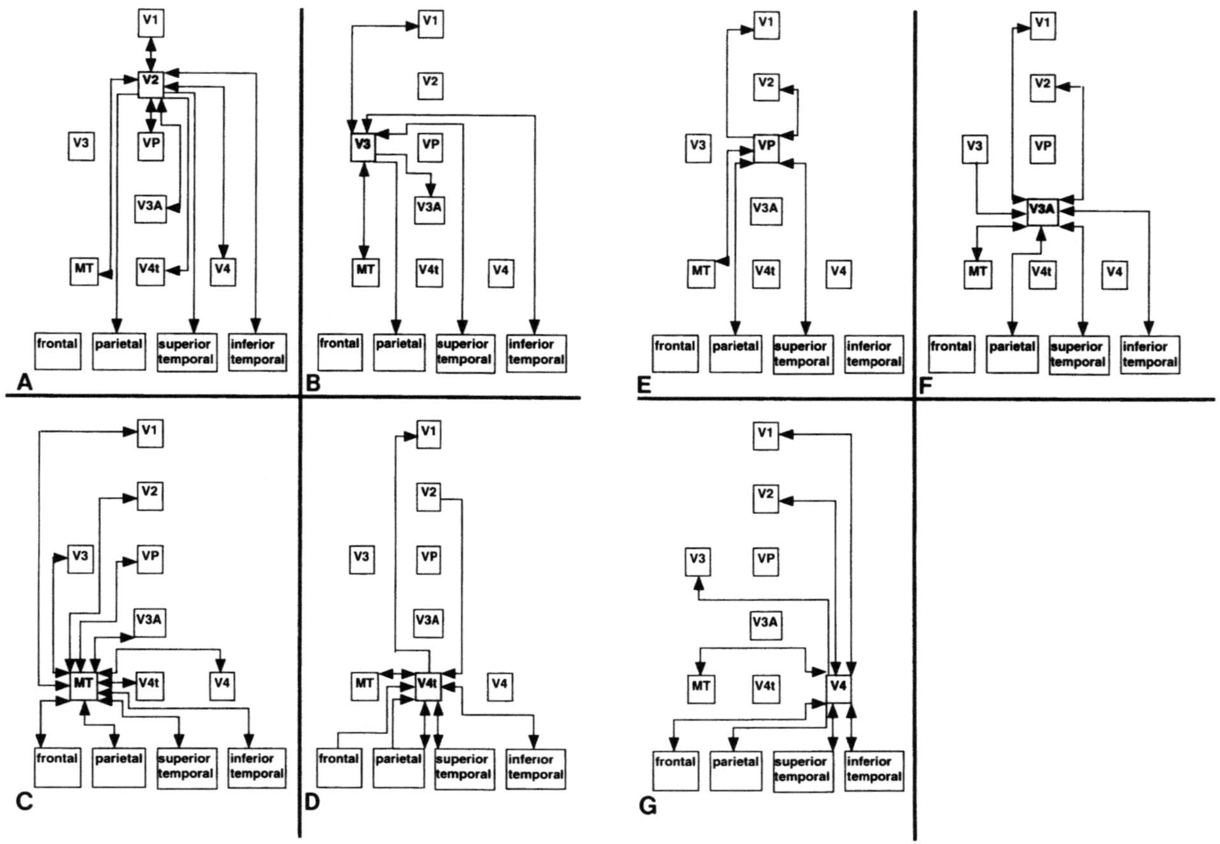

Figure 7. Schematic drawing of major intrahemispheric connections of V2(A), V3(B), V5/MT(C), V4t(D), VP(E), V3A(F), and V4(G) in macaque monkeys. This summary is based on data from Boussaoud *et al.* (1990), Casagrande and Kaas (1994), Colby *et al.* (1988), Cusick *et al.* (1995), Desimone and Ungerleider (1986), Distler *et al.* (1993), Doty (1983), Kennedy and Bullier (1985), Krubitzer and Kaas (1989), Maunsell and Van Essen (1983a), Nakamura *et al.* (1993), Newsome *et al.* (1986), Perkel *et al.* (1986), Rockland (1992a, b, 1994, 1995), Rockland and Pandya (1979), Rockland and Van Hoesen (1994), Rockland *et al.* (1994), Rosa *et al.* (1993), Shipp and Zeki (1989a, b, 1995), Spatz (1975, 1977), Steele *et al.* (1991), Tigges *et al.* (1981), Ungerleider and Desimone (1986a, b), Ungerleider and Mishkin (1979), Van Essen *et al.* (1981, 1986), Vogt [illegible]

V3 between V3/V3A (Zeki and Sandeman, 1976; Van Essen and Zeki, 1978) and V3/V4, and along the anterior border of VP between VP/V4. A broad termination zone of callosal fibers is found within V4 and V5/MT, which is larger than the representation of the vertical meridian (Maunsell and Van Essen, 1987; Newsome and Allman, 1980; Van Essen and Zeki, 1978). Thus, the termination zones of callosal fibers represent in many but not all cases the vertical meridian representation (Kennedy *et al.,* 1991; Newsome and Allman, 1980; Van Essen and Zeki, 1978; Van Essen *et al.,* 1982; Zeki, 1969, 1970, 1977b, 1978a, b), but the termination fields of interhemispheric fibers can include also representations of parts of the visual field with larger distances from the vertical meridian (Kennedy *et al.,* 1991). Area V2 sends homotopic callosal fibers to the V1/V2 border and heterotopic fibers to the region in and around V3 and V5/MT. The callosal connections from V2 to V1 avoid layer IV of V1 (Kennedy and Dehay, 1988; Kennedy *et al.,* 1986; Maunsell and Van Essen, 1987; Van Essen *et al.,* 1982).

4. Transmitter Receptors in Extrastriate Areas of Nonhuman Primates

4.1. Transmitter Receptors in Area V2

The border between V1 and V2 of the macaque and marmoset visual cortex is clearly visible by differences in mean densities and laminar distribution patterns of the receptors of most classical transmitters (Gebhard *et al.,* 1993; K. Zilles, unpublished observations).

Receptor autoradiography of *AMPA receptors* with [^{3}H]AMPA in macaques reveals the V1/V2 border by a sudden disappearance of the very low receptor density in layer IVB of V1 (Fig. 8A). The relatively high AMPA receptor density in layer II of V1 increases moderately in V2. The AMPA receptors occupy a somewhat broader part of the cortical width (layers II and upper III) in V2d and V2v than in the adjoining V1.

The *NMDA receptors* had highest densities in layer II, deeper layer IV, and layers V–VI of V2 (Rosier *et al.,* 1993).

Huntley *et al.* (1993) demonstrated *kainate receptors* with a monoclonal antibody. The highest numbers of stained perikarya were seen in layer II and the lower half of layer III. Layers IV–VI showed lower densities. Only the most proximal parts of the dendrites were stained. The staining in V1 was markedly weaker than in V2. This lower mean receptor density in V1 is corroborated by autoradiography of kainate receptors with [^{3}H]kainate (Fig. 8B; K. Zilles, unpublished observations). However, the laminar distribution pattern of autoradiographically demonstrated kainate receptors is different from that of the immunohistochemical observations. The kainate-binding sites reach their relatively highest densities both in layers II and V/VI of V2d and V2v. Low densities are found in layers III–IV. This difference between distribution patterns demonstrated by immunohistochemistry or by receptor autoradiography may be explained by the different structures (specific epitopes of receptor protein subunits versus binding site) visualized with the different techniques. The unknown

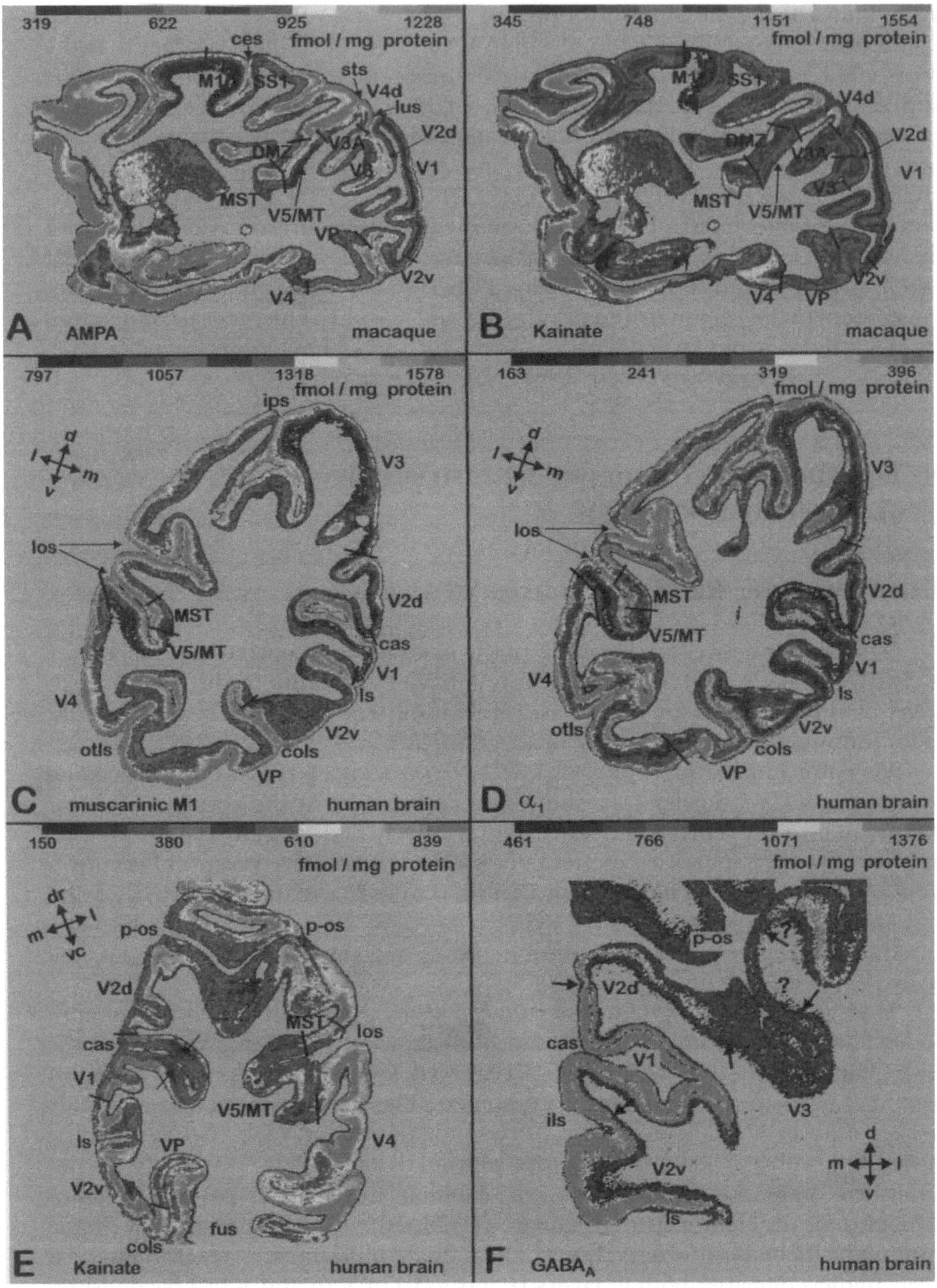

Figure 8. (A, B) Regional distribution of AMPA and kainate receptors respectively (demonstrated with [^{3}H]AMPA and [^{3}H]kainate binding; K. Zilles, unpublished observations), in adjacent parasagittal sections through the macaque cortex. The color scales show the receptor densities in fmol/mg protein. A considerable regional heterogeneity of the receptor distribution is visible, which allows the

dynamics of the transport of receptor proteins from the perikaryon to their final location in the membrane of cell bodies and/or neuronal processes, and the possible temporary local accumulation of the protein at specific places (Golgi cisternae, etc.) along the way from the place of synthesis (cell body) to the final location (dendrites, axon terminals), may also lead to these differences in distribution patterns of the structures visualized by the different techniques. The border between V2v and VP is characterized by an increase in [^{3}H]kainate binding-site density especially in the infragranular layers of VP.

Monoclonal antibodies against the *$GABA_A$ receptor* label punctate profiles in the neuropil and around cell bodies (Hendry *et al.*, 1990). The $GABA_A$ receptors (demonstrated with antibodies or [^{3}H]muscimol binding) are most densely packed in layers II–III, reach intermediate densities in layers I and IV, and show low densities in layers V–VI (Hendry *et al.*, 1990; Rakic *et al.*, 1988). *In situ* hybridization studies (Huntsman *et al.*, 1994) of seven different $GABA_A$ receptor subunits revealed a remarkably heterogeneous laminar distribution in V2. The [^{3}H]muscimol-binding sites clearly indicate the V1/V2 border both in macaques and marmosets by the sudden disappearence of the heavily labeled sublayer IVC, and a slightly lower mean density in V2 (Gebhard *et al.*, 1993; Hendry *et al.*, 1990; Rakic *et al.*, 1988; K. Zilles, unpublished observations).

All *muscarinic cholinergic receptors* can be visualized by labeling with [^{3}H]quinuclidinylbenzilate (QNB). The highest densities of these receptors are found in layer II, followed by intermediate densities in layers I and III–V and low densities in layer VI of the macaque V2 (Rakic *et al.*, 1988). A more selective labeling with [^{3}H]pirenzepine, predominantly of the muscarinic *M1* receptors, shows a similar distribution pattern, with the highest densities in layers II–IV and relatively low densities in layers I, V, and VI (Vannucchi and Goldman-Rakic, 1991; K. Zilles; unpublished observations). In contrast to the areal differences described for AMPA and kainate receptors (see above), the muscarinic M1 receptors show a much higher density in V1 than V2. Layer IVC and the stripe of Gennari are clearly visible in the M1 receptor imaging of V1, and the labeling of both layers stops abruptly at the V1/V2 border both in macaques and marmosets (Gebhard *et al.*, 1993; K. Zilles, unpublished observations). The border between

Figure 8. *(Continued)* delineation of numerous extrastriate areas, including V1, V2v, V2d, VP, V3, V3A, V4, V4d/V4t, V5/MT, MST, and the dorsomedial zone, DMZ. Further delineations are possible in frontal, prefrontal, and temporal regions. Abbreviations: ces, central sulcus; lus, lunate sulcus; M1, primary motor cortex; SS1, primary somatosensory cortex; sts, superior temporal sulcus. (C–F) The regional distribution of muscarinic (mainly) M1, α_1, kainate, and $GABA_A$ receptors, respectively (labeled with [^{3}H]pirenzepine, [^{3}H]prazosin, [^{3}H]kainate, and [^{3}H]muscimol; K. Zilles, unpublished observations) in adjacent coronal sections (C: M1; 8D: α_1) and an oblique (between coronal and horizontal cutting planes) section (E: kainate receptor) through the human occipital and parietal cortex. (D) Distribution of $GABA_A$ receptors at higher resolution. The color scales show the respective receptor densities in fmol/mg protein. Crossed arrows indicate the different orientations of the sections (d, dorsal; dr, rostrodorsal; l, lateral; m, medial; v, ventral; vc, ventrocaudal). A considerable regional heterogeneity of the receptor distribution is visible in all cases, which allows the delineation of various extrastriate areas. Abbreviations: cas, calcarine sulcus; cols, collateral sulcus; fus, fusiform gyrus; ils, intraligual sulcus; ips, intraparietal sulcus; los, lateral occipital sulcus; ls, lingual sulcus; otls, occipitotemporal sulcus; p-os, parietooccipital sulcus. Thin arrows in panel E point to layer IVB. For further details see text. A four-color reproduction of this figure appears following page xxiii.

V2 and VP or V3 cannot be clearly recognized in the M1 receptor autoradiography. Predominantly *M2* receptors can be shown with [^{3}H]oxotremorine-M binding in the occipital cortex of macaques and marmosets (Gebhard *et al.*, 1993; K. Ziles, unpublished observations). The difference in mean M2 receptor densities between V1 and V2 (high in V1, low in V2) is easily visible in both species. Highest densities are found in layers II–IVA and VC, lower densities in layers I, IVB, and V–VI.

Noradrenergic α_1-binding sites (labeled with [^{3}H]prazosin) reach the highest density in layer II, followed by an intermediate density in layer I and lower densities in layers III–VI of macaque V2 (Rakic *et al.*, 1988). The mean density of α_1 receptors appears to be slightly lower in V1 than V2 of macaques and marmosets (Gebhard *et al.*, 1993; Rakic *et al.*, 1988). The highest density of [^{3}H]clonidine-labeled *noradrenergic α_2-binding sites* is visible in layer III, intermediate densities are found in layers I–II and IV, and low densities in layers V–VI of V2 (Rakic *et al.*, 1988). The mean density of this binding site is lower in V2 than V1. *Noradrenergic β receptors* labeled with [^{125}I]iodopindolol are most densely packed in layer II, followed by layers I and III. Much lower densities are found in layers III–VI (Rakic *et al.*, 1988). Again, the mean density is lower in V2 than V1.

Dopaminergic D1 receptors reach highest densities in layers I–IIIA and intermediate densities in layers V–VI (Lidow *et al.*, 1991). A slightly different pattern was described in a recent immunohistochemical study (Smiley *et al.*, 1994), where the D1 receptors were found to be of highest density in layers I–II and V–VI, of intermediate density in layer III, and of very low density in layer IV. The *D2 receptors* are most densely packed in layer V and show intermediate densities in layers I–IIIA (Lidow *et al.*, 1991).

The highest density of *[^{3}H]serotonin-binding sites* was found in layers I–IIIA, whereas intermediate densities were found in layers IIIB–IV and low densities in layers V–VI of the macaque V2 (Parkinson *et al.*, 1989; Lidow *et al.*, 1989; Rakic *et al.*, 1988). In marmosets, an equal distribution of [^{3}H]5-HT (serotonin)-binding sites was found, with a lower mean density in V2 than V1 (Gebhard *et al.*, 1993). The [^{3}H]ketanserin-binding sites (with preference for *5-HT$_2$ receptors*) are most densely packed in layer III, show intermediate densities in layers I–II and IV–V, and show low densities in layer VI (Parkinson *et al.*, 1989; Lidow *et al.*, 1989; Rakic *et al.*, 1988). *Serotonin uptake sites* are most densely packed in layers I–II and IV (Lidow *et al.*, 1989).

Opiate receptors (μ- and δ-like) show an intermediate density in V2 compared with V1 (lower density) and V3 (higher density) of macaques (Lewis *et al.*, 1981). Since this increase continues in the temporal lobe, a remarkable association of opiate receptor densities with increasing functional complexity in the ventral stream can be observed.

The *neuropeptide Y receptors* decrease steadily from layer I to layer VI (Rosier *et al.*, 1990). The *somatostatin receptors* are highly concentrated in lower layer V and upper layer VI (Rosier *et al.*, 1991b). The *galanin receptors* also reach their highest densities in lower layer V and upper layer VI, but they have a second peak with intermediate density in layer IV (Rosier *et al.*, 1991a). The *cholecystokinin-binding sites* reach the highest densities in deeper layer III and layer IV and intermediate densities in deeper layer VI (Kritzer *et al.*, 1987).

The AMPA receptor distribution allows the identification of further borders (Fig. 8; K. Zilles, unpublished observations) between ventral V2 and VP by a slight increase in receptor density in the infragranular layers of VP, between dorsal V2 and V3/V3A by a slight increase of the receptor density in the most superficial layers, and between V3/V3A and V4 by a further increase in all layers. The border between V4t and V5/MT is striking by the sudden decrease in AMPA receptor density in V5/MT. The MST can be delineated from V5/MT by an increase in receptor density, and is also different from the rostrally adjoining dorsomedial zone (DMZ in Fig. 8) by its somewhat higher receptor density (Fig. 8).

By comparing the changes in AMPA and kainate receptors between V1, V2, and the rostrally adjoining areas, it seems that a slight but consistent increase in receptor density can be observed along the sequence from V1 through V2 to VP or V3. As with AMPA receptors (see above), V5/MT is clearly identifiable with [^{3}H]kainate binding by the lowest receptor density of all extrastriate areas (K. Zilles, unpublished observations).

5-HT_{1A} receptors show a similar laminar distribution in V3 as already described for V2. However, an overall increase in receptor density was reported at the V2/V3 border (Parkinson *et al.*, 1989).

In V5/MT a higher density of NMDA receptors was found in layer II, whereas the other layers show a more or less homogeneous and lower density of NMDA receptors (Rosier *et al.*, 1993). A decrease in density of 5-HT_1A receptors compared with surrounding regions was reported in a region which probably corresponds to V5/MT (Parkinson *et al.*, 1989).

5. Human Extrastriate Visual Cortex

5.1. Functional Specialization within Human Extrastriate Cortex

In contrast to the situation in nonhuman primates, where many observations about structural–functional relations in well-defined extrastriate areas are available, our knowledge concerning these relations in the human brain is still very limited. Most of the data about the *functional* aspects of the human extrastriate areas come from observations with the new imaging techniques. However, when reading those papers, including the short review to follow of results from functional imaging studies, one should be aware that the assignment of activation foci to particular architectonically defined areas is problematic. It is based on the assumption that architectonically defined areas are associated with macroscopic landmarks (e.g., sulci and gyri, the anatomical structures visible in positron emission tomography, [PET] and functional magnetic resonance imaging [fMRI] images in a relatively constant manner. This is not yet proven, and was doubted already by Brodmann (1909) (see also Fig. 2). Moreover, the interindividual variability in position and extent of architectonically defined cortical areas (Fig. 2) is not considered in presently used brain atlases. Atlases providing a probabilistic approach (Roland and Zilles, 1994) to localization of architectonic areas are necessary for a critical and reliable evaluation of such assignments.

Although this chapter is not planned as a review of PET and fMRI data on the human visual cortex, a short summary of functional aspects should be given in order to allow a comparison between the functions of defined extrastriate areas of nonhuman primates and those of the putatively equivalent areas in the human brain. PET and fMRI data of only those human extrastriate areas which have been architectonically defined with some reliability are briefly reviewed here. In addition, organizational aspects, such as the feedback projections from higher order to lower order visual areas, which have been intensively studied with functional and anatomical techniques in nonhuman primates, e.g., V2 or V1, can now be analyzed in the human brain. This general organizational aspect of the V1/V2 connectivity was demonstrated recently with fMRI in the human brain (Friston *et al.*, 1995).

Activation of *V2* was demonstrated with PET during shape discrimination, if shape was defined by texture boundaries (Gulyás and Roland, 1994b). This task is similar to illusory contour tasks, which also revealed a V2 activation in monkeys (Peterhans and Von der Heydt, 1989; Von der Heydt and Peterhans, 1989). Areas V1 and V2 and other extrastriate areas (not only V4) were activated by color discrimination tasks (Gulyás and Roland, 1991, 1994b) as already shown in monkeys (see Section 2.2). Area V2 is also active during visual recall of real objects, indicating comparable physiological mechanisms in perception and imagery (Le Bihan *et al.*, 1993; Roland and Gulyás, 1994, 1995). Apparently, V2 is not only involved in relatively simple perceptual functions, but is also activated—as part of a system of multiple cortical areas (McIntosh *et al.*, 1994)—by more complex perceptual, cognitive, and mnestic (e.g., face and word processing; Halgren *et al.*, 1994) tasks.

Human *V3* was mapped by fMRI as an area adjoining the dorsal border of V2 (Tootell *et al.*, 1995a) and the ventral half of the anterior border of dorsal V2 (Sereno *et al.*, 1995). The anterior border of V2 represents the horizontal meridian. Area V3 contains the lower visual field as in monkeys with a mirror-image representation of the visual field. The role of human V3 in vision is not completely clear. It may contribute to dynamic form perception (Zeki and Shipp, 1988; Zeki 1990c) and has a high contrast selectivity (Tootell *et al.*, 1995a), which was interpreted as further evidence for the classification of this area as part of the magnocellular dorsal stream. Stimulation by "optical flow" (streaming-like coherent motion of objects through the visual field; Gibson, 1950) compared with incoherent motion (directions of moving objects are randomized) involves three major occipitoparietal and occipitotemporal areas, as shown in a PET study by de Jong *et al.* (1994). One of these regions is located on the posterior surface of the cuneus. This region may be functionally comparable with the dorsal V3 area in monkeys, and it contains the lower vertical meridian, and may include also V3A (de Jong *et al.*, 1994). The same region is activated by comparing simple motions to a static pattern (Watson *et al.*, 1993; Zeki *et al.*, 1993). It can be activated by moving stimuli if there is a large V5/MT lesion (Shipp *et al.*, 1994). A putative V3A human homologue was mapped with fMRI as a region adjoining V3 laterally of the interhemispheric fissure on a frontal section (Tootell *et al.*, 1995a).

The other two regions activated by optical flow are located clearly outside V3, but will be considered here because of their activation by the same test paradigm. One of these regions outside V3 is located ventral to V5 at the occipitotemporal junction and lateral to the "color center" in the fusiform gyrus (de Jong *et al.*, 1994). Its position varies from the inferior occipital lobe to the lateral extension of the fusiform gyrus, depending on the individual surface anatomy. A distance of 10–20 mm is found between V5/MT and this region, which is also activated by illusory rotation motion (Zeki *et al.*, 1993). The best-studied area of the macaque showing activation by optical flow is the dorsal part of MST. However, the difference in topology between the human occipitotemporal area and the monkey MSTd rules out a clear comparison. The MSTd borders V5/MT in the monkey, but the human occipitotemporal optic flow area is clearly separated from V5/MT. Another candidate could be the TEO area of monkeys. The TEO receives input from V5/MT (Distler *et al.*, 1993), but it remains unknown whether TEO is engaged in optical flow perception. In conclusion, the monkey counterpart of the human occipitotemporal optic flow region is not identified, but MSTd is the most likely candidate. However, this would require the assumption of a different topographic organization of the MT–MST complex in humans compared with macaques.

The third region activated by optical flow is located on the posterior surface of the precuneus and as such it is part of the superior parietal lobule, where human BA 7 (Brodmann, 1909) or PE (Economo and Koskinas, 1925) is located. It has been concluded from lesion studies that the superior parietal lobule is implicated in visually guided reaching (Perenin and Vighetto, 1988). A functionally comparable macaque area is MIP (Colby and Duhamel, 1991), which receives input from V6/PO and the more dorsal V6A/POd (Colby *et al.*, 1988; Galetti *et al.*, 1991, 1993; Neuenschwander *et al.*, 1994), both located in the anterior bank of the parietooccipital sulcus. Thus, V6A/POd seems to be a candidate for an area comparable topographically and functionally to the parietal optical flow region in the human brain.

Human *VP* (= *V3v*) was mapped by fMRI as an area adjoining the anterior and lateral border of ventral V2 (Sereno *et al.*, 1995; Tootell *et al.*, 1995a). This border represents the horizontal meridian. Area VP contains the upper visual field as in monkeys with a mirror-image representation of the visual field.

Lesions of the lingual and fusiform gyri lead to achromatopsia (Damasio *et al.*, 1980; Meadows, 1974; Rizzo *et al.*, 1992; for review see Zeki, 1990b). PET, fMRI, and electrophysiological data from the same region showed activation by color (Allison *et al.*, 1993; Corbetta *et al.*, 1991; Gulyás and Roland, 1991, 1994b; Gulyás *et al.*, 1994a, b; Lueck *et al.*, 1989; Sakai *et al.*, 1995; Sereno *et al.*, 1995; Zeki *et al.*, 1991) and shape (Corbetta *et al.*, 1991). Thus, part of the lingual and fusiform gyri has been proposed as human correlate of the monkey area V4, which contains many color-, space-, and shape-activated neurons (Desimone and Schein, 1987; Desimone and Ungerleider, 1989; De Yoe *et al.*, 1994; Heywood *et al.*, 1992; Schiller and Lee, 1991; Walsh *et al.*, 1993; Zeki, 1973, 1977a, 1980c, 1983a, b, 1985, 1990a). The putative human *V4* is located mainly on the posterior part of the fusiform gyrus and abuts the rostral border of the putative human VP (Sereno *et al.*, 1995) as in the monkey. Knowledge about colors as attributes of objects is also represented in the posterior fusiform gyrus rostrally to the putative human V4 (Martin *et al.*, 1995). Medially of this region, a face recognition area was identified with electrophysiological techniques (Allison *et al.*, 1993, 1994). Additional color foci were described on the human dorsolateral occipital cortex (Corbetta *et al.*, 1991).

Since lesions of the macaque V4 did not produce profound and permanent color impairment (Heywood and Cowey, 1987; Schiller and Lee, 1991; Heywood *et al.*, 1992) as seen after lesions of the fusiform gyrus in human brains (Damasio *et al.*, 1980; Rizzo *et al.*, 1992), it is not completely clear whether the human V4 in the fusiform gyrus represents the equivalent of the macaque V4 (Heywood *et al.*, 1995; Meadows, 1974; Merigan, 1993; Ungerleider and Haxby, 1994; Walsh *et al.*, 1993; Wild *et al.*, 1985; Zeki, 1993). Furthermore, object- but not color-related activity was localized in a region of the human lateral occipital cortex "just abutting the posterior aspect of the motion-sensitive area MT/V5" (Malach *et al.*, 1995). This region, which was called LO, would be better compared with the macaque V4 area by its more lateral location. However, the object-related area could be alternatively also a posterior part of the monkey's PIT/TEO or TE areas if this part of the cortex is shifted to a more posterior position in the human than the monkey brain.

The ventral part of the human *V4* (= *V4v*) was mapped by fMRI (Sereno *et al.*, 1995) during phase-encoded retinal stimulation as an area adjoining the ventral half of the anterior border of VP. This border represents the vertical meridian, whereas the horizontal meridian is found at the anterior border of V4v. Area V4v contains the upper visual field as in monkeys with a non-mirror-image representation of the visual field.

Visual motion perception is specifically impaired following lesions or magnetic stimulation in the anterolateral part of the human occipital lobe (Barton *et al.*, 1995; Hotson *et al.*, 1994; Vaina, 1989; Zeki, 1991; Zihl *et al.*, 1983). This postlesional deficit makes the temporooccipital transition region a good candidate for the human *V5/MT* area. PET and fMRI studies have demonstrated that this area has a high sensitivity to contrast (15 times higher than in V1; Tootell *et al.*, 1995a), is strongly activated by moving visual stimuli, and is active during the production of visual motion aftereffects (Corbetta *et al.*, 1990, 1991; Dupont *et al.*, 1994; Lueck *et al.*, 1989; Probst *et al.*, 1993; Tootell *et al.*, 1993, 1995a, b; Watson *et al.*, 1993; Zeki *et al.*, 1991). With these techniques providing a sufficient local resolution, the region of interest could be pinned down to a location close to the level of the AC–PC line and posterior to the contact of the ascending limb of the inferior temporal sulcus with the lateral occipital sulcus (Watson *et al.*, 1993). The center of V5/MT varies considerably, by 27 mm in the left and 18 mm in the right hemisphere (Watson *et al.*, 1993). This is in agreement with anatomical data localizing V5/MT (Clarke, 1994b; Clarke and Miklossy, 1990; Tootell and Taylor, 1995; Zilles and Schleicher, 1993). A further visual motion-selective area was also identified in the posterior parietal cortex with PET (Watson *et al.*, 1993) and fMRI (Tootell *et al.*, 1995a), which may be identical with the cytochrome oxidase-rich area "PX" described by Tootell and Taylor (1995) in macaque and human brains.

Although V5/MT receives major input from V1 (directly or via V2) and

V5/MT projects back to V1 and V2 as demonstrated in the monkey (Cragg, 1969; Lund *et al.*, 1975; Shipp and Zeki, 1985, 1989a, b; Zeki, 1969, 1971a), it was shown in a PET study that human V5/MT, and in addition V3 and area 7 were active also in a completely hemianopic patient blinded by a lesion in area V1 (Barbur *et al.*, 1993). He could discriminate the direction of a moving bar and was consciously aware of the nature and direction of the stimulus. Thus human V5/MT may receive direct visual input without passing through V1. Input to V5/MT from the lateral geniculate body (Benevento and Yoshida, 1981; Fries, 1981; Yukie and Iwai, 1981) and indirectly via the superior colliculus and the pulvinar (Benevento and Fallon, 1975; Keating, 1980; Pasik and Pasik, 1982; Standage and Benevento, 1983) was observed in monkeys, and a direct input to the prestriate cortex was also reported in fetal human brains (Kostovic and Rakic, 1984). Moreover, it was found that experimental disconnection of V1 and V5/MT did not abolish the directional selectivity of V5/MT (Rodman *et al.*, 1989; Girard *et al.*, 1992), although some changes of electrophysiological responses of V5/MT neurons were found. Direct evidence for a component of directional motion discrimination bypassing human V1 comes from transcranial magnetic stimulation of V1 and V5/MT (Beckers and Zeki, 1995). The inactivation of V5/MT abolishes motion perception (akinetopsia), whereas V1 inactivation had only marginal effects and at a much longer delay. Thus "perceptually effective motion signals reach V5/MT before they reach V1" (Beckers and Zeki, 1995, p. 49).

Another motion area of the human brain with a somewhat different location and function compared with V5/MT was described by Orban *et al.* (1995). This area is activated much more by kinetic contours (contours defined by motion differences) than by luminance contours or uniformly moving random dots as typically found in V5/MT. A correlate of this area has not been identified in the monkey brain.

The naming of actions as possible attributes of an object activates an area immediately in front of the motion area V5/MT (Martin *et al.*, 1995).

Reports about the putative human MST are rare (Cheng *et al.*, 1995; Dupont *et al.*, 1994). Several cortical regions are activated by wide-field visual motion, including V1/V2 and surrounding areas in the cuneus and superior occipital gyrus (probably the human V3/V3A homologues), the frontal eye field (BA 8), parts of the most lateral premotor cortex (BA 6) and the superior parietal lobule (BA 7), and two regions in the lateral occipitotemporoparietal cortex. Cheng *et al.* (1995) reported that the more posterior of these two regions is also activated by small-field visual stimuli and corresponds probably to V5/MT. The more anterior field was activated bilaterally in the temporoparietal cortex. This field may be the human homologue of the MST area. It is located in the inferior parietal lobule and on the superior and middle temporal gyri. Anatomical and especially architectonic observations are lacking for this area.

Although the functional data from the human brain are not as detailed as the results from nonhuman primate brains, it is reasonable to adopt the hypothesis that there is a recognizably comparable organization of the extrastriate areas. We do not have enough evidence for establishing definite homologies between monkey and human extrastriate areas, but we have sufficient knowledge for comparing putatively equivalent areas between the species. This is an encouraging situation for the generation of working hypotheses to guide future functional and anatomical studies in the human extrastriate cortex.

5.2. Cyto- and Myeloarchitecture of Human Extrastriate Areas

Human *V2* is approximately 1–1.5 cm wide at the level of the foveal representation and about 1.8–3.4 cm at the level of presumptive intermediate eccentricities. The total surface area of V2 is approximately 1.7 times larger than that of V1 (Tootell and Taylor, 1995). Area V2 differs from V1 in the packing densities of neurons (V1: 40,000 neurons/mm^3; V2: 31,500 neurons/mm^3) and and glial cells (V1: 27,000 glial cells/mm^3; V2: 32,000 glial cells/mm^3) (Leuba and Garey, 1989), in the proportions of tissue volumes occupied by cell bodies or neuropil (more neuropil in BA 18) (Zilles *et al.*, 1986), and in the widths of layers (layers I, II, IV: BA 17 > BA 18; layer V: BA 17 < BA 18; Ziles *et al.*, 1986).

In myelin-stained sections, V2 can be distinguished easily from V1 (Vogt and Vogt, 1919; Sanides and Vitzhum, 1965a, b). The lateral boundary of V2 can be detected in the inferior part of the occipital cortex—the adjacent VP has a lighter staining—but not in the superior part (Clarke and Miklossy, 1990; Zilles and Schleicher, 1993). In flattened myelin-stained sections of human V2, alternating dense and light stripes were observed (Fig. 9); the respective stripes have a distance of 6–8 mm (Tootell and Taylor, 1995) and are much coarser and less frequent than the myelinated stripes in V1 (Table V). It is not clear whether the

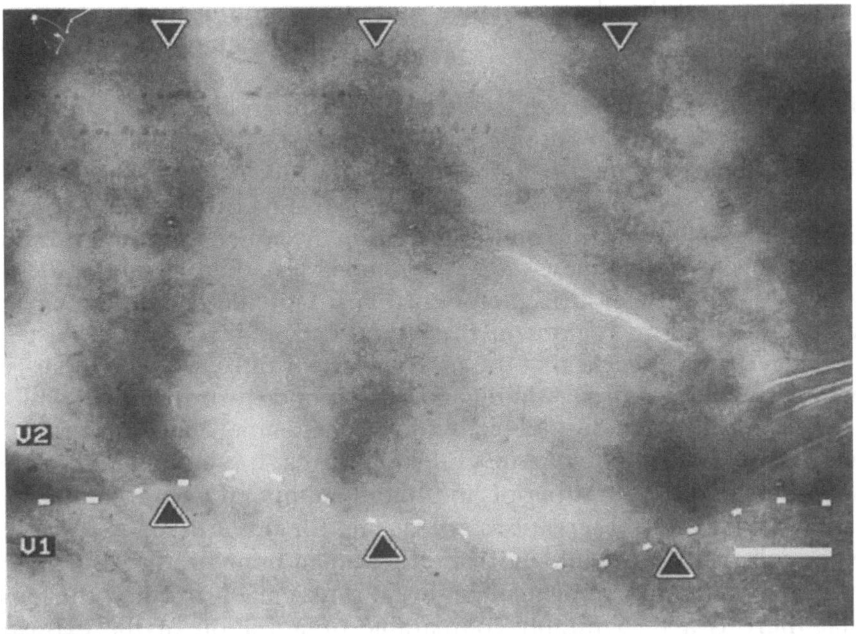

Figure 9. Myelin staining in human V2 as revealed by averaged images from four tangential sections through layers III–V. Dashed line indicates V1/V2 boundary. Bar = 2 mm. Note dark stripes running roughly perpendicular to the 17/18 boundary. From Tootell and Taylor (1995).

Table V. Intraareal Compartments[a,b]

	Squirrel monkey	Macaque	Human
V1 *blobs in CO stains:*			
Diameter (μm)	210 (1)	262–377 (2) 150–250 (10)	300–450 (3) 250–500 (4)
Center-to-center distance within row (μm)	500 (10)	450 (2) 550 (10)	500–800 (3)
Center-to-center distance between rows (μ)	500 (10)	350 (2) 350 (10)	600–1200 (4) 650–1050 (3)
Average density (1/mm²)	4.5 (10)	5.1 (3) 5 (10)	2–3 (3)
Blobs also darker than interblob in:			
GABA receptors	—	Yes (12)	Yes (12)
AChE	Yes (10)	Yes (10)	Yes (11)
Ocular dominance stripes, width (mm)	—	0.6–1.2 (8)	1.2—2.0 (7)
V2 *in CO stains:*			
Width of dark stripes (mm)	0.4–0.7 (9)	—	1–2.75 (13) 0.5–0.7 (4)
Width of pale stripes (mm)	—	—	1.5–2.75 (13) 0.5–0.7 (4)
Diameter of globular zones[c] (μm)	—	—	580–1400 (3)
Center-to-center distance between globular zones within stripes (μm)	—	—	850–1850 (3)
Center-to-center distance between stripes (mm)	1.0–1.5 mm (9)	1.5–2.5 (9)	1.5–2.3 (3)
CAT-301:			
Width of stripes (mm)	—	—	1–3 (5)
periodicity of thick and thick stripes (mm)	—	—	4–8 (5)
myelin stains:			
Width of stripes (mm)	0.4–0.7 (9)	—	6–8 (6)
Center-to-center distance between stripes (mm)	1.0–1.5 (9)	1.5–2.5 (9)	—
MT-V5: *CO*			
Interpatch distance in normals (mm)	—	0.85 (6)	0.7–1.4 (6)

[a]Measures are given as described in the quoted papers; they correspond to measures on histological sections, without corrections for shrinkage or deformation introduced during histological treatment.
[b]References: (1) Carroll and Wong-Riley (1984); (2) Wong-Riley and Carroll (1984); (3) Wong-Riley *et al.* (1993); (4) Clarke (1994b); (5) Hockfield *et al.* (1990); (6) Tootell and Taylor (1995); (7) Hitchcock and Kickey (1980), Horton and Hedley-Whyte (1984); Tootell and Taylor (1995); (8) Tootell *et al.* (1988a); (9) Tootell *et al.* (1983); (10) Horton (1984); (11) Horton and Hedley-Whyte (1984); (12) Hendry *et al.* (1994); (13) Burkhalter and Bernardo (1989).
[c]Patches in Hitchcock and Hickey (1980) and Horton and Hedley-Whyte (1984).

densely myelinated stripes in human V2 are colocalized with every other interstripe or represent either thick or thin stripes. The sizes of human V2 stripes and interstripes estimated in CO-stained sections (Burkhalter and Bernardo, 1989; Hockfield *et al.*, 1990) are not compatible with the distances between the densely myelinated stripes if every stripe or interstripe in CO-stained sections displayed a high myelin density. The most probable hypothesis based on CAT-301 staining of human V2 (Hockfield *et al.*, 1990) is that the CAT-301 (Table V) stripes are colocalized with thick CO-rich stripes.

Human *VP*, but not dorsal V3, can be easily distinguished from the surrounding cortex by cyto- and myeloarchitectonic characteristics alone (Clarke and Miklossy, 1990; Zilles and Schleicher, 1993). It lacks large pyramidal neurons in both layers III and V, whereas adjacent V2 has large pyramids in layer III and adjacent V4 in III and V. It is lightly myelinated, with predominance of fine fibers; both striae of Baillarger are absent.

Human *V4* can be distinguished from the adjacent V2 by the same architectonic criteria.

Human *V5/MT* is believed to correspond to a heavily myelinated area near the occipitotemporal junction on the lateral convexity (Clarke and Miklossy, 1990; Sereno and Allman, 1991; Tootell and Taylor, 1995; Ziles and Schleicher, 1993). In myelin stains, V5/MT is a relatively small oval area (2.0–1.5 × 1.4–1.0 cm) in the dorsal part of the inferior temporal sulcus and about 4.7 cm from foveal primary visual cortex V1 (Tootell and Taylor, 1995). Area V5/MT cannot be distinguished by *cyto*architectonic criteria alone from the surrounding cortex (Clarke and Miklossy, 1990).

5.3. Cytochrome Oxidase Staining as a Marker for Visual Areas and Intraareal Compartments

The cytochrome oxidase (CO) pattern in primary visual cortex of nonhuman primates has been described in great detail by Wong-Riley (1994) (see also Table V). Human V1 was shown to display a similar pattern of CO staining as in nonhuman primates (Horton and Hedley-Whyte, 1984; Burkhalter and Bernardo, 1989; Burkhalter *et al.*, 1993; Wong-Riley, 1993; Wong-Riley *et al.*, 1993; Clarke, 1994b). A dark band with a sharp lower and a gradual upper limit was found in layer IV and a system of patches in supragranular layers. In tangential sections these patches appeared oval (ca. 250 × 500 μm; Horton and Hedley-Whyte, 1984; Wong-Riley *et al.*, 1993; Clarke, 1994b) and tended to be organized in rows. Center-to-center distances within rows were reported to be 0.5–0.8 mm and center-to-center distances between rows 0.65–1.05 mm (Wong-Riley *et al.*, 1993). Average density of blobs was 2–3 blobs per mm^2 (Wong-Riley *et al.*, 1993). As in nonhuman primates, CO-dense blobs in human V1 were shown to stain also more intensely for acetylcholine esterase (Horton and Hedley-Whyte, 1984) and subunits of the $GABA_A$ receptor (Hendry *et al.*, 1994) than the interblob regions.

Human extrastriate visual cortex displays a distinct CO pattern that is different from that of V1 (Table V). Two basic patterns of CO pattern were described. First, a one-band pattern was found in most of the extrastriate occipital cortex; it was characterized by a dark band at midcortex, centered on layers III and IV with gradual upper and lower limits (Burkhalter and Bernardo, 1989; Wong-Riley *et al.*, 1993; Clarke, 1994b). The overall intensity and homogeneity of staining varied throughout the occipital cortex (Wong-Riley *et al.*, 1993; Clarke, 1994b). Second, a two-band pattern has been found in a restricted region of the posterosuperior precuneus; this region stained relatively lightly and had an even lighter band at midcortex, centered on layer IV (Clarke, 1994b).

Comparison of CO staining (Fig. 10) with cyto- and myeloarchitecture in

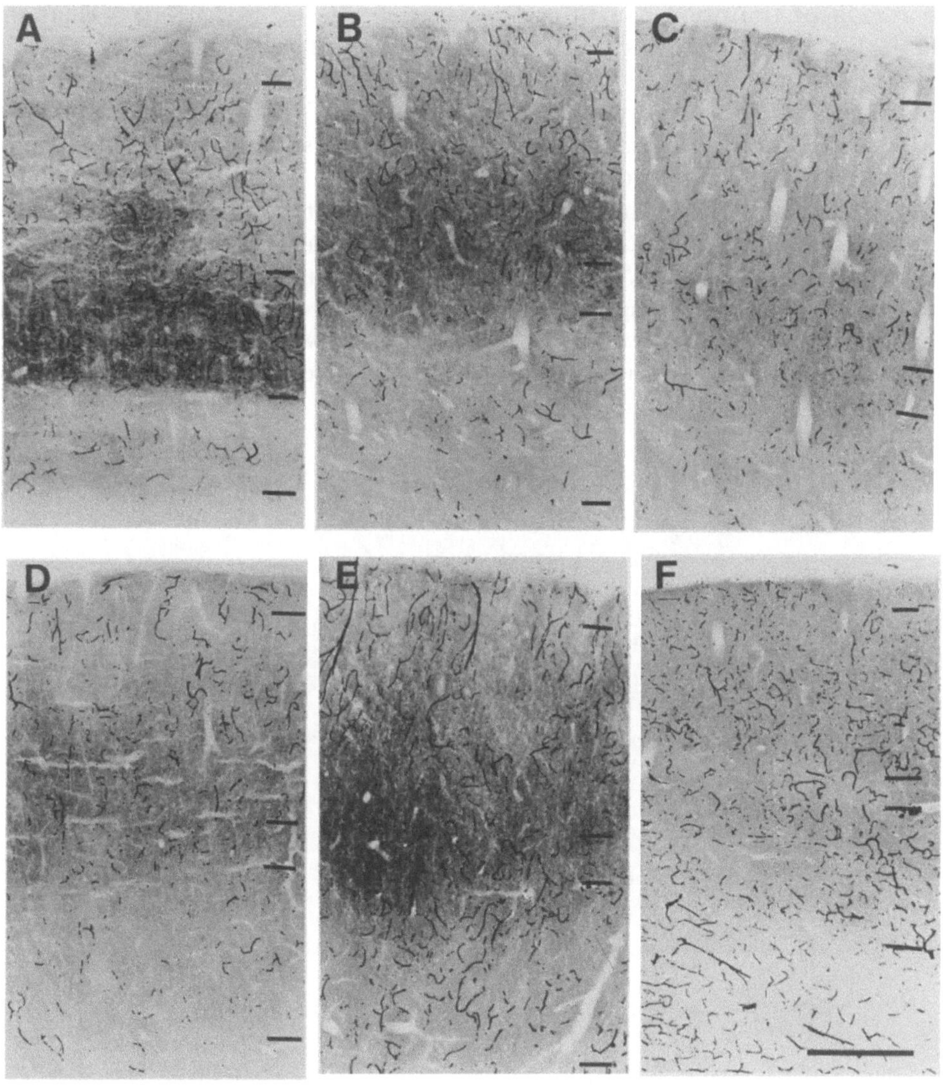

Figure 10. Photomicrographs showing cytochrome oxidase profiles in human (A) V1 (B) V2, (C) VP, (D) V4, (E) V5/MT, and (F) in a small region of the posterosuperior precuneus. Short lines indicate limits between layers I/II, III/IV, IV/V, and between layer VI and the white matter; bar = 0.5 mm. Note tangential variations of staining in V5/MT. From Clarke (1994b).

adjacent sections lead to the identification of characteristic CO patterns of several human extrastriate visual areas (Clarke, 1994b). All of the individually identified extrastriate areas displayed the one-band pattern of CO staining. In radial sections, the overall intensity of staining varied between areas: VP was very light, V2, V3, and V4 were darker, and V5/MT was very dark (Fig. 11). Furthermore, in V2, V3, V4, and V5/MT, but not VP, layer II tended to be darker than the infragranular layers. Changes in CO staining at myeloarchitectonically defined areal boundaries occurred gradually over distances ranging between 0.5 and 1.7 mm. The region of transition can be relatively large in comparison with the extent of the corresponding area (defined here as the region with unequivocal characteristics, i.e., without the region of transition). At the V2/VP boundary, the region of transition constituted 5–10% of V2 and 8–13% of VP; at the VP/V4 boundary, 7–12% of VP; and at the boundary of V5/MT, ca. 6% of V5/MT. Thus, a stripelike elongated area such as VP was surrounded by a region with an area as large as 25% of that of VP that showed transitional characteristics in CO stains. For an oval-shaped area such as V5/MT the transitional area may correspond to as much as 25%. In tangential sections of the flattened cortex, some of the areal differences in staining intensity were prominent and other characteristics, probably related to particular areas, became apparent (Tootell and Taylor, 1995). Thus, V5/MT could be identified as a dark, oval-shaped area measuring ca. 1 × 2 cm (Fig. 11). In some cases a crescent-shaped, patchy region surrounded V5/MT. Area V2 was identified as a "patchy," elongated area 1.8–3.4 cm wide. Other regions were identified anterolaterally to

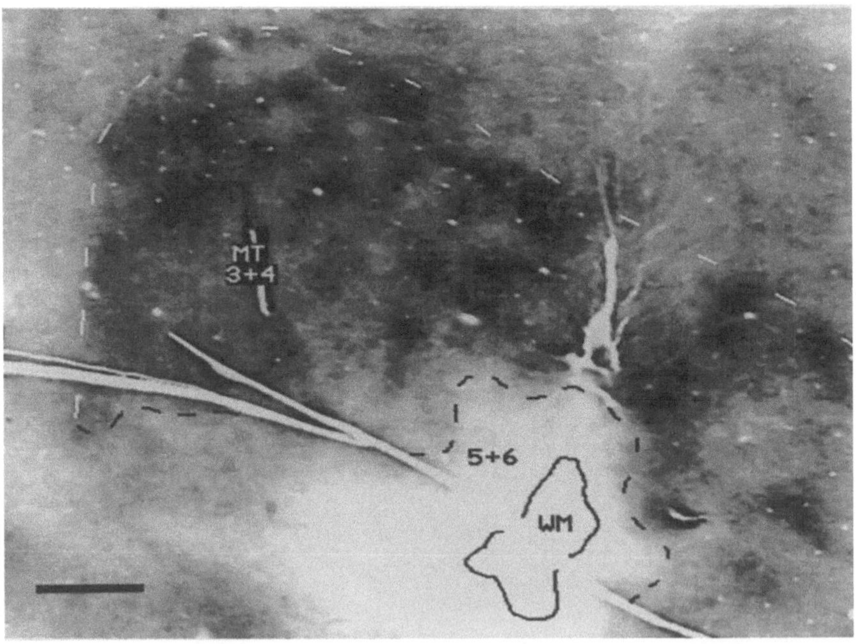

Figure 11. Cytochrome oxidase staining in tangential sections through layers V and VI of human V5. Bar = 2 mm. Note the darker staining than in the surrounding cortex (for radial profile compare Fig. 10). From Tootell and Taylor (1995).

V2. Area PX, posterior parietal area X, was a darkly staining area adjacent to upper V2; its size varied from 2.8 to 3.5 cm in the anteroposterior and from 1.5 to 2.0 cm in the lateromedial direction. Area VX, ventral area X, was a darkly stained area adjacent to lower V2. Areas PX and VX were proposed to correspond to areas V3 and V4, respectively; this correspondence could not, however, be established by other than positional criteria.

In two extrastriate areas, V2 and V5/MT, intraareal compartments were visualized with the CO stain. In V2, more or less patchy stripes were found to run perpendicularly to the V1/V2 boundary (Burkhalter and Bernardo, 1989; Wong-Riley *et al.,* 1993; Clarke, 1994b; Tootell and Taylor, 1995). The reported size of the stripes or of the patches constituting the stripes varied among studies. It is difficult to estimate whether these variations reflect genuine interindividual or regional differences or whether they result from different conditions in fixation and/or histological treatment. The dark and pale stripes (Table V). were estimated to be 1.0–2.75 and 1.5–2.75 mm wide, respectively (Burkhalter and Bernardo, 1989). Wong-Riley *et al.* (1993) described globular zones organized into stripes, the diameter of the globular zones being 580–1400 μm and center-to-center distances between stripes 1.5–2.3 mm. In yet another study, parts of dark and of pale stripes were found in the range of 0.5–0.7 mm (Clarke, 1994b). Patchy staining was found also in tangential sections through V5/MT; in normal subjects the interpatch distance was estimated at 0.7–to 1.4 mm (Fig. 11; Tootell and Taylor, 1995). A similar patchy pattern was observed in a case of bilateral age-related macular degeneration (Clarke, 1994b).

In conclusion, there are striking similarities in, but also considerable differences between, the patterns of CO staining of human and nonhuman primate extrastriate cortex. As in nonhuman primates, human V5/MT was proposed to correspond to a distinct dark area on the occipital convexity. This identification in human is supported by four lines of evidence: (i) the topology of the darkly stained region, the presumptive V5/MT, in relation to the other visual areas (Clarke, 1994b; Tootell and Taylor, 1995); (ii) its strong myelination (Clarke and Miklossy, 1990; Zilles and Schleicher 1993) shown in adjacent sections (Clarke, 1994b); (iii) the presence in the same region (of other cases) of a highly myelinated area darkly stained with the monoclonal antibody CAT-301 (Tootell and Taylor, 1995; DeYoe *et al.* 1990); and (iv) the persistence of very dark CO staining in presumptive V5/MT following a parvocellular deprivation (as present in age-related macular degeneration; Clarke, 1994b). Human area V2 was described as a region surrounding V1 with a patchy (Tootell and Taylor, 1995) or stripelike CO pattern (Wong-Riley *et al.,* 1993).

In contrast to clearly visible CO stripes in nonhuman primates, human V2 stripes appear patchy and irregular. Additional evidence for stripelike compartments in human V2 comes from two other observations. First, staining with monoclonal antibody against CAT-301, which probably labels preferentially magnocellular compartments in macaque visual system, showed a pattern of dark stripes 1–3 mm wide and 1.5–3 cm long, oriented perpendicular to the V1/V2 boundary (Table V; Hockfield *et al.,* 1990). The periodicity of the CAT-301-positive stripes varied between 4 and 8 mm. Second, parvocellular deprivation, such as present in age-related macular degeneration, was accompanied in human V2 by a complex pattern of dark (presumably magnocellular) and very light (presumably parvocellular) stripes that ran roughly perpendicular to the V1/V2 boundary; the width of the dark stripes varied between 1 and 3 mm and their periodicity was 3–5 mm (Clarke, 1994b).

6. Connectivity of Human Extrastriate Cortex

6.1. Thalamic Connections

Human thalamocortical connections are believed to be organized similarly as in nonhuman primates. Some of these connections were studied directly in human using one of the following three approaches. First, retrograde degeneration, i.e., neuronal loss, was evaluated in the different thalamic nuclei following focal hemispheric lesions (for review see Van Buren and Borke, 1972). Second, the radial distribution of thalamic terminals within the striate cortex was studied by means of the Nauta method (Mesulam, 1979; Miklossy, 1993). Third, metabolic consequences of small, discrete thalamic lesions were evaluated at the cortical level with the PET technique (e.g., for the connectivity between the posterior cingulate cortex and anterior thalamic nuclei; Clarke *et al.*, 1994). Only the first approach has yielded results relevant to the extrastriate cortex.

The human extrastriate visual cortex comprises occipital cortex outside BA 17, the posterior part of the parietal cortex, and the inferior part of the temporal cortex. These cortical parts are believed to be interconnected with two thalamic nuclei, the pulvinar and the nucleus lateralis posterior (adapted from Walker, 1938; or nucleus dorsocaudalis according to Hassler, 1959). The pulvinar and the nucleus lateralis posterior (adapted from Walker, 1938; or nucleus dorsocaudalis according to Hassler, 1959). The pulvinar and the nucleus lateralis posterior constitute the posterior and middle part of the lateral group of thalamic nuclei, pulvinar being the more posterior one (for a recent review, see, e.g., Creutzfeldt, 1995). The pulvinar is a large region which appears rather homogeneous in Nissl and myelin stains and hence is not easy to subdivide. Several types of subdivisions have been proposed, varying between one (Simma, 1954) and 11 parts (Hassler, 1959). We will adopt here the subdivision into four parts as proposed by Van Buren and Borke (1972): oralis, lateralis, medialis, and intergeniculatus, the latter corresponding to the part of the pulvinar situated between the medial and lateral geniculate nuclei.

As in nonhuman primates, there appears to be at least a rough topographical relationship between extrastriate visual cortex and thalamic nuclei in human (Fig. 12; Table VI). However, the detailed arrangements are far from clear. Some of the thalamocortical connections were convincingly documented, whereas for others contradicting reports were published by different authors (Table VI). The best evidence comes from relatively small cortical lesions which had been repeatedly shown to be associated with retrograde degeneration in the same thalamic subdivisions. This is the case for the connections from the inferior part of the nucleus pulvinaris medialis to the lingual and posterior fusiform gyri and from the nucleus pulvinaris oralis to the parietal operculum (Van Buren and Borke, 1972). Still convincing but less clear-cut evidence comes from cases with

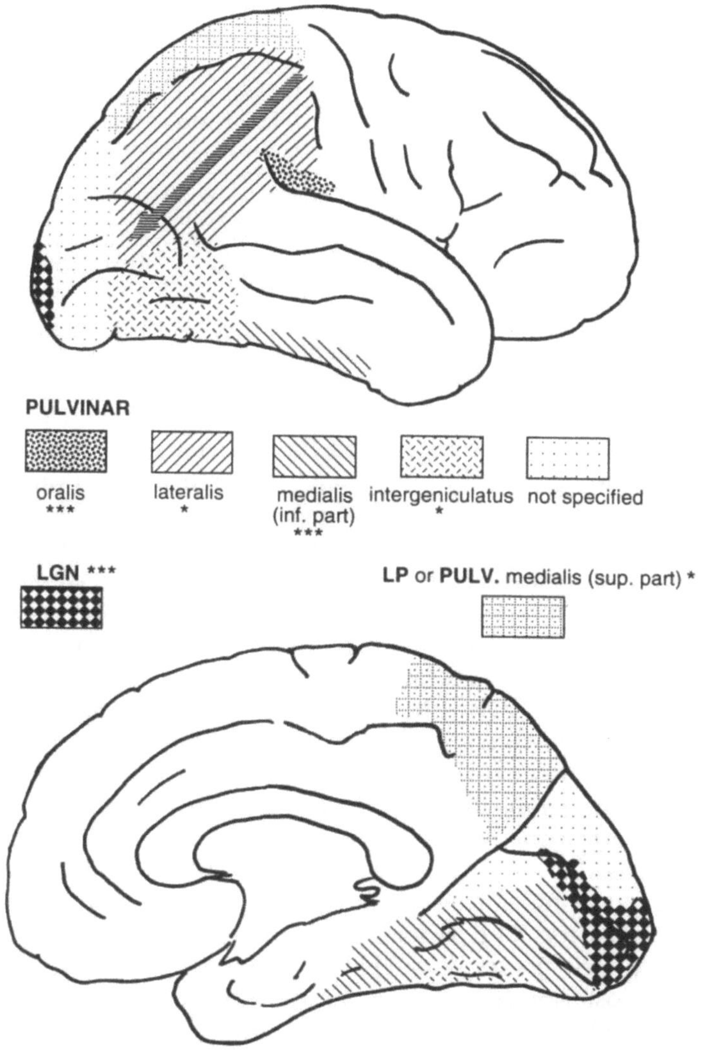

Figure 12. Topographical relationship between the human extrastriate cortex and thalamic nuclei as apparent from analysis of retrograde degeneration in thalamic nuclei following localized hemispheric lesions (for review and data see Van Buren and Borke, 1972). For details see text. The arrow on the parietal convexity marks the cortical projection of the anterior-to-posterior axis of the nucleus pulvinaris lateralis. The certainty of evidence is indicated with asterisks: ***solid evidence from cases with small, circumscribed lesions; **evidence from large lesions without reported contradictions in different studies; *evidence from large lesions with reported contradictions in different studies. Note that there seems to be a relatively simple mapping rule between the pulvinar and the cortex which it innervates: the nucleus oralis, which is the most anterior part of the pulvinar, is connected with the most anterior part of the cortical territory, nucleus lateralis with lateral part, nucleus medialis with medial parts, and nucleus intergeniculatus with posterolateral part.

Table VI. Thalamic Connections of Human Extrastriate Areas[a]

Visual area	Thalamic nucleus	Subdivision
V2	Pulvinar**	Unknown
dV3	Pulvinar**	Unknown
VP	Pulvinar***	Inferior part of pulvinaris medialis***
V4	Pulvinar***	Inferior part of pulcinaris medialis***
V5/MT	Pulvinar**	Pulvinaris lateralis* or intergeniculatus*

[a]Based on comparison of data on thalamocortical connections as presented in Fig. 11 and of the localization of human extrastriate areas. The certainty of evidence is indicated with asterisks: ***Solid evidence from cases with small, circumscribed lesions; **evidence from large lesions without reported contradictions in different studies; *evidence from large lesions with reported contradictions in different studies. For details see text.

large lesions, in particular if these were deep in the white matter. This type of lesion seemed to involve often the parietal and occipital convexity, which led to ambiguous results concerning its thalamic connections. The nucleus pulvinaris lateralis was proposed to be connected to the parietal cortex below the intraparietal sulcus including the supramarginal and angular gyri (Van Buren and Borke, 1972); other authors included also the lateral occipital cortex in the cortical territory of this nucleus (Clark and Russell, 1939). Analysis of cases with lesions placed more or less anteriorly on the parietooccipital convexity suggested the existence of a finer topographical organization; the anterior part of the nucleus pulvinaris lateralis was interconnected with the part of parietal convexity immediately behind the postcentral gyrus, the posterior part of the nucleus pulvinaris lateralis (Table VI) with the region at the occipitoparietotemporal junction (Van Buren and Borke, 1972). The least convincing evidence comes from subtractions of effects of large lesions. This led to the assumption that nucleus pulvinaris intergeniculatus is connected to the cortex at the lateral and possibly inferior part of the occipitotemporal junction (Van Buren and Borke, 1972). No unequivocal evidence has been shown for the thalamic connections of large parts of the occipital cortex (dotted in Fig. 12) and of the superomedial part of the parietal cortex. The parietal cortex above the intraparietal sulcus and in the interhemispheric fissure is widely believed to be interconnected with the nucleus lateralis posterior (e.g., Creutzfeldt, 1995), whereas Van Buren and Borke (1972) showed indirect evidence (subtraction of the effects of two large lesions) that it may be connected to the superior part of the nucleus pulvinaris medialis (Table VI). Finally, the lateral occipital cortex and the cuneus are most likely interconnected with the pulvinar, but it remains uncertain with which parts of it. As mentioned above, connections from the nucleus pulvinaris lateralis (Table VI) to the lateral occipital cortex have been proposed (Clark and Russell, 1939), but were dismissed by Van Buren and Borke (1972) because of uncertainties due to the very large lesions involved.

Thalamic connectivity of distinct areas in human can be extrapolated from a comparison of data summarized above (and Fig. 12) and the position of putative extrastriate visual areas (Fig. 5J). Human areas VP and V4 appear to be interconnected with the inferior part of pulvinaris medialis, and V5/MT with pulvinaris lateralis or intergeniculatus (Table VI). This type of connectivity bears some resemblence with that of macaque visual areas (Table IV). However, it is difficult to establish whether differences in connectivity reflect species differences or whether they are due to technical limits of human studies.

6.2. Interhemispheric Connections

Human visual callosal connections are relatively well understood as compared to those of other systems. Most of this understanding has been extrapolated from work on nonhuman primates, with only a limited amount of work done actually in human.

Human extrastriate cortex is believed to participate in interhemispheric connections. Anatomical evidence as to the *origin of these connections* comes from the analysis of brains with small lesions in which the presence of degenerating fibers within the corpus callosum (Van Valkenburg, 1908, 1913; Clarke and Miklossy, 1990) or in the (contralateral) target territories (Clarke and Miklossy, 1990) was demonstrated. Thus, the inferomedial part of the occipital cortex (Van Valkenburg, 1908, 1913; Clarke and Miklossy, 1990), the superolateral part of BA 19 (Clarke and Miklossy, 1990), and the posteroinferior part of the temporal cortex (Di Virgilio and Clarke, 1996) were shown to give rise to interhemispheric connections. The same is presumably true for the posterior part of the human parietal cortex, but there is not yet unequivocal evidence. In the published cases, the parietal lesions involved always the underlying white matter and interrupted, among other pathways, the tapetum, which had been shown to convey callosal fibers from the temporal lobe.

Very little is known about the neurons of origin of the visual callosal connections. Some of the visual callosal axons seem to belong to the large pyramidal neurons within OBγ of Economo and Koskinas (1925). These neurons have been reported lacking in acallosal patients (Shomura *et al.*, 1975). In the rhesus monkey, the equivalent pyramidal neurons disappear after section of the corpus callosum (Glickstein and Whitteridge, 1976) and have indeed been shown to project through the corpus callosum (Lund *et al.*, 1975). It is unknown whether other cortical layers and other neuronal types contribute also to callosal connections in human.

6.2.1. Pathways in the White Matter

Callosal axons form several distinct fascicles within the white matter of the occipital lobe (Van Valkenburg, 1913; Ludwig and Klinger, 1938; Clarke and Miklossy, 1990). Some callosal fibers travel in the forceps major (i.e., medially to the posterior horn of the lateral ventricle) and terminate mostly around the BA 17/18 boundary; others go through the tapetum (i.e., above and laterally to the posterior horn of the lateral ventricle) and enter mainly the cortex of the fusiform gyrus, although a few seem to pass in a separate bundle under the ventricle to reach the lower BA 17/18 boundary. Callosal fibers from the inferoposterior part of the temporal lobe are believed to join the tapetum and thus pass laterally and above the lateral ventricle, whereas those from the posterior parietal cortex travel dispersed through the parietal white matter toward the corpus callosum (Ludwig and Klinger, 1938). It is unknown to what extent human callosal fibers originating in a limited part of the cortex travel together in tight bundles in the white matter and to what extent then spread over large parts of the callosal pathway. In cat temporal cortex, callosal fibers originating in discrete parts of cortex first travel together in a tight bundle up to the ipsilateral lateral ventricle, then spread over large parts of the callosal pathway and intermingle with other callosal axons (Clarke *et al.*, 1995b). Although never actually shown in human, similar behavior of callosal axons from extrastriate visual cortex may explain why partial callosal lesions have to be relatively large to cause detectable neuropsychological deficits (e.g., Trevarthen, 1990).

6.2.2. Topographical Organization of Visual Fibers in the Human Corpus Callosum

Callosal axons originating or terminating in the human extrastriate cortex are believed to cross the midsagittal plane within the posterior half of the corpus callosum. Axons from the inferior part of the temporal cortex and from parts of the occipital convexity were also shown to cross in the anterior commissure (Nauta method; Di Virgilio and Clarke, 1996). It is generally assumed that there is a topographical arrangement of visual axons within the corpus callosum according to their origin (Fig. 13). However, the actual anatomical evidence is

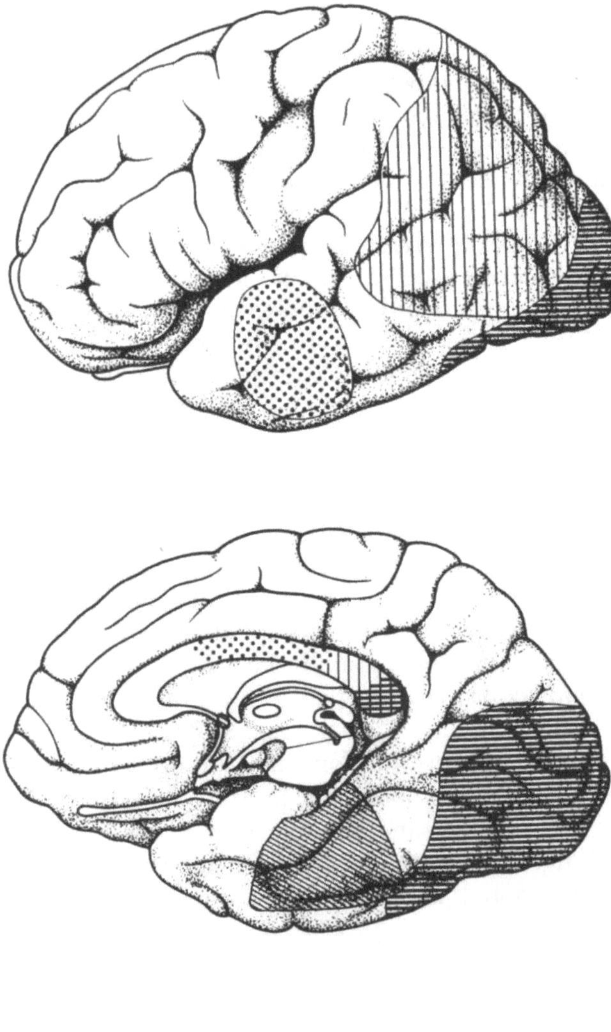

Figure 13. The visual pathway in the human corpus callosum and anterior commissure. Schematic representation of a lateral view of the left hemisphere; outline of the corpus callosum and anterior commissure underneath. Shading is used to indicated the passage of fibers in the corpus callosum; e.g., occipital cortex, lower part of the posterior splenium; posterior part of the parietal and temporal cortex, remaining part of the splenium, and posterior part of the body (references as in text). Compiled from the analysis of the corpus callosum (De Lacoste *et al.*, 1985) and the anterior commissure (Di Virgilio and Clarke, 1996) following circumscribed lesions. Note that the topography within the human telencephalic commissures may be much more complex than shown here (for the topography within the cat corpus callosum see, e.g., Payne and Siwek, 1991; Clarke *et al.*, 1995b).

limited, due to sensitivity problems of the applicable tracing techniques. Callosal axons from the occipital cortex cross in the splenium, most of them in its lower part; the degeneration of this pathway has been shown after small lesions with the Weigert–Pal stain (Dejerine and Dejerine-Klumpke, 1895; Van Valkenburg, 1908) and with the Nauta method (Clarke and Miklossy, 1990). Axons from the parietal and posterior temporal cortex cross in the splenium and possibly also in the posterior part of the body of the corpus callosum; signs of Wallerian degeneration (myelin pallor, loss of fibers, presence of phagocytes) were observed there after lesions in the corresponding parts of the hemispheres (de Lacoste *et al.*, 1985).

6.2.3. Distribution of Callosal Afferents—Visual Areas

The intracortical distribution of visual callosal afferents was studied in human using a modification of the Nauta technique for anterogradely degenerating axons (Clarke and Miklossy, 1990; Clarke *et al.*, 1995a). Human brains with unilateral occipital or occipitotemporal infarctions were silver-impregnated for degenerating axons, thereby revealing callosal afferents from the infarction to the intact, contralateral cortex.

Within the visual and visually related cortex the tangential distribution of callosal afferents is discontinuous; some regions receive callosal afferents, whereas others receive none or only very few. Within parts of the occipital and posteroinferior temporal cortex the callosal afferents form a bandlike pattern. A band of callosal afferents is found along the BA 17/18 boundary; callosal afferents terminate both in the adjacent parts of BA 17 and BA 18 (Fig. 14). The remaining part of BA 17 is almost devoid of callosal afferents. A 15- to 45-mm-wide stripe of BA 18 adjacent to the callosal band along the 17/18 boundary receives no or only very few callosal afferents; it is limited laterally by a region of callosal afferents (Fig. 15). On the flattened reconstructions of the posterior one-third of the human hemisphere, the bandlike distribution of callosal afferents is also apparent on the lateral part of the occipital lobe, the posterior part of the parietal cortex, and the posteroinferior part of the temporal cortex (Fig. 15). Some bands of callosal afferents are found in the vicinity of sulci, but there was no consistent relationship between sulci and callosal bands. Thus callosal bands were seen to cross both banks of a sulcus; furthermore, a callosal band did not always have the same relationship with a given sulcus in the hemispheres of different individuals (S. Clarke, unpublished observations).

The precise radial distribution of callosal afferents remains unknown in human. With the Nauta method, degenerating callosal axons are found mainly in the deep layers, and are scarce in the granular and supragranular layers both in the occipital (Clarke and Miklossy, 1990) and temporal and parietal cortices (Clarke *et al.*, 1995a). This bias is probably due to the limitations of the technique; in animal material, the Nauta technique has shown well the tangential distribution of callosal afferents, but only incompletely the radial extent of their terminal arborizations (Heimer *et al.*, 1967).

6.2.4. Topographical Organization of Visual Callosal Connections

Within the occipital cortex, there is a topographical relationship between the region of origin of callosal projections and the tangential extent of their

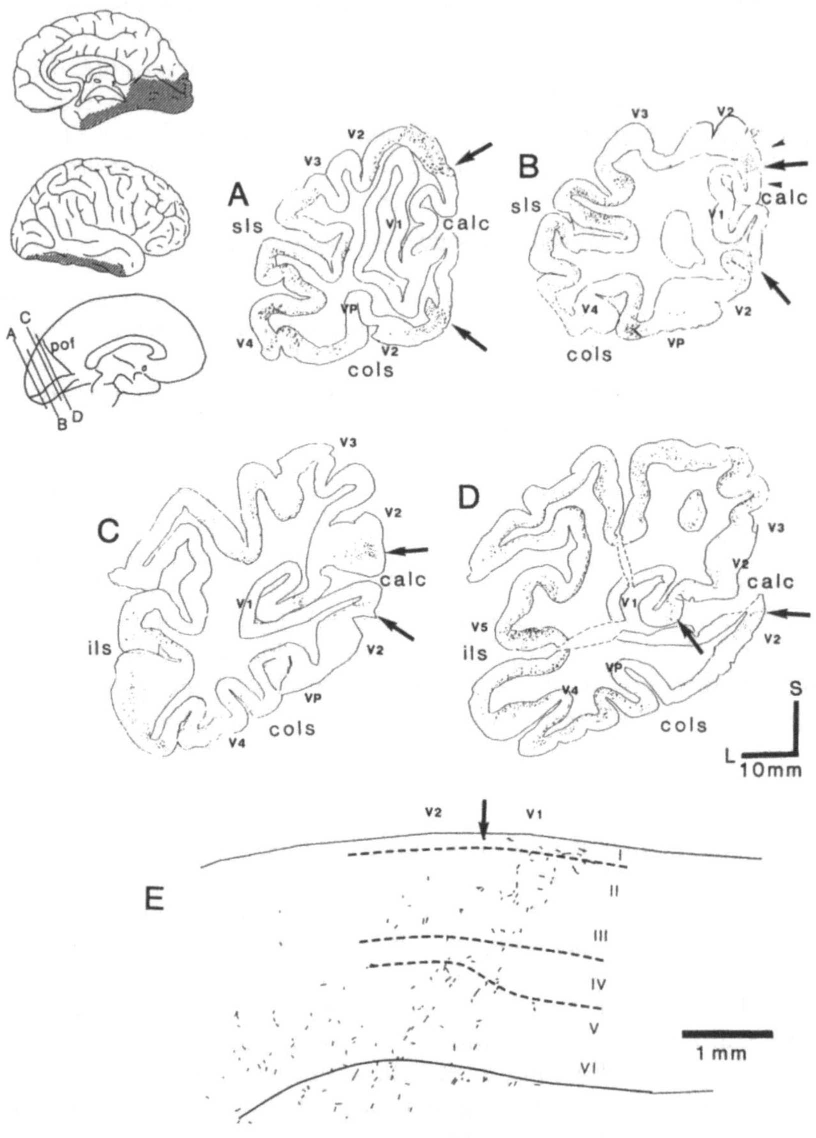

Figure 14. Distribution of visual callosal afferents in the human occipital cortex. Four coronal sections from the hemisphere contralateral to a large occipitotemporal infarction (hatched in inset) are represented by the section outline and the limit cortex/white matter. Dots indicate anterogradely degenerating fibers (as revealed by a modification of the Nauta method) which correspond to callosal afferents. Arrows point the the 17/18 boundary. Approximate positions of visual areas are indicated. Calc, calcarine fissure; cols, collateral sulcus; sls, superior lateral (occipital) sulcus; ils, inferior lateral (occipital) sulcus; ss, superior (occipital) sulcus. Note that the tangential distribution is discontinuous (A–D) and that callosal afferents are present in V1 within 1 mm of the V1/V2 boundary. Adapted from Clarke and Miklossy (1990).

terminations (Clarke and Miklossy, 1990). This becomes apparent when the output from small parts of the cortex is studied. A case with a small lesion in the upper part of BA 19 limited to about 1 cm^2 of the cortex was analyzed for the callosal and intrahemispheric (see Section 6.4.3) connections. For reasons summarized in the legend of Fig. 18, the lesion was most likely in a retinotopically organized visual area lying beyond area V3A, and more precisely in the lower quadrant representation within this area. Most of the callosal connections originating in the damaged cortex terminated in the symmetrical region of the opposite hemisphere, and only a few in other parts. Among the latter were regions that are known to contain the same retinotopic representation, such as the upper V1/V2 and the V3/V3A boundaries. No indication as to the topography of the callosal conenctions originating and terminating within the occipital cortex was found in cases with larger lesions. In particular, there seems to be no strict mediolateral relationship, i.e., the medial part of the occipital cortex does not project exclusively to the medial part of the opposite occipital cortex, nor the lateral to the lateral. There is not even a clear-cut difference between the distribution of callosal afferents in the posterior half of the occipital lobe after contralateral interruption of the forceps major alone or of the forceps major and the tapetum.

6.2.5. Homotopic and Heterotopic Visual Callosal Connections

It is often assumed that human callosal connections are homotopic and heterotopic, i.e., link symmetrical and nonsymmetrical regions in the two hemispheres, respectively. Data from visual callosal connections confirm this and give, for the first time, an idea about the range of heterotopic connections. As mentioned above, there are heterotopic callosal connections from the medial to the lateral part of the occipital cortex (Clarke and Miklossy, 1990). Further analysis has shown that V1 and V2 send callosal connections not only to the whole occipital cortex, but also to the posterior part of the temporal and parietal cortices, including regions proposed to be involved in specifically human functions such as the angular gyrus (Clarke *et al.*, 1995a; Fig. 16).

6.3. Visual Areas as Derived from the Pattern of Callosal Connections

In macaque monkeys, several boundaries between visual areas (defined as individual representations of the contralateral visual hemifield) are characterized by bands of callosal afferents and/or by changes in myeloarchitecture (for review see Knierim and Van Essen, 1992). Callosal afferents to the human visual cortex have been found to be distributed in stripes alternating with callosal-poor regions (see above; Clarke and Miklossy, 1990; Clarke *et al.*, 1995a). On this basis several boundaries between visual areas were determined in human (Figs. 5H–J, 14, and 17). Three of the visual areas identified in this way, V2, VP, and V4, differ in their cyto- and myeloarchitecture and can thus be localized even in brains where the callosal connections cannot be traced. Another area,

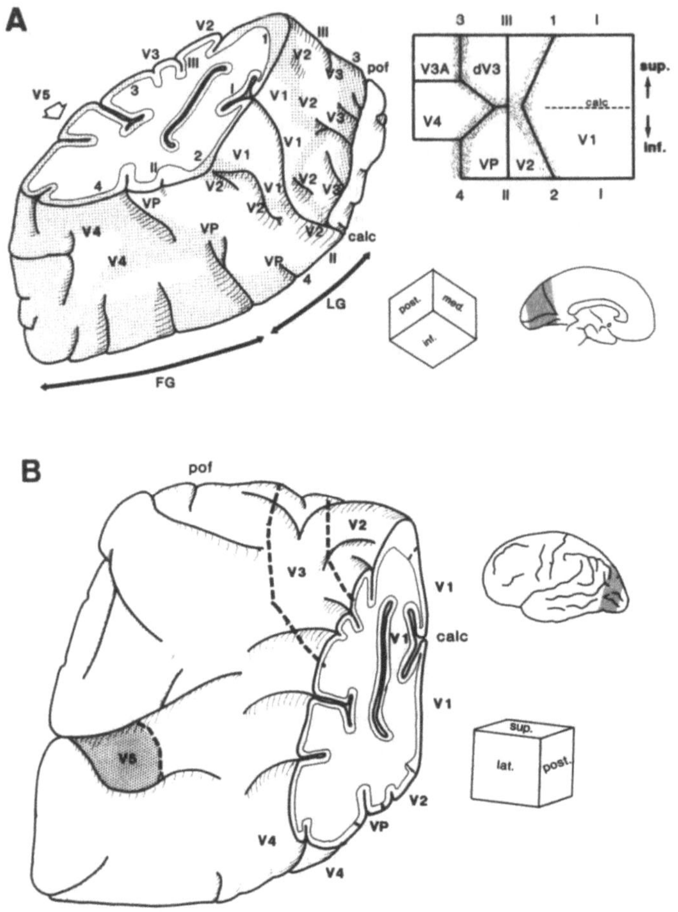

Figure 15. (A) Distribution of callosal afferents (stippled) and the resulting subdivision into visual areas as seen in a inferomedial view of a left occipital lobe (hatched in brain inset, orientation as shown on cube whose side = 10 mm). (B) The heavily myelinated region on the occipital convexity, corresponding to area V5/MT as seen on a lateral view of a left occipital lobe, (C) Distribution of callosal afferents (high density in dark stippling, low density in light) on a flattened reconstruction of the posterior one-third of the left hemisphere. Flattening paradigm developed by H. Drury and D. Van Essen. (A, B) Adapted from Clarke and Miklossy (1990); (C) from unpublished work by S. Clarke, N. Hadhikhani, D. Van Essen, and H. Drury.

V5/MT, can be identified because of its position in relation to other visual areas and because of its very heavy myelination.

6.4. Intrahemispheric Connections

6.4.1. Associative Pathways in the Hemispheric White Matter

The existence of large and distinct interhemispheric corticocortical bundles, i.e., the corpus callosum and the anterior commissure, is generally accepted and the full length of the implicated axons has been deduced from tracing studies even in human (see above). The situation is much less clear for intrahemispheric bundles. In 1895 Dejerine and Dejerine Klumpke summarized in their seminal book what remains the basis for current neurological models. They described five long association fascicles, identified by the dissection of the white matter and by microscopic observation of myelin-stained sections of normal and pathological human brains. Two of these fascicles are related to what we call today the extrastriate visual cortex. The *superior longitudinal fasciculus,* or *arcuate*

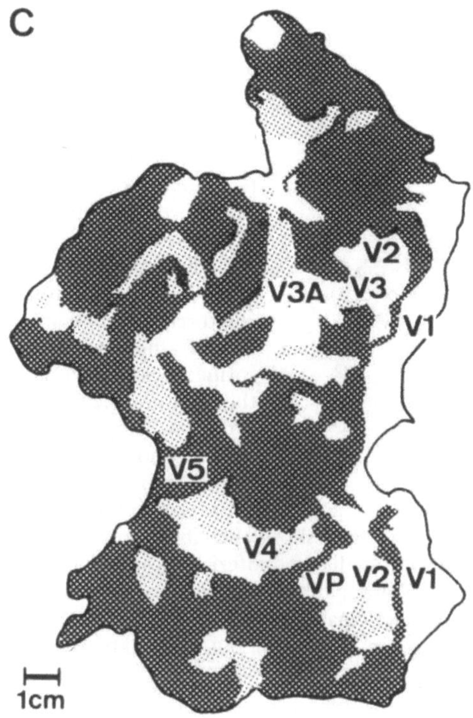

Figure 15. *(Continued).*

fasciculus, stretches between parts of the temporal and parietal cortices and the frontal convexity. When dissected, the arcuate fasciculus does not appear as a compact bundle except during its trajectory in the parietal operculum. It can be easily identified in histological sections, but its exact limits remain unclear. Dejerine and Dejerine Klumpke believed that fibers within the arcuate fasciculus did not run through its whole length, but that the fasciculus was composed of short association fibers. In myelin-stained sections from cases with lesions in appropriate regions, the degenerated axonal pathways could not be followed through the whole length of the fasciculus, but seemed to terminate in nearby gyri. Knowing, however, the limitations of degeneration studies based on myelin stains, we cannot exclude the existence of a small number of very long fibers. The *inferior longitudinal fasciculus* stretches between the occipital and temporal poles. Dejerine and Dejerine Klumpke (1895) stated that it contained fibers of different length. Tusa and Ungerleider (1985) reexamined the inferior longitudinal fasciculus in macaque and in human and stressed that, certainly in the first species, the inferior longitudinal fasciculus consisted of a series of short fibers.

6.4.2. Origin of Intrahemispheric Connections

Very little is known about the origin of intrahemispheric connections in human, but it is presumed to be similar to that in macaque monkeys. In particular, there are no data on the layers of origin and the neuronal types involved. Thus, in human, hierarchical organization of extrastirate visual areas cannot yet

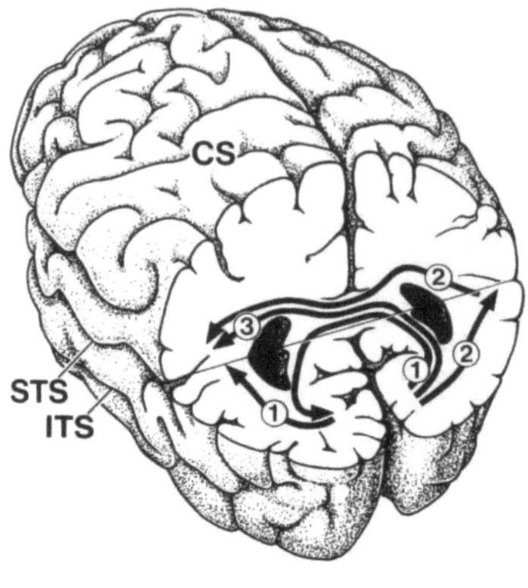

Figure 16. Callosal connections putatively involved in reading. One set of visual callosal connections underlying specifically human functions are those involved in reading. Dejerine (1892) reported a case of pure alexia following lesion in the left calcarine region and the splenium of the corpus callosum (Dejerine and Vialet, 1893). He interpreted this case as a disconnection of the right, noninjured calcarine region from the left angular gyrus, which he proposed to be involved in reading. There are three possible pathways by which visual information from the right calcarine region could reach the left angular gyrus: (1) A combination of a homotopic connection between the calcarine regions and association connection from the left calcarine region to the left angular gyrus; (2) a heterotopic connection from the right calcarine region to the left angular gyrus; and (3) a combination of an association connection from the right calcarine region to the right angular gyrus and a homotopic connection between the angular gyri. Neuropsychological evidence shows that ability to read does not rely critically on the first route; lesions limited to the left medial and inferior occipital lobe alone are not associated with reading deficits; a lesion of the callosal pathways needs to be associated (Binder and Mohr, 1992). The second route, i.e., a monosynaptic pathway from the right calcarine region to the left angular gyrus, does exist in human (Clarke *et al.*, 1995a), but it is unknown whether this route alone can sustain reading.

be deduced from the laminar pattern of connections (for macaque monkey see Felleman and Van Essen, 1991).

6.4.3. Tangential Extent of Intrahemispheric Connections

It is unknown which are the longest intrahemispheric corticocortical connections. It has long been supposed that occipitofrontal fibers may be the longest, but there has been no unequivocal evidence for their existence in human (see above). A study with the Nauta method showed that association connections can be longer than U-fibers (5 cm or longer; Clarke, 1994a). Association connections are most likely specific, since a small part of the extrastriate occipital cortex was shown to send connections of different density to different parts of the surrounding cortex (see Section 6.4.4; Clarke, 1994a).

6.4.4. Connectivity Between Visual Areas

Figure 17 summarizes anatomical evidence on connectivity between identified visual areas in human. A reciprocal connection was demonstrated between V1 and V2 (Burkhalter and Bernardo, 1989). The input from V1 to V2 was shown to terminate in layers III and IV, and the return projection in upper and lower layers. This laminar pattern of termination is compatible with a hierarchical relationship between V1 and V2, in which V2 represents a higher processing

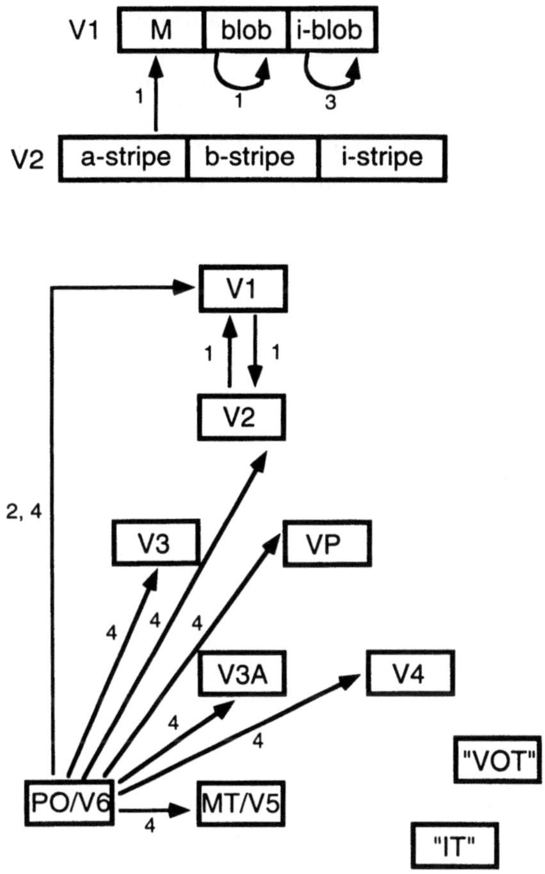

Figure 17. Identified visual areas, demonstrated connectivity and putative hierarchical organization in human. Macaque visual areas have been proposed to form a hierarchical model with a certain degree of parallel processing (Felleman and Van Essen, 1991). We do not know to what extent such a model is adapted for human visual areas. Certain characteristics of the macaque model have been found in human. In particular, there are selective connections between compartments within V1 and between V1 and V2 (Burkhalter and Bernardo, 1989; Burkhalter *et al.*, 1993) and there is a wide range of backprojections from higher order visual area (Clarke and Miklossy, 1990; Clarke, 1994a).

stage than V1. The backprojection from V2 to V1 may respect internal compartments; projections originating in selected parts of human V2 terminate in layer IVB in V1, whereas those originating in other parts do not. This is reminiscent of the magnocellular stream as described in macaque monkeys.

The connectivity of the superolateral part of BA 19 has been studied with the Nauta method for anterogradely degenerating axons (Clarke, 1994a). Circumstantial evidence suggested that the lesion lay anterolateral to visual areas V3 and V3A (Fig. 18), probably in the human equivalent of PO (Colby *et al.*, 1988) or V6 (Zeki, 1986). The infarcted part of cortex formed dense intrahemispheric connections in V1, V2, and V3 (and probably in a part of V4) and weaker connections in VP and V5/MT. Connections to V1 and V2 were mainly directed to the representation of the lower paramedian part of the visual field.

6.5. Intrinsic Connections

Intrinsic connections of the human extrastriate cortex were shown to be patchy. A small injection of the carbocyanine dye DiI into superficial V2 yielded patchy arrangement of anterograde labeling (Burkhalter and Bernardo, 1989). Terminal clusters, visible in layers II–V, were 0.3–0.5 mm wide and closely spaced (Fig. 19). The long intrinsic connections shown with the Nauta method in a case with a small lesion in the superolateral part of BA 19 had a columnar arrangement (Clarke, 1994a; Fig. 20). Thus, patchy arrangement of long intrinsic connections seems to be a general organizational principle of human extrastriate visual areas and perhaps more generally of the human cerebral cortex.

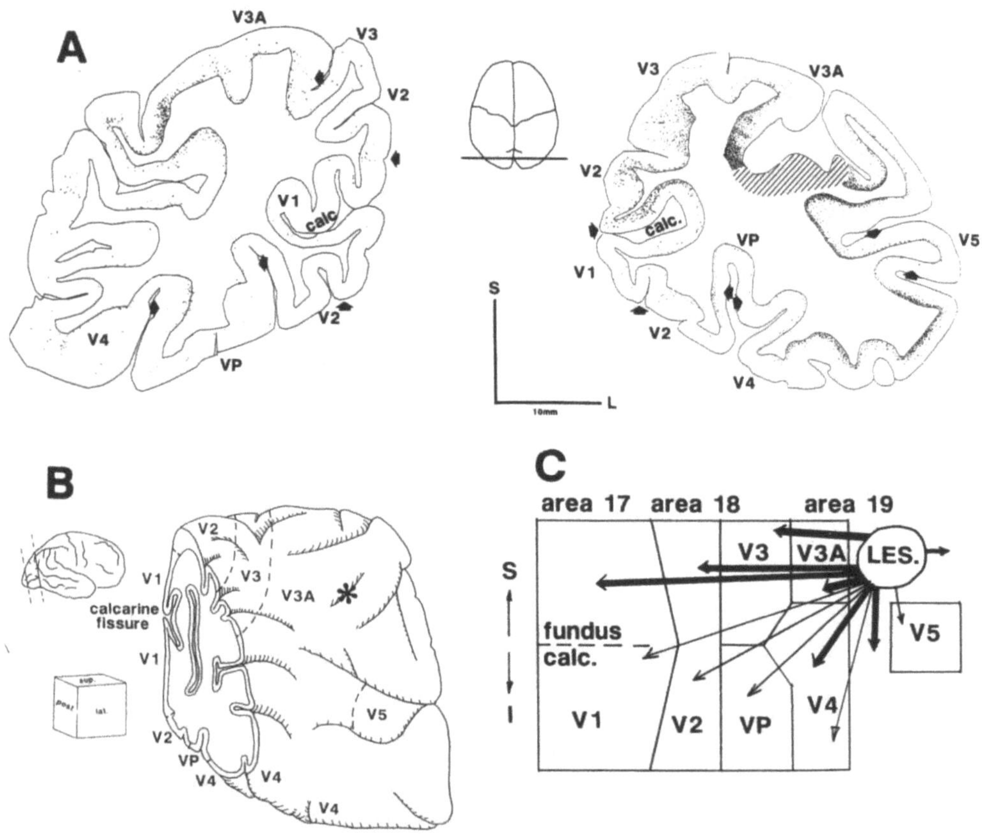

Figure 18. Human inter- and intrahemispheric connections of a small region in the putative homologue of macaque V6/PO, as demonstrated with a modification of the Nauta method. A small lesion lay in the superolateral part of area 19. In this part of the human occipital cortex, architectonic criteria alone are not sufficient for the identification of visual areas (Clarke, 1993). Indirect evidence suggests, however, that the lesion lay anterolateral to human areas V3 and V3A: (i) a comparison, based on macroscopic features of occipital lobes, of the site of lesion and the approximate positions of visual areas indicates that the lesion was likely to be anterior and lateral to V3 (C); and (ii) the lesion was in a part of area 19 giving rise to callosal projections, the distribution of which is partially shown in panel A. Callosal afferents were found in a band at the V1/V2 boundary and in another band lying more laterally at the V3/V3A boundary. The region with the densest callosal afferents was ca. 24 mm lateral to the lateral boundary of dorsal V3. Callosal afferents are likely to be strongest between homotopic parts of the visual cortex, which suggests that the infarcted region was lateral and anterior to V3A. An interesting possibility is that the present lesion lay within the human equivalent of macaque PO (Colby *et al.*, 1988) or V6 (Zeki, 1986).

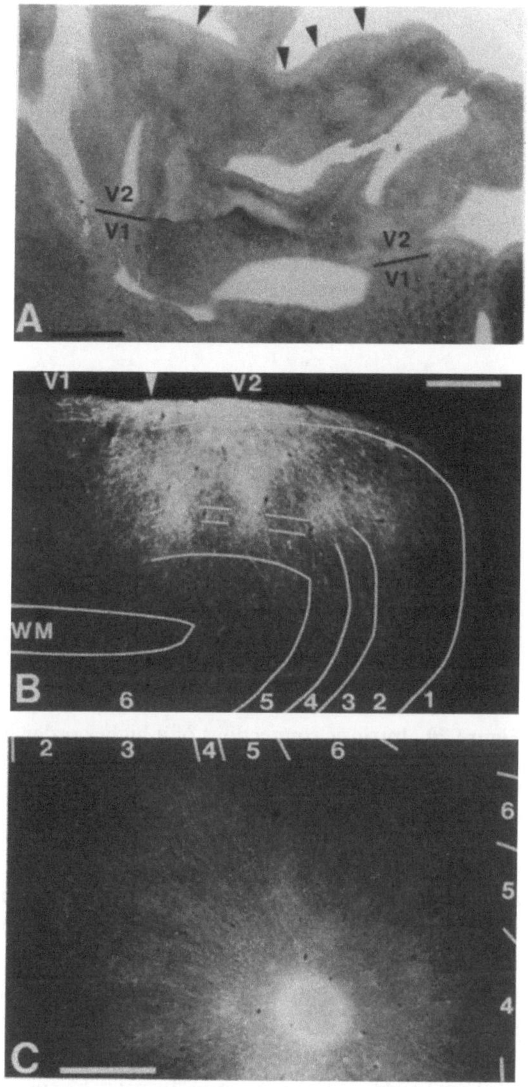

Figure 19. (A) Cytochrome oxidase stripes and (B, C) patchy distribution of intrinsic connections within human V2 (Fig. 5B from Burkhalter and Bernardo, 1989).

7. Transmitter Receptors in Extrastriate Areas of the Human Brain

Numerous observations on the regional and laminar distribution of transmitter receptors in the *human primary visual cortex* have been published (Burnet *et al.*, 1995; Cortés *et al.*, 1986, 1989; Dietl *et al.*, 1987; Hendry *et al.*, 1994; Jansen *et al.*, 1989; Pazos *et al.*, 1987a, b; Richards *et al.*, 1987; Young and Kuhar, 1979; Young and Penney, 1991; Zilles, 1991, 1992; Zilles and Schleicher, 1993, 1995; Zilles *et al.*, 1990, 1991, 1993). In contrast, only a few reports about receptors in V2 (Burnet *et al.*, 1995; Cortés *et al.*, 1989; Dietl *et al.*, 1987; Jansen *et al.*, 1989; Zilles and Schleicher, 1995; Zilles *et al.*, 1991, 1993) and no data for other

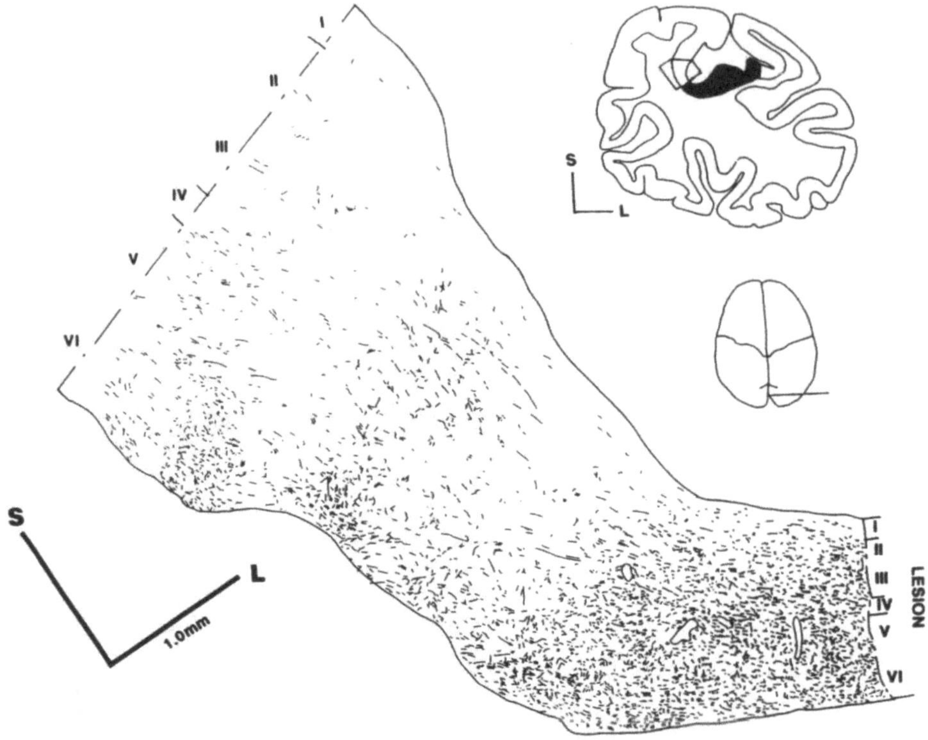

Figure 20. Intrinsic connections of a higher order human visual area (putative homologue of macaque V6/PO) as revealed with a modification of the Nauta method. The position of the charted lesion in relation to the lesion is indicated in the inset. Note the columnar arrangement of intrinsic connections. From Clarke (1994a).

anatomically or functionally defined extrastriate areas are available. This is probably due to the difficulty in identifying the precise localization and extent of extrastriate areas in Nissl-stained sections, which is the basis for the determination of the anatomical identity of cortical areas in receptor autoradiographic studies (Zilles and Schleicher, 1995). Moreover, receptor autoradiography in the human extrastriate cortex requires oversized cryostat sections through (preferably) unfixed, deeply frozen hemispheres, which is a difficult task. Despite these problems, the comparative analysis of receptor distribution patterns in macaque (where the parcellation of extrastriate areas has progressed much more than in human brains) and human visual cortices is a promising approach, because receptors are efficient markers for establishing neurochemical homologies between macaque and human extrastriate areas (see V5/MT below). Therefore, one of the aims of this section is the demonstration of the comparability of receptor distribution patterns in some extrastriate regions between macaque and human brains.

7.1. Receptors in V2

An abrupt change in laminar pattern is visible for most receptors at the V1/V2 border as already demonstrated in nonhuman primates (see Section 4.1).

Table VII. Relative Densities of Transmitter Receptors and Their Laminar Distributions in Human Areas V1 and V2[a]

	Area V1							Area V2					
Receptor[b]	I	II	III-IVA	IVB	IVC	V	VI	I	II	III	IV	V	VI
NMDA	−	++	+	−	+	−	−	−	++	++	−	−	−
AMPA	+	++	+	−	−	+	−	+	++	++	−	−	−
Kainate	+	++	−	−	−	++	++	−	++	+	+	++	++
$GABA_A$	−	++	+	−	++	−	−	−	++	++	++	−	−
Muscarinic binding sites	+	++	++	−	+	+	−	+	++	++	++	−	−
Muscarinic M1	−	++	++	−	+	+	−	−	++	++	++	−	−
Muscarinic M2	−	++	++	−	++	+	−	−	++	++	++	−	+
Nicotinic	−	−	−	−	+	++	++	+	+	−	−	++	++
$5\text{-}HT_{1A}$	++	++	+	−	−	−	−	++	++	+	−	−	−
$5\text{-}HT_2$	−	+	++	−	++	−	−	−	+	++	+	++	−
α_1	+	++	++	−	−	+	−	+	++	+	−	+	−
α_2	−	+	++	−	−	+	−	−	+	−	+	−	−

[a]K. Zilles (unpublished observations).
[b]Labeling of receptors: NMDA, [^{3}H]dizocilpine; AMPA, [^{3}H]AMMPA; kainate, [^{3}H]kainate; $GABA_A$, [^{3}H]muscimol; muscarinic binding sites, [^{3}H]*N*-methyl scopolamine; muscarinic M1, [^{3}H]pirenzepine; muscarinic M2, [^{3}H]oxotremorine-M; nicotinic, [^{3}H]cystisine; $5\text{-}HT_{1A}$, [^{3}H]8-OH-DPAT; $5\text{-}HT_2$, [^{3}H]ketanserin; α_1, [^{3}H]prazosin; α_2, [^{3}H]rauwolscine. Coding of relative densities: −, low density; +, intermediate density; ++, high density.

The laminar patterns of *[^{3}H]glutamate*-binding sites and *NMDA* receptors (labeled by [^{3}H]TCP and [^{3}H]glycine binding) show this border by a heavy labeling of layer IVC in V1, which is not visible in V2 (Jansen *et al.*, 1989). However, Jansen *et al.* (1989) reported that *AMPA* and *kainate* receptors did not reveal the V1/V2 border. In contrast to this finding, our own observations have shown that the laminar patterns and densities of AMPA (K. Zilles, unpublished observations) and kainate receptors (Fig. 8E) also change abruptly at the V1/V2 border. This observation was possible in contrast-enhanced autoradiographs. Layer IVB is prominent by itsvery low AMPA and kainate receptor density (Table VII). This low-density band stops abruptly at the V1/V2 border. Layer IVB is visible in Fig. 8E only in those parts of V1 where layer IVB reaches its largest width, and can thus be recognized at the low spatial resolution of Fig. 8E. At higher resolution, layer IVB is clearly recognizable in these ligand-binding studies (as in macaques, Fig. 8A, B) over the whole extent of human V1. Additionally, the mean density of kainate receptor is higher in the ventral part of V2 (V2v) than in V1. In contrast to the higher density in V2v, the dorsal part of V2 (V2d) does not show a clear difference in the mean receptor density from V1 (Fig. 8E). This low receptor density in layer IVB is also displayed by NMDA as well as nine other receptors (Table VII).

$GABA_A$ receptors show completely different laminar distributions between V1 and V2 (Zilles and Schleicher, 1995). Two maxima of receptor densities are visible in V1, whereas a unimodal distribution can be seen in V2 with the maximum in layers II–IV (Fig. 8F, Table VII). The dorsal (V2d) and ventral (V2v) parts of V2 contain different densities of $GABA_A$ receptors, with the higher density in V2v (Fig. 8F). Thus, two receptors–kainate (see above) and $GABA_A$—point to a neurochemical heterogeneity between V2v and V2d.

Muscarinic and nicotinic acetylcholine receptors have completely differing laminar patterns in both V1 and V2 (Table VII). Nicotinic receptors reach their highest densities in layers V–VI of both areas, whereas muscarinic binding sites are most densely packed in layers II–IVA (V1) and II–IV (V2). The regional distribution of muscarinic M1 receptors is shown in Fig. 8C.

Noradrenergic α_1 receptors reach a higher density in V2 than in V1 probably because of the higher density of the α_{1B} subtype (Zilles *et al.*, 1993). The α_1 receptors are most densely packed in layer II of V2. They reach intermediate densities in layers I, III, and V (Fig. 8D; Table VII). The α_2 differ from the α_1 laminar pattern in V2, especially by the higher density in layer IV V2 of (Table VII). The α_{2A} receptors ([^{3}H]idazoxan binding in the presence of unlabeled prazosin) show a lower density in V2 than in V1 (Zilles *et al.*, 1993). Layer VI reaches the highest density, followed by layers III and V. The lowest densities were found in layers I–II and IV. The V1/V2 border can be recognized not only by different laminar patterns in both areas, but also by the lower mean density of α_1 receptors in V1 than in V2 (Fig. 8D) and the lower density of α_2 receptors in V2 than in V1 (K. Zilles, unpublished observations).

Dopamine D_1 receptors (visualized with [^{3}H]SCH23390) were found with higher densities in V1 than in V2. The V1/V2 border is clearly recognizable by a higher density in V1 than in V2 and a layer IVB of V1 with low densities abruptly disappearing at the border (Cortés *et al.*, 1989). The laminar distribution of these receptors is rather homogeneous in V2. The D_2-binding sites seem to occur in extremely low densities in the neocortex and probably do not contribute to a recognition of different extrastriate areas (Camps *et al.*, 1989).

$5\text{-}HT_1$ binding sites are most densely packed in sublayer IVCβ of V1 and reach a higher mean density in V1 than in V2 (Pazos *et al.*, 1987a). The $5\text{-}HT_{1A}$ subtype, however, reaches highest densities in layers I–II of V1 and V2, with a rather similar distribution of this receptor subtype in both areas (Table VII; Pazos *et al.*, 1987a; our unpublished observations). $5\text{-}HT_2$ sites were demonstrated with [^{3}H]ketanserin binding in V1 and V2 (Pazos *et al.*, 1987b). Whereas the highest density was seen in layer IVC of V1, layers III and V showed the highest values in V2 (Table VII). The $5\text{-}HT_{2A}$-receptor mRNA is located with highest densities in layer IVC, with intermediate densities in layers II–IVA and with low densities in layer IVB of V1 (Burnet *et al.*, 1995). The density is higher in V1 than V2 and the V1/V2 border is clearly demarcated both in receptor autoradiographs and by *in situ* hybridization of mRNA (Burnet *et al.*, 1995; Pazos *et al.*, 1987b; our unpublished observations). The laminar distribution is more homogeneous in V2 than in V1. This pattern is closely correlated with the pattern revealed by receptor autoradiography with [^{3}H]ketanserin.

Cholecystokinin receptors reach highest densities in layers II to upper III and layer V of human V2 and lowest densities in lower layer III and layer IV (Fig. 4E in Dietl *et al.*, 1987). The V1/V2 border is clearly visible.

7.2. Receptors in Other Extrastriate Areas

Transmitter receptors reveal a considerable regional neurochemical heterogeneity of the extrastriate region (Fig. 8). Muscarinic M1 receptors show a lower mean packing density in V3 than in V2d, and a higher packing density in VP and

V4 than V2v (Fig. 8C). The V2d/V3 border is clearly recognizabel in $GABA_A$ receptor autoradiographs due to the lower packing density in V3 than in V2d (Fig. 8F). The most intersting result seems to be the delineability of a cortical region with a comparable location as previously described for the putative human V5/MT (Clarke and Miklossy, 1990; Watson *et al.*, 1993; Zilles and Schleicher, 1993). The delineability is based on the very low density of M1, α_1, and kainate receptors. This neurochemical chracteristic corresponds to data in the macaque V5/MT (Fig. 8A, B; K. Zilles, unpublished observations). An area of intermediate density can be delineated between V5/MT and the parietal cortex, which shows a relatively higher density (Fig. 8C–E). This area is located in the lateral occipital sulcus (los in Fig. 8C–E) and may correspond to the macaque MST.

In conclusion, receptor autoradiography is a promising tool for the parcellation of the human extrastriate region and for testing the "homologies" between cortical areas in humans and nonhuman primates.

8. Anatomical Studies as a Key to the Functional Organization of the Human Extrastriate Cortex

8.1. Concept of Visual Areas

An open question which needs further investigation is that of equivalence between areas in humans and in nonhuman primates. Although the same designations are used for macaque and human areas, the full extent of similarities and differences is not known.

Macaque extrastriate cortex has been shown to contain a large number of visual areas defined by three or more of the following criteria: (i) presence of retinotopic representation of the contralateral visual hemifield or of a part of it; (ii) functional specialization of neurons; (iii) specific effects of lesions; (iv) specific connectivity pattern; and (v) specific cyto-, myelo-, and/or chemoarchitecture (for review see Van Essen, 1985). Over 30 areas have been identified in macaque extrastriate cortex and 6–9 in human (Fig. 5). The definition of areas is based on less stringent criteria in human than in macaque. Two areas, V2 and VP, may fulfill three criteria: retinotopic organization (revealed by activation studies), particular pattern of callosal connections, and distinct architecture. However, there is no study which links retinotopic representation to the anatomical criteria in the same case. Another area, V4, may fulfill four criteria: functional specialization (revealed by certain activation studies), effects of lesions (leading to central achromatopsia), particular pattern of callosal connections, and distinct architecture. Again, there is no direct link between activation and anatomical studies. Furthermore, lesions leading to achromatopsia exceed by far putative area V4.

The size of single human extrastriate visual areas, as far as identified in anatomical material, is generally smaller than that of V1 (Table IV). Areas further away from V1 tend to be smaller than those close to V1; e.g., V5/MT is smaller than V2. When compared to macaque monkey, the decrease in surface between V1 and V5/MT is greater in human.

8.2. Modular Organization of Visual Areas

Some of the identified human extrastriate visual areas were shown to contain anatomically distinct modules. In V2 a pattern of stripes is visible in myelin and in CO stains (Burkhalter and Bernardo, 1989; Tootell and Taylor, 1995) and intrinsic connections are patchy (Burkhalter and Bernardo, 1989). In V5/MT the CO staining is patchy (Tootell and Taylor, 1995), and in a higher order area, a putative equivalent of macaque V6/PO, the intrinsic connections are clustered (Clarke, 1994b). Thus, distinct modules occur within early-stage visual areas, as they do in nonhuman primates.

8.3. Extrastriate Neuronal Network

Two models, not mutually exclusive, have influenced our understanding of information processing within the extrastriate cortex during the last decade.

Extrastriate visual areas of nonhuman primates can be grouped into two major families of areas with limited cross-link (Baizer *et al.*, 1991; Desimone and Ungerleider, 1989; DeYoe and van Essen, 1988; Kaas and Garraghty, 1991; Livingstone and Hubel, 1988; Maunsell, 1995; Maunsell and Newsome, 1987; McIntosh *et al.*, 1994; Mishkin *et al.*, 1983; Morel and Bullier, 1990; Ungerleider and Mishkin, 1982; Weller, 1988; Young, 1992; Zeki and Shipp, 1988): (i) a *ventral* stream from the occipital to the temporal cortex (V1, V2, VP, V4, inferior temporal cortex with areas PIT, CIT, and AIT) for pattern discrimination and object recognition ("What" system), and (ii) a *dorsal* stream from the occipital to the parietal cortex (V1, V2, V3d, V5/MT, MST, VIP, LIP, parietal cortex with area 7a) for the perception of relative spatial locations of objects and the visual guidance of movements toward objects ("Where" system). Recently, a third stream of areas (V1, V2, V3d, V5/MT, MST, and FST, anterior part of the superior temporal sulcus) has been suggested for the analysis of visual motion (Boussaoud *et al.*, 1990; Distler *et al.*, 1996).

Based on layers of their origin and termination, interareal connections were classified as feedforward, feedback, or within the same level of a hierarchical areal organization. Consequently, these connections can be interpreted as ascending, descending, or lateral according to the concept of *distributed hierarchical processing* (van Essen and Maunsell, 1983; Maunsell and van Essen, 1983a; Felleman and van Essen, 1991). This concept allows the occurrence of (i) various areas at each level of the hierarchy, (ii) feedforward and feedback connections between the areas and levels, (iii) skipping of intermediate levels for direct connections between very low and very high levels, (iv) parallel processing through multiple, functionally distinct channels, and (v) context-dependent shifts of one area to different hierarchical levels.

Both models were derived from anatomical and physiological work in *nonhuman primates.* Their validity for *human* cortex has rarely been tested. Partial validation comes from observations of brain-damaged patients (for review see, e.g., Grüsser and Landis, 1991). Tracing studies are rare in human, but three studies indicate that there seems to be a dorsal and a ventral pathway, although they may be less segregated than in macaque monkeys. First, the parvo- and magnocellular pathways remain relatively segregated in human V1 and V2 (Burkhalter and Bernardo, 1989). Second, a higher order visual area, most likely situated within the dorsal pathway, sends intrahemispheric connections to areas situated within the ventral pathway (Clarke, 1994a). Third, the inferior part of the occipitotemporal cortex, comprising probably several areas of the ventral pathway, sends interhemispheric connections to the posterior part of the parietal cortex (Clarke *et al.,* 1995a). The concept of distributed hierarchical processing may apply to the human visual cortex, but there is little evidence of a hierarchical order of human visual areas. Tracing methods that can be used in human material tend to be unreliable as to the precise extent of the termination fields and as to the layers of origin. However, there are indications that feedback projections to V1 and V2 exist from putative higher order areas (Fig. 17; Clarke, 1994a)

ACKNOWLEDGMENTS. Published and unpublished work by the authors reviewed in this chapter was supported by the Deutsche Forschungsgemeinschaft (SFB 194/A6), the Swiss National Science Foundation (31-36297.92, 3231-041607.94), the Human Frontier Science Program, and the European Community (HC&MP, BioMed, BioTech). We thank Meishu Qü, Marc Dammann, Renate Dohm, Brigitte Machus, and Kerstin Bisgiel for assistance in preparing the autoradiographs, Nicola Palomero-Gallagher for bibliographic assistance, and Christine Opfermann-Rüngeler for graphical work.

9. References

Albright, T. D., 1984, Direction and orientation selectivity of neurons in visual area MT of the macaque, *J. Neurophysiol.* **52:**1106–1130.

Allison, T., Begleiter, A., McCarthy, G., Roessler, E., Nobre, A. C., and Spencer, D. D., 1993, Electrophysiological studies of color processing in human visual cortex, *Electroenceph. Clin. Neurophysiol.* **88:**343–355.

Allison, T., Ginter, H., McCarthy, G., Nobre, A. C., Puce, A., Luby, M., and Spencer, D. D., 1994, Face recognition in human extrastriate cortex, *J. Neurophysiol.* **71:**821–825.

Allman, J. M., and Kaas, J. H., 1971, A representation of the visual field in the caudal third of the middle temporal gyrus of the owl monkey (*Aotus trivirgatus*), *Brain Res.* **31:**85–105.

Allman, J. M., and Kaas, J. H., 1975, The dorsomedial cortical visual area: A third tier area in the occipital lobe of the owl monkey (*Aotus trivirgatus*), *Brain Res.* **100:**473–487.

Allman, J. M., Kaas, J. H., and Lane, R. H., 1973, The middle temporal visual area (MT) in the bushbaby (*Galago senegalensis*), *Brain Res.* **57:**197–202.

Bailey, P., and von Bonin, G., 1951, *The Isocortex of Man,* University of Illinois Press, Urbana, IL.

Baizer, J. S., Ungerleider, L. G., and Desimone, R., 1991, Organization of visual inputs to the inferior temporal and posterior parietal cortex in macaques, *J. Neurosci.* **11:**168–190.

Baker, J. F., Petersen, S. E., Newsome, W. T., and Allman, J. M., 1981, Visual response properties of neurons in four extrastriate visual areas of the owl monkey (*Aotus trivirgatus*): A quantitative comparison of medial, dorsomedial, dorsolateral and middle temporal areas, *J. Neurophysiol.* **45:**397–416.

Baleydier, C., and Morel, A., 1992, Segregated thalamocortical pathways to inferior parietal and inferotemporal cortex in macaque monkey, *Visual Neurosci.* **8**(5)**:**391–405.

Barbur, J., Watson, J., Frackowiak, R., and Zeki, S., 1993, Conscious visual perception without V1, *Brain* **116:**1293–1302.

Barton, J. J. S., Sharpe, J. A., and Raymond, J. E., 1995, Retinotopic and directional defects in motion discrimination in humans with cerebral lesions, *Ann. Neurol.* **37:**665–675.

Beckers, G., and Zeki, S., 1995, The consequences of inactivating areas V1 and V5 on visual motion perception, *Brain* **118:**49–60.

Benevento, L. A., and Davis, B., 1977, Topographical projections of the prestriate cortex to the pulvinar nuclei in the macaque monkey: An autoradiographic study, *Exp. Brain Res.* **30**(2–3)**:**405–424.

Benevento, L. A., and Fallon, J. H., 1975, The ascending projections of the superior colliculus in the rhesus monkey (*Macaca mulatta*), *J. Comp. Neurol.* **160:**339–362.

Benevento, L. A., and Yoshida, K., 1981, The afferent and efferent organization of the lateral geniculo-prestriate pathways in the macaque monkey, *J. Comp. Neurol.* **203**(3)**:**455–474.

Betz, W., 1874, Anatomischer Nachweis zweier Gehirncentra, *Zentralbl. Med. Wiss.* **19:**193–195, 210–213, 231–234.

Binder, J. R., and Mohr, J. P., 1992, The topography of callosal reading pathways. A case–control analysis, *Brain* **115:**1807–1826.

Blümcke, I., and Celio, M. R., 1992, Parvalbumin and calbindin D-28k immunoreactivities coexist within cytochrome oxidase-rich compartments of squirrel monkey area 18, *Exp. Brain Res.* **92:**39–45.

Bolton, J. S., 1900, The exact histological localisation of the visual area of the human cerebral cortex, *Phil. Trans. R. Soc. Lond.* **193:**165–222.

Bonin, G. von, and Bailey, P., 1951, *The Isocortex of Man*, University of Illinois Press, Urbana, IL.

Born, R. T., and Tootell, R. B. H., 1992, Segregation of global and local motion processing in primate middle temporal visual area, *Nature* **357:**497–499.

Boussaoud, D., Ungerleider, L. G., and Desimone, R., 1990, Pathways for motion analysis: Cortical connections of the medial superior temporal and fundus of the superior temporal visual areas in the macaque, *J. Comp. Neurol.* **296:**462–495.

Boussaoud, D., Desimone, R., and Ungerleider, L. G., 1992, Subcortical connections of visual areas MST and FST in macaques, *Visual Neurosci.* **9**(3–4)**:**291–302.

Braak, H., 1977, The pigment architecture of the human occipital lobe, *Anat. Embryol.* **150:**229–250.

Braak, H., 1980, *Architectonics of the Human Telencephalic Cortex*, Springer, Berlin.

Brodmann, K., 1903, Beiträge zur histologischen Lokalisation der Grosshirnrinde. Zweite Mitteilung: Der Calcarinatypus, *J. Psychol. Neurol.* **2:**133–159.

Brodmann, K., 1905, Beiträge zur histologischen Lokalisation der Grosshirnrinde. Dritte Mitteilung: Die Rindenfelder der niederen Affen, *J. Psychol. Neurol.* **4:**177–226.

Brodmann, K., 1906, Beiträge zur histologischen Lokalisation der Grosshirnrinde. Fünfte Mitteilung: Über den allgemeinen Bauplan des Cortex pallii bei den Mammaliern und zwei homologe Rindenfelder im besonderen. Zugleich ein Beitrag zur Furchenlehre, *J. Psychol. Neurol.* **6:**275–400.

Brodmann, K., 1908a, Beiträge zur histologischen Lokalisation der Grosshirnrinde. VI. Mitteilung. Die Cortexgliederung des Menschen, *J. Psychol. Neurol.* **10:**231–246.

Brodmann, K., 1908b, Beiträge zur histologischen Lokalisation der Grosshirnrinde. VII. Mitteilung: Die cytoarchitektonische Cortexgliederung der Halbaffen Lemuriden, *J. Psychol. Neurol.* **10:**287–334.

Brodman, K., 1909, *Vergleichende Lokalisationslehre der Grosshirnrinde in ihren Prinzipien dargestellt auf Grund des Zellenbaus*, Barth, Leipzig.

Bullier, J., and Kennedy, H., 1983, Projection of the lateral geniculate nucleus onto cortical area V2 in the macaque monkey, *Exp. Brain Res.* **53:**168–172.

Burkhalter, A., and Bernardo, K. L., 1989, Organization of cortico-cortical connections in human visual cortex, *Proc. Natl. Acad. Sci. USA* **86:**1071–1075.

Burkhalter, A., Bernardo, K. L., and Charles, V., 1993, Development of local circuits in human visual cortex, *J. Neurosci.* **13:**1916–1931.

Burnet, P. W. J., Eastwood, S. L., Lacey, K., and Harrison, P. J., 1995, The distribution of 5-HT_{1A} and 5-HT_{2A} receptor mRNA in human brain, *Brain Res.* **676:**157–168.

Campbell, A. W., 1905, *Histological Studies on the Localisation of Cerebral Function*, Cambridge University Press, Cambridge.

Camps, M., Cortés, R., Gueye, B., Probst, A., and Palacios, J. M., 1989, Dopamine receptors in human brain: Autoradiographic distribution of D_2 sites *Neuroscience* **28:**275–290.

Carroll, E. W., and Wong-Riley, M. T. T., 1984, Quantitative light and electronmicroscopic analysis of cytochrome oxidase-rich zones in the striate cortex of the squirrel monkey, *J. Comp. Neurol.* **222:**1–17.

Casagrande, V., and Kaas, J. H., 1994, The afferent, intrinsic, and efferent connections of primary visual cortex in primates, in: *Cerebral Cortex,* Volume 10, *Primary Visual Cortex in Primates* (A. Peters and K. Rockland, eds.), Plenum Press, New York, pp. 201–259.

Cheng, K., Fujita, H., Kanno, I., Miura, S., and Tanaka, K., 1995, Human cortical regions activated by wide-field visual motion: An $H_2{}^{15}O$ PET study, *J. Neurophysiol.* **74:**413–427.

Clark, W. E., and Russell, W. R., 1939, Observations on the efferent connections of the centre median nucleus, *J. Anat.* **73:**255–262.

Clarke, S., 1993, Callosal connections and functional subdivision of the human occipital cortex, in: *Functional Organization of the Human Visual Cortex* (B. Gulyás, D. Ottoson, and P. E. Roland, eds.), Pergamon Press, Oxford, pp. 137–150.

Clarke, S., 1994a, Association and intrinsic connections of human extrastriate visual cortex, *Proc. R. Soc. Lond. B Biol. Sci.* **257:**87–92.

Clarke, S., 1994b, Modular organization of human extrastriate visual cortex: Evidence from cytochrome oxidase pattern in normal and macular degeneration cases, *Eur. J. Neurosci.* **6:**725–736.

Clarke, S., and Miklossy, J., 1990, Occipital cortex in man: Organization of callosal connections, related myelo- and cytoarchitecture, and putative boundaries of functional visual areas, *J. Comp. Neurol.* **298:**188–214.

Clarke, S., Assal, G., Bogousslavsky, J., Regli, F., Townsend, D. W., Leenders, K. L., and Blecic, S., 1994, Pure amnesia after unilateral lift polar thalamic infarct: Topographic and sequential neuropsychological and metabolic (PET) correlations, *J. Neurol. Neurosurg. Psychiatry* **57:**27–34.

Clarke, S., Van Essen, D., Hadjikhani, N., Drury, H., and Coogan, T., 1995a, Understanding human areas 18 and 37: Contribution of two-dimensional maps of visual callosal afferents, *Hum. Brain Mapping Suppl.* **1:**33.

Clarke, S., Ribaupierre, F., de, Bajo, V. M., Rouiller, E. M., and Kraftsik, R., 1995b, The auditory pathway in cat corpus callosum, *Exp. Brain Res.* **104:**534–540.

Colby, C. L., and Duhamel, J. R., 1991, Heterogeneity of extrastriate visual areas and multiple parietal areas in the macaque monkey, *Neuropsychologia* **29:**517–537.

Colby, C. L., Gattass, R., Olson, C. R., and Gross, C. G., 1988, Topographical organization of cortical afferents to extrastriate visual area PO in the macaque: A dual tracer study, *J. Comp. Neurol.* **269:**392–413.

Corbetta, M., Miezen, F. M., Dobmeyer, S., Shulman, G. L., and Petersen, S. E., 1990, Attentional modulation of neuronal processing of shape, color and velocity in humans, *Science* **248:**1556–1559.

Corbetta, M., Miezin, F. M., Dobmeyer, S., Shulman, G. L., and Petersen, S. E., 1991, Selective and divided attention during visual discrimination of shape, color and speed: Functional anatomy by positron emission tomography, *J. Neurosci.* **11:**2383–2402.

Cortés, R., Probst, A., Tobler, H. J., and Palacios, J. M., 1986, Muscarinic cholinergic receptor subtype in the human brain. II. Quantitative autoradiographic studies, *Brain Res.* **362:**239–253.

Cortés, R., Gueye, B., Pazos, A., Probst, A., and Palacios, J. M., 1989, Dopamine receptors in human brain: Autoradiographic distribution of D_1 sites, *Neuroscience* **28:**263–273.

Cragg, B. G., 1969, The topography of the afferent projections in the circumstriate visual cortex of the monkey studied by the Nauta method, *Vision Res.* **9:**733–747.

Creutzfeldt, O. D., 1995, *Cortex Cerebri. Performance, Structural and Functional Organization of the Cortex,* Oxford University Press, Oxford.

Cusick, C. G., Seltzer, B., Cola, M., and Griggs, E., 1995, Chemoarchitectonics and corticocortical terminations within the superior temporal sulcus of the rhesus monkey: Evidence for subdivisions of superior temporal polysensory cortex, *J. Comp. Neurol.* **360:**513–535.

Damasio, A., Yamada, T., Damasio, H., Corbett, J., and McKee, J., 1980, Central achromatopsia: Behavioral, anatomical and physiological aspects, *Neurology* **30:**1064–1071.

Dejerine, J., 1892, Contribution à l'étude anatomo-pathologique et clinique des différentes variétés de cécité verbale, *C. R. Hebd. Séances Mém. Soc. Biol.* **4:**61–90.

Dejerine, J., and Dejerine-Klumpke, A., 1895, *Anatomie des centres nerveux,* Rueff, Paris.

Dejerine, J., and Vialet, N., 1893, Contribution à l'étude de la localisation de la cécité verbale pure, *C. R. Hebd. Séances Mém. Soc. Biol.* **5:**790–793.

De Jong, B. M., Shipp, S., Skidmore, B., Frackowiak, R. S. J., and Zeki, S., 1994, The cerebral activity related to the visual perception of forward motion in depth, *Brain* **117:**1039–1054.

De Lacoste, M. C., Kirkpatrick, J. B., and Ross, E. D., 1985, Topography of the human corpus callosum, *J. Neuropathol. Exp. Neurol.* **44:**578–591.

Desimone, R., and Schein, S. J., 1987, Visual properties of neurons in area V4 of the macaque: Sensitivity to stimulus form, *J. Neurophysiol.* **57:**835–868.

Desimone, R., and Ungerleider, L. G., 1986, Multiple visual areas in the caudal superior temporal sulcus of the macaque, *J. Comp. Neurol.* **248:**164–189.

Desimone, R., and Ungerleider, L. G., 1989, Neural mechanisms of visual processing in monkeys, in: *Handbook of Neuropsychology*, Volume 2 (F. Boller, and J. Grafman, eds.), Elsevier, Amsterdam, pp. 267–299.

DeYoe, E. A., and Van Essen, D. C., 1985, Segregation of efferent connections and receptive field properties in visual area V2 of the macaque, *Nature* **317:**58–61.

DeYoe, E. A., and Van Essen, D. C., 1988, Concurrent processing streams in monkey visual cortex, *Trends Neurosci.* **5:**219–226.

DeYoe, E. A., Hockfield, S., Garren, H., and Van Essen, D. C., 1990, Antibody labeling of functional subdivisions in visual cortex: CAT-301 immunoreactivity in striate and extrastriate cortex of the macaque monkey, *Visual Neurosci.* **5:**67–81.

DeYoe, E. A., Felleman, D. J., Van Essen, D. C., and McClendon, E., 1994, Multiple processing streams in occipitotemporal visual cortex, *Nature* **371:**151–154.

Dietl, M. M., Probst, A., and Palacios, J. M., 1987, On the distribution of cholecystokinin receptor binding in the human brain: An autoradiographic study, *Synapse* **1:**169–183.

Distler, C., Boussaoud, D., Desimone, R., and Ungerleider, L. G., 1993, Cortical connections of inferior temporal area TEO in macaque monkey, *J. Comp. Neurol.* **334:**125–150.

Distler, C., Bachevalier, J., Kennedy, C., Mishkin, M., and Ungerleider, L. G., 1996, Functional development of the corticocortical pathway for motion analysis in the macaque monkey: A ^{14}C-2-deoxyglucose study, *Cerebral Cortex* **6:**184–195.

Di Virgilio, G., and Clarke, S., 1996, Human anterior commissure contains axons originating in the inferior part of the temporal lobe, *Experientia* **52:**A76.

Doty, R. W., 1983, Nongeniculate afferents to striate cortex in macaques, *J. Comp. Neurol.* **218:**159–173.

Dupont, P., Orban, G. A., De Bruyn, B., Verbruggen, A., and Mortelmans, L., 1994, Many areas in the human brain respond to visual motion, *J. Neurophysiol.* **72:**1420–1424.

Elliot Smith, G., 1907, A new topographical survey of the human cerebral cortex, being an account of the distribution of the anatomically distinct cortical areas and their relationship to the cerebral sulci, *J. Anat.* **41:**237–254.

Felleman, D. J., and Kaas, J. H., 1984, Receptive-field properties of neurons in middle temporal visual area (MT) of owl monkeys, *J. Neurophysiol.* **52:**488–513.

Felleman, D. J., and Van Essen, D. C., 1991, Distributed hierarchical processing in the primate cerebral cortex, *Cerebral Cortex* **1:**1–47.

Filimonoff, I. N., 1932, Über die Variabilität der Grosshirnrindenstruktur. Mitteilung II. Regio occipitalis beim erwachsenen Menschen, *J. Psychol. Neurol.* **44:**1–96.

Filimonoff, I. N., 1933, Über die Variabilität der Grosshirnrindenstruktur. Mitteilung III. Regio occipitalis bei den höheren und niederen Affen, *J. Psychol. Neurol.* **45:**69–137.

Fiorani, M., Jr., Gattas, R., Rosa, M. G. P., and Sousa, A. P. B., 1989, Visual area MT in the *Cebus* monkey: Location, visuotopic organization and variability, *J. Comp. Neurol.* **287:**98–118.

Flechsig, P., 1898, Neue Untersuchungen über die Markbildung in den menschlichen Grosshirnlappen, *Neuro. Centralblatt* **21:**977–996.

Flechsig, P., 1920, *Anatomie des menschlichten Gehirns und Rückenmarks auf myelogenetischer Grundlage*, Thieme, Leipzig.

Fries, W., 1981, The projection from the lateral geniculate nucleus to the prestriate cortex of the macaque monkey, *Proc. R. Soc. Lond. B* **213:**73–80.

Friston, K. J., Ungerleider, L. G., Jezzard, P., and Turner, R., 1995, Characterizing modulatory interactions between areas V1 and V2 in human cortex: A new treatment of functional MRI data, *Human Brain Mapping* **2:**211–224.

Galletti, C., Battaglini, P. P., and Fattori, P., 1991, Functional properties of neurons in the anterior bank of the parieto-occipital sulcus of the macaque monkey, *Eur. J. Neurosci.* **3:**452–461.

Galletti, C., Battaglini, P. P., and Fattori, P., 1993, Parietal neurons encoding spatial locations in craniotopic coordinates, *Exp. Brain Res.* **96:**221–229.

Gattass, R., and Gross, C. G., 1981, Visual topography of striate projection zone (MT) in posterior superior temporal sulcus of the macaque, *J. Neurophysiol.* **46:**621–638.

Gattass, R., Gross, C. G., and Sandell, J. H., 1981, Visual topography of V2 in the macaque, *J. Comp. Neurol.* **201:**519–539.

Gattas, R., Sousa, A. P., and Gross, C. G., 1988, Visuotopic organization and extent of V3 and V4 of the macaque, *J. Neurosci.* **8:**1831–1845.

Gebhard, R., Zilles, K., Schleicher, A., Everitt, B. J., Robbins, T. W., and Divac, I., 1993, Distribution of seven major neurotransmitter receptors in the striate cortex of the New World monkey *Callithrix jacchus, Neuroscience* **56:**877–885.

Gibson, J. J., 1950, *The Perception of the Visual World,* Houghton Mifflin, Boston.

Girard, P., Salin, P. A., and Bullier, J., 1992, Response selectivity of neurons in area MT of the macaque monkey during reversible inactivation of area V1, *J. Neurophysiol.* **67:**1437–1446.

Glickstein, M., and Whitteridge, D., 1976, Degeneration of layer II pyramidal cells in area 18 following destruction of callosal input, *Brain Res.* **104:**148–151.

Grüsser, O. J., and Landis, T., 1991, Visual agnosias and other disturbances of visual perception and cognition, in: *Vision and Visual Dysfunction,* Volume 12 (J. Cronly-Dillon, ed.), Macmillan Press, London.

Gulyás, B., and Roland, P. E., 1991, Cortical fields participating in form and colour discrimination in the human brain, *NeuroReport* **2:**585–588.

Gulyás, B., and Roland, P. E., 1994a, Binocular disparity discrimination in human cerebral cortex: Functional anatomy by positron emission tomography, *Proc. Natl. Acad. Sci. USA* **91:**1239–1243.

Gulyás, B., and Roland, P. E., 1994b, Processing and analysis of form, colour and binocular disparity in the human brain: Functional anatomy by positron emission tomography, *Eur. J. Neurosci.* **6:**1811–1828.

Halgren, E., Baudena, P., Heit, G., Clarke, M., and Marinkovic, K., 1994, Spatiotemporal stages in face and word processing. 1. Depth-recorded potentials in the human occipital and parietal lobes, *J. Physiol.* **88:**1–50.

Hammarberg, C., 1898, Studien über Klinik und Pathologie der Idiotie nebst Untersuchungen über die normale Anatomie der Hirnrinde, *Nova Acta Regiae Soc. Sci. Ups.* **III/17:**1–126.

Hardy, S. G., and Lynch, J. C., 1992, The spatial distribution of pulvinar neurons that project to two subregions of the inferior parietal lobule in the macaque, *Cerebral Cortex* **2**(3):217–230.

Hassler, R., 1959, Anatomy of the thalamus, in: *Einführung in die stereotactischen Operationen mit einem Atlas des menschlichen Gehirns* (G. Schaltenbrand, and P. Bailey, eds.), G. Thieme, Stuttgart.

Heimer, L., Ebner, F. F., and Nauta, W. J. H., 1967, A note on the termination of commissural fibers in the neocortex, *Brain Res.* **5:**171–177.

Heinze, G., 1954, Zytoarchitektonische Untergliederung der Area occipitalis, *J. Hirnforsch.* **1:**173–198.

Hendry, S. H. C., Jones, E. G., and Beinfeld, M. C., 1983, Cholecystokinin-like immunoreactive neurons in rat and monkey cerebral cortex make symmetrical synapses and have intimate associations with blood vessels, *Proc. Natl. Acad. Sci. USA* **80:**2400–2404.

Hendry, S. H. C., Hockfield, S., Jones, E. G., and McKay, R., 1984, Monoclonal antibody that identifies subsets of neurones in the central visual system of monkey and cat, *Nature* **307:**267–269.

Hendry, S. H. C., Jones, E. G., Hockfield, S., and McKay, R. D. G., 1988, Neuronal populations stained with the monoclonal antibody Cat-301 in the mammalian cerebral cortex and thalamus, *J. Neurosci.* **8:**518–542.

Hendry, S. H. C., Fuchs, J. L., de Blas, A. L., and Jones, E. G., 1990, Distribution and plasticity of immunocytochemically localized $GABA_A$ receptors in adult monkey visual cortex, *J. Neurosci.* **10:**2438–2450.

Hendry, S. H. C., Huntsman, M., Vinuela, A., Möhler, H., de Blas, A., and Jones, E., 1994, $GABA_A$ receptor subunit immunoreactivity in primate visual cortex: Distribution in macaques and humans and regulation by visual input in adulthood, *J. Neurosci.* **14:**2383–2401.

Hernandez-Gonzalez, A., Cavada, C., and Reinoso-Suarez, F., 1994, The lateral geniculate nucleus projects to the inferior temporal cortex in the macaque monkey, *NeuroReport* **5**(18):2693–2696.

Heywood, C. A., and Cowey, A., 1987, On the role of cortical area V4 in the discrimination of hue and pattern in macaque monkeys, *J. Neurosci.* **7:**2601–2617.

Heywood, C. A., Gadotti, A., and Cowey, A., 1992, Cortical area V4 and its role in the perception of color, *J. Neurosci.* **12:**4056–4065.

Heywood, C. A., Gaffan, D., and Cowey, A., 1995, Cerebral achromatopsia in monkeys, *Eur. J. Neurosci.* **7:**1064–1073.

Hitchcock, P. F., and Hickey, T. L., 1980, Ocular dominance columns: Evidence for their presence in humans, *Brain Res.* **182:**176–179.

Hockfield, S., Tootell, R. B. H., and Zaremba, S., 1990, Molecular differences among neurons reveal an organization of human visual cortex, *Proc. Natl. Acad. Sci. USA* **87:**3027–3031.

Hof, P. R., and Morrison, J. H., 1995, Neurofilament protein defines regional patterns of cortical organization in the macaque monkey visual system: A quantitative immunohistochemical analysis, *J. Comp. Neurol.* **352:**161–186.

Horton, J. C., 1984, Cytochrome oxidase patches: A new cytoarchitonic feature of monkey cortex, *Phil. Trans. R. Soc. Lönd. Biol.* **304:**199–253.

Horton, J. C., and Hedley-Whyte, E. T., 1984, Mapping of cytochrome oxidase patches and ocular dominance columns in human visual cortex, *Phil. Trans. R. Soc. Lönd. Biol.* **304:**255–272.

Hotson, J., Braun, D., Herzberg, W., and Boman, D., 1994, Transcranial magnetic stimulation of extrastriate cortex degrades human motion direction discrimination, *Vision Res.* **34:**2115–2123.

Hubel, D. H., and Livingstone, M. S., 1987, Segregation of form, color and stereopsis in primate area 18, *J. Neurosci.* **7:**3378–3415.

Huntley, G. W., Rogers, S. W., Moran, T., Janssen, W., Archin, N., Vickers, J. C., Cauley, K., Heinemann, S. F., and Morrison, J. H., 1993, Selective distribution of kainate receptor subunit immunoreactivity in monkey neocortex revealed by a monoclonal antibody that recognizes glutamate receptor subunits GluR5/6/7, *J. Neurosci.* **13:**2965–2981.

Huntsman, M. M., Isackson, P. J., and Jones, E. G., 1994, Lamina-specific expression and activity-dependent regulation of seven $GABA_A$ receptor subunit mRNAs in monkey visual cortex, *J. Neurosci.* **14:**2236–2259.

Jansen, K. L. R., Faull, R. L. M., and Dragunow, M., 1989, Excitatory amino acid receptors in the human cerebral cortex: A quantitative autoradiographic study comparing the distributions of [^{3}H]TCP, [^{3}H]glycine, L-[^{3}H]glutamate, [^{3}H]AMPA and [^{3}H]kainic acid binding sites, *Neuroscience* **32:**587–607.

Jen, L. S., and Zeki, S., 1984, High cytochrome oxidase content of the V5 complex of macaque monkey visual cortex, *J. Physiol. (Lond.)* **348:**23P.

Kaas, J. H., 1986, The structural basis for information processing in the primate visual system, in: *Visual Neuroscience* (J. D. Pettigrew, W. R. Levick, and K. J. Sanderson, eds.), Cambridge University Press, Cambridge, pp. 315–340.

Kaas, J. H., 1988, Changing concepts of visual cortex organization in primates, in: *Neuropsychology of Visual Perception,* Erlbaum, Hillsdale, NJ, pp. 1–32.

Kaas, J. H., 1989, Why does the brain have so many visual areas? *J. Cognitive Neurosci.* **1:**121–135.

Kaas, J. H., 1993, The organization of visual cortex in primates: Problems, conclusions, and the use of comparative studies in understanding the human brain, in: *Functional Organisation of the Human Visual Cortex* (B. Gulyás, D. Ottoson, and P. E. Roland, eds.), Pergamon Press, New York, pp. 1–12.

Kaas, J. H., and Garraghty, P. E., 1991, Hierarchical, parallel, and serial arrangements of sensory cortical areas: Connection patterns and functional aspects, *Curr. Biol.* **1:**248–251.

Kaas, J. H., and Krubitzer, L. A., 1991, The organization of extrastriate visual cortex, in: *Neuroanatomy of the Visual Pathways and their Development* (B. Dreher and S. R. Robinson, eds.), Macmillan, Houndsmills, UK, pp. 302–323.

Kaas, J. H., and Lin, C. S., 1977, Cortical projections of area 18 in owl monkeys, *Vision Res.* **16:**739–741.

Kaas, J. H., and Morel, A., 1993, Connections of visual areas of the upper temporal lobe of owl monkeys: The MT crescent and dorsal and ventral subdivisions of FST, *J. Neurosci.* **13:**534–546.

Kaas, J. H., and Preuss, T. M., 1993, Archontan affinities as reflected in the visual system, in: *Mammal Phylogeny, Placentals* (F. S. Szalay, *et al.*, ed.), Springer, New York, pp. 115–128.

Keating, E. G., 1980, Residual spatial vision in the monkey after removal of striate and preoccipital cortex, *Brain Res.* **187:**271–290.

Kennedy, H., and Bullier, J., 1985, A double-labeling investigation of the afferent connectivity to cortical areas V1 and V2 of the Macaque monkey, *J. Neurosci.* **5**(10)**:**2815–2830.

Kennedy, H., and Dehay, C., 1988, Functional implications of the anatomical organization of the callosal connections of visual areas V1 and V2 in the macaque monkey, *Behav. Brain Res.* **29:**225–236.

Kennedy, H., Dehay, C., and Bullier, J., 1986, Organization of the callosal connections of visual areas V1 and V2 in the macaque monkey, *J. Comp. Neurol.* **247:**398–415.

Kennedy, H., Meissirel, C., and Dehay, C., 1991, Callosal pathways and their compliancy to general rules governing the organization of corticocortical connectivity, in: *Neuroanatomy of the Visual Pathways and Their Development* (B. Dreher and S. R. Robinson, eds.), Macmillan, Houndsmills, UK, pp. 324–359.

Kleist, K., 1934, *Gehirnpathologie,* Barth, Leipzig.

Knierim, J. J., and Van Essen, D. C., 1992 Visual cortex: Cartography, connectivity and concurrent processing, *Curr. Opin. Neurobiol.* **2:**150–155.

Kondo, H., Hashikawa, T., Tanaka, K., and Jones, E., 1994, Neurochemical gradient along the monkey occipito-temporal cortical pathway, *NeuroReport* **5:**613–616.

Kostovic, I., and Rakic, P., 1984, Development of prestriate visual projections in the monkey and human fetal cerebrum revealed by transient cholinesterase staining, *J. Neurosci.* **4:**25–42.

Kritzer, M. F., Innis, R. B., and Goldman-Rakic, P. S., 1987, Regional distribution of cholecystokinin receptors in primate cerebral cortex determined by *in vitro* receptor autoradiography, *J. Comp. Neurol.* **263:**418–435.

Krubitzer, L. A., and Kaas, J. H., 1989, Cortical integration of parallel pathways in the visual system of primates, *Brain Res.* **478:**161–165.

Krubitzer, L. A., and Kaas, J. H., 1990a, Cortical connections of MT in four species of primates: Areal, modular and retinoptic patterns, *Visual Neurosci.* **5:**165–204.

Krubitzer, L. A., and Kaas, J. H., 1990b, Convergence of processing channels in extrastriate cortex of monkeys, *Visual Neurosci.* **5:**609–613.

Le Bihan, D., Turner, R., Zeffiro, T. A., Cuénod, C. A., Jezzard, P., and Bonnerot, V., 1993, Activation of human primary visual cortex during visual recall: A magnetic resonance imaging study, *Proc. Natl. Acad. Sci. USA* **90:**11802–11805.

Leuba, G., and Garey, L. J., 1989, Comparison of neuronal and glial numerical density in primary and secondary visual cortex of man, *Exp. Brain Res.* **77:**31–38.

Levitt, P., Rakic, P., and Goldman-Rakic, P., 1984, Region-specific distribution of catecholamine afferents in primate cerebral cortex: A fluorescence histochemical analysis, *J. Comp. Neurol.* **227:**23–36.

Lewis, M. E., Mishkin, M., Bragin, E., Brown, R. M., Pert, C. B., and Pert, A., 1981, Opiate receptor gradients in monkey cerebral cortex: Correspondance with sensory processing hierarchies, *Science* **211:**1166–1169.

Lidow, M. S., Goldman-Rakic, P. S., Gallager, D. W., and Rakic, P., 1989, Quantitative autoradiographic mapping of serotonin 5-HT_1 and 5-HT_2 receptors and uptake sites in the neocortex of the rhesus monkey, *J. Comp. Neurol.* **280:**27–42.

Lidow, M. S., Goldman-Rakic, P. S., Gallager, D. W., and Rakic, P., 1991, Distribution of dopaminergic receptors in the primate cerebral cortex: Quantitative autoradiographic analysis using [^{3}H]raclopride, [^{3}H]spiperone and [^{3}H]SCH23390, *Neuroscience* **40:**657–671.

Lin, C. S., Weller, R. E., and Kaas, J. H., 1982, Cortical connections of striate cortex in the owl monkey, *J. Comp. Neurol.* **211:**165–176.

Livingstone, M. S., and Hubel, D. H., 1983, Specificity of cortico-cortical connections in monkey visual system, *Nature* **304:**531–534.

Livingstone, M. S., and Hubel, D. H., 1984, Anatomy and physiology of a color system in the primate visual cortex, *J. Neurosci.* **4:**309–356.

Livingstone, M. S., and Hubel, D. H., 1987, Connections between layer 4B of area 17 and the thick cytochrome oxidase stripes of area 18 in the squirrel monkey, *J. Neurosci.* **7:**3371–3377.

Livingstone, M. S., and Hubel, D. H., 1988, Segregation of form, color, movement, and depth: Anatomy, physiology and perception, *Science* **240:**740–749.

Ludwig, E., and Klinger, J., 1938, *Noyaux et faisceaux du cerveau humain,* Georges Thomas, Nancy, France.

Lueck, C. J., Zeki, S., Friston, K. J., Dieber, M. P., Cope, P., Cunningham, V. J., Lammertsma, A. A., Kennard, C., and Frackowiak, R. S. J., 1989, The colour centre in the cerebral cortex of man, *Nature* **340:**386–389.

Lund, J. S., Lund, R. D., Hendrickson, A. E., Bunt, A. H., and Fuchs, A. F., 1975, The origin of efferent pathways from the primary visual cortex (area 17) of the macaque monkey as shown by retrograde transport of horseradish peoxidase, *J. Comp. Neurol.* **164:**287–304.

Lungwitz, W., 1937, Zur myeloarchitektonischen Untergliederung der menschlichen Area praeoccipitalis (Area 19 Brodmann), *J. Psychol. Neurol.* **47:**607–638.

Lysakowski, A., Standage, G. P., and Benevento, L. A., 1988, An investigation of collateral projections of the dorsal lateral geniculate nucleus and other subcortical structures to the cortical areas V1 and V4 in the macaque monkey: A double label retrograde tracer study, *Exp. Brain Res.* **69**(3)**:**651–661.

Malach, R., Reppas, J. B., Benson, R. R., Kwong, K. K., Jiang, H., Kennedy, W. A., Ledden, P. J., Brady, T. J., Rosen, B. R., and Tootell, R. B. H., 1995, Object-related activity revealed by functional magnetic resonance imaging in human occipital cortex, *Proc. Natl. Acad. Sci. USA* **92:**8135–8139.

Marcar, V. L., Xiao, D.-K., Raiguel, S. E., Maes, H., and Orban, G. A., 1995, Processing of kinetically defined boundaries in the cortical motion area MT of the macaque monkey, *J. Neurophysiol.* **74:**1258–1270.

Martin, A., Haxby, J. V., Lalonde, F. M., Wiggs, C. L., and Ungerleider, L. G., 1995, Discrete cortical regions associated with knowledge of color and knowledge of action, *Science* **270:**102–105.

Maunsell, J. H. R., 1995, The brain's visual world: Representation of visual targets in cerebral cortex, *Science* **270:**764–769.

Maunsell, J. H. R., and Newsome, W. T., 1987, Visual processing in monkey extrastriate cortex, *Annu. Rev. Neurosci.* **10:**363–401.

Maunsell, J. H. R., and Van Essen, D. C., 1983a, The connections of the middle temporal visual area (MT) and their relationship to a cortical hierarchy in the macaque monkey, *J. Neurosci.* **3:**2563–2586.

Maunsell, J. H. R., and Van Essen, D. C., 1983b, Functional properties of neurons in middle temporal visual area of the macaque monkey I: Selectivity for stimulus direction, speed, and orientation, *J. Neurophysiol.* **49:**1127–1147.

Maunsell, J. H. R., and Van Essen, D. C., 1987, Topographic organization of the middle temporal visual area in the macaque monkey. I: Representational biases and the relationship to callosal connections and myeloarchitectonic boundaries, *J. Comp. Neurol.* **266:**535–555.

McIntosh, A., Grady, C., Ungerleider, L., Haxby, J., Rapoport, S., and Horwitz, B., 1994, Network analysis of cortical visual pathways mapped with PET, *J. Neurosci.* **14:**655–666.

Meadows, J. C., 1974, Disturbed perception of colours associated with localized cerebral lesions, *Brain* **97:**615–632.

Merigan, W. H., 1993, Human V4? *Curr. Biol.* **3:**226–229.

Mesulam, M. M., 1979, Tracing neural connections of human brain with selective silver impregnation. Observations on geniculocalcarine, spinothalamic and entorhinal pathways, *Arch. Neurol.* **36:**814–818.

Meynert, T., 1872, Vom Gehirn der Säugethiere, in: *Handbuch der Lehre von den Geweben des Menschen,* Volume 2 (S. Stricker, ed.), Engelmann, Leipzig, pp. 694–808.

Miklossy, J., 1993, The geniculocalcarine pathway in man, and some putative visual areas involved in visuo-spatial attention, in: *Functional Organization of the Human Visual Cortex* (B. Gulyás, D., Ottoson, and P. Roland, eds.), Pergamon Press, Oxford, pp. 123–136.

Mishkin, M., Ungerleider, L. G., and Macko, K. A., 1983, Object vision and spatial vision: Two cortical pathways, *Trends Neurosci.* **6:**414–417.

Morel, A., and Bullier, J., 1990, Anatomical segregation of two cortical visual pathways in the macaque monkey, *Visual Neurosci.* **4:**555–578.

Nakamura, H., Gattass, R., Desimone, R., and Ungerleider, L., 1993, The modular organization of projections from areas V1 and V2 to areas V4 and TEO in macaques, *J. Neurosci.* **13:**3681–3691.

Neuenschwander, S., Gattass, R., Sousa, A. P. B., and Pinon, M. C. G. P., 1994, Identification and organization of areas PO and POd in *Cebus* monkey, *J. Comp. Neurol.* **340:**65–86.

Newsome, W. T., and Allman, J. M., 1980, Interhemispheric connections of visual cortex in the owl monkey, *Aotus trivirgatus,* and the bushbaby, *Galago senegalensis, J. Comp. Neurol.* **194:**209–233.

Newsome, W. T., Maunsell, J. H. R., and Van Essen, D. C., 1986, Ventral posterior visual area of the macaque: Visual topography and areal boundaries, *J. Comp. Neurol.* **252:**139–153.

Newsome, W. T., Britten, K. H., Salzman, C. D., and Movshon, J. A., 1991, Neuronal mechanisms of motion perception, *Cold Spring Harbor Symp. Quant. Biol.* **55:**697–705.

Orban, G. A., Dupont, P., De Bruyn, B., Vogels, R., Vandenberghe, R., and Mortelmans, L., 1995, A motion area in human visual cortex, *Proc. Natl. Acad. Sci. USA* **92:**993–997.
Parkinson, D., Coscia, E. C., and Daw, N. W., 1989, Identification and localization of 5-hydroxytryptamine receptor sites in macaque visual cortex, *Visual Neurosci.* **2:**515–525.
Pasik, P., and Pasik, T., 1982, Visual functions in monkeys after total removal of visual cerebral cortex, *Contrib. Sensory Physiol.* **7:**147–200.
Payne, B., and Siwek, D. F., 1991, The visual map in the corpus callosum of the cat, *Cerebral Cortex* **1:**173–188.
Pazos, A., Probst, A., and Palacios, J. M., 1987a, Serotonin receptors in the human brain. III. Autoradiographic mapping of serotonin-1 receptors, *Neuroscience* **21:**97–122.
Pazos, A., Probst, A., and Palacios, J. M., 1987b, Serotonin receptors in the human brain. IV. Autoradiographic mapping of serotonin-2 receptors, *Neuroscience* **21:**123–139.
Perenin, M.-T., and Vighetto, A., 1988, Optic ataxia: A specific disruption in visuomotor mechanisms. I. Different aspects of the deficit in reaching for objects, *Brain* **111:**643–674.
Perkel, D. J., Bullier, J., and Kennedy, H., 1986, Topography of the afferent connectivity of area 17 in the macaque monkey: A double-labelling study, *J. Comp. Neurol.* **253:**374–402.
Peterhans, E., and von der Heydt, R., 1989, Mechanisms of contour perception in monkey visual cortex. II. Contours bridging gaps, *J. Neurosi.* **9:**1749–1763.
Probst, Th., Plendl, H., Paulus, W., Wist, E., and Scherg, M., 1993, Identification of the visual motion area (area V5) in the human brain by dipole source analysis, *Exp. Brain Res.* **93:**345–351.
Rakic, P., Goldman-Rakic, P. S., and Gallager, D., 1988, Quantitative autoradiography of major neurotransmitter receptors in the monkey striate and extrastriate cortex, *J. Neurosci.* **8:**3670–3690.
Richards, J. G., Schoch, P., Häring, P., Takacs, B., and Möhler, H., 1987, Resolving $GABA_A$/benzodiazepine receptors: Cellular and subcellular localization in the CNS with monoclonal antibodies, *J. Neurosci.* **7:**1866–1886.
Rizzo, M., Nawrot, M., Blake, R., and Damasio, A., 1992, A human visual disorder resembling area V4 dysfunction in the monkey, *Neurology* **42:**1175–1180.
Rockland, K., 1992a, Laminar distribution of neurons projecting from area V1 to V2 in macaque and squirrel monkeys, *Cerebral Cortex* **2:**38–47.
Rockland, K., 1992b, Configuration, in serial reconstruction of individual axons projecting from area V2 to V4 in the macaque monkey, *Cerebral Cortex* **2:**353–374.
Rockland, K., 1994, The organization of feedback connections from area V2 (18) to V1 (17), in: *Cerebral Cortex,* Volume 10, *Primary Visual Cortex in Primates* (A. Peters, and K. Rockland, eds.), Plenum Press, New York, pp. 261–299.
Rockland, K., 1995, Morphology of individual axons projecting from area V2 to MT in the macaque, *J. Comp. Neurol.* **355:**15–26.
Rockland, K. S., and Pandya, D. N., 1979, Laminar origins and terminations of cortical connections of the occipital lobe in the rhesus monkey, *Brain Res.* **179:**3–20.
Rockland, K. S., and Pandya, D. N., 1981, Cortical connections of the occipital lobe in the rhesus monkey: Interconnections between areas 17, 18, 19 and the superior temporal sulcus, *Brain Res.* **212:**249–270.
Rockland, K. S., and Van Hoesen, G. W., 1994, Direct temporal-occipital feedback connections to striate cortex (V1) in the macaque monkey, *Cerebral Cortex* **4:**300–313.
Rockland, K., Saleem, K., and Tanaka, K., 1994, Divergent feedback connections from areas V4 and TEO in the macaque, *Visual Neurosci.* **11:**579–600.
Rodman, H. R., Gross, C. G., and Albright, T. D., 1989, Afferent basis of visual response properties in area MT of the macaque. I. Effects of striate cortex removal, *J. Neurosci.* **9:**2033–2050.
Roland, P. E., and Gulyás, B., 1994, Visual imagery and visual representation, *TINS* **17**(7)**:**281–287.
Roland, P. E., and Gulyás, B., 1995, Visual memory, visual imagery, and visual recognition of large field patterns by the human brain: Functional anatomy by positron emission tomography, *Cerebral Cortex* **5**(1)**:**79–93.
Roland, P. E., and Zilles, K., 1994, Brain atlases—A new research tool, *TINS* **17:**458–467.
Rosa, M. G. P., Soares, J. G. M., Fiorani, Jr., M., and Gattass, R., 1993, Cortical afferents of visual area MT in the *Cebus* monkey, possible homologies between New and Old World monkeys, *Visual Neurosci.* **10:**827–855.
Rosier, A. M., Orban, G. A., and Vandesande, F., 1990, Regional distribution of binding sites for

neuropeptide Y in cat and monkey visual cortex determined by *in vitro* receptor autoradiography, *J. Comp. Neurol.* **293:**486–498.

Rosier, A. M., Vandesande, F., and Orban, G. A., 1991a, Laminar and regional distribution of galanin binding sites in cat and monkey visual cortex determined by *in vitro* receptor autoradiography, *J. Comp. Neurol.* **305:**264–272.

Rosier, A. M., Leroux, P., Vaudry, H., Orban, G. A., and Vandesande, F., 1991b, Distribution of somatostatin receptors in the cat and monkey visual cortex demonstrated by *in vitro* receptor autoradiography, *J. Comp. Neurol.* **310:**189–199.

Rosier, A., Arckens, L., Orban, G., and Vandesande, F., 1993, Laminar distribution of NMDA receptors in cat and monkey visual cortex visualized by [^{3}H]-MK-801 binding, *J. Comp. Neurol.* **335:**369–380.

Sakai, K., Watanabe, E., Onodera, Y., Uchida, I., Kato, H., Yamamoto, E, Koizumi, H., and Miyashita, Y., 1995, Functional mapping of the human color centre with echo-planar magnetic resonance imaging, *Proc. R. Soc. Lond. B Biol. Sci.* **261:**89–98.

Sanides, F., and Vitzthum, H., 1965a, Zur Architektonik der menschlichen Sehrinde und den Prinzipien ihrer Entwicklung, *Dtsch. Z. Nervenheilk.* **187:**680–707.

Sanides, F., and Vitzthum, H., 1965b, Die Grenzerscheinungen am Rande der menschlichen Sehrinde, *Dtsch. Z. Nervenheilk.* **187:**708–719.

Sarkissov, S. A., Filimonof, I. N., Kononowa, E. P., Preobraschenskaja, I. S., and Kukuew, L. A., 1955, *Atlas of the Cytoarchitectonics of the Human Cerebral Cortex,* Medgiz, Moscow.

Schein, S. J., and Desimone, R., 1990, Spectral properties of V4 neurons in the macaque, *J. Neurosci.* **10:**3369–3389.

Schein, S. J., Marrocco, R. T., and de Monasterio, F. M., 1982, Is there a high concentration of color-selective cells in V4 of monkey visual cortex? *J. Neurophysiol.* **47:**193–213.

Schiller, P. H., and Lee, K., 1991, The role of the primate extrastriate area V4 in vision, *Science* **251:**1251–1253.

Sereno, M. I., and Allman, J. M., 1991, Cortical visual areas in mammals, in: *The Neural Basis of Visual Function* (A. G. Leventhal, ed.), Macmillan, London, pp. 160–172.

Sereno, M. I., Dale, A. M., Reppas, J. B., Kwong, K. K., Belliveau, J. W., Brady, T. J., Rosen, B. R., and Tootell, R. B. H., 1995, Borders of multiple visual areas in humans revealed by functional magnetic resonance imaging, *Science* **268:**889–893.

Shipp, S., and Zeki, S., 1985, Segregation of pathways leading from V2 to areas V4 and V5 of macaque monkey visual cortex, *Nature* **315:**322–325.

Shipp, S., and Zeki, S., 1989a, The organization of connections between areas VI and V5 in the macaque monkey visual cortex, *Eur. J. Neurosci.* **1:**309–332.

Shipp, S., and Zeki, S., 1989b, The organization of connections between areas V5 and V2 in macaque monkey visual cortex, *Eur. J. Neurosci.* **1:**333–354.

Shipp, S., and Zeki, S., 1995, Segregation and convergence of specialised pathways in macaque visual cortex, *J. Anat.* **187:**547–562.

Shipp, S., de Jong B. M., Zihl, J., Frackowiak, R. S. J., and Zeki, S., 1994, The brain activity related to residual motion vision in a patient with bilateral lesions of V5, *Brain* **117:**1023–1038.

Shomura, K. T., Ando, T., and Kato, K., 1975, Structural organization of "callosal" OBγ in human corpus callosum agenesis, *Brain Res.* **93:**241–252.

Simma, K., 1954, Die thalamocorticale Projektion beim Menschen, *Monatssehr. Psychiat. Neurol.* **127:**301–316.

Smiley, J. F., Levey, A. I., Ciliax, B. J., and Goldman-Rakic, P. S., 1994, D1 dopamine receptor immunoreactivity in human and monkey cerebral cortex: Predominant and extrasynaptic localization in dendritic spines, *Proc. Natl. Acad. Sci. USA* **91:**5720–5724.

Spatz, W. B., 1975, An efferent connection of the solitary cells of Meynert. A study with horseradish peroxidase in the marmoset *Callithrix, Brain Res.* **92:**450–455.

Spatz, W. B., 1977, Topographically organized reciprocal connections between area 17 and MT visual area of superior temporal sulcus, in the marmoset *Callithrix jacchus, Exp. Brain. Res.* **27:**559–572.

Spatz, W. B., and Kunz, B., 1984, Area 17 of anthropoid primates does participate in visual callosal connections, *Neurosci. Lett.* **48:**49–53.

Standage, G. P., and Benevento, L. A., 1983, The organization of connections between the pulvinar and visual area MT in the macaque monkey, *Brain Res.* **262:**288–294.

Steele, G. E., Weller, R. E., and Cusick, C. G., 1991, Cortical connections of the caudal subdivision of the dorsolateral area (V4) in monkeys, *J. Comp. Neurol.* **306:**495–520.

Takeuchi, Y., and Sano, Y., 1983, Immunohistochemical demonstration of serotonin nerve fibers in the neocortex of the monkey (*Macaca fuscata*), *Anat. Embryol.* **166:**155–168.

Tanaka, M., Lindsley, E., Lausmann, S., and Creutzfeld, O. D., 1990, Afferent connections of the prelunate visual association cortex (areas V4 and DP), *Anat. Embryol.* **181**(1)**:**19–30.

Tigges, J., Tigges, M., Anschel, S., Cross, N. A., Letbetter, W. D., and McBride, R. L., 1981, Areal and laminar distribution of neurons interconnecting the central visual cortical areas 17, 18, 19 and MT in the squirrel monkey (*Saimiri*), *J. Comp. Neurol.* **202:**539–560.

Tootell, R. B. H., and Born, R. T., 1990, Patches and direction columns in primate area MT, *Invest. Ophthalmol. Vis. Sci.* **31:**238.

Tootell, R. B. H., and Hamilton, S. L., 1989, Functional anatomy of the second visual area (V2) in the macaque, *J. Neurosci.* **9:**2620–2644.

Tootell, R. B. H., and Taylor, J. B., 1995, Anatomical evidence for MT and additional cortical visual areas in humans, *Cerebral Cortex* **5:**39–55.

Tootell, R. B. H., Silverman, M. S., DeValois, R. L., and Jacobs, G. H., 1983, Functional organization of the second visual cortical area of primates, *Science* **220:**737–739.

Tootell, R. B. H., Hamilton, S. L., and Silverman, M. S., 1985, Topography of cytochrome oxidase activity in owl monkey cortex, *J. Neurosci.* **5:**2786–2800.

Tootell, R. B. H., Hamilton, S. L., Silverman, M. S., and Switkes, E., 1988a, Functional anatomy of macaque striate cortex. I. Ocular dominance, binocular interactions and baseline conditions, *J. Neurosci.* **8:**1500–1530.

Tootell, R. B. H., Hamilton, S. L., and Switkes, E., 1988b, Functional anatomy of macaque striate cortex. IV. Contrast and magno-parvo streams, *J. Neurosci.* **8:**1594–1609.

Tootell, R. B. H., Born, R. T., and Ash-Bernal, R., 1993, Columnar organization in visual cortex in non-human primates and man, in: *Functional Organisation of the Human Visual Cortex* (B. Gulyás, D. Ottosson, and P. E. Roland, eds.), Pergamon, New York, pp. 59–74.

Tootell, R. B. H., Reppas, J. B., Kwong, K. K., Malach, R., Born, R., Brady, T. J., Rosen, B. R., and Belliveau, J. W., 1995a, Functional analysis of human MT and related visual cortical areas using magnetic resonance imaging, *J. Neurosci.* **15:**3215–3230.

Tootell, R. B. H., Reppas, J. B., Dale, A. M., Look, R. B., Sereno, M. I., Malach, R., Brady, T. J., and Rosen, B. R., 1995b, Visual motion aftereffect in human cortical area MT revealed by functional magnetic resonance imaging, *Nature* **375:**139–141.

Trevarthen, C., 1990, *Brain Circuits and Function*, Cambridge University Press, Cambridge.

Tusa, R. J., and Ungerleider, L. G., 1985, The inferior longitudinal fasciculus: A reexamination in humans and monkeys, *Ann. Neurol.* **18:**583–591.

Ungerleider, L. G., and Desimone, R., 1986a, Projections to the superior temporal sulcus from the central and peripheral field representations of V1 and V2, *J. Comp. Neurol.* **248:**147–163.

Ungerleider, L. G., and Desimone, R., 1986b, Cortical connections of visual area MT in the macaque, *J. Comp. Neurol.* **248:**190–222.

Ungerleider, L. G., and Haxby, J. V., 1994, "What" and "where" in the human brain, *Curr. Opin. Neurobiol.* **4:**157–165.

Ungerleider, L. G., and Mishkin, M., 1979, The striate projection zone in the superior temporal sulcus of *Macaca mulatta:* Location and topographic organization, *J. Comp. Neurol.* **188:**347–366.

Ungerleider, L. G., and Mishkin, M., 1982, Two cortical visual systems, in: *Analysis of Visual Behavior* (D. J. Ingle, M. A. Goodale, and R. J. W. Mansfield, eds.), MIT Press, Cambridge, MA, pp. 549–586.

Ungerleider, L. G., Desimone, R., Galkin, T. W., and Mishkin, M., 1984, Subcortical projections of area MT in the macaque, *J. Comp. Neurol.* **223**(3)**:**368–386.

Vaina, L., 1989, Selective impairment of visual motion interpretation following lesions of the right occipito-parietal area in humans, *Biol. Cybernet.* **61:**347–359.

Van Buren, J. M., and Borke, R. C., 1972, *Variations and Connections of the Human Thalamus. 1. The Nuclei and Cerebral Connections of the Human Thalamus,* Springer-Verlag, Berlin.

Van Essen, D. C., 1985, Functional organization of primate visual cortex, in: *Cerebral Cortex,* Volume 3, *Visual Cortex* (A. Peters and E. G. Jones, eds.), Plenum Press, New York, pp. 259–329.

Van Essen, D. C., and Maunsell, J. H. R., 1983, Hierarchical organization and functional streams in the visual cortex, *Trends Neurosci.* **6:**370–375.

Van Essen, D. C., and Zeki, S. M., 1978, The topographic organization of rhesus monkey prestriate cortex, *J. Physiol. (Lond.)* **277:**193–226.

Van Essen, D. C., Maunsell, J. H. R., and Bixby, J. L., 1981, The middle temporal visual area in the macaque monkey: Myeloarchitecture, connections, functional properties and topographic organization, *J. Comp. Neurol.* **199:**293–326.

Van Essen, D. C., Newsome, W. T., and Bixby, J. L., 1982, The pattern of interhemispheric connections and its relationship to extrastriate visual areas in the macaque monkey, *J. Neurosci.* **2:**265–283.

Van Essen, D. C., Newsome, W. T., Maunsell, J. H. R., and Bixby, J. L., 1986, The projections from striate cortex (V1) to areas V2 and V3 in the macaque monkey: Asymmetries, areal boundaries, and patchy connections, *J. Comp. Neurol.* **224:**451–480.

Vannucchi, M. G., and Goldman-Rakic, P. S., 1991, Age-dependent decrease in the affinity of muscarinic M1 receptors in neocortex of rhesus monkeys, *Proc. Natl. Acad. Sci. USA* **88:**11475–11479.

Van Valkenburg, C. T., 1908, Zur Anatomie der Projections- und Balkenstrahlung des Hinterhauptlappens sowie des Cingulum, *Monatsschr. Psychiat. Neurol.* **24:**320–339.

Van Valkenburg, C. T., 1913, Experimental and pathologico-anatomical researches on the corpus callosum, *Brain* **36:**119–165.

Vogt, O., 1903, Zur anatomischen Gliederung des Cortex cerebri, *J. Psychol. Neurol.* **2:**160–180.

Vogt, M., 1929, Über fokale Besonderheiten der Area occipitalis, *J. Psychol. Neurol.* **39:**506–510.

Vogt, C., and Vogt, O., 1919, Allgemeinere Ergebnisse unserer Hirnforschung. Vierte Mitteilung. Die physiologische Bedeutung der architektonischen Rindenfelderung auf Grund neuer Rindenreizungen, *J. Psychol. Neurol.* **25:**399–462.

Vogt Weisenhorn, D. M., Illing, R.-B., and Spatz, W. B., 1995, Morphology and connections of neurons in area 17 projecting to the extrastriate areas MT and 19DM and to the superior colliculus in the monkey *Callithrix jacchus*, *J. Comp. Neurol.* **362:**233–255.

Von der Heydt, R., and Peterhans, E., 1989, Mechanisms of contour perception in monkey visual cortex. I. Lines of pattern discontinuity, *J. Neurosci.* **9:**1731–1748.

von Economo, C., and Koskinas, G. N., 1925, *Die Cytoarchitektonik der Hirnrinde des erwachsenen Menschen,* Springer, Vienna.

Walker, A. E., 1938, *The Primate Thalamus,* Chicago University Press, Chicago.

Walsh, V., Carden, D., Butler, S. R., and Kulikowski, J. J., 1993, The effects of V4 lesions on the visual abilities of macaques: Hue discrimination and colour constancy, *Behav. Brain Res.* **53:**51–62.

Watson, J. D. G., Myers, R., Frackowiak, R. S. J., Hajnal, J. V., Woods, R. P., Mazziota, J. C., Shipp, S., and Zeki, S., 1993, Area V5 of the human brain: Evidence from a combined study using positron emission tomography and magnetic resonance imaging, *Cerebral Cortex* **3:**79–94.

Webster, M., Bachevalier, J., and Ungerleider, L., 1993, Subcortical connections of inferior temporal areas TE and TEO in macaque monkeys, *J. Comp. Neurol.* **335**(1):73–91.

Weller, R. E., 1988, Two cortical visual systems in Old World and New World primates, in: *Progress in Brain Research,* Volume 75 (T. P. Hicks and G. Benedek, eds.), Elsevier, Amsterdam, pp. 293–306.

Weller, R. E., and Kaas, J. H., 1983, Retinotopic patterns of connections of area 17 with visual areas V-II and MT in macaque monkeys, *J. Comp. Neurol.* **220:**253–279.

Weller, R. E., Wall, J. T., and Kaas, J. H., 1984, Cortical connections of the middle temporal visual area (MT) and the superior temporal cortex in owl monkeys, *J. Comp. Neurol.* **228:**81–104.

Wild, H. M., Butler, D., Carden, D., and Kullikowski, J. J., 1985, Primate cortical area V4 important for colour constancy but not wavelength discrimination, *Nature* **313:**133–135.

Wong-Riley, M. T. T., 1993, Cytochrome oxidase studies on the human visual system, in: *Functional Organisation of the Human Visual Cortex* (B. Gulyás, D. Ottoson, and P. E. Roland, eds.), Pergamon, New York, pp. 165–180.

Wong-Riley, M. T. T., 1994, Primate visual cortex. Dynamic metabolic organization and plasticity revealed by cytochrome oxidase, in: *Cerebral Cortex,* Volume 10, *Primary Visual Cortex in Primates* (A. Peters, and K. S. Rockland, eds.), Plenum Press, New York, pp. 141–200.

Wong-Riley, M., and Carroll, E., 1984, Cytochrome oxidase-rich zones in primate visual cortex, *Invest. Ophthalmol. Vis. Sci.* **25:**16–109.

Wong-Riley, M., Hevner, R., Cutlan, R., Earnest, M., Egan, R., Frost, J., and Nguyen, T., 1993,

Cytochrome oxidase in the human visual cortex; distribution in the developing and the adult brain, *Visual Neurosci.* **10:**41–58.

Yoshida, K., and Benevento, L. A., 1981, The projection from the dorsal lateral geniculate nucleus of the thalamus to extrastriate visual association cortex in the macaque monkey, *Neurosci. Lett.* **22**(2):103–108.

Young, M. P., 1992, Objective analysis of the topological organization of the primate cortical visual system, *Nature* **358:**152–155.

Young, W. S., and Kuhar, M. J., 1979, Autoradiographic localisation of benzodiazepine receptors in the brains of humans and animals, *Nature* **280:**393–395.

Young, A. B., and Penney, J. B., 1991, Bezodiazepine, GABA, and glutamate receptors in cerebral cortex, hippocampus, basal ganglia, and cerebellum, in: *Receptors in the Human Nervous System* (F. A. O. Mendelsohn and G. Paxinos, eds.), Academic Press, San Diego, pp. 9–47.

Yukie, M., and Iwai, E., 1981, Direct projection from the dorsal lateral geniculate nucleus to the prestriate cortex in macaque monkeys, *J. Comp. Neurol.* **201:**81–97.

Zeki, S. M., 1969, Representation of central visual fields in prestriate cortex of monkey, *Brain Res.* **14:**271–291.

Zeki, S. M., 1970, Interhemispheric connections of prestriate cortex in monkey, *Brain Res.* **19:**63–75.

Zeki, S. M., 1971a, Convergent input from the striate cortex (area 17) to the cortex of the superior temporal sulcus in the rhesus monkey, *Brain Res.* **28:**338–340.

Zeki, S. M., 1971b, Cortical projections from two prestriate areas in the monkey, *Brain Res.* **34:**19–35.

Zeki, S. M., 1973, Colour coding in rhesus monkey prestriate cortex, *Brain Res.* **53:**422–427.

Zeki, S. M., 1974, Functional organization of a visual area in the posterior bank of the superior temporal sulcus of the rhesus monkey, *J. Physiol.* (Lond.) **236:**549–573.

Zeki, S. M., 1975, The projection to the superior temporal sulcus from areas 17 and 18 in the rhesus monkey, *Proc. R. Soc. Lond. B Biol. Sci.* **193:**199–207.

Zeki, S. M., 1977a, Colour coding in the superior temporal sulcus of rhesus monkey visual cortex, *Proc. R. Soc. Lond. B Biol. Sci.* **197:**195–223.

Zeki, S. M., 1977b, Simultaneous anatomical demonstration of the vertical and horizontal meiridans in areas V2 and V3 of rhesus monkey visual cortex, *Proc. R. Soc. Lond. B Biol. Sci.* **195:**517–523.

Zeki, S. M., 1978a, Functional specialization in the visual cortex of the rhesus monkey, *Nature* **274:**423–428.

Zeki, S. M., 1978b, The third visual complex of rhesus monkey prestriate cortex, *J. Physiol.* (Lond.) **277:**245–277.

Zeki, S. M., 1978c, Uniformity and diversity of structure and function in rhesus monkey prestriate cortex, *J. Physiol.* (Lond.) **277:**273–290.

Zeki, S. M., 1980a, The response properties of cells in the middle temporal area (area MT) of owl monkey visual cortex, *Proc. R. Soc. Lond. B Biol. Sci.* **207:**239–248.

Zeki, S. M., 1980b, A direct projection from area VI to area V3a of rhesus monkey visual cortex, *Proc. R. Soc. Lond. B Biol. Sci.* **207:**499–506.

Zeki, S., 1980c, The representation of colours in the cerebral cortex, *Nature* **284:**412–418.

Zeki, S., 1983a, Colour coding in the cerebral cortex: The reaction of cells in monkey visual cortex to wavelengths and colours, *Neuroscience* **9:**741–765.

Zeki, S., 1983b, The distribution of wavelength and orientation selective cells in different areas of monkey visual cortex, *Proc. R. Soc. Lond. B Biol. Sci.* **217:**449–470.

Zeki, S., 1985, Colour pathways and hierarchies in the cerebral cortex, in: *Central and Peripheral Mechanisms of Colour Vision* (D. Ottoson, and S. Zeki, eds.), Macmillan, London, pp. 19–44.

Zeki, S., 1986, The anatomy and physiology of area V6 of macaque monkey visual cortex, *J. Physiol.* (Lond.) **381:**62P.

Zeki, S., 1990a, Colour vision and functional specialisation in the visual cortex, *Disc. Neurosci.* **6:** 7–64.

Zeki, S., 1990b, A century of cerebral achromatopsia, *Brain* **113:**1721–1777.

Zeki, S. M., 1990c, Parallelism and functional specialization in human visual cortex, *Cold Spring Harbor Symp. Quant. Biol.* **55:**651–661.

Zeki, S., 1991, Cerebral akinetopsia (visual motion blindness): A review, *Brain* **114:**811–824.

Zeki, S., 1993, *A Vision of the Brain,* Blackwell, Oxford.

Zeki, S. M., and Sandeman, D. R., 1976, Combined anatomical and electrophysiological studies on the boundary between the second and third visual areas of rhesus monkey visual cortex, *Proc. R. Soc. Lond. B Biol. Sci.* **194:**555–562.

Zeki, S., and Shipp, S., 1988, The functional logic of cortical connections, *Nature* **335:**311–317.

Zeki, S., and Shipp, S., 1989, Modular connections between areas V2 and V4 of macaque monkey visual cortex, *Eur. J. Neurosci.* **1:**494–506.

Zeki, S., Watson, J. D. G., Lueck, C. J., Friston, K. J., Kennard, C., and Frackowiak, R. S. J., 1991, A direct demonstration of functional specialization in human visual cortex, *J. Neurosci.* **11:**641–649.

Zeki, S. M., Watson, J. D., Frackowiak, R. S. J., 1993, Going beyond the information given: The relation of illusory visual motion to brain activity, *Proc. R. Soc. Lond. B Biol. Sci.* **252:**215–222.

Zihl, J., von Cramon, D., and Mai, N., 1983, Selective disturbance of movement vision after bilateral brain damage, *Brain* **106:**313–340.

Zilles, K., 1991, Codistribution of receptors in the human cerebral cortex, in: *Receptors in the Human Nervous System* (F. A. O. Mendelsohn, and G. Paxinos, eds.), Academic Press, San Diego, pp. 165–206.

Zilles, K., 1992, Receptors in the central nervous system. Neurotransmitter receptors in the forebrain: Regional and laminar distribution, in: *Histochemistry of Receptors,* (W. Graumann and J. Drukker, eds.), Fischer-Verlag, Stuttgart, pp. 229–240.

Zilles, K., and Schleicher, A., 1993, Cyto- and myeloarchitecture of human visual cortex and the periodical $GABA_A$ receptor distribution, in: *Functional Organization of the Human Visual Cortex* (B. Gulyás, D. Ottoson, and P. Roland, eds.), Pergamon Press, Oxford, pp. 111–121.

Zilles, K., and Schleicher, A., 1995, Correlative imaging of transmitter receptor distributions in human cortex, in: *Autoradiography and Correlative Imaging* (W. E. Stumpf and H. F. Solomon, eds), Academic Press, San Diego, pp. 277–307.

Zilles, K., Werners, R., Büsching, U., and Schleicher, A., 1986, Ontogenesis of the laminar structure in areas 17 and 18 of the human visual cortex. A quantitative study, *Anat. Embryol.* **174:**339–353.

Zilles, K., zur Nieden, K., Schleicher, A., and Traber, J., 1990, A new method for quenching correction leads to revisions of data in receptor autoradiography, *Histochemistry* **94:**569–578.

Zilles, K., Qü, M., Schröder, H., and Schleicher, A., 1991, Neurotransmitter receptors and cortical architecture, *J. Hirnforsch.* **32:**343–356.

Zilles, K., Qü, M., and Schleicher, A., 1993, Regional distribution and heterogeneity of a-adrenoceptors in the rat and human central nervous system, *J. Hirnforsch.* **34:**123–132.

16

Functional Organization of Human Visual Cortical Areas

BALÁZS GULYÁS

1. Introduction

The visual cortex in lower mammals and primates occupies a relatively large part of the neocortex, and this is one indication of the uniquely important role of the processing and analysis of visual information in the brain. This may, however, be too broad an observation, as the notion "visual cortex" is after all loosely defined. The classical anatomical definition (Cajal, 1899a, b) was that the visual cortex in several species, including human, monkey, and dog, has a special anatomical character; namely an intermediary white "line," stria, or layer visible to the naked eye (stria Gennari, described independently by Francesco Gennari, 1782, and Félix Vicq-d'Azyr, 1786). This has proved to be far too narrow a definition, as the part of the cortex characterized by this landmark includes only the primary visual area within the visual cortex. Clinical and anatomical studies around the turn of the century brought into the picture "the visual association cortex" (for review see Zeki, 1993; Kuljis, 1994; Rizzo, 1994; and the chapter by Gross, in this volume), the second visual area next to the striate cortex "constituted for the elaboration and interpretation of [visual] sensations" (Campbell, 1905). In the broadest sense, the neurobiological literature also includes cortical areas or corti-

BALÁZS GULYÁS • Division of Human Brain Research, Department of Neuroscience, Karolinska Institute, S-17177 Stockholm, Sweden.

Cerebral Cortex, Volume 12, edited by Rockland *et al.* Plenum Press, New York, 1997.

cal regions as part of the visual cortex if their main functions are related to the processing and analysis of visual information, whether they be located in the occipital lobe and its neighboring parietal and temporal cortices or at sites far away from it (e.g., frontal eye field; see, e.g., Drury *et al.*, 1996; and the chapters by Siegal and Read and by Schall, in this volume). No perfect consensus exists as to precisely what the extent of the visual cortex is and how to define it in the best way. One reason is that certain cortical areas may participate in various sensory or sensory–motor operations, for instance, in visuotactile or visuomotor "transformations." These areas may be a part of both the visual and sensory or visual and motor systems. Other "multimodal" areas may also show different functional features, depending upon task requirements.

As many of the techniques used to define visual cortical areas in monkeys cannot be used in human visual cortex research, and research with techniques appropriate for human cortex is still in its early phase, the exact definition of "human visual cortex" remains putative and a final answer is still to come.

The present chapter focuses on the functional mapping and functional organisation of visual cortical areas in humans.

2. Techniques Applicable to the Functional Exploration of Human Visual Cortex

Primate studies on the determination and localization of visual cortical areas can use a large battery of techniques, including quantitative and qualitative anatomical measurements (analysis of cyto-, myelo-, and receptor-architectural patterns), the analysis of cortical connection patterns, single-cell recordings (for purposes such as topographic mapping, exploration of functional properties of neurons, measuring delays), and psychophysical studies combined with well-controlled ablations or temporal suspension of cortical functions.

Whereas a number of anatomical methods can be used in both monkey and human studies, most of the techniques widely used in the exploration of the functional organization of monkey visual cortex cannot be used in human studies. Some of the most important methodologies, such as single-cell recording or ablations combined with pre- and postlesion psychophysics, essential for the determination of the functional properties of visual areas and their constituent neurons, are not present in the usual battery of research into the human visual cortex (Table I).

Complementing this repertoire, there are basically two main sources of information about the functional organization of human visual cortex: (1) clinical (neurological or neurosurgical) observations on patients with visual deficits, and (2) brain activation studies on normal subjects and patients, using functional imaging techniques, including positron emission tomography (PET) and functional nuclear magnetic resonance imaging (fMRI).

1.) Direct information about the functional organization of visual cortical areas in humans came in the past mainly from clinical (neurological or neurosurgical) observations: (a) the exploration of neurological patients with well-circumscribed anatomical lesions accompanied by well-defined visual deficits,

Table I. Techniques Used in the Exploration of Monkey and Human Visual Cortex

	Monkey	Human	Some characteristic examples for human studies
Anatomy			
Postmortem, including	√	√	
cytoarchitecture			Zilles (1990)
myelo-architecture			Tootell and Taylor (1995)
receptor architecture			Zilles (1991)
cytochrome oxidase			Clarke (1993), Wong-Riley (1993)
other stains			Tootell *et al.* (1995a, b)
intracortical connections			Burkhalter and Bernardo (1989), Burkhalter (1993)
callosal connections			Clarke and Miklossy (1990)
In vivo labeling + postmortem	√	—	
In vivo lesions + postmortem	√	√	Clarke and Miklossy (1990)
Lesions, ablations			
With both pre- and postlesion psychophysics	√	—	
With postlesion psychophysics	√	√	Zihl *et al.* (1983), Heywood *et al.* (1992)
Single-cell recording			
Topographic mapping	√	—	
Functional properties	√	—	
Temporal measurements	√	—	
Electrophysiological recording			
Surface recording (ECoG)	√	√	Allison *et al.* (1994)
Transcranial (EEG, VEP, MEG) temporal measurements	—	√	Buchner *et al.* (1994), ffytche *et al.* (1995)
Stimulation			
Direct	√	√	Brindley and Lewin (1968), Dobelle *et al.* (1979), Fried *et al.* (1982), Bak *et al.* (1990)
Transcranial	√	√	Meyer and Diehl, 1992
Optical imaging	√	—	
Functional brain imaging techniques (PET, fMRI)			
Metabolic mapping	(—)	√	
Receptor mapping	(—)	√	
Topographic mapping	(—)	√	Fox *et al.*(1986, 1987), Sereno *et al.* (1995), Tootell *et al.* (1995a,b), Belliveau *et al.* (1991), Corbetta *et al.* (1991), Gulyás and Roland (1991), Haxby *et al.* (1991), Zeki *et al.* (1991)
Functional mapping	(—)	√	

and (b) direct stimulation of visual cortex during neurosurgical interventions or under chronic conditions.

(a) Patient studies proved to be the essential way to explore the organization of striate cortex in the beginning of the century (Holmes and Horrax, 1919; Kuljis, 1994; Rizzo, 1994) and they were for decades the most important source for the localization of visual functions in strict anatomical context (see, e.g., Poppelreuter, 1917; Teuber *et al.*, 1960). Patient investigations combining psychophysical measurements and neuroimaging studies locating the damaged or nonfunctional brain regions are still an important source of information on human visual cortical organization.

Basically, there have been two major conclusions drawn from these cases. (i) Wounds localized to the pole as well as the cuneal and lingual parts of the occipital lobe revealed a strict topographic organization of the external visual field in the striate cortex (Holmes, 1945; Spalding, 1952a, b). (ii) Another major outcome of studies on patients with circumscribed anatomical lesions in their visual cortex was the accumulating body of evidence that in the extrastriate visual cortex partially separated modality-specific visual subsystems exist. A relatively large number of studies have dealt with selective functional loss of visual submodalities following lesions in the human brain related to color (cerebral achromatopsia) (Meadows, 1974a; Heywood and Cowey, 1987; Zeki, 1990), forms and faces (object agnosia, prosopagnosia; Benson and Greenberg, 1969; Meadows, 1974b; Landis *et al.*, 1988), motion (akinetopsia; Zihl, 1983; Vaina, 1989; McLeod *et al.*, 1989; Hess *et al.*, 1989; Regan *et al.*, 1992; Vaina, 1994), and stereovision (astereopsis; Holmes and Horrax, 1919; Richards, 1970; Blake and Cormack, 1979; Rizzo, 1989). Though most of these studies focus on one single visual submodality, the symptoms are usually accompanied by the disturbance of a number of visual functions, and the underlying anatomical lesions occupy rather large parts of the cortex and/or white matter. This fact limits the value of lesion studies in humans. However, these observations have already suggested that the processing and analysis of various visual cues take place in partially disparate networks of neuronal populations in extrastriate visual cortex before an integrated visual percept is generated in the brain.

(b) In addition to observations in lesion patients, the direct stimulation of visual cortex during neurosurgical operations or by implanted electrodes has also contributed to our understanding of the functional organization of human visual cortex. Penfield and Rasmussen's (1957) classic study on the localization of functions in the cerebral cortex explored the major features of visual cortical topography and, in broad lines, the functional division between striate and extrastriate visual regions. Although more detailed studies, using chronically implanted surface electrodes, have contributed to our knowledge on visual topography (e.g., Dobelle *et al.*, 1979) or localization of functions (e.g., Allison *et al.*, 1994), the scope of these techniques is limited and their role in visual cortex research has been dwarfed by the fast-developing functional imaging techniques.

2. The advance of functional imaging techniques, including positron emission tomography (PET) and functional nuclear magnetic resonance imaging (fMRI), has provided as with a unique opportunity to explore the functional organization of human visual cortex. In their most common form of application, both methodologies are based upon the fact that under normal circumstances regional cerebral blood flow (rCBF) changes are faithful indicators of the metabolic changes in synaptic levels which accompany neuronal activity (Roland,

1993). Changes in neuronal activities can therefore be localized and indirectly measured by measuring and localizing changes in rCBF with PET or fMRI, or, alternatively, those in blood oxygenation levels. The spatial resolution of these techniques, however, lags behind the spatial resolution of anatomical magnetic resonance imaging, and for this reason the functional images of the brain obtained by PET or fMRI should be complemented with high-precision anatomical images obtained with MRI, in order to combine anatomical and functional information. In addition to the aforementioned functional imaging with PET or fMRI, the application of these techniques in brain research in general and in human visual cortex research in particular is much wider. Similarly, with PET, a number of metabolic parameters can be quantitatively measured or the receptor architecture of the human brain can be mapped.

3. What the Monkey Brain Tells Us about the Human Visual Cortex

The main source of our knowledge about the functional organization of visual cortex is the primate brain, which has been studied extensively during the past decades. Over 30 visual cortical areas have been mapped and defined with respect to their anatomical and physiological features (Felleman and Van Essen, 1991; Drury *et al.*, 1996). Visual cortical areas in the macaque occupy over 50% of the cerebral cortex in this species.

Decades of research into the functional organization of the primate visual system have resulted in a number of guiding principles, models, and working hypotheses of how the human visual system deals with the processing of visual information at early (subcortical, striate, and early extrastriate) and higher levels. These principles, models, or hypotheses not only serve as a basis for research into the functional organization of the human visual system, but provide us with an impetus to explore the human brain as a challenge to find similarities or eventually differences between humans and other primates.

(i) The prime guiding hypothesis is that the visual cortex is subdivided into anatomically distinct cortical visual areas that each contribute to the processing and analysis of visual information. The number of visual areas in primates is large; in the macaque monkey, as mentioned above, it is over 30. Some visual areas, including areas V1, V2, and MT, can be found in each primate species so far explored; and it is therefore likely that these areas are evolutionarily conserved and exist in humans as well (Tootell *et al.*, 1995a). Indeed, ample anatomical and physiological evidence proves the existence of these areas in humans. It has also been shown that the proportion of the sizes of these areas to each other is similar in various species (Horton and Hoyt, 1991; Tootell *et al.*, 1995a).

In addition, there is compelling evidence that all primates have multiple visual areas and that all or most primates, including humans, share at least several visual areas in addition to V1, V2, and MT (Kaas, 1993), and the anatomical microorganisation of these areas and intracortical connections between these areas are similar in various monkey taxa. All monkey species examined show in many respects a similar organization of the visual cortex, although not of all individual visual areas.

(ii) A dominant hypothesis, supported by a wealth of anatomical and physiological data, is that there is a "division of labor" in the analysis and processing of visual information, at least at early levels in the visual system. In other words, the contribution of various visual areas to the analysis of various visual submodalities or visual cues, such as color, motion, or disparity, is different. An extreme position regarding this hypothesis is that there is "functional specialization" between visual areas: for instance, there is an area uniquely engaged with the analysis of color signals, whereas other areas are specialized for, e.g., the processing of visual motion information alone (see, e.g., Zeki *et al.*, 1991).

This functional division of labor is related to the anatomical subdivisions of the visual pathways, most notably to the parvocellular—magnocellular (P–M) pathways, already present at the retinal level. The analysis of movement is mostly confined to the magnocellular lateral geniculate nucleus (LGN), layers 4Ca–4B (V1), thick stripe (V2), area V5 (MT), parietal visual area pathway (see the chapters by Siegel and Read and by Schall in this volume). The analysis of form (high-resolution, static form perception) is related to the parvocellular LGN, layer 4Cb, interblob (V1), interstripe (V2), (areas 3, 4), inferotemporal cortex pathway. The analysis of color is processed by the parvocellular LGN, layer 4Cb, blob (V1), thin stripe (V2), area 4, interferotemporal visual areas pathway (Casagrande and Kaas, 1994). Binocular interactions are present in the parvocellular pathway—in both the interblobs and layer 4B of V1, in the interstripe and thick stripe regions of V2, and in V3, V4, and V5. This scheme, however, indicates only a strong tendency rather than a strict rule, since anatomical connections relaying "cross-talk" between the different functional streams are present (Morel and Bullier, 1990; Boussaoud *et al.*, 1990, Van Essen *et al.*, 1992). Nonetheless, cross-species comparisons indicate that the geniculocortical connections (M and P pathways) are similarly distributed in the various monkey taxa.

(iii) Anatomical studies, single-cell recordings, and behavioral observations indicate that at exstrastriate levels the functional division of labor in the visual system results in a marked separation of visual information processing regarding the characteristics of a particular object ("object vision") and its/their relation to other objects, each other, or the rest of the visual scene ("spatial vision"). Anatomically two major "functional pathways" or "processing streams" can be described: the occipitotemporal one ("ventral stream"), predominantly participating in object vision, and the occipitoparietal one ("dorsal stream"), engaged primarily with spatial vision (Ungerleider and Mishkin, 1982).

(iv) The revisiting of the above hypothesis, based mainly upon clinical observations in humans has led to a new interpretation of the "two functional pathways" or "processing streams," indicating that the occipitotemporal pathway mediates the perception of objects, whereas the occipitoparietal pathways deals with the practical-pragmatic aspects of visual information and mediates the control of actions directed to seen objects (Goodale and Milner, 1992; Milner and Goodale, 1993; Goodale, 1993).

(v) Single-cell recordings and lesion studies in monkeys, supported by functional imaging studies in humans, indicate another "division of labor" of visual information processing: whereas early visual areas (the striate and neighboring extrastriate cortex) are heavily engaged in the analysis of elementary visual information ("computational areas"), visual areas further away from the striate cortex are more involved in the internal representation of visual information ("representational areas") (Roland and Gulyás, 1994).

Monkey studies support the view that there are more similarities than differences between the organization of the visual system of different primate species, but cautions have also been raised. For instance, it is known that there can be differences in single-unit properties or topographic arrangements in presumably homologous areas in different species, and that species differences exist in color vision, contrast sensitivity, acuity, and motion selectivity. These facts must be accounted for in the extrapolation of monkey models to the human brain.

4. Human Visual Cortical Areas and Human Visual Cortex

Cortical areas and their number, location, and borders have been determined on the basis of multiple criteria, including cyto-, myelo-, and receptor-architectonic differences, topographic organization, functional properties of cells, cortical connectivity patterns, and psychophysical–behavioral changes after lesions.

In the human only some of the above criteria for determining visual cortical areas can be used. Differences in cyto-, myelo-, and receptor architecture can clearly define early-level visual cortical areas (chapter by Zilles and Clarke in this volume). The cortical connectivity patterns of these areas have been explored to a certain extent as well (Burkhalter and Bernardo, 1989). The exploration of patient cases with cortical lesions is of somewhat less importance in the determination of cortical areas, as the lesions usually are of traumatic or vascular origin and do not respect anatomical borders either between cortical areas or between gray and white matter. However, early lesion studies were essential in the determination of the topographic organization of striate cortex; and investigation of patients with well-defined lesions can still result in confirmative or even innovative findings about models of visual cortex organization in human. With the exception of a few intraoperative cases, single-cell recordings can hardly be performed in human subjects. Therefore the functional properties of individual neurons and the distribution of neurons with similar features within an area, as well as the mapping of intraareal topography with receptive field mapping (two important criteria for determining a cortical area in animal studies), are not in the repertoire of human visual cortex research.

Instead, novel physiological criteria for defining *visual cortical areas* are coming to the fore with the advance of brain activation studies using functional imaging techniques. These can even be said to replace criteria based upon single-cell recording. (i) The topographic representation of the external visual field in early visual cortical areas can be explored to some extent with PET, and can be mapped thoroughly and precisely with fMRI. (ii) On the basis of monkey models, one can test whether certain regions in the visual cortex (i.e., the proposed human analogs of primate visual areas) respond more vigorously than others to specific visual cues, such as color, contrast, or motion. (iii) Using computerized brain atlases, unified stereotactic convention, and intersubject averaging techniques, one can put data obtained in various anatomical and functional imaging studies in register. This provides us with more and more robust data on the anatomical-functional properties of cortical regions. Evidence about a given cortical *region* showing different anatomical–functional features from its neighboring regions can result in the definition of a visual cortical *area*.

Naturally, these functional criteria are complemented, especially at lower levels, with quantitative anatomical criteria, including cyto-, myelo-, and receptor-architectural ones (see, e.g., the chapter by Zilles and Clarke in this volume; Zilles, 1990;, 1991; Roland and Zilles, 1994). In this respect, the definition of *a visual cortical area* in human is the result of an iterative process which includes the combination of anatomical and functional techniques.

As far as the extent of the human cerebral cortex involved in straightforward visual activities is concerned, (i.e., "*the human visual cortex*"), no precise measurements have been made. A tentative early picture of the extent of human visual cortex is shown in Fig. 1. The figure is based upon a large series of functional imaging experiments in which we used various visual paradigms in human subjects, including visual perception, discrimination, recognition, learning, memory, and imagery tests. In each case, the subject's brain images were standardized with respect to shape and size (Roland *et al.*, 1993) and an experimental task–reference task subtraction was made in order to eliminate brain activation related to the nonvisual and nonspecific task components. As each resulting brain image was in standard format, the images were added to each other and in that way the brain regions of significant activations (Roland *et al.*, 1994) were superimposed. Only those regions are shown in the figure which, after the combination of the resulting images from the test series, form a continuum in the occipital lobe and its neighboring regions in the parietal and temporal lobes.

The tentative human visual cortex occupies entirely the occipital lobe and

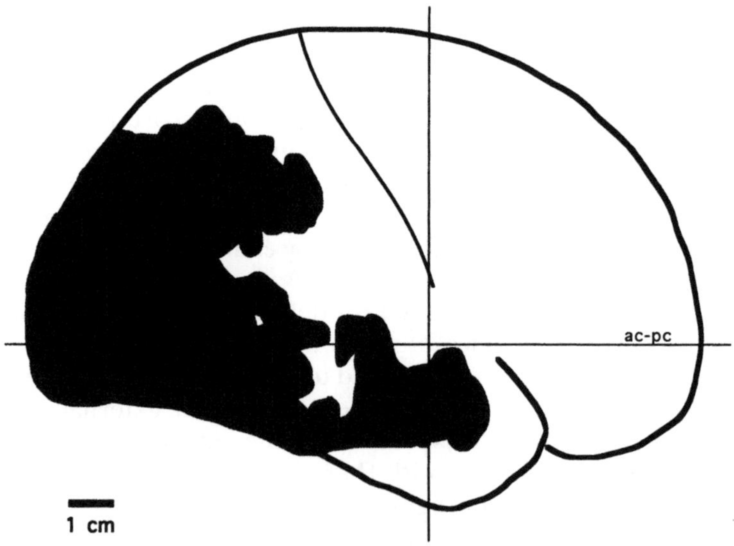

Figure 1. The tentative extent of human visual cortex. The figure is based upon 11 series of PET experiments (on the average, each involving data from 10 experimental subjects), related to visual perception, discrimination, recognition, learning, memory, and imagery processes, i.e., a wide scale of visual processes (Gulyás and Roland, 1991, 1993, 1994a, b, 1995; Gulyás *et al.*, 1994a, b; Roland and Gulyás 1993, 1995; as well as unpublished data by the present author). Talairach convention (Talairach and Tournoux, 1988), collapsed parasagittal view. The horizontal line (ac–pc) indicates the ac–pc plane, and the vertical line intersecting it indicates the Vca plane, the vertical plane of the Talairach stereotactic system. The central sulcus is shown.

enters into the posterior part of the parietal lobe as well as that of the temporal lobe, as indicated on a collapsed parasagittal projection of the brain, shown in Fig. 1. A smaller part of this tentative visual cortex region has been explored in detail for its anatomical structure, as shown in Fig. 2. However, detailed anatomical *and* functional information is available for a few early visual cortical areas, as indicated in this figure.

Further combined anatomical and functional studies are needed to define

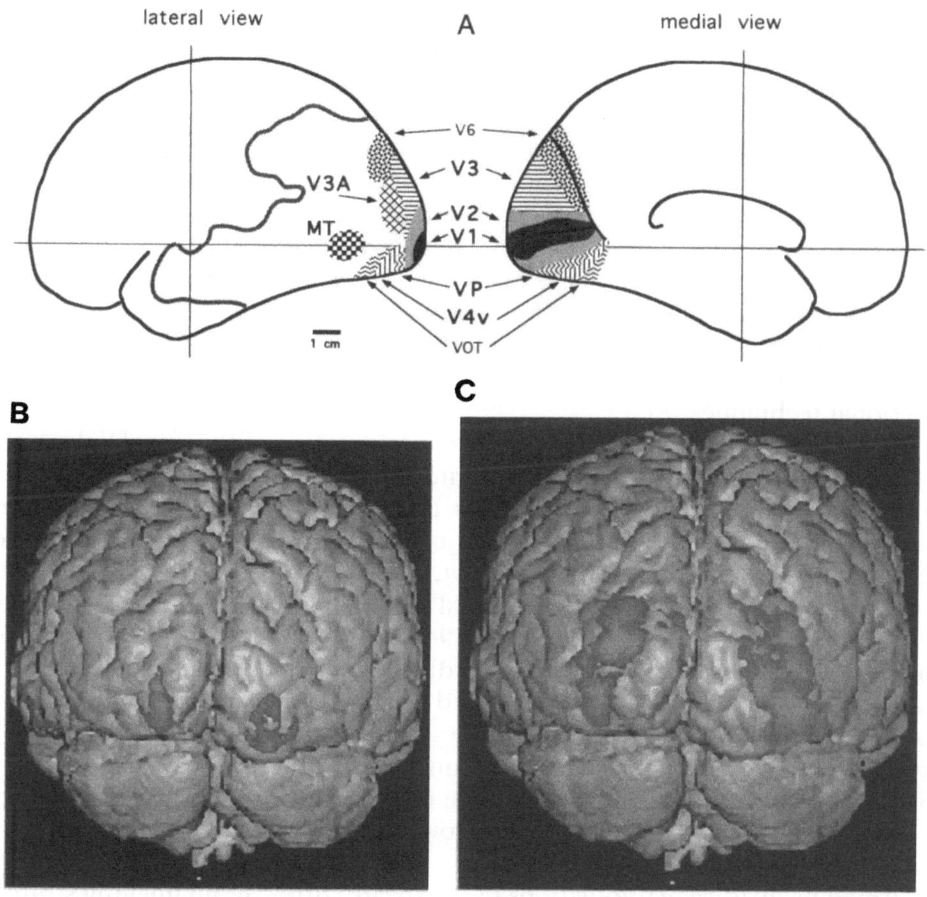

Figure 2. (A) The approximate arrangement of topographically organized visual cortical areas in the human brain (lateral and medial aspects). Talairach convention (as in Fig. 1). The extent of the visual cortex is indicated on the lateral view (hatched line), whereas the corpus callosum, the calcarine sulcus, and the parietooccipital sulcus are indicated on the medial view (solid lines). The figure is based upon data from Clarke and Miklossy (1990), Watson *et al.* (1993), Hadjikani (1995), Sereno *et al.* (1995), and the chapter by Zilles and Clarke in this volume. The notations in bold indicate areas that have been identified by both anatomical and functional techniques, whereas the smaller notations (not in bold) indicate areas that have only been identified with anatomical techniques. The location of the primary visual cortex (area V1) (B) and area V2 (C) on the posterior surface of the brain. The reconstruction is based on quantitative cytoarchitectonic data of five normal postmortem brains, the colored regions represent 50% or more overlap levels and are projected onto the normal MRI brain of the Human Brain Atlas (Roland *et al.,* 1994). Original data by Karl Zilles and Katrin Amunts; reconstruction made by Jonas Larsson. A four-color reproduction of Fig. 14 (B–C) appears following page xxiii.

visual cortical areas outside the region of early visual areas. On the basis of the monkey model of visual cortex, it can be expected that at least about 25–30 areas will be defined within the tentative borders of the human visual cortex displayed in Fig. 1 and outside the borders of the well-defined areas of V1, V2, V3, V3A, V4v, and MT/V5.

5. Topographic Organization of Early Visual Cortical Areas

Similar to the visual cortex of other vertebrates, the human visual cortex is centered around the striate cortex (or the primary visual cortex, area V1), a topographically well organized cortical area, containing the representation of the whole visual field. According to recent knowledge, the primary visual cortex is surrounded by five topographically organized extrastriate visual cortical areas: V2, V3, V3A, VP, and V4 (Clarke and Miklossy, 1990; Tootell *et al.*, 1995a, b; Zilles and Clarke, this volume). Outside these, no visual cortical areas with clear-cut topographic organization have been demonstrated; and with the exception of area V5 (MT), a visual area well characterized by its distinct anatomical features and functional properties, no further human visual cortical areas have been described with certainty and great precision using *both* anatomical and functional techniques.

The first detailed descriptions of the representation of the visual field in the human striate cortex go back to the beginning of the century, when wounded war veterans with occipital lobe deficits caused by penetrating missiles were examined and the first retinotopic maps of the human striate cortex were created (for references, see Horton and Hoyt, 1991; Kuljis, 1994; Rizzo, 1994; and Gross, this volume). Gordon Holmes' visual field map of the human striate cortex has become a textbook classic (Holmes, 1945) (Fig. 3A), despite two major inaccuracies: first, as precise anatomical studies have shown, the striate cortex is mainly buried in the calcarine sulcus and is not on the medial surface of the brain as extensively as indicated in Holmes figure, and second, Holmes severely underestimated the size of the macular and perimacular representation in the cortex, whereas he overestimated the size of the peripheral representation.

However, the Holmes map clearly shows the main features of the representation of the visual field in the primary visual cortex, and these have been confirmed by invasive manipulations in the striate cortex, brain imaging studies, as well as everyday clinical practice. The striate cortex of one hemisphere contains the representation of the contralateral visual hemifield. The fovea is represented in or close to the occipital pole, the far periphery close to the occipitoparietal sulcus; the upper part of the visual field is represented on the lower lips of the calcarine sulcus; whereas the lower part of the visual field is on the upper lip. The representation of the horizontal meridian runs along the base of the calcarine sulcus, whereas the representations of the lower and upper vertical meridians run parallel with the calcarine sulcus on its upper and lower lips, respectively, at the boundary between the striate cortex, area V1, and the first extrastriate area, area V2. The visual field representation in the striate cortex receives visual input from both eyes, with the exception of two segments: the "representation" of the contralateral eye's blind spot (only ipsilateral input) and that of the temporal crescent, the monocular C-shaped lateral part of the visual hemifield (only contralateral input). Recently, Holmes' map has been meticulously revised by Horton and Hoyt (1991) (Fig. 3B).

Recent functional brain imaging studies using PET and fMRI have basically confirmed the correctness of the Holmes' map. However, from measurements in the classic Holmes' map, the center of the macula is in or close to the occipital pole. This has not been confirmed by PET and fMRI studies, which systematically indicate a macular location anterior to the occipital pole (Fox *et al.*, 1986,

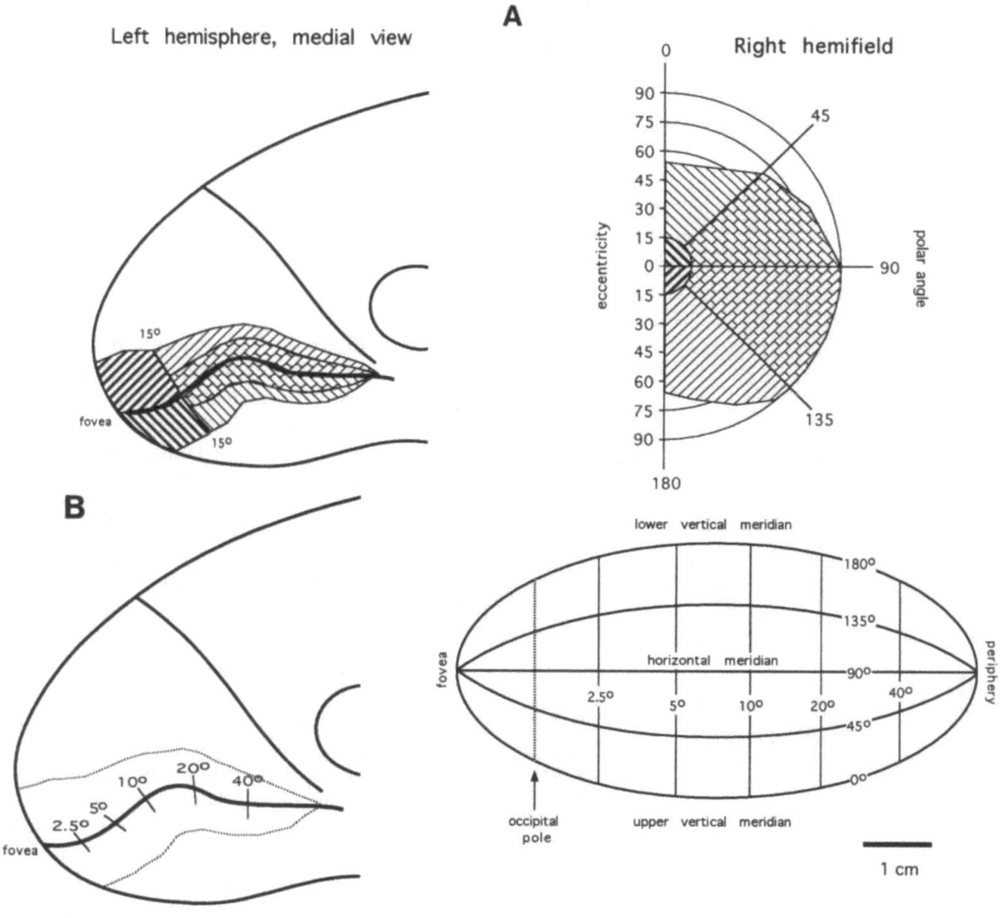

Figure 3. (A) The topographic arrangement of the external visual field in the striate cortex, according to Holmes (1945). (The figure has been redrawn after Fig. 5 in Holmes, 1945.) The contralateral hemifield (right panel) is represented topographically in the striate cortex of the left hemisphere (left panel). The foveal representation is close to the occipital pole, whereas the peripheral representation is close to the parietooccipital sulcus. (B) The topographic arrangement of the external visual field in the striate cortex, according to Horton and Hoyt (1991). (The figure has been redrawn after part of Fig. 8 in Horton and Hoyt, 1991, using Holmes' convention of visual field eccentricity and polar angle values.) The schematic topographic representation of the visual field in the folded area V1 (left panel) and in the flattened striate cortex (right panel). Note the significantly larger representation of the fovea as compared to Holmes' map.

A

B

C

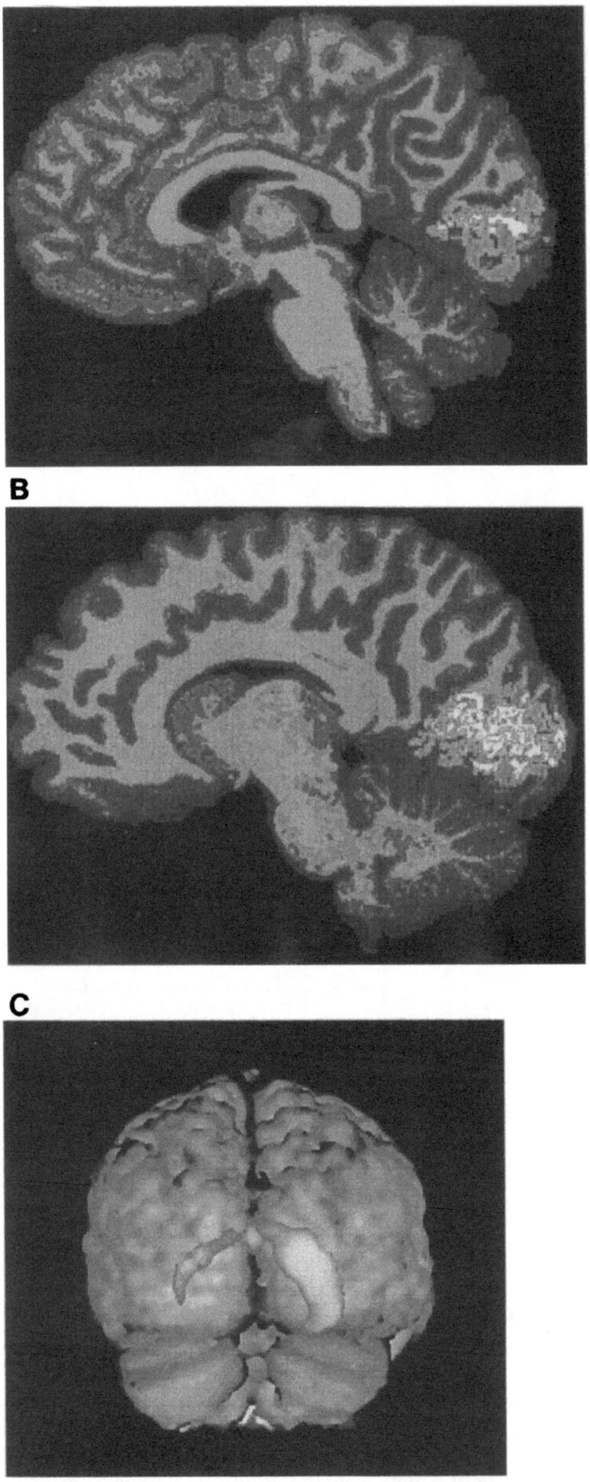

Figure 4. The anatomical variability of a visual cortical area: striate cortex. Five normal postmortem brains were anatomically standardized with the help of the Human Brain Atlas (Roland *et al.*, 1994). The images show overlaps between areas V1 of the five anatomically standardized brains. (A) Parasagittal slice made 10 mm from the midline in the left hemisphere (i.e., $x = -0$ mm), showing the largest overlaps. (B) Parasagittal slice made 5 mm from the midline in the left hemisphere ($x = -5$ mm), i.e., showing the overlaps approximately on the medial surface of the hemisphere. (Original cytoarchitectonic data were kindly provided by Karl Zilles and Torsten Schorman.) (C) The anatomical variability of the area V1/V2 border. The colored region projected onto the posterior surface of the brain represents the overlap between those parts of areas V1 and V2 that are not shown in Fig. 2C, i.e., that represent at the 50% overlap level the border between areas V1 and V2. Reconstruction made by Jonas Larsson. A four-color reproduction of this figure appears following page xxiii.

1987; Belliveau *et al.*, 1991; Gulyás and Roland, 1993, 1994b; Watson *et al.*, 1993; Gulyś *et al.*, 1996). The reason behind this apparent contradiction may be multiple. (i) There is a large anatomical variation of striate cortex borders between individuals, which is present even after anatomical standardization of individual brains. An example is shown in Fig. 4. Such variability regarding the borders of visual cortical areas is present in other cases as well (Watson *et al.*, 1993). (ii) In functional brain imaging studies, due to averaging procedures, the resulting activation peaks may not coincide with the activation peaks in the constituent individual images. Consequently, what functional brain activation studies show in this respect may only be a sector of the striate cortical representation of the macula.

Similar to the representation of the visual field in area V1, areas V2, V3, VP, V3A, and V4v have a topographically strictly organized visual field representation. Some studies indicate that other fields, such as MT (V5) or dorsal V4 (Sereno *et al.*, 1995) may also have at least some topography, although this is not universally endorsed (R. B. H. Tootell, personal communication).

The topographic organization of early extrastriate areas is orderly (see also the chapter by Zilles and Clarke in this volume). In area V2 the fovea is represented at the conjunction of the dorsal and ventral flanks of the area adjacent to the foveal representation of area V1 (Figs. 5 and 6). The dorsal flank contains the lower visual field (hemiquadrant) representation, whereas the lower flank of the area contains the upper field (upper hemiquadrant) representation. Along the border between areas V1 and V2 runs the vertical meridian, whereas the horizontal meridian runs along the border between areas V2 and V3. It is still a subject of investigation whether area V3 and area VP are confluent and form one continuous area adjacent to area V2 or there are two separated areas, a dorsal one, area V3, and a ventral, area VP. Anatomical studies argue for two separate areas (Hadjikani, 1995; and the chapters by Kaas and by Rosa in this volume), and functional brain activation studies further support this view (Sereno, 1995). In any case, areas V3 and VP together have a full representation of the external visual field, so that the lower field is represented in area V3 and the upper field is represented in VP. The fovea is represented at the edges of V3 and VP facing each other. (In fact, the foveal representations of areas V1, V2, V3, and VP are confluent.) The horizontal meridian runs between dorsal V2 and V3 (lower field) as well as ventral V2 and VP (upper field), whereas the vertical meridian runs between V3 and V3A (lower field) and VP and V4v (upper field), respectively. Area V3A has a complete representation of the contralateral hemifield (both the upper and lower quadrants are represented). The lower vertical meridian runs at the area's anterior border (neighboring area V3), whereas the upper vertical meridian runs at its posterior border (neighboring topographically less well organized visual areas). The foveal representation in area V3A is in its inferior edge, some millimeters superior to the foveal representations of areas V1, V2, V3, and VP. Area V4v has the upper field (upper contralateral quadrant) representation, with a foveal representation in its superior edge. The vertical meridian runs between the border of VP and V4v, whereas the horizontal meridian runs between V4v and less well-defined areas lateral and anterior to it. Functional imaging studies offer an opportunity to localize foveal representations in extrastriate visual areas, too, as shown in Fig. 7.

The representations of increasing eccentricities in the visual field are present at increasing distances away from the occipital pole, whereas the polar angles change in an orderly manner between the vertical and horizontal meridian representations in visual areas. The polarity of the polar angle representation indicates whether an area has a mirror image or a non-mirror image of the contralateral visual field. Areas V1 and V3/VP contain mirror-image representa-

A

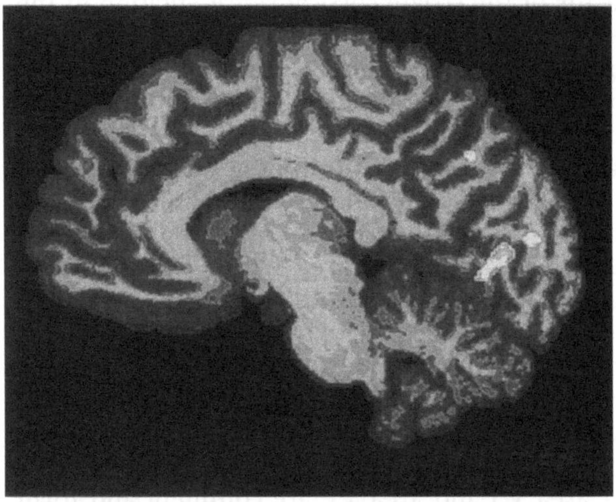

B

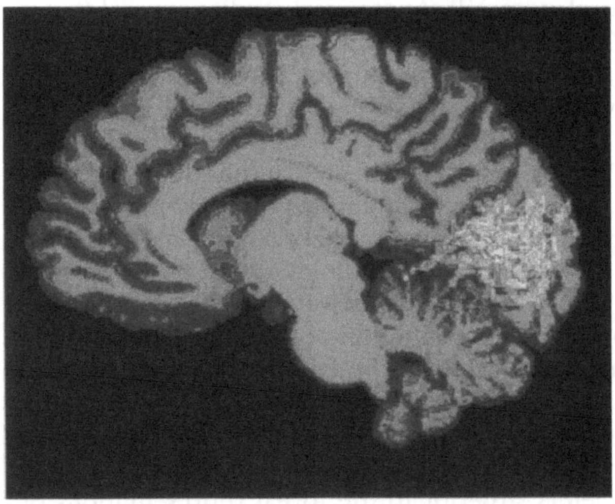

Figure 5. The anterior sector of the putative foveal representation in area V1 (population average). (A) The average representation of the fovea as determined by functional brain activation studies (region in red, covering parts of the upper and lower lips of the calcarine sulcus; parasagittal slice, $x = -7$ mm). In three test series, we measured with PET the regional cerebral blood flow (rCBF) changes in 33 normal volunteers who performed visual tests designed to activate the central representations in area V1. The tests included three color-related paradigms: (i) detection of wavelength changes under Ganzfeld viewing conditions (Gulyás *et al.*, 1997), (ii) discrimination of whole-field colored random dot patterns (Gulyás and Roland, 1994a), and (iii) reaction to changes in color and luminance of an LED placed in the center of gaze of the subjects (Kinomura *et al.*, 1995). As the cone density is highest in the macula and it rapidly decreases with eccentricity, color stimulation of the retinal sheet should almost exclusively stimulate the macular representations. The data obtained in the functional imaging studies were also processed and analyzed with the help of the Human Brain Atlas system (Roland *et al.*, 1994). Overlaps between regions activated in all visual tasks were created on the basis of averaged standardized images of the brains containing the significantly activated regions. Note that this region is anterior to the occipital pole wherein the foveal representation in the striate cortex is suggested by numerous studies. (B) This discrepancy can be explained by the fact that in an overlap image of the anatomically standardized striate cortex the maximum overlap region is also anterior to the occipital pole (for explanation, see Fig. 6; parasagittal slice, $x = -7$ mm; white region: five overlaps). This is mainly due to the fact that sulcal variability increases away from the center of the brain, i.e., more occipital sectors of the foveal representations in the striate cortex will result in fewer overlaps. A four-color reproduction of Fig. 5(B) appears following page xxiii.

tions, whereas areas V2, V3A, and V4v contain non-mirror-image representations.

It is strongly assumed that other visual areas outside the above ones may also contain at least partial representation of the external visual field. Dorsal area V4 may be one of these areas, as indicated by functional studies (Sereno *et al.*, 1995); however, others, on the basis of anatomical evidence, argue against the existence of a dorsal area V4 (Hadjikani, 1995).

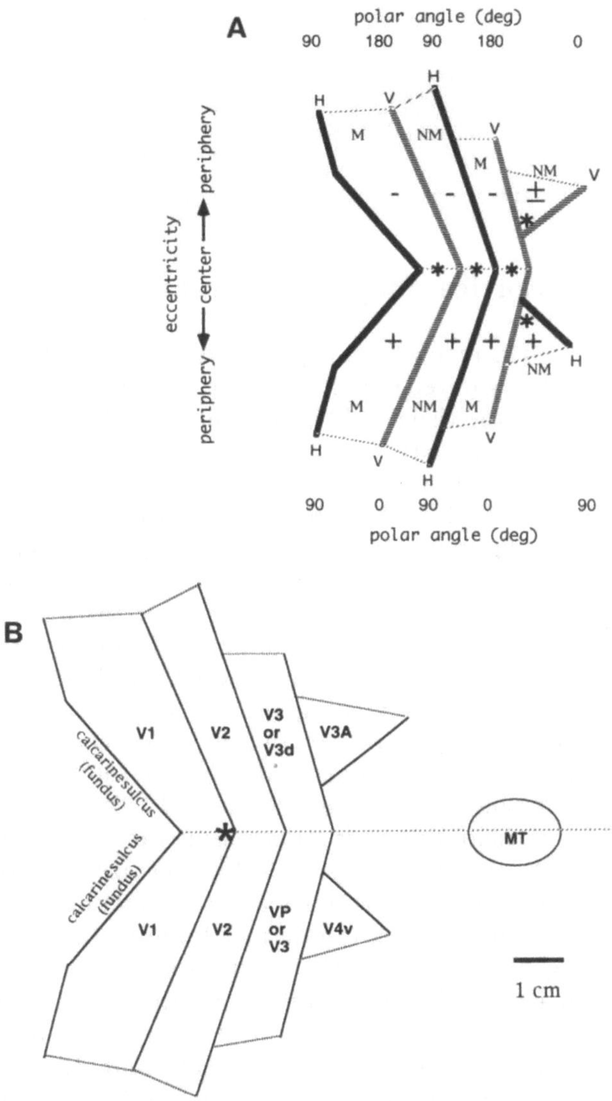

Figure 6. (A) Reconstruction of the relative arrangement of early human visual cortical areas in the flattened cortex, based upon data from Sereno *et al.*, (1995), Tootell *et al.*, (1996), and Kaas (1995). The striate cortex is cut into an upper and lower flank along the calcarine sulcus. The dotted horizontal line runs approximately along the ac–pc plane (z = 0 mm) of the Talairach coordinate system. However, whereas the distances are conserved, the directions in the figure cannot be related to the Talairach system, due to the flattening. The asterisk indicates the approximate position of the occipital pole. (B) Topographic representation of the visual field in early topographically organized visual cortical areas. Same convention as in Panel A. The foveal representations are indicated by asterisks; the representations of increasing eccentricities are further away from it in an orderly manner. In fact, the foveal representations of areas V1, V2, V3, and VP are confluent. The representation of the vertical meridians is indicated by dashed lines and V, whereas that of the horizontal meridians is indicated by solid lines and H. The representation of the upper visual field (i.e., contralateral upper quadrant) is indicated by a plus sign, the representation of the lower visual field (i.e., contra lateral lower quadrant) by a minus sign. Area V3A contains a full-field (i.e., contralateral hemifield) representation, indicated by ±. The polar angle changes in an orderly way between the vertical and horizontal meridian representations, i.e., between 180 and 90 deg in the lower field representations and between 0 and 90 deg in the upper field representations. Area V3A has a full representation of the contralateral field, and the polar angle changes between 180 deg (lower vertical meridian) and 0 deg (upper vertical meridian). Areas V1, V3, and VP contain a mirror-image representation (M) of the contralateral visual field, whereas V2, V3A, and V4v contain a nonmirror-image representation (NM).

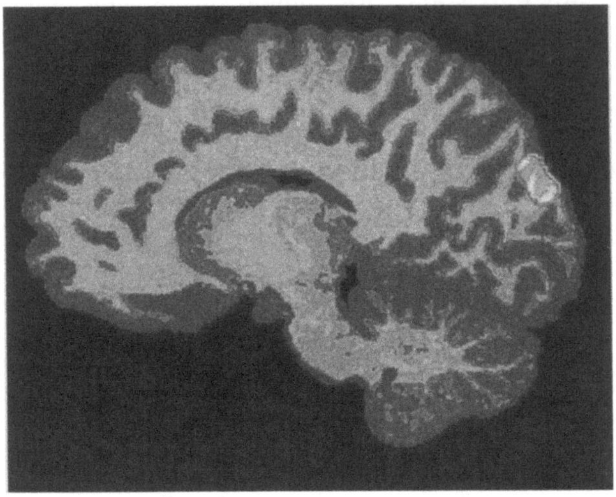

Figure 7. Functional localization of the foveal representation in putative area V3A. Using the same functional tests as described in legend to Fig. 5, the foveal representation in the putative area V3A is shown on a parasagittal slice (x = 12). As this region is lateral and superior to the confluent foveal representation of area V1 and its distance on the flat-mounted cortical surface from the previous region is very close to what Tootell *et al.*, (1996) describe for that of V3A, and it is strongly motion-sensitive (Larsson *et al.*, 1996), it may be identified as the putative foveal representation of area V3A.

6. Cortical Magnification Factors

The regular visuotopic representation of the external visual field in topographically organized visual cortical areas is disproportional and the cortical representation of the visual field is not uniform. The center of the visual field is overrepresented, whereas the peripheral visual field is underrepresented, so that the central representation of a given area in the central visual field occupies a larger area on the cortical surface than that of an equisized area in the peripheral visual field.

The anatomical basis for this disproportionality between the external (visual field) and central (visual cortex) representations is already present at the retinal level. In the mammalian visual system, there are at least three factors responsible for a systematically distorted input from the retina to higher levels in the visual pathways: (i) There is an unequal distribution of photoreceptors in the retina. The cone density is highest in the fovea and it sharply decreases with eccentricities, whereas the rod density, which is zero in the center of the fovea, sharply increases up to about 20 deg eccentricity, above which it decreases again with increasing eccentricities (Rodieck, 1973; Coren *et al.*, 1978). (ii) With increasing eccentricities there is a decreasing density of retinal ganglion cells (Perry *et al.*, 1984). (iii) The ganglion cell soma size increases with eccentricity, and, with increasing eccentricities a larger number of photoreceptor inputs impinge upon each ganglion cell (Stone and Johnston, 1981). Due to divergence of the retinogeniculate input, the strict visuotopic organization of the lateral geniculate nucleus is already distorted and the central (foveal) representations are overrepresented as compared to the peripheral representations (Sanderson, 1971; Malpeli and Baker, 1975; Connolly and Van Essen, 1984).

This disproportionality between between the external visual field and its central representation as a function of eccentricity is continuous across a topographically organized visual cortical area and is measured by the cortical magnification factor (CMF) (Fig. 8).

The linear CMF (lCMF) is the number of millimeters of cortex representing 1 deg of visual angle of the external visual field (millimeters of cortex/degree of visual field; mm/deg); the areal CMF (aCMF) expresses the relationship between

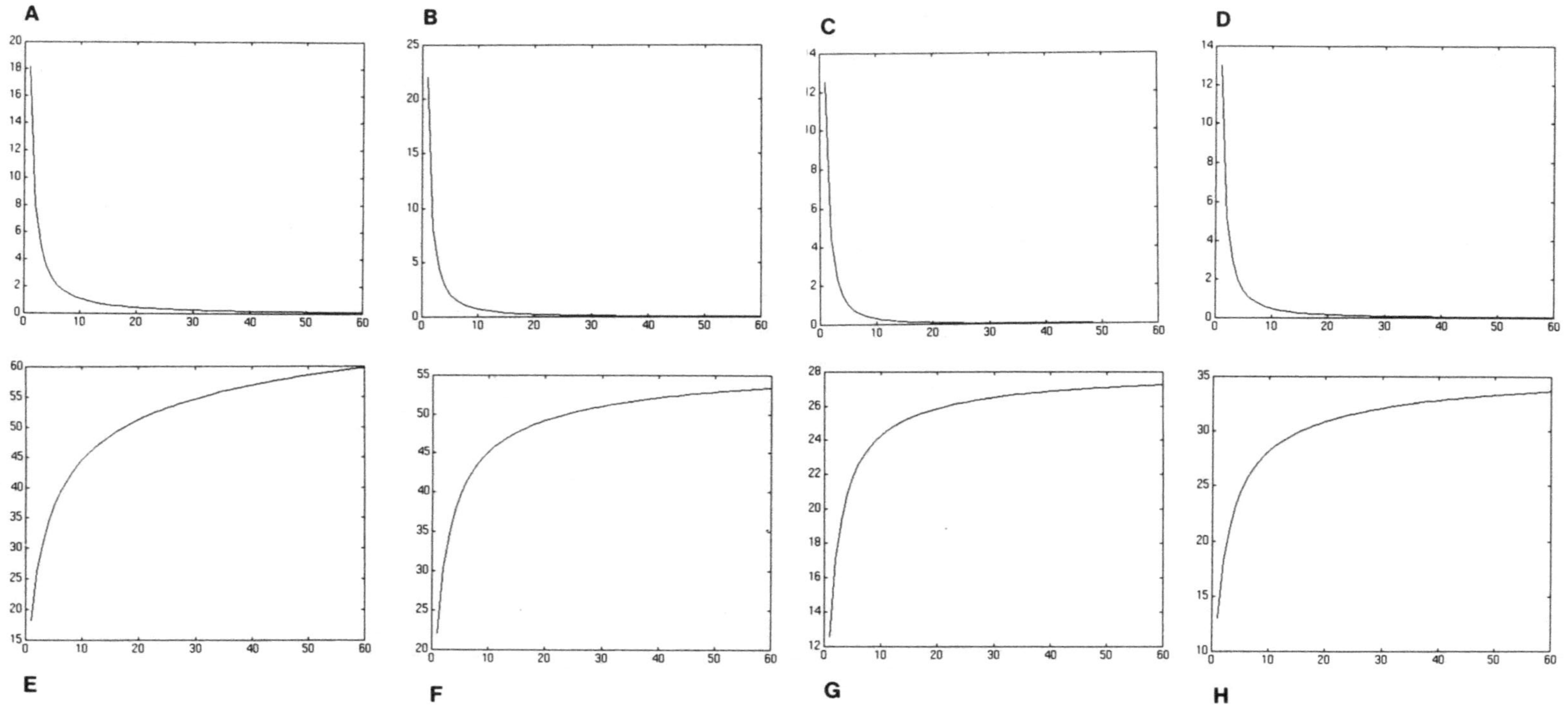

Figure 8. (A–D) The cortical magnification factor as a function of eccentricity in topographically organized areas (A) V1, (B) V2, (C) VP, and (D) V4v (based upon the equations of Sereno *et al.*, 1995). (E–H) The cortical mapping function as a function of eccentricity in areas (E) V1, (F) V2, (G) VP, and (H) V4v (based upon the equations of Sereno *et al.*, 1995).

Table II. Linear Cortical Magnification Factor (lCMF) in Human Foveal Striate Cortex

lCMF (mm/deg)	Technique[a]	Reference
15.1	a	Cowey and Rolls (1974, 1975)
11.5	e	Drasdo (1977)
5.27	b	Rovamo (1978)
7.75	b	Rovamo *et al.* (1978); Virsu and Rovamo (1979)
7.99	b	Rovamo and Virsu (1979)
6–7	f	Dobbele *et al.* (1979)
3.4	g	Fox *et al.* (1987)
20–25	—	Tolhurst and Ling (1988)
9.9	d	Horton and Hoyt (1991)
18.20	g	Sereno *et al.* (1995)

[a]For explanation, see list of techniques (a)–(f) in text.

the area of the cortex representing a sector of the visual field (mm^2/deg^2). The CMF is inversely proportional to eccentricity. Each topographically organized visual cortical area has its characteristic CMFs and CMF–eccentricity relationships.

The human lCMF in the striate cortex has been measured in various ways, including (a) comparative studies and extrapolation between monkeys and humans, (b) visual psychophysics, (c) clinical observations on migraine scotoma dimensions and their location in the visual field, (d) clinical observations on patients with well-localized cortical lesions, (e) electrophysiological studies, combined with psychophysics, (f) electrical stimulation of the cortex by surface electrodes, and (g) functional brain imaging studies, including positron emission tomography (PET) and functional magnetic resonance imaging (fMRI). These measurements have yielded rather large variations of values of the foveal lCMF (Table II).

On the basis of individual measurements, various expressions have been published for the lCMF as a function of eccentricity for both striate and topographically organized extrastriate visual cortical areas (Table III). It is also possible to demonstrate the relationship between the external visual field and its cortical representation by way of the cortical mapping function, expressing the cortical distance from the foveal representation as a function of eccentricities in the visual field (mm/deg).

Table III. Formulas Expressing the Relationship Between lCMF and Eccentricity in Human Striate Cortex and Topographically Organized Extrastriate Visual Cortical Areas[a]

Area	$lCMF_E$ (mm/deg)	Reference
V1	$11.5(1 + \text{const} \cdot E)^{-1}$	Drasdo (1977)
	$7.99(1 + 0.42E + 0.000055E^3)^{-1}$	Rovamo and Virsu (1984)
	$18.93(0.78 + E)^{-1.1}$	Tolhurst and Ling (1988)
	$17.3(0.75 + E)^{-1}$	Horton and Hoyt (1991)
	$20.05(0.08 + E)^{-1.26}$	Sereno *et al.* (1995)
V2	$25.19(0.09 + E)^{-1.53}$	Sereno *et al.* (1995)
VP	$18.28(0.24 + E)^{-1.75}$	Sereno *et al.* (1995)
V4v	$18.17(0.24 + E)^{-1.55}$	Sereno *et al.* (1995)

[a]E stands for a given eccentricity value in degrees.

7. Functional Properties of Early Visual Cortical Areas

In primate studies the functional features of a visual area are determined by analyzing the results of single-cell recordings and comparing them with those obtained in other visual areas. In addition, ablations with detailed psychophysical testing procedures contribute to our understanding of the overall physiological properties of visual cortical areas. These techniques, however, are not applicable to human visual cortex research. Instead, in recent years experiments with functional neuroimaging techniques have provided us with the opportunity to explore in detail the functional properties of human visual cortical areas. A prerequisite of these observations is, of course, the determination of the border of a visual cortical area either by anatomical techniques (e.g., postmortem anatomy combined with computerized brain atlases; Roland and Zilles, 1994) or by functional techniques (e.g., topographic determination of area borders; Sereno *et al.*, 1995). In most experiments aimed at the exploration of functional features of human visual cortical areas, observations in monkeys guided the design of the

Table IV. Some Functional Properties of Early Visual Areas

Area	Visual features	Selected references
V1	Sensitivity for photic stimulation	Belliveau *et al.* (1991)
	Disparity sensitivity	Gulyás and Roland (1994a, b); Savoy *et al.* (1993)
	Color sensitivity	Gulyás and Roland (1994a), Reppas *et al.* (1995), Gulyás *et al.* (1997)
	Orientation discrimination	Sergent *et al.* (1992)
	Spatial frequency discrimination	Sergent *et al.* (1992), Gulyás and Roland (1995)
V2	Disparity sensitivity	Gulyás and Roland (1994a, b)
	Color sensitivity	Gulyás and Roland (1994a), Gulyás *et al.* (1997)
	Orientation discrimination	Gulyás and Roland (1995)
V3	Disparity sensitivity	Gulyás and Roland (1994a, b)
	Color sensitivity	Gulyás and Roland (1994a), Gulyás *et al.* (1997)
	High-contrast sensitivity	Tootell *et al.* (1995c)
V3A	Disparity sensitivity	Gulyás and Roland (1994a, b), Savoy *et al.* (1995)
	Color sensitivity	Gulyás *et al.* (1997)
	Motion sensitivity	Tootell *et al.* (1996), Larsson *et al.* (1996)
	High-contrast sensitivity	Tootell *et al.* (1995c), Tootell *et al.* (1996)
V4 (V4v)	Color sensitivity	Zeki *et al.* (1991), Corbetta *et al.* (1991), Gulyás and Roland (1991, 1994b), Reppas *et al.* (1995)
	Form discrimination	Gulyás and Roland (1994b)
MT	Motion sensitivity	Tootell *et al.* (1995a, d)
	Motion from hue at isoluminance	ffytche *et al.* (1995), Tootell *et al.* (1995a), Larsson *et al.* (1996)
	Direction sensitivity	Tootell *et al.* (1995a, b), McKeefry *et al.* (1996)
	Motion aftereffect	Tootell *et al.* (1995b)
	High-contrast sensitivity	Tootell *et al.* (1995a, c)

experiments. It has also been the case that under certain circumstances the functional predictions about the functioning of a human visual area based upon observations in primates facilitated the determination of the borders of that area (e.g., area MT; Watson *et al.*, 1993; Tootell *et al.*, 1995a). A selective overview of functional properties of human visual cortical areas is given in Table IV.

8. Information Processing in Human Visual Cortex: Basic Visual Cues

Several physiological observations suggest that neuronal signals about elementary aspects of the visual scene, such as color, motion, and disparity, may be processed and analyzed by different functional pathways at early levels in the primate visual system (Ramachandran and Gregory, 1978; Zeki, 1974; Livingstone and Hubel, 1987; Schiller *et al.*, 1990; Schiller, 1993). Although there are interactions between the visual subsystems coding various visual cues, the underlying pathways are regarded as initially independent (Carney *et al.*, 1987; De Yoe and Van Essen, 1988).

At early cortical levels in the primate brain some cortical areas, by way of containing a large proportion of cue-sensitive neurons, exhibit a stronger involvement in the processing and analysis of selected visual cues, such as color or motion. Initial research with functional imaging techniques into the processing of elementary visual cues in human visual cortex focused on the mapping of visual cortical areas specifically activated by these selected visual cues.

8.1. Color-Sensitive Regions

As the retinal cone density is the highest in the center of the fovea and sharply decreases with retinal eccentricities, color is mainly represented in the macular and perimacular zones of the visual field, and the central representations of these zones in the visual cortex may indicate the center of the visual field in a number of cortical areas containing color-sensitive cells. It is therefore not surprising that several functional imaging studies aiming at the localization of color-sensitive cortical areas find activations in the primary visual cortex (V1) (Fig. 5) as well as in adjacent locations in the extrastriate visual cortex.

The centers of gravity of the activated fields in the calcarine cortex in the various studies are in very good agreement with each other; the Talairach and Tournoux (1988) coordinates are (*x, y, z;* right and left, respectively): Zeki *et al.* (1991): −4, −90, 0 (one confluent field); Gulyás and Roland (1994a): 7, −79, 3 and −8, −79, 1; Roland and Gulyás (1996; overlap of two color studies!): 5, −84, 2 and −4, −84, 2. These values are in very close agreement with the coordinate values of the macular representation of the visual field in area V1 as described by Fox *et al.* (1986, 1987) (*y, z* coordinates are −80, −3) and by Belliveau *et al.* (1991) (*x, y, z* coordinates are 8, −77, 8 and −7, −80, 8 for right and left, respectively) (Fig. 5).

The literature frequently assumes that "the color area" in humans is in the lingual gyrus (or in the lingual gyrus extending into the fusiform gyrus), and that this color area is the human analogue of area V4 in the monkey (Zeki *et al.*, 1991). Nevertheless, with the conspicuous exceptions of Verray (1888) and MacKay and Dunlop (1899), no clinical literature supports the existence of a single, well-circumscribed color area. Usually, lesions resulting in cerebral achromatopsia involve rather large sectors of the cortex in the occipital (and often the temporal and parietal) lobe(s) as well as extensive white matter parts. It is nevertheless impressive that basically all studies related to the mapping of cortical areas involved in color vision report that two regions symmetrically located in the

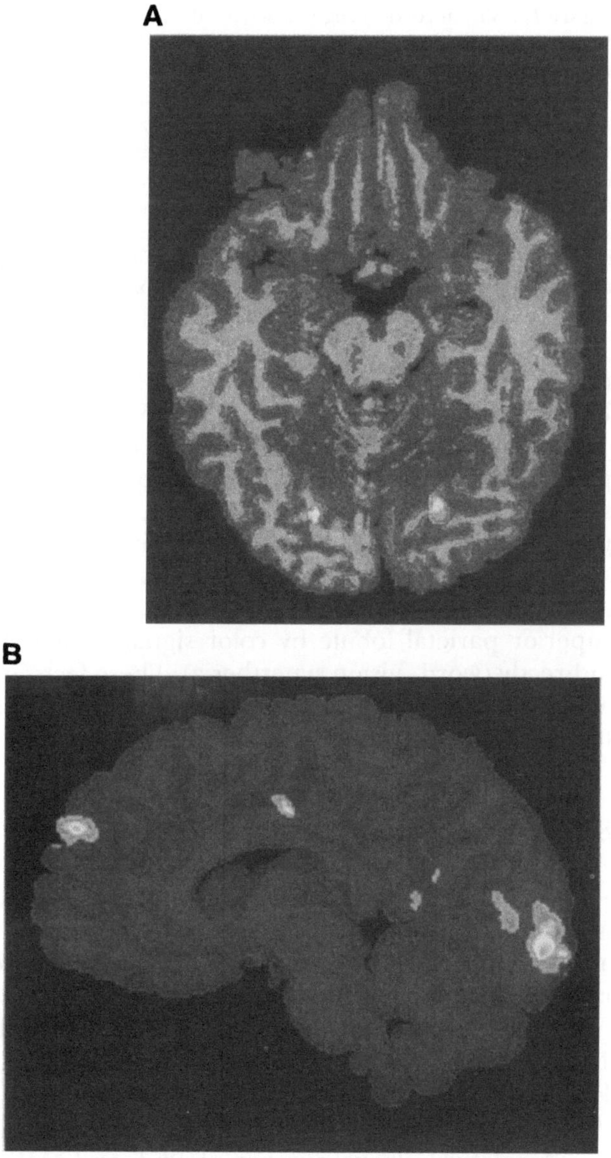

Figure 9. (A) Color-sensitive regions in the fusiform gyrus. Horizontal slice, $z = -15$ mm. Radiological convention (left hemisphere on the right side, right hemisphere on the left side). (From the experimental series by Gulyás and Roland, 1994a). This color-sensitive region in the fusiform gyrus has been described by a number of independent PET studies and is regarded as one of the most consistently activated regions engaged by various forms of color vision (Zeki *et al.*, 1991; Corbetta *et al.*, 1991; Gulyás and Roland, 1991, 1994a; Roland and Gulyás, 1996). (B) Disparity-sensitive regions. Parasagittal slice, $x = -6$ mm. (From the experimental series described by Gulyás and Roland, 1994a, b). Note the strong activation in the putative striate cortex. The images were processed by the HBA (Roland *et al.*, 1994) and activations are projected onto the standard MRI brain of the HBA.

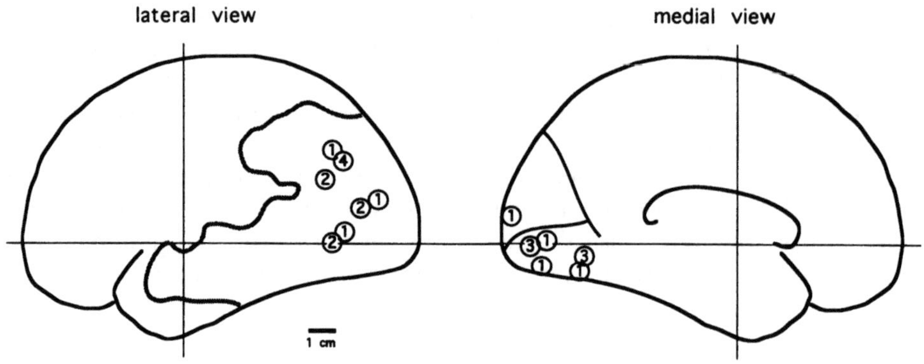

Figure 10. Visual cortical regions activated by color vision. Only those regions are indicated in this figure that are systematically activated under different experimental conditions using color stimulation. Only those studies were analyzed which contained appropriate stereotactic coordinates applicable for the present comparison. Centers of gravity: (1) Overlap regions between three different experimental series, described in Gulyás and Roland (1994a), Gulyás *et al.* (1996), as well as B. Gulyás, P. E. Roland, and J. Larsson, unpublished observations. (2) Zeki *et al.* (1991). (3) Corbetta *et al.* (1991). (4) Paulesu *et al.* (1995). Note that in addition to activation in the striate cortex and early extrastriate cortex that have been described by a number of studies, there is also activation in the posterior superior parietal lobule both by frank visual stimulations (1) and by indirect stimulation in word-color synesthetes (4). Same convention as in Fig. 1; lateral and medial views. The hatched line indicates the putative border of the visual cortex (Fig. 1).

fusiform gyri are active (Zeki *et al.,* 1991; Corbetta *et al.,* 1991; Gulyás and Roland, 1991, 1994a; Sakai *et al.,* 1995; Fig. 9A).

As expected on the basis of the cortical distribution of color-processing pathways originating in the retina, in addition to the activated color region in the striate cortex and the fusiform gyrus, a number of color-sensitive cortical regions are present in the extrastriate visual cortex (see Fig. 10) that may correspond to the macular representations in extrastriate visual cortical areas (Fig. 7). A most interesting finding is the systematic activation of a region in the posterior superior parietal lobule by color signals either directly (visual perception) or indirectly (word–vision synesthesia). These facts indicate that the processing and analysis of color information in the visual cortex engage a complex network of multiple cortical areas (Fig. 10).

8.2. Visual Motion-Sensitive Regions

Probably the most distinctive specialized visual cortical is area V5 or MT. It is present in all primate species explored. It has a unique myeloarchitecture, has an extremely high proportion of motion- and direction-sensitive neurons, and receives fast magnocellular input not only via the striate cortex, but also directly from the LGN (Benevento and Yoshida, 1981; Fries, 1981; Yukie and Iwai, 1981). The ablation of area MT results in a deficit of visual motion perception. For these distinctive properties area V5/MT is often termed "the motion area" (for review, see Zeki, 1991).

A relatively large number of PET and fMRI studies using visual motion paradigms (Zeki *et al.,* 1991; Watson *et al.,* 1993; Tootell *et al.,* 1995a), including even more sophisticated forms such as motion-from-hue at equiluminance (ffythce *et al.,* 1995; Larsson *et al.,* 1996), illusory motion (Zeki *et al.,* 1993), and motion aftereffect (Tootell *et al.,* 1995b) have aimed at the exploration in the human brain of this motion-sensitive area. The investigations systematically found activation in V1 and V5/MT, and, under a number of different circumstances, also activations in a number of extrastriate cortical regions (Orban *et al.,* 1995, and chapter this volume). The most important of these motion-sensitive regions outside V5/MT is area V3A (Tootell *et al.,* 1996; Larsson *et al.,* 1996), which is highly sensitive for motion signals (Fig. 11).

8.3. Disparity

Primate studies show the heavy involvement of striate and extrastriate cortices in stereoscopic processing (Poggio and Fischer, 1977; Poggio and Talbot, 1981; Felleman and Van Essen, 1987). It has therefore been expected that disparity signals would activate the human striate cortex as well as a number of early extrastriate areas. Indeed, both PET and fMRI studies clearly showed extensive activation of human central striate cortex and early extrastriate visual areas during both stationary (Gulyás and Roland, 1993; 1994a, b; Ptito *et al.,* 1993; Savoy *et al.,* 1993, 1995) and dynamic (de Jong *et al.,* 1994) depth stimulation. In addition to these regions, in stereovision there were fields of activation in the posterior superior parietal lobule (inferior parietal cortex) as well (Gulyás and Roland, 1993, 1994a, b; Savoy *et al.,* 1995), emphasizing the role of the occipitoparietal pathway in visuospatial processes (Fig. 12).

8.4. Visual Form

One of the most flexible features of the visual system is that it is capable of generating the same visual form by various visual cues or visual signals, such as

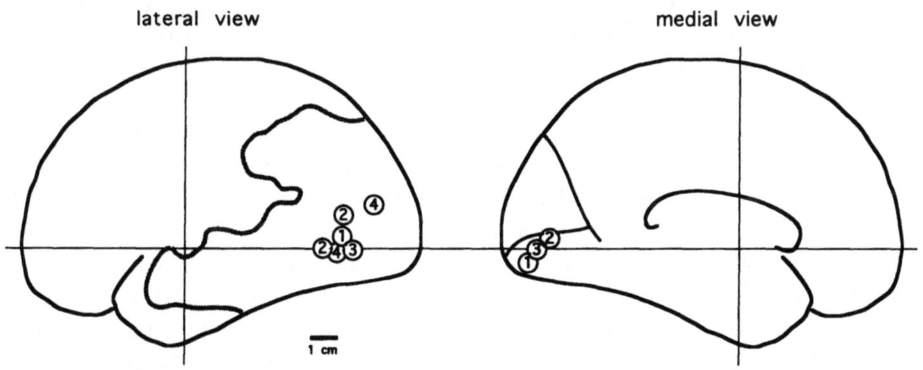

Figure 11. Visual cortical regions activated by visual motion. Same convention and procedures as described in Figure 10. (1) Zeki *et al.* (1991). (2) Corbetta *et al.* (1991). (3) Watson *et al.* (1993). (4) Larsson *et al.* (1996).

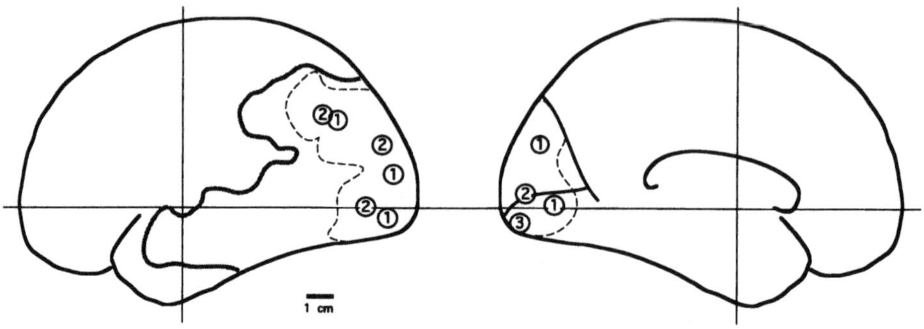

Figure 12. Visual cortical regions activated by disparity. Same convention and procedures as described in Fig. 10. (1) Gulyás and Roland (1994b). (2) Gulyás and Roland (1994a). (3) Ptito *et al.* (1993). The thin dashed line indicates the borders of the large confluent region activated by stereovision, as described in (2). For this reason, in the medial view figure, the centers of gravity belonging to (1) are within the region (dashed line) with center of gravity (2).

color, disparity, motion, orientation, luminance, or texture. It has been shown in monkeys that information from various form cues impinges upon the same form-sensitive cells in the inferior temporal cortex (Sáry *et al.*, 1993). It has been expected, however, that this is due to the early functional segregation of the visual pathways. Different visual signals, even if they finally generate the same visual form, are processed and analyzed by various subsegments of the visual system before an integrated form percept comes into existence. The generation of a simple visual form on the basis of elementary visual signals, including color, motion, disparity, texture, and luminance cues, has been the object of a number of PET studies, which have indicated that, indeed, the processing and analysis of various visual cues takes place in nonoverlapping parallel visual subsystems in the human brain (Gulyás *et al.*, 1994a, b; Gulyás and Roland, 1994a; (Figs. 13 and 14). Whereas most cortical regions participating in the generation of visual form on the basis of various visual cues are localized in occipitotemporal regions, some of them are localized in occipitoparietal regions. This indicates that there is no sharp division between the two visual subsystems, at least as far as basic form processes are concerned.

More complex ways of processing complex visual objects or forms, including their detection, discrimination, and recognition, have also been investigated. The learning and recognition of large-field complex visual patterns heavily involve both occipitotemporal and occipitoparietal visual cortical areas (Roland and Gulyás, 1995). In a similar manner, the discrimination of spatial frequency and orientation signals, basic constituents of complex visual patterns, engages cortical fields in both the occipitotemporal and occipitoparietal visual subsystems (Gulyás and Roland, 1995). On the other hand, operations with more specific complex forms, such as faces, tend to occupy almost exclusively occipitotemporal regions in the visual cortex (Sergent *et al.*, 1992; Haxby *et al.*, 1993, 1994, 1996; Fig. 15).

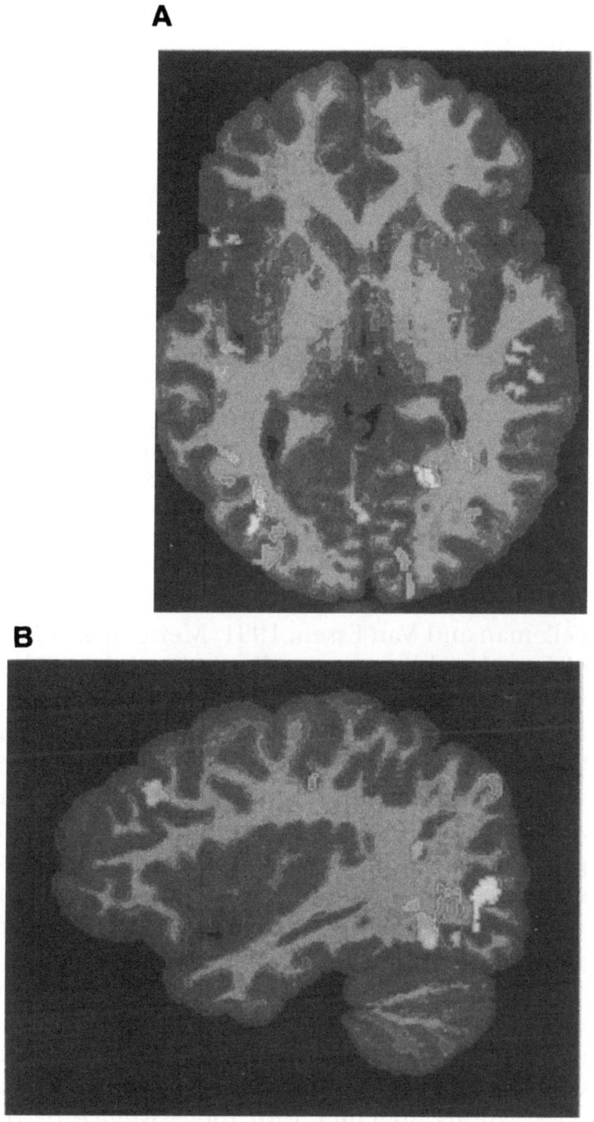

Figure 13. Parallel processing of visual form information generated by various visual cues. (A) Horizontal slice, $z = 2$ mm. HBA convention, as in Fig. 9 (Roland *et al.*, 1994). (B) Parasagittal slice, $x = -35$ mm. Note that there is only one overlap region in the left occipito-parietal sulcus; for the rest the regions are nonoverlapping. From the experimental series described in Gulyás *et al.* (1994a, b). A four-color reproduction of this figure appears following page xxiii.

8.5. Form Perception versus Visuospatial Operations in the Visual Cortex?

In the macaque monkey, anatomical studies have suggested the existence of two major distinct pathways or "processing streams," both originating in V1. Early single-cell recordings and behavioral observations indicated that one of these pathways, the ventral or occipitotemporal pathway, is involved in frank visual perception and object recognition, whereas the dorsal or occipitoparietal pathway deals primarily with visuospatial perception and visuomotor performance. This sharp separation within the visual system has been challenged in recent years, partly on the basis of robust interconnections between the two

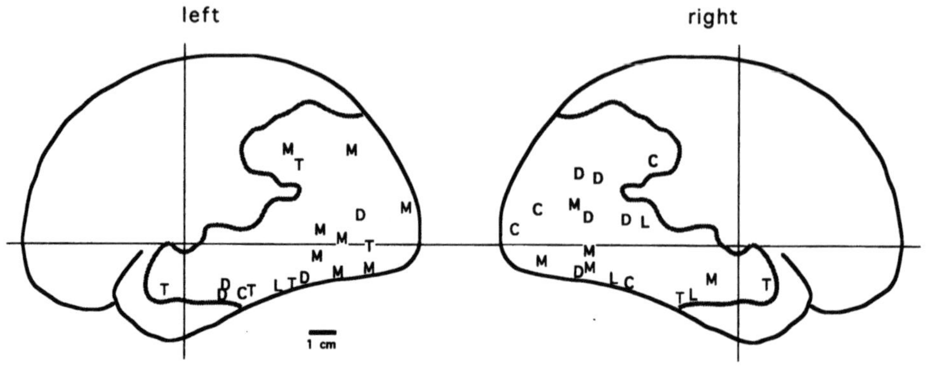

Figure 14. Visual cortical regions activated by visual form stimuli generated by various visual cues: C, color; M, motion; L, luminance; T, texture; D, disparity. Left and right hemispheres; the same convention and procedure as described in previous figures. Note the tendency of occipitotemporal localization, but also the involvement of occipitoparietal regions. Data taken from experiments described in Gulyás and Roland (1994a) and Gulyás *et al.* (1994a, b).

anatomical–functional streams (Morel and Bullier, 1990; Boussaoud *et al.*, 1990; Felleman and Van Essen, 1991, Merigan and Maunsell, 1993), partly on the basis of functional imaging studies indicating a widespread cortical activation in human visual cortex during visual tasks designed to activate one of the two functional subsystems (e.g., a simple object recognition task; e.g., Sergent *et al.*, 1992; Gulyás and Roland, 1994a; Gulyás *et al.*, 1994a, b). Still the hypothesis of the two visual subsystems or processing streams has in recent years provided considerable impetus for several brain activation experiments on normal subjects or psychophysical paradigms designed for patients with anatomically well circumscribed cortical lesions. Some of the functional imaging experiments support the separation between a dorsal visual subsystem, engaged mainly with visuospatial processes, and a ventral visual subsystem, engaged with object and form perception (Haxby *et al.*, 1991, 1993; Fig. 15) On the other hand, other studies show that a number of visual form-related stimuli cause widespread activations in the entire visual cortex, including parietal visual regions as well (Sergent *et al.*, 1992; Gulyás and Roland, 1994a; Gulyás *et al.*, 1994a, b; Haxby *et al.*, 1996). These observations, together with the existence of robust interconnections between parietal and temporal lobe regions in primates (Morel and Bullier, 1990;

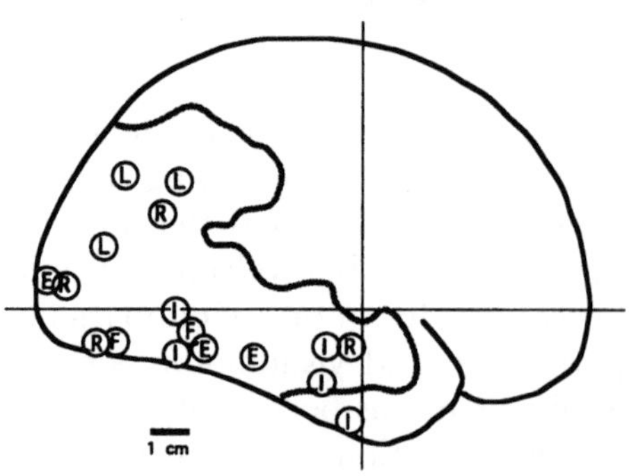

Figure 15. Visual cortical regions activated by a visuospatial and face-related perceptual visual tasks: L, location matching task (Haxby *et al.*, 1993); F, face matching task (Haxby *et al.*, 1993); I, face identity task (Sergent *et al.*, 1992); E, face encoding (Haxby *et al.*, 1996); R, face recognition (Haxby *et al.*, 1996). Same convention as in the previous figures; collapsed parasagittal view. Note the strong (but not pure) tendency for the localization of regions related to face-related visual tasks in the occipitotemporal visual pathway, and the localization of visuospatial-task-related cortical regions in the occipitoparietal pathway.

Boussaoud *et al.*, 1990), indicate that visual cortical operations involve a widespread network of cortical fields and in most cases are not limited to a visual subsystem.

9. Convergence and Divergence in the Human Visual System

Patient and functional imaging data on the organization of the human visual system, as well as primate studies using other methodologies, indicate that at the macronetwork level a relatively large number of visual cortical areas (or subsegments of areas) are involved in the processing and analysis of both elementary visual cues or submodalities and complex visual forms. It can be demonstrated that a part of this patchwork of areas or area segments are active in two or more visual tasks, and the prediction is that basically each subsegment of the visual cortex participates in various visual operations. The same visual percept can be generated by various visual cues, and the neuronal populations taking part in such operations can form an independent nonoverlaping patchwork of cortical areas or area segments.

These observations allow us to conclude that (i) the very same submodality is processed and analyzed by a number of cortical areas or area segments (Figs. 5, 7, 9–15), and (ii) the very same cortical area (or area segment) may be involved in the processing and analysis of a number of visual submodalities (Fig. 16). The very same visual percept can be generated by using different visual cues, each processed and analyzed by nonoverlapping cortical areas or area segments before the resulting information impinges upon the same cortical region (Fig. 13).

In short, there is (i) *divergence* and (ii) *convergence* of functional pathways

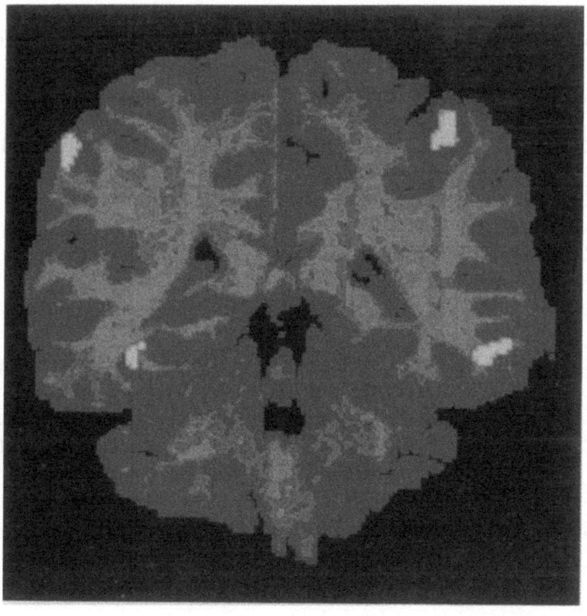

Figure 16. Convergence of various visual cues upon the same cortical regions. The regions present in symmetric positions in the parietal and temporal lobes are overlap regions between cortical fields activated by spatial frequency and orientation discrimination. Coronal slice, y = −39 mm. From the experimental series described in Gulyás and Roland (1995). HBA convention as in Figs. 9 and 13 (Roland *et al.*, 1994).

relaying visual cue-specific information in the human visual cortex. The detailed and complete processing and analysis of a visual scenery may require the concerted action of a large number of cortical areas or area segments in order to produce a unified and integrated visual percept in the human brain.

Acknowledgments. The author expresses his gratitude to Drs. Kathleen S. Rockland, Roger B. H. Tootell, and Per E. Roland for their valuable suggestions. The work was supported by the Karolinska Institute and the Swedish Medical Research Council.

10. References

Allison, T., Ginter, H., McCarthy, G., Nobre, A. C., Puce, A., Luby, M., and Spencer, D. D., 1994, Face recognition in human estrastriate cortex, *J. Neurophysiol.* **71:**821–825.

Bak, M., Girvin, J. P., Hambrecht, F. T., Kufta, C. V., Loeb, G. E., and Schmidt, E. M., 1990, Visual sensations produced by intracortical microstimulation of the human occipital cortex, *Med. Biol. Eng. Comp.* **28:**257–259.

Belliveau, J. W., Kennedy, D. N., McKinstry, R. C., Buchbinder, B. R., Weisskoff, R. M., Cohen, M. S., Vevea, J. M., Brady, T. J., and Rosen, B. R., 1991, Functional mapping of the human visual cortex by magnetic resonance imaging, *Science* **254:**716–719.

Benevento, L. A., and Yoshida, K., 1981, The afferent and efferent organisation of the lateral geniculo-prestriate pathways in the macaque monkey, *J. Comp. Neurol.* **203:**455–474.

Benson, D. F., and Greenberg, J. P., 1969, Visual form agnosia, *Arch. Neurol.* **20:**82–89.

Blake, R., and Cormack, R., 1979, Psychophysical evidence for monocular visual cortex in stereoblind humans, *Science* **203:**274–275.

Boussaoud, D., Ungerleider, L. G., and Desimone, R., 1990, Pathways for motion analysis, cortical connections of the medial superior temporal and fundus of the superior temporal visual areas in the macaque, *J. Comp. Neurol.* **296:**462–495.

Brindley, G. S., and Lewin, W., 1968, The sensations produced by electric stimulation of the visual cortex, *J. Physiol.* (Lond.) **196:**479–493.

Buchner, H., Weyen, U., Frackowiak, R. S., Romaya, J., and Zeki, S., 1994, The timing of visual evoked potential activity in human area V4, *Proc. R. Soc. Lond. B* **257:**99–104.

Burkhalter, A., 1993, Development of forward and feedback connections between areas V1 and V2 of human visual cortex, *Cerebral Cortex* **3:**487–476.

Burkhalter, A., and Bernardo, K. L., 1989, Organization of corticocortical connections in human visual cortex, *Proc. Natl. Acad. Sci. USA* **86:**1071–1075.

Cajal, S. R., 1899a, Notes on a structural study of the visual cortex of the human cerebrum, *Rev. Ibero-Am. Ciencias Medicas* **1:**1–14.

Cajal, S. R., 1899b, Studies on the human cerebral cortex (I): Visual cortex, *Rev. Trimestral Micrografica* **4:**1–63.

Campbell, A. W., 1905, *Histological Studies on the Localisation of Cerebral Function,* Cambridge University Press, Cambridge.

Carney, T., Shadlen, M., and Switkes, E., 1987, Parallel processing of motion and colour information, *Nature* **328:**647–649.

Casagrande, V. A., and Kaas, J. H., 1994, The afferent, intrinsic, and efferent connections of primary visual cortex in primates, in: *Cerebral Cortex,* Volume 10, *Primary Visual Cortex in Primates* (A. Peters and K. Rockland, eds.), Plenum Press, New York, pp. 201–259.

Clarke, S., 1993, Callosal connections and the functional subdivisions of the human occipital cortex, in: *Functional Organization of the Human Visual Cortex,* (B. Gulyás, D. Ottoson, and P. E. Roland, eds.), Pergamon Press, Oxford, pp. 137–149.

Clarke, S., and Miklossy, J., 1990, Occipital cortex in man, organization of callosal connections, related myelo- and cytoarchitecture, and putative boundaries of functional visual areas, *J. Comp. Neurol.* **298:**188–214.

Connoly, M., and Van Essen, D. C., 1984, The representation of the visual field in parvicellular and magnocellular layer of the lateral geniculate nucleus in the macaque monkey, *J. Comp. Neurol.* **226:**544–564.

Corbetta, M., Miezin, F. M., Dobmeyer, S., Shulman, G. L., and Petersen, S. E., 1991, Selective and divided attention during visual discrimination of shape, color, and speed, Functional anatomy by positron emission tomography, *J. Neurosci.* **11:**2383–2402.

Coren, S., Porac, C., and Ward, L. W., 1978, *Sensation and Perception,* Academic Press, New York.

Cowey, A., and Rolls, E. T., 1974, Human cortical magnification factor and its relation to visual acuity, *Exp. Brain Res.,* **21,** 447–454.

De Jong, B. M., Shipp, S., Skidmore, B., Frackowiak, R. S., and Zeki, S. 1994, The cerebral activity related to the visual perception of forward motion in depth, *Brain* **117:**1039–1054.

De Yoe, E. A., and Van Essen, D. C., 1988, Concurrent processing streams in monkey visual cortex, *Trends Neurosci.* **11:**219–226.

Dobelle, W. H., Turkel, J., Henderson, D. C., and Evans, J. R., 1979, Mapping the representation of the visual field by electrical stimulation of human visual cortex, *Am. J. Ophthalmol.* **88,** 727–735.

Drasdo, N., 1977, The neural representation of the visual space, *Nature* **266:**554–556.

Drury, H. A., Van Essen, D. C., Anderson, C. H., Lee, C. W., Coogan, T. A., and Lewis, J. W., 1996, Computerized mappings of the cerebral cortex: A multiresolution flattening method and a surface-based coordinate system, *J. Cognitive Neurosci.* **8:**1–28.

Felleman, D. J., and Van Essen, D. C., 1987, Receptive field properties of neurons in area V3 of macaque monkey extrastriate cortex, *J. Neurophysiol.* **57:**889–920.

Felleman, D. J., and Van Essen, D. C., 1991, Distributed hierarchical processing in the primate cerebral cortex, *Cerebral Cortex* **1:**1–47.

ffytche, D. H., Guy, C. N., and Zeki, S., 1995, The parallel visual motion inputs into areas V1 and V5 of human cerebral cortex, *Brain* **118:**1375–1394.

Fox, P. T., Mintun, M. A., Raichle, M. E., Miezin, F. M., Allman, J. M., and Van Essen, D. C., 1986, Mapping human visual cortex with positron emission tomography, *Nature* **323:**806–809.

Fox, P. T., Miezin, F. M., Allman, J. M., Van Essen, D. C., and Raichle, M., 1987, Retinotopic organization of human visual cortex with positron emission tomography, *J. Neurosci.* **7:**913–922.

Fried, I., Mateer, C., Ojemann, G., Wohns, R., and Fedio, P., 1982, Organization of visuospatial functions in human cortex: Evidence from electrical stimulation, *Brain* **105:**349–371.

Fries, W., 1981, The projection from the lateral geniculate nucleus to the prestriate cortex of the macaque monkey, *Proc. R. Soc. Lond. B* **213:**73–80.

Gennari, F., 1782, *De Peculiari Structura Cerebri,* Parma, Italy.

Goodale, M. A., 1993, Visual pathways supporting perception and action in the primate cerebral cortex, *Curr. Opin. Neurobiol.* **3:**578–585.

Goodale, M. A., and Milner, A. D., 1992, Separate visual pathways for perception and action, *Trends Neurosci.* **15:**20–25.

Gulyás, B., and Roland, P. E., 1991, Cortical fields participating in form and colour discrimination in the human brain, *NeuroReport* **2:**585–588.

Gulyás, B., and Roland, P. E., 1993, Visual cortical regions involved in stereovision, in: *Functional Organization of the Human Visual Cortex* (B. Gulyás, D. Ottoson, and P. E. Roland, eds.), Pergamon Press, Oxford, pp. 341–357.

Gulyás, B., and Roland, P. E., 1994a, Processing and analysis of visual sub-modalities of form, colour, and binocular disparity in the human brain: Functional anatomy by positron emission tomography, *Eur. J. Neurosci.* **6:**1811–1828.

Gulyás, B., and Roland, P. E., 1994b, Binocular disparity discrimination in human cerebral cortex: Functional anatomy by positron emission tomography, *Proc. Natl. Acad. Sci. USA* **91:**1239–1243.

Gulyás, B., and Roland, P. E., 1995, Visual cortical fields participating in spatial frequency and orientation discrimination: Functional anatomy by positron emission tomography, *Hum. Brain Mapping* **3:**133–152.

Gulyás, B., Heywood, C. A., Popplewell, D. B., Cowey, A., and Roland, P. E., 1994a. Visual form discrimination from colour or motion cues: Functional anatomy by positron emission tomography, *Proc. Natl. Acad. Sci. USA* **91:**9965–9969.

Gulyás, B., Roland, P. E., Heywood, C. A., Popplewell, D. B., and Cowey, A., 1994b, Visual form discrimination from luminance or disparity cues: Functional anatomy by positron emission tomography, *NeuroReport* **5:**2367–2371.

Gulyás, B., Roland, P. E., Larsson, J., Kinomura, S., Vidnyánszky, Z., Schormann, T., and Zilles, K., 1996, The macular representation in human striate cortex: A combined PET and cytoarchitectural study, *Neuroimage* **3**(3):S276.

Gulyás, B., Larsson, J., Amunts, K., Zilles, K., and Roland, P. E., 1997, Cortical regions in the human hair systematically participating in the processing and analysis of color, *Neuroimage* **5**:S2.

Hadjikani, N., 1995, The pattern of visual callosal connections in human extrastriate cortex, Thesis, University of Lausanne, p. 48.

Haxby, J. V., Grady, C. L., Horwitz, B., Ungerleider, L. G., Mishkin, M., Carson, R. E., Herscovitch, P., Schapiro, M. B., and Rapaport, S. I., 1991, Dissociation of object and spatial visual processing pathways in human extrastriate cortex *Proc. Natl. Acad. Sci. USA* **88**:1621–1625.

Haxby, J. V., Grady, C. L., Horwitz, B., Salerno, J., Ungerleider, L. G., Mishkin, M., Shapiro, M. B., and Rapaport, S. I., 1993, Dissociation of object and spatial vision processing pathways in human extrastriate cortex, in: *Functional Organization of the Human Visual Cortex* (B. Gulyás, D. Ottoson, and P. E. Roland, eds.), Pergamon Press, Oxford, pp. 329–340.

Haxby, J. V., Horwitz, B., Ungerleider, L. G., Maisog, J. M., Pietrini, P., and Grady, C. L., 1994, The functional organization of human extrastriate cortex: A PET-rCBF study of selective attention to faces and locations, *J. Neurosci.* **14**:6336–6353.

Haxby, J. V., Ungerleider, L. G., Horwitz, B., Maisog, J. M., Rapaport, S. I., and Grady, C. L., 1996, Face encoding and recognition in the human brain, *Proc. Natl. Acad. Sci. USA* **93**:922–0927

Hess, R. H., Baker, C. L., and Zihl, J., 1989, The "motion-blind" patient, low-level spatial and temporal filters, *J. Neurosci.* **9**:1628–1640.

Heywood, C. A., and Cowey, A., 1987, On the role of cortical area V4 in the discrimination of hue and pattern in macaque monkeys, *J. Neurosci.* **7**:2601–2617.

Heywood, C. A., Gadotti, A., and Cowey, A., 1992, Cortical area V4 and its role in the perception of color, *J. Neurosci.* **12**:4056–4065.

Holmes, G., 1945, The organisation of the visual cortex in man, *Proc. R. Soc. Lond. B* **132**:348–361.

Holmes, G., and Horrax, G., 1919, Disturbances of spatial orientation and visual attention, with loss of stereoscopic vision, *Arch. Neurol. Psychiatry* **1**:389–407.

Horton, J. C., and Hoyt, W. F., 1991, The representation of the visual field in human striate cortex, *Arch. Ophthalmol.* **109**:816–824.

Kaas, J. H., 1993, The organization of visual cortex in primates: Problems, conclusions, and the use of comparative studies in understanding the human brain, in: *Functional Organisation of the Human Visual Cortex* (B. Gulyás, D. Ottoson, and P. E. Roland, eds.), Pergamon Press, Oxford, pp. 1–11.

Kaas, J. H., 1995, Human visual cortex, *Current Biol.* **5**:1126–1128.

Kinomura, S., Larsson, J., Gulyás, B., and Roland, P. E., 1995, Attention activates the mid-brain reticular formation and thalamic interlaminar nuclei in man, *Science* **271**:512–515.

Landis, T., Regard, M., Bliestle, A., and Kleihues, P., 1988, Prosopagnosia and agnosia for noncanonical views, *Brain* **111**:1287–1297.

Larsson, J., Gulyás, B., and Roland, P. E., 1996, Discrimination of motion at isoluminance activates MT, V3 and superior parietal lobule in man, *Neuroimage* **3**(3):S285.

Lindsay, P. H., and Norman, D. A., 1977, *Human Information Processing,* Academic Press, New York.

Livingstone, M. S., and Hubel, D. M., 1987, Psychophysical evidence for separate channels for the perception of form, color, movement, and depth, *J. Neurosci.* **3**:3416–3468.

MacKay, G., and Dunlop, J. C., 1899, The cerebral lesions in a case of complete acquired colour-blindness, *Scott. Med. Surg. J.* **5**:503–512.

Malpeli, J. G., and Baker, F. H., 1975, The representation of the visual field in the lateral geniculate nucleus of *Macaca mulatta, J. Comp. Neurol.* **161**:569–594.

McKeefry, D. J., Watson, J. D. G., Frackowiak, R. S. J., and Zeki, S., 1996, Activity in human areas V1/v2, V3 and V5 during the perception of coherent and incoherent motion, *Neuroimage* **3**(3):S289.

McLeod, P., Heywood, C., Driver, J., and Zihl, J., 1989, Selective deficit of visual search in moving displays after extrastriate damage, *Nature* **339**:466–467.

Meadows, J. C., 1974a, Disturbed perception of colours associated with localized cerebral lesions, *Brain* **97**:615–632.

Meadows, J. C., 1974b, The anatomical basis of prosopagnosia, *J. Neurol.* **37**:489–501.

Merigan, W. H., and Maunsell, J. R. H., 1993, How parallel are the primate visual pathways? *Annu. Rev. Neurosci.* **16**:369–402.

Meyer, B. U., and Diehl, R. R., 1992, Examination of the visual system with transcranial magnetic stimulation, *Nervenarzt* **63:**328–334.

Milner, A. D., and Goodale, M. A., 1993, Visual pathways to perception and action, *Prog. Brain Res.* **95:**317–338.

Morel, A., and Bullier, J., 1990, Anatomical segregation of two cortical visual pathways in the macaque monkey, *Visual Neurosci.* **4:**555–578.

Orban, G. A., Dupont, P., De Bruyn, B., Vogels, R., Vandenberghe, R., and Mortelmans, L., 1995, A motion area in human visual cortex, *Proc. Natl. Acad. Sci. USA* **92:**993–997.

Paulesu, E., Harrison, J., Baron-Cohen, S., Watson, J. D. G., Goldstein, L., Heather, J., Frackowiak, R. S. J., and Frith, C. D., 1995, The physiology of coloured hearing. A PET activation study of colour-word synaesthesia, *Brain* **118:**661–676.

Pearlman, A. L., Birch, J., and Meadows, J. C., 1979, Cerebral color blindness, An acquired defect in hue discrimination, *Ann. Neurol.* **5,** 253–261.

Penfield, W., and Rasmussen, T., 1957, *The Cerebral Cortex of Man,* Macmillan, New York.

Perry, V. H., Oehler, R., and Cowey, A., 1984, Retinal ganglion cells that project to the dorsal lateral geniculate nucleus in the macaque monkey, *Neuroscience* **12:**1101–1123.

Poggio, G. F., and Fischer, B., 1977, Binocular interaction and depth sensitivity in striate and prestriate cortex of behaving rhesus monkey, *J. Neurophys.* **40:**1392–1405.

Poggio, G. F., and Talbot, W. F., 1981, Mechanisms of static and dynamic stereopsis in foveal cortex of the rhesus monkey, *J. Physiol.* **315:**469–492.

Poppelreuter, W., 1917, *Die Psychischen Schädigungen durch Kpfschuss im Kriege 1914/16,* Verlag von Leopold Voss, Leipzig.

Ptito, A., Zatorre, R., Petrides, M., Frey, S., Alivisatos, B., and Evans, A. C., 1993, Localization and lateralization of stereoscopic processing in the human brain, *NeuroReport* **4:**1155–1158.

Ramachandran, V. S., and Gregory, R. L., 1978, Does colour provide an input to human motion perception? *Nature* **275:**55–56.

Regan, D., Giaschi, D., Sharpe, J. A., and Hong, X. H., 1992, Visual processing of motion-defined form: Selective failure in patients with parietotemporal lesions, *J. Neurosci.* **12:**2198–2210.

Reppas, J. B., Tootell, R. B. H., and Rosen, B. R., 1995, Regional variation of color and luminance contrast sensitivity between and within human visual cortical areas, *Hum. Brain Mapping Suppl.* **1:**61.

Richards, W., 1970, Stereopsis and stereoblindness, *Exp. Brain Res.* **10:**380–388.

Rizzo, M., 1989, Astereopsis, in: *Handbook of Neuropsychology,* Volume 2 (F. Boller and J. Grafman, eds.), Elsevier, Amsterdam, pp. 415–427.

Rizzo, M., 1994, The role of striate cortex: Evidence from human lesion studies, in: *Cerebral Cortex,* Volume 10, *Primary Visual Cortex in Primates,* A. Peters and K. S. Rockland, eds., Plenum Press, New York, pp. 505–540.

Rodieck, R. W., 1973, *The Vertebrate Retina: Principles of Structure and Function,* Freeman, San Francisco.

Roland, P. E., 1993, *Brain Activation,* Wiley-Liss, New York.

Roland, P. E., and Gulyás, B., 1994, Visual imagery and visual representation, *Trends Neurosci.* **17:**281–287.

Roland, P. E., and Gulyás, B., 1995, Visual memory, visual imagery, and visual recognition of large field patterns by the human brain. Functional anatomy by positron emission tomography, *Cerebral Cortex* **5:**79–93.

Roland, P. E., and Gulyás, B., 1996, Assumptions and validations of statistical tests for functional neuroimaging, *Eur. J. Neurosci.,* in press.

Roland, P. E., and Zilles, K., 1994, Brain atlases—A new research tool, *Trends Neurosci.* **17:**458–467.

Roland, P. E., Levin, B., Kawashima, R., and Åkerman, S., 1993, Three-dimensional analysis of clustered voxels in ^{15}O-butanol brain activation images, *Hum. Brain Mapping* **1:**3–19.

Roland, P. E., Graufelds, C. J., Wählin, J., Ingelman, L., Andersson, M., Ledberg, A., Pedersen, J., Åkerman, S., Dabringhaus, A., and Zilles, K., 1994, Human Brain Atlas: For high-resolution functional and anatomical mapping, *Hum. Brain Mapping* **1:**173–184.

Rovamo, J. H., 1978, Receptive field density of retinal ganglion cells and cortical magnification factor in man, *Med. Biol.* **56:**97–102.

Rovamo, J., and Virsu, V., 1979, An estimation and application of the human cortical magnification facor, *Exp. Brain Res.* **37:**495–510.

Rovamo, J., and Virsu, V., 1984, Isotropy of cortical magnification and topography of striate cortex, *Vision Res.* **24:**283–286.

Sanderson, K. J., 1971, Visual field projection columns and magnification factors in the lateral geniculate nucleus of the cat, *Exp. Brain Res.* **13:**159–177.

Sakai, K., Watanabe, E., Onodera, Y., Uchida, I., Kato H., Yamamoto, E., Koizumi, H., and Miyashita, Y., 1995, Functional mapping of the human colour centre with echo-planar magnetic resonance imaging, *Proc. R. Soc. Lond. B* **261**(1360)**:**89–98.

Sáry, Gy., Vogels, R., and Orban, G. A., 1993, *Science* **260:**995–997.

Savoy, R. L., Kwong, K. K., and Cohen, M. S., 1993, Searching for stereopsis in humans using ultrafast functional MRI: Stimuli, analysis techniques, and preliminary data, *Soc. Neurosci. Abstr.* **19:**1500.

Savoy, R. L., Tootell, R. B. H., O'Craven, K. M., and Reppas, J. R., 1995, Cortical localization of fMRI response to stereo disparity, *Hum. Brain Mapping Suppl.* **1:**57.

Schiller, P. H., 1993, Parallel pathways in the visual system, in: *Functional Organization of the Human Visual Cortex* (B. Gulyás, D. Ottoson, and P. E. Roland, eds.), Pergamon Press, Oxford, pp. 43–58.

Schiller, P. H., Logothetis, N. K., and Charles, E., 1990, The role of the color-opponent and the broad-band channels in vision, *Visual Neurosci.* **5:**321–346.

Sereno, M. I., Dale, A. M., Reppas, J. B., Kwong, K. K., Belliveau, J. W., Brady, T. J., Rosen, B. R., and Tootell, R. B. H., 1995, Borders of multiple visual areas in humans revealed by functional magnetic resonance imaging, *Science* **268:**889–893.

Sergent, J., Ohta, S., and MacDonald, B., 1992, Functional neuroanatomy of face and object processing. A positron emission tomography study, *Brain* **115:**15–36.

Spalding, J. M. K., 1952a, Wounds of the visual pathway. Part I: The visual radiation, *J. Neurol. Neurosurg. Psychiatry,* **15:**99–109.

Spalding, J. M. K., 1952b, Wounds of the visual pathway. Part II: The striate cortex, *J. Neurol. Neurosurg. Psychiatry* **15:**169–183.

Stone, J., and Johnston, E., 1981, The topography of primate retina: A study of the human, bushbaby, and New- and Old-World monkeys, *J. Comp. Neurol.* **196:**205–223.

Talairach, J., and Tournoux, P., 1988, *Co-Planar Stereotaxic Atlas of the Human Brain,* Thieme, Stuttgart.

Teuber, H.-L., Battersby, W. S., and Bender, M. B., 1960, *Visual Field Defects After Penetrating Missile Wounds of the Brain,* Harvard University Press, Cambridge, MA.

Tolhurst, D. J., and Ling, L., 1988, Magnification factors and the organization of the human striate cortex, *Hum. Neurobiol.* **6:**247–254.

Tootell, R. B. H., and Taylor, J. B., 1995, Anatomical evidence for MT and additional cortical visual areas in humans, *Cerebral Cortex* **5:**39–55.

Tootell, R. B. H., Reppas, J. B., Kwong, K. K., Malach, R., Born, R. T. Brady, T. J., Rosen, B. R., and Belliveau, J. W., 1995a, Functional analysis of human MT and related visual cortical areas using magnetic resonance imaging, *J. Neurosci.* **15:**3215–3230.

Tootell, R. B. H., Reppas, J. B., Dale, A. M., Look, R. B., Sereno, M. I., Malach, R., Brady, T. J., and Rosen, B. R., 1995b, Visual motion aftereffect in human cortical area MT revealed by functional magnetic resonance imaging, *Nature* **375:**139–141.

Tootell, R. B. H., Reppas, J. R., and Rosen, B. R., 1995c, Functional analysis of human visual cortical areas V3, VP and V3A using magnetic resonance imaging, *Hum. Brain Mapping, Suppl.* **1:**62.

Tootell, R. B. H., Dale, A. M., Mendola, J. D., Reppas, J. B., and Sereno, M. I., 1996, fMRI analysis of human visual cortical area V3A, *Neuroimage* **3**(3)**:**S358.

Ungerleider, L. G., and Mishkin, M., 1982, Two cortical visual systems, in: *Analysis of Visual Behaviour,* (D. J. Ingle, M. A. Goodale, and R. J. W. Mansfield, eds.), MIT Press, Cambridge, MA, pp. 549–586.

Vaina, L. M., 1989, Selective impairment of visual motion interpretation following lesions of the right occipito-parietal area in humans, *Biol. Cybernet.* **61:**347–359.

Vaina, L. M., 1994, Functional segregation of color and motion processing in the human visual cortex: Clinical evidence, *Cerebral Cortex* **5:**555–572.

Van Essen, D. C., Anderson, C. H., and Felleman, D. J., 1992, Information processing in the primate visual system: An integrated systems perspective, *Science* **255:**419–423.

Verray, L., 1888, Hémiachromatopsie droite absolue, *Arch. Ophthalmol.* (Paris) **8:**289–301.

Vicq-d'Azyr, F., 1786, *Traité d'Anatomie,* Paris.

Virsu, V., and Rovamo, J., 1979, Visual resolution, contrast sensitivity, and the cortical magnification factor, *Exp. Brain Res.* **37:**475–494.

Watson, J. D. G., Myers, R., Frackowiak, R. S. J., Hajnal, J. V., Woods, R. P., Mazziotta, J. C., Shipp, S., and Zeki, S., 1993, Area V5 of the human brain: Evidence from a combined study using positron emission tomography and magnetic resonance imaging, *Cerebral Cortex* **3:**79–94.

Wong-Riley, M. T. T., 1993, Cytochrome oxidase studies on the human visual cortex, in: *Functional Organization of the Human Visual Cortex,* (B. Gulyás, D. Ottoson, and P. E. Roland, eds.), Pergamon Press, Oxford, 165–180.

Yukie, M., and Iwai, E., 1981, Direct projection from the dorsal lateral geniculate nucleus to the prestriate cortex in macaque monkeys, *J. Comp. Neurol.* **201:**81–97.

Zeki, S., 1974, Functional organization of a visual area in the posterior bank of the superior temporal sulcus of the rhesus monkey, *J. Physiol.* (Lond.) **236:**549–573.

Zeki, S., 1990, A century of cerebral achromatopsia, *Brain* **113:**1721–1777.

Zeki, S., 1991, Cerebral akinetopsia (visual motion blindness). A review, *Brain* **114:**811–824.

Zeki, S., 1993, The visual association cortex, *Curr. Opin. Neurobiol.* **3**(1993)**:**155–159.

Zeki, S., Watson, D. G., Lueck, C. J., Friston, K. J., Kennard, C., and Frackowiak, R. S. J., 1991, A direct demonstration of functional specialization in human visual cortex, *J. Neurosci.* **11:**641–649.

Zeki, S., Watson, J. D. G., and Frackowiak, R. S. J., 1993, Going beyond the information given: The relation of illusory visual motion to brain activity, *Proc. R. Soc. Lond. B* **252:**215–222.

Zihl, J., Von Cramon, D., and Mai, N., 1983, Selective disturbance of movement vision after bilateral brain damage, *Brain* **106:**313–340.

Zilles, K., 1990, Cortex, in: *The Human Nervous System* (G. Paxinos, ed.), Academic Press, San Diego, pp. 757–802.

Zilles, K., 1991, Codistribution of receptors in the human cerebral cortex, in: *Receptors in the Human Nervous System* (F. A. O. Mendelsohn and G. Paxinos, eds.), pp. 165–206.

17

The Color and Motion Systems as Guides to Conscious Visual Perception

S. ZEKI

1. Introduction

Is there one overall consciousness or are there many? Does each system have its own mechanism for conscious experience or is consciousness suprasystemic? Can each of the many visual areas that constitute the primate visual brain contribute directly to conscious experience, without the necessity for pre- or postprocessing by other visual areas, and especially by the primary visual cortex (area V1)? Such questions may have been asked before, but, with the discovery of functional specialization in the primate visual cortex (Zeki, 1974b, 1978a; Livingstone and Hubel, 1984a), they have become mandatory. In any case, there never has been a more propitious time for addressing them experimentally than the present. In this chapter, I do so with reference to the visual motion and color systems.

Our thinking about conscious visual experience has been dominated by the controversial syndrome of blindsight (Campion *et al.*, 1983; Fendrich *et al.*,. 1992; Celesia *et al.*, 1991; Ruddock, 1991; Weiskrantz, 1986, 1990), when subjects blinded by lesions to area V1 are said to be able to discriminate ("see") visual stimuli of which they have no conscious experience, with the implication that an intact and healthy V1 is necessary for conscious visual experience. Whatever the

S. ZEKI • Wellcome Department of Cognitive Neurology, University College London, London WC1E 6BT, England.

Cerebral Cortex, Volume 12, edited by Rockland *et al.* Plenum Press, New York, 1997.

merits of the different views about blindsight as a phenomenon, it is interesting to consider another phenomenon which stands, in a sense, diametrically opposed to blindsight. This second phenomenon, which has been called residual vision (Blythe *et al.*, 1987), is one in which subjects blinded by lesions in V1 are not only able to see, but also experience consciously what they have seen. The first and most extensive collection of cases of residual visual motion following damage to V1 ever assembled was published by George Riddoch in 1917 and sadly ignored by most since then. But other examples, some recent, have also been published and together they give a fascinating glimpse into how the visual brain functions.

2. Residual Vision

The motion system is much the better understood and it is perhaps more useful to begin with it. A definition of the residual vision for motion can be obtained by reference to the descriptions given by Riddoch (1917). His patients had suffered gunshot wounds during the Great War and, as a temporary army captain in the Royal Army Medical Corps, he had occasion to examine them. The tests were not ones which would be regarded as satisfactory today. This is partly due to the fact that his instruments, including those available for perimetric study, lacked the sophistication of today's, a fact recognized by Riddoch, who wrote of the "admitted crudeness of the perimetric method." It is also partly due to the fact that his patients had suffered injury in various parts of the Empire as well as on the battlefields of France. Riddoch was only able to examine them after they had been transferred to the Empire Hospital for Officers in London, frequently after a relatively long interval. Moreover, the location of the damage in the brain was never precise. It depended upon notes made by the surgeons after the operations or based on X-ray examinations, thus leading to descriptions which we would find incomplete and unsatisfactory today. In spite of these difficulties, Riddoch managed to characterize a syndrome which has far-reaching consequences for our understanding of conscious vision, although it is certain that he himself did not see the significance of his findings in this light. And because he could not give a plausible neurological explanation for the phenomenon, it was rapidly dismissed by Holmes and relegated to oblivion (see Zeki, 1991). But others (Blythe *et al.*, 1987; Mestre *et al.*, 1992; Barbur *et al.*, 1993) have more recently replicated Riddoch's findings and it is interesting to enquire whether a study of the neural pathways which are the basis of this syndrome will give us any insight into the more general phenomenon of conscious vision. It is not that the syndrome has become suddenly illuminating. Rather, recent anatomical and physiological findings would predict that such a syndrome should exist: the anatomical evidence shows that there is a direct pathway from the pulvinar that bypasses area V1 and leads to a prestriate cortical area that emphasizes motion (area V5) and the physiological evidence suggests that the characteristics of V5, directional motion selectivity, can be sustained by the direct pathway. Hence the only reason that there should not be a residual and conscious vision for motion would be the inability of the direct pathway to prestriate cortex to mediate conscious visual perception. The issue therefore revolves around the question of whether a pathway reaching the prestriate cortex without passing through V1 is sufficiently potent to mediate conscious vision.

2.1. Characteristics of Residual Vision

Riddoch studied 10 subjects and divided them into three groups. Of these, the most interesting from our point of view is the first, consisting of five patients. It is described as one "*Showing only perception of Movement in the Affected Field, in Addition to some Perception of Light.*"

All five patients were the victims of enemy action, none became unconscious immediately after being struck, and all save one soldiered on in the immediate aftermath of the injury. One was hit at Rawal Pindi in India, two at Gallipoli in Ottoman Turkey, one in France, and one in unrecorded territory. All were eventually evacuated to England and examined between 3 and 5 months later. They shared the following important features: (1) a complete dissociation between movement and object vision; (2) a substantial difference in the area of scotoma, as plotted using stationary versus moving objects; and (3) a conscious perception of visual motion.

2.1.1. The Dissociation of Motion Vision

All patients were able to detect the presence of motion without being able to characterize the other attributes of the stimulus. For patient 1, "The 'moving things' have no distinct shape, and the nearest approach to colour that can be attributed to them is a shadowy grey." For patient 2, "The 'moving something' had neither form nor colour. It gave him the impression of a shadow . . . the size, or the colour (black or white), of the test object make little difference." Patient 3 could detect the movement of feet in the street "though they had no shape"; patient 4 "declared he could distinguish no object . . . but he knew that something had moved through his blind field"; while patient 5 said, "They [the moving objects] don't appear to have any colour or shape. They look like shadows. Sometimes I can tell if the moving things are white." Thus one feature of the syndrome of residual motion vision is the ability to detect the presence of motion without being able to assign any other attribute to the moving object or stimulus.

2.1.2. The Scotoma for Moving and Stationary Objects

As might be expected from the above, there was a difference in the blind field when plotted with a stationary and a moving object. In some, as in case 1, the dissociation was complete, there being a "full field for movement." This evidently surprised Riddoch, because, using movement, "The nearest approach to a scotoma I could get by repeated examination was a feebler perception of movement in a small part of the upper quadrant about 20° from the fixation point; but the patient declared emphatically that it was never absolutely lost." In patient 3, the scotoma for movement improved considerably, but he still "cannot see stationary objects," while patient 4 who, after being wounded had no appreciation of movement in the blind fields, was found to have a nearly complete field for movement after 8 months, but the field "has not changed for the recognition of the object." A second feature of the syndrome, and one which follows from the first, is therefore the inequality of the scotoma when studied with stationary and moving objects. The inequality can be complete from the start, as in patient 1, or incomplete, some part of the field remaining blind when tested either way. Where there is a return of vision, movement vision may recover more quickly and extensively.

2.1.3. Conscious Perception of Visual Motion

A phenomenon rarely remarked on, though one which is of cardinal importance, is that patients are consciously aware of having seen movement. Riddoch himself did not make a special point of this, but made repeated references to it in his paper, almost certainly without realizing its true significance. In fact, its significance is a post-blindsight hindsight, so to speak, because the conscious dimension distinguishes the syndrome of residual motion vision so neatly from the syndrome of blindsight. Riddoch would no doubt have emphasized the conscious dimension if he had known about the phenomenon of blindsight. That the perception was a conscious one is only implicit in the fact that the patients made verbal reports to him to indicate that they had seen the movement. Thus he writes of "The frequency with which patients with restricted visual fields from occipital wounds . . . were immediately *conscious* of 'something' moving when the object was oscillated," of patient 1 in whom "the *consciousness* of 'something moving' kept up a continual desire to turn the head," and of patient 4, who "*knew* that something had moved through his blind field" (my emphases). Thus my definition of the syndrome would include a conscious awareness of the presence of motion.

It is nevertheless worth emphasizing that conscious awareness of the visual stimulus is restricted to movement and does not necessarily extend even to the direction of motion. All Riddoch's patients were "quite sure that neither shape nor colour can be attributed to [the movement]." But also, significantly, "The patients have great difficulty in describing the nature of the movement that they see: it is so vague and shadowy." Directional selectivity was not known at that time and Riddoch did not make a special point of examining it or discussing it. But the "vague and shadowy" motion that he speaks of need not exclude the ability to detect the direction of motion. Patient GY, an hemianopic patient with a syndrome similar to that described by Riddoch, and whom my colleagues and I have studied (Barbur *et al.*, 1993), was able to perceive consciously both the presence and the direction of motion to a high degree of accuracy, although he required rapidly displaced stimuli of high contrast to be able to do so. Patient 25 of Blythe *et al.* (1987) was also able to detect the direction of motion at high contrasts. Nevertheless the quality and degree of residual motion vision left after lesions of the striate cortex may vary somewhat, though it is severely abnormal. At one extreme, the conscious perception of motion and of its direction is relatively normal even though stimuli of high contrast must be used and even though not all directions of motion can be discriminated correctly or with equal ease. Our studies of patient GY showed that he had great difficulty in distinguishing the direction of oblique motion, reporting only the presence of a moving stimulus; in contrast, he saw with ease, and consciously, movement in the cardinal directions.

2.2. The Neural Basis of Residual Motion Vision

Riddoch was never very explicit in assigning his patients' capacities to a particular locus in the brain because the extent of the lesions could not be definitively ascertained. He wrote: "Operations had been performed in most cases at the casualty clearing stations or at the base hospitals; but as a rule some indication was given on the transfer sheet or on the medical history of what had been done." On the basis of these notes and on X-ray examination for bone fragments, he ascertained the area of damage and specified whether subcortical structures were involved (they were in 8 of the 10 patients). He nevertheless suggested implicitly that it is due to a sparing of tissue in the striate cortex itself. This was largely inferred from (a) the presence of a scotoma and (b) the position of the scotoma for the moving and the stationary tests, based on such knowledge of the visual projections to the cortex as was available at the time. Thus he writes that in "cases of hemianopic loss, where appreciation of movement returns, the recovery is never complete, a paracentral or central scotoma remaining as an end result, to represent the amount of irreparable damage done to the visual area," implying that the return of movement perception is due to damage that is not irreparable. Moreover, he considered that "in those cases in which no dissociation of the visual perceptions can be demonstrated, and in which no recovery of vision occurs, the damage has been extensively subcortical; but I cannot bring forward definite proof of this at present." It needs to be emphasized that, since the patients of Riddoch were consciously aware of having seen the movement, Riddoch was implicitly empowering the striate cortex, even a damaged striate cortex, with the capacity of conscious visual perception. This was in accord with the views of the time, when area V1 was considered to be the "sole" visual perceptive center and the cortex surrounding it, the association cortex, of associating received visual "impressions" with past visual "impressions" as well as with "impressions" derived from other senses. Both Henschen and Holmes considered that it was with the striate cortex that one "sees," Holmes writing more explicitly in his Ferrier Lecture of 1945 that the striate cortex is a "perceptive centre" (Holmes, 1945).

There are conditions of selective sparing where one can account for the syndrome in a plausible way, by supposing that a component of the striate cortex and its extension onto the relevant areas of the prestriate cortex is spared. One good example is the chromatopsia of carbon monoxide poisoning, which has been accounted for by supposing that the metabolically more active and hence more richly vascularized blobs and thin stripes of V1 and V2 are selectively spared the effects of hypoxia (Zeki, 1990a, 1993). But when the damage is due to physical disruption, it becomes far more difficult to account for the selective sparing of visual motion perception in terms of spared tissue in the damaged striate cortex or in V2.

In V1, directionally selective cells are largely confined to layer 4B and to upper layer 6 (Blasdel and Fitzpatrick, 1984; Dow, 1974; Hawken *et al.*, 1988; Orban *et al.*, 1986). They are predominantly located in the thick stripes of area V2, but our studies, as well as those of others, indicate that their proportions there are low (Levitt *et al.*, 1994; Peterhans and von der Heydt, 1993). To account for the syndrome of residual motion vision (especially when it is the consequence of gunshot wounds) in terms of spared tissue within V1, one would have to suppose that layer 4B and/or layer 6 are selectively spared and, moreover, that, even when spared, they can function more or less normally in the absence of the remaining tissue; to account for it in terms of spared tissue within V2, one would have to suppose that the thick stripes are selectively spared. On present evidence, this would seem somewhat improbable.

2.2.1. The Prestriate Cortex and Residual Motion Vision

An alternative would be to look at the cortex surrounding area V1. This cortex was long considered to be a single visual "association" cortex and the repository of "higher" visual functions. It was assigned the vague role of "interpreting" and "understanding" the visual image "received" by V1 (see Zeki, 1993, for a review), implying a separation between the cortical areas involved in "seeing" and "understanding." It is in fact made up of multiple visual areas. The most interesting of these in the context of residual motion vision is area V5, which is specialized for motion in both monkey (Dubner and Zeki, 1971; Zeki, 1974a) and human (Zeki *et al.*, 1991; Watson *et al.*, 1993; Tootell *et al.*, 1995b; Dupont *et al.*, 1994). Area V5 is more commonly referred to as MT, following the description of an area of that name in the owl monkey. But it is as well for all those who use that terminology to realize that the original definition of MT in the owl monkey (Allman and Kaas, 1971) was made on the basis of a mapping study alone, not a study of functional organization, and that that paper contains no statement whatsoever about directional selectivity or motion specialization of area MT or indeed a specialization for motion at all. In fact, such a specialization was only demonstrated in owl monkey area MT 10 years after its demonstration in macaque V5 (Zeki, 1980; Petersen *et al.*, 1980), and present evidence suggests that owl monkey MT may emphasize orientation as much as motion, quite unlike macaque V5 (Zeki, 1980; Malonek *et al.*, 1994)

2.2.2. The Visual Input to V5

The input to area V5 and to another prestriate area, V3, is dominated by a specialized group of retinal ganglion cells, well suited to register motion, and which reach layer 4B of V1 via the lower two magnocellular (magno- or M) layers of the subcortical visual center, the lateral geniculate nucleus (LGN). V5 and V3 receive their input from layer 4B of V1 and from specific subcompartments of area V2 (Shipp and Zeki, 1985, 1989b; De Yoe and Van Essen, 1985) with both of which V5 and V3 are reciprocally connected (Zeki and Shipp, 1988; Shipp and Zeki, 1989b). If V1 is destroyed, all these cortically relayed signals to V5 are lost. This relatively well studied input to V5 is supplemented by another, less well studied input that bypasses V1 and reaches it directly from subcortical visual centers such as the LGN (Fries, 1981) and the pulvinar (Standage and Benevento, 1983). The latter nucleus is connected with a midbrain visual center, the superior colliculus, which in turn receives input from the retina. It was indeed the superior colliculus that was considered to be the center mediating residual visual capacities in monkeys after removal of V1 (Pasik and Pasik, 1982; Keating, 1980). I, too, sought for an explanation in subcortical structures to account for the syndrome of residual motion vision (Zeki, 1991). But I now find such an explanation unsatisfactory, chiefly because residual motion vision has a conscious dimension, a fact not apparent from studying residual vision in animals.

This dual input to V5 naturally raises the question of which of the two sustains the characteristic of area V5, directional selectivity, and whether one can ascribe the syndrome of residual motion vision to the direct one, the one that bypasses V1. Recordings from area V5 in monkeys whose V1 had been inactivated (Rodman *et al.*, 1989; Girard *et al.*, 1992) and which consequently had no input from it into V5 have shown that V5 is not totally dependent upon V1 for its specialization. With such inactivation, the directional selectivity that characterizes V5 is not lost, but is only rendered less sharp. Assuming that there is a relationship between physiology and perception, it is tempting to enquire whether the residual motion vision that obtains after damage to V1 in the human, and which is characterized by a crude motion vision, reflects the capacities of the direct input to V5 and is crude because the sharpness of directional selectivity that is conferred by the indirect input, through V1, is lost. Of course, V5 is not the only area that receives a direct input from the pulvinar. Other areas, including V3, also receive such an input (Benevento and Rezak, 1976). Hence areas besides V5 may be involved in the crude but conscious motion vision. But even if activity restricted to V5 is not in itself a sufficient condition for conscious motion perception, can activity in a V5 disconnected from V1 lead to a conscious visual experience? The issue is of some interest because if activity in a V5 disconnected from V1 has a conscious dimension, the implication would be that individual specialized visual areas may be able to contribute to conscious, if crude, vision, without the need to act in concert with V1, either through pre- or postprocessing by the latter. In brief, V1, which, explicitly in the early literature and implicitly in the current theorizing about blindsight (Cowey and Stoerig, 1991), has been considered to be the only perceptually effective gateway into conscious vision, may not in fact be so.

2.2.3. Human V5

Obviously the best way of addressing the problem of whether a direct input into V5 is able to mediate conscious vision is to undertake a study in the human, where a verbal report is a quick and efficient way of establishing the conscious dimension to an experience. There would be two conditions for such a study: (a) the identification and localization of area V5 in the human brain and (b) access to a patient with a damaged V1, but an intact V5; if signals reach V5 in such a patient, then the clear implication is that there is a direct pathway to V5, just as in the macaque. Fortunately, the technique of positron emission tomography (PET) has enabled the identification of V5 (Zeki *et al.*, 1991; Watson *et al.*, 1993; Dupont *et al.*, 1994) (Fig. 1). Moreover, this has been confirmed by studies employing functional magnetic resonance imaging (fMRI) (Tootell *et al.*, 1995a). Perhaps reflecting the inevitable broad similarity between macaque and human brain, activation of area V5 in normals always entails a coactivation of other areas, either those like V1 and V2 which feed the main retinogeniculocortical signal to V5, or those like area V3 and the parietal cortex which share a common, M-dominated input with V5 (Watson *et al.*, 1993; Dupont *et al.*, 1994).

It is interesting to note that the lesion in patient LM (Zihl, 1980; Zihl *et al.*, 1991), which produced the syndrome of motion imperception or akinetopsia (Zeki, 1991), includes the territory of area V5. The patient cannot see the motion of objects. Closer examination reveals, however, that she is able to see motion provided it is very slow (about 2° sec^{-1}) and then only if the motion is in the cardinal directions, that is, vertical and horizontal (Hess *et al.*, 1989; Baker *et al.*, 1991; Shipp *et al.*, 1994). This is, in a sense, another kind of residual motion vision and we have argued that it is the motion that is mediated by the parts of her visual cortex that receive motion signals and are left intact by the lesion (Shipp *et al.*, 1994). Her motion vision consists therefore of what these areas are explicitly able to contribute, in their limited capacities, to motion perception in the absence of V5. This reinforces the view presented elsewhere that each area, or small set of areas, is able to contribute explicitly to visual perception according to its capacities, without channeling perception through one or a set of master areas (Zeki, 1993).

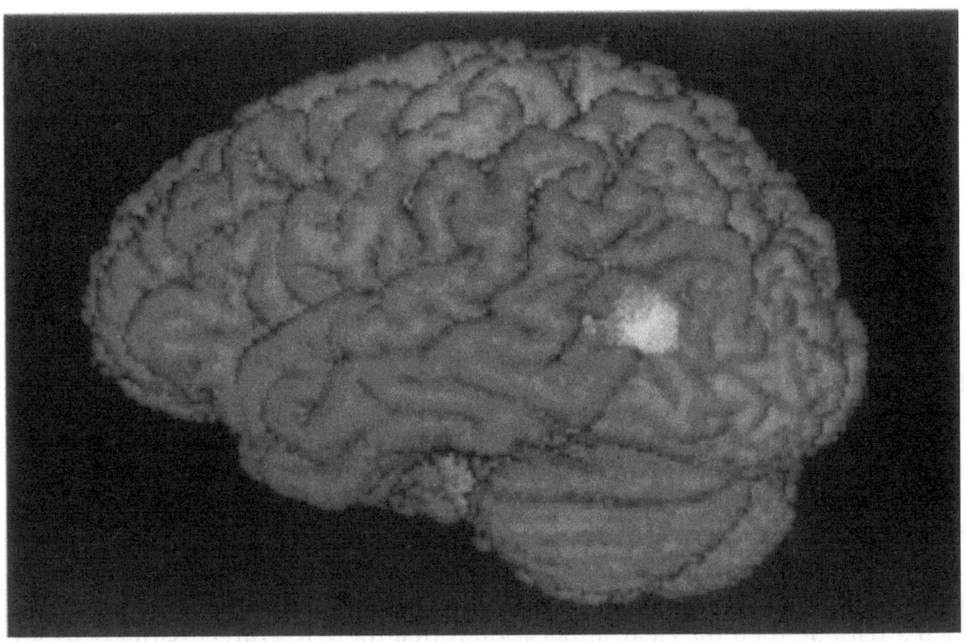

Figure 1. The location of human V5 on the surface of the human brain, as determined by a combination of PET and MRI. From Watson *et al.* (1993).

2.2.4. Patient GY

To ascertain whether the secondary input to V5 is capable of mediating a conscious vision of motion, all one needs to do is to find a patient with a damaged V1 and an intact V5, and who is consciously aware of motion presented to his or her blind field. In short, one needs to find a classical Riddoch patient and enquire whether his or her V5 is active when the patient perceives visual motion consciously. One would naturally not expect to see activity in V1 and V2 in such a patient, although one might expect to see some activation of area V3 and possibly of parietal cortex. This is the experiment that we embarked upon. Just such a patient, GY, became available to us. He had been unilaterally blinded by a lesion to his left V1, sustained in a car accident at age 7. The massive damage included the medial occipital lobe of the left hemisphere as well as the optic radiation on the left side, medially and less so laterally; but the occipital pole was spared. The consequence was a homonymous hemianopia with macular sparing. We studied him at age 36. It is perhaps worth describing briefly why I began studying GY. He had been extensively studied and described as a classic blindsight patient (see Weiskrantz, 1990), in other words, a patient who could discriminate visual stimuli but had no conscious awareness of having seen them. Here, I thought, was the opportunity of showing that, when he was able to discriminate without conscious awareness, the activity in his visual brain would be restricted to subcortical structures and would not involve the cerebral cortex. That, I thought, would be one big step, though not a conclusive one, toward demonstrating that conscious vision was the corollary of cortical activity. I had perhaps assumed too readily and unquestioningly that blindsight uniquely characterizes the syndrome of GY; although aware of Riddoch's work, I had perhaps supposed too readily that his were exceptional cases. It is almost certain that I had not been sufficiently impressed by Riddoch's repeated use of the word "conscious" in his description. But the experiment took an unexpected turn. I was surprised to find that, when we used high-contrast, rapidly moving stimuli, GY could report verbally, with 100% accuracy, whether a stimulus presented to his blind field was moving to the left or the right. He was perfectly conscious of what he saw, although what he saw was crude compared to what normals see; his vision, in the words of Riddoch, was "so vague and shadowy." I enquired anxiously of GY what he actually saw. His reply was precise. He told me that his vision through his blind field could be compared to what a normal person saw when he closed his eyes, looked at a source of light, and moved his hands in front. He could readily distinguish the cardinal axes of motion (vertical and horizontal), but had great difficulty in distinguishing the obliques. This unexpected conscious dimension to GY's residual vision thus changed the character of the experiment, though not the procedure; it now revolved around the question of whether there would be cortical activity, and in particular activity in V5, when GY consciously experienced the crude motion that he reported verbally with such accuracy, and whether therefore activity in V5, without parallel activity in V1, would accompany the conscious visual experience.

Using high-contrast, rapidly moving stimuli and measuring the activation in his brain with the technique of PET, we were able to show that areas V5 and V3,

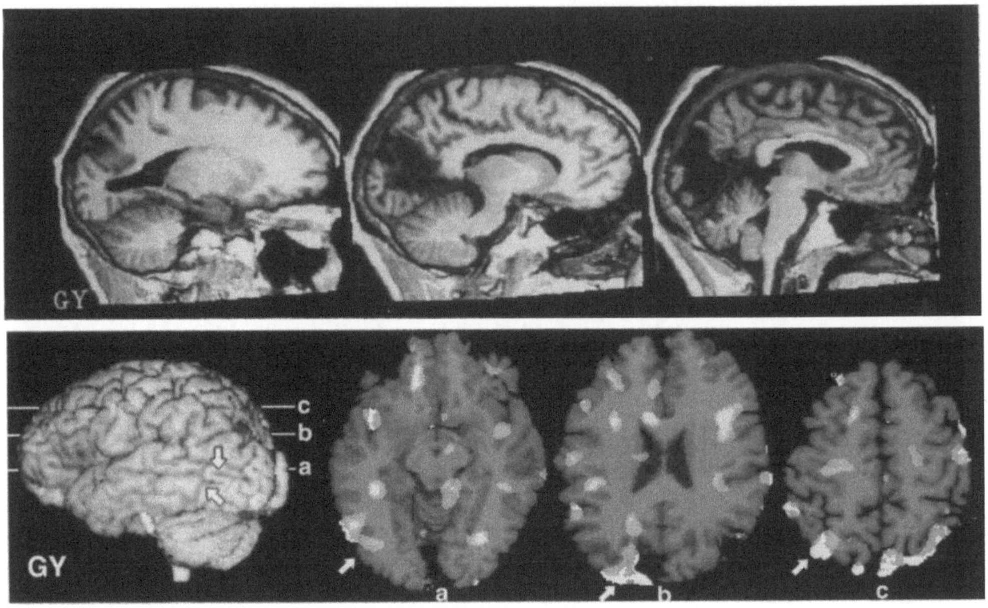

Figure 2. Top: Magnetic resonance images (MRI) of the brain of GY showing sagittal sections. The damage is quite extensive, with nearly complete destruction of medial occipital cortex and the underlying optic radiation, but sparing of the occipital pole. Bottom: Positron emission tomography (PET) data and MRI data have been coregistered and superimposed. On the left is a solid image of the brain. The horizontal lines indicate the corresponding levels of the transverse sections shown on the right. The arrows indicate (a) the activation of V5, (b) V3, and (c) Brodmann's area 7. From Barbur *et al.* (1993).

as well as the parietal cortex, but not V1, were active when patient GY was experiencing visual motion consciously (Barbur *et al.*, 1993; Fig. 2). One can conclude that a very crude but conscious perception of visual stimuli is possible without V1 and that activity in the specialized areas defined above is an essential part of the conscious process of perceiving visual motion. If Riddoch had had access to this information, he, too, would not have wished to interpret his syndrome in terms of spared tissue within V1. I may in fact be wrong to write in terms that imply that this capacity belongs uniquely to V5. The PET results show activation of V3 and parietal cortex as well and these, too, may be involved in GY's crude but conscious vision. I emphasize V5 more because of directional selectivity. What we can conclude from this experiment is that a crude but conscious vision is possible when specialized visual areas are active without parallel activity in V1, or without pre- or postprocessing by V1.

2.2.5. Conscious Perception of Optical Flow in the Blind

Nor should GY be thought of as a unique example. Just as fascinating is another patient who, 4 years after an extensive damage to his striate cortex, was able to perceive optical flow consciously in his blind field (Ceccaldi *et al.*, 1992; Mestre *et al.*, 1992). He could indeed perceive both the direction and the velocity of the flow field but, like GY, his capacity to do so was limited to fast movement,

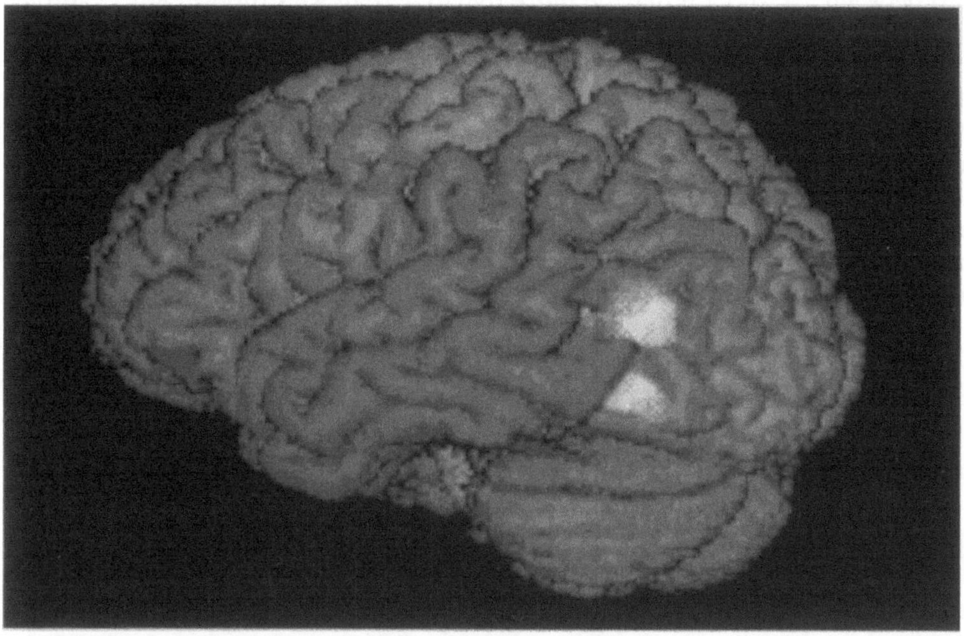

Figure 3. The location of human V5 on the surface of the human brain as determined by a combination of PET and MRI, and shown just below it, the estimated position of the area activated by an optic flow stimulus. Adapted from Watson *et al.* (1993) and de Jong *et al.* (1994).

an interesting point considered further below. No activation studies have been done with this patient, but it is likely that another, distinct visual area will be found to be active when he is so studied, perhaps in addition to V5, perhaps without parallel activity in V5. Our studies of optical flow in humans (de Jong *et al.*, 1994) have shown that an area close, or even contiguous, to V5, but lying ventral to it and encroaching upon the fusiform gyrus, becomes active when humans view such stimuli (Fig. 3). It is possible that such an area also receives a direct input that bypasses V1. Whatever the details, the capacities of this patient reinforce the view that activity in individual areas, or small group of areas, can lead to conscious visual perception without pre- or postprocessing by V1.

3. The Reciprocal Connections between V1 and V5

I have emphasized above that V5 can act autonomously of V1, since it is still active in the absence of V1 and can still mediate conscious perception of visual motion, perhaps on its own and perhaps in concert with the other areas that are active. It is results like this that have led me to propose that the individual areas of the prestriate cortex can each contribute explicitly to conscious visual perception, without the need for pre- or postprocessing by V1 or indeed other visual areas (Zeki, 1993). This is not the same thing as saying that, in the intact brain, these areas always act autonomously. Other results also suggest that the relationship between V1 and V5 may be more dynamic than we suppose. For example, when the change in cerebral blood flow (rCBF) of subjects who see motion in the rings of Leviant's *Enigma* figure is measured (Fig. 4), the largest increase is found in and around area V5, with no detectable increase in V1 or in V2 (Zeki *et al.*, 1993). This does not prove that V1 is inactive during the perception of phenomenal motion, since it may have been as active when subjects were looking at the Leviant figure as during its control, and thus the activity canceled out in the final images. But it does show that the balance of activity between V1 and V5 is not the same as in the condition when the same subjects see objective motion, when V1 and V2 show the highest activity. Such a result also draws attention to the possi-

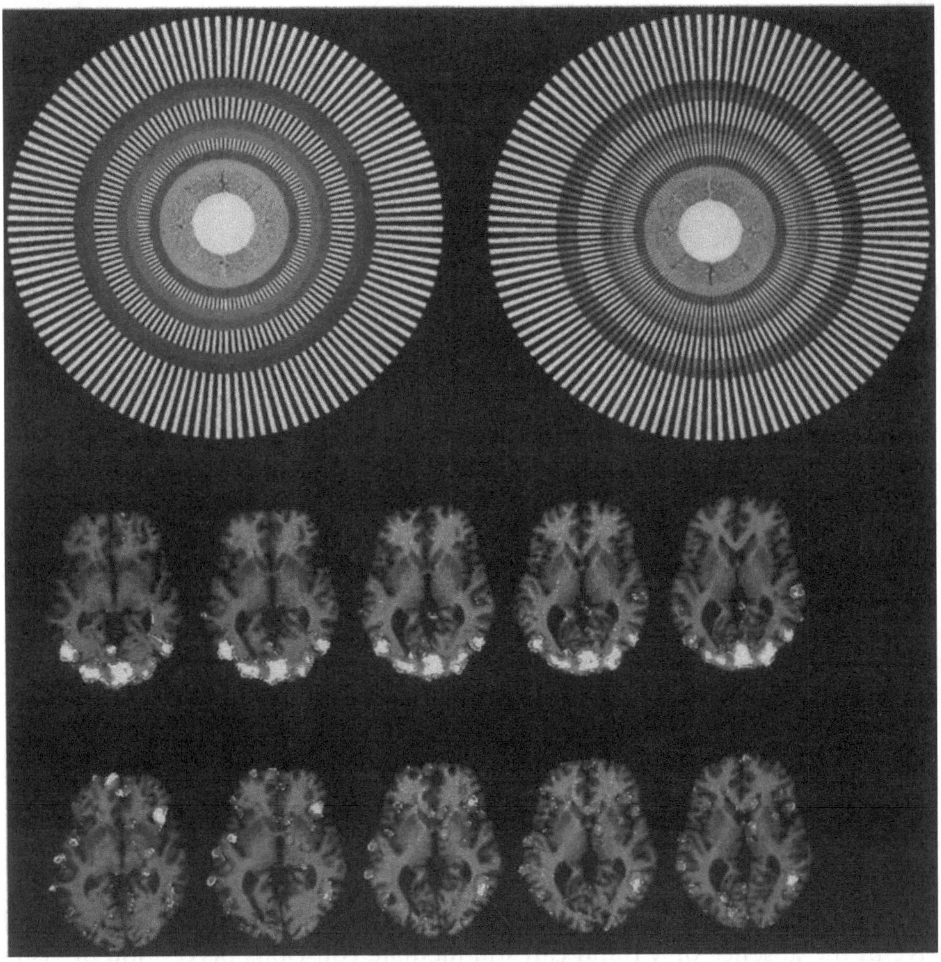

Figure 4. Top: The Leviant illusion. In the stimulus figure (left), many, but not all, normal subjects see rotatory motion confined to the rings. In the control figure (right), where the rings are broken by the radial spokes, the illusion of motion is much attenuated or even abolished. Middle: The activity of the human visual cortex when subjects view objective motion, shown in transverse PET and MRI coregistered sections (as in the study of Watson *et al.*, 1993). Bottom: The activity in the same brains as those of the middle row, to show the activity when the subjects saw the motion in the Leviant illusion.

bility that the visual areas may be capable of acting more autonomously than we generally suppose. The innocent, or those who deliberately wish to misread, or those who like to set up straw men that they can subsequently assail, may suppose from this that I am saying that V1 is not necessary for visual perception in general or visual motion perception in particular, or that, if necessary, it can be dispensed with with impunity, or that the quality of motion vision in the absence of V1 is as good as in its presence. None of these statements is true and I assert none. On the contrary, the quality of motion vision in the absence of V1 is poor, an observation implicit in the use of the term "residual." I am merely drawing attention to the fact that the relationship between V1 and the prestriate areas may not be quite so rigid and that the latter areas may be able to contribute more to conscious vision than we have been willing to suppose.

The Interaction between V1 and V5 During the Perception of Form-from-Motion

The poverty of vision without V1 may be due in part to the necessity for preprocessing by V1, but it may also be due in part to postprocessing in V1. Area V1 not only sends a powerful projection to V5, but receives a powerful projection back from it (Shipp and Zeki, 1989a; Krubitzer, and Kaas, 1990; Fig. 5). One characteristic of the latter is that, unlike the former, it is not patchy, but diffuse, and does not respect the territory of the cells in layer 4B projecting to V5, but also invades the territory of cells in layer 4B that project to V3 and to V2. Such a diffuse return projection is also the hallmark of the backprojection from V5 to V2 (Shipp and Zeki, 1989b; Krubitzer and Kaas, 1990; Fig. 5). It would be astonishing if such a powerful return projection were not put to some use. We have speculated elsewhere that it is one way to inform several subdivisions of V1 of the results of the operations undertaken by V5 (Zeki and Shipp, 1989). We have also suggested that such a feedback connection may be the kind of neural machinery needed to generate one attribute of vision from another (Zeki and Shipp, 1989). As an example, we chose structure from motion and have used it for a PET activation study (S. Zeki, R. S. J. Frackowiak, J. D. G. Watson, and K. J. Friston, The role of human area V1 in visual integration, unpublished results).

The rationale for this study is as follows. The oriented lines to which the cells of V3 respond so well (Zeki, 1978b; Burkhalter and Van Essen, 1986) can be generated from differences in luminance between the oriented lines and the background (*condition A*); they can also be generated if a texture moves in two different directions, say to the left in the zone defined by the oriented lines and to the right in the zone defined by the background (*condition B*). In our PET study we made the supposition that in condition A, the orientation-selective cells of V1 are stimulated first, followed by the orientation-selective cells of area V3, through the direct anatomical link connecting the two areas (Cragg, 1969; Zeki, 1969); in condition B the directionally selective cells of layer 4B are stimulated first, followed by an activation of area V5, which is then followed by an activation of V1 through the return connections from V5 to V1, followed lastly by an activation of the same orientation-selective cells of area V3. On the other hand, if one were to stimulate with uniform motion of the same textured stimulus alone (*condition C*), one should stimulate V1 and V5. Put more briefly, area V1 will be activated once in conditions A and C and twice in condition B. The paradigm for this study was developed in joint discussions between this laboratory and that of G. Edelman and it is fascinating to note that in their computer simulation of the paradigm the generation of form-from-motion becomes impossible if V5 is "lesioned" (Finkel and Edelman, 1989). What would happen in the human brain? If one were to use human subjects to view these three conditions, one should expect that the highest activity will be in V1, in condition B. Our results, using six subjects, showed that there was a highly significant increase in rCBF (and therefore activity) within areas V1 and V2 when subjects had been viewing the form-from-motion stimulus compared to the weighted

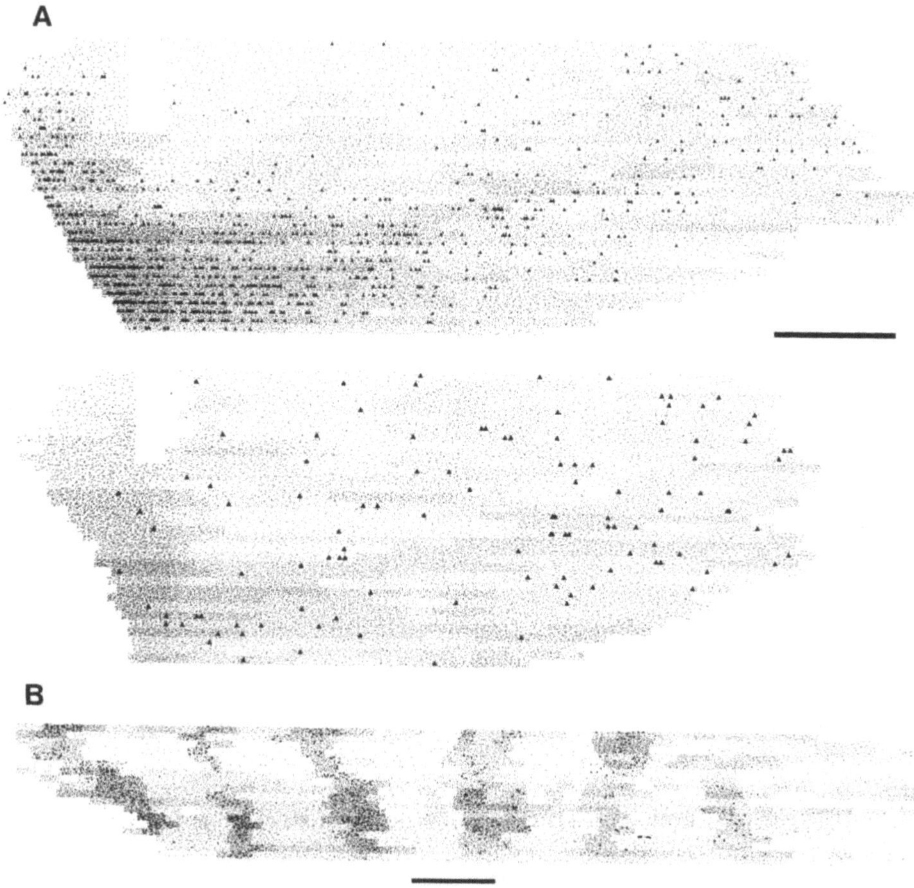

Figure 5. (A) Reconstruction from serial horizontal sections of the distribution of labeled cells (triangles) and labeled axonal terminals (stipple) in macaque striate cortex on the external surface of the occipital operculum following an injection of horseradish peroxidase–wheat germ agglutinin (HRP-WGA) into area V5. The upper reconstruction is from layer 4B and the lower one from layer 6. The scale bar is 5 mm. From Shipp and Zeki (1989a). (B) Computer-aided reconstruction of the distribution of labeled material in area V2 of the macaque monkey following an injection of HRP-WGA into area V5. The scale bar is 2 mm. From Shipp and Zeki, (1989b).

mean activity when they were viewing the motion stimulus alone and the form stimulus alone (Fig. 6). Although I choose to interpret this result to mean that motion is being extracted from the stimulus first, in V5, and the results of the operation thus performed in V5 are then relayed back to V1 to activate the V3-projecting cells there, I acknowledge that there may be other interpretations of this, in particular that the form-from-motion stimulus is a more potent one for the cells of V1 than either the form alone or the motion alone. But, whatever the exact interpretation, the results of this experiment, when compared to the results of our earlier experiment (Zeki *et al.*, 1991; Watson *et al.*, 1993), in which subjects viewed the movement of small squares in different directions, and the result of experiments in which subjects viewed Leviant's *Enigma,* all show that area V5 is dynamic in relationship with area V1 and that the strength of that relationship may be variable, depending upon the stimulus.

4. Dynamic Parallelism

The dynamic nature of the relationship between the two areas, V1 and V5, is not restricted to cortical connections alone. Other evidence, reviewed below, has also led us to the notion of dynamic parallelism (ffytche *et al.*, 1995; Zeki, 1995), by which we mean that the two pathways reaching V5 from the retina may be jointly active, or one may be active without the other, depending upon the nature of the stimulus. This study is based on the characteristics of the signals that reach V5 without passing through V1, and which lead to activity in V5. This activity has a conscious dimension, but without the participation of V1.

4.1. The Parallel Pathways to V1 and V5

A good idea of the functional contribution that the two parallel pathways to V5, one through V1 and the other through the superior colliculus and pulvinar, make is to compare the visual motion capacities of patient GY, with a damaged V1 and an intact V5, with the capacities of patient LM, who has the obverse damage, i.e., a damaged V5, but an intact V1 (Zihl *et al.*, 1983, 1991; Shipp *et al.*, 1994). We have found (Barbur *et al.*, 1993) that patient GY can only detect high-contrast, fast-moving ($>10°\ sec^{-1}$) stimuli. The studies that we and others have conducted on patient LM have shown that she, in comparison, can only detect very slowly moving stimuli ($<10°\ sec^{-1}$) (Hess *et al.*, 1989; Baker *et al.*, 1991; Shipp *et al.*, 1994). It is safe to assume, therefore, that one characteristic of the direct pathway to V5 is that it is better suited to signal fast motion. In fact, theoretical calculations and experimental observations show that the direct pathway is able to deliver signals at latencies of about 30 msec (Beckers and Zeki, 1995). The technique of transcranial magnetic stimulation, through which one can inactivate human visual areas briefly and selectively (Beckers and Hömberg, 1992), not only shows that V5 is more critical in the perception of fast signals, already implied by the clinical evidence, but also leads to the surprising conclusion that signals from fast-moving stimuli use a fast pathway to reach V5 before

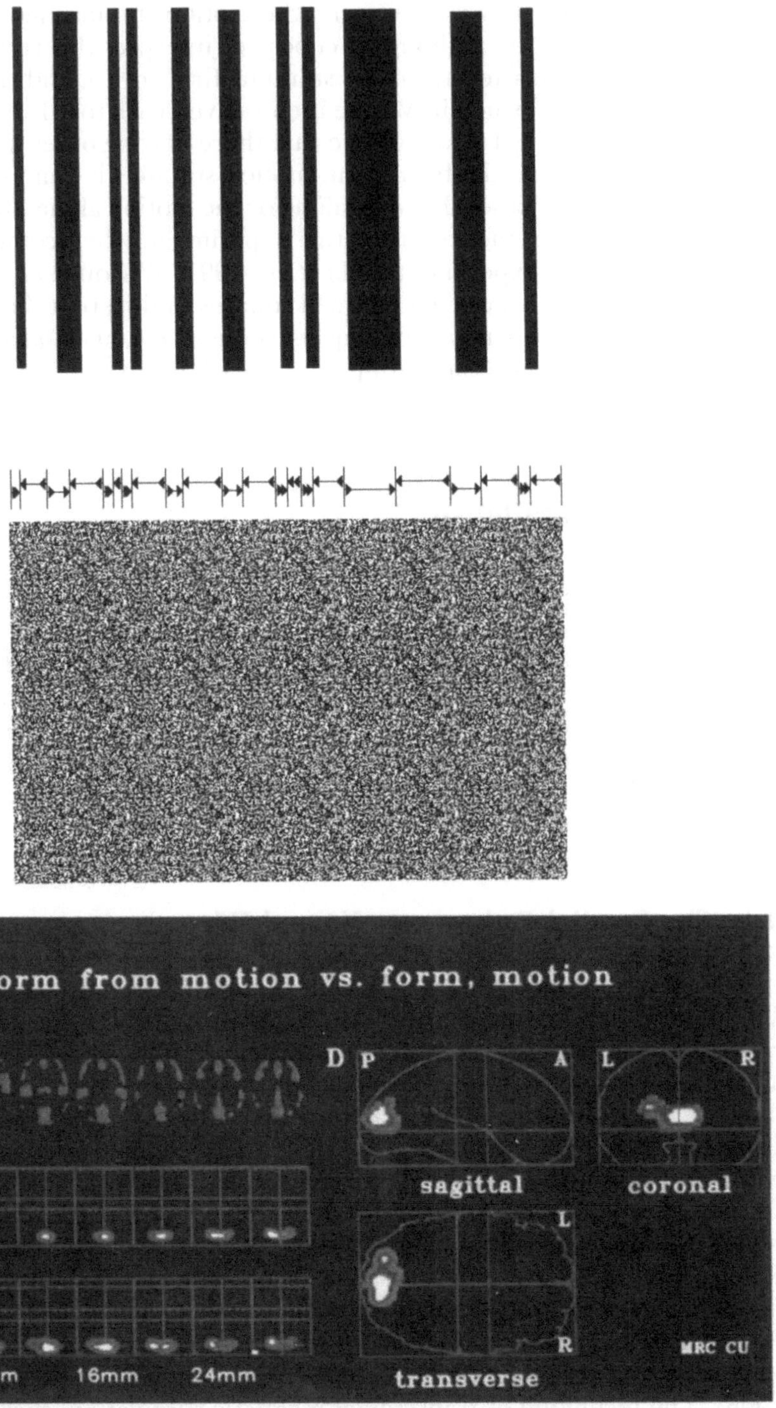

Figure 6. Top: The stimulus used in the form-from-motion PET study. The stimulus for "form" was a series of black-and-white vertical bars generated from luminance differences. The stimulus for "form-from-motion" was a moving textured sheet. There was a rightward-moving texture where there was a black line in the luminance stimulus, and where there was a white line there was a leftward-moving texture. A third stimulus was two superimposed, transparent moving sheets of

they reach V1 and that they reach human V5 at far higher speeds than had been supposed (Beckers and Zeki, 1995). The inference can be confirmed by direct measurements from the scalp of normal humans, using the electroencephalography (EEG) technique (ffytche *et al.*, 1995) This method, which gives latencies of 37 msec for the first arrival of signals, shows directly that signals from fast-moving stimuli reach V5 first; by contrast, those from slow-moving ones reach V1 first. It follows that activity in prestriate cortex occurs in parallel with or even precedes that in V1. Because the two parallel pathways can be simultaneously active or not, depending upon the stimulus, we speak of a dynamic parallelism (ffytche *et al.*, 1995; Zeki, 1995). It is also worth noting that, because the information on the fast input to V5 comes from normal humans, it follows that the activation of the fast, direct, pathway does not become operational only when a lesion compromises the more classic pathway. It is of course possible that, in the absence of V1, the fast pathway may gradually become more efficient. GY had had many years of visual experience, in fact about 20 years when we studied him. Riddoch, on the other hand, had examined his patients only a few months after the injuries. But the condition that he describes is so similar to the one characterizing GY that one must presume that a lengthy period of experience is not necessary for the intact pathway to exhibit its characteristics. The comparison of GY to Riddoch's patients also leads to another interpretation, namely that the many years of visual experience did not sufficiently enhance GY's visual capacity. It would seem that the visual cortex, and the direct input to the prestriate cortex, is not sufficiently labile to compensate for a loss so severe.

4.2. The Direct Measurement of Latencies in Visual Cortex

These general conclusions about the timing of arrival of signals in visual cortex and the parallel input to V1 and V5 are confirmed by the results of direct recordings from single cells in the cortex itself, which show that signals can reach the visual cortex in 20–30 msec in monkey (Maunsell and Gibson, 1992; Petersen *et al.*, 1988; Kawano *et al.*, 1994) and 30 msec in human (Ducati *et al.*, 1988; Wilson *et al.*, 1983). Recordings in monkey show that the earliest signals are picked up from V5, not V1 (Raiguel *et al.*, 1989). Such evidence may seem to fly in the face of earlier conclusions, reached from EEG studies, which have consistently failed to detect early motion-induced activity in prestriate cortex and supposed instead that visual cortex outside V1 may not receive motion signals at all (Van Dijk *et al.*, 1987; Maier *et al.*, 1987) or that it is always activated after V1, in sequential manner, with peak deflections occurring at intervals of 160–220 msec after delivery of the motion stimulus (Probst *et al.*, 1993; Drasdo *et al.*,

---

Figure 6. *(Continued)* texture, i.e., there were the same motions as the form-from-motion stimulus, but no structural component. Bottom: (A) The mean blood flow across all scans for all subjects. (B) The difference in blood flow across all subjects between the form-from-motion stimulus and an average of both controls. (C) A statistical parametric map, SPM(*t*), showing the significance of the difference in panel B. (D) The projection images showing the "hottest" pixels along the line of sight in the three views indicated. Adapted from Zeki (1993).

1993). In fact, there is no obvious disagreement; the fast component is only detected when fast-moving stimuli are used. Almost all previous EEG studies have used very slow motion (<6° sec^{-1}) and have thus not detected the fast component.

5. The Color System

The evidence presented above shows that conscious experience could be the correlate of activity in one or a small number of prestriate areas without the involvement of V1. It also raises the possibility that conscious experience could also be the correlate of activity in other visual areas, ones that emphasize other attributes of the visual scene. But one is naturally impelled to go beyond the demonstrations described above and ask more searching questions. Could we, for example, either produce experimentally or find clinical examples, of conditions in which all visual areas save one are destroyed? Would such an organism be able to see, and, if so, what? We might want to have an example in which only one system, consisting of a specialized area plus the pathways that feed it, is intact, with all else destroyed. What would vision be like in these conditions? What would the characteristics of the residual vision be if one of the areas concerned with a given attribute is destroyed, leaving earlier ones intact? We should also like to learn whether the general principles of residual vision derived from a study of the motion system are applicable to other kinds of vision. We can come closer to finding such examples if we look at the color system.

It is unfortunate that the color system is not quite as well studied, but there is at least one hint that there may be a residual vision for color in the absence of V1. The hint comes from the paper of Blythe *et al.* (1987) in which a patient blinded by a lesion in V1, was able to perceive, correctly and consciously, different colors. Henschen would have cared nothing for such a patient. In dismissing the results of Verrey (1888) and of MacKay and Dunlop (1899), he had written that a color center outside the primary visual cortex would be an absurdity, "for with the calcarine cortex destroyed and the cortex of that other gyrus [fusiform] intact, the patient would have to be absolutely blind and yet be able to see colors, which makes no sense" (Henschen, 1910). But apparently nonsensical facts are not uncommon in neurology and a very similar scenario occurs in the motion system; that scenario is no longer nonsensical, because we have a plausible anatomical explication for it. A similar scenario may not be so nonsensical for the color system either. Given that, in the monkey at least, color signals can reach the V4 complex without passing through V1, through the direct input from the lateral geniculate nucleus (Fries, 1981; Yukie and Iwai, 1981), it is plausible that the reported residual color vision may be traced to the operation of similar neural pathways as the one operating in the case of residual motion vision. But the evidence is scanty and the argument difficult to prove. There are, however, other interesting insights into conscious vision that a study of the color system is able to provide. Before looking at them it is useful to review the color pathways in the cerebral cortex in brief.

That we can speak of color pathways in the brain implies that they are to some extent segregated, and both anatomical and physiological evidence broadly support this notion. The Pa cells, better suited to register specific wavelengths, project through the upper four, parvocellular, layers of the LGN (the parvo- or P-layers) to V1 and eventually terminate in layers 2 and 3, where they divide into two subdivisions, ones that terminate in the cytochrome oxidase-rich puffs or blobs and others that terminate in regions in between, the interblob zones. Anatomical evidence shows that the input to layers 2 and 3 is not exclusively from the parvo system; the magno system makes a contribution, too, as do the W cells with their characteristic long latencies and large receptive fields (Hendry and Yoshioka, 1994; Lachica *et al.*, 1992). But physiological recordings show that substantial numbers of cells within the blobs tend to be wavelength-selective and unconcerned with the orientation, while the reverse situation obtains in the interblob zones (Tootell *et al.*, 1988; Livingstone and Hubel, 1984a). This being the domain of color vision, it would be surprising if there were a unanimity of views on such a parcellation. A contrary view has been presented by Lennie *et al.* (1990) and Leventhal *et al.* (1995). I find these latter views less convincing, if only because neither paper contains compelling anatomical evidence of the lack of relationship between the blobs and wavelength-selective cells: neither has a single photomicrograph of a cytochrome oxidase-stained section, let alone a photomicrograph of a penetration that has sampled from the blobs, or from the interblobs for that matter. Neither paper can therefore have the same status as the ones that have documented the anatomical evidence fully and found a correlation between physiology and cytochrome oxidase architecture. I also sense that Leventhal *et al.* (1995) are none too convinced of their own findings, for they hedge them with the disclaimer that "it remains possible that a parcellation of function by different cells in V1 will prove to be correct once additional information is obtained."

Area V1 has a systematic and topographically organized projection to V2 (Cragg, 1969; Zeki, 1969), an area with a distinctive cytochrome oxidase (CO) architecture consisting of three sets of stripes, the darkly staining thick and thin stripes, separated from each other by the more lightly staining interstripes (Fig. 7) (Livingstone and Hubel, 1982; Tootell *et al.*, 1983). The separate systems are maintained segregated in the projection from V1 to V2, the blobs of V1 projecting to the thin stripes of V2 and the interblobs to the interstripes (Livingstone and Hubel, 1984b). The thick stripes receive their input from layer 4B. Area V2, in turn, projects to V4 (among other areas), with both thin stripes and interstripes being involved (Fig. 8), while the thick stripes project to V3 and to V5 (Zeki, 1971; Zeki and Shipp, 1989; Shipp and Zeki, 1989b). Combined physiological and anatomical studies show that wavelength-selective cells in V2 are concentrated in the thin stripes (Hubel and Livingstone, 1987; De Yoe and Van Essen, 1985; Shipp and Zeki, 1985; Levitt *et al.*, 1994; Roe and Ts'o, 1995). Recordings from the V4 complex show that there is a high concentration of wavelength-selective cells there, and that even many of the orientation-selective cells respond better when the oriented line is of one wavelength or another (Zeki,

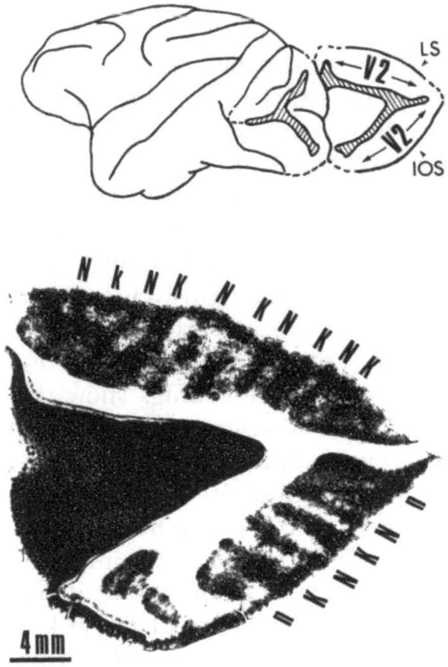

Figure 7. Top: Diagram of the left hemisphere of a macaque monkey brain, with the occipital operculum reflected to reveal the parts of area V2 occupying the posterior banks of the lunate and inferior occipital sulci (LS and IOS). If the operculum is flattened and sectioned tangentially (i.e., in the plane of the paper), the sections pass roughly parallel to the layers of V2 in both the LS and IOS. Bottom: An operculum section stained for the enzyme cytochrome oxidase to reveal the characteristic striped architecture of V2. K, k: Thick stripe; N, n: thin stripe, the pale stripes are known as interstripes (I). The distinction between thick and thin is easier for some stripes than for others; lowercase letters indicate some degree of uncertainty in the identification in regions of irregularity. From Zeki and Shipp (1989).

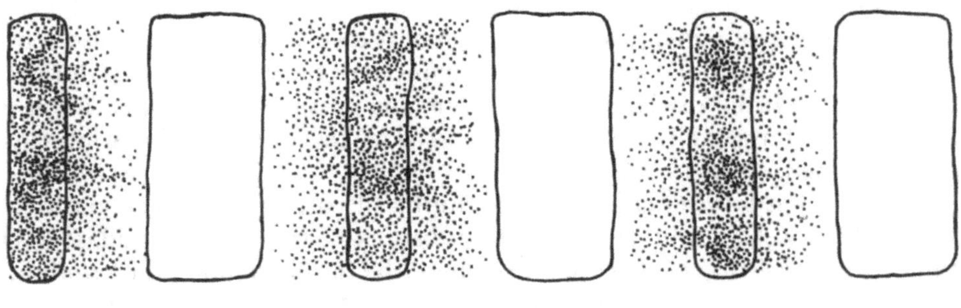

Figure 8. A schematic diagram of the two types of connectivity pattern between V2 and V4. In type 1 (above), there is one band of V4-efferent cells per cycle of cytochrome oxidase stripes, centered on the thin stripes. In type 2 (below), there are two bands, centered on the interstripes. From Zeki and Shipp (1989).

1974b, 1983; Desimone and Schein, 1987; Tanaka *et al.*, 1986). By the V4 complex I mean not only the area in the anterior bank of the lunate sulcus and the surface of the prelunate gyrus, but also the cortex that extends from there to the lateral one-third of the posterior bank of the superior temporal sulcus, cortex which also contains color cells (Zeki 1977). Moreover, since the callosal band that identifies the latter part of the V4 complex continues ventrally to appear in front of the inferior occipital sulcus (Zeki, 1970; Figs. 6 and 7 of Zeki, 1977), the V4 complex extends fairly ventrally and then medially and also includes at least the posterior part of the territory of the area known as TEO, as well as substantial parts of the ill-defined PIT. Boussaoud *et al.* (1991) obtained similar results to Zeki (1977), but traced the callosal connections more ventrally and medially and studied the topographic organization of the area that they call TEO in much greater detail. But the essential similarity of the two sets of results, together with the results of anatomical experiments which show that both divisions of the V4 complex, just like TEO, receive input from the thin stripes and interstripes of V2 (Zeki, 1971; Zeki and Shipp, 1989; Nakamura *et al.*, 1993), make it very likely that at least the posterior part of TEO is part of the V4 complex as defined by me earlier (see also Zeki, 1996). The V4 complex is therefore not quite as restricted as those unacquainted with the published evidence suppose it to be.

Finally, it appears that cells which are orientation-selective and those which lack such selectivity are each grouped together within the cortex of area V4 (Zeki, 1983). Whether the strongly wavelength-selective cells of V4 reside in subzones of high cytochrome oxidase content is not known.

Once again, this being the field of color vision, there is no unanimity of view on the physiology, although one has not heard dissenting voices of any note about the anatomical picture. In particular, Levitt *et al.* (1994) believe that their data on V2 "provide only equivocal support for the idea that the CO compartments capture the continued segregation of parallel functional pathways in the visual cortex." This conclusion finds little support from the anatomical and physiological evidence obtained in other laboratories and is not entirely supported by their own results (in particular the representative penetrations, shown in their Figs. 10 and 11). Moreover, this continued compartmentalization helps explain some of the syndromes that I enumerate below, which in turn help us better to understand the main theme developed in this chapter, that is, whether individual areas or systems specialized for particular attributes of vision can function more or less autonomously and result in conscious vision for that attribute.

5.2. Carbon Monoxide Poisoning

Perhaps one of the most fascinating syndromes in the clinical visual literature is the syndrome of carbon monoxide poisoning, consistently and unaccountably ignored since it was first described by Wechsler in 1933. Other reports have since confirmed the original description of Wechsler and there is no need to doubt its validity (see Zeki, 1990a, 1993, for a full review and bibliography). The syndrome comes nearest to being an example of the selective preservation of a functionally specialized system. To that extent, it aids us in answering the question: "Can a system [rather than an area], acting in isolation, yield a conscious experience of vision?" The answer must surely be "yes" because the syndrome is one in which, following carbon monoxide intoxication, patients become generally very debilitated visually while preserving a remarkably accurate vision for colors, even extending to the shades of a color. The weakness of the argument lies in the fact that the destruction of much of the cortex is largely inferred; it is possible that some cortex besides that devoted to color vision is preserved, though it can hardly be functioning normally, given the extreme visual incapacity of the subjects. Wechsler (1933) concluded that "The most probable answer is that certain parts of the cortex or the layers presumed to perceive and elaborate color vision escaped destruction or they recovered while the other visual areas did not," an extraordinary prediction for its time, which is no doubt why it was so universally ignored. A syndrome such as this becomes more difficult to account for if there is no neat functional compartmentalization in the cortex and if cells coding for color are randomly distributed throughout the visual cortex. A remarkable modern version of this syndrome has appeared in a recent elegant paper (Humphrey *et al.*, 1995). A patient had become virtually blind after a high-voltage electric shock had produced diffuse damage in his occipital lobes, not being even able to discriminate between two lines of orthogonal orientation. And yet, when subjected to the McCulloch paradigm, where the color of the afterimage is contingent upon the orientation of the lines, he could readily distinguish the color of the afterimage, though he could say nothing useful about the orientation of the lines that had been used to induce the colored afterimage.

I have interpreted these syndromes in the context of the color pathways that are centered on the cytochrome oxidase-rich subdivisions of areas V1 and V2 and their prolongation into the V4 complex (Zeki, 1990a, 1993). Zheng *et al.* (1991) have shown that the blobs are more richly vascularized than the interblobs, and my interpretation has been that this richer vascularization has cushioned them, and the wavelength-selective cells within them, from the effects of hypoxia. My interpretation may be wrong or at least one among others, but it is very difficult not to suspect that a pathway with some specially resistant properties is active in these cases, itself an argument for the selective preservation of a system. The syndrome has many interests, not least of which is that the patients are consciously aware of seeing the colors and can report them verbally and correctly. Hence this would suggest that a system acting more or less in isolation, while much else is destroyed or at least functioning highly abnormally, can mediate conscious vision. The patient of Humphrey provides another interesting conclusion, drawn by the authors themselves. Since the color of the afterimage is contingent upon the orientation of the bars and since the patient can report the color of the afterimage correctly like normals, without being able to say anything about the orientation of the bars, this may provide some evidence in support of the notion that the (orientation-selective) cells of V1 are contributing to perception, though not to conscious perception. The evidence may therefore provide some support for the hypothesis of Crick and Koch (1995) that we are not aware of what happens in V1. But nothing about the anatomy of the cortex in this patient and in others with widespread cortical damage is clear, nor is it certain that the basis of the perception is an interaction between color and orientation in V1. The probability that this is so seems high, given that the perceptual paradigm described by McCulloch can best be accounted for by the physiology of V1, and more specifically by the monocularity or the ocular dominance of many cells there (Hubel and Wiesel, 1977), in contrast to the prestriate areas, where ocular segregation or dominance is much less common (Zeki, 1979).

This evidence suggests that an entire system functioning on its own, without the support of other systems, can lead to conscious vision, though naturally a much degraded one.

The chromatopsia of carbon monoxide poisoning provides us with yet another intriguing problem. The form perception of such patients is often very highly degraded, while their color vision is surprisingly good (see, for example, Benson and Greenberg, 1969). One possible reason for this may be that the orientation-selective cells, presumed to mediate form perception, are much more vulnerable because they are in less richly vascularized territory; a corollary of this would be that the orientation-selective cells of the V4 complex also lie in metabolically less active subzones. This may well turn out to be true, but there is no evidence to support the conjecture. An alternative explanation would be that the color center is destroyed in such patients and their color vision is really the expression of the activity of wavelength-selective cells in V1 and V2 that have been spared the effects of hypoxia—a conjecture for which there is even less evidence, which is not to say that it may not turn out to be true. A good test would be to ask whether the color vision of such patients obeys the laws of color constancy, an experiment that has yet to be conducted. It is, finally, possible that the process of separating form and color has proceeded further in humans and that, instead of having subdivisions of wavelength- and orientation-selective cells in one and the same area, the assignment of the respective categories has been to two different areas in the human brain (Zeki, 1993).

Whatever the explanation, it raises the interesting question of the relationship between orientation selectivity and form perception and why it is that form vision can be preserved in cases of achromatopsia or color blindness of cerebral origin. Some might argue that form vision is commonly not preserved, indeed that achromatopsia is commonly accompanied by other disturbances. They are quite right. But the fact remains that there are examples, of which the patient of Oliver Sacks is one that I have studied myself, in which form is only minimally disturbed or not disturbed at all, and these cases are by far the more interesting ones. In response, one could argue that the form vision of such patients has not been adequately enough studied. Again, there is justice in this response, but it is perhaps worth emphasizing that form vision is commonly not studied in such detail precisely because it is largely intact and does not elicit a spontaneous complaint from the patient. At any rate, even if present, it is often not nearly as severe as achromatopsia or dyschromatopsia.

5.3. V4, Orientation Selectivity, and Form Analysis

The V4 complex is commonly considered to be a pivotal set of areas in form perception, in addition to their role in color perception. One reason for this is that it has orientation-selective cells (Zeki, 1975, 1983; Desimone and Ungerleider, 1986). It is a common assumption that orientation selectivity is the physiological precursor of form perception even though the precise relationship between the two is far from clear, either physiologically or psychologically. Indeed the connection between orientation selectivity and form perception is far more intuitive than physiological, but that intuition has much appeal and perhaps even considerable validity. Because oriented lines constitute elementary forms and are indeed the constituent elements of so many more complex forms, it is naturally tempting to consider them as a sort of 'building block' from which cells that code for more complex forms are generated. This was the original assumption of Hubel and Wiesel (1962) and that assumption has remained with us, sometimes explicitly but more often implicitly (see, for example, Desimone and Schein, 1987; Livingstone and Hubel, 1988). Hence it has become common practice to refer to small subdivisions of areas in which orientation-selective cells predominate, as well as to entire areas with heavy concentrations of orientation-selective cells, as being concerned with form (Livingstone and Hubel, 1988). The former is exemplified by the interblobs of area V1 and the thick stripes and interstripes of area V2, while a good example of the latter is the V4 complex.

The assumption may lead to the conclusion that the more tightly tuned a cell is for lines of a particular orientation, the more heavily it is involved in form perception; by extension, an area in which cells have sharper orientation tuning curves than another may be considered to be more heavily involved in form perception. In fact, by this reasoning, V4 does not do as well as V3 or V3A: our recordings show that, on the whole, the orientation-selective cells of the V4 complex are not as tightly tuned as the orientation-selective cells of V1, V2, or V3 (Fig. 9). One is forced to the conclusion that either orientation selectivity is not such a powerful indicator of a specialization for form, or that it is not the only indicator and that other factors, as yet unknown, are just as important. It is therefore a little surprising that the areas of the V3 complex have rarely been considered to be involved in form vision, although they are replete with orientation-selective cells (Burkhalter *et al.*, 1986; Zeki, 1978b). One reason for this may be that V3 and V3A do not project directly to the inferior temporal cortex (our unpublished results). But the extensive representation of orientation selectivity in both the V3 and the V4 complex may give us a clue as to why there has never been a clinical description of a total and specific loss of form vision. One possible explanation for this is that the form system based on V3 and V3A may be able to compensate for the form system based on the V4 complex, or vice versa. Such a suggestion receives support from the observation, made independently several times, that when agnosic patients are not able to recognize static objects, they are often able to do so when the same objects are set in motion (Bender and Feldman, 1972; Kertesz, 1979; Humphreys and Riddoch, 1987; see also Zeki, 1993, for a general review). In the chromatopsia of carbon monoxide poisoning, with the resulting much impoverished form vision, one may conjecture that all the orientation-selective cells and other form-detecting cells, whether in the V4 complex or elsewhere, are compromised because they are in metabolically less active territory. Whatever the merits of these speculative suggestions, the physiology and the pathology raise further interesting problems of parallelism, of how and under what conditions processing is switched from one area to another.

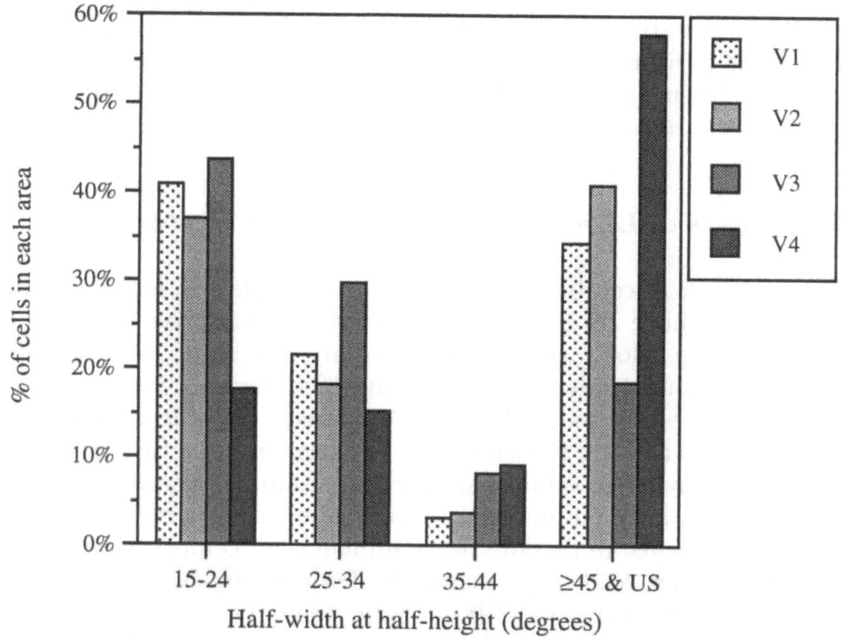

Figure 9. The distribution of cells in each of the visual areas V1, V2, V3, and V4 according to the width of their orientation tuning curve, as determined by computerized model fitting of the cells response profile to a moving, oriented bar. Cells showing any direction selectivity have not been included. Recent results from this laboratory. For V1, n = 93, for V2, n = 181, for V3, n = 135, and for V4, n = 119.

5.4. The Human Color Center

As stated above, it is very difficult to know whether the color system of carbon monoxide-poisoned patients operates because of an intact pathway extending to V4 or whether the perceptual capacities we observe in such patients are due to the expression of the physiological capacities of earlier levels of the color pathways, say V1 and V2. The phenomenon, having only been recently rediscovered, has not been extensively enough studied yet. But we could frame the question in a different way and ask how the system functions if one of its later components is damaged. The question revolves essentially around the role of the color center in the human.

Our human activation studies using color identified a color center located in the lingual and fusiform gyri, which we call human V4 (Lueck *et al.*, 1989; Zeki, *et al.*, 1991; see also Clarke and Miklossy, 1990, and Corbetta *et al.*, 1991). In fact, the identification of this zone as a color center was much aided by knowledge of the human pathology that leads to the syndrome of cerebral achromatopsia, a syndrome first described over a century ago and then conveniently buried until the results of experiments on the functional organization of monkey cortex became known (see Zeki, 1990a, 1993, for reviews). It was Meadows (1974) who, in modern times, first pointed out a consistent association between the syndrome and lesions in the fusiform and lingual gyri (see also Damasio *et al.*, 1980; Kölmel, 1988; Henaff and Michel, 1993). With color stimulation, activation is stronger in the fusiform gyrus (Zeki *et al.*, 1991) and, since V2 is likely to be situated in the lingual gyrus, the fusiform is the more likely site for the human color center (Zeki *et al.*, 1991).

5.5. The Effects of Lesions in the Human Color Center

The syndrome of cerebral achromatopsia is complex; in many patients, the achromatopsia is more appropriately described as a dyschromatopsia, a condition in which some colors, usually the greens and the blues, are more affected than others, though there have been cases in which the reds were more severely affected. In addition, in many (but not all) patients there is a recovery of the ability to see colors and the consequence of the lesion in the fusiform gyrus can only be detected by more subtle tests (see, for example, Kennard *et al.*, 1995). Rizzo *et al.* (1993) have emphasized variations in color perception in human achromatopsic patients. But lesions in the human color center, just like lesions in monkey V4 complex, lead to permanently elevated thresholds for wavelength discrimination (Heywood *et al.*, 1987; Vaina, 1994; W. Fries and S. Zeki, unpublished results). Moreover, the work of Walsh *et al.* (1993) in macaque and Kennard *et al.* (1995) in human show that in both species such damage leads to a fairly severe inability to "discount the illuminant" (Helmholtz), that is to say that the discrimination in both is made according to wavelength composition of the light reflected from a surface, not its color. Put more briefly, color constancy mechanisms are disrupted in both species.

5.6. The Implications of the Loss of Color Constancy Mechanisms

That human patients with lesions in the color center may lose that most essential property of the color system, the ability to discount the illuminant and assign a constant color to a surface, and that they use the dominant wavelength composition to classify objects according to color raises interesting problems with regard to conscious perception and categorization. It implies that there are hierarchies of classification and categorization, a conjecture well supported by previous physiological studies which show that the wavelength-selective cells of area V1 respond to the presence and intensity of light of their preferred waveband reflected from a surface in their receptive fields, and not to its color. By contrast, there exist cells in monkey V4 whose responses correlate with perceived color and are independent of the precise wavelength composition of the light reflected from the surface in their receptive fields (Zeki, 1983). It would therefore seem that the perceptual capacities of patients and monkeys with damage to cortical areas outside V1 and V2 that are intimately involved with color vision correlate better with the physiological properties of V1, and possibly of V2. From this one can deduce that, when one stage of the hierarchy is obliterated or partially destroyed, the previous stage can make the direct contribution to the conscious perception of the attribute in question and that that perception is better tailored to the physiology of the areas left intact by the damage (Zeki 1993). In fact, just such a view was proposed for color vision by Kulikowski *et al.* (1994).

6. Visual Consciousness as a Distributed Process: The Direct Contribution of Individual Areas to Conscious Vision

The evidence presented above, in particular the evidence relating to conscious visual motion in the absence of V1 and that relating to conscious color vision when the color system alone is properly functional, would seem to suggest that there are several consciousnesses, each one attached to a particular set of areas, and that visual consciousness itself is therefore a distributed process. With hindsight, that is what one might expect from a functionally specialized system. Moreover, the evidence shows that, provided a visual input is relayed to them, individual visual areas may contribute directly and explicitly to conscious vision according to their capacities, without the necessity for pre- or postprocessing by V1. It also shows that there can be a conscious dimension to the activity of the specialized visual areas, and that that dimension does not necessarily become available only after all the processing in all the visual areas is over. Just as integration is a multistage process (Zeki, 1990b, Zeki, 1993), so consciousness may operate at several different levels, each level being able to contribute explicitly to conscious vision, without further processing and in accordance with its capacities. Does this mean that the appropriate activity in any visual area, without parallel activity in V1, can lead directly to a conscious visual experience, however impoverished, and, by extension, that activity in V1 alone, without the prestriate cortex, can lead to a crude but conscious visual awareness? The answer to both these questions is uncertain. There are many visual areas in the primate brain and it is by no means certain that what obtains for motion will also obtain for, say, depth. It is also possible that, with any given visual area, the involvement of V1 through postprocessing will depend upon the stimulus. It is quite obvious that the kind of task that GY had to undertake is not critically dependent upon postprocessing by V1. By contrast, there is a good chance that a form-from-motion task will be so dependent.

Just as problematic is whether we are aware of what happens in V1. One critical test would be to test patients with specific defects in the prestriate cortex for the attributes that the damaged areas are specialized for and note whether the capacities of the patient correspond in any way to the physiology of V1 with respect to the same attribute. A good example is in the color domain. That the reaction of achromatopsic patients and monkeys with V4 lesions corresponds more to the physiology of wavelength-selective cells in V1, and that the patients should be fully aware of the nature of their reactions, would seem to imply that we can, under some conditions, be aware of what is happening in V1. On the other hand, there is no evidence that V2 was inactive in such patients, and the activity in it may contribute to both the perception and its conscious awareness. It is also possible that whether we are aware of what happens in V1 depends upon the task. It is reasonable to argue that the patient of Humphrey *et al.* (1995)

was not aware of the activity of the orientation-selective cells in V1, which determined the color of the afterimage. But was the color of the afterimage dependent upon cells outside V1 or was it dependent upon other V1 cells? If the latter, then the implication would be that we may be unaware of some, but not all, of what happens in V1. There thus remain many issues to resolve. What is certain is that the injection of consciousness into investigations of the visual cortex is likely to enliven the debate about the functions and functioning of the visual cortex more than ever before.

NOTE ADDED IN PROOF. This article was completed in November 1995, was revised in August 1996, and has not been updated since.

ACKNOWLEDGMENTS. I thank G. Wray for the work he has done on this manuscript. This work was supported by a grant from the Wellcome Trust.

7. References

Allman, J. M., and Kaas, J. H., 1971, A representation of the visual field in the caudal third of the middle temporal gyrus of the owl monkey (*Aotus trivirgatus*), *Brain Res.* **31:**85–105.

Baker, C. L., Hess, R. H., and Zihl, J., 1991, Residual motion perception in a "motion-blind" patient, assessed with limited lifetime random dots, *J. Neurosci.* **11:**454–461.

Barbur, J. L., Watson, J. D. G., Frackowiak, R. S. J., and Zeki, S., 1993, Conscious visual perception without V1, *Brain* **116:**1293–1302.

Beckers, G., and Hömberg, V., 1992, Cerebral visual motion blindness: Transitory akinetopsia induced by transcranial magnetic stimulation of human area V5, *Proc. R. Soc. Lond. B* **249:**173–178.

Beckers, G., and Zeki, S., 1995, The consequences of inactivating areas V1 and V5 on visual motion perception, *Brain* **118:**49–60.

Bender, M. B., and Feldman, M., 1972, The so-called 'visual agnosias', *Brain* **95:**173–186.

Benevento, L. A., and Rezak, M., 1976, The cortical projections of the inferior pulvinar and adjacent lateral pulvinar in the rhesus monkey (*Macaca mulatta*): an autoradiographic study, *Brain Res.* **108:**1–24.

Benson, D. F., and Greenberg, J. P., 1969, Visual form agnosia: A specific defect in visual discrimination, *Arch. Neurol. Psychiatry (Chicago)* **20:**82–89.

Blasdel, G. G., and Fitzpatrick D., 1984, Physiological organization of layer 4 in macaque striate cortex, *J. Neurosci.* **4:**880–895.

Blythe, I. M., Kennard, C., and Ruddock, K. H., 1987, Residual vision in patients with retrogeniculate lesions of the visual pathways, *Brain* **110:**887–905.

Boussaoud, D., Desimone, R., and Ungerleider, L. G., 1991, Visual topography of area TEO in the macaque, *J. Comp. Neurol.* **306:**554–575.

Burkhalter, A., Felleman, D. J., Newsome, W. T., and Van Essen, D. C., 1986, Anatomical and physiological asymmetries related to visual areas V3 and VP in macaque extrastriate cortex, *Vision Res.* **26:**63–80.

Burkhalter, A., and Van Essen, D. C., 1986, Processing of color, form and disparity information in visual areas VP and V2 of ventral extrastriate cortex in the macaque monkey, *J. Neurosci.* **6:**2237–2351.

Campion, J., Latto, R., and Smith, Y. M., 1983, Is blindsight an effect of scattered light, spared cortex, and near-threshold vision? *Behav. Brain Sci.* **6:**423–486.

Ceccaldi, M., Mestre, D., Brouchon, M., Balzamo, M., and Poncet, M., 1992, Autonomie deambulatoire et perception visuelle du mouvements dans un cas de cécité corticale quasi totale, *Rev. Neurol.* **148:**343–349.

Celesia, G. G., Bushnell, D., Toleikis, S. C., and Brigell, M. G., 1991, Cortical blindness and residual vision: Is the 'second' visual system in humans capable of more than rudimentary visual perception? *Neurology* **41:**862–869.

Clarke, S., and Miklossy, J., 1990, Occipital cortex in man: Organization of callosal connections, related myelo- and cytoarchitecture, and putative boundaries of functional visual areas, *J. Comp. Neurol.* **298:**188–214.

Corbetta, M., Miezin, F., Dobmeyer, S., Shulman, G. L., and Petersen, S. E., 1991, Selective and divided attention during visual discriminations of shape, color, and speed: Functional anatomy by positron emission tomography, *J. Neurosci.* **11:**2383–2402.

Cowey, A., and Stoerig, P., 1991, The neurobiology of blindsight, *Trends Neurosci.* **14:**140–145.

Cragg, B. G., 1969, The topography of the afferent projections in circumstriate visual cortex studied by the Nauta method, *Vision Res.* **9:**733–747.

Crick, F., and Koch, C., 1995, Are we aware of neural activity in primary visual cortex? *Nature* **375:**121–123.

Damasio, A., Yamada, T., Damasio, H., Corbett, J., and McKee, J., 1980, Central achromatopsia: Behavioural, anatomic, and physiologic aspects, *Neurology* **30:**1064–1071.

De Jong, B. M., Shipp, S., Skidmore, B., Frackowiak, R. S. J., and Zeki, S., 1994, The cerebral activity related to the visual perception of forward motion in depth, *Brain* **117:**1039–1054.

Desimone, R., and Schein, S. J., 1987, Visual properties of neurons in area V4 of the macaque: Sensitivity to stimulus form, *J. Neurophysiol.* **57:**835–867.

Desimone, R., and Ungerleider, L. G., 1986, Multiple visual areas in the caudal superior temporal sulcus of the macaque, *J. Comp. Neurol.* **248:**164–189.

DeYoe, E., A., and Van Essen, D. C., 1985, Segregation of efferent connections and receptive field properties in visual area 2 of the macaque, *Nature* **317:**58–61.

Dow, B. M., 1974, Functional classes of cells and their laminar distribution in monkey visual cortex, *J. Neurophysiol.* **37:**927–946.

Drasdo, N., Edwards, L., and Thompson, D., 1993, Models of the visual cortex based on visual evoked potentials, in: *Functional Organisation of Human Visual Cortex* (B. Gulyas, D. Ottoson, and P. Roland, eds.), Pergamon Press, Oxford, pp. 255–269.

Dubner, R., and Zeki, S. M., 1971, Response properties and receptive fields in an anatomically defined region of the superior temporal sulcus in the monkey, *Brain Res.* **35:**528–532.

Ducati, A., Fava, E., and Motti, E. D. F., 1988, Neuronal generators of the visual evoked potentials: Intracerebral recording in awake humans, *Electroenceph. Clin. Neurophysiol.* **71:**89–99.

Dupont, P., Orban, G. A., De Bruyn, B., Verbruggen, A., and Mortelmans, L., 1994, Many areas in the human brain respond to visual motion, *J. Neurophysiol.* **72:**1420–1424.

Fendrich, R., Wessinger, C. M., and Gazzaniga, M. S., 1992, Residual vision in a scotoma: Implications for blindsight, *Science* **258:**1489–1491.

ffytche, D., Guy, C. N., and Zeki, S., 1995, The parallel visual motion inputs into areas V1 and V5 of human cerebral cortex, *Brain* **118:**1375–1394.

Finkel, L. H., and Edelman, G. M., 1989, Integration of distributed cortical systems by reentry: A computer simulation of interactive functionally segregated visual areas, *J. Neurosci.* **9:**3188–3208.

Fries, W., 1981, The projection from the lateral geniculate nucleus to the prestriate cortex of the macaque monkey, *Proc. R. Soc. Lond. B* **213:**73–80.

Girard, P., Salin, P. A., and Bullier, J., 1992, Response selectivity of neurons in area MT of the macaque monkey during reversible inactivation of area V1, *J. Neurophysiol* **67:**1437–1446.

Hawken, M. J., Parker, A. J., and Lund, J. S., 1988, Laminar organization and contrast selectivity of direction selective cells in the striate cortex of the Old-World monkey, *J. Neurosci.* **8:**3541–3548.

Henaff, M. A., and Michel, F., 1993, A case of central achromatopsia: Possible implicit treatment of hue, *Invest. Ophthalmol. Vis. Sci.* **34:**745.

Hendry, S. H. C., and Yoshioka, T., 1994, A neurochemically distinct third channel in the macaque dorsal lateral geniculate nucleus, *Science* **264:**575–577.

Henschen, S. E., 1910, Zentrale Sehstörungen, in: *Handbuch der Neurologie* (M. Lewandowsky, eds.), Springer, Berlin, pp. 891–918.

Hess, R. H., Baker, C. L., and Zihl, J., 1989, The "motion-blind" patient: Low level spatial and temporal filters, *J. Neurosci.* **9:**1628–1640.

Heywood, C. A., Wilson, B., and Cowey, A., 1987, A case study of cortical colour 'blindness' with relatively intact achromatic discrimination, *J. Neurol. Neurosurg. Psychiatry* **50:**22–29.

Holmes, G., 1945, The Ferrier Lecture: The organization of the visual cortex in man, *Proc. R. Soc. Lond. B* **132:**348–361.

Hubel, D. H., and Livingstone, M. S., 1987, Segregation of form, color and stereopsis in primate area 18, *J. Neurosci.* **7:**3378–3415.

Hubel, D. H., and Wiesel, T. N., 1962, Receptive fields, binocular interaction and functional architecture in the cat's visual cortex, *J. Physiol. (Lond.)* **160:**106–154.

Hubel, D. H., and Wiesel, T. N., 1977, Ferrier Lecture. Functional architecture of macaque monkey visual cortex, *Proc. R. Soc. Lond. B* **198:**1–59.

Humphrey, G. K., Goodale, M. A., Corbetta, M., and Aglioti, S., 1995, The McCollough effect reveals orientation discrimination in a case of cortical blindness, *Curr. Biol.* **5:**545–551.

Humphreys, G. W., and Riddoch, J. M., 1987, *To See But Not to See: A Case Study of Visual Agnosia*, Erlbaum, Hillsdale, NJ.

Kawano, K., Shidara, M., Watanabe, Y., and Yamane, S., 1994, Neural activity in cortical area MST of alert monkey during ocular following responses, *J. Neurophysiol.* **71:**2305–2324.

Keating, E. G., 1980, Residual spatial vision in the monkey after removal of striate and preoccipital cortex, *Brain Res.* **187:**271–290.

Kennard, C., Lawden, M., Moreland, A. B., and Ruddock, K. H., 1995, Colour identification and colour constancy are impaired in a patient with incomplete achromatopsia associated with prestriate cortical lesions, *Proc. R. Soc. Lond. B* **260:**169–175.

Kertesz, A., 1979, Visual agnosia: The dual deficit of perception and recognition, *Cortex* **15:**403–419.

Kölmel, H. W., 1988, Pure homonymous hemiachromatopsia: Findings with neuroopthalmolgic examination and imaging procedures, *Eur. Arch. Psychiatr. Neurol. Sci.* **237:**237–243.

Krubitzer, L. A., and Kaas, J. H., 1990, Cortical connections of MT in four species of primates: Areal, modular, and retinotopic patterns, *Visual Neurosci.* **5:**165–204.

Kulikowski, J. J., Walsh, V., McKeefrey, D., Butler, S. R., and Carden, D., 1994, The electrophysiological basis of colour processing in macaques with V4 lesions, *Behav. Brain Res.* **60:**73–78.

Lachica, E. A., Beck, P., and Casagrande, V. A., 1992, Parallel pathways in macaque monkey striate cortex: Anatomically defined columns in layer III, *Proc. Natl. Acad. Sci. USA* **89:**3566–3570.

Lennie, P., Krauskopf, J., and Sclar, G., 1990, Chromatic mechanisms in striate cortex of macaque, *J. Neurosci.* **10:**649–669.

Leventhal, A. G., Thompson, K. G., Liu, D., Zhou, Y., and Ault, S. J., 1995, Concomitant sensitivity to orientation, direction, and color of cells in layers 2, 3, and 4 of monkey striate cortex, *J. Neurosci.* **15:**1808–1818.

Levitt, J. B., Kiper, D. C., and Movshon, J. A., 1994, Receptive fields and functional architecture of macaque V2, *J. Neurophysiol.* **71:**2517–2542.

Livingstone, M. S., and Hubel, D. H., 1982, Thalamic inputs to cytochrome oxidase-rich regions in monkey visual cortex, *Proc. Natl. Acad. Sci. USA* **79:**6098–6101.

Livingstone, M. S., and Hubel, D. H., 1984a, Anatomy and physiology of a color system in the primate visual cortex, *J. Neurosci.* **4:**309–356.

Livingstone, M. S., and Hubel, D. H., 1984b, Specificity of intrinsic connections in primate primary visual cortex, *J. Neurosci.* **4:**2830–2835.

Livingstone, M. S., and Hubel, D. H., 1988, Segregation of form, color, movement, and depth: Anatomy, physiology, and perception, *Science* **240:**740–749.

Lueck, C. J., Zeki, S. Friston, K. J., Deiber, M. P., Cope, P., Cunningham, V. J., Lammertsma, A. A., Kennard, C., and Frackowiak, R. S. G., 1989, The colour centre in the cerebral cortex of man, *Nature* **340:**386–389.

MacKay, G., and Dunlop, J. C., 1899, The cerebral lesions in a case of complete acquired colour-blindness, *Scott. Med. Surg. J.* **5:**503–512.

Maier, J., Dagnelie, G., Sperkreijse, H., and Van, D. B. W., 1987, Principal components analysis for source localization of VEPs in man, *Vision Res.* **27:**165–177.

Malonek, D., Tootell, R. B. H., and Grinvald, A., 1994, Optical imaging reveals the functional architecture of neurons processing shape and motion in owl monkey area MT, *Proc. R. Soc. Lond. B* **258:**109–119.

Maunsell, J. H. R., and Gibson, J. R., 1992, Visual response latencies in striate cortex of the macaque monkey, *J. Neurophysiol.* **68:**1332–1344.

Meadows, J. C., 1974, Disturbed perception of colours associated with localized cerebral lesions, *Brain* **97:**615–632.

Mestre, D. R., Brouchon, M., Ceccaldi, M., and Poncet, M., 1992, Perception of optical flow in cortical blindness: A case report, *Neuropsychologia* **30:**783–795.

Nakamura, H., Gattass, R., Desimone, R., and Ungerleider, L. G., 1993, The modular organization of projections from areas V1 and V2 to areas V4 and TEO in macaques, *J. Neurosci.* **13:**3681–3691.

Orban, G. A., Kennedy, H., and Bullier, J., 1986, Velocity and direction selectivity of neurons in areas V1 and V2 of the monkey: Influence of eccentricity, *J. Neurophysiol.* **56:**462–480.

Pasik, T., and Pasik, P., 1982, Visual functions in monkeys after total removal of visual cerebral cortex, *Contrib. Sensory Physiol.* **7:**147–200.

Peterhans, E., and von der Heydt, R., 1993, Functional organization of area V2 in the alert macaque, *Eur. J. Neurosci.* **5:**509–524.

Petersen, S. E., Baker, J. F., and Allman, J. M., 1980, Dimensional selectivity of neurons in the dorsolateral visual area of the owl monkey, *Brain Res.* **197:**507–511.

Petersen, S. E., Miezin, F. M., and Allman, J. M., 1988, Transient and sustained responses in four extrastriate areas of the owl monkey, *Exp. Brain Res.* **70:**55–60.

Probst, T., Plendl, H., Paulus, W., Wistl, E. R., and Scherg, M., 1993, Identification of the visual motion area (area V5) in the human brain by dipole source analysis, *Exp. Brain Res.* **93:**345–351.

Raiguel, S. E., Lagae, L., Gulyas, B., and Orban, G. A., 1989, Response latencies of visual cells in macaque areas V1, V2 and V5, *Brain Res.* **493:**155–159.

Riddoch, G., 1917, Dissociations of visual perception due to occipital injuries, with especial reference to appreciation of movement, *Brain* **40:**15–57.

Rizzo, M., Smith, V., Pokorny, J., and Damasio, A. R., 1993, Colour perception profiles in central achromatopsia, *Neurology* **43:**995–1001.

Rodman, R., Gross, C. G., and Albright, T. C., 1989, Afferent basis of visual response properties in area MT of the macaque. I. Effects of striate cortex removal, *J. Neurosci.* **9:**2033–2050.

Roe, A. W., and Ts'o, D. Y., 1995, Visual topography in primate V2: Multiple representation across functional stripes, *J. Neurosci.* **15:**3689–3715.

Ruddock, K. H., 1991, Spatial vision after cortical lesions, in: *Spatial Vision* (D. M. Regan, ed.), Macmillan, London, pp. 261–289.

Shipp, S., and Zeki, S., 1985, Segregation of pathways leading from area V2 to areas V4 and V5 of macaque monkey visual cortex, *Nature* **315:**322–325.

Shipp, S., and Zeki, S., 1989a, The organization of connections between areas V5 and V1 in macaque monkey visual cortex, *Eur. J. Neurosci.* **1:**309–332.

Shipp, S., and Zeki, S., 1989b, The organization of connections between areas V5 and V2 in macaque monkey visual cortex, *Eur. J. Neurosci.* **1:**333–354.

Shipp, S., de Jong, B. M., Zihl, J., Frackowiak, R. S. J., and Zeki, S., 1994, The brain activity related to residual motion vision in a patient with bilateral lesions of V5, *Brain* **117:**1023–1038.

Standage, G. P., and Benevento, L. A., 1983, The organization of connections between the pulvinar and visual area MT in the macaque monkey, *Brain Res.* **262:**288–294.

Tanaka, M., Weber, H., and Creutzfeldt, O. D., 1986, Visual properties and spatial distribution of neurones in the visual association area on the prelunate gyrus of the awake monkey, *Exp. Brain Res.* **65:**11–37.

Tootell, R. B. H., Silverman, M. S., DeValois, R. L., and Jacobs, G. H., 1983, Functional organization of the second cortical area in primates, *Science* **220:**737–739.

Tootell, R. B. H., Silverman, M. S., Hamilton, S. L., DeValois, R. L., and Switkes, E., 1988, Functional anatomy of macaque striate cortex. III. Color, *J. Neurosci.* **8:**1569–1593.

Tootell, R. B. H., Reppas, J. B., Kwong, K. K., Malach, R., Born, R. T., Brady, T. J., Rosen, B. R., and Belliveau, J. W., 1995, Functional-analysis of human MT and related visual cortical areas using magnetic-resonance-imaging, *J. Neurosci.* **15:**3215–3230.

Vaina, L. M., 1994, Functional segregation of color and motion processing in the human visual cortex: Clinical evidence, *Cerebral Cortex* **4:**555–572.

Van Dijk, B., Dagnelie, G., and Spekreijse, H., 1987, Motion onset–offset visual evoked potentials from rhesus visual cortex, in: *Evoked Potentials III: The Third International Evoked Potential Symposium* (C. Barber and T. Blum, eds.), Butterworths, London, pp. 205–212.

Verrey, D., 1888, Hémiachromatopsie droite absolue. Conservation partielle de la perception lumineuse et des formes. Ancien kyste hémorrhagique de la partie inférieure du lobe occipital gauche, *Arch. Ophthalmol. (Paris)* **8**:289–300.

Walsh, V., Carden, D., Butler, S. R., and Kulikowski, J. J., 1993, The effects of V4 lesions on the visual abilities of macaques: Hue discrimination and color constancy, *Behav. Brain Res.* **53**:51–62.

Watson, J. D. G., Myers, R., Frackowiak, R. S. J., Hajnal, J. V., Woods, R. P., Mazziotta, J. C., Shipp, S., and Zeki, S., 1993, Area V5 of the human brain: Evidence from a combined study using positron emission tomography and magnetic resonance imaging, *Cerebral Cortex* **3**:79–94.

Wechsler, I. S., 1933, Partial cortical blindness with preservation of colour vision: report of a case following asphyxia (carbon monoxide poisoning?), *Arch. Ophthalmol.* **9**:957–965.

Weiskrantz, L., 1986, *Blindsight*, Oxford University Press, Oxford.

Weiskrantz, L., 1990, The Ferrier lecture, 1989. Outlooks for blindsight: Explicit methodologies for implicit processes, *Proc. R. Soc. Lond. B* **239**:247–278.

Wilson, C. L., Babb, T. L., Halgren, E., and Crandall, P. H., 1983, Visual receptive fields and response properties of neurons in human temporal lobe and visual pathways, *Brain* **106**:473–502.

Yukie, M., and Iwai, E., 1981, Direct projection from the dorsal lateral geniculate nucleus to the prestriate cortex in macaque monkeys, *J. Comp. Neurol.* **201**:81–97.

Zeki, S. M., 1969, Representation of central visual fields in prestriate cortex of monkey, *Brain Res.* **14**:271–291.

Zeki, S. M., 1970, Interhemispheric connections of prestriate cortex in the monkey, *Brain Res.* **19**:63–75.

Zeki, S. M., 1971, Cortical projections from two prestriate areas, *Brain Res.* **34**:19–35.

Zeki, S. M., 1974a, Cells responding to changing image size and disparity in the cortex of the rhesus monkey, *J. Physiol. (Lond).* **242**:827–841.

Zeki, S. M., 1974b, Functional organization of a visual area in the posterior bank of the superior temporal sulcus of the rhesus monkey, *J. Physiol. (Lond.)* **236**:549–573.

Zeki, S. M., 1975, The functional organization of projections from striate to prestriate visual cortex in the rhesus monkey, *Cold Spring Harbor Symp. Quant. Biol.* **40**:591–600.

Zeki, S. M., 1977, Colour coding in the superior temporal sulcus of rhesus monkey visual cortex, *Proc. R. Soc. B* **197**:195–223.

Zeki, S. M., 1978a, Functional specialization in the visual cortex of the monkey, *Nature* **274**:423–428.

Zeki, S. M., 1978b, The third visual complex of rhesus monkey prestriate cortex, *J. Physiol. (Lond.)* **277**:245–272.

Zeki, S. M., 1979, Functional specialization and binocular interaction in the visual areas of rhesus monkey prestriate cortex, *Proc. R. Soc. Lond. B* **204**:379–397.

Zeki, S., 1980, The responses of cells in the anterior bank of the superior temporal sulcus in macaque monkeys, *J. Physiol. (Lond.).* **308**:85.

Zeki, S., 1983, Colour coding in the cerebral cortex: The reaction of cells in monkey visual cortex to wavelengths and colours, *Neuroscience* **9**:741–765.

Zeki, S., 1990a, A century of cerebral achromatopsia, *Brain* **113**:1721–1777.

Zeki, S., 1990b, A theory of multi-stage integration in the visual cortex, in: *The Principles of Design and Operation of the Brain* (J. C. Eccles and O. Creutzfeldt, eds.), Pontifical Academy of Science, Vatican City, pp. 137–154.

Zeki, S., 1991, Cerebral akinetopsia (visual motion blindness)—A review, *Brain* **114**:811–824.

Zeki, S., 1993, *A Vision of the Brain*, Blackwell, Oxford.

Zeki, S., 1995, The motion vision of the blind, *Neuroimage* **2**:231–235.

Zeki, S., 1996, Are areas TEO and PIT of monkey visual cortex wholly distinct from the fourth visual complex (V4 complex)? *Proc. R. Soc. Lond. B* **263**:1539–1544.

Zeki, S., and Shipp, S., 1988, The functional logic of cortical connections, *Nature* **335**:311–317.

Zeki, S., and Shipp, S., 1989, Modular connections between areas V2 and V4 of macaque monkey visual cortex. *Eur. J. Neurosci.* **1**:494–506.

Zeki, S., Watson, J. D. G., Lueck, C. J., Friston, K. J., Kennard, C., and Frackowiak, R. S. J., 1991, A direct demonstration of functional specialization in human visual cortex. *J. Neurosci.* **11**:641–649.

Zeki, S., Watson, J. D. G., and Frackowiak, R. S. J., 1993, Going beyond the information given: The relation of illusory visual motion to brain activity, *Proc. R. Soc. Lond. B* **252**:215–222.

Zheng, D., LaMantia, A. S., and Purves, D., 1991, Specialized vascularization of the primate visual cortex, *J. Neurosci.* **11:**2622–2629.

Zihl, J., 1980, "Blindsight": Improvement of visually guided eye movements by systematic practice in patients with cerebral blindness, *Neuropsychologia* **18:**71–77.

Zihl, J., Von Cramon, D., and Mai, N., 1983, Selective disturbance of movement vision after bilateral brain damage, *Brain* **106:**313–340.

Zihl, J., Von Cramon, D., Mai, N., and Schmid, C., 1991, Disturbance of movement vision after bilateral posterior brain damage. Further evidence and follow up observations, *Brain* **114:**2235–2252.

Index

Acetylcholine receptors, 725, 726
Acetylcholinesterase, 641, 653, 705
Achromatopsia, 67, 114, 701, 727, 763, 799, 801–802
Activity-driven plasticity, 143
Activity subcompartments, 304
Acuity thresholds, 181
Adrian, E. A., 40, 43
Aggregate receptive fields, 140, 143
Agnosia, visual, 33–34, 45
Alcmaeon, 6–7
Alexandrian science, 8–10
Alhazen, 36, 37
Allman and Kaas model, 98, 99, 100, 132, 145, 146, 166
All-or-none law, 40
Amacrine cells, 61
2-Aminophosphonobutyrate (APB), 68
AMPA, 695, 696, 699, 725
Amygdala, 47, 284, 495
  cortical connections, 285
  development, 645
  TE projections, 469
Analysis of visual scene, 58–87
  areal specificity
    models of, 63–67
    revisions and alternative models, 75–80
  extrastriate cortex, 181
  neuronal specificity
    models of, 60–63
    revisions and alternative models, 67–75
  topography and visual representation, 80–85
Anaxagoras, 6
Angular gyrus, 31–32, 35
Anisotropy, 147–148, 311
Antagonistic organization, 41
Antagonistic surround, 372, 398–404; *see also* Center-surround organization
  areal extent, 401, 404
  incidence, 398–401
  parameter selectivities, 401, 402, 403, 404
  receptive field shaping, 184
Anterior cingulate cortex, 530, 549, 643
Anterior superior temporal sulcus, 642
Anterior temporal area (AT), 156, 643, 646
Anterograde degeneration methods, 134–135
Anterolateral (AL) area, superior temporal sulcus, 169, 170
Antidromic activation, 208–209, 210, 559, 560
Antiphase kinetic boundaries, 392
Aotus, 107, 168; *see also* Owl monkey
Arab science, 36, 37
Arbors, *see also* Architecture
  layer 6 variability, 273
Architecture, 243–287, 688–692
  architectonic features/maps, 93, 97–98, 103, 130, 135–136, 675–680; *see also* Cytoarchitecture; Cytochrome oxidase; Myelin, myeloarchitecture
    Brodmann's areas, 675–680
    historical developments, 682–685
    MT and V2, 107–108
    striate and extrastriate cortex nomenclature, vii–viii, 674
    STS, 438 448–452
  computational, 245–277
    connectional weight, 258–264, 265
    intrinsic connections, 275–277, 278, 279
    laminar terminations of single axons, 245–250

Architecture (*cont.*)
computational (*cont.*)
postsynaptic population, 254–258
quantitative parameters, 264–273, 274, 275, 276
spatial configuration of single axons, 250–253
subtypes of connections, 273
correlation with function, 795
hierarchy, 277, 279–284
collateralized connections, 282–283
laminar patterns, 279–280
reciprocity, 280–282
human studies, 699–710, 727
comparison with other primates, 688–692
cyto- and myeloarchitecture, 704–706
cytochrome oxidase staining, 706–710
FEF, 548
functional specialization, 699–704
MT and satellites, 373–375
and object recognition: *see* TE
STP, 448–462
architectonic subdivisions of upper bank and fundus of STS, 448–452
TPO as separate cortical area, 461–462
TPO connections with high-order association cortex, 452–460
TPO subdivision physiological correspondences, 461–462
structure-function relationship: *see* Structure-function correspondence
V2 in macaque, 295–328
connectivity, 314–322
cytochrome oxidase, 296–297
function of multiple processing stages, 326–328
modular organization within stripes, 303–305
receptive field properties, stripes and, 297–302
relationship of structure and circuitry to perception, 322–326
topography, 305–314
Arcuate fasciculus, 718–719
Arcuate sulcus
area 8Ar, 530
FEF, 542, 543
smooth pursuit, 580
Area 6, 548
Area 7a, 181, 692; *see also* Inferior parietal lobule
cortical and subcortical connections, 502–504
FEF connectivity, 551
macaque model, 502, 503
Area 8, 545, 560, 674
Area 9, 605–610
Area 12, 588, 610–612
Area 13, 612–614
Area 15, 674
Area 17, 225, 722
Area 18, 100, 674, 722
Area 19, 100, 225, 674
Area 23, 674
Area 28, 674
Area 46, 605–610
Area F7, 530
Area IPa, 453
Areal organization, *see also* Theories of visual cortex organization; Visuotopic organization
cerebral cortex, 436
development, 640–641
early modern models, 98, 99
frontal cortex, 530
human studies, 749
STP: *see* Superior temporal polysensory area
Areal specificity
models of, 63–67
revisions and and alternative models, 75–80
Area MT: *see* V5/MT
Area VX, 709
Area X, 74, 709
Area Y, 74
Area Z, 74
Aristotle, 4, 9, 13, 16
Association cortex, 77, 94–95, 436
corticocortical axons, 212–214
lesions of, 499, 500
parietal, 48–51
temporal, 44–48
Associationism, 43
Association tasks, 443
Associative pathways, hemispheric white matter, 718–719
AT (anterior temporal area), 156
Attention, 76
architectonic mapping, 677
and IT neuron properties, 48
RF size modulation, 186–187
selection by, 492
visual, 50
Attention deficits, ssociational cortical lesions, 499, 500
Auditory areas, 95, 97, 243–244
connections to area TPO, 460
models of functional areas, 95
organization of, 190
Auditory association tasks, 443
Autopsy material, 745
Autoradiography, 724, 727
Avian studies, 179
Axonal arborization, 276
Axonal conduction time, 206–215
delays, 210–215
estimation of conduction delays, 207–209
minimizing cortical axon size, 206
specificity of for different connections and species, 209–210
Axon size: *see* Size, neuron

Bagshaw, 47
Baillarger, stripe of, 680, 682, 691
Bain, Alexander, 43
Banding, 94, 106–107
Barlow, Horace, 41, 62
Bartholin, Thomas, 16
Bartley, J. Howard, 40
Basal forebrain development, 645
Basal ganglia, 96
  oculomotor system, 542
  TE projections, 469
Baumgartener, G., 42
BDA, 281
Beaded specializations, 272
Beaded terminations, 270, 271, 275
Beck, Adolf, 38, 40
Behavior
  posterior temporal lobe and, 462
  RF size modulation, 186
  temporal organization of, 609
Behavioral studies
  MT/V5 in macaques, 404–421
    body movement guidance, 418
    extraction of 2D and 3D shape from motion, 418–421
    eye movement guidance, 412–418
    motion perception, 405–412
  STP cortex ablation, 442–443
Bell, Charles, 30
Bias-free maps, 134
Bidirectional flow of information, slow brain, 233
Binding activities, IT columns, 492–493
Binocular convergence, 32, 33
Binocular disparity
  contour definition, 339
  distribution of neurons across stripes, 352
  functional segregation, 337
Binocular interactions
  depth perception, 69, 70–71
  figure-ground segregation, 352–353
  human studies, 748, 765
  PLGN lesions and, 316
  registration, maximal RF shift, 187
Biocytin, 207, 251, 254, 255, 256, 257, 267, 268, 270, 271, 275
Biotinylated dextran amine (BDA), 244, 246, 247, 248, 281
Bipolar cell connections, 60, 61
Blind, residual vision, 781–787
Blindsight, 657, 666, 777, 779–787
Blobs, 94, 113, 296, 300, 301, 305, 335–336
  color signals, 795
  connectivity, 693, 694
  contour representation, 323
  hierarchical organization, 721
  neurotransmitters, 705, 706
  parallel processing of V1, V2, V4, and MT, 315
Blood flow: *see* Cerebral blood flow
Body movement
  cells responding to, 493
  MT/V5 in macaques, 418
Bolton, J. S., 43
Bolton–Campbell area, 43
Borders (object)
  color border cells, 300, 302
  contour perception, 340–343
Border tuft, 682
Boundaries (anatomic and functional), 71, 97, 99, 100; *see also* Architecture
  architectonic shifts at, 675
  cortical areas, 92–93
  electrophysiological visuotopic mapping, 133–134
  FSTd and DLa, 156, 157, 158
  intrinsic connections, 277
  laminar, 250
  MST-STP, 439
  topography, 307
  variability of, 178–179
  visuotopic maps, 143–144
Bouton architecture, 270, 271
Brain, W. R., 49
Brainstem
  motor error, 554
  saccadic eye movement, 540
  SEF connectivity, 587
Brightness, 7, 78
Broadband cells
  center-surround, 302
  multiple maps, 307
  receptive fields, 308
Broca, Paul, 25–26
Brodmann, K., 64, 680
Brodmann's areas, 46, 64, 91, 96, 97, 98, 100, 675–680
  architectonic maps, 675
  BA 4: *see* Motor cortex
  BA 6, FEF location in humans, 546, 547, 682
  BA 17, 682; *see also* Striate cortex
  BA 27 (presubiculum), 675
  BA 28 (entorhinal area), 675
  comparison of primate species, 686, 688
  Filimonoff subdivision correspondences, 679–680
  MT and V2, 105
  myelogentic subdivision correspondences, 685
  optical parakoniocortex model versus, 681–682
  sub-parcellation of, 682–685
  thalamic connectivity, 692
Brouillard, J. B., 26
Brown, S., 49
Bucy, L., 45–46
Burst neurons, saccadic eye movement, 540, 541
Bypass projections, 282, 284

Cajal, interstitial nucleus of, 548, 587
Cajal, Santiago Ramon y, 60, 61, 67, 80–81
Calbindin, 285, 689, 691
Calcar avis, 36
Calcarine cortex, 762
Calcarine sulcus, 36, 685, 752
Callithrix, 107, 150; *see also* Marmoset
Callosal connections
  anterograde degeneration methods, 134–135
  development, 644–645, 662–663
  extrastriate cortex
    caudal neocortex, 168
    DL subdivisions, 165, 166, 167
  human studies, 713–718, 722, 727, 745
  nonhuman primates, 693–605
  numbers of synapses, 259
  single-axon analysis in cat, 265–266
  timing of information transfer, 214–215
  visuotopic organization, 134–135, 171–172, 173
Callosal connection zone, 98
Campbell, Alfred, 43, 674, 675
Campbell model, 95, 96
Capuchin monkey, 138, 143, 146, 148, 149, 150
Carbon monoxide poisoning, 781, 797–799, 800
Cat, 40, 62, 98, 100, 108, 147
  architecture, 275
  callosal connections, 265–266
  development, 663
  second visual area, 144
  timing of information transfer
    callosal, 214–215
    chronaxies, 216
    conduction delay measurement, 211–212
    latencies, 221–222, 224–225
    superior colliculus, 215
CAT-301, 316, 689, 692
  human studies, 709
  intraareal compartments, 705
Catecholamines, 689
Caton, Richard, 40
Caudal dorsolateral area, 112, 114–115
Caudal extrastriate cortex, 112
Caudal STP, 362, 436, 445, 446, 448–449, 460
Caudal STS, 361, 439
Caudal TPO, 362, 436, 445, 446, 448–449, 450, 456, 458, 460
Caudate nucleus. 541, 542, 646
Cebus, 102, 104, 107, 111, 150, 306
Center of receptive field, 137
Center-surround organization, 68; *see also* Antagonistic surround
  color opponent field, 302
  FEF neuron movement field, 573
  on-off antagonism, 41
Central field representation of MT, 279–280
Cercopithecus, 677, 679, 686, 692; *see also* Macaques
Cerebellum
  architecture, 271
  FEF connectivity, 549
  oculomotor system, 542
  SEF connectivity, 586, 587
Cerebral blood flow, 746, 747, 788, 790
  cingulate cortex, 609
  FEF, 546, 547, 583
Cerebral cortex
  discovery of visual center, 31–42
  search for sensory areas in, 29–31
Cheirogaleus, 150, 171
Chemical markers, *see also* Architecture; Cytochrome oxidase
  circuitry, 207, 652
  development, 652–653
  visuotopic maps, 142–143
Chemoarchitecture; *see* Architecture
Chimpanzees, 542
Chirality, 133
Cholecystokinin, 689, 698, 726
Cholinergic receptors, 653, 696, 697–698
Chromaticity, contour definition, 339
Chromatopsia, 781, 797–799, 800
Chronaxie measurement, 216
Cingulate cortex, 614–616
  development, 643
  FEF connectivity, 549
  SEF connectivity, 587, 588
Cingulate gyrus, area 7a connections, 504
Classical receptive field (CRF), 372, 401
Claustrum, 321, 645
Clinical studies, 744, 745
CO: *see* Cytochrome oxidase
Coactivation patterns, 284
Coherent motion, 348–349. 350, 351
Coincidence detection, 219–220, 233
Collaterals
  architecture, 267, 269
    arbors, 277
    corticopulvinar neurons, 271
    feedback, 257
    feedforward axons, 253
    infragranular, 250
    terminal specializations, 275, 276
  infragranular layers, 247, 248, 251, 252, 253
  pyramidal cell connections, 491
Colliculopulvinocortical connections, 245
Colliding saccade, 557, 558
Color, 62, 63, 65, 69, 79, 71, 72, 75, 78
  architectonic mapping of function, 677
  contour definition, 339
  cross-modal interactions, 300
  2-D shape encoding in, 323
  functional segregation, 337
  functional transformations, 301

Color (*cont.*)
human studies, 701
area functional properties, 761
color system, 794–803
cortical regions activated by, 768
as distributed process, 802–803
information processing in color-sensitive regions, 762–764
parallelism, 791–794
residual vision, 778–787, 794–803
VP, 701–702
interleaved one step forward, two steps back mapping, 309
modular organization, 304, 305
parallel processing in V1, V2, V4, and MT, 315
plasticity, extrastriate lesions, 657
PLGN and, 316
receptive field types, 298–300
segregated processing, 316–317, 319, 320–322
specificity of functional interactions, 317–318
stimulus dependence, 319
subcompartments, 304
thin stripes, 310
V4 lesions and, 67, 691
Color border cells, 300, 302
Color borders, contour perception, 342–343
Color cells, 307, 308
Color constancy, loss of, 802–803
Color disparity cell, 300
Color domain, 311
Color pathways, 795–797
Color-selective cell pairs, 317–318
Color stripes, 297, 310, 311–312
Color tuning, 300
Columnar organization, 64, 65, 66, 76
IT: *see* Inferior temporal cortex, columnar organization
STP, 461
Columns, cortical representations in, 326
Commisural connectivity, development, 644–645; *see also* Callosal connections
Complex cells, 62
Complex objects and patterns
human studies, 766
MT/V5 in macaques, 388–398
Component-direction selective MT neurons, 388–390
Computational architecture, 245–277
connectional weight, 258–264, 265
intrinsic connections, 275–277, 278, 279
laminar terminations of single axons, 245–250
postsynaptic population, 254–258
quantitative parameters, 264–273, 274, 275, 276
spatial configuration of single axons, 250–253
subtypes of connections, 273
Computational areas, 264
architectonic mapping of function, 677
human studies, 748
Computational study, illusory contours, 348
Concurrency, hierarchy frameworks and, 284
Conditional motor tasks, 535
Conduction delays: *see* Delays
Conduction time: *see* Axon conduction time
Conduction velocity
FEF pyramidal cell activation, 560
magnocellular neurons, 265
Cones, 60, 61, 68, 86
Conformal mapping, 81
Connectional weight, 258–264, 265
Connections, 95, 96, 98; *see also* Architecture
definition of cortical areas, 93–94
development, 641–646, 662–663
frontal eye field, 548–552
human studies, 710–723
comparison with other primates, 692–695
hierarchical, 728
interhemispheric, 713–717
intrahemispheric, 718–721
intrinsic, 721–723, 724
thalamic, 710–713
visual areas as derived from callosal connections, 717–718
inferior parietal lobe, 502–504
and information transfer speed, 231
interhemispheric, 134–135
intersensory integration, 436; *see also* Superior temporal polysensory area
retinal cells, 60–63
STP-high-order areas, 448–462
architectonic subdivisions of upper bank and fundus of STS, 448–452
TPO association cortex, 452–460
TPO as separate cortical areas, 461–462
TPO subdivision physiological correspondences, 461–462
supplementary eye field, 585–594
timing of information transfer: *see* Timing of information transfer
V1-V2: *see* V1-V2
V2 in macaques, 314–322
color, form, and motion, 324–320
serial or parallel, 320–322
V5/MT in macaques, 363–367
Conscious perception: *see* Perception/perceptual processes
Contour, 300, 301, 318; *see also* Object vision
functional segregation, 337
human studies, 703
illusory, 353
position, orientation, size and attributes defin-ings, 473–476
relationship of cortical structure to function, 323, 325–326
V2 structure-function relationships, 338–343

Contrast, 68
contour perception, 339, 340–342
human studies, area functional properties, 761
MT and satellite receptive field, 379, 387–388
Contrast polarity, 379
Convergent factor, 253
Convergent input, 259
higher order receptive field property generation, 301
human studies, 769–770
ON and OFF cells, 69
V4, 76
Coordination of eye, head, forelimb movements, 549
Corpus callosum, 98, 109; *see also* Callosal connections; Commisural connectivity
development, 662
giant axons, 206
Correlograms, 317, 318, 319, 321, 322
Cortical blindness, 34
Cortical connections, 91; *see also* Architecture; Callosal connections; Corticocortical connections; Corticothalamic connections; Intracortical/intrahemispheric connections
area 7a, 502–504
intrinsic connection interactions, 275
reciprocity, 280
Cortical magnification, 307, 311–312
Cortical magnification factor, 142, 148, 758–761
Cortical retina, 43
Corticocortical connections, 259, 326; *see also* Intracortical/intrahemispheric connections
association cortex, 212–214
computational properties, 267
neurofilament-containing neurons, 276–277
single axon spatial configuration, 250–253
timing of information transfer: *see* Timing of information transfer
Corticogeniculate axons: *see* Geniculocortical system
Corticospinal pathways, 285
Corticotectal neurons, 560
Cortico-thalamo-cortical circuits, 283
Countermanding paradigm, 564
Cragg model, 98, 99
Craniotopic representation, 594
Creutzfeldt, O., 42
Critchley, Malcolm, 49
Cross-correlation analysis, 315
conduction delay, 207, 208
retinogeniculate axons, 210–211
timing of information transfer, 218–219
Crossed lesion paradigm, 47
Cross-modal interactions
color oriented cells of V2, 300
integration of sensory information, 436
Cross-species comparisons, 748
Cuneus, 423
Cytoarchitecture, 46, 64, 130–131, 675, 675–680, 745
CO staining comparisons, 707–708
development, 641, 642
historical developments, 682–685
human studies, 704–706
anatomical variability, 754
extrastriate cortex, 704–706
FEF, 548
topography, 751
parietal lobe, 501–502
SEF borders, 589
TE: *see* TE
visuotopic analyses, 135–136
Cytochrome oxidase, 64, 75, 81, 103, 110, 111, 113, 147–150, 461
correlation with function, 795
development, 652, 664–665
dorsolateral cortex, 114
human studies, 745
extrastriate cortex, 706–710
intrinsic connections, 722, 723
intraareal compartments, 705
intrahemispheric connections, 693, 694
PX, 702
single axon terminations, 251
V1
and conduction latencies, 226–227
V1-V2 and MT connections, 363–365
V2, 106–107, 335, 336, 337, 688–689
and conduction latencies, 226–227
macaque, 296–297; *see also* Architecture, V2 in macaque
in rhesus monkey, 340
V1-V2 and MT connections, 363–365
V2-V4 connectivity patterns, 795, 796
V5/MT, 106–107, 691–692
V5/MT surrounding areas, 156
ventrolateral cortex, 169

DA (dorsoanterior area), 163
Darkschewitsch, nucleus of, 548
DARRP-32, 652
Darwin, Charles, 26, 27
Da Vinci stereopsis, 353
DDD characteristic, 387
Delayed crossing effect, 218
Delay period activity, prefrontal cortex neurons, 606
Delays
axonal conduction time, 210–215
axon size and, 206
estimation of, 207–209
Democritus, 6
Dendritic arbors, 273; *see also* Architecture
Dendritic spines, 270

Densely myelinated zone of MST, 361, 362
Dentate nucleus, 549
2-Deoxyglucose, 135, 296, 304, 649, 652
Depth, 66, 67, 69, 70–71, 70, 71, 76
  segregated processing, 316–317, 319
  spatial cues, 506
  specificity of functional interactions, 317–318
  thick stripes, 310
Depth cells, 298
Descartes, R., 37, 38
Development, 639–666
  comparison with other mammals, 662–663
  and IT neuron properties, 48
  myelination, 179
  normal, 640–653
    connectivity, 641–646
    neurochemical markers, 652–653
    neuronal activity, 646–652
    prenatal specification of areas, 640–641
  plasticity in, 660–662
  plasticity in adulthood, 653–660
    damage-induced, 654–658
    experience or learning-induced, 658–660
  recovery after injury or damage, 665–666
  sequence of development and functional streams, 664–665
Developmental hypotheses of visuotopic organization, 176–177
de Viessens, Raymond, 18
DI (dorsointermediate area), 113
Diamidino yellow, 261, 263, 264
DiI dye studies, 721–722
Diogenes, 6
Direction, 62, 376–383
  functional segregation, 337
  human studies, area functional properties, 761
  STP, 442
Directional tuning, MT neurons, 377, 380–381, 382, 388, 389
Direction-selective cells
  human studies, 764
  localization in layers 4b and 6, 782
  V5, 786
Direction selectivity
  for expansion/contraction or rotational movement, 391, 393–398
  MT neurons, 360, 361, 377–380, 381, 382, 383, 387
    antagonistic surround in MT and MSTd, 402
    pattern and component direction selective, 388–390
  V3, 690
  V4, 690
  visual pursuit cells, 412
Discontinuities, 309, 310, 313–314
  contour definition, 339
  DM of marmoset, 162
  field, 141–142, 152, 158, 178, 179, 180
  map, 141, 142–143, 152, 162, 178
  significance of, 178, 179, 180
Disparity, 72
  contour definition, 339
  cross-modal interactions, 300
  distribution of neurons across stripes, 352
  functional segregation, 337
  functional transformations, 301
  human studies
    area functional properties, 761
    basic visual cues, 765
    cortical regions activated by, 768
  interleaved one step forward, two steps back mapping, 309
  modular organization, 304, 305
  MT and satellite receptive field, 387
  parallel processing in V1, V2, V4, and MT, 315
  spatial cues, 506
Disparity cells, 208, 300, 302
Disparity contour, 323
Disparity domains, 311, 323
Disparity stripes, 297, 311
Disparity types, 304
Distributed processes
  FEF, 577
  hierarchical, 181, 728, 729
  optic flow, 512
  visual consciousness as, 803–804
Diurnal habit
  relative area occupied by visual functions, 686
  visuotopic organization for, 146
Divergence, 260, 769–770
DL: *see* Dorsolateral area
DM: *see* Dorsomedial area
Dopamine D1 receptors
  development patterns, 652
  motor learning task, 596
  prefrontal cortex, 608
  V2, 698, 725, 726
Dorsal area (D)
  galago, 165
  V2, 143–144
Dorsal extrastriate cortex, 143–144
Dorsal FST (FSTd), 110, 115–116, 156, 157, 158
Dorsal MST: *see* Medial superior temporal visual area, dorsal
Dorsal (occipitoparietal) pathway, 51
  development, 649
  hierarchical organization and, 728
  human studies, 728, 748, 768
  plasticity, 657
  PLGN and MLGN and, 316
  pulvinar connections, 441
  spatial vision, 366

Dorsal (occipitoparietal) pathway (*cont.*)
STP
connections, 440–441
directional selectivity information, 442
visuomotor control, 492
Dorsal pontine nucleus, 503
Dorsal postarcuate premotor cortex (PMd), 609
Dorsal prelunate sulcus, 502, 503
Dorsal premotor area, 530, 535–540
Dorsal V3, 108–110, 113, 161, 163
Dorsal-ventral division of inferior temporal cortex, 101
Dorsal zone (DZ), 155
Dorsoanterior area (DA), 163
Dorsointermediate area (DI), 113
Dorsolateral area (DL), 99, 100, 102, 104, 113, 165
anterior (DLa), 156, 157, 158, 165, 166, 167
caudal, 112, 114–115, 165, 167
continuation as VP, 117
intermediate, 165, 166, 167, 169, 171
MT and V2, 106
posterior, 165, 166, 167, 169, 171
retrograde labeled feedforward neurons, 261, 262, 263
rostral, 110, 114–115, 165
visuotopic organization
cytochrome oxidase reactive regions, 169
subdivisions of, 165–167
Dorsolateral prefrontal cortex, 527
areas 46 and 9, 605–610
Dorsomedial area (DM), 99, 100, 102, 112, 113
connectivity, intrahemispheric, 693, 694
theories of organization, 110–111
V3 connections, 113
visuotopic organization, 158–165, 178
Dorsomedial zone (DMZ)
FEF connectivity, 551
transmitter receptors, 699
Double-step saccade test, 554, 555, 556, 557, 558
du Laurens, Andre, 13
du Petit, Francois Pourfour, 19–20, 21, 22
Durer, Albrecht, 12
Dynamic form analysis, 351
Dynamic parallelism, 791–794
Dyschromatopsia, 799

Ego motion: *see* Movements
Egypt, 2–4
Elaborate cells, 483
Electroencephalography, 40, 745, 793–794
Electron microscopy, 207, 254–257
Electrophysiology, 131, 133; *see also* Single-cell physiology; Single-unit electrophysiology
conduction delay estimation, 208–209
and development of extrastriate cortex, 646–649, 650, 651
galago, 150
history, 50–51
Electrophysiology (*cont.*)
human studies, 745
inferior parietal lobe function, 505–516
extraretinal signals, 514–516
visual properties, 505–514
IT columnar organization, 486
MT and V2, 105–106
SEF, 589–594
STP, 436
End-stopped cell, 302
functional segregation, 337
V1, 323
End-stopped receptive fields, 342
Entorhinal area, 675
Epigenetic mechanisms, 143
Erasistratus, 8, 9–10, 11, 19, 22
Evolution, 26–29, 177
Excitatory disparity cells, 298
Excitatory postsynaptic potentials (EPSPs)
latency measurement, 207
TE, 491
timing of information transfer, 218–219
Excitatory receptive fields, visuotopic order and, 136
Expansion/contraction (E) cells, 391, 393–398
Experience
and IT neuron properties, 48
plasticity, 658–660
Express saccades, 564
External taenia, 685
External visual field, 755
Extrastriate cortex, 64–65
development, 641, 653
dorsal and ventral processing pathways, 51
early modern theories of organization, 98–100
FEF connectivity, 549
models of functional areas, 95, 96
neurochemical markers, 653
processing in, 95
reorganization after damage, 657–658
V2: *see* V2
Extrinsic connections, 276, 277, 279
Eye movement, 187; *see also* Frontal lobe visuomotor areas
architectonic mapping of function, 677
area 7a connections and, 504
FEF connections controlling, 549
inferior parietal lobule, 504
maintenance of spatial constancy, 553–559
MT/V5 in macaques, 412–418
oculomotor signals
feedback control, 553
FEF connectivity, 540–542, 549, 562
inferior parietal lobule, 514–516
plasticity, extrastriate lesions, 657
RF size modulation, 186
SEF stimulation and, 589–594
stabilization during, 368, 383
STP and, 441, 442–443, 444

Eye position
maintenance of spatial constancy, 553–559
and posterior parietal neuron response, 50
SEF signal, 594
ventral premotor cortex, 532
visual response modulation, 514–515, 517–518, 519

F1, 530
Faces, 62
human studies, 700
cortical regions processing, 768
FEF, 546
SEF connectivity, 587
IT activation, 487, 488, 489
STP and, 443
Face-selective cells, 48, 232, 493
Feature detection model, 69–70, 71
Feature extraction, 69, 181, 418–421
Feature space, TEO-TE connections, 484
Feedback connections, 76, *see also specific areas, connections, and processes*
architecture: *see* Architecture
development, 642
higher-order area connections, 729
human studies, 728, 729
oculomotor system, 553
STP, 460
STS, 455
TE, 493
timing of information transfer: *see* Timing of information transfer
V1-V2 and MT connections, 363–365
Feedforward connections
architecture: *see* Architecture
development, 642
human studies, 728
STS to TPO subdivision, 462
timing of information transfer: *see* Timing of information transfer
FEF: *see* Frontal eye fields
Ferrier, David, 31–32, 34, 35, 36, 44, 48–49
Field discontinuities, 141–142, 152, 158, 178, 179, 180
Figure-ground segregation, 343–353
motion cues, 348–352
occlusion cues, 343–348
from occlusion cues, 346–348
stereoscopic cues, 352–353
Filimonoff, I. N., 674, 675, 679–680
First-order transformations, 179
MT, 150–151, 152
visual field, 140, 141
Fissures, 104
Fixation, FEF activity, 561
Fixation neurons, 514, 600
Flechsig, Paul, 36, 37, 42, 685
Flechsig's areas, 685
Flicker, 67, 73, 74, 423, 425
Fluorens, Pierre, 24–25, 29
Fluorescence studies
connectivity
architecture, 258
development, 642
V1 and V2 parallel processing, 321
Fluorescent labels, 244
Flying fox, 150, 156
Form, 677; *see also* Object vision; Shape
achromatopsia, 799
architectonic mapping of function, 677
cross-modal interactions, 300
dynamic analysis, 351
functional transformations, 301
human studies
area functional properties, 761
basic visual cues, 765–769
regional activation, 768
V1-V5 interactions in perception of form from motion, 789–791
V4, orientation selectivity and form analysis, 799–801
interleaved one step forward, two steps back mapping, 309
modular organization, 304, 305
pale stripes, 310
processing, degree of segregation of, 320–322
STP and, 443
V2 oriented cells, 300
Form-from-motion perception, 792
Fovea
magnification factor, 146
marmoset, 169
Foveal receptive fields, FEF pyramidal cells, 560
Foveal representations
human studies, 753, 755, 756
MST, 412
of MT, 361, 373
Freiburg group, 42
Freud, Sigmund, 33
Fringe area, 682
Fritsch, C. J., 28, 29, 30, 31
Frontal eye fields (FEF), 442, 443, 540–584, 548
area 7a connections, 502, 504
connectivity, 548–552
function in humans, 582–583
location
in humans, 545–548
in nonhumans, 542–545
maintenance of spatial constancy, 555, 557
oculomotor system, 540–542, 542
plasticity, extrastriate lesions, 657
saccade production, 558–569
motor signal, 559–563
superior colliculus, 566–569
trigger signal, 563–566
SEF connections, 587–589

Frontal eye fields (FEF) (*cont.*)
sensory function, 569–580
auditory responses, 569–571
saccade target selection, 571–580
slow pursuit eye movements, 580–582
timing of information transfer, 229, 232
fast brain, 229, 230
interconnections and synchronization of firing, 231
Frontal lobe/frontal cortex
comparison of visual areas for primate species, 686, 687
connectivity
area 7a, 503
intrahemispheric, 694
development, 664
connectivity, 643
myelination, 646
eye movement generation, 544
FEF in humans, 546, 547, 548
plasticity, experience/learning-induced, 659
Frontal lobe visuomotor areas, 110, 527–616
cingulate cortex, 614–616
frontal eye field, 540–584
connectivity, 548–552
function in humans, 582–583
location in humans, 545–548
location in nonhumans, 542–545
oculomotor system, 540–542
saccade production, 552–569
sensory function, 569–580
slow pursuit eye movements, 580–582
organization of frontal cortex, 529–530
postarcuate premotor cortex, 530–540
dorsal, 535–540
ventral, 531–535
prefrontal cortex, 604–614
dorsolateral, areas 46 and 9, 605–610
inferior orbital, area 13, 612–614
ventrolateral, area 12, 610–612
supplementary eye field, 584–601
connectivity, 585–594
neuronal properties, 594–601, 602
saccade production, 601, 603–604
supplementary motor area, 584–585
FST: *see* Fundal area of superior temporal sulcus
Fulton, John, 47–48, 49
Functional binding, 323
Functional mapping
comparison of primate species, 686, 687
FEF projection to superior colliculus, 566–569
human studies, 745, 747, 750
Functional models, 95, 96
Functional organization, *see also* Theories of visual cortex organization; Visuotopic organization
anatomic studies as key to, 727–729
Functional organization (*cont.*)
development, 664–665
frontal cortex, 529–530
in humans
convergence and divergence in, 769–770
cortical magnification factors, 758–761
early visual areas, 761–762
information processing, 762–769
monkey brain and, 747–749
specialization of visual cortex, 699–704
technique, 744–747
topographic organization, 752–758
visual cortical areas and visual cortex, 749
processing streams, 440–445; *see also* Dorsal pathway; Ventral pathway
structure-function correspondence: *see* Structure-function correspondence
V2, 335–354
Functional segregation
blobs-interblobs and stripes, 300–301
form domain, 300
Fundal area of superior temporal sulcus (FST), 52, 102, 115–116, 436, 442
connectivity, 366
FEF, 550, 551
processing streams, 366
SEF, 588
thalamic, 693
TPO, 441
dorsal subdivision of, 110, 115–116, 156, 157, 158
plasticity, 655
receptive field size, 138
STS, 438, 440, 450, 451
architectonic subdivisions, 449
frontal projections, 454
posterior parietal projections to TPO subdivisions, 458
prefrontal terminals, 455
theories of organization, 115–116
ventral, 115–116
Fusiform gyrus, 701, 702, 763, 764
achromatopsia, 701
architectonic mapping, 677
callosal connections, 713
color center, 801–802
connectivity, 710, 711, 713
optical flow perception in blind, 787

GABA, 184, 257
blobs, 705, 706
development patterns, 652
human studies, 725–726, 727
V2, 696, 697, 725–726
Gain field, 515, 516
Galago, 11
maps, 686, 687, 688
MT and V2, 105, 106, 107

Galago (*cont.*)
visuotopic organization, 143–144, 150
dorsomedial complex, 163, 164
interhemispheric connections, 171–172
lateral and ventral areas, 168
Galanin receptors, V2, 698
Galen, 10–11, 12, 13, 29–30
Gall, Franz Joseph, 16, 20, 23–24, 25–26, 28, 29, 30, 31, 63, 74
Ganglion cells, 60, 61, 86
Gap effect, 564
Gap task, 565
Gating
FEF, 577
preparatory set cell function, 598
Gaze, 531
FEF and, 548, 549
frontal cortex, 546, 547, 648
Gaze angle, 533, 535
Gaze field, 514
Geniculate, 62
connectivity, 367
LGN: *see* Lateral geniculate nucleus
MGN, 504, 710, 711
timing of information transfer, 217
Geniculocortical system, 253, 267–268
single axon analysis, 264
single axon spatial configuration, 250, 251
timing of information transfer, 215, 217
Geniculoextrastriate pathway development, 662, 663
Gennari, Francesco, 22–23, 743
Gennari, stripe of, 22–23, 36, 682
Geometry of arbors, 253
Giant axons, 206
Global maps, 181
Global model of visuotopy, 145
Global orientation, 436
Global processing, MT and V2, 108
Glutamate, 725
Golgi, Camillo, 60
Gorilla, 542
Gradient-like organization, 285
Gradients, visuotopic, 145, 176
Grandmother cells, 507
Granit, Ragnar, 40
Granule cells, 181, 679
Gratiolet, P., 36
Gratiolet's radiation, 36
Group selection theory, 181
Grusser, O. J., 42

Haller, Albrecht von, 20, 21–22, 25
Hand movement, cells responding to, 493
Hartline, Halden Keffer, 41, 61–62
Head, Henry, 75
Head movement
FEF connections controlling, 549
stabilization during, 368, 383
Head position, maintenance of spatial constancy, 553–559
Hemianopia, 32, 780–781
Hemifields, 93, 93, 94, 105, 106, 107
Hemispatial neglect, 500
Henschen, Salomon, 36
Herodotus, 3
Herophilos, 8–9, 10
Heterotopic callosal connections, 717
Hierarchy/hierarchical organization and processing, 62, 277, 279–284
architectural subdivisions of visual cortex, 675–680
area 7a connections, 503
cat cortical areas, 225
collateralized connections, 282–283
color constancy mechanism, 802–803
historical development of concepts, 675–682
human studies, 719–720, 721, 728, 729
inferior parietal lobule, 502
laminar patterns, 279–280
other connections, 283–284
reciprocity, 280–282
STS, 462
timing of information transfer, 230–231
High contrast stimuli, 68
High-order areas, 93–94, 436
architectonic mapping, 677
STP connections, 448–462
High-order processing
association cortex: *see* Superior temporal polysensory area
development, neurochemical markers, 653
receptive field properties, 314
Hippocratic doctors, 7–8, 16
Histochemical stains, architecture, 688–692
Histology, 93; *see also* Architecture
architectural subdivisions of visual cortex, 675–680
cytochrome oxidase, 296–297
History, 673–685
cyto-, myelo- and pigmentoarchitectonic parcellations, 682–685
discovery of visual center in cerebral cortex, 31–42
Ferrier versus Munk and Schaffer, 31–36
microelectrode studies, 40–42
neurophysiological studies, 38–40
Panizza, 31
striate cortex established as visual cortex, 36–38
early modern era, 23–31
Broca, 25–26
Fluorens, 24–25
Gall, 23–24
search for sensory areas, 29–31
Egypt, ancient, 2–4
extrastriate cortex, 42–51

History (*cont.*)
first hierarchical concepts, 675–682
Greek science, 4–11
medieval, 11–14
myelogenetic studies, 685
Renaissance to early modern era, 14–23
Gennari, 22–23
Swedenborg, 20–21
Vesalius, 14–15
Von Haller, 21–22
Willis, 16–20
STP region, 435–440
early revision of extent, 438–440
early studies, 437–438
early views on organization, 435–437
theories of organization
early maps, 94–98
early modern theories, 98–100
two or more cortical visual systems, 51–52
visuosensory versus visuopsychic regions, 674–675
Hitzig, E., 28, 29, 30, 31
Holmes, Sir Gordon, 37, 38, 49
Holmes' map, 752, 753
Homonymous hemianopia, 32
Homotopic callosal connections, 717
Horizontal connections, V2, 303
Horizontal meridian, 145, 158
field discontinuity, 180
human studies, 752
V2, 146–147
visual representation in dorsal V2, 306, 307
zero, 114
Horseradish peroxidase, 244, 245, 258, 560
Horseshoe pattern of parastriate, 685
5-HT receptors, 698, 699
Hubel, David, 1, 40, 41–42, 43, 62, 64, 98
Hubel-Wiesel model, 64, 65
Human cortex, 107, 181
architecture, 699–710
cyto- and myeloarchitecture, 704–706
cytochrome oxidase staining, 706–710
functional specialization, 699–704
areal organization of visual cortex, 436
associational cortical lesions, 499, 500
comparisons with nonhuman primates, 685–692
architecture, 688–692
connectivity, 692–695
maps, 686–688
transmitter receptors, 695–698
connectivity, 710–723
comparison with other primates, 692–695
interhemispheric, 713–717
intrahemispheric, 718–721
intrinsic, 721–723
thalamic, 710–713
visual areas as derived from callosal connections, 717–718

Human cortex (*cont.*)
extrastriate area equivalents, 686
frontal eye field
function, 582–583
location, 545–548
functional organization of visual areas
anatomic studies as key to, 727–729
convergence and divergence in, 769–770
cortical magnification factors, 758–761
functional propertied of early visual areas, 761–762
information processing, 762–769
monkey brain and, 747–749
technique, 744–747
topographic organization, 752–758
visual cortical areas and visual cortex, 749
historical concepts and early maps, 673–685
cyto-, myelo- and pigmentoarchitectonic parcellations, 682–685
first hierarchical concepts, 675–682
myelogenetic studies, 685
visuosensory versus visuopsychic regions, 674–675
inferior parietal lesions, 500–501; *see also* Inferior parietal lobule
maps
Brodmann's areas, 686, 688
comparison with other species, 686, 687
extent of areas, 689
MT and V2, 105, 106, 108
parietal lesions in, 49
plasticity, V1 lesion studies, 656
posterior temporal lobe, 462
transmitter receptors, 723–727
comparison with other primates, 695–698
in V2, 724–726
V4, 114
ventral premotor cortex, 531
Hume, David, 43
Huxley, T. H., 36
Hyperacuity, 181, 186
Hypercolumns, 64, 70–71, 313
Hypercomplex neurons, 62
Hypermetamorphosis, 45

ibn al-Haytham, 36, 37
Ice cube model, 64, 65
Illusory contour cells, 300, 301, 302
Illusory contours, 337, 343–346, 353
Illusory motion, 765
Illusory rotation, 701
Imhotep, 3
Inferior longitudinal fasciculus, 719
Inferior occipital gyrus, V4, 690
Inferior occipital lobe, 701
Inferior occipital sulcus, V2, 796

Inferior parietal (IP) lobule, 499–520
behavioral methods, 504–505
electrophysiological studies, 505–516
extraretinal signals, 514–516
visual properties, 505–514
macaque, connections in, 502–504
macaque model, 501–502
mathematical model for spatial representation, 516–519
neurological findings, 500–501
STS connections, 460
VIP, 103
plasticity, 655
theories of organization, 117–118
timing of information transfer, 213, 214
Inferior temporal cortex, 47, 64, 101, 102, 103, 173, 181, 362, 365, 367, 528
area 7a connections, 504
columnar organization, 469–495
binding activities in distant columns, 492–493
changeability of selectivity in adult, 489–490
face neurons in STSa, 493
object recognition by activities distributed over, 494–495
optical imaging of, 486–489
position, orientation, size and attributes defining contours, 473–476
TE, 477, 478, 479–480
TE, afferents to, 477, 480–485
TE, function in, 490–492
TE, stimulus selectivity of cells, 470–473
connectivity
FEF, 549, 550, 551
FSTv, 116
intrahemispheric, 694
pulvinar, 441
STS, 440, 443
thalamic, 692
development, 665, 666
connectivity, 642–643, 645
electrophysiological studies, 647, 648, 649, 650, 651
myelination, 646
neurochemical markers, 652–653
DLc and, 114–115
dorsal (ITCd), 116
hierarchical organization, 721
mapping, area definition problems, 362, 686
plasticity
experience/learning-induced, 658, 659
extrastriate lesions, 657
learning or experience-dependent, 659
reorganization after damage, 658
V1 lesion studies, 655
posterior: *see* Posterior inferotemporal area
Inferior temporal cortex (*cont.*)
preferential localization serotonenergic terminations, 285
prosimians, 173
STP and, 445
timing of information transfer, 229
visuotopic analyses, 135–136
Information processing: *see* Processing; *specific areas and functions*
Infragranular layer
collaterals, 247, 248, 250, 251, 252, 253
development, 642
pyramidal cell somas, 256
Inhibitory disparity cells, 298
Inhibitory neurons, response selectivity establishment, 232
Inhibitory postsynaptic potentials (IPSP), coincidence detection, 220
Inhibitory surround: *see* Antagonistic surround; Center-surround organization
Inouye, Tatsuji, 37
Input convergence: *see* Convergent input
Integration
intersensory: *see* Superior temporal polysensory area
visual, V1-V5 interactions, 789–790
Integration time
chronaxie measurement, 216
response selectivity establishment, 232
at resting membrane potential, 217–218
Interareal interactions
human studies, 728–729
parallel processing, 320–322
Interarea transfer times, 228
Interblobs, 94, 297, 300, 301, 305, 795
color signals, 795
connectivity, intrahemispheric, 693, 694
contour representation, 323
parallel processing of V1, V2, V4, and MT, 315
Interhemispheric connections: *see* Callosal connections
Interindividual variability
architecture, 680
in humans, 699, 700
human visual cortical area, 754, 755
Interlaminar S layers, 321
Interleaved architecture, 310, 309, 323
Intermediate fields (developmental stages), 685
Intersensory integration: *see* Superior temporal polysensory area
Interstitial nucleus of Cajal, 548, 587
Interstripes, 795
color vision, 796, 797
connectivity, intrahemispheric, 693, 694
Intraareal compartments, 705
Intraareal information processing, 211–212
Intraareal lateral interactions, 178

Intracortical/intrahemispheric connections, 314
DM, 693, 694
FEF, 549–551, 552
higher-order area connections, 729
human studies, 718–721, 729, 745
nonhuman primates, 693
SEF, 587
Intralaminar nuclei, 441, 549
Intrinsic connections, 275–277, 278, 279, 326
architecture, 275–277, 278, 279
human studies, 721–723, 724
Inverse summation test, 372, 383
IPa
FEF connectivity, 550, 551
MT and satellites, area definition, 362
STS
architectonic areas, 438
posterior parietal projections to TPO subdivisions, 458, 459, 460
prefrontal terminals, 455
timing of information transfer
latency, 229
IPL: *see* Inferior parietal lobule
IT: *see* Inferior temporal cortex, columnar organization; TE cytoarchitectonic area
IVC layer, development, 665

Jackson, J. H., 29
James, William, 33, 35
Jung, Richard, 42, 62
Juxtastriate areas, 98, 99

Kaes-Bechterew, stripe of, 681
Kainate, 695–697, 699, 725, 727
Kepler, J., 36
Kinesthetic fields, 97
Kinetic boundaries, antiphase, 392
Kinetic contours, MT, 703
Kinetic occipital area, 423
Kluver, Heinrich, 45–46, 75
Konorski map, 97–98
Koskinas, G. N, 677, 682
K system, slow brain, 233
Kuffler, Stephen, 41

Lagged inputs, slow brain, 233
Lamarck, J.-B., 26
Laminar boundaries, 250
Laminar patterns, 66, 76, 209
anterograde transport of MT-injected label, 364
architecture, 244
feedforward and feedback connections, 257, 258
hierarchy levels, 279
human studies, 719–720
STP, 461
terminations of horizontal collaterals, 277
V1-V2 border region, 724–725
Laminar position, and receptive field size, 137, 138
Laminar-specific differences in spatial mapping, 251
Laminar terminations, single axons, 245–250
Language, 531
Lashley, Karl, 46, 49, 75
Latency, 220–231
cat cortex, 224–225
direct measurement in visual cortex, 793–794
FEF, 577
antidromic activation, 560
pyramidal cell activation, 560
MT responses, 376
primates, 225–231
retina and LGN, 221–224
saccade, 564, 583
stimulus-evoked activity, 600–601
temporal scatter, 245
V1-V2, 219–220, 321
Lateral area visuotopic organization, 165–172
Lateral connections, 76, 244
hierarchy levels, 279, 728
human studies, 728
inter-areal interactions, 178
retina, 71
STS, 455
and visuotopic organization, 179
Lateral geniculate nucleus, 63, 66, 69, 367
color signals, 794, 795
connectivity, 692–693
development, 646
human striate cortex, 710, 711
retinogeniculate axons, 210–211
V5 input, 703, 783
development, 641, 646, 665
human studies, 748
motion perception, 764
striate cortex, 710, 711
inactivation studies, 320–321
plasticity
early lesion effects, 662
V1 lesion studies, 656
single axon analysis, 264–265
timing of information transfer, 215, 221–224, 225, 226
Lateral geniculate nucleus, dorsal
axon classes, 210–211
connectivity, 646, 692
development, 641, 646, 663, 665
oculomotor system, 541
plasticity
early lesion effects, 662
V1 lesion studies, 656
Lateral interparietal area (LIP), 117–118
area 7a inputs, 502
connectivity, 441
area 7a, 502, 503, 504
FEF, 550, 551, 552
SEF, 587, 588
thalamic, 692

Lateral interparietal area (LIP) (*cont.*)
fuscin staining, 684–685
macaque models, 501–502
maintenance of spatial constancy, 559
oculomotor system, 541
plasticity, 655
RF position shifts, 187
theories of organization, 117–118
timing of information transfer
fast brain, 230
latency, 229
Lateral occipital cortex, V4 correlates, 702
Lateral occipital sulcus, 727
Lateral subdivision of MST, 156, 157, 158, 361, 439
Lateral visual cortex myelination, 646
Lateromedial area (LM) feedback connections, 257
Laurens, Andre du, 13
Layers, 674; *see also* Architecture
architecture, 244
synaptic, timing of information transfer, 228
Learning-induced plasticity, 658–660
Leonardo da Vinci, 12
Leucoagglutin: *see Phaseolous vulgaris* leucoagglutin
Leviant illusion, 788, 791
Limbic structures, 97, 504
Limes parastriatus gigantocellularis, 679
Line orientation, 62
Lines defined by motion, 348–349. 350, 351
Lingual fusiform gyrus, 710
Lingual gyrus, 114, 763
achromatopsia, 701
color center, 801–802
LIP: *see* Lateral interparietal area
Lipofuscin, 684–685
Lissauer, H., 33
Locus coeruleus, 503, 504, 645
Luminance, 63, 70, 71, 72, 304
architectonic mapping, 677
contour definition, 339
cortical regions activated by, 768
2-D shape encoding in domains, 323
oriented lines, 789
Lunate sulcus, 272, 642, 796
Lungwitz, prestriate areas of, 682, 683–684

Macaques, 44, 101, 102, 104, 107, 110, 111, 112, 113, 114, 462; *see also* Cercopithecus
architecture, 690
intraareal compartments, comparative, 705
topography, 306
V2, 689
callosal axon size, 206
connectivity, 692–693, 722
development, 642, 645
interhemispheric, 206, 214–215
intrahemispheric, 719, 721
thalamic, 712
V2, 314–322
Macaques (*cont.*)
development, 642, 645, 647, 648, 649, 650, 651
dorsolateral cortex, 114, 115
FEF localization, 548
frontal cortex, 529
FST, 116
human comparisons
behavioral methods, 504–505
connections, 502–504
electrophysiological studies, 505–516
hierarchical organization and, 728–729
inferior temporal organization: *see* Inferior temporal cortex, columnar organization
intraareal compartments, 705
maps
comparison with other species, 686, 687
extent of areas, 689
MST, 727
neurotransmitters, 724
optical flow activation of dorsal MST, 701
point image size in, 181
prefrontal cortex, 609
saccade representation, 546
stripes, 147
timing of callosal information transfer, 214–215
upper temporal lobe areas, 116
V1-V2 border transmitter receptors, 695–698
V1-V5 interactions, 790
V2, 295–328, 796
connectivity, 314–322
cytochrome oxidase, 296–297
function of multiple processing stages, 326–328
modular organization within stripes, 303–305
receptive field properties, stripes and, 297–302
relationship of structure and circuitry to perception, 322–326
topography, 305–314
V4, 702, 802
V5/MT and V2, 105, 106, 107, 782
visuotopic organization, 128, 138, 144, 146
dorsomedial complex, 161, 163
lateral and ventral areas, 165, 168
MT, 151, 152–154
surrounding areas of MT, 155
TEO, 17
VP and VA, 116–117
Macula, location of, 753
Macular degeneration, 709–710
Macular representation, 762, 764
Magendie, Francois, 30n.
Magnetic resonance imaging (MRI), 699, 700, 700, 744, 744, 745, 746, 747, 749, 753, 765
FEF in humans, 546
motion-sensitive regions, 764–765
MT imaging, 421–425
residual vision studies, 784, 785, 788
V4v, 702
VP studies, 701

Magnification factors, 81, 148
  cortical, 142, 148, 758–761
  foveal, 146
  in V2, 311–312
  in V5/MT, 157–158
Magnocellular layers, 63
Magnocellular LGN, 66, 78, 79, 108, 782–783
Magnocellular MGN, area 7a connections, 504
Magnocellular-parvocellular pathways
  development, 665
  human studies, 728–729, 747–748
  inputs to V2, 336
  processing streams, 367
  segregation of inputs, 315–317
  single axon analysis, 264–265
  timing of information transfer, 223
Magnocellular ventroanterior nucleus, FEF connectivity, 549
Malphigi, Marcello, 15–16
Map discontinuities, 141, 142–143, 152, 162, 178
Maps, 81; *see also* Architecture; Theories of visual cortex organization; Visuotopic organization
  in humans, 745; *see also* Humans, functional organization of visual areas
  visuotopic, 131–133
Map singularities, 142
Marmoset, 112, 113
  FEF, 542
  maps, comparison with other species, 686, 687
  MT and V2, 106, 107
  receptive field segregation, 150
  transmitter receptors, V1-V2 border, 695–698
  visuotopic organization, 146
    dorsomedial complex, 160–161
    lateral and ventral areas, 165, 166, 167–168, 169–171
    MT surrounding areas, 155, 156–158
Mathematical model, spatial representation, 516–519
Matthews, Bryan, 40
McCulloch model, 96
M channel
  latencies, 226, 227
  slow brain, 233
  timing of information transfer, 230
MD: *see* Mediodorsal nucleus
Medial area (M), 99, 100, 102, 110, 111–113, 233
Medial geniculate nucleus (MGN), 504, 710, 711
Medial intraparietal area, 117
Medial longitudinal fasciculus, 548
Medial superior temporal visual area (MST), 52, 66, 103, 110, 115–116, 362, 442
  architecture, 692
  area definition, 362
  border with polysensory cortex, 439
  Brodmann's areas, 677
Medial superior temporal visual area (MST) (*cont.*)
  caudal, 366
  connectivity, 366
    area 7a, 502, 503
    FEF, 550, 551
    SEF, 587
    thalamic, 693
  convergence of inputs, 301
  dorsal, 361, 439, 701
    MT and satellites, area definition, 362
    optical flow activation of, 701
    STP organization, 436
    STS frontal projections, 454
  human correlate of, 727
  human studies, 703
  lateral subdivision, 156, 157, 158, 361, 439
  MT and satellites, area definition, 362
  optical flow activation, 701
  plasticity, 655
  processing streams, 366
  receptive field size, 138
  STS, 450, 451
    architectonic subdivisions, 449
    frontal projections, 454
    prefrontal terminations, 456
  theories of organization, 115–116
  timing of information transfer, 213, 214
    fast brain, 229
    interconnections and synchronization of firing, 231
    latency, 229
  TPO connections, 441
  transmitter receptors, 696
  ventral, 361, 362
  zone between cSTP, 460
Mediodorsal nucleus
  FEF connectivity, 549
  oculomotor system, 541. 542
Mediolateral gradient of visuotopy, 145
Memory, 531
  architectonic mapping of function, 677
  area 7a connections, 504
  human studies, 700
  prefrontal cortex, 608
Memory saccade task, 565
Mental imagery, 245
Mescaline, 45
Metabolic activity
  development of extrastriate cortex, 649, 652
  development studies, 649, 652
Metabolic mapping, 135, 745, 746
Meynert cells, 267
Microcircuitry, 244
Microelectrode mapping, 50–51, 104; *see also* Single-cell physiology; Single-unit electrophysiology
  DM, 111
  early, 40–42

Microelectrode mapping (*cont.*)
V3 and dorsal V3, 109
V4, 113–114
Microsaccadic eye movements, 187
Middle temporal area (MT)
posterior, 157
architectonic subdivisions, 449
connections, 366
STS prefrontal terminations, 456
V5: *see* V5/MT
Middle temporal crescent, 115–116, 155, 165, 166
Midget system, 62, 63, 66, 67, 68, 69, 70, 76, 79
Mill, John Stuart, 43
Minicolumns, 491, 492
Mirror-image representations, 100, 145, 152, 153
Missl stains, 710
Mixed visual field sign, 157
MLGN: *see* Magnocellular LGN
Modality specific areas, 436
Modular organization, *see also* Architecture
human studies, 728
TPO subdivisions as, 461
V2, 147, 148, 303–305
Monaminergic connections
development patterns, 652
numbers of synapses, 259
Morphological method, axon diameter-conduction delay estimation, 207
Morphometric analysis, 244
Mosaic transformations, 141, 143
Motion (object), 52, 62, 63, 65, 67, 70, 71, 73, 74, 76; *see also* Disparity
analysis of components of, 71
architectonic mapping, 677
coherent, lines defined hy, 348–349. 350, 351
contour definition, 339
development, 652
FEF pyramidal cells as, 560
figure-ground segregation cues, 348–352
functional interactions, specificity of, 317–318
functional segregation, 337
human studies, 701
area functional properties, 761
cortical regions activated by, 768
dynamic parallelism, 791–794
information processing in motion-sensitive regions, 764–765
motion vision, dissociation of, 779
MT, 703
residual vision, neural basis of, 781–787
scotoma for moving and stationary objects, 779–780
V1-V5 interactions in perception of form from motion, 789–791
V5/MT, 702
lines defined by, 337
MST, FST and, 284
Motion (object) (*cont.*)
MT and satellite receptive field, 67, 284, 376–388
contrast and wavelength, 387–388
direction, 376–383
general response properties, 367–376
in macaques, 405–412
orientation and disparity, 386–387
speed, 383–386
transparent stimuli, 392
optic flow selectivity, 510–514
processing, 66, 366
parallel in V1, V2, V4, and MT, 315
segregated, 316–317, 319
residual vision, 778, 779
conscious perception of, 78–781
scotoma for movement and stationary objects, 779–780
shape extraction from, 418–421
as spatial cue, 507–514
STP, 441, 444
structured, 512
thick stripes, 298
Motion: *see* Motor areas; Movements
Motion flow, 509–514
Motion onset latency, 376
Motor areas, 30, 527, 528, 530; *see also* Frontal lobe visuomotor areas
maps of visual cortical areas for primate species, 686, 687
supplementary eye field, 584–585
timing of information transfer, 218
Motor control, dorsal pathway, 492
Motor error, 553–554, 555, 558
Motor learning task, 596
Motor signal, FEF, 559–563
Motor system, 285
Mountcastle, Vernon, 50, 51
Movement cells, 560
Movements, *see also* Frontal lobe visuomotor areas
architectonic mapping of function, 677
dorsal premotor cortex and, 535
FEF connections controlling, 549
MT/V5 in macaques, 412–415
perfrontal cortex activity related to, 606
processing pathways, 748
ventral premotor cortex and, 531–535
MST: *see* Medial superior temporal visual area
MTc (middle temporal crescent), 115–116, 155, 165, 166
Muller, Johannes, 11, 29–31
Multimodal systems, *see also* Superior temporal polysensory area
in humans: *see* Humans, functional organization of visual areas
IPa and PGa, 550
STP, in infant macaques, 647

Munk, Hermann, 32–33, 34, 35, 36, 64, 94
Muscarinic receptors
  human studies, 725, 726–727
  V2, 696, 697–698
Myelin, 110, 179
  anterior bank of STS, 361
  architecture (myeloarchitecture), 271, 675–680, 683–684, 688–692, 745
    comparison of visual cortical areas for primate species, 686, 687
    CO staining comparisons, 707–708
    historical developments, 682–685
    human studies, 704–706
    mapping, 680–681
    OA, OB, OC area correspondences, 677, 678–679
  development, 646, 685
  feedback axons, 270, 271
  FEF connectivity, 550
  pulvinar, 710
  retinogeniculate pathway, 210–211
  timing of information transfer, 230
  upper temporal lobe areas, 115
  V2, 108
  V3, 690
  V4, 691
  V5/MT, 108, 360, 691–692
    area definition, 362
    in humans, 718
Myers model, 98, 99

Nauta method, 715, 720, 722
Nemesius, 13
Neocortex, 96, 284
  development, 653
  timing of information transfer, 218–219
Network size, V1-V2 interactions, 321–322
Neurochemical gradients, 285; *see also* Neurotransmitters
Neurofilaments, 276–277, 689
Neuronal group selection theory, 177
Neuronal specificity, 64
  models of, 60–63
  revisions and alternative models, 67–75
Neuron size: *see* Size, neuron
Neuropeptide Y receptors, 698
Neurophysiology, *see also* Electrophysiology; Neurotransmitters
  history, 38–40
  MT/V5 in macaques, 367–404
    antagonistic surround, 398–404
    general response properties, 367–376
    RF, responses to complex stimuli, 388–398
    RF, simple parameter selectivities, 376–388
  STP functional heterogeneity within upper bank of STS, 443–445
  structure-function relationship: *see* Structure-function correspondence
Neurophysiology (*cont.*)
  TPO subdivision physiological correspondences, 461–462
Neurotransmitters, 491, 723–727
  blobs, 705, 706
  development of, 652
  human studies, 745
    comparison with other primates, 695–698
    in V2, 724–726
  localization, 285
  V2, 689–690, 724, 725, 726
Newton, Isaac, 32, 33, 37
Nissl stain, 688–692, 724
NMDA receptors, 491, 699
NMR: *see* Magnetic resonance imaging
Nocturnal habit
  relative area occupied by visual functions, 686
  visuotopic organization for, 146
Nomenclature, 165, 191, 674
Noradrenergic neurons
  development of, 652
  localization, 285
  V2, 698, 725, 726
Nuclear magnetic resonance: *see* Magnetic resonance imaging
Nuclei, thalamic, 587, 646, 693, 710, 711, 712
  development, 646
  oculomotor system, 540, 541, 548
Nucleus of Darkschewitsch, 548
Nucleus X, 586, 587
Numbers/numerals, 677
Numbers of neurons, V1, V2, V4, 259
Numbers of synapses, and connectional strength, 259

OA (area peristriata), 43, 674, 677, 678–679
OB (area parastriata), 43, 674, 677, 678, 679, 682, 713
Object-centered representation, 600
Object vision, 436; *see also* Faces; Inferior temporal cortex, columnar organization; Shape
  activities distributed over brain, 494–495
  architectonic mapping of function, 677
  connections, 366
  face neurons, 493
  FEF connections controlling, 549
  human studies, 748
  IT
    position, orientation, size and attributes defining contours, 473–476
    stimulus selectivity of TE cells, 470–473
  ventral pathway, 366, 528
  ventral subsystem, 768
Oblique motion, 781, 785
OC (area striata), 674, 677, 678, 679

Occipital cortex, 35, 51, 75, 77, 83, 84; *see also* Extrastriate cortex; Striate cortex
 area 7a connections, 503
 comparison of visual areas for primate species, 686, 687
 functional areas, 95
 human studies, 752
  color processing areas, 763
  connections, 713–718, 722
  extent of visual cortex, 750
 lateral, 702
 models of functional areas, 95, 96
 tripartition of, 675–682
 ventrolateral cortex borders, 169
Occipital parakoniocortex, 681–682
Occipitoparietal (dorsal) pathways: *see* Dorsal pathway
Occipitoparietal sulcus, 767
Occipitotemporal connections
 development, 642
 gradient-like organization, 285
 nonadrenergic terminations, 285
Occipitotemporal cortex
 comparison of primate species, 686, 687
 human studies
  complex object/form processing, 766
  higher-order area connections, 729
Occipitotemporal (ventral) pathways: *see* Ventral pathway
Occipitotemporal region, 48
Occipitotemporal sulcus, 249
Occipitotemporal transition (TEO), 48, 135, 173
 afferents to TE, 480–481, 484
 anterograde label of V2 feedback axons, 273
 arbors, 253
 architecture
  laminar patterns, 280
  reciprocity, 282
  secondary bypass projections, 282
  single axon terminations, 246
  single axon terminations in TEO, 249
 development, 642, 643, 645, 646, 663, 665
 FEF connectivity, 551, 552
 human correlates in lateral occipital cortex, 702
 human studies, 701
 MT and satellites, area definition, 362
 plasticity, V1 lesion studies, 655
 pulvinar connections, 441
 SEF connectivity, 588
 TE connections, 477, 480–485, 493
 thalamic connectivity, 692, 693
 timing of information transfer, 228, 230
 V1 and V2 sources of common input, 321
 ventral pathway, 469; *see also* Ventral pathway
 visuotopic organization, 172–173, 174–175
Occlusion, figure-ground segregation cues, 346–348
Ocular dominance columns, 81, 251, 304
 blobs, 705, 706
 interleaved maps, 309, 310
Ocular dominance hypercolumn, 313
Oculomotor signals, *see also* Frontal lobe visuomotor areas
 feedback control, 553
 FEF connectivity, 540–542, 549, 562
 inferior parietal lobule, 514–516
OFF response, latency scatter, 226–227
OFF system, 62, 68, 69, 70
One-area, one-function hypothesis, 64, 76
One step forward, two steps back mapping, 309
ON/OFF system, 62, 69
ON response, latency scatter, 226–227
ON system, 62, 68, 69, 70
Operculum, 710
Opiate receptors, 698
Opponent vector neurons, 508–509
Optical flow
 conscious perception of in blind, 786–787
 human studies, 701
 inferior parietal neuron response, 510–514, 515–516
 MT and satellite responses, 391, 393–398
 perception of in blind, 786–787
Optical imaging, 304, 308, 311
 human studies, 745
 IT columnar organization, 486–489
Optic chiasm latencies, 223
Optic flow components (OFC), 393
Optic nerve single cell recordings, 61–62
Optic radiation, 213
Optokinetic nystagmus (OKN), 368, 415, 418
Orangutan, 542, 679, 688
Orbitofrontal area
 development, neurochemical markers, 653
 timing of information transfer, 229
Organology, 24n.
Orientation, 69, 70, 71, 72, 75, 311
 color and, in V1, 798–799
 functional segregation, 337
 MT, 782
 parallel processing in V1, V2, V4, and MT, 315
 STP selectivity, 443, 444
 V4 and, 799–801
Orientation columns, 305
Orientation-selective neurons, 41–42, 64
 broadband cell pairs, 317
 form analysis, 799–801
 MT and satellite receptive field, 386–387
 multiple maps, 307
 STP, 443, 444
 V3, 690, 789
 V4, 690
 wavelength of oriented line, 795
Oriented broadband cells, 307, 308
Oriented cells, 300, 302, 317–318

Oriented color cells, 308
Oriented lines, 789
Orofacial representation, 531
Orthodromic action potential, 223, 265–266
Overlapping receptive fields, 142
Overlapping representation, TE columnar organization and, 492
Owen, Sir Richard, 15, 36
Owl monkey, 98, 99, 100, 103, 110, 111, 112, 113, 439; *see also Aotus*
  architecture, topography, 306
  comparison with other species, 686, 687
  dorsolateral cortex, 114
  dorsomedial complex, 158–160, 163
  FEF, 542, 545
  lateral and ventral areas, 165, 166
  MT, 105, 106, 107, 132, 150, 359, 782
  plasticity, 658
  upper temporal lobe areas, 115
  visuotopic organization, 128, 146, 171
  V2, 105, 106, 107
  VP and VA, 116–117

Pale stripes, 304; *see also* Stripes
  cortical magnification, 311–312
  form processing, 310
Panizza, Bartolomeo, 31
Parahippocampal gyrus, area 7a connections, 504
Parahippocampal region
  development of connectivity, 643
  polysensory cortex, 436
Paralaminar nucleus, FEF connectivity, 549
Paralimbic areas, development, 653
Parallel activation and plasticity of extrastriate areas, 663
Parallel processing, 326, 327
  human hierarchical organization, 721
  in humans, 767
  human studies, 728
  V1, V2, 320–322
  V1, V2, V4 and MT, 315
Parameter selectivity: *see* Selectivity; *specific parameters (color, direction, etc.)*
Parasol system, 62, 63, 66, 67, 68, 69, 70, 76, 79
Parastriate area, 43, 95, 674
  architectonic studies, 684–685
  models of functional areas, 95
  OA, OB, OC area correspondences, 677, 678–679
Parietal association cortex, 48–51
Parietal cortex, 44, 51, 79, 80, 97, 423, 495; *see also* Inferior parietal lobule
  area 7a connections, 503
  comparison of visual areas for primate species, 686, 687
  connectivity
    FEF, 549, 550, 551
    intrahemispheric, 694
Parietal cortex (*cont.*)
  development
    myelination, 646
    neurochemical markers, 653
  dorsal pathway: *see* Dorsal pathway
  human studies, 768
    callosal connections, 713–718
    color processing areas, 763, 764
    extent of visual cortex, 750, 751
    higher-order area connections, 729
    residual motion, 786
  theories of organization, 117–118
  timing of information transfer, 230
  visuotopic remnants, 136
Parietal neglect syndrome, 500
Parietal operculum, 710
Parietal processing, STP and, 440
Parietooccipital area (PO), 111–112, 161, 177
  anterior, 103
  connectivity, 722
    FEF, 550, 551
    SEF, 588
    thalamic, 692
  hierarchical organization, 721
  human studies, 701, 721
    areas activated by different visual cues, 768
    complex object/form processing, 766
  intrinsic connections, 724
  mosaic organization in, 143
  PIP, 117–118
  plasticity, 655
Parietooccipital sulcus, 423, 767
Pars reticulata, 600
  oculomotor system, 542
  SEF connectivity, 586, 587
Parvalbumin, 448, 449, 450–451, 452, 453, 688, 691, 692
Parvocellular area, 63
Parvocellular deprivation, 709–710
Parvocellular input, development, 665
Parvocellular LGN, 66, 78, 79, 795
Parvocellular mediodorsal nucleus, FEF connectivity, 549
Parvocellular pathways: *see* Magnocellular-parvocellular pathways
Patches
  functional interactions with V1 and V2, 315
  intrinsic connections in humans, 722, 723
  size of, 253
  stripe connections, 303
  termination of pyamidal neurons, 277, 278
  V5/MT, 691–692
Pattern, 67, 78
  MT and satellite responses, 388–390
  slow brain, 233
  STP and, 443
Pattern-direction selective MT neurons, 388–390
P channels, latencies, 226, 227

Perception/perceptual processes, 86, 748, 768
color system, 794–803
carbon monoxide poisoning, 797–799
color pathways, 795–797
lesions, effects of, 802
loss of color constancy mechanism, 802–803
V4, orientation selectivity, and form analysis, 799–801
dynamic parallelism, 791–794
motion: residual vision, 778–787
characteristics of, 779–781
neural basis, 781–787
V1-V5 connections, 787–791, 791–793
visual consciousness as distributed process, 803–804
Perceptual binding, 318, 322–323, 324
Peripheral representation, human studies, 753, 755
Peripheral vision, 81, 111
Perirhinal complex, 469, 495, 644
Peristriate areas, 95
models of functional areas, 95
OA, OB, OC area correspondences, 677, 678–679
Peristriate cortex, 43
PG, 503, 692; *see also* Area 7a
PGa
FEF connectivity, 550, 551
STS architectonic areas, 438, 449
PGm, 503
PHA-L: *see Phaseolous vulgaris* leucoagglutin
Phaseolous vulgaris leucoagglutin, 207, 244, 249, 252, 257, 266, 267, 269, 272, 273, 274, 483
Photopic vision, 67
Physiological neuronography, 96
Physiological studies: *see* Electrophysiology; Neurophysiology
Piccolomini, Archangelo, 15
Pigmentoarchitectonic studies, 684–685
PIP (posterior intraparietal area), 117–118
Piriform cortex, 277
PIT: *see* Occipitotemporal transition; Posterior inferotemporal area
Plaids, 388–390
Plasticity
in adulthood, 653–660
damage-induced, 654–658
experience or learning-induced, 658–660
recovery after injury or damage, 665–666
retention of infant pattern by high-order elements, 653
developmental changes, 652, 660–662
Plasticity of representation, 186
Plasticity of RF, 188–189
Plato, 6–7
PLGN: *see* Parvocellular LGN
PMLS
development, 663
timing of information transfer latency, 225
PO: *see* Parietooccipital area
Point image
versus point set, 312–313
size of, 136–140
Point image size, 187
Point set, 139, 312–313, 313
Polar angle, 154, 756
Polarity, MT and satellite receptive field, 379
Polymodal neurons: *see* Multimodal systems
Polysensory areas, 136, 439; *see also* Superior temporal polysensory area
Pons, FEF connectivity, 548, 561, 562
Pontifical neuron, 62, 64
Pontine reticular formation
FEF connectivity, 548
saccadic eye movement, 540
SEF connectivity, 587
Population, neuronal, 259
Population coding, 183, 494
Poseidonus, 13
Positron emission tomography (PET), 419, 744, 745, 746, 747, 749, 750, 753, 765, 766
connectivity studies, 710
FEF, 546, 547, 583
form-from-motion study, 792
motion-sensitive regions, 764–765
MT/V5, 702
plasticity
early lesion effects, 662
V1 lesion studies, 656
residual vision studies, 783–784, 788
ventral premotor cortex, 531
VP studies, 701
Postarcuate cortex
FEF connectivity, 549
movements elicited by sensory stimuli, 597
premotor, 530–540
dorsal, 535–540
ventral, 531–535
Posterior association cortex, 44
Posterior cingulate, area 7a connections, 503
Posterior fusiform gyrus, 710
Posterior inferotemporal area (PIT), *see also* Occipitotemporal transition
human correlates in lateral occipital cortex, 702
STS
architectonic subdivisions, 449
posterior parietal projections to TPO subdivisions, 458
prefrontal terminals, 455
Posterior intraparietal area (PIP), 117–118
Posterior parietal area (PP), 99, 115, 117–118
FEF connectivity, 549, 550, 551
lesions, 49, 50
polysensory cortex, 436, 439, 445–446
projections to TPO subdivisions, 460
pulvinar connections, 441

Posterior parietal area (PP) (*cont.*)
STP and, 443, 445
STS fundus connections, 440
visuotopic organization, 159
Posterior parietal area 7a, 181
Posterior temporal cortex
humans, 462
maps, definition problems, 686
Posteromedial (PM) area, 169, 170
Postsynaptic connections
laminar patterns, 280
single axon terminations, 248
Postsynaptic population, 254–258
Pourfour du Petit, Francois, 19–20, 21, 22
PP: *see* Posterior parietal area
Prearcuate FEF, 548
Precentral gyrus, FEF in humans, 546
Precuneus, 701
Prefrontal cortex, 495, 527, 604–614
area 7a, 502
dorsolateral, areas 46 and 9, 605–610
FEF connectivity, 549
inferior orbital, area 13, 612–614
projections to superior temporal cortex, 462
temporal modulation in, 598
TE projections, 469
timing of information transfer, 229
ventrolateral, area 12, 610–612
Prelunate gyrus, 321
Prelunate sulcus, 503
Premotor cortex, 527, 528, 530–540
dorsal, 535–540
dorsal prearcuate (PMd), 609
and gaze behavior, 548
movements elicited by sensory stimuli, 597
temporal modulation in, 598
ventral, 531–535
Preoccipital area (OA 19), 675
Presaccadic movement cell, 560
Prestriate areas, 144
FEF connectivity, 545
lesions, 46
of Lungwitz, 682
subdivisions, 682
Presubiculum, 675
Pretectum, 69, 367
Pribham, Karl, 46, 47, 50
Primary motor cortex, 527, 530
Primary sensory areas, 436
Primary visual cortex: *see* V1
Primordial fields, 685
Probabilistic sampling, 186
Probability fields, 185
Processing, 69, 95, 528
bidirectional flow in slow brain, 233
computational architecture: *see* Computational architecture
connection systems in, 283–284
Processing (*cont.*)
dorsal and ventral streams, 51; *see also* Dorsal pathway; Ventral pathway
early models, 94–96
human studies, 728, 729, 762–769; *see also* Humans, functional organization of visual areas
intra-areal, 211–212
in MT-V5 and satellites: *see* V5/MT, processing in macaques
multiple stages, function of, 326–328
parallel: *see* Parallel processing
selectivity of response, establishment of, 232
serial, 326
arbor mapping and modeling, 266
cat cortical areas, 225
timing of information transfer, 230
V2, 320–322
visuotopic organization, 136
Processing fields, 95–97
Processing streams, 51–52; *see also* Dorsal pathway; Ventral pathway
FEF connectivity, 549–550
human studies, 748, 767–769
segregated, 316–317, 549–550
STP region, 440–445
behavioral studies after ablation of STP cortex, 442–443
dorsal stream cortical connections, 440–441
neurophysiological evidence for functional heterogeneity within upper bank of STS, 443–445
pulvinar connections, 441–442
STS and connections, 366
Progressions, receptive field, 309
Projection focus, divergent feedback connections, 282
Projection neurons, 268
collateralization, 283
numbers of, 259, 260
Projection patterns, 98, 99
Proprioception
area 7a connections and, 504
stabilization during motion, 368, 383
Psychic blindness, 33, 45, 51, 94
P system, slow brain, 233
Pulvinar, 320, 321, 367
connectivity, 692
area 7a, 503
cortical connections, 259
development, 645
FEF, 549
patchy, 303
STP, 441–442
development, 640–641, 645
human studies, 710, 711, 712
oculomotor system, 541, 542
plasticity, V1 lesions, 656–657

Pursuit cells, 373, 412–418
Pursuit eye movements, 361, 553, 580–582, 657
Putamen, 593
PX, 709
Pyramidal cells
  architecture, 257, 267
  area OB, 679
  callosal integrity and, 713
  collaterals, 277, 491
  connections, 254–256, 257, 491
  extrastriate cortex, 675
  FEF, 545, 560
  modules, 181, 182, 185
  neurochemical gradients, 285
  terminations, 279
  total number of synapses, 259
  V2, 689
Pyramidal tract
  antidromic latency, 211–212
  timing of information transfer, 218

Quasi-hierarchy, 279

Radial model, 65
Radial motion, 508–509
Raphe nuclei, 645
Rat, 75, 147, 257, 491
Readiness potential, 596
Reading, 677
Receiver operating characteristics (ROC), 578–579
Receptive field, 41, 71
  excitatory, visuotopic order and, 136
  FEF pyramidal cells, 560
  MT and satellites, 371–372, 373
    antagonistic surround, 398–404
    complex stimuli responses, 388–398
    contrast and wavelength, 387–388
    direction of motion, 376–383
    optic flow stimuli, 391, 393–398
    orientation and disparity, 386–387
    plaid stimuli, 388–390
    speed of motion, 383–386
    transparent stimuli, 390–391, 392
  MT/V5 in macaques
    responses to complex stimuli, 388–398
    simple parameter selectivities, 376–388
  parallel progressions, 308
  point image versus point set, 312–313
  position, dynamics of, 186–189
  progressions, 308, 309
  at stripe borders, 309
  topography, 307
  types of, 301, 302
  V2 in macaques, 297–302
  visual pursuit cells, 373
Receptive field eccentricity, 341, 543
Receptive field heterogeneity, 182, 183, 184, 185, 186
Receptive field overlap
  MT, 373–374
  specficity of functional interactions, 317–318
Receptive field plasticity, 188–189
Receptive field polar angle, 169
Receptive field scatter, 154, 185
  and point image size, 136–140
  and spatial resolution, 181–186
  topography, 307
Receptive field shaping, 184, 185
Receptive field shifting, 187–188
Receptive field size
  local lesioning effects, 189, 190
  MSTd, 361
  MT, 374
  MT and satellites, area definition, 362
  and point image size, 136–140
  and spatial resolution, 181, 181–186
  V2 versus V1, 311
Receptive field spatial heterogeneity, 185
Receptor architecture, 745
Receptor mapping: *see* Neurotransmitters
Reciprocal asymmetry, V1-V2 and MT connections, 363–364
Reciprocal connections, cat cortical areas, 225
Recognition: *see* Inferior temporal cortex, columnar organization
Regional cerebral blood flow: *see* Cerebral blood flow
Reisch, G., 12, 14
Relays, 228
Renaissance, 12, 37
Repetitive transformations, 141, 143
Residual vision, 778–787, 778
  characteristics of, 779–781
  neural basis of residual motion vision, 781–787
Resolution limits, 181, 182
Response selectivity, 232
Reticular formation, 587
  development, 645
  FEF connectivity, 548
  saccadic eye movement, 540
Reticular nucleus, 693
Retina, 68–69, 69, 71, 783
  connections of, 60–63, 66
  superior collicus inputs, 541
  timing of information transfer, 221–224
Retinal disparity, 298
Retinal error, 553
Retinal ganglion cells, 62
Retinocentric representations, FEF, 557
Retinogeniculate pathway, axon classes, 210–211
Retinothalamo-extrastriate projections, 657, 662
Retinotopy, 91, 98
  area 7a neuron, 515, 516, 519
  DI, 113
  FEF innervation, 550
  human studies, 727, 752

Retinotopy (*cont.*)
MT, 373
MT and V2, 106
TE, 493
V3, 690
V4, 113–114
Retroinsular area, 423
Rhesus monkey, 690, 713; *see also* Macaque
Rhinal sulcus, 469, 495, 644
Riddoch, George, 778
Rindenblindheit, 34
Rods, 60, 61, 68, 86
Rostral DL, 114–115
Rostral STS, 366
Rostral superior colliculus, 600
Rostral TPO (TPOr), 436, 445, 446, 449, 451, 452, 458
Rotation, 701
Rotation (R) cells, 391, 393–398
Ruysch, Frederik, 16

Saccades, 187, 556; *see also* Frontal lobe visuomotor areas
FEF, 546, 553
frontal cortex areas, 544
frontal eye field, 552–569
oculomotor system, 540–541
SEF and, 589–594, 601, 603–604
STP role, 442–443
supplementary eye field, 601, 603–604
target selection, 571–580
ventromedial and dorsolateral representation, 545
Samiri, 107; *see also* Squirrel monkey
Sampling fallacy, 78
Scanning eye movements, 573, 574, 575, 576, 577
Schäfer, Edward Albert, 34–36, 49
Scotoma
moving and stationary objects, 779–780
residual motion vision, 781
Scotopic vision, 67, 78
Search condition, 573, 574
Sechenov, I. M., 38
Secondary projections, 282, 284
Secondary sensory areas, 436
Second-order transformations, 158
split horizontal meridian in V2, 307
ventrolateral cortex, 169
visual field, 140, 141–142, 145
Second visual area: *see* V2
Seelenblindheit, 33
SEF: *see* Supplementary eye field
Segregate and generalize strategy, 323
Segregated processing
degree of strictness, 316–317
FEF connectivity, 549–550
Selection by attention, 492
Selective covert attention, 186
Selectivity
antagonistic surround, 401, 402, 403, 404
changeability of in adult, 489–490
color, 317–318
direction, *see* Direction selectivity
feature, 48, 232, 493
MT/V5 in macaques, 376–388
orientation: *see* Orientation-selective neurons
TE cells, 470–473
Selectivity index (SI), 393
Self-organizing feature algorithm, 180, 181
Sensory areas, 29–31, 45, 674–675, 744; *see also* Parietal cortex; Superior temporal polysensory area; *specific sensory functions*
congruent borders, 93
in humans: *see* Humans, functional organization of visual areas
integration of sensory information, 436
somatosensory area, 49, 95, 97
organization of, 190
reorganization after damage, 657–658
Sensory function, frontal eye field, 569–580
Sequential behavior, frontal cortex role, 609–610
Serial activation, arbor mapping and modeling, 266
Serial flow of information, 326
Serial organization
timing of information transfer, 230
V2, 320–322
Serial processing, cat cortical areas, 225
Serotonin
development patterns, 652
localization, 285
V2, 689, 698, 725, 726
Shape, 66, 67, 76, 78; *see also* Form; Object vision
architectonic mapping, 677
extraction from motion, 418–421
human studies, 701
Shape, receptive field, 182, 183, 184, 185
Silent region, classical receptive field: *see* Antagonistic surround
Silver degeneration technique, 243
Silver stain, 60
Single-arbor analysis, 253, 260
Single-axon analysis, 264–267
Single-cell recordings, 44, 64
human studies, 745, 747
IT columnar organization, 486, 487
IT cortex, 48
optic nerve, 61–62
Single-step saccade test, 556
Single-unit electrophysiology, 304
anterior temporal cortex development, 647
motor error, 558, 559
patterns of functional interactions, 315
Singularities, map, 142

Size, neuron
  geniculocortical axons, 267–268, 269
  layer 6 variability, 273
  minimizing, 206–208
  timing of information transfer, 223
S layers, interlaminar, 321
Slow brain, 233
Slow pursuit eye movements, frontal eye field, 580–582
SMA
  movements elicited by sensory stimuli, 597
  SEF connectivity, 588
SMI-32, 448, 449, 450, 451, 452, 453, 689, 690, 691
Smith, G. Elliott, 36, 43, 95, 96, 674, 675, 676, 680
Smith Papyrus, 1–3
Smooth pursuit, 442, 444, 504
Somatosensory area, 49, 95, 97
  organization of, 190
  reorganization after damage, 657–658
Somatosensory band, models of functional areas, 95
Somatostatin receptors, 698
Spatial analysis, 66, 652
Spatial configuration of single axons, 250–253
Spatial constancy, maintenance of, 553–559
Spatial detail, 181–182
Spatial distribution, frontal cortex neurons, 552
Spatial filters, 69
Spatial frequency
  midget system, 62, 67, 68, 69, 70
  motion direction tuning, 380–381
Spatial heterogeneity, receptive field, 182, 183, 184, 185, 186
Spatial interpolation, 186
Spatial mapping, computer injected current, 265–266
Spatial representation, inferior parietal lobule, 516–519
Spatial resolution
  receptive field size and scatter, 181–186
  RF size and, 181
Spatial vision, 51
  architectonic mapping, 677
  connections, 366
  dorsal pathway, 528
  FEF connections controlling, 549
  human studies, 701, 748, 767–769
  STP and, 443, 444
  visual cues, 505–514, 767–769
Spatiotemporal activation, arbor mapping and modeling, 266–267
Specificity
  areal, 63–67, 75–80
  axonal conduction time, 209–210
  neuronal, 60–63, 64, 67–75
Speed of motion, MT and satellite receptive field, 379, 383–386
Spencer, Herbert, 29
Sperry, Roger, 41
Spike-triggered averaging, 208, 211–212, 219
Spiny stellate cells, 267
Spiny terminations, 270, 271
Spiral motion, 396
Spot cells, 299, 301, 302, 308
Spurzheim, John Gaspar, 24, 28, 29, 63, 74
Squirrel monkey, 320, 439; *see also* Samiri
  architecture
    intraareal compartments, comparative, 705
    topography, 306
    V2, 688–689
  dorsolateral cortex, 114
  FEF, 542
  intraareal compartments, 705
  maps, comparison with other species, 686, 687
  visuotopic organization, ITc, 173
Stationary stimuli, visual cues, 506–507
Stellate cells, 267
Stereopsis, 63, 65
  figure-ground segregation cues, 352–353
  human studies, 765
  PLGN lesions and, 316
Stereoscopic cues, figure-ground segregation, 353
Stereoscopic depth perception, 69, 70–71
Stereotyped circuitry, 326
Stereotyped functional transformations, 314
Stimulus selectivity: *see* Selectivity; specific parameters
STP: *see* Superior temporal polysensory area
Stria Gennari, 743
Striate area, *see also* OC; V1
  lipofuscin stained structure, 685
  models of functional areas, 95
  OA, OB, OC area correspondences, 677, 678–679
Striate cortex, 35, 40, 41–42, 43, 51, 68; *see also* OC; V1
  callosal connections, 98
  development, metabolic studies, 649
  early microelectrode studies, 40–42
  early neurophysiological studies, 38–40
  establishment as visual cortex, 36–38
  human studies, 752–753, 752, 753
    disparity information, 765
    motion perception, 764
  maps, comparison with extrastriate cortex area, 686
  models of organization, 64–66
  MT input, 365–367
  residual vision, 781
  STP and, 441–442
Striate receptive area, 98
Striatum
  SEF connectivity, 585–586
  TE projections, 469

Stripe of Baillarger, 680, 682, 691
Stripe of Gennari, 22–23, 36, 682
Stripe of Kaes-Bechterew, 681
Stripes, 147–150, 297, 310
  color vision, 796, 797
  contour representation, 323
  functional segregation, 300–301, 337
  hierarchical organization, 721
  inputs to, 335–336
  interleaved one step forward, two steps back mapping, 309
  MT connections, 365
  multiple maps, 307, 309
  parallel processing of V1, V2, V4, and MT, 315
  parallel receptive field progressions, 308
  receptive field progression at borders, 309
Structure: *see* Architecture
Structure-function correspondence, 267, 727–729
  FEF, 560, 562
  histological features and, 297
  IT
    TE columnar organization, 486–489
    TPO, 461–462
  V2 in awake monkey, 338, 339, 340, 341
  V2 in macaque, 322–326
    contour perception, 323, 325–326
    perceptual binding across modalities, 322–323, 324
  visuotopic organization hypotheses, 176, 177
Strychninization, 96, 97
STS: *see* Superior temporal sulcus
Subangular gyrus, 685
Subcortical area, myelogenesis, 685
Subcortical connections
  area 7a, 502–504
  development, 640–641
  FEF, 548–549
  SEF, 585–587
Substantia nigra, 600
  FEF connectivity, 549
  oculomotor system, 541, 542
  SEF connectivity, 586, 587
Subsystems, 51–52
Subunit processing, 63
Superior colliculus, 66, 69, 365–366, 367, 600
  development, 646
  FEF connectivity, 549, 560, 562, 566–569
  FEF pyramidal cell activation, 560
  oculomotor system, 542
  plasticity, 656
  saccade generation, 541, 553, 554–555, 557
  SEF connectivity, 586, 587, 588, 601, 603
  STP and, 441–442
  timing of information transfer, 215
  V5 input, 783
  visual afferents and visual inputs, 541
Superior longitudinal fasciculus, 718–719
Superior parietal lobule, 701
Superior temporal cortex
  connectivity
    development, 642
    intrahemispheric, 694
    projections from prefrontal cortex, 462
  medial: *see* Medial superior temporal visual area
Superior temporal gyrus, projections to TPO subdivisions, 459, 460
Superior temporal motion area, 151
Superior temporal polysensory area, 52, 136, 181, 435–462
  anterior, 436, 447–448, 503
  area 7a connections, 503, 504
  caudal (CSTP; TPOc), 362, 436, 445, 446, 456, 458
    architectonic area revision, 448–449, 450
    TPO correspondences, 460
  definition, location, and extent of, 435–440
    early revision of extent, 438–440
    early studies, 437–438
    early views on organization, 435–437
  development
    connectivity, 642
    electrophysiological studies, 647, 648, 651
  medial area: *see* Medial superior temporal visual area
  plasticity, 654, 655, 656
  posterior, 436, 447–448
  processing streams, 440–445
    behavioral studies after ablation of STP cortex, 442–443
    dorsal stream cortical connections, 440–441
    neurophysiological evidence for functional heterogeneity within upper bank of STS, 443–445
    pulvinar connections, 441–442
  revision based on chemoarchitecture and connections with high-order areas, 448–462
    architectonic subdivisions of upper bank and fundus of STS, 448–452
    TPO as separate cortical areas, 461–462
    TPO connections with high-order association cortex, 452–460
    TPO subdivision physiological correspondences, 461–462
  SEF connectivity, 588
  subparcellation of, 445–448
    caudal and rostral STP, 445, 446
    PP and STP proper, 445–446
    STPp and STPa, 447–448
    TPO connectional and architectonic subdivisions, 446–447
  visual cortical inputs to, 440–441
Superior temporal projection zone, 115

Superior temporal sulcus (STS), 359, 360, 366, 436
anterior bank of, 361, 493
architectonic mapping, 677
area 7a connections, 503
caudal, 361, 439
development
connectivity, 642
metabolic studies, 649
face-selective cells, 232, 493
processing streams, 366
SEF connectivity, 587
thalamic connectivity, 692
timing of information transfer, 229
V4, 690
Supplementary eye field (SEF), 530, 541, 544, 547, 584–601
connectivity, 585–594
FEF connectivity, 549, 552, 587–589
neuronal properties, 594–601, 602
saccade-related activity, 594–600
stimulus-evoked activity, 600–601, 602
oculomotor system, 542
saccade production, 601, 603–604
stimulus-evoked activity, 600–601, 602
supplementary motor area, 584–585
Supra-areal topography, 143, 179
Supragranular layers, latency, 224
Supragranular projections, connectivity, 642
Supramodal spatial deficit, 50
Surround inhibition: *see* Antagonistic surround; Center-surround organization
Swedenborg, Emmanuel, 20–21
Swirl model, 65–66
Synaptogenesis, 646, 652
Synchronization, 245, 492
arbor mapping and modeling, 266
hierarchy frameworks and, 284
timing of information transfer, 231
Synchronized firing, coincidence detection, 220
Synthetic areas, 181

TA, 229
TAa, 438, 449
Tactile learning deficit, 50
Talairach stereotactic system, 750
Talbot, S.A., 40, 44, 64
Tapetum, 713
TAs, 458
TE (cytoarchitectonic area), 362, 449, 458, 477, 478, 479–480; *see also* Inferior temporal cortex, columnar organization
afferent organization, 477, 480–485
arbors, 253
architecture
laminar patterns, 280
single axon terminations, 246
development, 642–643, 645, 646, 665
FEF connectivity, 551, 552
TE (cytoarchitectonic area) (*cont.*)
function in, 490–492
human correlates in lateral occipital cortex, 702
pulvinar connections, 441
SEF connectivity, 588
stimulus selectivity of cells, 470–473
thalamic connectivity, 692, 693
timing of information transfer, 228, 230
ventral pathway, 469, 470
TEa, 362, 449, 458
Tectopulvinar route, 656
TEd, 470
TEm, 362, 449, 458
Temporal association cortex
Temporal cortex, 44, 51, 66, 79, 80, 97
area 7a connections, 504
Brodmann's areas, 677
callosal connections, 713
comparison of visual areas for primate species, 686, 687
development
connectivity, 643
myelination, 646
neurochemical markers, 652
FEF connectivity, 549, 550, 551
human studies
color processing areas, 763
extent of visual cortex, 750, 751
slow brain, 233
STP: *see* Superior temporal polysensory area
theories of organization, 115–116
timing of information transfer, 230
ventral pathway: *see* Ventral pathway
ventral posterior (TVP), 173
visuotopic remnants, 136
Temporal field, 101
Temporal processing, *see also* Timing of information transfer
hierarchy frameworks and, 284
human studies, 745
prefrontal cortex, 609–610
STP and, 440
Temporal (timing) properties
computer injected current, 26
MLGN and, 316
MT and satellite responses, 376
parasol system, 67–68
saccade initiation, 598
slow brain, 233
Temporal (timing) scatter, 245
Temporal (timing) summation, 219–220
10 millisecond per unit area rule, 228
TEO: *see* Occipitotemporal transition
Terminations
architecture, intrinsic connections, 277, 278, 279
developmental stages, 685
human studies, 729
laminar, 245–250

Terminations (*cont.*)
preferential localization nonadrenergic and serotonergic, 285
specialization, 270–275, 276, 285
Texture, 67, 71, 72, 78
contour definition, 339
human studies, cortical regions activated by, 768
Texture domains, 2-D shape encoding in, 323
TF (cytoarchitectonic area)
area 7a connections, 503, 504
visually responsive and unresponsive sectors, 172
Thalamus, 69, 283, 285, 448, 449, 692–693; *see also* Pulvinar
collateralization, 283
development, 645–646
FEF connectivity, 549
human studies, 710–713
neocortical connections, 284
nonhuman primates, 692–693
oculomotor system, 541, 542
plasticity, 656–657
SEF connectivity, 586–587
terminal specializations, 270, 271
timing of information transfer, 215
TPO connections, 441
Thales, 5–6
Theories of visual cortex organization, 91–119
areas, modules, overlaps, 92–94
areas of third tier versus V3 and V4, 108
DI, 113
DLr and DLc, 114–115
DM, 110–111
early maps, 94–98
early modern theories, 98–100
M, 111–113
parietal areas: LIP, VIP, PIP, and VPP, 117–118
recent maps, 100–104
upper temporal lobe areas, MTc, FSTd, FSTv, and MST, 115–116
V2 and MT, 104–108
V4, 113–114
Vp and Va, 116–117
Thick stripes: *see* Stripes; Cytochrome oxidase
Thin stripes: *see* Stripes; Cytochrome oxidase
Timing of information transfer, 205–233
fast brain, 230
latencies to and between areas, 220–231
cat cortex, 224–225
primates, 225–231
retina and LGN, 221–224
magnocellular neuron conduction velocity, 265
rapidity of processing, 231–233
structural constraints, 206–220
axonal conduction time, 206–215
neuronal integration time, 215–220
synchronization, 245
temporal scatter, 245
Titchener, E. B., 43
Topographic organization, 80–85, 436
callosal connections, 715–716
FEF connectivity, 549
human studies, 745, 752–758
MT and satellites, 373–375
V2 in macaques, 305–314
TPO
architectonic subdivisions, 438, 446–447, 449
connectivity, 446–447
association cortex, 452–460
thalamic projections, 441
visual cortical inputs to, 440–441
physiological correspondences, 461–462
as separate cortical area, 461–462
timing of information transfer, 229
TPOc, 362, 436, 445, 446, 448–449, 450, 456, 458, 460
TPOi, 449, 450, 458
TPOr, 436, 445, 446, 449, 451, 452, 458, 460
Transcortical conduction, 283
Transformations of visual field, 140–144
Transitional V4, 155
Translation
MSTd neuron sensitivity to, 382
rotation-selective MTSd cell response, 393, 395
Transmitter receptors: *see* Neurotransmitters
Transparent stimuli, MT and satellite responses, 390–391, 392
Tree shrew, 150
Trigger features, frontal cortex neurons, 528
Trigger signal, FEF, 563–566
True Blue labeled neurons, 262, 263; *see also* Fluorescence studies
TS, timing of information transfer, 229
Tuned-excitatory obligatory binocular cells, 298
Turing, Alan, 73
Two-dimensional position test, 373, 382
Two steps forward, one step back mapping, 310

Upper visual field, V4, 690

V1 (primary visual area), 66, 76, 79, 91, 92–93, 94, 98, 111, 113, 700
architecture, 244, 267, 269, 270, 273
extent of area, 689
intrinsic connections, 277
numbers of neurons in, 259
secondary bypass projections, 282, 283
transitions at borders, 144
blob and interblob zones, 94
CO, 296–298
color pathways, 795
connectivity
color pathways, 795
human studies, 729
interhemispheric, 693
intrahemispheric, 693, 694
V5 input, 782, 782, 783, 784

V1 (primary visual area) (*cont.*)
contour representation, 323
development, 660–662, 664–665
early modern models, 98, 99
human studies, 747
blindsight, 777
callosal connections, 717
chromatopsia, 798–799
color constancy mechanism, 802
color sensitivity, 761, 762
conscious vision, contributions to, 803–804
cortical magnification factors, 759, 760
cytochrome oxidase profiles, 707, 708
functional organization, 727, 761
hierarchical organization, 721
higher-order area connections, 729
intraareal compartments, comparative, 705
motion perception, 765
MT projections, 703
neurotransmitters, 725
residual motion vision, 782
residual vision, 785
topography, 751, 752
visual field representation, 755, 756
interleaved one step forward, two steps back mapping, 309
magnification factor, 157
mental imagery, 245
and MT: *see* V1-V5/MT
oculomotor system, 541
orientation tuning curve width, 801
perceptual processing of form-from-motion, 789
plasticity, 654–657
receptive field plasticity, 188–189
receptive field positions, 187
receptive field scatter, 138–139
receptive field size, 311
retrograde labeled feedforward neurons, 261, 262, 263
RF position shifts, 187
subcompartments, 305
TEO projections to, 173
TE projections, 469, 493
timing of information transfer
fast brain, 229
information content, 232
interconnections and synchronization of firing, 231
latencies, 225–228
latency, 229
relays between TE, 228
transmitter receptors, 696
visuotopic maps, 144
V1–V2, 146–147, 317
arbors, 253
V1–V2 (*cont.*)
architecture
extent of areas, 689
feedforward arbors, 252
laminar patterns, 279–280
numbers of projection neurons, 260
single axon terminations, 245, 246
conduction delays, 213
connectivity, interhemispheric, 693
cytochrome oxidase staining, 297
development, 641
dynamics, 318–320
human studies, 703
connections, 722
cyto- and myeloarchitecture, 704
cytochrome oxidase profiles, 709–710
MRI, 700
neurotransmitters, 724–725
and MT, 363–364
reflection of visual field across, 311
specificity of, 317–318
stripes, modular organization, 303–304
timing of information transfer
latencies, 223
latency difference, 228
prediction of onset latency difference, 219–220
transmitter receptors, 695
V1 lesion effects, 148
visual representation in dorsal V2, 306, 307
V1-V5/MT, 367
color perception, 787–793
single axon terminations, 245, 246
timing of information transfer, 228
V2 (secondary visual area), 66, 76, 79, 91, 98, 104–108, 112, 461
anterograde label of feedback axons to V1, 274
architecture, 252, 266, 269, 270, 273, 688–690
extent of areas, 689
intrinsic connections, 277, 278
numbers of neurons in, 259
numbers of projection neurons, 260
secondary bypass projections, 282, 283
architecture in macaque, 295–328
connectivity, 314–322
cytochrome oxidase, 296–297
function of multiple processing stages, 326–328
modular organization within stripes, 303–305
receptive field properties, stripes and, 297–302
relationship of structure and circuitry to perception, 322–326
topography, 305–314
bands, 94
color vision, 300, 796, 797

V2 (secondary visual area) (*cont.*)
connectivity, 796
interhemispheric, 693
intrahemispheric, 693, 694
thalamic, 692
contour representation, 323
development, 664, 665, 666
connectivity, 641–642, 645
metabolic studies, 649
myelination, 646
neurochemical markers, 653
DI connections, 113
DLr and, 115
dorsolateral cortex, 114
early modern models, 98, 99
form, color, depth information, 310
functional organization in awake monkey, 335–354
figure-ground segregation mechanisms, 343–353
methodology, 336–338
visual contours, 338–343
human studies, 700, 747
callosal connections, 717
chromatopsia, 799
color constancy mechanism, 802
connections, 722, 723
conscious vision, contributions to, 803
cortical magnification factors, 759, 760
cyto- and myeloarchitecture, 704, 706
cytochrome oxidase profiles, 707, 708–709
functional organization, 727
functional properties, 761
hierarchical organization, 721
higher-order area connections, 729
intraareal compartments, comparative, 705
MT projections, 703
neurotransmitters, 724–726, 726, 727
thalamic connectivity, 712
topography, 751, 752
transmitter receptors, 724–726
visual field representation, 755, 757
visuotopic organization, 727
interleaved one step forward, two steps back mapping, 309
modular organization, 305
MT visuotopic maps, 132
orientation tuning curve width, 801
plasticity, 655
receptive field types, 302
retrograde labeled feedforward neurons, 261, 262, 263, 264
single axon terminations, 245, 246
TE projections, 493
theories of organization, 104–108
timing of information transfer, 213, 230
fast brain, 229

V2 (secondary visual area) (*cont.*)
timing of information transfer (*cont.*)
information content, 232
interconnections and synchronization of firing, 231
latencies, 225–229
transmitter receptors, 695–698
V1 interactions: *see* V1-V2
ventral pathway, 469
visuotopic organization, 144–150
anisotropy, 147–148
discontinuities, 178
transitions at borders, 143–144
V2d, visual representation in, 306
V2–V1
development, 642
feedback connections, 315
V2–V3, 179
architecture, 690
transmitter receptors, 699, 727
V2–V4
arbors, 253
numbers of projection neurons, 260
V3, 98, 100, 101, 102, 103, 112
architecture, 277, 690
connectivity, 366
human studies, 722
interhemispheric, 693, 695
intrinsic, 277
V5 input, 782, 783, 784, 785–786
dorsal (dV3), 161, 163
FEF connectivity, 551
human studies, 700, 703
connections, 722
cyto- and myeloarchitecture, 706
functional properties, 761
hierarchical organization, 721
neurotransmitters, 726
orientation-selective cells, 789
residual motion, 786
residual vision, 785
topography, 751, 752
visual field representation, 755, 756
VP region, 701–702
maps, extent of areas, 689
orientation-selective cells, 800
orientation tuning curve width, 801
PIP, 117–118
plasticity, 655
theories of organization, 108–110
timing of information transfer, 230
transmitter receptors, 696, 699
V1 and V2 sources of common input, 321
V3A, 161, 163, 170, 352, 423
architecture, 690
human studies, 700, 703
connections, 722
foveal representation, 758

V3A (*cont.*)
human studies (*cont.*)
hierarchical organization, 721
topography, 751, 752
visual field representation, 755, 757
orientation-selective cells, 800
transmitter receptors, 696
V3d, 112
thalamic connectivity, 712
theories of organization, 108–110
V3S, plasticity, 655
V3v: *see* VP
V3–V4, 690
V4, 66, 67, 74, 76, 77, 78, 79, 80, 101, 102, 103, 112
afferents to TE, 480–481, 484
architecture, 252, 266, 269, 690–691
intrinsic connections, 277
laminar patterns, 280
numbers of neurons in, 259
numbers of projection neurons, 260
secondary bypass projections, 282
single axon terminations, 249
color patch, 315
color responses, 802
color signals, 794
color vision, 796, 797
connectivity, 366, 796
interhemispheric, 695
intrahemispheric, 693, 694
patchy, 303
thalamic, 692
convergence of inputs, 301
development
connectivity, 641–642
metabolic studies, 649
FEF connectivity, 550, 551
hierarchical view, 320
human studies, 727
callosal connections, 717
chromatopsia, 799
connections, 722
cyto- and myeloarchitecture, 706
cytochrome oxidase profiles, 707, 708
functional properties, 761
hierarchical organization, 721
neurotransmitters, 727
orientation selectivity and form analysis, 799–802
thalamic connectivity, 712
visual field representation, 755, 757
human VP as correlate of, 701–702
inputs from V2, 323
MT connections, 364
orientation patch, 315
orientation selectivity, and form analysis, 799–801
orientation tuning curve width, 801
V4 (*cont.*)
plasticity, 655
experience/learning-induced, 659
extrastriate lesions, 657
V1 lesion studies, 655
TEO and PIT, 796
TE projections, 493
theories of organization, 108, 113–114
timing of information transfer, 227–228
information content, 232
relays between V1 and TE, 228
transitional, 155
transmitter receptors, 696
V1 and V2 sources of common input, 321
ventral pathway, 469
visuotopic organization, 165, 169, 170, 171
wavelength-selective cells, 795
V4i, 458
V4t, 169, 170
connections, 366
STS
architectonic subdivisions, 449
frontal projections, 454
prefrontal terminals, 455
V4v, human studies, 702
cortical magnification factors, 759, 760
functional properties, 761
topography, 751, 752
visual field representation, 755, 757
V5 (areal MT), 151
V5/MT, 52, 66, 74, 78, 79, 80, 91, 102–103, 104–108, 110, 151, 442
architecture, 244, 691–692
collaterals, 250, 252
extent of areas, 689
intraareal compartments, 705
intrinsic connections, 277
laminar patterns, 279–280
secondary bypass projections, 283
borders, 701
Brodmann's areas, 677
CAT 301 staining, 316
columnar organization, 326
connectivity
interhemispheric, 695
intrahemispheric, 693, 694
patchy, 303
single axon terminations, 245, 246
thalamic, 692
convergence of inputs, 301
development, 665
connectivity, 642
electrophysiological studies, 649, 651
metabolic studies, 652
myelination, 646
DLr connections, 114
early modern models, 98, 99
FEF connectivity, 550, 551

V5/MT (*cont.*)
functional imaging in humans, 421–425
hierarchical view, 320
human studies, 700, 701, 702–703, 747
callosal connections, 717–718
comparison with other primates, 701
cyto- and myeloarchitecture, 706
cytochrome oxidase profiles, 707, 708, 709
functional organization, 727
functional properties, 761
hierarchical organization, 721
latency measurement, 793
localization, 783–784
motion perception, 764–765
neurotransmitters, 727
residual vision, 782–785, 786
thalamic connectivity, 712
topography, 751, 752
visual field representation, 755
visual input to, 782–783
inputs from V2, 323
lesions, 77
MLGN and, 316
motion perception, 67
motion processing, 351
organization
surrounding areas, 154–158
theories of, 104–108, 115–116
visuotopic, 150–154
plasticity, 654–656
experience/learning-induced, 659
extrastriate lesions, 657
learning or experience-dependent, 659
receptive field scatter, 138–139
SEF connectivity, 588
STP-MST border, 439
STP organization, 436
STS
architectonic subdivisions, 449
frontal projections, 454
posterior parietal projections to TPO subdivisions, 458
prefrontal terminals, 455
prefrontal terminations, 456
TEO projections to, 173
timing of information transfer, 213, 230
fast brain, 229, 230
interconnections and synchronization of firing, 231
latency difference, 228
transmitter receptors, 696, 699
V1 and V2 sources of common input, 321
V1-V5 connections, 787–793
visuotopic organization, 132, 135–136
first-order transformations of visual field, 140–141
interhemispheric connection patterns, 168
lateral and ventral areas, 165, 166, 167, 170
V5/MT, processing in macaques
areas, definitions of, 360–363
behavioral studies, 404–421
body movement guidance, 418
extraction of 2D and 3D shape from motion, 418–421
eye movement guidance, 412–418
motion perception, 405–412
connections, 363–365
general response properties, 367–376
nature of moving stimulus, 367–370
receptive field, 371–372, 373
temporal aspects, 376
topographical architecture, 373–375
variability of, 375–376
input to, nature of, 365–367
physiological studies, 367–404
antagonistic surround, 398–404
general response properties, 367–376
RF, responses to complex stimuli, 388–398
RF, simple parameter selectivities, 376–388
V6
connectivity, 722
human studies
hierarchical organization, 721
topography, 751
intrinsic connections, 724
V6A, 701
VA: *see* Ventroanterior nucleus
Ventral area 6, 531
Ventral area visuotopic organization, 165–172
Ventral-dorsal division of inferior temporal cortex, 101
Ventral extrastriate cortex: *see* V2
Ventral FST: *see* FSTv
Ventral intraparietal area (VIP), 117–118
Ventral MST, 361, 362
Ventral (occipitotemporal) pathway (VOT), 51, 172
development, 649, 665
electrophysiological studies, 649
human studies, 728, 748, 768
hierarchical organization and, 721, 728, 729
topography, 751, 752
IT organization: *see* Inferior temporal cortex, columnar organization
MT and, 367
object vision, 366
PLGN and MLGN and, 316
pulvinar connections, 441
TE projections, 469
Ventral posterior inferior nucleus, area 7a connections, 504
Ventral posterior intraparietal area (VPP), 102, 103, 110, 117–118
Ventral premotor cortex, 531–535

Ventroanterior nucleus (VA), 112
FEF connectivity, 549
oculomotor system, 541, 542
theories of organization, 116–117
visuotopic organization, 165, 166, 167, 169, 170
Ventrolateral prefrontal cortex, 527
Ventroposterior (VP) area, 99, 100, 101, 102, 103, 112
architecture, 690
DL continuation as, 117
human studies, 701–702
callosal connections, 717
cortical magnification factors, 759, 760
cyto- and myeloarchitecture, 706
cytochrome oxidase profiles, 707, 708
functional organization, 727
hierarchical organization, 721
thalamic connectivity, 712
topography, 751, 752
visual field representation, 755, 756
visuotopic organization, 727
macaque, 169
maps, extent of areas, 689
theories of organization, 116–117
transmitter receptors, 696, 699
visuotopic organization, 165, 166, 167, 169, 170
Ventroposterior area plus VA, 110
Vertical meridian, 171
DLp and DLi, 165
field discontinuity, 180
MT, 151, 152
visual representation in dorsal V2, 306, 307
Vervet, 306; *see also* Macaque
Vesalius, 10, 12, 14–15
Vestibular area, 423
Vestibular function, 504, 594
Vestibulo-ocular reflex (VOR), 368, 504
VF, 172
Vicq d'Azyr, Felix, 20, 23, 743
Viessens, Raymond de, 18
Vinci, Leonardo da, 12
VIP, 103
plasticity, 655
theories of organization, 117–118
timing of information transfer, 213, 214
Visual agnosia, 33–34, 45
Visual animated objects, 97
Visual areas
concept of, 727–728
cytochrome oxidase markers, 706–710
definition of, 749–752
human studies, derivation from callosal connections, 717–718
intraareal compartments, 705
modular organization of, 728
STP inputs, 440–441
Visual association cortex: *see* Association cortex
Visual auditory association tasks, 443
Visual center, 49
Visual field
architectonic mapping, 677
human studies, 753, 755–758
Visual field sign, 133
Visual hemifields, 93
Visual learning, 78
Visual pursuit, 361, 553, 580–582, 657
Visual pursuit cells, 373, 412–418
Visual representation, 80–85
Visual search condition, 573, 574
Visuoauditory areas: *see* Auditory areas
Visuomotor areas: *see* Motor areas
Visuomotor control, dorsal pathway, 492
Visuopsychic area, 43, 95, 674–675; *see also* Extrastriate cortex
Visuosensory regions, 674–675; *see also* Striate cortex; *specific sensory functions*
Visuosomatosensory area, 95
Visuospatial processes, *see* Spatial processes
Visuotactile operations, 744
Visuotopic organization, 40, 44, 127–191, 306–307
areas and maps in different species, 144–176
DM complex, 158–165
lateral and ventral areas, 165–172
MT, 150–154
MT surrounding area, 154–158
TEO, 172–173, 174–175
V2, 144–150
human studies, 727, 758
mapping methods, 131–133
multiplicity of, 176–181
nonvisuotopically organized areas, 135–136
receptive field position, dynamics of, 186–189
receptive field size, scatter, and point image size, 136–140
receptive field size, scatter, and spatial resolution, 181–186
types of representation and topographic transitions between areas, 140–144
Vogt, C., 674, 680–681, 683–684, 685
Vogt, O., 674, 675, 680–681, 683–684, 685
Vogt's cytoarchitectural areas, 548
Von Economo, C., 43, 91, 95, 96, 101, 674, 675, 676, 677, 680, 682, 713
Von Economo model, 95, 96
von Haller, Albrecht, 20, 21–22, 25
VOT: *see* Ventral (occipitotemporal) pathway
VP: *see* Ventroposterior area
VPP, 102, 103, 110, 117–118
VTF, 172
VX, 709

Wald, George, 40
Wavelength discrimination, 387–388, 802
Wavelength-selective cells, 795
W cells, 66, 69, 79, 215, 223
W channels, 62
Weigert-Pal stain, 715
Wheat germ agglutinin-horseradish peroxidase (WGA-HRP), 245, 460
White matter
callosal pathways in humans, 713–714
intrahemispheric associative pathways, 718–719
Willis, Thomas, 16–19, 20

X area, 74, 709
X cells, 221–222, 223, 225
X channels, 62

Y area, 74
Y cells, 215, 221, 222, 223, 225
Y channels, 62
Yerkes Primate Center, 46, 49
Y group, 504

Z area, 74
Zero horizontal meridian, 114

MIX
Papier aus verantwortungsvollen Quellen
Paper from responsible sources
FSC® C105338

If you have any concerns about our products,
you can contact us on
ProductSafety@springernature.com

In case Publisher is established outside the EU,
the EU authorized representative is:
Springer Nature Customer Service Center GmbH
Europaplatz 3, 69115 Heidelberg, Germany

Printed by Libri Plureos GmbH
in Hamburg, Germany